15th International Conference on Micro and Nanotechnology for Power Generation and Energy Conversion Applications (PowerMEMS 2015)

Journal of Physics: Conference Series Volume 660

Boston, Massachusetts, USA
1 - 4 December 2015

ISBN: 978-1-5108-1814-9
ISSN: 1742-6588

Printed from e-media with permission by:

Curran Associates, Inc.
57 Morehouse Lane
Red Hook, NY 12571

Printed by Curran Associates, Inc. (2016)

For permission requests, please contact the Institute of Physics
at the address below.

Institute of Physics
Dirac House, Temple Back
Bristol BS1 6BE UK

Phone: 44 1 17 929 7481
Fax: 44 1 17 920 0979

techtracking@iop.org

Additional copies of this publication are available from:

Curran Associates, Inc.
57 Morehouse Lane
Red Hook, NY 12571 USA
Phone: 845-758-0400
Fax: 845-758-2634
Email: curran@proceedings.com
Web: www.proceedings.com

TABLE OF CONTENTS

Wednesday, December 2, 2015

PLENARY PRESENTATION I

SESSION W1A: ELECTRET-BASED ENERGY HARVESTING

We have developed an energy harvester that is specifically desired for impulse acceleration of infrastructure vibrations such as sudden motion at railway bridges. The energy harvester, based on potassium-ion-electret on the sidewalls of 1.8um-gap comb drives, generated 64uAp-p current during low impulse acceleration. This current was high enough to light a LED brightly.

We developed for the first time a MEMS electrostatic vibration energy harvester (e-VEH) with corona-charged vertical electret having a very large bandwidth due to the presence of nonlinear elastic stoppers. At 2.0 gRMS, for a bias voltage of 21 V the device harvests more than 0.1 μW within 45~ 444 Hz, and more than 0.4 μW within 147~444 Hz. With a bias of 46 V, the bandwidth for 0.9 μW is 45~432 Hz, and for 2.1 μW is 146~432 Hz. The maximum RMS harvested power is 6.6 μW at 428 Hz.

We have developed a soft electret gel for vibrational energy harvesters. The soft electret material was obtained by solidifying ionic liquid and immobilizing cations (+) on the surface. When a piece of soft electret gel is sandwiched between a pair of electrodes, large amount of charge is induced in an electrical-double-layer capacitor (1.0-10 μFcm^{-2}) appearing at the interface of the electrode and the ionic liquid gel. By retracting the electrode repeatedly, we obtained the current output of a several $\mu A_{p-p}/cm^{-2}$ stably.

SESSION W1B: MAGNETIC SYSTEMS

W1B-01 HYBRID ELECTROMAGNETIC AND ELECTROSTATIC MICROMACHINED SUSPENSION WITH ADJUSTABLE DYNAMICS pg. 19

K. Poletkin[1], Z. Lu[1], U. Wallrabe[1], and V. Badilita[2]

[1]University of Freiburg, GERMANY and [2]Karlsruhe Institute of Technology (KIT), GERMANY

We present a new hybrid micromachined contactless suspension (CS) based on combining electromagnetic inductive and electrostatic actuation. The stiffness components are dynamically adjusted during the operation phase using a series of integrated electrodes. We experimentally demonstrate vertical linear positioning of a disk-shaped conductive proof mass in a range from 30 to 200μm, controlled tilting about two orthogonal axes in the horizontal plane ranges from ±1 to ±4 degrees, controlled oscillation about the vertical axis with angular displacement of 37° at 1.5Hz frequency. To demonstrate dynamical stiffness adjustment, we experimentally show that the angular stiffness component is increased by a factor of two at 100μm levitation height.

W1B-02 SIMULATION AND EXPERIMENTAL VALIDATION OF A SELECTIVE MAGNETIZATION PROCESS FOR BATCH-hPATTERNING MAGNETIC LAYERS pg. 24

C. Velez, W.C. Patterson, and D.P. Arnold

University of Florida, USA

This work pertains to a selective magnetization method for batch-patterning magnetic poles into magnetic layers, recently developed by our group. This method is important to the PowerMEMS community for expanding the use of permanent magnetic materials in MEMS via the ability to create complex magnetic field patterns. In this paper, we show, for the first time, systematic experimental validation of a simulation tool that models the process.

W1B-03 FERROFLUID-BASED STRETCHABLE MAGNETIC CORE INDUCTORS pg. 29

N. Lazarus and C.D. Meyer

US Army Research Laboratory, USA

We demonstrate a stretchable magnetic core inductor based on liquid metal traces around a ferrofluid core. Using ferrofluid allows a higher inductance density without affecting the mechanical properties of the surrounding silicone. The low frequency inductance was found to increase from 255 nH before fill to 390 nH after fill with ferrofluid, an increase of 52%. The inductor was also shown to survive uniaxial strains of up to 100%.

SESSION W2A: ADDITIVE MANUFACTURING FOR ENERGY STORAGE

W2A-01 PRINTABLE ON-CHIP MICRO BATTERY FOR DISPOSAL BIO-SENSING DEVICE pg. 34

T. Tsukamoto and S. Tanaka

Tohoku University, JAPAN

This paper reports an on-chip micro battery which can be fabricated by printing method. The battery consists of screen-printed Mg and AgCl thick films. The open circuit voltage as high as 1.58 V and the maximum power as high as 1.4 mW were obtained. A demonstration device self-powered by the on-chip battery worked in a 0.14 M NaCl solution, and digital data transmission from the device was confirmed.

W2A-02 LAYER-BY-LAYER FULLY PRINTED Zn-MnO_2 BATTERIES WITH IMPROVED INTERNAL RESISTANCE AND CYCLE LIFE pg. 39

B. Kim, R. Winslow, I. Lin, K. Gururangan, J. Evans, and P.K. Wright

University of California, Berkeley, USA

This research created direct layer-by-layer printed zinc-based secondary batteries with an ionic liquid-based gel polymer electrolyte to power micro- and meso-scale devices. Fully printed cells have demonstrated average discharge capacities of 0.548 mAh/cm^2, energy densities of 8.20 mWh/cm^3, and specific energies of 2.46 mWh/g with some cells achieving over 1000 cycles without catastrophic failure. Layer-by-layer printed devices exhibited decreased DC internal resistance and longer cycle life over previous mechanically assembled cells.

W2A-03 RAPID FABRICATION OF SUPERCAPACITOR ELECTRODES USING BIO-NANOSCAFFOLDS IN CAPILLARY MICROFLUIDICS pg. 44

F. Zang, S. Chu, K. Gerasopoulos, J.N. Culver, and R. Ghodssi

University of Maryland, USA

We report the utilization of capillary microfluidics to rapidly create nanostructure-patterned electrodes for energy storage applications. Using open-channel capillary photoresist microfluidics, virus bio-nanoscaffolds are autonomously delivered on gold electrodes, forming a dense and porous bio-nanoscaffold layer. The performance enhancement was evaluated in a NiO supercapacitor electrode system, where the galvanostatic charge/discharge cycle showed a 3.6-fold increase in areal capacitance for the nanostructured electrode compared to the planar electrode.

SESSION W2B: ENERGY HARVESTING I

An EM low frequency vibration energy harvester is newly developed based on dual halbach array which is suspended in two magnetic springs. Each halbach array concentrates the magnetic flux lines on one side of the array while suppressing the flux lines on the other side. Dual halbach array allows the concentrated magnetic flux lines to interact with the same coil in a way where maximum flux linkage occurs. The prototype device offers 33.4μWcm-3 average power density which is much higher than recently reported EMEH.

This paper presents experiments on how to approach the physical limits on power from vibration energy harvesting under displacement-constrained operation. A harvester with voltage-control of system stiffness is used for the purpose. Power saturation is shifted to higher accelerations by load optimization and tunable electromechanical coupling. The obtained effectiveness is beyond 60% under optimization. This work shows a first demonstration of reaching optimal power in intermediate acceleration-range between two extremes of maximum efficiency and maximum power transfer.

We design, model and fabricate a novel MEMS device for low-frequency, low-g vibration energy harvesting. The bi-stable buckled beam structure based new design is aiming to address the remained challenge of applying the energy harvesting technology to the real environment – working at low frequencies (<100Hz) and low g's (<1g). An electromechanically coupled lumped model has been formulated with Lagrangian equations and solved by harmonic balance method. A macro-scale prototype has been built for validation. A MEMS prototype has been fabricated and tested, showing wide bandwidth below 150Hz at 0.2g.

PLENARY PRESENTATION II

SESSION W3A: POWER ELECTRONICS

This paper presents the design and simulation results of a self-starting electrical interface circuit for parametric frequency increased vibrational energy harvesters. Due to the inherent ring down of frequency up-converting harvesters, an active rectifier is interfaced to a buck-boost stage in order to decouple the load and always provide an optimal impedance to the transducer. The circuit provides an efficiency between 75% - 93% for input voltages in the range of 0.5 - 2V. The minimum input voltage and power are 800mV and 3μW respectively.

Journal of Physics: Conference Series **660** (2015) 011002 doi:10.1088/1742-6596/660/1/011002

W3A-02 INTERFACE CIRCUIT FOR VIBRATION ENERGY HARVESTING WITH ADJUSTABLE BIAS VOLTAGE pg. 71

J. Wei[1], E. Lefeuvre[1], H. Mathias[1], and F. Costa[2]
[1]University Paris Sud, FRANCE and [2]University Paris Est, FRANCE

We develop a new interface circuit for electrostatic vibration energy harvesting with adjustable bias voltage. An electronic switch is used to modify the circuit configuration so that the harvested energy increases the voltage across a biasing capacitor. Decrease of this biasing capacitor voltage occurs naturally due to the circuit imperfections. Such a control of the bias voltage enables to adjust the amount of energy converted by the variable capacitor on each cycle.

W3A-03 FULLY INTEGRATED ULTRA-LOW VOLTAGE STEP-UP CONVERTER WITH VOLTAGE DOUBLING LC-TANK FOR ENERGY HARVESTING APPLICATIONS pg. 76

H.M.P.C. Jayaweera, W.P.M.R. Pathirana, and A. Muhtaroĝlu
Middle East Technical University Northern Cyprus Campus, TURKEY

We design, fabrication, and testing of a novel integrated interface topology for ultra-low voltage step up converter in 0.18 µm CMOS technology without the use of off-chip components.Fully integrated center-tap differential inductors are introduced in proposed LC oscillator design to enhance the efficiency of the system by eliminating the traditional clock buffer circuits used to drive the charge-pump.

SESSION W3B: CATALYSIS, COMBUSTION, AND THRUSTERS

W3B-01 Not published

FLAME PROPAGATION OF N-BUTANE/AIR MIXTURE IN A 2.5mm GAP CONSTANT VOLUME CHAMBER

H. Su[1], L. Jiang[2], H. Cao[1], Z. Nian[1], X. Li[2], and D. Zhao[2]
[1]Zhengzhou University, CHINA and [2]Chinese Academy of Sciences, CHINA

We have developed a visual constant volume bomb to study the micro-scale flame propagation. By spark ignition, under certain experimental conditions, it can be observed to flame propagation speed is far slower than the speed of flame propagation under normal scale. Uneven heat and mass diffusion and hydrodynamic reasons caused flame oscillation which is very easily observed.

W3B-02 FLEXIBLE WIRELESS WALL TEMPERATURE SENSOR FOR UNSTEADY THERMAL FIELD pg. 81

M. Lee, K. Morimoto, and Y. Suzuki
University of Tokyo, JAPAN

We report the development of a flexible wireless wall temperature sensor and its performance evaluation in an unsteady thermal field. The sensor is made of thermally stable polyimide and can be fitted onto a curved surface. A 1 mm-sized sensing resistor is additionally sputtered on the copper coil to improve the spatial resolution of the sensor. It is demonstrated that the temperature measurement uncertainty of ±6.4 °C has been achieved with the measurement time interval as small as 2.48 ms.

W3B-03 EFFECT OF MICRO COOLING CHANNELS ON A HYDROGEN PEROXIDE MONOPROPELLANT MICROTHRUSTER PERFORMANCE pg. 86

J. Huh and S. Kwon
Korea Advanced Institute of Science and Technology (KAIST), SOUTH KOREA

We design and fabricate the liquid micro thruster with regenerative cooling channels, to relieve thermal shock on the micro structure. Photosensitive glass is main material for the micro thruster fabrication and platinum catalyst is inserted to the catalyst bed of the micro thruster for propellant decomposition. Two micro thruster with and without channels are used for the experimental test. The results shows lower thermal shock for the regenerative cooling channel added micro thruster. However, chamber temperature and pressure is also reduced with excessive cooling effect.

Thursday, December 3, 2015

PLENARY PRESENTATION III

TPB-03 ROBOTIC INSECTS: MANUFACTURING, ACTUATION, AND POWER CONSIDERATIONS pg. 91
Robert Wood
Harvard University, USA

SESSION T4A: POWER MANAGEMENT AND CONTROL OF HARVESTER SYSTEMS

T4A-01 MICROWATT POWER CONSUMPTION MAXIMUM POWER POINT TRACKING CIRCUIT USING AN ANALOGUE DIFFERENTIATOR FOR PIEZOELECTRIC ENERGY HARVESTING pg. 92
Z.J. Chew and M. Zhu
University of Exeter, UK

This paper reports the design, implementation, and testing of a newly proposed analogue maximum power point tracking (MPPT) circuit with microwatt power consumption for piezoelectric energy harvesting. The proposed MPPT circuit uses an analogue differentiator and comparator to monitor the changes in the voltage induced by the PEH and enables a DC-DC converter for energy transfer at maximum power point (MPP) with efficiency of 81% to 99%, without using sensing and control circuits and microcontroller, that are in traditional MPPT scheme.

T4A-02 A SELF-POWERED HYBRID ENERGY SCAVENGING SYSTEM UTILIZING RF AND VIBRATION BASED ELECTROMAGNETIC HARVESTERS pg. 97
H. Uluşan, K. Gharehbaghi, Ö. Zorlu, A. Muhtaroğlu, and H. Külah
Middle East Technical University, TURKEY

This study presents a novel hybrid energy scavenger system that combines the power generated simultaneously by a vibration-based (<10Hz) electromagnetic harvester and a UHF band (900 MHz) RF harvester. The novel scavenger interface uses a power management circuit in 180 nm CMOS technology to step-up and to regulate the combined output. The system volume is 6.5 cm^3. The hybrid system behaves as a typical battery, and keeps the output voltage stable at 3 V up to 18 µW of output power.

T4A-03 SELF-SUFFICIENT ELECTRONIC CONTROL FOR NONLINEAR, FREQUENCY TUNABLE, PIEZOELECTRIC VIBRATION HARVESTERS pg. 102
S. Heller, S. Neiss, M. Kroener, and P. Woias
University of Freiburg - IMTEK, GERMANY

We develop the first self-sufficient nonlinear harvester system which is based on a ultra-low power electronic control. For the implementation of a closed loop control, the phase shift between the excitation and the system response is used as control variable. Depending on the speed and frequency of changes in the excitation different control concepts were implemented and tested to optimize the power output of the system.

T4A-04 ELECTROSTATIC VIBRATION ENERGY HARVESTER USING AN ELECTRET-CHARGED MEMS TRANSDUCER WITH AN UNSTABLE AUTO-SYNCHRONOUS CONDITIONING CIRCUIT pg. 107
A. Karami[1], P. Basset[2], and D. Galayko[1]
[1]Sorbonne Universités, FRANCE, and [2]Université Paris-Est Marne-la-Vallée, FRANCE

We show through experimental results that MEMS electrostatic vibration energy harvester charged with electret can benefit in some situations of the use of recently reported unstable auto-synchronous conditioning circuits, instead of primitive or simple rectifier conditioning circuits that have been extensively used with electret electrostatic devices.

SESSION T4B: PUMPS AND ACTUATORS

We present the design, fabrication, and demonstration of a set of 2 mm^2, high-force actuators that combine piezoelectric in-plane actuators with MEMS-enabled scissor mechanisms (motion amplifiers) to simultaneously produce out-of-plane forces of >4 mN and out-of-plane displacements of >4 um. The actuators utilize a new fabrication approach to achieve 15 times smaller area than previous generation of devices, while offering greater actuator displacement and force per unit area of device footprint.

We have designed, fabricated, and characterized an electrostatic ion pump architecture that is compatible with novel applications such as chip-scale, cold-atom interferometry systems. This design improves upon previous designs to increase the ionization probability using a new arrangement and structure of electrodes. Electrons from a nanostructured field emitter array impact ionize the gases within the chamber; the ions are then gettered by a titanium electrode, creating vacuum.

We develop, simulate and test a MEMS out-of-plane (vertical) translational lead-zirconate-titanate (PZT) Archimedes' spiral arm bimorph actuator, which offers large deflection and large force in a compact, mechanically and thermally robust structure that is readily fabricated by commercial PZT manufacturing processes, simply connected to its voltage supply, and readily integrated into more complex MEMS systems. To characterize the unique behavior of spiral-arm actuators, two types of actuators with different electrode patterns (arms are half actuated or fully actuated) were simulated, fabricated, and tested.

We propose, fabricate and validate a novel ECF (electro-conjugate fluid) micropump with TPSEs (triangular prism and slit electrode pair) by electroforming process using newly developed self-aligned micro molds in order to improve alignment accuracy and alleviate fabrication difficulty. Compared with the traditional MEMS process, the concept is to utilize back-side UV light and electrode pattern on the glass wafer for patterning thick photoresist, instead of using ordinary front-side UV light and photomasks.

SESSION T5A: ENERGY HARVESTING II

This paper reports a membrane-based MEMS piezoelectric vibration energy harvester with incorporating design approaches to achieve broader band response than a comparable device with a classical topology. A key drawbacks of a plain membrane compared to cantilevers, is the strain neutralisation of regions of high strain energy through housing the centred proof mass. Experimental results reported up to two folds of power enhancement, over plain membrane, when subjected to band-limited white noise.

This paper reports the design, prototyping and experimental validation of a human-machines interface (HMI) integrated into a glove with energy harvesting from the motion of fingers. The device is addressed to industrial applications where the interaction with machines is restricted by safety procedures, to medical applications, design tools and virtual reality field.

T5A-03 SCALABLE FABRICATION OF TRIBOELECTRIC NANOGENERATORS FOR COMMERCIAL APPLICATIONS pg. 142

L. Dhakar[1], X. Shan[2], Z. Wang[2], B. Yang[3], F.E.H. Tay[1], C.-H. Heng[1], and C. Lee[1]
[1]National University of Singapore, SINGAPORE, [2]Singapore Institute of Manufacturing Technology (SIMTech), SINGAPORE, and [3]Shanghai Jiao Tong University, CHINA

We develop a triboelectric mechanism based large size energy harvesting device using roll-to-roll ultraviolet embossing. These large size devices can be embedded in the indoor floorings, pathways and roads, to harvest mechanical impact energy from human and vehicle movement. This work is aimed towards developing fabrication flow and processes for large volume manufacturing of triboelectric nanogenerators for commercial applications.

SESSION T5B: THERMAL, NUCLEAR, AND COMBUSTION SYSTEMS

T5B-01 ALKALI METAL BASED MICRO COMBUSTION USING GRAPHENE MICRO-VALVE TRIGGER pg. 147

A. Ruyack, V. Gund, K. Camera, S. Ardanuç, C. Ober, and A. Lal
Cornell University, USA

We report an arrayable micro-scale electrical to thermal energy amplification approach utilizing a graphene based micro-valve trigger and an alkali metal fuel source. Using this approach we can react nanoliter droplets of rubidium with ambient air on demand, producing electrical to thermal energy gains of seventy five times. By arraying the fuel into discrete packets, we can achieve temporal control of the fuel source, allowing for tailored peak and average power output.

T5B-02 PROTOTYPE OF RADIOISOTOPE THERMO-PHOTOVOLTAIC SYSTEM USING PHOTONIC CRYSTAL SPECTRAL CONTROL pg. 151

X. Wang, W.R. Chan, V. Stelmakh, M. Soljačić, J.D. Joannopoulos, I. Celanović, and P.H. Fisher
Massachusetts Institute of Technology, USA

This work reports the design, simulation and measurement results of a portable high-efficiency radioisotope thermophotovoltaic system (RTPV) powered by one plutonia fuel pellet. Two dimensional photonic crystal is used to realize spectral control. Our RTPV has the potential to reach higher efficiency and specific power than conventional RTGs and meet the energy demands of systems requiring long lifetime and low maintenance such as spacecrafts and remote sensors.

T5B-03 Not published

NUMERICAL STUDY ON FLAME PROPAGATION OF N-BUTANE/AIR MIXTURE IN A 2.0MM GAP MICRO CONSTANT VOLUME CHAMBER

L.Q. Jiang, H. Su, L.G. Zhao, and D.Q. Zhao
Chinese Academy of Sciences, CHINA

We used a 2D axisymmetric model to simulate the characteristics of n-butane/air flame propagation in a Φ35mm × 2.0mm disk-like gap micro constant volume chamber at 0.1Mpa and 300K. Three kinds geometry of flamefront (i.e convex, plat and concave flames) exist along the direction to unburned mixture. The velocity of flame propagation in the micro chamber is not uniformed and is lower than that in a conventional chamber bomb.

PLENARY PRESENTATION IV

TPB-04 NANOENGINEERED SURFACES FOR THERMAL ENERGY CONVERSION pg. 156

B. Bhatia, D.J. Preston, D.M. Bierman, N. Miljkovic, A. Lenert, R. Enright, Y. Nam, K. Lopez, N. Dou, J. Sack, W.R. Chan, I. Celanović, M. Soljačić and **Evelyn N. Wang**
Massachusetts Institute of Technology, USA

SESSION T6A: RF AND HIGH-CURRENT SYSTEMS

T6A-01 ADVANCES IN VERTICAL SOLID-STATE CURRENT LIMITERS FOR INDIVIDUAL FIELD EMITTER REGULATION IN HIGH-DENSITY ARRAYS pg. 162

F.A. Hill and L.F. Velásquez-García
Massachusetts Institute of Technology, USA

This paper reports the design, fabrication, and characterization of improved solid-state elements intended for individual regulation of field emitters part of high-density arrays. The work demonstrates a high-yield, CMOS fabrication process that creates silicon vertical ungated field-effect transistors (FETs) that achieve current saturation with ~3.5 V bias voltage while limiting the current to ~ 6μA (4 A/cm2). Experimental characterization of arrays of FETs with different array size demonstrates the same per-FET value. Finite element simulations of the device predict a saturation voltage close to the experimental value and a saturation current within a factor of two of the experimental value.

T6A-02 WIRELESSLY POWERED MICRO-TRACER ENABLED BY MINIATURIZED ANTENNA AND MICROFLUIDIC CHANNEL pg. 167

G. Duan, X. Zhao, H.R. Seren, C. Chen, and X. Zhang
Boston University, USA

A miniaturized antenna was fabricated and integrated with a commercialized RFID chip to form a micro-tracer. The micro-tracer was wirelessly powered and interrogated through near field coupling with reader antennas embedded under a microfluidic channel, which can manipulate large number of tracers simultaneously by in channel focusing and separation. The total size of the micro-tracer is 2mm by 1mm, with a 380μm by 380μm integrated miniaturized antenna, optimized at the frequency of 840MHz. The wireless power transfer and real-time communication was demonstrated as the micro-tracer flows inside the microfluidic channel carried by pump oil.

T6A-03 WATT-LEVEL WIRELESS POWER TRANSMISSION TO MULTIPLE COMPACT RECEIVERS pg. 172

A. Garraud, D.J. Munzer, M. Althar, N. Garraud, and D.P. Arnold
University of Florida, USA

We report an electrodynamic wireless power transmission system using a 450 Hz magnetic field to transmit watt-scale power levels to multiple compact receivers. A single 3.0 cm^3 receiver achieves 1.25 W power transfer at 1 cm (0.35 W/cm^3 power density), a 12x increase in power density compared to previous research. Moreover, we demonstrate simultaneous recharging of two wearable devices, using two receivers located in arbitrary positions and orientations.

SESSION T6B: ENGINEERED MATERIALS FOR ENERGY HARVESTING

T6B-01 PIEZOELECTRIC CELLULAR MICRO-STRUCTURED PDMS MATERIAL FOR MICRO-SENSORS AND ENERGY HARVESTING pg. 177

A. Kachroudi[1,2,3], S. Basrour[1,2], L. Rufer[1,2], and F. Jomni[3]
[1] University Grenoble Alpes, FRANCE, [2] CNRS, FRANCE, and
[3] Université de Tunis El Manar, TUNISIE

This paper reports a novel low-cost fabrication process of a charged cellular micro-structured polydimethylsiloxane (PDMS) material referred as piezo-electret or ferro-electret for micro-sensors applications. The dielectric spectra reached on these structures exhibit a high piezoelectric longitudinal coefficient d_{33} of 350pC/N. A mechanical characterization method proves the reliability of this material for low-frequencies applications around 100Hz.

T6B-02 OPTIMIZATION OF A PDMS STRUCTURE FOR ENERGY HARVESTING UNDER COMPRESSIVE FORCES pg. 182

J. Shi[1], D. Zhu[1], Z. Cao[2], and S.P. Beeby[1]
[1] University of Southampton, UK and [2] Coventry University, UK

This paper reports the optimization, fabrication and testing of an optimized PDMS ferroelectret material for energy harvesting applications. The optimized ferroelectret structure was fabricated by PDMS molding using 3D-printed plastic moulds. To quantify the applied compressive forces and its frequency, an electrodynamic test instrument was used to apply compressive forces to the material.

T6B-03 FABRICATION AND CHARACTERIZATION OF NANO/MICRO TEXTURED ELECTRET TO AVOID ELECTROSTATIC STICTION AND ENHANCE ITS SURFACE POTENTIAL pg. 187

M. Suzuki[1], M. Shimokizaki[1], T. Takahashi[1], Y. Yoshikawa[2], and S. Aoyagi[1]
[1]Kansai University, JAPAN and [2]ROHM Co., Ltd., JAPAN

A novel SiO2 electret having micro/nano-scale-textured surfaces was fabricated and characterized its surface potential. The textures were formed by lithography and wet etching using hydrofluoric acid. The width/pitch of each pattern is 0.2 μm/0.6 μm or 1.0 μm/2.0 μm. Both initial value and stability of surface potential of textured electret was better than that of non-textured electret. On the other hand, the stiction force between textured electret and movable electrode was smaller than that between non-textured electret and electrode.

Friday, December 4, 2015

PLENARY PRESENTATION V

FPB-05 HISTORY AND MODERN APPLICATIONS OF NANO-COMPOSITE MATERIALS CARRYING GA/cm² CURRENT DENSITY DUE TO A BOSE-EINSTEIN CONDENSATE AT ROOM TEMPERATURE PRODUCED BY FOCUSED ELECTRON BEAM INDUCED PROCESSING FOR MANY EXTRAORDINARY NOVEL TECHNICAL APPLICATIONS pg. 191

Hans W.P. Koops
HaWilKo GmbH, GERMANY

SESSION F7A: ENGINEERED MATERIALS FOR ENERGY SYSTEMS

F7A-01 FLEXIBLE SUPERCAPACITOR BASED ON MnO_2 COATED LASER CARBONIZED ELECTRODES pg. 195

R. Rahimi, M. Ochoa, W. Yu, and B. Ziaie
Purdue University, USA

This paper reports on a simple and low-cost technique using a CO2 laser engraver to create high-surface-area carbon nanoparticles by pyrolyzing the polyimide layer of a commercially-available Cu/Kapton substrate. This strategy offers improvements over previous laser-assisted supercapacitor efforts by offering three distinctive advantages: (1) improved electrical contact between the carbon nanoparticles and the bottom current collector to reduce the internal resistance; (2) an increase of 55% in specific capacitance with a MnO2/carbon hybrid structure; and (3) more compact, stackable architecture with low interconnection resistance.

F7A-02 MICROSTRUCTURED SILVER SURFACES PRODUCED BY FREEZE CASTING FOR ENHANCED PHASE CHANGE HEAT TRANSFER pg. 200

G.J. Gouws and N. Shortt
Victoria University of Wellington, NEW ZEALAND

The fabrication of porous silver microstructured surfaces on copper substrates by a modified freeze casting process is described. These surface layers display a hierarchical porous structure that can be controlled by process parameters such as nanoparticle loading, polymer concentration or the freeze rate. Due to the large surface area, void structure and high thermal conductivity these structures have potential in energy transfer applications, and their use in enhancing heat transfer during the nucleate phase of pool boiling is evaluated.

F7A-03 BIO-NANOTEXTURED HIGH ASPECT RATIO MICROPILLAR ARRAYS FOR HIGH SURFACE AREA ENERGY STORAGE DEVICES pg. 206

S. Chu, K. Gerasopoulos, and R. Ghodssi
University of Maryland, USA

This paper presents fabrication and characterization of high aspect-ratio (AR) bio-nanotemplated hierarchical NiO supercapacitor electrodes. The hierarchical structure is based on self-assembly of Tobacco mosaic viruses (TMVs) on high AR micropillar arrays. Enhanced assembly of the bio-nanoparticles was achieved by increasing TMV solution accessibility into the deep microcavities of the pillar arrays. Electrochemical characterization of the hierarchical NiO supercapacitor electrodes revealed a 25-fold increase in charge capacity compared to flat NiO with excellent cycle stability over 1500 charge/discharge cycles at 2mA/cm2

SESSION F7B: THERMAL SYSTEMS

A. Arnaud[1,2], J. Boughaleb[1,3,4], S. Monfray[1], F. Boeuf[1], O. Cugat[2], and T. Skotnicki[1]
[1]STMicroelectronics, FRANCE, [2]University of Grenoble, FRANCE, [3]Institut National Des Sciences Appliquees, FRANCE, and [4]Commissariat à l'Energie Atomique (CEA), FRANCE

The bimetallic strip heat engines (BSHE) are devices exploiting the first-order phase transition of thermo-mechanically bistable membranes to turn heat fluxes generated by thermal gradients into mechanical energy. This strain energy can be converted into electric charges by means of electromechanical transducers. In this paper we model the coupling between a BSHE and a piezoelectric transducer and we evaluate the theoretical performances of these engines when they describe a Stirling cycle.

R. Chutani[1], F. Formosa[2], M. de Labachelerie[1], A. Badel[2], and F. Lanzetta[1]
[1]Universite de Bourgogne Franche-Comte, FRANCE and [2]Universite Savoie Mont Blanc, FRANCE

This work details the clean room wafer-level microfabrication of hybrid fluidic membranes (HFMs) that would be suitable working in quite severe conditions of temperature and pressure. The concept of HFMs consists of two thin RTV silicone based membranes, which encapsulate an incompressible fluid. The development of such HFMs is to provide a technological solution for micro engines by replacing piston with HFMs. The latter will help to avoid friction losses and leaks due to the cylinder-piston gap which are detrimental to the engine operation.

H. Nakajima, S. Oh, A. Ueno, K. Morimoto, and Y. Suzuki
University of Tokyo, JAPAN

For precise thermal control in satellites under varying internal heat dissipation and thermal boundary condition, we propose a high-fill factor MEMS radiator enhanced by the near-field effect. We have successfully fabricated a prototype with parylene shared springs. A fill factor as high as 89 % has been achieved. It is found that at the ON state,the diaphragm temperature is increased from 71.4 degrees Celsius to 93.8 degrees Celsius, showing up to 70 % enhancement in the radiation heat flux.

SESSION F8A: ENERGY HARVESTING III

J. Oxaal, M. Hella, and D.-A. Borca-Tasciuc
Rensselaer Polytechnic Institute, USA

A MEMS vibration energy harvester with gap-closing interdigitated electrodes is designed for and tested on HVAC air ducts. A two level stopper system is implemented which utilizes cantilever beams and a film of parylene deposited on the electrode sidewalls, defining the absolute minimum gap between the electrodes and providing electrical insulation. The device operates in impact mode when tested on the ducts with an RMS acceleration of 155 mgRMS. The peak power is 12nW (0.6 nW RMS) with a PSD of 6.9e-11 W/Hz at 240 Hz.

J.M. Chen[1], L. Bu[1], W.Y. Xu[1], B.J. Xu[1], and L. Song[2]
[1]China University of Geosciencs, CHINA and [2]State Grid Jibei Electric Power Co., Ltd., CHINA

This paper reports the design, fabrication and testing of a novel vibro-impact type triboelectric energy harvester. Strong dynamic nonlinearities are measured for this vibro-impact system, and wideband frequency response under diverse structural parameters are analyzed. The proposed device is applied in two large amplitude scenarios, and generates maximal peak-to-peak voltage of 18V in foot swinging condition and 45V in arm swinging condition.

J. Nakano[1], K. Komori[1], Y. Hattori[1], and Y. Suzuki[2]
[1]Hokuriku Electric Industry Co. Ltd., JAPAN and [2]University of Tokyo, JAPAN

This paper reports the development of MEMS rotational electret energy harvester (EH) for human motion. With a large-scale proof-of-concept model, power output of 50 μW has been obtained at a very low rotational speed of 1 Hz. Optimal design of MEMS EH is also proposed, and a rotational structure with embedded ball bearing has been successfully developed using 1.5 mm-thick Si substrates. Experimental data of the prototype electret generator with asymmetric added-mass will be presented in the conference.

Pendulum structures are used for converting translational into rotational motion. This work investigates three different structures as well as their geometrical parameters with respect to their ability to couple human leg motion into a rotational inductive energy harvesting system. Multiple-mass setups are considered both in system simulations and experiments. An average output power larger than 20 mW was achieved under real-world conditions using a load resistance equal to the coil resistance.

SESSION F8B: FUEL CELLS

This work is aimed in the fabrication and operation of a paper-based membraneless nanofluidic fuel cell. The advantage of this device lies in the use of adhesive polyester as the support film and as the hook-and-loop tape, offering simplicity, easy to operate, and a significant decrease in cell dimensions. The enhancement of several aspect such as surface area, oxygen source, stream rate and distance between electrodes resulted in a current and power density of 600 mA cm-2 and 132 mW cm-2.

This paper reports the design, fabrication of a miniaturized PEM fuel cell combined microreactor with self regulated hydrogen mechanism. Miniaturized PEM fuel cell comprises of screen printed PEM fuel cell assembly and on board hydrogen storage medium; in order to extend the run time of the system hydrogen production microreactor is combined. Microreactor produces hydrogen based on catalytic hydrolysis of NaBH4; platinum catalyst used in this work is fabricated using pulsed electrodeposition and dealloying technique. Produced hydrogen from the microreactor is self regulated passively by hydrophobic surface based gas venting.

A nanofluidic fuel cell with flow through electrodes was used to investigate the role of pH in the cell performance using ethanol as fuel and two Pd nanoparticles, commercially available and synthesized using ionic liquids. The use of an ionic liquid as reaction medium allows to obtain nanoparticles by a clean pathway, which combined with the technology of fuel cells can be an attractive green solution to supply energy requirements of small devices.

Solid oxide fuel cells (SOFCs) are known as one of the possible next generation energy systems; however, lowering the operating temperature lower than 600C is indispensable to enhance their application fields. In order to solve this issue, we proposed a new SOFC structure using a thin film electrolyte to reduce its resistance at lower temperatures. In this study, we will present a recent progress on our SOFCs using a proton conductive thin film as an electrolyte.

SESSION F9A: FLUID-BASED HARVESTERS

F9A-01 A MINIATURE RADIAL-FLOW WIND TURBINE USING PIEZOELECTRIC TRANSDUCERS AND MAGNETIC EXCITATION pg. 266

H. Fu and E.M. Yeatman
Imperial College London, UK

This paper presents a miniature radial-flow wind turbine that converts wind energy into electricity using piezoelectric transducers and magnetic excitation. It is the first realization of a piezoelectric radial-flow wind turbine considering compactness, reliability and the low start-up wind speed into the design. A 159 μW power output was obtained with a 270 kΩ load. The device starts working at 3.5 m/s and keeps operating down to 1.84 m/s after start-up.

F9A-02 A MEMS TURBINE PROTOTYPE FOR RESPIRATION HARVESTING pg. 271

U. Goreke[1], S. Habibiabad[1], K. Azgin[1], and M.I. Beyaz[2]
[1]Middle East Technical University, TURKEY and [2]Antalya International University, TURKEY

The design, manufacturing, and performance characterization of a MEMS-scale turbine is reported. The turbine is designed for integration into a respiration harvester that can convert normal human breathing into electrical power through electromagnetic induction. The turbine is packaged and tested at respiration flow rates between 5-25 lpm. The highest rotational speed was measured to be 9.84 krpm at 25 lpm, resulting in 8.96 mbar pressure drop across the device and 370mW actuation power.

F9A-03 ENERGY HARVESTING FROM HUMAN MOTION USING FOOTSTEP-INDUCED AIRFLOW pg. 275

H. Fu[1], R. Xu[2], K. Seto[2], E.M. Yeatman[1], and S.G. Kim[2]
[1]Imperial College London, UK and [2]Massachusetts Institute of Technology, USA

We present an in-shoe energy harvester converting foot-strike energy into electricity to power wearable or portable devices. A flexible air pump is employed to convert the vertical foot motion into airflow, and a miniature wind generator is developed to generate electricity at high frequencies. This design overcomes the difficulties of shoe energy harvesting, including the limited vertical distortion of shoes and the low human-motion frequency. A 6 mW power output was obtained with a 4.9 Ω load.

F9A-04 OPTIMIZATION OF MICROFLUIDIC BASED ELECTROMAGNETIC ENERGY HARVESTER FOR SHOE INSOLES pg. 280

M.M. Rahman[1], R. Atkin[2], and H. Kim[1,2]
[1]University of Utah, USA and [2]SOLFIRE, USA

This paper reports the improved performance in both simulation and experimental results of the 4th generation microfluidic-based energy harvester where the power density has been increased by 6.89 times to achieve higher power density in comparison to the previous versions. Such improvement was achieved by optimizing device geometry (4 channels in series with 6 magnets per channel), resulting in one of the highest power densities among human-body-induced vibration based energy harvesters (PZ [2], EM [3], ES [4]). The fabricated device was tested by constructing a custom testing setup that mimics the periodic acceleration of human foot movements. The final device produced total power of 455.77mW from a volume of 20x3.74x0.75cm3, resulting in a power density of 8.13mW/cm3.

SESSION F9B: ENERGY STORAGE

F9B-01 HIERARCHICAL CELLULOSE-DERIVED CARBON NANOCOMPOSITES FOR ELECTROSTATIC ENERGY STORAGE pg. 285

V. Kuzmenko[1], A.M. Saleem[1,2], A. Bhaskar[1], H. Staaf[1], V. Desmaris[1,2], and P. Enoksson[1]
[1]Chalmers University of Technology, SWEDEN and [2]Smoltek AB, SWEDEN

In this study, new freestanding hierarchical carbon nanocomposite materials are evaluated as electrodes for energy storage. These electrodes consist of carbon nanotubes (CNTs) with a 1-20 nm tube diameter deposited on top of cellulose-derived carbon nanofibers (CNFs) with a 50-250 nm fiber diameter. CNF matrix has great mechanical and electrochemical stability along with valuable mesoporosity, while CNTs contribute to material surface area and electrical conductivity, which is beneficial for supercapacitor high energy performance.

F9B-02 A LIQUID-ACTIVATED TEXTILE BATTERY FOR WEARABLE BIOSENSORS pg. 290

X. Liu and P.B. Lillehoj
Michigan State University, USA

We demonstrate the first liquid-activated textile battery which can achieve a voltage of 1.3 V. Details and experimental data on the design, fabrication and characterization of our Ag-Al textile battery are presented, which offers a unique platform for on-demand power generation for the detection of aqueous samples.

doi:10.1088/1742-6596/660/1/011002

K. Hoeppner[1], M. Ferch[1], A. Froebe[2], R. Gernhardt[2], R. Hahn[2], P. Mackowiak[1], B. Mukhopadhyay[2], S. Roder[1], I. Saalhofen[2], and K.-D. Lang[1,2]
[1]Berlin Institute of Technology, GERMANY and [2]Fraunhofer IZM, GERMANY

We report the fabrication and testing of silicon-integrated lithium ion secondary micro batteries with interdigitally arranged side-by-side electrodes. Cavities separated by narrow silicon spacers were etched into Si, passivated and provided with thin film current collectors. Slurries of the electrode materials, $Li_y(Ni_{1/2}Co_{1/5}Mn_{3/10})O_2$ as the cathode and $Li_4Ti_5O_{12}$ as the anode, were filled into the cavities by micro dispensing. The fabricated batteries were operated with 1M LiPF6 in EC:DMC as the electrolyte and feature a rate capability of > 10C and a linear capacity loss rate of <0.1 % per cycle over 30 full cycles.

H. Jung, K. Gerasopoulos, X. Zhang, and R. Ghodssi
University of Maryland, USA

This work presents an in situ operando approach that allows characterization of both stress and structural evolutions in Li-ion battery electrodes using a MEMS platform. New capabilities of our platform are also highlighted, allowing visual observation of crystal phase-dependent structural changes in the electrode, with an improved interference pattern monitoring method, yielding higher sensitivity compared to a previously reported technique. Simultaneous characterization of the stress and structural changes are achieved in a single experiment, showing excellent agreement with literature reports.

PLENARY PRESENTATION VI

Gang Chen
Massachusetts Institute of Technology, USA

Poster Presentations

SCIENCE AND ENGINEERING

Catalysis & Combustion

M.H. Ervin, B. Isaacson, and L.B. Levine
US Army Research Laboratory, USA

Porous silicon (PSi) is being investigated as an on-chip energetic material. When an oxidizer is added to PSi, a very energetic combustion reaction can be obtained. We have developed an electrochemical (galvanic) PSi etch process that will allow the integration of PSi with preexisting devices which may utilize the mechanical force, gases, light, or heat released by the PSi reaction. Proof-of-principle energy harvesting of the combustion reaction energy by an off-chip macroscale piezo cantilever has been demonstrated.

PB-02 Not published
STUDY ON COMBUSTION CHARACTERISTIC OF NON-PREMIXED HYDROGEN MICRO-JET-FLAME
J. Zhang, X. Li, H.L. Yang, L.Q. Jiang, X.H. Wang, and D.Q. Zhao
Chinese Academy of Sciences, CHINA

The combustion characteristics of non-premixed hydrogen micro-jet-flames with coflow air were studied numerically in present work with the emphasis on the interaction between flame and the solid tube. The variations of the maximum flame temperature and flame height with the Reynolds number in a range from 0.56 to 481 were obtained. The effect of the radiation heat loss on the combustion characteristics, and the effect of different coflow airs on the quenching velocities were also discussed.

Energy Conversion Physics & Devices

This work presents the design and characterization of a first-of-a-kind millimeter-scale thermophotovoltaic (TPV) system using a metallic microburner, photonic crystal (PhC) emitter, and low-bandgap photovoltaic (PV) cells. We report on the integration of the microburner and PhC in a complete TPV system: we present measurements and simulations of electrical power generation of TPV systems with and without the PhC emitter.

We model a new kind of vibration transducer with autoparametric capabilities for vibration-based energy harvesters. First a special class of 1D vibration transducers are introduced with strong nonlinear springs that show large resonances and capable for an effective energy transfer but not capable of autoparametric resonance. Then a new 2D class of vibration transducers are presented having strong nonlinear character and well suited for autoparametric resonance vibrations.

An electrical circuit equivalent approach has been set up allowing elementary microengine components to be modeled. They cover gas channels/chambers thermodynamics, viscosity and thermal effects, mechanical structures and electromechanical transducers. We propose new developments taking into account scaling effects to establish models suitable for microengines. They are based on simplifications derived from the comparison of the hydraulic radius with respect to the viscous and thermal penetration depths respectively.

We develop a new set of equations that defines the internal resistances in bi-conductive membrane (BCPM) fuel cells and allows optimization of its critical dimensions and power density has been developed. Based on these equations, a design approach is proposed and fabrication steps that limit the power density of the final device are identified. To validate the model, the results from the developed equations are compared with the experimental results of Hamel et al.

This paper reports the design, fabrication and characterization of porous silicon based super capacitor electrodes for hybrid and/or monolithic integration with energy harvesters. The supercapacitor technology reported here is based on hybrid nanomaterial utilizing ALD (atomic layer deposition) coated nano-porous silicon. Such electrodes exhibit excellent electrical properties and stability in different electrolytes - superior in comparison other Si-based approaches. The stability and volumetric densities of energy (1.3 mWh/cm3) and power (214 W/cm3) compare also favorably to those of carbon based approaches. Proof-of-concept on-chip integration is also demonstrated.

This paper reports the design, fabrication and characterization of a flexible fabric solid-state electrical double layer supercapacitor. The proposed device was fabricated using a spray coated fabric substrate to impregnate the textile with carbon particles to form the supercapacitor electrodes. These were combined with a non-hazardous gel electrolyte and achieved high capacitance.

This paper reports multiphysics simulations (COMSOL) of relatively low conductive cathode oxide materials in nanoarchitectures that operate within the appropriate potential range (cut-off voltage 2.5 V) at 3 times the C-rate of micron scale thin film materials while still accessing 90% of material. This paper also reports a novel anode fabrication of Germanium sputtered on a Cu nanotube current collector for lithium-ion batteries. Germanium on Cu nanotubes is shown to alleviate the effect of volume expansion, enhancing mechanical stability at the nanoscale and improved the electronic characteristics for increased rate capabilities.

Engineered & Nanostructured Materials for Energy Conversion

We develop c-axis tilted AlN films on Si(100) substrates in the aim of high output power vibration energy harvesters (VEHs). Tilted c-axis AlN film is predicted that show higher electrical electromechanical coupling (k31) than normal c-axis AlN, leading to higher output power. The tilted-angle of c-axis was controlled by substrate temperature.

In this study, we succeeded in improving surface potential of electret based on inorganic-organic nanocomposite by using fluoropolymer named "CYTOP" and silica nanoparticles. The silica nanoparticles were mixed in the CYTOP before curing or stacked between two CYTOP layers. The surface potentials of CYTOP mixed or stacked with silica nanoparticles were higher than those of a control electret made of pure CYTOP. Comparing two types of electrets, the performance of stack-typed electret is better than that of mix-typed electret.

Herein we describe the fabrication and characterization of piezoelectric composites consisting of $BaTiO_3$ nanoparticles, double wall carbon nanotubes (DWCNT's) and PDMS. By the careful addition and mixing of arc discharge grown DWCNT's into a $BaTiO_3$/PDMS mixture we investigate the effect that varying amounts of DWCNT's will have on the piezoelectric D33 constant on composite films. Piezoelectric constants are measured using a Berlincourt type piezoelectric measurement tool. Final composite structure is verified via scanning electron microscopy.

In the present work, we report the synthesis and evaluation of a methanol tolerant PtAg/C bimetallic catalyst for the oxygen reduction reaction. Electrodeposits of Pt and Ag were carried out by chronoamperometry technique. The PtAg particles synthesized exhibit properties of tolerance to methanol up to 5 M. This cathode is intended to be used in a high-load methanol microfluidic fuel cell or even in a single stream fuel cell.

We report the design, fabrication, and characterization of 2D photonic crystal (PhC) thermal emitters for a millimeter-scale hydrocarbon TPV microgenerator as a possible replacement for batteries in portable microelectronics, robotics, etc. In our TPV system, combustion heats a PhC emitter to incandescence and the resulting radiation is converted by a low-bandgap TPV cell. The PhC tailors the photonic density of states to produce spectrally confined thermal emission that matches the bandgap of the TPV cell, enabling high heat-to-electricity conversion efficiency. The work builds on a previously developed fabrication process to produce a square array of cylindrical cavities in a metal substrate. We present ongoing incremental improvements in the optical and thermo-mechanical properties, the fabrication process, and the system integration, as recently combined with fabrication using novel materials, such as sputtered coatings, to enable a monolithic system.

Fabrication for Energy Systems

This paper reports the fabrication of a rotary MEMS bearing structure for examining the properties of Carbon Nanotube (CNT)-based rotating bearings. Applications of the device are explored.

Textile-based sensor and piezoelectric contributed recently to smart textile development especially focused on medical applications. In this work, the development of a textile-based piezoelectric material has been addressed based on a method compatible with large area manufacturing at low-cost. We first report on the actuation of paper cantilevers using pure crystals demonstrating the efficient and well-known piezoelectric character of the pure material: Rochelle salt. Then the fabrication of the composite material is described. Finally, the converse piezoelectric effect of the composite material is demonstrated.

This paper reports the design, fabrication, and testing of an electromagnetic energy harvester fabricated by using the mature and low-cost rigid-flex printed circuit board (PCB) technology. The rigid boards are used for mechanical frames and coil winding whereas the flexible polyimide is used for mechanical springs and shuttle mass platform on which a permanent magnet is attached. The harvester showed a strong nonlinearity due to the spring hardening effect, thus increasing its operation bandwidth.

A cost effective bottom-up process for the fabrication of micro thermoelectric generators (μTEGs) was developed. It is based on a novel fabrication method involving two selectively dissolvable photoresists and galvano-static electrodeposition of thermoelectric materials. The characterization of a 71 thermocouples μTEG will be presented accompanied with some data from simulations.

In this paper we present a piezoelectric based vibrational energy harvesting device (PEH) which could be integrated into a biocompatible package to power implantable sensor or therapeutic medical devices. The presented architecture is a serial bimorph laminated with ultra-thinned (ranging from 15μm to 100μm) outer PZT "skins" that could operate at a "very low frequency", below 25Hz typically. The core process flow is disclosed (see Figure 1) and detailed with regards to cost and reliability.

We present the fabrication and evaluation of integrated on chip micro-scale supercapacitors. The cells strings to be ultimately attractive and viable for electronic applications must demonstrate the performance uniformity for the individual micro-EDLCs. One of the important aspect of this issue is the reproducibility and yield of the micro-EDLCs fabricated in a one process flow.To address this issue, we performed few characterizations techniques among them: electrochemical Cyclic voltammetry (CV) and impedance spectroscopy (IS).

Power Electronics

This paper presents design and simulation of a power management circuit for MEMS electrostatic energy harvesters with significantly small capacitances in order to make it applicable to highly miniaturized generators. The interface circuitry is based on a flyback technique and automatically detects the switching instants for optimal power transfer through the flyback path without using any external control signals. The need to power the circuitry by extracted power only and the limited available power makes low power consumption of the circuit a main design driver.

Heat & Mass Transfer in Energy Systems

We fabricated in-plane thermoelectric micro-generator on a free-standing Silicon nitride thin film substrate, and evaluated the performance of the generator. The measured temperature difference in the generator was 22.3 K at 373K due to its high thermal resistance of the structure. The calculated exergy of the thermoelectric device is up to 7 %. The efficiency is much higher than that of a conventional type thermoelectric module. It is possible to get higher performance by the thermal design, which is a more simple way than an improvement of ZT.

Tribology

This paper reports a comprehensive analysis of the performance for gas-lubricated bearing-rotor system in silicon microengine. The generalized lubrication model and dynamics model are developed and the effects of micro fabrication defects, viscous friction force and temperature on both steady-state and dynamic characteristics of bearing system are investigated. The results can serve as a useful tool for the design of silicon microbearing-rotor systems.

APPLICATION

Energy Harvesting Devices & Systems

This paper demonstrates an entirely new, parallel-beam architecture for passively-switched energy harvesting and tailors the beams' interaction to vary performance. The harvester's volume is 5 times smaller than previous version; however, it generates power over the entire range from 0.2 g to 2 g and from 5 Hz to 20 Hz.

PA-25 A LOW-FREQUENCY HYBRID ENERGY HARVESTER WITH HIGH OUTPUT PERFORMANCE pg. 416

Y. Xia, W. Liu, T. Chen, Z. Yang, P. Wang, H. Liu, and L. Sun
Soochow University, CHINA

We demonstrated a low-frequency hybrid energy harvester with high output performance. Under a low driving frequency, an electromagnetic harvesting structure is able to impact with high-frequency PVDF harvesting cantilevers with large amplitude. The output performance can be improved significantly by both the electromagnetic harvesting structure and the PVDF harvesting cantilevers. For an excitation acceleration of 0.5g, the maximum output voltages of 1.9 V for electromagnetic part and 3V for PVDF cantilever are obtained, respectively, at resonant frequency of 23.6 Hz.

PB-26 A NON-LINEAR 3D PRINTED ELECTROMAGNETIC VIBRATION ENERGY HARVESTER pg. 420

P. Constantinou and S. Roy
Tyndall National Institute, IRELAND

This paper reports the design, modeling, fabrication, and detailed testing of a non-linear electromagnetic vibration energy harvester using additive manufacturing techniques, which is poised to be a disruptive technology and has the potential impact of revolutionizing manufacturing methodologies. The frequency response of the system is non-linear, due to magnetic force interactions, and is that of a softening spring type. A model for the system is presented and corresponds favorably to measurements. The produced prototype has a demonstrated power density of 0.4 mW/cm3

PA-27 A SILICON DISK WITH SANDWICHED PIEZOELECTRIC SPRINGS FOR ULTRA-LOW FREQUENCY ENERGY HARVESTING pg. 425

J. Lu, L. Zhang, T. Yamashita, R. Takei, N. Makimoto, and T. Kobayashi
National Institute of Advanced Industrial Science and Technology (AIST), JAPAN

In this paper, a micromachined silicon disk with piezoelectric springs was successfully developed with ultra-low resonant frequency of 15.36~42.42Hz and Q-factor of 39~55. Footprint size of the device was less than half of piezoelectric cantilever, while the device can scavenge reasonably high power of 0.5μW@0.1g.

PB-28 A VIBRATION POWERED WIRELESS MOTE ON THE FORTH ROAD BRIDGE pg. 430

Y. Jia[1,2], J. Yan[2], T. Feng[2], S. Du[2], P. Filder[2], K. Soga[2], C. Middleton[2], and A.A. Seshia[2]
[1]University of Chester, UK and [2]University of Cambridge, UK

This paper reports the development, lab-based experimentation and field-site testing (at the Forth Road Bridge in Scotland) of a packaged miniaturised macro-scale electromagnetic vibration energy harvester, which is responsive to both directly excited resonance and auto-parametric resonance for the purpose of operational bandwidth enhancement. The harvester was able to recover up to >1 mW of average raw AC power from bridge vibration and sustain the duty-cycled transmission of a wireless mote.

PA-29 ACOUSTIC ENERGY TRANSMISSION IN CAST IRON PIPELINES pg. 435

M.E. Kiziroglou, D.E. Boyle, S.W. Wright, and E.M. Yeatman
Imperial College London, UK

In this paper we propose acoustic power transfer as a method for the remote powering of pipeline sensor nodes. A theoretical analysis of acoustic power propagation in metal structures is defined, and the effectiveness of the proposed method experimentally validated. Analysis of the results allows the identification of the main engineering challenges in the implementation of an optimized and practical acoustic-powered wireless monitoring system.

PB-30 ALL DISPENSER PRINTED FLEXIBLE 3D STRUCTURED THERMOELECTRIC GENERATORS pg. 440

Z. Cao, J.J. Shi, R.N. Torah, M.J. Tudor, and S.P. Beeby
University of Southampton, UK

We developed a 3D structured thermoelectric generator using dispenser printing technology. Different structures have been investigated in this study, While the performances have been presented. 3D dispenser printing shows potential to fabricate devices for energy harvesting application.

PA-31 AN IMPROVED SWITCHING CONTROL LAW FOR THE OPTIMIZED SYNCHRONOUS ELECTRIC CHARGE EXTRACTION CIRCUIT pg. 445

W. Liu[1], A. Badel[2], F. Formosa[2], C. Liu[1], and G. Hu[1]
[1]Southwest Jiaotong University, CHINA and [2]Université Savoie Mont Blanc, FRANCE

This paper presents an improved switching control law for the Optimized Synchronous Electric Charge Extraction technique. The new control law presented here is to let the switches act ahead or after the maximal point with a phase tuning value. Low load dependency and bandwidth widening effect are two most expected advantages for wideband energy harvesting.

This paper presents an analysis of three different magnet configurations to achieve magnetic plucking for frequency up-converting harvesters based on the distributed Gilbert model for permanent cube magnets. Each configuration has a unique actuation mechanism which provides advantages in certain real-world applications.

This article presents a design of a nonlinear vibration energy harvester which has widened resonance band with the same level of the peak performance compared with that of a linear harvester. To this end, a pair of mechanical stoppers and a pair of repulsive magnets were introduced. It was found that the simultaneous use of the stoppers and magnets can widen the resonance band while maintaining the peak level of the resonance curve of the original linear harvester.

A batteryless self-powered wireless transmitting system has been developed that sends the signal from a drum pad to a synthesizer. The power generated by a piezoelectric generator functions both as the "Play" signal for the synthesizer and as the power source for the transmitter. Experimental results for an electronic drum without any connecting wires fully demonstrated the feasibility of self-powered 800 us latency wireless transmission.

In this work, crystalline Si nanowires (Si NWs) in combination with different fabrication techniques and a vertical device architecture have been proposed to develop an all-silicon nanostructured thermoelectric generator. To produce such device, a suitable vertical integration of Si NWs on patterned microstructures, which define the thermoelectric legs of the generator, has been achieved by bonding top and bottom silicon structures through nanowires. The process has been proved to be a feasible approach that employs a regrowth process of the nanowires for bonding purposes.

This paper introduces a prototype energy harvester device with integrated MEMS capacitive plate and two impact oscillators for transferring energy from low frequency structural vibration with varying mechanical spectra to a vibration of a high resonance frequency cantilever. The use of the two impact oscillators not only harvested energy at low frequencies but also had demonstrated exceptionally sufficient and optimum dynamic responses to a broad frequency bandwidth between 13 Hz and 39 Hz, the bandwidth covering most residual vibrations in structures and systems.

We presents an approach to study the reliability of energy harvesting micro-devices in autonomous systems and more particularly active medical implants. The structure is a cantilever and will be submitted to a low frequency mechanical force. Piezoelectric transduction converts mechanical energy provided by heartbeat into electrical energy. Knowing the electromechanical characteristics of the structure is made possible through its analytical modeling and FEM simulations as a function of different types of excitation. To study the reliability and durability of the structure we propose to establish an accelerated test bench, massively parallel, where the environment and stimuli can be precisely controlled over a wide period of time. This will permit the characterization of potential failure modes through a regular review by mechanical and / or electrical system.

PB-38 DESIGN AND MANUFACTURE OF PERPENDICULAR BI-STABLE CANTILEVER FOR VIBRATIONAL ENERGY HARVESTING ON THE BASIS OF STOCHASTIC RESONANCE pg. 479

M. Kawano, Y. Zhang, R. Zheng, K. Nakano, and B. Kim
University of Tokyo, JAPAN

We designed and fabricated a new vibrational energy harvester that can convert low frequency vibrations to electrical energy over a wide frequency band. To realize the vibrational energy harvester, we proposed to use a remarkable phenomenon called stochastic resonance, which makes amplitude of bi-stable oscillators increase. Then, we modeled a vibrational energy harvester that has double well potential system and numerical simulations were implemented. In accordance with simulation results, an energy harvester was designed and fabricated.

PA-39 DYNAMIC STUDY FOR PERFORMANCE IMPROVEMENTS OF A THERMO-MECANICALLY BISTABLE HEAT ENGINE pg. 484

J. Boughaleb[1,2,4], A. Arnaud[1], S. Monfray[1], P.J. Cottinet[3], S. Quenard[2], G. Pitone[4], F. Boeuf[1], D. Guyomar[4], and T. Skotnicki[1]
[1]STMicroelectronics, FRANCE, [2]Commissariat à l'Energie Atomique (CEA), FRANCE,
[3]Institut National des Sciences Appliquées (INSA), FRANCE, and [4]Delta-concept, FRANCE

This paper focuses on a dynamic study of a thermal energy harvester . We propose here a dynamic model of the device, simulations results and experimental validation by using a complete set of bimetallic strips characterized mechanically.

PB-40 ELECTRICAL CHARACTERIZATION OF A BUCKLING THERMAL ENERGY HARVESTER pg. 489

E. Trioux[1], L. Rufer[1], S. Monfray[2], T. Skotnicki[2], P. Muralt[3], and S. Basrour[1]
[1]Université Grenoble-Alpes, FRANCE, [2]CNRS, FRANCE, [3]STMicroelectronics, FRANCE, and
[4]École Polytechnique Fédérale de Lausanne (EPFL), SWITZERLAND

This paper presents the electrical characterization of a novel concept of micro thermal energy harvesters based on thermally activated bilayer buckling structures. A piezoelectric layer is directly integrated into the bistable structure. The buckling is first induced mechanically by a tip to avoid the eventual pyroelectric charges. Then the thermal bucking is characterized and the output power of different geometries is measured.

PA-41 ELECTROSTATIC ENERGY HARVESTER UTILIZING HIGH DENSITY OF ELECTRODE FOR HIGHER OUTPUT POWER pg. 493

K. Minami, N. Miwatani, K. Kanda, T. Fujita, and K. Maenaka
University of Hyogo, JAPAN

The improvement of output energy from the electret type vibration energy harvester is demonstrated. By using high-density doubled number of electrodes and optimizing the electrode width and pitch, the output power of 26 uW was obtained at constant vibration with 400 ump-p, 350 Hz from the numerical calculation, which is as five times large as previous work.

PB-42 ENERGY HARVESTER USING CONTACT-ELECTRIFICATION OF MAGNETIC FLUID DROPLETS UNDER OSCILLATING MAGNETIC FIELD pg. 498

D. Kim and K.-S. Yun
Sogang University, SOUTH KOREA

We propose an energy harvester using ferrofluid droplets that deform in an oscillating magnetic field. The proposed device consists of top and bottom plates with a conducting electrode coated with a hydrophobic layer and water-based ferrofluid droplet. The contact area between the ferrofluid and the solid surface is changed according to the magnetic field applied by a magnet, which generates AC output power. We demonstrate that the modulation of ferrofluid droplets can be easily and effectively controlled by remote actuation from external magnetic field.

PA-43 EXPERIMENTAL VERIFICATION OF LOAD RESISTANCE SWITCHING FOR GLOBAL STABILIZATION OF HIGH-ENERGY RESPONSE OF A NONLINEAR WIDEBAND ELECTROMAGNETIC VIBRATION ENERGY HARVESTER pg. 501

T. Sato, A. Masuda, and T. Sanada
Kyoto Institute of Technology, JAPAN

This paper presents an experimental verification of a self-excitation control of a resonance-type vibration energy harvester with a Duffing-type nonlinearity which is designed to perform effectively in a wide frequency range. An experimental prototype harvester was designed and fabricated to conduct validation tests. It was shown that the numerical and experimental results agreed very well, and the highest-energy solutions were successfully stabilized globally.

PB-44 FABRICATION OF SCALABLE INDOOR LIGHT ENERGY HARVESTER AND STUDY FOR AGRICULTURAL IoT APPLICATIONS pg. 506

M. Watanabe[1], A. Nakamura[1], A. Kunii[2], K. Kusano[2], and M. Futagawa[3]
[1]LAPIS Semiconductor Co., Ltd., JAPAN, [2]LAPIS Semiconductor MIYAGI Co., Ltd., JAPAN, and [3]Shizuoka University, JAPAN

We have developed scalable indoor lighting energy harvester for IoT application which especially using for agricultural environmental sensing. Optical harvester device fabrication processes were using MEMS and printing hybrid manufacturing process for low cost and for scalable characteristics. We have evaluated input power and output power scalability under several environmental conditions.

PA-45 FINITE ELEMENT ANALYSIS OF COMBINED MAGNETOELECTRIC-ELELECTRODYNAMIC VIBRATION ENERGY CONVERTER pg. 511

S. Bradai[1,2], S. Naifar[1,2], and O. Kanoun[1]
[1]Technische Universität Chemnitz, GERMANY and [2]University of Sfax, TUNISIA

This work reports the design and optimization of a novel combined vibration energy harvester based on the use of electrodynamic and magnetoelctric principle for reliable and powerful system within a challenging size. Finite element analysis for the combined device is developed to evaluate the convertor. The system consists on the use of magnetoelctric transducer placed between two magnets and surrounded with a coil. Rectangular geometry for the magnetoelctric transducer is selected.

PB-46 FREQUENCY UP CONVERSION APPROACH TO SCAVENGE MECHANICAL ENERGY FROM AN ELECTROMAGNETIC DIGITAL ACTUATOR pg. 516

L. Yan[1], A. Badel[1], L. Petit[2], and F. Formosa[1]
[1]Universite Savoie Mont Blanc, FRANCE and [2]Université Technologique de Compiègne, FRANCE

This work reports the design and optimization of a novel combined vibration energy harvester based on the use of electrodynamic and magnetoelectric principle for reliable and powerful system within a challenging size. Finite element analysis for the combined device is developed to evaluate the convertor. The system consists on the use of magnetoelectric transducer placed between two magnets and surrounded with a coil. Rectangular geometry for the magnetoelectric transducer is selected.

PA-47 IMPROVED THERMAL ISOLATION OF SILICON SUSPENDED PLATFORMS FOR AN ALL-SILICON THERMOELECTRIC MICROGENERATOR BASED ON LARGE SCALE INTEGRATION OF Si NANOWIRES AS THERMOELECTRIC MATERIAL pg. 521

L. Fonseca[1], I. Donmez[1], M. Salleras[1], C. Calaza[1], G. Gadea[2], J.D. Santos[2], A. Morata[2], and A. Tarancon[2]
[1]Centre Nacional de Microelectronica (CNM-CSIC), SPAIN and
[2]Institut de Recerca en Energia de Catalunya (IREC), SPAIN

We have designed special suspended microplatforms as a part of Silicon compatible planar thermoelectric microgenerators. Bottom-up grown silicon nanowires will bridge such platforms to the surrounding silicon bulk rim acting as thermoelectric material. In order to maximize the temperature developed between both areas, any other support bridging structures (usually auxiliary silicon beams) are rethought as low conductance thin film membranes with a sieve-like design that allows fabricating them with a short additional wet anisotropic etch step.

PB-48 MAXIMIZING OUTPUT POWER IN A CANTILEVERED PIEZOELECTRIC VIBRATION ENERGY HARVESTER BY ELECTRODE DESIGN pg. 526

S. Du, Y. Jia, and A. Seshia
University of Cambridge, UK

This paper reports on the theory and experimental verification underpinning optimization of the active electrode area of a cantilevered piezoelectric vibration energy harvester in order to maximize output power. The results show that increasing active area does not always increase output power; on the other hand, output power may be reduced if the region corresponding to low mechanical strain is covered.

PA-49 MEMS BASED NONLINEAR MONOSTABLE ELECTROMAGNETIC VIBRATIONAL ENERGY HARVESTER FOR WIDER BANDWIDTH pg. 531

D. Mallick[1], A. Amann[2], and S. Roy[1]
[1]Tyndall National Institute, IRELAND and [2]University College Cork, IRELAND

This work presents the design, fabrication and characterization of a nonlinear MEMS Electromagnetic Vibrational Energy Harvester (VEH) with wide operational bandwidth. The nonlinearity is introduced in the proposed device through the stretching strain of the specially designed fixed-fixed beam arms, which gives rise to the wider bandwidth of 57 Hz at 1g. The device is characterized at different level of integration to report a comparative study. This work also discusses an electrical actuation mechanism to overcome the fundamental limitation of multi-stability of nonlinear VEH at micro-scale.

PB-50 MEMS NARROW GAP ELECTROMAGNETIC HARVESTER WITH MITIGATION OF CURVATURE DISTORTION pg. 536

K. Yamaguchi, T. Fujita, K. Kanda, and K. Maenaka
University of Hyogo, JAPAN

The narrowing air-gap of electromagnetic energy harvester is one of the most important parameter to improve the generated energy. In this paper we demonstrate a compensation method for residual stress on the NdFeB sputtered magnetic film on Si wafer. By using both-side sputtering method, we successfully reduced the residual stress from 144 MPa to 9.40 MPa.

PA-51 Not published
MICRO ELECTROMAGNETIC VIBRATION ENERGY HARVESTER BASED ON COMBINED FREE/RIGID IMPACT MOTION FOR LOW FREQUENCY - LARGE AMPLITUDE OPERATION

A. Haroun[1,2], I. Yamada[1], and S. Warisawa[1]
[1]University of Tokyo, JAPAN and [2]Cairo University, EGYPT

We present a micro electromagnetic vibration energy harvester (FIMG) that can work effectively at low frequencies with analytical and experimental study. It simply consists of a permanent magnet mass that allowed to move freely inside a tube carrying an electrical coil directly connected to the vibration source and closed at both ends with two hard end stops. The proposed harvester shows a significant output power at low frequency - large amplitude vibration. A prototype of D9L12 cylindrical size can generate 91.3μW at 2.5Hz and 9.87 ms^{-2}. In addition, it has a simple construction that allows size minimization.

PB-52 MULTILAYER FERROELECTRET-BASED ENERGY HARVESTING INSOLE pg. 541

Z. Luo[1], D. Zhu[1,2], and S.P. Beeby[1]
University of Southampton, UK and [2]Coventry University, UK

This paper reports a flexible energy harvesting insole made of multilayer ferroelectrets, and demonstrates that this insole can sufficiently power wireless transmission. The insole can produce more than 100 μJ of energy from every step of walking. For every 3 to 4 footsteps, the transmitter gains energy from the insole and is able to send an 8-bit wireless signal to its receiver, which is 6 to 8 meters away from the source.

PA-53 PASSIVELY-SWITCHED, NON-CONTACT ENERGY HARVESTER FOR BROAD OPERATIONAL RANGE AND ENHANCED DURABILITY pg. 546

W.Z. Zhu and C. Livermore
Northeastern University, USA

This paper presents the first simulation and demonstration of passively-switched dynamics in a multi-element harvester in which the beams interact by non-contact (magnetic) means rather than by mechanical impact.

PB-54 PIEZO-MAGNETIC ENERGY HARVESTING FROM MOVEMENT OF THE HEAD pg. 551

A. Delnavaz and J. Voix
École de Technologie Supérieure, CANADA

We design, optimize and test the piezo-magnetic energy harvester that is capable of converting non-harmonic movements of the human head into electricity. The device is composed of a doubly-clamped piezoelectric fiber composite (PFC) beam and two permanent magnets. The PFC beam comprises unidirectionally aligned piezoceramic micro-fibers with interdigitated electrodes. The device can operate with the small-amplitude and low-frequency movements of the head for which other energy harvesting solutions are rarely efficient. Also, the proposed energy harvester can inconspicuously be integrated with the glasses.

PA-55 PIEZOELECTRIC ENERGY HARVESTING FROM HEARTBEAT VIBRATIONS FOR LEADLESS PACEMAKERS pg. 556

M.H. Ansari and M.A. Karami
University at Buffalo, USA

In this research, we study energy harvesting from heartbeat vibrations using fan-folded piezoelectric beams. The generated energy from the heartbeat can be used to power a leadless pacemaker. The proposed device does not incorporate magnets and is thus Magnetic resonance imaging (MRI) compatible. The required power for a pacemaker is about 1 μW. It is shown that our proposed device generates an order of magnitude more power than the nominal power needed for a leadless pacemaker. The small size of the energy harvester and sufficient output power of the device are the advantages of this new design. The size of the device is 2 $cm \times 0.5 cm \times 1 cm$ (1 cc in volume).

PB-56 PIEZOELECTRIC VIBRATION ENERGY HARVESTER USING INDIRECT IMPACT OF SPRINGLESS PROOF MASS pg. 561

S. Ju and C.-H. Ji
Ewha Womans University, SOUTH KOREA

This paper presents an impact based piezoelectric vibration energy harvester using freely movable spherical proof mass and MFC (Macro Fiber Composite) beams as piezoelectric cantilevers. When external vibration is applied, a metal sphere moves freely along channel and impacts both ends of cavity, which induces the vibration of MFC beams and generates electric power. Device having the form-factor of a wristwatch has been designed and tested. Moreover, proof mass made of different materials has been tested to improve the long-term reliability and audible noise level.

PA-57 REMOTE POWER TRANSFER USING MAGNETO-ELECTRIC DEVICES pg. 566

K. Sinha[1] and M. Tabib-Azar[1,2]
[1]University of Utah, USA and [2]University of California, Berkeley, USA

Coils are used to inductively power implantable devices.We report a MEMS device consisting of a piezoelectric beam integrated with a permanent magnet foot-mass used to convert acoustic and magnetic signals to electricity. The device has a high input impedance (as opposed to coils) and can be miniaturized aggressively to below 100 micron linear dimensions. The piezoelectric beams have much higher quality factors(Q) larger than 1000 while coils have low Qs (~ 20) and the harvesting efficiency is proportional to Q.

PB-58 SILICON MEMS BISTABLE ELECTROMAGNETIC VIBRATION ENERGY HARVESTER USING DOUBLE-LAYER MICRO-COILS pg. 571

P. Podder, P. Constantinou, D. Mallick, and S. Roy
Tyndall National Institute, IRELAND

This work reports the development of a MEMS bistable electromagnetic vibrational energy harvester(EMVEH) consisting of a silicon-on-insulator (SOI) spiral spring, double layer micro-coils and NdFeB magnets. Furthermore, with respect to the spiral silicon spring based VEH, several different micro-coil topologies and dimensions have been investigated to determine the optimal coil dimensions. In order to achieve a wide frequency bandwidth a bistable nonlinearity is introduced through a repulsive interaction of fixed and moving permanent magnets.

PA-59 SMART DESIGN SELFTUNING PIEZOELECTRIC ENERGY HARVESTER INTENDED FOR GAS TURBINES pg. 576

L.G.H. Staaf, E. Köhler, M. Soeiro, P. Lundgren, and P. Enoksson
Chalmers University of Technology, SWEDEN

We report a design, simulation and measurement results of a piezoelectric harvester with self-tuning for wider bandwidth and coupled piezoelectric cantilevers to maintain a high power output by extended strain distribution. The harvester is intended for powering sensors on gas turbines and has certain conditions on size, temperature and frequency. Two cantilevers are clamped and coupled at each ends, hence using all available piezoelectric capacity to optimize the power output. To extend the bandwidth, self-tuning is introduced by a sliding weight.

PB-60 STUDY ON TIRE-ATTACHED ENERGY HARVESTER FOR LOW-SPEED ACTUAL VEHICLE DRIVING pg. 581

Y. Zhang, R. Zheng, T. Kaizuka, D. Su, and K. Nakano
University of Tokyo, JAPAN

This study proposes a tire-attached energy harvester, in which a cantilever beam pasted piezoelectric film and magnets with the same polarity are fabricated as a bistable vibrating system, for low-speed actual-vehicle driving. As the wheel rotates, the energy harvester is subjected to the noise produced from the interaction between the paved road and the rotating tire, and tangentially gravitational force as a periodic input can be applied to achieve the occurrence of stochastic resonance. Stochastic resonance can significantly stimulate the response of the bistable vibrating system, and therefore enhance the energy harvesting efficiency.

PA-61 VIBRATION ENERGY HARVESTER WITH BI-STABLE CURVED BEAM SPRING OFFSET BY GRAVITATIONAL ACCELERATION pg. 586

K. Yamamoto[1], T. Fujita[1], A. Badel[2], F. Formosa[2], K. Kanda[1], and K. Maenaka[1]
[1]University of Hyogo, JAPAN and [2]Universite Savoie Mont Blanc, FRANCE

We developed MEMS bi-stable spring for vibration energy harvester (VEH), which consists of intrinsically curved shape spring and gravitational acceleration. By applying the gravitational acceleration, the curved beam is offset to the gravity direction. It will make more symmetrical bi-stable motion and the symmetrical ratio is improved from 3.3 to 65.4 %.

PB-62 WHICH IS BETTER, ELECTROSTATIC OR PIEZOELECTRIC ENERGY HARVESTING SYSTEMS? pg.590

A.D.T. Elliott[1], L.M. Miller[2], E. Halvorsen[3], P.K. Wright[4], and P.D. Mitcheson[1]
[1]Imperial College London, UK, [2]Alphabet Energy, USA, [3]Buskerud and Vestfold University College, NORWAY, and [4]University of California, Berkeley, USA

This paper answers the often asked, and until now inadequately answered, question of which MEMS compatible transducer type achieves the best power density in an energy harvesting system. The results presented allow an engineer to choose an optimal transduction mechanism as a function of harvesting operating frequency, acceleration and device size.

Fuel Cells, Reactors, Combustors & Heat Engines

PA-63 A NOVEL STUDY OF THE KINETICS OF EXTERNAL HIERARCHICAL NANOSTRUCTURES IN METHANOL FUEL CELL pg.595

M. Al-Halhouli, J. Kieninger, O. Yurchenko, and G. Urban
University of Freiburg, GERMANY

We investigate the kinetics, catalytic activity and mass-transport of a layer of porous electrode that consists of external hierarchical nanostructures through 2D COMSOL simulations of an anode of a direct methanol fuel cell (DMFC). Investigations include the influence of nanostructures morphology on total catalytic activity and specific activity based on pore accessibility as well as catalytic efficiency of nanostructures. Additionally, Thiele modulus equation was applied inside the simulations to support and provide a well understanding of the results.

PB-64 STRETCHABLE GLUCOSE BIOFUEL CELL WITH WIRINGS MADE OF MULTIWALL CARBON NANOTUBES pg. 600

Y. Fujimagari and Y. Nishioka
Nihon University, JAPAN

Recently, enzymatic biofuel cells that use glucose in a human body to produce electricity have been of special interest. In this study, we fabricated of a glucose biofuel cell with flexible and stretchable electrodes using MWCNT on a PDMS substrate. The maximum power of 0.46 μW at 96.5 mV that corresponds with the power density of 1.52 μW/cm2 was realized by introducing a glucose solution of 100 mM at room temperature.

PA-65 EVALUATION OF ALCOHOL DEHYDROGENASE AND ALDEHYDE DEHYDROGENASE ENZYMES AS BY-ENZYMATIC ANODES IN A MEMBRANELESS ETHANOL MICROFLUIDIC FUEL CELL pg. 606

J. Galindo-de-la-Rosa[1], N. Arjona[2], L.G. Arriaga[3], J. Ledesma-García[1], and M. Guerra-Balcázar[1]
[1]Autónoma de Querétaro, Querétaro, MEXICO, [2]Centro de Investigación y Desarrollo Tecnológico en Electroquímica, MEXICO, and [3]Centro de Investigación y Desarrollo Tecnológico en Electroquímica, MEXICO

We develop a membraneless enzymatic microfluidic fuel cell, which operates using ethanol as fuel. Alcohol dehydrogenase (DH) enzyme was selected as catalyst, and altogether with aldehyde DH, were used as bioanodes to enhance the ethanol electro-oxidation reaction. The study of the enzymatic array (combined, vertically and horizontally separated) resulted in the highest cell performance to date reported corresponding to the combined enzymatic electrode: 7.01 mW cm-2.

PB-66 Not published

INVESTIGATION OF SILICON-BASED AIR-BREATHING MICRO DIRECT METHANOL FUEL CELLS WITH DIFFERENT ANODE FLOW FIELDS

Y.J. Tuo, Y.F. Zhang, P. Zhang, and X.W. Liu
Harbin Institute of Technology, CHINA

We develop air-breathing μ-DMFCs with different anode flow fields to improve the cell performances. From the results obtained in this work, it can be concluded that the single-serpentine flow field structure parameters demonstrated in the μ-DMFC is also applicable to the greater DMFCs or the PEMFCs running on the anode.

Energy Management, System Integration & Energy Aware Systems

PA-67 A SOLAR CELL POWERED ADAPTIVE CHARGING CIRCUIT FOR CMOS INTEGRATED MICRO FUEL CELLS pg. 611

C. Moranz[1], H. Ghafarian[1], K. Ylli[2], and Y. Manoli[1,2]
[1]University of Freiburg, GERMANY and [2]Hahn-Schickard, GERMANY

In this work an integrated CMOS circuit is presented which automatically maximizes the charging current for rechargeable chip integrated micro fuel cell accumulator arrays. These accumulators comprise a fuel cell for delivering energy to systems and a hydrolysis cell for recharging the integrated hydrogen storage. The charging current supplied by small solar cells which are projected to be part of the chip package is continuously monitored and an interface circuit is adjusted in order to maximize the charging current.

PB-68 MICROWATT-POWERED, LOW-COST, PRINTED GRAPHENE OXIDE HUMIDITY SENSORS FOR DISTRIBUTED NETWORK APPLICATIONS pg. 616

A.P. Taylor[1] and L.F. Velásquez-García[2]
[1]Edwards Vacuum, USA and [2]Massachusetts Institute of Technology, USA

This paper reports the design, fabrication, and characterization of novel conductometric humidity sensors that employ an ultrathin film of graphene oxide (GO) nanoflakes as transducing element. The GO film is deposited at atmospheric pressure and at a temperature slightly above room temperature using a home-built electrospray printer. In dynamic humidity tests conducted at atmospheric pressure, the sensor tracks the response of a commercial sensor and reacts to changes in humidity in less than 500 ms. There is a quadratic dependence of the relative humidity on the sensor resistance for the studied relative humidity range between 3% and 63%, with more than a three-fold change in resistance over the range. The power consumption of the sensor is less than 30 μW while drawing 7 μA, and less than 15 μW while drawing 5 μA. Our devices are promising candidates for deployment in a distributed sensor network due to their low cost, small size, and low power consumption.

RF Energy Conversion

PA-69 DESIGN OF AN INDUCTIVE POWER TRANSFER SYSTEM WITH FLEXIBLE COILS FOR BODY-WORN APPLICATIONS pg. 621

L.R. Clare[1], S.G. Burrow[1], B.H. Stark[1], N.J. Grabham[2], and S.P. Beeby[2]
[1]University of Bristol, UK and [2]University of Southampton, UK

This paper describes the high level design of an IPT (Inductive Power Transfer) system for bodyworn electronics, and in doing so investigates the challenges for an IPT system that arise specifically inthis scenario. Principally, these are: highly variable coil coupling through time-varying miss-alignmentand coil separation; a requirement that one or both of the coils must be wearable and thus flexible andwashable; and proximity to the human body introducing limits on the maximum EM field.

PB-70 NEAR FIELD WIRELESS POWER TRANSFER USING CURVED RELAY RESONATORS FOR EXTENDED TRANSFER DISTANCE pg. 626

D. Zhu[1,3], L. Clare[2], B.H. Stark[2], and S.P. Beeby[1]
[1]University of Southampton, UK, [2]University of Bristol, UK, and [3]Coventry University, UK

This paper investigates performance of a near field wireless power transfer system that uses curved relay resonator coils to extend transfer distance. Near field wireless power transfer can transfer energy over a relatively short distance which is on the same order of dimensions of the coupled coils. Previous work has shown that the energy transfer distance can be increased using flat relay resonators. This research focuses on using the curved relay coils to extend energy transfer distance. The concept can be used in e-textiles applications.

Thrusters & Propulsion Systems

PA-71 A MEMS-BASED SOLID PROPELLANT MICROTHRUSTER ARRAY FOR SPACE AND MILITARY APPLICATIONS pg. 631

A. Chaalane[1,2], R. Chemam[1], M. Houabes[1], R. Yahiaoui[3], A. Metatla[4], B.Ouari[5], N. Metatla[6], D. Mahi[7], A. Dkhissi[8,9], and D. Esteve[10]
[1]University of Annaba, ALGERIA, [2]ESIEE-Paris, FRANCE, [3]FEMTO-ST, FRANCE, [4]University of Skikda, ALGERIA, [5]University of Tlemcen, ALGERIA, [6]University of Mila, ALGERIA, [7]University of Laghouat, ALGERIA, [8]Ghent University, BELGIUM, and [9]LAAS-CNRS, FRANCE

Since combustion is a simple way to achieve large quantities of energy from a small volume, we develop a MEMS based solid propellant microthruster array for small spacecraft and micro-air-vehicle applications. A large range of thrust force is obtained by varying chamber and nozzle geometry parameters in one step of DRIE. The use of polysilicon initiator deposited on thin membrane ensures a better heat transfer to the propellant and thus a short response time and low energy consumption. Experimental tests of ignition and combustion employing home made propellants are reported and discussed.

ADDITIONAL

Fluidic

PB-72 ON DESIGNING LOW PRESSURE LOSS WORKING SPACES FOR A PLANAR STIRLING MICROMACHINE pg. 636

M.-A. Hachey[1], É. Léveillé[1], L.G. Fréchette[1], and F. Formosa[2]
[1]Université de Sherbrooke, CANADA and [2]Université de Savoie Mont Blanc, FRANCE

Low pumping loss working spaces were developed for a novel planar free-piston Stirling microengine. The complexity of its flow geometry and its expected high operating frequency (kHz) are difficult to analyze. Thus, experimental research was favored. We describe the design process used to shape the exchangers and how static pressure was measured along a steady unidirectional flow and analytically treated to determine pumping loss. It was surmised to be an efficient way of immediately depicting pumping loss issues in both directions for what would be a transient pulsating flow.

Magnetic

PA-73 ENERGY HARVESTER FOR USE ON NEARBY CURRENT-CARRYING CONDUCTORS pg. 641

D.-S. Nguyen, Z. Wu, and R. White
University of California, Berkeley, USA

We design, model and test a new concept of electret-based energy harvester for use on nearby an electrical conductor carrying an AC current. The experimental results of an energy harvester, using external bias voltage, verify the working principle of the harvester. The RMS output voltage sharply increase when the current amplitude gets 36 A - 40 A and continuously increase when the bias voltage varying from 5V-67V.

PB-74 MITIGATION OF INTERFACIAL SILICIDE REACTIONS FOR ELECTROPLATED CoPt FILMS ON Si SUBSTRATES pg. 645

O.D. Oniku and D.P. Arnold
University of Florida, USA

We report the influence of film thickness on the material properties of electroplated CoPt permanent magnets. It was observed that the magnetic properties decrease substantially for films below 2 μm. This effect was determined to be a consequence of metal-silicide reactions at the substrate interface. Subsequently, a TiN diffusion-barrier layer was shown to inhibit the silicide reaction and thereby maintain strong properties in micron-thick layers.

Piezoelectric

PA-75 A HIGHLY EFFICIENT SIMULATION TECHNIQUE FOR PIEZOELECTRIC ENERGY HARVESTERS pg. 650

R. Ardito, A. Corigliano, and G. Gafforelli
Politecnico di Milano, ITALY

We present a new computational technique which is aimed at obtaining fast and accurate simulations of piezoelectric beams, used in inertial energy harvesting MEMS.A refined, yet simple, model is proposed with reference to the multi-physics problem of piezoelectric energy harvesting by means of laminate cantilevers. The main objective is to retain a simple structural model (Euler-Bernoulli beam), with the inclusion of effects connected to the actual three-dimensional shape of the device.

THE 15TH INTERNATIONAL CONFERENCE ON MICRO AND NANOTECHNOLOGY FOR POWER GENERATION AND ENERGY CONVERSION APPLICATIONS

PowerMEMS Boston 2015

MASSACHUSETTS

DECEMBER 1 - 4

Technical Digest

Conference Chairs:

Carol Livermore
Northeastern University

Luis Fernando Velásquez-García
Massachusetts Institute of Technology

SPONSORED BY

TRF

Northeastern University

Massachusetts Institute of Technology

The 15th International Conference on Micro and Nanotechnology for Power Generation and Energy Conversion Applications (PowerMEMS 2015)

Boston, Massachusetts USA
December 1 - 4, 2015

Conference Chairs:
Carol Livermore - Northeastern University, USA
Luis Fernando Velásquez-García - Massachusetts Institute of Technology, USA

Journal of Physics: Conference Series **660** (2015) 011001 doi:10.1088/1742-6596/660/1/011001

WELCOME

Greetings, and welcome to Boston, MA and PowerMEMS 2015 – the 15th International Conference on Micro and Nanotechnology for Power Generation and Energy Conversion Applications!

The objective of PowerMEMS 2015 is to catalyze innovation in micro- and nano-scale technologies for the energy domain. The scope of the meeting ranges from basic principles, to materials and fabrication, to devices and systems, to applications. The many applications of Power MEMS range from the harvesting, storage, conversion and conditioning of energy, to integrated systems that manage these processes, to actuation, pumping, and propulsion.

Our Conference aims to stimulate the exchange of insights and information, as well as the development of new ideas, in the Power MEMS field. Our goal is to allow the attendees to interact and network within our multidisciplinary community that includes professionals from many branches of science and engineering, as well as energy, policy, and entrepreneurial specialists interested in the commercialization of Power MEMS technologies.

Since the first PowerMEMS in Sendai, Japan in 2000, the Conference has grown in size, reputation, impact, and technical breadth. This continuing growth is evident in this year's technical program, which includes an increasing number of papers on nanomaterials, additive manufacturing for energy systems, actuators, energy storage, harvesting strategies and integrated energy harvesting systems, for example.

This year's technical program is highlighted by six plenary talks from prominent experts on piezoelectrics, robotic insects, thermoelectrics, photovoltaics, nanocomposite cathodes, and thermal energy conversion systems. The contributed program received a large number of abstract submissions this year, 169 in total. After careful review by the 34-member Technical Program Committee, a total of 135 papers were selected for presentation. The 60 contributed oral presentations are arranged in two parallel sessions. The 75 posters are arranged in a "two-in-one" poster session in which the poster session time is divided in two; half the posters will be presented during each half-session, allowing the poster presenters to also browse the posters during the poster session. Posters will remain up during the meeting, so please feel free to peruse them at your leisure. The Proceedings will be visible and accessible through IOP after conclusion of the Conference.

We hope to maximize interaction among participants and stimulate lively discussion through the technical program and social events. For the first time at the Conference, a great many sponsors are making their presence known at PowerMEMS through tabletop displays, which will catalyze engaging conversations between our industrial partners and the research community. Also, for the first time at PowerMEMS, we will host a Rump Session – an informal event in which Conference participants engage in honest and fun discussions of topics of relevance to the community while strengthening bonds within the community.

This meeting is made possible by many generous contributions of time, effort, and financial support. Thanks are due to the Technical Program Committee for their intensive efforts in reviewing abstract submissions, and to the International Steering Committee for their advice and support. We are grateful to Preferred Meeting Management Inc. for their many and key contributions to the management and organization of our Conference. Finally, we gratefully acknowledge the organizational and financial support provided for this meeting by the Transducers Research Foundation, Northeastern University, the Massachusetts Institute of Technology, and our exhibitors and other supporters. We hope that you find our Conference enjoyable and fruitful, and that you enjoy your stay in the Boston, MA area.

Sincerely,

Carol Livermore
Conference Co-Chair

Luis Fernando Velásquez-García
Conference Co-Chair

Conference Officials

Conference Co-Chairs

Carol Livermore .... *Northeastern University, USA*
Luis Fernando Velásquez-García .... *Massachusetts Institute of Technology, USA*

PowerMEMS School Chair

Hanseup Kim .... *University of Utah, USA*

Technical Program Committee

Miguel A. Aleman-Arce .... *Instituto Politécnico Nacional, MEXICO*
David Arnold .... *University of Florida, USA*
Skandar Basrour .... *Grenoble Alpes Université, FRANCE*
Philippe Basset .... *Université Paris-Est, FRANCE*
Mustafa Beyaz .... *Antalya International University, TURKEY*
Danick Briand .... *Ecole Polytechnique Fédérale de Lausanne (EPFL), SWITZERLAND*
Steve Burrow .... *University of Bristol, UK*
Jan Dziuban .... *Wroclaw University of Technology, POLAND*
Alper Erturk .... *Georgia Institute of Technology, USA*
Luc Fréchette .... *Université de Sherbrooke, CANADA*
Takayuki Fujita .... *University of Hyogo, JAPAN*
Einar Halvorsen .... *Buskerud and Vestfold University College, NORWAY*
Andrew Holmes .... *Imperial College London, UK*
Yoshihiro Kawahara .... *University of Tokyo, JAPAN*
Hiroki Kuwano .... *Tohoku University, JAPAN*
Amit Lal .... *Cornell University, USA*
Jeff Lang .... *Massachusetts Institute of Technology, USA*
Janet Ledesma-García .... *Universidad Autónoma de Querétaro, MEXICO*
Carol Livermore .... *Northeastern University, USA*
Yiannos Manoli .... *University of Freiburg, GERMANY*
Koji Miyazaki .... *Kyushu Institute of Technology, JAPAN*
Michael Renaud .... *Holst Centre-Imec, THE NETHERLANDS*
Paul Ronney .... *University of Southern California, USA*
Shad Roundy .... *University of Utah, USA*
Tomonori Seki .... *OMRON Corporation, JAPAN*
Herbert Shea .... *Ecole Polytechnique Fédérale de Lausanne (EPFL), SWITZERLAND*
Junichiro Shiomi .... *University of Tokyo, JAPAN*
Michael Short .... *Massachusetts Institute of Technology, USA*
Yuji Suzuki .... *University of Tokyo, JAPAN*
Shuhei Takahashi .... *Gifu University, JAPAN*
Luis Fernando Velásquez-García .... *Massachusetts Institute of Technology, USA*
Mike Waits .... *US Army Research Laboratory, USA*
Xiaohong Wang .... *Tsinghua University, CHINA*
Peter Woias .... *University of Freiburg, GERMANY*

ACKNOWLEDGEMENTS

We gratefully acknowledge, at the time of printing this publication, the financial contributors to the Conference from the following:

Massachusetts Institute of Technology

TRF Transducer Research Foundation

Northeastern University

American Elements
DJ DevCorp
Draper
Edwards Vacuum
Femtotools
Konishiyasu Co., Ltd.
MEMS Industry Group
MEMS Journal, Inc.
MEMS and Nanotechnology Exchange
MiQro Innovation Collaborative Center (C2MI)
Multi-Functional Integrated System Technology Center (MIST Center)
Piezo Systems, Inc.
TECNISCO, LTD.
Tousimis
ULVAC Technologies

Peer review statement

All papers published in this volume of *Journal of Physics: Conference Series* have been peer reviewed through processes administered by the proceedings Editors. Reviews were conducted by expert referees to the professional and scientific standards expected of a proceedings journal published by IOP Publishing.

Piezoelectric MEMS for energy harvesting

Isaku Kanno

Kobe University, 1-1 Rokkodai-cho, Nada-ku, Kobe 659-8501, Japan

E-mail : kanno@mech.kobe-u.ac.jp

Abstract. Recently, piezoelectric MEMS have been intensively investigated to create new functional microdevices, and some of them have already been commercialized such as MEMS gyrosensors or miropumps of inkjet printer head. Piezoelectric energy harvesting is considered to be one of the promising future applications of piezoelectric MEMS. In this report, we introduce the deposition of the piezoelectric PZT thin films as well as lead-free KNN thin films. We fabricated piezoelectric energy harvesters of PZT and KNN thin films deposited on stainless steel cantilevers and compared their power generation performance.

1. Introduction

Recently, multifunctionality of ferroelectric materials have attracted attention in the field of microelectromechanical systems (MEMS) because they can give new and superior sensing or actuating performances by integrating ferroelectric thin films into the simple micro-structures. Among a variety of functionality in ferroelectric materials, piezoelectricity is the most important properties for the sensors and actuators in MEMS because the piezoelectric materials have superior electromechanical coupling characteristics. In order to integrate piezoelectric materials into MEMS, the piezoelectric materials should be prepared in thin-film form to be microfabricated by photolithography. The most popular piezoelectric materials are $Pb(Zr,Ti)O_3$ (PZT), and the deposition and characterization of PZT thin films have been intensively studied. Recently, some of piezoelectric MEMS composed of PZT films have been developed as commercial products, such as micropumps of inkjet printer heads or MEMS gyro sensors. Vibration energy harvesters (EHs) is one of the promising applications in piezoelectric MEMS because piezoelectric thin films enable to generate relatively large electric power by making the simple micro- or millimeter-scale unimorph cantilevers. However, to develop the high energy efficient and reliable piezoelectric MEMS EHs, several technical issues still remain; deposition of the well-crystallized piezoelectric thin films, and furthermore, precise measurement of piezoelectric properties of thin films. Environmental impact of lead in PZT is also a serious problem for the practical use of the piezoelectric MEMS, and the lead-free piezoelectric ceramics including thin films have been investigated intensively.

2. Deposition of piezoelectric PZT thin films

In the piezoelectric MEMS, PZT-based ferroelectric thin films are the most popular materials used because of their large piezoelectric properties compared with non-ferroelectric thin films such as AlN or ZnO. The required specification of the deposition of the PZT thin films are 1) thickness more than 1 μm, 2) uniformity of the film-thickness on large Si wafers (> 6 inches), 3) low-cost and high deposition rate. Several methods have been studied such as chemical solution deposition (CSD; sol-gel deposition), chemical vapor deposition (CVD), pulsed laser deposition (PLD), and rf-sputtering. Those methods can produce well-crystallized PZT thin films with high piezoelectric coefficient, however, each of them has advantage and disadvantage, especially from the viewpoint of mass-production of piezoelectric MEMS. CVD is another candidate for PZT deposition. This method has advantage for the excellent step coverage which is strongly required for semiconductor memories (FeRAM). However, the source materials of PZT are relatively expensive so that it is not suitable for the deposition of micrometer-thick PZT films. PLD is also a popular method for high-quality PZT deposition. The main drawback of PLD is the deposition area with uniform thickness by the conventional PLD was as small as a square centimeter, however, the PLD system for mass production has recently developed.

CSD and rf-sputtering are the most popular deposition processes of the PZT thin films for MEMS applications. CSD method can produce PZT thin films without expensive deposition apparatus such as vacuum system. However, the thickness of the layer by single spin coating is typically as thin as ~0.1 μm, and the thick PZT films more than 1 μm needs multi-spin coating with each pre-baking process. Therefore, the repetition of spin coating and baking processes needs long time to accomplish the total deposition. On the other hand, rf-magnetron sputtering is a practical deposition method of PZT films on large Si wafers with high deposition rate more than 1 μm/h. Initial cost of the sputtering apparatus is expensive, however, the cost of the sputtering targets is lower than the other deposition methods. If the deposition condition is optimized, high-quality PZT thin films can be stably produced on a variety of substrates.

Recently lead-free piezoelectric materials have been intensively studied to mitigate environmental impact of the piezoelectric devices. A variety of lead-free piezoelectric materials such as $BaTiO_3$(BT), (Bi,Na)TiO_3(BNT), and $BiFeO_3$, (K,Na)NbO_3 (KNN) have been studied. Among these lead-free materials, KNN shows the highest piezoelectric properties and their Curie temperature is also comparable with those of PZT. For the lead-free piezoelectric MEMS, we have successfully prepared KNN thin films on Si substrates by rf-magnetron sputtering and demonstrated that the KNN thin films have large piezoelectric properties comparable with the PZT thin films [1].

3. Lead-free piezoelectric thin-film EHs of stainless-steel cantilevers

3.1. Fabrication process of piezoelectric thin-film EHs

Piezoelectric MEMS EHs usually have simple unimorph cantilever structures composed of piezoelectric PZT thin films on Si substrates. In order to maximize the vibration amplitude, the resonance frequency of the cantilevers should be adjusted to environmental vibrations. Considering the brittleness of Si-based unimorph cantilevers, the fatal damages such as cracking or breaking of the cantilever are inevitable under the continuous and unregulated vibration. Furthermore, the resonance frequency of conventional MEMS EHs is usually more than 1 kHz because of the small size of the resonators, although the frequency of environmental vibrations is generally less than 200 Hz.

In order to solve these issues, we fabricated piezoelectric thin-film EHs using thin metal cantilevers to improve the toughness of the

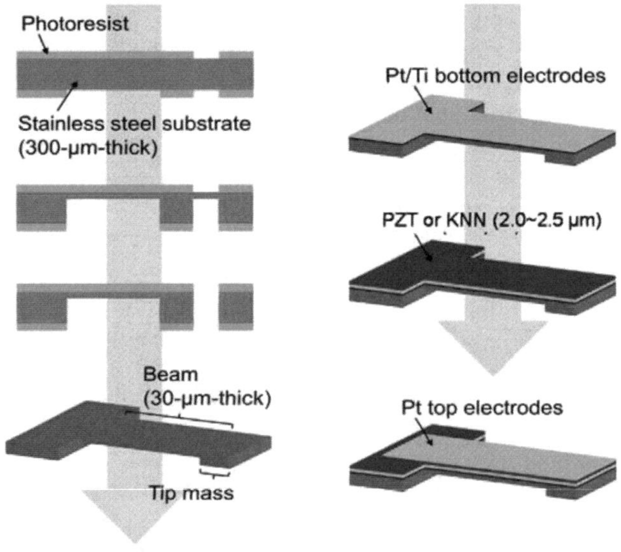

Figure 1. Fabrication process of piezoelectric MEMS energy harvesters of PZT- or KNN-thin films on stainless steel cantilevers.

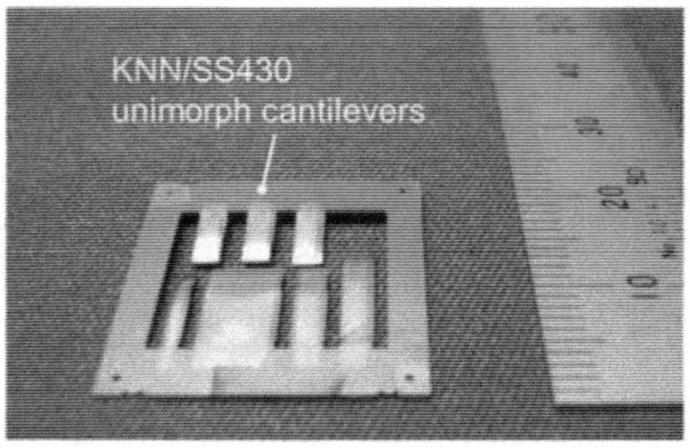

Figure 2. Image of lead-free KNN MEMS energy harvesters.

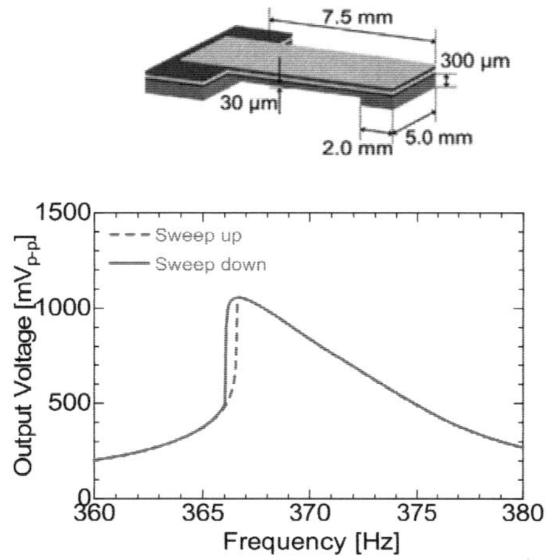

Figure 3. Schematic of cantilever and Frequency response of PZT MEMS energy harvesters.

resonators. The fracture toughness K_{Ic} of stainless steel is more than fifty times larger than that of Si. Therefore, the resonance frequency can be easily reduced even with the heavy tip mass on the thin cantilevers.

We deposited both of PZT and lead-free $(K,Na)NbO_3$ (KNN) thin films on the microfabricated metal cantilevers by rf-magnetron sputtering and compered their power generation performance as piezoelectric MEMS energy harvesters. The fabrication process is shown in Fig. 1. First, a photoresist pattern was formed on both sides of a 300-μm-thick stainless steel (SS430) plate to prepare the cantilever structure with tip mass and fixed end. After that, two-step spray etching process was conducted using ferric chloride solution and 30-μm-thick cantilevers were fabricated. The thickness of the tip mass and fixed end was 300 μm. Prior to PZT or KNN deposition, Pt/Ti bottom electrodes were deposited on the cantilevers. Successively, the PZT and KNN thin films were deposited by rf-magnetron sputtering onto the Pt/Ti-coated cantilevers. During the deposition of the piezoelectric thin films, the cantilevers were heated up to around 650°C, and a mixed gas Ar/O_2 was introduced into sputtering chamber. The thickness of the PZT and KNN films was 2.0~2.5 μm. Finally the Pt top electrodes were deposited through a shadow mask. Direct deposition of the piezoelectric thin films on the microfabricated cantilever could simplify the fabrication of the piezoelectric MEMS EHs. Fig. 2 shows images of the piezoelectric EHs with the KNN thin films on stainless steel cantilevers. We could obtain the flat unimorph cantilevers.

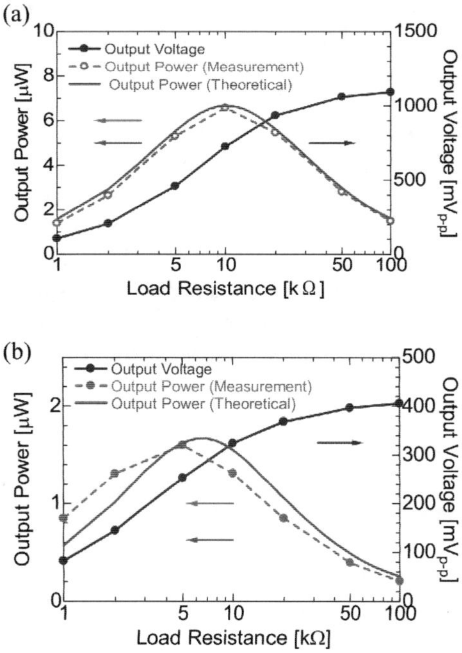

Figure 4. Output voltage and power as a function of load resistance: (a) PZT, and (b) KNN MEMS EHs.

3.2. Power generation characteristics

We examined the power generation performance of the PZT and KNN MEMS-EHs with the same dimensions of the cantilever (length : 7.5 mm, width : 5 mm) [2]. We mounted the EHs on the vibration exciter, and the top and bottom electrodes of the piezoelectric thin films were connected to a load resistance. Fig. 3 shows the frequency response of output voltage of PZT thin-film EHs in an open circuit state at the acceleration of 10 m/s^2. A clear peak of the output voltage appeared at a resonance frequency of 367 Hz, and the Q value was measured as 80 by the half-power bandwidth method. In addition, the non-linear vibration due to the softening spring effect was observed. This non-linear resonance is attributed to large deflection of the thin stainless steel cantilever at the resonance. Similar non-linear vibration was observed for KNN thin-film EHs.

Table 1. Comparison of the PZT- and KNN-EHs

	KNN	PZT
Dimensions [mm^2]	7.5 × 5.0	
Tip mass [mg]	25	
Relative dielectric constant (ε_r)	409	325
Dielectric loss (tanδ)	0.56	0.02
Peak output voltage [mV$_{p-p}$]	423	1114
Resonance frequency [Hz]	393	367
Q value [-]	90	80
K^2 [× 10^{-3}]	2.34	4.66
Optimum load resistance [kΩ]	5	10
Maximum output power [μW]	1.6	6.7
Power density [μW/mm^3]	0.41	1.7

Fig. 4 shows the output voltage and power of PZT- and KNN-EHs as a function of load resistance. The measurement was performed at the resonance frequency under the acceleration of 10 m/s^2. The averaged output power of the PZT-EHs reached 6.7 μW at the optimum load resistance of 10 kΩ, while those of the KNN-EHs were 1.6μW and 5 kΩ, respectively. Comparison of the KNN and PZT EHs is listed on Table 1. The generalized electromechanical coupling factor of K^2 was derived by the theoretical calculation [3]. The output power as well as K^2 of KNN-EHs were lower than those of the PZT-EHs in this experiment. Lower output power of KNN-EHs is attributed by large dielectric loss (tanδ) of the KNN thin films on stainless cantilevers. It means that the optimization of the deposition condition of the KNN thin films on the stainless steel would improve the output performance.

4. Summary

Piezoelectric thin films have been started to be used in the practical MEMS applications, and the progress of the deposition process and measurement technologies improve the piezoelectric properties and their stability of the thin films. EHs are one of the promising applications of piezoelectric MEMS which would enable small-size EHs with high efficiency. We fabricated piezoelectric thin-film EHs using stainless steel cantilevers and evaluated their power generation efficiency. PZT and lead-free KNN thin films were directly deposited on the microfabricated stainless-steel cantilevers by rf-sputtering, and we confirmed that the high energy efficiency as well as excellent toughness under the large vibration.

References

[1] I. Kanno, T. Ichida, K. Adachi, H. Kotera, K. Shibata, and T. Mishima, 2012 *Sens. Actuators A* **179** 132

[2] Y. Tsujiura, E. Suwa, F. Kurokawa, H. Hida, K. Suenaga, K. Shibata, and I. Kanno, 2013 *Jpn. J. Appl. Phys.* 52 09KD13

[3] M. Renaud, K. Karakaya, T. Sterken, P. Fiorini, C. Van Hoof, and R. Puers, 2008 *Sens. Actuators A*, **145–146** 380

Impulse-Excited Energy Harvester based on Potassium-Ion-Electret

H Ashizawa[1], H Mitsuya[1], K Ishibashi[1], T Ishikawa[1], H Fujita[2], G Hashiguchi[3] and H Toshiyoshi[2]

[1]SAGINOMIYA SEISAKUSHO, Inc., Saitama, Japan
[2]Institute of Industrial Science, the University of Tokyo, Tokyo, Japan
[3]Reserch Institute of Electronics, Shizuoka University, Shizuoka, Japan

E-mail: h-ashizawa@saginomiya.co.jp

Abstract. We have developed an energy harvester that is specifically desired for impulse acceleration of infrastructure vibrations such as sudden motion at railway bridges. The energy harvester based on potassium-ion-electret on the sidewalls of 1.8-μm-gap comb electrodes generated a 64 $\mu A_{p\text{-}p}$ current during low impulse acceleration, which was large enough to light a green LED.

1. Introduction

Trillion-sensor universe, where people annually consume more than a trillion sensors for comfortable life requires vast amount of energy wireless sensors operation and data transfer. Therefore, energy harvesting is a crucial component to avoid periodical change of battery as many as a trillion unit. Sensors for social infrastructures require all-day monitoring, which cannot be fulfilled by the solar cells, and hence it makes vibrational energy harvesters very attractive. There are three types of vibrational energy harvesters: piezoelectric, electromagnetic and electrostatic. Only the electrostatic type can efficiently harvest energy at low frequency vibrations (< 100 Hz) because the spring can be designed separately from the power generation regions.

Electret, a dielectric material storing semi-permanentl electric charges, has been attracting increasing attention for energy harvesters in recent years. High-performance electrostatic MEMS energy harvesters usually have a pair of opposing silicon electrodes with narrow gaps (< 2 μm) for high-output power. There are several methods to store electric charges in the electret films, such as corona discharge [1], ion implantation [2], electron beam injection [3], molecule ionization by X-ray radiation [4], and potassium ion-doping during thermal oxidation [5]. Amongst of these, electrets on the sidewall of narrow gaps can be made only by the potassium ion-doping Using this technique, we achieved high power generation during a low impulse acceleration due to extremely large force factor.

2. Concept and Device

2.1. Power generation method

The energy harvester device (illustrated in Figure 1) developed in this work has vertical comb-electrodes that were coated with potassium-ion-electrets. The harvester mechanism is consisted of

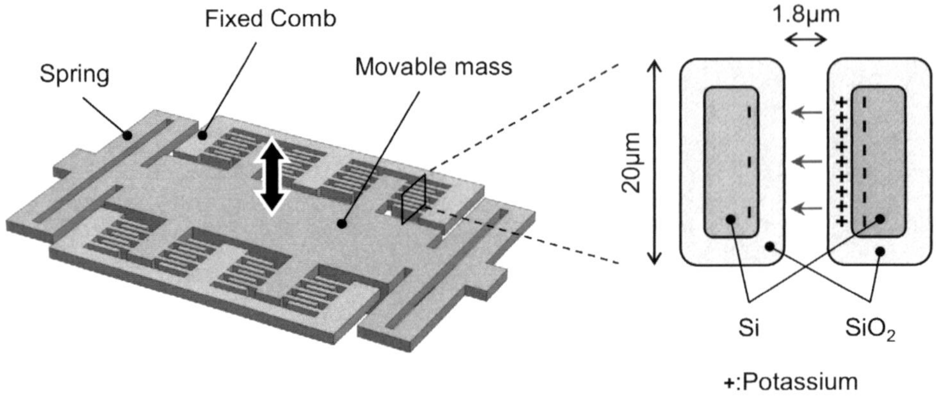

Figure 1. *Schematic of the electrostatic vibrational energy harvester.*

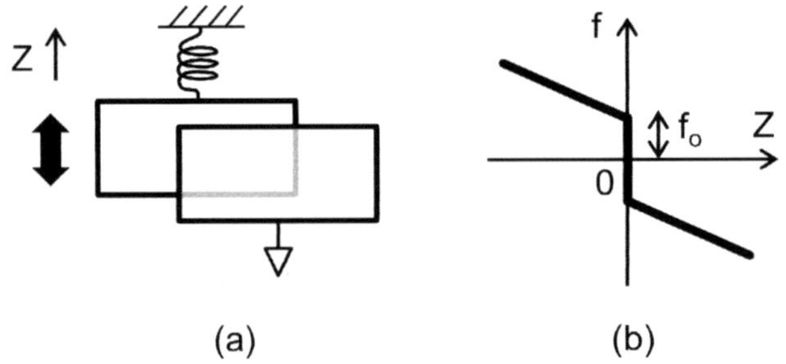

Figure 2. *Force (f) versus displacement of movable mass (Z).*

fixed combs and movable combs (with mass *m*) suspended springs on a silicon-on-insulator (SOI) wafer. The springs are made in the device layer, while the movable mass is in the handling layer to achieve relatively compliant springs and a large mass for efficient vertical motions at lower frequencies. A certain force level (f_o) is required to start the vertical motion (illustrated in Figure 2a) due to the strong electrostatic attraction that is built in between the electret combs at the initial position as schematically illustrated in the force curve in Figure 2b. Due to these characteristics, this device cannot use the mechanical resonance under small vibrations but it is more suitable for impulse acceleration.

In a case of power generation from a vibration source with stable frequency and phase, a resonance type energy harvester would have been more beneficial. However, we focused on the development of a impulse-type power generator because the environments surrounding the sensor network are rather rich in mechanical vibrations of random frequency and phase.

2.2. Fabrication and Device

The energy harvester was developed by by conventional photolithography and deep-RIE (DRIE) processes on an SOI wafer. Then, the whole device was thermally oxidized at 980°C in a wet environment by N_2-bubbling an aqueous solution of KOH to form a silicon oxide film resulted in a uniform inclusion of potassium ions in the silicon oxide film (Figure 3). During this process, the areas protected by the 100-nm-thick silicon-nitride (Si_3N_4) film by low-pressure chemical vapor deposition

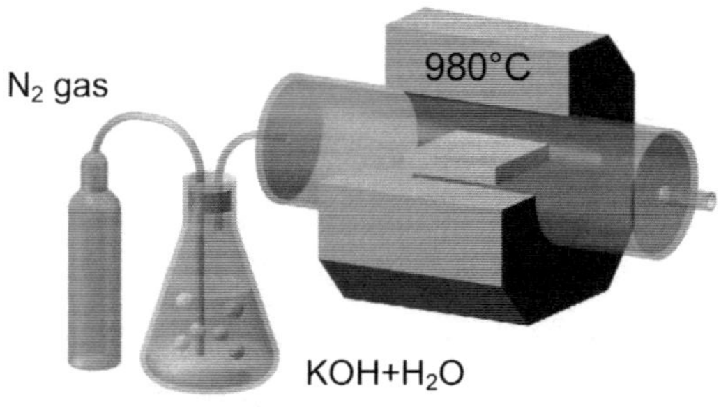

Figure 3. *Thermal oxidation.* by bubbling a stream of KOH solution.

Figure 4. *Bias-temperature procedure.*

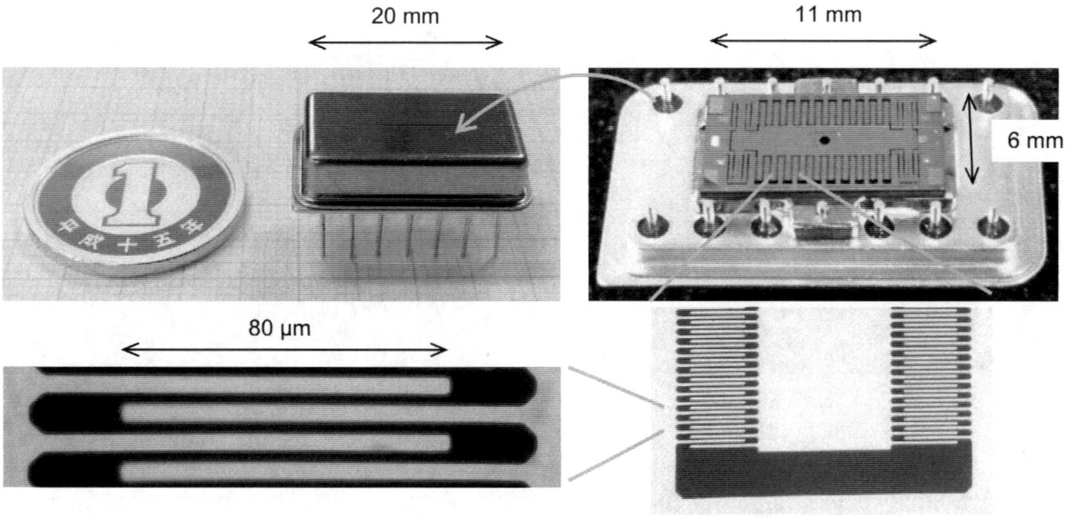

Figure 5. *Photographs of the developed energy harvester.*

(LPCVD) are free from the oxidation, and later used as electrical pads after stripping the nitride film. Finally, the included ions were electrically activated at approximately 500°C with 100-V-bias voltage to form 100-V- electret films between the comb electrodes (Figure. 4) [5]. The chip (11mm x 6mm) has 4000 pairs of 5-μm-width comb electrodes. The overlap between the coms is 80 μm with across a 1.8 μm gap.

3. Experimentals

For demonstration, the device was hermetically sealed in a nitrogen-filled package (12mm x 20mm, Figure 5). The packaged device was mechanically excited with consecutive impulses obtained by sliding a screw bolt on the package surface (Figure. 6). The output voltage at a 1-MΩ-load resistance and the short circuit current were 34.6 V_{p-p} and 64.2 μA_{p-p} respectively (Figure 7). The device provides considerably larger current than the previous devices reported elsewhere [6].

The output impedance was measured by the impulse acceleration using a voice coil. The output power peaked at a 0.6-MΩ-load resistance (Figure 9). Thanks to this very low output impedance and the large current, we could directly connect the harvester to a green LED and successfully light the diode brightly by sliding the bolt (Figure 8).

Typically, a power obtained from an energy harvester is once charged, and it is intermittently discharged. So a large output current is more important than just large output power. Therefore, our energy harvester based on potassium ion electret is very superior in practice.

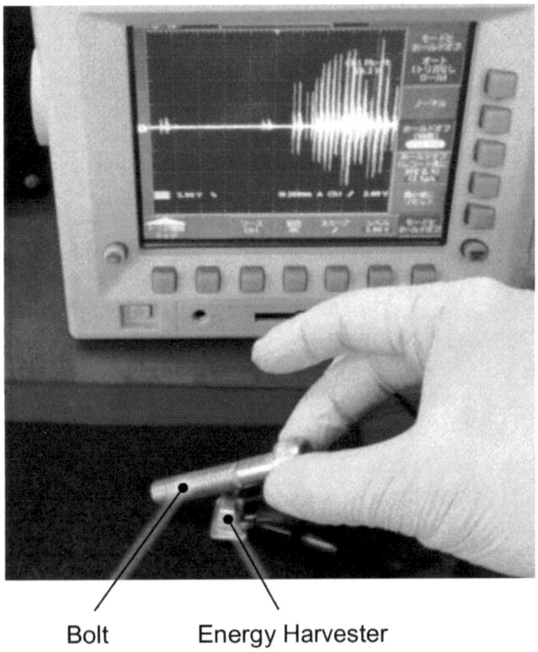

Figure 6. *Energy harvester device in operation.*

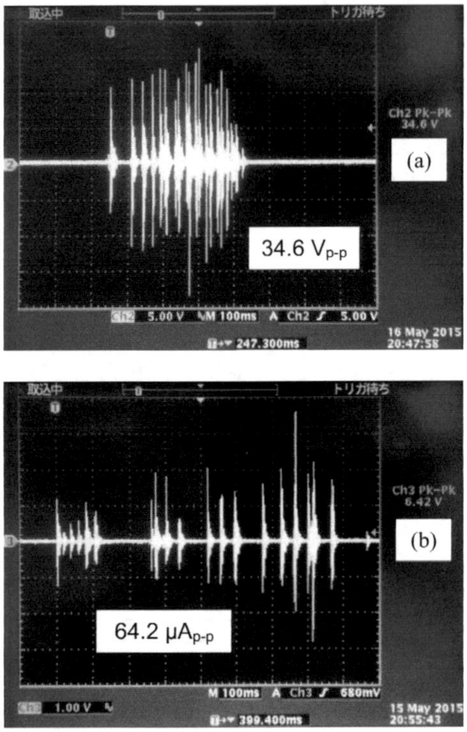

Figure 7. *Waveforms generated by impulses. (a) Output voltage of maximum 34.6 V_{p-p} for 1 MΩ load resistance. (b) Short current of 64.2 μA_{p-p}.*

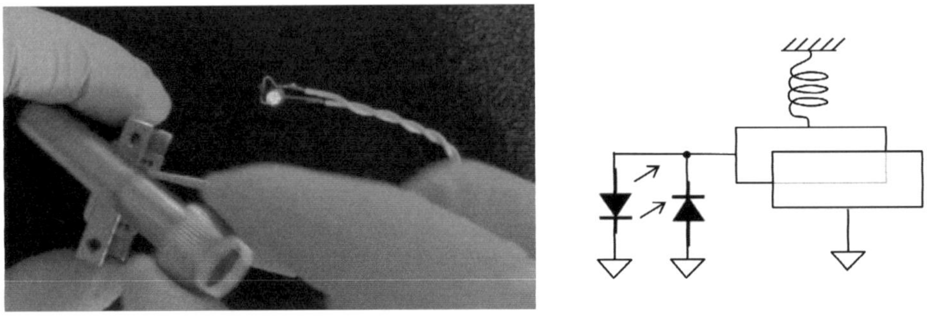

Figure 8. *A green LED directly driven by the energy harvester.*

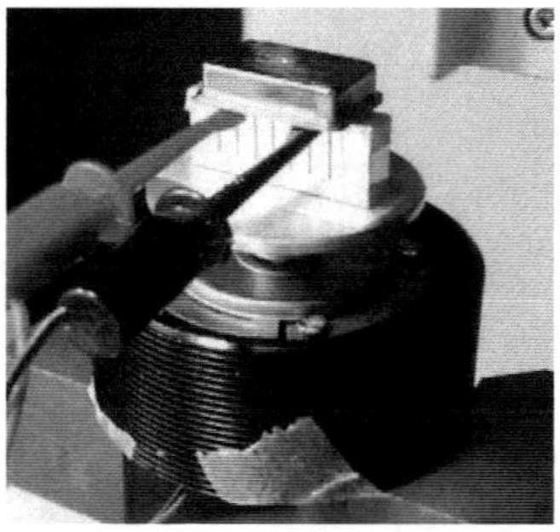

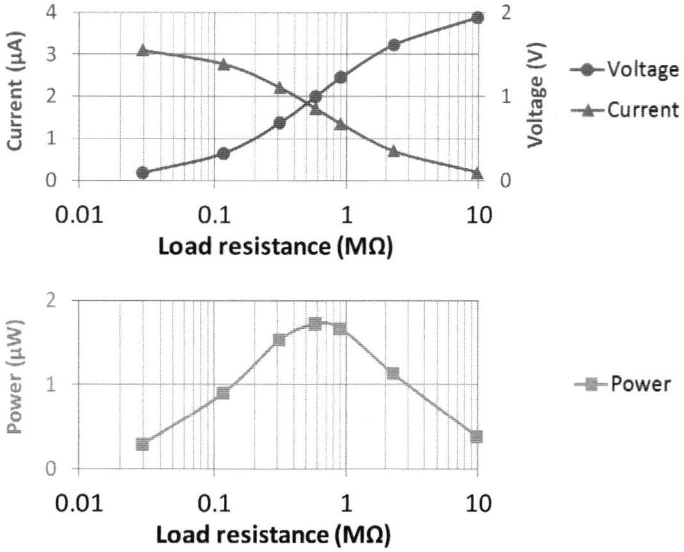

Figure 9. *Output voltage and current as a function of load-resistance (upper). Output power as a function of load-resistance (lower).*

4. Conclusion

In this study, we have developed an energy harvester based on potassium-ion-electret featuring (1) considerably large current (64.2 $\mu A_{p\text{-}p}$) and (2) very low output impedance (0.6 MΩ) owing to the superb performance of our potassium-ion-electret. The proposed technology allows us to develop independent energy source for sensor networks such as infrastructures that requires all-day monitoring.

5. Acknowledgments

This work is supported by the New Energy and Industrial Technology Development Organization (NEDO).

References

[1] Altafim R C A, Giacometti J M and Janiszewski J M 1991 *Proc. 7th International Symposium on Electrets (ISE 7)* (Berlin, Germany) 267-271

[2] Mescheder U, Urbanovic P, Müller B and Baborie S 2008 *Proc. PowerMEMS 2008+microEMS2008* (Sendai, Japan) 501-504

[3] Günther P 1992 *Sensors and Actuators* A **32** 357-360

[4] Hagiwara K, Honzumi M, Goto M, Tajima T, Yasuno Y, Kodama H, Kidokoro K, Kashiwagi K and Suzuki Y 2009 *Proc. PowerMEMS 2009* (Washington DC, USA) 173-176

[5] Sugiyama T, Aoyama M, Shibata Y, Suzuki M, Konno T, Ataka M, Fujita H and Hashiguchi G 2011 *Appl. Phys. Express.* **4** 114103-1-3

[6] Misawa K, Sugiyama T, Hashiguchi G, Fujita H and Toshiyoshi H 2015 *IEEE MEMS 2015* (Estoril, Portugal) 1071-1074

Low-Frequency MEMS Electrostatic Vibration Energy Harvester With Corona-Charged Vertical Electrets and Nonlinear Stoppers

Y Lu[1], F Cottone[1,2], S Boisseau[3], D Galayko[4], F Marty[1], P Basset[1]

[1]Université Paris-Est / ESYCOM / ESIEE Paris, France
[2]Now at University of Perugia, NIPS Laboratory, Department of Physics, Italy
[3]CEA, Leti, Minatec Campus, Grenoble, France
[4]UPMC-Sorbonne Universités / LIP6, CNRS, France

E-mail: yingxian.lu@esiee.fr

Abstract. This paper reports for the first time a MEMS electrostatic vibration energy harvester (e-VEH) with corona-charged vertical electrets on its electrodes. The bandwidth of the 1-cm^2 device is extended in low and high frequencies by nonlinear elastic stoppers. With a bias voltage of 46 V (electret@21 V + DC external source@25 V) between the electrodes, the RMS power of the device reaches 0.89 μW at 33 Hz and 6.6 μW at 428 Hz. The -3dB frequency band including the hysteresis is 223~432 Hz, the one excluding the hysteresis 88~166 Hz. We also demonstrate the charging of a 47 μF capacitor used for powering a wireless and autonomous temperature sensor node with a data transmission beyond 10 m at 868 MHz.

1. Introduction

It is widely admitted that low-frequency vibrations naturally exist in the environment and that they offer a potential resource of energy to power electronic devices or sensors. However, the efforts are always faced with the difficulty to capture low-frequency vibration energy with MEMS-based VEHs. Confined by the miniature size of the MEMS VEHs, the maximum displacement of the movable part is limited within the void space in the device, and the mass in the resonant structure is limited by its volume, which both limit the power of the device in low frequency. By using nonlinear soft springs, the operating frequency bandwidth is expanded at low frequencies [1], but the use of polymers in the device adds complexity to the fabrication process.

In this work we present a MEMS e-VEH with a large frequency bandwidth, effective from 50 Hz and obtained by the use of nonlinear external stoppers in silicon. In addition, although only soft x-ray and UV charging had been proposed as charging methods for MEMS vertical electrets [2], we demonstrate that it is also possible with a classical point-grid-plane triode Corona charging method.

2. Device Description and Fabrication

A simplified schematic of the prototype is shown in Figure 1. A moving silicon proof mass (400 μm-thick) is connected to fixed ends by linear serpentine springs. On both ends of the movable mass, stoppers standing against clamped-clamped elastic beams have been added [3]. At sufficient acceleration levels, the stoppers on the ends of the movable mass hit the elastic beams, which deforms them; this results in a low-frequency and a wideband response of the device.

The fabrication process is derived from [4]. The silicon layer is patterned with an Al mask using a DRIE process, and is anodicly bonded with a glass substrate. A layer of parylene is deposited all over the device. The parylene layer on the movable part is negatively charged with the corona charging setup while the parylene layer on the fixed electrodes is not charged.

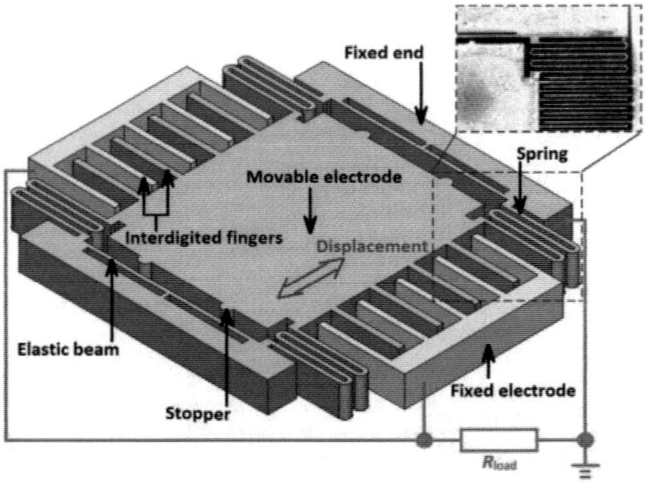

Figure 1. Simplified schematic of the silicon layer of the electrostatic energy harvester. The nonlinear spring system consists of the linear serpentine springs and the nonlinear elastic beams standing against stoppers.

3. Device Characterization

3.1. Built-in voltage measurement

In order to measure the built-in bias voltage (V_{bias}) provided by the electret, we propose to use the circuit shown in Figure 2(a). The saturation value of the voltage across the storage capacitor C_v is given by:

$$V_{store} = V_{bias}\left(1 - \frac{C_{min}}{C_{max}}\right)$$

where C_{min} and C_{max} are the minimum and the maximum values of the variable capacitance C_{var}. The saturation value of V_{store} is 11 V as given by the measurement shown in Figure 2(b), and the measurement of capacitance C_{var} indicates a minimum of 50 pF and a maximum of 105 pF respectively. Thus, the internal voltage provided by the electret is about V_{bias} = 21 V.

However, the device could stand for much higher bias voltage. In order to fully demonstrate the capability of the prototype, we also performed experiments with an additional DC bias at 25 V in series with the prototype to investigate its performance with a higher voltage bias (up to 46 V).

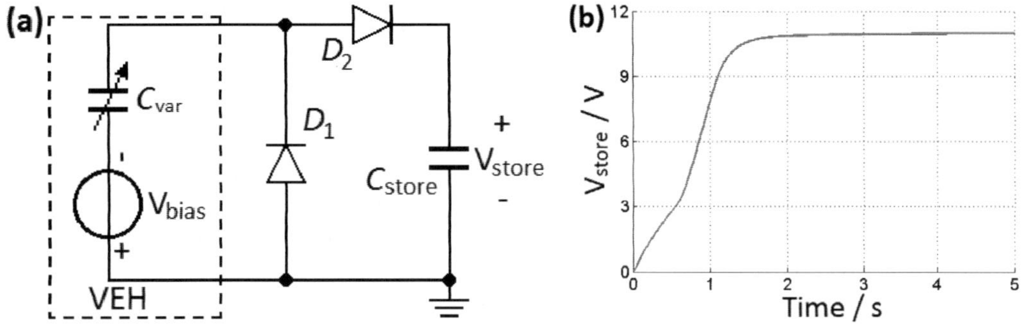

Figure 2 (a). Circuit for built-in voltage measurement of the electret layer, where C_{store}=1.1 nF; (b). Evolution of V_{store} at 2.0 g_{RMS}, 145 Hz.

3.2. Characterization of the performance in frequency domain

The device is connected in series with a load resistor of 6.65 MΩ and is subjected to power measurement with frequency sweeping. The device is firstly tested at 0.1 g_{RMS} for the natural frequency measurement f_0, which is observed at f_0 = 104 Hz.

Figure 3 gives the RMS power against the frequency at higher accelerations (0.5 and 2.0 g_{RMS}). Typical spring stiffening effects and hysteresis are observed, which is caused by the impacts between the stoppers and the nonlinear elastic beams.

With the bias voltage of 21 V (electret alone), the maximum output power at 0.5 g_{RMS} is 0.48 μW, and the one at 2.0 g_{RMS} reaches 1.3 μW. If the hysteresis is excluded, the maximum power at 0.5 g_{RMS} is 0.2 μW, while it reaches 0.4 μW at 2.0 g_{RMS}. An output power of at least 0.1 μW can be harvested within the 45~450 Hz frequency band including the hysteresis frequency span.

With a higher bias voltage of 46 V (electret at 21 V and external DC bias at 25 V), both the power and the bandwidth of the device are improved. The power reaches 1 μW at 0.5 g_{RMS} just below the frequency span of hysteresis, and reaches 2.1 μW at 161 Hz. At 2.0 g_{RMS}, the power reaches 6.6 μW at 428 Hz, and the power before reaching hysteresis is 2.4 μW. Within the frequency range of 29~432 Hz, the output power is superior to 0.6 μW. Finally, the -3 dB frequency bands with and without hysteresis are 223~432 Hz and 88~166 Hz, respectively.

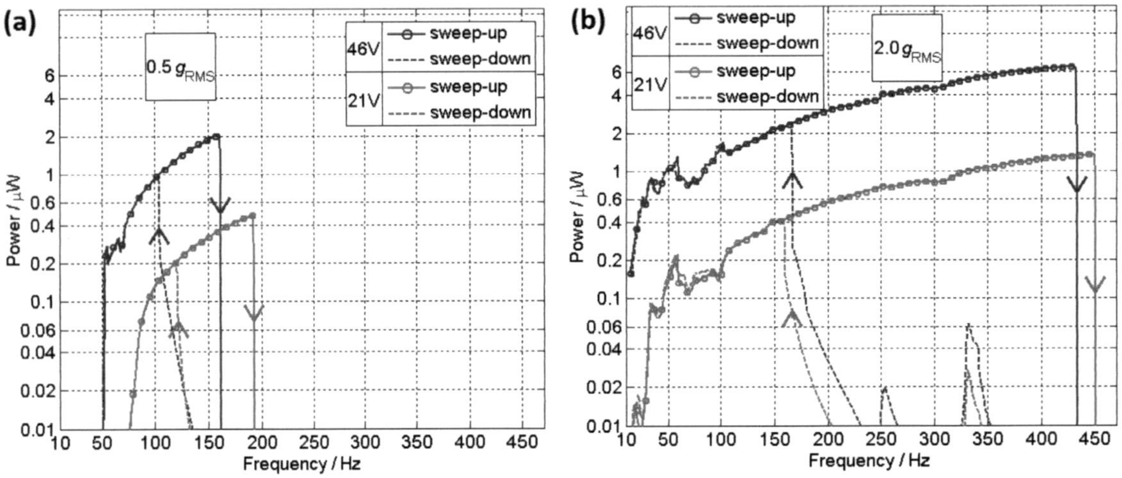

Figure 3. RMS output power vs. frequency for bias voltages of 21 V (electret alone) and 46 V (electret@21V + DC external bias@25V) at 0.5 g_{RMS} (a) and 2.0 g_{RMS} (b), R_{load}=6.65 MΩ.

3.3. Powering sensors and wireless data transmission

The power supply of a wireless temperature sensor node supplied by our energy harvester is shown in Figure 4(a). The output power of the device without any additional DC bias (electret alone – V_{bias} = 21 V) is rectified with a diode bridge, and stored in a 47-μF capacitor. This capacitor powers a low-power sensor node made of a temperature sensor, a MSP430 microcontroller and a RF chip working at 868 MHz (CC430). A low-power Schmitt trigger (Switch module) [5] is used to detect when enough energy has been stored in the capacitor to supply the wireless sensor node. The Switch module reads the voltage across C_s (U_{Cs}) and closes the switch to supply the wireless sensor node as soon as U_{Cs} reaches 3.8 V. The switch is re-opened when U_{Cb} falls below 2.8 V in order not to uselessly discharge C_s. The power consumption of the switch module is lower than 40 nA at 3 V. A photograph of the experimental setup is also shown in Figure 4(b).

Figure 4(c) shows the voltage evolution during a series of data transmission experiment. The storage capacitor C_s is charged with the e-VEH prototype working at 300 Hz with an acceleration of

2.0 g_{RMS}. The initial voltage across C_s rises from 0 V to 3.8 V in 7.2 min. This corresponds to a stored energy of 334 µJ; then, the average harvested power stored in C_s is 0.77 µW. U_{Cs} = 3.8 V is detected by the Schmitt trigger which closes the switch (PMOS transistor) to supply the wireless sensor node. The wireless sensor wakes up, performs its temperature measurement and sends the information wirelessly. This operation occurs twice each time the sensor node is turned on. The voltage drop for a measurement and a transmission up to 10 m is 0.7 V, which corresponds to an energy consumption of 102 µJ (155 µJ for two successive measurements and RF emissions). Finally, it takes about 2 min to recharge C_s from 2.8 V to 3.8 V, which corresponds to an average harvested power of 1.27 µW. Then, a new measurement cycle (2 measures and 2 data emissions) restarts… and so on.

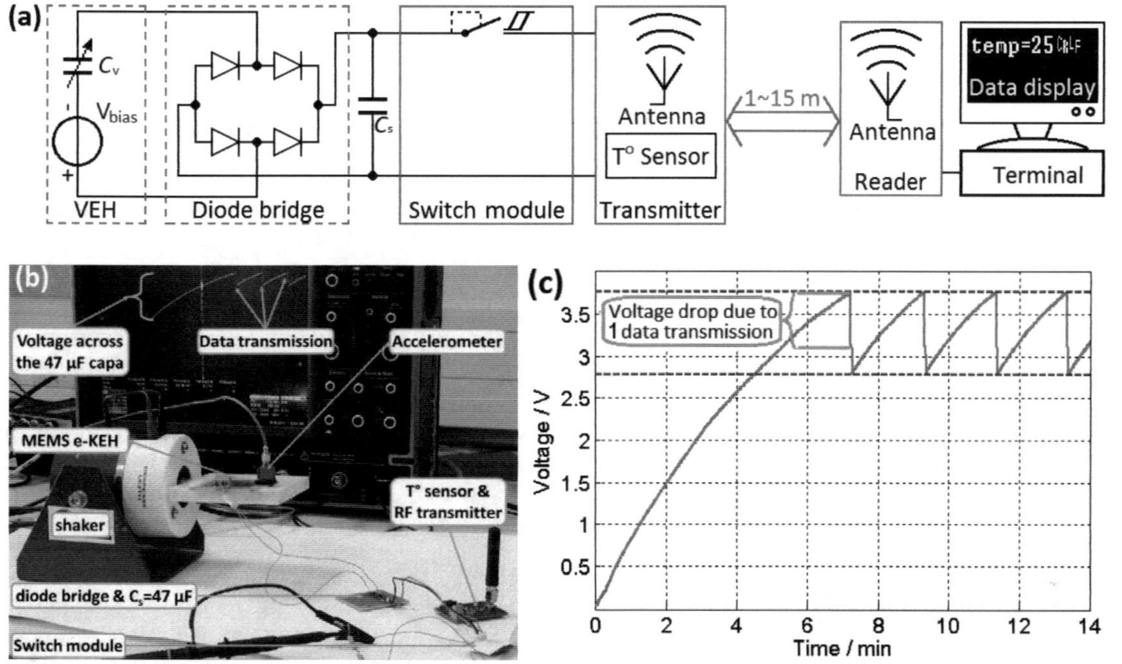

Figure 4 (a). Schematic and (b). photograph of the experiment of data transmission; (c). voltage evolution on C_s with time during the experiment.

4. Conclusion

We have reported on a low-frequency and large bandwidth 1-cm² silicon-based e-VEH prototype with Corona-charged vertical electrets and nonlinear external stoppers. The device has been connected to a diode-bridge capacitor circuit to supply a wireless sensor node based on a CC430 chip with an emission of a temperature measurement every 2 minutes.

Acknowledgments

This work is sponsored by the *SATT-IDFINNOV*.

References

[1] D. Miki, M. Honzumi, Y. Suzuki, et al, *Proc. of IEEE MEMS'10*, pp. 176-179, 2010
[2] Y. Suzuki, *IEEJ Trans. Electrical & Electronic Eng.* **6** pp 101-111, 2011
[3] F. Cottone, P. Basset, F. Marty, et al, *Proc. of IEEE MEMS'14,* pp. 385-388, 2014
[4] Basset P, Galayko D, Cottone F, et al. „*J. Micromech. & Microeng.*, **24** 035001, 2014
[5] STMicroelectronics/CEA Patent pending

Soft Electret Gel For Low Frequency Vibrational Energy Harvesters

H Mitsuya[1], S Ono[2], K Miwa[2], M Ataka[3], H Toshiyoshi[3], and H Fujita[3]

[1]SAGINOMIYA SEISAKUSHO, INC., 535 Sasai, Sayama, Saitama, 350-1395, Japan
[2]Central Research Institute of Electric Power Industry, 2-6-1 Nagasaka, Yokosuka, Kanagawa, 240-0196, Japan
[3]CIRMM, IIS, the University of Tokyo, 4-6-1 Komaba, Meguro, Tokyo, 153-8505, Japan

E-mail: hiro-mitsuya@saginomiya.co.jp

Abstract. A soft electret material was obtained by solidifying ionic liquid in a polymer network and immobilizing cations (+) on the surface. When a piece of soft electret gel is sandwiched between a pair of electrodes, a large amount of charge is induced in an electrical-double-layer capacitor (1.0-10 μF/cm^2) appearing at the interface of the electrode and the ionic liquid gel without an external voltage source. By retracting the electrode repeatedly, we obtained a current output of a few μA_{p-p}/cm^2 stably.

1. Introduction

There has been intensive investigation on MEMS vibrational energy harvesters due to their high applicability to IoT (Internet of Things) [1]. These inertial harvesters can be divided into three categories: piezoelectric, electromagnetic and electrostatic. Compared to the bulky electromagnetic type and limited-frequency-ranged piezoelectric type, electrostatic energy harvesters have several advantages. In particular, separating the spring and the power generation regions provides better design options for optimized performance. Although low frequency vibration (<50 Hz) is abundant in our daily life, e.g. motions by human body, infrastructure and transportation vehicles, most studies using electrostatic energy harvesters have been aiming at high frequency, e.g. vibrations generated by compressors or engines. The mismatch seems to stem from the fact that harvesters based on conventional MEMS technology are too fragile to achieve high-output power (>100 μW) at low frequency [2]. Therefore, to harvest energy at low frequencies while maintaining the advantages of electrostatic energy harvesters, it is necessary to develop a new method for capacitive power generation.

A triboelectric nanogenerator can obtain high-output power from low-frequency human motion [3]. However, compared to conventional devices, this type of device requires large external mechanical energy input such as pushing on a buzzer or stepping on the floor. In order to overcome this limitation, we employ a novel energy harvesting technique that changes the contact area between the soft electret gel having immobilized cations on the surface and the electrode. In this study, we demonstrated contact/retract operation of the electrode as a verification of our new power generation principle. This work improves the harvester based on the electrical double layer of ionic liquid reported at PowerMEMS 2014 [4].

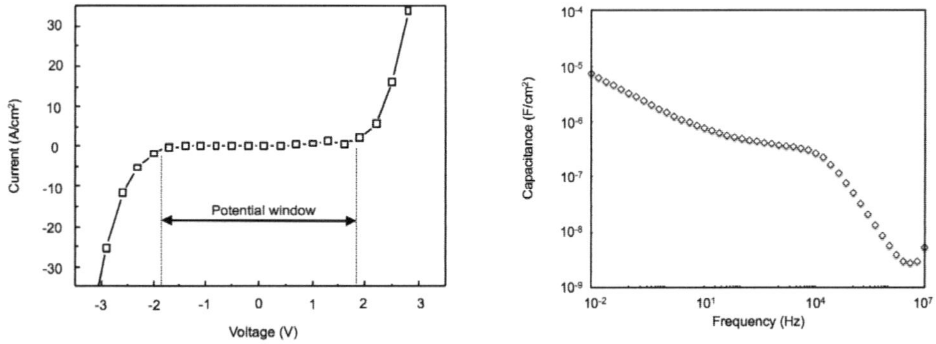

Figure 1. (a) The potential window of ionic liquid. (b) The capacitance of ionic liquid vs frequency.

2. Concept

2.1. Characteristics of the ionic liquid

Ionic liquids are composed of cations (+) and anions (-) with no other diluting solvent, and a wide variety of ionic pairs can be combined as desired. These ionic liquids have various unique characteristics, e.g. extremely low vapor pressure, resistance to high temperature, and the formation of an electrical double layer when a voltage applied across the material. Within these characteristics, we focused on the formation of an electrical double layer within the applied voltage range called the potential window (**Figure 1a**). An ionic liquid works as an insulator in the potential window, as it forms a 1 nm-thick electrical double layer on the interface between the ionic liquid and the electrode. Thus, it is capable of generating quite high capacitances in the order of 10 $\mu F/cm^2$ (**Figure 1b**).

2.2. Principle of power generation

The capacitance is utilized as variable capacitor whose contact area changes as the electrode is moves. In this way output current was obtained with the motion; this result was already reported in PowerMEMS 2014. However, this method has two primary problems: (1) Electrostatic attraction between the electrode and the ionic liquid prevented the change in contact area that is necessary for power generation and (2) external voltage source is needed to form the electrical double layer. Therefore, this work is the further extension of the previous study on utilizing the electrical double layer of ionic liquid while mitigating the aforementioned limitations.

Soft electret material was obtained by solidifying ionic liquid and immobilizing cations on the surface. When a piece of the soft electret gel is sandwiched between a pair of electrodes, large amount of charge are induced in an electrical-double-layer capacitor appearing at the interface of the electrode and the ionic liquid gel. By retracting the electrode repeatedly, we obtained the output-current (**Figure 2**).

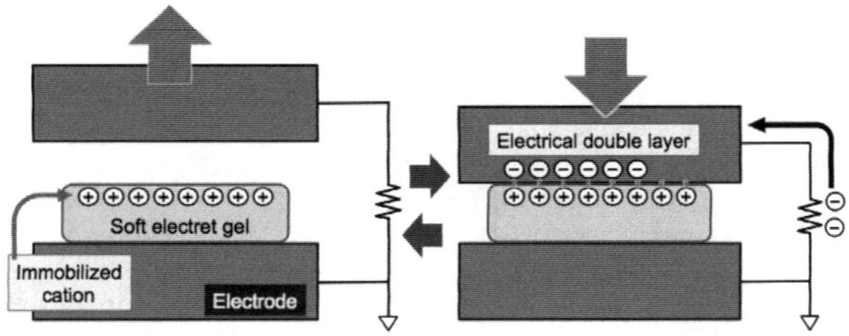

Figure 2. Schematic of principle of power generation by soft electret gel: Electrical-double-layer capacitance appearing at the interface of the electrode and the soft electret gel.

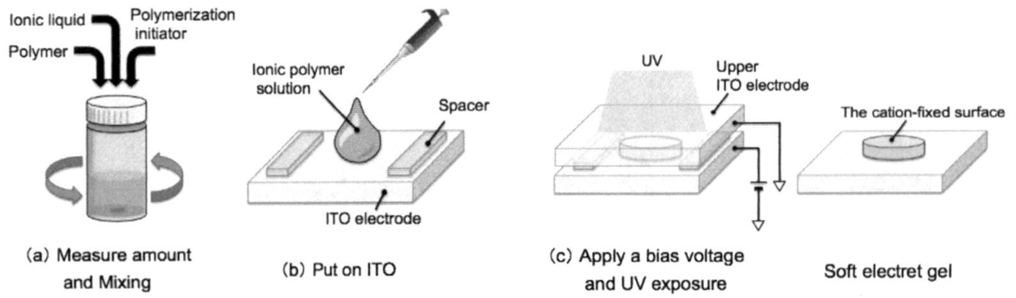

Figure 3. Schematic view of the soft electret gel manufacturing method: Cations of ionic liquid and fluid polymer have the same polymerizable functional group

3. Experimental section

3.1. Materials and manufacturing method

The soft electret gel consists of three materials: a base material, an ionic liquid and an initiator (**Figure 3a**). The base material is a fluid polymer with a polymerizable functional group. Accordingly, the ionic liquid has a cation with the same functional group for binding to the polymer. The initiator allows polymerization after mixing the appropriate amounts of constituent materials and exposing to UV light. First, we mixed and put the ionic polymer solution between a pair of transparent ITO (Indium Tin Oxide) electrodes with spacers to define the height of the gel (**Figure 3b**). We then applied a bias voltage within the solution's potential window (2 V DC) in order to form an electrical double layer. Finally, we exposed the sample to UV light (total about 45 min) for immobilizing cation ions on the surface of the gel (**Figure 3c**) while keeping the bias voltage.

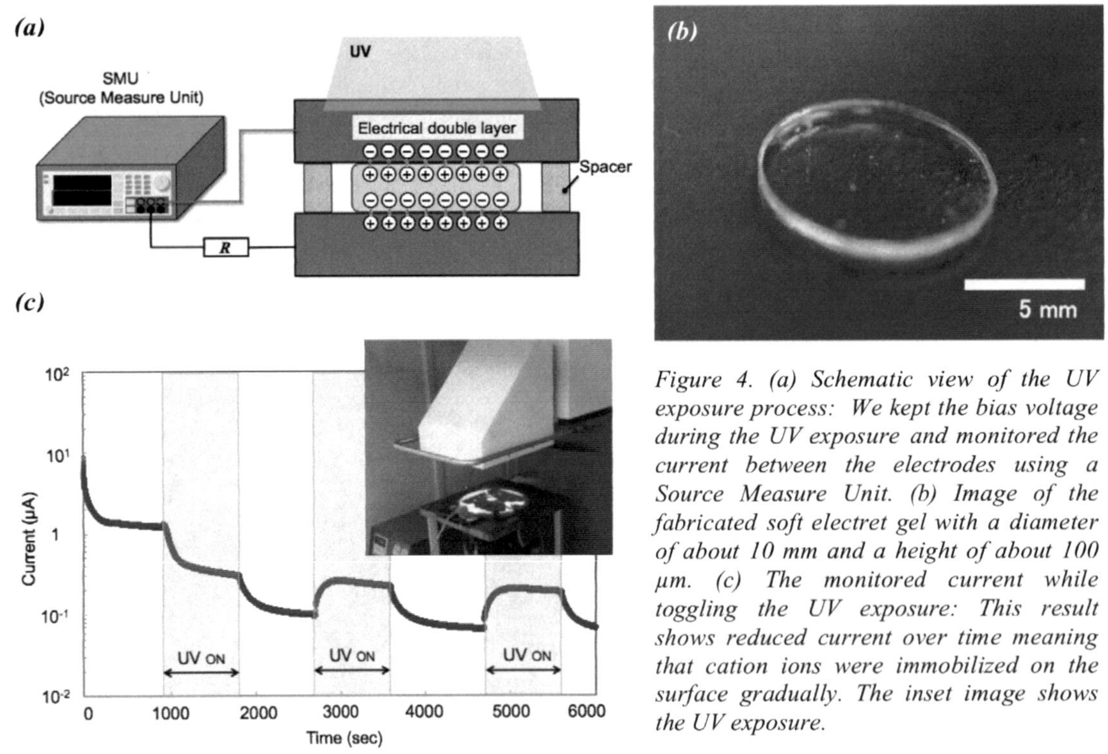

Figure 4. (a) Schematic view of the UV exposure process: We kept the bias voltage during the UV exposure and monitored the current between the electrodes using a Source Measure Unit. (b) Image of the fabricated soft electret gel with a diameter of about 10 mm and a height of about 100 µm. (c) The monitored current while toggling the UV exposure: This result shows reduced current over time meaning that cation ions were immobilized on the surface gradually. The inset image shows the UV exposure.

3.2. Immobilized cations

Through these processes, we monitored the current between the electrodes using a Source Measure Unit (SMU; Keysight B2900A, **Figure 4a**). **Figure 4b** shows the monitored current during cycles with and without UV exposure. This curve shows that the current reduced over time settling down to 0.1 $\mu A/cm^2$, which means that cations were immobilized on the surface gradually. The soft electret gel was fabricated with a diameter of 10 mm and a height of 100 μm (**Figure 4c**).

3.3. Power generation principle verification

We prepared a pair of ITO electrodes and placed a sample of the soft electret gel between them (**Figure 5a**). Then, we used a Digital Multi Meter (DMM; Agilent 34410A) and LabVIEW-setup to monitor the output current between the electrodes (**Figure 5b**). The output current was measured using the 10 MΩ setting of DMM. When the gel surface with immobilized cations touched the ITO electrode, we obtained an output current.

4. Result and discussion

The output current peaked when we touched the cations-fixed surface with the electrode (**Figure 5c**). We measured up to 2 $\mu A_{p\text{-}p}/cm^2$ from the soft electret gel by using a DMM and LabVIEW-setup.

Compared to previous study (PowerMEMS 2014), we were able to achieve an equivalent output current of 2 $\mu A_{p\text{-}p}/cm^2$ without using any external bias voltage. This means the soft electret gel works as an alternative to the external power supply. In this study, we fabricated a relatively stiff sheet-like gel of about 10mm diameter and about 100 μm thickness. The shape, dimension and stiffness of the gel can be modified as desired allowing for optimization of the output current from a device.

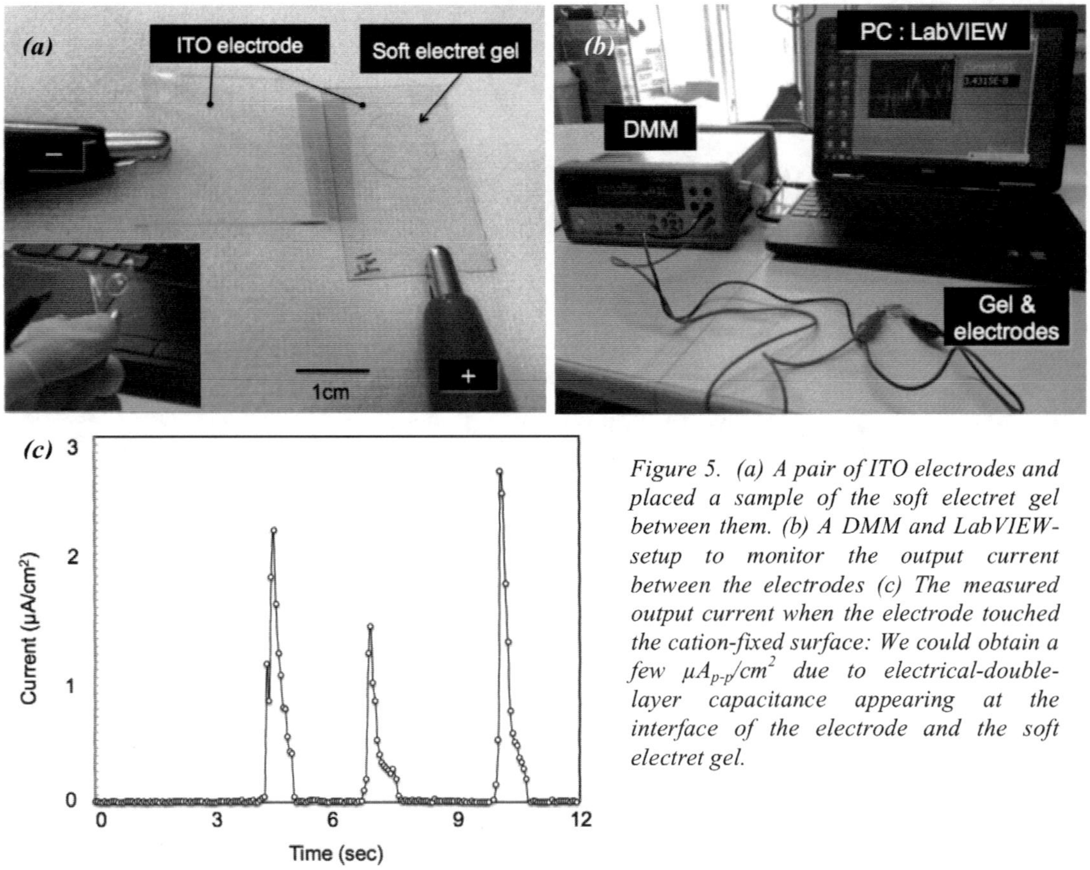

Figure 5. (a) A pair of ITO electrodes and placed a sample of the soft electret gel between them. (b) A DMM and LabVIEW-setup to monitor the output current between the electrodes (c) The measured output current when the electrode touched the cation-fixed surface: We could obtain a few $\mu A_{p\text{-}p}/cm^2$ due to electrical-double-layer capacitance appearing at the interface of the electrode and the soft electret gel.

5. Conclusion

In this study, we have developed a soft electret gel with immobilized cations on the surface. This technique provides robust devices by minimizing fragile regions. Unlike harvesters obtained by the conventional MEMS technology, the soft electret gel method shows superior performance at the low frequency range with low mechanical input power due to the characteristics of the ionic liquid based insulating material between the electrodes. Moreover, eliminating the external bias voltage to form an electrical double layer for high capacitance is crucial for low power applications. As a result, the proposed technology leads to a robust energy harvester at the low frequency range for low power applications such as wearable devices based on human motion. Currently, we are developing a MEMS energy harvesting device using this novel electret gel.

6. Acknowledgements

This work is partly supported by the New Energy and Industrial Technology Development Organization (NEDO).

References

[1] Atzori L , Iera A, and Morabito G, 2010 *Comput. Networks*, **54** 15 2787–2805
[2] Shen D, Park J H, Ajitsaria J, Choe S Y, Wikle H C and Kim D J 2008 *J. Micromech. Microeng.* **1** 055017-24
[3] Meng B, Tang W, Too Z, Zhang X, Han M, Liu M and Zhang H, 2013 *Energy Environ. Sci.* **6** 3235
[4] Yamada S, Mitsuya H, and Fujita H, 2014 *Proc. PowerMEMS 2014* (Hyougo, Japan) 55

Hybrid electromagnetic and electrostatic micromachined suspension with adjustable dynamics

K Poletkin[1], Z Lu[1], U Wallrabe[1] and V Badilita[2]

[1] Laboratory for Microactuators, Department of Microsystems Engineering - IMTEK, University of Freiburg, Freiburg im Breisgau D-79110, Germany

[2] Institute of Microstructure Technology, Karlsruhe Institute of Technology, Eggenstein-Leopoldshafen D-76344, Germany

E-mail: vlad.badilita@kit.edu, k.poletkin@gmail.com

Abstract. This paper introduces a novel design for a hybrid micromachined contactless suspension, whose operation is based on combining electromagnetic inductive and electrostatic actuation. Wirebonded microcoils provide the electromagnetic inductive actuation, while electrodes patterned on a Si wafer provide electrostatic control. The coil structure and the electrode structure are independently designed and fabricated, and are finally assembled into one device by flip-chip bonding. We demonstrate vertical linear positioning of an aluminium disk-shaped proof mass in a range from 30 to 200 μm based on the coil structure. The electrode structure is employed to dynamically adjust the stiffness components during the operation of the suspension, to control the tilting in a range from $\pm 1^\circ$ to $\pm 4^\circ$, as well as to control the oscillation about the vertical axis with a displacement of 37° at about 1.5 Hz frequency.

1. Introduction

This work is continuing the research efforts in our group in the field of micromachined inductive suspensions [1–3] and relates to a more general interest in the micro-engineering community to develop miniaturised actuators such as micro-gyroscopes and micro-motors, by completely eliminating the physical contact between the moving parts of these devices. Many of these devices where the technological solution involves some sort of mechanical contact such as spring structures, hinges, ball bearings, face a series of drawbacks: limited displacements, decrease in sensitivity or in the delivered output force. In order to overcome these limitations, levitation has been employed as a solution to eliminate mechanical contact, therefore to completely eliminate friction and wear in miniaturised sensors and actuators in a series of applications such as as micro-inertial sensors [4–7], frictionless micro-bearings [1, 8], bistable switches [9], micro-accelerator [10], linear-micro-actuator [11], and nano-force sensor [12].

Moreover, the combination of electromagnetics and electrostatics opens new avenues in terms of device capabilities and performance. If levitation and rotation of micro-objects have been demonstrated in a relative straightforward manner only exploiting electromagnetic levitation, by integrating electrodes for electrostatic actuation in the same device, Kraft et al. [10] have achieved an additional feature, i.e., accelerating micro-objects along a predefined path. Improved energy performance of a contactless suspension based on combining diamagnetic and inductive principles has been demonstrated as opposed to employing diamagnetic or inductive suspension separately [13].

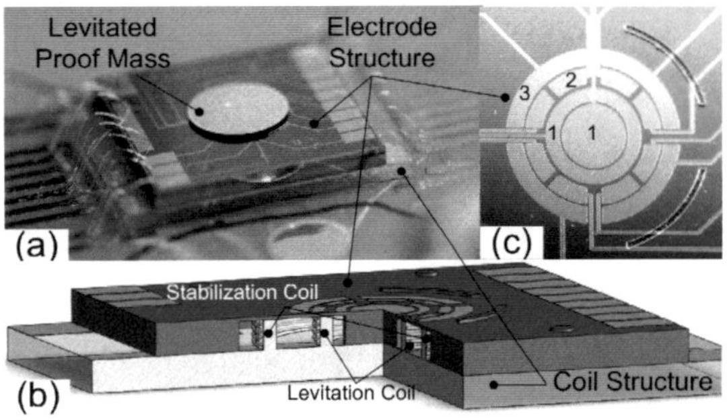

Figure 1. The hybrid suspension: a) fully assembled device with the Al proof mass during operation; b) 3D schematic of the device identifying the coil and the electrode structure; c) top view of the electrode structure identifying each set of electrodes: 1-suspending, 2-rotating and 3-tilting electrodes.

We have previously reported micromachined inductive suspensions based on electromagnetic levitation using 3D microcoils obtained by means of an automatic wirebonder. We have extensively characterised the performance of these suspensions [2] and we have developed an analytical model to predict the behaviour as a function of design parameters and input values [1]. In this work, in addition to electromagnetic induction, we employ electrostatics by integrating a series of electrodes in the suspension structure in order to have direct control on the stiffness components of the device. Significantly increased operational capabilities, i.e., linear and angular positioning with adjustable dynamics is also demonstrated herewith.

2. The hybrid suspension

2.1. Design and fabrication

The device reported here combines electromagnetic induction and electrostatic actuation and a comprehensive visual description is given in Figure 1. For each actuation principle a dedicated structure is designed, namely a coil structure for electromagnetic induction actuation, and an electrode structure for the electrostatic actuation. The coil structure is similar to those reported previously [14] and consists of two coaxial solenoidal microcoils obtained by automatic wirebonding. These coils are wound around 450 μm tall SU-8 structures defined by UV photolithography. Electrical contacting of the coils is realised via CrAu pads defined on the Pyrex substrate by metallisation, UV patterning and wet etching. In this work the diameters of the two coils are: 2000 μm and 3800 μm. As explained in the next section, the role of the inner coil is to provide levitation of the proof mass, while the role of the outer coil is to provide stabilisation of the proof mass during the levitation. In this work the inner (levitation) coil has 20 windings and the outer (stabilisation) coil has 12 windings. The wire used here is 25 μm insulated gold wire.

The electrode structure is fabricated on a Si wafer by sputtering CrAu (20/150nm), followed by UV photolithography and wet etching. Subsequently, the substrate wafer is etched by deep reactive ion etching (DRIE) from the backside in order to accommodate the coil structure. In the final stage, the electrodes are supported by a silicon membrane with a thickness of 40 μm. The function of these electrodes is to control the vertical and angular actuation of the proof mass.

In the final fabrication step the electrode structure is mounted on top of the coil structure and the final device is assembled by flip-chip bonding having the following dimensions: 9.4 mm x 7.4 mm x 1 mm. For this work we have used a disk-shaped aluminium proof mass with a diameter of 3.2 mm. Throughout this work we will refer to the very recent results obtained with a disk-shaped aluminium proof mass with a diameter of 2.8 mm [3] and we will place both data sets in the context of the theoretical model developed and reported in [1]. In this way we demonstrate that the experimental data is confirming our theoretically-generated stability map presented in [1].

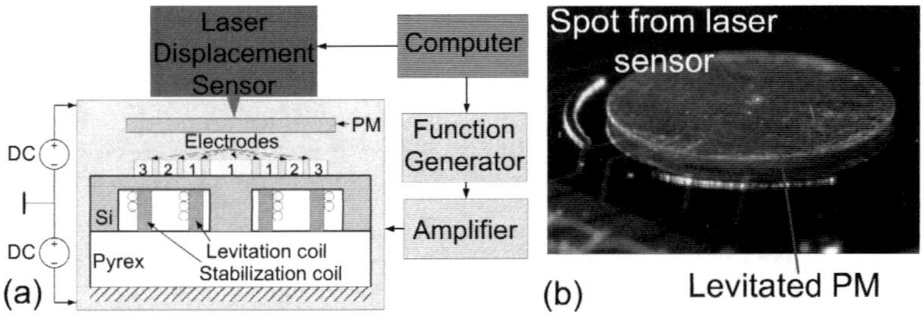

Figure 2. The experimental setup: a) schematic and measurement chain; b) Snapshot of the levitating proof mass during the experiment.

2.2. Operation of the device

The two coaxial solenoidal coils are biased through a current amplifier (LCF A093R) with square wave AC signals having a phase-shift of 180°, thus providing stable levitation of a circular conductive proof mass. The typical rms current values used in this paper range between 100 and 130 mA for a frequency of 9 MHz, whereas the levitation height varies from 30 to 180 μm, having the top of the electrode structure as reference. A function generator (Arbstudio 1104D) was used to modulate the amplitude and the frequency of the coil currents. A schematic of the experimental setup together with the measurement chain is presented in Figure 2.

The ensemble of the electrode structure fulfils several functions: modulates the vertical actuation, controls the tilting about two horizontal axes, as well as the oscillation about the vertical axis. When central electrodes ("1" - in Figure 1c and Figure 2a) are energised, the levitation height and the dynamics of the proof mass are modulated while maintaining the same amount of current in the coils. The set of electrodes "2" operate according to the variable capacitance principle and determine the angular oscillation of the proof mass when a voltage is applied. We control the tilt of the proof mass with respect to two orthogonal axis in the horizontal plane by biasing the set of electrodes "3".

A laser distance sensor (LK-G32) with a resolution of 10 nm was used to measure the linear displacements of the proof mass. The red dot visible in Figure 2b represents the spot from the laser distance sensor.

3. Interpretation of experimental results

The dynamics of the suspension is governed by the stiffness components. For a complete characterisation of the suspension capabilities, the stiffness components must be measured. To this end, the linear displacements have been recorded simultaneous with the voltages applied on the respective electrodes.

Figure 3a shows the vertical displacement of the proof mass as a function of the applied voltage, along with the vertical force. The electrostatic force between two plates that are oppositely charged is given by the product between the field produced by one of the plates and the charge on the other. In terms of actual physical dimensions and applied voltage, the electrostatic force is given by:

$$F = \epsilon_0 \epsilon_r \cdot \frac{A}{4} \cdot \frac{V^2}{h^2} \quad (1)$$

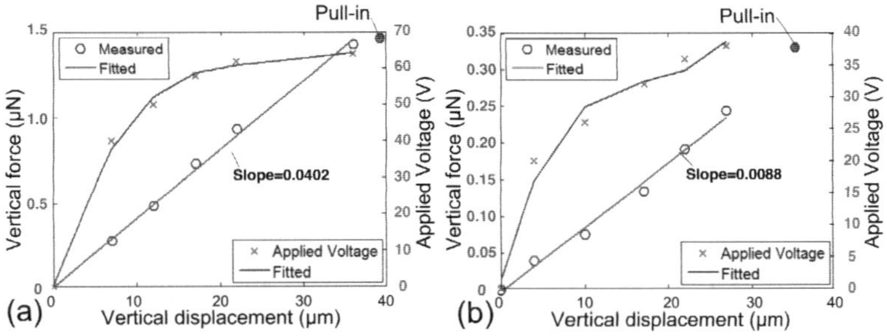

Figure 3. Experimental linear displacements and voltage values for the evaluation of stiffness components: (a) for central electrodes "1"; (b) for outer electrodes "3".

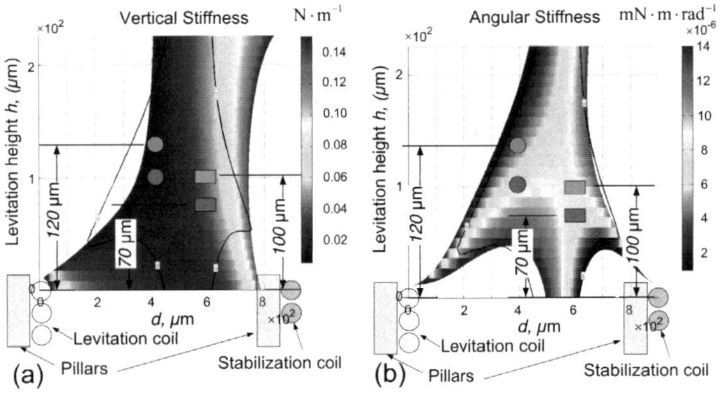

Figure 4. Mapping of the distribution of stiffness components within the stability domain (according to the model in [1]): (a) vertical stiffness; (b) angular stiffness. Circular/rectangular marks correspond to the proof mass with 2.8/3.2 mm diameter. Red/orange corresponds to the case when electrodes "1" are biased/not biased.

where the free space electrical permittivity is ϵ_0, the relative permittivity for air is ϵ_r, V is the applied voltage, A is the overlapping area between the electrodes and the proof mass and h is the distance between the plates, in this case the vertical displacement given in Figure 3. The value for the overlapping area A for electrodes "1" is $8.0 \cdot 10^{-7} m^2$, and for electrodes "3" is $4.3 \cdot 10^{-7} m^2$. The measurements presented in Figure 3, combined with the analytical model developed and presented in [1] yield the values for the vertical and angular stiffness components, which are included in Table I, column I, for the proof mass with a diameter of 3.2 mm used in this paper. As predicted by the theoretical model [1], we observe a consistent increase in stiffness compared to the experimental values reported in [3] when using a smaller proof mass, with a diameter of 2.8 mm.

In Figure 4 we provide a contextual overview by consolidating the experimental results obtained in this work using a proof mass with 3.2 mm diameter, as well as the results reported previously [3] using a lighter proof mass with 2.8 mm diameter, on the same map of stiffness components elaborated using our analytical model presented in [1]. Integrating the electrode structure offers the possibility to modulate the vertical positioning of the proof mass not only by changing the current in the coils, but also by applying a voltage to the electrodes, in this

Table 1. Measurement and modelling.

	Proof mass	3.2 mm	Proof mass	2.8 mm *
Measurement no.	I	II	III	IV
Measured stiffness				
Angular, $[Nm \cdot rad^{-1}]$	$1.1 \cdot 10^{-8}$	$0.7 \cdot 10^{-8}$	$0.6 \cdot 10^{-8}$	$2.1 \cdot 10^{-8}$
Vertical, $[N \cdot m^{-1}]$	0.040	0.040	0.026	0.026
Modelled stiffness**				
Angular, $[Nm \cdot rad^{-1}]$	$0.8 \cdot 10^{-8}$	$0.5 \cdot 10^{-8}$	$0.4 \cdot 10^{-8}$	$0.9 \cdot 10^{-8}$
Vertical, $[N \cdot m^{-1}]$	0.042	0.042	0.024	0.024
Parameters				
Levitation height, [μm]	100	70	120	100
Voltage on electrodes "1", [V]	0	54	0	42
Coil current, [mA]	115	115	115	115

* reported in [3]
** developed in [1]

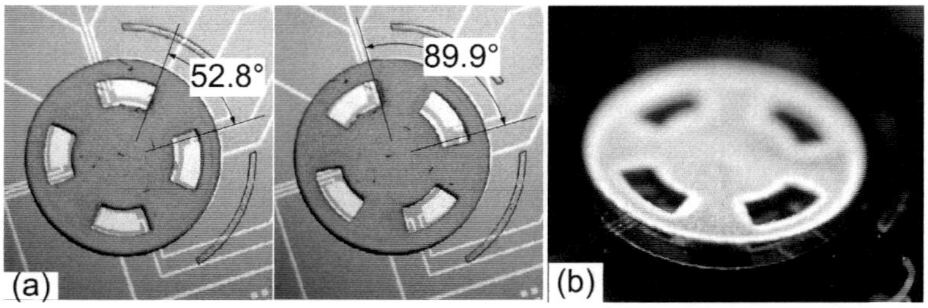

Figure 5. a) Angular oscillation in the horizontal plane around the vertical axis; b) Snapshot of the levitated carved proof mass executing the oscillations.

case the electrodes labeled "1". Using the electrostatic force generated while applying a voltage on electrodes "1", the proof mass can be maintained at a certain height even when the current in the levitation coil is increased, therefore changing the angular stiffness of the respective proof mass. This is demonstrated by applying a voltage of 54 V on electrodes "1" and recording a levitation height of 70 μm, as opposed to 100 μm for the same current in the coil structure, but no voltage applied to the electrodes.

The analytical model presented in [1] predicts a very slight change in the vertical stiffness when electrodes "1" are biased and the current in the coils is maintained constant. This is confirmed by the experimental results obtained here, as well as in [3], and reported in Table I: the vertical stiffness components both with and without applying the additional electrostatic force are 0.040 $N \cdot m^{-1}$ and 0.026 $N \cdot m^{-1}$ for the 3.2 mm and 2.8 mm diameter proof mass, respectively. The analytical model also predicts that the variation of the angular stiffness depends on the diameter of the proof mass. For the 2.8 mm diameter proof mass the angular stiffness increases upon biasing electrodes "1" ($0.9 \cdot 10^{-8}$ versus $0.4 \cdot 10^{-8} Nm \cdot rad^{-1}$), whereas for the 3.2 mm diameter proof mass the angular stiffness decreases upon biasing electrodes "1" ($0.5 \cdot 10^{-8}$ versus $0.8 \cdot 10^{-8} Nm \cdot rad^{-1}$). This excellent agreement between the theoretical predictions and the experiment is synthetically presented in Figure 4.

Figure 5 shows the operation of the device upon biasing electrodes "2" with a voltage of ±10V, thus creating two phases. The proof mass oscillates with 37° amplitude at 1.5 Hz frequency.

4. Conclusion

This work consolidates our studies on inductive suspensions and together with the results reported recently [3] confirms the predictions of our linear analytical model presented in [1].

Acknowledgments

KP acknowledges support of the Alexander von Humboldt Foundation (Grant RUS-1149346 STP). VB acknowledges support from German Research Foundation (Grant BA 4275/2-1).

References

[1] Lu Z, Poletkin K, den Hartogh B, Wallrabe U and Badilita V 2014 *Sensor. Actuat. A-Phys.* **220** 134–143
[2] Lu Z, Poletkin K, Wallrabe U and Badilita V 2014 *Micromachines* **5** 1469–1484 ISSN 2072-666X
[3] Poletkin K, Lu Z, Wallrabe U and Badilita V 2015 *Microelectromechanical Systems, Journal of* **in Press** DOI: 10.1109/JMEMS.2015.2469211
[4] Han F, Sun B, Li L and Wu Q 2015 *IEEE Sensors J.* **15** 209 – 217
[5] Su Y, Xiao Z, Ye Z and Takahata K 2015 *IEEE Electron Device Lett.* **36** 393–395
[6] Nakamura S 2005 MEMS inertial sensor toward higher accuracy & multi-axis sensing *Proc. 4th IEEE Conf. Sensors* pp 939–942
[7] Shearwood C, Ho K, Williams C and Gong H 2000 *Sensor. Actuat. A-Phys.* **83** 85–92
[8] Coombs T, Samad I, Ruiz-Alonso D and Tadinada K 2005 *IEEE Trans. Appl. Supercond.* **15** 2312–2315
[9] Dieppedale C, Desloges B, Rostaing H, Delamare J, Cugat O and Meunier-Carus J 2004 Magnetic bistable micro-actuator with integrated permanent magnets *Proc. IEEE Sensors* vol 1 pp 493–496
[10] Sari I and Kraft M 2015 *Sensor. Actuat. A-Phys.* **222** 15–23
[11] Ruffert C, Gehrking R, Ponick B and Gatzen H H 2006 *IEEE Trans. Magn.* **42** 3785–3787
[12] Abadie J, Piat E, Oster S and Boukallel M 2012 *Sensor. Actuat. A-Phys.* **173** 227–237
[13] Liu K, Zhang W, Liu W, Chen W, Li K, Cui F and Li S 2010 *Microsyst. Technol.* **16** 431–439
[14] Lu Z, Jia F, Korvink J, Wallrabe U and Badilita V 2012 Design optimization of an electromagnetic microlevitation system based on copper wirebonded coils *2012 Power MEMS* (Atlanta, GA, USA) pp 363 – 366

Simulation and experimental validation of a selective magnetization process for batch-patterning magnetic layers

C Velez, W C Patterson and D P Arnold

Interdisciplinary Microsystems Group, Department of Electrical and Computer Engineering, University of Florida, FL, 32611, United States

Email: camilovelez@ufl.edu

Abstract. This work reports experimental validation of a simulation tool that models a selective magnetization process for batch-patterning magnetic poles into magnetic layers. The magnetic patterning process is simulated via the finite-element method to solve the magneto-quasi-static Maxwell equations using the measured nonlinear magnetic properties of the materials (magnetization curves). The simulation results are compared to experimental measurements of the stray fields of selectively magnetized substrates prepared under different processing conditions. The fields are measured by a magneto-optical imaging technique approximately 50 μm above the surface. The simulations are shown to predict the general trends observed in the experimental results.

1. Introduction

Permanent magnetic materials are widely used in MEMS sensors, actuators, etc., but typically such materials are restricted to a single magnetization direction, i.e. the entire magnet is "poled" in only one direction. In contrast, at the macroscale, much more complex, three-dimensional magnetic field arrangements are typically used to create desirable magnetic field patterns, such as Halbach arrays. These arrangements generally enable stronger magnetic fields and/or field gradients, which enhance device end effect performance. The ability to create complex magnetic field patterns with sub-millimeter or even micrometer feature sizes in MEMS could significantly enhance the utility of magnetic materials, and also enable new microscale device architectures.

Previous attempts for creating complex magnetic structures (patterning) with sub-millimeter feature size began with machining permanent magnets, individually magnetizing, and assembly them [1–3]. Due to the high attractive forces between magnetic poles the assembly is troublesome, together with the lack of batch processing make this approach not very scalable to large manufacturing scales. A newer solution is the inkjet-print of micro-magnets [4], but this approach still requires additional permanent magnets to guaranty a certain field orientation. The solution presented here is to magnetically pattern permanent magnet layers by "writing" magnetic poles of desired shape and orientation. Three mechanisms for magnetic micro-patterning have been mainly used. First, poles can be created by generating spatially varying magnetic fields using current-carrying conductors [5–7], but scalability below 1 mm is an issue. Second, thermomagnetic patterning [8,9], where magnetization is altered using heat from laser irradiation enabling magnetic features in the range of 50-100 μm, but limited to the surface of the material. Lastly, magnetic fields can be shaped using soft-magnetic magnetizing heads to selectively magnetize poles into magnetic layers [10–14].

The selective magnetization process using shaped magnetizing heads has been shown to be widely adaptable to different pole sizes (microns to millimeters), materials (ultra-thin magnetic layers, electroplated thick films, bulk rare earth materials), and geometries (simple stripes to complex shapes). We have also introduced a simulation approach for modeling the process and the resultant magnetic structures [15]. In this paper, systematic and quantifiable comparisons are made between the simulation predictions and experimental results of the process, specifically the stray fields of selectively magnetized substrates fabricated under different process conditions (varying reversal magnetic field amplitudes). The magnetic stray fields of the patterned layer are measured by a magneto optical imaging technique [16].

2. Selective magnetization process description

To imprint magnetic pole patterns in the substrate, the selective magnetization method reported in [11] is used. First, the substrate is uniformly pre-magnetized out-of-plane (upward) using a 6 T pulsed magnetic field. Then a soft magnetic "magnetization mask" is placed in contact with the substrate, and a reversal (downward) magnetic field pulse is applied. The high permeability of the magnetization mask concentrates the magnetic flux in the poles of the mask, thereby flipping the magnetization down in those areas. In the other regions (between the poles underneath the air gaps of the mask), where the fields are weaker, the magnetization remains in the original upward orientation.

Figure 1 shows the magnetizing mask used in this work and the resulting pole pattern of a selectively magnetized substrate. The mask has a simple stripe pattern and is conventionally machined from AISI 1018 mild/low carbon steel (relative permeability μ_r=102) with the feature size (line/space) of 1 mm and feature depth of 0.6 mm (Figure 1). The magnetic substrate for selective magnetization is a flexible iron oxide material with 0.57 mm thickness, remanence of B_r=180 mT, and coercivity of H_{ci}=125 kA/m ($\mu_0 H_{ci}$=0.16 T).

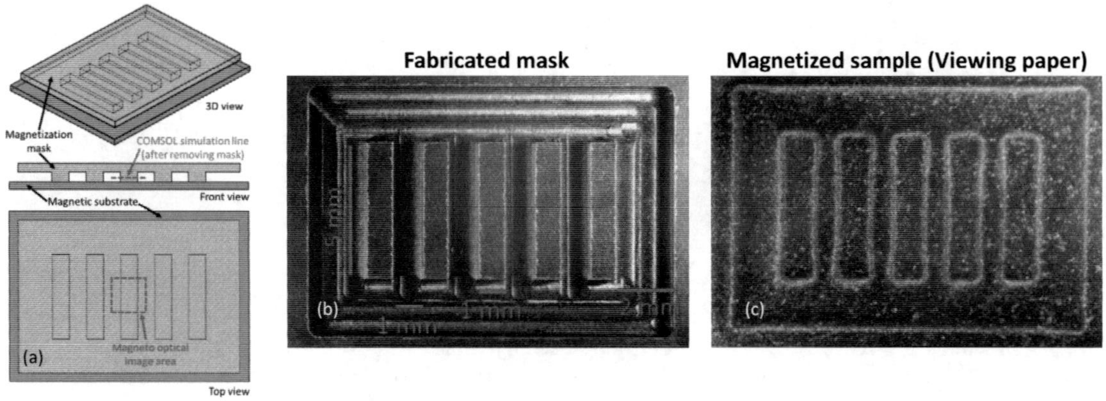

Figure 1. (a) Schematic of the magnetization mask (gray) over magnetic substrate (orange). Red dashed lines represent the simulation line and the images taken using magneto optical images at 50 μm of the surface (shown late). (b) Magnetization mask. (c) Magnetized pattern on flexible iron oxide observed using magnetic viewer paper.

3. Selective magnetization simulation

The selective magnetization process is modeled using COMSOL Multiphysics using a 2D simulation. The finite-element method is used to solve the magneto-quasi-static Maxwell equations, following the methods reported in [15]. The experimentally measured nonlinear magnetic properties of the substrate and mask materials (magnetization curves) are used in the simulations.

Simulations are divided in two consecutive and dependent steps: first, the application of the reversal magnetic field (Figure 2, top row), followed by second, the stray field generated by the remanent magnetization in the substrate after selective magnetization (Figure 2, bottom row). In the first step, the substrate is initially considered uniformly magnetized up, and the simulation models the fields during the peak of the downward selective reversal pulse. The result of this first step is then used to calculate the conditions for second step. Following Samwel, et al. [17], the remanent magnetization at every point in the substrate is calculated using the DCM (dc-demagnetization measurement) for the material. The DCM curve relates the peak field from step one to the resultant residual magnetization used in step two. Step two of the simulation computes the stray fields for the selectively magnetized substrate (with the magnetizing mask no longer present).

The sensitivity of the magnetization process with respect to the reversal B field amplitude is evident in simulations shown in Figure 2, comparing reversal fields of 100 mT, 400 mT, and 700 mT. Figure 3 shows the profile of the predicted B_z stray fields at 50 μm above the surface of the substrate for the same three cases. For the 400 mT case, the simulations indicate that it is possible to selectively reverse the substrate through its entire thickness, as compared to the thermomagnetic patterning [8,9], which was limited to ~1 μm in depth. This case also yields the strongest magnetic stray field contrast (Figure 3). It is evident that when the reversal field is too weak (100 mT), the substrate is not patterned (direction of the magnetization doesn't change), and only surface domains will be altered producing very weak remanent fields. In the opposite case, when a very high reversal field (700 mT) is applied, the entire substrate is reversed, not only the sections underneath the mask. These two extreme cases suggest there exists an ideal (optimal) reversal magnetic field at which the substrate sections underneath the mask reverse without affecting the magnetizations of the remaining substrate. Consequently, a parametric series of simulations and experiments are conducted, varying the reversal B field from 50 to 1100 mT in order to determine the optimal field (further discussed below).

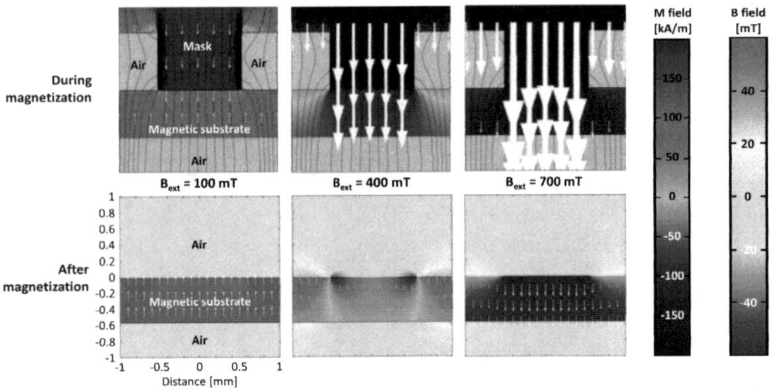

Figure 2. Selective magnetization process simulation (during magnetization – top row) and stray B field generated by the magnetic substrate after selective magnetization (after magnetization – bottom row). The M field inside the magnetic substrate and the mask (the shape of the mask is drawn for illustration).

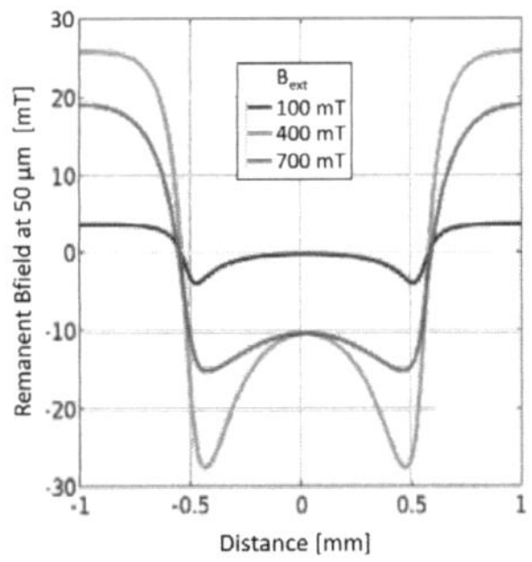

Figure 3. Simulation of the stray magnetic field 50 μm above the substrate for three reversal magnetic fields.

4. Selective magnetization and magneto-optical imaging measurements

Magnetization fields from 100 mT to 1000 mT are applied using a pulse magnetizer on independent samples. Magneto-optical images of each sample are taken [16] using a 250 mT saturation magneto-optical indicator film at 50 μm above the surface of the substrate. Figure 4a presents an example 3D representation of the field pattern after magnetization, and Figure 4b presents the average measurement of the B_z field across the feature line (average of 1240 profiles). Figure 4c shows the average B_z field (measured at 50 μm) generated by samples experimentally magnetized at 88 mT, 411 mT, and 926 mT. The values B_{min} and B_{max} defined in this figure correspond to the fields measured in the center of the mask pole and air gap, respectively.

Figure 5 compares B_z field simulation and experimental values. The maximal values of B_{max} and B_{min} were 45 mT and -23 mT, measured on sample magnetized with 411 mT. The results show the simulations accurately predict the general trends of the fields. However, there is some discrepancy in the absolute field magnitudes; the simulations underpredict the measurements. Discrepancies in the magnitude of the B fields between simulation and measurements can be attributed to experimental inaccuracies in the magneto-optical measurements; underestimation of the B fields generated by the mask during simulations; or oversimplifications in the simulation model (e.g. only permits magnetization in the z-direction).

As an overall figure of merit for the magnetization process, the magnetic contrast is defined as [18]: $C=(B_{max}-B_{min})/(B_{max}+B_{min})$. Figure 6 shows a comparison between the B field contrast from simulations and the experimental measurements. These results confirm the existence of an optimal external magnetic field value at which the contrast is maximized (~500 mT in this case). A contrast difference between simulations and measurements is on the order of 29%.

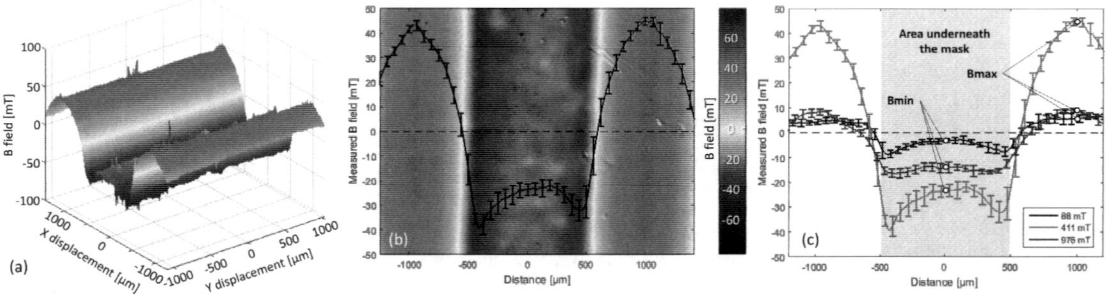

Figure 4. Magneto-optical images of the selectively magnetized sample (at 411 mT) taken at 50 µm above the surface (a) Magneto optical topographical view of the B_z field (b) Magneto optical image of the B_z field and average profile along the x axis. (c) B_z field profiles for samples magnetized with different external magnetic fields.

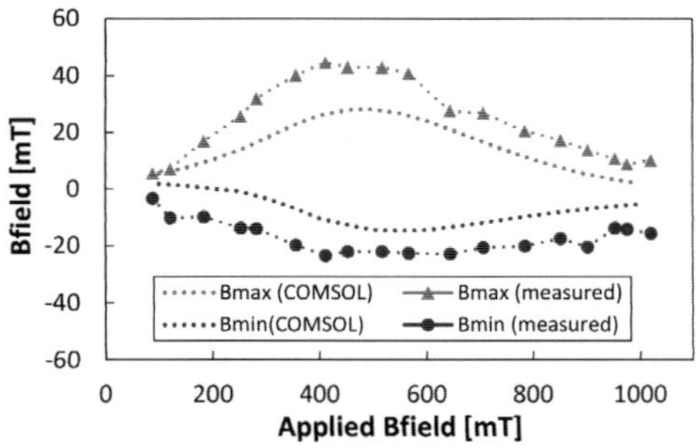

Figure 5. B_{max} and B_{min} measured/simulated 50 µm above the substrate for different reversal fields during magnetization.

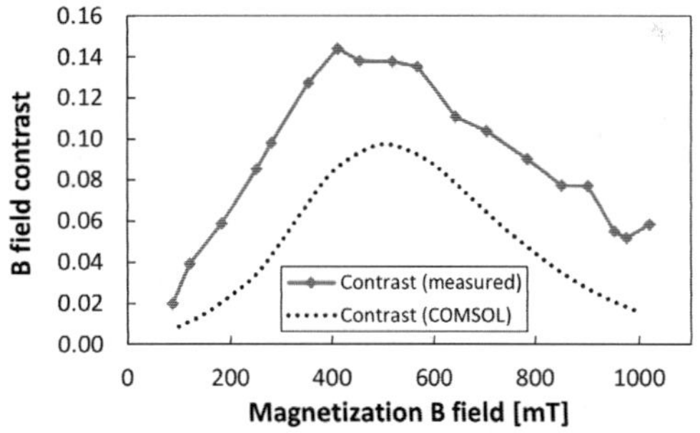

Figure 6. B-field contrast 50 µm above the substrate for different reversal fields during magnetization.

5. Conclusions

In summary, this paper demonstrates good capabilities of the simulation tool for predicting the magnetic stray fields for samples magnetized with various reversal fields. The simulations and experiments also prove that an ideal reversal magnetic field exists, and should be applied in order to maximize the magnetic field contrast on the magnetized sample. Ongoing efforts seek to improve and further validate the simulation tool over different process variables (dimensions, materials, etc.), such that it can be used to predict the stray fields from different magnetized layers under different process conditions.

References

[1] Yamaguchi K, Fujita T, Tanaka Y, Takehira N, Sonoda K, Kanda K and Maenaka K 2014 *J. Phys. Conf. Ser.* **557** 012033
[2] Herrault F, Yen B C, Ji C, Spakovszky Z S, Lang J H and Allen M G 2010 *J. Micromechanical Syst.* **19** 4–13
[3] Peterson B a, Patterson W C, Herrault F, Arnold D P and Allen M G 2012 *PowerMEMS* pp 2–5
[4] Chen P, Huang Y-Y, Bhave G, Hoshino K and Zhang X 2015 *Ann. Biomed. Eng.*
[5] Riley C D, Jewell G W and Howe D 1998 The Design and analysis of axial field multipole impulse magnetizing fixtures *7th Jt. MMM-Intermag Conf. Abstr. (Cat. No.98CH36275)* **7112** 1–4
[6] Töpfer J and Christopher V 2004 *Sensors Actuators, A Phys.* **113** 257–63
[7] Bishop D J, Jin S, Kim J and Ramirez A G 2000 *US pattent #: 6,124,650*
[8] Dumas-Bouchiat F, Zanini L F, Kustov M, Dempsey N M, Grechishkin R, Hasselbach K, Orlianges J C, Champeaux C, Catherinot a. and Givord D 2010 *Appl. Phys. Lett.* **96** 3–6
[9] Osman O, Toru S, Dumas-Bouchiat F, Dempsey N M, Haddour N, Zanini L F, Buret F, Reyne G and Frénéa-Robin M 2013 *Biomicrofluidics* **7** 0–11
[10] Arnold D P, Das S, Park J, Zana I, Lang J H and Allen M G 2006 *Ieee Trans. Microelectromechanical Syst.* **15** 1351–63
[11] Oniku O D, Ryiz P V., Garraud A and Arnold D P 2014 *J. Appl. Phys.* **115** 38–41
[12] Oniku O, Regojo R, Kaufman Z, Patterson W C and Arnold D P 2013 *IEEE Trans. Magn.* **49** 4116–9
[13] Gilles P A, Delamare J and Cugat O 2002 *J. Magn. Magn. Mater.* **242-245** 1186–9
[14] Zana I, Herrault F, Arnold D and Allen M 2005 *Tech. Dig. 5th Int. Workshop on Micro and Nanotechnology for Power Generation and Energy Conversion Apps. PowerMEMS 2005* (Tokyo, Japan) pp 0–3
[15] Oniku O D, Garraud A, Shorman E E, Patterson W C and Arnold D P 2014 *Tech. Dig. Solid-State Sensors, Actuators, and Microsystems Workshop (Hilton Head 2014)* (Hilton Head Island, South Carolina) pp 187–90
[16] Patterson W C, Garraud N, Shorman E E and Arnold D P 2015 *Rev. Sci. Instrum.* **86** 094704
[17] Samwel E O, Bissell P R and Lodder J C 1993 *J. Appl. Phys.* **73** 1353–9
[18] Peli E 1990 *J. Opt. Soc. Am. A.* **7** 2032–40

Acknowledgments

The authors acknowledge Robin Carroll, Dr. Alexandra Garraud, Nicolas Garraud, and Ololade Oniku for valuable technical discussion.

Ferrofluid-based Stretchable Magnetic Core Inductors

N Lazarus[1] and C D Meyer
Sensors and Electron Devices Directorate
US Army Research Laboratory, Adelphi, MD, USA

E-mail: nathan.lazarus2.civ@mail.mil

Abstract. Magnetic materials are commonly used in inductor and transformer cores to increase inductance density. The emerging field of stretchable electronics poses a new challenge since typical magnetic cores are bulky, rigid and often brittle. This paper presents, for the first time, stretchable inductors incorporating ferrofluid as a liquid magnetic core. Ferrofluids, suspensions of nanoscale magnetic particles in a carrier liquid, provide enhanced magnetic permeability without changing the mechanical properties of the surrounding elastomer. The inductor tested in this work consisted of a liquid metal solenoid wrapped around a ferrofluid core in separate channels. The low frequency inductance was found to increase from 255 nH before fill to 390 nH after fill with ferrofluid, an increase of 52%. The inductor was also shown to survive uniaxial strains of up to 100%.

1. Introduction

The development of soft, stretchy electronics in recent years has opened up a range of new applications such as biomedical monitoring, soft robotics and wearable computing [1]. A major challenge has been the creation of high performance stretchable inductors for wireless power and data transmission. The rigid magnetic materials common in inductors design are not easily translated to the stretchable domain. Previous stretchable inductors have instead been built in non-magnetic elastomers such as silicone or polyurethane, resulting in poor inductance density. In previous work, the first stretchable magnetic core inductor based on soft silicone loaded with magnetic particles was demonstrated [2]. The resulting magnetic composites had permeabilities as high as 2.9, while retaining the ability to stretch up to 100% without permanent damage. One disadvantage of this approach, however, is that the resulting composite has different mechanical properties from the original elastomer. The embedded particulate serves to crosslink neighbouring polymer chains, similar to vulcanized rubber, resulting in increased elastic modulus. Although the composites demonstrated stretchability, the greater mechanical stiffness could complicate interfacing with softer surfaces such as human skin.

In this work, the fabrication and testing of a magnetic core stretchable inductor using a ferrofluid as a core material is presented. Ferrofluids are stable suspensions of magnetic particles, typically with particle size on the order of 10 nm to prevent settling due to gravity [3]. Since the core is liquid, the mechanical behaviour of the surrounding elastomer is unchanged unlike in the case of the particle loaded composite. The stretchable inductor traces are made using a room temperature liquid metal in fluidic channels [4], allowing millimeter-scale cross sections for low resistance while

[1] To whom any correspondence should be addressed.

retaining stretchability. The resulting magnetic core inductor was found to have 52% higher inductance after ferrofluid fill, and retained functionality up to 100% uniaxial strain.

2. Fabrication

A fabrication process for creating a solenoid inductor based on molding silicone using a two-part 3D printed mold was developed in [5]. A modified version (Fig. 1) was used to create three layers of fluidic channels (top and bottom layers filled with liquid metal to form the solenoid, and a middle layer filled with ferrofluid). The process begins by filling a three-part mold with precursors of a soft silicone (Ecoflex 00-30, Smooth-on) (Fig. 1 (a)). The molds used in the process are shown in Fig. 2. As in the original two layer process, the top and bottom molds are used to define the solenoid inductor, while a pedestal defined in the bottom mold sets the placement of a 3D printed bar that defines the ferrofluid core. After the silicone is cured completely at 85°C on a hotplate, it is removed from the mold (Fig. 1 (b)). Since the molds are rarely perfect in contact, a thin layer of silicone remains in the via holes, requiring coring to form holes using a syringe tip.

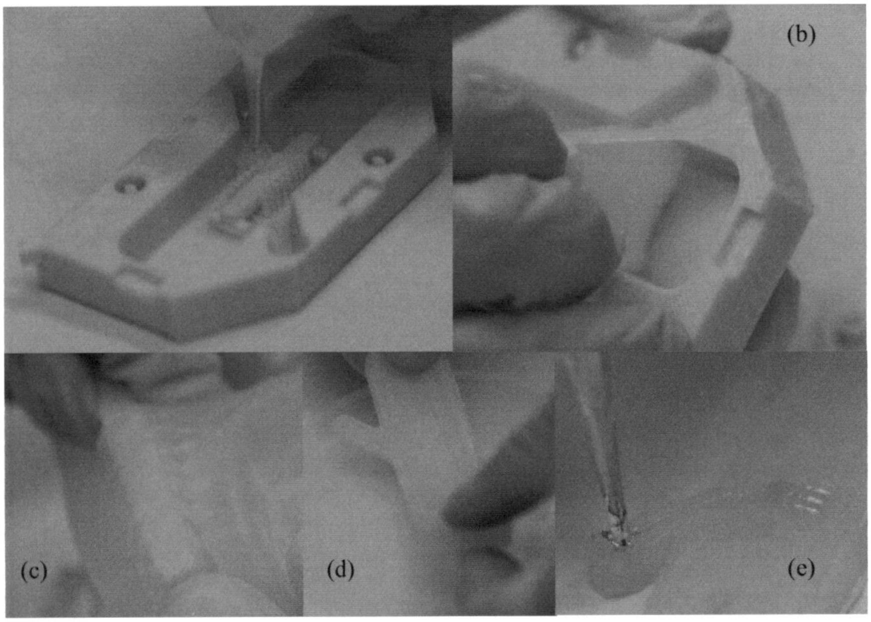

Figure 1. Fabrication process: (a) pour silicone elastomer into mold, (b) cure completely and remove, (c) pull out bar defining ferrofluid core (d) seal top and bottom to partially cured silicone and (e) fill with liquid metal and ferrofluid

At this stage, the bar defining the void for the ferrofluid remains embedded within the silicone. Since silicone does not adhere well to the 3D printed polycarbonate, it is then possible to slide the embedded piece out of the silicone (Fig. 1 (c)), a technique previously demonstrated using low friction fibers to define channels in [6]. The bottom channels are then sealed by laying the molded piece onto a pool of Ecoflex 00-30 that has been partially cured for one hour and fifteen minutes at room temperature to minimize the possibility of flowing into the channels (Fig. 1 (d)), a process that is then repeated to seal the top channels as well. Inlet and outlet holes are then cored out using a narrow gauge syringe tip, and the two channels are filled with galinstan and ferrofluid (Ferrotec EFH-1) for the electrical trace and core respectively. The holes are then sealed using drops of silicone. The final fabricated inductor after fill is shown in Fig. 3. Electrical connection for testing is made using copper wires at the inlet and outlet ports, which are in mechanically isolated regions (the tabs extending off the sides of the main inductor test piece) to minimize effects of stretching on the contacts.

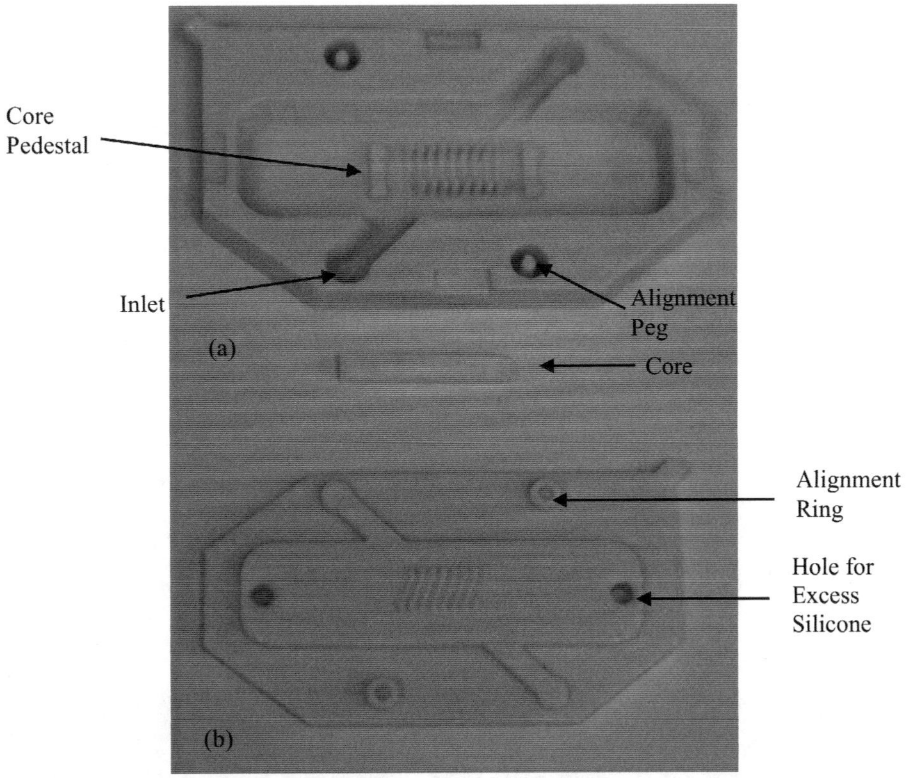

Figure 2. (a) Bottom/core and (b) top molds

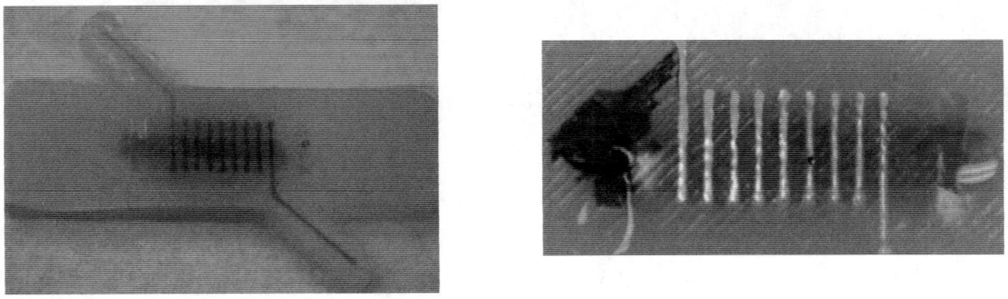

(a) (b)

Figure 3. Inductor (a) filled with colored water to verify fluidic isolation and (b) after final fill with liquid metal and ferrofluid

3. Testing

The solenoid inductor fabricated in this work consisted of 8.5 turns, with traces one millimeter in width and thickness. The effective core length was defined to be 21 mm, with total core width and thickness of 6 mm and 7 mm respectively. Due to the need for a layer of silicone to isolate the ferrofluid from the channels containing the liquid metal, the ferrofluid portion of the core is only 4 mm by 5 mm, filling 48% of the total core area. The inductor frequency response (Fig. 4) was measured with an HP 4294A impedance analyser before and after filling with the ferrofluid EFH-1 (Ferrotec, $\mu_r \approx 3.6$). The low frequency inductance was found to be 255 nH before fill, and rose to 390 nH after being filled with ferrofluid, an increase of 52%.

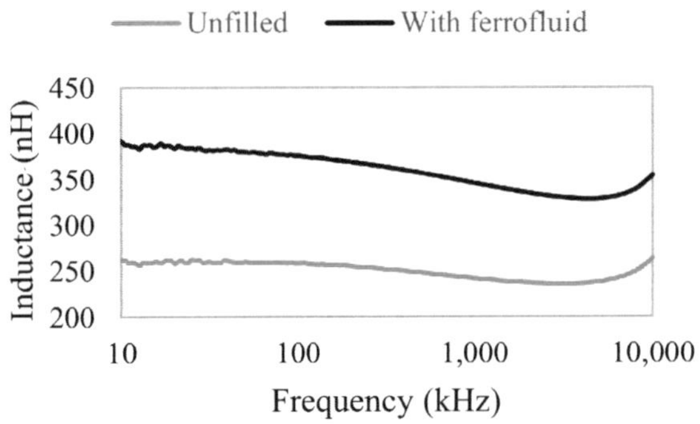

Figure 4. Freque ferrofluid

Both the ferrofluid-filled inductor and a second stretchable inductor with a solid silicone core were also characterized mechanically using a custom uniaxial strain testing setup, consisting of 3D printed clamps designed to fit on a standard bench vise. The test piece is clamped on each end, with a raised zigzag pattern in the clamp designed to bite into the silicone and minimize sliding. The bench vise is then used to adjust the clamp position along the threaded rod, while the position is measured using calipers to determine the applied strain. Fig. 5 shows the ferrofluid inductor during testing up to 100% uniaxial strain.

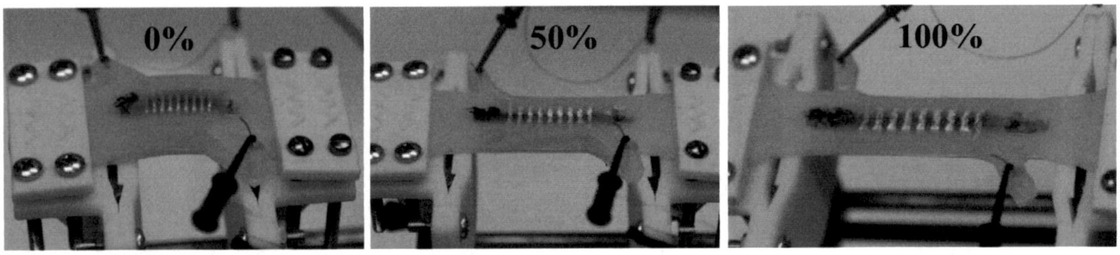

Figure 5. Ferrofluid inductor strain testing

The inductance of a coil of wire is dependent on the geometry, with both the self inductance of individual traces as well as the coupling to neighboring traces varying with strain. A model for the effects of stretching on a solenoid was previously developed in [5], finding that, for applied strain parallel to the core, the inductance declines rapidly with strain as the core length increases and the core cross sectional area decreases. This behaviour is evident in the silicone-cored inductor as shown in Fig. 6. For the ferrofluid inductor, the drop in inductance was even faster, with the inductance asymptotically converging to the non-magnetic behavior for higher strains. This is likely due to the drop in core cross section, as well as to the overall flattening of the solenoid for large strains. A similar drop in effective permeability was also found for the magnetic-loaded polymer based stretchable inductors in [2].

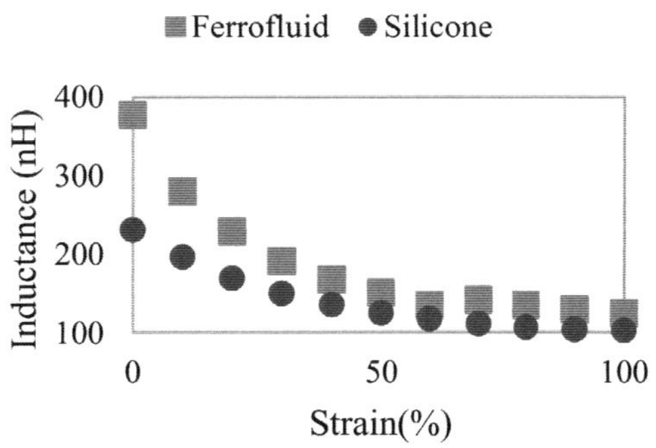

Figure 6. Uniaxial str

4. Conclusion

A magnetic core stretchable inductor was demonstrated using ferrofluid within a liquid metal solenoid. The ferrofluid was found to significantly enhance the inductance density of the inductor, increasing the total inductance by 52% after ferrofluid fill. The inductor was tested up to 100% uniaxial strain without damage, although the effect of the ferrofluid dropped rapidly with higher mechanical strains. This makes the device particularly valuable for strain sensing, one of the original applications for liquid metal inductors [4], since the ferrofluid results in a larger change in inductance with strain. Since most biomedical applications require use only up to about 30% strain, the stretchability of skin over much of the human body, this is also a useful improvement for creating an inductor for wireless power or data transmission in a stretchable system.

5. References

[1] Rogers J A, Someya T, and Huang Y 2010 *Science* **327** 1603
[2] Lazarus N, Meyer C D, Bedair S S, Slipher G A, and Kierzewski I M 2015 *ACS Appl. Mater. Interfaces*, **7** 10080
[3] Zahn M 2001 *J Nanoparticle Research* **3** 73
[4] Fassler A and Majidi C 2013 *Smart Mater. Struct.* **22** 055023
[5] Lazarus N, Meyer C D, Bedair S S, Nochetto H and Kierzewski I M 2014 *Smart Mater. Struct.* **23** 085036
[6] Park Y and Wood R J, 2013 *Proc. IEEE Sensors 2013* 1

Printable on-chip micro battery for disposal bio-sensing device

Takashiro Tsukamoto and Shuji Tanaka

Department of Bioengineering and Robotics, Tohoku University
6-6-01 Aramaki Aza Aoba Aoba-ku Sendai-shi, Miyagi, 980-8579, Japan

t_tsuka@mems.mech.tohoku.ac.jp

Abstract. This paper reports an on-chip micro battery which can be fabricated on a CMOS chip. The battery consists of screen-printed Mg and AgCl films. An open circuit voltage of 1.58 V, the maximum power of 1.4 mW and a total stored energy of 400 mJ were obtained from anode and cathode electrodes of about 3 × 3 mm^2. Demonstration devices self-powered by the on-chip battery worked in a 0.14 M NaCl solution, and digital data transmission from the devices using optical and electrical field methods were confirmed.

1. Introduction

A disposal bio-sensing device is expected to be used in a point of care testing system such as rapid disease inspection in developing countries and home health care system. In these systems, inspection cost should be as small as possible. For this purpose, a disposal bio-sensing chip was proposed [1], in which recycling cost can be minimized. In this system, a small sensor chip without post-process was used to make the chip cost low. However, the on-chip micro coil for power feed and data transfer limits the miniaturization, i.e. the cost reduction was limited. Therefore, in this paper, we proposed an on-chip micro battery which can be printed on the sensor chip. To simplify the system, a single cell battery should generate sufficient voltage for a CMOS chip. We chose Mg as an anode material and AgCl as a cathode material. A chip size can be reduced due to high power density of the battery, which reduces the chip cost. In addition, wireless communication method implemented in the chip does not require any electric wirings connected to the chip, which drastically reduces the post-process cost.

2. Disposal bio-sensor chip

Figure 1 shows the schematic of a disposal bio-sensor chip using an on-chip battery. The Hall sensor array and signal processor are fabricated by a standard CMOS process. Antibodies specifically bind to the inspection target (antigen) are fixed on the sensor array. A on-chip battery consists of Mg and AgCl films is placed on a CMOS chip. Bio-sensing is usually done in an aqueous solution such as saline, which can be also used for an electrolyte of the battery with some salt. This technique enables long-term storage and simple structure. Two types of wireless communication methods are proposed to transfer the inspection result: optical and electric field methods. The optical communication type device (Dig. 1(a)) transfers the digital data using light emission or color change. For this purpose, LEDs, organic ELs or electrochromic devices are fabricated on the chip. The electric field communication type device (Fig. 1(b)) induces the electric potential distribution in the electrolyte which can be sensed by external pick-up electrodes.

Figure 2 shows the inspection procedure. First, the sensor chip is immersed in an inspection solution, then the test sample (urine, blood etc.) are introduced. If the test sample contains an inspection target (antigen), the targets are fixed on the sensor chip by antigen-antibody coupling. Then micro beads made of magnetic material combined with antibodies which also specifically bind to the target antigen are introduced. Thus, the magnetic beads are fixed on the chip again using antigen-antibody coupling. If the magnetic beads exist on the sensor ship, the perpendicular magnetic field is induced by the applied magnetic field parallel to the surface, which can be detected by the hall sensors. The sensor chip is activated by adding salt such as NaCl or NH_4Cl to the solution, then it counts the number of magnetic beads and send the result by optical or electrical methods.

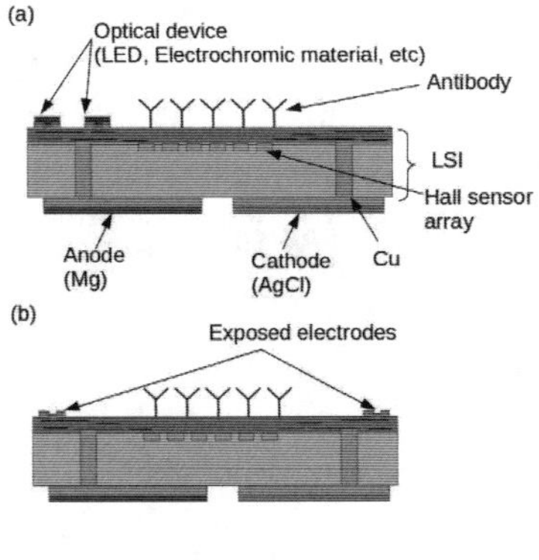

Figure 1. Structure of the disposal bio-sensor chip using (a) optical and (b) electric field data transfer.

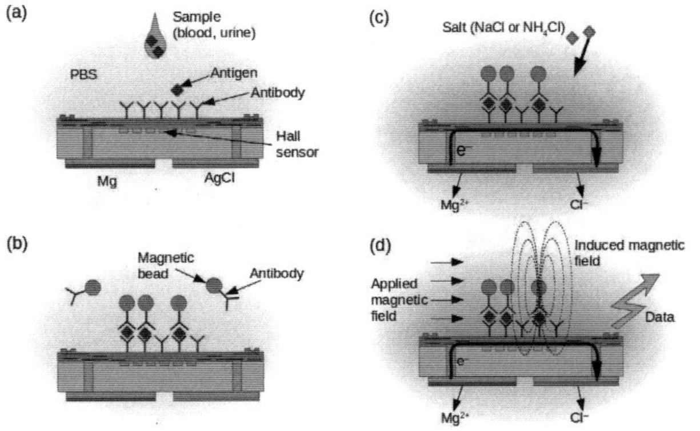

Figure 2. inspection procedure. (a) sensor chip and test sample are immersed to the solution. (b) Antibody coated magnetic beads and (c) salt is added. (d) Magnetic beads are detected by Hall sensor and the result is transferred.

3. Printable on-chip micro battery

3.1. Fabrication procedure

Several methods for deposition of the Mg film such as vacuum evaporation [2] and electroplating [3] were already reported. However, these methods take long time for thick film deposition. Therefore, in this paper, we proposed a printing method to deposit thick films for both Mg and AgCl electrodes. Figure 3 shows the fabrication process of the printable battery. First, current collection electrodes made of Cu were fabricated on the device. A stencil mask made of 250 μm polypropylene paper was prepared by a cutting plotter, and aligned to the Cu pattern. An Ag/AgCl paste (ALS Co., Ltd) was then screen-printed on the Cu electrode. The thickness of the printed paste can be controlled by that of stencil mask. After removing the mask and drying the paste at room temperature, another stencil mask was aligned and a Mg paste, which was prepared by mixing 0.05 g of Mg powder (Wako Pure Chemical Industries, Ltd) with 0.06 g of Ag dispersed conductive epoxy resin (Chemtronics), was printed and cured at 70°C. Finally, unnecessary area was covered by an UV curable resin. Figure 4 shows the schematic of the test sample. The area of both electrodes was 3 × 3 mm^2.

3.2. Performance of on-chip micro battery

The fabricated test battery was evaluated by a setup as shown in Fig. 4(b). 0.14 M NaCl solution was used as the electrolyte. Various load resistors were connected to the battery and the applied voltage was measured by a digital multimeter (34410A, Agilent Technologies Inc.). Figure 5 shows the measured performance of the battery. The open circuit voltage, the internal resistance and the maximum output power were 1.58 V, 520 Ω and 1.4 mW, respectively.

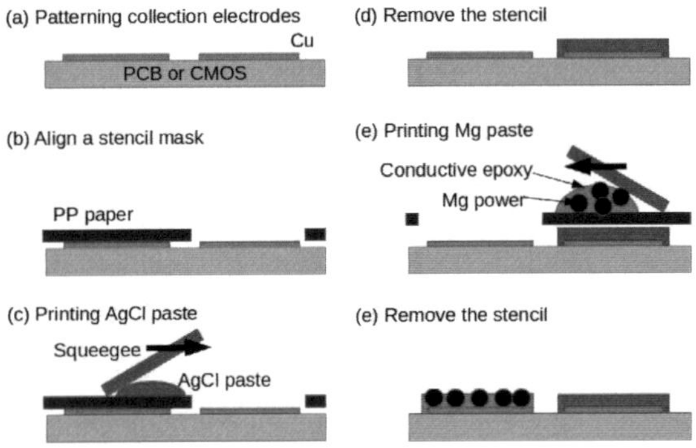

Figure 3. Fabrication procedure.

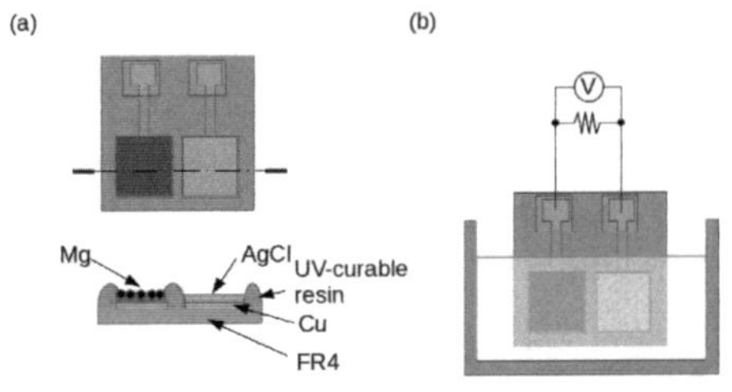

Figure 4. (a) Structure of the test device and (b) measurement equipment.

Figure 6 shows the time dependency of the output voltage using the 3.3 kΩ load resistor. The output voltage was stable as long as about 650 s and the total energy obtained by the battery was as high as 400 mJ, which is sufficient for the disposal application.

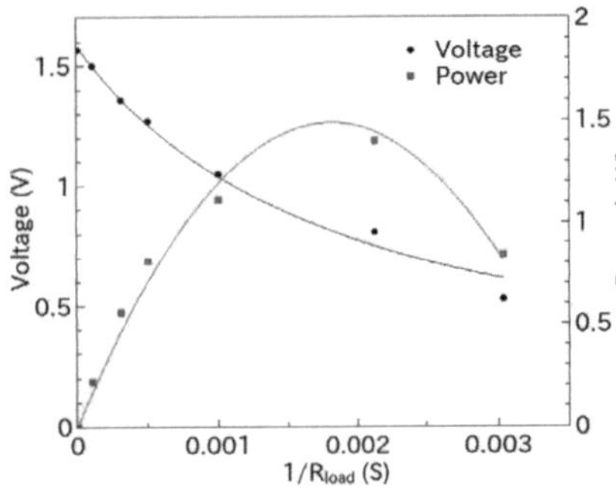

Figure 5. Output performance of the on-chip battery.

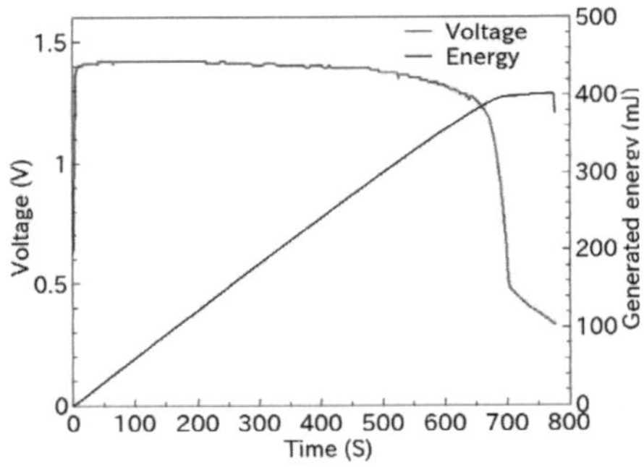

Figure 6. Long time operation and generated total energy.

Figure 7 and 8 show the scanning electron micrograph (SEM) and energy dispersive X-ray spectrum (EDS) of the anode and cathode electrodes, respectively. Before power generation, the Mg particles on the anode electrode were covered by Ag particles which came from the conductive epoxy resin (Fig. 7(a) and 8(a)), and the cathode electrode was covered by Ag and AgCl (Fig. 7(b) and 8(b)). After power generation, Mg was eroded and cracks and voids were formed, and the Ag particles on the Mg were partly removed (Fig. 7(c)). From the EDS analysis, the atom ratio between Mg and O was about 29:71 (Fig. 8(c)), which implies not only main reaction

$$Mg \rightarrow Mg^{2+} + 2e^-$$

but also side reaction

$$Mg + 2H_2O \rightarrow Mg(OH)_2 + H_2$$

occurred. In fact, the gas formation on the anode was clearly observed. On the other hand, cathode was covered by pure Ag (Fig. 8(d)), which means almost all of AgCl was reduced to Ag as

$$AgCl + e^- \rightarrow Ag + Cl^-.$$

Figure 7. SEM of (a) anode and (b) cathode before power generation, and (c) anode and (c) cathode after power generation.

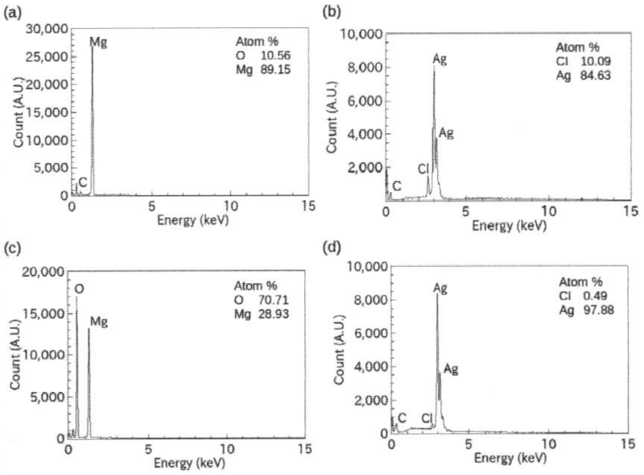

Figure 8. EDS of (a) anode and (b) cathode before power generation, and (c) anode and (c) cathode after power generation.

4. Demonstration device

A CMOS chip requires both long time and high cost for development. Therefore, in this paper, we used commercially available CMOS device combined with the developed on-chip battery to confirm the operation of the system. Two types of demonstration devices were fabricated and tested: optical and electric field data communication types.

Figure 9 shows the schematics of these demonstration devices. A microprocessor (PIC12LF1501, Microchip Technology Inc.) was used as the CMOS device. LEDs and communication electrodes were used for optical and electric field data transfer, respectively. A printed circuit board (PCB) was used as a base board and these devices were soldered on it. The developed on-chip battery was printed on the backside of the PCB. Unnecessarily area was covered by the transparent UV-curable resin. The PCB size was about 5 × 7 mm^2.

Figure 10 shows the operation result of the optical type demonstration device powered by the on-chip battery in 0.14 M NaCl solution. Two LEDs were used as clock and data signals. On and Off of each LED could be clearly observed by a standard CMOS camera, and the digital data could be decoded.

Figure 11 shows the experimental setup for the the electric field communication through the electrolyte. Two pick up electrodes were inserted in the solution about 2 mm away from each communication electrode on the device. Figure 12(a) is the applied potential to the communication electrode, which was directly measured by a standard probe with an external power supply. Before the valid data, start bits with digital data of "101010" was sent to determine the baud rate and the start timing. Figure 12(b) shows the received signal using pick up electrodes. Due to a high pass filter formed by electric double layer, the measured signal showed positive and negative sharp peaks at rising and falling edges of the sending digital data. After passing through the digital filter, the digital signal was restored as shown in Fig. 12(c). The sampling timing was determined by the start bits, then the digital data was successfully decoded.

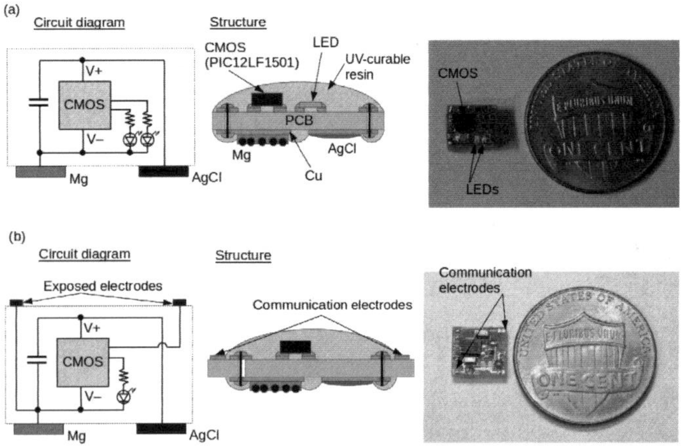

Figure 9. Structure of demonstration device using (a) optical and (b) electrical data transfer.

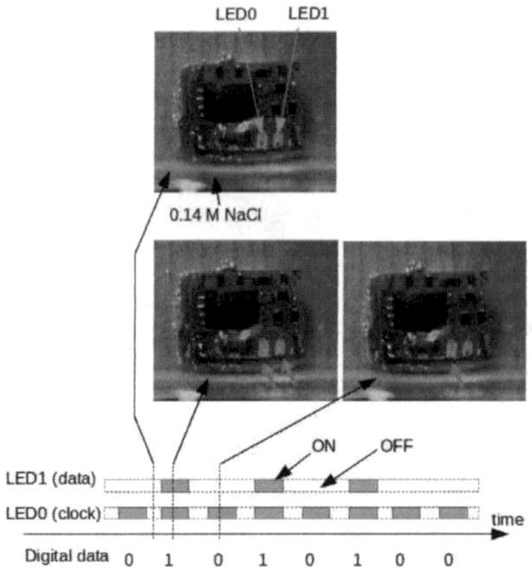

Figure 10. Optical signal generated by the demonstration device.

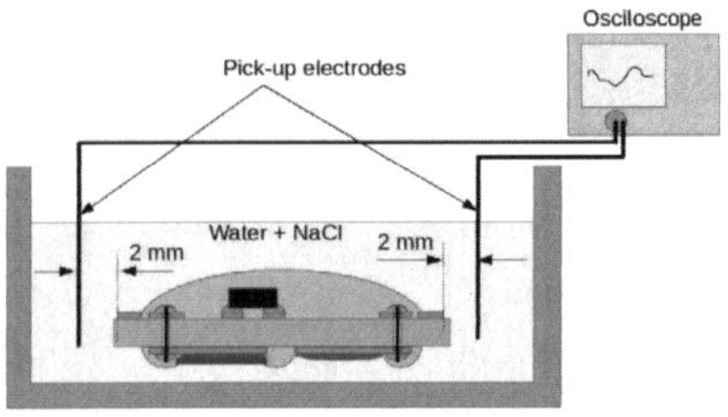

Figure 11. Experimental setup for the electric field data transfer.

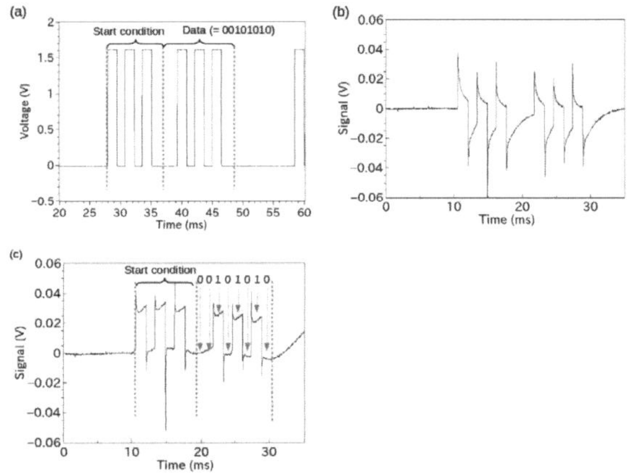

Figure 12. (a) sending and (b) received signal. (c) Restored signal using digital processing.

5. Summary

A printable on-chip battery for a disposal bio-sensor chip was developed and evaluated. The open circuit voltage of 1.58 V, maximum power of 1.4 mW and total stored energy of 400 mJ were obtained using 3 × 3 mm^2 electrodes. Demonstration devices using commercially available CMOS device were developed and the digital data was successfully transferred by optical and electrical methods. The proposed printable on-chip battery can be easily fabricated on any devices including CMOS chips at low cost. It is good candidate for the disposal bio-sensor as well as an ingestion sensor.

Acknowledgments

This research was partly supported by Center of Innovation Science and Technology based Radical Innovation and Entrepreneurship Program.

References

[1] T. Ishikawa *et al.*, *Proc. 32nd Sensor Symposiu,* 2013, 5PM3-PSS-87 (in Japanese).
[2] F. Sammoura *et al.*, *Sens. Actuators, A*, **111**, pp. 79-86, 2004.
[3] M. Tsang *et al.*, *Proc. MEMS 2014*, Jan. 2014, pp. 358-361.

Layer-by-layer fully printed Zn-MnO_2 batteries with improved internal resistance and cycle life

B Kim[1], R Winslow[1], I Lin[2], K Gururangan[2], J Evans[2] and P Wright[1]
[1]Department of Mechanical Engineering, University of California, Berkeley, California, 94720, USA
[2]Department of Materials Science and Engineering, University of California, Berkeley, California, 94720, USA

E-mail: bernard.kim@berkeley.edu

Abstract. This research created direct layer-by-layer printed zinc-based secondary batteries with an ionic liquid-based gel polymer electrolyte to power micro- and meso-scale devices. The use of a gel polymer electrolyte composed of [BMIM][Otf] ionic liquid, ZnOtf salt, and PVDF-HFP polymer binder enabled direct layer-by-layer printing of functional cells. The effects of additive printing methods on cell discharge capacity, cycle life, and internal resistance are discussed. Fully printed cells have demonstrated average discharge capacities of 0.548 mAh/cm^2, energy densities of 8.20 mWh/cm^3, and specific energies of 2.46 mWh/g with some cells achieving over 1000 cycles without catastrophic failure. Layer-by-layer printed devices exhibited decreased DC internal resistance and longer cycle life over previous mechanically assembled cells.

1. Introduction

As ionic liquids have grown in use, investigations have been carried out on their applicability in rechargeable batteries [1]. In particular, advancements in zinc-based chemistries with ionic liquid electrolytes have proven the feasibility of creating rechargeable zinc cells through additive manufacturing methods to power micro- and meso-scale devices [2, 3, 4].

Previous work has investigated the feasibility of using gel polymer electrolytes based on the ionic liquid 1-butyl-3-methylimidazolium trifluoromethanesulfonate ([BMIM][Otf]) and the salt zinc trifluoromethanesulfonate (ZnOtf) in Zn-MnO_2 secondary battery systems by mechanically assembling cells, but has not investigated the effects of layer-by-layer printing on cell performance [5, 6, 7]. This work improves upon past prototypes by characterizing the effects of fully printing cells and by utilizing a direct layer-by-layer printing process to improve cell performance and cycle life.

2. Experimental Methods

Individual inks were made for the cathode, anode, and electrolyte, the compositions of which are listed in Table 1. The cathode is composed of an MnO_2- and carbon-based ink developed for this work. The anode is composed of either zinc foil, for half printed cells, or a Zn and carbon ink similar to the cathode ink, for fully printed cells. The gel polymer electrolyte is composed of PVDF-HFP permeated with the ionic liquid and salt solution ([BMIM][Otf] and ZnOtf). For all inks, PVDF-HFP was first dissolved in NMP to create a gel. Once the gel was

Table 1. Compositions of electrode and electrolyte inks. All materials were measured in wt% with additional NMP added to tune rheology.

Cathode	Anode	Electrolyte
86% MnO_2	84% Zn	67% ZnOtf:[BMIM][Otf] (1:6.5 mass ratio)
6% Acetylene Black	7% Acetylene Black	–
8% PVDF-HFP	9% PVDF-HFP	33% PVDF-HFP

fully homogenized, the respective active materials and fillers were added, and additional NMP was added to tune the rheology (for the cathode and anode inks). The inks were thoroughly mixed and sonicated to ensure adequate homogenization and particle dispersion.

Cells were manufactured using stencil printing and doctor blade coating. A 100 μm thick PET sheet with 1 cm^2 square holes was used as the stencil. To fabricate cells, the cathodes were first stencil cast into 1 cm^2 squares on stainless steel foil and dried in a gravity oven at 80°C. The GPE was cast with a doctor blade over the cathodes in multiple passes with increasing blade offset distances, drying partially between layers. Once the GPE was fully dried, the anode was either stencil cast directly on top of the GPE and dried (for fully printed cells), or 1 cm^2 squares of zinc foil were placed on top (for half printed cells). Throughout the casting process, the cells were maintained at approximately 80°C by adhering the substrate to an aluminum block, which acted as a thermal sink and maintained film temperatures constant. After the entire cell was dried, the cells were peeled from the foil, cut out with a razor blade, and sealed in stainless steel 2032 coin cells (MTI Corp) (Figure 1). Half printed cells were manufactured with thick (200 μm) GPE layers, and fully printed cells were manufactured with both thick and thin (70 μm) GPE layers.

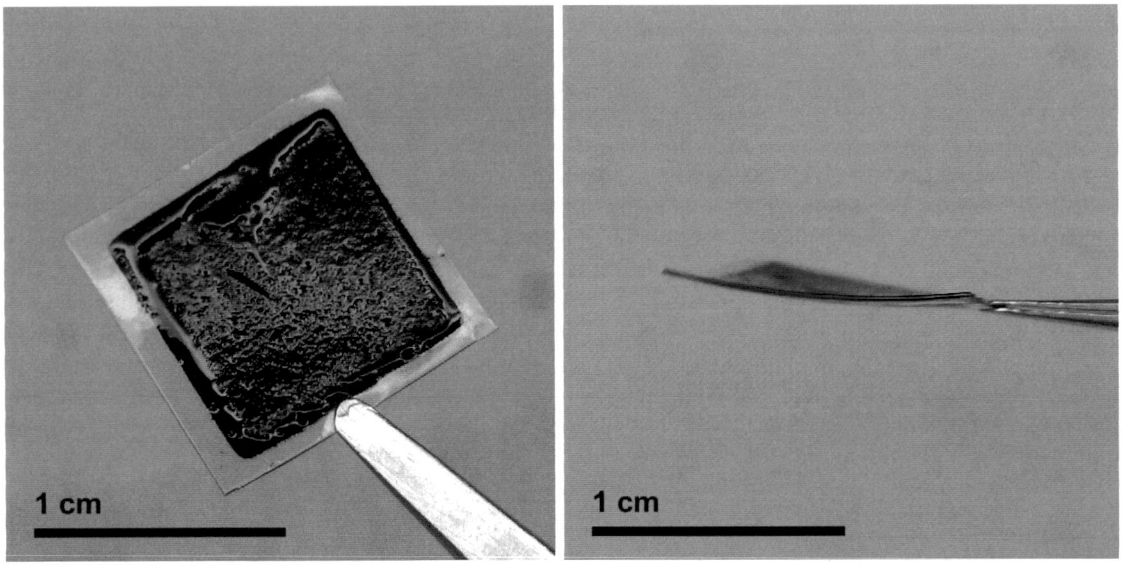

Figure 1. Left: front view of fully printed cell. *Layers from back to front: printed MnO_2 cathode, gel polymer electrolyte, printed Zn anode*; Right: side view of fully printed cell.

Mechanically assembled cells were manufactured by stencil printing the cathode in 1 cm^2 squares, casting the GPE in flat circular disks about 500 μm thick, cutting out 1 cm^2 squares of zinc foil, and arranging them in a sandwich structure in a coin cell, with additional electrolyte added at the interfaces for wetting.

The cells were cycled on a Neware battery tester with a constant current charging profile and were cycled from 1.0-1.8V (Table 2). Cell performance data was collected from the Neware testing software. SEM images were taken with a Hitachi TM-1000 Tabletop Scanning Electron Microscope. Samples were prepared by cutting cells in half with a razor blade and mounting them on 90° stages to examine the cross section.

Table 2. Cycling profile for a single step

	Step Name	Parameters
1	Constant Current Charge	125 μA from 1.0-1.8V
2	Constant Current Discharge	-125 μA from 1.8-1.0V
3	Rest	3 minutes

3. Results & Discussion

3.1. Interfacial Cohesion of Printed Components

SEM images of fully printed cells with thick and thin GPE layers illustrate the interfacial cohesion between printed components (Figure 2). The interfaces between the electrodes and GPE are continuous and free from any delamination. As a consequence of the processing parameters, the polymer structure in each additional layer is allowed to mix with the previous layer, forming a continuous structure with minimal interfacial discontinuities and improving cell performance.

Figure 2. Cross sections of fully printed cells with thick and thin GPEs. *Left: fully printed, thick GPE cell; bottom: fully printed, thin GPE cell*

3.2. Discharge Capacity

Figure 3 presents the discharge capacities of the first 100 cycles for all cells. Figure 4 presents the discharge capacities of the first 250 cycles for both fully printed cells. The mechanically assembled and half printed cells were cycled until failure, but testing was terminated before the fully printed cells reached failure.

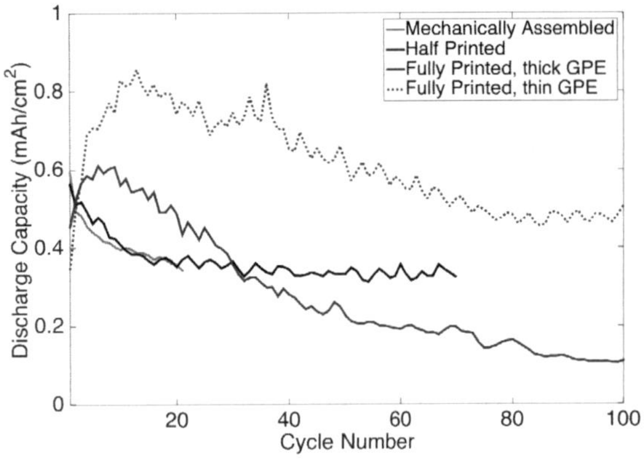

Figure 3. Discharge capacities of first 100 cycles for all cells. Only mechanically assembled and half printed cells were cycled until failure.

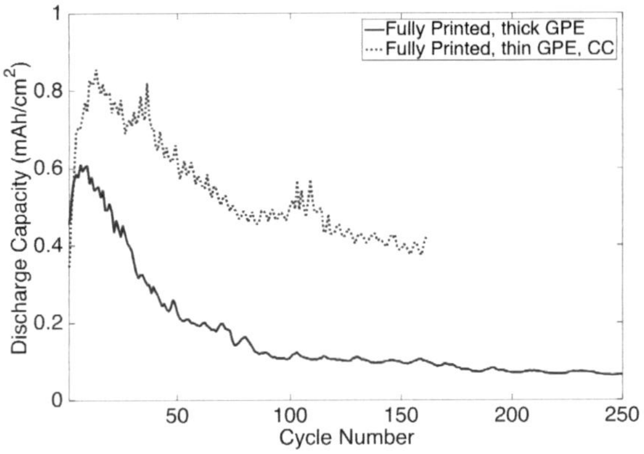

Figure 4. Discharge capacities of first 250 cycles for fully printed cells. Cells were cycled until tests were terminated.

Both fully printed cells demonstrated higher discharge capacities compared to the mechanically assembled cell for the same number of cycles, while the half printed cell demonstrated an equal discharge capacity in the same period. The fully printed, thin GPE cell exhibited the highest discharge capacity of all cells. Averaged over the data collected, the cell had a discharge capacity of 0.548 mAh/cm^2, energy density of 8.20 mWh/cm^3, and specific energy of 2.46 mWh/g. However, the half printed cell sustained a more stable discharge capacity over its lifetime compared to the fully printed cells.

The half printed and mechanically assembled cells exhibited continuously decaying discharge capacities, but the fully printed cells experienced a "break in" period where the discharge capacity initially increased before decaying. Further work is needed to provide confirmation, but this is likely due to the concentration of zinc at the electrolyte interface in the printed anode increasing with initial replating cycles, as opposed to the bulk zinc foil anode which already provided the highest concentration of zinc possible. The faster rate of decay in discharge capacity for the fully printed cells compared to the other two may also be a result of the composite nature of the printed zinc anode.

3.3. Cycle Life

While all half printed and fully printed cells demonstrated improved cycle life compared to mechanically assembled cells, fully printed cells exhibited about an order of magnitude improvement in cycle life (Figures 3 and 4). Previously, mechanically assembled cell failure was consistently accompanied by delamination between the cathode and electrolyte layers, but the interfacial cohesion between those two components in printed cells prevents delamination from occurring. Fully printed cells may have longer cycle lives than half printed cells due to similar interfacial cohesion issues at the anode, but additional experiments are necessary to confirm.

Figure 5 presents the full cycle life of a fully printed, thick GPE cell. Testing was terminated beyond 1000 cycles. While the discharge capacity decayed to a level too low for useful work, the cell did not reach a point where it was no longer able to cycle, in contrast to mechanically assembled and half printed cells.

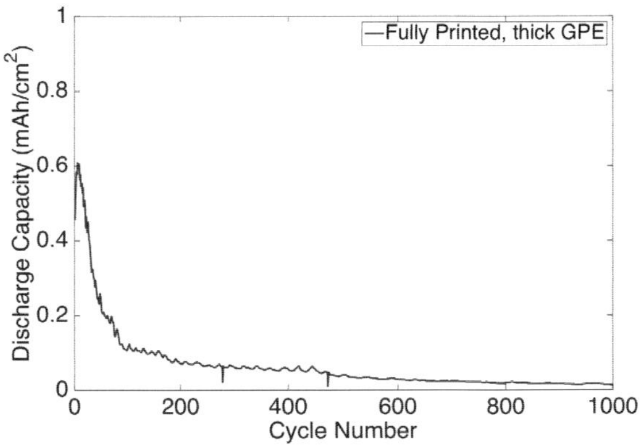

Figure 5. Cycle life of fully printed, thick GPE cell. Cycle life exceed 1000 cycles without cell failure.

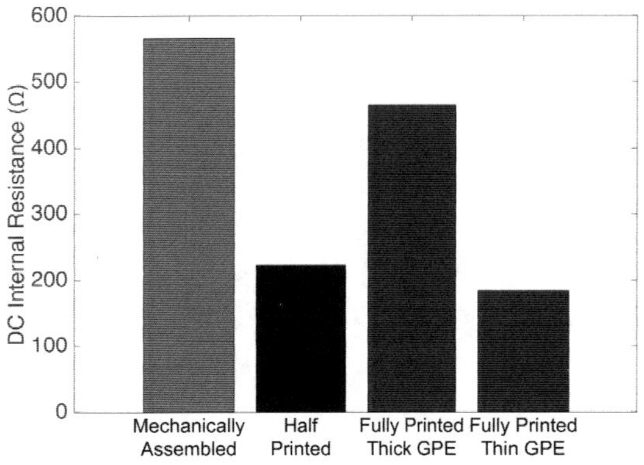

Figure 6. Average internal resistances of all cells at rest.

3.4. Internal Resistance

Figure 6 presents the average DC internal resistances of all cells while at rest between testing cycles. The internal resistances of half printed and fully printed cells are lower than the internal resistance of mechanically assembled cells. While the internal resistance of fully printed thick GPE cells is the highest among all printed cells, the mechanically assembled and half printed cells contained a zinc foil anode, which has lower resistance than the printed zinc anode. However, the fully printed thin GPE cells exhibited the lowest internal resistance of all printed cells, which may be attributed primarily to the difference in GPE thickness between cells as well as improved interfacial cohesion (Figure 2). This lower internal resistance therefore corresponds to higher discharge capacity, as the lower resistive losses due to the thinner GPE layer result in better cell performance (Figure 3).

4. Conclusion

Layer-by-layer fully printed Zn-MnO_2 secondary cells with an ionic liquid-based gel polymer electrolyte were successfully manufactured and cycled. Fully printed cells showed significant improvements in discharge capacity, cycle life, and internal resistance over previously mechanically assembled cells. These improvements are in large part due to the increased interfacial cohesion between layers afforded by direct layer-by-layer printing. Future experiments will optimize ink compositons and process parameters to improve performance metrics and will scale up printing to produce larger cells.

References

[1] Armand M, Endres F, MacFarlane DR, Ohno H and Scrosati B 2009 *Nature Materials* **8** 621-9
[2] Liu Z, El Abedin SZ and Endres F 2014 *J. Solid State Electrochem.* **18** 2683-91
[3] Simons TJ, MacFarlane DR, Forsyth M and Howlett PC 2014 *Chemelectrochem* **1** 1688-97
[4] Kumar GG and Sampath S 2003 *Solid State Ionics* **160** 289-300
[5] Ramesh S and Lu SC 2012 *J. App. Polymer Sci.* **126** E484-92
[6] Winslow R, Wu CH, Wang Z, Kim B, Keif M, Evans J and Wright P 2013 *J. Phys.: Conf. Series* **476**
[7] Ho CC, Evans JW and Wright PK 2010 *J.* Micromech. Microeng. **20**

Rapid fabrication of supercapacitor electrodes using bio-nanoscaffolds in capillary microfluidics

F Zang[1,2], S Chu[1,2], K Gerasopoulos[2], J N Culver[3] and R Ghodssi[1,2,*]
[1] Department of Electrical and Computer Engineering, University of Maryland, College Park, MD 20742, USA
[2] Institute for Systems Research, University of Maryland, College Park, MD 20742, USA
[3] Institute for Bioscience and Biotechnology Research and Department of Plant Science and Landscape Architecture, University of Maryland, College Park, MD 20742, USA

[*] E-mail: ghodssi@umd.edu

Abstract. This paper reports the utilization of capillary microfluidics to rapidly create nanostructure-patterned electrodes for energy storage applications. Using patterned photoresist as open-channel capillary microfluidics, *Tobacco mosaic* virus (TMV) bio-nanoscaffolds suspended in solution are autonomously delivered onto planar gold electrodes over a 1 cm^2 area. The TMVs assemble on the electrode and form a dense bio-nanoscaffold layer due to enhanced evaporation-assisted assembly in the open-channel capillary microfluidic device within an hour. The TMV structures are coated with Ni/NiO through electroless plating and thermal oxidation to form supercapacitor electrodes. The galvanostatic charge/discharge cycle showed a 3.6-fold increase in areal capacitance for the nanostructured electrode compared to planar structures.

1. Introduction

The needs for efficient storage of energy from intermittent sources are driving the development of high efficiency energy storage devices such as battery and supercapacitors [1]. Supercapacitors, which preserve electrical energy through electrochemical reaction, are attractive among various energy storage systems for their higher power and energy densities compared to batteries and conventional dielectric capacitors, respectively [2]. Similarly, nanostructured materials offer significant advantages compared to their bulk counterparts such as improved mechanical stability, faster ion/electron transfer kinetics, higher surface area and higher sensitivity. These properties make them ideal candidates to be used as electrode materials for redox-based supercapacitors to achieve higher energy density and higher power capability.

Novel functionalities can be engineered on biological molecules such as proteins, DNA and viruses. These mutated biological molecules preserve the nanostructures of nature while exhibiting tailored surface attachments. The *Tobacco mosaic* virus (TMV) is a cylindrical plant virus measuring 300 nm in length with a diameter of 18 nm. It can be genetically modified to include cysteine residues for enhanced binding affinity. These TMV mutants possess the ability to self-assemble onto various

[*] To whom any correspondence should be addressed.

substrate surfaces and the capability of being metallized as electrodes. TMVs have previously been demonstrated as promising high surface area nanoscaffolds to self-assemble on electrodes for high energy density microbatteries and supercapacitors [3, 4]. However, the TMV self-assembly process was limited by the long assembly time (3-18 hours), low utilization rate of virus samples, and difficulty in localization, preventing its rapid implementation in microdevices.

In this work, a TMV bio-nanoscaffold supercapacitor electrode was rapidly patterned though the integration of capillary microfluidics in the microfabrication process. The integration of capillary microfluidics leverages the autonomous sample delivery of a capillary device as well as the self-assembly of TMVs, and enables rapid coating of TMV nanoscaffolds in specific areas through room-temperature evaporation with optimal utilization of nanoscaffold samples. Subsequently, metallization with Ni from solution and thermal oxidation are easily performed on the TMV structures to form supercapacitor electrodes. The electrochemical performance of the electrodes is tested through galvanostatic charge/discharge cycle.

2. Methods

2.1. Supercapacitor electrodes

The bio-nanoscaffolded supercapacitor electrode leverages the high aspect ratio characteristic and the self-assembly capability of genetically modified TMV, and is aimed at a significant improvement in areal capacity compared to planar electrodes.

Figure 1. Optical and scanning electron microscopy images showing the TMV nanostructured electrode

The optical and scanning electron microscopy (SEM) images in figure 1 show the morphology of the TMV nanostructured electrode, where Ni/NiO coated TMV nanostructures are patterned in the center of the electrode, forming an active electrode footprint of $1cm^2$. The active electrode area is connected to the surrounding Cr/Au contact pad. As shown in the SEM image, the thin film Ni/NiO coating on the TMVs preserves the rod-like structures of the nanoscaffolds, increasing the active surface area compared with a planar electrode. The 8 μm-wide nanoscaffolded electrodes were separated by 2 μm-wide gaps left over from the capillary microfluidics-assisted fabrication process.

2.2. Capillary microfluidics-assisted nanoscaffolded electrode fabrication

Open-channel capillary microfluidics featuring autonomous liquid sample delivery and enhanced surface evaporation have been demonstrated as a suitable platform for rapid nanostructure assembly in microdevices [5]. The fabrication of the proposed supercapacitor electrodes utilizes a similar capillary microfluidics-assisted process to rapidly pattern nanostructures on a large electrode area with high precision.

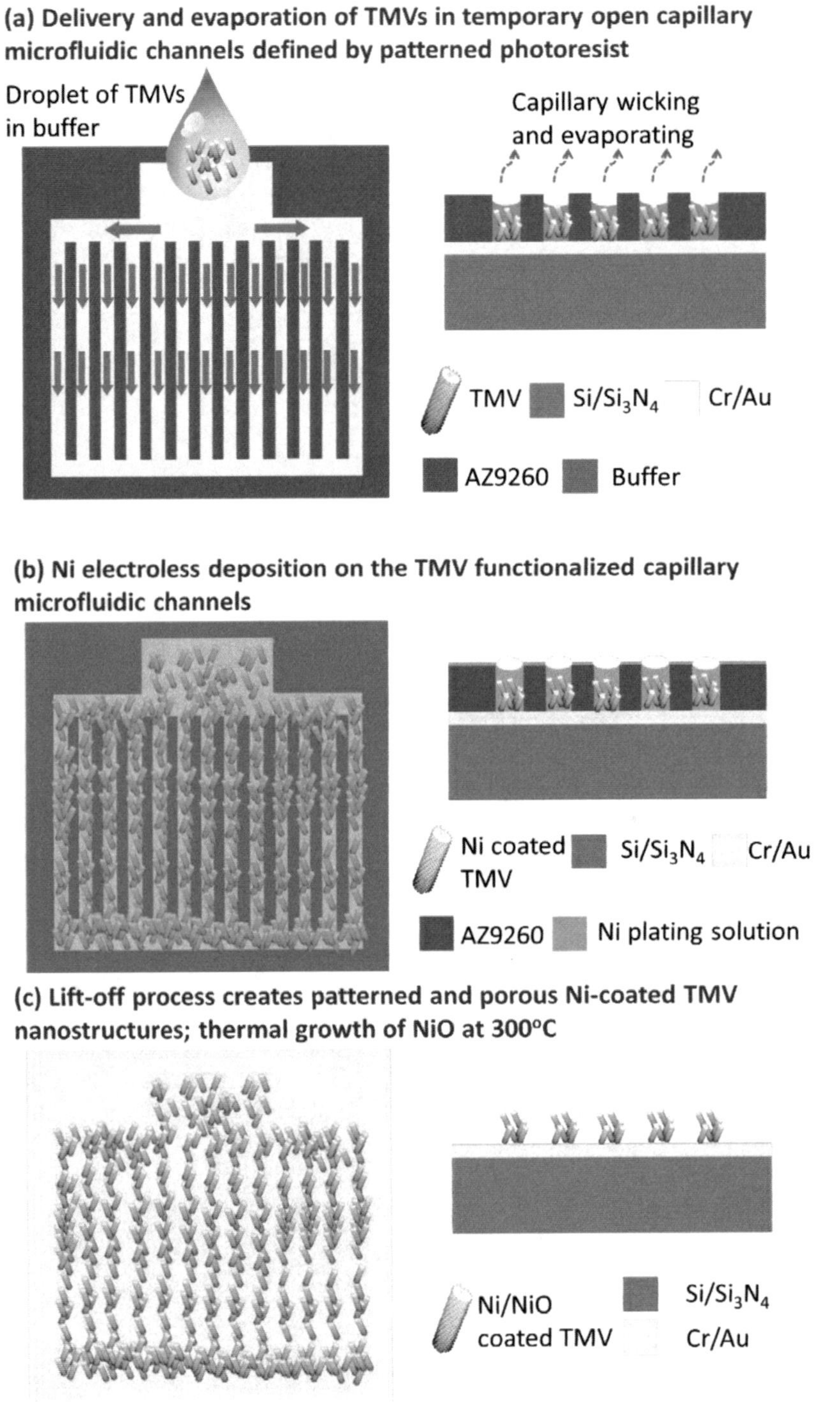

Figure 2. Fabrication process flow of creating TMV nanostructured electrode using capillary microfluidics and one-step photolithography

The microfabrication process flow to create a TMV-nanostructured electrode using temporary open capillary microfluidic patterns is demonstrated in figure 2. One-step photolithography was performed to create microstructured strips with 8 μm of thickness and 2 μm of width on AZ9260 photoresist. The surface was rendered hydrophilic by exposing the surface to O_2 plasma at 50 W for 60 seconds. The gaps between photoresist strips then served as the open capillary microfluidic channels. TMV nanoscaffolds suspended in deionized water were autonomously delivered on the capillary channel patterns from one end of the electrode over the whole active electrode surface due to capillary effects. The capillary microfluidic layer has its top surface exposed to the air, allowing for the rapid evaporation of TMV solution and accelerating TMV assembly on electrodes (figure 2a). The TMV nanoscaffolds were attached on the active electrode area and concentrated in the capillary microfluidic channels after one hour of room-temperature buffer evaporation. The electrode surface was then metalized though electroless plating of Ni, creating Ni-coated TMV-nanostructured electrode (figure 2b). The successive photoresist stripping in acetone removed the temporary capillary microfluidic channels. The NiO layer was formed on the nanostructured Ni surface in a furnace at 300 °C over 2 hours (figure 2c).

2.3. Equipment and test setup

The performance of the fabricated Ni/NiO nanoscaffolded supercapacitor electrodes was tested using redox cycling in a three-electrode setup. The test setup utilized the fabricated Ni/NiO working electrodes, commercial Ag/AgCl reference electrode (CH Instruments, TX, USA) and 2 cm^2 of platinum foils as the counter electrode. The constant current density charge and discharge tests were performed in 2 M KOH solution with a VSP-300 potentiostat (Bio-Logic Science Instruments SAS, France).

3. Results and discussions

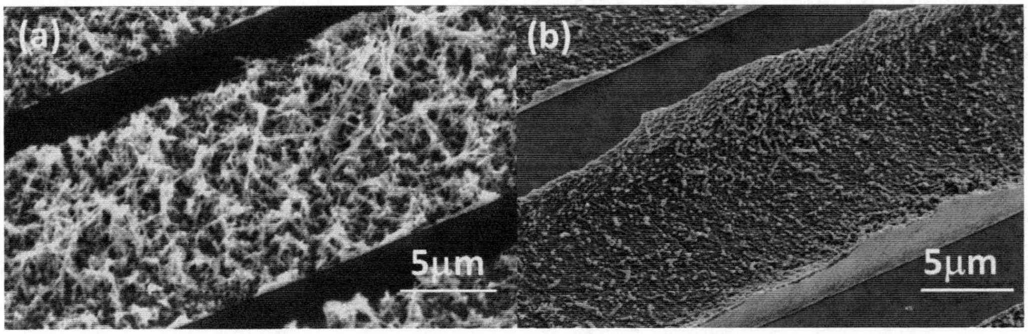

Figure 3. SEM images showing TMV nanoscaffolded Ni/NiO electrodes fabricated by evaporation of (a) 0.2 mg/ml and (b) 2 mg/ml of TMVs in capillary microfluidic channels

The SEM images in figure 3 show the metallized TMV-nanoscaffolded electrode patterns fabricated in the capillary microfluidics. The SEM images in figure 3a shows the morphology of nanostructured Ni/NiO electrode fabricated using the TMV nanoscaffolds in 50 μl of DI water at the concentration of 0.2 mg/ml. After autonomous delivery and the subsequent evaporation of the TMV solution, the electrode surface demonstrates clear porous structures composed of TMV nanorods in random orientations. However, as the TMV concentration further increases (2 mg/ml), the nanorods are closely stacked during evaporation, forming a thick film with much lower porosity after metallization (figure 3b). This would likely yield a decreased areal capacitance. Therefore, to ensure an enhanced surface area of the supercapacitor electrode, the TMV suspension with the concentration of 0.2 mg/ml is adopted to create porous TMV Ni/NiO electrode in the following electrochemical tests.

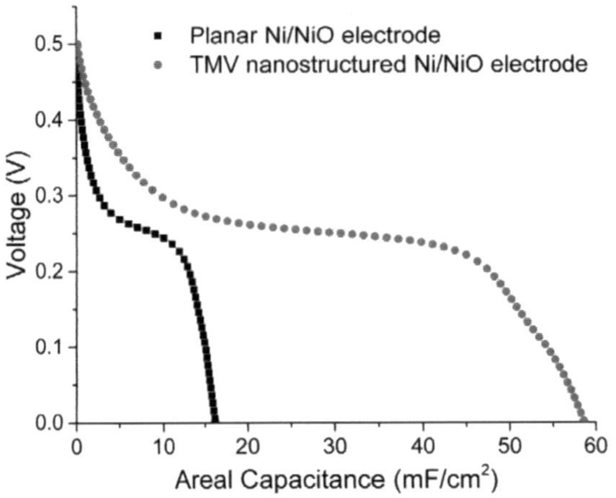

Figure 4. Discharge characteristics of the Ni/NiO coated planar and TMV nanostructured electrodes

The electrochemical performance of the NiO supercapacitor electrodes was tested by charge and discharge cycles at constant current of 500 μA (active electrode area of 0.8 cm^2). Figure 4 shows the discharge characteristics of both the planar and TMV nanostructured Ni/NiO electrodes at the 30th cycle. The TMV nanostructured electrode showed an approximate 3.6-fold increase in areal capacitance compared to the planar Ni/NiO electrode, validating the significant increase in the active surface area of NiO using the nanostructured electrode fabricated in capillary microfluidics. The capacitance increase is lower than the reported 7-fold increase [4]. This is caused by the occupation of microfluidic channel sidewalls on chip which results in the reduction of active electrode area, and the possible degradation in porosity of TMV layer during evaporation.

4. Summary

This is the first demonstration of using capillary microfluidics as an enabling technology to rapidly fabricate nanostructure-patterned electrodes for energy applications. The TMV bio-nanoscaffolds were autonomously delivered in temporary photoresist-patterned open capillary microfluidic channels, and functionalized the planar electrode with patterned high-surface-area nanostructures within an hour. This technology enables higher areal capacitance within a small footprint, efficiently utilizes minimum amount of nanostructure samples to functionalize a device surface, and will contribute to rapid fabrication of nanostructured energy storage devices in symmetric/asymmetric configurations using one-step photolithography.

Acknowledgments

The work was supported by the Army Research Office Biochemistry Program and National Science Foundation Nanomanufacturing Program. The authors would like to acknowledge the staff in Maryland NanoCenter for the support in fabrication process.

References

[1] Zhao X, Sanchez B M, Dobson P J and Grant P S 2011 *Nanoscale* **3** 839-855
[2] Jiang H, Ma J and Li C 2012 *Adv. Mater.* **24** 4197-4202
[3] Gerasopoulos K, Pomerantseva E, McCarthy M, Brown A, Wang C, Culver J N and Ghodssi R 2012 *ACS Nano* **6** 6422-6432
[4] Chu S, Gerasopoulos K and Ghodssi R 2015 *Proc. IEEE MEMS 2015* 1118-1121
[5] Zang F, Gerasopoulos K, McKinzie K, Culver J N and Ghodssi R 2015 *Solid-State Sensors, Actuators and Microsystems (TRANSDUCERS) 2015* 548-551

A low frequency vibration energy harvester using dual Halbach array suspended in magnetic springs

M Salauddin, M A Halim and J Y Park[1]
Micro/Nano Devices and Packaging Lab., Dept. of Electronic Engineering, Kwangwoon University, 447-1 Wolgye-dong, Nowon-gu, Seoul, 139-701 Korea.

E-mail: jaepark@kw.ac.kr

Abstract. An electromagnetic (EM) low frequency vibration energy harvester is newly developed based on dual Halbach array which is suspended in two magnetic springs. Each Halbach array concentrates the magnetic flux lines on one side of the array while suppressing the flux lines on the other side. Dual Halbach array allows the concentrated magnetic flux lines to interact with the same coil in a way where maximum flux linkage occurs. With the goal of higher power generation in low amplitude and low frequency vibrations, the magnetic structures (both the dual Halbach array and the magnetic springs) were optimized in terms of operating frequency and power density. A prototype was fabricated and tested. It is capable of delivering maximum 1.09mW average power to 44Ω optimum load at 11Hz resonant frequency and 0.5g acceleration. The prototype device offers 33.4μWcm^{-3} average power density which is much higher than recently reported electromagnetic energy harvesters.

1. Introduction

With current development in microelectronic technology, the dimension and the power expenditure of electronic components and integrated electronic systems have been meaningfully decreased. These formations qualify applications requiring high mobility and long term freedom. Generally permanent energy alternatives such as batteries have the limitations of limited lifespan, chemical hazards and sometimes larger size compare with micro-electromechanical system (MEMS) devices. Energy harvester is a possible substitute of batteries that converts the existing environmental energy into usable electric energy. The ambient energy scavenging, as known as energy harvesting or power harvesting, is the process of obtaining usable energy from natural and human-made sources that surround us in the everyday environment. General ambient sources are solar, acoustic noise, thermal, biochemical, radio waves, and vibration etc. Commonly used vibration based energy harvesting techniques are electromagnetic [1], electrostatic [2], and piezoelectric [3]. The electromagnetic energy harvesting technique offers high energy conversion efficiency and very low frequency operation due to the easy mechanical resonator composition of the device [4-5].

[1] To whom any correspondence should be addressed.

A suitable electromagnetic vibration energy harvester can be used to power various consumer electronics such as mobile phones, wrist watches, audio devices, hearing aids and implanted biomedical devices. Energy harvesting from low frequency vibration is attractive due to its availability around the ambient environment. But significant power generation at low frequency vibration is challenging because power flow decreases with the decrease in frequency. Moreover, designing a spring-mass system suitable for low frequency vibration energy harvesting is difficult. In an electromagnetic vibration energy harvester, the resonance frequency depends on the spring characteristics and magnet characteristics such as grade, structure, and mass. The selection of spring (material, structure and shape) plays an essential role in the performance of the harvester [6]. In order to address low frequency harvester design challenge, EM energy harvesters using magnetic spring was proposed in [7]. This paper is mainly focused on Halbach array based electromagnetic energy harvester. A Halbach array is an arrangement of magnets which increases the magnetic field in one side of the array, minimizing the magnetic field to nearly zero on the other side of the array [8]. Generally Halbach array can enhance the magnetic field intensity and reduce the overall volume of the electromagnetic energy harvester [9]. Halbach array in an EM energy harvester increases the magnetic flux density as well as the power generation at low frequencies which was addressed in [10,11] instead of a single magnet.

2. Harvester design and fabrication

Figure 1 shows the schematic structure of the proposed magnetic spring based low frequency vibration energy harvester using dual Halbach array. The electromagnetic low frequency energy harvester is made of dual Halbach array, magnetic springs and three copper coils. The dual Halbach array structure is suspended between two magnetic springs. Each Halbach array is formed using seven block ($10\times5\times2mm^3$) magnets. Dual array structure is made by attaching the Halbach array to a hollow rectangular shape polycarbonate structure. The air gap between two Halbach arrays is 14 mm and the volume of single Halbach array is ($35\times10\times2mm^3$). Each magnetic spring is formed by a magnet ($5\times5\times2mm^3$) fixed at the end cover of the prototype and a magnet ($5\times5\times2mm^3$) fixed at the side of the array structure in a repulsive manner. A bobbin containing 3 series connected coils (each 400 turns) is placed in the middle of the hollow Halbach array structure. Two polycarbonate spacers ($10\times5\times2mm^3$) is placed at both sides of Halbach array so that the flux lines of the magnets attached to the Halbach array frame do not interact with those of the spring magnets fixed to the Halbach array frame. In order to determine the position of each Halbach array and two other magnets with the hollow rectangular Halbach array structure, finite element simulation was done so that no significant magnetic interaction occurs among the magnets as shown in figure 2. All the components were assembled to make the prototype energy harvester and tested to observe its output performance as shown in figure 3. Each magnetic spring is formed by two spring magnets facing similar poles. During experiment, vibration was applied in horizontal direction in order to

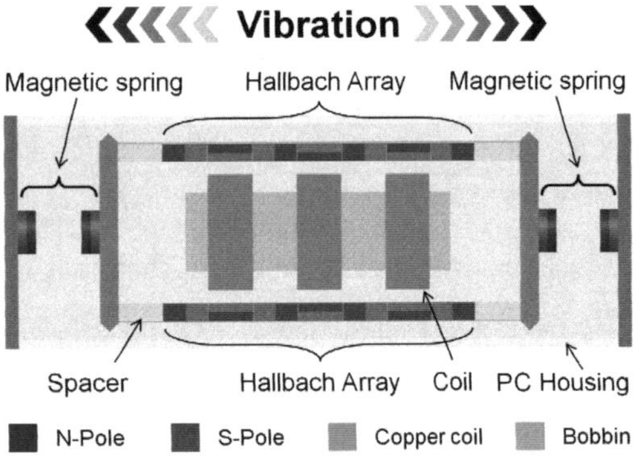

Figure 1. Schematic structure of the proposed dual Halbach array based electromagnetic energy harvester.

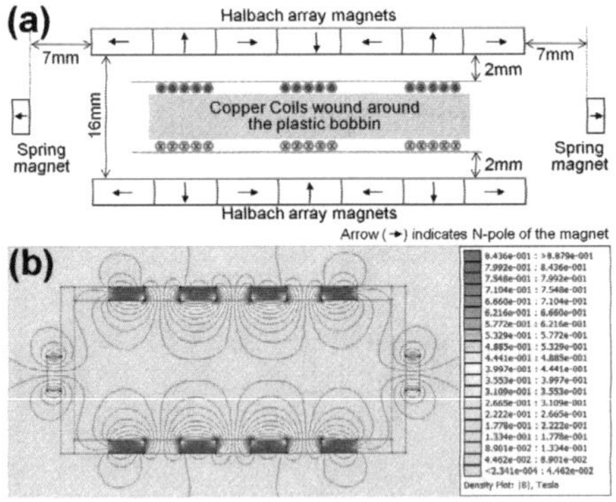

Figure 2. Finite element simulation by FEMM: (a) Schematic of simulation parameters and (b) magnetic flux lines.

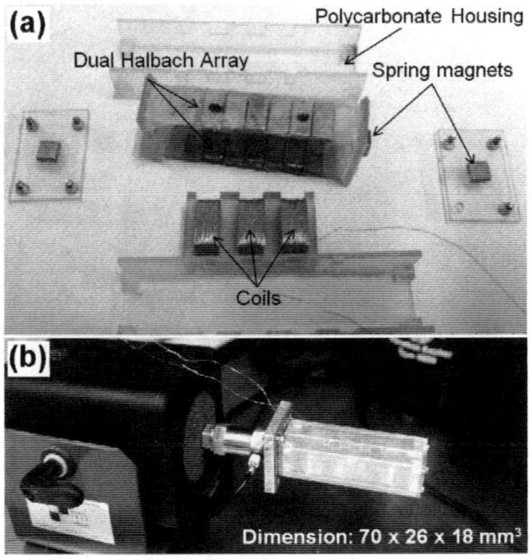

Figure 3. Photographs of the energy harvester prototype: (a) Components before assembling and (b) assembled prototype mounted on the shaker.

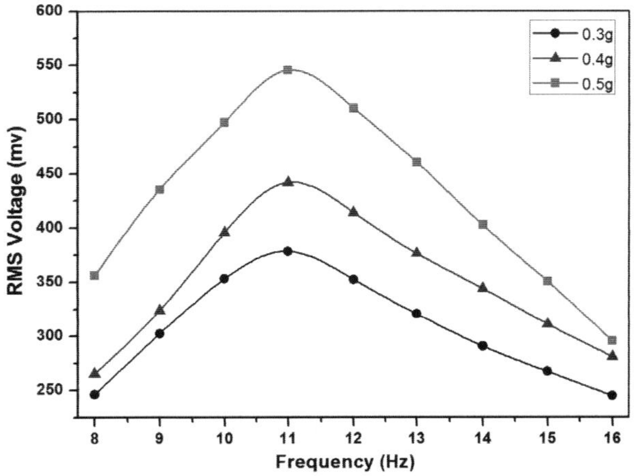

Figure 4. Frequency response of the dual Halbach array based EM energy harvester under various base excitations (accelerations).

reduce the gravity effect on the Halbach array structure. The dual Halbach array based generator structure has been modelled using finite element simulation by FEMM in order to realize the forces created by spring magnets and to determine the Halbach array position and flux linkage with the coil. Figure 2(b) shows the results of FEMM simulation of the corresponding Halbach array generator structure of figure 2(a), showing magnetic flux density at 2mm distance is 0.31T. According to Faraday's Law of electromagnetic induction, the induced open circuit e.m.f voltage (time varying) $V_{oc}(t)$ is given by

$$V_{oc}(t) = -N\frac{d(\int \vec{B}.d\vec{A})}{dt}\dot{z}(t) = NBl\dot{z}(t) \quad (1)$$

where $(\int \vec{B}.d\vec{A})$ is the net magnetic flux through the differential component area $d\vec{A}$ of the magnet coil assembly. N is the number of coil turns, B is the magnetic flux density, l is the coil length across the magnetic flux lines, and $\dot{z}(t)$ is the relative velocity between magnet and coil. The output power delivered to the load connected to the coil terminals at the resonance condition is [12]

$$P = \frac{\omega_n^2 Y_0^2 (NBl)^2}{8\xi^2(R_L + R_C)} \quad (2)$$

where, ξ is the damping ratio, R_L is the load resistance and R_C is the coil resistance. According to maximum power transfer theorem, maximum power is delivered when the load resistance R_L is equal to the coil resistance R_C.

3. Experimental results and discussion

The fabricated dual Halbach array energy harvester prototype was mounted on an electrodynamic vibration exciter (TMS 2004E) connected to a function generator (Agilent 33250A) in conjunction with a power amplifier (TMS 2004E21) which provide sinusoidal excitation of various frequencies and accelerations to the harvester prototype. The amplitude of the input vibration was monitored by a reference accelerometer (Type 8305; Brüel &Kjær) attached to the base of the vibration exciter, connected to a measuring amplifier (Type 2525; Brüel & Kjær). A digital storage oscilloscope (Tektronix TDS5052B) was connected to the harvester outputs in order to measure and record its output response.

Figure 4 shows the frequency response of the energy harvester prototype under various input accelerations. Results show that the device has a resonant frequency of 11Hz at which the output voltage is maximum. At resonance, 378.3mV, 441.7mV and 545mV RMS voltages are obtained while vibrated at 0.3g, 0.4g, and 0.5g accelerations, respectively. Figure 5 plots the measured RMS output voltage across the load resistances and the average power delivered to the loads at a resonant frequency of 11Hz under various acceleration of vibration. The voltage across the load increases with the increase in the load resistance values. However, a maximum average power of 447μW, 726μW, and 1093μW were delivered to the optimum load resistance of 44Ω at 0.3g, 0.4g, and 0.5g accelerations, respectively. As the acceleration increases, the output load voltages and power increases linearly. The generated average power is experimentally equal to V_{RMS}^2/R_L ; where V_{RMS} is the RMS load voltage across the load R_L.

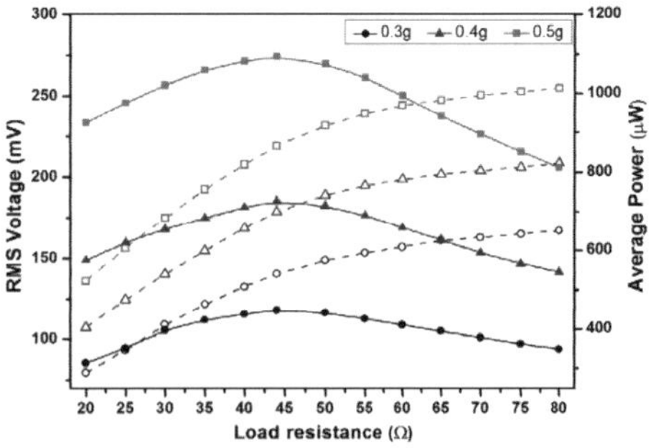

Figure 5. RMS load voltage (dotted lines) and average output power (solid lines) against load resistances at 11Hz resonance frequency.

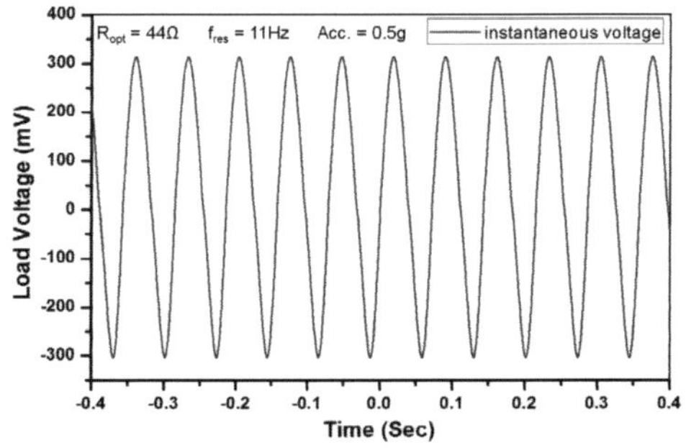

Figure 6. Instantaneous voltage waveform across the optimum load resistance at 11Hz resonant frequency and 0.5g acceleration.

Figure 6 shows the instantaneous voltage (619mV peak-peak voltage) waveform generated by the harvester prototype across the optimum load resistance 44Ω at 11Hz resonance condition and 0.5g acceleration. The device shows much better output performance than a number of recently reported electromagnetic energy harvesters in terms of both output power and average power density as shown in Table 1. This device is not properly optimized so that this device resonance frequency is 11Hz and other reported electromagnetic energy harvesters have the resonant frequency below 10Hz which were intended to be used in human-body-induced motion applications. In the next stage of the work, proper optimization will be done by improved design to reduce the resonant frequency so it can be used in human-body-induced motion applications. Moreover, reducing the volume and friction of the Halbach array structure with the housing of the harvester is in our consideration. The dual Halbach array generator presented here has been shown to generate meaningful power from low frequency fitful behaviour. The magnetic springs in this dual Halbach array structure has several benefits such as reliability, simple structure and easy to use at resonance frequency conditions.

Table 1. Performance comparison with recently reported EM energy harvesters.

Reference number	Operating condition	Acceleration(g) g=9.8 ms^{-2}	Average Power (μW)	Average Power Density (μWcm^{-3})
[7]	8Hz	0.04g	14.55	2.85
[13]	4.6Hz	2g	110	15.4
[14]	5.8Hz	2g	103.55	5.4
[15]	10Hz	1g	13.6	2.85
This work	*11Hz*	0.5g	*1093*	*33.4*

4. Conclusions

The proposed dual Halbach arrays energy harvester device has been designed, fabricated, and tested. Arranging magnets in Halbach array configuration is impressive because the flux density increases on one side of the array and reduces to nearly zero on the other side which in turn, reduces overall volume of the harvester. The goal of this work is to increase the power at low amplitude and low frequency vibrations by optimizing the magnetic spring structures. Our fabricated prototype offers maximum of 1.09mW average power to 44Ω optimum load at a resonant frequency of 11Hz and input acceleration of 0.5g. The prototype shows much better output performance than other recently reported electromagnetic energy harvesters in terms output power and average power density. The next stage of this work will include further optimization of the prototype to be used in energy harvesting from different human-body-induced motions such as hand-shaking, walking, running, cycling etc.

Acknowledgements

The authors are grateful to acknowledge the Basic Science Research Program (2013R1A1A2A 10064810) and the Pioneer Research Centre Program (2010-0019313) through the National Research Foundation of Korea (NRF) funded by the Ministry of Science, ICT and Future Planning, Korea.

References

[1] Beeby S P, Torah R N, Tudor M J, Glynne-Jones P, O'Donnell T, Saha C R and Roy S, 2007 *J.Micromech. Microeng.* **17** 1257–65
[2] Nguyen S D and Halvorsen E, 2011 *J. Microelectromech. Syst.* **20** 1225-1227
[3] Elfrink R, Kamel T M, Goedbloed M, Matova S, Hohlfeld D, Andel Y V and Schaijk R V, 2009 *J.Micromech. Microeng.* **19** 094005
[4] Liu H, Qian Y and Lee C, 2013 *Sens. Actuators A* **204** 37–43
[5] Cheng S and Arnold D P, 2010 *J. Micromech. Microeng.* **20** 025015
[6] Lee J M H, Yuen S C L, Li W J and Leong P H W, 2003 *IEEE Int.Symp.Circuit Syst.* **4** 876-879
[7] Saha C R, O'Donnell T, Wang N, and McCloskey P, 2008, *Sens. Actuators A*, **147** 248-253
[8] Zhu D, Beeby S, Tudor J and Harris N, 2012 *Smart Mater. Struct.* **21** 075020
[9] Zhu D, Beeby S, Tudor J, and Harris N, 2013 *Sens. Actuators, A*, **203** 11-19
[10] Liu H, Qian Y, Wang N, and Lee C, 2013 *Proc. MEMS,* 829-832
[11] Wang Y J, Chen C D, Sung C K, and Li C, 2012 *J. Intel. Mat. Syst. Struct.*, **23** 933-943
[12] Awaja N, Sood D, Vinay T, 2009 *J. sensors & Transducers* **103** 109-121
[13] Halim M A and Park J Y, 2014 *J. appl. Phys.* **115** 094901
[14] Halim M A and Park J Y, 2015 *Sensors Actuators A* **229** 50-58
[15] Galchev T, Kim H, and Najafi K, 2011 *J. Microelectromech. Syst.* **20** 852-866

Experiments on power optimization for displacement-constrained operation of a vibration energy harvester

Binh Duc Truong, Cuong Phu Le and Einar Halvorsen

Department of Micro- and Nanosystem Technology, Buskerud and Vestfold University College, Campus Vestfold, Raveien 215, 3184 Borre, Norway

E-mail: `Einar.Halvorsen@hbv.no`

Abstract. This paper presents experiments on how to approach the physical limits on power from vibration energy harvesting under displacement-constrained operation. A MEMS electrostatic vibration energy harvester with voltage-control of the system stiffness is used for this purpose. The power saturation problem, when the proof mass displacement reaches maximum amplitude for sufficient acceleration amplitude, is shifted to higher accelerations by use of load optimization and tunable electromechanical coupling k^2. Measurement results show that harvested power can be made to follow the optimal velocity-damped generator also for a range of accelerations that implies displacement constraints. Comparing to the saturated power, the power increases 1.5 times with the optimal load and an electromechanical coupling k^2=8.7%. This value is 2.3 times for a higher coupling k^2=17.9%. The obtained system effectiveness is beyond 60% under the optimization. This work also shows a first demonstration of reaching optimal power in the intermediate acceleration-range between the two extremes of maximum efficiency and maximum power transfer.

1. Introduction

Microelectromechanical systems (MEMS) vibration energy harvesting (VEH) is a potential approach to autonomously supply power for wireless sensors. When the harvested power is sufficient to operate the electronic system, use of inconvenient batteries can be eliminated. A typical VEH design is a spring-mass system. The proof mass motion induced by ambient vibrations leads to energy conversion based on either of three basic mechanisms: piezoelectric, electromagnetic and electrostatic [1, 2]. For low-loss resonant harvesters at the microscale, displacement-constrained operation is easily encountered because of the limited space available on chip. Beyond a critical acceleration sufficient to drive the proof mass displacement to its maximum amplitude, the power is saturated [3, 4] and left increasingly far below the theoretical power bound for displacement-limited operation [5]. This bound is conservatively approximated by the optimal velocity-damped generator (VDRG) whose power increases linearly with acceleration. The question is therefore how to avoid saturation and further improve power for acceleration amplitudes beyond the critical value? Utilization of transducing end-stops in our previous impact-device concept [6] somewhat overcomes this problem, but it demands complex optimization of the device design to further increase power in the impact regime. In addition, for a two-port harvester operating under displacement-constrained operation it has been shown that

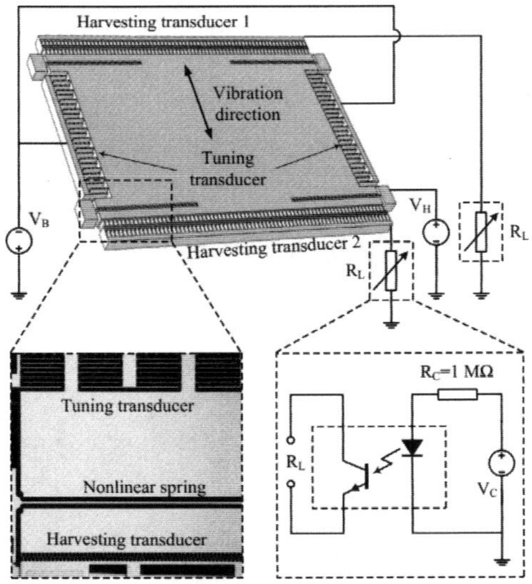

Figure 1. Key features of the device design, sketch of the electrical setup with the load tuning control. A close-up view of device fabricated using the SOIMUMPS process with the device layer thickness of t = 25 μm.

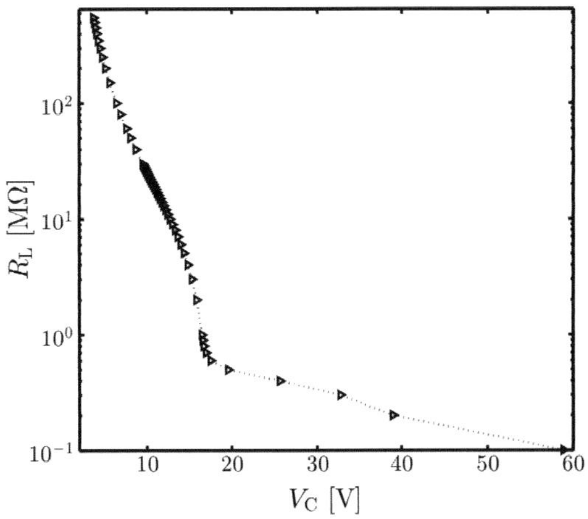

Figure 2. Characterization of the opto-coupling resistor under driving of the voltage source V_C.

it is ultimately optimal to maximize efficiency [7] which is equivalent to maximizing electrical damping. Hence, maximum efficiency and maximum power transfer (unconstrained motion) constitute two extremes, but the intermediate optimization cases have not been studied.

In this work, we investigate an approach to optimize power in both damping-limited regime and displacement-limited regime. Harvester power is then maximized in the intermediate range. The approach is motivated by electrical damping being the control parameter in optimization of the VDRG [2]. We here consider load-resistance optimization and adjustment of the electromechanical coupling. Both these factors directly affect the electric damping and are used to keep the displacement amplitude at the limit for the VDRG. It should be noted that a resistive load such as we use here can emulate a buck-boost converter that has no input filter capacitor [8]. It is therefore much more than an experimental convenience. For the experiments we employ a previous large-frequency-tuning-range resonator device [9] as a MEMS electrostatic vibration energy harvester. The system stiffness and the electromechanical coupling can be adjusted by an applied voltage. This, together with an electrically controllable load resistance let us explore the optimization problem.

2. MEMS device

Figure 1 shows key features of the electrostatic device. The harvesting transducers are two anti-phase comb-drive capacitance structures with a nominal capacitance C_0=0.47 pF. The proof mass is suspended by four single beams. The restoring force is designed to have hardening nonlinearity. The device stiffness is tuned by a bias control V_T=V_H+V_B of the tuning transducer which is a gap-closing capacitor structure. When the electrodes of the harvesting transducers are short-circuited, the net force is approximated by

$$F_{\text{net}}(x) = (k_{\text{m}} - \frac{C_{\text{T}}}{g^2}V_{\text{T}}^2)x + (\frac{k_{\text{m}}}{l^2} - \frac{2C_{\text{T}}}{g^4}V_{\text{T}}^2)x^3 \quad (1)$$

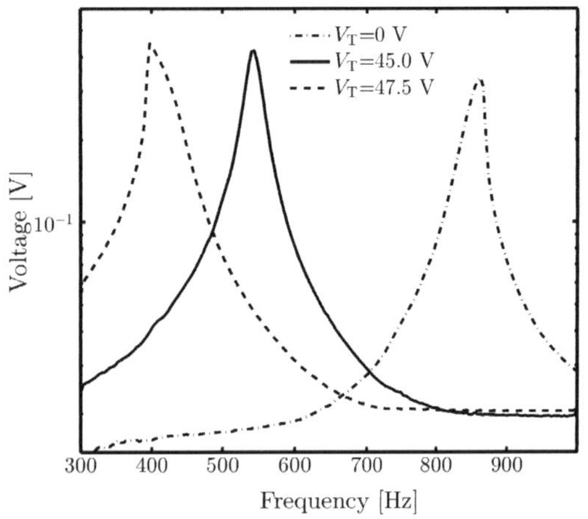

Figure 3. Measured frequency responses for increase of the tuning voltage V_T at a small acceleration A= 0.21 g and the load R_L=20 MΩ.

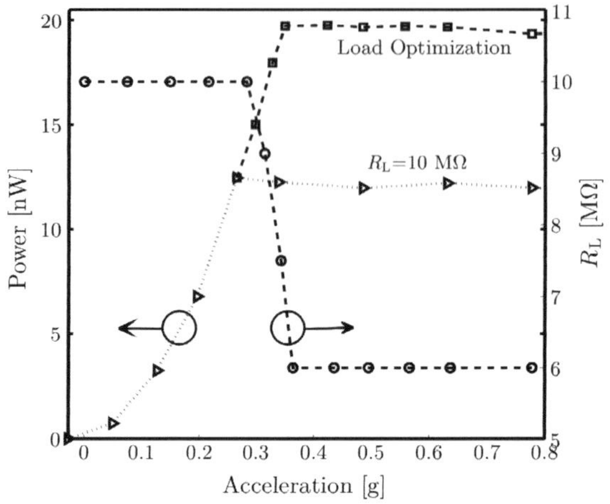

Figure 4. Measured power in damping-limited regime and displacement-limited regime with corresponding optimal load for V_T=45.0 V.

where k_m is the linear stiffness of the restoring force, $l=\sqrt{\frac{735}{512}}w$, w is the beam width, g and C_T are the nominal gap and the nominal capacitance of the tuning transducer. The generated power is obtained by connection of the fixed-electrode of the harvesting transducers to the variable load R_L. To avoid cumbersome resistor changes, the load used in the experiment is the opto-coupling resistor V0617A with low coupling capacitance. The load value R_L is set by a voltage V_C and a series resistance R_C. Figure 1 also displays a close-up view of the device, which is fabricated using the SOIMUMPS process with a device layer thickness of t=25 μm. The nonlinear spring and a part of all transducers are shown in the optical micrograph. By design, the maximum amplitude of the proof mass displacement X_{max}=5.5 μm is defined by rigid end-stops. Further details of the device parameters can be found in [9].

3. Measurements

The opto-coupling resistor is characterized under control of the voltage V_C and a series resistance R_C=1 MΩ as shown in figure 2. The optical coupling between the diode emitter and the phototransistor leads to the variable resistance that is high at low V_C and vice-versa. The load can be adjusted from 500 MΩ down to 100 KΩ when V_C varies from 3.7 V to 58.9 V. This characterization is used in all measurements of the device. The bias voltage for the harvesting transducers is chosen as V_H=45.0 V. Figure 3 shows the measured frequency responses at the small acceleration regime when the proof mass displacement is still below the limit X_{max}. The hardening effect due to the nonlinear spring is evident for V_T=0.0 V with a center frequency 831.5 Hz. The effective stiffness of the system reduces with increase of the bias V_T of the tuning transducer, giving higher output voltages and lower center frequencies. The system response is roughly linear for V_T=45.0 V, which gives a center frequency f_c=531.5 Hz. The device exhibits softening effects for V_T=47.5 V. The critical voltage that causes pull-in instability is estimated to $V_{cr} \sim 50$ V. We now use V_T = 45.0 V as a case for load optimization in both the damping-limited regime and the displacement-limited regime.

Figure 4 shows measured powers and corresponding optimal load for each RMS acceleration amplitude at the center frequency f_c. In the damping-limited regime, the optimal load is R_L=10 MΩ in this case. Keeping this resistance while varying the acceleration, the power saturates at

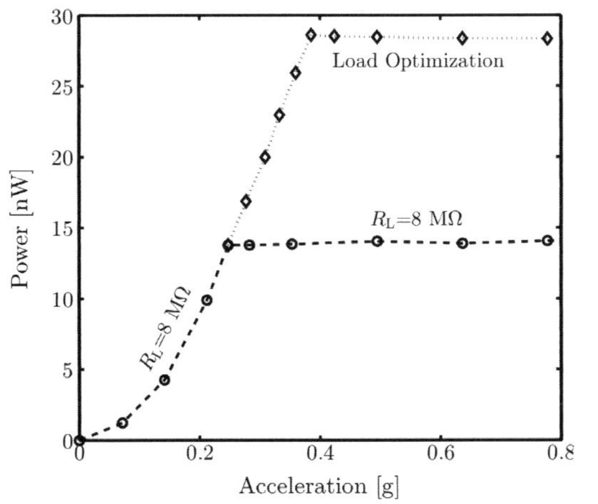

Figure 5. Measured power both with and without load optimization for V_T=47.5 V.

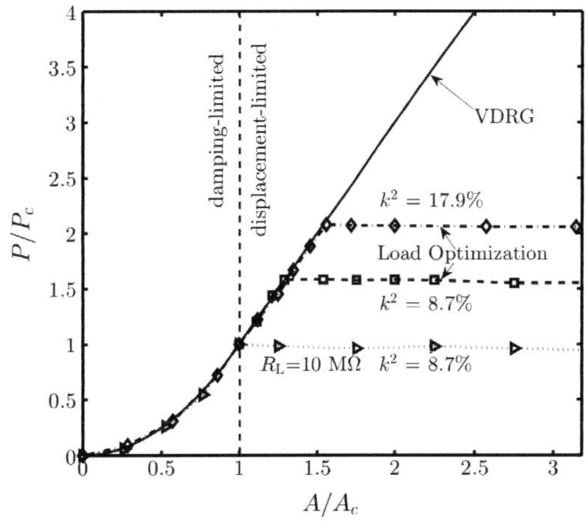

Figure 6. Comparison of the maximum power to the optimal velocity-damped generator VDRG for $k^2 = 8.7\%$ (V_T=45.0 V)and $k^2 = 17.9\%$ (V_T=47.5 V).

P_c=12.4 nW for accelerations larger than a critical value A_c=0.28 g. We interpret this as the proof mass displacement reaching the maximum X_{max} at A_c and that it hits the end-stops for $A > A_c$. However, the power can be improved for $A > A_c$ by adjusting the load separately for every acceleration amplitude. The measured result shows that the power can be improved for acceleration amplitudes between $A > A_c$ and 0.36 g where the power looks approximately linear in A. For $A > 0.36$ g the power saturates at 19.7 nW. All corresponding optimal loads and accelerations can be found in figure 4. The optimal load only varies between A_c and 0.36 g and is again constant for $A > 0.36$ g. Note that no attempt is made to avoid proof mass impacts on the end-stops. Therefore the power-acceleration curve here is sginificantly different from the theoretical result in [5]. It is reasonable to interpret the optimized value R_L=6.0 MΩ as the resistance value that gives maximum electrical damping. Increase of the tuning voltage V_T leads to a reduced net stiffness and higher electromechanical coupling k^2. This is advantageous to further increase the harvested power under displacement-constrained operation. Figure 5 shows the measured power for V_T=47.5 V when driven at the center frequency f_c=400.0 Hz. The device now reaches a higher critical power of P_c=13.8 nW at a lower A_c=0.25 g, compared to the previous case of V_T=45.0 V. With load optimization for $A > A_c$, the power continues to increase to a maximum of 28.6 nW at A=0.38 g. This power about 1.5 times better than the previous case.

In the linear approximation, the anti-phase comb-drive harvesting transducers are equivalently converted to a two-port model because of decoupling of the common and differential modes [10, 11]. For a displacement-limited two-port device, one can show that load optimization can boost the harvested power to a maximum value of

$$P_{max} \approx \frac{1}{2} k^2 Q_m P_c \qquad (2)$$

where the mechanical quality factor Q_m is estimated from the full bandwidth at half maximum of the open-circuit frequency response and k^2 is the electromechanical coupling of the linear two−port model, evaluated by

$$k^2 = 1 - \frac{f_r^2}{f_{ar}^2} \qquad (3)$$

Table 1. Measured electromechanical coupling factor and figure of merit k^2Q_m at A=0.03 g.

Tuning voltage	Electromechanical coupling k^2	k^2Q_m
V_T= 45.0 V	8.7%	3.2
V_T= 47.5 V	17.9%	4.7

where f_r is the resonant frequency measured with short-circuited output and f_{ar} is the resonant frequency measured with open-circuited output. Table 1 shows the measured coupling k^2 and a corresponding figure of merit k^2Q_m. The achieved power is compared to the maximum possible power of the optimal velocity-damped generator under displacement-constrained operation in figure 6. The power and the acceleration are normalized by the factors P_c and A_c respectively because P/P_c is a universal function of A/A_c for all VDRGs. The comparison in the damping-limited regime shows that the power obtained with load optimization closely approaches the optimal VDRG up to a maximum acceleration beyond A_c. In the displacement-limited regime, the range of accelerations where the optimum can be followed depends on k^2Q_m as is seen by comparing the maximum power $1.54P_c$ at A=$1.28A_c$ for k^2=8.7% to the maximum power $2.07P_c$ at A=$1.55A_c$ for k^2=17.9%. The corresponding estimates of maximum power from (2) are respectively $1.6P_c$ and $2.4P_c$ giving best correspondence with the lowest-coupling configuration which is also the most linear one. It is noteworthy that even though the high-coupling configuration exhibited clear softening nonlinearities, it follows the optimal VDRG curve as closely as the other alternative until its maximum is reached.

4. Conclusion

Electrical damping is controlled through load resistance and electromechanical coupling to maximize power for a vibration energy harvester under displacement-constrained operation. The measured power closely follows the optimal VDRG even between the two extremes of unconstrained proof mass motion and displacement-limited operation with saturated power. The load optimization makes a gradual transition between these two extremes which we can think of respectively as maximum power-transfer and maximum efficiency. For displacement-limited operation, there are significant improvements in power from increasing electromechanical coupling even for a high-coupling device.

Acknowledgment

This work was supported by the Research Council of Norway through Grant no. 229716/E20.

References

[1] T. Starner, *IBM Systems J.,* **35** 618 (1996)
[2] P. D. Mitcheson et. al., *Proceedings of the IEEE,* **96** 1457 (2008)
[3] M. S. M. Soliman et. al., *J. Micromech. Microeng.,* **18** 115021 (2008)
[4] D. Hoffmann, B. Folkmer and Y. Manoli, *J. Micromech. Microeng.,* **19** 094001 (2009)
[5] E. Halvorsen e.t al., *J. Phys.: Conf. Ser.* **476** 012026 (2013)
[6] C. P. Le, E. Halvorsen, Søråsen O and Yeatman E M, *J. of Int. Mat. Syst. Struc.,* **23** 1409 (2012)
[7] M. Renaud et. al., *J. Micromech. Microeng.,* **22** 105030 (2012)
[8] R. D'hulst, T. Sterken, R. Puers, G. Deconinck and J. Driesen, *IEEE Trans. Indust. Electro.,* **57** 4170 (2010)
[9] C. P. Le and E. Halvorsen, *Proceedings of the Transducers & Eurosensors XXVII,* 1352 (2013)
[10] F. Peano and T. Tambosso, *J. Microelectromech. Syst.,* **14** 429 (2005)
[11] H. A. C. Tilmans, *J. Micromech. Microeng.,* **6** 157 (1996)

Low-Frequency, Low-G MEMS Piezoelectric Energy Harvester

R Xu, S G Kim

Department of Mechanical Engineering, Massachusetts Institute of Technology, Cambridge, MA 02139, USA

Email: rzxu@mit.edu

Abstract. This paper reports the design, modeling and fabrication of a novel MEMS device for low-frequency, low-g vibration energy harvesting. The new design is based on bi-stable buckled beam structure. To implement the design at MEMS scale, we further proposed to employ residual stress in micro-fabricated thin films. With an electromechanical lumped model, the multi-layer beam could be designed to achieve bi-stability with desired frequency range and excitation amplitude. A macro-scale prototype has been built and tested to verifies the prediction of the performance enhancement of the bi-stable beam at low frequencies. A MEMS scale prototype has been fabricated and tested to verify the frequency range at low excitation amplitude. The MEMS device shows wide operating frequency range from 50Hz to 150Hz at 0.2g without external proof mass. The same device with external proof mass has lower frequency range (< 10Hz) with boosted deflection amplitude.

1. Introduction

Energy harvesting at MEMS scale will promisingly advance exciting applications such as large wireless sensor networks, Internet of Things etc. due to its high power generation efficiency and small form factor. However, the mismatches between the operating conditions of current MEMS energy harvesters and the characteristics of the ambient vibrations have long hindered wide applications of this technology. For instance, our previous MEMS device based on nonlinear resonance has achieved high output power of 45μW and wide bandwidth of >20%, while the operation frequency and vibration amplitude are still high (>1000Hz at 4g) [1]. Our new design is aiming to address the remained challenges of applying the energy harvesting technology to the real environment – working at low frequencies (<100Hz) and low g's (<1g).

Bi-stable oscillator based energy harvesters have been investigated for widening the bandwidth in recent years' work. Reported designs have achieved bi-stability utilizing mechanisms such as magnetic force [2, 3], mechanical compression [4, 5], and pre-shape [6]. The magnets are not suitable for MEMS device assembly; the pre-shaped curved beams vibrate in plane and are not compatible with piezoelectric materials spin-coating; and manually applying compression is not feasible for MEMS devices. Moreover, most of the bi-stable nonlinear resonance based devices are in macro scale [2-5]. To implement the bi-stable oscillator at MEMS scale, we have proposed to utilize the inherent property of micro-fabricated thin films – residual stress. Basing on the lumped model we have built, the bi-stable oscillator could be designed to achieve buckling and desired frequency and excitation amplitude range.

2. Modeling and Design

2.1. Modeling

An electromechanical coupled lumped model has been developed for analyzing the new design. The model is based on a clamped-clamped rectangular beam, which consists of multi-layer thin films of different material with residual stresses. Two governing equations for the mechanical and electrical domain respectively have been derived using *Lagrangian* equations [7]:

$$m\ddot{w}+k_{L}w+k_{N}w^{3}+b\dot{w}+C_{N}wV_{N}+C_{L}V_{L}=F \tag{1}$$

$$C_{0}\left(\dot{V}_{L}+\dot{V}_{N}\right)+\frac{V_{L}+V_{N}}{R}=I_{L}+I_{N} \tag{2}$$

where w, k_L, k_N, b, C_L, C_N and F are the beam center deflection, linear and nonlinear stiffness' of the beam, the mechanical damping coefficient, the linear and nonlinear electromechanical coupling coefficients, and the external excitation force respectively; $I_L=C_L\dot{w}$ and $I_N=C_N w\dot{w}$ are two parts of the electrical current generated by piezoelectric element through coupling; the induced voltages on the electrical port are written in separate parts V_L and V_N, since they come from two parts of the current respectively and have different frequencies due to different coupling; R is the load resistance and C_0 is the internal capacitance of the piezoelectric element. These lumped parameters are functions of the device dimensions and material properties. The nonlinear differential equations are further solved analytically using harmonic balance method, so that the dynamic responses of the post-buckling beam are obtained. The softening and stiffening responses corresponding to the so-called intra-well and inter-well oscillations have been simulated as shown in figure 1(a). The large-amplitude snapping of inter-well stiffening response increases the velocity of the beam oscillation significantly which enhances the power generation at low frequencies.

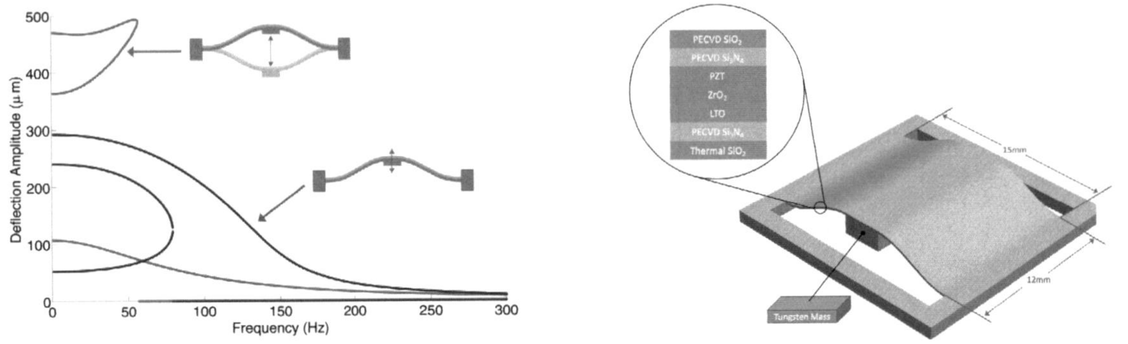

(a) (b)

Figure 1. (a) Two modes of oscillations are simulated for the macro-scale prototype at 3g. Blue and red lines are the softening response and stiffening response respectively. (b) Schematic of the MEMS device being fabricated. Beam thickness and buckling are exaggerated.

2.2. Design

The design of the MEMS-scale bi-stable energy harvesters takes into account the general compatibility with micro-fabrication as well as the PZT spin-coating process. A doubly clamped multi-layer beam design therefore has been adopted. Furthermore, the residual stress as an intrinsic property in micro-fabricated thin films is a critical parameter and should be carefully characterized and designed. The overall effect of the residual stress in different layers is compressing the beam to buckling and inducing bi-stability. Thermal dioxide, LPCVD or PECVD silicon nitride could have compressive stresses, which can also be tuned by changing the gas ratio or plasma frequency [8]. After measuring the residual stress in each material, the thicknesses of the stack of thin films can thus be designed and controlled knowing the deposition rates, to make the doubly clamped beam buckle to the designed level and make the stress distribution symmetric with respect to the neutral axis. The geometry and proof mass are also designed to match the targeting frequency range and g range. The MEMS device that has been designed is shown in the schematic in figure 1(b). It is projected to operate at low frequencies (50-200Hz) at low vibration amplitudes (<1g).

3. Prototypes and Testing

3.1. Macro-Scale Prototype

A macro-scale prototype has been built to test the design concept (figure 2(a)). The tested prototype has an aluminum sheet (76mmx20mmx0.4mm) as the elastic substrate, which is pre-compressed to buckle and then clamped at both ends. A proof mass of 16.5 grams is clamped at the center of the beam. Two PZT fiber based energy harvesters (Smart Materials Corp.) were attached on the substrate on both sides of the proof mass to generate electrical power. The energy harvesters were directly connected to a load resistor of 2MΩ for power measurement. The device is tested on an electromagnetic shaker, which is monitored by an accelerometer and controlled in real time by *Labview*. The generated power was measured at various frequencies at fixed excitation amplitude to generate the frequency response.

(a)

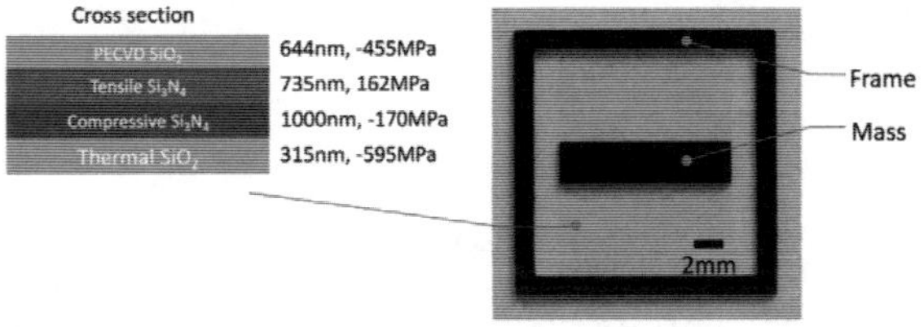

(b)

Figure 2. (a) A photo of the macro prototype mounted on the electromagnetic shaker. The clamped-clamped beam consists of an aluminum sheet of 76mmx20mmx0.4mm as the elastic substrates and two PZT fiber-based energy harvester patches. The center proof mass is 16.5g. (b) A photo of the fabricated MEMS prototype with a schematic of the beam composition. The beam is 15mmx12mm with backside silicon mass (3mmx12mmx530μm).

3.2. MEMS-Scale Prototype

To verify the frequency range and excitation amplitude range of the MEMS device, a simplified design has been fabricated first (figure 2(b)). The MEMS device leaves out the piezoelectric layer so that the fabrication cycle has been reduced significantly, while the device is similar enough to the original design to test the dynamic response. The MEMS device is smaller than a quarter coin, having a beam of 15mmx12mm (LxW). The beam composition is similar to the real design with four layers. The PECVD Si_3N_4 is the elastic substrate since its residual stress is compressive and can be easily altered in a wide range. The tensile Si_3N_4 is introduced to emulate the tensile PZT layer in the real device and achieve stress balance. Backside DRIE was used to release the beam and define the silicon proof mass on the backside of the beam (3mmx12mmx530μm). An optional tungsten proof mass of 0.24g could be attached on the silicon proof mass as an external mass. The fabricated device was tested with an electromagnetic shaker and a laser vibrometer. The shaker was excited harmonically with sweeping frequencies, so that the deflection of the beam could be measured in frequency domain.

4. Results

4.1. Macro-Scale Prototype

The power measurement at various frequencies and fixed excitation amplitude was done for both mono-stable and bi-stable configurations at high g. It should be noted that the two configurations are essentially the same in terms of the materials, with the only difference that the bi-table configuration is the beam with pre-compression to buckling. The testing was carried out at 3g (figure 3(a)). It is clear that the bi-stable system outputs higher power in a wide frequency range. The frequency response of the bi-stable system shifts to lower frequencies proves the enhancement of the output at lower frequencies of the bi-stable system.

The bi-stable configuration has also been tested at lower g (1g) by first sweeping frequency up and then sweeping down slowly (figure 3(b)). Since the input excitation does not provide enough energy to overcome the energy barrier, the bi-stable system oscillates around one of its equilibria (intra-well). The hysteresis clearly shows the characteristic softening response of the oscillation and verifies the model.

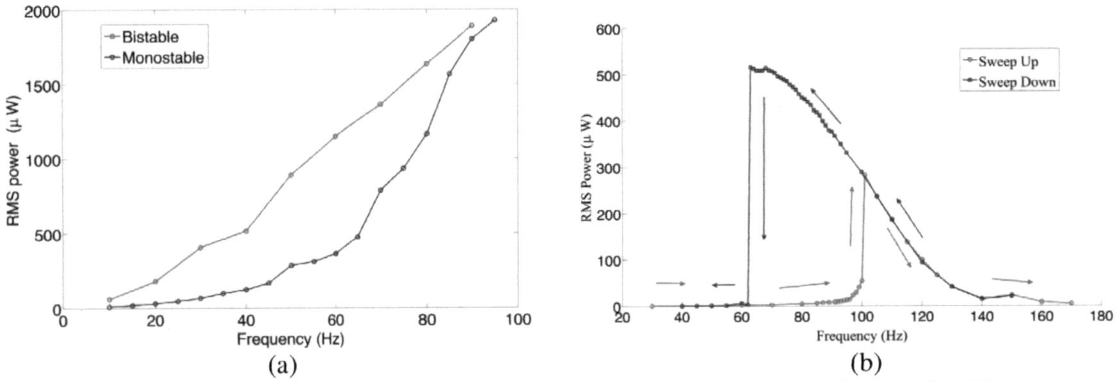

Figure 3. (a) RMS power responses of mono-stable and bi-stable configurations at 3g. The frequency cannot be set higher so that the jump-down has not been reached due to the limitation of the shaker. (b) RMS power responses of the bi-stable configurations at 1g shows softening response.

4.2. MEMS-Scale Prototype

The device shown in figure 2(b) was firstly tested without external mass by sweeping the frequency forward and backward (figure 4(a)). The excitation amplitude was monitored to have a maximum of 0.2g and g values close to it in a wide frequency range, and lower g's at low frequency range. The center and the frame of the device was scanned during sweeping to measure the center deflection at each frequency. During the forward frequency sweep, the beam center deflection increases gradually and jumps down at 140Hz. Backward frequency sweep was also done, and the system jumps up at 100Hz. The spring stiffening response has a wide bandwidth below 150Hz.

An external tungsten proof mass of 0.24g was attached to the backside silicon mass of the device to go through the same frequency sweep testing (figure 4(b)). It can be noticed that the frequency response shifted to much lower frequencies due to the heavier mass, and the large-amplitude region is below 10Hz. The deflection amplitude is also significantly increased compared to the same device without external proof mass. It is reasonable to infer the device with piezoelectric material will have a boosted power output due to the increased mass.

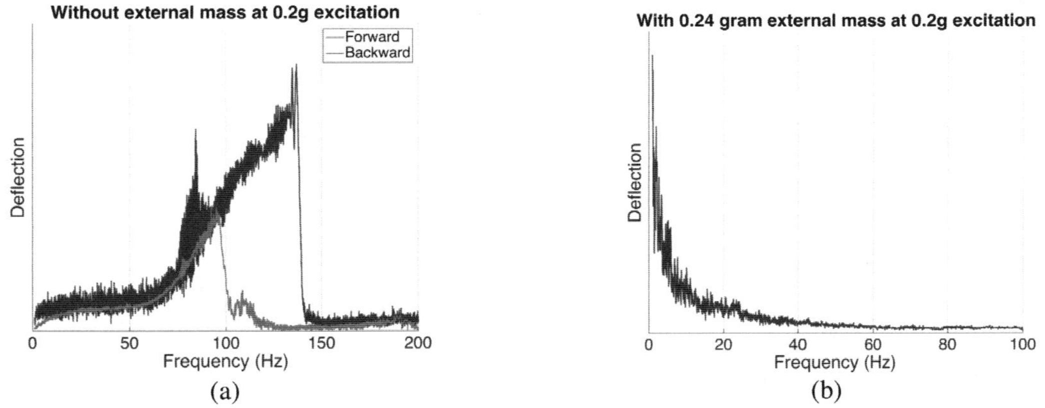

Figure 4. (a) Frequency response of the MEMS device without external proof mass. (b) Frequency response of the MEMS device with 0.24g external proof mass.

Currently, the full energy harvester with embedded PZT is under fabrication, and the results will be reported in near future.

5. Conclusions

Bi-stable buckled beam on MEMS scale has been proposed, designed and fabricated for low-frequency, low-g vibration energy harvesting. The analytical model derived from *Lagrangian* equation and solved with harmonic balance method presents the stiffening and softening frequency response for bi-stable system at high and low excitation amplitude respectively. With the model considering the multi-layer beam structure with residual stress embedded, we could design the MEMS device to achieve buckling and desired frequency and excitation amplitude range.

A macro-scale prototype has been built and tested, and verified the bi-stable system has higher power at lower frequencies than the mono-stable system at high g. At low g, the prototype display the characteristic spring softening response as simulated by the analytical model. A MEMS device prototype has also been fabricated and tested using the laser vibrometer. Test results show that both the devices with and without external proof mass meet our design targets, with high-amplitude responses below 150Hz frequency at 0.2g.

Acknowledgement
The authors would like to thank the MIT-SUTD International Design Center for funding this research.

References

[1] Hajati A and Kim S G 2011 *Appl. Phys. Lett.* **99** 083105.
[2] Mann B P and Owens B A 2010 *J. Sound Vib.* **329** 1215–26.
[3] Erturk A and Inman D J 2011 *J. Sound Vib.* **330** 2339–53.
[4] Cottone F, ea al 2012 *Smart materials and structures* **21** 035021.
[5] Jung S M and Yun K S 2010 *Appl. Phys. Lett.* **96** 111906.
[6] Qiu J, Lang J H and Slocum A H 2004 *J. Microelectromechanical Syst.* **13** 137–46.
[7] Xu R and Kim S G 2015 10th Energy Harvesting Workshop *Preprint* (Blacksburg, VA).
[8] Rossi C and Scheid E 1998 *J. of Vacuum Science & Technology* **16** 2003-07.

Can Integrated Micro-Optical Concentrator Technology Revolutionize Flat-Plate Photovoltaic Solar Energy Harvesting?

Michael W. Haney
Advanced Research Projects Agency – Energy
U.S. Department of Energy
1000 Independence Ave., SW
Washington, DC 20585

E-mail: Michael.Haney@hq.doe.gov

Abstract: The economies-of-scale and enhanced performance of integrated micro-technologies have repeatedly delivered disruptive market impact. Examples range from microelectronics to displays to lighting. However, integrated micro-scale technologies have yet to be applied in a transformational way to solar photovoltaic panels. The recently announced *M*icro-scale *O*ptimized *S*olar-cell *A*rrays with *I*ntegrated *C*oncentration (*MOSAIC*) program aims to create a new paradigm in solar photovoltaic panel technology based on the incorporation of micro-concentrating photo-voltaic (μ-CPV) cells. As depicted in Figure 1, *MOSAIC* will integrate arrays of micro-optical concentrating elements and micro-scale PV elements to achieve the same aggregated collection area and high conversion efficiency of a conventional (i.e., macro-scale) CPV approach, but with the low profile and mass, and hopefully cost, of a conventional non-concentrated PV panel. The reduced size and weight, and enhanced wiring complexity, of the *MOSAIC* approach provide the opportunity to access the high-performance/low-cost region between the conventional CPV and flat-plate (1-sun) PV domains shown in Figure 2. Accessing this portion of the graph in Figure 2 will expand the geographic and market reach of flat-plate PV. This talk reviews the motivation and goals for the *MOSAIC* program. The diversity of the technical approaches to micro-concentration, embedded solar tracking, and hybrid direct/diffuse solar resource collection found in the *MOSAIC* portfolio of projects will also be highlighted.

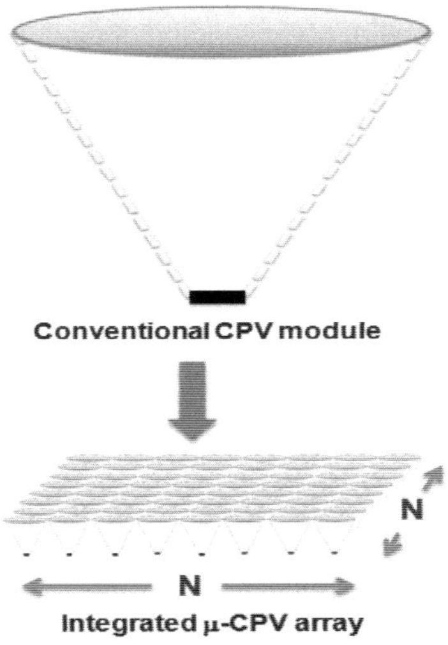

Figure 1. Schematic depiction of a conventional CPV module vs. the baseline *MOSAIC* micro-concentrator array approach.

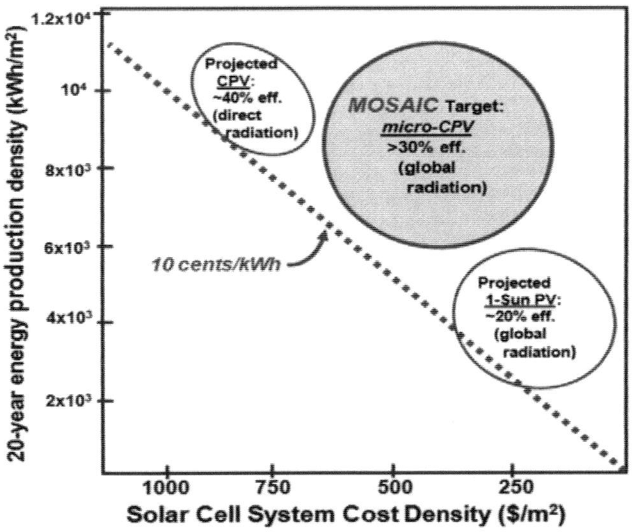

Figure 2. Long-term solar energy harvesting performance density vs. cost density. *MOSAIC* technologies will provide the opportunity to move into the low-cost/high-performance region (i.e., above the 10 cents/kWh line) in the "white space" between the projected conventional CPV and flat-plate PV domains.

A fully autonomous power management interface for frequency up-converting harvesters using load decoupling and inductor sharing

M Tom, O Paul and T Galchev

Microsystem Materials Laboratory, IMTEK, University of Freiburg
Georges-Köhler-Allee 103, 79110, Freiburg, Germany

Email: tzeno.galchev@imtek.de

Abstract. This paper presents the design and simulation results of a self-starting electrical interface circuit for Parametric Frequency Increased Generator (PFIG) vibrational energy harvesters. A significant challenge with these types of harvesters is the ring down of the transducer, introducing an exponential decrease in efficiency when interfaced with standard passive rectifiers. This work aims to address these challenges by interfacing a standard active rectifier to a buck-boost stage in order to decouple the load and always provide an optimal impedance to the transducer. Because PFIG harvesters have two outputs, an inductor sharing technique allows the hardware to be reused. The active buck-boost power management circuit is designed for XFAB's 0.35 μm 2P5M CMOS process. It can provide an efficiency between 75% - 93% for input voltages in the range of 0.5 - 2 V, reaching its maximum efficiency at 1.8 V. The active power consumption of the power converter is 800 nW. The system can start from as low as 800 mV of input voltage and 3 μW of input power.

1. Introduction

Advances in semiconductor technology over the past decade have led to sensor systems with increased functionality while the overall power consumption has been reduced. Additionally, improvements in low-power electronics and power management techniques for energy harvesters are predicted to lead to energy autonomous stand-alone sensor systems. One area of interest is in developing autonomous wearable and implantable systems. Human motion is a potential source of renewable energy, however it is particularly challenging to harvest due to its variability. Previously, a vibration harvester architecture was introduced, called the Parametric Frequency Increased Generator (PFIG) [1-2], particularly suited toward converting low-frequency and variable amplitude vibrations. However, due to the nature in which PFIGs operate, specialized power management techniques are needed which can deal with the alternating and decaying output signals. Furthermore, as energy harvesters are scaled down, reduced electromechanical coupling leads to lower output voltages and power levels. Therefore, future power management techniques must be able to address these technical issues while consuming nearly negligible power.

This paper discusses the design and simulation results of an electrical interface circuit for extracting power from PFIG harvesters. This type of harvester has two outputs, and therefore an architecture was developed that allows for energy from both outputs to be extracted, stored, and managed effectively while sharing and reusing hardware components in order to keep the overall system size small. One of the main tasks of the power management system is to decouple the load from the harvester in order to allow energy transduction to take place in an optimized way regardless of the variable load presented to the system. Both piezoelectric and electromagnetic transducers can be approximated as having a resistive source impedance while in resonance, which means that they will deliver optimal power to a matched resistive load [3]. This is also true in the case of an unforced ring-down, as is the case with PFIG harvesters. In this work, the behavior of a matching load resistor is emulated using a non-inverting discontinuous conduction mode buck-boost converter. While the circuit in this paper is designed for interfacing to a piezoelectric PFIG, the same concept can be applied to electromagnetic implementations.

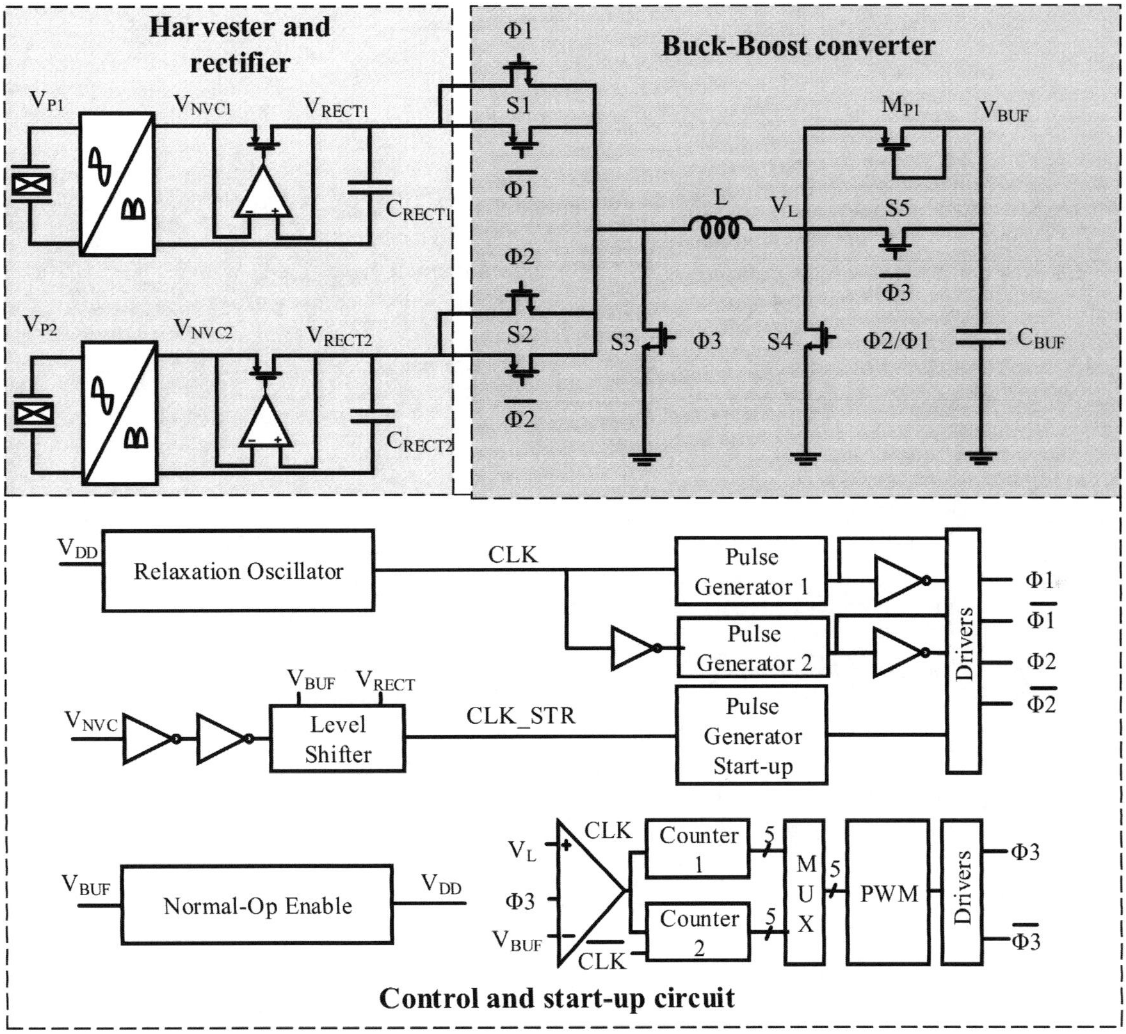

Figure 1. Block diagram of the PFIG power management interface

2. System architecture

The architecture of the proposed piezoelectric harvesting system is described in figure 1. The PFIG has two transducers and therefore the electrical circuit should be able to extract and combine the energy from both. As illustrated in figure 1, two separate active rectifiers are attached to the output of each of the two transducers. They are based on the design presented in [6] and their job is to convert the initially alternating voltage to DC. A single buck-boost converter is then shared between the two transducers. Digital control is used to time-share the buck-boost converter. Since, the duty cycle of the converter is very small, the rising and falling edges of the clock signal are used to switch between the two harvester sources.

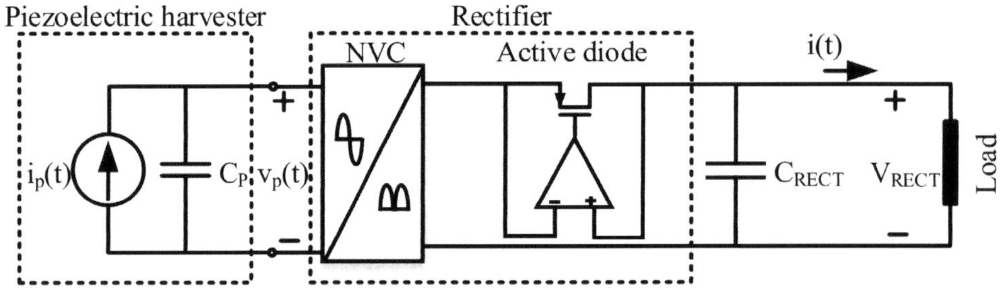

Figure 2. Piezoelectric transducer connected to an active rectifier and load.

2.1. Piezoelectric harvester model

Piezoelectric transducers can be approximately modeled in the electric domain as an ideal current source in parallel to a parasitic capacitance C_P and resistance R_P as shown in figure 2 [3]. Generally, R_P is very large compared to the impedance of C_P and therefore it is neglected. When connected to a rectifier, the maximum rectified voltage, V_{RECT}, occurs during no-load scenario and is equal to the open-circuit voltage, V_{OC}, given by $V_{OC} = I_P / \omega C_P$, where, I_P is the amplitude of the alternating current and ω is the frequency of vibration. It has been shown [4] that peak power is extracted, when V_{RECT} is half of V_{OC} and this maximum power corresponds to:

$$P_{RECT_{max}} = \frac{C_P \omega V_{OC}^2}{2\pi} \qquad (1)$$

2.2 Rectifier

The rectifiers are implemented by combining a negative voltage converter with an active diode as previously shown in [6]. In this way the voltage drop is reduced to a few millivolts, compared with the large voltage drop (> 0.25 V) associated with a diode implementations. This is very important for miniaturized implementation, where the output voltages are typically small.

2.3 Buck-boost converter and optimal power extraction

The DC-DC converter is used to optimize V_{RECT} and thereby track the maximum power point of each PFIG transducer [3, 5]. Operating the buck-boost converter in a discontinuous conduction mode creates the impedance match. The time averaged input resistance, R_{IN}, can be derived by formulating an expression for the average input current, I_{IN} (which is the inductor current I_L averaged over one switching cycle) and diving it by the input voltage V_{RECT} leading to the following formulation:

$$R_{IN} = \frac{2Lf_{CLK}}{D^2} \qquad (2)$$

As can be observed from equation 2, the input impedance depends only on the inductor value L, the duty cycle D, and the switching frequency f_{CLK}, and is independent of input or output voltage. The proposed buck-boost converter has five switches *S1*-*S5*. Switches *S1* and *S2* are used to extract energy from the rectified output of each of the PFIG transducers, while the remaining switches are shared. In order to improve efficiency at low voltages, switches *S1* and *S2* are implemented by a parallel combination of NMOS and PMOS transistors. Switches *S3* and *S4* are implemented using NMOS transistors and switch *S5* is realized using PMOS.

2.4 Control circuit

The converter control unit uses a digital control loop in-order to minimize the active power consumption [5]. The control loop has a 5 bit up/down counter, whose value determines the on-time of switch *S5*. In-order to achieve zero current switching, the counter always tries to settle to the optimum value by checking whether the present pulse value is higher or lower than optimal. This decision is made by a clocked comparator, which monitors node voltages V_L and V_{BUF}.

2.5 Clock generator

An on-chip clock for synchronization is generated using an RC relaxation oscillator. The signal is generated by the constant charging and discharging of a capacitor through two reference current sources. The relaxation oscillator designed in this work generates a clock of 2.5 kHz. The clock generator consumes less than 150 nW during active operation.

2.6 Start-up circuit

In order to achieve fully autonomous operation, the harvester should have a start-up mechanism to initially charge the buffer capacitor, C_{BUF} from a completely discharged state. The start-up circuit selectively uses switch *S4*, MOSFET diode M_{P1}, and inductor L to create a boost converter, which in-turn charges C_{BUF}. The clock signal for the start-up procedure (CLK_STR) is generated directly from the mechanical input source using the arrangement shown in figure 1.

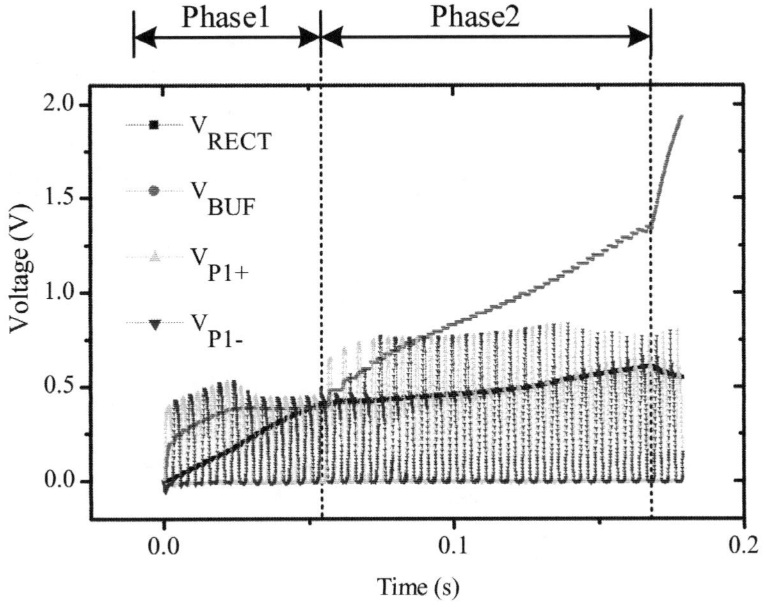

Figure 3. Transient plot showing the start-up of the system when coupled to a PFIG. The following parameters are used: $I_P = 25$ µA, $C_P = 25$ nF, $f_P = 200$ Hz, $C_{RECT} = 1$ µF, $P_{IN\_MAX} = 2.4$ µW, $L = 62$ µH, $C_{BUF} = 100$ nF.

3. Results

The architecture described in figure 1 was designed and simulated for the 0.35 µm CMOS process offered by XFAB. Each piezoelectric output of the PFIG is modeled to have $C_P = 25$ nF and an operating frequency of 200 Hz. The start-up process can be explained in two phases. In phase 1, the buffer capacitor, C_{BUF}, initially charges passively through a diode. In phase 2, once V_{BUF} reaches the MOSFET threshold voltage, charging continues using the start-up boost-converter described in section 2.6. This operation continues until V_{BUF} reaches 1.2 V after which normal operation begins. During the start-up phase, all of the control circuits are disabled. Start-up transients are shown in figure 3, illustrating the two phases. With this arrangement, the harvester was able to start from input voltages as low as 800 mV with as little as 3 µW of input power.

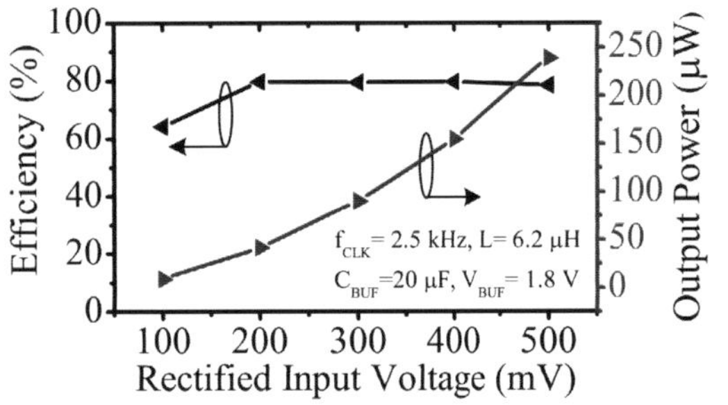

Figure 4. Efficiency of the buck-boost converter when coupled to a low-output impedance electromagnetic system.

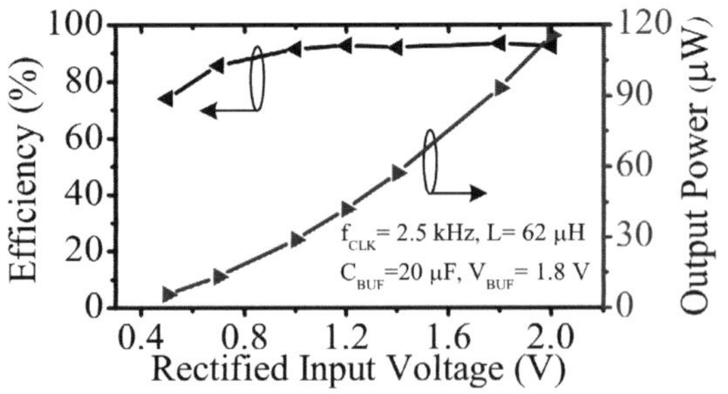

Figure 5. Efficiency of the buck-boost converter when coupled to a high-output impedance piezoelectric system.

The efficiency of the buck-boost converter during normal operation was analyzed by varying the input voltage between 0.5 - 2 V. Figures 4-5 show plots of power efficiency as a function of rectified input voltage, where the first range is more appropriate for low-impedance electromagnetic harvesters (figure 4) and the second for higher voltage and higher output impedance piezoelectric implementations (figure 5). The active power consumption of the converter is as low as 800 nW.

Table 1: Performance comparison of previously published work

Reference	Impedance matching	Converter efficiency	Rectified input voltage
This work	**Yes**	**75% - 93%**	**0.5 - 2.0 V**
[7]	No	58% - 86% (end to end)	> 0.8 V
[8]	Yes	65% - 79%	1.5 - 5.0 V
[5]	Yes	75% - 85%	2.0 - 5.0 V
[9]	Yes	40% - 60%	0.6 - 2.0 V

4. Conclusion

A fully autonomous electrical interface circuit was developed for Parametric Frequency Increased Generators in XFAB's 0.35 μm CMOS process. The designed buck-boost converter was able to achieve efficiencies in the range of 75% - 93% for input voltages in the range of 0.5 - 2 V, reaching its maximum efficiency at 1.8 V. The active power consumption of the power converter is 800 nW. The system can start from as low as 800 mV of input voltage and 3 μW of input power.

Acknowledgements

The research leading to these results has received funding from the European Union's Seventh Framework Program (FP7/2007-2013) under grant agreement n°331333.

References

[1] T. Galchev, H. Kim, and K. Najafi, *IEEE JMEMS*, vol. 20, no. 4, pp. 852-66, 2011.
[2] T. Galchev, E. Aktakka, and K. Najafi, *IEEE JMEMS,* vol. 21, no. 6, pp. 1311-20, 2012.
[3] G. Ottman, H. Hofmann, and G. Lesieutre, *IEEE TPEL*, vol. 18, no. 2, pp. 696–703, 2003.
[4] G. Ottman, H. Hofmann, A. Bhatt, and G. Lesieutre, *IEEE TPEL*, vol. 17, pp. 669–76, 2002.
[5] Y. K. Ramadass and A. P. Chandrakasan, *IEEE JSSC*, vol. 45, no. 1, pp. 189–204, 2010.
[6] A. S. Herbawi, O. Paul, and T. Galchev, *IEEE Sensors Conference*, pp. 1-4, 2013.
[7] E. E. Aktakka, K. Najafi, *IEEE JSSC*, vol. 49, pp. 2017–2029, 2014
[8] S. Bandyopadhyay, A. P. Chandrakasan, *IEEE JSSC*, vol. 47, no. 9, pp. 2199–2215, 2012.
[9] I. M. Darmayuda *et al.*, *IEEE Trans. Circuit Syst. II*, vol. 59, no. 9, pp. 578–82, 2012.

Interface Circuit for Vibration Energy Harvesting with Adjustable Bias Voltage

J Wei[1a], E Lefeuvre[a], H Mathias[a], F Costa[b]

[a]IEF – CNRS, Univ. Paris Sud, Orsay, France
[b]SATIE – CNRS, Univ. Paris Est Créteil, France

E-mail: jie.wei@u-psud.fr

Abstract. This paper presents a new interface circuit for electrostatic vibration energy harvesting with adjustable bias voltage. An electronic switch is used to modify the circuit configuration so that the harvested energy increases the voltage across a biasing capacitor. Decrease of this biasing capacitor voltage occurs naturally due to the circuit imperfections. Such a control of the bias voltage enables to adjust the amount of energy converted by the variable capacitor on each cycle. This feature can be used to optimize the mechanical damping induced by the energy conversion process in order to maximize the harvested power. Another feature of this interface circuit is that it is capable to get high bias voltage whatever the battery voltage with low energy loss.

1. Introduction

Energy harvesting is a promising way of generation of electrical supply for powering miniature device, like those used in wearable electronics, medical implants and wireless sensor networks. Mechanical vibration is a common and well known energy source. Mechanical vibration energy harvesting concept consists in converting the energy of the parasitic ambient mechanical vibration into the useful electrical energy. Among several types of Vibration Energy Harvester (VEH), such as electromagnetic, piezoelectric, electrostatic, etc, Electrostatic Vibration Energy Harvesters (e-VEH) appears to be promising for miniaturization.

An e-VEH is actually based on a variable capacitor whose moving electrode is actuated by ambient vibrations. Such electromechanical device can be profitably miniaturized using MEMS technology. One of the main challenges of e-VEH implementation is the design of the interface circuit, whose power efficiency is a critical issue. Recently, several interface circuits without switches and inductive components have been proposed [1-4], whose simplicity enables efficient standalone implementation.

However, as shown by Dhorziev et al., these circuits do not intrinsically optimize the system performances: there exists an optimal power supply voltage at which output power is maximum [5]. Such optimization can be performed using Yen's circuit [6], but in a limited way because of the limited achievable voltage range.

The interface circuit presented here can theoretically generate voltages across the variable capacitor without upper limit, enabling to get high bias voltages whatever the battery voltage. Dragunov et al.

[1] To whom any correspondence should be addressed.

[2] and De Queiroz [3] previously proposed interface circuits able to generate high bias voltage. However, their circuits include zener diodes as voltage stabilizer, which induce significant energy losses and lead to very poor efficiency. Contrary to those circuits, the proposed circuit has very high power efficiency. Moreover, its adjustable bias voltages can be used for real-time optimization of the harvested power, respectively with the variations of the ambient vibration characteristics.

2. Theoretical analysis of the proposed interface circuit

The new interface circuit for e-VEHs, shown in Figure 1, was derived from our previous work [4]. It is composed of biasing capacitor C_{bias}, a diode rectifier (D1and D2), an electronic switch SW and a multiplier circuit which consist of capacitors C1 and C2, and diodes D3, D4, D5. The input of the circuit is connected to the mechanically-actuated variable capacitor C_{var}, and the generated energy is stored in a rechargeable battery with V_{DC} voltage, which could be replaced by a capacitor or a supercapacitor.

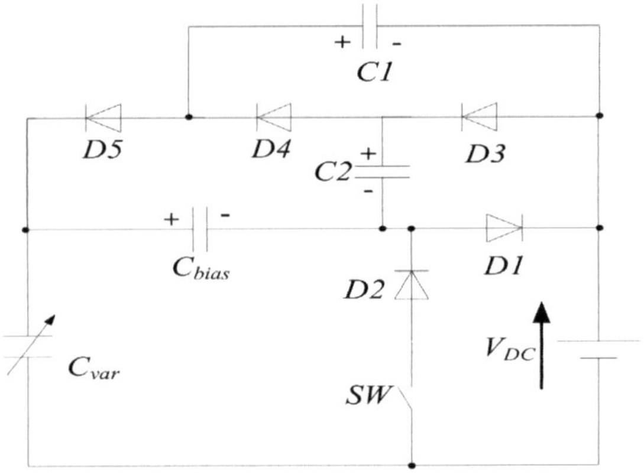

Figure 1. Schematic diagram of the proposed interface circuit

The operation of the circuit can be divided into two different modes. SW open: the variable capacitor is used for charging the biasing capacitor C_{bias}. SW closed: the variable capacitor is used for charging the battery. The following analysis is made assuming a steady variation of the capacitance C_{var} between a maximum value C_{max} and a minimum value C_{min}. Imperfections of the components are neglected.

2.1. SW open: charging of the biasing capacitor

Due to the diodes D3, D4 and D5, the variable capacitor is initially charged to voltage V_{DC}. When C_{var} decreases, the diodes D1 and D4 are conducting and the biasing capacitor is charged. The capacitors C1 and C2 are charged via the diode D4. The charging process ends when C_{var} reaches its minimum. Then, C_{var} increases and the diodes D1 and D5 become conducting. An electric current flows from the battery to C_{var} via D1 and C1. When C_{var} gets to its maximum, the voltage across C_{var} is greater than V_{DC} due to the previously charged C1. These different steps are repeated during the periodic variation of C_{var}, and the voltage across C_{bias} is progressively increased.

The charge-voltage diagram of the variable capacitor is shown in Figure 2. The diagram begins at point 1 with C1, C2 and C_{bias} initially discharged. The first QV cycle passes successfully at points 1, 2, 3, 4 and the second cycle passes at points 4, 5, 6, 7, 8 etc. The increase of the voltage across C_{bias} is represented by 4 to 5, 8 to 9 etc. At the end of each cycle, the charge Q of the variable capacitor increases. Hence, when the switch SW is open, the circuit may theoretically generate voltage values across the variable capacitor without upper limit.

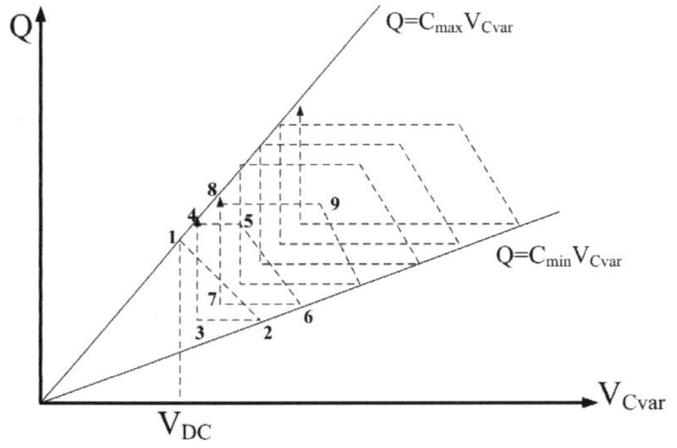

Figure 2. Theoretical charge-voltage diagram of the variable capacitor for the proposed circuit with SW opened

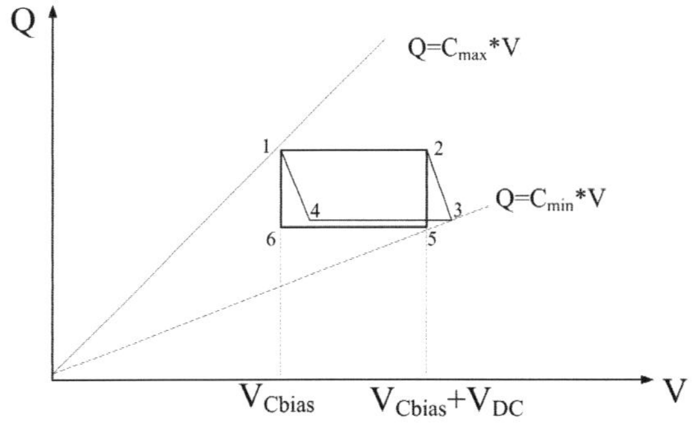

Figure 3. Theoretical charge-voltage diagram of the variable capacitor for the proposed circuit with SW closed

2.2. SW closed: charging of the battery

The charging of the battery may occur when SW is closed. In steady state operation, the voltage V_{Cbias} across the capacitor $C_{bias}$is greater or equal to $2*V_{DC}$. If V_{Cbias} is smaller than $2*V_{DC}$ and SW is closed, the charging of C_{bias} continues until V_{Cbias} reaches $2*V_{DC}$ and then the charging of the battery begins.

The charge-voltage diagram of the proposed circuit is shown in Figure 3. Due to D2, the voltage across C_{var} is greater or equal to V_{Cbias}. Between points 1 and 2, all the diodes are blocking and the charge of the C_{var} keep constant. Between points 2 and 3, D1 is conducting, meanwhile an electric current flows from C_{var} to the battery through C_{bias}. Between points 4 and 1, D2 is conducting, meanwhile an electric current flows from C_{bias} to C_{var}. Hence, the energy conversion cycles pass successively at points 1, 2, 3, 4 and 1. Assuming that the voltage variation C_{bias} is negligible, which is the case if C_{bias}>>C_{var}, the charge-voltage diagram tends to the rectangle passing at points 1-2-5-6. In this case, the harvested energy per cycle, with SW closed, can be expressed as (1).

$$W = V_{DC} V_{Cbias} C_{min} \left[\frac{C_{max}}{C_{min}} - \left(1 + \frac{V_{DC}}{V_{Cbias}}\right) \right] \quad (1)$$

This equation shows that the minimum value of the capacitance ratio C_{max}/C_{min} for which energy can be harvested tends to 1 for high values of V_{Cbias}. Our previous circuits with multicell voltage multiplier enabled also to reduce the limit of the capacitor ration C_{max}/C_{min}. However, the number of cells of the multiplier determined a fixed ratio between V_{Cbias} and V_{DC}, making thus impossible to adjust continuously and independently V_{Cbias}. More precisely, the voltage across the variable capacitor of previous circuit varied between $N*V_{DC}$ and $(N+1)*V_{DC}$ with a N-cell multiplier, as depicted in Figure 4. It did not intrinsically optimize the system performances: previous works [5] showed that there exits an optimal power supply voltage at which output power is maximum. The proposed circuit enables to solve this problem. Since V_{Cbias} can be adjusted to any value greater than $2V_{DC}$, the e-VEH can be operated under an optimal bias voltage by controlling the switch SW.

One should note that in practice, imperfections of the circuit components progressively decrease the voltage $V_{Cbias.}$ We experimentally observed exponential voltage decrease with time constant of a few minutes. Thus, in practice V_{Cbias} can be adjusted in both directions. If dynamic tuning of the voltage across C_{bias} requires faster decrease, this can be obtained by inducing extra losses in the circuit. For instance, we experimentally decreased the discharge time constant to a few tenths of seconds by adding a 5 GΩ resistor in parallel with Cbias, this without visible reduction of the circuit efficiency.

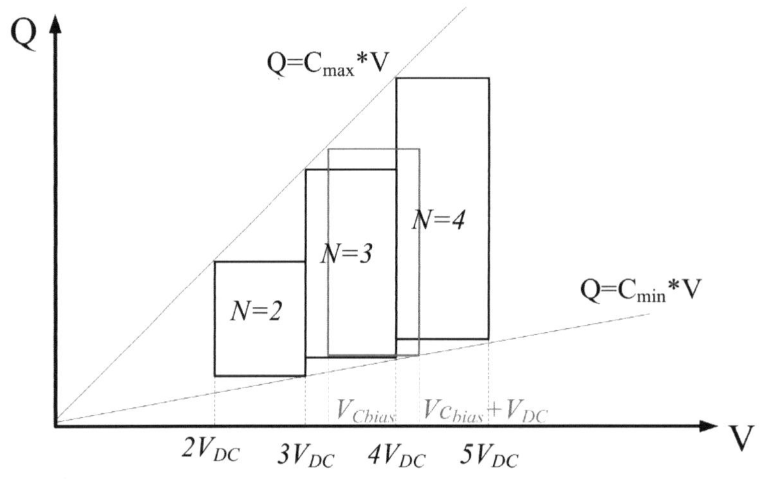

Figure 4. Charge-voltage diagram of the variable capacitor of the circuit using a multicell voltage multiplier (in black) and for the proposed circuit (in red)

3. Spice Simulation

To study the influence of the imperfection of electric components, the proposed circuit was modelled using LTSPICE software. The variable capacitor was varied from 25pF to 125 pF at 20 Hz frequency. The biasing capacitor was modelled by a 4.7 nF capacitor paralleled with a 2.10^{10} Ω resistor. The reverse current of the diodes was set to 10 pA and the forward drop voltage was set to 0.4 V. The ON and OFF states of the electronic switch SW were modelled by 1 Ω and 10^{10} Ω resistors respectively. The battery voltage V_{DC} was set to 5 V.

The circuit waveforms with periodic switching are shown in Figure 5. After 8 seconds initial charging of C_{bias}, SW was periodically turned ON and OFF with 8 s ON-time and 12 s OFF-time. One can clearly see that the output power is not null with SW closed, and that the power is increased when V_{Cbias} is larger. When SW is open, V_{Cbias} is increased. These simulation results confirmed the theoretical analysis of the proposed circuit. At the point where V_{Cbias}=25V, the simulated output power is 155 nW. This value represents 94% of the 165 nW theoretical power calculated using (1).

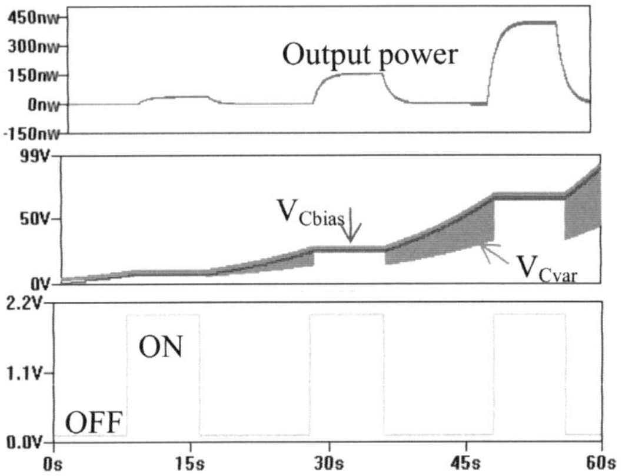

Figure 5. Simulated voltage of V_{Cbias} ,V_{Cvar} and power and SW control vs time

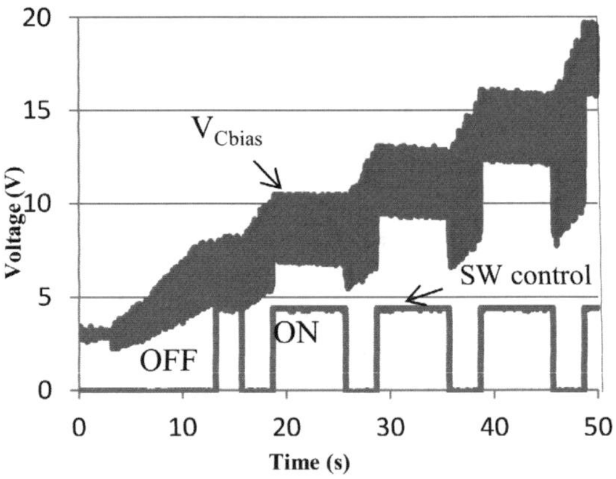

Figure 6. Experimental waveforms of V_{Cvar} and SW control signal vs time

4. Experimental results

The experimental validation was carried out using a rotating variable capacitor driven by a DC motor, with nearly 25 pF for C_{min} and 125 pF for C_{max}. The DC motor speed was 120 rpm, so the variable capacitor was varied at 20 Hz frequency. Low reverse current PAD5 diodes were used for *D1*, *D2* and *D3*. C_{bias} was a 4.7 nF capacitor. Due to the variable capacitor used, the generated power was not sufficient to power an electronic switch and its control electronics. Indeed, based on results presented by Dudka et al. [7], we estimated that the electronic switch and its integrated control circuit would consume about 1 μW. For this experiment, we emulated the electronic switch SW using an electromechanical relay controlled by a signal generator.

The experimental waveforms of the control signal of SW and V_{Cvar} voltage are shown in Figure 6. One can observe that these waveforms are quite similar to the simulation result of Figure 5 and confirm that the circuit operates as expected in both configurations.

5. Conclusion

This new interface circuit for electrostatic energy harvesters has a continuously adjustable bias voltage. It enables to generate high bias voltage whatever the battery voltage with very low energy loss. This degree of freedom can be used to optimize the mechanical damping induced by energy conversion process, in order to maximize the harvested power. Our future work will focus on the control principle and on the design of the integrated control circuit

6. Reference

[1] De Queiroz A C M, Domingues M. The doubler of electricity used as battery charger. Circuits and Systems II: Express Briefs, IEEE Transactions on, 2011, 58(12): 797-801.
[2] Dragunov V and Dorzhiev V. Electrostatic vibration energy harvester with increased charging current. Journal of Physics: Conference Series 476 (2013) 012115.
[3] De Queiroz A C M. Electrostatic Energy Harvesting Without Active Control Circuits. 2014 IEEE 5th Latin American Symposium on Circuits and Systems, 2014, 1-4.
[4] Lefeuvre E, Risquez S, Wei J, et al. Self-Biased Inductor-less Interface Circuit for Electret-Free Electrostatic Energy Harvesters. Journal of Physics: Conference Series. 2014, 557: 012052.
[5] Dorzhiev V, Karami A, Basset P et al. MEMS electrostatic vibration energy harvester without switches and inductive elements. Journal of Physics: Conference Series 557 (2014) 012126.
[6] Yen B C, Lang J H. A variable-capacitance vibration-to-electric energy harvester. Circuits and Systems I: Regular Papers, IEEE Transactions on, 2006, 53(2): 288-295.
[7] A. Dudka, D. Galayko, and P. Basset, "Design of Controller IC for Asynchronous Conditioning Circuit of an Electrostatic Vibration Energy Harvester", IEEE International Conference on Green Computing and Communications, pp. 660-663, Besancon, November 2012.

Acknowledgments

This work is supported by the "IDI 2014" project funded by the IDEX Paris-Saclay, ANR-11-IDEX-0003-02

Fully Integrated Ultra-Low Voltage Step-up Converter with Voltage Doubling LC-Tank for Energy Harvesting Applications

H M P C Jayaweera[1], W P M R Pathirana[1], Ali Muhtaroğlu[1,2]

[1]Sustainable Environment and Energy Systems
[2]Dept. of Electrical-Electronics Engineering
Middle East Technical University Northern Cyprus Campus
Kalkanli, Guzelyurt, Mersin 10, Turkey

E-mail: {pradeep.jayaweera, manula.pathirana, amuhtar}@metu.edu.tr

Abstract. This paper reports the design, fabrication, and validation of a novel integrated interface circuit for ultra-low voltage step up converter in 0.18 μm CMOS technology. The circuit does not use off-chip components. Fully integrated centre-tap differential inductors are introduced in the proposed LC oscillator design to achieve 38% area reduction compared to the use of four separate inductors. The efficiency of the system is hence enhanced through the elimination of clock buffer circuits traditionally utilized to drive the step-up converter. The experimental results prove that the system can self-start, and step 0.25 V up to 1.7 V to supply a 46 μW load with 15.5% efficiency. The minimum validated input voltage is 0.15 V, which is boosted up to 1.2 V under open circuit conditions.

1. Introduction

Due to recent developments in remote and embedded microsystems working with only micro- to milli-Watts of power, energy harvesting micro-power generators have become key enablers in electronic industry. However, small micro-power generators used to harvest energy from the ambient environment are limited in capacity. DC-DC boost converters are necessary to attain a sufficiently high voltage from ultra-low voltage output of the energy harvesters, such as thermoelectric micro-modules used in wearable computing. Charge-pumps have been commonly used in such miniaturized applications, for which on-chip integration of step-up circuits is a critical system design requirement.

Typical requirements from on-chip charge-pumps include maximized efficiency to enable more features at the load, self-starting operation with minimum or no external component support, and ultra-low input voltage range (~200 mV) – often much lower than the threshold voltage of the low cost MOSFET technologies [1] [2]. Most of the integrated charge-pumps utilized in memories, liquid crystal display drivers and control systems for motor drivers consist of non-inductive circuits [3]. Inefficiency of on-chip inductors is often perceived as an inhibitive design problem. Hence, conventional clock generator circuits, such as the ones presented in [4] and [5], employ ring oscillators with buffer circuits in the final stage to sustain edge rates and high drive currents required to switch the charge-pump capacitors. However, dynamic power consumed by large CMOS clock buffers can be a significant portion of the total power budget available from micro-power generators.

In this paper, validation data from an alternative clock generator design is presented. The circuit replaces digital oscillators with a low cost voltage-doubling LC tank implemented in standard 0.18 μm CMOS technology with on-chip inductors, eliminates large CMOS buffers, and thus provides high step-up ratio at higher efficiency and larger output power capacity compared to the alternatives. The resulting step-up converter, illustrated in Fig. 1, is capable of self-starting, and boosting an input voltage as low as 0.15 V. Section 2 presents the design and analysis of the clock generation circuit. The charge-pump design, also utilized in a number of previous designs, is briefly introduced in Section 3. Experimental results from the test chip under targeted load conditions are provided in Section 4. Finally, Section 5 highlights conclusions from this work.

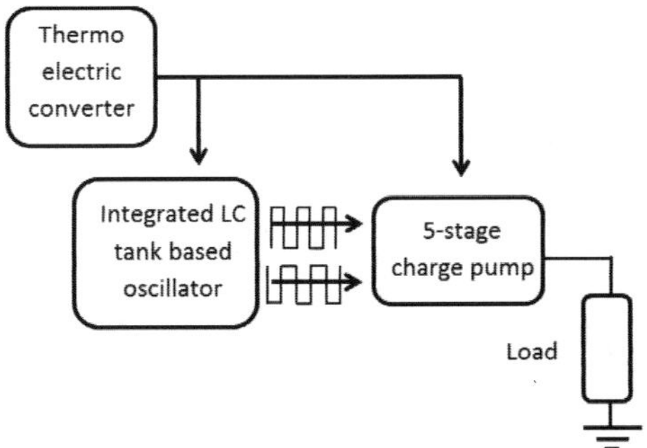

Figure 1. Block diagram of the proposed system.

2. Clock Generation Circuit

The clock generator is required to operate with an ultra-low voltage supply in a self-starting, efficient mode to provide two out-of-phase clock signals. Use of digital CMOS buffers in conventional ring oscillators increases dynamic power dissipation, and reduces efficiency. LC oscillators, which are more suitable for out-of-phase differential clock generation, have traditionally been avoided due to potentially high cost associated with on-chip inductor design. In this work an LC tank [10] based oscillator topology was enhanced, and implemented in standard (low cost) 180 nm CMOS technology. The circuit model for the LC tank based oscillator is depicted in Fig. 2(a). R_p is the parasitic resistance of the LC tank, and $-R_a$ is the negative resistance used to compensate R_p, in order to generate an undamped system. The cross-coupled NMOS pair; shown in Fig. 2(b) provides the negative resistance ($-R_a$) for the LC tank as describe in [10].

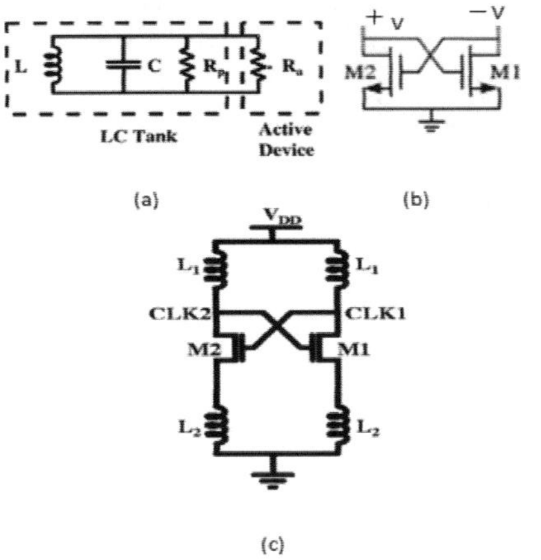

Figure 2. (a) Schematic of LC oscillator, (b) Cross coupled NMOS pair, (c) LC tank based oscillator

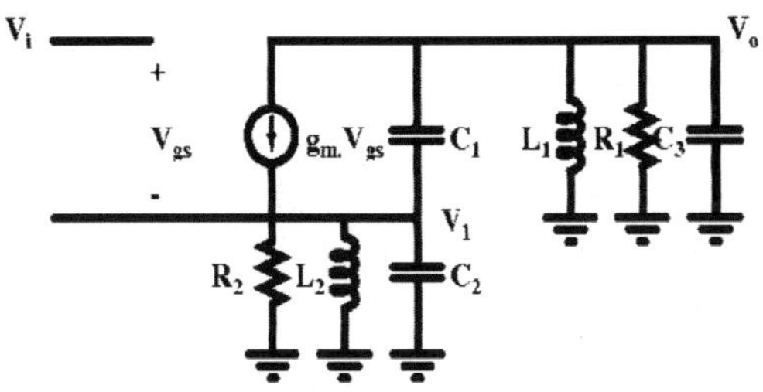

Figure 3. Half circuit model of the oscillator.

The frequency of the proposed LC oscillator is given by Equation (1):

$$f_0 = \frac{1}{2\pi\,(LC)^{1/2}} \tag{1}$$

where f_0 is the frequency of the LC tank based oscillator, C (C_{tank}) is the parasitic capacitance of the NMOS M1 and M2, and L (L_{tank}) is the inductance of the inductor shown in Fig. 2(c). The circuit parameters can calculate using the transient analysis for the half circuit model as shown in Fig. 3. According to the Kirchhoff's low, the current at node V1 can written as,

$$\frac{V_1}{R_2} + \frac{V_1}{L_2 s} + V_1 C_2 s - (V_2 - V_1)C_1 s = g_m V_{gs} \tag{2}$$

$$\frac{V_0}{L_1 s} + \frac{V_0}{R_1} + V_0 C_3 s - (V_1 - V_0)C_1 s = -g_m V_{gs} \tag{3}$$

By asserting V_0/V_i = -1 (Barkhausen criteria) to sustain oscillation, and considering the real part, the final equation can derive as follows:

$$a\omega^4 + b\omega^2 + c = 0 \tag{4}$$

where,

$$a = (C_1L_1R_1)(C_2L_2R_2) + (C_3L_1R_1)(C_2L_2R_2) + L_2R_2L_1R_1C_1C_3$$

$$b = -(C_3L_1R_1R_2 + L_1L_2 + C_2L_2R_2R_1 - g_mL_1R_1L_2 + C_1L_1R_1R_2 + g_mL_1R_2L_2 + C$$

$$c = R_2R_1$$

The inductors L_1 and L_2 are implemented as two center tap differential inductors to minimize the cost and chip area. To obtain 1 GHz frequency for the ultra-low voltage range, the inductances are selected as L_1=15 nH and L_2 = 1.6 nH. The quality factor and self-resistance of the inductor L1 are 5.8, 13.6 Ω, and L2 inductor is 3.5, 2.96 Ω respectively. The spiral Inductor Assistant software for Sonnet is used to design the layout of the inductor and 3D planner electromagnetic field solver software is used to model the inductor. The inductors are designed using UMC 0.18μm CMOS technology with 19 μm wide and 20KA Aluminum top Metal 6 and Metal 5 layers in order to generate relatively high quality factor values. The total area for the center-tap differential inductors is 0.61 mm^2. R_2 and C_2 are the parasitic resistance and capacitance of L_2. R_1 and C_3 are the parasitic resistance and capacitance of L_1. C_1 is the parasitic capacitance of the NMOS. For the clock frequency range of 1 GHz, R_1 is around 2240 Ω and R_2 is 77 Ω. The calculated transconductance of the NMOS is approximately 5 mA/V. C_1, C_2 and C_3 values are 100 fF, 0.4 pF, and 1.68 pF respectively. The oscillation frequency is derived as:

$$f_o = \frac{1}{2\pi\sqrt{LC}} = \frac{1}{2\pi\sqrt{15 \times 10^{-9} \times 1.68 \times 10^{-12}}} = 1GHz$$

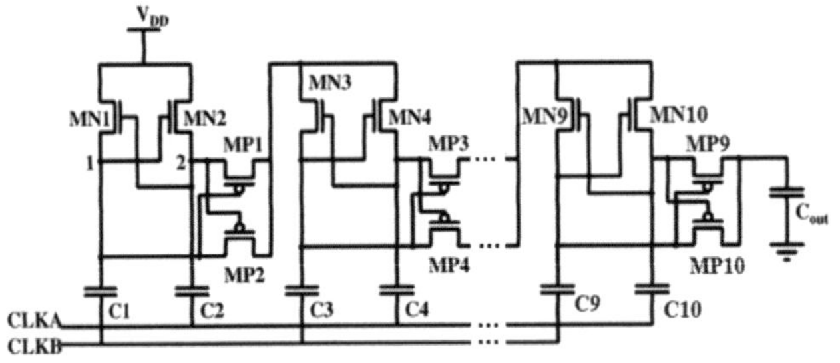

Figure 4. Charge-pump circuit with cross connected NMOS[9]

IOP Publishing
doi:10.1088/1742-6596/660/1/012017

3. Charge-Pump Circuit

Dickson charge-pump [6] and its derivatives [7], [8] are commonly used for on-chip voltage step-up. An ultra-low voltage derivative was proposed by [9] as shown in Fig. 4. A five stage version of this design has been implemented in this work with suitable MOSFET parameters to minimize power dissipation and current backflow. Further details on the operation of this design can be found in [9].

4. Experimental Results

The system has been designed in UMC 0.18μm CMOS technology, and SONNET 3D Planner Electromagnetic Field Solver Software is used to characterize the centre tap inductors for the LC tank based oscillator. The circuit with LDO regulation stage occupies 0.88 mm^2 of layout area, although regulation mode has not been utilized in this work. The full chip layout of the step-up converter is shown in Fig. 5.

The fabricated charge-pump design has been validated using 30 pF capacitive load and various load resistance values. The measured system efficiency curve is depicted in Fig. 6 against varying load resistance with 0.25 V input. The maximum efficiency is achieved for a load resistance of 60 kΩ. The proposed system can convert 0.25 V to 1.1 V with 30 kΩ load resistance, as depicted in Fig. 7. For 60 kΩ load resistance, the system can achieve 1 V output level at 0.2 V input, as shown in Fig. 8. The experimental results demonstrate the system can self-start, and step 0.25 V up to 1.7 V to supply a 46 μW load with 15.5% efficiency. The minimum validated input voltage is 0.15 V, which is boosted up to 1.2 V under open circuit conditions. At ultra-low voltage range, the proposed step-up converter is smaller in size, and has higher integration with improved efficiency compared to the alternatives [11], [12], [13] respectively in 0.18 μm technology.

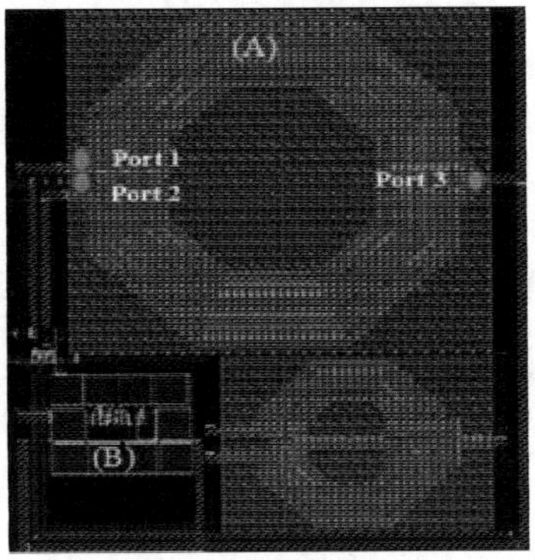

Figure 5. Die layout of the fabricated system.

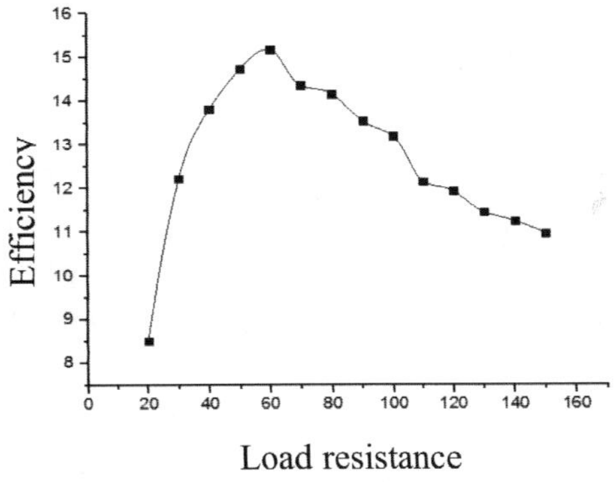

Figure 6. System Efficiency vs. load resistance for 0.25 V input voltage.

5. Conclusion

An ultra-low voltage step-up integrated charge-pump circuit topology with voltage doubling LC tank for interfacing micro-power generators has been designed, fabricated, and validated using 0.18μm standard CMOS technology, with 0.88 mm^2 area. The center tap differential inductors in the LC tank achieve 38% area saving compared to the usage of four distinct inductors. The system can self-start as low as 0.15 V, and generate an out-of-phase clock pair with amplitude 0.30 V at this input voltage. 0.25 V can be stepped up to 1.7 V with input load resistance of 60 kΩ and maximum efficiency of 15.5%. Therefore, a step-up ratio of 1:6.8 is achieved at highest efficiency. At 160 kΩ load resistance and 0.25 V input, the effective step-up ratio increases to 1:9.

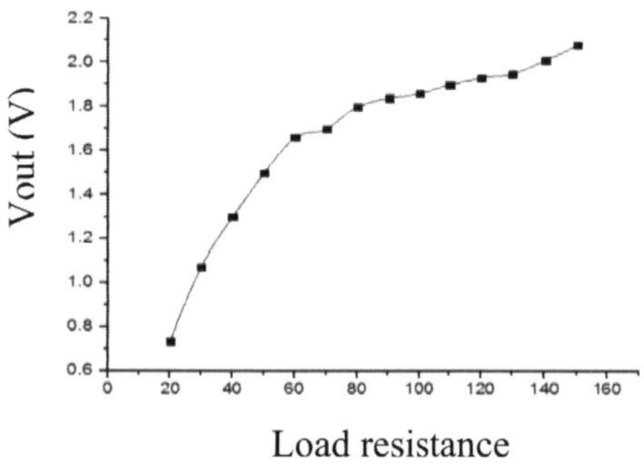

Figure 7. System unregulated output voltage with varying load resistance for 0.25 V input voltage.

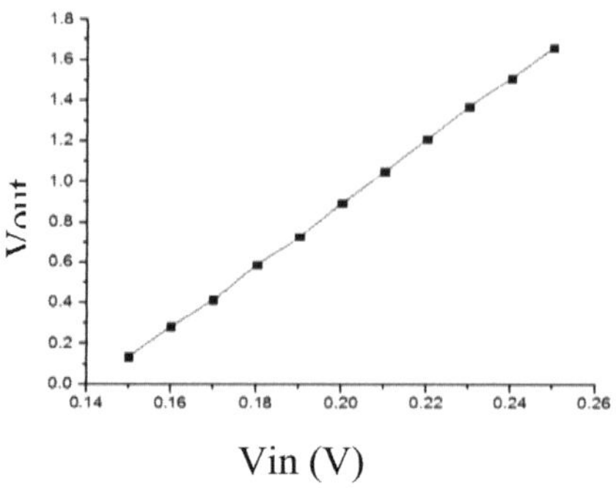

Figure 8. The variation of output voltage with input voltage for 60 kΩ load resistance.

References

[1] Y. K. Ramadass and A. P. Chandrakasan 2011 *A battery-less thermoelectric energy harvesting interface circuit with 35 mV startup voltage* IEEE J. Solid-State Circuits, vol.46, no.1, pp.333-341.

[2] P. H. Chen, et al 2012 *Startup techniques for 95 mV step-up converter by capacitor pass-on scheme and Vth-tuned oscillator with fixed charge programming IEEE J.* Solid-State Circuits, vol. 47, no. 5, pp. 1252- 1260.

[3] T. Tanzawa 2013 *Chapter 1 System Overview and Key Design Considerations*, in On-chip High-Voltage Generator Design, Springer,New York.

[4] W.P.M. R. Pathirana and A. Muhtaroglu 2013 *Low voltage fully integrated DC-DC converter for self-powered temperature sensor,* presented at the 19th International workshop on Thermal Investigation of ICs and System, Berlin, Germany.

[5] Kim, J., Mok, P. K., & Kim, C 2015 *A 0.15 V Input Energy Harvesting Charge Pump With Dynamic Body Biasing and Adaptive Dead-Time for Efficiency Improvement.* Solid-State Circuits, IEEE Journal of, 50(2), pp. 414-425.

[6] J. F. Dickson 1796. IEEE Journal of Solid-State Circuits, vol. 1, pp. 374-378.

[7] X. Wang, D. Wu, F. Qiao, P. Zhu, K. Li, L. Pan, and R. Zhou 2009 *A high efficiency CMOS charge pump for low voltage operation*, in IEEE 8th International Conference on ASIC, pp. 320–323.

[8] J.-T. Wu and K.-L. Chang, 1998 *MOS charge pumps for low-voltage operation*, IEEE J. Solid-State Circuits, vol. 33, no. 4, pp. 592–597.

[9] S. A. Bhalerao, A. V. Chaudhary, and R. M. Patrikar, 2007 *A CMOS low voltage charge pump*, in VLSI Design, 2007. Held jointly with 6th International Conference on Embedded Systems., 20th International Conference on, pp. 941–946.

[10] K. Kwok and H. C. Luong, 2005 *Ultra-low-Voltage high-performance CMOS VCOs using transformer feedback*, IEEE J. Solid-State Circuits, vol. 40, no. 3, pp. 652–660.

[11] Peng, Huan, et al 2014 *CMOS startup charge pump with body bias and backward control for energy harvesting step-up converters.* Circuits and Systems I: Regular Papers, IEEE Transactions on 61.6, pp 1618-1628.

[12] J. Kim, P. K. Mok, and C. Kim, 2014 *23.1 A 0.15 V-input energy-harvesting charge pump with switching body biasing and adaptive dead-time for efficiency improvement*, in Solid-State Circuits Conference Digest of Technical Papers (ISSCC), 2014 IEEE International, pp. 394–395.

[13] G. Bassi, L. Colalongo, A. Richelli, and Z. Kovacs-Vajna 2012 "*A 150mV-1.2V fully-integrated DC-DC converter for Thermal Energy Harvesting*," in 2012 International Symposium on Power Electronics, Electrical Drives, Automation and Motion (SPEEDAM), pp. 331–334.

Flexible Wireless Wall Temperature Sensor for Unsteady Thermal Field

Minhyeok Lee, Kenichi Morimoto and Yuji Suzuki

Department of Mechanical Engineering, The University of Tokyo,
7-3-1 Hongo, Bunkyo-ku, Tokyo, 113-8656, Japan

E-mail: mlee@mesl.t.u-tokyo.ac.jp

Abstract. We present a novel flexible wireless wall temperature sensor with high spatio-temporal resolution and its performance evaluation in an unsteady thermal field. A base part of the sensor is made of thermally-stable polyimide and the copper films. Using a Si hard mask fabricated by standard lithography and DRIE process, 1 mm-sized sensing resistor is sputtered on the copper coil. We enhance the time response for each measurement by reducing the frequency sweeping points. It is shown that the accuracy of the present temperature measurement is in acceptable range for most combustion studies, based on a series of error-estimation analyses. The temperature measurement uncertainty of ± 6.4 °C has been achieved with the measurement time interval as small as 2.48 ms.

1. Introduction

The wall temperature measurement for unsteady thermal fields is one of the major issues in combustion studies [1]. Contact thermometry using thermocouple has been widely used, but it requires wiring and physical contact that would easily introduce external disturbances [1, 2]. As non-contact thermometry, infrared pyrometer and thermographic phosphor are also available, but existing techniques require optical access, which could be problematic in combustion studies [1-4].

To overcome such limitations, we have proposed a wireless wall temperature sensor using TCR (Temperature Coefficient of Resistance) changes of Au thin film [5, 6]. The prototype sensor is successfully fabricated with standard MEMS process and the measured resonant frequency shows a quadratic increase in response to an increase of the temperature. However, the spatial resolution of the temperature measurement is as large as the diameter of the sensor coil, i.e., 10 mm, and its temporal resolution has not yet been examined.

In the present study, we propose a novel flexible wireless wall temperature sensor with improved spatio-temporal resolution.

2. Design of the wireless temperature sensor

The present sensor consists of the following elements: a planar spiral coil, a capacitor, and a resistor with temperature dependence. The sensor coil is inductively coupled with the read-out coil, and the circuit impedance changes in response to the resistance change of the resistor on the sensor. One advantage of this approach is that the sensor consists of passive devices only without any semiconductors, so that it is applicable to high temperature applications. Figure 1 shows the equivalent

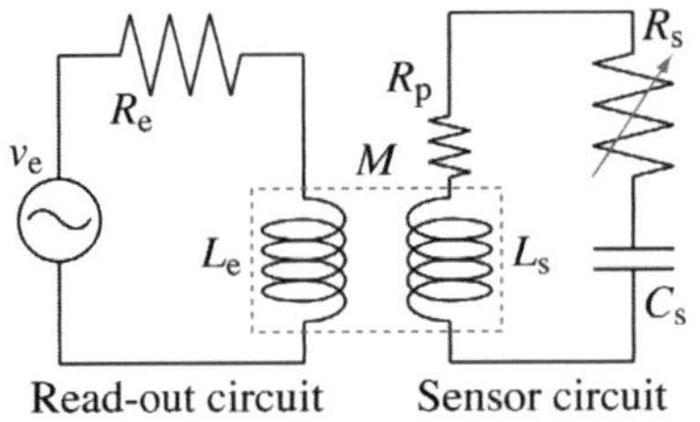

Figure 1. The equivalent circuit model for the wireless wall temperature sensor (right) and the read-out coil (left).

$$\angle Z_e(\omega) \cong \arctan\left\{\frac{(1-k^2)L_s^2C_s^2\omega^4 + \left\{(R_s+R_p)^2C_s^2 + (k^2-2)L_sC_s\right\}\omega^2 + 1}{k^2(R_s+R_p)L_sC_s^2\omega^3}\right\}, \quad (1)$$

circuit model for the sensor and its coupled read-out circuit, under an external driving voltage v_e. Based on the circuit analysis, the phase angle of the read-out circuit impedance $\angle Z_e(\omega)$ can be derived as equation (1) [5, 6], where L, C, R, and ω represent the inductance, the capacitance, the resistance, and the angular frequency, respectively. Subscripts e and s denote the external read-out circuit and the sensor. R_p is the parasitic resistance due to the sensor coil, and the coupling coefficient k is determined by the self-inductance of the two coils and their mutual inductance M, which is highly dependent on their geometric relationship. The resonant frequency is obtained from the extremum condition of equation (1), and its shift is determined by the sensor resistance and thus the temperature.

For the coupled series LCR circuit, the Q factor depends both on the resistance and the mutual inductance M. To achieve a higher Q factor for higher accuracy in the resonant frequency, the sensor resistance should be minimized and the mutual inductance should be maximized under constraint of the device. In our previous study [5, 6], however, due to large sheet resistance of the sensor coil with sputtered Au films, the parasitic resistance of the coil itself was large, making integration of the sensor resistor prohibited. Therefore, the sensor coil was used as the sensor resistor, which significantly deteriorates the spatial resolution of the temperature measurement.

In order to suppress the unwanted effect of the parasitic resistance, thickness of the sensor coil should be on the order of 10 μm. In the present study, we employ a Cu-laminated polyimide film as the sensor substrate. Note that polyimide, which has been widely used for the flexible printed circuits (FPC), is a material with suitable properties to be used in combustion fields: high thermal durability up to 400 °C, chemical stability, and flexibility.

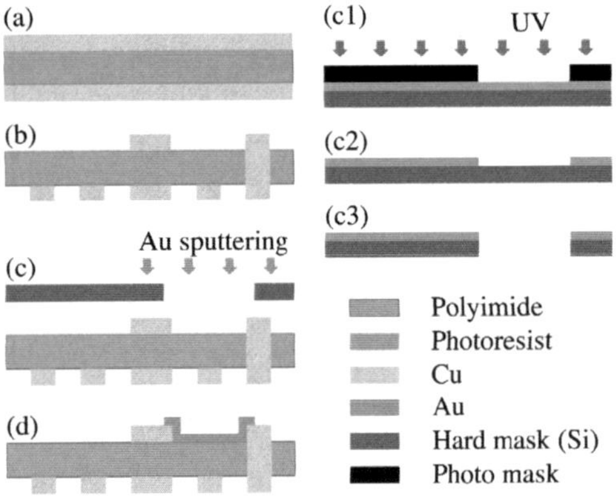

Figure 2. Sensor fabrication process: Firstly, Cu layers on the polyimide film are patterned into the electrodes and the planar spiral coil. A Si hard mask (figure 3a) is made through the photolithography and Deep-RIE process. Then, thin Au film is sputtered through the hard mask to deposit a sensing resistor as shown in figure 3b.

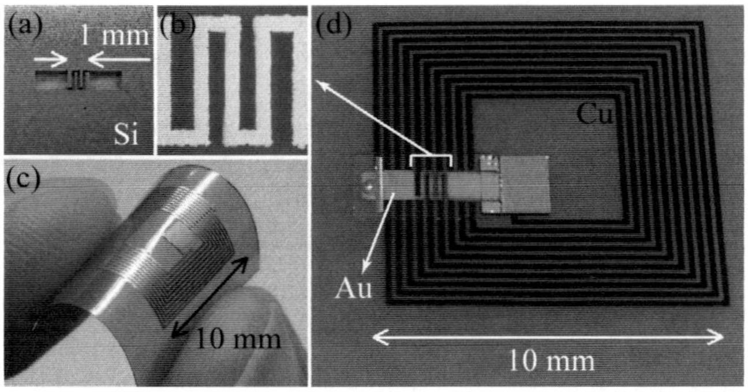

Figure 3. Photograph of (a) the Si hard mask, (b) the sensing resistor, and (c), (d) the fabricated polyimide-based wireless wall temperature sensor. The dimensions of the sensing resistor are 1 mm × 1 mm.

3. Fabrication of the sensor

Figures 2a-d show the fabrication process of the present sensor using a 12.5 μm-thick polyimide substrate. Firstly, 18 μm-thick Cu layers laminated on both sides of the polyimide film are patterned into the electrodes for the capacitor and the planar coil. A contact hole is drilled and electroless Au plating is performed for an electrical connection between the top and bottom Cu layers. Then, a Si hard mask for the sensing resistor is microfabricated by a standard lithography and etch-through process with DRIE (MUC-21 ASE-Pegasus, STS) as shown in figure 3a. Finally, an approximately 100 nm-thick Au film with a Cr adhesion layer is sputtered through the hard mask to deposit a sensing resistor (figure 3b), and to compose the sensor as a closed circuit.

Figure 3d shows a photograph of the successfully fabricated prototype sensor with a 1 mm sensing resistor. The measured resistance of the sensing resistor and the designed value of the parasitic resistance are respectively 130 Ω and 1 Ω. Note that the dimension of the resistor can be further reduced to as small as 50 μm for better spatial resolution.

4. Improvement of temporal resolution

When determining the resonant frequency, the sensor is inductively coupled with the read-out coil and its impedance is measured using a network analyzer. In our previous studies, the frequency sweeping required long time, so that the average time for the measurement remained as an order of 100 ms. Thus we try to shorten the measurement time by reducing the number of frequency sweeping points. As shown in figure 4, the measured phase angle data with different scanning points are fitted with

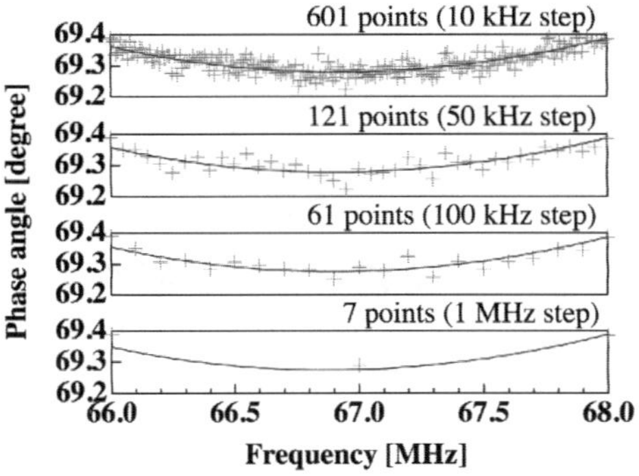

Figure 4. Phase angle of impedance versus the frequency for different number of sweeping points. The graphs are magnified in the vicinity of resonant frequency.

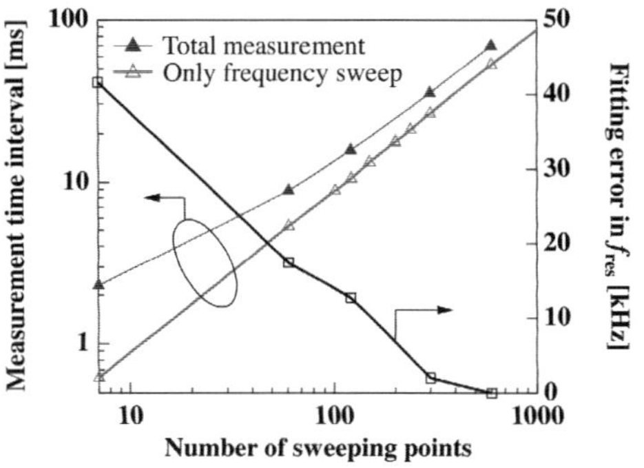

Figure 5. Measurement time interval and fitting error in resonant frequency versus number of sweeping points. The total measurement time includes not only the sweeping time but also time required for data transmission.

equation (1). With decreased number of data points, a fitting error on the resonant frequency is increased. The fitting error and the measurement time interval are plotted as a function of the number of sweeping points in figure 5. Note that the measured data with 601 points are used as a reference. When we decrease the number of sweeping frequency to 7 points, the sweeping time takes approximately 0.6 ms. When the time required for data-transfer is included, it corresponds to 2.3 ms. Due to the increase of fitting error, the uncertainty of the temperature measurement is also increased. However, the uncertainty around 100 °C is estimated to be ± 5.2 °C that should be acceptable for most combustion studies.

5. Performance evaluation in unsteady field

Figure 6 shows an experimental setup for the performance evaluation in an unsteady thermal field. The sensor is fixed on a plate and the read-out coil is attached on the other side with a distance of 1.6 mm. The read-out coil is connected to a network analyzer (ZNB20, Rohde & Schwarz) through a coaxial cable, and the wall surface temperature is simultaneously monitored with a K-type thermocouple once in every 100 impedance measurements. Both the network analyzer and the data acquisition unit are connected to a host PC through Ethernet for the remote control and the data transfer. VISA (Virtual Instrument Software Architecture) is used as an application programming interface with SCPI (Standard Commands for Programmable Instruments) commands.

To provide an unsteady thermal field to the wall surface, the sensor is suddenly approached to a 2

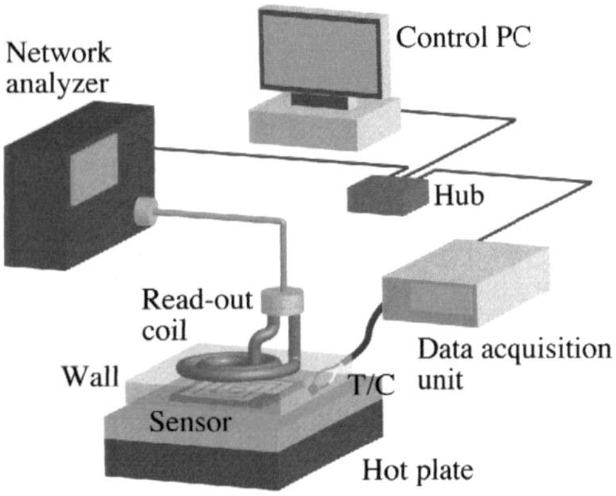

Figure 6. Schematic of the experimental setup for the measurement of unsteady temperature changes.

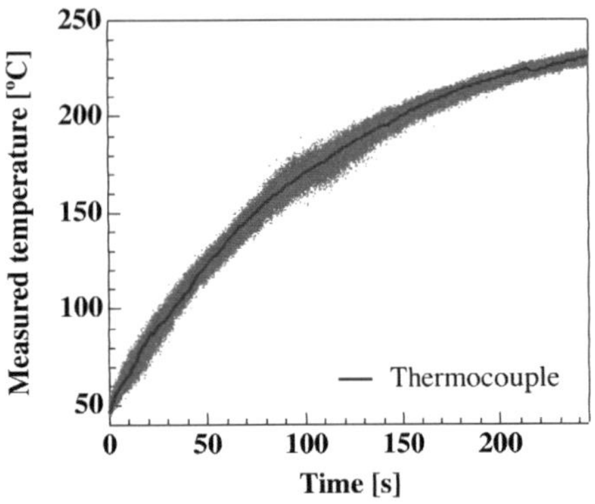

Figure 7. Temporal variation of the measured temperature converted from the resonant frequency along with the reference temperature measured with the thermocouple.

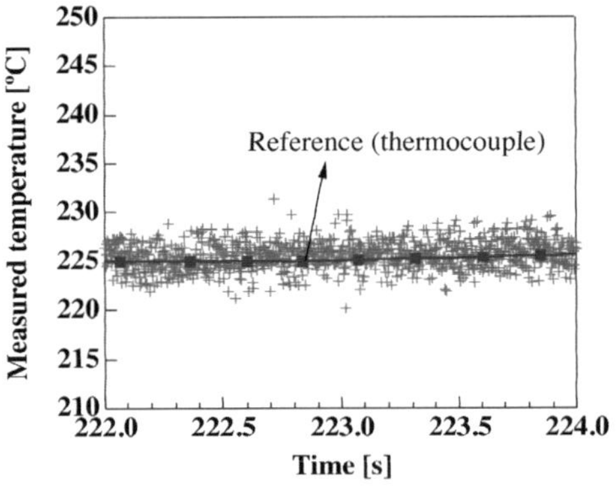

Figure 8. Measurement result which is magnified in the vicinity of 225 °C.

mm distance from a hot plate, which is heated up at 350 °C beforehand. The measurement is performed until surface temperature reaches 230 °C. The measured resonant frequency data are converted to the temperature and plotted as a function of the elapsed time as shown in figure 7. The measured data in the vicinity of 225 °C are magnified as shown in figure 8. The average time for each measurement is 2.48 ms, and the uncertainty of the temperature measurement is found to be ± 6.4 °C with the 95% coverage. By improving the measurement method, it is expected that the temporal resolution less than 1 ms can also be achieved.

6. Conclusions

We have developed a wireless wall temperature sensor with high spatio-temporal resolution for the use in combustion studies. The sensing system includes the sensor composed by a LCR circuit and the inductively-coupled read-out coil. The sensor is fabricated on the Cu-laminated polyimide film, and an additional 1 mm-sized sensing resistor is formed by standard MEMS process. The parasitic resistance of the coil is markedly suppressed by using the thick Cu layers, so that the additional sensing resistor dominates the whole sensor resistance. The fabricated sensor shows high flexibility, so that it can be fitted well on a curved surface. The sensor performance has been evaluated in an unsteady thermal field. Uncertainty in the temperature measurement of ± 6.4 °C with the mean measurement time interval of 2.48 ms has been achieved.

7. References

[1] Dreizler A and Böhm B 2015 *Proc. Combust. Inst.* **35** 37
[2] Childs P R N, Greenwood J R and Long C A 2000 *Rev. Sci. Instrum.* **71** 2959
[3] Fuhrmann N, Schild M, Bensing D, Kaiser S A, Schulz C, Brübach J and Dreizler A 2012 *Appl. Phys. B Lasers Opt.* **106** 945
[4] Someya S, Okura Y, Munakata T and Okamoto K 2013 *Int. J. Heat Mass Transf.* **62** 382
[5] Lee M, Kawahara Y, Morimoto K and Suzuki Y 2014 *J. Phys. Conf. Ser.* **557** 012077
[6] Lee M, Morimoto K and Suzuki Y 2015 *SAE Paper* 2015-01-1997

Acknowledgments

This work is partially supported by JSPS Grant-in-Aid for Exploratory Research (No. 26630060) and the Cross-ministerial Strategic Innovation Promotion Program (SIP) from JST. ML is supported through the Leading Graduates Schools Program, “Global Leader Program for Social Design and Management,” by MEXT Japan. Photo-mask is made using The University of Tokyo VLSI Design and Education Center (VDEC)’s 8-inch EB writer F5112+VD01 donated by ADVANTEST Corporation.

Effect of micro cooling channels on a hydrogen peroxide monopropellant microthruster performance

Jeongmoo Huh and Sejin Kwon
Department of Aerospace Engineering, Korea Advanced Institute of Science and Technology, 373-1 Guseong-dong, Yuseong-gu, Daejeon, 305-701, Republic of Korea

E-mail: trumpet@kaist.ac.kr

Abstract. In this paper, a hydrogen peroxide monopropellant microthrusters with and without regenerative micro cooling channels were fabricated and performance test results were compared to determine cooling effect of the regenerative micro cooling channels. Photosensitive glass was used as microfabrication material, which is cost-effective for MEMS fabrication process. Nine photosensitive glasses was integrated using UV and thermal bonding and composed the microthrusters. 90wt% hydrogen peroxide was used both as monopropellant and cooling fluid. For hydrogen peroxide decomposition, catalyst was fabricated and inserted into the microchamber. Platinum was used as the catalyst active material and γ-alumina was used as catalyst support. Experimental testing was conducted to determine effect of the cooling channels and the chamber pressure, temperature and surface temperature were measured. The performance test results showed that it was possible to relieve the thermal shock of the micro thruster structure by as much as 64% by adding regenerative micro cooling channels on both sides of the microthruster chamber. However, the chamber pressure and temperature decreased by regenerative cooling channels due to excessive cooling effects.

1. Introduction

For nanosatellites applications, many microthrusters has been developed using solid propellant propulsion. Solid propellant thruster has advantages of system simplicity, which is desirable for micro scale thruster operation, however, difficulties of throttling and re-ignition of the solid propellant thruster showed the limitation for nanosatellites attitude control, orbit compensation and orbit transfer. A monopropellant thruster is one alternative for nanosatellite applications, with system simplicity than that of bipropellant and possibility of throttling and re-ignition, which is difficult to function using solid propellant thruster. For these reasons, lots of microthruster was developed by monopropellant propulsion[1-3].

The microfabrication material with low thermal conductivity is desirable to conserve heat energy of the microthrustser, which has enhanced heat energy loss, due to large surface to volume ratio. Ceramics and glass are good fabrication material with low thermal conductivity and several previsou work shows the results of the microfabrication using these material. HTCC(high-temperature-co-fired ceramic) was used as an microfabrication material by Cheah et al.[4]. He found higher efficiency of the electric power consuming microthruster with HTCC than silicon, due to energy conservation effect of the fabrication material. LTCC(low-temperature co-fired ceramic) and glass were used also for microthrusters[3, 5]. Occasionally, however, brittle characteristics of the materials is a difficult challenge to overcome. Monopropellant microthruster was successfully fabricated using LTCC by Wu et al.[5], however, the crack occurred at the combustor wall and thrust level decreased. Glass was used by An[3] for micro reactor fabrication, and he found cracking of the micro chamber wall. These results showed the insufficient thermal strength of the ceramic and glass, which can be improved avoiding sudden temperature change and unevenly distributed thermal stress.

In this paper, a liquid microthruster with and without regenerative cooling channels were fabricated and to determine the effect of the cooling channels on the microthruster, which can be used for thermal shock handling. Experimental test was conducted and compared for the two microthrusters, and cooling channels effect on a liquid microthruster performance was studied.

2. Microthruster fabrication

The microthrusters were designed for 50mN class thrust generation, referring required microthruster thrust for nanosatellites operation[6]. Regeneratively cooled microthruster was designed and fabricated using nine glass wafers, and the components drawing is shown in figure 1. For similar heat capacity, uncooled microthruster was also designed and fabricated using nine glass wafers, and the drawing of the components is shown in figure 2. Using MEMS fabrication process, glass wafers were UV exposed using the pattern of designed microthruster profiles. Due to UV exposed parts of the glass is etched 20 times faster than unexposed parts by hydrofluoric acid, exposed glass can be selectively etched for the thruster profile. After polishing of the glass surfaces for thermal bonding, Nine glasses were integrated by both thermal bonding and UV bonding and composed the microthruster. Before all the wafers bonded, platinum/alumina catalyst was inserted into the catalyst with 100 mesh size sieve as catalyst holder, which catalyst was fabricated by the procedure of loading, calcination, and reduction. The integrated microthruster is shown in figure 3.

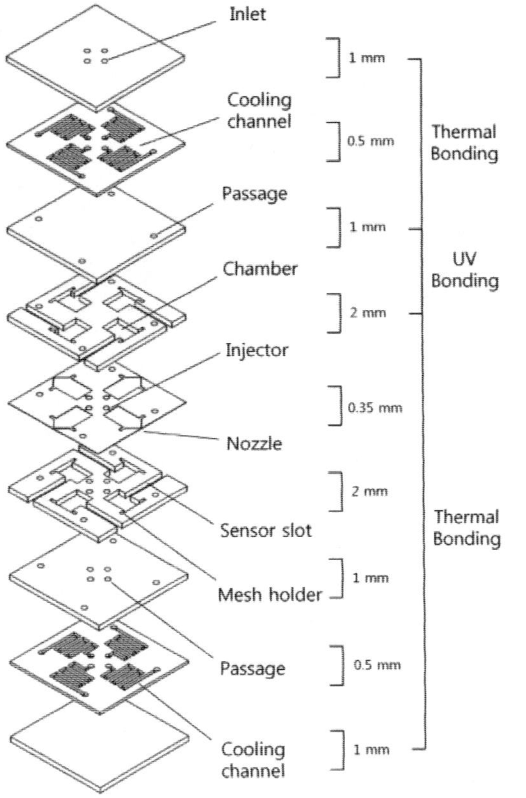

Figure 1. Regeneratively cooled microthruster components

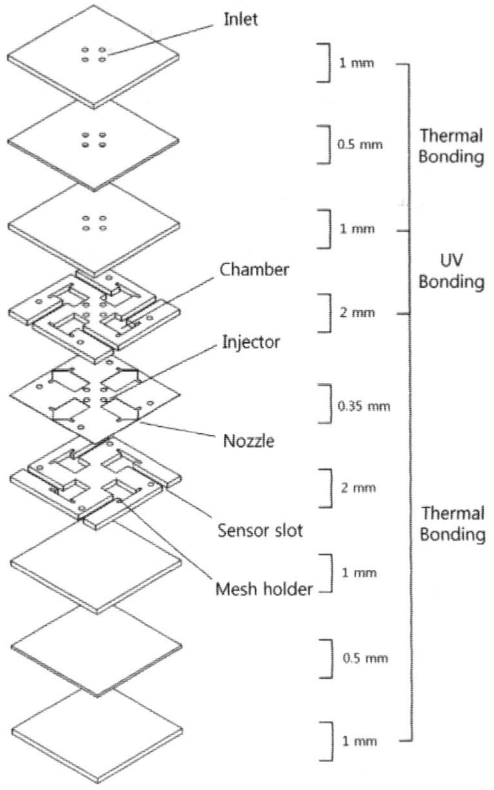

Figure 2. Uncooled microthruster components

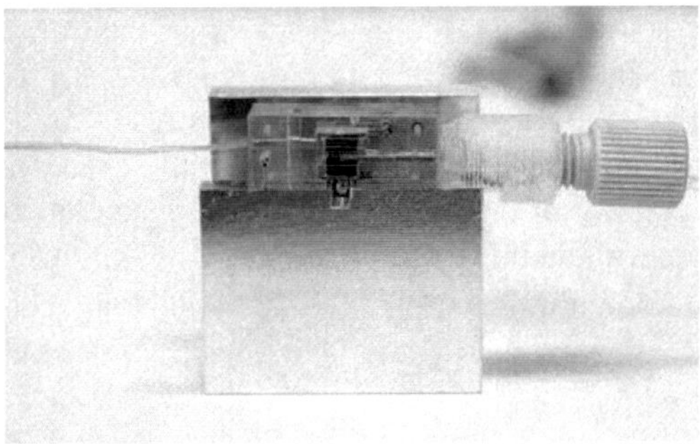

Figure 3. Integrated thruster installed at the thruster stand for performance test

3. Performance test and discussion

Experimental test was conducted with 90wt% hydrogen peroxide as a propellant, propellant feeding system and data acquisition device. Syringe pump and Teflon tubing composed propellant feeding system. The performance test results are shown in figure 4 and 5. Figure 4 shows the temperature differences for the chamber and surface of the microthruster. The chamber temperature was 323.0 °C for the cooled microthruster, and 582.6 °C for the uncooled microthruster. The surface temperature was 95.8 °C and 137.1 °C for cooled and uncooled microthruster, respectively. Figure 5 shows the chamber pressure differences between cooled and uncooled microthruster. The chamber pressure was 1.9 bar and 2.3 bar for the cooled and uncooled microthruster, respectively. Lower surface temperatures were desirable and it was intended for relaxation of thermal stresses stemming from abrupt temperature variations, but the lower chamber pressure and temperature are undesirable for thrust generation. The lower values for the pressure chamber might be explained by excessive effects from the cooling channels, which were experimentally determined for robustness during the wafer surface polishing process of the microfabrication procedure. The lower chamber temperature might cause a lower propellant decomposition efficiency by the catalyst, which varies exponentially depending on the temperature according to the Arrhenius equation.

Another factor we were interested in was the characteristic velocity. The characteristic velocity, also known as c*, is defined as:

$$c^* = P_c A_t / \dot{m} \tag{1}$$

where P_c is chamber pressure, A_t is nozzle throat area and $\dot{m}$ is mass flow rate. It may also be expressed in the following form for choked flow:

$$c^* = \frac{\sqrt{kRT_c}}{k[2/(k+1)]^{(k+1)/(2k-2)}} \tag{2}$$

where k is specific heat ratio, R is the gas constant, and T_c is chamber temperature. To find the degree of energy released from the propellant in a general rocket engine, c* efficiency is often used, which is the ratio of the actual value of c* derived from the equation (1) with the measured P_c to theoretical value of c* from the equation (2) using the estimated adiabatic temperature T_c. This is reasonable because the chamber temperature in the macroscale is very nearly the adiabatic temperature of the propellant. In the microscale, it is difficult to reach the adiabatic temperature due to excessive heat energy loss. Therefore, in this paper, as a denominator for the c* efficiency calculation, we used the actual c* determined from the measured chamber temperature, instead of the theoretical c* determined from the estimated chamber temperature. In other words, the ratio of c* derived from the equation (1) with the measured P_c to the c* from the equation (2) with the measured T_c was used. This describes how much chamber pressure was generated versus the temperature generated in the chamber. The performance results showed that the average maximum c* ratio at each pulse was 56.1% for the regeneratively cooled microthruster and 48.9% for the uncooled microthruster. This result may indicate that even though the temperature in the chamber was lower, the chamber pressure against the chamber temperature was relatively higher by as much as 7.2% with regenerative cooling channels than without. It may demonstrate the effectiveness of regenerative cooling even for non-optimized cooling channel design conditions.

The rise times for the pressure and temperatures were also estimated and were defined as the time for the values to reach 90% of their maximum from an initial state at each pulse. The results showed that it took an average of approximately 2.5 sec for the chamber pressure to reach 90% of its maximum for the regeneratively cooled microthruster and 3.0 sec for the uncooled microthruster. The rise time of the temperature was approximately 5.1 sec and 15.6 sec, respectively, for the chamber and the surface in the regeneratively cooled microthruster. In the uncooled microthruster, it was 3.5 sec and 7.5 sec, respectively, for the chamber and the surface. Despite the fact that the pressure rise time was lower and all of the maximum temperatures were lower, all of the temperature rise times were longer for the microthruster with the regenerative cooling channels than without. Therefore, lower thermal shock occurred with cooling channels. Using the definition of thermal shock as a temperature variation with respect to time, the maximum thermal shock of the microthruster chamber was approximately 63.9°C/sec for the regeneratively cooled microthruster and 167.5°C/sec for the uncooled microthruster. Additionally, the thermal shock on the microthruster surface was 6.2°C/sec and 18.2°C/sec with and without regenerative cooling channel, respectively. These results showed that the thermal shock was relieved by approximately 64% by adding regenerative cooling on the micro liquid monopropellant thruster. Figure 6 shows thermal shock the microthruster suffering at each pulse mode start, and approximately three times relieved thermal shock by regenerative cooling in the microthruser.

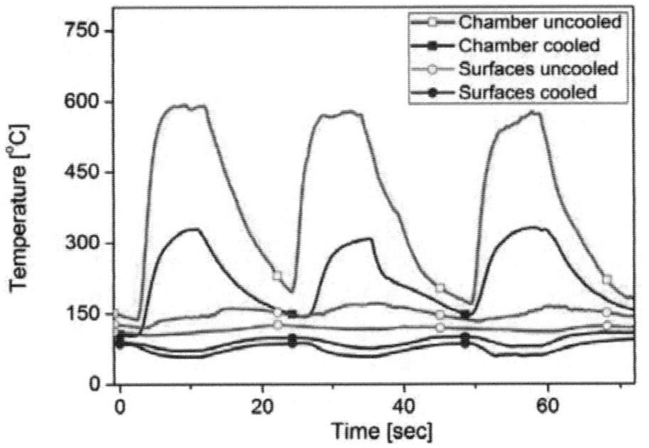

Figure 4. Temperature variations on three pulse operations

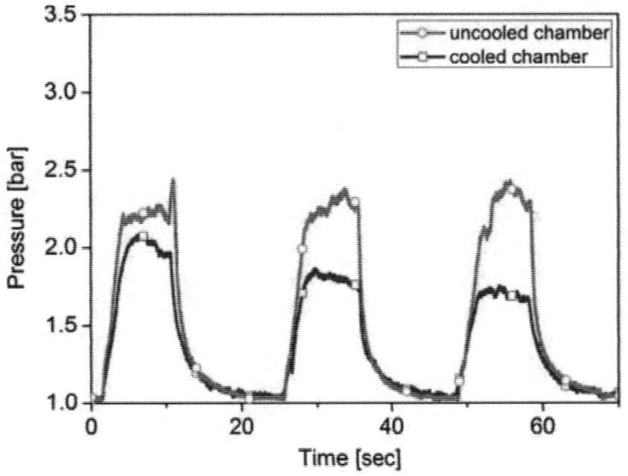

Figure 5. Chamber pressure variations on three pulse operations

Table 1. Summary of the microthruster experimental test results

Micro thruster test results	With channel	Without channel
Chamber presure[+]	1.9 bar	2.3 bar
Chamber temperature[+]	323.0 °C	582.6 °C
Surface temperature[+]	95.8 °C	137.1 °C
c* ratio[#]	56.1%	48.9%
Rising time of chamber pres	2.5 sec	3.0 sec
Rising time of chamber temp	5.1 sec	3.5 sec
Rising time of surface temp	15.6 sec	7.5 sec
Thermal shock in chamber	63.9 °C /sec	167.5 °C /sec
Thermal shock on surface	6.2 °C /sec	18.2 °C /sec

+ Average of the maximum values for each pulse
#Ratio of measured pressure c* to measured temperature c*

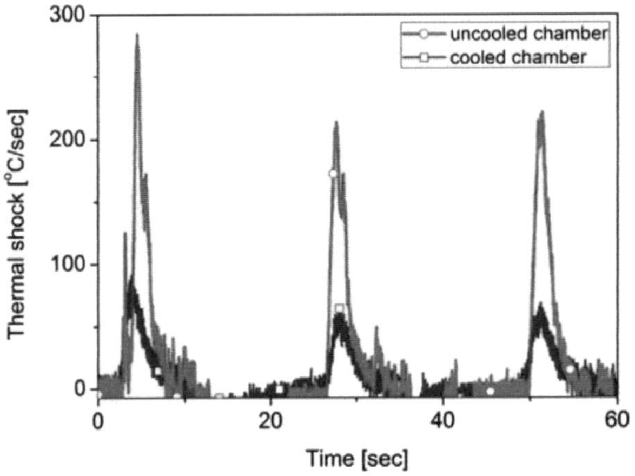

Figure 6. Thermal shocks of the cooled and uncooled monopropellant microthruster at each pulse start

4. Conclusion

Liquid monopropellant microthrusters with and without regenerative cooling channels have been fabricated and performance were compared for handling the thermal shock for microthrusters. Microthrusters with and without cooling channels were successfully fabricated and tested in this work. The performance results showed that despite the lower chamber and surface temperature in the regeneratively cooled microthruster, it took more time for the temperature to reach its maximum value, which reduced the chamber and surface temperatures variations with respect to time. Therefore, the thermal shock on the microthruster structure was relieved successfully, as much as 64% by adding regenerative cooling channels to the micro monopropellant thruster. Unexpectedly, the chamber pressure and temperature yielded poor results for the regenerative cooling channel, which may be accounted for by the excessive cooling effect of the channels and a lower propellant decomposition efficiency of the catalyst in the lower temperature condition. Additional work on cooling channel optimization to consider variations in the propellant decomposition efficiency of the catalyst as a function of temperature is expected to yield improved performance of the cooling channels.

Acknowledgment

This work was conducted at High-Speed Vehicle Research Center of KAIST with the support of Defense Acquisition Program Administration (DAPA) and Agency for Defense Development (ADD).

References

[1] Wu M H and Yetter R A 2009 Lab on a chip 9 910-6
[2] Huh J and Kwon S 2014 Journal of Micromechanics and Microengineering 24 9
[3] An S 2006 (Deajeon, South Korea: KAIST)
[4] Cheah K H and Low K S 2015 Journal of Micromechanics and Microengineering 25 10
[5] Wu M H and Lin P S 2010 Journal of Micromechanics and Microengineering 20 085026
[6] Janson S W, Helvajian H, Hansen W W and Lodmell L J 1999 In: The 2nd Int. Conf. on Integrated Micro Nanotechnology for Space Applications, (Pasadena, CA: The Aerospace Corporation)

Journal of Physics: Conference Series **660** (2015) 012021 doi:10.1088/1742-6596/660/1/012021

ROBOTIC INSECTS: MANUFACTURING, ACTUATION, AND POWER CONSIDERATIONS

Robert Wood

Charles River Professor of Engineering and Applied Sciences
Harvard John A. Paulson School of Engineering and Applied Sciences
Wyss Institute for Biologically Inspired Engineering
Harvard University, USA

Abstract

As the characteristic size of a flying robot decreases, the challenges for successful flight revert to basic questions of fabrication, actuation, fluid mechanics, stabilization, and power - whereas such questions have in general been answered for larger aircraft. When developing a robot on the scale of a housefly, all hardware must be developed from scratch as there is nothing "off-the-shelf" which can be used for mechanisms, sensors, or computation that would satisfy the extreme mass and power limitations. With these challenges in mind, this talk will present progress in the essential technologies for insect-like robots with an emphasis on multi-scale manufacturing methods, high power density actuation, and energy-efficient power distribution.

Microwatt power consumption maximum power point tracking circuit using an analogue differentiator for piezoelectric energy harvesting

Z J Chew and M Zhu
College of Engineering, Mathematics and Physical Sciences, University of Exeter, Exeter, EX4 4QF, UK

E-mail: z.j.chew@exeter.ac.uk

Abstract. A maximum power point tracking (MPPT) scheme by tracking the open-circuit voltage from a piezoelectric energy harvester using a differentiator is presented in this paper. The MPPT controller is implemented by using a low-power analogue differentiator and comparators without the need of a sensing circuitry and a power hungry controller. This proposed MPPT circuit is used to control a buck converter which serves as a power management module in conjunction with a full-wave bridge diode rectifier. Performance of this MPPT control scheme is verified by using the prototyped circuit to track the maximum power point of a macro-fiber composite (MFC) as the piezoelectric energy harvester. The MFC was bonded on a composite material and the whole specimen was subjected to various strain levels at frequency from 10 to 100 Hz. Experimental results showed that the implemented full analogue MPPT controller has a tracking efficiency between 81% and 98.66% independent of the load, and consumes an average power of 3.187 μW at 3 V during operation.

1. Introduction

Energy harvesting technology which converts energy from ambient environment into electrical energy is a promising solution to provide supply energy for fully autonomous systems such as wireless sensor nodes (WSNs) because power density harvested from the surroundings such as vibration, solar and temperature can be more than 100μW/cm^3 [1]. In parallel, with the advancement of low-power electronic devices, which power consumption is only a few milliwatts [2, 3], energy harvesting powered WSNs are therefore viable. Vibration-based energy harvesting is of particular attractive because vibration can occur almost everywhere either naturally from the environment or induced by means of other sources. Piezoelectric energy harvesters (PEHs) are most commonly used transducers to scavenge vibration energy because of their simple structures, rigidity, self-contained power generation capabilities and attractive high energy density [1, 2, 4]. However, power harvested by PEH is intermittent, in the form of AC and variable in magnitude of current and voltage. Therefore, energy storage and rectification circuity are needed for most applications. Furthermore, power harvested by PEH is very sensitive to connected electrical loads. If the electrical load mismatched the impedance of PEHs, the power delivered to the load would be a tiny fraction of power that the PEH is capable of harvesting. Therefore, an effective power management circuit is needed for PEHs to deliver the energy harvested to the electrical load.

Traditional power management circuits for PEHs use a rectifier to convert the AC voltage to the DC voltage and store it in a capacitor. This simplest form of electronics interface is not effective for most applications because it is not adaptive to variations from either the ambient vibration subjected by PEHs or the connected electrical load. From literature of [5, 6], it is known that PEHs exhibit linear electrical characteristic in which the maximum power transfer occurs at half the rectified open-circuit voltage V_{OC} of the PEH as shown in figure 1. With this simple electrical relationship of PEHs, an MPPT scheme by tracking the half the rectified open-circuit voltage V_{OC} of PEHs, through using an analogue differentiator and comparators proposed in this paper, could be used instead of the traditional MPPT methods using microcontroller. This leads to a low power overhead of this circuit, allowing energy to be drawn directly from a PEH for operation, and eliminating the need of sensing circuitry external power supply or start-up circuit. Without all these additional circuits and power hungry controller, power consumption of the MPPT controller can be greatly reduced. A prototype was implemented and tested over a wide range of vibrational frequencies and strain levels across the PEH. Experimental results have proven the feasibility of the proposed method.

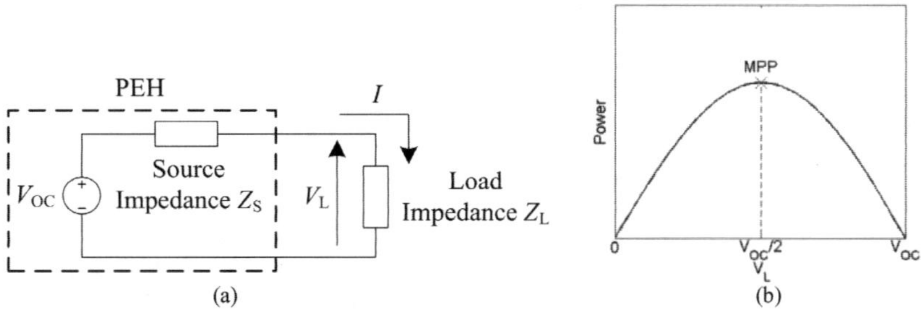

Figure 1. (a) Illustration of PEH as a linear circuit and (b) its maximum power point occurring at half of the open-circuit voltage.

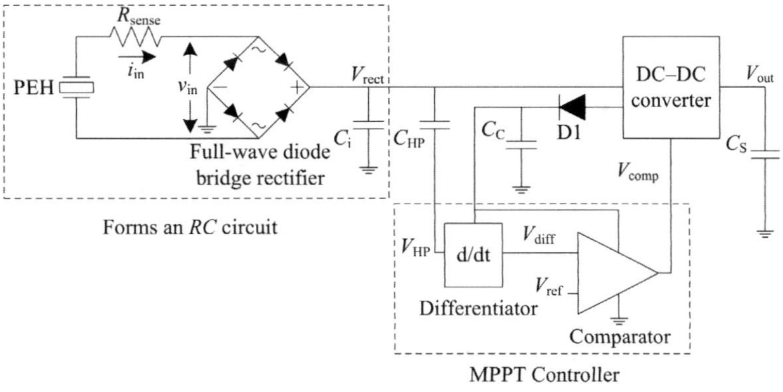

Figure 2. Schematic of the proposed MPPT circuit.

2. MPPT Using Differentiator

The proposed MPPT controller is shown in figure 2 and the operation is explained using figure 2 to 4. The MPPT circuit is used in conjunction with a rectifier and a DC–DC converter to manage the energy harvested from a PEH at its maximum power point (MPP). The voltage after the rectifier V_{rect} is monitored using a low-power operational amplifier-based differentiator. The theoretical background of this proposed method is based on the voltage charging profile of capacitor in an RC circuit. The DC–DC converter can be seen as open-circuited when it is disabled because no current can be drawn by it. In figure 2, before the MPP is reached, the circuit is effectively an RC circuit formed by the PEH, rectifier and smoothing capacitor because there is no current path through the DC–DC converter which has been disabled. The smoothing capacitor C_i will eventually be charged up to V_{OC} of the PEH if the

DC–DC converter is always disabled. This means V_{OC} of the PEH can be determined using the proposed circuit when the DC–DC converter is always disabled, that is, a simple *RC* circuit. In the *RC* circuit, the capacitor can be charged up very fast in an almost linear fashion initially but the increment rate of the voltage across the capacitor will gradually become slow as V_{rect} is increasing towards V_{OC}. A high-pass filter is used to pass through the fast changing signal and block the slow changing signal. The input voltage V_{HP} into the differentiator is the resultant voltage of feeding V_{rect} through a high-pass filter that attenuates the slow changing V_{rect} after V_{rect} is equal to $V_{OC}/2$ as shown in figure 3. V_{HP} gradually increases and peaks at $V_{OC}/2$. It should be highlighted that the input of the differentiator and C_{HP} forms a high-pass filter. After the peak, V_{HP} will decrease since the slow changing V_{rect} has been attenuated. Differentiation of V_{HP} is equal to zero at the maxima. Therefore, $V_{OC}/2$ can be determined by differentiating V_{HP}. $V_{OC}/2$ is reached when the derivative of V_{HP} is equal to zero. When V_{rect} reaches $V_{OC}/2$ of the PEH, output og the differentiator V_{diff} will drop below V_{ref} which subsequently triggers the comparator to send a signal pulse V_{comp} to enable the DC–DC converter. Energy will be transferred from the PEH at its MPP to an energy storage device. V_{rect} will decrease once the energy transfer begins. After V_{diff} has dropped below V_{ref}, the DC–DC converter will be disabled and the energy transfer process stops. C_i will be charged up again by the PEH until V_{rect} reaches $V_{OC}/2$ of the PEH and the energy transfer process begins again.

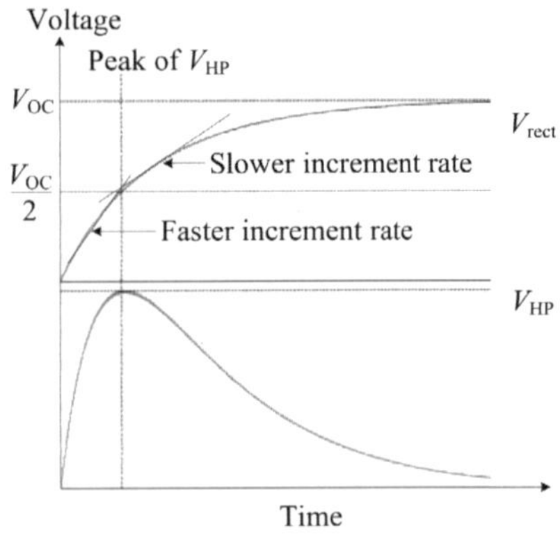

Figure 3. Illustration of the MPP occurring at $V_{rect}= V_{OC}/2$ at the peak of V_{HP}.

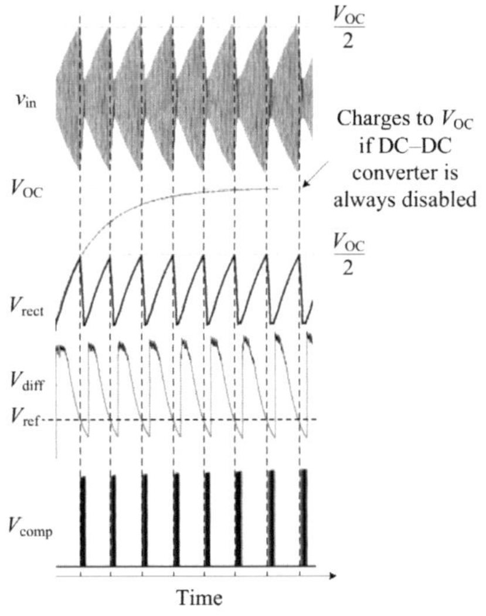

Figure 4. Operation of the proposed MPPT circuit

3. Experiment

A macro-fiber composite (MFC) bonded on a composite material using adhesive glue is used as the PEH in this experiment. An Instron E10000 ElectroPuls dynamic and fatigue test machine was used as the vibration source. Different vibrational frequencies up to 100 Hz and three sets of mechanical loadings with peak-to-peak strain levels of 349.99 µε, 696.70 µε, and 927.24 µε, applied by Instron machine on the MFC PEH were performed. The MPPT efficiency η_{MPPT} of the MPPT circuit can be defined as the ratio of the harvested power P_{in} flowing into the power management circuit at the FB rectifier to the power P_{max} dissipated by a perfectly matched load directly connected to the PEH:

$$\eta_{MPPT} = \frac{P_{in}}{P_{max}} \tag{1}$$

Measurements were made across the input of the FB rectifier to get voltage v_{in} and across a 20 Ω resistor R_{sense} for the current i_{in} generated from the PEH as shown in figure 2. The root mean square (RMS) value of input voltage and current waveform is calculated from the measured data using and the product of the root mean square of v_{in} and i_{in} gives the average harvested power P_{in} into the power management circuit. Measurements of V_{diff}, V_{rect}, V_{ref}, and V_{comp} were also made using an oscilloscope to verify the operation of the circuit. The power consumption of the MPPT, P_{MPPT}, is an important characteristic for the proposed scheme in this paper. To determine the power consumption P_{MPPT} of the core of the controller formed by the differentiator and comparators, the current and voltage were measured between D1 and C_C.

4. Results

Figure 5 shows the measured voltages of V_{diff}, V_{rect}, V_{ref}, and V_{comp}. The measured result agrees with the simulation result. V_{diff} gradually becomes smaller and falls below V_{ref}. This caused V_{comp} to become HIGH, enabling the buck converter for energy transfer. V_{rect} decreases sharply when the energy transfer occurs and gradually increases after the process ends. The energy transfer process and switching of the electronics switches caused large spikes at V_{ref}. Powers, P_{in} and P_{max}, obtained from the tests are shown in figure 6. It can be seen that the power harvested from the PEH using the proposed circuit closely matches the power obtained using manual resistive matching in the tested conditions. This prototyped MPPT controller shows efficiency of around 81 % to 98.66 %, which is calculated based on equation (1). This peak efficiency is better than the MCU-based MPPT controller [7]. The power consumption P_{MPPT} of the MPPT controller is shown in figure 7. Initially, there is a large inrush current into C_C because the capacitor is empty. The current eventually became very small when C_C has been fully charged up to around 3.3 V. When C_C is fully charged up and the controller begins to operate, the current consumption is about the same as shown in the inset. By looking into the inset of large scaled current consumption, it can be observed that the peak current during operation is about 10 μA and the average current consumption of the controller is 1.011 μA during operation. The supply voltage to the controller is relatively stable without significant drop during operation. This indicates that the current consumption of the circuit is indeed very low. Therefore, the average power consumption of this control circuit is only 3.178 μW, which is 100 times less than MPPT controller using microcontroller and 16 times less than MPPT controller using combination of analogue circuits and FPGA respectively [7, 8].

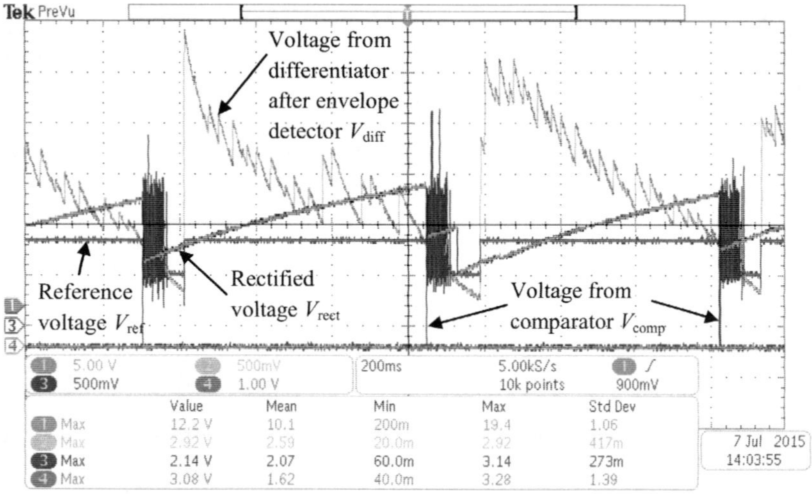

Figure 5. Measured waveforms of V_{diff}, V_{rect}, V_{ref}, and V_{comp} of the implemented MPPT circuit

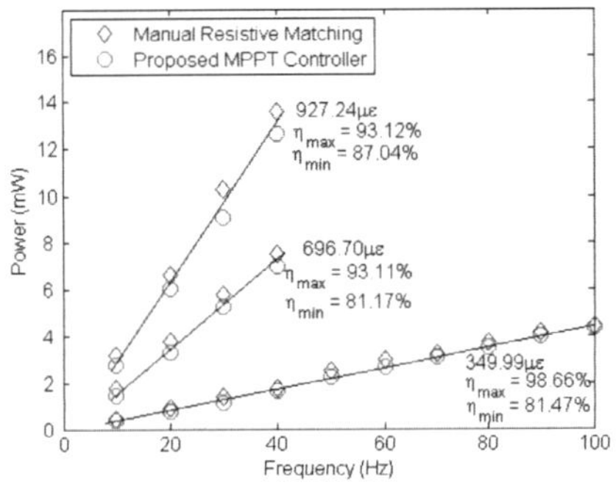

Figure 6. Efficiencies of the proposed MPPT.

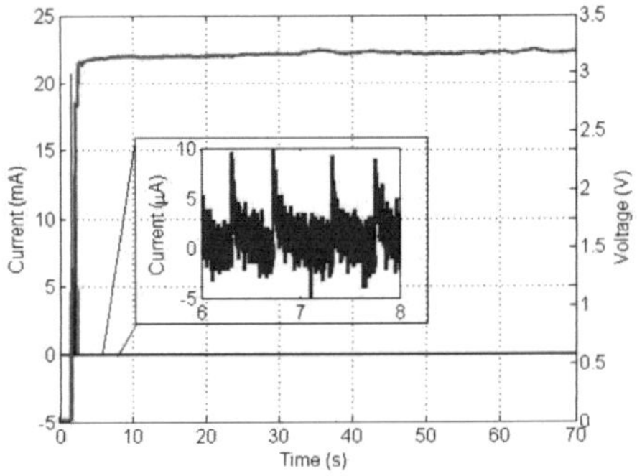

Figure 7. Power consumption of the circuit.

5. Conclusion

A microwatt power consumption MPPT circuit was presented and demonstrated for harvesting energy from an MFC piezoelectric transducer at its maximum power point under various vibrational frequencies and excitation forces. By exploiting the *RC* response of a charging capacitor and linear electrical characteristics of the piezoelectric element, an MPPT scheme based on open-circuit voltage has been successfully implemented to find the MPP of the PEH using differentiator. This innovation has greatly simplifies the MPPT controller and reduces the power consumption of the circuit. Without MCU or FPGA which are usually off-the-shelf products, the proposed circuit can be easily fabricated as a monolithic IC. With the low power requirement and reduced size in the form of IC, this circuit can find applications in other areas such as MEMS energy harvesting.

Acknowledgments

The authors would like to thank Innovate UK and EPSRC for their sponsorship of this project.

References

[1] Roundy S, Wright P K and Rabaey J 2003 A study of low level vibrations as a power source for wireless sensor nodes *Computer Communications* **26** 1131-44

[2] Steven R A and Henry A S 2007 A review of power harvesting using piezoelectric materials (2003–2006) *Smart Materials and Structures* **16** R1 - R21

[3] Yu H, Zhou J, Deng L and Wen Z 2014 A vibration-based MEMS piezoelectric energy harvester and power conditioning circuit *Sensors* **14** 3323-41

[4] Zhu M, Worthington E and Njuguna J 2009 Analyses of power output of piezoelectric energy-harvesting devices directly connected to a load resistor using a coupled piezoelectric-circuit finite element method *IEEE Transactions on Ultrasonics, Ferroelectrics and Frequency Control* **56** 1309-17

[5] Ottman G K, Hofmann H F, Bhatt A C and Lesieutre G A 2002 Adaptive piezoelectric energy harvesting circuit for wireless remote power supply *IEEE Transactions on Power Electronics* **17** 669-76

[6] Tabesh A and Fréchette L G 2010 A Low-Power Stand-Alone Adaptive Circuit for Harvesting Energy From a Piezoelectric Micropower Generator *IEEE Transactions on Industrial Electronics* **57** 840-9

[7] Kong N and Ha D-S 2012 Low-Power Design of a Self-powered Piezoelectric Energy Harvesting System With Maximum Power Point Tracking *IEEE Transactions on Power Electronics* **27** 2298-308

[8] Elliott A D T and Mitcheson P D 2014 Piezoelectric energy harvester interface with real-time MPPT *Journal of Physics: Conference Series* **557** 012125 (6 pp)

A Self-Powered Hybrid Energy Scavenging System Utilizing RF and Vibration Based Electromagnetic Harvesters

H Uluşan[1,*], K Gharehbaghi[1], Ö Zorlu[2], A Muhtaroğlu[3], and H Külah[1,2]

[1] METU, Department of Electrical and Electronics Engineering, Ankara, Turkey
[2] METU-MEMS Research and Applications Center, Ankara, Turkey
[3] METU Northern Cyprus Campus, Department of Electrical and Electronics Engineering, Güzelyurt, Mersin 10, Turkey

E-mail: hulusan@metu.edu.tr

Abstract. This study presents a novel hybrid system that combines the power generated simultaneously by a vibration-based Electromagnetic (EM) harvester and a UHF band RF harvester. The novel hybrid scavenger interface uses a power management circuit in 180 nm CMOS technology to step-up and to regulate the combined output. At the first stage of the system, the RF harvester generates positive DC output with a 7-stage threshold compensated rectifier, while the EM harvester generates negative DC output with a self-powered AC/DC negative doubler circuit. At the second stage, the generated voltages are serially added, stepped-up with an on-chip charge pump circuit, and regulated to a typical battery voltage of 3 V. Test results indicate that the hybrid operation enables generation of 9 μW at 3 V output for a wide range of input stimulations, which could not be attained with either harvesting mode by itself. Moreover the hybrid system behaves as a typical battery, and keeps the output voltage stable at 3 V up to 18 μW of output power. The presented system is the first battery-like harvester to our knowledge that generates energy from two independent sources and regulates the output to a stable DC voltage.

1. Introduction

Energy harvesters use environmental sources such as solar, thermal, vibration and RF energy to generate electrical energy. The decreasing power demand of the new generation integrated circuits allow the use of micro-harvesters as energy sources in microsystems as an alternative to batteries. However, each micro-harvester by itself is typically limited in power generation capacity, and output voltage level. Hence, hybrid systems are of high interest to effectively harvest multiple sources while generating sufficient voltage and power levels for an increased number of applications [1], [2].

There are several attempts in literature to build hybrid energy harvesters. System presented in [3] utilizes RF and vibration based piezoelectric harvester structures to generate continuous power for a wireless sensor by switching between the two harvester outputs. Another hybrid structure, which uses RF and EM harvesters for structural health monitoring, utilizes the harvested power to generate two discrete and independent outputs, one per harvesting mode [4]. These structures are useful when enough power is scavenged from either harvester to support the load by itself, but do not provide the means to effectively couple the two harvester outputs.

---

* Middle East Technical University, Department of Electrical and Electronics Engineering, Ankara, Turkey

In this paper a hybrid structure is presented, which adds and boosts the simultaneously scavenged power from RF and EM harvesters to support higher loads at the output. The structure consists of a positive and a negative dual rail rectification for the AC signals scavenged from RF and EM harvesters, and a low voltage DC/DC converter to boost and stabilize the output at the desired value. The interface circuit has been designed, fabricated at UMC 180 nm CMOS technology, and validated. The organization of this paper is as follows: Section II briefly describes the proposed circuit topologies. Section III presents the obtained test results of the hybrid harvester system. Finally, conclusions are summarized in Section IV.

2. Interface Circuit Description

Figure 1 presents the block diagram of the proposed hybrid system that scavenges the low frequency (< 20Hz) vibrations and UHF RF signals, and combines them to drive higher loads. The vibrations are harvested by an in-house EM energy harvester optimized to obtain best performance at 2x2x2 cm^3 of volume[5]. The RF harvester operates at 900 MHz frequency. The scavenged power from RF and EM modules is converted into positive and negative polarities respectively, with a common reference. Continuous current flow is therefore obtained from positive to negative terminals, and no external component is needed to combine the generated powers. The rectified low voltage differential is boosted through a charge pump, and is regulated to a desired voltage (3 V) in order to use the hybrid harvester as a supply to common ICs.

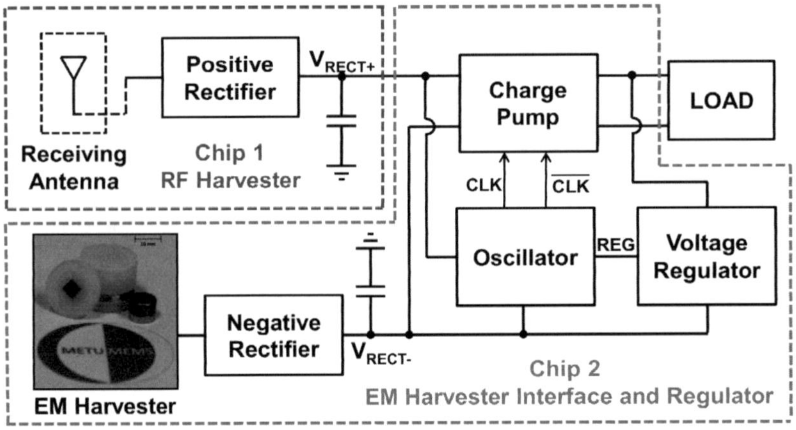

Figure 1. Hybrid energy harvester structure.

2.1. RF and EM Rectifier Circuits

The signal from the RF harvester is rectified, and stepped up using a 7-stage threshold compensated rectifier to generate a positive DC voltage. Bias voltage required by the MOSFET at each rectification stage is supplied by the source terminal of the right adjacent transistor, as shown in Figure 2 [6]. The configuration compensates for the voltage drop at each stage to enhance step-up performance. A dummy MOSFET-capacitor pair (M_d and C_d) in the last stage minimizes voltage fluctuation with load.

The harvested voltage at the EM side is rectified to a negative potential with a self-powered AC/DC doubler circuit depicted in Figure 3, which utilizes active diodes to increase current capacity, and to minimize the forward bias voltage drop. However, active diodes require a power supply unlike passive counterparts. The comparators in the active diodes are powered internally by a passive AC/DC positive doubler, and a negative quadrupler constructed with diode-connected MOSFETs. The positive doubler with passive diodes generates the required gate to source voltage to improve the load current delivered by the active diode. Furthermore, the negative quadrupler with passive diodes generates a negative supply required to precisely turn off the active diodes, thus minimizing losses at the output.

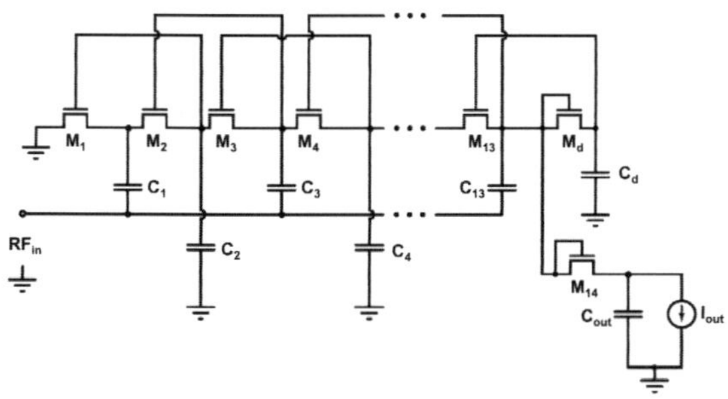

Figure 2. 7-stage threshold compensated RF rectifier [6].

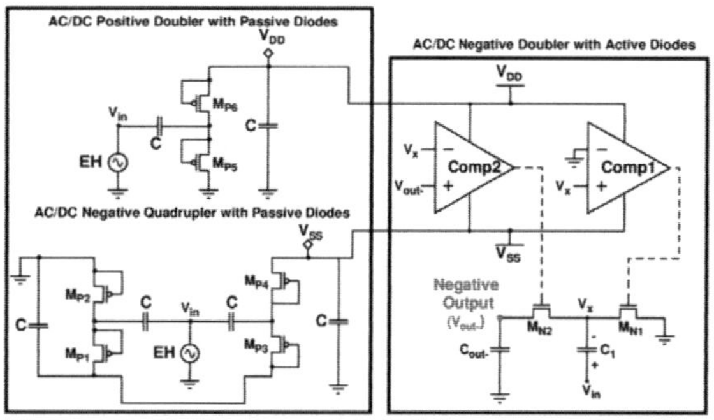

Figure 3. Self-powered AC/DC negative doubler.

Operation principle of the circuit in Figure 3 is as follows: When the input voltage is positive and V_x potential is higher than GND, the comparator Comp1 turns the M_{N1} transistor ON. As a result of this, the storage capacitor C_1 between the input and V_x node is charged up to negative peak voltage of the input. When V_x falls below ground potential, the switch M_{N1} is turned OFF, and a negative charge is stored on the capacitor C_1. Similarly, when the input is negative and the V_x potential is below negative output voltage V_{out-}, switch M_{N2} is turned ON, and current flows from V_{out-} to V_x since V_{out-} has lower potential compared to V_x. When V_{out-} falls below V_x, the switch is turned OFF and a doubled negative voltage is stored on the output storage capacitor C_{out-}.

2.2. DC/DC Converter and Voltage Regulator

In the next stage of the circuit, the rectified dual rail output from EM and RF harvesters is fed to a DC/DC boost converter that utilizes a 4-stage cross-connected charge pump [7]. The charge pump is suitable for low voltage applications, and in contrast to the inductor based DC/DC converters, it is suitable for on-chip CMOS integration. The non-overlapping clocks required by the charge pump are generated by a low power current-starved ring oscillator in Figure 4. The frequency of the oscillator is controlled through the biasing of the current starving transistors. The bias generator circuit includes a PMOS transistor operating in triode region (M_1), and helps to determine the bias current of the oscillator. The voltage regulator, shown in Figure 5, provides bias feedback based on the output voltage to modulate oscillator frequency through current starving MOSFETs, and hence adjusts the pumped current. The regulator consists of a voltage divider, a reference (V_{REF}) generator [8], and a comparator to detect the error between the divided output voltage and the reference.

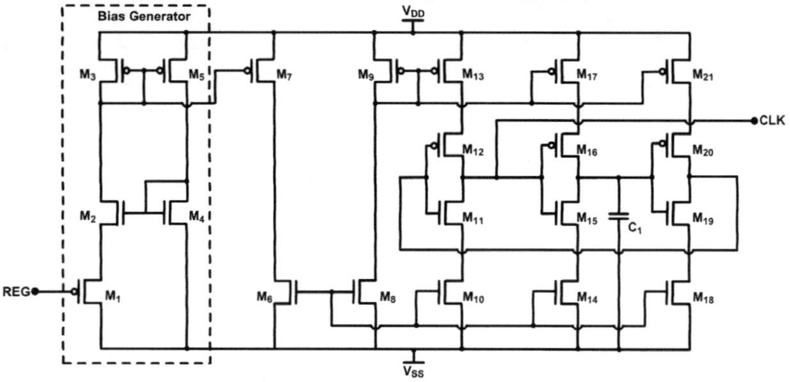

Figure 4. Current-starved ring oscillator with PTAT bias generator.

The operation of the voltage regulator circuit is as follows: The oscillator is pushed to maximum frequency until the output voltage (V_{OUT}-V_{SS}) reaches the desired value (3 V), at which point the divider output reaches V_{REF}, and the comparator readjusts the frequency of the oscillator.

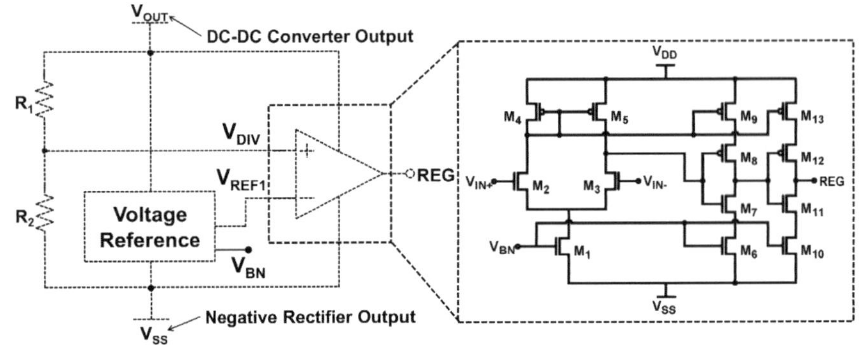

Figure 5. Block level representation of the voltage regulator circuit.

In the comparator circuit, NMOS input transistors are utilized to make the input common mode close to low voltages since the circuit compares the voltage reference and divider which are both at low levels. The circuit is composed of three stages, which enhances the precision and reduces the ripple at the output. The bias of the subthreshold comparator (V_{BN}) is also generated from the same reference circuit, which reduces power dissipation and sensitivity to supply voltage.

3. Test Results and Discussion

The hybrid interface circuit is designed and implemented in UMC 180 nm CMOS technology. Figure 6 shows the die micrograph of the implemented CMOS harvester chips. RF circuit is isolated from the remaining components on a different die in order to separate its substrate and minimize the body effect on the MOSFETs. The only off-chip components of the system are storage capacitors of the EM rectifier and resistors at the voltage divider, which can be integrated on the CMOS chip in a future stepping for cost reduction.

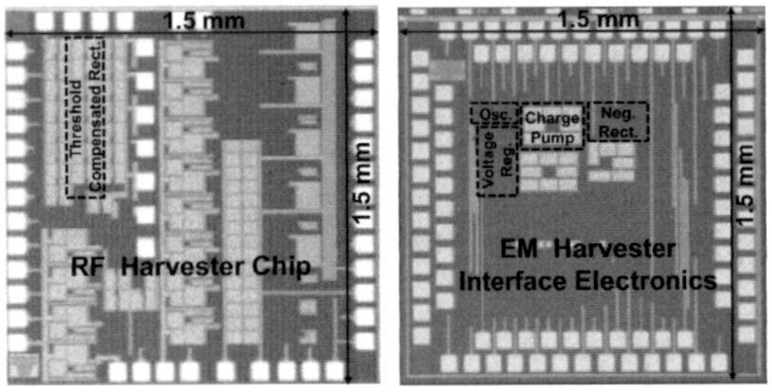

Figure 6. Micrograph of the RF and EM harvester chips.

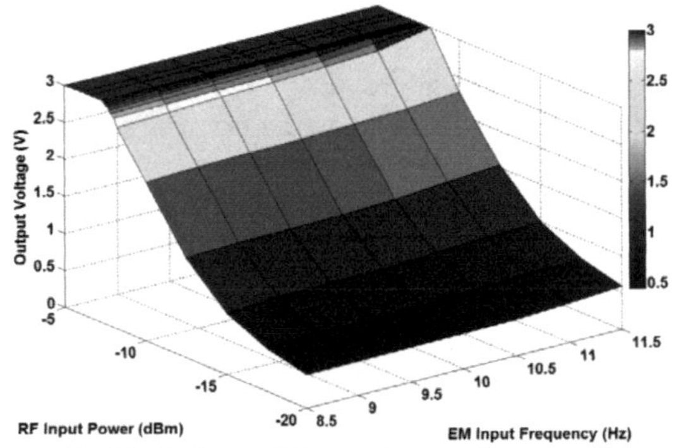

Figure 7. Variation of the output voltage with respect to the RF input power and EM harvester vibration frequency.

The hybrid system has been tested at different vibration frequencies and RF input power levels. EM harvester has been stimulated with 1g peak-to-peak acceleration at 8.5-11.5 Hz input frequency range. -20 dBm to -5 dBm input power has been applied to the RF harvester. Figure 7 presents variation of the output voltage with various input conditions at 1 MΩ load. As the RF input power increases, the output voltage ramps up, and attains the desired voltage (3V). It is observed that the required RF power for obtaining 3 V output varies with vibration frequency, and both harvester sources contribute to the generated power and voltage. Figure 8 shows the input and output voltage waveforms when the EM harvester is stimulated with 1g vibration at 10 Hz, and RF harvester input is

-5 dBm at 900 MHz. The hybrid system successfully delivers a stable 3 V source with a ripple voltage less than 30 mV to 1 MΩ load. The low level ripple at the output shows the accuracy and sensitivity of the voltage regulator. The voltage regulator rapidly responds to the output voltage variation and tunes the boosting ratio of the charge pump. Figure 9 presents the current driving capability of the hybrid system when the EM harvester is stimulated with 1g vibration at 10 Hz, and RF harvester input is -5 dBm at 900 MHz. As depicted in the figure, the battery-like operation is verified at 3 V for a wide range of loads up to 18 μW of output power.

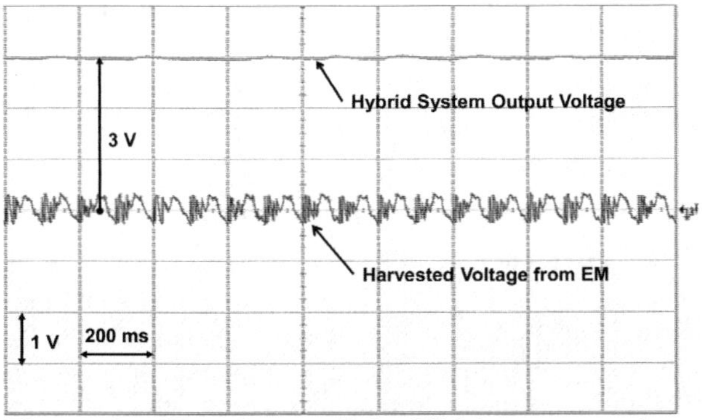

Figure 8. Waveforms of the EM harvester voltage and the generated 3 V hybrid system output voltage at 10 Hz and 1g vibration (RF input power: -5 dBm).

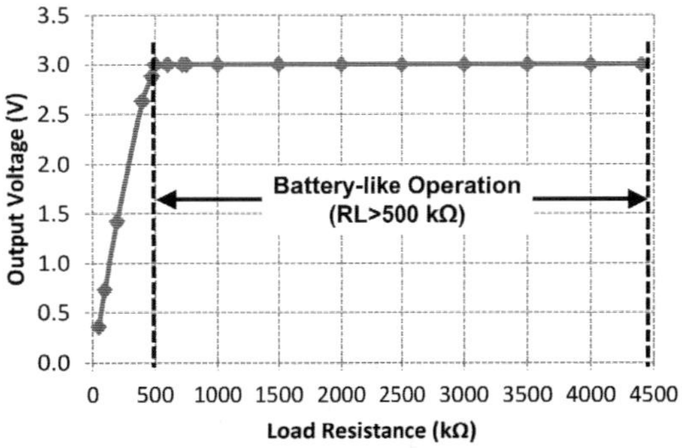

Figure 9. Variation of the output voltage with respect to the load resistance at 10 Hz and 1g vibration (RF input power: -5 dBm).

4. Conclusion

A self-powered and autonomous hybrid energy harvesting structure which utilizes the low frequency ambient vibrations and UHF RF signals is presented in this paper. The low frequency ambient vibrations are harvested by a small sized in-house EM harvester, and are rectified negatively by a self-powered autonomous rectifier. In contrast, the harvested RF signals are rectified by a 7-stage threshold compensated rectifier to a positive DC voltage. The rectified positive and negative voltages are then added serially by using the ground node as a reference. At the next stage, an on-chip DC/DC converter and a voltage regulator respectively boosts and stabilizes the rectified dual rail voltage to the desired 3 V output. The test results showed that the hybrid system provides 3 V stable output with 18 μW of output power for a wide range of input conditions while both of the harvesting sources contribute to the output.

5. Acknowledgemnet

This work is supported by Ministry of Science, Industry and Technology (SAN-TEZ) Project No: 1375, Turkey.

References

[1] Muhtaroğlu A, Yokochi A and Von Jouanne A 2008 *J. Power Sources* **178** 2–6.
[2] Sang Y, Huang X, Liu H, and Jin P 2012 *IEEE Transactions on Magnetics* **48** 4495–4498.
[3] Huang C and Chakrabartty S 2011 *IEEE ISCAS* (Rio de Janeiro: Brasil) p 2473–2476.
[4] Galchev T et al. 2011 *Proceedings of SPIE* **7983** 1–9.
[5] Ulusan H et al. 2015 *Eurosensors 2015* (Freiburg: Germany) **120** p 337-340.
[6] Papotto G, Carrara F, and Palmisano G 2011 *IEEE JSSC* **46** 1985–1997.
[7] Bhalerao S A, Chaudhary A V, and Patrikar R M 2007 *VLSID'07* (Bangalore: India) p 941–946.
[8] Seok M, Kim G, Blaauw B, and Sylvester D 2012 *IEEE JSSC* **47** 2534–2545.

Self-sufficient electronic control for nonlinear, frequency tunable, piezoelectric vibration harvesters

S Heller, S Neiss, M Kroener, P Woias

Laboratory for Design of Microsystems, IMTEK, University of Freiburg, Georges-Koehler-Allee 102, 79110 Freiburg, GERMANY

E-mail: simon.heller@imtek.uni-freiburg.de

Abstract. Research in vibration energy harvesting focuses increasingly on nonlinear harvesters. In comparison to linear harvesters they show an inherent larger bandwidth through hardening or softening effects and higher conversion efficiency. A further increase of the bandwidth and thus a higher energy yield can be achieved by controlled tuning of such a nonlinear system. In this paper a self-sufficient tuning control electronic, which is directly powered by the harvester, is presented.

1. Introduction

In the last few years the research in the field of vibration energy harvesting has focused on nonlinear systems [1-3]. As illustrated in figure 1, these systems show an inherently larger bandwidth due to their nonlinear frequency response which leads to a higher conversion efficiency over a wide frequency band. Therefore, in contrast to linear systems, no trade-off between the output power and the bandwidth is necessary.

However, nonlinear systems show a characteristic hysteresis whereby the harvester is either located in a preferable high oscillation orbit or in a low oscillation orbit depending on the change of the excitation (indicated by the arrows in figure 1).

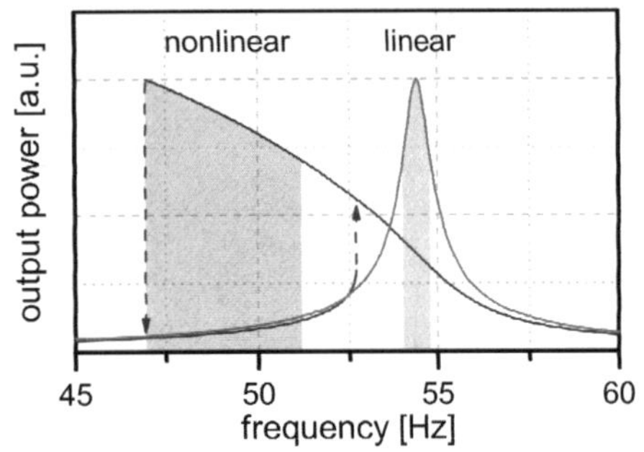

Figure 1. Typical frequency response of a nonlinear (blue) and a linear system (red).

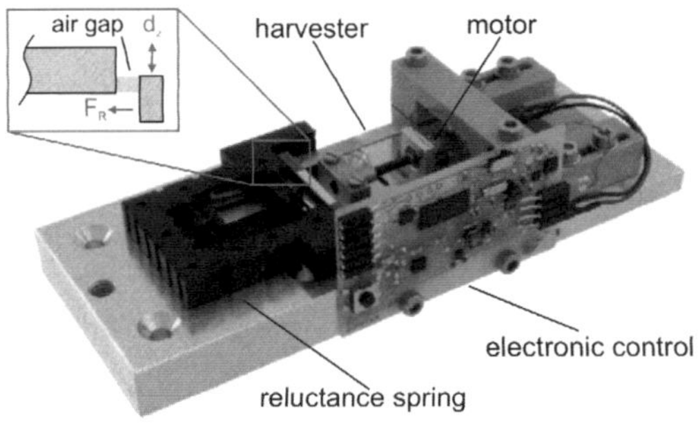

Figure 2. Photo of the nonlinear harvester and the control circuit.

The transition points between both orbits depend on the harvester's resonance frequency and degree of nonlinearity. Therefore, in order to be able to reach the high oscillation orbit and to further increase the bandwidth, the resonance frequency has to be made tunable by an electronic control. Within the research on linear harvesters, this task is accomplished by using the phase shift between the excitation signal and the system response [4]. However, [5, 6] use controls based on a look-up table in which the optimal working point for every excitation frequency is stored. The disadvantage of this open-loop control is its inability to react to interferences such as the degradation of the piezo ceramics which could cause a shift of the resonance frequency. Only [5] addresses this problem and proposes a combination of an open-loop and a closed-loop control. The control is based on a look-up table but the values stored there are verified at longer intervals by a measurement of the phase shift.

2. Nonlinear vibration harvester

The starting point of the presented work is the nonlinear vibration harvester depicted in figure 2. The nonlinear behavior is produced by a patented reluctance spring [7] which is leading to a nonlinear restoring force F_R on the seismic mass in function of the deflection d_z. For the adjustment of the resonance frequency the gap width between the seismic mass of the harvester and the reluctance spring will be changed. This is realized by an electric motor which is coupled to the reluctance spring by a spindle gear.

Figure 3 shows the frequency responses for different gap widths between 0.5 mm and 1.5 mm at an excitation level of 5 m/s^2. Over a gap width change of 1 mm a resonance frequency between 53.6 Hz (1.5 mm) and 63.9 Hz (0.5 mm) is achieved. The relative hysteresis width varies between 4.3% and 16% of the nonlinear resonance frequency depending on the gap width. The corresponding phase responses recorded with an infinite load resistance (see figure 4), prove that even for nonlinear systems the point of maximal amplitude occurs at a phase shift between excitation and system response of -90°.

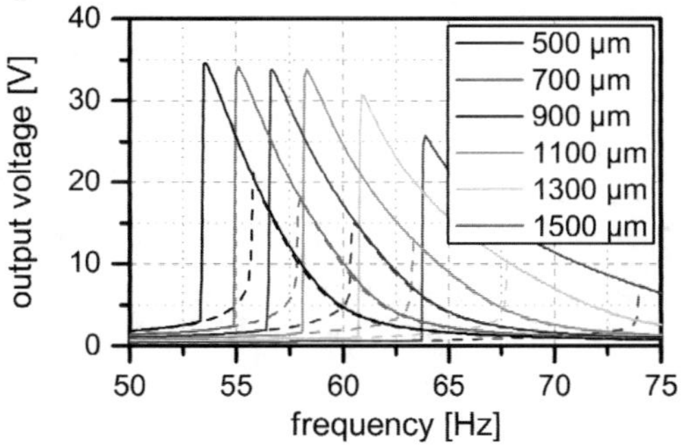

Figure 3. Frequency responses for different gap widths at an excitation level of 5 m/s^2.

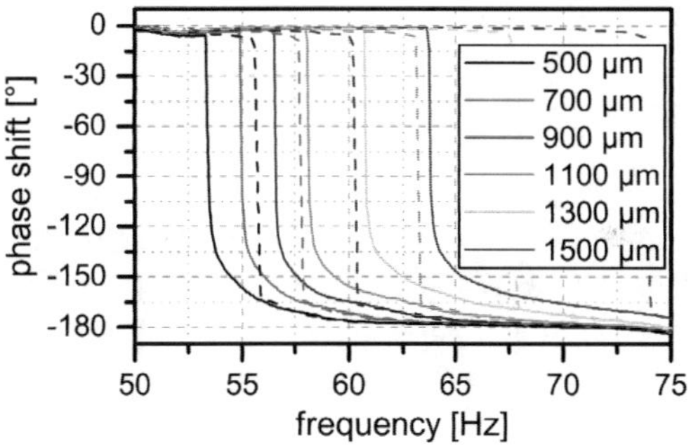

Figure 4. Phase responses for different gap widths at an excitation level of 5 m/s^2.

3. Control concepts

3.1. Tuning concept

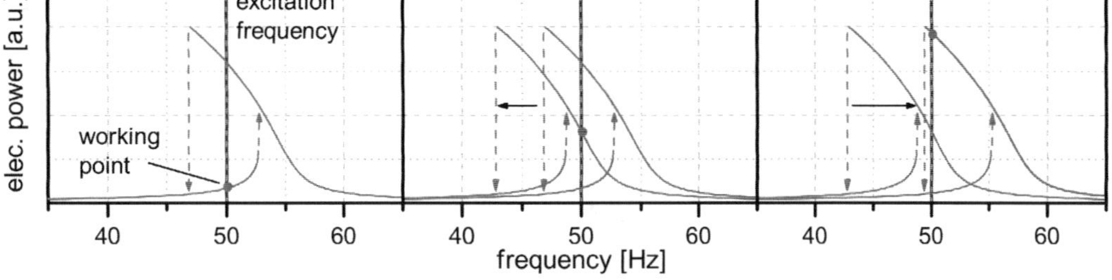

Figure 5. Tuning concept for transitions from the low oscillation orbit to the high oscillation orbit.

For every frequency inside the hysteresis band of the frequency response; two stable states exist as previously described: one in the high oscillation orbit and one in the low oscillation orbit. To enable the transition from the low oscillation orbit into the high oscillation orbit, a tuning concept as illustrated in figure 5 was developed: Starting from the low oscillation orbit (left), the nonlinear resonance frequency (respectively the stiffness of the system) has to be increased until the working point jumps from the low oscillation orbit to the high oscillation orbit (middle). Afterwards, the resonance frequency has to be decreased by reducing the stiffness of the system to reach the optimal working point (right).

3.2. Continuous monitoring
Due to an insufficient stability of the system at the point of maximum output voltage, a definition of the working point at a phase shift of -90° is not practical. A better way is the definition of a phase window, which corresponds to the bandwidth of the harvester and allows for sufficient stability while granting high-enough output power to enable the self-sufficient supply of the harvester. Moreover, the number of tuning steps can be dramatically reduced by this approach. For an unloaded harvester the phase window extends from -135° to -150°. Due to the electrical damping the phase response shift is reduced, so that the practical phase window extends from -75° to -100° in case of an electrical load.
The control loop starts with the measurement of the actual phase shift. When the phase shift is outside the phase window, the gap width is adjusted in accordance to the measured phase shift. To check the set gap width, the phase shift is measured again. This procedure repeats until it is close to the optimal working point. If this is the case the control electronic is powered down for 30s in favour of a higher output power. Afterwards, the procedure is restarted.

3.3. Continuous monitoring
The drawback of the previously presented control concept is it measures the phase shift at fixed points in time. Therefore, depending on the temporal change of the excitation, the phase shift is measured too often or not often enough. In both cases this leads to higher power consumption. Ideally, the phase shift is measured only when a change in the excitation occurs. Therefore, a second concept was developed which uses an additional ultra-low power accelerometer to measure the excitation level during the sleep phase. The output voltage of the harvester is used to monitor the excitation frequency. If one of these quantities changes over a defined tolerance, the microcontroller wakes up immediately and initiates a measurement of the phase shift.

4. Control electronic
For the realization of the two control concepts the self-sufficient control electronic depicted in figure 6 was developed. The energy and information flows between the different parts are shown in figure 7.

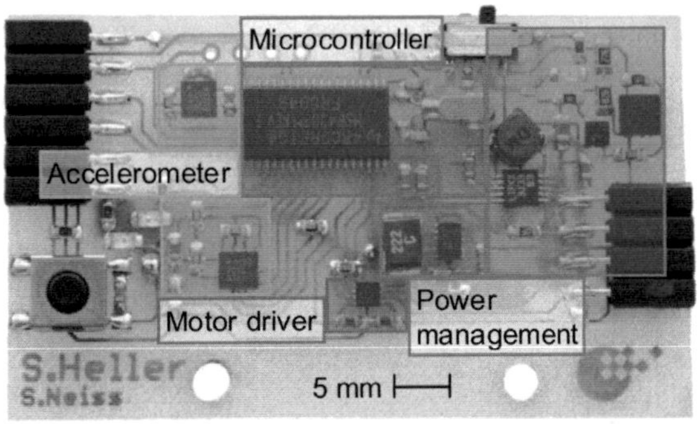

Figure 6. Photo of the circuit board with the control electronic.

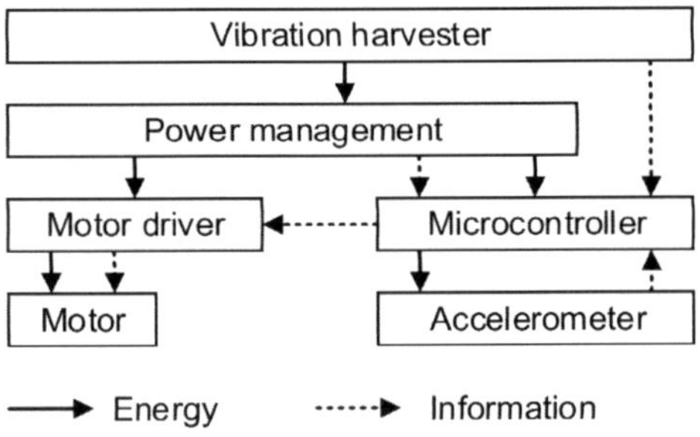

Figure 7. Block diagram of the electronic control.

The core of the system is a low-power microcontroller (MSP430FR5949, *TI*) which calculates the phase shift using the output signal of an analog accelerometer (KXTC-4100, *Kionix*) and the output voltage signal of the harvester. Furthermore, the microcontroller is used for the control of the motor (RE6, *Maxon Motor*) via a motor driver (DRV8838, *TI*). For the realization of the control concept with continuous monitoring an ultra low power accelerometer (ADXL362, *Analog Devices*) is used.
An efficient power management is needed for the self-sufficient operation of the control. First, a combination of a bridge rectifier and a buck converter (LTC3588-1, *Linear Technology*) produces a constant output voltage of 3.3V from the harvesters AC output voltage. This voltage is used to charge a super cap (GW209F, *CAP-XX*) with a capacity of 0.14 F. The electric motor, as well as the motor driver is powered at 3.3V, whereas the microcontroller and the sensors are powered at 2V. This has the advantage that the power consumption can be reduced further. To produce the 2V supply voltage a further high-efficient buck-converter (TPS62740, *TI*) is used.

5. Results

5.1. Frequency sweep

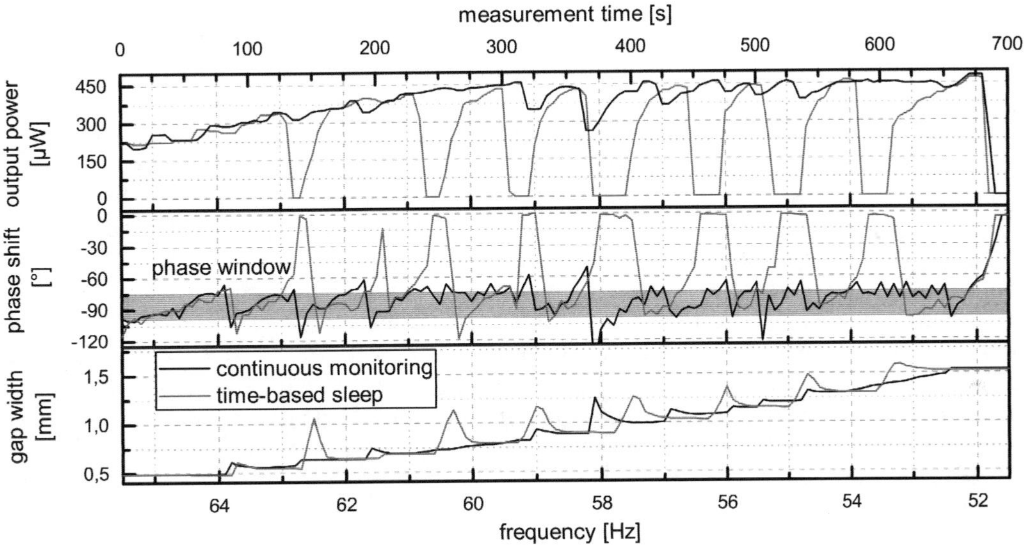

Figure 8. Comparison of the two control concepts in the working area for a frequency drift from 80 Hz to 50 Hz by means of the output power, the phase shift and the gap width at an excitation level of 5 m/s^2.

To characterize the control concepts, the control performance at a slow, continuous sweep of the excitation frequency of 0.1 Hz/5 s at a constant excitation level of 5 m/s^2 was investigated. The gap width, the output power of the harvester and the phase shift were recorded.
For an up-sweep no significant differences between the two concepts were found. The working point stayed most of the time inside the phase window, the gap width decreased gradually from 1.5 mm to 0.5 mm while an average output power of the harvester of 350 μW was achieved.
Substantial differences appear for the down-sweep (see figure 8). With the concept of the time-based sleep the harvester falls into the low oscillation orbit while sleeping whereby the output power drops to 0 μW. To get back into the high oscillation orbit after waking up, first the gap width has to be reduced and afterwards increased, as described in section 3.1 and easily seen at the spikes in the course of the gap width in figure 8. With the concept of the continuous monitoring in contrast, the harvester stays all the time in the high oscillation orbit. Thus, the energy yield for the whole sweep is increased by 39% to 307 mJ. Moreover the energy consumption of the actuator is reduced by 49% to

139 mJ. Taking into account the higher power consumption for continuous monitoring the net output power is 114 mJ whereas for the time-based sleep concept it is -72 mJ.

5.2. Recovery time
To compare the harvester system to other linear and nonlinear systems, the recovery time at an excitation level of 5 m/s^2 was measured. The recovery time was defined as the time period needed to harvest the energy which is required for a tuning process after a jump in the excitation frequency over the whole working area (52.5 Hz to 64.5 Hz). As the results in table 1 show, the recovery time for the concept of continuous monitoring is 30% longer in comparison to the concept of time-based sleep due to higher power consumption. In spite of a longer adjustment travel for a jump from the upper to the lower border of the working area, the recovery time is 36% less which can be explained by a lower mechanical damping at lower frequency and thereby a higher output power of the harvester.

Table 1. Recovery time for the two different control concepts at an excitation level of 5 m/s^2.

	52.5 Hz to 64.5 Hz	64.5 Hz to 52.5 Hz
Time-based sleep	518 s	375 s
Continuous monitoring	662 s	497 s

6. Control electronic
The presented results verify that a self-sufficient closed-loop control of a nonlinear vibration harvester based on the measurement of the phase shift between excitation and system response of the harvester is possible. At an excitation level of 5 m/s^2 an average output power of 350 μW is measured which leads to a recovery time of the system between 375 s and 662 s depending on the used control concept and the direction of the change in the excitation frequency.
The choice of the optimal control concept mainly depends on the temporal change of the excitation. For a constant excitation over a long period of time the concept of time-based sleep is preferable because of the lower power consumption. However, in case of a frequently changing excitation the concept of continuous monitoring is better suited because the fall into the low oscillation orbit can be avoided and therefore a significant amount of energy needed for the actuation can be saved.

Acknowledgements
We gratefully acknowledge the financial support from the German Research Association (DFG) within the Research Training Group (GRK) 1322 "Micro Energy Harvesting".

References
[1] M. Ferrari et al.: *Improved energy harvesting from wideband vibrations by nonlinear piezoelectric converters*, Sen. Actuat. A-Phys., 162 (2010), pp. 422 – 431.
[2] B. Mann et al.: *Uncertainty in performance for linear and nonlinear energy harvesting strategies*, J. Intell. Mater. Syst. Struct., 23 (2012), pp 1451 – 1460.
[3] D. Zhu et al.: *Strategies for increasing the operating frequency range of vibration energy harvesters: a review*, Maes. Sci. Technol., 21 (2010), pp. 022001.
[4] C. Peters et al.: *A closed-loop wide range tunable mechanical resonator for energy harvesting systems*. Jour. of Micormech. and Microeng., 19 (2009), pp. 094004.
[5] C. Eichhorn et al.: *A smart and self-sufficient frequency tunable vibration energy harvester*. Jour. of Micromech. and Microing., 21 (2011), pp. 104003.
[6] I. N. Ayala-Garcia et al.: *A tunable kinetic energy harvester with dynamic over range protection*. Smart Mat. and Struc. 19 (2010), pp. 115005.
[7] S. Neiss et al.: *Reluctance Springs for nonlinear Energy Harvesting Generators*. Digest Tech. Papers PowerMEMS'12, 2012, pp. 153 – 156.

Electrostatic vibration energy harvester using an electret-charged mems transducer with an unstable auto-synchronous conditioning circuit

Armine Karami[1], Philippe Basset[2], Dimitri Galayko[1]
[1]Sorbonne Universités, UPMC Univ Paris 06, CNRS, LIP6, Paris, France
[2]ESIEE-Paris, Université Paris-Est Marne-la-Vallée, Noisy-le-Grand, France

E-mail: armine.karami@lip6.fr

Abstract. This paper reports for the first time experiments using an electrostatic vibration energy harvester comprised of a low voltage electret-charged MEMS transducer joined to an unstable auto-synchronous conditioning circuit with rectangular charge-voltage characteristic, also known as the Bennet's doubler conditioning circuit. The experimental results show that the electret voltage, even if of low value, can be used as the necessary pre-charge for these type of electrostatic vibration energy harvesters. Also, the use of such a conditioning circuit with a low-voltage electret capacitive MEMS tranducer instead of the previously-reported conditioning circuits with direct connection to load or through a rectifier, can be advantageous in terms of maximal harvested power for a low-voltage electret, showing up to 95% higher converted power.

1. Introduction

In the recent years, extensive research has been done on electret-based electrostatic vibration energy harvesters (e-VEHs). High-voltages electret have enabled the realization of e-VEHs with relatively high harvested power density from relatively high-gap and/or in-plane moving electrode capacitive transducers [1]. These electret e-VEHs have been experimentally tested using either a direct connection to an impedance-matched load – also known as the primitive conditioning circuit –, or simple conditioning circuits consisting of a diode bridge connected to an impedance-matched load, with or without the use of a smoothing capacitor.

However, electrets may be subject to degradation resulting in the decrease of their built-in voltage over time. This problem is strongly linked to the materials and the technology used to implement the charged electret layer. As the electret voltage decreases, so does the converted power if using the previously-mentioned conditioning circuits.

Notwithstanding this decrease of the electret built-in voltage, recently, a new class of conditioning circuits was discovered: inspired from the "Bennet's doubler of electricity", they allow an exponential self-increase of their biasing, and thus of the converted power, at the small cost of a low pre-charge voltage. In the electrical domain, these circuits show no saturation of the harvested power over time of operation. Moreover, these circuits are inductorless and do not require any external command for synchronization [2, 3].

This paper presents the first experiment combining the Bennet's doubler conditioning circuit, which will be called the unstable auto-synchronous conditioning circuit in the rest of the paper,

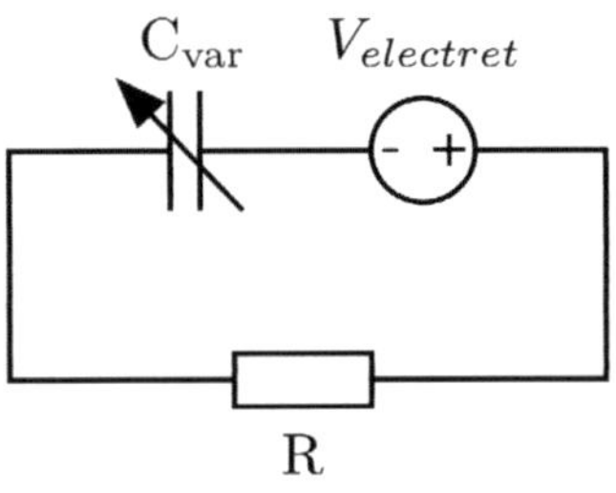

Figure 1. The primitive conditioning circuit

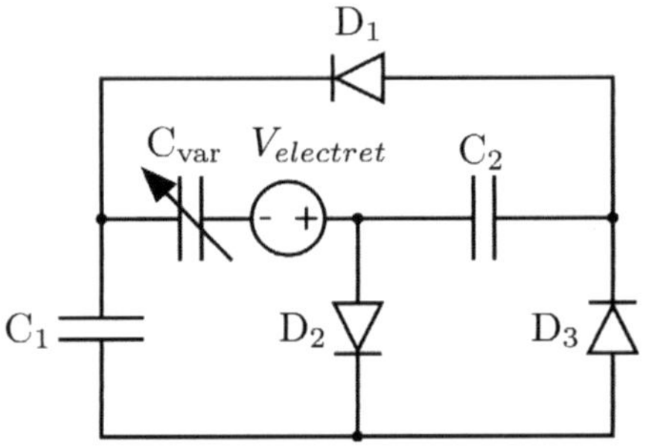

Figure 2. The used unstable auto-synchronous conditioning circuit. The electret charged transducer is modeled by C_{var} and $V_{electret}$

to an electrostatic transducer with a weakly charged electret layer. The goal of the experiment is to prove that an electret, even of low voltage value, can be used as the necessary pre-charge for e-VEHs using unstable auto-synchronous conditioning circuits. In order to appreciate the advantage of this configuration, an experiment where the same electret capacitive transducer is associated to a primitive conditioning circuit is also carried out. The maximum harvested powers with each one of the conditioning circuits are then compared. This comparison can be extended to variations of the primitive conditioning circuit that have been extensively used with electret e-VEHs [1].

2. Presentation of the e-VEH and previous experiments

2.1. Presentation of the electret MEMS capacitive tranducer

The MEMS device used in the experiment reported in this paper has a very similar geometry to the device presented in [4]. The device was charged with an electret layer using the corona charging technique. In the electrical circuit domain, the electret is electrically modeled by the voltage source $V_{electret}$ in series with the transducer, as depicted in Fig. 1 (joined to the primitive conditioning circuit) and Fig. 2 (joined to the unstable conditioning circuit). After charging, this voltage was measured at 22 V. However, because of the electret layer degradation over time, several weeks after the charging, the voltage $V_{electret}$ was measured at 5.3 V. All the experiments presented in this paper were carried out with this lower electret voltage.

2.2. Presentation of the unstable auto-synchronous conditioning circuit

The unstable auto-synchronous conditioning circuit derived from the Bennet's doubler electrostatic engine, is depicted in Fig. 2. The circuit operates as a charge pump, which shows an exponential increase of the voltages across the circuit's capacitors, if studied in the electrical domain. Indeed, the periodic variation of biased C_{var} induces the conversion of an amount of energy that is proportional to the square of C_{var} bias, and that can be computed from the rectangular-shaped QV conversion cycle of the circuit. A part of this energy is reinvested on C_{var} to increase its bias, hence implementing an avalanche mechanism of increase of bias across the transducer and the fixed capacitors of the circuit, and thus of converted power. The remaining part of the converted energy is splitted between the circuit's fixed capacitors. The total converted power also depends on the transducer extreme capacitance values ratio C_{max}/C_{min}. More details on the circuit's operation are given in [5] for an electret-free transducer. The global operation of the circuit is not modified by the use of an electret charged transducer. In the present paper, the capacitor values were chosen as $C_1 = 100$ nF and $C_2 = 2$ nF.

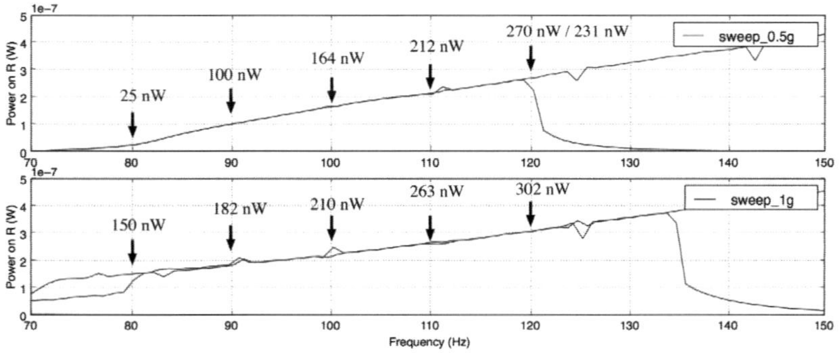

Figure 3. Results of experiment 1: results for the e-VEH with the primitive circuit excited under frequency-swept external accelerations of 0.5g rms and 1g rms. The measured quantity is the power dissipated on the resistance, plotted over the frequency of the input acceleration

3. Experimental results and discussion

3.1. Presentation of the experiments

3.1.1. Experiment 1 The electret MEMS transducer device was first tested in experiment 1 using the primitive conditioning circuit, depicted in Fig. 1. For this experiment, the transducer was submitted to a slow swept frequency sinusoidal input acceleration, between 70 Hz and 150 Hz, at two different rms values: 0.5 g and 1 g. For each amplitude of excitation, the optimal resistance for the primitive conditioning circuit was used, after an experimental measurement of its value.

Throughout the sweep, for each frequency of the input, the power dissipated on the load resistance R was measured. This measurement also gives an idea of the resonator's mechanical frequency response under the different accelerations, affected by mechanical nonlinearities as described in [7]. Note that the slow sweeping speed allows to consider that the dissipated power values on R were obtained in the steady-state mode of operation of the circuit for each frequency of the input. The results of experiment 1 are depicted in Fig. 3.

3.1.2. Experiment 2 Then, the transducer was tested in experiment 2 using the unstable auto-synchronous conditioning circuit, depicted in Fig. 2. To compare the maximal harvested power with the experiment 1, the frequencies of the input were chosen in the frequency band obtained in experiment 1. In experiment 2, the transducer was submitted to input accelerations at those fixed frequencies, of the same amplitudes as those used in experiment 1.

At each cycle of the circuit operation, the converted power is splitted between the circuit's capacitors and the transducer proportionally to their capacitance value. Because the capacitances values were chosen as $C_1 >> C_2 >> C_{max}$, the harvested power defined as the variation of the energy on C_1 throughout one cycle of operation is a good approximation of the total converted power. The evolution of the harvested power over time is measured in experiment 2, through the measurement of the voltage and energy over time across C_1.

Before the start-up ($t = 0$ in the plots), all nodes were short-circuited to the ground, thus nulling the initial voltages across the fixed capacitors and annhilating any possible external pre-charge. The results of experiment 2 are depicted in Fig. 4 and Fig. 5.

3.2. Results and discussion

First, note that in Fig. 4 and 5, no exponential increase of the voltage across the circuit's capacitors is observed. In fact, it does occur at the very beginning of each experiment, but the low maximum over minimum capacitance ratio of the used transducer compared to the transducer used in [6] immediately results in the observed saturation phenomenon, which is of

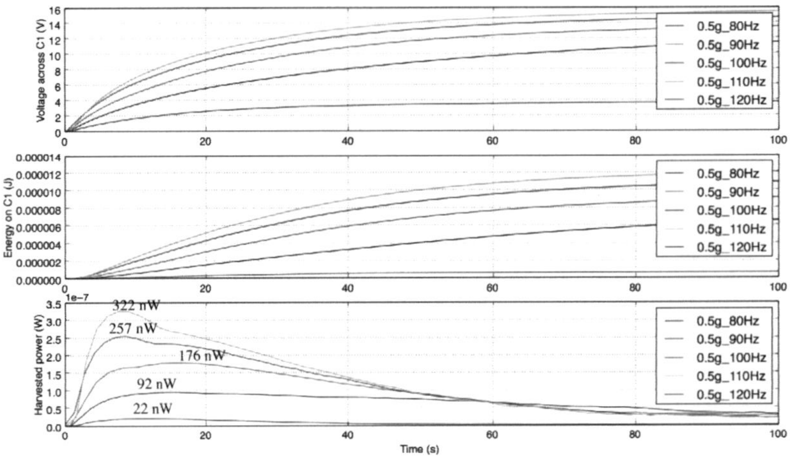

Figure 4. Results of experiment 2: e-VEH with the unstable auto-synchronous circuit, excited under different fixed frequencies external accelerations, of 0.5g rms amplitude. Measured voltage and energy across the largest fixed capacitor, and harvested power, all plotted over time

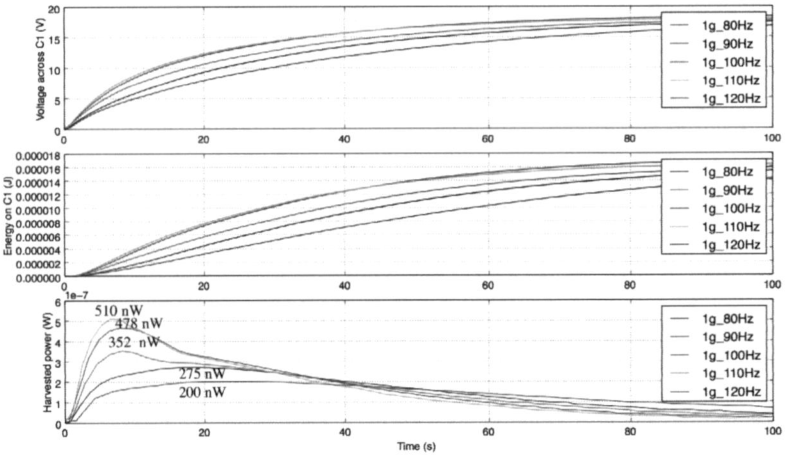

Figure 5. Results of experiment 2: e-VEH with the unstable auto-synchronous circuit, excited under different fixed frequencies external accelerations, of 1g rms amplitude. Measured voltage and energy across the largest fixed capacitor, and harvested power, all plotted over time

the same origin as in [6], namely the electromechanical coupling-induced resonator bandwidth shift. Also, the results show that the highest harvested power figures were obtained when the transducer was submitted to input accelerations of frequencies slightly lower than the frequency of the peak of amplitude in the resonator's response, for the same qualitative reasons discussed in [6].

For both of the tested amplitudes, there exists a frequency of input acceleration for which the maximal power point reached with the unstable auto-synchronous conditioning circuit is higher than the power dissipated on the primitive circuit's resistance. At 110 Hz of 1g amplitude input, the results with the unstable auto-synchronous conditioning circuit yield a maximal converted power up to 95% superior to the converted power with the primitive conditioning circuit for the same input. Thus, under specific input conditions and with the tested device, the use of the unstable auto-synchronous conditioning circuit is advantageous over the use of the primitive conditioning circuit.

Qualitative arguments about the operation of the system with each of these circuit can be used to explain the differences between the converted power, given the low value of the electret voltage for the device used in the experiments. Indeed, the converted power for the primitive conditioning circuit roughly increases with the square of the electret voltage, whereas the unstable auto-synchronous conditioning circuit uses its voltage as a pre-charge to reach a maximal power point at a higher voltage. Furthermore, the advantage of the unstable conditioning circuit configuration over the primitive circuit is transitive to diode-bridge rectifier-based conditioning circuits such as the circuit used in [8]. Indeed, the maximal converted power for these circuits is also proportional to the square of the electret voltage.

Still, the lack of a quantitative theoretical model that fully describes the dynamics of the electromechanically-coupled system makes it hard to determine, in general, which one of the conditioning circuits will have better performances, in terms of maximal harvested power. Indeed, the circuit's conditioning schemes are different and result in different impact of the electromechanical coupling, leading to different system dynamics, and thus harvested power. Hence, under given design constraints, there is currently no other mean than numerical simulation and/or experimental testing to formally chose a conditioning circuit over the other.

4. Conclusion

This work experimentally validated that an electret can be used as the necessary pre-charge source for an e-VEH using the unstable auto-synchronous conditioning circuit, also called the Bennet's doubler. Secondly, the experiments also show that in the presence of a lowly-charged electret, the use of this unstable auto-synchronous conditioning circuit can be advantageous over the use of simpler primitive or rectifier-based conditioning circuits. The maximal converted power was up to 95% higher with the unstable auto-synchronous conditioning circuit compared to the primitive conditioning circuit. The use of the unstable auto-synchronous conditioning circuit is thus a potential solution to the problem of the slow discharge of the electret layer.

References

[1] Suzuki Y, Miki D, Edamoto M, Honzumi M 2010 "A MEMS electret generator with electrostatic levitation for vibration-driven energy-harvesting application" *J. Micromech. Microeng.* **20** 104002

[2] Queiroz A C M and Domingues M 2011 "The doubler of electricity used as battery charger" *IEEE Trans. Circuits Syst. II, Exp. Briefs* **58** 797-801

[3] Lefeuvre E, Risquez S, Wei J, Woytasik M and Parrain F 2014 "Self-biased inductor-less interface circuit for electret-free electrostatic energy harvesters" *Journal of Physics: Conference Series* **557** 012052

[4] Basset P, Galayko D, Paracha A, Marty F, Dudka A and Bourouina T 2009 "A batch-fabricated and electret-free silicon electrostatic vibration energy harvester" *J. Micromech. Microeng.* **19** 115025

[5] Dorzhiev V, Karami A, Basset P, Dragunov V and Galayko D 2014 "MEMS electrostatic vibration energy harvester without switches and inductive elements" *Journal of Physics: Conference Series* **557** 12126-30

[6] Dorzhiev V, Karami A, Basset P, Marty F, Dragunov V and Galayko D 2015 "Electret-free micromachined silicon electrostatic vibration energy harvester with the bennets doubler as conditioning circuit" *IEEE Electron Device Lett.* **36** 183-5

[7] Basset P, Galayko D, Cottone F, Guillemet R, Blokhina E, Marty F and Bourouina T 2014 "Electrostatic vibration energy harvester with combined effect of electrical nonlinearities and mechanical impact" *J. Micromech. Microeng.* **24** 035001

[8] Miki D, Suzuki Y, Kasagi N 2009 "Effect of nonlinear external circuit on electrostatic damping force of micro electret generator" *Solid-State Sensors, Actuators and Microsystems Conference (TRANSDUCERS) International Conference on* 636-9

A high-force, out-of-plane actuator with a MEMS-enabled microscissor motion amplifier

Xin Xie[1] and Carol Livermore[2]

[1]Department of Electrical and Computer Engineering, Northeastern University, Boston, MA 02115 USA
[2]Department of Mechanical and Industrial Engineering, Northeastern University, Boston, MA 02115 USA

x.xie@neu.edu, livermore@neu.edu

Abstract: The design, fabrication, and demonstration of a set of 2 mm^2, high-force actuators that combine piezoelectric in-plane actuators with MEMS-enabled scissor mechanisms (motion amplifiers) to simultaneously produce high out-of-plane forces and large out-of-plane displacements are presented. The microscissor motion amplifier employs two layers of lithographically-patterned SU-8 microstructure laminated with a thin film of structural polyimide and adhesive to form hinges. Performance is optimized by varying layer thickness and adhesive types. Measured displacements of >3 μm and measured forces of >5 mN are observed, corresponding to a displacement per unit area of 1.6 $\mu m/mm^2$ and a force per unit area of 2.6 mN/mm^2. Cyanoacrylate adhesive provides superior performance to silicone adhesive, with larger force output. Thicker polyimide hinges provide smaller displacement but greater force than thinner polyimide hinges. These powerful, compact actuators have significant potential for high-force applications like tactile displays and micropumps.

1. Introduction

Providing information to those who are blind or have low vision is critical for enhancing mobility, situational awareness, education, and more. Text-based tactile information delivery can be effective, rapid, and private; examples include Braille and the Optacon [1,2], both of which deliver text through the motion of few centimeter scale piezoelectric bending beam actuators. Conveying graphical information that cannot be expressed as text through tactile stimuli is highly desired, but it is challenging to create actuators that are compact enough to provide high resolution graphical information while still having large force and displacement and being robust enough to be well sensed by users. Existing refreshable tactile graphical display technology includes electroactive polymer (EAP) actuators [3,4], piezoelectric bending beams [5], and other MEMS-enabled actuators [6,7]. EAP actuators are small in size, enabling a few millimeter pitch, but they have large actuation times on the order of seconds. Piezoelectric bending beams must be large enough to achieve the necessary forces, making integration into compact systems a challenge. MEMS devices offer a route to compactness, but engineering MEMS for high force and/or large displacement can be a challenge.

To address these challenges, tactile elements that magnify small in-plane vibrations into the vertical direction using scissor amplifiers were introduced in [8]. Although these tactile actuator elements were at the 30 mm^2 scale, the ultimate goal of this type of actuator is to achieve high resolution displays (similar to the 1 mm^2 resolution of human fingerpads) with rapid refreshability and the ability to convey spatial patterns, motions and rhythms for diverse information delivery [8]. The present research employs a new design and process, modeled on that of [9], to reduce the size (area/element) of this technology by 15X relative to [8]. The present tactile elements employ two layers of lithographically-patterned SU-8 microstructure laminated with a thin film of polyimide and adhesives to form hinged microscissor motion amplifiers. The microscissors are integrated with piezoelectric extensional actuators so that the microscissors convert the actuators' in-plane motions into larger out-of-plane vibrations (figure 1). This approach has potential not only for tactile displays but also for other applications in which high force is desired, such as micropumps.

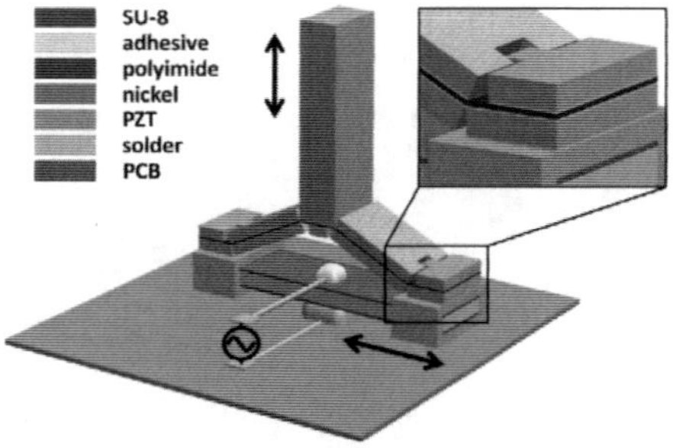

Figure 1. Schematic diagram of a single actuator for a tactile array. The actuator comprises an in-plane PZT actuator with a laminated, micro-fabricated scissor amplifier to convert small in-plane vibrations into 10X larger out-of-plane vibrations.

2. Design and Fabrication

The microscissor requires both flexible elements (for the hinges) and rigid elements (for the scissor linkages). The flexible elements are made of polyimide film, with a Young's modulus of about 2 GPa. The polyimide extends along the entire length of the scissor. Because the structure requires bending to be localized in the four hinges as in figure 1, the linkages are made more rigid by the lamination of photolithographically-defined SU-8 elements onto the polyimide film in the regions that are not supposed to bend. After the completed microscissor is mounted on the extensional piezoelectric actuator, contraction and expansion of the actuator bend the hinges so that the center of the scissor rises and falls. The offset, interleaved hinge structure leaves ample clearance for the rigid linkages during bending. Unlike in [8], the present hinge structure shortens the hinge length, maximizing the space available for longer linkages that provide larger vertical displacement.

The rigid elements for the linkages (figure 2) are 200 μm thick, 3 mm long, and 0.67 mm wide, corresponding to a 2 mm^2 actuator. Both the upper and lower rigid structures are patterned in SU-8 2150 by a single layer photolithography process. The SU-8 is patterned on top of a 15 nm layer of Omnicoat to enable subsequent release from the substrate. A temporary "carrier tape" made of polyimide is adhered to the SU-8 structures, and the Omnicoat is then dissolved to release the SU-8 from the wafer. The carrier tape preserves the relative positions of the SU-8 elements. The upper and lower SU-8 structures are laminated above and below the thin polyimide hinge layer with two layers of adhesive to form the microscissor as shown in figure 3. The carrier tape is then removed. The polyimide hinges are 30 μm long, 670 μm wide, and either 25 μm, 50 μm, or 125 μm thick, depending on the thickness of the polyimide film that is used. Two adhesives are used. The first is a 50 μm thick low-modulus silicone adhesive that comes laminated on both sides of a polyimide tape. The second is an approximately 5 μm thick layer of cyanoacrylate adhesive that is applied manually to the SU-8 parts before they are laminated onto a plain polyimide tape. The thickness of the cyanoacrylate is determined by measuring the thickness of the polyimide/SU-8 composite structure both with and without the adhesive layer; the volume of cyanoacrylate is also well controlled to prevent excess adhesive. The resulting microscissor is shown in figure 4a before the hinges are bent. Two hollow, rectangular SU-8 clamps are microfabricated separately for use in the actuator assembly process.

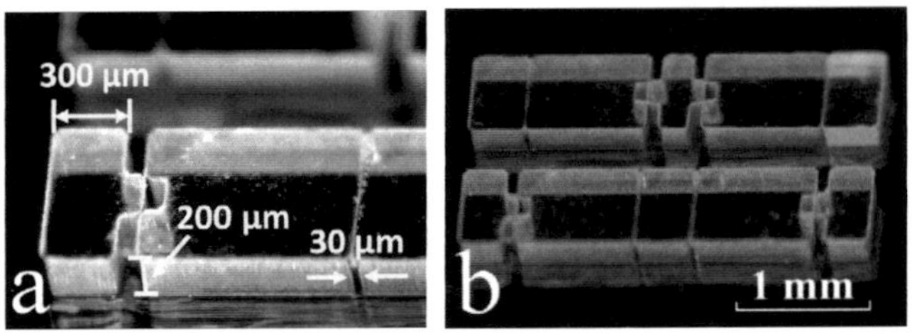

Figure 2. Optical micrographs of (a) the 200 µm thick SU-8 structures (rigid elements of the actuator) with 30 µm gaps for hinge bending, and (b) the upper and lower hinge structures implemented in SU-8 for a 3 mm x 0.67 mm micro-scissor.

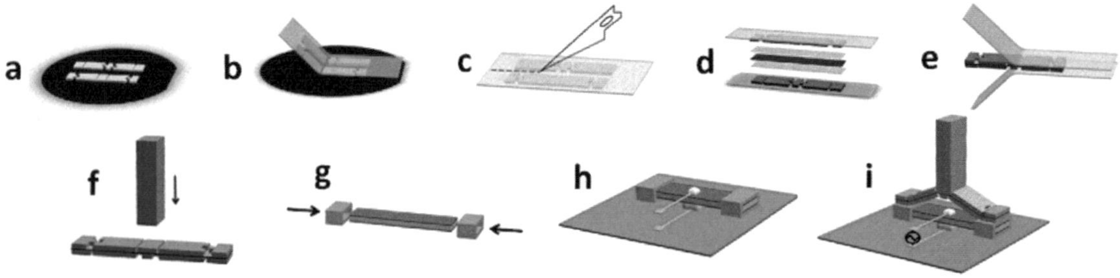

Figure 3. Fabrication process flow of a single actuator for a tactile array. a) Pattern SU-8 on wafer. b) Transfer SU-8 from wafer to polyimide carrier tape. c) Separate upper and lower SU-8 parts. d) Laminate polyimide film, adhesives and both sets of SU-8 parts. e) Peel off the polyimide carrier tape. f) Attach the tactile interface pin. g) Attach the SU-8 clamps (fabricated separately) to the PZT extensional actuator. h) Solder PZT to printed circuit board. i) Adhere scissor to PZT to form actuator.

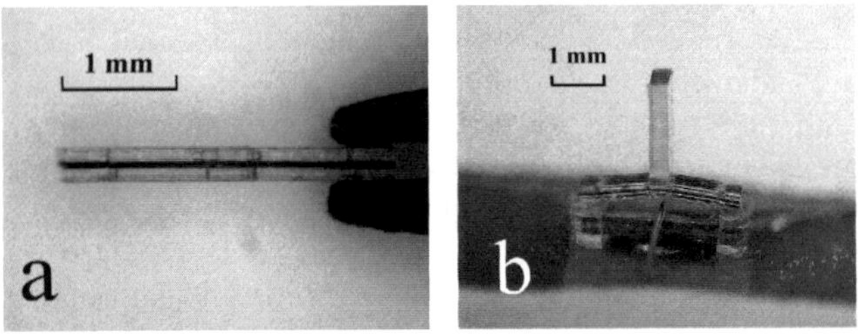

Figure 4. Optical micrographs of (a) an assembled SU-8 microscissor with total thickness of 430 µm, and (b) an assembled single tactile element with electrical connections.

The two clamps are mechanically registered to the ends of commercially-available, y-poled PZT actuators with dimensions of 3 mm x 0.67 mm x 0.38 mm with cyanoacrylate applied in the gap between clamps and PZT to ensure strong bonding. The scissors are then mounted on the clamps. The overall lengths of the two microscissors are designed to be longer than the length of the PZT actuator. When the microscissor is attached to the PZT, the hinges form a starting (unactuated) angle of 9° between the scissor linkage and the PZT. The clamps serve as an interface between the actuator and the scissor and provide clearance for electrical connections. The center of the actuator is soldered to a printed circuit board, creating electrical connection and mechanical support. The top of the PZT is grounded via a soldered connection. Because the present actuators are demonstrated for tactile displays, an SU-8 pin is integrated on the scissor's central element to create the sensing interface. Figure 4b shows a fully-assembled actuator with the scissor mounted on a PZT extensional actuator.

3. Experiments and Results

The individual actuators are mounted on a testing stage under an optical stereomicroscope so that the deformations of the scissor are visible in the microscope's viewing plane. A square wave bipolar AC voltage with a frequency of 0.5 Hz is output by a function generator, amplified, and applied to the device. The voltage amplitude is increased in 10 V steps from a starting value of 50 V to the maximum value of 170 V. The maximum peak voltage is below the PZT's maximum allowable voltage and reflects the limits of the test electronics rather than the device. The tip of the actuator moves under the applied voltage, and a microscope camera captures two or three images of the actuator in its maximum and minimum positions for each voltage. The actuator's displacement is measured by pixel counting in the captured images, in which each pixel is 0.45 μm. Displacement values are recorded for voltage values on which the pixel count increments by a full pixel, avoiding the need for interpolation.

To measure the amplitude of the force that the actuators apply when they vibrate in contact with a rigid surface, the individual actuators are mounted on a second testing stage that is in turn mounted to the grips of a mechanical tester (Instron 5943). The mechanical tester applies a compressive preload of 40 mN to the sensing pin when zero voltage is applied to the actuator. A square wave bipolar voltage with a frequency of 10 Hz is applied to the actuator, and its peak amplitude is increased from 50 V to 170 V in increments of 10 V. The resulting forces are measured by the mechanical tester's load cell.

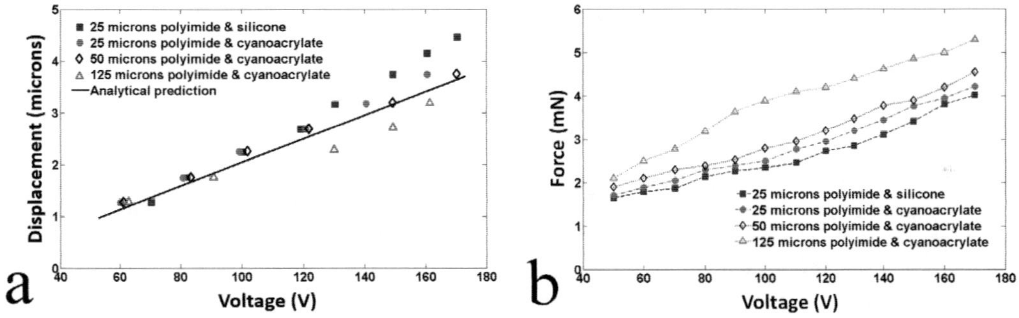

Figure 5. Plots of (a) the measured vibrational amplitude and (b) the measured force output by the 2 mm^2 tactile elements with various adhesives and layer thicknesses vs. applied voltage.

The measured peak-to-peak displacement and the measured force are plotted in figure 5. The solid line represents the displacement predicted analytically for the simplified case of an ideal microscissor with pinned hinges [8]. The dashed lines with markers represent measured data with different thicknesses of polyimide and different adhesives. The results show that the cyanoacrylate provides larger force output than the silicone adhesive does. Increasing the polyimide thickness at a given level of applied voltage typically decreases the displacement while increasing the measured force.

The actuator with a 125 μm thick polyimide hinge adhered by cyanoacrylate adhesive has a maximum displacement of 3.3 μm and a maximum force of 5.3 mN, corresponding to a displacement per unit of actuator area of 1.6 μm/mm^2 and a force per unit of actuator area of 2.6 mN/mm^2. This compares favorably with the results of [8], which yielded a displacement and force per unit of actuator area of 0.3 μm/mm^2 and 1.6 mN/mm^2, respectively. Its force output also compares favorably with [10-12], which yielded forces per unit of actuator area of 1.1 mN/mm^2, 0.9 mN/mm^2, and 0.6 mN/mm^2, respectively. Furthermore, the actuator's performance may be optimized for different applications by varying the layer thicknesses.

4. Conclusion

The design, fabrication, and operation of a set of compact tactile actuators that have a 15X smaller footprint than previous generations of actuators [8] but still achieve displacements of > 3 μm and forces of > 5 mN have been demonstrated. The maximum peak voltage of 170 V applied here is below the PZT's maximum allowable voltage, indicating that the maximum measured displacement and force are lower bounds on the actuator's performance. The measured performance corresponds to values of force per unit area and displacement per unit area that are 5.3 times and 1.6 times those of [8], respectively. The current actuators' performance is enabled by the use of a laminated hinge architecture that maximizes the length available for the scissor linkages in a compact area. The performance of the actuators is shown to depend on the details of the microscissor design. Varying the thickness of the polyimide hinge layer provides an effective means of tuning the actuators' force and displacement, and laminating the layers with a thin layer of cyanoacrylate adhesive instead of a more compliant silicone adhesive optimizes force output at the cost of a slight decrease in displacement. Overall, however, the measured displacements are close to predictions based on an ideal pinned-hinge scissor. Unlike conventional MEMS actuators, which are commonly used for applications like switches and optical devices, these powerful yet compact actuators have significant potential for high-force applications like tactile displays and micropumps.

Acknowledgements
This work was supported by the Samsung Think Tank Team. Microfabrication was performed in the George J. Kostas Nanoscale Technology and Manufacturing Research Center at Northeastern University.

References

[1] Nakajima M, Suzuki K, Su H R, Tani Y and Morita A 1991 Braille cell US Patent Specification 5226817.
[2] Bliss J C, Katcher M N, Rogers C H and Shepard R P 1970 Optical-to-tactile image conversion for the blind IEEE Transactions of Man-Machine Systems 11 58 - 65
[3] Mazzone A, Zhang R, and Kunz A 2003 Novel actuators for haptic displays based on electroactive polymers 2003 *Proc. of the ACM Symp. on Virtual reality software and technology* (New York, NY) pp 196-204
[4] Bolzmacher C, Hafez M, Khoudja M B, Bernardoni P and Dubowsky S 2004 Polymer-based actuators for virtual reality devices *Proc. of SPIE* (Bellingham, WA) **5385** pp 281- 9
[5] Pasquero J and Hayward V 2003 STReSS: A practical tactile display system with one millimeter spatial resolution and 700 Hz refresh rate *Proc. of Eurohaptics 2003* Dublin, Ireland 94 – 110
[6] Kitamura N, Chim J and Miki N 2013 Micro-needle electrode array for electro tactile display *Technical Digest* of *the Transducers* (Barcelona, Spain) pp 106 – 7
[7] Torras N et al. 2013 Nematic opto-mechanical actuators for the fabrication of refreshable tactile systems *Technical Digest of the Transducers* (Barcelona, Spain) pp 1691-4
[8] Xie X, Zaitsev Y, Velásquez-García L F, Teller S J, and Livermore C 2014 *Journal of Micromech. Microeng.* 24, no. 12 (2014): 125014.
[9] Sreetharan P S, Whitney J P, Strauss M D and Wood R J 2012 *Journal of Micromech. Microeng.* 22.5 (2012): 055027.
[10] Felder J, Lee E, and DeVoe D L 2015 *Journal of MEMS* **24** 4 pp 896-903
[11] Yosefi G, Mirzakuchaki S, Raissi F, and Afrang S 2014 Design and Analysis of a High Force, Low Voltage and High Flow Rate Electro-Thermal Micropump *Micromachines* **5** 4 pp 1323-1341.
[12] Yang G H, Kyung K U, Srinivasan M A, and Kwon D S 2007 Development of quantitative tactile display device to provide both pin-array-type tactile feedback and thermal feedback *Conf. of EuroHaptics* pp. 578-579

Miniature Electrostatic, High-Vacuum Ion Pump Architecture Using A Nanostructured Field Emission Electron Source

A Basu[1] [a], M A Perez[b], L F Velásquez-García[2] [a]
a) Microsystems Technology Laboratories, Massachusetts Institute of Technology, 77 Massachusetts Avenue, Cambridge MA 02139, USA
b) ColdQuanta, Inc., 3030 Sterling Circle, Boulder, CO 80301, USA.

Abstract. We report a field emission-based, electrostatic ion pump architecture for generation of high vacuum within a small chamber that is compatible with miniaturized cold-atom interferometry systems. The design increases the ionization probability using a helical electron collector. To create vacuum, electrons from a nanostructured field emitter array impact-ionize the gas molecules within the chamber; then, the ions generated are gettered by a negatively charged annular-shaped titanium ion collector. A proof-of-concept pump prototype was developed and characterized using a 200 cm^3 stainless steel vacuum chamber. The pressure inside the chamber was observed to decrease from 7.8×10^{-7} Torr to 7.2×10^{-7} Torr as the bias voltage on the ion collector was varied from -100 V to -1000 V while the emission current was kept constant at approximately 3.2 μA. The functional form of the experimental pump characteristics is in agreement with a proposed reduced-order model.

1. Introduction

Cold-atom interferometry of alkali atoms can be harnessed to implement a variety of high-precision sensors and timing devices such as atomic clocks, gyroscopes, accelerometers, magnetometers and gravimeters [1]. These systems require ultra-high vacuum (UHV, pressure < 10^{-9} Torr) to operate; therefore, chip-scale atomic sensors and timing devices require miniaturized UHV pumps resilient to alkali metal vapours that consume power at levels compatible with device portability.

In a macro-sized chamber UHV-level vacuum can be maintained using a conventional magnetic ion pump, which consists of an array of positively charged cylinders (anode), a flat negatively charged Ti plate (ion collector), and large permanent magnets that generate high magnetic fields (~ 0.2 T) within the pump volume [2] to 'trap' the electrons and achieve efficient impact ionization. While scaled-down versions of magnetic ion pumps have been reported [3], these are incompatible with miniaturized cold-atom interferometry systems because (i) a reduction in the pump size causes the required threshold magnetic field for electron trapping to increase [4], and (ii) the larger magnetic field associated with a miniaturized ion pump can disrupt operation of the cold-atom interferometry systems [5]. Non-evaporable getter (NEG) pumps are used in some cold-atom interferometry systems, e.g., commercial chip-scale atomic clocks; however, NEG pumps are unable to pump noble gases such as He and N_2 that are present in the chamber, and they inefficiently pump alkali vapours.

In this work we report an improved magnet-free ion pump design compatible with chip-scale cold-atom interferometry devices. The pump design is shown to operate at orders of magnitude lower

1 anirbasu@mit.edu
2 Velasquez@alum.mit.edu

pressure compared to the pump described in [6]. In addition, the pump speed, characterized by the pumping time constant τ, is shown to be much larger than the pump speed of our previous pump design reported in [7], which also was implemented in a 200 cm^3 chamber.

2. Device Design

The proposed field emitter array (FEA)-based magnet-free ion pump architecture is shown in Figure 1. In this pump design, a helical electron collector pulls the electrons toward itself, forcing them to first travel beyond the height of the electron collector, to then get pushed back due to the electrostatic mirror effect of the annular-shaped ion collector. Therefore, the trajectory of the electrons is significantly increased compared to a pump design with parallel-capacitor electrode configuration, such as the one described in [8], augmenting the probability of impact ionization.

Figure 2 shows a picture of the implemented proof-of-concept pump prototype assembled on a 2.75" CF flange with four electrical feedthroughs. The FEA chip is placed on top of a glass substrate. Two spring clips, one for the chip substrate and the other for the gate electrode pad, are used to provide mechanical stability as well as electrical connection between the chip and the feedthroughs. The two other electrical feedthroughs of the flange were connected to the ion and electron collectors via copper wires. The ion collector is made of titanium to efficiently getter ions, while the electron collector is made of a refractory metal to facilitate being heated to high temperatures when a current is passed through it during baking of the apparatus, thus reducing the outgassing of the electrode when the pump is in operation. The chamber is connected through an all-metal valve to an external support pump system that is not shown in the picture.

The FEA (Figure 3) consists of 62,500 nano-sharp silicon tips, each surrounded by a self-aligned gate electrode; the fabrication of the FEA is described elsewhere [9]. The nanometer-sharp tips generate high surface electric fields (>3×10^7 V/cm) at low bias voltage (< 40 V), causing electrons to quantum-tunnel to vacuum [10]. In a previous publication we showed that these FEAs do not degrade in the presence of Rb vapour [7].

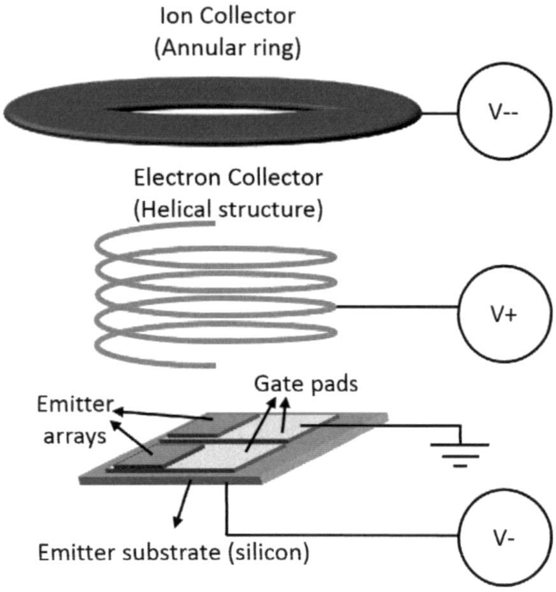

Figure 1. Schematic of the FEA-based, magnetic-less ion pump architecture.

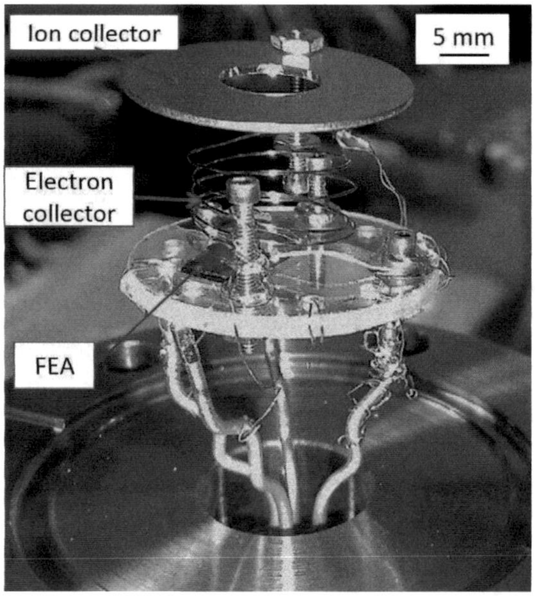

Figure 2. Picture of the FEA-based ion pump showing the electron and ion collectors.

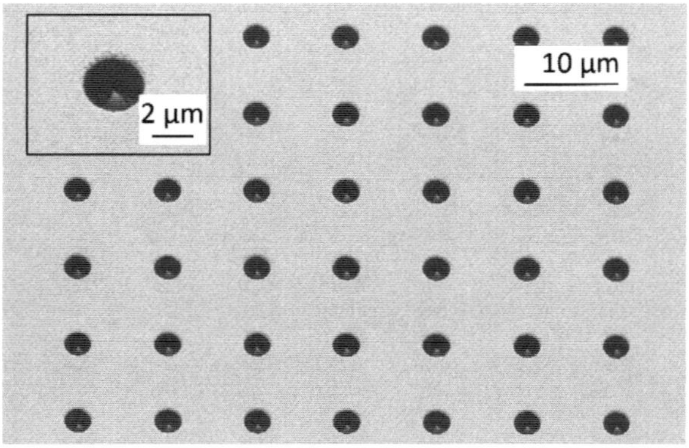

Figure 3. SEM of a section of the active area of the FEA chip, showing the nano-sharp tips surrounded by gate electrode apertures.

3. Device Modelling

Finite-element simulations using the commercial software COMSOL Multiphysics were conducted on a 3-D model of the proof-of-concept pump prototype. The helical electron collector has an outer diameter of 1.5 cm and a wire thickness of 0.4 mm; the ion collector is an annular structure with 2.5 cm outer diameter and 0.8 cm inner diameter, while the active area of the field emission electron source is a square 4 mm wide. The initial kinetic energy of the electrons was assumed to be 100 eV, which corresponds to a bias voltage between the chip substrate and the gate electrode equal to 100 V. Figure 4 shows the trajectories of 100 electrons evenly spread across the top surface of the electron source. In this simulation, the ion collector was biased at -1000 V, while the electron collector was biased at +1000 V. The simulation results show that the trajectory of the electrons is significantly increased compared to the parallel capacitor case, which results in an increased ionization probability.

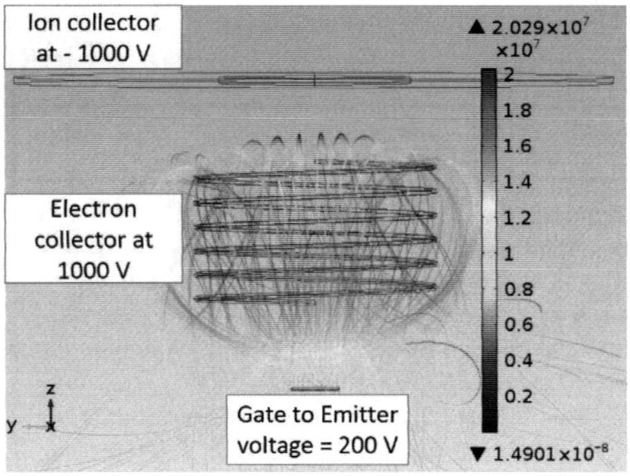

Figure 4. Finite element analysis of the trajectories of the electrons emitted by the FEA in the proposed ion pump design. The scale bar shows the speed of the electrons in m/s.

The theoretical relationship between chamber pressure as a function of time, volume of the chamber, and emitted electron current can be derived from first principles starting from the ideal gas law; the ion current is given by the electron impact ionization model. It can be shown that the chamber pressure P as a function of time t is equal to

$$P(t) = [P_o - P_f]e^{-t/\tau} + P_f \quad (1)$$

where P_o is the chamber pressure at $t = 0$, P_f is the ultimate pressure of the chamber ($t \gg \tau$), and τ is the characteristic pumping time constant, which is a function of chamber volume, electron current, length of the trajectory of the electrons, and ionization cross section of the gas species. Given the functional dependence of the chamber pressure versus time predicted by (1), the semi-log plot of the negative of the time derivative of the pressure versus time describes a straight line with slope equal to $-1/\tau$, i.e.,

$$ln\left[-\frac{dP(t)}{dt}\right] = ln\left[\frac{P_o - P_f}{\tau}\right] - \frac{t}{\tau} \quad (2)$$

4. Experimental Results

4.1 FEA Characterization

The FEA chip was characterized in a UHV chamber maintained at 5×10^{-10} Torr pressure. The current versus voltage (*I-V*) characteristics of the device are shown in Figure 5; the electron source turns-on with 32 V of bias voltage and emits over 1 mA with 99.8% transmission through the gate when biased at 117 V. Figure 6 shows the Fowler-Nordheim (*F-N*) plot of the data shown in Figure 5, i.e., $ln\ (I/V^2)$ vs. $1/V$ where I is the transmitted (collected) current and V is the bias voltage; the data describe a straight line, demonstrating that the current is field emitted. Using modelling reported in [9], from the slope of the F-N plot the radius of the tips is estimated at 1.73 nm, in agreement with SEM metrology.

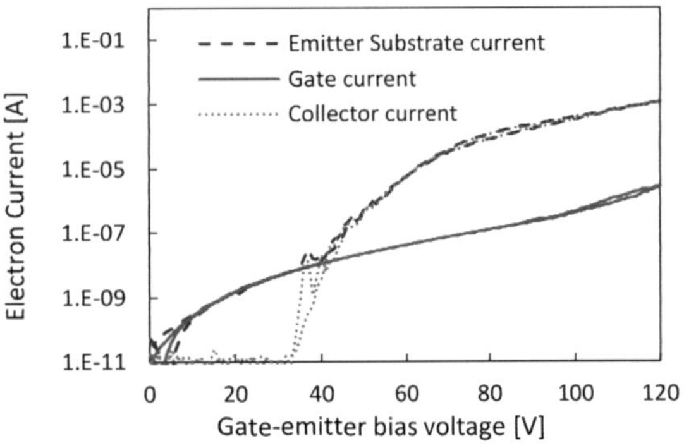

Figure 5. Current-voltage characteristics of the silicon FEA. The current is dominated by gate leakage until ~40 V, after which is mostly transmitted by the gate electrode.

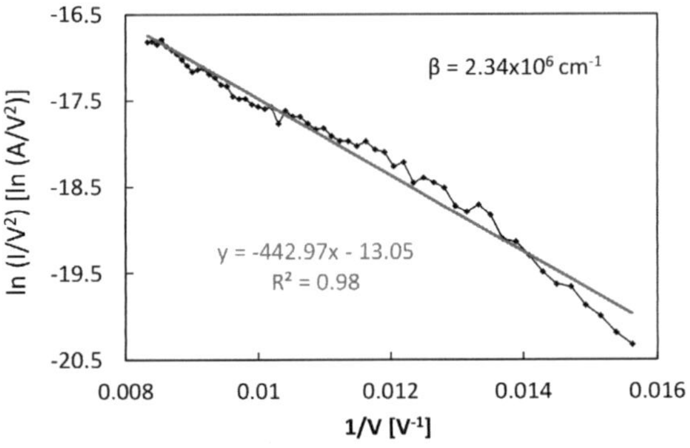

Figure 6. Fowler-Nordheim plot of the collected electron current from the FEA chip. A field factor β equal to 2.34×10^6 cm^{-1} is obtained from the slope of the plot.

4.2 Pump Characterization

Figure 7 shows the pressure versus time data collected with the proof-of-concept pump prototype. The data were acquired starting at the point when pressure is at 7.3×10^{-7} Torr and the ion collector was biased at -1000 V. The chamber pressure rises to 7.8×10^{-7} Torr when the bias voltage on the ion collector is set to -100 V, and it is seen to fall when the voltage is set to -1000 V with the lowest observed pressure being 7.2×10^{-7} Torr. Throughout the test, there was no variation in the emission current observed, showing that pumping was due to the ion collector voltage alone and not due to other factors such as reduced electron scrubbing as observed in [7]. Figure 8 shows the semi-log plot of the minus time derivative of the pressure versus time during pump-down, with the horizontal axis denoting the time since the beginning of each pump-down cycle. Each data point in the plot represents an average of the minus time derivative of the pressure considering all pump-down cycles. The slope of the linear fit of the data estimates the experimental pumping time constant at about 161 seconds.

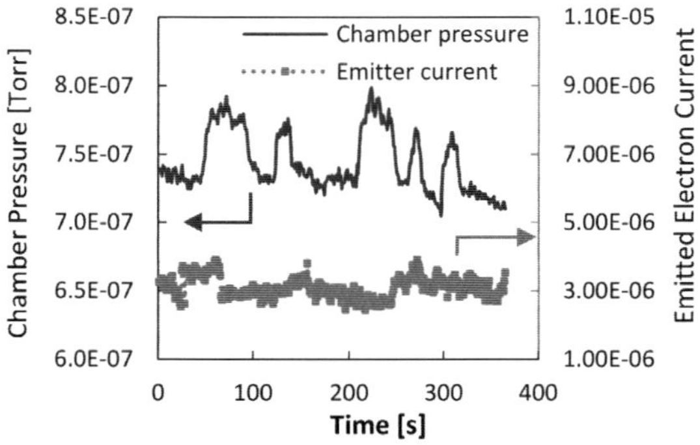

Figure 7. Pressure versus time. Pressure falls when the ion collector is biased at -1000 V, and it rises when the voltage is set to -100 V.

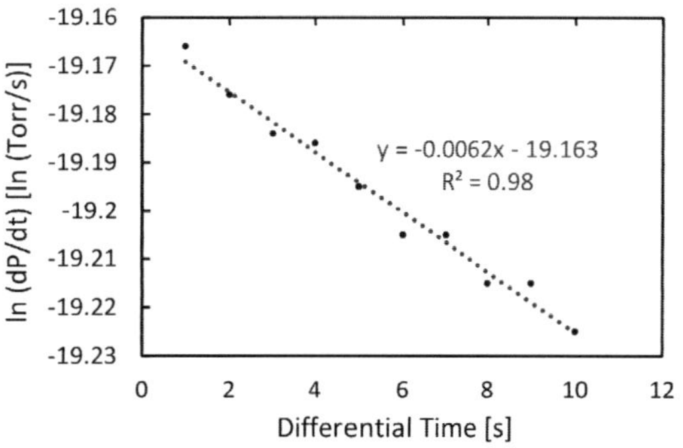

Figure 8. Semi-log plot of the negative of the time derivative of the chamber pressure vs. time. From the slope of the linear fit $\tau = 161.2$ s.

5. Conclusion

An improved proof-of-concept magnet-free pump prototype was implemented in a 200 cm^3 stainless steel vacuum chamber. The architecture consisted of a silicon FEA chip as electron source, a helical electron collector made of a refractory metal, and an annular ion collector made of titanium. Characterization of the FEA chip demonstrates that the electrons produced are field emitted. Characterization of the ion pump prototype shows that the pressure inside the chamber rapidly decreases from 7.8×10^{-7} Torr to 7.2×10^{-7} Torr as the bias voltage on the ion collector is varied from -100 V to -1000 V at a constant electron current of 3.2 μA, achieving a vacuum level that is orders of magnitude below the best reported values from a miniaturized magnetic-free ion vacuum pump. A reduced-order model of the pump dynamics predicts the functional form of the pumping data.

Acknowledgments

The fabrication of the nanostructured field emission electron source was done by Arash Fomani from MIT and was carried out in the Microsystems Technology Laboratories, MIT. We would like to thank Steve Hughes from ColdQuanta and David Johnson from Draper Laboratories for helpful discussions. This work was funded by the Defence Advanced Research Projects Agency / Microsystems Technology Office (DARPA/MTO) award number 022995-001 (program manager Robert Lutwak). Any opinions, findings, and conclusions or recommendations expressed in this publication are those of the authors and do not necessarily reflect the views of the US Government and therefore, no official endorsement of the US Government should be inferred.

6. References

[1] Kitching J, Knappe S, and Donley E A 2011 *IEEE Sensors Journal* **11** 1749
[2] Welch K 2001 *Capture pumping technology* (Elsevier)
[3] Green S R, Malhotra R, and Gianchandani Y B 2013 *J. Microelectromech. Syst.* **22** 309
[4] Audi M and De Simon M 1987 Vacuum **37** 629
[5] Rushton J A, Aldous M, and Himsworth M D 2014 *e-print arXiv:1405.3148*
[6] Grzebyk T, and Górecka-Drzazga A 2013 *J. Micromech. Microeng.* **23** 015007
[7] Basu A, Perez M A, Velasquez-Garcia L F 2015 *Proc. 18th Int. Conf. on Solid-State Sensors, Actuators and Microsystems* (Anchorage, AK) pp1021-1024
[8] Fomani A A, Velásquez-García L F, and Akinwande A I 2014 *Proc. 27th Int. Vacuum Nanoelectronics Conference* (Engelberg, Switzerland) pp. 210-211.
[9] Basu A, Swanwick M E, Fomani A A, and Velásquez-García L F 2015 *J. Phys. D: Appl. Phys.* **48** 25501
[10] Gomer R 1961 *Field emission and field ionization* (Cambridge, MA: Harvard University Press)

Out-of-Plane Translational PZT Bimorph Actuator with Archimedes' Spiral Actuating Tethers

Chenye Yang, Sanwei Liu and Carol Livermore

Department of Mechanical and Industrial Engineering, Northeastern University, Boston, MA 02115 USA

E-mail: c.yang@neu.edu, livermore@neu.edu

Abstract. The design, finite element analysis (FEA), and experimental characterization of a MEMS out-of-plane (vertical) translational lead-zirconate-titanate (PZT) bimorph actuator supported on Archimedes' spiral tethers are presented. Two types of bimorph actuators with different electrode patterns (with spiral tethers half actuated or fully actuated) are designed and fabricated. Both designs are fabricated by commercial processes and are compatible with integration into more complex MEMS systems. Finite element analysis (FEA) was used to analyze and predict the displacements of both types of actuators. The deflections of both fully-actuated and half-actuated devices were measured experimentally to validate the design. At an applied voltage of 110V, the out-of-plane deflections of the actuators with half-actuated and fully-actuated tethers were measured at about 17 μm and 29 μm respectively, in good agreement with FEA predictions of 17.1 μm and 25.8 μm. The corresponding blocking forces are predicted as 10 mN and 17 mN by FEA.

1. Introduction

Piezoelectric actuators can produce high forces and operate at high speed, but their native displacements are relatively small. Larger displacements can be engineered into piezoelectric actuators at the expense of some of their output force, for example through the use of piezoelectric bimorph benders. Thin film, microfabricated piezoelectric bending actuators are classic examples of trading force for increased displacement [1-4]. For applications in which large displacement from a small-footprint actuator is the primary requirement, thin film piezoelectric actuators can provide a superior solution. For applications in which actuator force is also important, thin film piezoelectric bending actuators are less attractive. In addition to the actuators' small force, the lithographic processing required to fabricate their complex, multi-layer thin film and released structures increases their cost.

For actuators in which a compromise between force and displacement is paramount, more macroscale approaches can offer satisfactory performance. Examples include fully macroscale actuators with mechanisms for displacement amplification, such as drum actuators [5-6] and cymbal and moonie actuators [7-9]. However, their large sizes limit their integration into microsystems. The actuator of [10] offers a smaller, partially microfabricated alternative, with tens of mN of force and displacements of up to 11 μm. However, the actuator in [10] is manually assembled from individual parts, limiting its manufacturability. For higher force actuators, their ability to be integrated into small-scale systems can be limited by their size or by their complexity of assembly.

This article presents a piezoelectric MEMS out-of-plane (vertical) actuator that addresses the coupled challenges of high force, high displacement piezoelectric actuators via an approach that can be manufactured efficiently and integrated with other MEMS parts at the die scale to create more complex systems. The present devices are translational lead-zirconate-titanate (PZT) Archimedes' spiral tether bimorph actuators. The actuators are fabricated by a commercial PZT manufacturing process, and their design offers large deflection and large force in a compact, mechanically and thermally robust structure that has accessible, easy to fabricate electrical connections. To characterize the unique behavior of spiral-tether actuators, two types of actuators with different electrode patterns (half-actuated tethers and fully-actuated tethers) are simulated, fabricated, and tested.

2. Design and Fabrication

The goal of the actuator design is to maximize the out-of-plane displacement and force while minimizing the lateral dimensions and maintaining the simplest possible fabrication process. Some trade-offs must be made between displacement and force; the present actuator is designed for a gas-phase microfluidic application and requires a minimum displacement of 25 μm. One seemingly straightforward approach would be to support a moveable central island on a set of straight actuating tethers, similar to [3]. Despite its simplicity, this design has drawbacks. The primary drawback lies in the challenges of achieving a compact device; the tethers must have sufficient length to enable the necessary displacement, but tether length directly translates into the lateral device dimensions for the case of straight tethers. A secondary drawback lies in the stress concentrations, which are magnified by the sharp corners of a purely straight tether architecture; stress concentrations may be moderated by rounding out the corners. Supporting a moveable central island on a set of folded flexures offers greater compactness, but providing electrical connections such that alternating regions of the actuating flexures will curve up and down to achieve large net displacements increases the fabrication complexity of the device [2, 11]. Stress concentrations also pose a challenge for folded flexures.

An alternative that can potentially achieve compactness while avoiding stress concentrations and maintaining simple wiring for straightforward fabrication is to use curved actuating tethers. Curved tethers can be tightly packed, minimizing wasted space, and their naturally rounded shape minimizes stress concentrations and the chance of failure. For these reasons, a central moveable plate supported on curved actuating tethers is chosen for the present device. In particular, an Archimedes' spiral shape is selected for the actuating tethers because it offers three key advantages. First, the spiral-supported actuator maximizes the length of the actuating tethers in a constrained space; second, the smooth Archimedes' spiral minimizes stress concentration; and third, the Archimedes' spiral guarantees uniform spacing between adjacent actuating tethers. To obtain large out-of-plane displacement in both the upward and downward directions, an X-poled bimorph is selected.

When a voltage is applied across the electrodes and the actuator's outer edge is constrained, the central island of the actuator moves up and down, depending on the voltage's polarity. If one were to design the spiral-supported actuator based on the physics of straight bimorph bending beams, one seemingly promising method would be to actuate the outer halves of the tethers (closest to the circular supporting frame) with one polarity and the inner halves of the tethers (closest to the central island) with the opposite polarity, similar to [2]. This approach is not chosen, for two reasons. First, the physics of curved actuators are not the same as the physics of straight actuators. For curved bimorphs, torsion plays an important role in generating net displacement, and it will be seen through finite element analysis (FEA) that actuating the halves with opposite polarity is ineffective. Second, even if that configuration were beneficial, it would greatly complicate wiring. Two alternatives are considered instead. In the first, only the outer half of the spiral (nearest the supporting frame) is actuated; this approach would also generate net displacement in a straight actuator, similar to [1, 3]. In the second, the full length of the spiral tether is actuated; this approach would generate zero displacement in a straight actuator but offers net displacement here because of the effects of torsion.

Two types of X-poled PZT bimorph actuators are fabricated as shown in figure 1. Each actuator comprises three actuating tethers shaped as Archimedes' spirals, a central island, and a circular supporting frame. The electrical and mechanical structures of both types of actuators are defined by laser cutting of nickel-coated PZT plates by a commercial vendor, Piezo Systems, Inc. Through-plate cutting is used to pattern the PZT features, and shallow cutting is used to pattern the electrode on one side of the actuator. The reverse-side electrode covers the full actuator. In figures 1a and 1c, the patterned electrode extends over only the outer half of the spiral tethers, whereas in figures 1b and 1d, the patterned electrode extends over the full length of the tethers. The material used to fabricate the actuators is PZT-5A4E, which guarantees a high Curie Point (350 °C) for compatibility with moderate temperature applications. Table 1 shows the actuator dimensions.

Table 1. Actuator dimensions

Property	Notation	Value
Archimedes' spiral constant ($r = a\theta$)	$\boldsymbol{a}$	0.768 mm
Thickness of single PZT layer	$\boldsymbol{t}$	0.135 mm
Width of single tether	$\boldsymbol{w}$	1.129 mm
Diameter of actuator	$\boldsymbol{d}$	19 mm
Electrode starting angle	$\boldsymbol{\theta_1}$	Full-actuated: 1.990 rad Half-actuated: 6.301 rad
Electrode ending angle	$\boldsymbol{\theta_2}$	8.988 rad

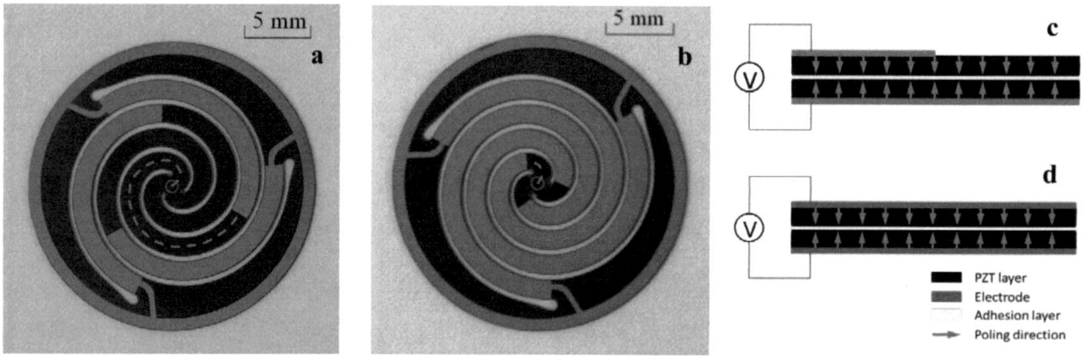

Figure 1. Photographs of Archimedes' spiral bimorph actuators with a) half-actuated tethers and b) fully-actuated tethers, and schematic cross-sections through c) a single half-actuated tether (as marked in a) and d) a single fully-actuated tether (as marked in b).

3. Finite Element Analysis

The behaviors of devices with fully-actuated and half-actuated tethers are modeled and simulated using finite element analysis (FEA) in Abaqus. Voltages are input, and the out-of-plane (vertical) translational displacement of the central island is predicted. The outer edge of the circular supporting frame is considered to be fully constrained, with zero rotation and zero deflection. Static voltages of 30 to 110 V are applied in increments of 10 V to drive each actuator. The thicknesses of the electrodes and of the adhesion layer that connects the bimorph's layers are assumed to be negligible. The C3D20E element (a 20-node quadratic piezoelectric brick element) is used in the mesh.

Figure 2 plots the FEA displacement contour at a voltage of 110 V for actuators with fully-actuated tethers and for actuators with half-actuated tethers. The maximum predicted out-of-plane deflections of the central island are 25.8 μm and 17.1 μm for the fully-actuated and half-actuated devices, respectively. The blocking forces for half-actuated and fully-actuated devices are calculated at 10 mN and 17 mN, respectively. For comparison, the out-of-plane deflection of the central island is predicted to be just 9 μm when the outer and inner halves of the tethers are actuated with opposite polarities as described above. The much smaller deflection under actuation conditions that would be optimal for a straight beam underscores the differences between the bending-driven physics of straight actuating beams and the combined physics of partially bending-driven and partially torsion-driven bending of curved actuating beams.

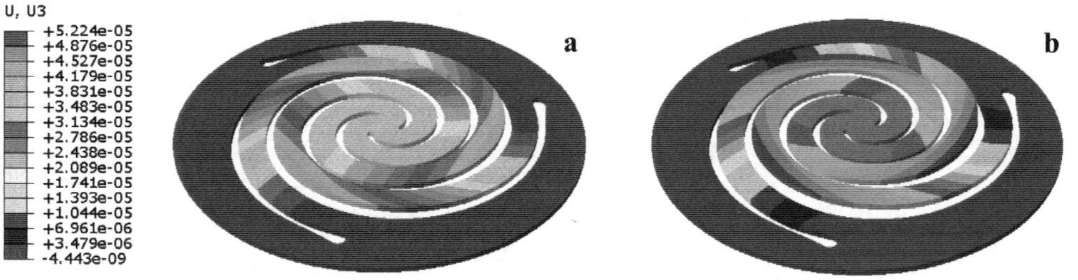

Figure 2. Displacement contour plots obtained by FEA using Abaqus at a voltage of 110V showing a) an actuator with half-actuated tethers and b) an actuator with fully-actuated tethers

4. Experiments

The deflections of both fully-actuated and half-actuated devices are measured. Each device is mounted on three pieces of pre-cut, 50 μm thick tin-plated copper foil tape that are evenly distributed around the edges of the actuators. The pieces of tape are soldered together to provide a common voltage. The tape provides electrical contact to the reverse-side electrodes and acts as a mechanical spacer above the underlying surface, ensuring that the center of the actuator can move both up and down. The tape and the actuators are constrained by epoxy on a 3D printed platform to approximate fully constrained boundary conditions. DC voltages of 30 V to 110 V are applied in 10 V increments with a DC power supply (MASTECH® HY5005E-2) and a high voltage amplifier (WMA-005, Falco Systems). The displacements of the actuators' central islands are obtained by measuring the heights at which a micrometer (Newport® 461 Series) achieves electrical contact with a metal tip glued on the central island. Three repeated displacement measurements are made at each voltage for each type of actuator. The average of the three measurements is recorded as the deflection for that voltage and actuator.

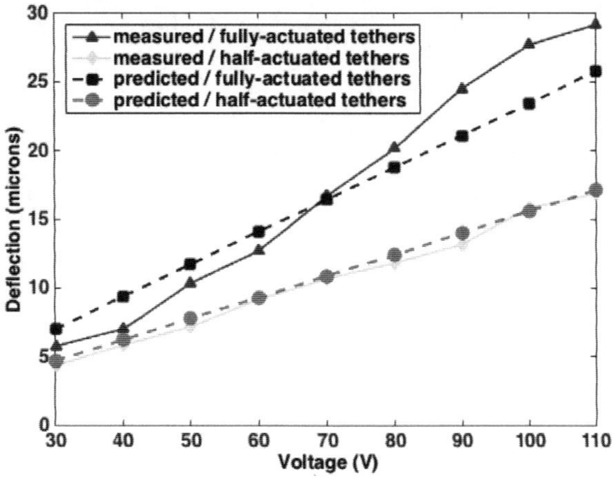

Figure 3. Measured and predicted out-of-place displacement for actuators with both half-actuated tethers and fully-actuated tethers.

Figure 3 plots FEA and experimentally-measured results for deflection versus actuating voltage for both half-actuated and fully-actuated devices. The results show that for both actuators, the displacements increase monotonically with the applied voltage, as predicted by the simulations. There is close agreement between FEA and experimental results for the actuator with half-actuated tethers, whereas the device with fully-actuated tethers shows reasonable agreement but with some deviation between FEA and experimental results. The fully-actuated device consistently offers greater displacement than the half-actuated device. The device with fully-actuated tethers deflects about 0.235 μm /V, whereas the device with the half-actuated tethers deflects about 0.155 μm /V.

5. Conclusion

Archimedes' spiral-supported piezoelectric bimorph actuators were shown to produce up to 29 μm of deflection and 17 mN blocking forces under unipolar actuation. The largest deflections and blocking forces were produced by devices in which uniform actuation was applied over the full length of the spiral tethers, in contrast to the behaviour of straight bimorph actuators. The high performance of the fully-actuated devices is attributed to the torsion generated in the spiral tethers. The experimental results are in good agreement with FEA predictions. The devices have a simple planar architecture for straightforward integration into more complex MEMS systems, and their fabrication via a commercial process provides an effective actuator solution that does not require access to fabrication facilities.

Acknowledgements

This work was supported by DARPA MTO under grant W31P4Q-10-1-0005. This article does not necessarily reflect the position or policy of the US government. The authors thank Dick Kensley for his help in fabricating the actuators at Piezo Systems, Inc.

References

[1] Choi J, Qiu Z, Rhee C, Wang T and Oldham K 2014 A three-degree-of-freedom thin-film PZT-actuated microactuator with large out-of-plane displacement *J. Micromech. Microeng.* **24** 075017

[2] Aktakka E E, Peterson R L and Najafi K 2013 A 3-DOF piezoelectric micro vibratory stage based on bulk-PZT/silicon crab-leg suspension *IEEE 26th Int. Conf. on Micro Electro Mechanical Systems (Taipei)* pp 576-9

[3] Qiu Z, Pulskamp J S, Lin X, Rhee C, Wang T, Polcawich R G and Oldham K 2010 Large displacement vertical translational actuator based on piezoelectric thin films *J. Micromech. Microeng.* **20** 075016

[4] Polcawich R G, Pulskamp J S, Judy D, Ranade P, Trolier-McKinstry S and Dubey M 2007 Surface micromachined microelectromechanical ohmic series switch using thin-film piezoelectric actuators, *IEEE Trans. Microw. Theor. Tech.* **55** 2642-54

[5] Sun C L, Lam K H, Chan H L W, Zhao X Z and Choy C L 2006 A novel drum piezoelectric-actuator *Appl. Phys.* A **84** 385-9

[6] Sun C L, Lam K H, Lu S G, Chan H L W, Zhao X Z and Choy C L 2007 Effect of geometry on the characteristics of a drum actuator *J. Intell. Mater. Syst. Struct.* **18** 1077-82

[7] Kim J H, Kim S H and Kwak Y K 2003 Development of a piezoelectric actuator using a three-dimensional bridge-type hinge mechanism *Rev. Sci. Instrum.* **74** 2918-24

[8] Haller D, Paetzold A, Goldin N, Neiss S, Goldschmidtboeing F, Nitsche W, King R and Woias P 2011 Cymbal type piezo-polymer-composite actuators for active cancellation of flow instabilities on airfoils *Proc. IEEE Transducers Conf. (Beijing)* pp 494-7

[9] Newnham R E, Dogan A, Xu Q C, Onitsuka K, Tressler J and Yoshikawa S 1993 Flextensional "Moonie" actuators *Proc. IEEE Ultrason. Symp.* **1** 509-13

[10] Xie X, Zaitsev Y, Velásquez-Garcia L F, Teller S J, Livermore C 2014 Scalable, MEMS-enabled, vibrational tactile actuators for high resolution tactile displays *J. Micromech. Microeng.* **24** 125014

[11] Fang H, Liu L and Ren T 2006 Modeling and design optimization of large-deflection piezoelectric folded cantilever microactuators *IEEE Trans. Ultrason. Ferroelec. Freq. Contral* **53** 237-40

ECF micropump fabricated by electroforming with novel self-aligned micro-molding technology

D Han[1,4], J W Kim[2], S Yokota[2], K Edamura[3]

[1] Interdisciplinary Graduate School of Science and Engineering, Tokyo Institute of Technology, J3-12, 4259 Nagatsuta-cho, Midori-ku, Yokohama 226-8503, Japan
[2] Precision and Intelligence Laboratory, Tokyo Institute of Technology, J3-12, 4259 Nagatsuta-cho, Midori-ku, Yokohama 226-8503, Japan
[3] New Technology Management Co., Ltd., 2-9-1-306 Higashi-shinkoiwa, Katsushika-ku, Tokyo 124-0023, Japan

E-mail: han.d.ab@m.titech.ac.jp

Abstract. This paper proposes and presents a novel ECF (electro-conjugate fluid) micropump with TPSEs (triangular prism and slit electrode pair) fabricated by electroforming process using newly developed self-aligned micro molds. ECF is a kind of functional and dielectric fluid. ECF micropump is based on the principle of ECF jet, which is a powerful and active jet flow generated between electrodes immerged in ECF, when high DC voltage is applied to the electrodes. Our previous research experimentally demontrated that the ECF micropump had high power density thanks to the 2D-integraton (serialized integration and paralleled integration) of our proposed MEMS fabrication method by using micro-molding and electroplating. Moreover, it was also proved that higher aspect ratio of TPSEs by the multilayer fabrication process resulted in higher flow rate of the ECF micropump. However, the multi-layer fabrication has demerit to require precise alignment that is time-consuming and extremely difficult to be met. In order to improve alignment accuracy and alleviate fabrication difficulty, this paper proposes a novel self-aligned MEMS fabrication process for high aspect ratio TPSEs. The ECF micropump by this newly-proposed MEMS process was successfully fabricated and the feasibility was proved by experimentally investigating output performance of the ECF micropump.

1. Introduction

Micropumps as critical power sources play a significant role in increasing numbers of microfluidic fields such as micro liquid-forced cooling devices, micro total analysis systems (μTAS) and micro dosing systems [1]. Among them, microfluidic actuators, especially those with higher output power on the further microscale, recently have come to the fore and attracted considerable attention worldwide, which are undoubtedly in greatly need of micropumps having high power density. For example, Yoshida et al proposed a 3-DOF microactuator with built-in fluid inertia (FI) micropump and electro-rheological (ER) microvalves for in-pipe working micromachines of about 10 mm in diameter. This FI micropump with a volume of 2.3 cm^3 generated output fluid power up to 0.22 W for water pumping. That is to say, it had the output power density of 95.7 mW/cm^3 [2]. As mentioned in that paper, the volumetric power density is recognized as one of crucial performance index for micropumps, in micro

[4] To whom any correspondence should be addressed.

hydraulics. Although a huge number of micropumps with various driving mechanisms have been developed as power sources for micro hydraulics, there are few micropumps to reach high power density with over a few hundred mW/cm^3 [3]

In order to achieve the aforementioned goal, we pay attention to electro-conjugate fluid (ECF) as a promising candidate for the driving mechanism. ECF is a kind of functional and dielectric fluid. An ECF micropump is based on the principle of ECF jet (figure 1), which is a powerful and active jet flow generated between electrodes immerged in ECF, when high DC voltage is applied to the electrodes. This phenomenon is defined as ECF effect [4]. Compared with other kinds of micropumps, the ECF micropump has the following advantages: (a) direct pumping mechanism (The energy is converted directly from electric energy into kinetic energy of the fluid by ECF effect.); (b) simple structure (Except for ECF and some electrode pairs, nothing else is needed for pumping.); (c) No mechanically movable parts (all the electrode pairs are fixed, resulting in low noise and vibration.). Therefore, the ECF micropump is preferable for micro hydraulic systems, such as micro fingers, hands, manipulation systems.

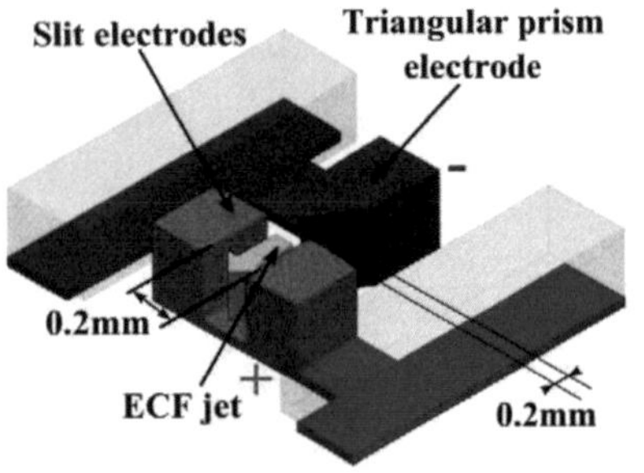

Figure 1. ECF jet by TPSE.

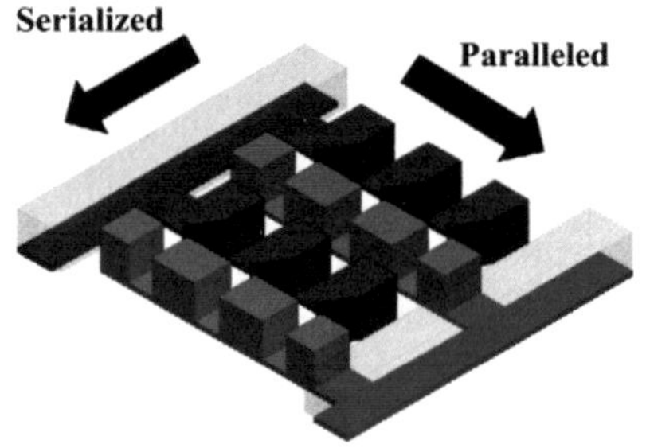

Figure 2. 2D-integration of TPSEs.

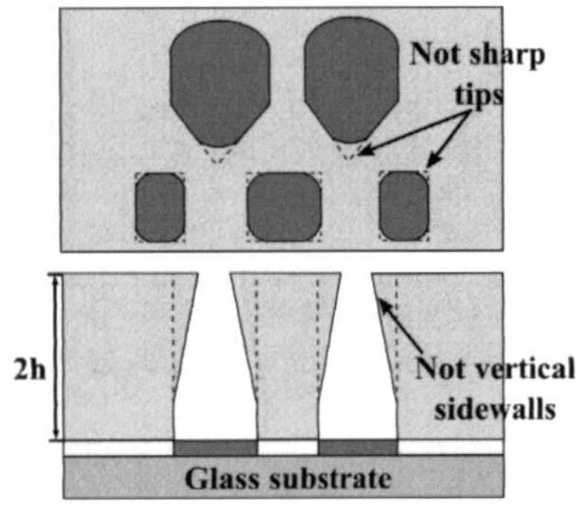

Figure 3. Deformed tip shape (top view) and sidewall profile (cross sectional view) of TPSE mold.

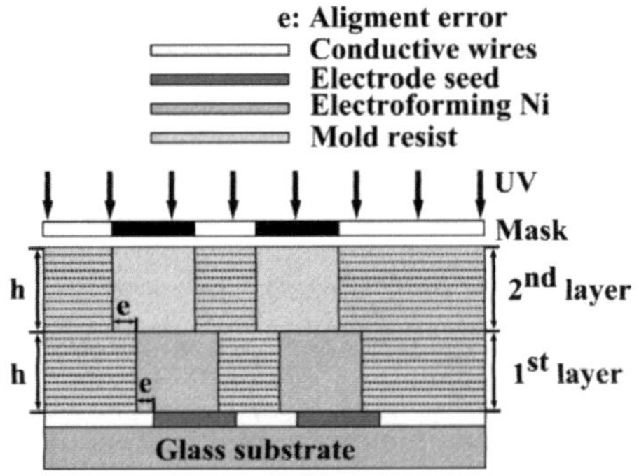

Figure 4. Misalignment of the multi-layer fabrication process.

In our previous research, it was proved that the ECF micropump consisting of TPSEs had higher power density, thanks to the 2D-integration (in series and in parallel) of TPSEs (figure 2). Moreover, we also proved that the output flow rate was increased by higher TPSE height [5]. For purpose of higher aspect ratio of TPSEs, a multilayer MEMS process based on our single layer process was proposed. However, there were some limitations. In the single layer fabrication of the relatively thick photoresist, the residual stress in cross-linking and swelling in mold exposure and developing could result in deformed TPSEs mold structures, including not sharp tips and not vertical sidewalls, as illustrated in figure 3. In the case of multi-layer fabrication of manual alignment operation, precise alignment is time-consuming and extremely difficult to be met, as shown in figure 4.

In order to improve the electrode profile and alignment accuracy of the TPSEs, this paper proposes a novel self-aligned micro molding technology for the ECF micropump with TPSEs. The concept of the novel self-aligned fabrication process is demonstrated in section 2. The proposed novel fabrication process in detail is described in section 3. The fabrication of ECF micropumps and characteristic experiments are mentioned in section 4. The conclusions are summarized in section 5.

2. Concept

In order to solve the problems that the previous fabrication processes are confronted with, we develop the ECF micropump fabricated by electroforming with novel self-aligned micro-molding technology. The concept is depicted in figure 5. The seed electrode pattern is nontransparent for UV light, while the glass substrate and negative photoresist KMPR (MicoChem Corp.) are both transparent. Therefore we can utilize back-side UV light and electrode seed pattern on the glass wafer for patterning thick photoresist, instead of using ordinary front-side UV light and mold photomask. After the bask-side UV exposure, the self-alignment is automatically realized between the mold pattern and the electrode seed pattern. The proposed self-aligned fabrication is also suitable for multi-layer fabrication process. For example, for the 2nd mold layer, the grown Ni electrode pattern in the 1st mold layer can be used as mold photomask for patterning the micro mold of 2nd layer. After the bask-side UV exposure, the self-alignment is also automatically achieved between the micro mold of 1st layer and that of 2nd layer, as shown in figure 6.

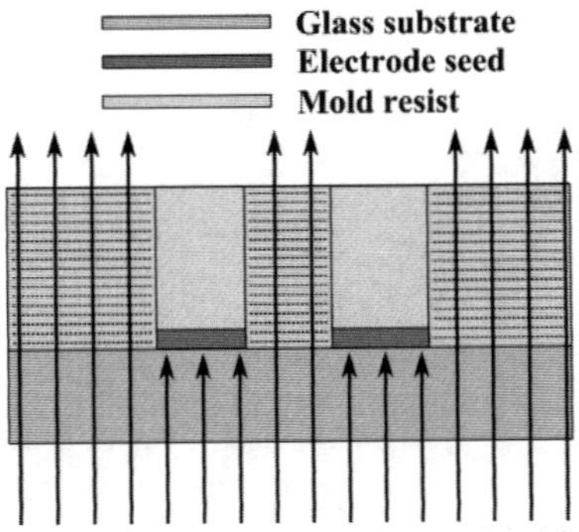

Figure 5. Concept of novel self-aligned fabrication for single layer.

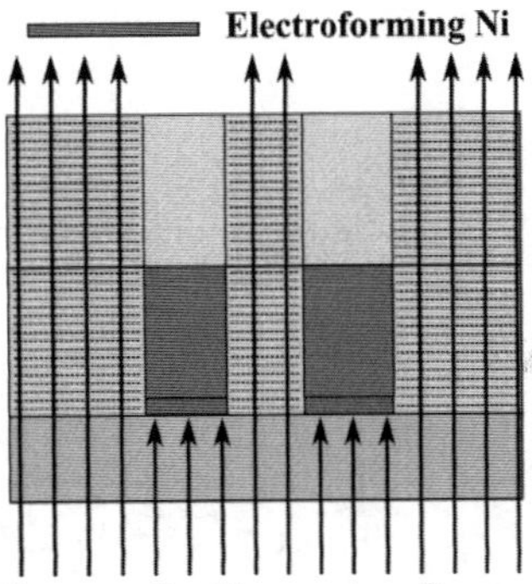

Figure 6. Concept of novel self-aligned fabrication for multi-layer.

3. Novel fabrication process

The proposed novel self-aligned fabrication process for ECF micropump with TPSEs is depicted in figure 7. First, Au/Ti are deposited and patterned as conductive seed layer, which is composed of electrode seed and conductive wires (single-line-pattern wires and parallel-lines-pattern wires). The electrode seed is used for positioning and formation of TPSE micro mold. The parallel-lines-pattern wires (line width = 10 μm, space = 39 μm) between the electrode seed, together with the single-line-pattern wires, are both utilized to confirm conductivity of electrode seed in the following electroplating. Then the thin negative photoresist (KMPR) with 5 μm thickness is formed to cover the whole surface except for the conductive seed layer for TPSE structures. Thick spin-coated KMPR for TPSEs is exposed and patterned by back-side UV exposure through transparent glass substrate on which the electrode seed pattern plays a role in forming micro-mold holes, so that self-aligned mold pattern is realized. By using this micro-mold, The TPSEs are formed by nickel electroplating. By selectively removing KMPR, the TPSE structures of nickel with height 440 μm were successfully fabricated and were coated with gold. Finally, another thick photoresist SU-8 (MicroChem Corp.) is patterned for the micro fluidic channels. The SEM picture of fabricated TPSEs and fluidic channels are displayed in figure 8.

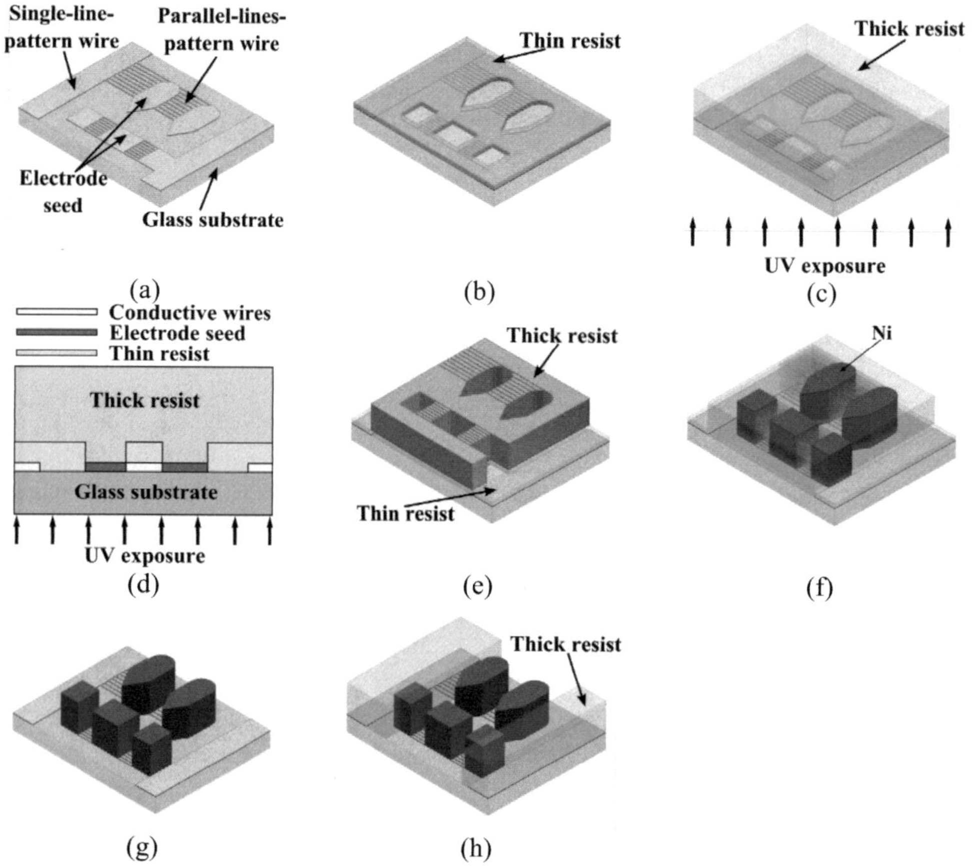

Figure 7. Proposed novel self-aligned fabrication process: (a) Seed layer patterning; (b) Insulation by cover; (c) Back-side UV exposure; (d) Cross sectional of back-side UV exposure; (e) Mold developing; (f) Ni electroplating; (g) Mold removal; (h) Channel forming.

4. Experiments and discussions

The fabricated ECF micropump (figure 8) occupies the volume of $30.6\times7.3\times1$ mm^3 and adopts 3 paralleled TPSEs and 10 serialized ones. The ECF micropump is packaged with a based plate, a bottom silicon rubber sheet, a top silicon rubber sheet and a cover plate.

Then the characteristic experiments of the fabricated ECF micropump are conducted. The output pressure is measured by the calibrated pressure sensor and the electric balance is adopted to measure the mass of ECF flowing out in specific time. FF-909EHA2 (New Technology Management Co., Ltd.) is used as the working fluid. Based on the experimental results illustrated in figure 9, the output pressure and output flow rate are both increased with higher DC voltage. The maximum output pressure and flow rate are 88 kPa and 411 mm^3/s respectively at the applied voltage of 4 kV.

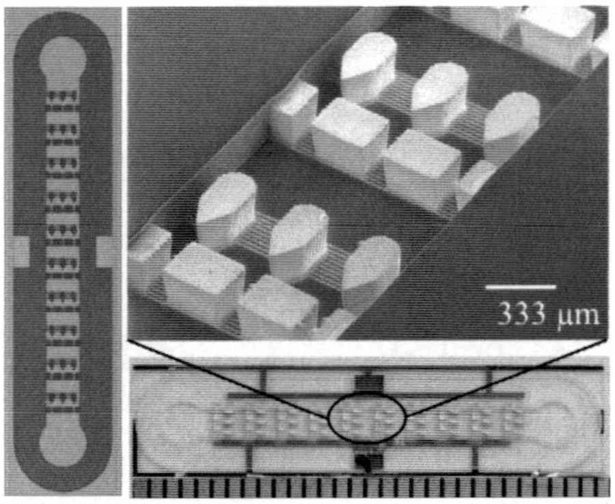

Figure 8. Design model (left) and fabricated prototype (right) of ECF micropump.

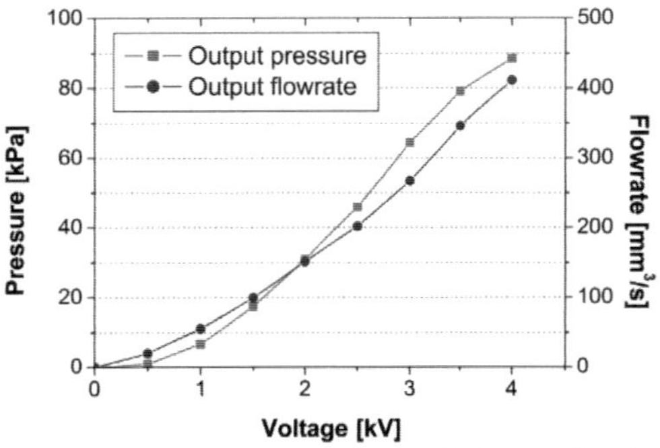

Figure 9. Output characteristics of the fabricated ECF micropump.

5. Conclusions

To improve electrode profile and alignment accuracy, we proposed the novel self-aligned micro-molding technology for ECF micropump with TPSEs, successfully fabricated the prototype, and experimentally proved the feasibility of the novel fabrication method. As future work, higher-aspect-ratio TPSEs will be realized by the multi-layer the back-UV multilayer process, in order to improve the output density of ECF micropumps.

References

[1] Iverson B D and Garimella S V 2008 *Microfluid. Nanofluidics.* **5** 145-174
[2] Yoshida K, Muto T, Kim J W and Yokota S *Proc. 11th Int. Conf. on Mechatronics Technology* (Ulsan, Korea, 5-9 November 2007) pp 450-455
[3] Laser D J and Santiago J G 2004 *J. Micromech. Microeng.* **14** R35
[4] Abe R, Takemura K, Edamura K and Yokota S 2007 *Sensor. Actuat.* A. **136** 629-637
[5] Kim J W, Suzuki T, Yokota S and Edamura K 2012 *Sensor. Actuat.* A. **174** 155-161

Cantilevers-on-membrane design for broadband MEMS piezoelectric vibration energy harvesting

Yu Jia[1,2], Sijun Du[2] and Ashwin A Seshia[2]

[1]Faculty of Science and Engineering, Thornton Science Park, University of Chester, Chester CH2 4NU, UK

[2]Department of Engineering, University of Cambridge, Trumpington Street, Cambridge CB2 1PZ, UK

E-mail: yu.jia.gb@ieee.org

Abstract. Most MEMS piezoelectric vibration energy harvesters involve either cantilever-based topologies, doubly-clamped beams or membrane structures. While these traditional designs offer simplicity, their frequency response for broadband excitation are typically inadequate. This paper presents a new integrated cantilever-on-membrane design that attempts to both optimise the strain distribution on a piezoelectric membrane resonator and improve the power responsiveness of the harvester for broadband excitation. While a classic membrane-based resonator has the potential to theoretically offer wider operational frequency bandwidth than its cantilever counterpart, the addition of a centred proof mass neutralises its otherwise high strain energy regions. The proposed topology addresses this issue by relocating the proof mass onto subsidiary cantilevers and integrates the merits of both the membrane and the cantilever designs. When experimentally subjected to a band-limited white noise excitation, up to approximately two folds of power enhancement was observed for the new membrane harvester compared to a classic plain membrane device.

1. Introduction

For the last decade, there has been a convergence of vibration energy harvesting (VEH) and microelectromechanical systems (MEMS) technologies [1] in an attempt to eventually realise integrated implementations of VEH and IC platforms for sensor systems. Amongst the different types of miniaturised mechanical-to-electrical transduction mechanisms, electromagnetism does not scale well at dimensions typical of MEMS devices, while electrostatic generators generally demonstrate poor power densities [2]. While there has been a growing trend of adapting electret transducers [3, 4], piezoelectric films remain the most popular choice for MEMS VEH to date [2, 5, 6, 9].

In terms of topologies, cantilever-based designs are by far the most employed structure [2, 5, 6] due to their simplicity, high responsiveness and the ability to house a proof mass near the free end without compromising the high strain energy regions of the active piezoelectric transducer [7, 8]. Membrane designs have been previously investigated in the field [2, 9] and provide the theoretical potential to provide wider frequency bandwidth. However, notable power or bandwidth enhancements from these designs have not been demonstrated yet [10].

One of the main drawbacks of the membrane topology, in comparison to cantilever-based designs, is the strain neutralisation of active piezoelectric transduction regions with the addition of an effective proof mass. Figures 1 and 2 illustrate a scenario where a significant area of otherwise high strain energy region of a circular disk membrane is sacrificed to house a centred proof mass. Since a given piezoelectric harvester relies on strain-induced charge generation across its active transduction area, strain optimisation is of paramount importance for power optimisation. Furthermore, the additional clamping conditions in a membrane structure yield lower compliance; thus, the membrane requires a higher excitation to manifest the same level of mechanical strain when compared to a cantilever beam.

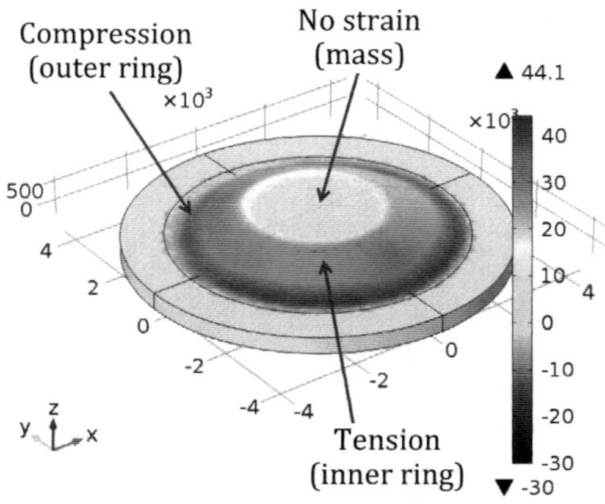

Figure 1: COMSOL model of a classic circular disk membrane (7 mm diameter) and a centred circular suspended proof mass (3 mm diameter).

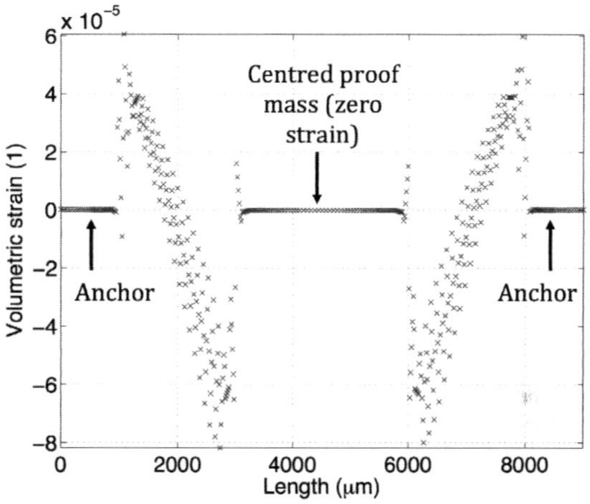

Figure 2: Strain distribution along the diameter of the classic circular disk membrane with an acceleration loading of 100 g.

2. Design and simulation

A new structural topology to improve the responsivity towards broadband excitation is proposed in figure 3. Instead of placing a single proof mass at the membrane centre, the masses are distributed on subsidiary cantilevers that extend outwards from the centre. Although regions of membrane are still sacrificed to accommodate the design complexity, the constant high strain regions of the membrane core are retained (figure 4). Additionally, the subsidiary cantilevers themselves possess high strain regions. Similar to a classic membrane structure, the membrane portion of the new design is composed of two opposing strain regions: the anchor strains near the clamped end and the bending strain near the centre. The bending strain of cantilevers align with that of the membrane core, thereby simplifying electrode design.

Table 2 summarises a simulated and calculated example where the device is subjected to 100 g of acceleration loading on the proof mass. In the simulation, 0.5 μm thick aluminium nitride (AlN) is assumed as the piezoelectric layer. It can be seen that the design complexity of the new membrane topology resulted in smaller total active piezoelectric area. However, as a result of higher compliance of the etched membrane structure as well as the larger effective proof mass, the average induced strain is several folds higher.

The theoretical maximum power extractable from the simulated strain response was calculated by computing the average electric charge generated by equation 1 and the power extractable across an ideal impedance given by equation 2 [8].

$$q = d_{31}\varepsilon_{av}Ea_{pz} \tag{1}$$

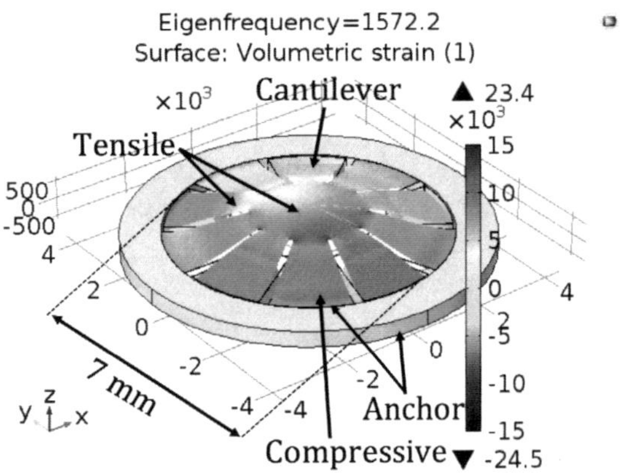

Figure 3: COMSOL model of the integrated cantilever- membrane design with the proof mass distributed on subsidiary cantilevers.

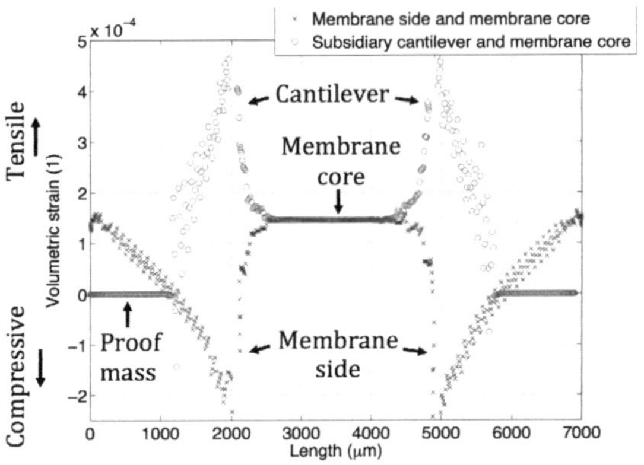

Figure 4: Strain distribution of the integrated cantilever membrane design for 100 g acceleration loading. The centre of the membrane experiences notable strain.

Table 1: Accumulated strain across the active piezoelectric area when subjected to an acceleration loading of 100 g, corresponding to maximum theoretically achievable power amplitude assuming optimal impedance matching for all constituent piezoelectric regions. Av. strain represents average strain across a particular active piezoelectric region.

Parameter	Plain membrane		Cantilever-on-membrane	
	Area (m^2)	av. strain (1)	Area (m^2)	av. strain (1)
Membrane anchor strain	1.88E-5	2.02E-5	7.85E-6	1.03E-4
Membrane bending strain	1.26E-5	4.07E-5	5.23E-6	1.86E-4
Membrane core bending strain	n/a	n/a	7.07E-6	1.29E-04
Cantilever bending strain	n/a	n/a	3.98E-6	2.62E-4
Proof mass (kg)	6.56E-6		8.45E-6	
Natural freq. 01 mode (Hz)	1478		1572	
Sum of area × av. strain (m^2)	8.91E-10		3.73-9	
Charge generated (C)	5.88E-10		2.46E-9	
Theoretical peak power (W)	6.41E-7		1.56E-5	

$$P = \frac{\omega h_p q^2}{\varepsilon_0 \varepsilon_r a_{pz}} \quad (2)$$

where, q is the charge generated, d_{31} is the piezoelectric charge constant in the 31 mode, ε_{av} is the average induced strain, E is the elastic modulus, a_{pz} is the active piezoelectric area, P is the peak power, ω is the frequency, h_p is the thickness of the piezoelectric layer, ε_0 is the permittivity of free space and ε_r is the dielectric constant of the piezoelectric material.

It can be seen from table 2, that given the same design area, the new membrane has the potential to attain a significantly higher peak power. However, this theoretical value is only achievable if all the various active piezoelectric regions are optimally matched in impedance. In practice, due to the complexity of the structure, it is difficult to simultaneously extract maximum energy from both the membrane and the subsidiary cantilevers.

3. Experiment and result

MEMS devices were fabricated using a 0.5 μm AlN on 10 μm doped silicon on insulator process. Un-etched regions of 400 μm thick substrate silicon were used as suspended proof masses. Figure 5 shows a MEMS harvester chip attached to a leadless chip carrier (LCC). The bottom of the LCC was hollowed out, through laser cutting, in order to allow unrestricted travel of the proof masses and minimal film-squeeze damping. Figure 6 presents a closer view of the device. The piezoelectric layer and top metallisation are not integrated on top of the proof masses.

Figure 5: Photograph of the MEMS chip (12 mm by 12 mm) on a leadless chip carrier. The backside of the chip carrier was etched out by laser micro-machining to minimise film-squeeze damping within the package.

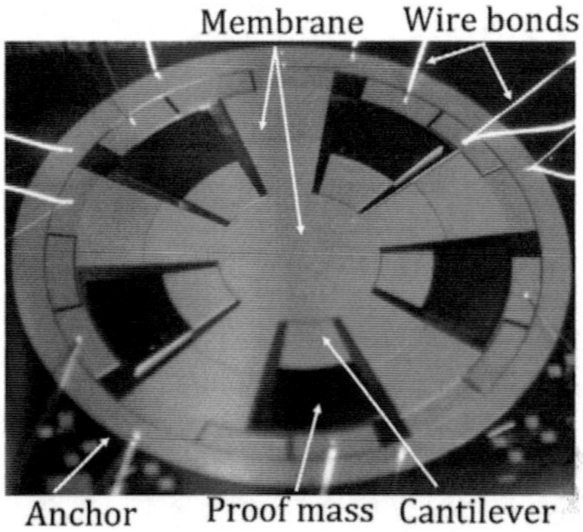

Figure 6: Micrograph of the cantilever-on-membrane VEH (active diameter: 7 mm).

The membrane diameter is 7 mm and the core diameter is 3 mm, while the end mass on each subsidiary cantilever takes up 60% of the cantilever length. While individual electrode zones were segmented during layout design, all transduction regions with the same strain polarity were connected together (off chip) for experimental simplicity. This however, came at the cost of suboptimal impedance matching and power extraction from the harvester. A comparable classic membrane device was also fabricated (7 mm membrane diameter and 3 mm centred mass diameter), which had a less complicated electrode connectivity.

In addition to the 01 mode, the first transverse modes of the subsidiary cantilevers within the new membrane structure can also be employed (table 2) to open up additional operational frequency bands. When driven with 1 g of acceleration at the 01 resonant mode, the plain membrane recorded $\sim$1.2 μW peak power while the new topology device yielded 2.2 μW. The power output of the new topology was 'diluted' by sharing the electrode connection with the cantilevers, which were minimally responsive at the 01 mode frequency range. The subsidiary cantilevers had a much higher compliance than the membrane and produced up to 0.5 μW (each) prior to fracture when driven beyond 0.5 g of acceleration at resonance. On the other hand, the classical membrane was more robust.

Table 2: Frequencies for measured classic membrane and new membrane devices. Modes where strain cancellation takes place are excluded.

Mode of interest	Plain membrane	New membrane
Membrane 01 mode	1200	1400
Subsidiary cantilevers 1 to 5, transverse mode	n/a	280, 290, 300, 310, 320

Figure 7 compares the average power output of the new membrane structure with the classic membrane structure when subjected to band-limited white noise from 10 Hz to 2 kHz. The experimentally matched load resistance ranged between 50 kΩ to 100 kΩ. However, these values represent a compromise due to the impedance mismatch of the piezoelectric regions on the membrane and the cantilevers.

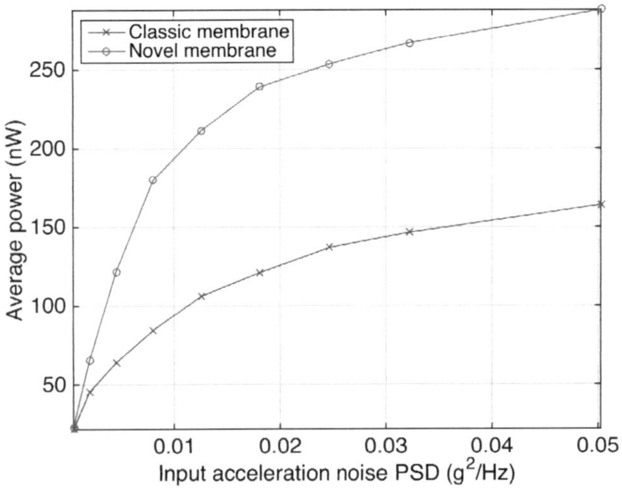

Figure 7: Measured average power for a classic membrane device and a new cantilevers-on-membrane device, subjected to band-limited white noise (10 Hz to 2 kHz).

Up to approximately two folds of power enhancement was observed between 0.005 g^2/Hz and 0.02 g^2/Hz of band-limited white noise. While subsidiary cantilevers readily fractured when driven beyond 0.5 g at resonance, such mechanical failure were not observed when subjected towards the band-limited white noise excitation of up to 0.05 g^2/Hz with peaks at 10 g.

4. Conclusion

A new cantilever-on-membrane topology for MEMS piezoelectric vibration energy harvesting is proposed. Instead of positioning the proof mass at the centre of a classic plain membrane, the masses are distributed onto subsidiary cantilevers. This enabled the harvesting of strain energy at the centre core of the membrane, which would otherwise be neutralised. Furthermore, additional frequency bands were introduced from the subsidiary cantilevers, thus, making it more responsive to broadband excitation. Simulation results suggest the theoretical potential to recover over an order of a magnitude higher peak power for a given acceleration loading, under ideal conditions. Experimentally, the new membrane harvester recorded up to two times higher power output than a plain membrane when subjected to band-limited white noise vibration.

Acknowledgement

This work was supported by the Engineering and Physical Sciences Research Council [EPSRC grant number: EP/L010917/1].

References

[1] A Harb, 2011, Energy harvesting: State-of-the-art, *Renewable Energy*, **36**, 2641-54
[2] S Priya and D Inman, 2009, *Energy Harvesting Technologies*, (New York: Springer US)
[3] F. Peano and T. Tambosso, 205, *J. Microelectromech. Syst.*, **14**(3), 429?35
[4] Y. Suzuki, D. Miki, M. Edamoto, and M. Honzumi, 2010, *J. Micromech. Microeng*, **20**(10)
[5] R. van Schaijk, R. Elfrink, T. Kamel, and M. Goedbloed, 2008, *IEEE Sensors*, 45?48
[6] R Andosca *et al.*, 2012, *Sens. Actuators A.*, **178**, 76-87
[7] Y Jia and A A Seshia, 2014, *J. Phys. Conf. Ser.*, **557**
[8] Y Jia and A A Seshia, 2015, *Microsyst. Technol.*
[9] S. Yoo, J. Kim, S. Park, C. Jang and H. Jeong, 2014, *J. Korean Phys. Soc.*, **64**(5), 706-709
[10] D. Zhu, M.J. Tudor and S.P. Beeby, 2010, *Meas. Sci. Technol.*, **21**(2)

Optical HMI with biomechanical energy harvesters integrated in textile supports

G De Pasquale[1], SG Kim[2], D De Pasquale[1]

[1] Department of Mechanical and Aerospace Engineering, Politecnico di Torino, Corso Duca degli Abruzzi 24, Torino, Italy
[2] Mechanical Department, Massachusetts Institute of Technology, 77 Massachusetts Avenue, Cambridge, Boston, USA

E-mail: giorgio.depasquale@polito.it, sangkim@mit.edu, daniele.depasquale.87@gmail.com.

Abstract. This paper reports the design, prototyping and experimental validation of a human-machine interface (HMI), named GoldFinger, integrated into a glove with energy harvesting from fingers motion. The device is addressed to medical applications, design tools, virtual reality field and to industrial applications where the interaction with machines is restricted by safety procedures. The HMI prototype includes four piezoelectric transducers applied to the fingers backside at PIP (proximal inter-phalangeal) joints, electric wires embedded in the fabric connecting the transducers, aluminum case for the electronics, wearable switch made with conductive fabrics to turn the communication channel on and off, and a LED. The electronic circuit used to manage the power and to control the light emitter includes a diodes bridge, leveling capacitors, storage battery and switch made by conductive fabric. The communication with the machine is managed by dedicated software, which includes the user interface, the optical tracking, and the continuous updating of the machine microcontroller. The energetic benefit of energy harvester on the battery lifetime is inversely proportional to the activation time of the optical emitter. In most applications, the optical port is active for 1 to 5% of the time, corresponding to battery lifetime increasing between about 14% and 70%.

1. Introduction

The function of human-machine interfaces (HMI) is to transfer the analog commands imposed by the user's body to machines. The majority of devices and systems developed are introducing some limitations to the user's comfort and motion freedom because of the presence of cables or large batteries. Additionally, tedious and long calibration procedures are often required. The first HMI glove prototypes were proposed in the 1970s and 1980s, such as the MIT-LED glove and the Sayre Glove [1]. The Data Glove [2] (1982) was based on plastic tubes and light sensors to measure and store joint angles. The most recent devices were commercialized starting from the 1990s for virtual modeling, motion analysis and medical applications (CyberGlove [3], Humanglove [4], StrinGlove [5], etc). The HMI gloves proposed in the past have some common limitations: reduced comfort/performances due to wire communications, long calibration procedures, resistance to hand motions, and power supply based on cables or batteries. The glove introduced here, called GoldFinger, overcomes some of these problems by harvesting biomechanical energy from hands, by integrating advanced materials, by miniaturizing components, and by coupling the interface to standard textile supports.

The most diffused applications of HMI gloves are manufacturing, design and 3D modeling, robotics, medicine and health care, wearable systems management, sign language understanding and data management. In the manufacturing field, hand motion can replace computer peripherals to manipulate virtual objects. Practical examples are represented by gloves used to select interior furniture of cars through virtual reality programs (Daimler-Benz), to simulate maintenance operations on aircrafts (Boeing) [6], or to test the performances of astronauts and pilots [7]. Other applications are in the medical field, where gloves support rehabilitation of hand capabilities as motion, grip strength, and temperature sensitivity [8].

The invasiveness of HMI gloves generally represents a limitation to user's hand motion and can reduce the portability of the device. Additionally, the fabric support may alter the sensors precision and accuracy, and the global measurement performances. The energy supply is usually provided through external cables or heavy batteries that additionally limit the user mobility. The next generation of HMI should include innovative features as high integration levels of sensors and electric connections with the fabric, energetic autonomy, good wearability and comfort of use, wireless communication and low cost.

In this paper, the design, prototyping and experimental validation of a HMI glove are presented. The device communicates by means of an optical port (LED) and integrates an energy harvesting system able to increase the battery charge with the fingers motion. The electro-mechanical conversion properties of the transducers were preliminary validated. Then, the electric power of the harvesting system installed to the glove was measured during the fingers opening-closing motion. The battery capacitance (160 mAh) can supply the optical port continuously for approximately 104 hours (including additional 11% power consumption due to circuitry losses). The energetic benefit of energy harvester on the battery lifetime is inversely proportional to the activation time of the optical emitter. In most applications, the optical port is active for 1 to 5% of the time, corresponding to battery lifetime increasing between about 14% and 70% [9].

2. Device description

The HMI glove, named GoldFinger, is represented in Fig. 1. The main features of the design are the energetic efficiency of the system, the integration of components (piezoelectric transducers, electric conductors and electric controls), the miniaturization of electronic components (including the power management circuit and the storage battery). An aluminum case situated on the back side of the glove is used to protect the electric parts against dust, wear and impacts. This kind of package is characterized by small dimension (55x46x10 mm^3), strength and lightness. The technical fabric of the glove provides elasticity and resistance to wear and water. Fabric-embedded wires are used to electrically connect the piezoelectric transducers to the circuit. The conductive wires (Fabrickit, electrical resistance = 0.3 Ω/m) have almost the same flexural deformability of the fabric.

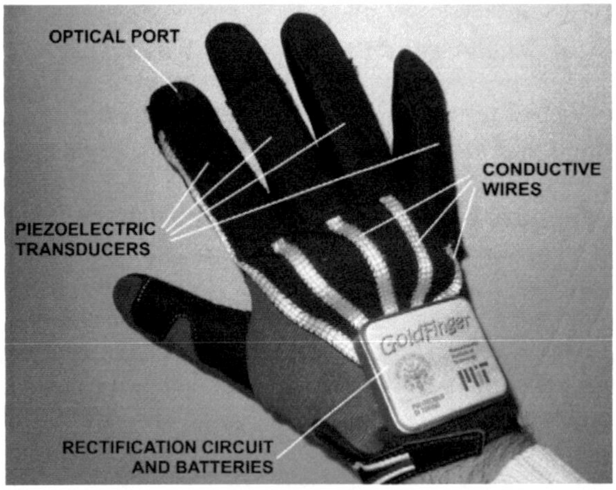

Figure 1. Functional components of GoldFinger.

The optical port is controlled by the user through an electric switch integrated in the glove and made with conductive metalized nylon fabric (Statex, surface resistance < 0.02 Ω, abrasion resistance = 500.000 cycles, temperature range = –30 / +90 °C, thickness = 0.1 mm, specific mass = 77 g/m^2), instead of using discrete electrical components. The conductive fabric is composed of a nylon support covered by three Sn/Cu/Ag layers. The optical port is made with a high efficiency LED L-383IDT (Kingbright, size = 9.6x5x2.5 mm^3, forward voltage = 1.7 V, forward current = 2.5 mA, wavelength = 627 nm, capacitance = 15 pF).

The functional scheme of the device is reported in Fig. 2. The hand analog command sequence is sent to the receiver through the light emitter mounted on the fingertip. The optical receiver is an analog-to-digital converter (ADC) represented by a camera. Through the software interface, the light path and light on/off flashing sequence is tracked and converted in Cartesian coordinates and the microcontroller is programmed accordingly.

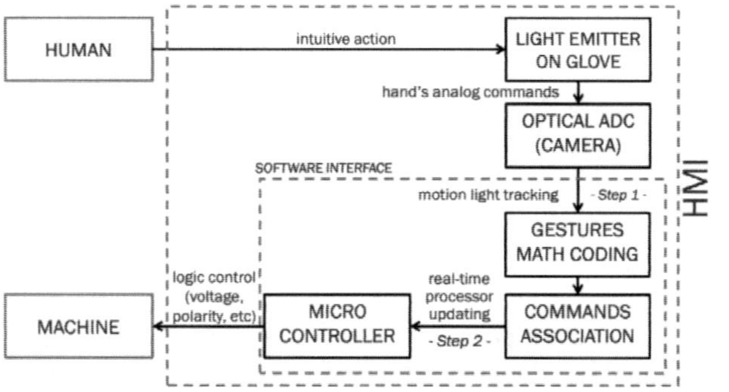

Figure 2. The device workflow.

3. Biomechanical energy harvesting

The harvesting system is made with four couples of piezoelectric transducers (Physik Instrumente, PIC252) integrated into the fabric in correspondence with the fingers joints, and in particular to the proximal-interphalangeal (PIP) joints of the 2-5 fingers. The transducer components are reported in Fig. 3a: they include the piezoelectric foil with two metalized surfaces connected to the electrodes and the polymeric package. The properties of the transducer are reported in Tab. 1. Laboratory tests also revealed good conversion efficiency and reliability under repetitive load cycles. The control circuit used to manage the harvested power and to control the optical port is reported in Fig. 3b.

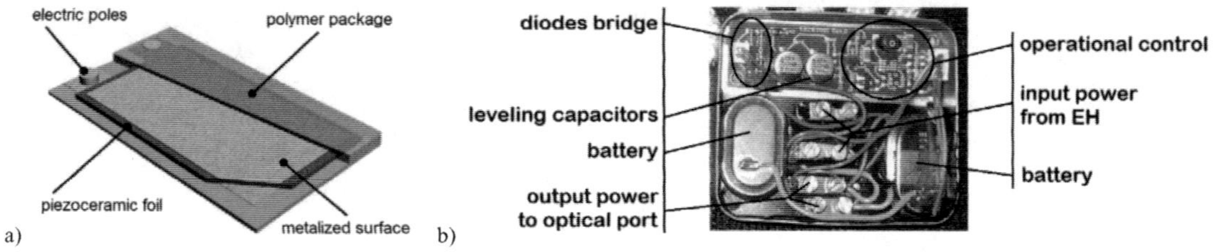

Figure 3. The piezoelectric transducer (a) and the rectification and storage circuit (b).

4. Software interface

The software interface works in two steps, as shown in Fig. 2. In Step 1, the LED position is detected by filtering the environmental light and by selectively tracking the wavelength corresponding to the LED color. Depending to the light path and to the on/off switching, the command message is composed and converted to Cartesian coordinates. The analog optical instructions are then translated by the software to codified logic instructions that have been previously associated with the gestures. In

Step 2, the real-time programming of the microcontroller is provided by modifying the parameters of the flow of execution instructions.

The area of detectable movements is approximately 1 m^2, situated in the range 0.5-1 m from the receiver objective. The optical reader used is a camera that can record images in color (i.e., 24-bit RGB with 3 channels and 8 bits/channel) at 15 fps with a resolution of 640x480 dpi. Figure 4 shows the image captured by the camera (left) and the digitally processed image generated by the interface software during Step 1 (right). The actual position and status (i.e., on/off) of the light emitter is detected and saved by two Cartesian coordinates.

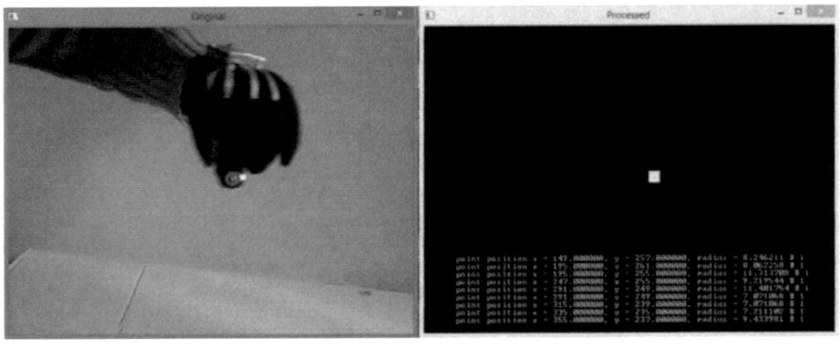

Figure 4. Software interface for light tracking and analog-to-digital commands conversion.

5. Validation tests

The electro-mechanical properties of the piezo transducers were preliminary validated in the cantilever mode by using a TIRA TV51120 electromechanical shaker. The output current and power were measured at variable resistance, and the resistance $R_{opt} = 216$ kΩ maximizing the power is obtained.

The following measurements were conducted with the transducers applied to the glove in their final configuration; more in detail, two transducers are coupled in opposite configuration for each finger to increase the output current. Finally, the glove has been equipped with four transducer couples and is then worn to run the tests, consisting in opening-closing fingers for 10 s. The output parameters of the open circuit voltage (V_0) and of voltage difference (V) across R_{opt} are measured and stored during the fingers motion. The average instant power generated by the glove is

$$\overline{P}_{inst}(t)=\frac{P_{cumul}(t)}{t}=\frac{\sum_{T=0}^{t}P(T)\cdot\Delta t}{t},$$

where P_{cumul} is the cumulative power obtained as the sum of the instant power over time with t_j the sampling times. The measured values of the cumulative power and of the average instant power are reported in Fig. 5.

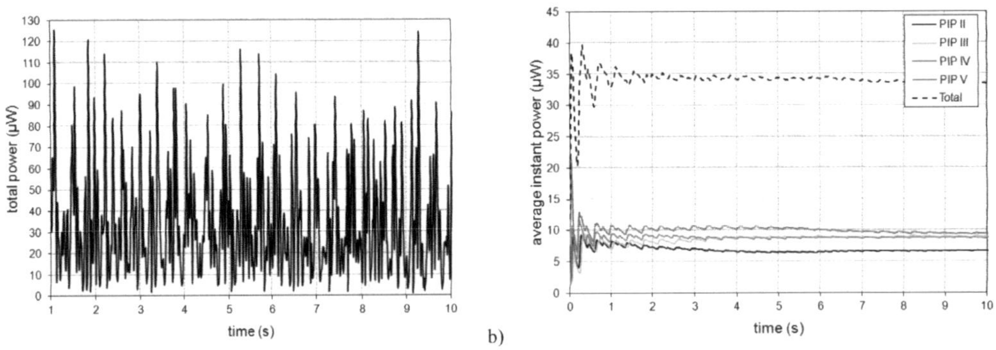

a) b)

Figure 5. Measured total output power generated by the glove (a) and average instant power (b).

From the experimental results, the total average power generated is 31.9 μW (6.6, 8.5, 8.5 and 8.3 μW for each finger, respectively). By considering the power consumption of the optical port (4.25 mW), the operative duty cycle of the system is 133.2. The power dissipated by the conditioning and leveling circuit is about 0.5 mW. Then, by including all of the sources of electric power consumption, the system duty cycle is approximately 146.5. With the reported duty cycle and without using batteries, the light emitter could be active for 30 s per hour. Then, the energetic performance of the harvester is suitable for increasing the battery charge and/or reducing the battery size. In particular, by using the GoldFinger glove instead of the gloves normally worn by operators, the lifetime of the battery of the GoldFinger is increased by the percentage reported in Tab. 2.

Table 1. Properties of piezoelectric transducers.

Property	Value	Dimension
length	61	mm
width	35	mm
thickness	0.4	mm
bending radius	12	mm
piezo layer thickness	100	μm
Young's modulus	16.4	GPa
capacitance	150	nF
temperature range	-20 / +180	°C
mass	2.1	g

Table 2. Duty cycle as function of the optical port activation.

Optical port active time period (%)	System lifetime (hours)		Lifetime increase (%)
	Battery only	Battery and energy harvesters	
1	10473	34450	69.6
5	2094	2433	13.9
10	1047	1126	6.96
20	523.6	542.5	3.48
50	209.4	212.4	1.39
100	104.7	105.5	0.70

6. Conclusions

GoldFinger demonstrates the applicability of biomechanical energy harvesters to HMI and the possibility to increase the battery lifetime through the conversion of biomechanical energy. The prototype, characterized by small power consumption, small, compact and fabric-integrated components, is also improved by the wireless communication of the optical port. The wearability, comfort, low stiffness to bending and small battery size can sensitively increase the added value of this kind of devices for the future. The overall duty cycle provided by the integrated piezoelectric transducers is approximately 146.5 and it is able to increase the battery lifetime between about 14% and 70% in those applications where the optical port is used between 1% and 5% of the time.

References

[1] D. J. Sturman, D. Zeltzer, "A survey of glove-based input," *IEEE Comput. Graph. Appl.*, vol. 14, no. 1, pp. 30-39, 1994.

[2] Digital data entry glove interface device, by G. Grimes, AT&T Bell Laboratory, Murray Hill, NJ. (1983). *Patent* U.S. 4 414537.

[3] P. Coiffet, G.C. Burdea, *Virtual reality technology*. New York, Wiley, 2003.

[4] L. Dipietro, A. M. Sabatini, P. Dario, "Evaluation of an instrumented glove for hand-movement acquisition," *J. Rehabil. Res. Dev.*, vol. 40, no. 2, pp. 179-190, 2003.

[5] T. Kuroda, Y. Tabata, A. Goto, H. Ikuta, and M. Murakami, "Consumer price data-glove for sign language recognition," in *Proc. Int. Conf. Disabil., Virtual Reality Associated Technol.*, 2004, pp. 253-258.

[6] R. Steffan, U. Schull, and T. Kuhlen, "Integration of virtual reality based assembly simulation into CAD/CAM environments," in *Proc. Annu. Conf. IEEE Ind. Electron. Soc.*, 1998, vol. 4, pp. 2535-2537.

[7] K. U. Doerr, H. Rademacher, S. Huesgen, W. Kubbat, "Evaluation of a low-cost 3D sound system for immersive virtual reality training system," *IEEE Trans. Vis. Comput. Graph.*, vol. 13, no. 2, pp. 204-212, 2007.

[8] G. C. Burdea, P. L. T. Weiss, D. Thalmann, "Guest editorial special theme on virtual rehabilitation," *IEEE Trans. Naural Syst. Rehabil. Eng.*, vol. 15, no. 1, 2007.

[9] G. De Pasquale, S.G. Kim, D. De Pasquale, "GoldFinger: wireless human–machine interface with dedicated software and biomechanical energy harvesting system", *IEEE/ASME T. Mech.*, 2015.

Scalable fabrication of triboelectric nanogenerators for commercial applications

Lokesh Dhakar[1,2], Xuechuan Shan[3], Zhiping Wang[3], Bin Yang[4], Francis Eng Hock Tay[2,5], Chun-Huat Heng[1] and Chengkuo Lee[1,6]*

[1] Department of Electrical and Computer Engineering, National University of Singapore, 4 Engineering Drive 3, Singapore 1175762
[2] NUS Graduate School for Integrative Sciences and Engineering, Centre for Life Sciences (CeLS), 28 Medical Drive, Singapore 117456
[3] Singapore Institute of Manufacturing Technology (SIMTech), 71 Nanyang Drive, Singapore 638075
[4] Department of Micro/NanoElectronics, Shanghai Jiao Tong University, Shanghai 200240, China
[5] Department of Mechanical Engineering, National University of Singapore, 9 Engineering Drive 1, Singapore 117575
[6] NUS (Suzhou) Research Institute, Suzhou, Jiangsu, China 215123

E-mail: elelc@nus.edu.sg

Abstract. Harvesting mechanical energy from irregular sources is a potential way to charge batteries for devices and sensor nodes. Triboelectric effect has been extensively utilized in energy harvesting devices as a method to convert mechanical energy into electrical energy. As triboelectric nanogenerators have immense potential to be commercialized, it is important to develop scalable fabrication methods to manufacture these devices. This paper presents scalable fabrication steps to realize large scale triboelectric nanogenerators. Roll-to-roll UV embossing and lamination techniques are used to fabricate different components of large scale triboelectric nanogenerators. The device generated a peak-to-peak voltage and current of 486 V and 21.2 μA, respectively at a frequency of 5 Hz.

1. Introduction

As the non-renewable sources of energy are depleting at a fast pace, generating green and sustainable energy is a huge challenge. Harvesting mechanical energy from various sources in environment is a potential solution to generate electrical power at a micro and macro scale. Triboelectric mechanism has become a popular approach for mechanical energy harvesting owing to its high power output, easy fabrication and wide choice of materials [1, 2] .

In any triboelectric nanogenerator (TENG) device, at minimum one out of two triboelectric layers comprises of a polymer film as they can hold charges for long duration due to their insulating properties. This is important as these embedded charges on polymer films are utilized to convert mechanical energy into electrical energy by changing the gap between two triboelectric layers. To improve the performance of triboelectric mechanism using polymer films, it is also important to pattern these polymer films to improve the effective contact area and triboelectric charge separation [3,

4]. The current fabrication processes used to fabricate triboelectric nanogenerators are not scalable as they are limited by the chamber or substrate size. This poses a bottleneck towards the large scale fabrication and commercialization of these devices. In this paper, we propose a scalable fabrication process for large size TENG devices. The polymer films are fabricated by patterning using roll-to-roll ultraviolet (UV) embossing process. This process is used to fabricate very large size samples of patterned polyethylene terephthalate (PET) films. The device electrode was fabricated by laminating large size copper film on a liquid crystal polymer (LCP) film substrate. These components are then assembled together to realize large size device for mechanical energy harvesting using triboelectric mechanism.

2. Device design and fabrication

2.1. Design

The device uses triboelectric effect to generate charges by contact and separation between two triboelectric layers. The first triboelectric layer comprises of patterned PET film whereas the second triboelectric layer is fabricated by laminating large size copper sheet to a LCP substrate. The two materials are chosen as triboelectric pair as patterned PET film has a higher tendency of accepting electrons as compared to copper when the two triboelectric layers come in contact due to external mechanical force. Spacers at regular intervals between the two triboelectric layers to maintain the gap to avoid sagging of the large size films together hindering the contact separation process.

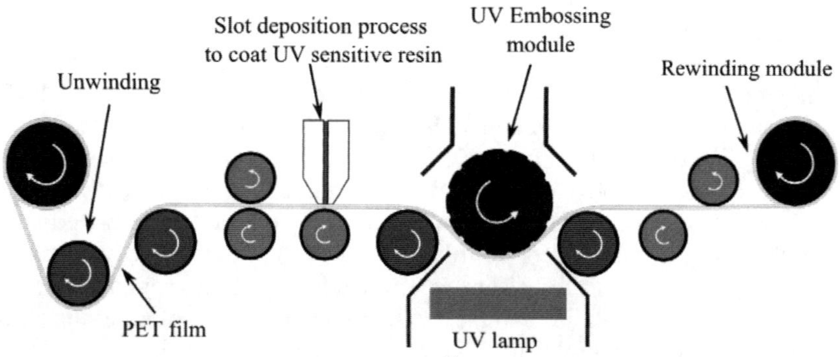

Figure 1. Schematic diagram of the roll-to-roll UV embossing setup for large size PET film

2.2. Fabrication

The large size patterned PET film used as first triboelectric layer was fabricated using roll-to-roll UV embossing [5]. The schematic of fabrication setup for roll-to-roll UV embossing is shown in Figure 1. The setup can be divided into four parts. First is the unwinding module which supplies the substrate PET film. The second part is the coating module which deposits UV curable resin on PET film using slot deposition process. Following the coating module is third part, which is UV embossing module for patterning microstructures on PET film. This module uses a patterned roller to transfer the required patterns and uses UV lamp placed beneath the embossing roller to cure the transferred patterns on the PET film. The fourth part is rewinding module that provides web tension for separating the embossed PET film from the embossing roller and then rewinds for collecting the embossed PET film. An optical image of a patterned PET film with line patterns (line pitch: 500 μm) using roll-to-roll UV embossing is shown in Figure 2. Large patterned films were cut into size of 40 cm x 40 cm size sheets. These patterned PET sheets were used to increase the contact area between the triboelectric layers during contact electrification.

For the second triboelectric layer, an 18 μm thick copper film was attached on top of 50 μm thick LCP substrate using lamination. To realize the complete device, 1 mm thick spacers were assembled on the patterned PET film in order to create a gap the two triboelectric layers as shown in Figure 2.

The copper film was then assembled on top of spacers. As the PET film has a large size, the film tends to sag down permanently onto the copper electrode which hinders the contact separation mechanism required for charge generation for mechanical energy harvesting. Therefore it is important to assemble spacers at regular spacing between copper and patterned PET triboelectric layer to help the large patterned polymer film rebound back to its original position after it is subjected to mechanical force. A schematic of the complete device is shown in Figure 2.

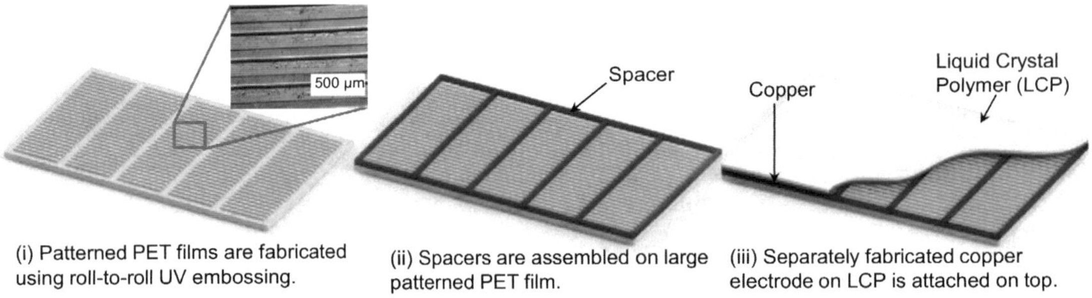

Figure 2. Assembly of the large scale triboelectric nanogenerators

2.3. Working mechanism

The schematic for operating mechanism of LS-TENG is shown in Figure 3. Both the triboelectric layers are initially uncharged before coming into contact with each other. As an external mechanical force is applied on the device due to different activities like walking, palm tapping etc., both the triboelectric layers come in contact with each other. This results into transfer of electrons from the copper film to patterned PET film due to copper's higher tendency to donate electrons relative to PET. The charges on the patterned PET film are preserved due to its insulating properties. And the charges on conductive copper film can flow through the load resistor connected to the ground as electric potential changes at the copper electrode. As the applied force is released, the patterned PET film with the negative charges separates from the copper film. This results into an increase in the potential at the copper electrode. This leads to flow of electrons from the reference ground electrode towards the copper electrode. As the patterned PET film reaches the maximum point of separation between the two triboelectric layers, electrostatic equilibrium is reached and there is no flow of electrons in the load resistor connected between electrode and ground. Thereafter, the patterned PET film again approaches towards the copper film due to periodic mechanical force. This leads to the flow of electrons in the load resistor in opposite direction. The flow of electrons in the load resistor during the approaching and separating motion of triboelectric layers can be used to power an electrical load.

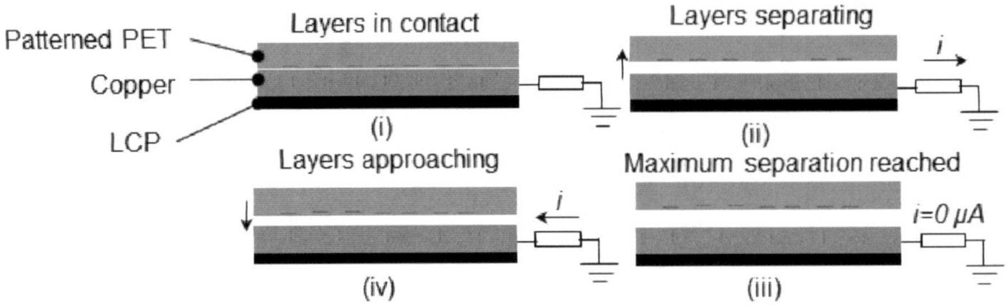

Figure 3. Operating principle of the TENG

3. Results and discussion

The fabricated device was tested under different conditions to observe the device characteristics. The device was also demonstrated for practical applications for different situations. The voltage characteristics were measured using a 100 MΩ probe with DSOX3034A oscilloscope. The current characteristics were measured using a low noise SR570 current pre-amplifier. The testing results are discussed in the following sections.

3.1. Device characteristics at different operating frequencies

The large size device was tested using hand tapping to measure the generated voltage and current. The voltage and current output were tested at different tapping frequencies from 1 Hz to 5 Hz. Both voltage and current output were observed to increase as the frequency of applied force was increased as shown in Figure 4. The peak-to-peak voltage increased from 232 V to 486 V as the frequency increased from 1 Hz to 5 Hz. The peak-to-peak current increased from 7.2 μA to 21.2 μA as the frequency increased from 1 Hz to 5 Hz. The increase in voltage output with increased frequency is due to increased value of applied force as the tapping frequency increased. The higher applied force led to increased contact area between the triboelectric layers resulting in higher voltage output. The increase in the current output can be explained by the combined effect of the increased force and the reduction of the cycle time for the charges to flow in the external circuit as the tapping frequency increased.

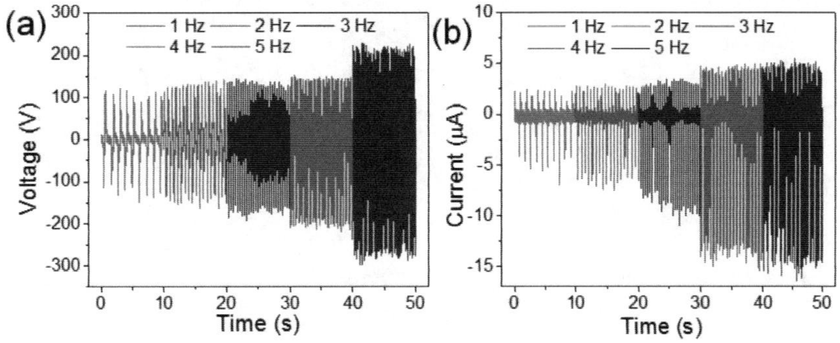

Figure 4. (a) Voltage and (b) current output of TENG at different frequency of the applied force

3.2. Applications

The device was tested for different practical scenarios to demonstrate its applicability. The device was tested using a normal sized football bouncing on the large sized device. The device generated a peak-to-peak voltage of 78 V due to the ball bouncing on the large sized device as shown in Figure 5a. The device was also assembled on the back of a chair to harvest energy from reclining human motion as shown in Figure 5b. As the reclining angle increased from 5° to 20°, the peak-to-peak voltage

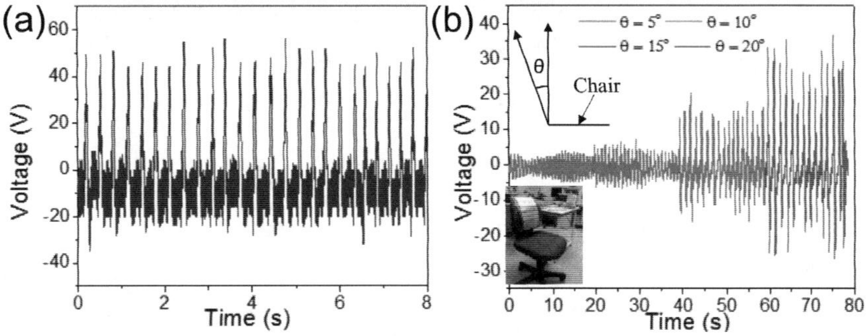

Figure 5. Voltage output of TENG (a) using a bouncing football (b) assembled on chair with different inclination angles

increased continuously as shown in Figure 5b. The increase in reclining angle resulted into higher impact force which in turn led to higher output voltage. The assembled device at the back of chair was also tested to harvest mechanical energy from irregular motion of the human subject while working on a desktop computer. A commercial 3-axis accelerometer ADXL325 was used to measure the activity level of the human subject as shown in Figure 6. The TENG device generated peak-to-peak voltage and current of up to 85 V and 0.75 μA, respectively during high activity as compared to 35 V and 0.3 μA during low activity.

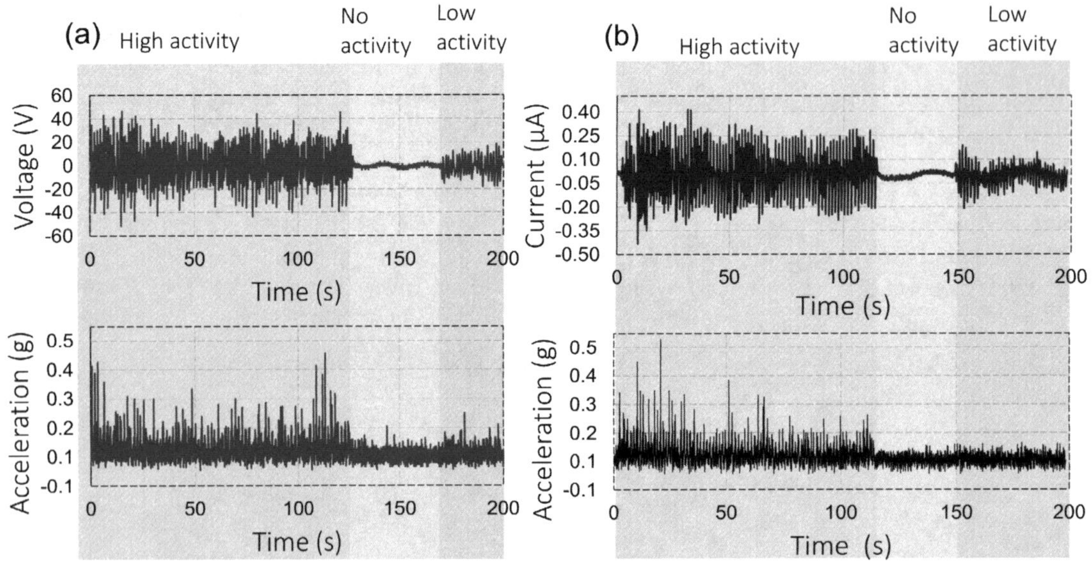

Figure 6. (a) Voltage and (b) current generated by irregular human motion

4. Conclusion

We have developed and demonstrated a process flow for large scale fabrication of TENGs with high throughput. Roll-to-roll UV embossing has been utilized to fabricate large size patterned polymer films. Using scalable fabrication processes, TENGs can be manufactured for commercial applications at a large scale. The device was tested different scenarios to demonstrate the practical application of the large scale device. The device generated a device generated a peak-to-peak voltage and current of 486 V and 21.2 μA, respectively.

Acknowledgements

This work is supported NRF2011 NRF-CRP001-057 Program 'Self-powered body sensor for disease management and prevention-orientated healthcare' (R-263-000-A27-281) from the National Research Foundation (NRF), Singapore.

References

[1] F.-R. Fan, Z.-Q. Tian, and Z. Lin Wang 2012 *Nano Energy* **1** 328-334.

[2] B. Meng, W. Tang, Z.-h. Too, X. Zhang, M. Han, W. Li and H. Zhang 2013 *Energy & Environmental Science* **6** 3235-3240.

[3] L. Dhakar, F. Tay, and C. Lee 2014 *Journal of Micromechanics and Microengineering* **24** 104002.

[4] F.-R. Fan, L. Lin, G. Zhu, W. Wu, R. Zhang, and Z. L. Wang 2012 *Nano letter* **12** 3109-3114.

[5] Z. Zhong and X. Shan *Journal of Micromechanics and Microengineering* **22** 085010.

Alkali Metal Based Micro Combustion Using Graphene Micro-valve Trigger

A Ruyack[1], V Gund[1], K Camera[2], S Ardanuç[1], C Ober[2] and A Lal[1]

[1]SonicMEMS Laboratory, School of Electrical & Computer Engineering, Cornell University, Ithaca, NY, USA
[2]Ober Group, Department of Material Science & Engineering, Cornell University, Ithaca, NY, USA

E-mail: arr68@cornell.edu

Abstract. Controllable combustion of alkali metals offers a pathway to high energy density, high power density scalable thermal systems. The volatile nature of these materials however, has limited their incorporation into useable devices. Recent findings in low trigger power MEMS one-shot valves offer a viable pathway forward for creating systems in which to exploit the unique properties of alkali metals as thermal energy sources. By miniaturizing and combining these fuels with such a valve, and incorporating them with a polymer substrate, a micro-packet of combustible metal can be created. Here we report such a device utilizing a graphene based micro-valve trigger and a rubidium metal fuel source to realize an arrayable micro-scale electrical to thermal energy amplification scheme. Using this approach we can react nanoliter droplets of rubidium with ambient air on demand, producing electrical to thermal energy gains of seventy five times. By arraying the fuel into discrete packets, we can also achieve temporal control of the fuel source, allowing for tailored peak and average power output.

1. Introduction

Alkali metals are an energy dense fuel source that inherently have high power density due to their low ionization energy and thus high reactivity. This makes these elements particularly interesting for energy applications. Their reactivity, however, makes the processing, storage and actuation of these fuels in a controlled manner difficult. As a result, the properties of these materials are not typically exploited.

Recent advances in MEMS one-shot valves have offered a pathway forward for integrated alkali metal systems by providing low input-power actuation sources with a leak-free seal [1]. These valves provide oxidation free storage for containment and are a perfect trigger mechanism for actuating fuels utilizing ambient oxygen for combustion. Our approach employs such a device as a key portion of a novel, arrayable and scalable integrated thermal power generation system employing alkali metal fuel, micro-valve and a polymer layer acting as a seal and characterization mechanism.

2. Device Concept

Due to the reactive nature of the proposed fuels, temporal control of the fuel reaction is imperative for exploitation of the unique properties of these materials. In order to solve this issue, we divide the fuel into tiny packets that are placed into the micro-valve and can be triggered on demand to tailor peak and average power output. The basic cell of our device is comprised of three parts: a rubidium droplet (the fuel) contained within a graphene micro-valve (the trigger/barrier) sealed with a polymer film on the backside (structural layer/seal) bonded by heating. Figure 1 shows the assembly and overlying structure of such a device. The operation of the device begins with a current pulse, which fractures the graphene valve and allows ambient air to react with the rubidium. The heat generated by this reaction vaporizes the underlying polymer, which provides a mechanism for quantifying the amount of heat generated. Figure 2 gives a theoretical analysis of the energy delivered or required for each system component as a function of cell dimension.

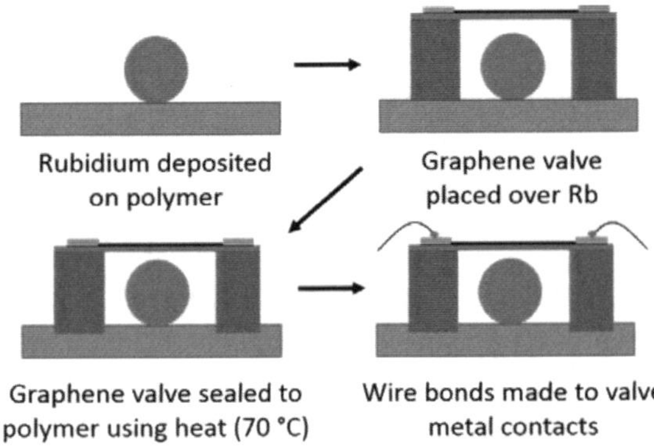

Figure 1. Fabrication schematic of micro-valve alkali metal cell for energy conversion.

3. Alkali Fuel Source

Alkali metal fuel sources offer the potential for extremely high specific power densities under oxidation, while maintaining excellent energy densities. For this device, we focused on rubidium as our fuel as volumetric energy density and reactivity of the alkali metals generally increases with increasing period. Furthermore, rubidium is pyrophoric and spontaneously ignites in air, a behavior which lower period alkalis do not possess, making an ignition source unnecessary.

When Rb fuel is allowed to spontaneously ignite in air, it has a specific power density of about 2 MW/kg, dwarfing conventional supercapacitors (around 15 kW/kg) by about two orders of magnitude [2]. Its energy density of about 550 W-hr/kg beats current lithium ion technology, winning by 2-5 times [3]. However, the fact that these fuels rely on spontaneous chemical reactions means that power delivery lasts only about one second, making long-term applications difficult. Figure 3 shows the short timescale energy release of a 100 μL Rb droplet.

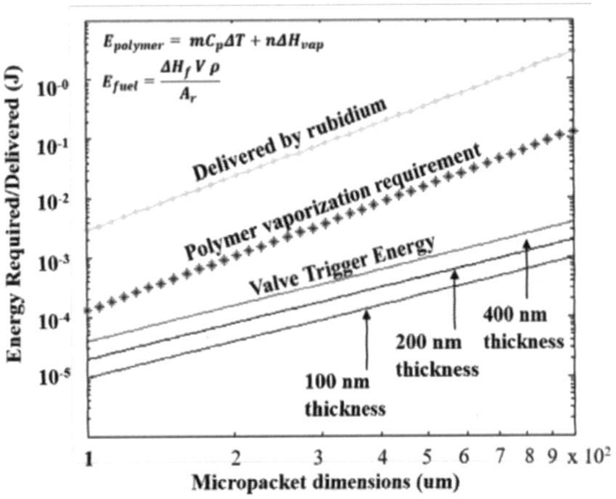

Figure 2. Energy requirement or delivery for each part of the system. Polymer thickness taken to be 10% of micro-packet dimension. Delivered energy far outweighs the energy input requirement. For a detailed analysis of the valve energy see [1]. Polymer energy is calculated from a molar heat of vaporization of 110 kJ/mol and heat capacity of 1.2 J/g-K. Rubidium energy is calculated from the enthalpy of formation of Rb_2O (-169.5 kJ/mol, worst case scenario).

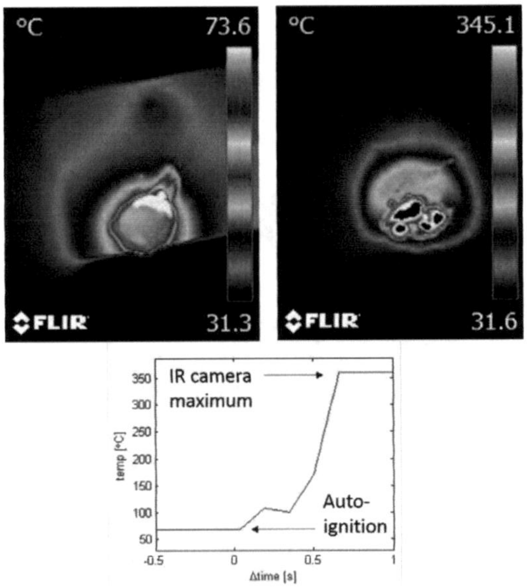

Figure 3. Top: IR images separated by 0.3s showing before (left) and after (right) auto-ignition of Rb. Bottom: Corresponding IR data for a point on the Rb surface. We see ignition timescales on the order of 0.5-1.0 s for droplets 100 nL – 100 uL.

4. Graphene Micro-valve Trigger

Rubidium needs to be packaged and remain in inert atmosphere until triggered and exposed to ambient air. Our triggering approach follows previously reported work on graphene-based one-shot micro-valves [1] that use significantly low input electrical energy for thermo-mechanical cleaving. We have shown 100% reliability for triggered actuation of >100 valves. Valves were fabricated in the same method as previously reported, producing valves 750 x 750 μm having approximately 360 nm thick silicon nitride membranes with Ni electrodes and patterned graphene. These valves have been shown to be leak-free and the fabrication processes for these devices allows for extensive variation, enabling scaling of valve size and density for tailoring of application. Figure 4 shows a valve before and after valve triggering.

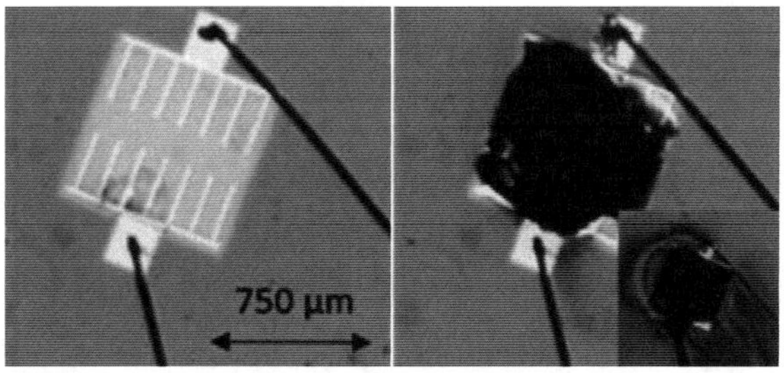

Figure 4. Left: Top view of graphene micro-valve with integrated Rb and sealed with polymer. The Rb is under the low set of electrode fingers (off-center). Right: Graphene micro-valve with Rb after triggering. Inset: 4 minutes after triggering showing discoloration.

5. Polymer Based Substrate and Characterization Platform

On the backside of the valve we utilize a blended polycarbonate polymer to realize both mechanical rigidity for use as a backing layer as well as low vaporization temperature for power generation characterization. The polymer also acts as the backside seal for the device, allowing for the fuel to be protected from ambient air until triggered via graphene valve.

The blended polycarbonate used in this study was a 30/70 blend of a vaporizable polycarbonate (VPC) synthesised based on [4] and commercially available polypropylene carbonate (PPC). This VPC offers low vaporization temperature but is limited to thin spin coated films [5]. Blending the VPC with commercial PPC allows for thick films approximately 125 μm thick to be drop cast which can act as packaging to our system while still having excellent decomposition properties.

Furthermore, the vaporizable nature of our polymer allows for characterization of our thermal energy release system. When actuated, the heat produced by the fuel will vaporize the polymer substrate. Through optical microscopy and profilometry we can determine the amount of removed polymer. Combining this with the decomposition properties of the polymer obtained using commercial thermogravimetric analysis, we have a means of quantifying energy release on the small scale of our devices. Our blended polymer typically show a decomposition temperature of ~ 180 °C.

6. Device Assembly

All assembly takes place in an inert atmosphere devoid of oxygen (argon glove box with O_2 concentration < 1 ppm). Rubidium (99.75% metals basis) is stored under mineral oil until deposited. Deposition was accomplished by heating the Rb in oil to above the melting temperature and transferred via a resistively heated zero-dead volume syringe. Volumes of Rb droplets ranging from tens of nL to hundreds of μL can be prepared in this manner.

Rb droplets were transferred to polymer substrates in this manner. The graphene valve is then placed over the droplet and the whole system moved to a hot plate heated to below, but near the polymer glass transition temperature (typically 70 °C). This softens the polymer enough to adhere the MEMS valve to the surface in order to seal the system. A silicon spacer is used to prevent the polymer from sticking to the hot plate. Once sealed, the devices are placed in sealed containers and transferred from the glovebox for testing. The devices remain under argon until wire bonding and subsequent actuation.

7. Device Testing

Tested devices were 750x750 μm and consume 140 mW for about 14 ms before triggering. These can fit 500 nL of Rb, which can release up to 1.53 J upon oxidation. This implies a theoretical maximum

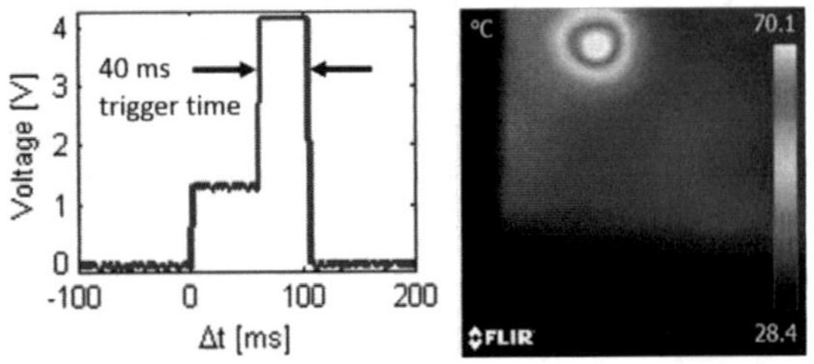

Figure 5. Left: Triggering voltage waveform showing short timescale valve actuation. Right: IR image from the backside of the polymer showing heat generation immediately after valve triggering.

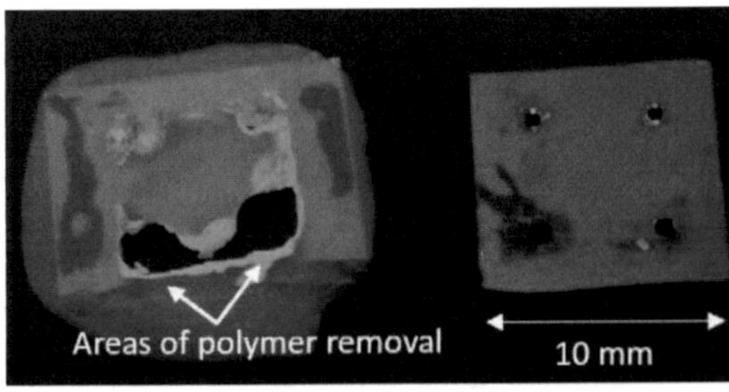

Figure 6. Left: Tested device showing polymer removal due to rubidium heat generation. Right: Corresponding 2 x 2 valve array.

energy gain of nearly 800 times. Figure 5 shows the waveform used to trigger the valve and corresponding temperature increase as seen from the backside of the polymer via IR camera. Tested devices, shown in Figure 6, demonstrate polymer etching. These devices were heated to 65 °C before triggering in order to melt the Rb and aid in spontaneous ignition. This higher than ambient initial temperature was accounted for in calculation. Calculations show up to 9.6% of the theoretical Rb energy is transferred to the polymer, corresponding to 75x electrical to thermal energy gain. This demonstrates successful sealing, actuation and heat generation of our proposed device. Loss mechanisms to the air and valve structure are significant, and future experimentation will focus on removing these loss mechanisms by creating a fully polymer system.

8. Conclusions

While successful storage and actuation of rubidium fuel has been demonstrated, we foresee various improvements and advancements with the current system. Moving from rubidium to caesium is expected to increase power and energy density, while varying valve size and density will allow for on-demand control of energy and power output. Integration of thermoelectrics poses an interesting application for a low electrical energy input to high energy output system and finally, a fully polymer system could pave the way for total device transience and totally vanishing electronics.

9. Acknowledgements

This work was performed in part at the Cornell NanoScale Facility, a member of the National Nanotechnology Infrastructure Network, which is supported by the National Science Foundation (Grant ECCS-0335765). This work also made use of the Cornell Center for Materials Research Shared Facilities which are supported through the NSF MRSEC program (DMR-1120296). Funding for this project was provided by the DARPA VAPR program.

References

[1] V. Gund, et al., "Graphene one-shot micro-valve: Towards vaporizable electronics," *Proc. of 28th IEEE International Conference on Micro Electro Mechanical Systems*, 2015.

[2] Guoping Wang, et al., "A Review of Electrode Materials for Electrochemical Supercapacitors," *Chem. Soc. Rev.*, 2012, 41, pp 797–828

[3] Haisheng Chen, et al., "Progress in electrical energy storage system: A critical review," *Prog. Nat. Sci.*, 19 (3), 2009, pp 291-312.

[4] F. M. Houlihan, et al., "Thermally Depolymerizable Polycarbonates. 2. Synthesis of Novel Linear Tertiary Copolycarbonates by Phase-Transfer Catalysis", *Macromolecules*, 19 (1), pp 13-19, 1986.

[5] V. Gund, et al., "Multi-modal Graphene Poplymer Interface Characterization Platform For Vaporizable Electronics," *Proc. of 28th IEEE International Conference on Micro Electro Mechanical Systems*, 2015.

Prototype of radioisotope thermophotovoltaic system using photonic crystal spectral control

X Wang[1, 3], W R Chan[3], V Stelmakh[1, 3], M Soljacic[2], J D Joannopoulos[2, 3], I Celanovic[3] and P H Fisher[2, 3]

[1]Electrical Engineering Department, Massachusetts Institute of Technology, Cambridge, MA, USA
[2]Physics Department, Massachusetts Institute of Technology, Cambridge, MA, USA
[3]Institute of Soldier Nanotechnologies, Massachusetts Institute of Technology, Cambridge, MA , USA

Email: xiawaw@mit.edu

Abstract. This work describes a prototype of a small-size radioisotope thermophotovoltaic (RTPV) system with the two-dimensional metallic photonic crystal emitter and the low bandgap TPV cell. The project demonstrates the simulation and measurement of a system powered by an electrical heat source that mimics the radioisotope fuel pellet. The photonic crystal and the polished Ta3%W substrate are both used as the emitting surfaces to demonstrate the benefits of spectral control. The rest of the system is thermally insulated to increase the overall system efficiency. The photonic crystal emitter demonstrates four times more output power over a flat metal emitter from the 1 cm^2 TPV cell. With more cell areas, better TPV cells and improved insulation design, the system is expected to reach an efficiency of 7.8%.

1. Introduction

Nuclear power systems are the most promising way to provide long term electrical power for long duration when providing additional energy input, fuel or maintenance is impossible. The most common way to use the nuclear energy is to use the heat released during the radioactive decay. Plutonia, which is an alpha emitter with high decay heat, is fabricated into fuel pellets and packaged into the rectangular General Purpose Heat Source (GPHS) containing four pellets [1]. Thermal-based radioisotope powered systems have been developed and utilized since the 1960s and have demonstrated numerous safe and reliable usages in space and land missions such as Apollo, Cassini-Huygens, etc. [2].

Over the past 50 years, various kinds of radioisotope powered thermal-to-electric conversion mechanisms have been investigated: the thermal-electric generators (RTG) [3], Stirling engines [4], alkali-metal thermal-to-electric converter (AMTEC) [5] and radioisotope thermophotovoltaic systems (RTPV) [6] [7]. RTGs are most widely used and have provided reliable sources for spacecrafts and remote sensors for decades, but the efficiency is limited by the thermoelectric material to ~6.6%. Stirling engines can achieve an efficiency over 30%, but the moving piston creates reliability concerns. AMTEC systems are chemical heat engines with a sodium or potassium working fluid that have demonstrated an efficiency of 20%, but the specific power of the systems is limited to ~6 W/kg. In an RTPV system, radioisotope decay heats an emitter to incandescence, and the resulting thermal radiation is converted to electricity by a low bandgap TPV cell as shown in figure 1. The energy conversion elements are solid-

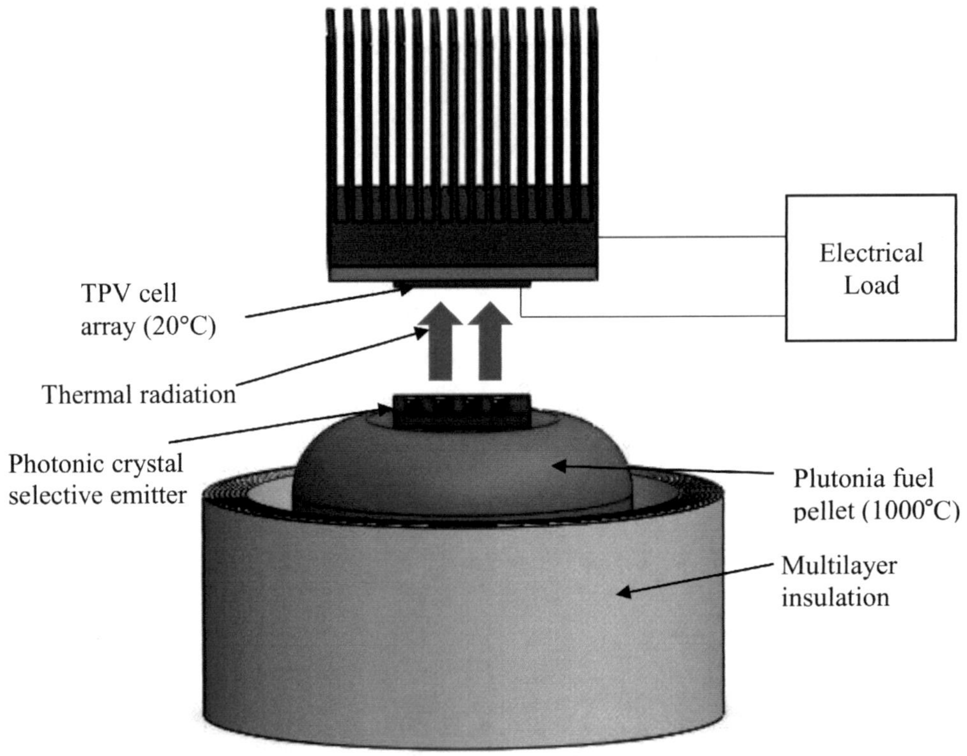

Figure 1. Radioisotope thermophotovoltaic system using photonic crystal spectral control.

state, light-weight components that make RTPV the most promising technology to achieve high efficiency, high energy density, and high reliability.

In an RTPV system, spectral control is needed to modify the thermal emission spectrum to match the TPV cell. Ideally, the emitter has high emission for the photons above the cell bandgap to generate more electricity and low emission for the low energy photons in far infrared to reduce waste heat. With recent advances in photonic crystals, a micro-fabricated metallic photonic crystal patterned with periodic holes can shape the spectrum of the hot emitter to approach the ideal emitter [8] [9] [10]. Spectral control using photonic crystals has demonstrated many successful results in other TPV systems, such as the propane powered micro-reactor [11], solar TPV system [12], etc.

2. System construction and characterization

In the current prototype, the heat source was an electrical button heater enclosed in a cylindrical Inconel case to mimic the plutonia fuel pellet. The heater was supported through a hollow Inconel tube, which was back supported by a stainless steel plate fixed on the vacuum flange. The selective emitter was brazed on top of an extruded stand.

The emitter was a 1×1 cm^2 Ta3%W photonic crystal comprised of a square array of cylindrical cavities fabricated by semiconductor process. The geometry (cavity radius, depth and lattice period) of the photonic crystal determines the resonant modes which correspond to the cutoff wavelength of the emission spectrum. For photons above the bandgap, the deep holes on the photonic crystal make it a good emitter. For photons below the bandgap, the emitter appears to be a polished reflective metal. The photonic crystal gives the emission approaching a blackbody emitter in the useful region, but does not

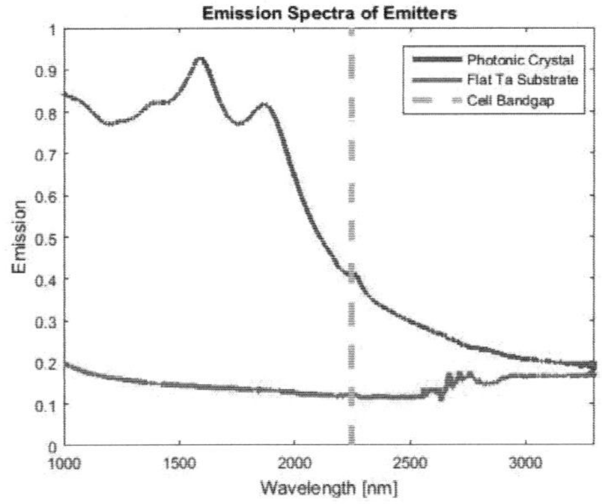

Figure 2. Measured spectra of photonic crystal and flat metal emitter at room temperature.

emit much waste heat produced by long wavelength photons. The measured spectra of the photonic crystal and the flat metal substrate at room temperature is shown in figure 2.

Thermal radiation from the emitter illuminates the TPV cells. We used an array of two InGaAsSb cells with a bandgap of 0.55 eV (2250 nm) with a total area of 1 cm^2 [13]. The distance between the emitter and the cell was adjusted to be ~1.2 mm,

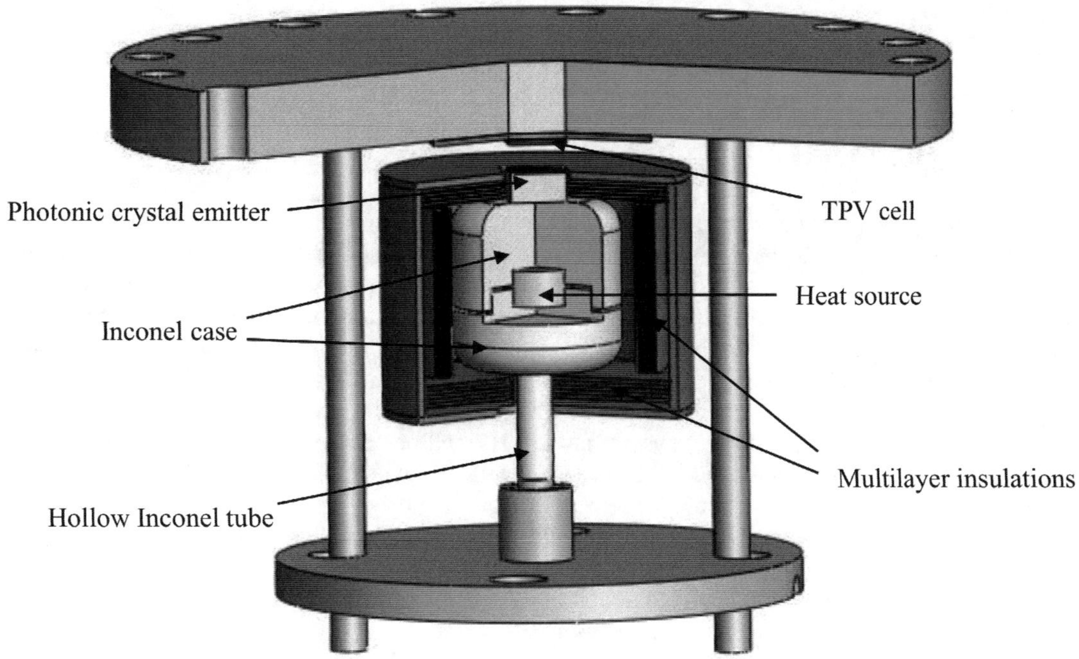

Figure 3. Illustration of RTPV system prototype.

The system configuration is shown in figure 3. In order to improve the system efficiency, all surfaces of the heater except the emitter were insulated by multilayer insulation layers (MLI) made with copper foil and zirconia powder. MLI layers were spirally wound around the heater side. Patterned MLI layers with cutouts for the emitter and support tube were placed on the front and back surfaces of the heater.

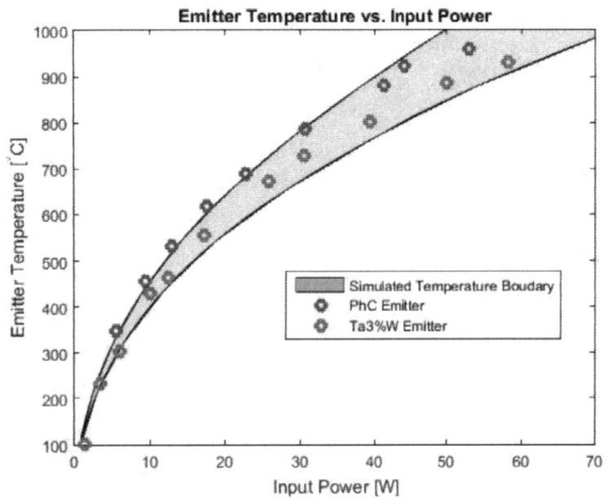

Figure 4(a). Measured emitter temperature as function of input power.

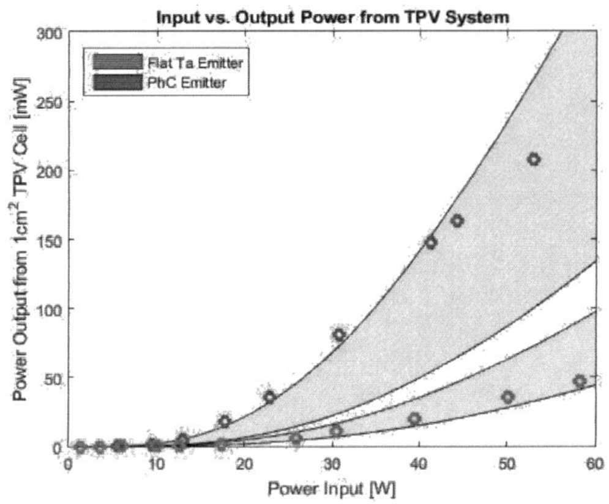

Figure 4(b). Input power vs. output power from 1 cm^2 TPV cell.

The heater and MLIs were placed inside a polished copper case only leaving the emitter area to be exposed.

The system was fired inside a vacuum chamber with a pressure in the range of 10^{-5} Torr. The TPV cell was maintained at 20°C by a chiller outside of the chamber. The generated electrical power and the required heat input were measured up to 950°C for both the photonic crystal and flat metal emitters. The reflection spectra of the emitters were measured before and after the firing to verify that there was no degradation.

The temperature of the heater as a function of the input power with both emitters did not show much difference as shown in figure 4(a). The only difference in the two system was the 1 cm^2 emitter area that was very small comparing to the whole heater. The system with the photonic crystal emitter showed four times more output power in comparison to the flat metal emitter as shown in figure 4(b) because of the increased in-band emissivity compared to unstructured metal.

3. System modelling and improvements

The overall simulation was done by COMSOL and Matlab Livelink. The heat transfer through the complex geometry was simulated using COMSOL Heat Transfer Module. The heat flux through the multilayer insulation layers (MLI), the photocurrent and the maximum power output from the TPV cell were computed using Matlab nonlinear solver and optimization functions. Since many material parameters are temperature dependent, the simulation was performed based on the best and worst case material parameters.

One major limitation of the current system is the TPV cell performance. The current TPV cell only has 0.6 internal quantum efficiency above the cell bandgap, where a quantum efficiency of 0.9 can normally be achieved [13]. In addition, the thermal insulation of the system can also be further improved. Copper foils are used in the current setup to shield the radiation loss. The system performance can be much more improved by utilizing gold foils which are more reflective in infrared regime.

A realistic RTPV system with 8 cm^2 TPV cell insulated with gold foil was estimated in figure 5. The system based on one fuel pellet can potentially output 5.5W power with 70W of thermal input power, reaching an efficiency of 7.8%. Due to the cylindrical shape of the fuel pellet, the curved side surfaces of the heater is not utilized efficiently. In the RTPV system powered by the General Purpose Heat Source (GPHS), all surfaces can potentially be used efficiently as the emitters to reach a higher efficiency. The weight for the RTPV system will be estimated in future work.

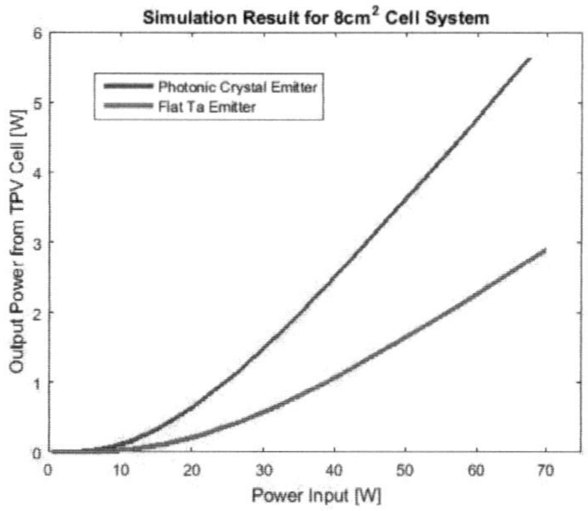

Figure 5. Simulation result for an 8 cm^2 cell system insulated with gold foil.

4. Conclusions

The project demonstrated the simulation and measurement results of an RTPV prototype system using photonic crystal spectral control to realize high efficiency. Comparing to other types of thermal-based radioisotope power sources, RTPV system has the advantages of reaching higher efficiency than the thermal-electric generators. The absence of moving parts or working fluids also makes it an ideal power source in unattended regions for several decades of battery lifetime.

References

[1] Franco-Ferreira E and Goodwin G 1997 *Platin.Met.Rev.* **41(4)** 154-63
[2] Angelo J and Buden D 1985 *Space Nulcear Power* (Florida: Orbit Book Company)
[3] Rinehart G 2001 *Prog. Nucl. Energy* **39** 305-19
[4] Chan J, Wood J and Schreiber J 2007 *AIP. Conf. Proc.* **880** 615
[5] Shock A, Mukunda M, Or C, Kumar V and Summers G 1995 *Acta Astron.* **37** 21-57
[6] Koudelka R D, Murray C S, Fleming J G, Shaw M J, Teofilo V and Alexander C 2006 *AIP Conf. Proc.* **813** 545
[7] Wolford D, Chubb D Clark E, Pal A, Scheiman D and Colon J 2009 *Emittance Measruements Relevent to a 250Wt Class RTPV Generator for Space Exploration* NASA/TM-2009-215619
[8] Rinnerbauer V, Ndao Sidy, Yeng Y, Senkevich J, Jensen K and Joannopoulos J 2013 *J. Vac. Sci. Tecchnol. B.* **31(1)** 011802
[9] Stelmakh V, Rinnerbauer V, Geil R, Aimone P, Senkevich J, Joannopoulos J, Soljacic M and Celanovic I 2013 *Appl.Phys,Lett* **103** 123903
[10] Rinnerbauer V, Yeng Y, Chan W, Senkenvich J, Joannopoulos J, Sojacic M and Celanovic I 2013 *Opt. Express* **21(9)** 11482-91
[11] Chan W, Bermel P, Pilawa-Podgurski R, Marton C, Jesen K, Senkevich J, Joannopoulos J, Sojacic M and Celanovic I 2013 *Proc. Natl. Acad. Sci. U.S.A* **110** 5309-13
[12] Lenert A, Bierman D, Nam Y, Chan W, Celanovic I, Soljacic M and Wang E 2014 *Nat. Nanotechnol.* **9** 126-30
[13] Chan W, Huang R, Wang C, Kassakian J, Joannopoulos J and Celanovic I 2010 *Sol. Energ.Mat.Sol.* **94** 509-14

Nanoengineered Surfaces for Thermal Energy Conversion

Bikram Bhatia, Daniel J. Preston, David M. Bierman, Nenad Miljkovic, Andrej Lenert, Ryan Enright, Youngsuk Nam, Ken Lopez, Nicholas Dou, Jean Sack, Walker R. Chan, Ivan Celanović, Marin Soljačić and Evelyn N. Wang

Massachusetts Institute of Technology, Cambridge, MA, USA

E-mail: enwang@mit.edu

Abstract. We provide an overview of the impact of using nanostructured surfaces to improve the performance of solar thermophotovoltaic (STPV) energy conversion and condensation systems. We demonstrated STPV system efficiencies of up to 3.2%, compared to ≤1% reported in the literature, made possible by nanophotonic engineering of the absorber and emitter. For condensation systems, we showed enhanced performance by using scalable superhydrophobic nanostructures *via* jumping-droplet condensation. Furthermore, we observed that these jumping droplets carry a residual charge which causes the droplets to repel each other mid-flight. Based on this finding of droplet residual charge, we demonstrated electric-field-enhanced condensation and jumping-droplet electrostatic energy harvesting.

1. Introduction

All thermal energy conversion processes are governed by the three modes of heat transfer: conduction, convection and radiation. Nanoengineered surfaces can influence each of these three modes resulting in efficiency enhancements in a variety of energy conversion systems. This work focuses on the application of nanoengineered surfaces in two areas: solar thermophotovoltaic (STPV) energy conversion energy conversion and condensation heat transfer.

Existing approaches used to generate power from sunlight use either photovoltaics, in which sunlight directly excites electron–hole pairs in a semiconductor, or solar–thermal, in which sunlight drives a mechanical heat engine. Photovoltaic power generation is intermittent and typically only exploits a portion of the solar spectrum efficiently, whereas the intrinsic irreversibilities of small heat engines make the solar–thermal approach best suited for utility-scale power plants. There is, therefore, an increasing need for hybrid technologies, such as STPV, that can combine photovoltaics and solar-thermal to harness the entire solar-spectrum.

Unlike STPV which is a nascent solar energy conversion approach, condensation is a ubiquitous process often observed in nature and harnessed in many industrial processes such as power generation, desalination, thermal management, and building environmental control. Recent advances in surface engineering have offered new opportunities to enhance condensation heat transfer by drastically changing the wetting properties of the condenser surface. Specifically, the development of superhydrophobic surfaces has been pursued to enhance condensation heat transfer, where the low droplet surface adhesion and small droplet departure sizes increase the condensation heat transfer coefficient. Additionally, more detailed insights on droplet interactions on these surfaces can lead to new opportunities for a wide variety of possible applications including self-cleaning, anti-icing, and energy harvesting [1].

2. Solar thermophotovoltaic energy conversion

A solar thermophotovoltaic device converts sunlight into thermal emission tuned to energies directly above the photovoltaic bandgap using a hot absorber–emitter. By doing so, solar thermophotovoltaics promise to achieve high efficiency, by harnessing the entire solar spectrum; scalability and compactness, because of their solid-state nature; and dispatchablility, owing to the ability to store energy using thermal or chemical means. However, efficient collection of sunlight in the absorber and spectral control in the emitter are particularly challenging at high operating temperatures which has limited previous experimental demonstrations of this approach to conversion efficiencies around or below 1% [2]. STPV systems with structured surfaces with tailored spectral properties were theoretically predicted to achieve efficiencies exceeding 40%, however no such experimental demonstration has been reported in the literature. We developed a full solar thermophotovoltaic device consisting nanoengineered absorber and emitter surfaces that enabled us to reach experimental efficiencies of 3.2% [3].

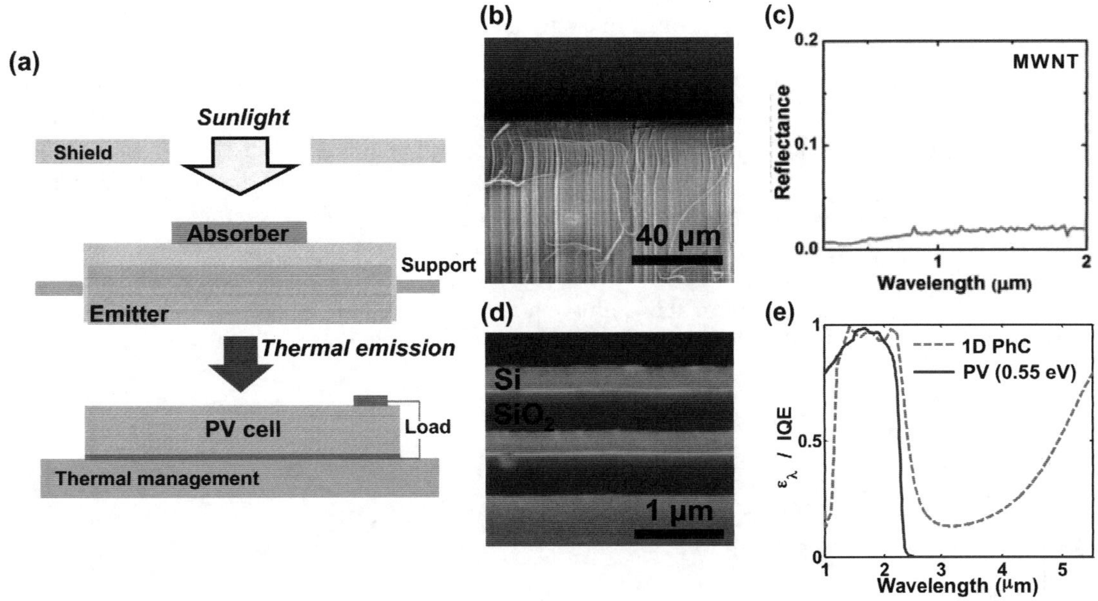

Figure 1. (a) Schematic of our vacuum-enclosed devices composed of an aperture/radiation-shield, an array of MWNTs as the absorber, a 1D PhC, a 0.55 eV-bandgap photovoltaic cell (InGaAsSb) and a chilled water cooling system. (b) SEM cross-section of the MWNTs. (c) Measured optical reflectance of the MWNT absorber. (d) SEM cross-section of the 1D PhC showing the alternating layers of silicon and SiO_2. (e) Measured emittance of the PhC emitter (red) and internal quantum efficiency of the InGaAsSb PV cell (blue) plotted as a function of wavelength. (Adapted from [3])

Figure 1a shows a schematic of our STPV device that integrates a multiwalled carbon nanotube (MWNT) absorber and a one-dimensional Si/SiO_2 photonic-crystal (PhC) emitter. Thermal resistance between the absorber and emitter is minimized by integrating the absorber and emitter on the same conductive silicon substrate such that heat is effectively delivered to the emitter via thermal spreading. In addition, we varied the absorber–emitter areas to tune the energy balance of the device. The absorber area is reduced with respect to the planar area of the sample, so we supply enough heat for the absorber–emitter to reach optimal emitter temperature by increasing the level of irradiance and leveraging the high absorptance of the nanotube array. Emitter-to-absorber area ratios (AR) >1 also reduced the area for remissive losses from the nearly blackbody nanotube array surface, thus boosting thermal efficiency. To reduce parasitic radiative losses, we metallized the sides of the silicon substrate and inactive area around the nanotube absorber with tungsten, a relatively low-emissivity, high-temperature material, and incorporated a high-reflectivity silver-coated shield (Figure 1a) to recycle this parasitic radiation back to the device.

Vertically aligned carbon nanotubes were chosen as the solar absorber because of their high-temperature stability in vacuum and their nearly ideal absorptance, crucial for absorbing highly concentrated irradiance at elevated emitter-to-absorber area ratios. As shown in Figure 1b, the as-grown nanotubes are 10–15 nm in outer diameter and 80–100 μm tall, with a ±0.5 μm variation in height at the tips. The broad-spectrum absorptance of the nanotube array in this study exceeds 0.99 (Figure 1c), consistent with previous reports for similar nanotube array geometries.

The multilayer Si/SiO_2 structure of the 1D photonic crystal, composed of five alternating layers of silicon and SiO_2 (Figures 1d,e), improves the spectral matching between the emittance of the emitter and the internal quantum efficiency of the InGaAsSb photovoltaic cell (E_g = 0.55 eV). These materials were chosen for ease of fabrication and high-temperature compatibility with the silicon substrate. The layer thicknesses were optimized via a constrained global optimization of the product of efficiency and power density.

Figure 2 shows the thermal-to-electrical conversion efficiency ($\eta_t\eta_{tpv}$) plotted as a function of input solar irradiance H_s. Overall, the highest conversion efficiency we measured was 3.2+0.2% using an AR = 7 device, which is three times greater than that obtained in previous STPV experiments. This was achieved using a compact design at substantially lower levels of optical concentration (~750 times), which enables higher optical efficiencies. Significant enhancements in efficiency relative to a greybody absorber–emitter (ε = 0.5) were achieved through the use of (1) a 1D PhC for improved spectral performance of the emitter and a vertically aligned MWNT array for nearly ideal solar absorptance (a twofold contribution to the improvement in $\eta_t\eta_{tpv}$) and (2) optimization of the active emitter-to-absorber area ratio (an additional twofold improvement).

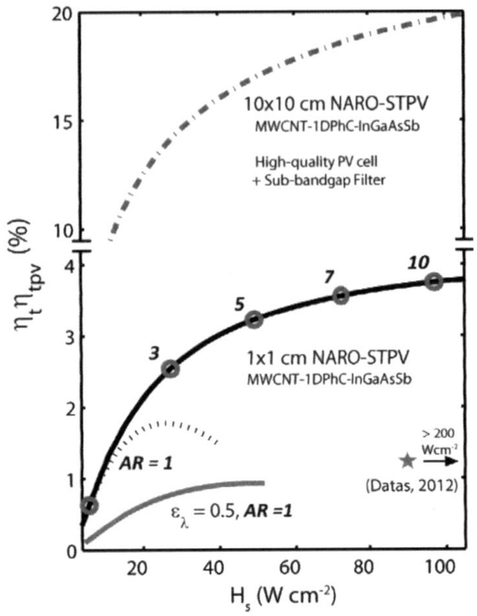

Figure 2. Relative improvements in efficiency and near-term predictions for nanophotonic STPVs. Conversion efficiency $\eta_{th}\eta_{pv}$ as a function of a solar irradiance H_s. Red symbols show experimental results with devices different AR (values specified next to symbols) and solid lines represent modeling results. Efficiencies approaching 20% were predicted with a scaled-up (10×10 cm^2) STPV device utilizing a high-quality 0.55 eV photovoltaic module with a sub-bandgap reflector. (Adapted from [3])

Significant improvements are predicted as the device planar area is increased, since parasitic losses are reduced drastically. Moreover, the efficiency can be further enhanced through improvements in low-bandgap photovoltaics, better spectral control and higher-temperature operation [4]. Re-optimizing nanophotonic absorber–emitter surfaces, *e.g.* using 2D PhC to better match photovoltaic bandgaps, as well as using higher-temperature photonic-crystal materials such as refractory metals could further lead to performance enhancement [5]. The efficiency improvements demonstrated in this work, as well as the promising predictions using a validated model, suggest the viability of nanophotonic STPVs for efficient and scalable solar energy conversion.

3. Jumping-droplet-enhanced condensation

When droplets coalesce on a superhydrophobic nanostructured surface, the resulting droplet can jump from the surface due to the release of excess surface energy. If designed properly, these superhydrophobic nanostructured surfaces can not only allow for easy droplet removal at micrometric length scales during condensation but also promise to enhance heat transfer performance. However, the rationale for the design of an ideal nanostructured surface as well as heat transfer experiments demonstrating the advantage of this jumping behavior are limited. Our work shows that silanized copper oxide surfaces (Figure 3a) created via a simple fabrication method can achieve highly efficient jumping-droplet condensation heat transfer [6].

Figure 3c compares the condensation heat transfer coefficients for conventional filmwise, flooded and dropwise condensing Cu surfaces with jumping-droplet condensation on silanized CuO superhydrophobic surfaces. It is experimentally demonstrated that a 25% higher overall heat flux and 30% higher condensation heat transfer coefficient can be achieved using jumping-droplet condensation compared to conventional dropwise condensing copper (Cu) surfaces at low super-saturations ($S < 1.12$). These CuO surfaces offer ideal condensation behavior in terms of emergent droplet morphology and coalescence dynamics and a significant enhancement in heat transfer performance when compared to state-of-the-art condensing surfaces. Furthermore, the chemical-oxidation-based CuO fabrication process provides a simple and readily scalable method to create superhydrophobic condensation surfaces that can sustain droplet jumping behavior. Accordingly, these surfaces are attractive for applications such as atmospheric water harvesting and dehumidification where the heat fluxes are relatively low and droplets can be maintained in a highly mobile state.

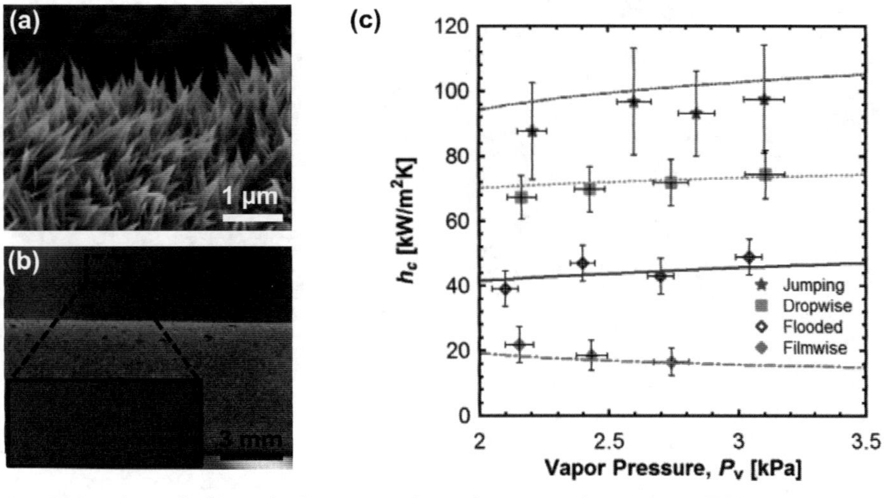

Figure 3. (a) Side view field emission scanning electron microscopy (FESEM) image of a CuO surface with no silane. (b) Jumping-droplet superhydrophobic condensation on a nanostructured CuO tube. (c) Experimental and theoretical steady state condensation coefficient (h_c) as a function of saturated vapor pressure (P_v) for tube surfaces undergoing filmwise, dropwise, flooded, and jumping condensation. (Adapted from [6])

While this droplet jumping phenomenon has been studied on a range of surfaces, past work has neglected electrostatic interactions and assumed charge neutrality of the droplets. We show that jumping droplets on a variety of superhydrophobic surfaces, including copper oxide, zinc oxide, and silicon nanopillars, gain a net positive charge that causes them to repel each other mid-flight [7]. The charge is determined experimentally by observing droplet motion in a uniform electric field. The results show that surfaces with identical coatings, showed identical charge trends irrespective of the surface structure or surface finish which indicates that the charging of the jumping droplets occurs at the solid–liquid interface, rather than after departing from the surface. We conclude that the mechanism for the charge accumulation is associated with the formation of the electric double layer at the droplet-coating interface and subsequent charge separation during droplet jumping governed by the fast time scales of droplet coalescence.

One application of this charging phenomenon is further enhancement of condensation heat transfer by preventing droplet reversal and return to the condenser surface due to the presence of vapor flow towards the surface, which increases the drag on the jumping droplets. This effect limits the possible heat transfer enhancement because larger droplets form upon droplet return to the surface that impede heat transfer until they can be either removed by jumping again or finally shedding *via* gravity. By characterizing individual droplet trajectories during condensation on superhydrophobic nanostructured copper oxide surfaces, this vapor flow entrainment is shown to dominate droplet motion for droplets smaller than $R \approx 30\ \mu m$ at moderate heat fluxes ($q'' > 2\ W/cm^2$) [8]. Subsequently, electric-field-enhanced (EFE) condensation is demonstrated, whereby an externally applied electric field prevents jumping droplet return due to the positive charge obtained by the droplets upon jumping (Figure 4). As a result, with scalable superhydrophobic CuO surfaces, a 50% higher overall condensation heat transfer coefficient is demonstrated compared to a jumping-droplet surface with no applied field for low supersaturations (<1.12).

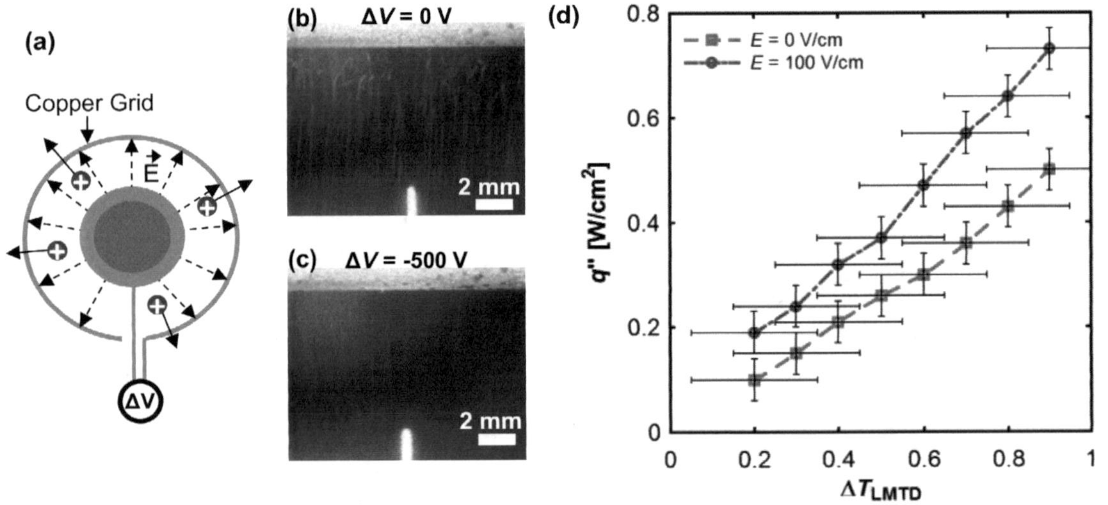

Figure 4. Electric-field-enhanced droplet removal. (a) Schematic of EFE condensation. The outer copper grid is biased negative relative to the condensing tube, creating an electric field and attracting jumping droplets away from the surface and preventing droplet return due to vapor flow entrainment. Long exposure time image (40 ms) of water vapor condensate ion on a superhydrophobic CuO tube with a copper electrode located beneath with (b) zero bias voltage having significant droplet-droplet interactions and return to the surface against gravity (pointing down) and (c) 500 V bias (electrode negative, tube ground). The image shows the concept of electric-field-enhanced (EFE) condensation with no droplet return to the surface and significant attraction of jumping droplets away from the surface ($P_v = 2700 \pm 75$ Pa, $S \approx 1.04$). (d) Experimental steady-state overall surface heat flux (q□) as a function of log mean water-to-vapor temperature difference (ΔT_{LMTD}) for tube surfaces undergoing jumping droplet condensation and EFE condensation. (Adapted from [8])

Another application of charged jumping droplets is use of these droplets for electrostatic energy harvesting. Here, the charged droplets jump between superhydrophobic copper oxide and hydrophilic copper surfaces to create an electrostatic potential and generate power during formation of dew under atmospheric conditions. Power densities of ~0.06 nW/cm^2 were demonstrated, which, in the near term, can be improved to ~1 $\mu W/cm^2$ [9]. This work demonstrated a surface engineered platform that is low cost and scalable for atmospheric energy harvesting and electric power generation. These applications of charged jumping droplets offer new avenues for improving the performance of self-cleaning and anti-icing surfaces as well as thermal diodes, and may also provide a competitive mode of energy harvesting from temperature gradients.

4. Summary

In summary, we presented an overview of the impact of using nanostructured surfaces to improve performance of solar thermophotovoltaic energy conversion and condensation systems. Our STPV system, consisting of a MWNT absorber and 1D PhC emitter with optimized nanophotonic properties, achieved experimental efficiencies of up to 3.2% in comparison to ≤1% reported in the literature. For condensation systems, we employed scalable superhydrophobic CuO nanostructures, among others, that can achieve significant improvements in the condensation coefficients *via* jumping-droplet condensation and electric-field-enhanced condensation modes. Furthermore, these nanostructured surfaces hold promise for several other applications including atmospheric energy harvesting, electric power generation, anti-icing and self-cleaning.

References

[1] Preston D J 2014 **S.M. Thesis** Electrostatic Charging of Jumping Droplets on Superhydrophobic Nanostructured Surfaces: Fundamental Study and Applications *Department of Mechanical Engineering* Massachusetts Institute of Technology

[2] Datas A and Algora C 2012 Development and experimental evaluation of a complete solar thermophotovoltaic system *Prog. Photovoltaics* **21** 1025-39

[3] Lenert A, Bierman D M, Nam Y, Chan W R, Celanovic I, Soljacic M and Wang E N 2014 A nanophotonic solar thermophotovoltaic device *Nature Nanotech.* **9** 126-30

[4] Lenert A, Nam Y, Bierman D M and Wang E N 2014 Role of spectral non-idealities in the design of solar thermophotovoltaics *Opt Express* **22** A1604-A18

[5] Rinnerbauer V, Lenert A, Bierman D M, Yeng Y X, Chan W R, Geil R D, Senkevich J J, Joannopoulos J D, Wang E N, Soljacic M and Celanovic I 2014 Metallic Photonic Crystal Absorber-Emitter for Efficient Spectral Control in High-Temperature Solar Thermophotovoltaics *Adv. Energy Mater.* **4**

[6] Miljkovic N, Enright R, Nam Y, Lopez K, Dou N, Sack J and Wang E N 2013 Jumping-Droplet-Enhanced Condensation on Scalable Superhydrophobic Nanostructured Surfaces *Nano Lett.* **13** 179-87

[7] Miljkovic N, Preston D J, Enright R and Wang E N 2013 Electrostatic charging of jumping droplets *Nature Communications* **4**

[8] Miljkovic N, Preston D J, Enright R and Wang E N 2013 Electric-Field-Enhanced Condensation on Superhydrophobic Nanostructured Surfaces *ACS Nano* **7** 11043-54

[9] Miljkovic N, Preston D J, Enright R and Wang E N 2014 Jumping-droplet electrostatic energy harvesting *Appl. Phys. Lett.* **105**

Advances In Vertical Solid-State Current Limiters For Individual Field Emitter Regulation In High-Density Arrays

Frances A Hill and Luis F Velásquez–García*

Microsystems Technology Laboratories, Massachusetts Institute of Technology, 77 Massachusetts Ave., Cambridge, MA, 01239, United States

* Email: Velasquez@alum.mit.edu

Abstract. We report the design, fabrication, and characterization of improved solid-state elements intended for individual regulation of field emitters part of high-density arrays. We demonstrate a high-yield, CMOS compatible fabrication process of single-crystal, vertical, ungated, n-type silicon field-effect transistors (FETs); each device behaves as a current source when is biased at a voltage larger than its drain-source saturation voltage. An ungated FET in saturation connected in series to a field emitter can compensate for the wide variation in current-voltage characteristics of the field emitters due to the tip radii spread present in any field emitter array, which should result in emitter burn-out protection, larger array utilization, and smaller array emission non-uniformity. Using 1-2 Ω·cm single-crystal n-Si wafers, we fabricated arrays of 25 μm tall vertical ungated FETs with 0.5 μm diameter that span two orders of magnitude of array size. Experimental characterization of the arrays demonstrates that the current is limited with > 3.5 V bias voltage to the same ~6 μA (6 $A.cm^{-2}$) per-FET value. Finite element simulations of the device predict a saturation voltage close to the experimental value and a saturation current within a factor of two of the experimental value.

1. Introduction

Field emission is the quantum tunneling of electrons into vacuum from the surface of metallic or semiconducting materials due to the presence of a high electrostatic field (> 3×10^7 V/cm) that narrows the potential barrier that traps the electrons within the material, i.e., workfunction [1]; a field emitter with a nanoscaled tip radius is capable of generating such electrostatic fields at a low (< 100 V) voltage [2],[3]. Field emitter electron sources are an attractive alternative to thermionic cathodes because they can be switched faster [4], work at a lower vacuum [5], consume less power, and produce electron beams with lower energy spread and higher brightness. Field emitter arrays (FEAs) are implemented to increase the total current from a field emission cathode and have been used in field emission displays [6], high-frequency amplifiers, gas ionizers [7],[8], X-ray sources [9],[10], multi-electron beam lithography, and other vacuum microelectronic and nanoelectronic devices. However, field emission current has an exponential dependence on the emitter tip radius, and, unfortunately, arrays of nano-sharp emitters can have associated tip diameter distributions with long tails [11]. Variations of tip diameters across an array of emitters lead to non-uniform emission. At a given bias voltage the duller tips might not be active while the sharper tips are already burned-out, which can cause severe array subutilization.

Researchers have proposed technologies to regulate the current emitted by FEAs, including p-n junction diodes operated in reverse bias [12], linear resistors [14], and CMOS transistors [15].

However, these approaches are not ideal because they cannot attain high enough current per tip, are not power efficient, or only regulate clusters of emitters that still present emission non-uniformity within the cluster. Our group pioneered the use of vertical ungated field-effect transistors (FETs), i.e., high aspect-ratio, single-crystal, n-type silicon columns, that are operated in saturation to limit the current of each field emitter [16]. The technology takes advantage of velocity saturation of carriers and channel pinch-off in a semiconductor to implement a current source-like ballast component that occupies the footprint of the emitter, leaving unaltered the emitter density of the FEA [17]. We previously demonstrated FEAs with 1 μA per emitter and 1 A.cm^{-2} current density with this structure [18]; more recently, we reported characterization of a scaled-down version of the vertical limiters with per-emitter saturation currents up to 0.1 μA [19]. In this work, we report ungated FETs that achieve current saturation with ~3.5 V bias voltage and that limit the current to ~ 6 μA (6 A.cm^{-2}).

2. Device Design and Modeling

Extensive axisymmetric process simulations of high aspect-ratio, single-crystal silicon columns were conducted using the SILVACO software (Silvaco International, Santa Clara, CA). Using the process simulation results, the electrical response of the devices was simulated and the parameters of the transistor, i.e., drain-source saturation current I_{dss}, drain-source saturation voltage V_{dss}, linear conductance g_{lin} (inverse of the drain-source resistance for bias voltages significantly lower than V_{dss}), and output conductance g_{out} (inverse of the drain-source resistance for bias voltages significantly higher than V_{dss}) were obtained.

In a first set of simulations, the cross-section of the device is circular with a 0.5 μm diameter, the channel length was set at 25 μm, and the doping concentration was varied between 10^{12} and 10^{16} cm^{-3}; the transistor parameters obtained are shown in Figure 1. Through variation of the doping concentration only, the same structure can deliver a saturation current between 20 pA and 14 μA with a saturation voltage between 0.06 V and 7.15 V. In addition, the output resistance is at least two orders of magnitude larger than the linear resistance, evidencing the current source-like behavior of the ungated FET when biased at voltages larger than V_{dss}, which results in a better capability of the ungated FET to regulate current compared to a linear resistor. For a doping concentration of 4×10^{16} cm^{-3} (resistivity ~ 1-2 Ω.cm), I_{dss} is equal to 3.0 μA and V_{dss} is equal to 3.7 V.

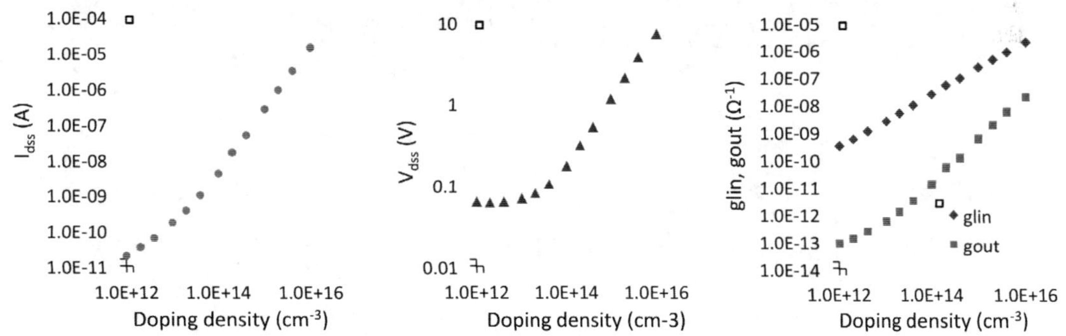

Figure 1. Finite element modeling results of the saturation current vs. doping (left), saturation voltage versus doping (center), and linear/output conductance vs. doping (right) for an ungated FET with a 0.5 μm diameter cross-section and channel length equal to 25 μm.

In a second series of simulations, we varied the channel length between 10 μm and 80 μm for an ungated FET with a 0.5 μm diameter cross-section made of n-Si with a doping concentration equal to 2×10^{15} cm^{-3}; the transistor parameters obtained are shown in Figure 2. There is great modulation in the saturation current for variation on the channel length, but there is only a moderate variation in the saturation voltage of the transistor, with minimal changes in V_{dss} for channel length above 40 μm.

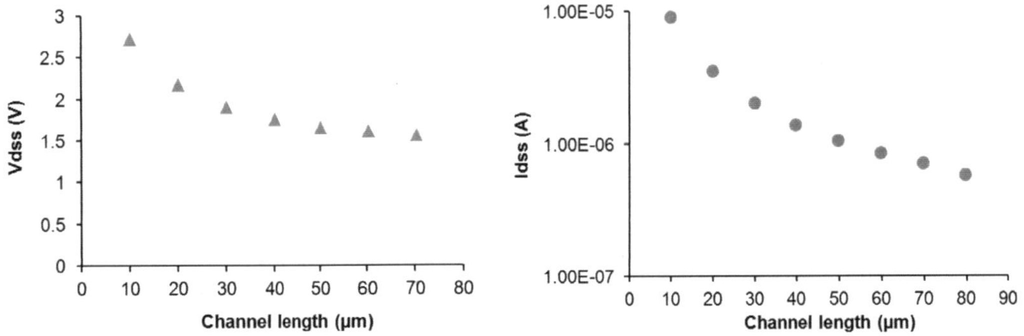

Figure 2. Finite element modeling results of the saturation voltage (left) and saturation current (right) versus channel length for an ungated FET made of n-Si with ~10 Ω.cm (2×10^{15} cm^{-3} doping concentration).

3. Device Fabrication

Arrays of vertical ungated FETs were made using single-crystal, n-type, silicon wafers with a resistivity equal to 1-2 Ω·cm. First, 25 μm-tall columns with 1 μm diameter cross-section are patterned on the substrate with deep reactive-ion etching, and the silicon around the columns is recessed by 2 μm so that the tops of the columns are taller than the surrounding silicon (Figure 3(a)). The columns are then thinned to a 0.5 μm diameter cross-section using wet oxidation (Figure 3(b)). Next, the gaps around the columns are filled-in with a 1.5 μm thick TEOS film (Figure 3(c)). After that, the oxide on the tops of the columns is etched back using plasma etching, which exposes the silicon core (Figure 3(d)). Finally, a 1.5 μm-thick aluminium film on top of a 10 nm-thick titanium film is sputtered on the top surface of the wafer, patterned, and sintered to form ohmic contacts (Figure 3(e)). Scanning electron microscope (SEM) images of the fabricated chip with a close-up of array of a 11 by 11 ungated FET array are shown in Figure 4.

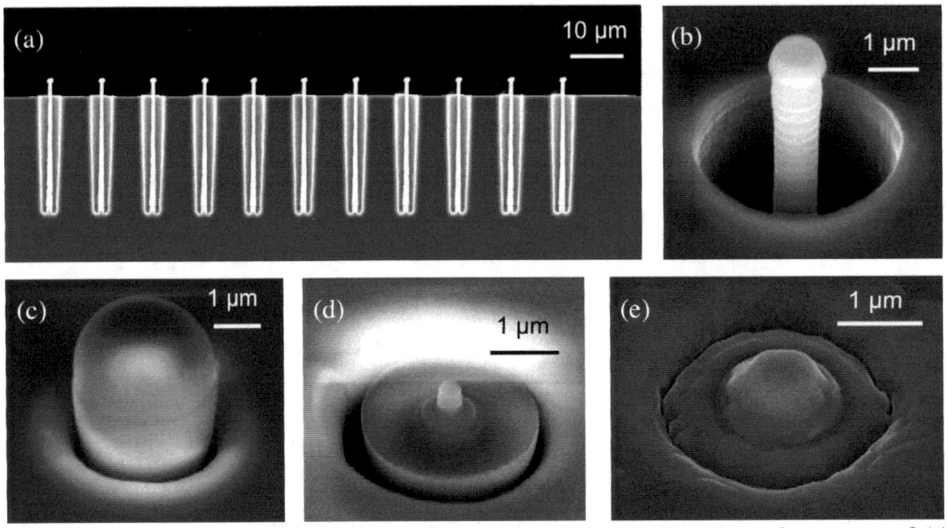

Figure 3. Selected SEM images of the fabrication process: (a) cross-section of an array of 11 high aspect-ratio silicon columns, each of which will form the channel of an ungated FET; (b) close-up of a silicon column after thinning; (c) close-up of a silicon column after the cavity filling-in; (d) close-up of a silicon column after the oxide on top has been removed, leaving the silicon intact; (e) metallized top.

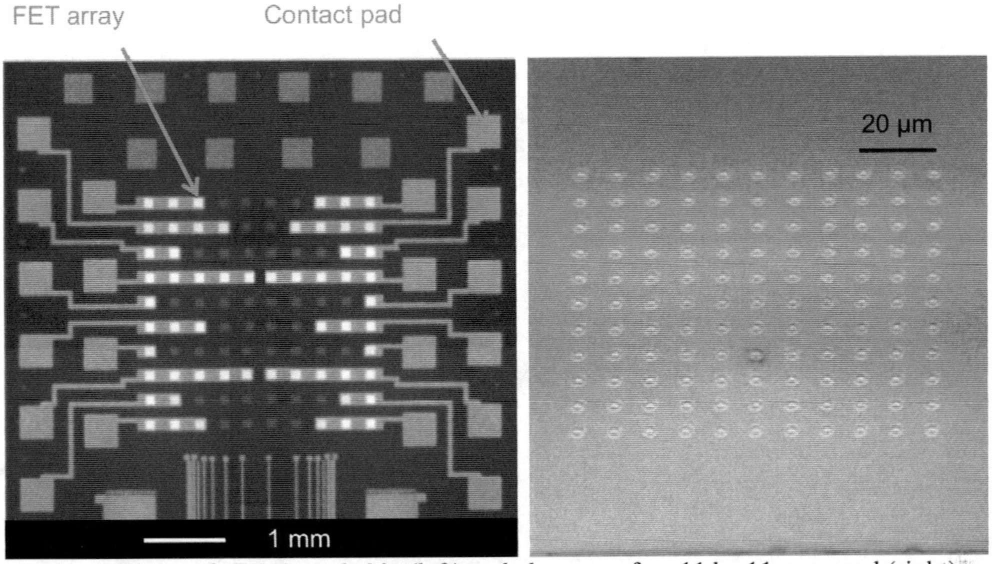

Figure 4. Fabricated chip (left) and close-up of an 11 by 11 array pad (right).

4. Device Characterization

Current-voltage *(I–V)* characteristics of arrays of ungated FETs with different array size were obtained using a probe station using an Agilent 4156C semiconductor parameter analyser; arrays of 1, 4, 16, 64 and 121 ungated FETs were characterized. During the measurements, the substrate (hence the source end of the channel) was held at 0 V, and a positive voltage V_{ds} was applied to the drain contact (the top of the structure). The results of the experiments are shown in Figure 5; in the figure, we assumed that the FETs part of the same array operated uniformly, that is, we divided the measured current by the number of FETs present in the array to calculate the per-ungated FET current-voltage characteristics. For $V_{ds} < 0.5$ V, the characteristics of the ungated FETs are linear; for V_{ds} between 0.5 V and about 3.5 V, there is a transition regime between the linear operation (no channel pinch-off) and saturated operation (channel pinched-off); for $V_{ds} > 3.5$ V, the ungated FET behaves like a current source, that is, it delivers a certain current while having associated a high impedance, resulting in small variation in the current for a significant variation in V_{ds}. The average saturation voltage is ~3.5 V, which is very close to the expected value from the finite element simulations; the average saturation current per ungated FET is 6 μA, which is within a factor of two from the value predicted by the finite element simulations.

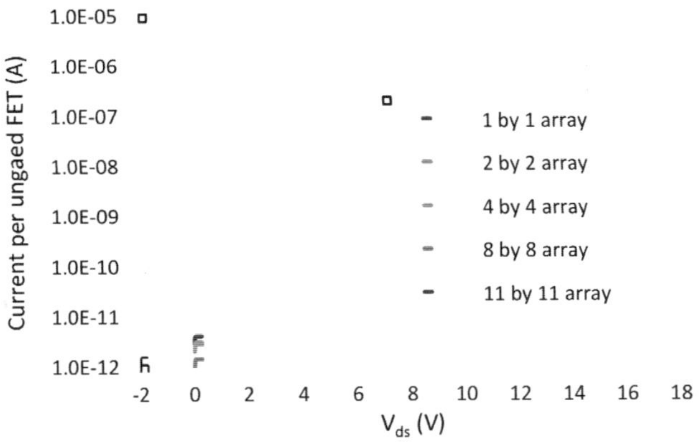

Figure 5. Per-ungated FET current-voltage characteristics from arrays of ungated FETs spanning a range of between 1 and 121 elements. The data demonstrates uniform array operation.

5. Conclusions

We demonstrated improved solid-state ballasting elements intended for individual regulation of field emitters part of high-density arrays. The devices are single-crystal, vertical, ungated, n-type silicon FETs that behave as current sources when biased at a voltage larger than their drain-source saturation voltage. Using 1-2 Ω·cm single-crystal n-Si wafers, we fabricated arrays of 25 μm tall vertical ungated FETs with 0.5 μm diameter cross-section that spanned two orders of magnitude of array size. Experimental characterization of the arrays demonstrates that the current is limited with > 3.5 V bias voltage to the same ~6 μA (6 $A{\cdot}cm^{-2}$) per-FET value. Finite element simulations of the device predict a saturation voltage close to the experimental value and a saturation current within a factor of two of the experimental value.

Acknowledgements

The devices were made at the Microsystems Technology Laboratories of the Massachusetts Institute of Technology. This work was funded in part by the Defense Advanced Research Projects Agency / Microsystems Technology Office, grant N66001-11-4204 (Program Managers A I Akinwande and D Palmer, Principal Investigator R Gupta), subaward 219877 through the Massachusetts General Hospital. Any opinions, findings, and conclusions or recommendations expressed in this publication are those of the authors and do not necessarily reflect the views of the US Government and therefore, no official endorsement of the US Government should be inferred.

[1] Gomer R 1961 *Field Emission and Field Ionization* (Cambridge, MA: Harvard University Press)
[2] Spindt C A 1968 *J. Appl. Phys.* **39** 3504
[3] Dvorson L, Sha G, Kymissis I, Hong C-Y, Akinwande A I 2003 *IEEE Trans. Electron Dev.,* **50** 2548–2558
[4] Swanwick M E, Keathley P D, Fallahi A, Krogen P R, Laurent G, Moses J, Kärtner F X, Velásquez-García L F 2014 *Nano Letters* **14** 5035-5043
[5] Fomani A A, Akinwande A I, Velásquez-García L F. 2013 *J. Phys. Conf. Ser.* **476** 012014
[6] Vaudiane P, Meyer R 1991 *Proc. Int. Electron Devices Meeting (IEDM 1991),* 197-200.
[7] Velásquez-García L F, Gassend B L P, Akinwande A I 2010 *J. Microelectromech. Syst.* **19** 484-493
[8] Chen, L-Y, Velásquez-García L F, Wang X, Teo K, Akinwande A I 2007 *Proc. Int. Electron Dev. Meeting (IEDM 2007),* 843-846
[9] Cheng S, Hill F A, Heubel E V, Velásquez- García L F 2015 *J. Microelectromech. Syst.* **24** 373 – 383
[10] Basu A, Swanwick M E, Fomani A A, Velásquez-García L F 2015 *J. Phys. D: Appl. Phys.* **48** 225501
[11] Nilsson L, Groening O, Kuettel O, Groening P, Schlapbach L 2002 *J. Vac. Sci. Technol. B* **20** 326–337
[12] Liu K, Chiang C-J, Heritage J P 2006 *J. Appl. Phys.* **99** 034502
[13] Fursey G 1996 *Appl. Surf. Sci.* **94/95** 44-59
[14] Vaudiane P, Meyer R 1991 *IEDM Tech. Dig.* 197–200
[15] Hong C-Y, Akinwande A I 2005 *IEEE Trans. Electron Devices* **52** 2323–2328
[16] Velásquez-García L F, Adeoti B, Niu Y, Akinwande A I 2007 *Proc. IEEE Int. Electron Dev. Meeting (IEDM 2007),* 599-602
[17] Velásquez-García L F, Guerrera S, Niu Y, Akinwande A I 2011 *IEEE Trans. Electron Dev.* **58** 1775 – 1782
[18] Velásquez-García L F, Guerrera S, Niu Y, Akinwande A I 2011 *IEEE Trans. Electron Dev.* **58** 1783 – 1791
[19] Guerrera S, Velásquez-García L F, Akinwande A I 2012 *IEEE Trans. Electron Dev.* **59** 2524-2530

Wirelessly powered micro-tracer enabled by miniaturized antenna and microfluidic channel

G Duan, X Zhao, H R Seren, C Chen and X Zhang[1]
Department of Mechanical Engineering, Boston University, Boston, MA, USA

E-mail: xinz@bu.edu

Abstract. A miniaturized antenna, 380μm by 380μm in size, was fabricated and integrated with a commercialized passive RFID chip to form a micro-tracer, whose size was 2mm by 1mm in total. The micro-tracer was wirelessly powered and interrogated by a single layer spiral reader antenna through near field coupling. To maximize the working distance, the resonant frequency of micro-tracer and reader antenna were matched at 840MHz. Due to the ultra small size of the tracer antenna, power transfer efficiency decreased dramatically as the distance between tracer antenna and reader antenna increased, thus the working distance of the micro-tracer was limited within 1mm. To achieve massive operation of the micro-tracer, a microfluidic platform was fabricated with in channel focusing and separation. Acrylic sheets were laser cut to define the channel and cover structure, then bonded together layer by layer with a glass substrate, on which reader antenna was integrated. Pump oil was used as the fluidic media carrying the micro-tracer flowing inside the microfluidic channel. The wireless power transfer and real-time communication was demonstrated with the micro-tracer flowing above the reader antenna, as the ID of the micro-tracer was retrieved and displayed on a computer screen.

1. Introduction

Thanks to CMOS and microelectromechanical systems (MEMS) technologies, integrated circuit (IC) and physical sensor's feature size has been successfully reduced to micrometre (μm) scale [1, 2]. This enabled the possibility of integrated microsensors. With miniaturized size, microsensors can play important roles in various areas, such as health and environmental monitoring, homeland security, and natural resource exploration, among others. However, for a functional sensor, especially in micrometer range, wireless communication and battery charging are always required, thus an integrated antenna is needed. Nevertheless, the radiation efficiency of small antennas follows the Chu-Harrington limit [3], setting a minimum size for the far-field antenna design for a given frequency. Therefore, near field wireless communication stands out as a viable option.

In our study case, to achieve successful size miniaturization, near field with higher power density and less demand of antenna dimension is utilized. Due to the short propagation distance of near field, sensors, that are designed to operate at distant locations, are usually dispatched to the target environment and then collected back to retrieve stored information [4]. In this paper, a 380μm by 380μm double layer antenna was designed, fabricated and integrated with a commercialized RFID chip [5] to form a micro-tracer 1mm by 2mm in size. The micro-tracer can work alone or be integrated with sensors and a battery to form a microsensor. A spiral reader antenna was designed and fabricated

[1] To whom any correspondence should be addressed.

to work with the micro-tracer. Wireless communication and power transfer was successfully realized at 1mm distance. A microfluidic channel platform with focusing and separation functions was designed to manipulate multiple miniature tracers.

2. System design and fabrication

2.1. Antenna design

A commercialized RFID chip was chosen to form the micro-tracer with an impedance of 18.6-j171.2 at the frequency of 915MHz. Due to the negative imaginary portion, the RFID could be treated as a capacitance in series with a resistance. A double layer spiral antenna was designed as an inductance to match the impedance of the RFID chip to maximize the power transfer efficiency. Antennas with different side lengths were designed and simulated in CST Microwave Studio to find the optimized antenna dimension. The simulation results are shown in table 1. Identical line width (10μm) and number of turns (4) are applied in the simulation. From the simulation results, it can be seen that the impedance match is realized at a side length between 360μm and 380μm.

Table 1. The simulation result of micro-tracer antennas with different side length.

	Type1	Type2	Type3	Type4
Side length (μm)	340.0	360.0	380.0	400.0
Imaginary portion	140.8	162.0	181.8	203.1
Inductance (nH)	24.5	28.2	31.6	35.3

The reader antenna was designed as a single layer spiral antenna with a coupling loop as shown in figure 1. The RF signal fed into the coupling loop excites the inner spiral antenna at the resonance frequency, at which maximum magnetic field is generated The resonance frequencies of the antennas can be accurately designed by varying the spiral coil geometry (e.g. size, number of turns, track and gap widths, etc.), and changing the coupling ratio between the coupling loop and the spiral. This is clearly demonstrated in figure 2, in which the reflection coefficient (S11) were plotted for various antennas with resonant dips.

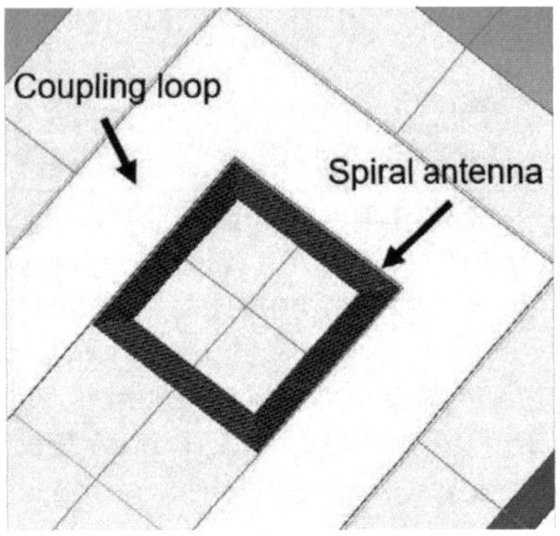

Figure 1. Reader antenna structure

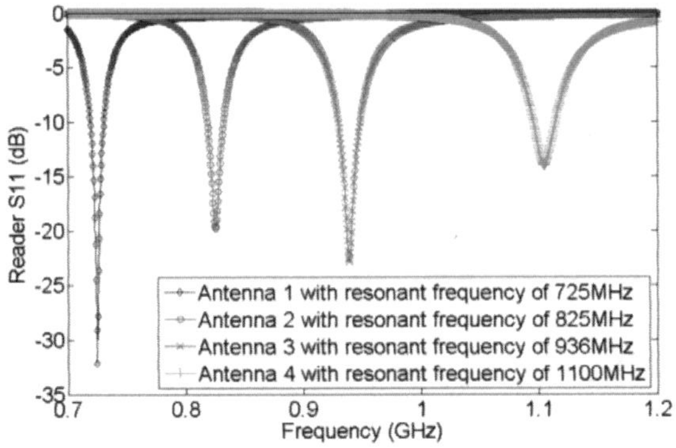

Figure 2. S11 of different reader antenna designs

2.2. Antenna fabrication and chip integration

The fabrication process of the antennas is illustrated in figure 3. The reader and tracer antenna were fabricated on a Pyrex glass substrate with 500μm thickness. A thin layer of copper was evaporated as the seed layer on top of the substrate. Then, 5μm thick photoresist was spun and patterned to define the structure of the first layer of the antenna, after which electroplating was applied to grow the first layer of the spiral to 5μm thickness. The wafer was soaked in acetone to remove the photoresist.

Subsequently, the seed layer was removed by wet etching. At this point, in the case of the reader antenna, the structure was complete while in the case of the micro-tracer antenna, another spiral structure was required. SU8, as the separation and supporting layer, was photo-defined to form the vias, which were filled by electroplating to connect the first and second layer. Following the same flow as the first layer fabrication, the second layer of the antenna was fabricated as shown in steps 8 through 12 in figure 3. Finally, a thin layer of silicon nitride was sputtered as an insulation and protection layer. The SEM picture of a fabricated tracer antenna is shown in figure 4. Figure 4 (a) shows that the two layers were connected by a via in the middle, while the pad on the left was connected to the bottom layer. The spiral structure can be seen clearly in figure 4 (b). After antenna fabrication, the antenna and RFID chip was bonded together with silver epoxy to form a micro-tracer.

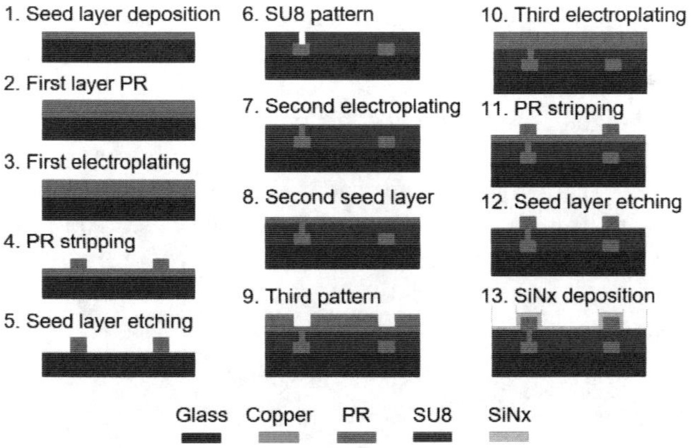

Figure 3. Fabrication process of the micro-tracer antenna

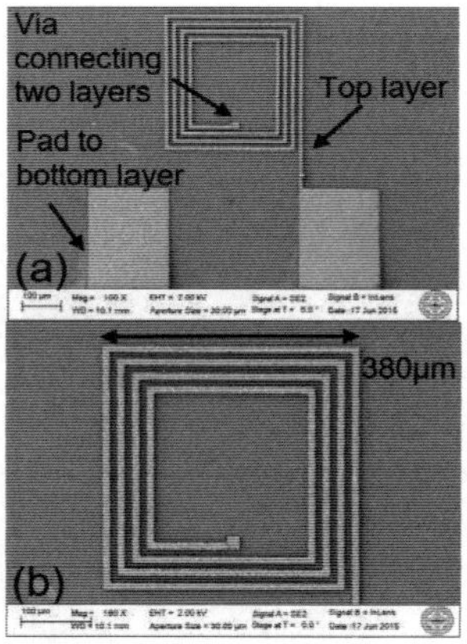

Figure 4. SEM picture of micro-tracer antenna.

3. Microfluidic channel design and fabrication

3.1. Microfluidic channel design

A microfluidic channel with cross section of 2mm in width and 1.2mm in height was designed to manipulate multiple micro-tracers. Pump oil was chosen as the fluidic media to transport the sensors flowing inside the channel. Considering large numbers of micro-tracers may arrive at the microfluidic channel simultaneously, in order to avoid clogging, a focusing structure was designed as shown in figure 5 (a). When multiple sensors arrive at the funnel structure, the two side inputs begin pumping at the same time with a high velocity in order to create a focusing effect for the main stream. Thus, the micro-tracers are lined up such that only a single micro-tracer passes the main channel at a time.

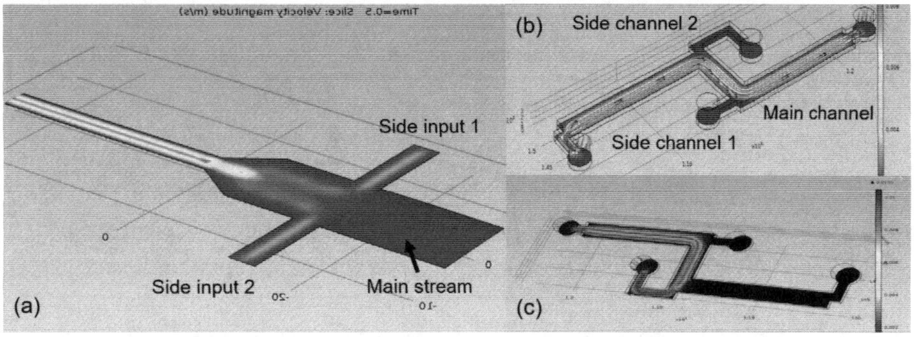

Figure 5. Microfluidic channel design. (a) Focusing structure to line up micro-tracers. (b) and (c) Separation structure to prevent reading failure due to collision

If two micro-tracers are so close that they reach the reading range of the reader antenna simultaneously, data collision may cause reading failure. A separation unit was fabricated to guarantee enough time delay between two deliveries as shown in figure 5 (b) and (c). When one of the micro-tracers reaches side channel 1, the main channel stops pumping while side channel 1 begins to pump. Thus, the newly arrived sensor is forced to the second sidewall with others keeping their position. Next, the side channel 1 stops pumping, while side channel 2 starts to pump and only a single micro-tracer passes the microfluidic channel. In this fashion, by controlling the time delay between two operation circles, an adequate interval delivery time can be guaranteed.

3.2. Microfluidic channel fabrication
The microfluidic channels were composed of three layers: the first layer was a glass substrate, on which reader antennas were fabricated, the second layer was the structure of the channel, and the third layer was the channel cover. The second and third layers were fabricated by laser cutting an acrylic sheet. As the thickness of the acrylic sheet was 0.5mm, two layers were stacked together to form the channel layer. The three layers were bonded together with epoxy adhesive. A syringe was connected to the microfluidic channel through a soft pipe, which was inserted into the connection ports; the flow speed inside the channel was controlled by a syringe pump. The reader antennas were wire bonded to SMA connectors, which were connected to the reader circuit through co-axial cable. The fabrication process is illustrated in figure 6 (a) and the fabricated microfluidic channel is shown in figure 6 (b).

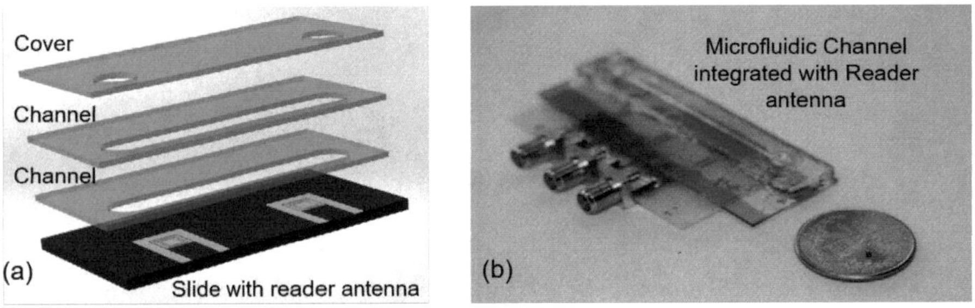

Figure 6. (a) Microfluidic channel fabrication process. (b) Picture of the microfluidic channel.

4. Experiment and result
After micro-tracer integration and microfluidic channel fabrication, we performed the in-channel real-time communication experiment. The experimental setup is shown in figure 7 (a). The reader antenna

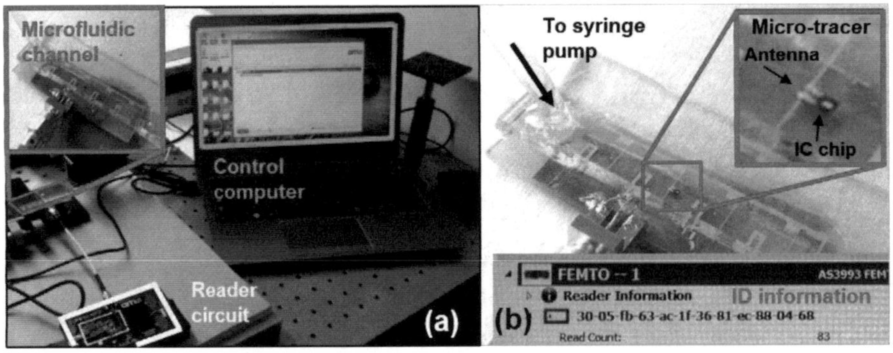

Figure 7. (a) Test setup. (b) RFID information reading out

was embedded below the microfluidic channel, and connected to the reader circuit through a co-axial cable, which was powered and controlled by a computer using a USB cable. A micro-tracer was placed inside the microfluidic channel and pump oil was employed as the carrying media to transport the micro-tracer within the microfluidic channel. A syringe pump was connected to one end of the microfluidic channel to control the flow speed, while the other end of the channel was connected to an oil reservoir. As the oil flowed, the micro-tracer travelled across the reader antenna. When the communication link was complete, the ID information of the RFID was retrieved and displayed on the computer screen as shown in figure 7 (b).

5. Conclusion

We developed a prototype of integrated microfluidic channels to interrogate micro-tracers using near field RF communication technique. A miniaturized antenna was designed and optimized to match its impedance with a RFID IC chip within a limited chip area (< 400µm×400µm). After the integration of the antenna and RFID IC, wireless communication was demonstrated in the microfluidic channel. Beyond the specific application developed in this work, the micro-tracer may also be employed as a communication unit for a variety of MEMS sensors.

Acknowledgments

This work was supported by the Advanced Energy Consortium: http://www.beg.utexas.edu/aec/ Member companies include Petrobras, Repsol, Statoil, Shell, and Total.

References

[1] Dai H, Nobuyuki S, Kazuyoshi T, Akio S and Daisuke R 2007 ULSI Devices: Current Status and Future Prospects of Research and Development *Hitachi Review.* **56** 25–33
[2] Chen L, Chang J, Hsu C and Cheng W 2009 Fabrication and performance of MEMS-based pressure sensor packages using patterned ultra-thick photoresists *Sensors* **9** 6200–18
[3] Wheeler H 1975 Small antennas *IEEE Trans. Antennas and Propag.* **AP-23** 462–469
[4] Seren H R, Zhao X, Chen C, Wang C and Zhang X 2015 Enabling a microfluidic RFID readout system via miniaturization and integration *J. Microelectromech. Syst.* **24** 395–403
[5] SL3ICS1002/1202 UCODE G2XM and G2XL 2012 *Product data sheet* Rev. 3.7-9

Watt-level wireless power transmission to multiple compact receivers

A Garraud, D J Munzer, M Althar, N Garraud, D P Arnold
Interdisciplinary Microsystems Group, Department of Electrical and Computer Engineering, University of Florida, Gainesville, FL, 32611, USA

E-mail: agarraud@ufl.edu

Abstract. This paper reports an electrodynamic wireless power transmission (EWPT) system using a low-frequency (300 Hz) magnetic field to transmit watt-scale power levels to multiple compact receivers. As compared to inductively or resonantly coupled coils, EWPT facilitates transmission to multiple non-interacting receivers with little restriction on their orientation. A single 3.0 cm^3 receiver achieves 1.25 W power transmission with 8 % efficiency at a distance of 1 cm (350 mW/cm^3 power density) from the transmitter. The same prototype achieves 9 mW at a distance of 9 cm. Moreover, we demonstrate simultaneous recharge of two wearable devices, using two receivers located in arbitrary positions and orientations.

1. Introduction

Over the past ten years, an alternative concept of recharging method for electronics devices, has gained interest: the wireless power transmission (WPT) [1,2]. The main advantage is to remove the burden of having to plug mobile electronic devices (laptops, smart phones, tablets, wearables) [3], to ease the recharge of electric vehicles [4], or to avoid surgical operations for battery replacement in biomedical implants [5]. Most of the research has focused on the electromagnetic coupling between two coils to wirelessly transfer power, namely inductive WPT [1,2]. In this approach, a transmitter coil produces a time-varying magnetic field that directly induces a voltage and current in a receiver coil (via Faraday induction) placed at a remote distance.

We have recently presented a different approach, named electrodynamic wireless power transmission (EWPT) [6–8]. This method relies on the use of a small permanent magnet in the receiver to achieve WPT. The transmitter produces a time-varying magnetic field, thereby inducing a magnetic torque/force on the magnet in the receiver. This coupling leads to the magnet motion, generating a voltage in the receiver windings, and if a load is connected, the EWPT is achieved. The system is operated at much lower frequencies (100–1000 Hz) than inductively or resonantly coupled systems (0.1–10 MHz) [1]. The lower frequencies facilitate higher safety margins for magnetic field amplitudes [9,10] and better penetration through electrically conductive media (metal, human body, seawater, …). This enables wireless power delivery in a crowded environment, such as an office space, an automobile, or to a medical implant within the human body. Furthermore, unlike resonant-type WPT systems, there is little interference between receivers.

Various concepts of magnetic-to-mechanical-to-electrical wireless power transfer have been previously demonstrated, the earliest by magnetically exciting mechanical resonance in a receiver and then using either electrodynamic [6–8] or piezoelectric [11] transduction to convert this mechanical motion into electrical power. A system employing a cantilever-based oscillator yielded a power transmission of 0.15 mW at 2.2 cm with 12 % efficiency [6]. A different prototype utilized a torsional resonator, achieving 3.1 mW at 1 cm (equivalent to a power density of 0.14 mW/cm^3), and 0.13 mW at 7 cm [7]. In these coil-based receivers, the inductive coupling between the transmitter coil and the receiver windings was shown to have a minor influence in the proposed EWPT mechanism [7]. An alternative design using a piezoelectric cantilever demonstrated a transmission efficiency up to 3 % [11].

A more recent variation on the EWPT concept was based on the continuous rotation of the permanent magnet rather than a resonant-based system. Under steady-state operation, the rotating magnet acted as a synchronous machine, supplying electrical power to an external load [8]. We demonstrated EWPT of up to 99 mW over a 5 cm distance at 0.5% efficiency (power density of 31 mW/cm^3) and 5 mW over 20 cm. This non-resonant approach enables power transmission over a range of frequencies, as opposed to only at the resonant frequency of the receiver.

This paper presents the recent improvements in the EWPT using the continuously rotating magnet system, eventually reaching a watt-level power transmission at close distance. The design of the transmitter coil has been modified in order to achieve a larger B-field at a lower input power. In parallel, the receiver architecture has been revised to enhance the flux change and increase the voltage output. These changes have enabled the implementation of DC rectifying electronics that can then lead to the recharge of wearable devices.

Section 2 first describes the working principle of EWPT using a continuously rotating magnet and it introduces the experimental setup used to perform EWPT along with the basic operation. Section 3 discusses the experimental results. Conclusions are provided in section 4.

2. Methods and Materials

2.1. EWPT Working Principle

The EWPT system reported here includes two major parts: a coil-based transmitter and a receiver system featuring a permanent magnet rotor that can continuously rotate within receiver windings, as illustrated in Figure 1.a. The transmitter coil is a pancake-shaped coil with 200 turns of 16 AWG copper wire, having an outer diameter of 15 cm, an inner diameter of 2.5 cm, and a length of 6 mm (L=3 mH, R=0.8 Ω) as illustrated by Figure 1.b. A low-carbon steel sheet (thickness=1 mm) is placed on the backside of the transmitter to enhance the magnetic flux lines on the transmitter frontside, therefore increasing the magnetic flux density in direction of the receiver. As presented in Figure 1.c, the receiver system comprises a diametrically magnetized, cylindrical, N42 grade Nd–Fe–B permanent magnet rotor (K&J Magnetics, D36DIA, diameter=4.8 mm, length=9.5 mm) that rotates on a ferrofluid bearing within a glass tube. The tube is inserted into a 3D-printed structure and wound with a single coil of 600 turns of 36 AWG copper wire. The final size is: width=17 mm, length=17 mm, height=11 mm, for a total active volume of 3 cm^3.

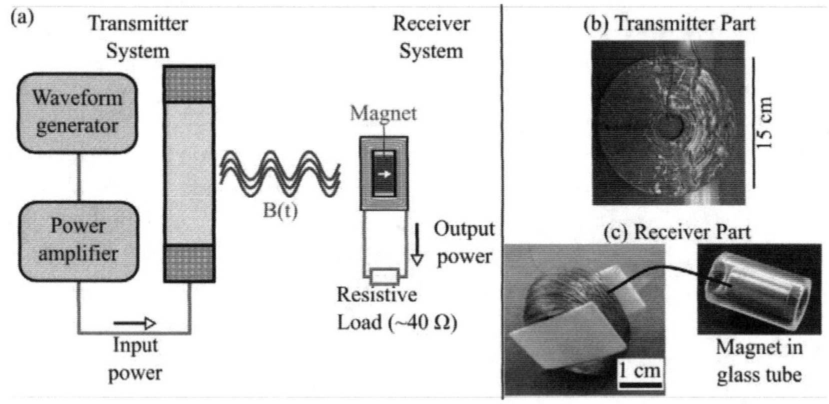

Figure 1: (a) Setup description of the electrodynamic wireless power transmission; (b) Picture of the coil-based transmitter; (c) Picture of the magnet in the glass tube that is part of the receiver.

To wirelessly transmit power, an AC current is injected into the transmitter coil, which results in the generation of a sinusoidal magnetic flux density B(t). When the receiver system is placed nearby, the transmitted B-field magnetically couples with the receiver magnet, and forces synchronous rotation of the permanent magnet. The magnet motion produces a time-varying magnetic flux in the receiver windings, hence a voltage. If an electrical load is connected to the two terminals of the receiver windings, output power is delivered.

2.2. Experimental Setup

The EWPT system is characterized using the following setup. The initial input waveform is generated by a USB-powered Analog Discovery interface (Digilent Inc.) driven by a custom-made Matlab code. The signalis supplied, as an AC voltage, to a Crown XLS2500 linear power amplifier. The resulting output is an AC current input for the transmitter coil, monitored by a Tektronix TCP312A current probe connected to a Tektronix TCPA300 current probe amplifier. The transmitter voltage is measured with two 10x oscilloscope probes, each connected to one terminal. An Agilent DSO-X-2004A 4-channel oscilloscope is used to monitor the input current and voltage to the transmitter, and to subsequently calculate the input power. The output receiver voltage, measured across the receiver load, is measured with the Analog Discovery board, and the signal is later post-processed to extract the RMS voltage as a function of the transmitter frequency. The resistance of the receiver windings is independently measured with a Keithley 2000 multimeter, using the 4-wire configuration. The load resistance is matched to the coil resistance for optimal power transfer. For a given AC excitation of the transmitter (frequency, amplitude), the resultant B-field spatial distribution is measured with a Lakeshore XHMMA-1482 axial Hall probe connected to a Lakeshore 475DSP gaussmeter.

2.3. Basic Operation

The frequency is swept from a low frequency, 40 Hz—where the receiver locks into synchronous, steady-state rotation—up to a point where the magnet stays rotating continuously (200–400 Hz). If the frequency is kept increasing, at a certain point, the magnet will lose synchronization with the transmitted B-field, resulting in no more magnet rotation. This upper frequency is called the maximal frequency and the output power at that frequency, the maximal output power. The frequency where synchronization is lost represents the "pull out torque" for a synchronous machine—the point where the magnetic torque acting on the magnet equals the sum of the opposing mechanical and magnetomotive torques [12].

3. Results and Discussion

First, the output power as a function of the transmitted B-field is characterized along the central axis of the transmitter coil. Next, the on- and off-axis wireless transmissions are measured. Last, successful recharge of commercial wearable devices is demonstrated.

3.1. Output Power vs B-Field

After placing the receiver system 1 cm away from the transmitter along its central axis, a sinusoidal current is injected in the transmitter coil. The resistive load is adjusted to 50 Ω. The amplitude of the input current is varied from 0.6 A to 2.2 A, corresponding to an input power from 0.3 W to 15 W, respectively. It results in a variation of the B-field at the receiver location from 3.2 mT_{pk} to 13 mT_{pk}. The resulting maximal output power is measuring as a function of the transmitted B-field and the results are presented in Figure 2.a, showing a quadratic relationship, as previously reported [8]. A maximum output power of 1.25 W is achieved at 13 mT_{pk}. The transmission efficiency is extracted as the ratio of the maximal output power to the input power, and it is reported in Figure 2.b. It is found as 9.3 ± 1.1 % for all tested B-fields.

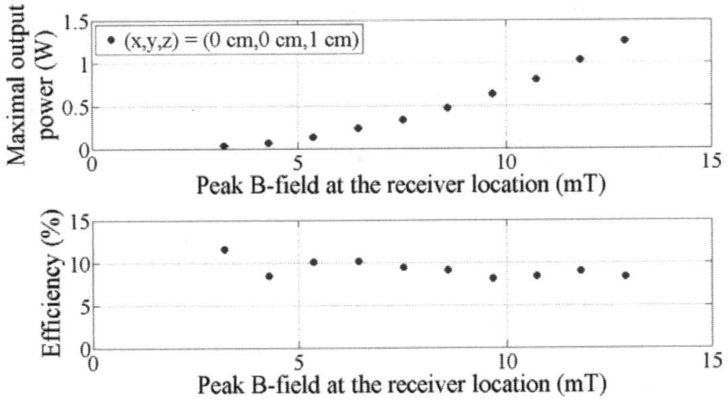

Figure 2: (a) Maximal output power and (b) transmission efficiency, as a function of the peak B-field at the receiver location.

3.2. On- and Off-Axis EWPT

For the subsequent experiments, the input current is fixed at 2.2 A (15 W). The distribution of the output power as a function of the on-axis position is performed and the results are presented in Figure 3. Each point represents the average value of three measurements, with the error bars too small to be noticed. Here, 1.25 W of power is wirelessly transmitted at 1 cm, while 9 mW is transmitted at 9 cm. This reduction in power is a consequence of the decay of the magnetic flux density with distance. A linear fit to the semilog function results in a slope of (-0.27).

The off-axis output power distribution is also measured, and the results are presented in Figure 4. The receiver is moved in one direction off axis, 1 cm above the transmitter, and the output power is extracted at each position. Each point represents the average value of three measurements, with the error bars too small to be noticed. At 7 cm off-axis (~transmitter radius), an output power transmission down to 8 mW is measured.

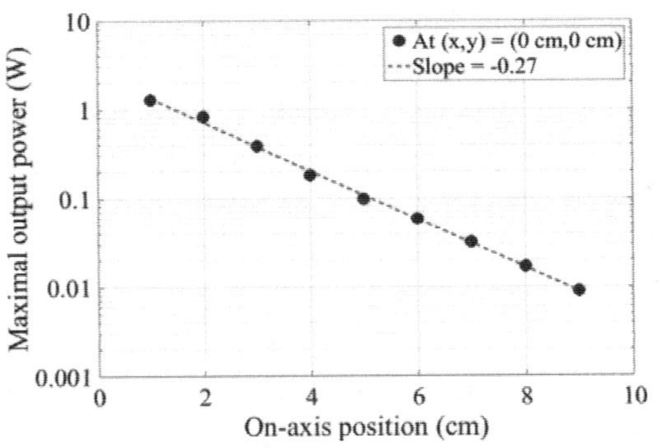

Figure 3: Maximal output power, as a function of the on-axis position.

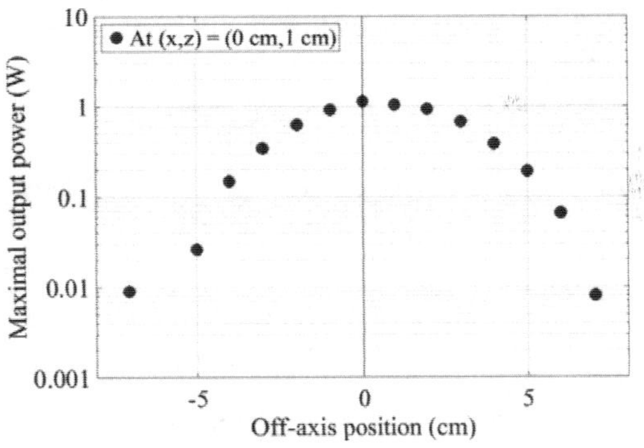

Figure 4: Maximal output power, as a function of the off-axis position.

3.3. Recharging of Commercial Devices

The output AC signal can be rectified into a DC signal using a diode bridge rectifier. A 5 V regulator (LM7805) is then implemented to provide a constant 5 V DC signal, necessary to recharge most conventional wearable devices, as described in Figure 5. Two printed-circuit boards are made and then used with two different receivers to power two different wearable devices simultaneously: an iPod Shuffle media player and a Garmin Vivosmart health monitoring watch. While simultaneously charging the two devices, as presented in Figure 6, at a frequency of 300 Hz, the DC power supplied to each device is measured to be 250 mW and 350 mW, respectively. It should be noted that the receivers are distributed at no particular location on the transmitter surface, nor the receivers windings are put in a particular orientation with respect to the transmitter coil.

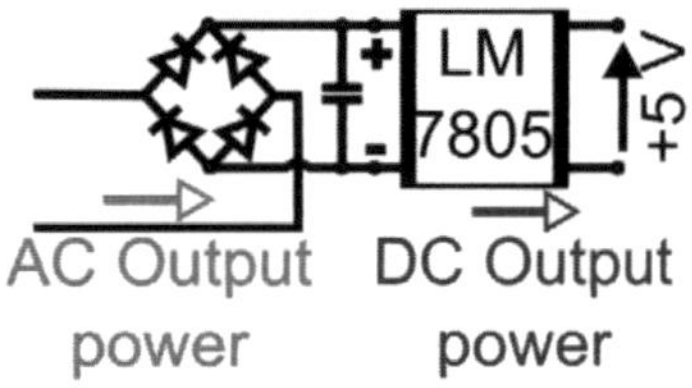

Figure 5: AC to DC conversion electrical circuit.

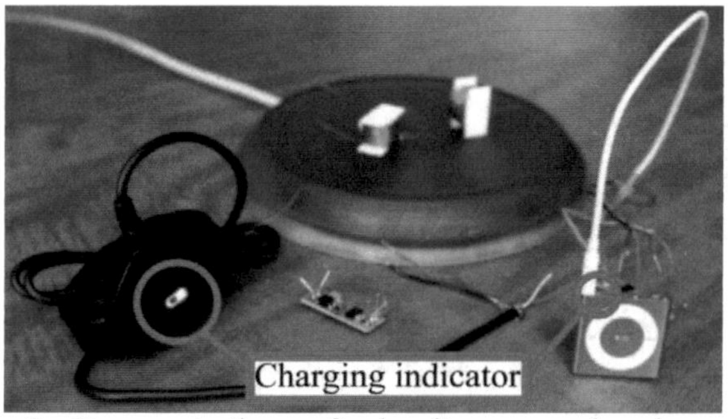

Figure 6: Demonstration of simultaneous charging of two commercial wearable devices: an iPod Shuffle and a Garmin Vivosmart health monitoring watch using two receivers.

4. Conclusions

In this paper, an EWPT system using the continuous rotation of a permanent magnet in the receiver is discussed. Compared to previous works, a significant improvement in the output power delivered to a resistive load is achieved, reaching up to 1.25 W, with a power density up to 350 mW/cm^3 and an efficiency of 8 %. The on- and off-axis power deliveries from the transmitter, as well as powering multiple receivers at the same time, are demonstrated. The successful charging of two wearable devices placed in arbitrary positions and orientations demonstrates the potential application of this EWPT mechanism in consumer electronic products.

Acknowledgments
This work was supported by the US Army Communications-Electronics Research, Development and Engineering Center (CERDEC) through the General Technical Services sub-contract GTS-S-13-226.

References

[1] Kurs A, Karalis A, Moffatt R, Joannopoulos J D, Fisher P and Soljacic M 2007 *Science* **317** 83–6
[2] Wei X, Wang Z and Dai H 2014 *Energies* **7** 4316–41
[3] Rim C T and Chris Mi C 2015 *IEEE Trans. Power Electron.* **30** 6015–6
[4] Musavi F and Eberle W 2014 *IET Power Electron.* **7** 60–6
[5] Kim J-D, Sun C and Suh I-S 2014 *2014 IEEE Wireless Power Transfer Conference* (IEEE) pp 166–9
[6] Challa V R, Mur-Miranda J O and Arnold D P 2012 *Smart Mater. Struct.* **21** 115017
[7] McEachern K M and Arnold D P 2013 *J. Phys. Conf. Ser.* **476** 012004
[8] Garraud A, Jimenez J, Garraud N and Arnold D P 2014 *J. Phys. Conf. Ser.* **557** 675–9
[9] C95.6 2002 *IEEE Standard for Safety Levels With Respect to Human Exposure to Electromagnetic Fields, 0-3 kHz*
[10] C95.1 2010 *IEEE Standard for Safety Levels With Respect to Human Exposure to Radio Frequency Electromagnetic Fields, 3 kHz to 300 GHz*
[11] Paprotny I, Waterbury A, Challa V, Xu Q, Arnold D P, Sanders S and Wright P K 2012 *Tech. Dig. 12th PowerMEMS 2012* (Atlanta, GA, USA) pp 263–6
[12] Chapman S J 2002 *Electric Machinery and Power System Fundamentals* (McGraw-Hill Higher Education)

Piezoelectric cellular micro-structured PDMS material for micro-sensors and energy harvesting

A Kachroudi[1,2,3], S Basrour[1,2], L Rufer[1,2] and F Jomni[3]

[1] Univ. Grenoble Alpes, TIMA Laboratory, F-38031 Grenoble, France
[2] CNRS, TIMA Laboratory, F-38031 Grenoble, France
[3] Université de Tunis El Manar, Faculté des Sciences de Tunis, Laboratoire Matériaux, organisation et propriétés (LMOP), 2092, Tunis, Tunisie

E-mail: achraf.kachroudi@imag.fr

Abstract. This paper reports a novel low-cost fabrication process of a charged cellular micro-structured polydimethylsiloxane (PDMS) material referred as piezo-electret or ferro-electret for micro-sensors applications. The dielectric spectra reached on these structures exhibit a high piezoelectric longitudinal coefficient d_{33} of 350pC/N. A mechanical characterization method proves the reliability of this material for low-frequencies applications around 100Hz.

1. Introduction

Piezoelectric materials are widely used in the micro-sensors and the actuators field. Depending on the desired applications, specific properties like piezoelectric parameters, the patterning process, the mechanical stiffness and the thermal stability may be taken into consideration and properly adjusted. During the last fifteen years, the implementation of the materials referred as piezo-electrets or ferro-electrets have opened new opportunities in the sensors and actuators design due to their mechanical flexibility and piezoelectric properties. Ferro-electrets or piezo-electrets are composite materials, which are presented in the matrix form containing internal voids organized in a regular or random way. These structures become piezo-electrically functional once opposite charge carriers are implanted in the inner voids' surfaces. These charges are responding to quasi-permanent macro-dipoles, the piezoelectric activity arises from the changes of these internal dipole moments resulting from the mechanical or electrical stresses in these materials. The piezoelectric materials have been already implemented in many applications. In fact, they were used in artificial muscles, microphones, loudspeakers, in shoes for converting mechanical energy into electrical energy, for flexible array position detectors (touchpads) and stretchable large array pressure sensors. In order to control the piezoelectric properties of this structures type, a fabrication process for controlling the shape, the size and the spatial distribution of the voids in the polymer matrix is required. These structures usually exhibit a high piezoelectricity in the thickness direction characterized by a large piezoelectric coefficient d_{33}. This constitutes an advantage for applications requiring electro-mechanical transduction in the thickness direction. The thickness piezoelectric coefficient d_{33} can be easily determined through their direct or converse piezoelectric response. In this paper, we propose a novel low-cost fabrication process of the charged cellular micro-structured polydimethylsiloxane (PDMS) material for ferro-electret micro-sensors and energy harvesting applications. We proposed the dielectric spectroscopy as a powerful tool to test the performance of the PDMS ferro-electret and a mechanical method exhibiting the reliability of such structures for low-frequencies applications.

2. Sample preparation

The polydimethylsiloxane (PDMS) used in this work was prepared using the Sylgard 184 Silicone Elastomer Kit from Dow Corning. The kit is composed of two elements: the pre-polymer (dimethyl-vinyl terminated silicone base polymer) and the curing agent. These elements were mixed at *1:10* weight ratio recommended by the supplier and also used by other authors [1]. The obtained mixed uncured PDMS was degassed in a vacuum desiccator. To develop the cellular micro-structured PDMS material, we proposed a low-cost method that enables us to control the size and the shape of the micro-cavities introduced in the PDMS polymer. The process was carried out in three steps. Firstly, the molds for the micro-structured layers were made using a photosensitive dry film (Riston® MM540, Dupont) bonded to the substrate and a photo-mask, the package was inserted into the UV curing machine (Krub, Kloé) to be exposed to *50%* of the maximum power during *20s* for the photolithography. Then, the photosensitive film was developed using a *1%* anhydrous Sodium Carbonate solution (Na_2CO_3). The obtained molds are made up of cylindrical pads with a height of $40\mu m$, a diameter of $100\mu m$ and a pitch of $150\mu m$ (Figure 1). The micro-structured $40\mu m$ thick layers were obtained by spin-coating the PDMS on the molds that had already been made. Bulk $55\mu m$ thick layers were also obtained by the same technique. The thickness of the PDMS layers was checked with a mechanical sensor (Mitutoyo). The micro-structured PDMS material is presented in the form of a sandwich of three layers, two bulk layers of $55\mu m$ thickness separated by a micro-structured layer of $40\mu m$ thickness (Figure 2). The three layers were bonded together using oxygen plasma treatment for *2s*. Our structures were considerably minimized compared to those implemented by previous work [2].

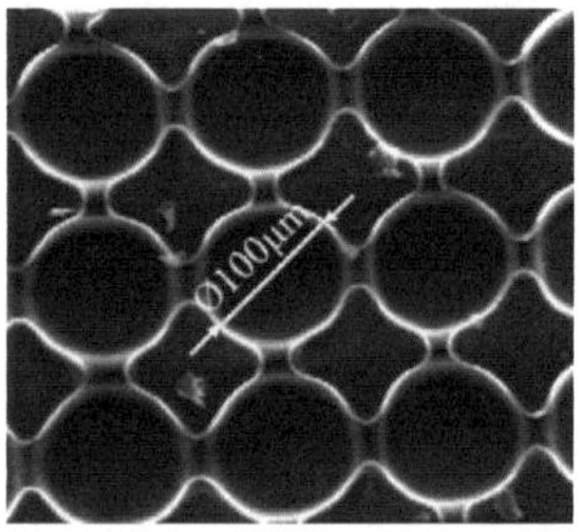

Figure 1. SEM top view of the micro-structured layer with bottom bulk layer.

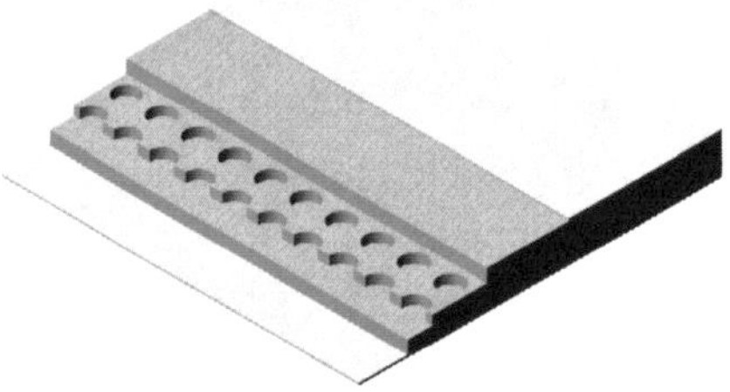

Figure 2. Schematic illustration of the micro-structured PDMS material with top and bottom gold electrodes.

3. Electrical charging

In this study, we used the contact charging, which is the most promising technique for the electrical charging of such structures that can be performed with simple equipment. This technique consists of applying a quasi-static voltage directly across the gold electrodes that have been deposited in advance onto the sample surfaces. The applied electric field that is estimated by the Paschen's law, triggers the electric breakdown into the micro-cavities to generate the macro-dipoles. We chose to charge our structures by applying a quasi-static triangular voltage across the sample electrodes with amplitudes between *1kV* and *4kV* with a frequency of *0.5Hz* for *15min*. The electrical charging equipment is composed of a function generator [Keithley, 3390, 50MHz] coupled with a High Voltage amplifier [Spellman, SL60] in order to provide the needed amplitudes. The amplifier has a breakdown detector to protect the equipment and the samples. The electrical charging cell is provided with electrodes of smaller areas than these of the samples. The cell is installed in a thermalized chamber.

4. Results and discussion

4.1. Dielectric-resonance spectra

The method of dielectric-resonance spectra relies on the converse piezoelectricity of the sample. In fact, the structures deform upon an application of a voltage and vibrate periodically if they are subjected to an ac-voltage. If the ac-frequency coincides with a structure mechanical vibration mode, a resonance will be detected. Dielectric properties of the poled cellular micro-structured PDMS ferroelectret material were reached with the dielectric analyzer [Novocontrol, BDS 20]. Samples were placed in a vacuum chamber to avoid parasitic effects and interferences during the measurements. The temperature was controlled by the temperature programmer [Linkam, TMS94]. All measurements were made under inert gas (Nitrogen) to avoid the oxidation phenomenon at high temperatures. Samples were subjected to an alternative voltage of $1V_{rms}$ in the frequency range *100Hz* to *1MHz* and in the temperature range from *-25°C* to *85°C* by a step of *10°C*. The electroded samples were brought into contact with two measuring points that could vibrate freely with the applied sinusoidal voltage, with only minimal effects from spring-loaded electrode contacts. Figures 3 and 4 represent the frequency-temperature map of the dielectric loss C'' of the non-charged and charged at *25°C* cellular micro-structured PDMS material, respectively.

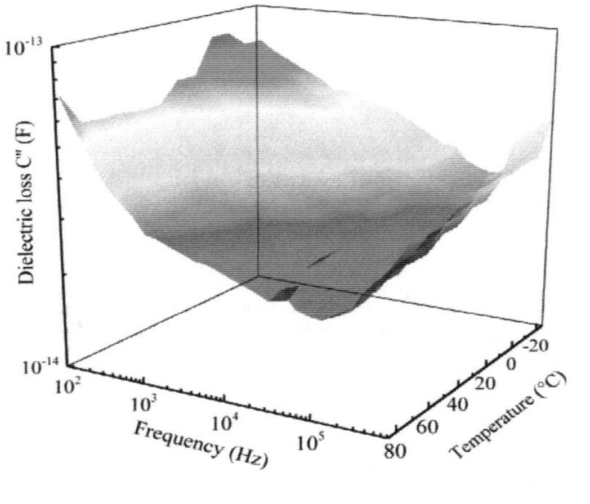

Figure 3. Frequency-temperature map of the dielectric loss of the non-charged micro-structured PDMS material.

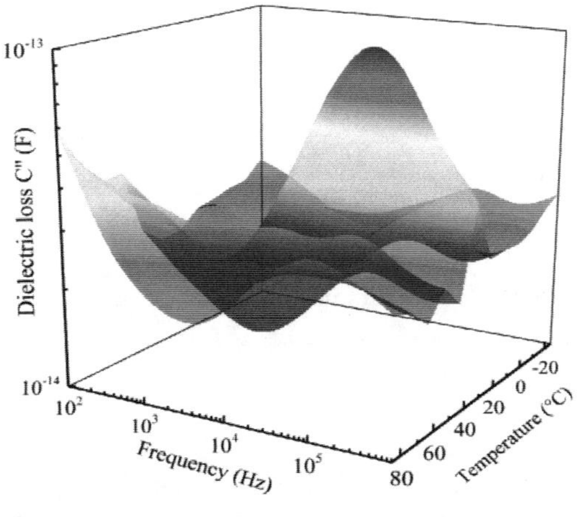

Figure 4. Frequency-temperature map of the dielectric loss of the charged micro-structured PDMS material at *25°C*.

We can observe the resonant phenomena that appear in Figure 4 compared to Figure 3. The dielectric resonance loss C'' spectrum of the piezoelectric micro-structured material near the thickness-extension (TE) mode anti-resonance frequency can be fitted by a least square fit method through the imaginary part of Equation 1 [3, 4].

$$C(f) = \frac{\varepsilon_0 \varepsilon_r A}{h} \frac{1}{1 - k_{33}^2 \dfrac{\tan(\pi f / 2 f_p)}{(\pi f / 2 f_p)}} - iC_{loss} \qquad (1)$$

Where ε_0 is the permittivity of the free space and ε_r is the relative permittivity of the sample, A and h are the electroded sample area and the thickness, respectively. k_{33} is the complex coupling factor and f_p is the complex anti-resonance frequency of the TE mode. C_{loss} is related to the loss that can occur in the material other than those related to the resonance phenomena. f_p is related to the complex elastic stiffness c_{33} and the sample mass density ρ through Equation 2.

$$f_p = \frac{1}{2h}\sqrt{\frac{c_{33}}{\rho}} \qquad (2)$$

The coupling factor k_{33}, is given by Equation 3.

$$k_{33}^2 = \frac{d_{33}^2 c_{33}}{\varepsilon_0 \varepsilon_r} \qquad (3)$$

f_p, k_{33}, and C_{loss} are determined via the fit for each temperature; then, c_{33} and d_{33} were calculated via Equations 2 and 3, respectively. The temperature dependence of these quantities is shown in Figures 5 and 6.

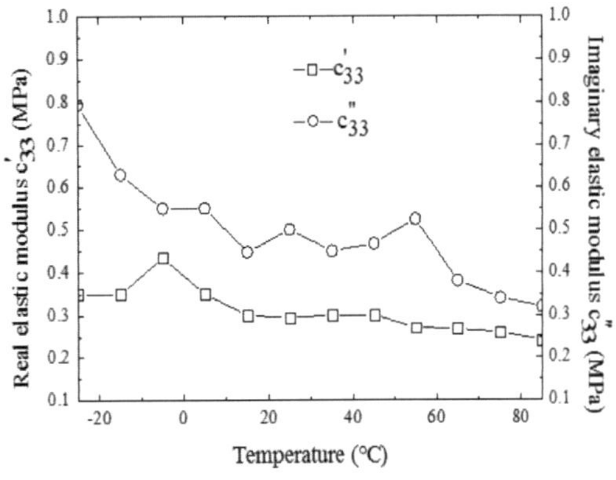

Figure 5. Temperature dependence of the real and imaginary stiffness c'_{33} and c''_{33}.

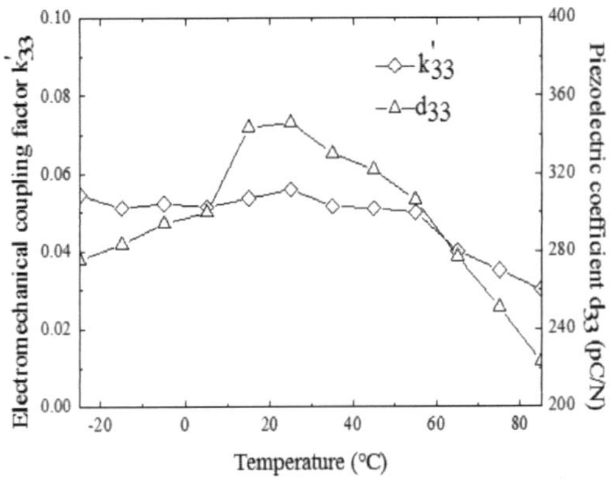

Figure 6. Temperature dependence of the coupling factor k'_{33} and the piezoelectric coefficient d_{33}.

The micro-structured PDMS material shows a low constant elastic stiffness in the temperature range between *-25°C* and *85°C* (Figure 5). However, the low real part of the elastic stiffness c'_{33} is accompanied by an important imaginary part c''_{33} even exceeding the real part. The electromechanical coupling factor k'_{33} represents a maximum value of *0.055*, this value matches the maximum value of the piezoelectric coefficient d_{33} of *350pC/N* at *25°C*. This temperature corresponds to the temperature of the electrical charging process. Moreover, a high compliance modulus of the PDMS, increased by the structure morphology, boosts the piezoelectric coefficient d_{33} and thus the piezoelectric effect of the material. When samples are heated and attain *60°C*, d_{33} decreases considerably compared to its maximum value due to the thermally stimulated loss of internally implanted charge carriers. However, the cellular micro-structured material is still piezo-electrically operating over a wide temperature range from -25°C to 85°C.

4.2. Electromechanical piezo-electret sample testing

The experimental setup used for the electromechanical testing of our structures is similar to that used in work published by Hillenbard and Sessler [5], and Kressmann [6]. The micro-structured PDMS piezo-electret material connected with a wideband charge amplifier (B&K 2634) and loaded with an additional mass of *m=0.1kg* on its top surface is actuated with the shaker (Data Physics-V20) in a frequency range from *10Hz* to *200Hz*. The shaker acceleration is set to $a=0.4m.s^{-2}$. When the piezo-electret sample is under excitation, two forces act on the sample: the static force, f_s, from gravitational acceleration g and the dynamic force, f_d, related to the acceleration a imposed by the shaker to the vibrating mass given by Equation 4.

$$f_s = mg;\ f_d = ma \tag{4}$$

The dynamic force is responsible for the generation of charge carriers on the electrode surface and retrieves the response by the charge amplifier (Figure 7).

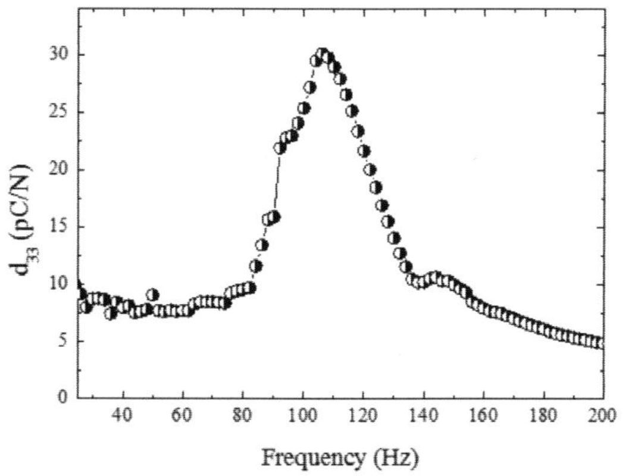

Figure 7. Voltage-to-input force frequency response function of piezo-electret PDMS material.

Figure 7 shows a force sensitivity of *30pC/N* at the vicinity of *110Hz*, which can head towards a use of such micro-structured ferro-electret PDMS material as a micro-sensor in a low-frequency range. The resonance-determined d_{33} is much larger than that obtained from the electromechanical testing. This can be explained by the different measurement conditions used in these two techniques. In fact, for the electromechanical testing method, in addition to the dynamic force used to calculate d_{33}, the sample is loaded with a static force in order to be fixed. Therefore, as the cellular material is compressed, its elastic modulus increases, which decreases the d_{33}. This fact is valid especially for this type of samples presenting a very low elastic moduli. Contrarily, during the resonance determination, the sample is free-standing, and can vibrate freely without any extra load.

5. Conclusion and future work

In this work, a novel low-cost method for implementation of cellular micro-structured ferro-electret PDMS material was proposed. This process allows us to control the shape and the size of the cellular structures. The piezoelectric properties of such structures, investigated using the dielectric resonance spectroscopy, result in high piezoelectric coefficient d_{33}=*350pC/N*. An additional electromechanical testing shows a force sensitivity of *30pC/N* at a frequency of *110Hz* to confirm the reliability of such structures as micro-sensors for low-frequency applications. In the future, we will improve the piezoelectric response of these miniaturized structures for energy harvesting.

6. References

[1] Johnston I D, Mccluskey D K, Tan C K L and Tracey M C 2014 *J. Micromechanics Microengineering* **24** 035017
[2] Shi J, Zhu D and Beeby S P 2014 *J. Phys. Conf. Ser.* **557** 012104
[3] Mellinger A, Wegener M, Wirges W and Gerhard-Multhaupt R 2001 *Appl. Phys. Lett.* **79** 1852–4
[4] Fang P, Holländer L, Wirges W and Gerhard R 2012 *Meas. Sci. Technol.* **23** 035604
[5] Hillenbrand J and Sessler G M 2004 *IEEE Trans. Dielectr. Electr. Insul.* **11** 72–9
[6] Kressmann R 2001 *J. Appl. Phys.* **90** 3489–96

Optimization of a PDMS structure for energy harvesting under compressive forces

J Shi[1], D Zhu[2], Z Cao[1] and S P Beeby[1]

[1] Electronics and Electrical Engineering Group, Electronics and Computer Science University of Southampton, Southampton, SO17 1BJ, UK
[2] School of Computing, Electronics and Maths, Faculty of Engineering, Environment and Computing, Coventry University, Coventry, CV1 5FB

E-mail:js4g11@soton.ac.uk

Abstract. This paper reports the optimisation of a PDMS structure for energy harvesting applications. The PDMS structure was optimised using ANSYS simulation. The fabrication processes were also optimised for maximum PDMS ferroelectret energy harvesting performance. The optimised PDMS structure was fabricated using a 3D-printed plastic mould. The ANSYS simulation and experiment results demonstrated the variation in energy harvesting performance of the PDMS ferroelectret depending upon the different void geometry. The optimised operating frequency is determined by the geometry of the voids inside ferroelectrets and the Young's modulus of the PDMS. The peak voltage and energy generated per strike under a compressive load was obtained experimentally as a function of force applied and frequency. The experimental maximum peak output power of an optimised PDMS structure with an area of 2×2 cm^2 was 0.46 μW across an optimum load resistance of 21 MΩ under a square wave force of 800 N and 17 Hz.

1. Introduction

Ferroelectrets are thin films of polymer foams which can store electric charges in its internal voids, presenting strong piezoelectric-like properties after electric charging [1]. They are able to generate an electric signal during compressive and bending deformations. Due to its high piezoelectric properties, ferroelectrets have already been utilized extensively in sensor or transducer applications, such as microphones, headphones and loudspeakers [2-4].

With the development in the field of wearable electronics, there is a demand for energy supplies, more specifically renewable, long life time energy sources. Kinetic energy harvesting technology that gathers energy from human motion, such as running and walking or ambient vibration, such as machinery vibration, has received growing attention over the last decade. This technology enables wearable devices to be operated using the power that is harvested from human activities. A common approach to harvest energy from human motion is utilizing piezoelectric effect that converts mechanical energy to electrical energy. Commonly, investigated piezoelectric materials for kinetic energy harvesting are lead zirconate titanate (PZT) and polyvinylidene fluoride (PVDF). Even though PZT has relatively high piezoelectric charge constant d_{33}, its high Young's modulus (63 GPa) is unsuitable for wearable applications. On the other hand, PVDF has low Young's modulus (2.9 GPa), which makes suitable for wearable energy harvesting. But, its d_{33} is an order of magnitude lower than that of PZT. Comparing with these two piezoelectric materials, ferroelectrets are light, thin, soft material and have a relatively high d_{33} that is comparable to PZT. These properties make it possible for using ferroelectrets as the active material to harvest energy from human motion and power wearable electronic devices.

The piezoelectric like properties of ferroelectrets has already been theoretically analysed [5, 6]. However, most ferroelectrets are formed by a modified film blow and extrusion process for producing the desired cellular structures. The results of this fabrication method are usually ill-controlled in both individual void geometry and overall cellular structure. Furthermore, the existing modelling methods to approximate the piezoelectric like properties of these materials and structures are simplified models that do not consider individual void geometry and void distribution. We have develop a model that predicts the variation in performance of PDMS ferroelectret with different void geometries [7].This paper presents the optimization, fabrication and testing of an optimized PDMS ferroelectret material for energy harvesting applications.

2. Operating principle

As illustrated in Figure 1, the structure consists of top and bottom electrodes, and two void layers inside. When a force is applied to a charge-implanted cellular film, the thickness of the film, as well as the induced charge density on the top and bottom electrodes will change, which results in the piezoelectric-like effect. An analytical model has been developed to investigate the material properties of the composite structures, with key parameters estimated by ANSYS simulation. Equation (1) and (2) show the relationships between its piezoelectricity and the composite structures.

$$d_{33} = n(n+1)\varepsilon_1 \sigma tr_1 (1 - sr_1)^2 [sr_1[(n+1) + n\varepsilon_1 tr_1]^2 c_{33}]^{-1} \quad (1)$$

$$d_{31} = n(n+1)\varepsilon_1 \sigma tr_1 (1 - sr_1)(sr_2 - 1)\gamma [n(n+1)\varepsilon_1 \sigma tr_1 (1 - sr_1)(sr_2 - 1)\gamma]^{-1} \quad (2)$$

Where ε_1 is the relative dielectric constant of PDMS; n is the number of void layers; c_{33} is the elastic modulus of PDMS; σ is the charge density on void surface; γ is the Poisson's ratio; x_1 and x_2 are the thickness of solid and void layers, respectively; sr_1 and sr_2 are dimensions A_2 divided by A_1 and A_4 divided by A_3, respectively; tr_1 is dimension x_2 divided by x_1.

According to Paschen's law [8], the breakdown voltage for void is described by the equation

$$V_b = apx_2[\ln(pd) + b]^{-1} \quad (3)$$

Where V is the breakdwon voltage; p is the gas pressure in the void; x_2 is the void height; a and b are constants which depend upon the composition of the gas.

Results from the model are presented in Figure 4 and Figure 6.

3. Fabrication processes and experiment

3.1. *Fabrication process for PDMS ferreoelectret*

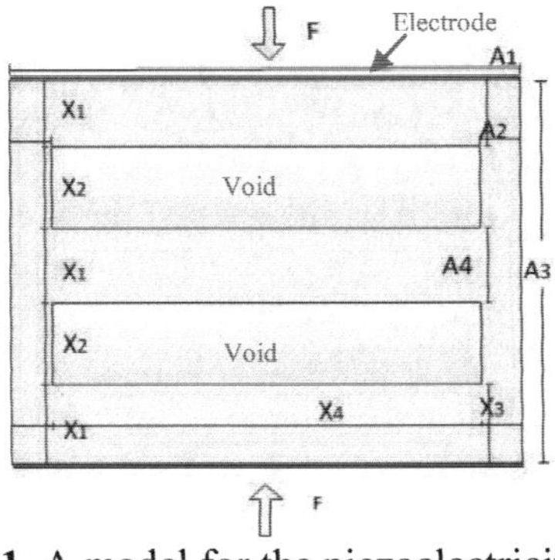

Figure 1. A model for the piezoelectricity of a charge-implanted composite microstructure

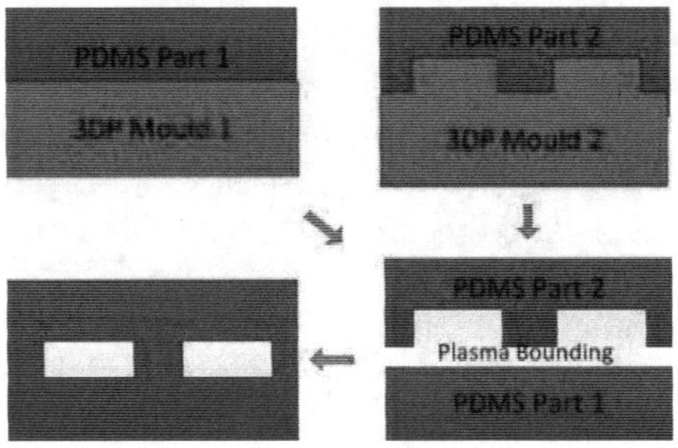

Figure 2. Schematic of fabrication processes

The PDMS ferroelectrets used in the test is fabricated using a moulding process as illustrated in Figure 2. The plastic moulds were 3D- printed by a Connex350 TM 3D printing system (Stratasys, MN, USA). The VerClear^{TM} was selected as the mould material because it is transparent, which makes it easy to observe the entire fabrication process. Because the 3D printable material inhibits PDMS polymerization and the printed structure tends to warp to some extent when it is removed from the printer, the resulting 3D-printed mould is not immediately suitable for PDMS casting. The moulds must be baked in an oven at 80 °C for 24 hours. Next, the moulds were placed and glued using a drop of degassed non-polymerized PDMS to a piece of clear glass, with its base in contact with the PDMS film. After this step, the glass-backed moulds were exposed to a silane vapour for 1 hour to be coated with a thin layer of trichloro(1H,1H,2H,2H-perfluorooctyl)silane (Sigma Aldrich, MO, USA) with promote of the PDMS.

Once the plastic moulds are fabricated, liquid PDMS and curing agent (Sylgard 184 from Dow Corning, MI, USA) were mixed at a 10:1 weight ratio and this mixture was degassed in a vacuum desiccator. The degassed PDMS was poured into the glass-backed mould which were again degassed and then baked at 80 °C for 1 hour. When the degassed PDMS was poured into the mould, the fluid level in the mould was monitored to ensure the degassed PDMS would not overflow. This step is used to control the thickness of the PDMS. After detaching the polymerized PDMS from the moulds, an oxygen plasma treatment was applied on the patterned surface of one PDMS layer and the smooth surface of the other PDMS layer (Femto Asher, Diener, Germany, 30 S at 35-40W) and these were then bonded together. The bonded PDMS was then baked in oven at 80 °C for 1 hour. A 10 N force was applied to the sample during the bake to improve the bonding effect, and also recover the original hydrophobic PDMS surface chemistry. The image of a fabricated PDMS ferreoelectret is shown in Figure 3. After the fabrication, the sample was charged using a corona charging method. For this experiment, corona charging was applied at room temperature and normal atmospheric pressure. A corona-tip voltage of -25 kV and a charging time of 2 mins were employed.

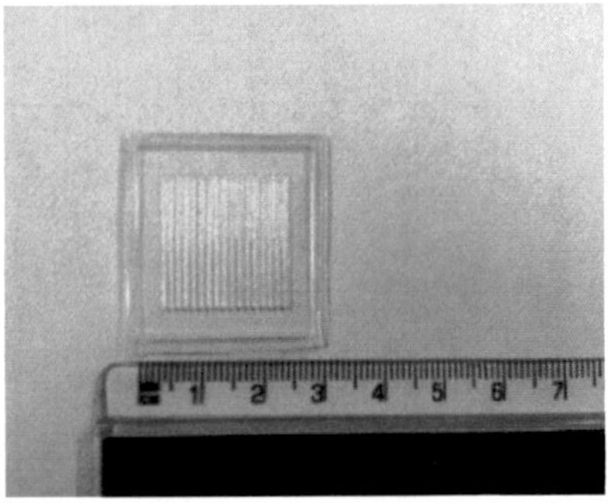

Figure 3. Image of ferroelectret material

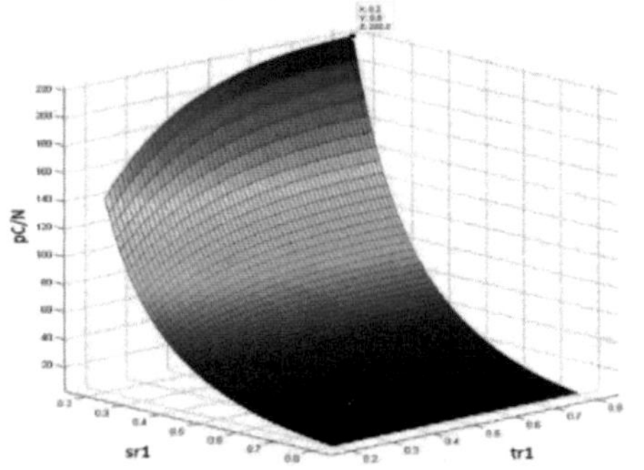

Figure 4. Analytical rectangle model results varying with the size of voids [7]

3.2. Testing

To investigate how the properties of the material are affected by the void structure, three different structures were designed, basing upon the maximum precision of the 3D printer, the restrictions of fabrication process and analytical results (Figure 4). From the Figure 4, the analytically modelled piezoelectric properties of the PDMS ferroelectret are improved with an increase of the void area and the optimized structure was achieved with a ratio tr_1=0.8 for a single void layer of PDMS ferroelectret material [7].The detailed structural parameters are shown in table 1. To quantify the applied forces on the test samples, an Instron electrodynamic instrument (ElectroPuls E1000, Instron Ltd) was used. The maximum dynamic force this equipment can produce is 1000 N and it can provide several types of compressive force including square, sinusoidal and triangular wave forces. The frequency of the compressive forces applied on the test samples was varied from 0 to 100 Hz. These parameters can be modified in the control software.

The output voltages of the PDMS ferroelectret under square wave compressive force were recorded using an oscilloscope. In order to calculate the output energy from the voltage signal, the measurement outputs were transferred from the oscilloscope to a computer and analysed. In addition, the material was also characterized by measuring the output voltage pulses, whilst scanning the load resistance from 1MΩ to 51MΩ.

Table 1. The dimensions for three designed void structures

	Tr_1	Sr_1	Thickness (Solid layer)	Thickness (Void layer)	Width (Void)	Width (Void gap)	Natural frequency (ANSYS)
Structure 1	0.8	0.4	0.4mm	0.32mm	0.75mm	0.5mm	24Hz
Structure 2	0.8	0.4	2mm	1.6mm	0.75mm	0.5mm	21Hz
Structure 3	0.8	0.5	2mm	1.6mm	0.625mm	0.625mm	22Hz

4. Results and Discussion

The output voltage of the ferroelectret PDMS film under compressive force takes the form of a voltage pulse as shown in Figure 5. From this figure, the maximal peak voltage was measured after 1hour and 100 hours after poling and were 0.9 V and 0.48 V, respectively. Thus device voltage is reduced by around 47% after 99 hours. This high decay ratio is related to the poor charge stability at the PDMS/air interface.

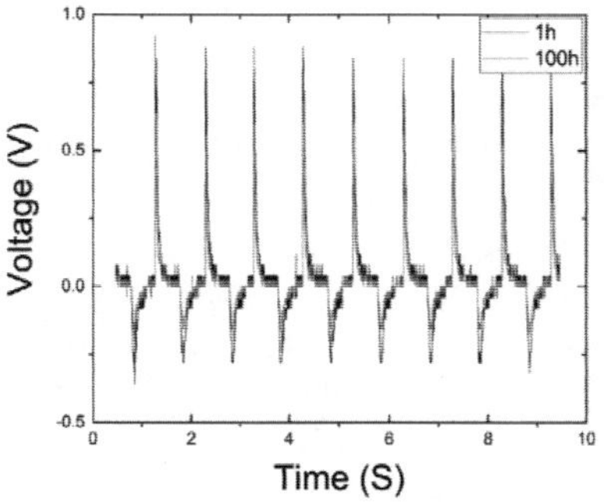

Figure 5. Measured voltage output of structure 1 from a 800 N compressive force, 1 Hz frequency, and 21 MΩ load resistance

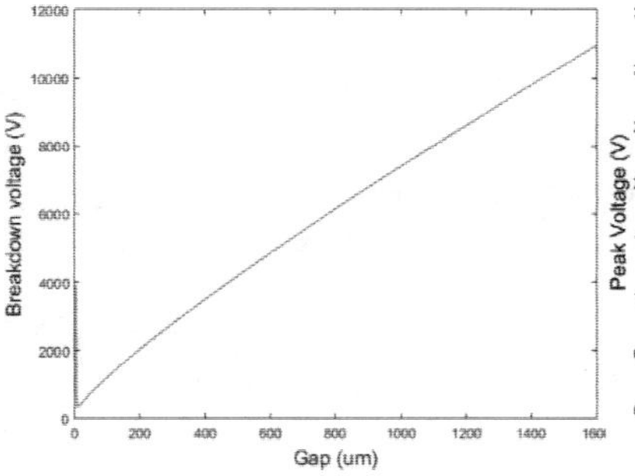

Figure 6. Analytical minimum breakdown voltage as function of the void height

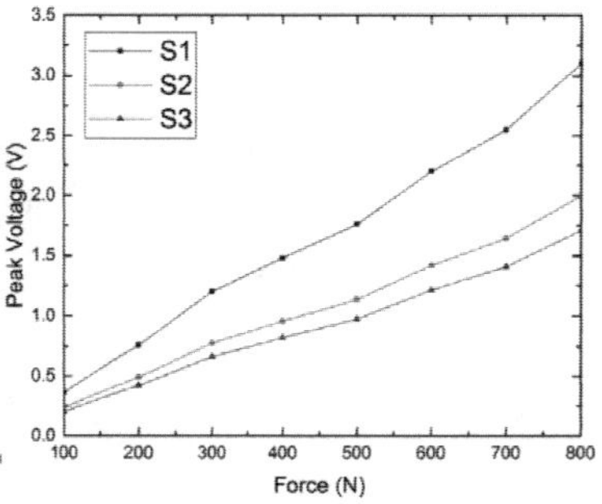

Figure 7. Measured voltage output of from 100 to 800 N compressive force at 17 Hz and 21 MΩ load resistance

Figure 6 shows the minimum required breakdown voltage to achieve polarization for different void height. With the void thickness height, ferroelectret requires a higher voltage to polarize. The thinner ferroelectret can achieve a higher surface charge density than thicker ferroelectet based on the same poling voltage resulting in higher piezoelectric-like properties.

Figure 7 shows the peak output voltage at different applied forces with the frequency of 17 Hz and load resistance of 21 MΩ for these three samples. The output voltage steadily increases with increasing applied force and the maximum peak voltage is achieved at 800 N for these samples.

Figure 8 shows the peak output voltage of the three samples at different frequencies. The output peak voltage initially increases with increasing applied frequency and then start to decline at 17Hz. The optimized operating frequencies are closed to the first natural frequencies which are defined by the geometry and stiffness of materials. The thinner film have higher optimized operating frequency while the lower effective area ratio can reduce the frequency.

Figure 9 shows the characteristic I-P curves of the three PDMS films under 800 N compressive forces with the frequency of 17 Hz, as a function of load resistance. From the figure, it can be observed that the maximum peak power of 0.46 μW can be achieved from the structure 1 PDMS ferroelectret on 21 MΩ load resistance at 17 Hz. These results indicated that the lower effective area ratio and thinner PDMS ferreoelectet can produce more output power. Both theoretical and experimental results indicate that a thinner PDMS ferroelectret will also further enhance the charge long term reliability and charge density.

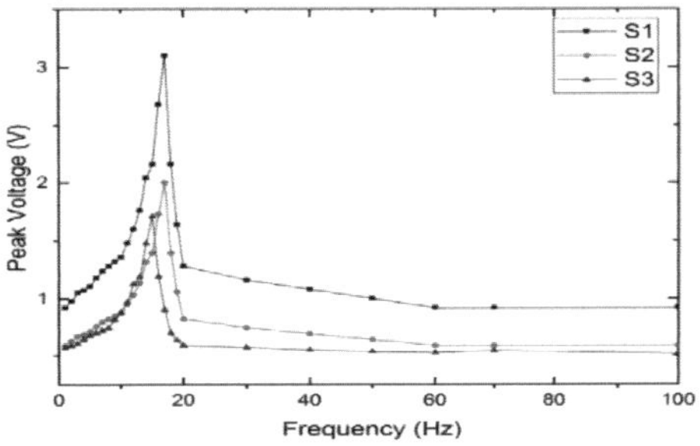

Figure 8. Measured voltage output under 800 N compressive force at different frequency, and 21 MΩ load resistance

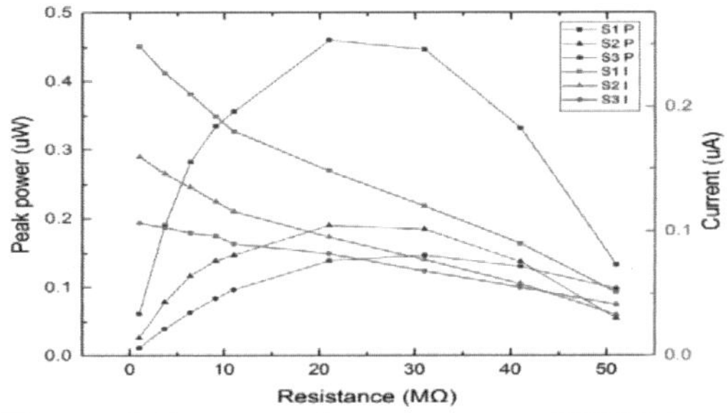

Figure 9. Power-resistance trace of PDMS material under 800 N compressive force and 17 Hz force frequency

5. Conclusion

In this work, the simulation, fabrication and testing of PDMS ferroelectrets with a variety of dimensions has been investigated. The energy harvesting properties of PDMS ferroelectret are improved with an increase in void area and thinner total thickness as predicted by the model. In addition, the operating frequencies can significantly affect the output power. The optimized operating frequency is achieved at the natural frequency which is determined by the geometry of material. This needs to be taken into consideration when designing the power electronics for the energy harvester. The optimized PDMS ferroelectret with the size of 2 cm x 2 cm generates a maximum of 3.1 V, 0.46 μW when connected to a 21 MΩ resistive load under a square compressive force of 800 N at 17 Hz. This equals a power density of 0.12 μW/cm^2. Future work will focus on improving the charge density, stability and integrated in wearable devices.

Acknowledgements

This work was performed under the SPHERE IRC funded by the UK Engineering and Physical Sciences Research Council (EPSRC), Grant EP/K031910/1. Prof Beeby would also like to acknowledge his EPSRC Fellowship, Grant EP/1005323/1

References

[1] G. M. Sessler, "Charge distribution and transport in polymers", *IEEE Trans. Dielectr. Electr. Insul,* Vol. 4, No.5, pp. 614-628, 1997.

[2] S. Bauer, R. Gerhard-Multhaupt, and G.M. Sessler, "Ferroelectrets: Soft electroactive foams for transducers," *Physics Today* vol. 57, no. 2, pp. 37-43, 2004

[3] M. Paajanen, "The cellular polypropylene electret material – Electromechanical properties," *D. Tech. Dissertation, VTT Publications,* Vol. 436, Espoo, 2001.

[4] Michael Wegener and Werner Wirges "Optimized Electromechanical Properties and Applications of Cellular Polypropylene - a New Voided Space-Charge Electret Material" *The Nano-Micro Interface* (H.-J. Fecht, M. Werner, Editors) (WILEY-VCH, Weinheim, 2004) pp. 303-317 (2004).

[5] J. Hillenbrand and G. M. Sessler, "Piezoelectricity in cellular electret films", *IEEE Trans. Dielect. Electr. Insul.*, vol. 7 (4), pp. 537-542, 2000.

[6] M. Paajanen, J. Lekkala, and H. Välimäki, "Electromechanical modeling and properties of the electret film EMFi", *IEEE Trans. Dielect. Electr. Insul.*, vol. 8 (4), pp. 629-636, 2001.

[7] Shi. J, Zhu. D and Beeby. S.P. "An investigation of PDMS structures for optimized ferroelectret performance". *Journal of Physics: Conference Series* , 557, 012104. 2014.

[8] Lieberman, M. A., & Lichtenberg, A. J. "Principles of plasma discharges and materials processing." *MRS Bulletin, 30,* 899-901. 1994

Fabrication and Characterization of Nano/Micro Textured Electret to Avoid Electrostatic Stiction and Enhance Its Surface Potential

M Suzuki[1], M Shimokizaki[1], T Takahashi[1], Y Yoshikawa[2], and S Aoyagi[1]

[1] Faculty of Engineering Science, Kansai University, 3-3-35 Yamate-cho, Suita, Osaka564-8685, Japan
[2] ROHM Co., Ltd., 21 Saiin Mizosaki-cho, Ukyo-ku, Kyoto 615-8585, Japan

E-mail: m.suzuki@kansai-u.ac.jp

Abstract. An SiO_2 electret with nano/micro scaled texture was developed to enhance generation power of our previously developed electrostatic vibration energy harvester. The texture on SiO_2 was formed by lithography and wet etching process. The width/pitch of the each texture is 0.2 μm/0.6 μm or 1.0 μm/2.0 μm, respectively. Both initial value and stability of surface potential of textured electret was better than that of control sample. It was also found that the smaller texture enhance these properties. On the other hand, the stiction force using textured electret was smaller than that using control electret.

1. Introduction

Recently, vibrating energy harvesters (VEHs) which generate electrical power from environmental vibration have attracted considerable attention as one of the power supplies stable for wireless sensor networks [1-5]. Since frequency of environmental vibration is usually on the order of a few to several tens of hertz, electrical power generated by electrostatic VEH using an electret are generally higher than that generated by an electromagnetic VEH using a magnet [3]. Electret is dielectric material in which electric charges are stored semi-permanently.

We previously developed a VEH, as shown in Fig. 1 [4, 5]. This VHE consists of a fixed electret on bade electrode and an insulating plate which is suspended by vertically vibrated spring acts as a counter electrode. In this device, distance of air gap between the electret and the insulating plate is periodically changed according to vibration of the spring; and the output becomes maximum level when the insulating plate is closest to the electret. Surface potential of electret is most impotent characteristic for the VEH because the output power is theoretically proportional to square of it. However, the insulating plate are stuck to the electret by electrostatic force when the surface potential was exceeds a certain amount. This problem is generally called "stiction." To overcome these problems, we tried to develop a novel electret via addition of nano/micro scald texture on its surface, which has both high-surface potential and low stiction force. In this paper, we present the fabrication method and characteristics of the developed electret. The output power of the VEH using developed electret is also shown.

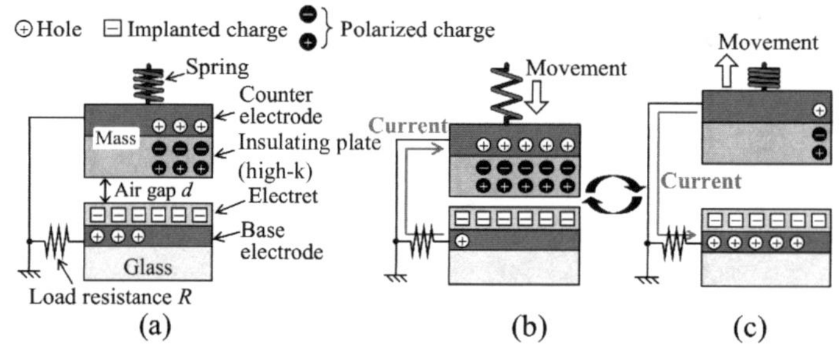

Figure 1. (a) Schematic of developed VEH in which the counter (upper) electrode and insulating are vertically vibrated by external ambient vibration. (b) When the upper electret and insulating plate approach to the electret, the insulating plate was polarized by electric potential difference from the electret, and it attracts electric charge. (c) When the counter electrode and insulating plate move away from the electret, electric charge is attracted to the base (lower) electrode. The movement of the electric charge is changed to electric energy by the load resistance.

2. Fabrication of nano/micro textured electret

Process flow for fabricating nano/micro textured electret is shown in Fig. 2. At first, a low resistive silicon substrate (p-type, (100) orientation, resistivity < 0.01 Ω·cm, 500 μm in thickness) was thermally oxidized in vapour of deionized water at 1,000°C to fabricate SiO_2 film. SiO_2 is one of most famous electret material. In this study, two-types of SiO_2 film with different thickness were fabricated, one of which is 0.8 μm and the other of which is 1.6 μm. Then, arrays of square shaped electron beam (EB) resist were formed on the surface of SiO_2 (0.8 μm in thickness) by EB lithography, of which the width and pitch are 0.2 μm and 0.6 μm, respectively. As the same time, arrays of square shaped electron photoresist were formed on the surface of SiO_2 (1.6 μm in thickness) by photolithography, of which the width and pitch are 1.0 μm and 2.0 μm, respectively. Schematic of fabricated SiO2 electret with nano/micro scaled texture. Since drawing speed of the EB lithography is much slower than that of photolithography, EB lithography was used only to form the nano scaled texture. Next, the SiO_2 films were etched using hydrofluoric acid diluted by DI water at 10%. The etching depth of SiO_2 is equal to with of EB/photoresist. Next, EB/photoresist was removed, followed by wiring to the silicon substrate. Total area of the nano/micro texture is 225 mm^2 (15 mm × 15 mm), as shown in Fig. 3. Finally, electrons were implanted to the textured SiO_2 by corona discharge method to form electret. The setup for corona discharge is illustrated in Fig. 3. SEM images of textured SiO_2 are shown in Fig. 4. Control samples (non-textured SiO_2 electret, 0.8 μm in thickness) were also fabricated.

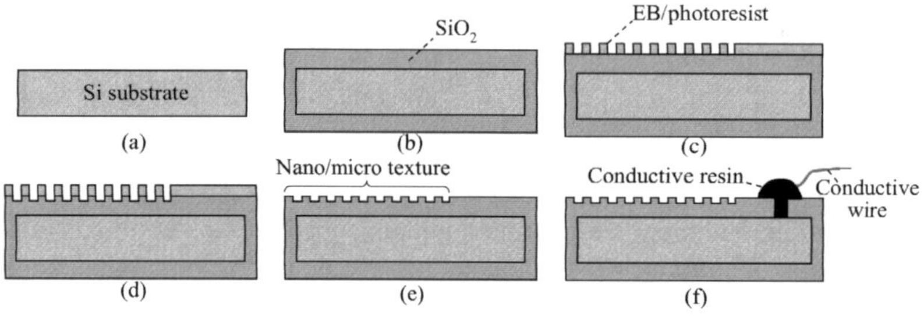

Figure 2. Fabrication flow of the nano/micro textured electret. (a) Cleaning of a low conductive silicon substrate. (b) Thermal oxidation of the silicon substrate to form SiO_2 film. (c) EB lithography or photolithography to form nano/micro scaled arrays. (d) SiO_2 etching by diluted HF acid, in which the arrays made of EB/photoresist was used as etching mask. (e) Removal of EB/photoresist. (f) Wiring to the silicon substrate using conductive resin.

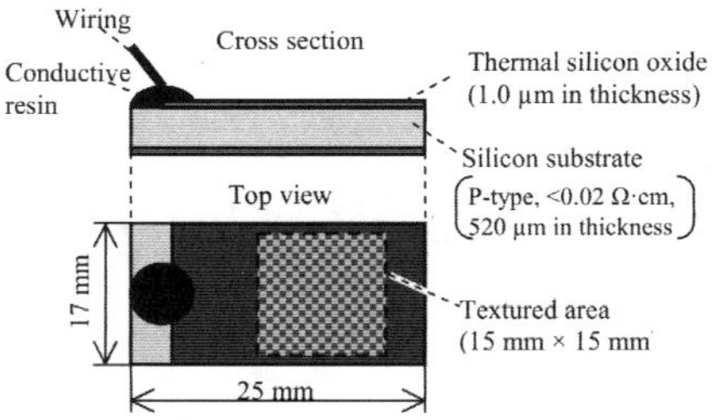

Figure 3. Schematic of fabricated SiO_2 electret with nano/micro scaled texture.

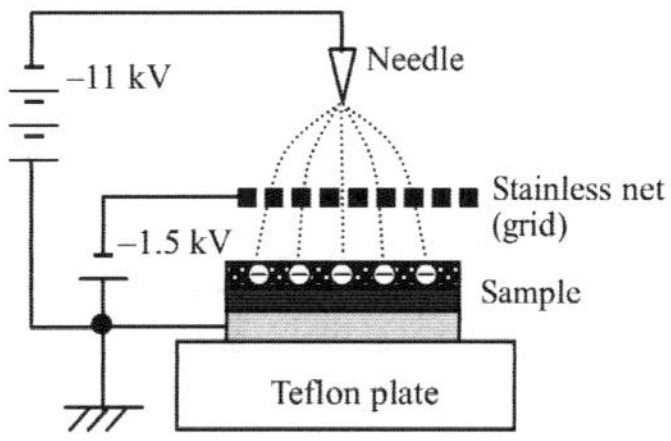

Figure 4. Schematic of setup for corona discharge to implant electron to the nano/micro textured SiO_2 film.

3. Measurement result and discussion

3.1. Surface potentials and those stability

The surface potential of fabricated electret was measured by a surface electrometer. Initial surface potential of the electrets and those time degradation are shown in Fig. 6. These results indicate that both initial value and its stability of the textured electrets are better than those of control sample. It was also found that the smaller texture enhance these advantage.

3.2. Stiction force

The setup for measuring stiction force between the high-*k* insulating plate under counter electrode and electret on base electrode is shown in Fig. 7. In this setup, the electret is fixed to an electronic balance. First, SiO_2 electret was contacted with high-k plate with counter electrode. Then the electret was lifted off. The electrostatic attractive force of the moment, in which the two parts leaved, was measured. This measurement was repeated 50 times for each sample. The measurement results are shown in Fig. 8. This graph shows that the stiction force between textured electret and the insulator is smaller than that between control electret and the insulator. Additionally, it was found that the surface potential of textured electret was hardly decreased by the measurement of stiction force (it means 50 times repetition of contact and release), as shown in Table 1.

3.1. Discussion

From the results obtained in these investigations, we propose a model of nano/micro textured electret shown in Fig. 9. In this model, implanted charges are mainly trapped near electret surface. Since surface area of the textured electret is larger than that of non-textured electret, the amount of electrical charge kept in the textured electret is larger than that kept in the non-textured electret.

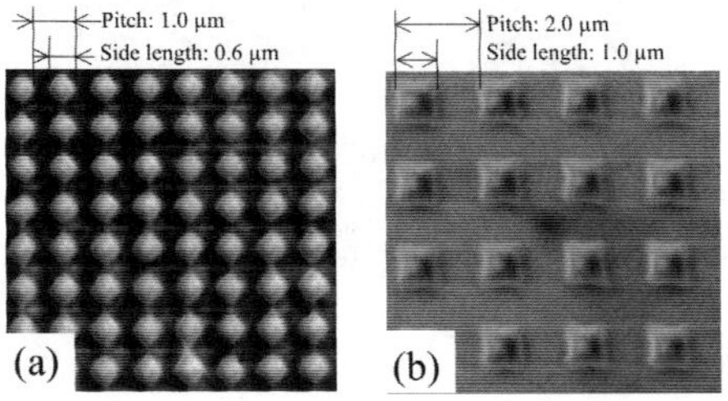

Figure 5. SEM images of surface of textured electrets, fabricated using (a) EB lithography or (b) photolithography.

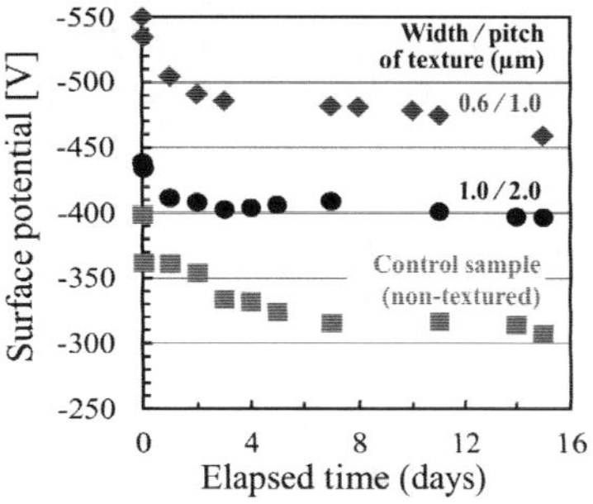

Figure 6. Time degradation of surface potential of textured and control (non-textured) electrets made of SiO_2.

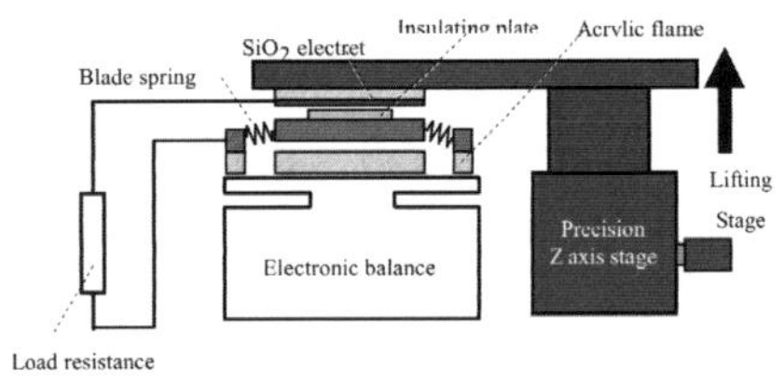

Figure 7. Schematic of setup for stiction force measurement.

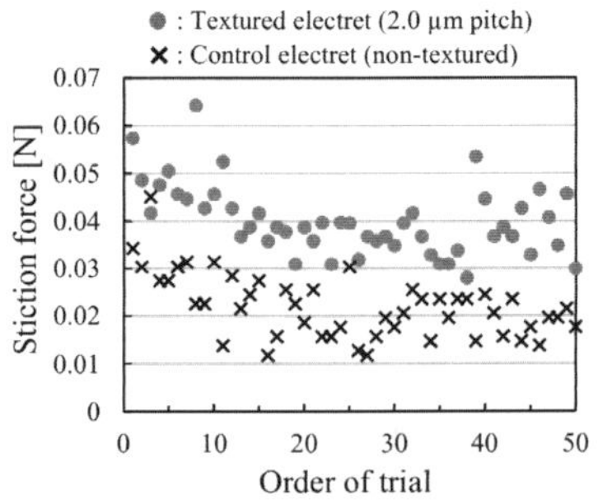

Figure 8. Measured stiction force between electret and insulating plate with counter electrode.

Table 1. Summary of the measured stiction force between electret and insulating plate with counter electrode, and surface potential before and after testing.

	Textured electret	Control electret
Side length / Pitch	2.0 μm / 1.0 μm	NA
Stiction force	22 mN	40 mN
Surface potential before testing	-500 V	-330 V
Surface potential after testing	-420 V	-240 V

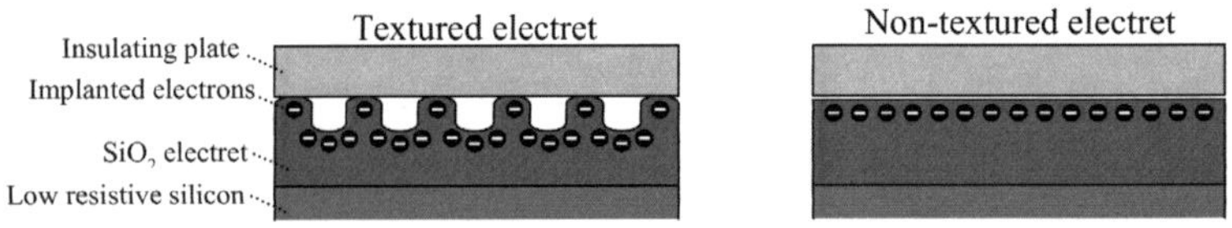

Figure 9. Proposed mode of (a) textured electret which has both high surface potential and low stiction force, and (b) non-textured electret.

Next, we also consider stiction force of the textured electret using the proposed model. The textured electret contacts with the insulating plate with counter electrode only at its top, as shown in Fig. 9(a). Therefore, the stiction force becomes relatively smaller because the electric attraction is inversely proportional to square of distance between two objects. On the other hand, whole surface of control electret contacts with the counter electrode, as shown in Fig. 9(b). Therefore, electrostatic attraction between electret and the insulating is probably lowered by formation of texture to the electret.

4. Acknowledgments

This work was supported in part by a grant of "Strategic Research Foundation Grant-aided Project for Private Universities": Matching Fund Subsidy of MEXT (Ministry of Education, Culture, Sport, Science, and Technology, Japan), 2015-2019. This work was supported in part by JSPS (Japan Society for the Promotion of Science) KAKENHI (26870728).

5. References

[1] S. P. Beeby, M. J. Tudor, and N. M. White 2011 *Meas. Sci. Technol.*, **17**, 175-195
[2] K. Matsumoto, K. Saruwatari, Y. Suzuki 2011 *in Digest Tech. PowerMEMS'11*, 134-137
[3] T. Tsutsumino, *et al.* 2006 *in Digest Tech. IEEE Internat. Conf. MEMS'06* 98-101
[4] T. Takahashi, *et al.* 2012 *in Digest Tech. MEMS'12* 1265-1268
[5] T. Takahashi, *et al.* 2013, *in Digest Tech. MEMS'13* 877-880

History and modern applications of nano-composite materials carrying GA/cm² current density due to a Bose-Einstein Condensate at room temperature produced by Focused Electron Beam Induced Processing for many extraordinary novel technical applications.

Hans W. P. Koops
CEO of HaWilKo GmbH, Ober-Ramstadt, Germany

E-mail: hans.koops@t-online.de, www. hawilko.com

Abstract The discovery of Focused Electron Beam Induced Processing and early applications of this technology led to the possible use of a novel nanogranular material "Koops-GranMat®" using Pt/C and Au/C material, which carries at room temperature a current density > 50 times the current density which high TC superconductors can carry. The explanation for the characteristics of this novel material is given. This fact allows producing novel products for many applications using Dual Beam system having a gas supply and X,Y,T stream data programming and not using GDSII layout pattern control software. Novel products are possible for energy transportation, -distribution,-switching, photon-detection above 65 meV energy for very efficient energy harvesting, for bright field emission electron sources used for vacuum electronic devices like amplifiers for HF electronics, micro-tubes, 30 GHz to 6 THz switching amplifiers with signal to noise ratio >10(!), THz power sources up to 1 Watt, in combination with miniaturized vacuum pumps, vacuum gauges, IR to THz detectors, EUV- and X-Ray sources. Since focusing electron beam induced deposition works also at low energy, self-cloning multibeam-production machines for field emitter lamps, displays, multi-beam - lithography, - imaging, and - inspection, energy harvesting, and power distribution with switches controlling field-emitter arrays for KA of currents but with < 100 V switching voltage are possible. Finally the replacement of HTC superconductors and its applications by the Koops-GranMat® having Koops-Pairs at room temperature will allow the investigation devices similar to Josephson Junctions and its applications now called QUIDART (Quantum interference devices at Room Temperature). All these possibilities will support a revolution in the optical, electric, power, and electronic technology.

Contents
The lecture reports on many subjects.

Discovery of Focused Electron Beam Induced Processing [1].
Early applications of this technology [2] for Micro-data recording [3] and high resolution Projection Lithography[4].

Discovery of the novel nanogranular material in 1994, now called " Koops-GranMat®", which carries at room temperature a current density up to 3 GA/cm² [5].

Explanation of the novel nanogranular material "Koops-GranMat®" having a Bose Einstein Condensate at room temperature due to the deposition process [6] and the materials growth process.
A recently published paper by L. Butov and Remeika, from UCSD allowed to calculate the temperature at which the Bose Einstein Condensate in Koops-GranMat® can form [7].

Koops-GranMat® is not a superconductor, but carries 100 times higher current density than those.
In High TC superconductors the repelling force of 2 electrons is balanced by the antiparallel spins with their attractive power [8] and which are in general composed from large crystalline compounds, which contain many acoustic and optical phonons. Superconductivity is only achieved by cooling the material to the temperature, where the resistance drops from a finite value to zero (TC) to allow the formation of the Cooper-Pairs. Koops-Pairs are formed in the condensate of Koops-GranMat®. This, however, is free from optical phonons due to geometry quantization in the nanocrystals , even at room temperature [9].

Novel products for energy transportation, -distribution, and -switching are recently presented [10] and patented [11]. Photonic crystals can be built with this deposition method to be used as WDM Filters for the optical net [12].
Very sensitive photon-detection at room temperature above 65 meV energy allows IR cameras without a cooling device, and also very efficient energy harvesting. At red light Koops-GranMat® has an efficiency which is 3 times that which Si solar cells have in the entire white spectrum.

If applicability for powerful electron sources for nano-electronics is requested, all heated cathodes are excluded, diamond delivers at room temperature sufficient current for radiation insensitive electronics. Koops-GranMat® surpasses all other characteristics by very high currents and current density, and, which is most important, low required extraction voltage. Maximum currents of 1 mA from one tip are obtained at 24V (Au/C) and 70 V(Pt/C). Such voltages can be handled in nano-electronics wiring patterns with 1 µm pitch [13].
Bright field emission electron sources are used for miniaturized vacuum electronic devices, a miniaturized vacuum pump and vacuum gauges, like an "Orbitron" pump [14].
Amplifiers for HF electronics, micro-tubes, 30 GHz to 6 THz switching amplifiers with signal to noise ratio (S/N) < 10(!) are possible, since the cathode to Grit capacitance is 0.24 AF (240 FF)
Very powerful THz sources with a high Signal to Noise ratio are possible.[15].

Table 1: Characteristics for IR-sources in the 0.2 to 10 THz range

Frequency THz	Wavelength µm	Resonator E0 µm	1 Electron pulse at 100 V travels µm = IR window µm	S/N at 1 mA DC
0.2	1500	750	15	173
0.5	600	300	6	77,5
1	300	150	3	54
5	60	30	0,6	24,6
10	30	15	0,3	17

EUV- and X-Ray sources can make use of the material which cannot get hot due to the nanocrystalline metal/carbon compound structure used for the cathode and the anode. There only Rutherford scattering can act and escaping electrons can be retarded not to heat the tube wall [16].

Focused electron beam induced deposition works also using low energy electrons and allows by FEBIP Self-cloning the multibeam-production of SEM's for massive parallel beam machines [17],[18]
Other mass producible objects are: Field emitter lamps and displays, photosensitive areal devices for energy harvesting, power distribution with switches controlling field-emitter arrays for KA of currents but with < 100 V switching voltage.

"Nano Granular material" protects Koops-GranMat® and other material combinations and their many applications forms. Applicant Dr. Hans W. P. Koops containing a) European Patent Application No. 12 183 564.9 (KOO 7532 EP -. Katscher, Habermann patent attorneys Darmstadt), b) National applications: China, USA, Germany under process [19]
Finally: Koops-GranMat® having Koops-Pairs at room temperature and it's applications can replace all uses of HTC superconductors cooled to 40K . E.g. cooled SQUIDS by at Room Temperature operating QUIDART (Quantum Interference Device At Room Temperature), Magnets, Q-Bit-computers etc. Patents filed 2014 in Germany.

Koops-GranMat® is discovered in 1994 at FTZ German Telekom Research Centre as a novel material which is capable to replace cooled superconducting materials, and can revolutionalize energy transport, detection, Field-Emitter lamps and displays, GHz to THz switching electronics, THz radiation power sources, and X-Ray sources.

Conclusions

The applications of Koops-GranMat® can revolutionalize the electronic technology in a dramatic way, with the capability of 100 times of what super-conducting materials can do, but with the advantage to allow miniaturized tube electronics for GHz to THz applications and powerful miniaturized X-ray tubes which operate at room temperature and have no need for expensive cooling systems.

Acknowledgments

The author appreciates the cooperation and advice of Dr. Hiroshi Fukuda, Hitachi High Technologies Tokyo, Japan and his theoretical support, and is grateful to Dr. J. Sellmair, Sellmair Nano-electrotechnics GmbH, DE, and to the RAITH GmbH, DE, for experimental support. He thanks Prof. H. L. Hartnagel, IMP TU Darmstadt, DE, for metrology support and for discussions on THz radiation applications, Dr. U. Koops, his son, DE, for chemistry discussions, and also Dr. H. Ito, Hitachi High Technologies, Japan, for supports.

References

[1] R. Speidel Optik 23 125, 1965/66

[2] G. Heide Z. Angewandte Physik 15 (1963) 116, P. Holl Optik 30 (1970) 116.

[3] G. Möllenstedt, R. Schief, et al. (1968). "Electron Optical Micro Recorder Controlled by Television." Optik 27(7)

[4] H. Koops, "Verwendung der Kondensor-Objektiv-Einfeldlinse zur elektronenoptischen Mikrominiaturisation", Optik 29,(1969) 119 – 121, and H. Koops, "Elektronenoptische Herstellung von Transmissionsgittern mit 100 nm Gitterkonstante für weiche Röntgenstrahlung", Dissertation Tübingen 1971

[5] H.W.P. Koops, J. Kretz, M. Rudolph, M. Weber, G. Dahm. K.L. Lee, "Characterization and application of materials grown by electron beam induced deposition", Jpn. J. Appl. Phys Vol. 33 (1994) 7099-7107.

[6] H.W.P. Koops, A. Kaya, M. Weber,"Fabrication and Characterization of Platinum Nanocrystalline Material Grown by Electron-beam Induced Deposition", J. Vac. Sci. Technol. B 13(6) Nov/Dec (1995) 2400-2403

[7] H. Koops, H. Fukuda, "Giant current density via indirect exciton orbit overlapping in polarized nanogranular materials" J. Vac. Sci. Technol. B 33, 02B108 (2015),

http://dx.doi.org/10.1116/1.4904732

[8] B. Steele, Cornell Chronicle July 28, **(2014)** , and J. C. Séamus Davis, Cornell, S. Avici et al. *Nature Comm.* 5, Article number: 3845, 22 May **(2014)**, and D.S. Inosov, et al. Nature Physics Vol.6. 178, March **(2010)**

[9] Patton K. R. & Geller, M. "Phonons in a nanoparticle mechanically coupled to a substrate", Phys. Rev. B 67, 155418 (2003).

[10] Hans W.P. Koops "Focused electron beam induced processing, a technology to develop and produce miniaturized electron-, IR, THz-, X-ray sources, high resolution detectors and sensors for IR- and X-ray tomography" Invited paper Proc. Eurosensors XXVI, September 9-12, **2012**, Kraków, Poland, Procedia Engineering Volume 47, 2012, Pages 1460–1465

[11] Hans W.P. Koops "Nano Granular Materials (NGM) material for sheets of high refractive index, for high current density carrying wires, sheets and IR to X-ray electromagnetic radiation sensitive photon detector and process for making the same" EP 12 183 564.9, 7. September 2012

[12] H. W. P. Koops, O. E. Hoinkis, et al. (2001). "Two-dimensional photonic crystals produced by additive nanolithography with electron beam-induced deposition act as filters in the infrared." Microelectronic Engineering 57-8: 995-1001.

[13] F. Floreani, H.W. Koops, W. Elsäßer Concept of a miniaturised free-electron laser with field emission source, Nuclear Instruments and Methods in Physics Research A 483 (2002) 488-492

[14] Koops, Hans Wilfried Peter, Orbitron-Pumpe, DE000010241549B4 05.09.2002 HaWilKo GmbH. a) DE 102 41 549, b) EU 03 015 579.0,

[15] Dr. Hans WP Koops applicant "A device for generating THz radiation with free electrons". a) EU 10 170 823.8 b) PCT / EP 2011/062803

[16] Applicant Dr. Hans W.P. Koops "Miniaturized X-ray Tube with Koops-GranMat® cathode, anode, and decelerator" AZ 102015001440.4 , Filed to DPMA Munich 9.2.2015

[17] Hans W.P. Koops "Method and apparatus for the production of particle beam systems" a) DE No. 02 103 794.7, b) EP 04704570.3, c) US 10 / 543.175, and also for the production of products for energy distribution-switches, lithography, imaging, and inspection

[18] Hans W.P. Koops, HaWilKo GmbH Ober-Ramstadt, Germany," Fast lithography for dyes, how to speed up defect review, and precision photo mask repair" Invited paper LSITS Nov. 2012 Osaka, Japan

[19] Hans W. P. Koops a) European Patent Application No. 12 183 564.9 b) China, USA, DE

Flexible supercapacitor based on MnO_2 coated laser carbonized electrodes

Rahim Rahimi [1,2], Manuel Ochoa [1,2], Wuyang Yu [1,2], and Babak Ziaie [1,2]

[1]Birck Nanotechnology Center, Purdue University, West Lafayette, IN 47907, USA
[2]School of Electrical and Computer Engineering, Purdue University, West Lafayette, IN 47907, USA

E-mail: bziaie@purdue.edu

Abstract. This paper presents a facile, low-cost approach for fabrication of flexible hybrid carbon/ MnO_2 pseudo-capacitors. The highly porous carbon electrodes of the supercapacitor are fabricated by laser pyrolysis of polyimide-laminated copper sheet and subsequently coated with a uniform thin layer of MnO_2. The porous laser carbonized polyimide provides a high effective surface area for the MnO_2 coating, resulting in an increase of 55% in the electrochemical performance of the supercapacitor. The fabricated device exhibits a specific capacitance of $7.1mFcm^{-2}$ at a scan rate of $40mVs^{-1}$. Moreover, the copper backing film provides a proper electrical contact to the high surface area carbon/ MnO_2 composite for stability under mechanical deformation and low internal resistance.

1. Introduction

The recent push for more human-friendly, wearable electronics creates a demand for portable energy storage units that are lightweight and flexible [1]. Supercapacitors are a promising candidate to complement or replace conventional rechargeable batteries due to their high power density and multi-cycle lifetime [2]. Electrochemical double-layer capacitors (EDLC), in particular, are suitable for use in flexible electronics due to light weight and reliability, but they are ultimately limited by surface area [3]. Recently, researchers have explored the use of carbon-based materials such as graphene and CNT as active electrodes in EDLCs due to their extraordinarily high surface area, chemical inertness, and desirable electrical properties [4]. Their storage performance can be further improved by incorporating metal oxides as pseudo-capacitive materials that undergo fast reversible redox reactions; such as MnO_2, RuO_2, and Fe_2O_3 [5]. Despite their enhanced capacitive performance, the synthesis processes for obtaining graphene and CNT are complex and require access to cleanroom facilities. Recently, several reports have shown more economical approaches using graphene oxide reduction [6] and high temperature pyrolization of organic materials [7], which unfortunately suffer from decomposition of GO over time and high temperature processes.

As an alternative approach, several groups including our own have explored the use of laser carbonization of commercial polymeric material to fabricate micro/nano carbon structures for various sensing [8] and energy storage application [9]. Herein, we extend this method to the fabrication of a simple and low-cost carbon/MnO_2 pseudo-capacitor, created by laser pyrolization of the polyimide layer of a commercially-available Cu/Kapton substrate followed by uniform *in situ* generation of MnO_2 on the carbonized polyimide, Figure 1. This strategy brings improvements over previous laser-assisted supercapacitor efforts by offering three distinctive advantages: (1) improved electrical contact between the carbon nanoparticles and the bottom current collector to reduce the internal resistance; (2) an increase of 55% in specific capacitance with a MnO_2/carbon hybrid structure; and (3) more compact and stackable architecture with low interconnection resistance.

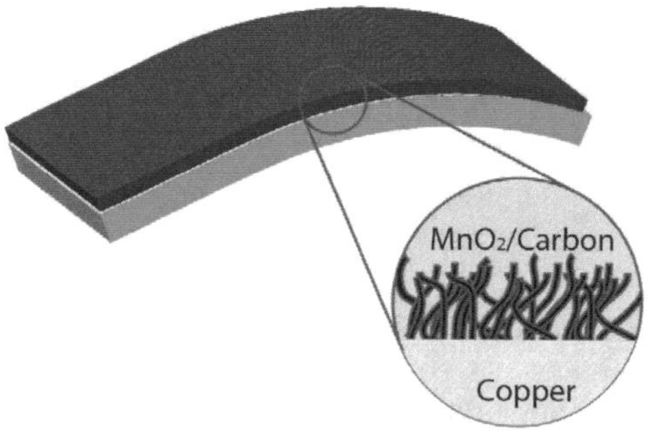

Figure 1. Schematic illustration of the flexible MnO_2 coated porous carbon supercapacitor electrode. The highly porous laser carbonized polyimide is connected to the bottom copper collector. Circular inset shows a schematic of the hybrid MnO_2/porous carbon.

2. Fabrication

Figure 2 illustrates the fabrication process of the flexible supercapacitor. For the fabrication process a commercial polyimide-laminated copper sheet (Pralux AC, DuPont ®) with a polyimide thickness of 9 μm and copper thickness of 12 μm is used as the precursor for carbonization and current collector, respectively. The polymer and copper surfaces are first cleaned with DI water followed by isopropyl alcohol (IPA). To define 15 mm × 15 mm copper pads, the copper is masked and subsequently etched. The mask is defined by screen printing a layer of acrylic ink onto the copper side of the film using a laser-patterned tape as a stencil (pattern defined by a CO_2 laser engraver, Universal Laser Systems, Inc., Scottsdale, AZ). After screen printing, the tape is carefully removed and the ink is fully cured by annealing on the hotplate at 80 °C for 10 min. The exposed copper is then removed by a wet etching process in a commercial copper etchant solution (RadioShack®) at 80 °C, continuously agitated for 15 min. The ink mask is subsequently removed by immersing the film in acetone. Next, the polymer surface is cleaned in a solution of IPA/DI with a 1:1 ration and sonicated for 10min, followed by rinsing with DI water and drying at 80 °C for 2 hours.

The conductive carbon patterns are directly pyrolyzed onto the polyimide side of the film using the CO_2 laser engraver at the optimal setting (7.5 W power, 1.4 ms^{-1} speed) that was previous reported by our group [8]. This process pyrolizes the polyimide film into highly porous carbon composed of micro/nano carbon structures throughout the thickness of the polyimide film and forms an electrical contact with the bottom metal collector. During the laser carbonization process the polyimide-laminated copper is attached to a PET sheet with double-sided tape for handling rigidity and for keeping the film surface flat during processing. The entire laser carbonization process is performed under ambient conditions.

The carbon/MnO_2 composite is prepared by precipitating MnO_2 microparticles in the carbonized patterns in pH 7 conditions. For this process the carbonized film is immersed in a beaker with a 0.1 M $KMnO_4$ solution. Then a solution of 0.1 M Na_2SO_4 (Alfa Aesar) is poured into the above mention solution (1:1 v/v), which is agitated and maintained at 80 °C for 5 min. The thickness of the MnO_2 coating can be adjusted by adjusting the immersion time. Previous studies have shown that in neutral pH solutions, permanganate ions (MnO_4^-) can be reduced to MnO_2 on various nano carbon surfaces (e.g. CNT, graphene, and nano petals) via the following reaction [10]:

$$MnO_4^- + 3C + H_2O \leftrightarrow 4MnO_2 + CO_3^{2-} + 2HCO^{3-} \quad (1)$$

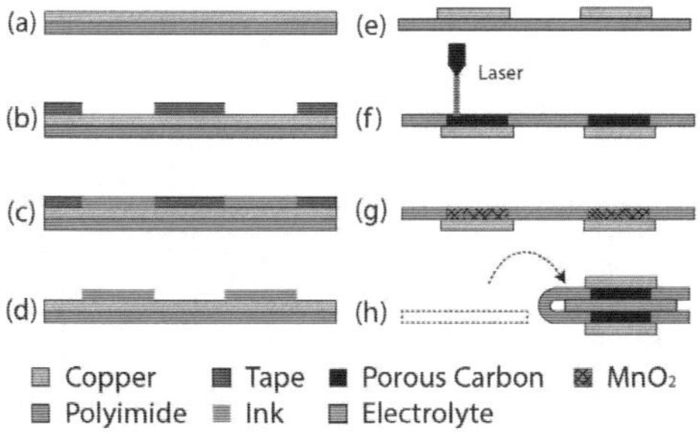

Figure 2. Fabrication process of flexible supercapacitor using laser pyrolization of polyimide. (a-d) screen printing insulating ink; (e) wet etching copper; (f) laser carbonizing polyimide sheet; (g) coating the carbonized polymer with MnO_2; (h) fold the electrodes with electrolyte.

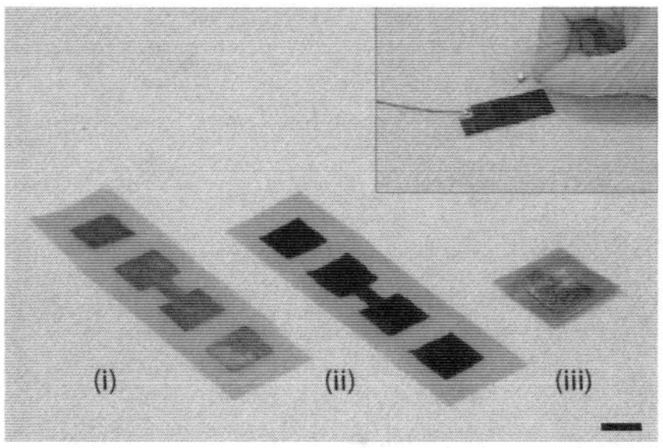

Figure 3. Photograph of assembly process of two stacked supercapacitors, connected in series. (i) Wet etched copper pads; (ii) laser carbonized polyimide with MnO_2 coating; (iii) the final folded and assembled supercapacitors. Inset shows electrical connection between porous carbon and copper pads. Scale bar: 10mm.

The samples are then rinsed in deionized water and annealed on a hotplate at 100 °C for 2 hours. In order to test the performance of the capacitor, the substrate is folded with a filter paper used as a separator between the two electrodes and a 3 M aqueous solution of KCl used as the electrolyte. Figure 3 shows a photograph of the final fabricated device with the intermediate fabrication steps of two stacked supercapacitors electrically connected in series. Figure 3i illustrates the electrical connection pads after the copper etching process. Figure 3ii shows the porous carbon pads after the laser carbonization process of the polyimide film. Figure 3iii shows the final assembled supercapacitor in series with an effective area of 15 × 15 mm^2. The size of the device can be easily modified by changing the area of the laser carbonized pattern on the polyimide surface. The high conductivity and electrical connection between the laser carbonized patterns and the copper pads is qualitatively demonstrated in the inset in Figure 3, wherein an LED is lit up.

3. Characterization and results

3.1. Qualitative characterization

High intensity laser treatment of the polyimide can increase the temperature of the polymer to over 1000 K. This high temperature results in explosive phase changes including boiling, vaporization, and carbonization of the polyimide, which results in micro/nano structures in the laser ablated areas. In addition, laser treatment alters the polymer surface energy, making it super hydrophilic. Contact angle measurements show the high hydrophilicity and capillary action of the carbonized micro/nano structure after laser treatment, Figure 4. This high level of hydrophilicity eliminates the need for surface functionalizing the carbonized polymer in concentrated acidic solutions.

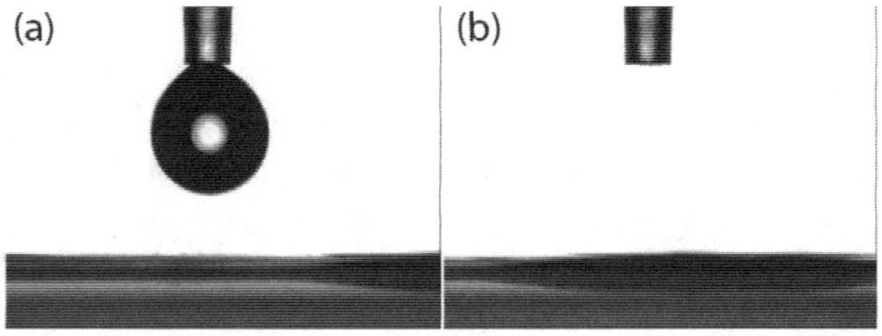

Figure 4. Goniometric measurements show the hydrophilicity of laser carbonized polyimide. (a) Water droplet before depositing on the surface. (b) Water droplet flattens on the surface (super hydrophilic)

Scanning electron microscopy (SEM) was used for investigating the surface morphology of the carbonized structures before and after coating with MnO_2. Figure 5a-b reveals that carbonized surface consists of multiple parallel lines of aligned highly porous carbon micro/nano structures. This alignment is due to multiple back-and-forth passes of the laser beam. The laser used in this process has a spot size 60 μm, therefore large-area ablations require multiple sweepings of the laser over the target area, resulting in these aligned structures. A magnified view of the carbonized traces reveals high aspect ratio features on the surface with height < 1 μm. The voids between carbon patterns provide the catalyst reactant solutions with easy access to the carbon surface for reduction. Figure 5c shows the uniform coating of MnO_2 on the carbonized surface after 5 min of catalyst deposition, resulting in carbon/ MnO_2 composite electrodes. The coating is thin but uniform and does not clog the porous morphology of the carbon. Catalyst thickness and its agglomeration increases with the immersion time, resulting in higher internal resistance of the capacitor. A cross-section image (Figure 5d) confirms the carbon porosity as well as the conformal coating of MnO_2 on carbon while maintaining electrical contact with the laminated copper film.

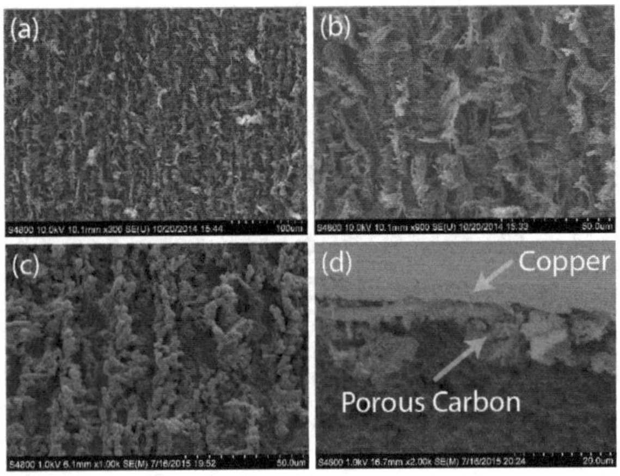

Figure 5. SEM of supercapacitor electrodes. (a-b) images of laser pyrolized of polyimide. (c) MnO_2 coated onto porous carbon electrode. (d) Cross-section image of the carbon traces showing contact between carbonized material and copper sheet.

3.2. Quantitative characterization

Figure 6a shows cyclic voltammetry scans of the carbon/MnO_2 composite at scan rates ranging from 40 mV s^{-1} to 500 mV s^{-1}. As evident by the quasi-rectangular CV plots, the device shows an electrical double-layer capacitive and pseudocapacitance behavior for scan rates up to 500 mVs^{-1}. The specific capacitance, C, achieved from the CV plots can be calculated from [7]:

$$C = \frac{I_{avg}}{V_s A} \tag{2}$$

where I_{avg} is the average current over the positive and negative voltage sweeps, V_s is the voltage scan rate, and A is the surface area of the electrode in contact with the electrolyte. The specific capacitance is calculated to be 7.1 mF cm^{-2} at a scan rate of 40 mV s^{-1} and drops to 6 mF cm^{-2} at 500 mV s^{-1}, Figure 7. Figure 6b compares the electrochemical behavior of device before and after the MnO_2 coating. The plot shows a unique increase of 55 % in the specific capacitance of carbon/ MnO_2 composite electrodes. These results show a 50% improvement over previously reported laser-fabricated super capacitors [9]. Figure 6c shows a near-ideal capacitive performance for the galvanostatic charge/discharge curves of the carbon/ MnO_2 composite electrodes. The charge/discharge curves display a small potential drop, indicating a low internal resistance in the device. The measured internal resistance is less than 2 Ω cm^2, which arises from the intimate electrical contact between the carbon/ MnO_2 composite and the copper film. In order to probe the performance of the device under mechanical stress, the device was subjected to conformal wrapping around a glass cylinder with a 5 mm radius of curvature and shown to have excellent retention of capacitance, Figure 6d.

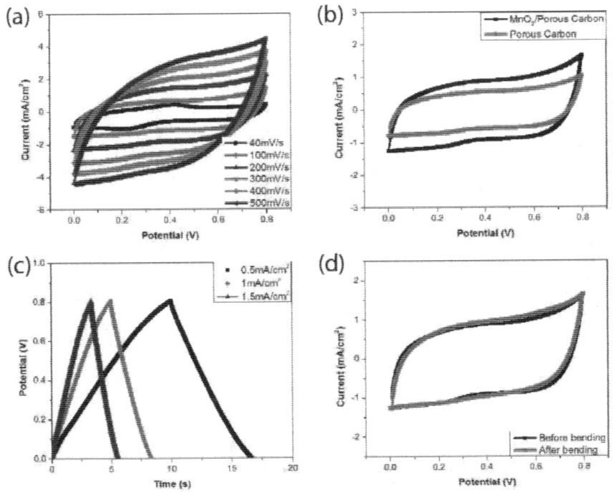

Figure 6. Cyclic voltammetry and galvanostatic charge/discharge characterizations. (a) CV of MnO_2/porous carbon at a variety of scan rates. (b) CV curves with and without MnO_2 coating with scan rate of 500mV/s. (c) Galvanostatic charge discharge curves at different current densities. (d) Comparison of the CV curves before bending and under bending with scan rate of 500 mV/s.

4. Conclusion

We demonstrated a new carbon/MnO_2 composite flexible supercapacitor by using simple low-cost laser pyrolysis of polyimide and redox deposition MnO_2 in a solution-based process. The conformal coating of MnO_2 on the carbonized micro/nano structures provides an increase in the specific capacitance of the device (up to 55%). It is anticipated that this technology can be further extended to other metal oxides pseudo-capacitors and other flexible electronic applications.

References

[1] Rahimi R, Ochoa M, Yu W and Ziaie B 2014 A sewing-enabled stitch-and-transfer method for robust, ultra-stretchable, conductive interconnects *J. Micromechanics Microengineering* **24** 095018

[2] Zeng W, Shu L, Li Q, Chen S, Wang F and Tao X-M 2014 Fiber-Based Wearable Electronics: A Review of Materials, Fabrication, Devices, and Applications. *Adv. Mater.* 5310–36

[3] Kang Y J, Chung H, Han C-H and Kim W 2012 All-solid-state flexible supercapacitors based on papers coated with carbon nanotubes and ionic-liquid-based gel electrolytes *Nanotechnology* **23** 289501

[4] Chen T and Dai L 2014 Flexible supercapacitors based on carbon nanomaterials *J. Mater. Chem. A* **2** 10756

[5] Xue M, Xie Z, Zhang L, Ma X, Wu X, Guo Y, Song W, Li Z and Cao T 2011 Microfluidic etching for fabrication of flexible and all-solid-state micro supercapacitor based on MnO2 nanoparticles. *Nanoscale* **3** 2703–8

[6] Le L T, Ervin M H, Qiu H, Fuchs B E and Lee W Y 2011 Graphene supercapacitor electrodes fabricated by inkjet printing and thermal reduction of graphene oxide *Electrochem. commun.* **13** 355–8

[7] Hsia B, Kim M S, Vincent M, Carraro C and Maboudian R 2013 Photoresist-derived porous carbon for on-chip micro-supercapacitors *Carbon N. Y.* **57** 395–400

[8] Rahimi R, Ochoa M, Yu W and Ziaie B 2015 Highly Stretchable and Sensitive Unidirectional Strain Sensor via Laser Carbonization. *ACS Appl. Mater. Interfaces*

[9] In J Bin, Hsia B, Yoo J-H, Hyun S, Carraro C, Maboudian R and Grigoropoulos C P 2015 Facile fabrication of flexible all solid-state micro-supercapacitor by direct laser writing of porous carbon in polyimide *Carbon N. Y.* **83** 144–51

[10] Lee S W, Kim J, Chen S, Hammond P T and Shao-Horn Y 2010 Carbon nanotube/manganese oxide ultrathin film electrodes for electrochemical capacitors *ACS Nano* **4** 3889–96

Microstructured silver surfaces produced by freeze casting for enhanced phase change heat transfer

G J Gouws [1] and N Shortt
School of Engineering and Computer Science, Victoria University of Wellington, Wellington, New Zealand

E-mail: gideon.gouws@vuw.ac.nz

Abstract. Microporous silver surface layers were formed on copper substrates by means of a modified freeze casting method. The structure of such layers is the result of the templating action of the ice crystals and layers were found to contain a hierarchical porous structure. Three different pore morphologies were present in the microstructure with pore sizes ranging from approximately 0.5 to 200 μm. The application of these surface structures was found to considerably enhance the heat flux during the nucleate phase of pool boiling, with heat fluxes up to five times higher from a microporous surface compared to a bare copper substrate. Bubble formation and departure was found to be significantly different on the two types of surfaces, with smaller bubbles formed with a high density on the microporous surface. The enhancement in heat flux by these structures is most likely due to the combined effect of an increased surface area with high thermal conductivity, an increase in nucleation sites for bubble formation as well as effective wicking from micropores to sustain bubble growth and departure.

1. Introduction

Materials with a porous microstructure play an important role in enhancing many energy related processes and technologies such as batteries, fuel cells, supercapacitors and catalysts [1]. These structures not only provide a large surface area, but potentially also increase the density of reaction sites to augment the energy transfer process. In the case of a phase change heat transfer process such as pool boiling, the use of porous or structured surfaces is the primary mechanism that can be used to enhance the heat flux or to delay the onset of a critical heat flux [2]. Several methods for the fabrication of such surfaces can be found in the literature, ranging from sintering of small particles to self-assembly or microfabrication [3].

This paper presents the fabrication of porous silver microstructured surfaces on copper substrates by a modified freeze casting process. The technique is a relatively simple and inexpensive method to produce a hierarchical porous surface structure with control over pore dimensions and alignment. In addition to the pore structure, these layers also have a large surface area and high thermal conductivity and thus have the potential for application in many different energy transfer and storage processes. The efficacy of these surfaces in enhancing heat transfer during the early nucleate phase of pool boiling will be evaluated.

[1] To whom any correspondence should be addressed.

Freeze casting has gained popularity over the last decade as a method for preparing porous microstructures with a high degree of pore alignment [5,6]. The basic process as illustrated in figure 1 consists of making a colloidal suspension of nano- or micro particles in a solvent (typically water) together with dispersants and binders. This solution is then subjected to a directional freeze, which causes the nucleation of ice crystals (a). These ice crystals grow and propagate through the colloid in the form of hexagonal platelets (b). Any impurities will be rejected by the advancing ice front, and should be trapped in spaces between the ice platelets. At completion of the freeze process, the ice crystals can be removed by sublimation, leaving a porous structure as defined by the templating of the ice crystals (c). The remaining structure can then be annealed to burn off the binder, sinter the particles and produce a denser and mechanically robust structure (d). The resultant pore morphology can be controlled to a degree by a variation of process parameters such as the freeze rate, particle size and shape or the chemical nature of binders and additives.

Using variations of the basic freeze casting method, many workers have shown the preparation of porous ceramic, metal or polymer structures. Applications of material produced by the method have mostly focussed on potential biomaterials such as tissue or bone scaffolds [7,8], drug delivery scaffolds [9] or ultrasound tissue phantoms [10]. The potential uses of these structures in porous electrode materials and in fuel cells have also been suggested [11].

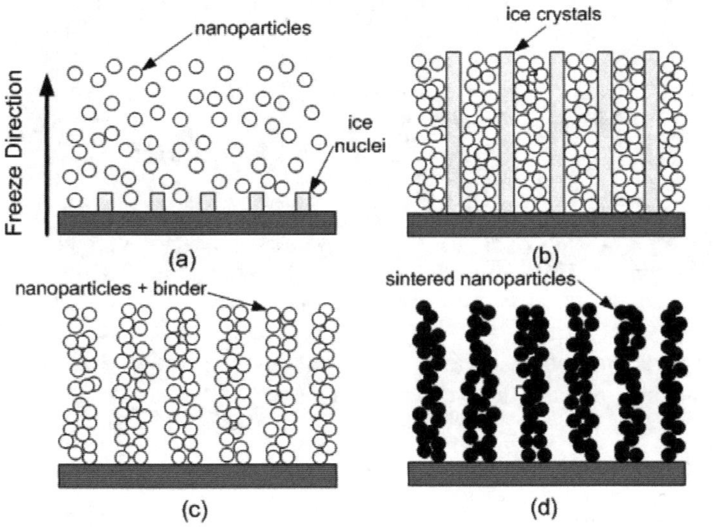

Figure 1: The idealised formation of a porous structure by a freeze cast process. A colloidal suspension of nanoparticles is subjected to a temperature gradient to initiate freezing (a) and the growing ice crystals reject any impurities (b). The ice can be removed in a sublimation step (c) after which an anneal step is used to sinter the nanoparticles and improve mechanical strength (d).

2. Experimental Method

Silver nanoparticles with diameters between 100 – 500 nm were used as the starting material. Colloidal suspensions were made by slowly dispersing the nanoparticles in a 1% solution of polyvinyl alcohol in water, producing mass loadings of nanoparticles from 10% to 40%. Vortex mixing and ultrasonic disruption were used to prevent particle aggregation and promote dispersion.

Substrates in the form of copper discs with a typical thickness of 2.5 mm and a diameter of 25 mm were machined, degreased in acetone and isopropanol and given a final etch in acetic acid to remove surface oxides. A measured volume of the silver nanoparticle colloidal solution was then dispensed on the substrate and the substrate placed on top of a copper cold finger. A small hole in the base of the substrate allowed access to a thermocouple for temperature measurement during the freeze cast step as well as later during heat transfer tests. The temperature of the cold finger was controlled by Peltier thermoelectric elements and a controlled ramp from room temperature to -40 °C was induced in the cold finger.

At the conclusion of the freeze step, samples were removed from the cold finger and placed in a vacuum freeze drier for a period of ~ 48 hours, after which samples were annealed in a tube furnace under inert gas flow for 4 hours at a temperature of 500 °C.

The efficacy of the films in enhancing heat transfer during the nucleate phase of pool boiling was investigated by mounting the samples on the top of a copper heat flux finger that was heated from the bottom by cartridge heaters. The sample was then fitted into the bottom of a glass boiling vessel containing saturated, pre-boiled deionised water at atmospheric pressure as the working liquid. By measuring the temperature profile along the heat flux finger as well as the surface temperature of the copper substrate, the heat flux as a function of excess temperature ΔT_{ex} ($\Delta T_{ex} = T_{surface} - T_{saturation}$) was calculated for values of ΔT_{ex} in the range 0 – 10 °C. A camera was fitted at the top of the test system enabling viewing of the sample surface during the process.

Figure 2: A copper substrate with Ag nanoparticle colloid on a cold finger during the freeze step. Freezing proceeds from the outside inwards and the position of the freeze front can be seen as the circular ring.

Figure 3: The appearance of the porous silver layer at the completion of the anneal step to remove the polymer binder and sinter the Ag nanoparticles.

3. Material Microstructure

The appearance of the silver surface after annealing is shown in figure 3, while the detailed microstructure structure is presented at different length scales in figure 4. It is found that the structure increases in density from the outside perimeter of the substrate to the centre, i.e. in the direction of the freeze front. This is consistent with particles and other impurities being ejected by the advancing ice front and eventually concentrating in the centre of the sample. This effect is least noticeable in samples with high Ag particle mass loadings or when very high freeze rates were used and may be due to the fact that in these cases the moving freeze front was unable to reject particles, leading to a more even structure over the substrate.

Studying the microstructure in more detail in the SEM (figure 4) allows the observed pores to be broadly divided into three types based on their shape and size:

Type A: Relatively large (~ 100 to 200 μm) irregular shaped pores that are relatively randomly distributed through the structure (figure 4(a)).

Type B: Elongated porous channels separated by nanoparticle walls, with channels typically 40 to 80 μm in length and approximately 10 μm wide (figure 4(b)). The walls consists of sintered nanoparticles as was ejected by the ice crystals and have a typical thickness of 5 to 10 μm. These channels and walls are found to be aligned over small areas (0.2 to 0.5 mm^2) of the sample.

Type C: Micropores form within the nanoparticle walls, as these walls are not densely packed. These pores have typical dimensions of 0.5 – 5 μm. Figure 4(c)

The observed pore structure was found to be weakly dependant on the freeze rate, with higher freeze rates leading to the preferential formation of Type A and C pores at the expense of Type B. This is to be expected, as high freeze rates will produce smaller ice crystals with fast moving fronts which will destroy the periodic nature of the Type B pores. It can also be assumed that the circular tool marks on the substrate acts as nucleation points for ice platelets and as such it will be difficult to produce structures with better alignment of the walls and channels without more elaborate surface preparation.

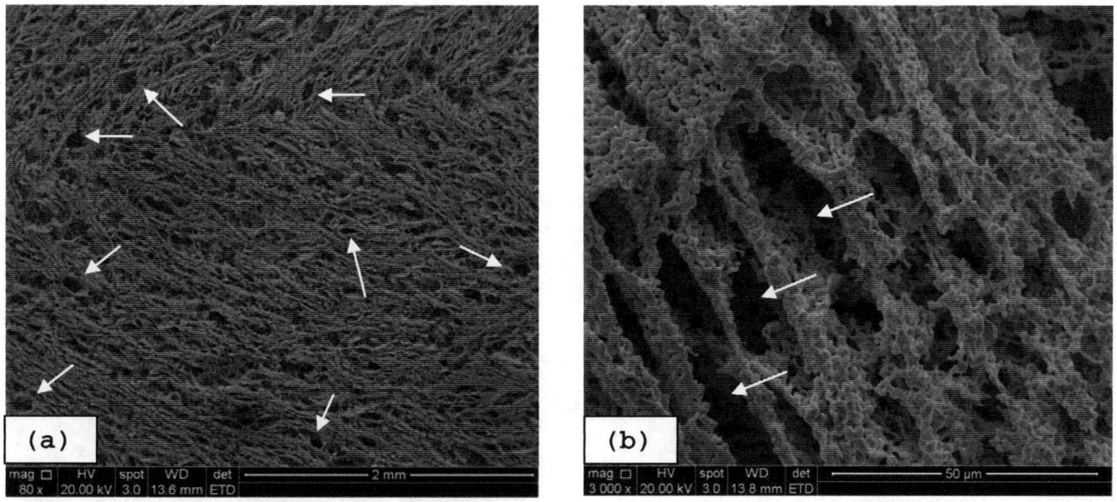

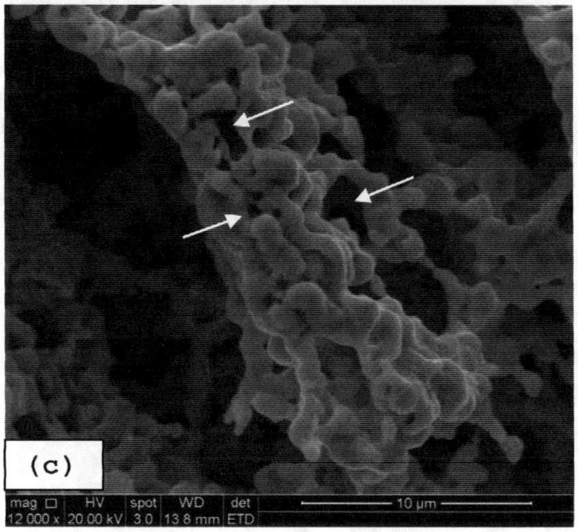

Figure 4: The microstructure of the freeze cast surface at different length scales, showing the appearance of three types of pores. Large irregular pores can be seen in (a), while channel-like pores between the nanoparticle walls can be seen in (b). The nanoparticle walls themselves contain micropores between the particles (c).

4. Boiling heat transfer measurements

The results from heat flow during pool boiling on the porous silver surface were compared to that from on a standard Cu substrate with no surface layer (Figure 5). Very little difference in the heat flux from the different samples were noted at low values of excess temperatures ($\Delta T_{ex} < 2$ °C). For the bare copper substrate, the heat flux increased relatively slowly until $\Delta T_{ex} \sim 5.5$ °C, after which a sharp increase in the heat flux is observed. In the case of a microporous layer present on the surface, this departure from the initial low value of heat flux occurs at much lower values of ΔT_{ex}; ranging from approximately ΔT_{ex} = 2.5 °C for a 10% Ag mass loading to ΔT_{ex} = 1.8 °C for a 40% Ag mass loading. It is interesting to note that the performance of samples with a mass loading of 10 – 30 % Ag are comparable, but the 40% mass loading has significantly better heat transfer. At an excess temperature of ΔT_{ex} = 8 °C the first three samples produced a heat flux that was approximately 3.3 times that of the copper substrate, while the 40% silver sample had a heat flux 5.3 times that of the copper.

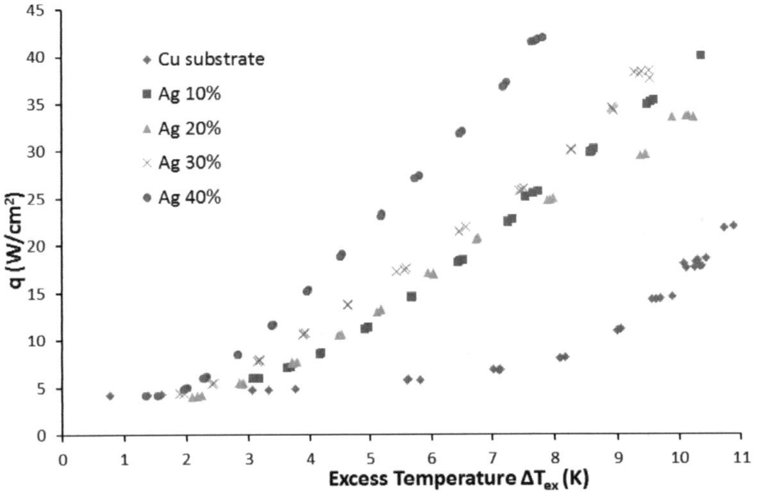

Figure 5: The heat flux from porous surfaces with different Ag mass loadings as a function of excess temperature ΔT_{ex} compared to that observed on a bare copper substrate.

The formation of bubbles and their departure was found to be significantly different on the two types of surfaces at these low values of excess temperature. Bubbles that form on the machined copper surface at ΔT_{ex} of ~ 2 °C (figure 6) had an average diameter of 0.41 ± 0.16 mm, while at similar temperatures on the porous silver surfaces (figure 7) a mean bubble diameter of 0.28 ± 0.08 mm was observed. The density of bubbles was also significantly higher on porous surface, with a typical density of 24 bubbles/cm^2 on the copper surface compared to approximately 80 bubbles/cm^2 on the porous surfaces. In the latter case, low contrast between bubbles and surface makes clear identification harder and together with the smaller size of the bubbles it may well lead to an underestimation of bubble density. Significant bubble departure from the microstructured surface also occurs well in advance from that on the copper surface, in agreement with the point of increase in the heat flux observed for the different samples.

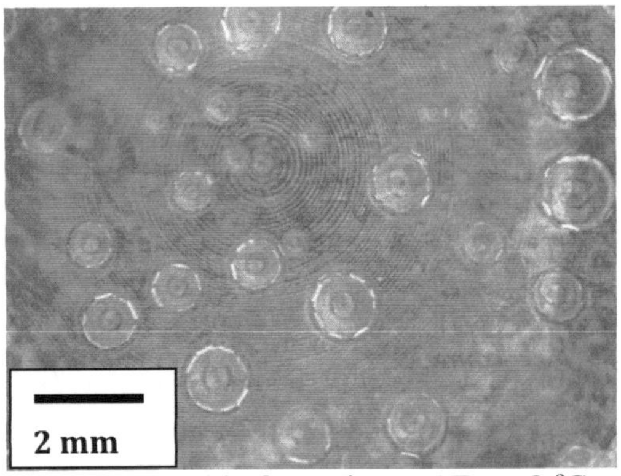

Figure 6: Bubble formation at ΔT_{ex} ~ 2 °C on a bare copper substrate.

Figure 7: Bubble formation at ΔT_{ex} ~ 0.8 °C on a microporous Ag surface.

The different types of pores in the structure may potentially each play a role in contributing to the enhanced heat flux. Both the Type A and B pores will enhance the number of potential nucleation sites. Due to their channel shape with a high aspect ratio, the Type B pores may also contribute to an upward squirt or jet effect during bubble departure as was suggested by Wang [12]. This has the potential to further enhance fluid convection during the release of bubbles. The micropores in the walls may play a role in feeding liquid to the growing bubbles and sustain the process, while at the same time the high thermal conductivity of the silver nanoparticles as well as the associated high surface area of the porous structure will be effective in transferring heat from the copper surface into the fluid.

5. Summary

Porous surface structures were prepared from silver nanoparticles using a modified freeze cast method. Three different types of pores could be identified, with pore sizes spanning three orders of magnitude. The surfaces were evaluated for the enhancement of heat transfer during the early nucleate stage of pool boiling and heat fluxes up to five times higher than that from a standard machined copper surface was obtained. Bubble densities increased on the microporous surfaces, together with significantly higher departure rates at lower values of excess temperature. The roles played by different types of pores from the structure in enhancing the heat flux needs further study in order to understand the mechanisms of heat transfer enhancement and optimise the microstructure.

References

[1] Li Y, Fu Z and Su B 2012 Hierarchically Structured Porous Materials for Energy Conversion and Storage *Adv. Funct. Mater.* **22** 4634

[2] Attinger D, Frankiewicz C, Betz A R , Schutzius T M, Ganguly R, Das A, Kim C and Megardis C M 2014 Surface Engineering for Phase Change Heat Transfer: A Review *MRS Energy and Sustainability* **1** E4

[3] Patil C M and Kandlikar S G 2014 Review of the Manufacturing Techniques for Porous Surfaces Used in Enhanced Pool Boiling *Heat Transfer Engineering* **35(10)** 887

[4] Lu Y and Kandlikar S G 2011 Nanoscale Surface Modification Techniques for Pool Boiling Enhancement – A Critical Review and Future Directions *Heat Transfer Engineering* **32(10)** 827

[5] Deville S, Saiz E, Nalla R K and Tomsia A P 2006 Freezing as a Path to Build Complex Composites *Science* **311** 515

[6] Deville S 2008 Freeze-Casting of Porous Ceramics: A Review of Current Achievements and Issues *Advanced Engineering Materials* **10(3)** 155

[7] Pawelek K M, Husmann A, Best S M and Cameron R E 2014 Ice-templated structures for biomedical tissue repair: From physics to final scaffolds *Applied Physics Reviews* **1** 021301

[8] Deville S 2010 Freeze-Casting of Porous Biomaterials *Materials* **3** 1913

[9] Gutiérrez M C, García-Carvajal Z Y, Jobbágy M, Rubio F, Yuste L, Rojo F, Ferrer M L and Del Monte F 2007 Poly(vinyl alcohol) Scaffolds with Tailored Morphologies for Drug Delivery and Controlled Release *Adv. Funct. Mater.* **17** 3505

[10] Dawson A, Harris P, Gouws G J 2010 Anisotropic Microstructured Poly(Vinyl Alcohol) Tissue-Mimicking Phantoms *IEEE Transactions on Ultrasonics, Ferroelectrics and Frequency Control* **57(7)** 1494

[11] Chen Y, Bunch J, Li T, Mao Z and Chen F 2012 Novel functionally graded acicular electrode for solid state cells fabricated by the freeze-tape-casting process *Journal of Power Sources* **213** 93

[12] Wang X S, Wang Z B and Chen Q Z 2010 Research on Manufacturing Technology and Heat Transfer Characteristics of Sintered Porous Surface Tubes *Advanced Materials Research* **97-101**, 1161

Bio-nanotextured high aspect ratio micropillar arrays for high surface area energy storage devices

S Chu[1,2], K Gerasopoulos[2] and R Ghodssi[1,2,*]
[1] Department of Electrical and Computer Engineering, University of Maryland, College Park, MD 20742, USA
[2] Institute for Systems Research, University of Maryland, College Park, MD 20742, USA

*E-mail: ghodssi@umd.edu

Abstract. This paper presents fabrication and characterization of bio-nanotextured hierarchical nickel oxide (NiO) supercapacitor electrodes. The hierarchical electrode structure is created through self-assembly of *Tobacco mosaic viruses* (TMVs) on high aspect-ratio micropillar arrays. Enhanced assembly of the bio-nanoparticles was achieved by increasing TMV solution accessibility into the deep microcavities of the pillar arrays. Electrochemical characterization of the hierarchical NiO supercapacitor electrodes revealed a 25-fold increase in charge capacity compared to a planar NiO, and demonstrated excellent cycle stability over 1500 charge/discharge cycles at 2 mA/cm^2. This study leverages the unique bio-nanoscaffolds for small scale energy storage devices through further optimization of the hierarchical structures and wetting techniques for significant improvements in micro/nano scale energy storage devices.

1. Introduction

Micro- and nanoscale materials have received significant attention for developing electrical energy storage devices such as Lithium-ion batteries, fuel cells, and supercapacitors. The unique advantages of using materials exhibiting these length-scales comes from their ability to facilitate fast electron/ion transfer kinetics with densely packed high surface-area-to-volume structures, which enable high energy and power density performances through efficient use of energy storage materials [1].

TMV is a biological nanomaterial characterized by its high aspect-ratio structure. Previously, TMVs have been genetically modified to express cysteine residues on their outer most surface to enhance self-assembly and metal binding properties [2]. Subsequent TMV metallization and integration with thin film deposition techniques (atomic layer deposition, sputtering, etc.) enables the demonstration of bio-scaffolds for various energy storage devices [3,4]. The robust metallized TMVs coated with thin film energy storage materials yield a core (metal)/shell (active material) nanoelectrode structure ensuring continuous electrical connectivity between the two electrode components while maintaining the scaffold's nanoscale structure.

* To whom any correspondence should be addressed.

To further increase the electrode surface area, the TMV nanoscaffolds were integrated with gold micropillar arrays to create hierarchical structures. This is one of the most effective ways to create high surface area electrodes, taking advantage of building blocks with different length-scales [5]. However, TMV self-assembly suffers from limitations on high surface area microstructures, showing reduced bio-nanoparticle surface coverage on the vertical and bottom surfaces of high AR geometries [6]. In this work, critical geometric parameters of the microstructures that govern the virus assembly morphology were characterized, thus enabling the use of optimized bio-templated hierarchical architectures. Electrochemical performance of the final hierarchical electrodes was evaluated as NiO supercapacitor electrodes.

2. Methods

2.1. Design and fabrication of micropillar array electrodes

A set of four different micropillar array electrodes was fabricated to study the spacing AR (SAR) dependence of the TMV self-assembly process. As illustrated in figure 1a, a silicon substrate is patterned with an image reversal photoresist (AZ-5214), and etched using deep-reactive-ion-etching (DRIE). The etched wafer was diced into electrodes with dimension of 1×1.7 cm^2 (1×1 cm^2 pillar array area), and both sides of the electrode were passivated with PECVD Si_3N_4 (500 nm at 300 °C) followed by sputter deposition of Ti (30 nm)/Au (120 nm). All pillars were 70 μm tall and 7 μm in diameter (10:1 AR) (figure 1b), but the spacing between the pillars was varied from 7 to 21 μm corresponding to a SAR of 10:1 to 3.3:1 (figure 1c).

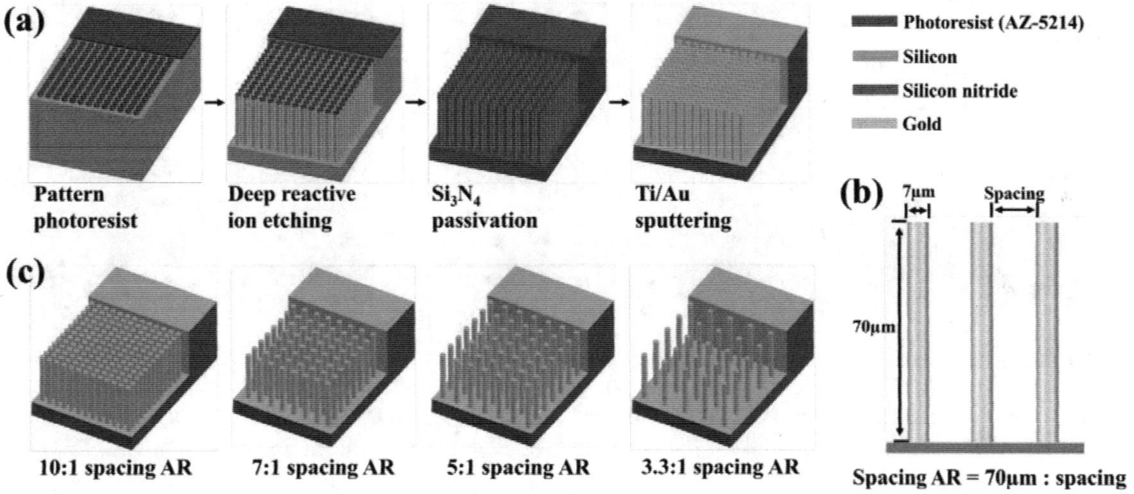

Figure 1. Schematics of micropillar array electrodes structures and fabrication process flow. (a) Fabrication process of micropillar array electrodes. (b) Cross-sectional view of the pillar array. All pillars were 10:1 AR while the SAR varied from 10:1 to 3.3:1, and (c) schematic of the four electrodes with different SAR for varied solution accessibilities.

2.2. TMV nanotexturing on high AR geometries

Three different approaches were tested for nanotexturing the micropillar array electrodes – (1) typical, previously published TMV self-assembly and metallization protocol [5], (2) pre-treatment of the electrodes using isopropyl alcohol (IPA), and (3) ultrasonication after the immersion of the electrode into TMV solution. The typical process involves overnight self-assembly of TMVs after immersion of Au-coated electrode into 0.2 mg/ml TMV solution in 0.1 M phosphate buffer (pH 7). Subsequently, an electroless Ni metallization process follows initial Pd surface activation. For the IPA treatment, the micropillar array electrodes were rinsed and immersed in IPA for 5 minutes followed by the typical

TMV self-assembly and metallization. For ultrasonication, the electrodes were immersed in TMV solution and placed in an ultrasonic bath for 5 minutes prior to overnight self-assembly and metallization. The IPA treatment was expected to increase hydrophilicity of surfaces with hydroxyl groups, which can lead to enhanced wettability of the array structure. The ultrasonication is intended to facilitate TMV diffusion into the microcavities between pillars by agitating the virus solution within the array structure.

2.3. Electrochemical characterization
The enhanced energy storage performance of the bio-nanotextured hierarchical structures was characterized using a NiO supercapacitor system. The NiO supercapacitor electrodes were prepared by annealing Ni-coated hierarchical electrodes at 300 °C for two hours in a box furnace (details of the NiO electrode preparation and characterization have been published elsewhere [7]). The electrodes were characterized using galvanostatic (2 mA/cm^2) charge/discharge cycles in 2 M KOH aqueous electrolyte in a three-electrode electrochemical cell (working: hierarchical NiO, counter: Pt foil, reference: Ag/AgCl).

Figure 2. Cross-sectional SEM images of micropillar array electrodes after typical TMV self-assembly and electroless Ni metallization process (scale bars are 50 μm otherwise specified). (a-d) SEM images taken at the very edge of the four electrodes, and (e-h) the images taken at the electrode center area after cleaving.

3. Results and analysis

3.1. TMV coverage on high AR micropillar arrays
The cross-section scanning electron microscope (SEM) images in figure 2 show results from the typical TMV self-assembly and Ni metallization process on the micropillar array electrodes. As clearly shown in the images (figure 2a-d), uniform nanotexturing of the vertical sidewalls is observed on the pillars located at the edge of all four electrodes. However, on the pillars located in the center of the electrode, a gradual reduction of virus surface coverage towards the bottom of the pillars is observed as the SAR decreases (figure 2e-g). Uniform TMV assembly was only achieved only on pillar arrays with 3.3:1 SAR (figure 2h). This, combined with the TMV assembly profile from the edge towards the center of the pillar arrays (figure 3, 7:1 SAR), is evidence that the reduced TMV assembly density can likely be attributed to the surface tension of the TMV solution on the structurally hydrophobic pillar

arrays, thereby reducing surface wettability at the base of the pillar arrays that are located in the center of the electrode.

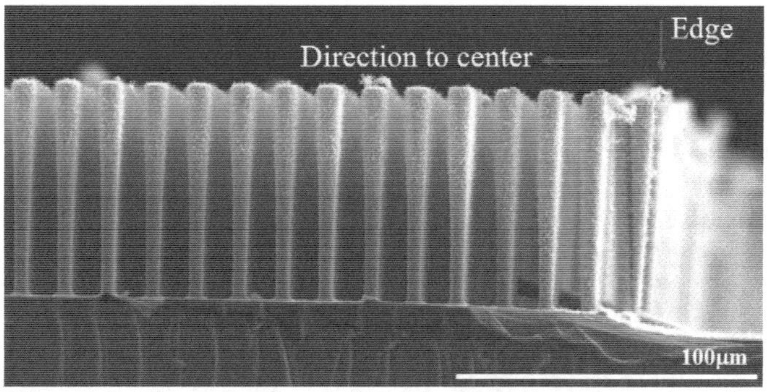

Figure 3. Cross-sectional SEM image taken from the edge of the center-cleaved electrode (7:1 SAR) after typical TMV self-assembly and Ni metallization protocol.

The cross-sectional SEM images shown in figure 4 are from pillar arrays (7:1 SAR) located at the center of electrodes treated with ultrasonication (figure 4a) and IPA (figure 4b,c,d). When compared with figure 2f for the same SAR, an obvious improvement of the virus coating is observed from both methods. A homogeneous TMV coating is achieved at the bottom of pillar sidewalls, while TMV was hardly observed in figure 2f. Low uniformity along the pillar sidewalls, showing less TMV at the bottom (figure 4c) compared to more complex clusters toward the top (figure 4d) – especially for the case of IPA treatment – shows the need for further study and improvements. For the electrochemical characterization experiments, IPA treatment method was selected due to localized breakdown of the micropillars during the ultrasonication.

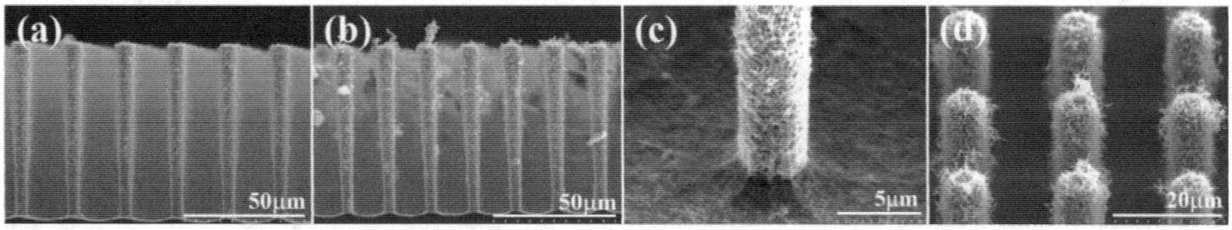

Figure 4. Cross-section SEM images taken from the center area of 7:1 SAR electrodes after the TMV and Ni coating process using (a) ultrasonication and (b) IPA treatment. The angled SEM images taken from the (c) top and (d) bottom of the electrodes treated with IPA.

3.2. Electrochemical performance of the hierarchical electrodes

Electrochemical characterization results of the hierarchical NiO electrodes are shown in figure 5. Based on the results above, micropillar array electrodes with 5:1 SAR were used to ensure a highly uniform coverage of the microstructures with TMV, and to be able to characterize the surface area enhancement factor from the microstructures under that assumption. Figure 5a shows the areal capacitance of the hierarchical NiO electrodes measured from 1500 galvanostatic (2 mA/cm^2) charge/discharge cycles. This shows excellent cycle stability with coulombic efficiency remaining above 99 % for the entire cycle duration. When comparing the areal capacitance between the hierarchical and planar NiO electrodes, a significant improvement from 18mF/cm^2 to 450mF/cm^2 is achieved. The 25-fold increase in areal capacitance is accompanied by highly stable NiO charge/discharge characteristics up to at least the 1500th cycle (figure 5b). Considering the calculated 3.5 surface area enhancement factor from the micropillar array, the remaining factor of 7 due to nanotexturing alone compares favourably with previously reported enhancement factors on flat surfaces in the range of 7~9.

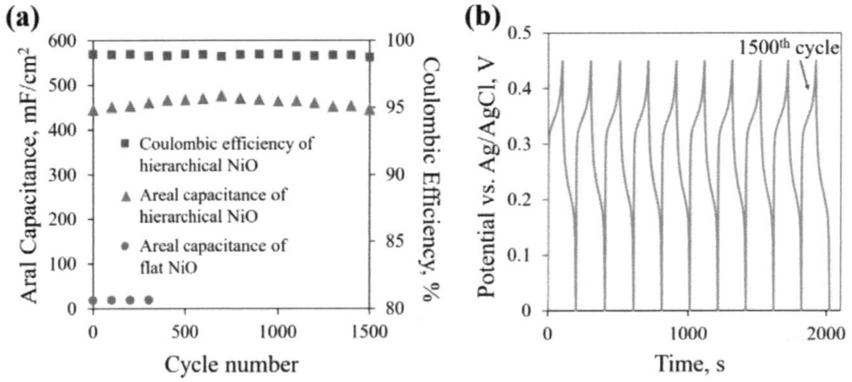

Figure 5. Electrochemical characterization of hierarchical NiO electrode. (a) Areal capacitance and coulombic efficiency for 1500 galvanostatic charge/discharge cycles at 2 mA/cm^2. (b) Stable galvanostatic charge/discharge plot of last 10 cycles.

4. Summary

This work studies the essential parameters of micropillar array geometries which govern the TMV nanotexturing quality over high surface area microstructures. It was found that there is a trade-off between SAR of micropillar arrays vs uniform TMV coating. This is attributed to structural hydrophobicity of densely packed micropillar arrays resulting in limited wetting of TMV solution into deep microcavities. IPA treatment for enhanced wettability resulted in increased surface coverage of TMVs on the high SAR array structures. However, uniformity of the coverage along the pillar sidewalls still requires further optimization and improvement. The electrochemical characterization results suggest a great advantage of this unique bio-nanoscaffolds for small scale energy storage devices with significant improvements in energy storage performance.

Acknowledgement

This research was funded by Nanostructures for Electrical Energy Storage, an Energy Frontiers Research Center funded by the U.S. Department of Energy, Office of Science, Office of Basic Energy Sciences. The authors would like to thank Professor James N. Culver for providing access to their lab for TMV purification, and the staff of the FabLab at the Maryland Nanocenter for providing access to their clean-room and spectroscopy/imaging facilities.

5. References

[1] Arico A S, Bruce P, Scrosati B, Tarascon J-M, and Schalkwijk W V 2005 *Nat. Mater.* **4** 366-377
[2] Royston E, Ghosh A, Kofinas, P, Harris M T, and Culver J N 2008 *Langmuir* **24** 906-912
[3] Gnerlich M, Ben-Yoav H, Culver J N, Ketchum D R, and Ghodssi R 2015 *J. Power Sources* **293** 649-656
[4] Chen X, Gerasopoulos K, Guo J, Brown A, Wang C, Ghodssi R, and Culver J N 2011 *Adv. Funct. Mater.* **21** 380-387
[5] Gerasopoulos K, Pomerantseva E, McCarthy M, Brown A, Wang C, Culver J, and Ghodssi R 2012 *ACS Nano* **6** 6422-6432
[6] Gerasopoulos K, McCarthy M, Banerjee P, Fan X, Culver J N, and Ghodssi R 2010 *Nanotechnology* **21** 055304 (11pp)
[7] Chu S, Gerasopoulos K, and Ghodssi R 2015 *Proc. IEEE MEMS 2015* 1118-1121

Electrical performances of pyroelectric bimetallic strip heat engines describing a Stirling cycle

A Arnaud[1,2], J. Boughaleb[1,3,4], S Monfray[1], F Boeuf[1], O Cugat[2], T Skotnicki[1]

1 ST Microelectronics, F38926 Crolles, France
2 G2Elab, Univ. Grenoble Alpes, F38000 Grenoble, France
2 CNRS, Univ. Grenoble Alpes, F38000 Grenoble, France
3 LGEF, INSA Lyon, F69621 Villeurbanne
4 CEA-LITEN, F38054 Grenoble

E-mail: arthur.arnaud@st.com

Abstract. This paper deals with the analytical modeling of pyroelectric bimetallic strip heat engines. These devices are designed to exploit the snap-through of a thermo-mechanically bistable membrane to transform a part of the heat flowing through the membrane into mechanical energy and to convert it into electric energy by means of a piezoelectric layer deposited on the surface of the bistable membrane. In this paper, we describe the properties of these heat engines in the case when they complete a Stirling cycle, and we evaluate the performances (available energy, Carnot efficiency...) of these harvesters at the macro- and micro-scale.

Introduction

The development of bimetallic strip heat engines has been pursued as an alternative to the Seebeck thermoelectric generators based on the properties of semiconductors like bismuth tellurides, or to the conventional heat engines using fluids as a heat transfer medium. Proofs of concept have been reported in [1-5]. These devices exploit the sudden displacement of a bimetallic membrane switching from a critical equilibrium position to a stable one, to harvest a part of the thermal energy flowing through the bimetal and to convert it into kinetic energy. The existence of a hysteretic behavior, consequence of their bistability, enables to cycle beams between a hot source and a cold one (Fig.1a,b), causing then an astable behavior that can be modeled like a switched thermal capacitor [6]. They can be coupled with a piezoelectric transducer deposited on the surface of the bistable membrane in order to transform the strain energy of the membrane into electric energy. Up to now, the modeling of these harvesters has been limited to the study of the properties of the thermal snap-through and of the efficiency of the thermo-mechanical transduction performed by the bimetallic strip [7]. In this paper we address the issue of modeling the performances of the thermo-mechano-electrical transduction realized by a piezoelectric bimetallic strip heat engine. Given that thermo-mechanically bistable beams can be fabricated at the macroscale and at the microscale (Fig.1a-b), we evaluate the efficiency of these devices whatever their sizes. Our model is based on the architecture represented in Fig.1d, where we assume that the membrane behaves like a plate strip.

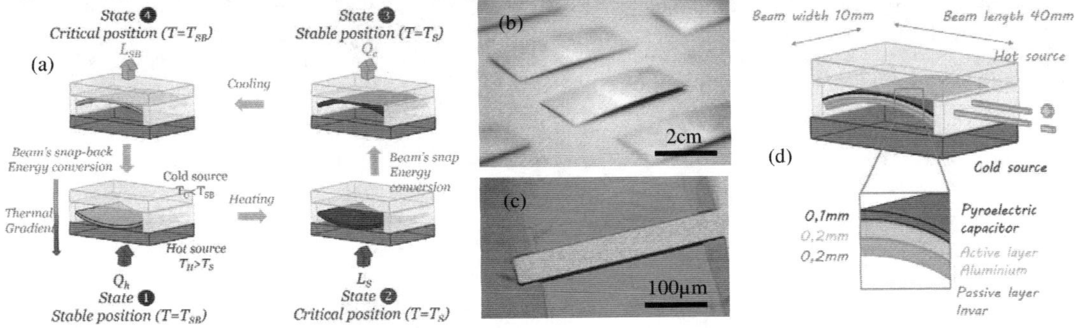

Figure 1. (a) Principle of the heat engine. (b) Bistable NC4-Invar bimetallic membrane at the macroscale. (c) Bistable Al-SiON beams at the microscale. (d) Simplified representation of the piezoelectric harvester studied in this paper.

1. Modeling of the Stirling heat engine

The modeling of the thermo-mechanical instability of simply-supported initially-buckled composite beams was presented in [8], where the use of the Galerkin method enabled to demonstrate that the equilibrium was described by a third-order polynomial of the beam's deflection amplitude. In this paper, we extend this method to the case of partially-clamped beams, according to the technique presented in [7-9], in order to predict the behavior of the heat engine at the microscale. The boundary conditions are modeled by a finite torsional stiffness γ varying from 0 (simply-supported beams) to ∞ (clamped-clamped beams). As explained in [7], the choice of a rigidity verifying $\gamma.L=36.7$ is enough to describe a clamped-clamped beam and authorizes first-order buckling modes. Whatever the value of the clamp stiffness, the beam's shape at the equilibrium can be approximated by (1a) where k depends on the γ according to (1b)

$$w(x) = w_1.\left(cos(k.x) - cos\left(k.\frac{L}{2}\right)\right) \tag{1a}$$

$$k.cos\left(k.\frac{L}{2}\right) + \gamma.sin\left(k.\frac{L}{2}\right) = 0 \tag{1b}$$

The equilibrium is then described by (2a), where n_e is the beam's axial stiffness, i_e the beam's bending stiffness, n_α the beam thermal expansion, m_α the asymmetry of the thermal expansion. The effect of the residual stress is modeled by two parameters, F_o being the total residual stress and M_o the asymmetry of the residual stress. Thus the thermo-mechanical bistability appears only if the effective residual stress P_o exceeds the Euler buckling load (2b).

$$n_e.\frac{(k^2.L - k.sin(k.L))^2}{8.L}.w_1^3 + \frac{k^2.L - k.sin(k.L)}{2}.(i_e.k^2 - P_o).w_1 + 2.k.sin\left(k.\frac{L}{2}\right).(M_o - m_\alpha.T) = 0 \tag{2a}$$

$$P_o = n_\alpha.\frac{M_o}{m_\alpha} - F_o \geq i_e.k^2 \tag{2b}$$

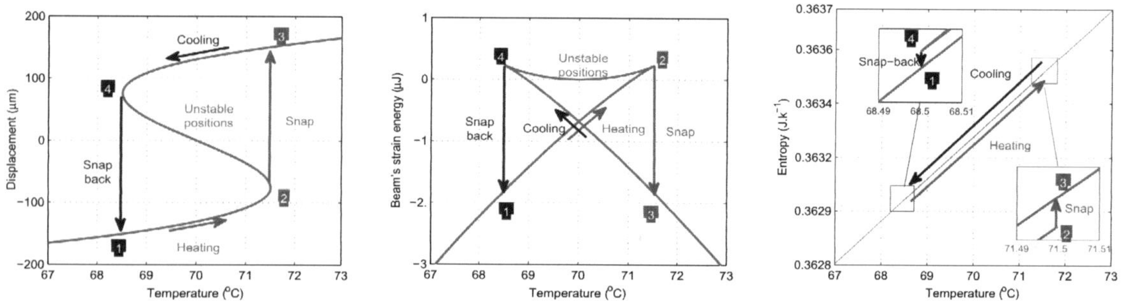

Figure 2. (a) Hysteretic behavior of a Al-Invar beam in the diagram (w_1,T). (b) Evolution of the mechanical energy of the bistable Al-Invar beam. (c) Evolution of the entropy of the bistable Al-Invar beam.

The beam's thermal hysteresis and the beam's deflection amplitude w_c of the unsnapped state at T_S can be found by deriving (2a). The evolutions of the deflection, total strain energy and entropy of a bistable Al-Invar bimetallic beam ($L=4cm$, $v=1cm$, $t_1=t_2=200\mu m$) covered by a PZT layer ($t_p=100\mu m$) are represented in Fig.2. In Fig.2b we observe that the variations of the energy during the beam's snap are weak because it corresponds to the sum of the energy variations of every layers, which see positive or negative variations of their strain energy due to internal stresses either tensile or compressive.

$$\Delta T = 2.\frac{\sqrt{(k^2.L - k.sin(k.L)).L}}{k.sin\left(k.\frac{L}{2}\right)}.\sqrt{\frac{(P_o - i_e.k^2)^3}{27.n_e.m_\alpha}} \tag{3a}$$

$$w_c = \sqrt[3]{\frac{4.L.m_\alpha.sin\left(k.\frac{L}{2}\right).\Delta T}{k.(k.L - sin(k.L))^2.n_e}} \tag{3b}$$

We now describe the electrical behaviour of a piezoelectric layer of thickness t_p deposited on the bistable bimorph beam, either on the passive layer (low thermal expansion material) or the active layer of the bistable membrane (high thermal expansion material). The constitutive equation of this layer is (4).

$$d\mathcal{D}_3 = p_3^{\mathcal{S}}.dT + e_{31}.d\mathcal{S}_1 + \varepsilon_{33}^{\mathcal{S}}.d\mathcal{E}_3 \tag{4}$$

In weak coupling, the effect of the stress induced by the polarization of the piezoelectric material is neglected: the equilibrium equation (2a) is not modified and the occurrence of the thermal snap-through does not depend on the electric field inside the piezoelectric layer. By integrating (4), we find the electrical equation of the piezoelectric bimetallic strip heat engine, linking the electric charge Q_e carried by the piezoelectric layer to its output voltage U_p, the temperature of the system and the deflection of the beam (5). Knowing that w_1 depends on the temperature (see (2a)), we can interpret the effect of the thermomechanical instability like a tertiary nonlinear pyroeletric effect.

$$dQ_e = \mathcal{C}_p.\left(-d\mathcal{U}_p + \frac{t_p.p_3^S}{\varepsilon_{33}^S}.dT + \frac{t_p.e_{31}}{L.\varepsilon_{33}^S}.\left(\frac{k^2.L - k.sin(k.L)}{2}.w_1 + 2.(z_p - z_o).k.sin\left(k.\frac{L}{2}\right)\right).dw_1\right) \quad (5)$$

In this paper, we only study the performances of the Stirling thermodynamic cycle due to its high theoretical efficiency and to the possibility to easily implement it with a SECE power management circuit [10,11]. The Stirling cycle generically describes a cycle made of two isochoric transformations (heating and cooling of the engine) and two isothermal transformations (phases of compression and expansion). In the case of a pyroelectric heat engine, the extensive parameter is the electric charge: the isochoric transformations are replaced by open-circuited transformations and the isothermal transformations by electrical charges and discharges of the piezoelectric capacitor. We consider three different working scenarios, depending on how and when the harvester is polarized. Given that the piezoelectric layer can be deposited on the active layer or on the passive layer of the bistable beam, six different Stirling cycles are presented in Fig.3 in the case of the previous Al-Invar-PZT beam. In our study, we only consider static equilibrium positions and we neglect the voltage peaks observed when the beam snaps and snaps back.

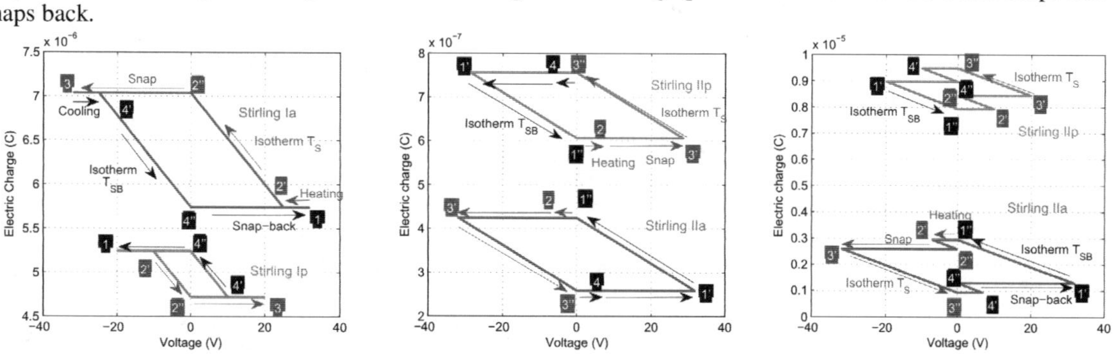

Figure 3. (a) Diagram (Q,V) of the Stirling cycles Ia and Ip. (b) Diagram (Q,V) of the Stirling cycles IIa and IIp. (a) Diagram (Q,V) of the Stirling cycles IIIa and IIIp.

In the first scenario (Fig.3a), the "isochoric" heating of the bistable mover happens between state 4 and 2, meaning that the bimetallic beam is open-circuited before its unsnaps and until before it snaps. Similarly the "isochoric" cooling of the device happens between 2 and 4. The isothermal charge and discharge of the piezoelectric capacitance are performed in 2 and 4. *Ia* represents the case of a piezoelectric layer deposited on the active layer of the beam (Al) and *Ip* the case of a layer deposited on the passive layer of the beam (Invar). For the cycles *IIa* and *IIp* (Fig.3b), the "isochoric" heating and cooling are realized from 1 to 3 and from 3 to 1. The isothermal transformations are realized in 1 and 3. The last Stirling cycle (*IIIa* and *IIIp*) is a combination of the two other cycles: the piezoelectric capacitance is open-circuited from 1 to 2, then it is quickly discharged in 2. The composite beam snaps in open circuit from 2 to 3 and is discharged in 3. We show that the available electric energy obeys to (6a) for the cycles *Ia* and *Ip*, to (6b) for *IIa* and *IIp* and to (6c) for *IIIa* and *IIIp*.

$$W_e = C_p.\left(\frac{p_3^S.t_p}{\varepsilon_{33}^S}.\Delta T - 4.\frac{e_{31}.t_p}{\varepsilon_{33}^S}.(z_p - z_o).k.sin\left(k.\frac{L}{2}\right).w_c\right)^2 \quad (6a)$$

$$W_e = C_p.\left(\frac{p_3^S.t_p}{\varepsilon_{33}^S}.\Delta T + 8.\frac{e_{31}.t_p}{\varepsilon_{33}^S}.(z_p - z_o).k.sin\left(k.\frac{L}{2}\right).w_c\right)^2 \quad (6b)$$

$$W_e = C_p.\left(\left(\frac{p_3^S.t_p}{\varepsilon_{33}^S}.\Delta T\right)^2 + \left(\frac{3.e_{31}.t_p}{4.\varepsilon_{33}^S}.(k^2.L - k.sin(k.L)).w_c\right)^2 + \left(2.\frac{e_{31}.t_p}{\varepsilon_{33}^S}.(z_p - z_o).k.sin\left(k.\frac{L}{2}\right).w_c\right)^2 + 4.\frac{e_{31}.p_3^S.t_p^2}{{\varepsilon_{33}^S}^2}.(z_p - z_o).k.sin\left(k.\frac{L}{2}\right).w_c.\Delta T\right) \quad (6c)$$

In the next parts we compare the performances of these Stirling engines at the macroscale and microscale, and we study the influence of the thermo-mechanical instability and the impact of the piezo-/pyroelectric properties of the piezoelectric layer on the efficiency of the energy harvester.

2. Performances at the macroscale

In order to find the most efficient Stirling cycle, we plot in Fig.4a the available electric energy as a function of the thermal hysteresis of an Al-Invar-PZT beam whose thermal hysteresis is centered on the medium value 70°C. This figure confirms the observations made on Fig.2 where the area enclosed by the working cycle is higher for *IIp*. The energy produced by *IIp* progressively decreases because for highly unstable beams, the stress is tensile when the beam is snapped and unsnapped, thus reducing the electric charge variation. By comparing the amount of energy produced by the harvester with the energy produced by a classical pyroelectric engine based on the pyroelectric effect only, we can see the potential benefit of using a bistable beam to create a pseudo-tertiary pyroelectric effect to boost the performance of the generator, and to harvest heat from thermal gradients instead of heat from temporal variation of the temperature. For a 3K hysteresis, the available energy of *IIp* is 75.6μJ, corresponding to an energy surface density of 18.9μJ.cm^{-2}. According to [3], the cycling frequency of a harvester with a membrane working with $T_{SB}=67°C$ and $T_S=70°C$ is of around 0.25Hz, meaning that a piezoelectric harvester with a 3K-hysteresis beam could deliver up to 18.9μW, enough to supply a Wireless Sensor Node [12].

Given that the thermo-mechano-electrical coupling is weak, the dependence of the heat capacitance on the mechanical stress and on the piezoelectric effect is weak, and the latent heat is negligible. In that conditions, the heat supplied by the hot source can be approximated by

$$Q_h \approx C_v.\Delta\mathrm{T} \tag{7}$$

We thus define the thermo-electrical efficiency of the device by (8a), and the Carnot efficiency of the heat engine by (8b). Given that our model does not take into account the thermal transfers inside the harvester, the efficiency (8a) is normalized by the ratio of the thermal hysteresis to the snap temperature instead of the true hot source temperature and cold source temperature. The formula (8b) thus gives the maximal Carnot efficiency but corresponds to the working conditions that generate the lowest electrical power.

$$\eta_{abs} = \frac{W_e}{C_v.\Delta T} \tag{8a}$$

$$\eta_{rel} = \frac{W_e}{C_v.\Delta T^2}.\left(\frac{M_o}{m_\alpha} + \frac{\Delta T}{2} + 298K\right) \tag{8b}$$

Fig.4b. plots the evolution of these two efficiencies as functions of the thermal hysteresis for the cyle IIp. It shows that after reaching a minimal value around 2K, the efficiency increases as the thermal hysteresis widens. However, the rapid fall of the Carnot efficiency for highly hysteretic devices means that the transformation becomes less reversible. This performances degradation is mainly linked to the efficiency of the thermo-mechanical transduction according to [7]. Typical Carnot efficiencies of around 0.4% are obtained with a 3K hysteresis. In order to quantify the impact of the piezoelectric properties on the efficiency of the heat engine, the available energy and the Carnot efficiency of various piezoelectric materials are represented as a function of the electro-thermal coupling factor defined in [11]. The fact that PZT gives better results than PVDF shows that the performances of the heat engine are more linked to the piezoelectric coupling than to the pyroelectric effect. Knowing that the electro-thermal coupling factor represents the Carnot efficiency of a classical pyroelectric generator, we can see that the pyroelectric bimetallic strip heat engine can be more efficient than a classical pyroelectric generator when PZT is used as an electro-mechanical transducer.

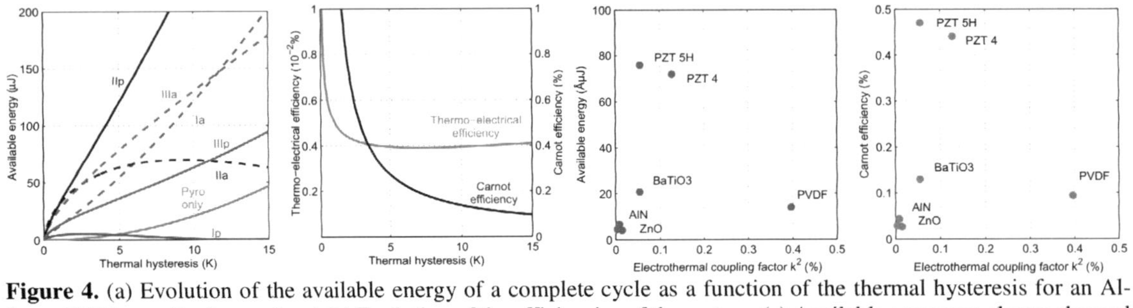

Figure 4. (a) Evolution of the available energy of a complete cycle as a function of the thermal hysteresis for an Al-Invar-PZT beam (L=4cm v=1cm) (b) Evolution of the efficiencies of the system. (c) Available energy vs electro-thermal coupling factor [11] of various piezoelectric materials for a 3K hysteresis. (c) Carnot efficiency vs electro-thermal coupling factor.

3. Efficiency of the harvester at the microscale

Since miniaturized beams are fabricated by successive depositions of materials layers, the beams edges are can be considered as clamped to the substrate and the beams cannot continue being modeled as simply-supported beams:

the clamp stiffness is therefore taken equal to $\gamma=36.7/L$ [7]. In these simulations, we model the performances of the stacks PZT-Al-SiO_2 and Al-SiO2 PZT ($L=400\mu m$, $v=100\mu m$, $t_{Al}=2\mu m$, $t_{SiO2}=2\mu m$ $t_{PZT}=1\mu m$), corresponding to scaling factor of 100 compared to the macro-harvester. Fig.5a first represents the evolution of the energy generated by the Stirling cycle *IIp* as a function of the instability of the beam represented by its thermal hysteresis. In order to quantify the impact of the torsional stiffness, the performances of a simply-supported beam with a single capacitor are compared with those of a clamped-clamped beam with a single capacitor. The significant decline of the generated energy (factor 15 for a hysteresis of 3K) is due to the coexistence of tensile stresses and compressive stresses inside the piezoelectric layer that the charges created by the tertiary non-linear pyroelectric effect. The energy is generated by the pyroelectric effect. By using three series piezoelectric capacitors, we can harvest the electric charges due to the tertiary pyroelectric effect instead of the classical pyroelectric effect, explaining the efficiency reduction observed in Fig.5a (69pJ for a CC beam instead of 100pJ for a SS beam) and Fig.5b (0.57% instead of 0.83% of Carnot efficiency). To finish we compare the impact of the piezoelectric properties on the performances of the harvester. Because they are made of weakly pyroelectric materials, bistable beams with GaN, AlN, ZnO layers are not significantly impacted by the scale reduction and produce few picojoules of energy per cycle (Fig.5c). In our study, PZT-based harvesters gives the best performances and can produce energy densities of around $0.2\mu J.cm^{-2}$ for a hysteresis of 3K.

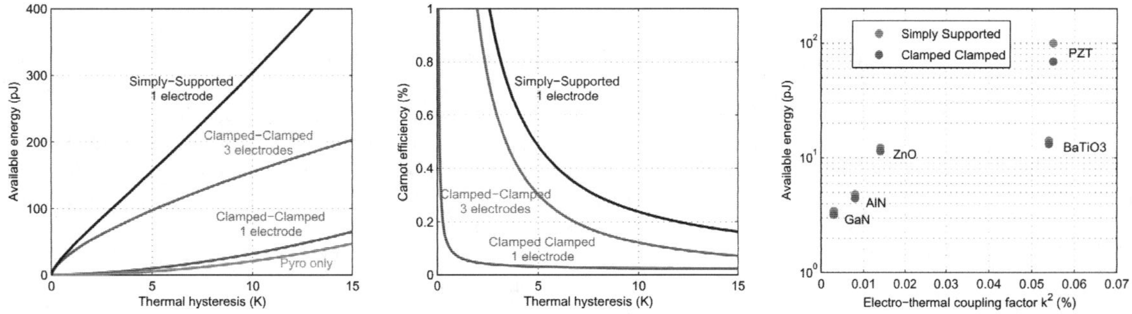

Figure 5. (a) Electrical energy per cycle produced by an Al-SiO2-PZT beam at the microscale as a function of the thermal hysteresis. (b) Carnot efficiency of the beam at the microscale. (c) Comparison of various piezoelectric beams (L=400µm v=100µm t=5µm) at the microscale for a hysteresis of 3K.

4. Conclusion

In this paper, we presented the results of the modeling of pyroelectric bimetallic strip heat engine. By using the Galerkin method, we found simple laws describing the non-linear behavior of thermo-mechanically bistable beams and its impact on the electric behavior of the generator and we defined it like a tertiary non-linear pyroelectric effect. We then studied the performances of the energy harvester when it describes a Stirling cycle and we show that energy densities of around $20\mu J.cm^{-2}$ can be obtained at the macroscale with a PZT-based heat engine (hysteresis of 3K), corresponding to a Carnot-relative efficiency of 0.4%, superior to the efficiency of a classical pyroelectric heat engine. This shows that the use of non-linear energy harvesting techniques can also be used to increase the performances of pyroelectric heat engines.

Acknowledgements

The project was partially funded by the French Inter-Ministerial Fund (FUI) through HEATec project.

5. References

[1] Boisseau S *et al* 2013, *Smart Mater. Struct.* **22**, 025021
[2] Arnaud A *et al* 2013, *J. Phys.: Conf. Ser.* **476**, 012062
[3] Boughaleb J *et al* 2014, *Smart Mater. Struct.* Submitted article
[4] Ravindran S K T *et al* 2011, *Smart Mater. Struct.* 99, 104102
[5] Puscasu O *et al* 2012, *Proc. Electron Devices Meeting (IEDM),* 12.5.1-12.5.4
[6] Boughaleb J *et al* 2014, *Sens. actuators A*, Submitted article
[7] Arnaud A *et al* 2014, *J. Micromech. Microeng.* Accepted article
[8] Arnaud A *et al* 2015, *New Circuits and Systems Conference (NEWCAS), 2015 IEEE 13th International*, pp.1-4
[9] Michael A and Kwok C Y 2006, *J. Micromech. Microeng.* **16**, 2034
[10] Olsen R B *et al* 1985, *J. Appl. Phys.* **58**, 4709
[11] Sebald G *et al* 2008, *IEEE Trans Ultrason. Ferroelectr. Freq. Control*, **55**,3 pp.538-551
[12] Boisseau S *et al* 2015, *New Circuits and Systems Conference (NEWCAS), 2015 IEEE 13th International*, pp.1-4

Microfabrication of hybrid fluid membrane for microengines

R Chutani[1], F Formosa[2], M de Labachelerie[1], A Badel[2] and F Lanzetta[1]
[1]FEMTO-ST, Besancon, Universite de Bourgogne Franche-Comte, France
[2]Laboratoire SYMME, Universite Savoie Mont Blanc, Annecy, France

E-mail: ravinder.chutani@gmail.com

Abstract. This paper describes the microfabrication and dynamic characterization of thick membranes providing a technological solution for microengines. The studied membranes are called hybrid fluid-membrane (HFM) and consist of two thin membranes that encapsulate an incompressible fluid. This work details the microelectromechanical system (MEMS) scalable fabrication and characterization of HFMs. The membranes are composite structures based on Silicon spiral springs embedded in a polymer (RTV silicone). The anodic bonding of multiple stacks of Si/glass structures, the fluid filling and the sealing have been demonstrated. Various HFMs were successfully fabricated and their dynamic characterization demonstrates the agreement between experimental and theoretical results.

1. Introduction

Microengines based on the pistons mostly suffer from friction losses and leaks due to the cylinder-piston gap which is detrimental to the engine operation [1]. Therefore, membranes are preferred when microminiaturizing. However their natural frequency is a critical aspect for optimization, since low values matching the characteristics times of the micromachine cycle are required [2]. In this paper, we propose and demonstrate an alternative solution to the piston, which is called a Hybrid Fluidic Membrane (HFM), to provide a technological solution for low frequency microengines.

1.1 Geometry description

The architecture of the HFM is based on the work of [3], which demonstrates the possible use of fluid-membrane interaction to achieve membranes with a low resonance frequency. It consists in an incompressible fluid confined between two thin membranes. The incompressible fluid is used as an inertial mass which allows a drastic reduction of the resonance frequency to match dynamic requirements of microengines. The height of the structure can be easily tuned to provide more or less inertia effect. Additionally, a spiral spring can be added to the membrane, thus allowing stiffness control. This additional structure is also used to aim at an optimized flexural shape to increase the swept volume. One other advantage of this architecture is the uncoupling between stiffness and sealing. The latter can be achieved using soft material such as RTV silicone.

The chosen fluid has to fulfil the trade-off between high density to enhance the fluid inertia effect and low viscosity to limit the mechanical dissipations. Figure 1a describes the HFM geometry as well as its geometrical parameters.

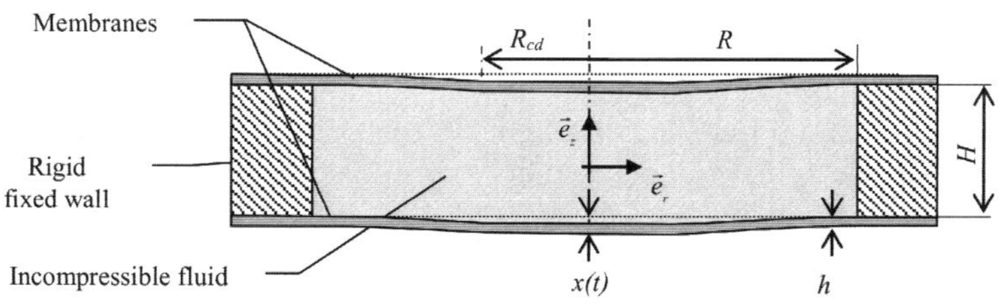

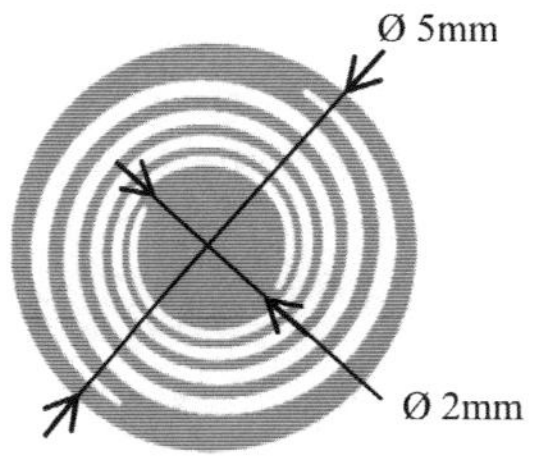

a) b)

Figure 1. a) HFM architecture and geometrical parameters, b) spiral spring

The 5 mm (= 2R) diameter spiral spring geometry is shown in Figure 1b. It consists in four intertwined logarithmic spiral shapes defining two planar springs. It is worth noting that their width vary proportionally with the spiral angle to aim at uniform stress distribution along the structure. A rigid 2 mm diameter central disk increases the swept volume compared to a simple plane membrane.

Table 1. Geometrical and material parameters.

	RTV Silicone membrane	Fluid	Frame
Geometry	$h = 200$ µm $R_{cd} = 1$ mm		1.6 mm $< H <$ 6 mm $R = 2.5$ mm
Material	$E = 1.5$ MPa $\nu = 0.49$ $\rho_m = 1100$ kg/m^3	Vaseline $\rho_f = 800$ kg/m^3 Glycerine $\rho_f = 1200$ kg/m^3	

1.2 Analytical modelling approach

An analytical fluid-structure model was proposed in [3]. The modeling strategy is based on the Ritz method to obtain approximate solutions for the fluid motion and the mechanical diaphragm deformation amplitude. Additionally, the study is limited to axisymmetric in phase motions of the membranes (*i.e.* first eigenmode of the HFM) and the fluid is assumed to be inviscid.

Each membrane is considered as a plain annular diaphragm and an additional linear spring. Consequently, a simple model can be established and while capturing the complexity of the spiral spring behaviour in a single stiffness parameter. It is also important to notice that FEM can be used to design the spiral spring to aim at the desired stiffness value (Figure 2a).

The flexural deformation *w(r)* of the diaphragm is the first natural mode shape obtained through modal analysis of the annular diaphragm and is expressed from Bessel's functions [4]. The accuracy of the approximate shape is deemed well enough as can be seen in Figure 2b when compared to FEM result Figure 2a. The dynamic equilibrium of the HFM is obtained from the energy method applied to the global structure (top and bottom membranes and fluid). From the model, the resonance frequency of the HFM can be analytically expressed provided that the material properties and central disk diameter are set.

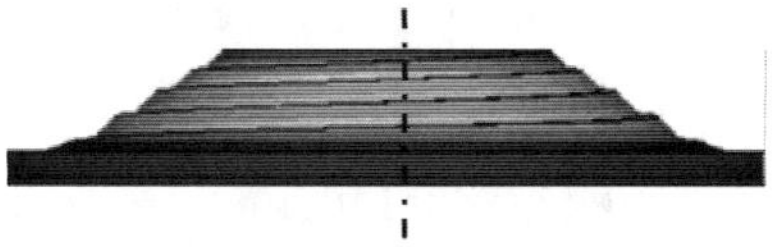

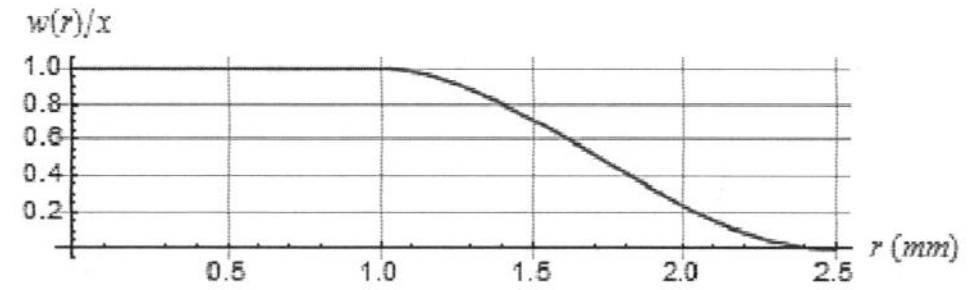

a) b)

Figure 2. a) FEM model of the membrane, b) Deformed shape used for the fluid-structure HFM model.

The effects of HFM height, type of fluid or additional stiffness can be easily studied. Figure 3 presents the evolution of the natural frequency with respect to the chamber height (*H*) for the two different fluids of Table 1. In Red and blue lines are the natural frequency curves for Glycerine and Vaseline respectively whereas the initial membrane natural frequency is plotted in dashed line. A reduction ratio up to 1/3 can be obtained as shown in Figure 3b.

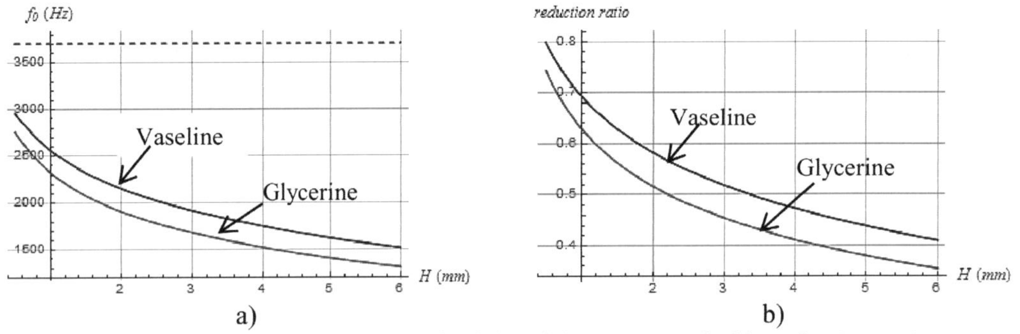

Figure 3. a) Theoretical effect of height of the HFM on f_0 , b) Reduction ratio.

2. Fabrication and realization of HFMs

The wafer-level fabrication approach of HFMs is constrained: all elements should withstand the required temperature for the machine operation and for the fabrication process itself. This includes high temperature (≥ 250°C) anodic bonding (AB) steps which ensure hermitical sealing of wafers. RTV silicone elastomer, which can withstand 340°C, is used as the membrane material [5]. Glycerine and low viscosity Vaseline oil having boiling point ≈ 300°C are used for the fluid. They have been selected as incompressible fluid for the HFM.

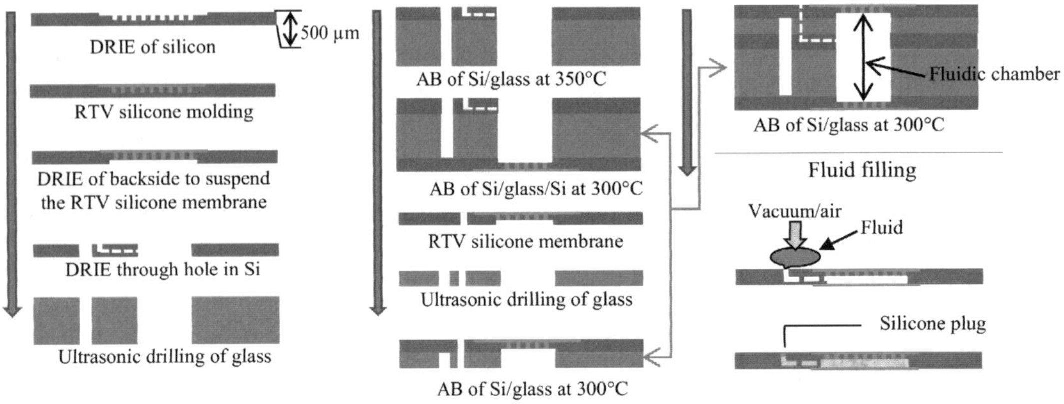

Figure 4. Microfabrication steps from free suspended RTV silicone membrane to HFM structures.

The developed fabrication process flow is described is in Figure 4. It starts with the patterning of 150µm deep silicon planar spring structure using DRIE. RTV silicone moulding allows hermetical membrane to be obtained (Figure 5a). Besides, it allows a drastic improvement of the Si spring robustness. The Si spirals are embedded in RTV silicone with an additional elastomer layer of 50 µm thickness. The latter is then etched from back side to suspend the embedded planer spring based RTV silicone membrane. In the next step, the RTV silicone membrane wafer was anodically bonded to the central thick multiple Si/glass based frame. The HFM structures are then filled with a fluid using the wafer level filling method described in Figure 4. The wafer-level filling process of the chambers was performed in a vacuum chamber, allowing the suppression of air bubbles left inside the chambers to guarantee the dynamic characteristics of the HFM. Once the chamber connecting membranes were filled with fluid through a thin channel (500 μm diameter), HFM chips were diced. HFM with two different chamber heights (HFM1 and HFM2, with 1.6 mm and 6 mm chamber height respectively) were successfully fabricated (Figure 5b).

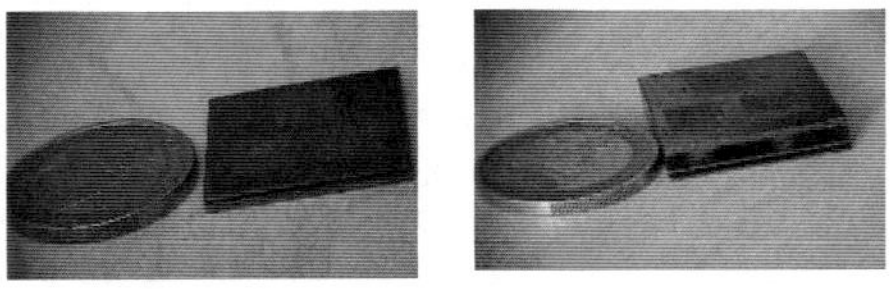

HFM 1 HFM 2

a) b)

Figure 5. a) Silicone spiral spring plus RTV silicone membrane, b) HFM structures.

3. Experimental measurements, results and discussions

3.1 Dynamic characterization

Dynamic characterizations of HFMs were performed using a MEMS analyser (Figure 6a), in which a laser vibrometer measures the center disk displacement when the structure is submitted to low-level dynamic excitation using a PZT buzzer. Figure 6b shows the experimental spectral response for a membrane alone and a HFM filled with Vaseline oil, which shows the effective reduction of the HFM resonance in comparison to the membrane alone.

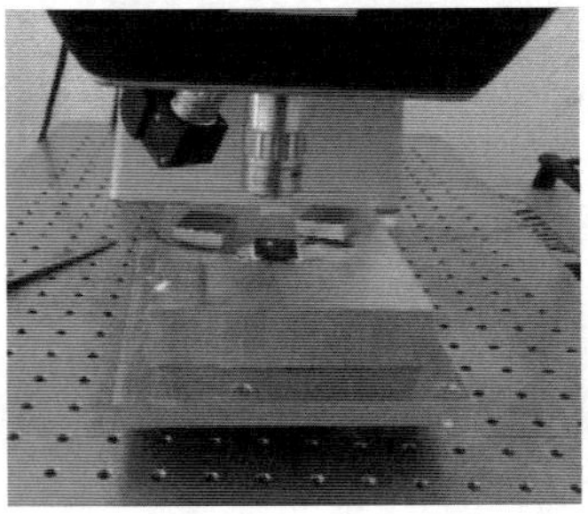

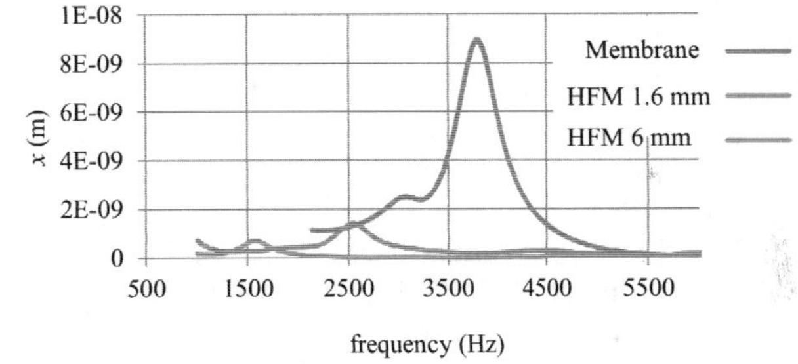

a) b)

Figure 6. a) MEMS analyzer b) Experimental resonance frequencies.

Table 2 sums up the experimental results on selected samples.

Table 2. Experimental frequencies.

	Membrane alone	HFM 1.6 mm	HFM 6 mm
Frequency (Hz)	3793, 3605, 3716	Vaseline : 2504, 2546, 2357 Glycerine: 2162, 2131, 2265	Vaseline 1570

3.2 Model validation

Figure 7 presents theoretical as well as experimental results. The plain red and blue curves are theoretical results obtained for Glycerine and Vaseline when updating the stiffness value of the model using the mean value of the measured frequencies of the membranes. Red and blue dots lines represents the scattering of the theoretical results based on the maximal and minimal measured frequencies and deduced stiffness. Though some dispersions are observed, the experimental results for the HFMs are in good agreement with the theoretical results. The maximal discrepancy between experimental and theoretical approaches is 6.6 %.

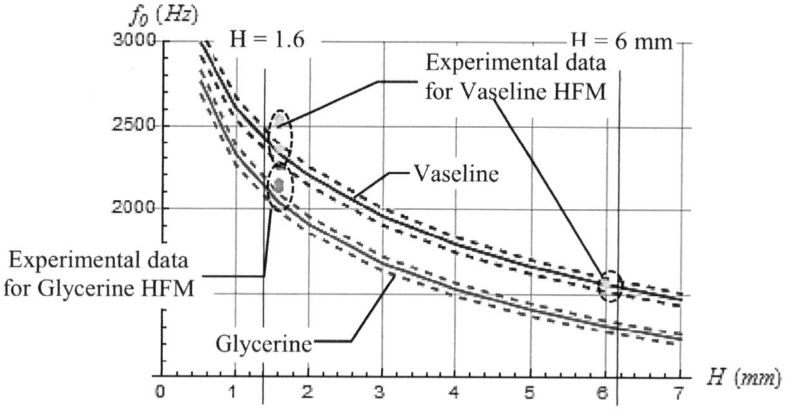

Figure 7. Experimental and theoretical result for f_0.

4. Conclusion

We have demonstrated the fabrication of low frequency MEMS membrane using the concept of HFM. In addition, characterizations of the fabricated HFM membranes show the expected dynamic features, namely the reduction of the membrane resonance frequency. The choice for materials and processes have been made targeting oscillating engines which operates with a moderate temperature of 200°C. Additional developments will aim at repeatability improvement, the evaluation of quality factor and the large amplitude behavior study. The integration of an electromechanical transduction will be a next step towards a fully integrated component for micro-engine concepts.

Acknowledgement
This study has been funded by the French National Research Agency (ANR) under grant no. ANR-12-SEED-0005-01 (MISTIC Project) and it was also partly supported by the French RENATECH network and its FEMTO-ST technological facility.

References

[1] S. K. Chou, W. M. Yang, K. J. Chua, J. Li, and K. L. Zhang, "Development of micro power generators – A review," *Appl. Energy,* vol. 88, no. 1, pp. 1–16, Jan. 2011.
[2] H. K. Bardaweel, M. J. Anderson, R. F. Richards, and C. D. Richards, "Optimization of the dynamic and thermal performance of a resonant micro heat engine," *J. Micromechanics Microengineering*, vol. 18, no. 10, p. 104014, 2008.
[3] F. Formosa, A. Badel, and H. Favrelière, "Development of low frequency, insulating thick diaphragms for power MEMS applications," *Sensors Actuators A Phys.*, vol. 189, pp. 370–379, Jan. 2013.
[4] S. Timoshenko, S. Woinowsky-Krieger, and S. Woinowsky, *Theory of plates and shells*, New York: Wiley-Interscience, 1959.
[5] Permatex® High-Temp Red RTV Silicone; http://www.permatex.com.

Development of Near-Field-Enhanced High-Fill-Factor MEMS Radiator with Shared Spring

H Nakajima, S Oh, A Ueno, K Morimoto and Y Suzuki
Department of Mechanical Engineering, The University of Tokyo, 7-3-1 Hongo, Bunkyo-ku, Tokyo, Japan
E-mail: nakajima@mesl.t.u-tokyo.ac.jp

Abstract. For precise thermal control in satellites under varying internal heat dissipation and thermal boundary condition, we propose a high-fill factor MEMS radiator enhanced by the near-field effect. We have successfully fabricated a prototype with parylene shared springs and achieved a fill factor of as high as 89 %. It is found that at the ON state, the diaphragm temperature is increased from 58.0 °C to 106.4 °C, showing 144 % enhancement in the radiation heat flux.

1. Introduction

For precise thermal control in satellites under varying internal heat dissipation and/or irradiation of sunlight, active radiation control has become an important technical challenge [1, 2]. We previously proposed a MEMS-based active radiator that can change the emissivity by controlling the distance between the substrate and diaphragms with the electrostatic force [3]. The heat flux ratio at the ON/OFF states has been measured for their early prototype, and up to 42 % heat flux enhancement was obtained due to the increased thermal-contact conductance [4]. The fill factor (the ratio of the actual radiation area to the projection area of the device) was higher than that of convectional devices with louvers or shutters [5], but still limited to 61 %. Compared to conventional thermal control devices like bulky heat exchangers, the proposed MEMS radiator is advantageous for next-generation small-scale satellites due to its light weight and low energy consumption.

In the present study, we redesign the MEMS radiator with shared springs for significant improvement of the fill factor. In addition, we reduce the gap between the diaphragm and the substrate at the "ON" state, in order to activate the near-field thermal radiation and thus to obtain further enhancement of the heat flux ratio. We design the structural parameters of the present device based on mechanical considerations and heat transfer analyses. With the present proof-of-concept experiments, we demonstrate the heat-flux enhancement by the near-field radiation effect for the first time.

2. Design of the present MEMS radiator

Figure 1 shows the principle of the electrostatically-driven MEMS radiator. At the "OFF" state, the gap between the diaphragm and the substrate is as large as several micrometers, and the radiation heat flux is determined by the Stephan-Boltzmann law. At the "ON" state, on the other hand, the gap becomes smaller than the infrared wavelength responsible for the thermal radiation, so that the heat flux is enhanced by the near-field effect. Thus the near-field thermal radiation [6, 7] is expected to enhance the heat transfer from inside the satellites to the space.

Figure 2 shows the schematic of the present MEMS radiator. Thermally-durable parylene (diX-HR, KISCO) is used for the structural material for reducing parasitic heat conduction along the spring and

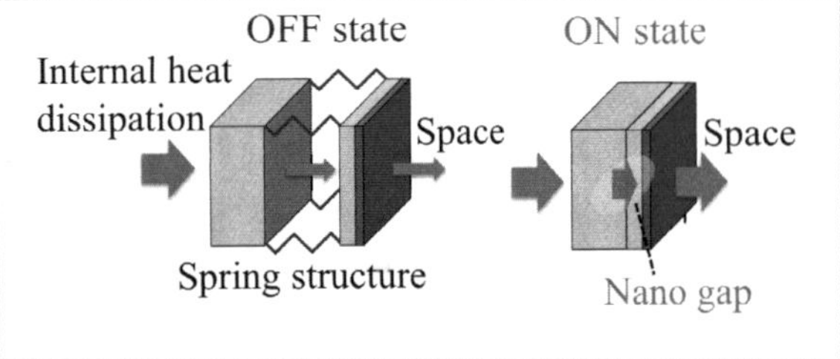

Figure 1. Principle of the MEMS radiator. At the "OFF" state, the thermal resistance is mainly determined by the conventional radiation. At the "ON" state, the gap between the diaphragm and the substrate becomes smaller than the nominal wavelength of thermal radiation, and thus the thermal resistance is drastically reduced due to the near-field thermal radiation effects.

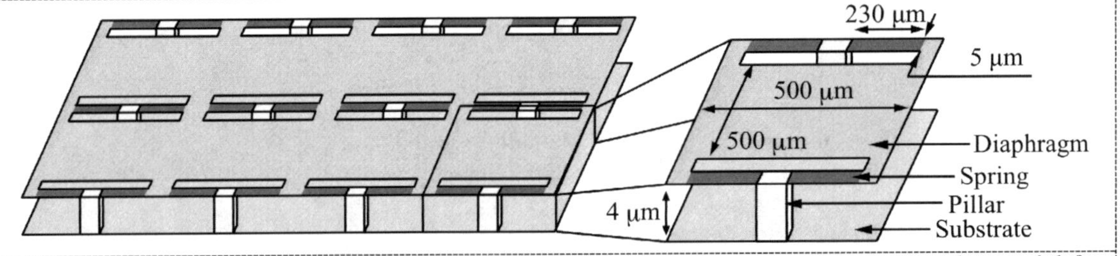

Figure 2. Schematic of the present MEMS radiator. Parylene is used as the structural material for reducing parasitic heat conduction along the spring and lowering the driving voltage. The parylene springs are shared among neighboring diaphragms to increase the fill factor. Compared to our previous prototype with independent diaphragms [4], the fill factor is increased from 61 % to 89 %.

for lowering the driving voltage. Parylene has low Young's modulus and low thermal conductivity, and is a proven material for the space applications [4]. The structure is basically the same as the Ueno & Suzuki's prototype [4], and the improved points are as below: 1) The parylene springs are shared among adjacent diaphragms in order to further improve the fill factor. 2) The insulating layer at the lower side of diaphragms is removed, and the thickness of the insulating layer on the substrate is decreased to 50 nm in order to reduce the gap between the substrate and the diaphragm at the ON state for the near-field effects. 3) Titanium is adopted as a top electrode material because of its low thermal conductivity, which prevents the heat conduction along the springs and improves the heat flux ratio.

The radiator performance is greatly influenced by the spring length and width. In the present design, the dimension of the springs are determined for satisfying the following three requirements: a) the driving voltage is smaller than 10 V, b) the resonance frequency is higher than 2 kHz, and c) the heat flux ratio between the ON/OFF states is maximized. For requirement a), the driving voltage is estimated from a parallel-plate capacitor model [4]. The driving voltage corresponds to the pull-in voltage, with which the diaphragm is snapped down to the bottom substrate due to the electrostatic force between the top Ti and bottom Au electrodes. For requirement b), high resonance frequency is assumed for durability against vigorous vibration during the launch of the satellite. We estimate the resonance frequency based on a spring-mass model, where the polymer springs are assumed as multi-layered cantilever beams.

For requirement c), the heat-flux ratio between the ON/OFF states is estimated from heat transfer analyses. We have performed a series of numerical simulations based on a finite volume method [8]. The amount of the heat flux from a diaphragm to the outer space is estimated for different spring length and width at the ON/OFF states. The heat balance equation for the diaphragm with the temperature T, the density ρ, the heat capacity C_p, and the thickness of Δz can be expressed as:

$$\rho C_p \frac{\partial T}{\partial t} = q_x + q_y + \frac{q_{far} - q_{space} + q_{con} + q_{near}}{\Delta z}, \tag{1}$$

where the heat-flux terms on the R.H.S. of equation (1) are respectively given by:

$$q_x = \lambda \frac{d^2T}{dx^2},\ q_y = \lambda \frac{d^2T}{dy^2} \text{ : Two-dimensional heat conduction in the diaphragm,} \tag{2}$$

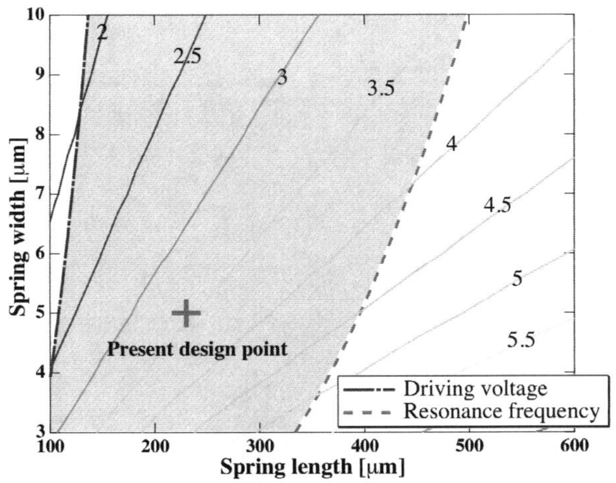

Figure 3. Contours of the heat flux enhancement ratio for different spring lengths and widths. The blue dashed line corresponds to the lower limit of spring length for the driving voltage < 10 V, while the red broken line sets the upper limit of the spring length for the resonant frequency > 2 kHz. By considering practical limitation of the micro fabrication process, the spring length and the width are respectively determined as 230 μm and 5 μm, which corresponds to 241 % heat flux enhancement.

$$q_{far} = \frac{\varepsilon_1 \varepsilon_2 \sigma (T_{wall}^4 - T^4)}{\varepsilon_1 + \varepsilon_2 - \varepsilon_1 \varepsilon_2} \text{ : Far-field thermal radiation inside the gap,} \quad (3)$$

$$q_{space} = \varepsilon_2 \sigma (T^4 - T_{\mathrm{inf}}^4) \text{ : Far-field thermal radiation to the space environment,} \quad (4)$$

$$q_{con} = \frac{T_{wall} - T}{R_{contact} A} \text{ : Contact thermal conductance,} \quad (5)$$

$$q_{near} = \varepsilon_N \sigma (T_{wall}^4 - T^4) \text{ : Near-field thermal radiation between the nano-gap.} \quad (6)$$

Here, λ and σ are the thermal conductivity and the Boltzmann constant, respectively. ε_1 and ε_2 are the emissivity of the substrate and the diaphragm, respectively. In the present analysis, the near-field heat flux between Ti and Au electrodes for the 100 nm gap is calculated using the previous model [6, 7], and the effective emissivity ε_N due to the near-field thermal radiation is computed. T_{wall} and T_{inf} denote the given temperatures at the substrate surface and for the surrounding at infinity, respectively. $R_{contact}$ and A represent the contact thermal resistance [5] and the projected area of the diaphragm.

The heat flux from the substrate to the space at the OFF state is determined with the far-field radiation, the heat conduction through springs, and the heat conduction in the diaphragm. At the ON state, the near-field radiation between the nano-gap and the contact heat conduction are also involved. A second-order central differencing scheme with an Euler explicit method is employed, and the steady-state solution for the temperature distribution on the diaphragm is obtained.

Figure 3 shows the contours of the heat flux enhancement ratio for different spring lengths and widths. The substrate and surrounding temperatures are 125 °C and 27 °C respectively. The blue dashed line corresponds to the lower limit of the spring length for the driving voltage less than 10 V, while the red broken line sets the upper limit of the spring length for the resonance frequency higher than 2 kHz. Then the spring length and width should be chosen from the region between the two lines. By considering practical limitation of the MEMS-based fabrication process, in the present study, the spring length and width are determined as 230 μm and 5 μm, respectively. The fill factor of the present design is as high as 89 %. With the present geometry, 241 % heat-flux enhancement is expected.

3. Fabrication process

Figure 4 shows the fabrication process of the present device. Patterned Ti electrodes are covered with the parylene (diX-HR, KISCO) structures with etching holes. The whole process is summarized as: 1) Thermal oxidation of a Si wafer for electrical insulation, 2) Deposition/patterning of the bottom Cr/Au/Cr electrodes, 3) Sputtering of 50 nm-thick SiO_2 layer as another insulating layer. The thickness is determined in order to obtain the near-field effects. 4) Patterning of 0.2 μm-thick poly-Si as the release layer, 5) Patterning photoresist sacrificial layer, 6) Deposition of the top Ti electrode, 7) deposition of 4 μm-thick diX-HR and its patterning with etching holes as the structural layer. The etching holes are arranged on the diaphragms in order to remove the sacrificial layer smoothly. Note

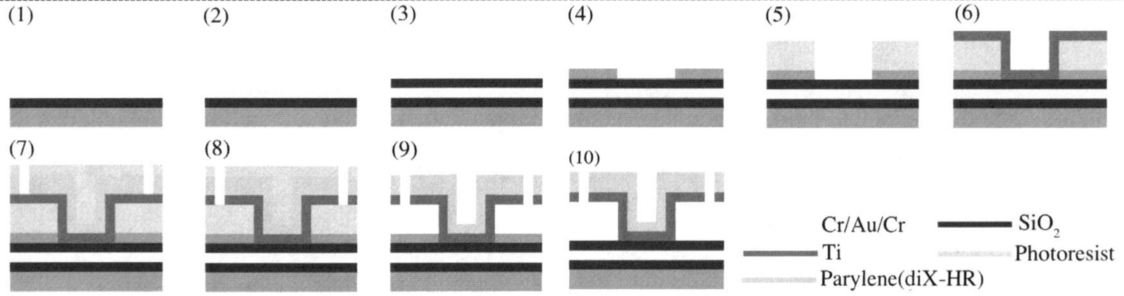

Figure 4. Fabrication process of the present device.

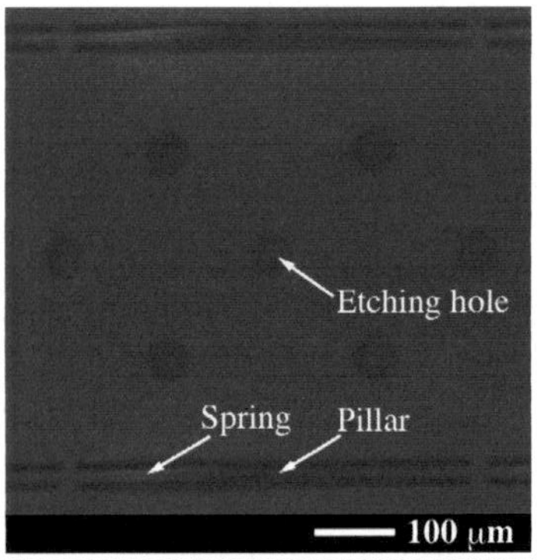

Figure 5. SEM image of a unit diaphragm on the present device.

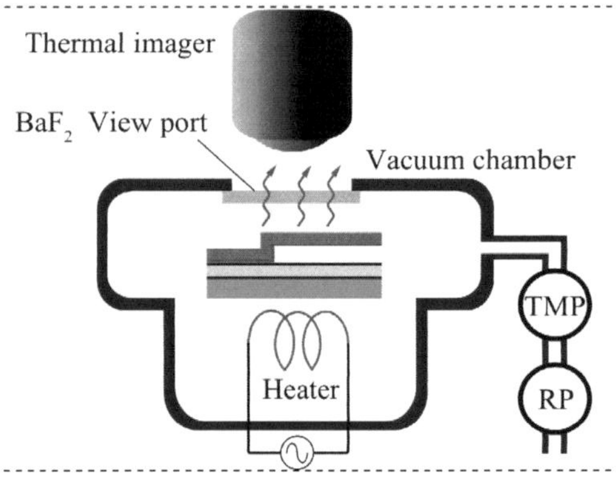

Figure 6. Experimental setup for thermal radiation measurements. The device is mounted on a hot chuck inside a vacuum chamber kept at 2 ~ 8 × 10^{-3} Pa, so that heat conduction across the gap is negligible.

that these etching holes occupy only about 5 % of the total surface area. 8) Patterning the top Ti electrode, 9) Removal of the photoresist sacrificial layer, 10) Etching of the Si sacrificial layer with XeF_2. Finally, as shown in Figure 5, a 22 × 22 array of the unit diaphragm is formed on a single chip.

4. Thermal radiation measurement

Figure 6 shows the experimental setup for thermal radiation measurements. The device is mounted on a hot chuck inside a vacuum chamber, where the chamber pressure is kept at 2 ~ 8 × 10^{-3} Pa. Thus the heat conduction across the gap is negligible, because the pressure is low enough to achieve the free molecular flow regime. The heater temperature is kept at 126 °C, and the surface temperature of the device is measured using a thermal imager (Apiste, FSV-1200) with a spatial resolution of ~250 μm. In the present study, the infrared-light absorption of the BaF_2 window of the vacuum chamber and the emissivity of the surface materials are compensated. Firstly, through a heating experiment using a sample with a black-body paint of known emissivity, the transmission rate of BaF_2 is estimated as 0.95. Secondly, a Si chip with a 200 nm Ti film below a 4 μm-thick parylene layer, which is the same composition as the diaphragms, was prepared, and the emissivity was measured by monitoring the surface temperature with a thermocouple. It is found that the emissivity is 0.316 at 106 °C and 0.391 at 58 °C. Somewhat large change of emissivity in a narrow temperature range is partially attributed to uncertainty of the present uncooled thermal imager at lower temperature.

Unfortunately, the present device could not be driven by the applied voltage. This is partially due to the failure of the insulating layer, but further study is needed for clarifying the reason and for achieving the active switching of the ON/OFF states. In the present study, in order to evaluate the thermal performance of the present radiator, the ON state is attained through the permanent stiction of

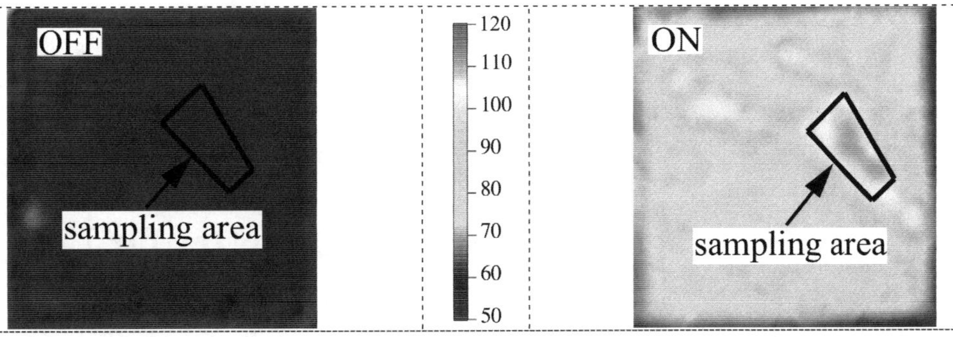

Figure 7. Thermal images of the present MEMS radiator for the ON/OFF states. The surface temperature is increased from 58.0 °C to 106.4 °C at the ON sate, corresponding to 144 % increase in the radiation heat flux.

the diaphragm to the substrate by applying a volatile liquid on the device.

Figure 7 shows the thermal images of the present MEMS radiator for the ON/OFF states. At the OFF state, the temperature distribution is quite uniform, and the surface temperature averaged over the sampling area (shown in the figure) is 58.0 °C. On the other hand, at the ON state, a non-uniform temperature distribution with a high-temperature area is obtained. This is attributed to the fact that the permanent stiction is achieved only at the area, which is verified by observation under microscope. The surface temperature averaged over the sampling area is increased to 106.4 °C at the ON state. This increase of the surface temperature corresponds to 144 % enhancement in the radiation heat flux. Without the near-field thermal radiation, the enhancement in the radiation heat flux is estimated in the present numerical analysis to be at most 46 %, which corresponds to the previous condition [4]. Therefore, it can be concluded that the near-field thermal radiation plays a dominant role in the drastic increase of the radiation heat flux of the present radiator. Since the numerical analysis shows 241 % heat flux enhancement, further improvement of the heat flux can be expected.

5. Conclusion

We have developed a high-fill-factor MEMS radiator enhanced by the near-field thermal radiation effect. We have fabricated our radiator prototype with shared springs and achieved a fill factor of as high as 89 %. It is demonstrated from proof-of-concept experiments that the surface temperature is increased from 58.0 °C at the OFF state to 106.4 °C at the ON state, which corresponds to 144 % enhancement in the radiation heat flux, showing realization of the near-field thermal radiation.

6. References

[1] Williams A and Palo E S 2006 *Proc. SPIE* **6221** 622108
[2] Demiryont H and Moorehead D 2009 *Sol. Energy Master. Sol. Cells* **93** 2075
[3] Ueno A and Suzuki Y 2011 *Transducers '11* (Beijing) 2654
[4] Ueno A and Suzuki Y 2014 *Appl. Phys. Lett.* **104** 093511
[5] Farrar D, Schneider W, Osiander R, Champion L J, Darrin G A, Douglas D and Swanson D T 2002 *ITHERM 2002* (San Diego) 1020
[6] Ueno A and Suzuki Y 2014 *Proc. Int. Heat Transfer Conf.* IHTC15-9636
[7] Narayanaswamy A and Chen G 2009 *Nano Lett.* **9** 2909
[8] Patankar SV 1980 *Numerical Heat Transfer and Fluid Flow* (New York: McGraw Hill)

Electrostatic MEMS vibration energy harvester for HVAC applications

J Oxaal[1], M Hella[2], D-A Borca-Tasciuc[1]

[1] Department of Mechanical, Aerospace, and Nuclear Engineering, Rensselaer Polytechnic Institute, Troy, NY, USA
[2] Department of Electrical, Computer and Systems Engineering, Rensselaer Polytechnic Institute, Troy, NY, USA

E-mail: borcad@rpi.edu

Abstract. This paper reports on an electrostatic MEMS vibration energy harvester with gap-closing interdigitated electrodes, designed for and tested on HVAC air ducts. The device is fabricated on SOI wafers using a custom microfabrication process. A dual-level physical stopper system is implemented in order to control the minimum gap between the electrodes and maximize the power output. It utilizes cantilever beams to absorb a portion of the impact energy as the electrodes approach the impact point, and a film of parylene with nanometer thickness deposited on the electrode sidewalls, which defines the absolute minimum gap and provides electrical insulation. The fabricated device was first tested on a vibration shaker to characterize its resonant behavior. The device exhibits spring hardening behavior due to impacts with the stoppers and spring softening behavior with increasing voltage bias. Testing was carried out on HVAC air duct vibrating with an RMS acceleration of 155 mg_{RMS} and a primary frequency of 60 Hz with a PSD of $7.15{\cdot}10^{-2}$ g^2/Hz. The peak power measured is 12nW (0.6 nW RMS) with a PSD of $6.9{\cdot}10^{-11}$ W/Hz at 240 Hz (four times of the primary frequency of 60 Hz), which is the highest output reported for similar vibration conditions and biasing voltages.

1. Introduction

Improving energy efficiency of buildings is of critical importance as power consumption has increased significantly in both private and public building sectors [1]. Building energy usage makes up 40% of the US energy consumption and is one of the largest worldwide polluters; for instance in 2005 U.S. buildings contributed 9% of the world's carbon dioxide emissions, ranking third worldwide as a source of carbon emission [2]. To address this, an increasing level of research effort is being dedicated to developing building automation systems (BAS) to better manage energy expenditures. Developing autonomous energy sources to power wireless sensors of BAS is another facet of this effort, as batteries are perceived as nuisance and added maintenance cost by potential BAS users due to their relatively short lifetime. In this context, this paper investigates vibration electrostatic MEMS vibration energy harvesters designed to work on HVAC ducts in a large building.

Vibrations are of particular interest for energy harvesting as they are ubiquitous in buildings, especially in HVAC systems which have air ducts that span the building interior, and they have ample potential for harvesting. The MEMS harvester designed here uses interdigitated electrodes with a gap closing configuration to form a variable capacitor. The energy harvested with such structure depends on the difference between and ratio of maximum and minimum capacitance, as well as frequency of vibration and applied bias. While the minimum capacitance is when the mobile electrode is equally distanced from each fixed electrode that is set by initial device geometry, maximum capacitance is controlled here by displacement limiting physical stoppers. The physical stopper consists of two different stopping mechanisms that act in conjunction: first, when approaching the fixed electrode, the moving electrode makes contact with cantilever beams connected to the non-moving portion of the device that absorb a portion of the impact energy (figure 1). However, the electrode continues to move in the same direction, albeit decelerating. Next the final contact is with a film of parylene-C with nanometer thickness deposited on the electrode sidewalls, which defines the absolute minimum gap and also provides electrical insulation (figure 1). The thin film stopper can provide much smaller electrode separation gap as compared to microfabricated physical stoppers which are defined by lithography and etching, a process that limit the resolution (and minimum gap) to 1 μm or larger [3–6]. Thus, the power output in the presence of the dual-level stopper is expected to increase significantly. The devices are fabricated with silicon-on-insulator (SOI) wafers using a custom microfabrication process. They are tested first on a shaker table to characterize the dynamic response, followed by performance characterization when mounted on HVAC ducts in the laboratory.

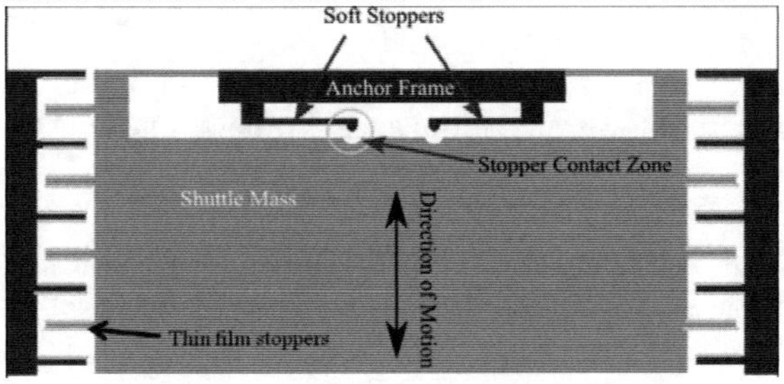

Figure 1. Dual-level stopper system in a gap closing electrostatic harvester.

2. Vibration study

Before discussing harvester design, fabrication, and testing, a few design considerations are presented first. As mentioned above, the devices investigated here are intended to work on HVAC systems. Thus, a thorough vibration study was conducted on the HVAC duct work in the authors' laboratory in order to understand the frequency spectrum available. An Analog Devices ADXL335 3-axis accelerometer was mounted to a 1 inch square stainless steel plate and magnetically attached to 16 different locations on the air ducts. In regard to the orientation of the accelerometer with the air duct, the z-axis is always normal to the duct wall, the y-axis parallel to the air flow direction, and x-axis perpendicular to the flow direction.

The vibration data was analyzed using the power spectral density function (PSD) to find the frequency and power content of the vibrations. The Signal Processing Toolbox of Matlab contains several PSD estimate functions and the `pmtm` function was chosen in this analysis for its high quality estimation. The results of the PSD for all axis and locations tested are shown in figure 2. All together the most prominent frequencies were 45, 60, 65, and 120 Hz. The axis normal to the duct wall (z-axis) contained the highest amplitude of vibrations by two orders of magnitude. The average PSD amplitude at these peaks is on the order of 10^{-4} g^2/Hz. This amplitude is on the lower end of what others have reported measurements of [7] or driven devices with [4,6,8]. Based on these results, the devices were design to resonate at 65 Hz. Spring softening from electrostatic forces and spring hardening from stopper impacts will broaden the resonant bandwidth to where acceleration PSD is greatest (45-120 Hz).

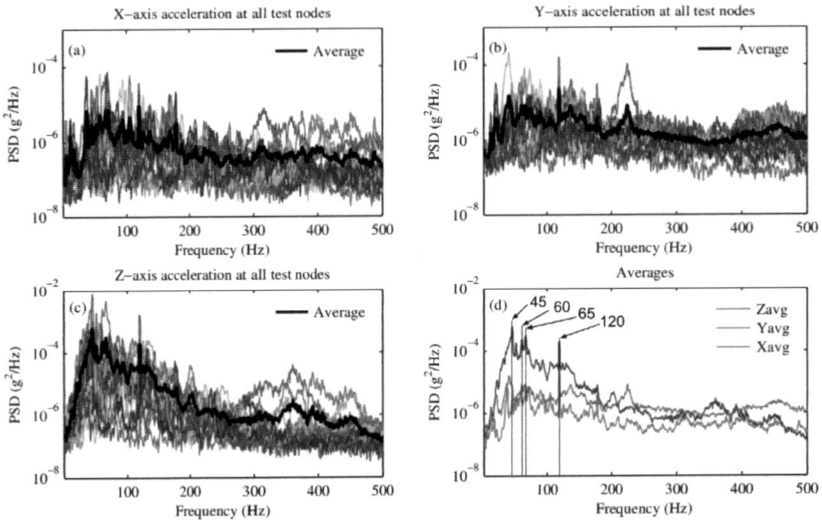

Figure 2. (a) X, (b) Y, (c) Z and (d) average PSD as function of frequency for different locations of the accelerometer on HVAC duct.

3. Design and fabrication

The device was fabricated on SOI wafer with 200 μm device layer and 350 μm handle layer using a custom microfabrication process developed at Cornell Nanoscale Science and Technology Facility (CNF). As mentioned above, it is designed to resonate 65 Hz frequency range, where the maximal amount of vibration power is available in the HVAC ducts. Gap-closing interdigitated electrodes with 22 μm nominal gaps form the variable capacitor. The shuttle mass is suspended from four corners with double 4-fold flexure beams. Device dimensions and other properties are shown in table 1. The fabrication process is illustrated in figure 3a and a completed device is shown in figure 3b. The deposition of thin film stopper (parylene) was carried out with a shadow mask (step 8), after the devices were cleaved. The thickness of parylene was 200 nm, as determined afterward with a profilometer. Following this the devices where wire bonded and packaged in a similar manner as reported in [9].

Table 1. Device properties and dimensions

Property	Symbol	Value	Unit
Die size		13x14	mm x mm
Shuttle area size		9.3 x 8.4	mm x mm
Shuttle mass	m	90	miligrams
Electrode overlap length	L_f	400	μm
Electrode width	W_f	8	μm
Nominal gap	g_0	22	μm
Number of electrodes	n	336	
Spring beam width	w_s	10	μm
Spring beam length	l_s	1670	μm
Total stiffness	k	14.6	N/m^2

4. Results

First the static capacitance of the device was measured using an LCZ meter. The nominal capacitance was found to be 30.7 pF, close to the analytical estimation of 25 pF. Next, determination of the frequency response, including the effect of the biasing voltage, was carried out. Figure 4 shows the results of a frequency up-sweeps conducted at 20 mg acceleration with the device connected to a series resistance of 3 MΩ. The output shows an increase in the nonlinear spring softening as the bias voltage increases (left shifting peaks), and spring hardening due to mechanical impact with stoppers (right slanting plateau with drop off effect.

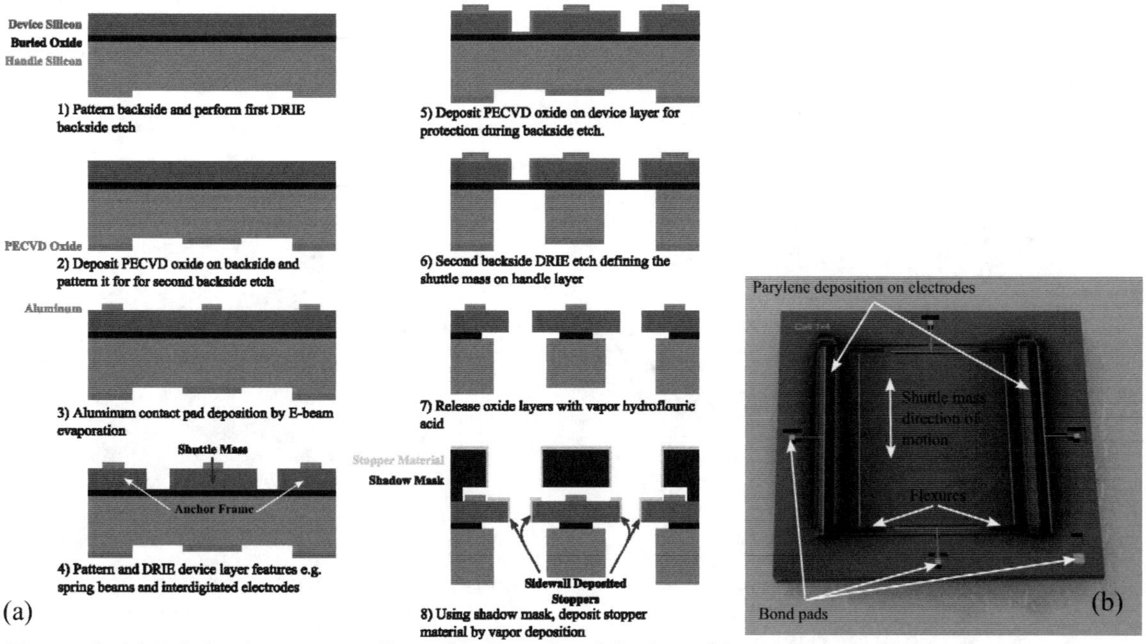

Figure 3. (a) Fabrication process flow. (b) Completed device with parylene on electrodes.

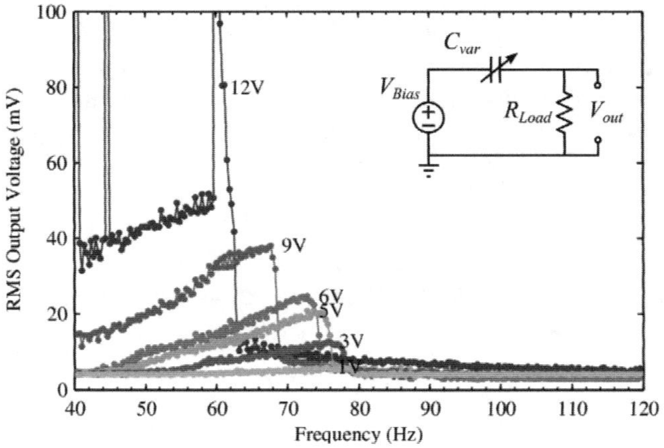

Figure 4. Voltage across load resistor during vibration frequency up sweeps at 20 mg RMS amplitude and with various DC bias voltages.

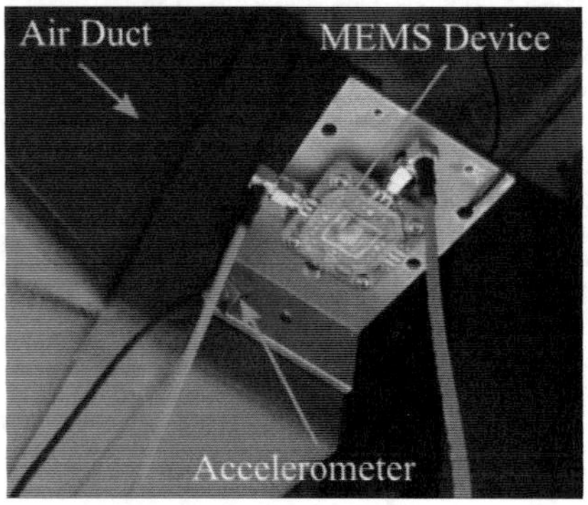

Figure 5. MEMS device being tested on HVAC air duct.

Next, measurements were taken with the device on HVAC ducts hanging from the ceiling of the lab. An L-bracket was used to mount the harvester on the duct sidewall with its orientation normal to the duct wall where the vibration amplitude is maximum (figure 5). The raw data from the acceleration and voltage measurements is shown in figure 6. The peak power for a device biased at 15 V is 12 nW (0.6 nW RMS) at an RMS acceleration of 155 mg. The power spectral density (PSD) of the device is greatest at 240 Hz (6.9e-11 W/Hz) as shown in figure 7. This frequency corresponds to four times the acceleration peak, 60 Hz and 7.15e-2 g2/Hz, indicating that the device is operating in impact mode with the stoppers. Theoretically the devices were expected to reach significantly higher power, but the performance was significantly decreased due to a steep sidewall profile of the electrodes which prevented the devices from reaching a uniform minimum gap, reducing the maximum achievable capacitance. The sidewall profile was due to limitations associated with the etch tool. As an example, under static conditions, the trench for the nominal gap started with a width of 22 μm on top and reduced to 16 μm at the bottom of the 200 μm etch. Regardless, even with these imperfections, the device performed well, or better, when compared to similar harvesters working under the same range of broadband vibrations reported in literature [3,4,6,8].

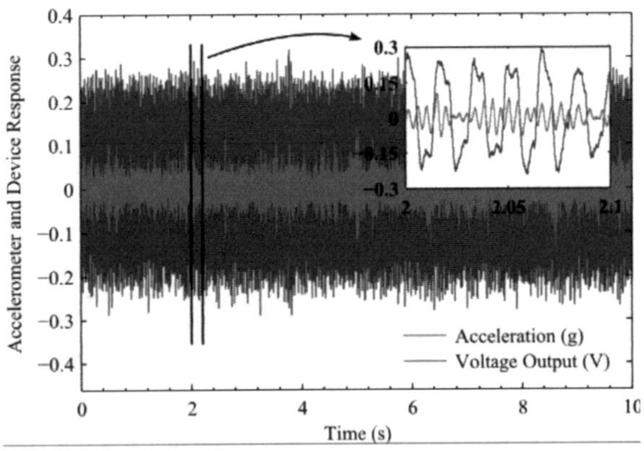

Figure 6. Raw data from the device biased at 15 V and accelerometer attached to the HVAC duct.

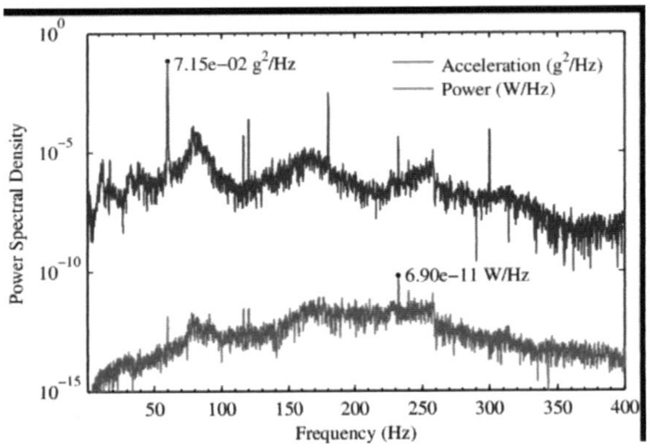

Figure 7. PSD of acceleration and device voltage when on HVAC duct with 15 V bias.

5. Conclusion.

Energy harvesting from ambient vibration is of interest for powering sensors in building automation systems. In this work an electrostatic energy harvester using a gap-closing, interdigitated variable capacitor structure was designed for HVAC vibration power harvesting and tested. To maximize the power output of the device, a dual-level stopper system was incorporated, fixing the minimum gap in the nanometer range. When tested on HVAC air ducts of a laboratory in a large building the devices reached 12 nW of peak power, one of the highest reported power levels for similar devices working under equivalent conditions.

Acknowledgment

This work was supported by New York State Energy Research Development Authority (NYSERDA) Award No. 0925733

References

[1] Vullers R J M, van Schaijk R, Doms I, Van Hoof C and Mertens R 2009 S. S. Elec. **53** 684–93
[2] Estrin D, Culler D, Pister K and Sukhatme G 2002 IEEE Pervasive Comput. **1** 59–69
[3] Le C P and Halvorsen E 2012 J. Micromech. Microeng. **22** 074013
[4] Le C P, Halvorsen E, Søråsen O and Yeatman E M 2013 Smart Mater. Struct. **22** 075020
[5] Chiu Y and Tseng V F G 2008 J. Micromech. Microeng. **18** 104004
[6] Basset P, Galayko D, Cottone F, Guillemet R, Blokhina E, Marty F and Bourouina T 2014 J. Micromech. Microeng. **24** 035001
[7] Miller L M, Halvorsen E, Dong T and Wright P K 2011 J. Micromech. Microeng. **21** 045029
[8] Nguyen S D, Halvorsen E and Jensen G U 2013 J. Microelectromechanical Syst. **22** 892–900
[9] Oxaal J, Foster D, Hella M and Borca-Tasciuc D-A 2015 J. Micromech. Microeng **25** 105010

Vibro-Impact Type Triboelectric Energy Harvester for Large Amplitude and Wideband Applications

J M Chen[1], L Bu[1], W Y Xu[1], B J Xu[1], L Song[2]
[1]School of Information Engineering, China University of Geosciences, Beijing, China
[2]State Grid Jibei Electric Power Co., Ltd., Beijing, China

E-mail: lingbu@cugb.edu.cn

Abstract. This paper reports the design, fabrication and testing of a novel vibro-impact type triboelectric energy harvester. The dynamics of vibro-impact converts external vibration to large contact force for triboelectric power generation. Strong nonlinearities are measured for this vibro-impact system, and wideband frequency response under diverse structural parameters are analyzed. The proposed device is applied in two large amplitude scenarios, and generates maximal peak-to-peak voltage of 18V in foot swinging condition @2Hz 30cm, and maximal peak-to-peak voltage of 45V in arm swinging condition during running @5Hz 40cm.

1. Introduction

Energy harvesters such as piezoelectric, electromagnetic and electrostatic harvesters have been extensively explored in recent years [1-3]. The output power of energy harvesters is typically around the order of 1~100μW, which is still stringent for most sensor applications. It is necessary to further boost output power of energy harvesters.

Triboelectric energy harvesters are brought forward by Wang Z L et al. [4], and are able to generate high output voltage (up to the order of 100V) and power (up to the order of 10mW) [5]. The basic principle of triboelectric energy harvesters is that when two materials contact each other, electrons transfer through external load between the materials due to different electronegativity. So far, triboelectric energy harvesters have been utilized in lightening 100 LEDs [6], fabric of clothes [7], mobile phone battery charger [8], and etc. The high output voltage and power of triboelectric energy harvesters are usually achieved under pressing, rubbing, and bending conditions [9], which requires large contact force. But in real environment such as bridge vibration and human body movement, contact force is small while vibrations are abundant. To utilize triboelectric energy harvester in these scenarios, it is necessary to convert vibration to large contact force.

This paper presents a novel vibro-impact type triboelectric energy harvester to generate high output voltage and power under small contact force environment. The dynamics of vibro-impact introduce strong non-linearity into device performance over a wide range of frequency. The proposed device is more suitable for large amplitude vibrations, and can be utilized in human body wearable applications.

2. Concept

The vibro-impact type triboelectric energy harvester consists of two facing electrodes with frictional surface and one inner rotor. The inner rotor is springless, and is capable of free movement between electrodes. When horizontal vibration occurs, the inner rotor collides with the frictional surface of one electrode. The impact generates large contact force between rotor and frictional surface, and the difference in electronegativity generates transferring electrons through external load in the process of contact and separation. As horizontal vibration continues, the inner rotor collides with the other electrode, and triboelectric power is produced in a similar way.

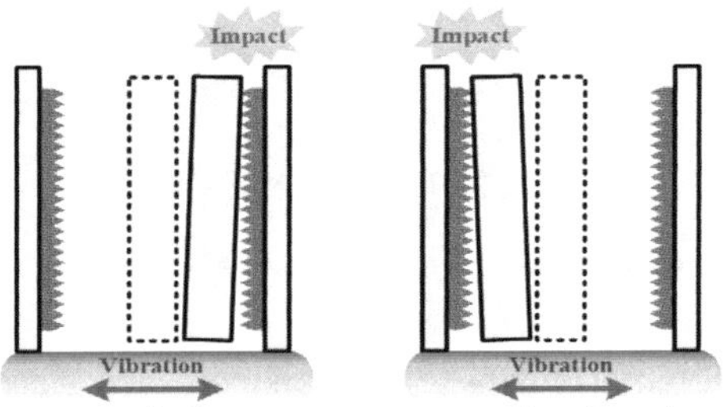

Figure 1. Concept of vibro-impact type triboelectric energy harvester.

3. Device Fabrication

The frictional surface of the device is PDMS pyramid arrays, the fabrication process of which is shown in Figure 2. 1000Å thermal SiO_2 is first grown on 4 inch silicon wafer, and then 1000Å Si_3N_4 is deposited using LPCVD on both sides. After spin coating and patterning 1μm photoresist, the Si_3N_4/SiO_2 double layer on one side is dry etched, so that etching windows with the dimension of 300μm×300μm are formed. The remaining Si_3N_4/SiO_2 double layer serves as the mask for bulk silicon etching, and the etching thickness is approximately 210μm. After the formation of silicon cup, PDMS (Sylgard 184, Dow Corning, polymer: curing agent=10:1 mass ratio) is poured in, cured at 45℃, and peeled off. The thickness of PDMS is controlled by calculating total mass, and is approximately 0.4mm from the cone to the base. The fabricated PDMS pyramid arrays are shown in Figure 3(a).

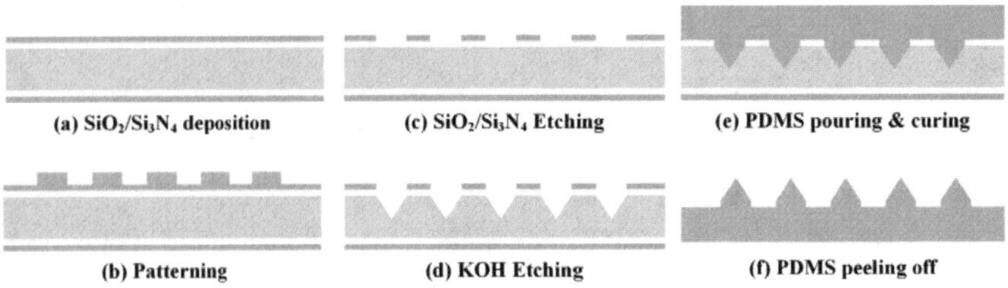

Figure 2. Fabrication process flow.

Then, PMMA spacer is formed using Computer Numerical Control (CNC) machining of the transparent acrylic material. The outside and inside dimensions of the PMMA spacer are 20mm×20mm and 12mm×12mm, respectively. The two electrodes and inner rotor are made of stainless steel. The dimensions of the electrodes and inner rotor are 20mm×20mm and 10mm×10mm, respectively. PDMS pyramid arrays are stuck directly to the electrodes due to the viscidity of silicon rubber, and are cut into 11mm×11mm pieces so as to fit inside the PMMA spacer. The assembly part of the device is shown in Figure 3(b). Finally, two electrodes with frictional surface and PMMA spacer is stuck together to form a complete sandwich structure, with the inner rotor freely placed inside. Different structural parameters are adopted: spacer height g=3mm, 5mm, 10mm; rotor thickness d=1mm, 3mm, 8mm. The complete device is shown in Figure 3(c).

Figure 3(d) shows the testing rig, in which the device is clamped between two PMMA fixtures and is tested under horizontal vibration. The PMMA fixtures are fixed to the electromagnetic shaker, and the acceleration and frequency is controlled by function generator. A 3-axis accelerometer is also mounted in the front of the PMMA fixture to record excitation acceleration. Generated electrical signals are recorded using oscilloscope.

Figure 3. (a) PDMS pyramid arrays; (b) parts for assembly; (c) photo of entire device; (d) testing rig.

4. Results and Discussion

Figure 4 shows the measured voltage-time traces generated on a single electrode. The device is of structural parameters as: g=3mm, d=1mm, and vibration frequency is 20Hz. As acceleration increases from 1.2g, 1.6g, to 1.8g, maximum voltage increases from 3.3V, 4.1V, to 6.5V. Increased vibration acceleration results in stronger impacts and larger contact force, and thus increases triboelectric output voltage. But the number of impacts per cycle decreases as acceleration increases. The number of impacts is shown by the number of voltage peaks per cycle, which is less than one per cycle @1.8g, one per cycle @ 1.6g, and two per cycle @1.2g. Because the inner rotor is freely moving and impacting, the number of impacts is correlated not only to external vibration parameters, but also to the relative position and pitch attitude of the inner rotor. It is speculated that, for 1.2g acceleration, the inner rotor might collide with the testing frictional surface at two different edges consecutively due to different pitch attitude. For the relatively large acceleration of 1.8g, it is possible that the relative position of the inner rotor is close to the opposite frictional surface, so that less than one impact per cycle is recorded.

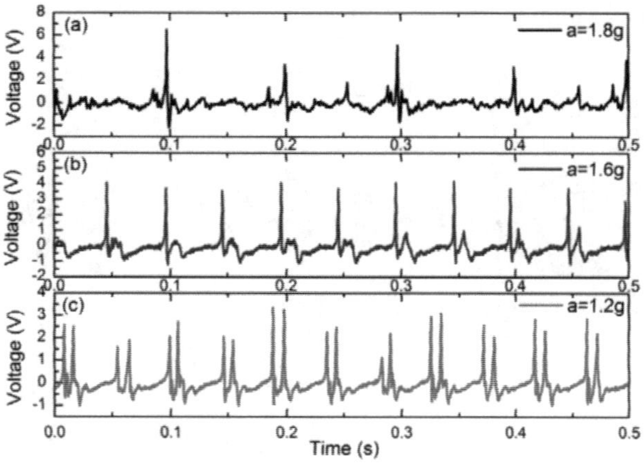

Figure 4. Voltage-time traces @20Hz and at diverse accelerations. (a) less than one impact per cycle @1.8g; (b) one impact per cycle @1.6g; (c) two impacts per cycle @1.2g.

The frequency response of the device also exhibits strong nonlinearity. As is shown in Figure 5(a), at fixed spacer height g=10mm, smaller rotor thickness d results in higher output voltage at lower frequency range, and vice versa. Figure 5(b) shows that, at fixed rotor thickness d=1mm, an optimal spacer height *g*=5mm exists for maintaining high output voltage at lower frequency range and wider bandwidth. In Figure 5(c), rotors are of the same free length (approximately 1.2mm, considering the height of PDMS pyramid arrays). The device with *g*=5mm and *d*=3mm generates highest output voltage in 25~30Hz frequency range, while the device with *g*=3mm and *d*=1mm generates highest output voltage in the remaining test frequency range. In all six different traces shown in Figure 5, the device with *g*=5mm and *d*=1mm functions optimally in the <25Hz low frequency range. Despite low output voltage, the device with g=10mm and *d*=8mm exhibits flattest frequency response and widest bandwidth. Such dynamic characteristics are very different from the traditional spring-mass-damper energy harvester, and implies the potential diverse application of the devices in low frequency, wideband, and selected frequency range in accordance to different structural parameters.

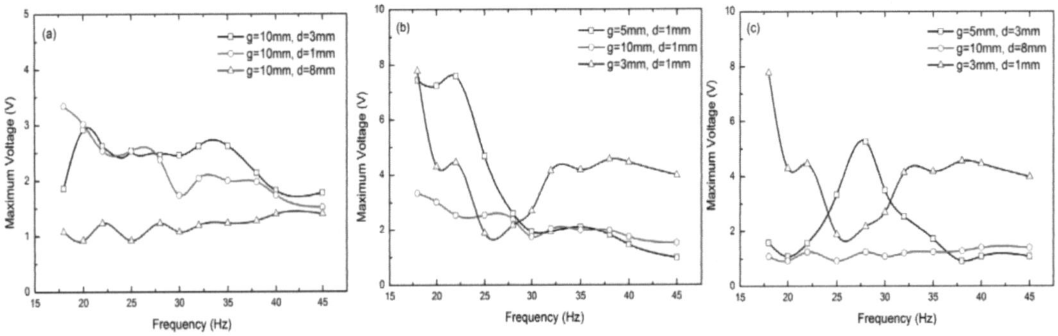

Figure 5. Frequency responses of devices (a) with the same spacer gap but diverse rotor thicknesses; (b) with the same rotor thickness but diverse spacer gaps; (c) with the same free length but diverse volumes.

5. Application

The proposed device (*g*=3mm and *d*=1mm) is applied in large amplitude human motion environment. Figure 6 shows the foot swinging condition, and the device is mounted in the rear part of the shoe, as is shown in Figure 6(a). The knee of the tester is almost still while the foot is swinging back and forth at around 2Hz frequency and around 30cm amplitude. Figure 6(b) shows the measured output voltage, with maximal peak-to-peak voltage of 18V.

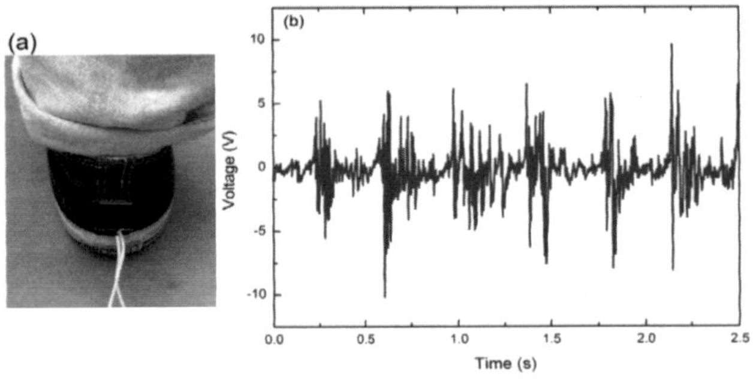

Figure 6. (a) Device mounted in the rear part of the shoe. (b) Measured output voltage.

Figure 7 shows the arm swinging condition in the running process, and the device (*g*=3mm and *d*=1mm) is placed on the arm, as is shown in Figure 7(a). Limited by the wiring length, the tester runs in situ during test, and the swing of the arm is approximately 5Hz in frequency and 40cm in amplitude. Figure 7(b) shows that the device generates maximally 45V peak-to-peak voltage. Such results prove the suitability of proposed device in large amplitude, low frequency and wideband applications.

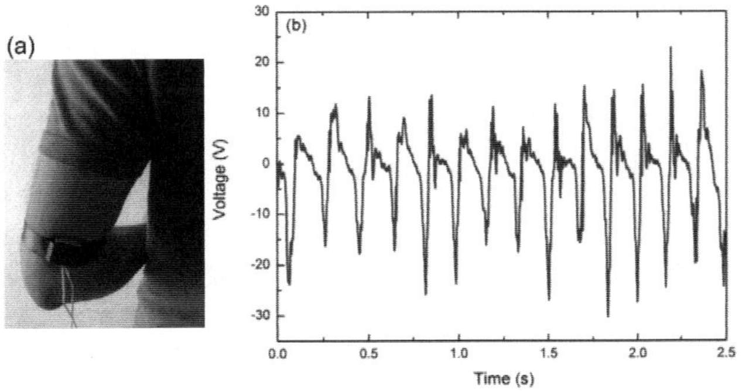

Figure 7. (a) Device mounted on the arm, in running condition. (b) Measured output voltage.

Conclusion

A novel vibro-impact type triboelectric energy harvester is presented, which effectively converts external vibration to large contact force. The conversion is achieved through vibro-impact dynamics, which is of strong non-linearity. Fabricated device exhibits diverse number of impacts per cycle in voltage-times traces due to different accelerations, and exhibits diverse low frequency, wideband, and selected frequency range in frequency response curves due to different structural parameters. The proposed device is applied in wearable applications. In foot swinging condition @2Hz 30cm, maximal peak-to-peak voltage reaches 18V. In arm swinging condition @5Hz 40cm, the device generates maximally 45V peak-to-peak voltage. These results prove the viability of proposed device in large amplitude and wideband applications.

Acknowledgements

This work is supported by Fundamental Research Funds for Central Universities (Grant No. 2-9-2014-010) of China.

References

[1] Aktakka E E, Peterson R L, Najafi K. A CMOS-compatible piezoelectric vibration energy scavenger based on the integration of bulk PZT films on silicon. *Proceedings of IEEE International Electron Devices Meeting (IEDM'10)*, 2010, pp. 31.5.1-31.5.4.

[2] Galchev T, Hanseup Kim, Najafi K. Micro power generator for harvesting low-frequency and nonperiodic vibrations. *J. Microelectromech. S.*, 2011, 20(4), pp. 852-866.

[3] Boisseau S, Despesse G, Ricart T, et al. Cantilever-based electret energy harvesters. *Smart Mater. Struct.*, 2011, 20(10), 105013(11pp).

[4] Wang Z L. Triboelectric Nanogenerators as New Energy Technology for Self-Powered Systems and as Active Mechanical and Chemical Sensors. *ACS Nano*, 2013(7), pp. 9533-9557.

[5] Wang S H, Lin L, Wang Z L. Triboelectricnanogeneratorsasself-powered active sensors. *Nano Lett*, 2012(12), pp. 6339-6346.

[6] Meng B, Tang W, Zhang X, et al. Self-powered flexible printed circuit board with integrated triboelectric generator. *Nano Energy*, 2013(2), pp.1101-1106.

[7] Lee S, Ko W, Oh Y, et al. Triboelectric energy harvester based on wearable textile platforms employing various surface morphologies. *Nano Energy*, 2015(12), pp.410-418.

[8] Zhu G, Chen J, Zhang T J, et al. Radial-arrayed rotary electrification for high performance triboelectric generator. *Nat. Commun.* 2014(5), pp.3426.

[9] Dhakar L, Tay F E H, Lee C. Skin based flexible triboelectric nanogenerators with motion sensing capability. *Proceedings of International Conference on Micro Electro Mechanical Systems (MEMS'15)*, 2015, pp.106-109.

MEMS Rotational Electret Energy Harvester for Human Motion

J Nakano[1], K Komori, Y Hattori, and Y Suzuki[2]
[1]Hokuriku Electric Industry Co. Ltd., 3158 Shimookubo, Toyama-shi, Toyama 932-2292, Japan
[2]Department of Mechanical Engineering, The University of Tokyo, 7-3-1 Hongo, Bunkyo-ku, Tokyo 113-8656, Japan
E-mail: ysuzuki@mesl.t.u-tokyo.ac.jp

Abstract. This paper reports the development of MEMS rotational electret energy harvester (EH) for capturing kinetic energy of human motion. Optimal design method of rotational electret EH is proposed by considering both the rate of overlapping-area-change and the parasitic capacitance. A rotational MEMS electret EH with embedded ball bearing has been successfully developed. Up to 3.6 μW has been obtained at 1 rps rotation with an early prototype.

1. Introduction

Output power of resonance-type vibration energy harvesters (VEH) is limited by the velocity-damped resonance generator (VDRG) limit [1], and it is not a straightforward process to design VEH with large output power from human motion due to its very low vibration frequency. On remedy is to enlarge the vibration amplitude of mass [2, 3], but the dimension of the device in the vibration direction should be enlarged. Another approach is to use a moving "rotor" without tethers [4-6], but its advantage over the resonance generators in terms of the output power remains unclear.

On the other hand, rotational generator is widely used as the perpetual power source of wristwatches [7], and prototypes with electromagnetic [8] and piezoelectric [9] principles are also proposed for capturing both linear and rotational accelerations. A gyroscopic device is also proposed [10]. At low rotational speed, the electret generator should have higher output power than electromagnetic or piezoelectric ones [11]. The objective of the present study is to develop a MEMS rotational electret EH prototype.

2. Design of Rotational MEMS Electret Generator

Figure 1 shows the concept of MEMS rotational electret EH. Fan-shaped electret and guard electrodes are patterned on the rotor, and interdigitated electrodes are formed on the stator for collecting the induced charges. The rotor is connected to the stator with a miniature ball bearing, and can be rotated freely around its center. The rotor has a eccentric added mass, which makes the center of gravity of the rotor off-centered.

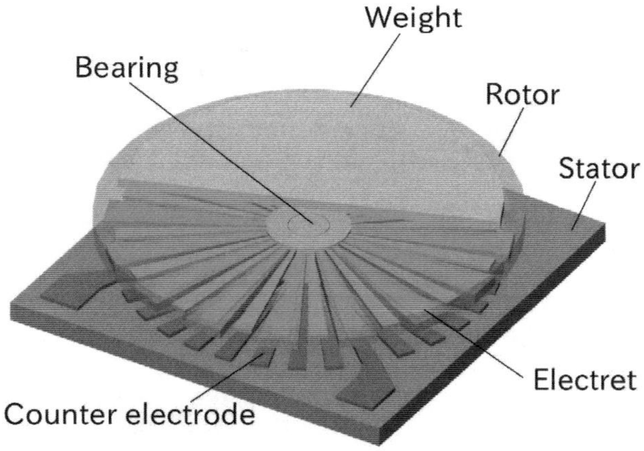

Figure 1. Schematic of rotational electret energy harvester.

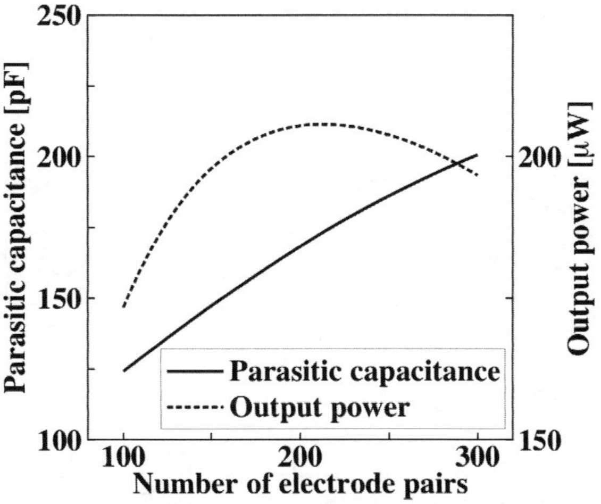

Figure 2. Parasitic capacitance and output power versus the number of poles N for a rotational electret energy harvester.

In order to maximize its output power, number of poles N and the gap between the interdigitated electrodes δ should be optimized. With increasing N, the rate of change of the overlapping area between electret and electrode is increased. On the other hand, with decreasing N, the parasitic capacitance C_p is increased, which deteriorates the generator performance [12, 13]. Similarly, with increasing δ, effective electrode area is decreased, while C_p is increased with decreasing δ.

In the present study, in order to determine the optimum N, we employ the analytical formula for the output power of rotational electret generator including the effect of C_p [12]. For C_p, we employ the analytical formula for interdigitated electrodes [14]. On the other hand, δ at the innermost radius is chosen as 30 μm based on the requirement of the present MEMS process.

In this analysis, the rotational speed, the outer radius of the rotor, the innermost diameter of the electrodes, the air gap between the top and bottom substrates, and the surface charge density are assumed to be 1 Hz, 20 mm, 9 mm, 100 μm, and 1 mC/m^2, respectively. The relative permittivity of the stator substrate is assumed to be 3.2, since a thick SU-8 layer is formed on the stator as explained below. The external load is optimized for each N. Figure 2 shows the parasitic capacitance C_p and the output power versus number of poles N. It can be seen that C_p is increased with N, and the output power is maximized at N=213, where the output power of 206 μW can be obtained. Based on this analysis, N is chosen as 220, by which the output power is unchanged.

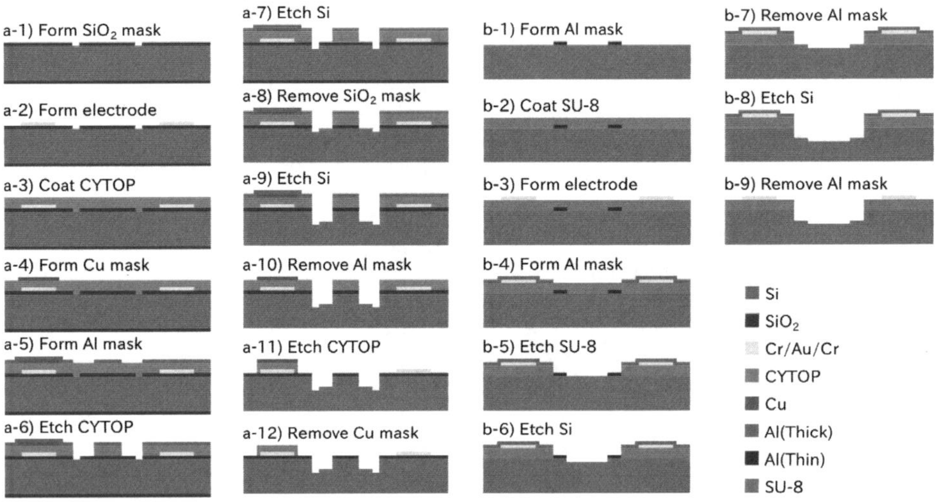

Figure 3. MEMS process of a) Rotor and b) Stator for a 40 mm-diameter rotational electret energy harvester using a ball bearing.

a) *b)*

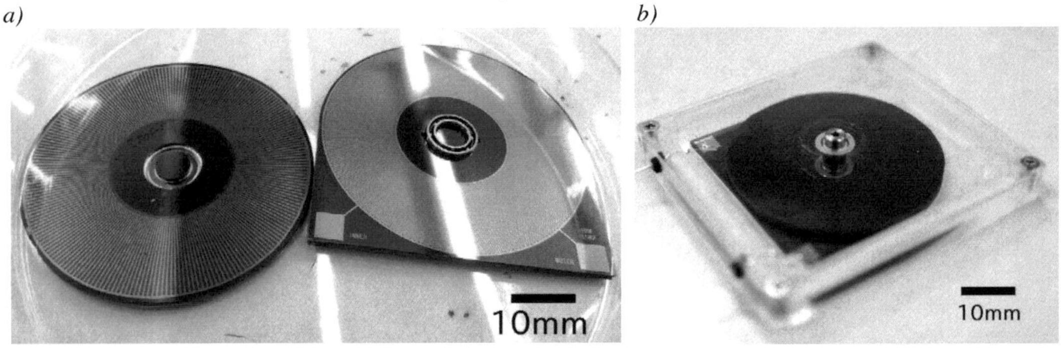

Figure 4. MEMS rotational electret energy harvester. A ball bearing is successfully installed its housing etched into the Si substrate. a) Rotor and stator substrate with a ball bearing, b) Rotational energy harvester stored in a plastic package.

3. Fabrication Process

Figure 3 shows MEMS fabrication process using 1.5 mm-thick Si wafers with thermal oxide. On the rotor side, the SiO_2 insulation layer is patterned for the first mask of the bearing housing, and Cr/Au/Cr interdigitated electrodes are patterned on the SiO_2 layer. Then, 15 μm-thick CYTOP EGG (Asahi Glass) electret layer is formed, and then Cu hard mask for CYTOP and Al mask for bearing housing are patterned. The center part of CYTOP is removed with O_2 plasma. This is followed by two-step DRIE for forming the stepped bearing housing; the first mask is the bottom SiO_2, and the second mask is the Al layer. Finally, CYTOP is patterned with O_2 plasma and the Cu mask is stripped.

On the stator side, an Al mask is patterned for the bearing housing, and 100 μm-thick SU-8 is formed. Thickness of SU-8 is determined for minimizing C_p [13]. Then, Cr/Au/Cr interdigitated electrodes and another Al hard mask are patterned on the SU-8 layer. The center part of SU-8 is removed with O_2/SF_6 plasma. This is followed by two-step DRIE for forming the stepped bearing housing; the first mask is the bottom Al, and the second mask is the top Al layer. Finally, the Al mask is stripped.

Figure 4 shows the rotor and stator substrates successfully mirofabricated. A 2 mm-high miniature ball bearing is successfully buried into the Si wafers, and the rotor can be smoothly rotated with very low friction. The measured value of C_p is around 200 pF, which is slightly higher than the present estimate of 176 pF.

4. Experimental Results

In the present early prototype, the gap between the rotor and the stator is unfortunately non-uniform, and around 100 μm. This is because the ball bearing is not well-fitted to the housing due to the rough

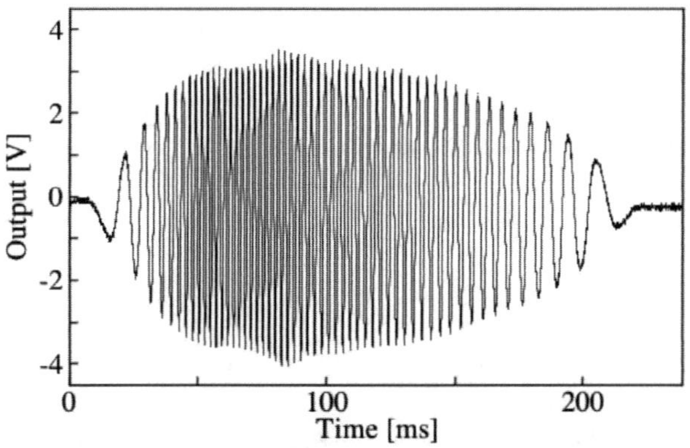

Figure 5. Transient output voltage after applying impulse rotational force to the rotor.

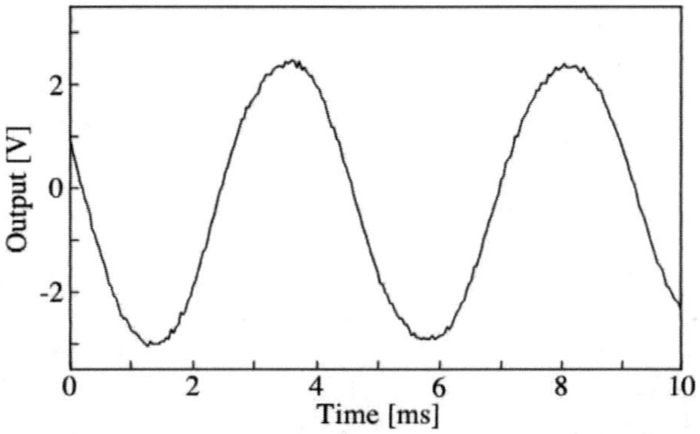

Figure 6. Output voltage at around 1 rps, which corresponds to the output power of 3.6 μW. External load is 1 MΩ, which is lower than the optimum load of 3.3 MΩ.

bottom surface of the housing. In addition, the surface potential is about -160 V, which is much lower than the target value of the surface potential is -700~-800 V. This lower surface potential could be solved by using a thicker SiO_2 insulation layer and/or improved design of the interdigitated electrode suppressing discharge.

Figure 5 shows the output voltage with an impulse rotational force across 1 MΩ load resistance. The waveform in Fig. 5 with about 1 rps is extracted, and re-plotted in Fig. 6. The peak-to-peak voltage amplitude is about 5.5 V, which corresponds to 3.6 μW power generation. The experimental data are in reasonable agreement with the simulation result of 4.1 μW. With the optimum load of 3.3 MΩ, the output power in simulation is 7.2 μW. Since the output power is proportional to the surface voltage of the electret squared, the output power in accordance with the designed value of 200 μW can be obtained with the present generator.

5. Conclusion

MEMS rotational electret energy harvester (EH) for capturing kinetic energy of human motion is developed. Optimal design of number of poles is proposed by considering both the rate of overlapping-area-change and the parasitic capacitance. A prototype with embedded ball bearing has been successfully developed. Up to 3.6 μW has been obtained at 1 rps rotation with the early prototype.

Acknowledgement

Photo-mask is made using The University of Tokyo VLSI Design and Education Center (VDEC)'s 8-inch EB writer F5112+VD01 donated by ADVANTEST Corporation.

References

[1] P. D. Micheson, T. C. Green, E. M. Yeatman, and A. S. Holmes, "Architectures for Vibration-driven Micropower Generators," J. Microelectromech. Syst., Vol. 13, pp. 429-440 (2004).
[2] C. R. Saha, T. O'Donnell, N. Wang, and P. McCloskey, "Electromagnetic Generator for Harvesting Energy from Human Motion," Sensors Actuators, A, Vol. 145, pp. 248-253 (2008).
[3] Y. Naruse, N. Matsubara, K. Mabuchi, M. Izumi, and K. Honma, "Electrostatic Micro Power Generator from Low Frequency Vibration such as Human Motion," J. Micromech. Microeng., Vol. 19, 094002 (2009).
[4] H.-W. Lo, and Y.-C. Tai, "Parylene-based Electret Power Generators," J. Micromech. Microeng., Vol. 18, 104006 (2008).
[5] M. E. Kiziroglou, C. He, and E. M. Yeatman, "Rolling Rod Electrostatic Microgenerator," Trans. IEEE, Ind. Electr., Vol. 56, pp. 1101-1108 (2008).
[6] B. J. Bowers, and D. P. Arnold, "Spherical, Rolling Magnet Generators for Passive Energy Harvesting from Human Motion," J. Micromech. Microeng., Vol. 19, 094008 (2009).
[7] Y. Iijima, "Microgenerator for the Wrist Watch," PowerMEMS 2000, pp. 44-52 (2000).
[8] E. Romero, M. R. Neuman, and R. O. Warrington, "Rotational Energy Harvester for Body Motion," Proc. IEEE MEMS 2011, Cancun, pp. 1325-1328 (2011).
[9] P. Pillatsch, E. M. Yeatman, and A. S. Holmes, "A Piezoelectric Frequency Up-converting Energy Harvester with Rotating Proof Mass for Human Body Applications," Sensors Actuators, A, 206, pp. 178-185 (2014).
[10] T. Ishii, J. Iwasaki, and H. Hosaka, "Study on Dynamics Characteristics of Gyroscopic Power Generator," Proc. SPIE, Vol. 6794, 67941O-1 (2007).
[11] Y. Suzuki, "Recent Progress in MEMS Electret Generator for Energy Harvesting," IEEJ Trans. Electr. Electr. Eng., 6, pp. 101-111 (2011).
[12] U. Bartsch, C. Sander, M. Blattmann, J. Gaspar, and O. Paul, "Influence of Parasitic Capacitances on the Power Output of Electret-based Energy Harvesting Generators," Proc. PowerMEMS 2009, Washington D.C., pp. 332-335 (2009).
[13] R. Chen and Y. Suzuki, "Suspended Electrodes for Reducing Parasitic Capacitance in Electret Energy Harvesting Devices," J. Micromech. Microeng., 23, 125015 (2013).
[14] R. Igreja, and C. J. Dias, "Analytical Evaluation of the Interdigital Electrodes Capacitance for a Multi-layered Structure," Sensors Actuators, A, 112, pp. 291-301 (2004).

Investigation of Pendulum Structures for Rotational Energy Harvesting from Human Motion

K Ylli[1,3], D Hoffmann[1], A Willmann[1], B Folkmer[1], Y Manoli[1,2]
[1] Hahn-Schickard, Wilhelm-Schickard-Str. 10, 78052 Villingen-Schwenningen, Germany
[2] Fritz Huettinger Chair of Microelectronics, Department of Microsystems Engineering – IMTEK, University of Freiburg, Germany

E-mail: Klevis.Ylli@hahn-schickard.de

Abstract. Energy Harvesting from human motion as a means of powering body-worn devices has been in the focus of research groups for several years now. This work presents a rotational inductive energy harvester that can generate a sufficient amount of energy during normal walking to power small electronic systems. Three pendulum structures and their geometrical parameters are investigated in detail through a system model and system simulations. Based on these results a prototype device is fabricated. The masses and angles between pendulum arms can be changed for the experiments. The device is tested under real-world conditions and generates an average power of up to 23.39 mW across a resistance equal to the coil resistance of the optimal pendulum configuration. A regulated power output of the total system including power management of 3.3 mW is achieved.

1. Introduction

Body-worn devices are becoming increasingly popular as they can perform a wide range of tasks from fitness and health tracking to ambient assisted living or simply as fashion accessories. The power supply issue however has not been satisfactorily solved to this date as the devices require periodic recharging. Energy harvesting (EH) from human motion offers a potential solution to this problem. EH devices are continuously getting smaller and can thus be integrated more easily into clothing items.

However, previously reported devices generate average powers in the low milliwatt range [1–3] and many devices are not fully characterised under real-world walking conditions [4]. Additionally, previous rotational devices have used a single eccentric mass to couple external motion into the rotating system [4–6]. In contrast this work analyses the influence of the geometrical parameters of the three-armed pendulum on the energy that can be coupled into the system and consequently the energy that can be provided for an electronic system.

The first part of this work focuses on the theoretical analysis of three pendulum structures and their ability to couple human leg motion into the rotational system. Of the three investigated structures the most feasible setup with the highest power output is chosen and a prototype is fabricated. This device consists of three pendulum arms with a proof mass attached to each of them and an adjustable angle between the arms. Different angular positions are characterised experimentally and the proof mass is varied.

[3] To whom any correspondence should be addressed.

2. Theoretical analysis

2.1. External acceleration conditions

Before developing the system model, a range of measurements was performed to obtain the accelerations that occur at different body locations and varying walking speeds. Three subjects walked on a treadmill at speeds of 4, 6, 8 and 10 km/h. This data is used as an excitation source in the simulations.

In order to compare the different body locations (an exemplary acceleration curve is shown in Figure 1 (a)) the RMS acceleration was calculated from the acceleration data. Figure 1 (b) indicates that the lower body locations might be the most suitable positions for energy harvesting from low-frequency, large-amplitude motions. A system model is developed in order to estimate the power output for all body locations.

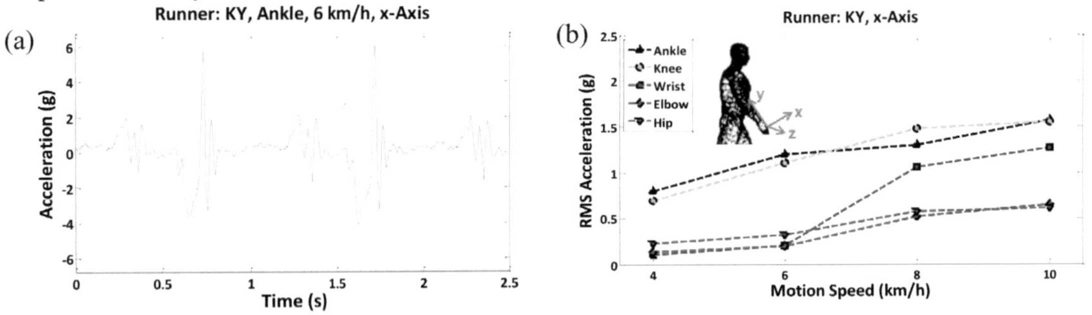

Figure 1. (a) Acceleration along the x-axis at the ankle for subject KY on a treadmill at 6 km/h. (b). RMS accelerations calculated from acceleration measurements on a treadmill at different body locations and walking speeds. The insert shows the axis orientation with the x-axis pointing forward at standstill

2.2. System model

In this work pendulum structures with three pendulum arms are investigated. The length of the pendulum arms, the angle between the arms, as well as the proof mass attached to each arm can be varied. In the first of three configurations, PEH1, the angle between pendulum arms is fixed. In the second configuration PEH2 the arms are connected by springs, thus the angle between arms can change dynamically during operation. Lastly, in PEH3 the angle is fixed, but the proof masses are spring mounted, i.e. the length of the pendulum arms can vary dynamically during operation. The schematic of each configuration is shown in Figure 2.

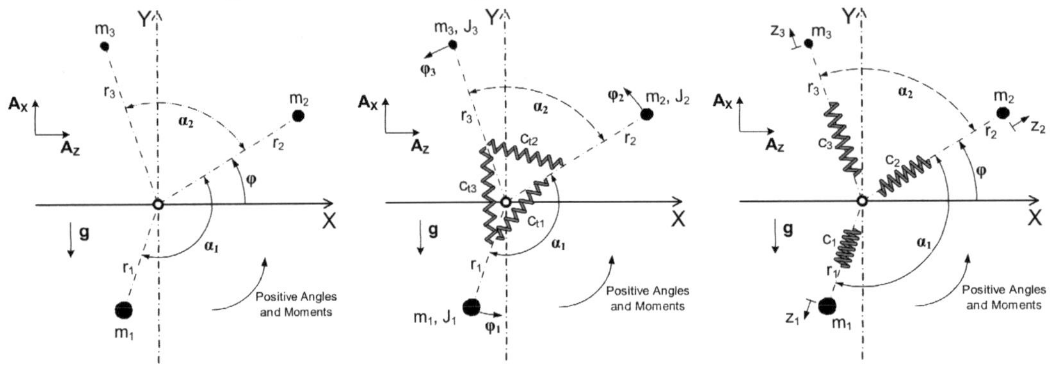

Figure 2. Schematic of the three-armed pendulum. From left to right: PEH1-PEH3. Relevant parameters are shown. The theoretical analysis also includes the dynamic variation of radii and angles during rotation in the case of PEH2 and PEH3 respectively.

The equations of motion are derived for each configuration, based on the moments acting on the pendulum arms. As an example, the equations for PEH1 are given below.

$$J\ddot{\varphi} = -c_r\dot{\varphi} + \sum M_g + \sum M_{Ax} + \sum M_{Az} \quad (1)$$

$$\sum M_g = -g\, m_1\, r_1 \cos(\varphi - \alpha_1) - g\, m_2\, r_2 \cos(\varphi) - g\, m_3\, r_3 \cos(\varphi + \alpha_2) \quad (2)$$

$$\sum M_{Ax} = A_x\, m_1\, r_1 \cos(\varphi - \alpha_1) + A_x m_2\, r_2 \cos(\varphi) + A_x m_3\, r_3 \cos(\varphi + \alpha_2) \quad (3)$$

$$\sum M_{Az} = -A_z m_1\, r_1 \sin(\varphi - \alpha_1) - A_z m_2\, r_2 \sin(\varphi) - A_z m_3\, r_3 \sin(\varphi + \alpha_2) \quad (4)$$

$$J = m_1 r_1^2 + m_2 r_2^2 + m_3 r_3^2 \quad (5)$$

2.3. Simulation results

In order to find the optimal geometry for each PEH, the RMS number of rotations per second "OmegaRMS" is introduced. It is used as a measure of how much motion is coupled into the rotational system and serves as the target function of an optimisation algorithm before the transduction mechanism is modelled.

Using the system model for each PEH, an optimization is performed with the following boundary conditions: The radii are limited from 5 mm to 25 mm and the proof masses from 5 g to 25 g per pendulum arm with a step size of 1 mm and 1 g respectively. The angles α_1 and α_2 are static in the case of PEH1 and PEH3 and can range from 0° to 180° in steps of 5°. Hence, in the case of $\alpha_1 = \alpha_2 = 0°$ a single-arm pendulum is calculated. Figure 3 (a) shows the optimal design for one particular motion speed of PEH1 and it can be seen that each design is only optimal for one motion speed(each line is above all other lines at only one speed). Figure 3 (b) compares the optimal design of PEH1 and PEH3 for the second runner: DH. It can be noticed, that PEH3 shows a minor advantage in the average number of rotations over PEH1. However, the PEH3 optimal design found by the optimisation algorithm is "unfeasible" with respect to the spring expansion which reached values larger than 50 cm from the starting point of 2.5 cm. The optimal design of PEH2 was identified as an unfeasible design, because the angle between pendulum arms reached large values (effectively winding the spring around the rotational axis).Consequently, the spring-less pendulum PEH1, is found to be the most feasible design with the highest power output that can be implemented as a human motion harvester.

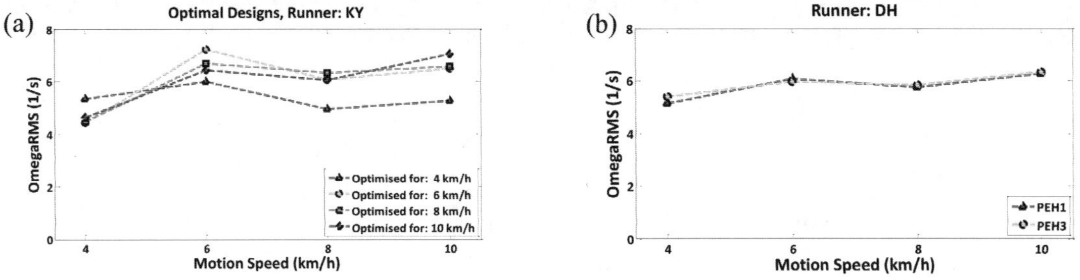

Figure 3. (a) Comparison of the optimum designs calculated for each particular motion speed for subject KY. Each line represents one design. (b) Comparison of the average number of rotations per second for PEH1 and PEH3 and subject DH.

Focusing on PEH1, an electro-magnetic transduction mechanism is modelled according to Figure 4. It consists of 8 segment magnets with a 45° arc each, which when placed next to each other (in alternating polarisation) form a ring. The magnets are linked to the pendulum and perform the same rotation. A total of 8 coils are placed in a way that all coils observe a change in flux density at the same time, as the magnets pass across them. Table 1 shows the simulated maximum average power output at the most suitable location on the lower body (ankle) and upper body (wrist). The geometrical boundaries used to find the designs that generate the shown power output are: pendulum arm lengths no longer than 25 mm and a combined proof mass of 75 g. For both test subjects the highest power output is simulated with acceleration data obtained from the ankle which is the only body location considered henceforth. Although each power output in Table 1 is based on a separate optimal design calculated for that particular walking condition, the tendency of the system parameters remains constant throughout the simulations. The best designs tend to have the maximum proof mass of 25 g per pendulum arm and the maximum allowed arm length of 25 mm. The optimum magnet height is calculated to be 5 mm and the optimal coil height ranges between 3 and 3.5 mm. The angles between pendulum arms range between 0° and 10°, indicating that the largest power output can be achieved when the three pendulum arms and thus the three masses are overlaid to effectively form a classic single-armed pendulum. This can be explained by the fact that in this case, the systems' centre of gravity is the furthest from the rotational axis and consequently the external acceleration can act on the largest possible lever.

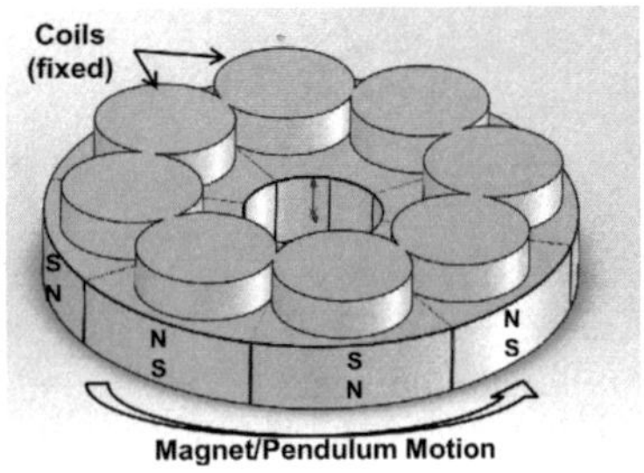

Figure 4. Schematic of transduction mechanism. Fixed coils placed opposite rotating magnets

Table 1. Simulated maximum average power output for different conditions and maximum pendulum arm lengths of 25 mm and a total proof mass of 75 g.

	Runner KY		Runner DH	
	4 km/h	6 km/h	4 km/h	6 km/h
Ankle P_{RMS} [mW]	10.46	16.4	6.62	9.11
Wrist P_{RMS} [mW]	0.36	2.51	0.16	0.19
Ankle ω_{RMS} [1/s]	3.55	4.37	2.77	3.26
Wrist ω_{RMS} [1/s]	0.63	1.73	0.4	1.06

3. Experimental evaluation

An adjustable prototype was fabricated for the experimental verification of the previously calculated best theoretical designs for PEH1 (Figure 5 (a)). The device consists of three pendulum arms with proof masses of 26 g each (high-density wolfram alloy, 18 g/cm³).The angle between pendulum arms can be adjusted in steps of 15° and the proof masses are removable. Segment magnets with an inner radius of 5 mm, outer radius of 19 mm, height of 5 mm and arc of 45° according to Figure 4 were used. The coils have inner and outer diameters of 3 and 10 mm and a height of 3 mm (~80Ω/coil).

A total of 7 setups was characterised with two test subjects on a treadmill. The parameters of each setup as well as the average power output across a resistance equal to the total coil resistance are shown in Table 2. The pendulum arm length is 25 mm for all setups.

Table 2. Average power output for the prototype device based on PEH1 and seven different configurations as measured during treadmill runs with two subjects.

Setup	Geometrical Parameters					P_{RMS} [mW], Runner: DH		P_{RMS} [mW], Runner: KY	
	α_1	α_1	m_1	m_2	m_3	4 km/h	6 km/h	4 km/h	6 km/h
1	0°	0°	26 g	26 g	26 g	8.47	23.39	9.04	22.16
2	0°	180°	26 g	26 g	26 g	3.55	5.04	2.91	4.54
3	90°	90°	26 g	26 g	26 g	3.84	5.09	6.16	8.41
4	15°	15°	26 g	26 g	26 g	9.11	16.92	9.09	19.79
5	0°	0°	26 g	26 g	0 g	10.77	10.87	7.19	12.87
6	0°	0°	26 g	0 g	0 g	8.92	11.13	5.98	11.8
7	90°	0°	26 g	26 g	0 g	7.34	9.14	10.24	12.88

Additional measurements were performed to determine the net average power into a supercapacitor by using a power management circuit consisting of a full-bridge rectifying circuit, an overvoltage protection, a 0.25 F supercapacitor and hysteresis regulation as presented in [7]. A total of 213 s was required to charge the capacitor from 0 V up to 3 V where the regulated voltage output is enabled by the hysteresis control and the consumer system can start operating. The voltage at the capacitor (measured during fast outdoor walking) is shown in Figure 5 (b). The average consumed power at which a regulated voltage output of 3V can be maintained is 3.3 mW.

(a)

(b)

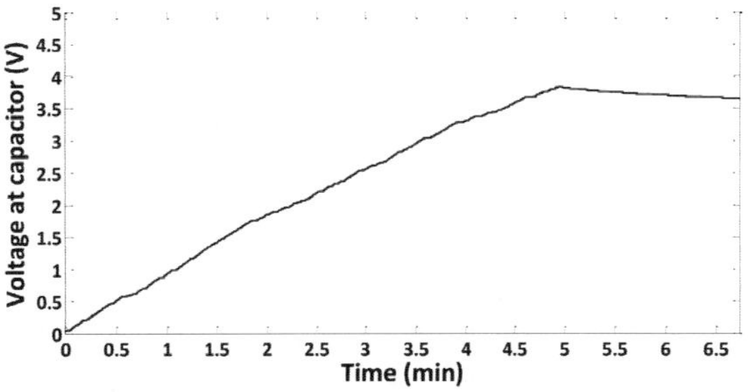

Figure 5. (a) Prototype device attached at the ankle with two of three proof masses mounted on the pendulum arms. (b) Voltage at a 0.25 F supercapacitor being charged during fast outdoor walking (approximately 6 km/h, runner: DH)

4. Conclusion

The previous sections confirmed that the classic one-armed pendulum is the most suitable of the three investigated pendulum structures. The analysed prototype pendulum harvester can easily be attached to the ankle using a Velcro strip and the housing is of a size that can easily be worn hidden under the pants with its external dimensions of 60x60x40 mm^3 (HxLxW). The net average power of 3.3 mW that is provided is sufficient for wearables like a fitness tracker or an indoor-navigation system.

Currently ongoing improvements focus on a smaller device size and particularly a smaller width. In addition, modifications to the magnetic circuit are expected to allow a reduction of the device weight, while maintaining the current high power output level. This is advantageous in terms of wearing comfort as the user is less likely to be disturbed by a reduced eccentric proof mass.

References

[1] K. Ylli, D. Hoffmann, A. Willmann, P. Becker, B. Folkmer, and Y. Manoli, “Energy harvesting from human motion: exploiting swing and shock excitations,” *Smart Mater. Struct,* vol. 24, no. 2, p. 025029, 2015.

[2] D. Hoffmann, B. Folkmer, and Y. Manoli, “Human Motion Energy Harvester for Biometric Data Monitoring,” *J. Phys.: Conf. Ser,* vol. 476, p. 012103, 2013.

[3] D. Carroll and M. Duffy, “Modelling, design, and testing of an electromagnetic power generator optimized for integration into shoes,” *Proceedings of the Institution of Mechanical Engineers, Part I: Journal of Systems and Control Engineering*, vol. 226, no. 2, pp. 256–270, 2012.

[4] P. Pillatsch, E. M. Yeatman, and A. S. Holmes, “A wearable piezoelectric rotational energy harvester,” *BSN 2013*, pp. 1–6, 2013.

[5] E. Romero, M. R. Neuman, and R. O. Warrington, “Rotational energy harvester for body motion,” *MEMS 2011*, pp. 1325–1328.

[6] D. Spreemann, Y. Manoli, B. Folkmer, and D. Mintenbeck, “Non-resonant vibration conversion,” *J. Micromech. Microeng,* vol. 16, no. 9, pp. S169–S173, 2006.

[7] D. Hoffmann, A. Willmann, B. Folkmer, and Y. Manoli, “Tunable Vibration Energy Harvester for Condition Monitoring of Maritime Gearboxes,” in *PowerMEMS*, 2014, p. 012099.

A bendable and compactdevice for low-power application

E. Ortiz-Ortega[1], M. P. Gurrola[2], L. G. Arriaga[2] and J. Ledesma-García[1*]

[1]División de investigación y Posgrado, Facultad de Ingeniería, Universidad Autónoma de Querétaro, Cerro de las Campanas S/N, C.P. 76010, Querétaro, México.

[2]Centro de Investigación y Desarrollo Tecnológico en Electroquímica S.C., Parque Tecnológico Querétaro Sanfandila, C.P. 76703, Querétaro, México.

E-mail: janet.ledesma@uaq.mx

Abstract. In this work we present a novelmodel of a membraneless microfluidic fuel celldevice based on adhesive polyester film fabricated by a non-sophisticated technique at room temperature and reached high performance. This concept overcomes the concerns about the reliability, versatility and weight of fuel cell devices for portable energy applications.Current densities until500 mA cm^{-2} and power densities about90mW cm^{-2} were achieved using formic acid as fuel and air and dissolved oxygen as oxidant.

1. Introduction

The miniaturization trend of technology is continuing in a variety of products, such as cell phones, laptops, music players, and so on. The advantages of the current devices increased processing speed and more executable concurrent tasks; however, battery technology remains a challenge because it does not meet the continuous energy demand of today's devices. In this sense, small energy conversion devices such as co-laminar or membraneless micro/nanofluidic fuel cells were created to supply the energy demand of small portable electronic devices as an alternative power source to batteries, which have a limited lifetime. Recently, more sophisticated applications for this type of devices have been visualized in the medical area to satisfy the energy requirements of pacemakers, glucometers and other biosensors. However, some barriers remain and are related to reproducibility, lifetime, size/weight and flexibility. Co-laminar microfluidic fuel cells have evolved to solve some of these aspects. Initially, micro-fabrication consisted of modifying hard materials such as poly-(methyl methacrylate), glass slides and polydimethylsiloxane using the hot-press method or photolithography [1-3]. Recent studies are focused on finding a small, lightweight and flexible energy conversion microdevice. Tominaka [4] used a flexible polymer substrate named cyclo-olefin polymer (COP) to fabricate a bendable fuel cell without using pumps. Some others [5] described alternatives based on thin and flexible films and different fabrication techniques. With this method, biofuel cells based on paper, such as filter paper, were reported [6]. Other works related to the use of filter paper with a cathode gas-diffusion layer or HB-pencil-stroked graphite were reported to fabricate a biofuel cell [7] or a self-pumping air-breathing membraneless fuel cell [8-9], respectively. In general, sophisticated fabricated techniques, complex assembled process, low performance and rigid devices are the issue to overcome.

In this work, we report the construction and evaluation of a flexible membraneless fuel cell based on adhesive polyester film, which is adhered to paper foils to assemble and disassemble the device, where the fuel (formic acid) and oxidant (oxygen from air) flow-through the electrodes for energy conversion.

2. Experimental Section

2.1. Construction of membraneless air-breathing device.

Thedevicewas constructed using flexible materials and non-sophisticated technique. The cell operation principle is derived from a design previously reported [10]. Adhesive white polyester (ARcare® 8259, Adhesives Research, thickness of 200 μm) was employed as support film and hook-and-loop tape,and home-made silicone elastomer film (Silastic®, Dow Corning, prepared using an Elcometer® Film Applicator with a final thickness of 600 μm) as sealing and cell channel structure, both materials are responsible to provide flexibility and versatility to the device. Commercial carbon nanofoam (Marketech Inc.) coated with commercial electrocatalyst materials (Pt/C, ETEK 30 wt% and Pd/C, ETEK 20 wt%) as flow-through and air-breathing electrode, finally aluminium foil was employed as electric contact. The four elements integrate an inexpensive and versatile (Figure 1a).

Two supporting films were made by adhesive white polyester, the top plate include a window that allows the access of oxygen from the air and bottom plate with two inlets for the reagents and one outlet for the reaction sub-products. The cell channel structure made by silastic allows the reagents that were located in the middle of electrode provide a homogeneous distribution; both supporting films and channel structure were fabricated using a Silhouette® cutting plotter. The electric contact and 3D electrodes were cut by hand and the electrocatalyst materials were incorporated by spray technique employing an air brush until the metal loading was 0.6 mg, the electrode dimensions are15 mm in length, 2 mm in widthand 0.1 mm in height (geometrical area of 0.015 cm^2). The total weight and volume of the device were 603.7 mg and 0.62 cm^3, respectively.

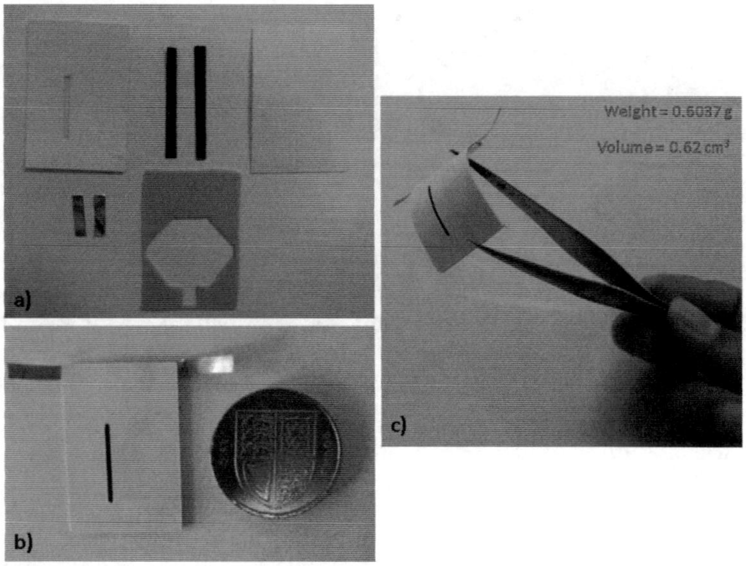

Figure 1 Photographs of the cell a) components of flexible device, b) comparison with size of one pound, c) flexibility of membraneless device.

2.2. Evaluation of membraneless air-breathing device performance.
The fuel and oxidant streams were fed through the nanoporous electrodes with scan rate of 6 ml h^{-1} using syringe pumps (NE-4000, New Era Pump Systems Inc) at 0.5M Formic acid and combined O_2-satured solution and O_2 from the air dissolved in 0.5 M H_2SO_4 solutions, respectively. Were evaluated the performance, stability and lifetime

The stability test was carried out by a chronoamperometry at maximum power voltage (0.44 V) employing 1.5 ml of the fuel. The electrochemical impedance was employed to measure the total resistance of the device after several tests.

3. Results and discussion

3.1. Evaluation of membraneless air-breathing device performance.
The polarization and power density curves (Figure 2a) were the same after the evaluation of stability employing 1.5 ml of fuel (Figure 2b), reached 0.9 V and around 580 mW cm^{-2}. On the other hand, the current density was constant throughout the test indicating a good stability of the membraneless device due there were no poisoning by reaction products in according with the polarization curves too.

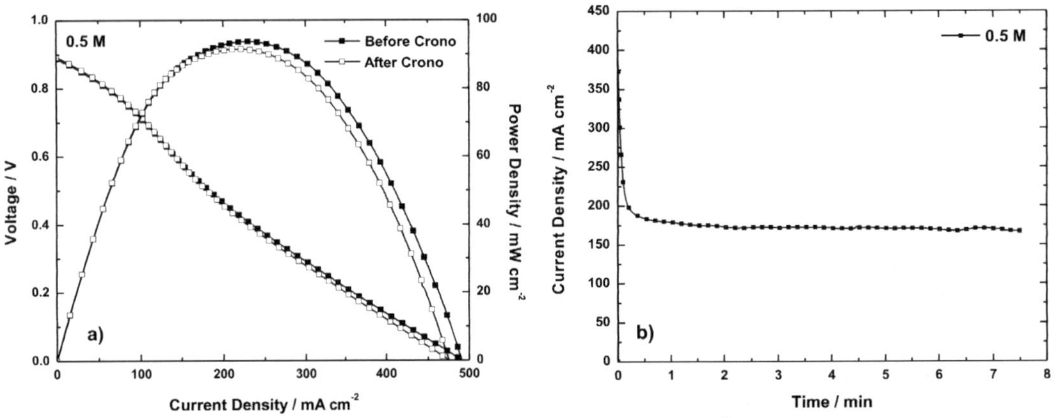

Figure 2a) Polarization and power density curves at 6ml h^{-1} of flow rate at 0.5M formic acid, before and after the stability test, and b) Stability curve at 0.44 V for 1.5 ml of the fuel.

Is important to emphasize why the concept of hook-and-loop tape based, due to either of the supporting films can be easily removed and re-attachment again at several times. The cell was assembled and disassembled between each test, were made several polarization curves and chronoamperograms; the figure 3 shows as the cell was physically after the ten, twenty five and forty seven times of disassembled and can be observed that the resistance of device was maintained within a range of 0.5 and 1 ohm cm^{-2} which indicates the cell can be used over and over again. The advantages of assembled and disassembled were the accumulated bubbles CO_2 were eliminated and if necessary can change any element of the cell, in this way the device gains versatility.

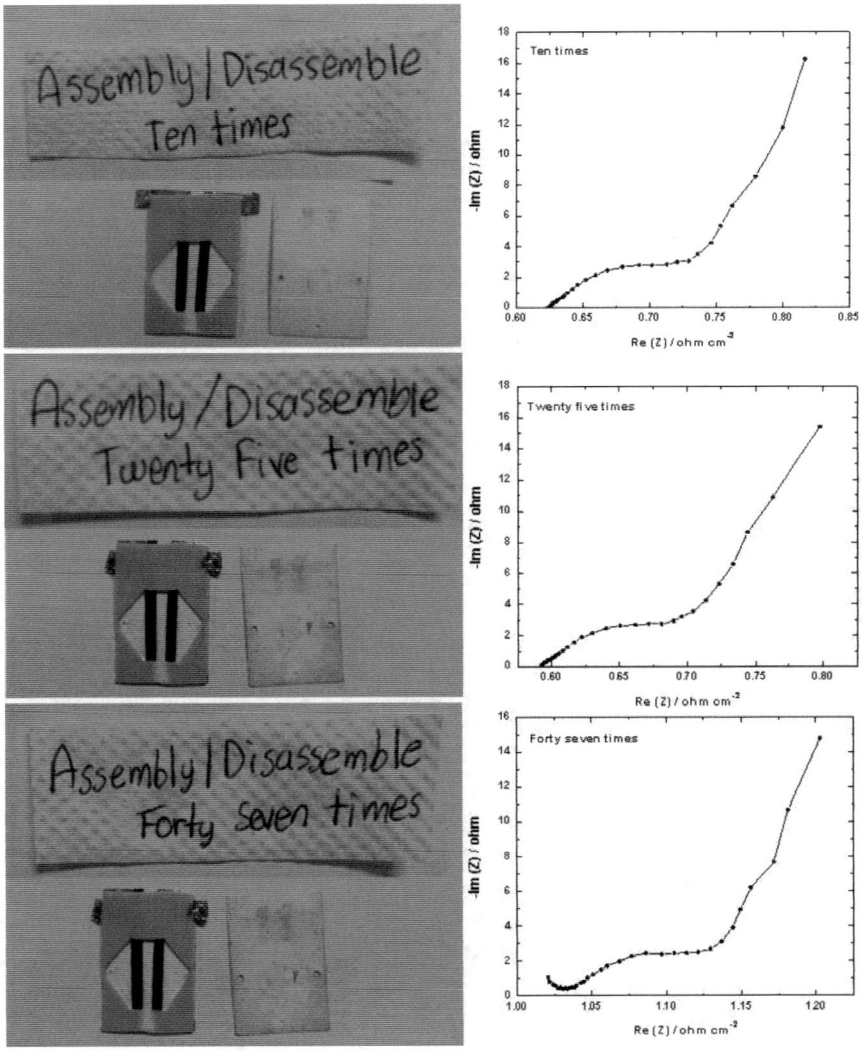

Figure 3 Impedance spectra after 10, 25 and 47 cycles of assembly/disassembly of the tape-based membraneless fuel cell.

Finally, the membraneless air-breathing device was connected to a microelectronic system contained a booster that raised the voltage from 0.83 V to 3 V (Figure 4a). The energy produced by the device was sufficientto light a LED (Figure 4c) while the fuel and oxidant were injected. This showed that a flexible device coupled with microelectronic system can employ in a future to supply energy to low power device employing low concentration and clean energy.

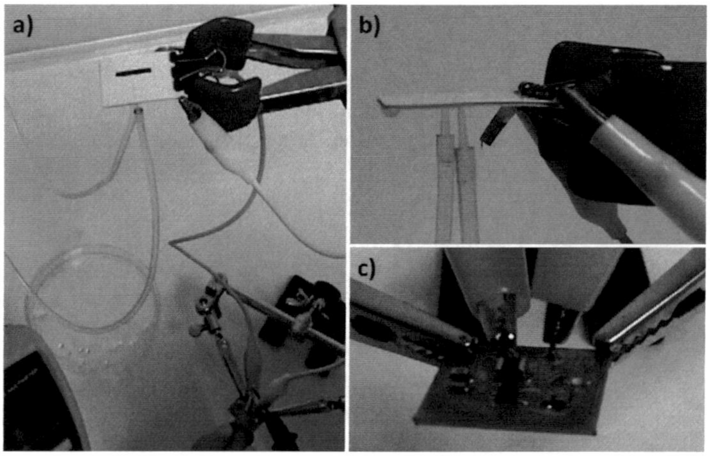

Figure 4 a) membraneless air-breathing device connected to microelectronic system with a LED b) lateral photograph of membraneless air-breathing device c) microelectronic system with a LED lighting.

4. Conclusions

In summary, these results shows a small, thin, lightweight, flexible, reusable, easy to assemble and to fabricate, and high performance at low fuel concentration and metal load, membraneless air breathing device. In this sense, the device shows a great potential and is the first step to create an energy source for portable electronic devices or medical applications in the near future of low-power.

References

[1] Dector A, Esquivel JP, González MJ, Guerra-Balcázar M, Ledesma-García J, Sabaté N, Arriaga LG. Formic acid microfluidic fuel cell evaluation in different oxidant conditions. Electrochim Acta 2013;92:31-35.

[2] Morales-Acosta D, Rodríguez HG, Godinez LA, Arriaga LG. Performance increase of microfluidic formic acid fuel cell using Pd/MWCNTs. J Power Sources 2010;195:1862-1865.

[3] Kjeang E, Michel R, Harrington DA, Djilali N, Sinton D. A Microfluidic Fuel Cell with Flow-Through Porous Electrodes. J Am Chem Soc 2008;130:4000-4006.

[4] Tominaka S, Nishizeko H, Mizuno J, Osaka T. Bendable fuel cells: on-chip fuel cell on a flexible polymer substrate. Energy Environ Sci 2009;2:1074-1077.

[5] Focke M, Kosse D, Müller C, Reinecke H, Zengerle R, Stetten FV. Lab-on-a-Foil: microfluidics on thin and flexible films. Lab Chip 2010;10:1365-1386.

[6] Zhang L, Zhou M, Wen D, Bai L, Lou B, Dong S. Small-size biofuel cell on paper. Biosens Bioelectron 2012;35:155-159.

[7] Narváez-Villarrubia CW, Lau C, Ciniciato GPMK, Garcia SO, Sibbett SS, Petsev DN, Babanova S, Gupta G, Atanassov P. Practical electricity generation from a paper based biofuel cell powered by glucose in ubiquitous liquids. Electrochem Commun 2014;45:44-47.

[8] Arun RK, Halder S, Chanda N, Chakraborty S. A paper based self-pumping and self-breathing fuel cell using pencil stroked graphite electrodes. Lab Chip 2014;14:1661-1664.

[9] Esquivel JP, Del Campo FJ, Gomez de la Fuente JL, Rojas S, Sabaté N. Microfluidic fuel cells on paper: meeting the power needs of next generation lateral flow devices. Energy Environ Sci 2014;7:1744-1749.

[10] Ortiz-Ortega E, Goulet MA, Lee JW, Guerra-Balcázar M, Arjona N, Kjeang E, Ledesma-García J, Arriaga LG. A nanofluidic direct formic acid fuel cell with a combined flow-through and air-breathing electrode for high performance. Lab Chip 2014;14:4596-4598

Design and fabrication of miniaturized PEM fuel cell combined microreactor with self-regulated hydrogen mechanism

A. Balakrishnan[1*], M. Frei[2], S. Kerzenmacher[2], H. Reinecke[1], C. Mueller[1]
[1]Laboratory for Process Technology, [2]Laboratory for MEMS Applications
IMTEK-Department of Microsystems Engineering, University of Freiburg,
Georges-Koehler-Allee 103, 79110 Freiburg, Germany
[*]E-mail: arvind.balakrishnan@imtek.de

Abstract. In this work we present the design and fabrication of the miniaturized PEM fuel cell combined microreactor system with hydrogen regulation mechanism and testing of prototype microreactor. The system consists of two components (i) fuel cell component and (ii) microreactor component. The fuel cell component represents the miniaturized PEM fuel cell system (combination of screen printed fuel cell assembly and an on-board hydrogen storage medium). Hydrogen production based on catalytic hydrolysis of chemical hydride takes place in the microreactor component. The self-regulated hydrogen mechanism based on the gaseous hydrogen produced from the catalytic hydrolysis of sodium borohydride ($NaBH_4$) gets accumulated as bubbles at the vicinity of the hydrophobic coated hydrogen exhaust holes. When the built up hydrogen bubbles pressure exceeds the burst pressure at the hydrogen exhaust holes the bubble collapses. This collapse causes a surge of fresh $NaBH_4$ solution onto the catalyst surface leading to the removal of the reaction by-products formed at the active sites of the catalyst. The catalyst used in the system is platinum deposited on a base substrate. Nickel foam, carbon porous medium (CPM) and ceramic plate were selected as candidates for base substrate for developing a robust catalyst surface. For the first time the platinum layer fabricated by pulsed electrodeposition and dealloying (EPDD) technique is used for hydrolysis of $NaBH_4$. The major advantages of such platinum catalyst layers are its high surface area and their mechanical stability. Prototype microreactor system with self-regulated hydrogen mechanism is demonstrated.

1. Introduction

Though there are existing research understanding in the field of low temperature Proton Exchange Membrane (PEM) fuel cell for small scale portable or stationary applications, the main reason which hinders its commercialization is the fuel source (hydrogen, H_2). PEM fuel cell driven with H_2 paves the way for an energy source which can operate as a primary source or as secondary sources in combination with batteries. Speaking about the low temperatures fuel cell systems there are acidic and alkaline fuel cell systems. PEM fuel cells belong to the acidic fuel cell category where the redox reaction of hydrogen-oxygen happens at the anode and cathode forming electricity, water and heat. Protons are transported through the PEM layer and the electrons are transported across the external circuit. H_2 driven PEM fuel cell system provides many advantage over batteries in terms of increased energy density, instant recharge eco friendliness. As mentioned before the H_2 fuel storage is still an existing challenge from an engineering stand point. There are various H_2 storage techniques available such as, compressed hydrogen gas (700 bar), liquefied hydrogen (approximately -200 °C), adsorption on metal organic frameworks (MOFs), reformation of hydrocarbons and metal-chemical hydrides. The possibilities of using the first two options are ruled out for portable applications due to its practical feasibility. The above mentioned other possibilities are extensively researched from both engineering and material stand points. Recently many works have been reported with on-board hydrogen generation as reliable alternates for the hydrogen fuel storage for PEM fuel cells which can be adapted in portable applications. The on-board hydrogen generation is based on metal hydrides and chemical hydrides. The metal hydride (also intermetallic compounds and alloys) represent the materials that can form metal-hydrogen compounds. This attributed to very specific lattice structure and the electronegativities of the element. The hydrogenation (adsorption) and dehydrogenation (desorption) process are reversible making them very attractive for portable applications. It is know that hydrogen actively reacts with metals at elevated temperature forming hydrides and these metals are called transitions metals. Palladium is well-known for its hydrogen adsorption-desorption behaviour at ambient temperature and low pressures of hydrogen (1 bar) forming palladium hydride which is represented as $PdH_{0.7}$. Other intermetallic compounds and alloys such as TiFe, $CeNi_3$, $TiMn_2$ and $LaNi_5$ are also been studied in order to adapt them for low temperature portable applications. The term chemical hydride is general representation which is further sub classified as following [1]

- binary saline hydrides: NaH, CaH_2, MgH_2
- aluminium-based complex hydrides: $LiAlH_4$, $NaAlH_4$
- boron-based complex hydrides: $LiBH_4$, KBH_4, $NaBH_4$

One of the most preferred chemical hydride is $NaBH_4$ due to its high stability at ambient atmosphere, capability to produce hydrogen at ambient conditions, relatively low exothermic in nature when compared to other hydrides and its eco friendliness. The hydrolysis reaction of $NaBH_4$ is shown in equation (1). The stoichiometric amount of hydrogen that can be generated according to equation (1) is 10.8 mass%. The self-hydrolysis of $NaBH_4$ is controlled by increasing the pH of the solution and then using an active catalyst to reinitiate the hydrolysis reaction.

$$NaBH_4 + 2H_2O \rightarrow NaBO_2 + 4H_2 + \left(-300\frac{KJ}{mol}\right) \quad (1)$$

The other mentioned hydrides are very unstable at ambient air which has significant risk in handling. It is worthwhile mentioning one other hydride which has gained significant interest; MgH_2 due to its stability at ambient air. A novel approach was presented with PEM fuel cell integrated pallidum based hydrogen storage and an air breathing cathode fabricated on a silicon substrate [2]. The difference between the conventional PEM fuel cells and the chip integrated fuel cell accumulator is the absence of the gas diffusion layers and flow fields, which eventually leads to a simplified system design and easy fabrication. One other major advantage of such system is the ability to integrate CMOS for regulation of the output voltage. Figure 1 shows a fully fabricated chip integrated fuel cell accumulator.

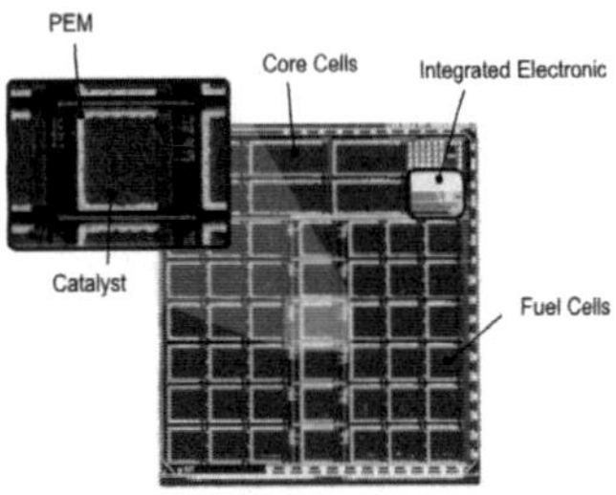

Figure 1. Chip integrated fuel cell accumulator with on-board palladium based H_2 storage and integrated CMOS electronics.

In the above mentioned system the PEM layer and the cathode layers were screen printed whereas the pallidum layer was realized by sputtering and electroplating process. Major drawback of this system is the limited run time of the system which is due to the amount of hydrogen stored in the palladium. Simply increasing the layer thickness of bulk palladium causes irreversible damage to the substrate due to the volumetric expansion of palladium on hydrogen adoption. In order to overcome this drawback an approach was presented in [3], where the bulk palladium layer was replaced by a palladium-polymer composite (chip integrated fuel cell accumulator with palladium-polymer composite H_2 storage will be hereon depicted as miniaturized PEM fuel cell). Using this approach the challenge related to the volumetric expansion of bulk palladium is overcome and simultaneously an increased hydrogen storage volume is achieved thereby an increased run time of the system. For further increasing the run time of the miniaturized PEM fuel cell system it was combined with the hydrogen production reactor (on-board range extender) [4]. The hydrogen was produced by the catalytic hydrolysis of alkaline $NaBH_4$ solution. The challenging aspect of the miniaturized PEM fuel cell combined on-board range extender system was the controllability of hydrogen production and its regulation. There are numerous approaches presented in the literature to generate hydrogen on-board using chemical hydrides and then stream the produced hydrogen into the flow fields of the PEM fuel cell after passing through a gas-liquid phase separator (regulation is done by conventional approach; external pumps and valves driven by parasitic energy) [5]. The hydrogen supply to the fuel cell is controlled by pumping speed of chemical hydride solution on to the catalyst surface. Our motivation is to address this challenge in the miniaturized PEM fuel cell combined microreactor system by using passive components; hydrophobic surface based gas venting technique [6].

1.1. System concept

In this work we present the design, fabrication process chain of miniaturized PEM fuel cell combined microreactor system with self-regulation hydrogen mechanism. A schematic view of the system design is shown in Figure 2. The system consists of two components (i) fuel cell component and (ii) microreactor component. The fuel cell component represents the miniaturized PEM fuel cell system (combination of screen printed fuel cell assembly and an on-board hydrogen storage medium).

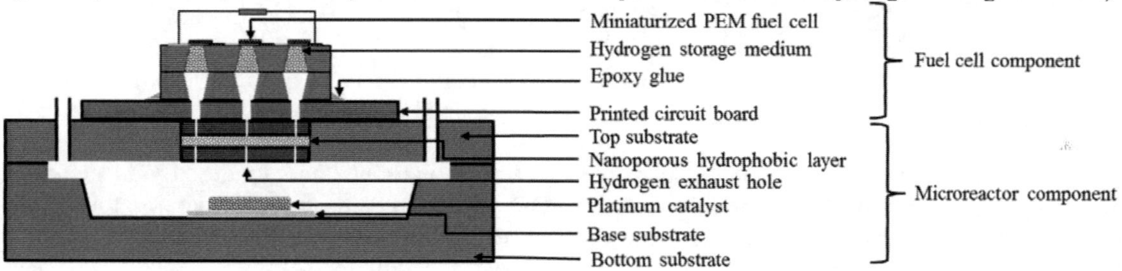

Figure 2. Schematic view of miniaturized PEM fuel cell combined microreactor with integrated self-regulation hydrogen mechanism

Hydrogen production based on catalytic hydrolysis of chemical hydride takes place in the microreactor component which comprises of top and bottom substrate (fabricated by micromilling on a polymer substrate). The top substrate houses the hydrogen exhaust hole; through which the hydrogen is transported outside the reactor. The bottom substrate houses a porous platinum layer. The gaseous hydrogen produced from the catalytic hydrolysis of sodium borohydride ($NaBH_4$) accumulates (bubbles) at the hydrogen exhaust holes. The built up hydrogen bubbles at the surface of hydrogen exhaust holes collapse when the pressure inside the hydrogen bubble overcomes the burst pressure of the hydrogen exhaust holes. This collapse causes a surge of fresh $NaBH_4$ solution onto the catalyst surface leading to the removal of the reaction by-products ($NaBO_2$) formed at the catalyst active sites of the catalyst.

2. Experimental

The system consists of two major components; (i) fuel cell component and the (ii) microreactor component. The fabrication of the fuel cell component involves both clean room and back end process. The polymer substrate based microreactor component is realized by micromilling.

2.1. Fabrication and characterization

The fuel cell component comprises of the miniaturized PEM fuel cell and the printed circuit board. A detailed description about the fabrication of the miniaturized PEM fuel cell with palladium-polymer composite hydrogen storage medium is presented elsewhere [4]. The microreactor component consists of top and bottom substrates; top substrate houses a sandwich of hydrogen exhaust holes and nanoporous hydrophobic layer. The hydrogen exhaust holes are realized by high aspect ratio dry etching (ICP) on silicon substrate followed by hydrophobic layer coating. The sandwich of the two silicon chips and the nanoporous hydrophobic layer were glued together resulting in one chip; hydrogen exhaust hole. The bottom substrate consists of the platinum catalyst layer deposited on to a base substrate. Initially two base substrates were adapted; nickel foam and CPM. Platinum catalyst was deposited by standard electroplating technique but there were considerable drawbacks such as, bad coverage of the 3 dimensional surface area of the base substrate, mechanical stability of the deposited platinum on to the base substrate (violent hydrogen production during the catalytic hydrolysis rips off the platinum from the base substrate causing a degradation in hydrogen production behavior of the system when used continuously, these results are not presented here). In search of solution for the above mentioned drawbacks the novel technique of electropulsed deposition and dealloying (EPDD) technique was adapted for nickel foam, CPM in addition to the ceramic plate base substrate [7]. Major advantages of such platinum catalyst layers are high surface area and high mechanical stability and the definition of the coating area, which is difficult to achieve using conventional electroplating. Figure 3 shows all three base substrates coated with platinum by EPDD.

Figure 3. Base substrates coated with platinum layer by EPDD technique; the left most in the figure is nickel foam followed by CPM and the ceramic plate. The dark shade (colour version) at the centre of each sample is the deposited platinum.

Since ceramic plate was nonconductive in nature it was first patterned and sputtered with platinum thin film followed by the EPDD process. Same EPDD process parameters were applied for all the three base substrates. SEM micrographs of all the 3 base substrates are shown in Figure 4. The high voltage used for SEM micrographs were 5 KV, the magnification for the nickel foam and CPM samples are 100 X and for the ceramic plate was 1000 X. This difference is attributed to less visibility at 100 X magnification on the ceramic plate.

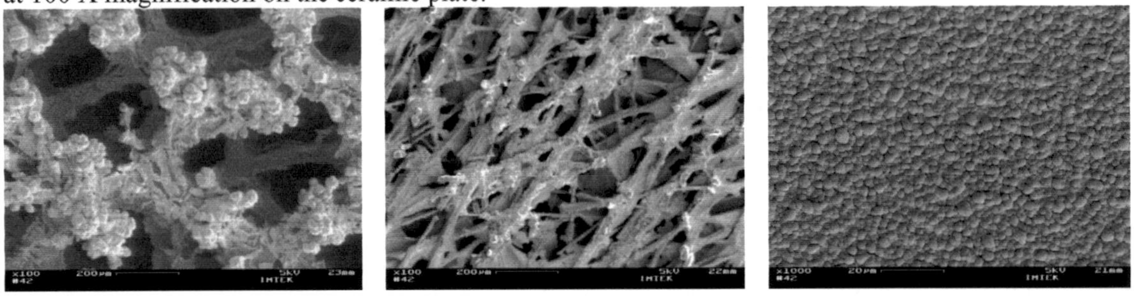

Figure 4. SEM micrographs of the base substrates deposited with platinum by EPDD technique. The left most image is nickel foam, middle is the CPM and the right most is the ceramic plate.

The proof of concept and functionality of microreactor with self-regulated hydrogen mechanism is demonstrated by a prototype microreactor; small reactor volume with single hydrogen exhaust hole

(500 μm diameter) coated with PTFE layer. Hydrogen concentration sensor was used to measure the produced hydrogen from the microreactor (CarboSen 1.000 from LAMTEC GmBH). The $NaBH_4$ stock solution was inserted into the prototype microreactor system with the nickel foam coated platinum catalyst and the change in hydrogen concertation was recorded with respect to time.

3. Result & discussion

The hydrogen produced from the catalytic hydrolysis of 800 μl of 0,7 wt% $NaBH_4$ and 4 wt% NaOH solution is shown in figure 5. The reactor initially produced hydrogen at a consistent rate followed by a pulsed release of H_2. The peaks can be observed in the figure 5 which represent the unique behaviour of the self-regulated hydrogen mechanism of the prototype microreactor. When the measurement was repeated similar peaks were observed, which represent the self-regulated hydrogen release. This kind of H_2 release behaviour is advantageous for the miniaturized PEM fuel cell system. The pulsed H_2 could be further enhanced by implemented the silicon based hydrogen exhaust holes.

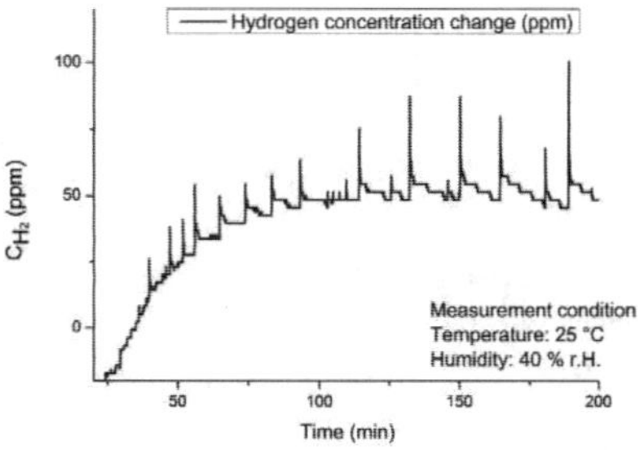

Figure 5. Functionality of the microreactor with integrated hydrogen self-regulation mechanism was tested by producing hydrogen by catalytic hydrolysis of 800 μl of stock solution. The microreactor was placed inside the measurement chamber and the change in hydrogen concentration inside the measurement chamber was measured with the hydrogen conc. sensor (ppm) with respect to time.

4. Conclusion & outlook

Design and fabrication of the miniaturized PEM fuel cell system combined microreactor with self-regulation hydrogen mechanism is presented. The prototype testing of the microreactor system is demonstrated with the self-regulation hydrogen mechanism. EPDD plated platinum on a ceramic base substrate is used for the hydrolysis of $NaBH_4$. Work in progress includes the prototype testing and characterization of the miniaturized PEM fuel cell combined microreactor system (see Figure 6) and the optimization of EPDD deposited platinum for hydrolysis reaction of alkaline $NaBH_4$. Advantages of miniaturized PEM fuel cell combined microreactor system include increased hydrogen storage possibilities, instant refilling and on-board energy conversion; this paves a platform for the development of new generation of energy sources for various applications.

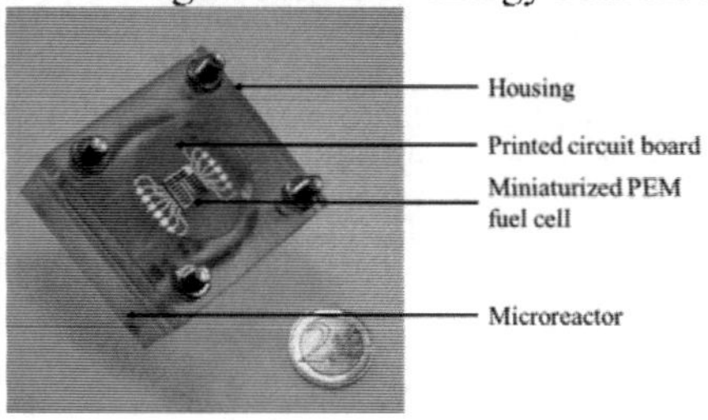

Figure 6. Miniaturized PEM fuel cell combined microreactor system. This system is the realization of the schematic diagram shown in figure 1. Microreactor and the housing is fabricated with PMMA (transparent and stable against alkaline solutions)

5. Reference

[1] A. Zuettel, Hydrogen as a Future Energy Carrier, Weinheim: Wiley-VCH, 2008
[2] G. Erdler et al., Sensors and Actuators A 132 pp. 331–336, 2006.
[3] A. Balakrishnan at al., Advanced Materials Research, pp. 904-912, 2013.
[4] A. Balakrishnan et al., Journal of Physics: Conference Series 557, 012049, 2014.
[5] T.Kim et al., International journal of Hydrogen Energy, vol. 37, pp. 2240-2446, 2012.
[6] D.D.Meng et al., Journal of Micromechanics and Microengineering, vol. 16, pp. 419-424, 2006.
[7] K. Christian et al., Journal of Power Sources, vol. 242, pp. 255-263, 2013.

Effect of pH in a Pd-based ethanol membraneless air breathing nanofluidic fuel cell with flow-through electrodes

C A López-Rico[1,4], J Galindo-de-la-Rosa[2], J Ledesma-García[2], L G Arriaga[1], M Guerra-Balcázar[2], N Arjona[3]*

[1] Centro de Investigación y Desarrollo Tecnológico en Electroquímica, 76703, Querétaro, México.
[2] División de Investigación y Posgrado, Facultad de Ingeniería, Universidad Autónoma de Querétaro, 76010, Querétaro, México.
[3] Centro de Investigación y Desarrollo Tecnológico en Electroquímica, Unidad Tijuana, 22444, Baja California, México.
[4] Tecnológico Nacional de México, Instituto Tecnológico de Tijuana, Apdo. Postal 1166 Tijuana, 22000, Baja California, México

E-mail: wvelazquez@cideteq.mx & noe.arjona@yahoo.com.mx

Abstract. In this work, a nanofluidic fuel cell (NFC) in which streams flow through electrodes was used to investigate the role of pH in the cell performance using ethanol as fuel and two Pd nanoparticles as electrocatalysts: one commercially available (Pd/C from ETEK) and other synthesized using ionic liquids (Pd/C IL). The cell performances for both electrocatalysts in acid/acid (anodic/cathodic) streams were of 18.05 and 9.55 mW cm^{-2} for Pd/C ETEK and Pd/C IL. In alkaline/alkaline streams, decrease to 15.94 mW cm^{-2} for Pd/C ETEK and increase to 15.37 mW cm^{-2} for Pd/C IL. In alkaline/acidic streams both electrocatalysts showed similar cell voltages (up to 1 V); meanwhile power densities were of 87.6 and 99.4 mW cm^{-2} for Pd/C ETEK and Pd/C IL. The raise in cell performance can be related to a decrease in activation losses, the combined used of alkaline and acidic streams and these high values compared with flow-over fuel cells can be related to the enhancement of the cathodic mass transport by using three dimensional porous electrodes and two sources of oxygen: from air and from a saturated solution.

1. Introduction

Membraneless microfluidic fuel cells (MMFCs), are a type of fuel cells which operate with both, liquid fuel and oxidant under a co-laminar regime, flowing tangentially, and forming a natural interface between them. This interface carry out a similar function than typical exchange ion membranes. [1] The advantages of these MMFCs over traditional fuel cells are therefore, a decrease of ohmic losses because the absence of a physical barrier, the possibility to mix electrolytic solutions of different pH, [2] and the advantages of usage of all kind of liquid fuels (i.e. methanol, ethanol, glycerol, vanadium, glucose, etc.) and oxidants (hydrogen peroxide, potassium permanganate). [3] Thermodynamically, the ethanol electro-oxidation reaction occurs at 0.09 V *vs.* NHE in an acidic pH and at -0.74 V in alkaline pH [4]. Moreover, the oxygen reduction reaction in acidic and alkaline take place at 1.23 and 0.4 V *vs.* NHE, respectively. Their use in a fuel cell as fuel and oxidant can thermodynamically result, in cell voltages from 1.17 V to 2 V.

Some features to improve in MMFCs are as well as in fuel cells, the synthesis of highly active electrocatalysts, development of supports with high surface area, and the decrease of mass transport limitations [5]. For the synthesis of electrocatalysts, the use of ionic liquids as "all-in-one" reaction medium: solvent, reducing and stabilizer agent, can allow to obtain nanoparticles by a clean pathway [6] and they could follow most of the twelve principles of green chemistry [7]. Which combined with the clean technology of membraneless micro/nanofluidic fuel cells, can be an attractive green solution to supply energy requirements of small devices. In this work, Pd/C nanoparticles from a commercially available source (ETEK) as well as synthesized using an "all-in-one" ionic liquid, were used as anodic electrocatalysts in a membraneless nanofluidic fuel cell (MNFC). This novel MNFC operates using three dimensional flow-through electrodes, and combines two sources of oxygen: from air and from an saturated aqueous electrolyte. The aim of this work, was to evaluate -for the first time- the effect of changing the nature of the electrolyte (in terms of pH) in the cell performance (voltage, current density and power density) using two different anodic electrocatalysts.

2. Experimental

2.1. Synthesis of Pd/C

Pd/C nanoparticles were easily synthesized using 2-hydroxy ethylammonium formate ionic liquid as a green chemistry "all-in-one" solvent. Briefly, 50 mg of Na_2PdCl_4 (Sigma-Aldrich, 98%) were dissolved in the ionic liquid and sonicated by 10 minutes. It is important to denote, that a dark color was observed 3 minutes after the addition of the metallic salt precursor. After that, 150 mg of Vulcan carbon (XC-72, Cabot®) were added and the sonication was retained another 30 minutes. The mixture was preserved under refrigeration (at -20°C) for 24 h to ensure a complete reaction. Finally, the solution was centrifuged at 4000 rpm for 30 minutes and the resulting powder was washed several times with deionized water.

2.2. Membraneless nanofluidic fuel cell operation.

Pd/C ETEK and Pd/C IL were deposited on carbon nanofoam by spray coating as well as Pt/C (ETEK, 30 wt. %) used as the cathodic electrocatalyst, maintaining a total mass content of 0.3 mg of electrocatalyst. The membraneless nanofluidic fuel cell (Fig. 1) was patterned and fabricated using a computer numerical control, and was composed of i) two plates of poly-methyl methacrylate (PMMA, Fig. 1b). ii) A silicone-based film (Silastic®) which acted not only as gasket but also as pattern to place the porous electrodes (Fig. 1a). And iii) the carbon nanofoam (Marketech® International Inc.) slides as three dimensional electrodes. The geometrical area of electrodes was of 0.015 cm^2.

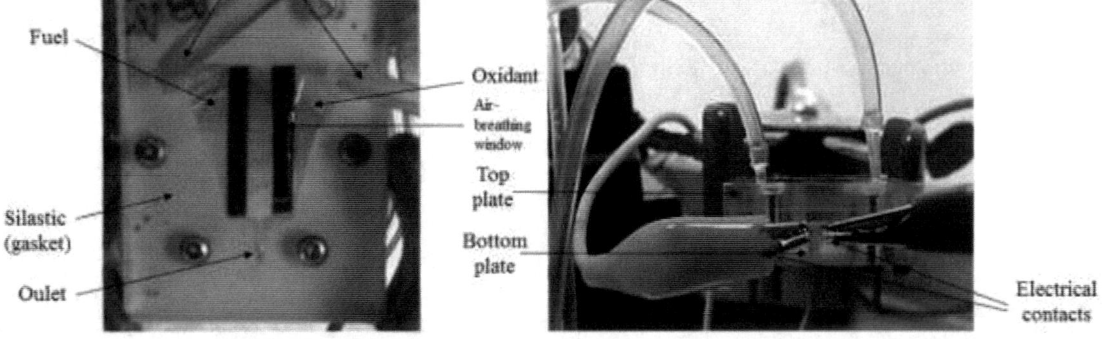

Figure 1. Components of the membraneless nanofluidic fuel cell

The cell design herein presented take advantage of the high surface area of carbon nanofoam (450 $m^2\ g^{-1}$) in order to enhance the electrocatalytic activity. Ethanol (Denatured, 96%, J.T. Baker) was used as fuel at three concentrations: 0.5, 1 and 1.5 M and an oxygen-saturated solution (4.3 U.A.P. Praxair) combined with oxygen from air was used as oxidant (open window of 0.22 cm^2). Both streams were prepared using 0.3 M KOH (87%, J.T. Baker) as the electrolyte for an alkaline fuel cell and 0.5 M H_2SO_4 (98.5%, J.T. Baker) for the acidic fuel cell. Also, an alkaline anodic stream and an acidic cathodic stream were combined. The streams were pressure-driven using a Harvard apparatus syringe pump (PHD Ultra Syringe Pump).

3. Results and discussion

X-ray diffraction patterns are shown in Figure 2a. The (111), (200), (220), (311) and (222) crystallographic planes were identified in both, Pd/C ETEK and Pd/C IL, and are characteristics of a face-centred cubic structure. Crystallite sizes were determined using the Scherrer´s equation and lattice parameters applying the Bragg's law (Table 1). Average crystallite sizes were of 8.5 and 4.5 nm for Pd/C IL and ETEK, respectively.

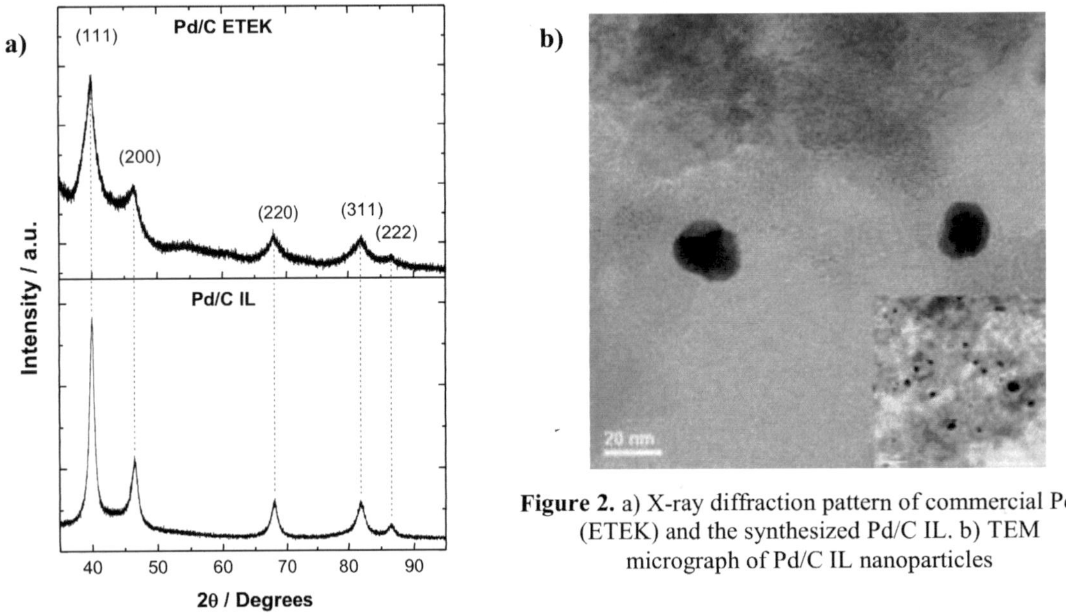

Figure 2. a) X-ray diffraction pattern of commercial Pd/C (ETEK) and the synthesized Pd/C IL. b) TEM micrograph of Pd/C IL nanoparticles

TEM micrographs of the synthesized Pd/C nanoparticles through the "all-in-one" 2-hydroxy ethylammonium formate ionic liquid are shown in Fig. 2b. Particles with sizes of 22 nm were found. Agglomeration of particles also occurred as a consequence of a weak stabilization by the ionic liquid. This could be related to the multiple tasks that the ionic liquid plays, reducing its capability to reduce and stabilize at the same time. TEM image of Pd/C ETEK can be found elsewhere [8].

Table 1. Summary of crystallite sizes, particles sizes, and mass content determined by XRD, TEM and TGA, respectively.

Material	Crystallite size		Lattice	Particle	Mass content /
	Plane	Size (nm)	parameter (Å)	size (nm)	(%wt.)
Pd/C (IL)	(111)	9.678	3.9024	22	74% XC-72
	(200)	8.524			
	(220)	7.636			26% Pd
	(311)	8.022			
Pd/C (ETEK)	(111)	4.098	3.8918	6	70% XC-72
	(200)	5.635			
	(220)	4.326			30% Pd
	(311)	3.902			

An analysis of effect of ethanol concentration was made in the three fuel cell configurations (alkaline, acidic, and alkaline anodic stream/acidic cathodic stream). For space purposes, in Figure 3 only results for Pd/C IL and Pd/C ETEK in the alkaline NFC are shown. The higher cell performances (1.5 M ethanol) in each fuel cell configuration are summarized in Figure 5.

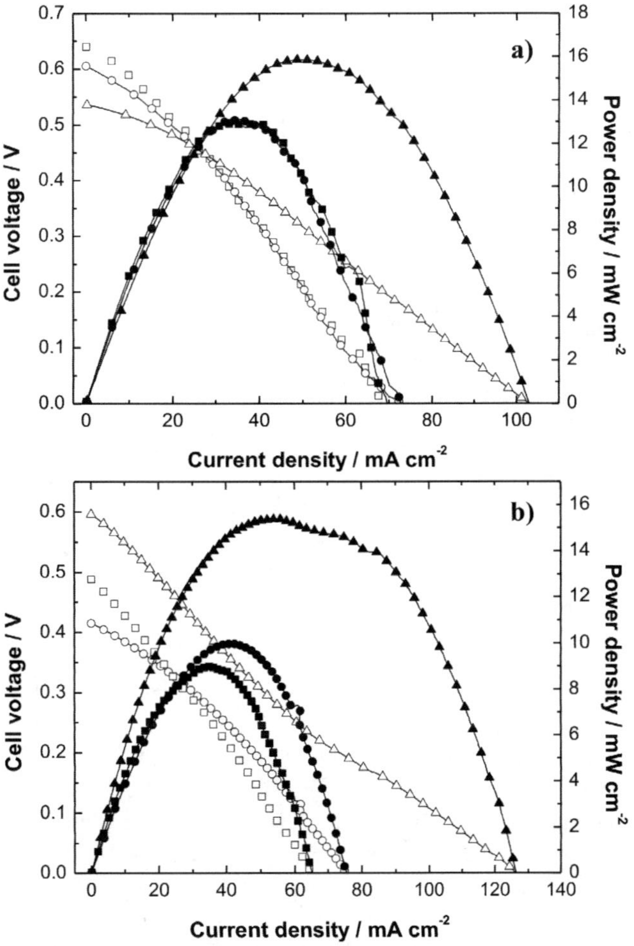

Figure 3. Polarization and performance curves for a) Pd/C ETEK and b) Pd IL as function of concentration 0.5 (squares), 1 (circles) and 1.5 M (triangles) in an alkaline ethanol NFC

Pd/C ETEK showed better performance at 0.5 M ethanol (Fig. 3a). However, Pd/C IL improved its performance with the increase in the ethanol concentration (Fig. 3b). The higher performance was achieve using Pd/C IL with 1.5 M ethanol showing a cell voltage of 0.6 V, a current density of 126 mA cm^{-2} and a power density of 15.4 mW cm^{-2} (Fig. 3b, triangles). The cell performance of a complete alkaline nanofluidic fuel cell (NFC) is showed in Figure 4a. The results obtained for the acidic NFC are showed in Figure 4b and the combined NFC in Figure 4c. It was observed from figures 5a and 5b that these materials showed different behaviour in both, alkaline and acidic medium. Pd/C IL showed better performance in alkaline than in acidic and could be related to the enhancement of reaction kinetics at higher pH [1]. On the other hand, Pd/C ETEK showed similar performances in alkaline and acidic media. In the combined MNFC, the use of alkaline anodic stream and the acidic cathodic stream resulted in cell voltages of 1.11 and 1.03 V for Pd/C ETEK and Pd/C IL. Furthermore, this combination enhanced the power density, obtaining values of 87.66 and 99.42 mW cm^{-2}, respectively.

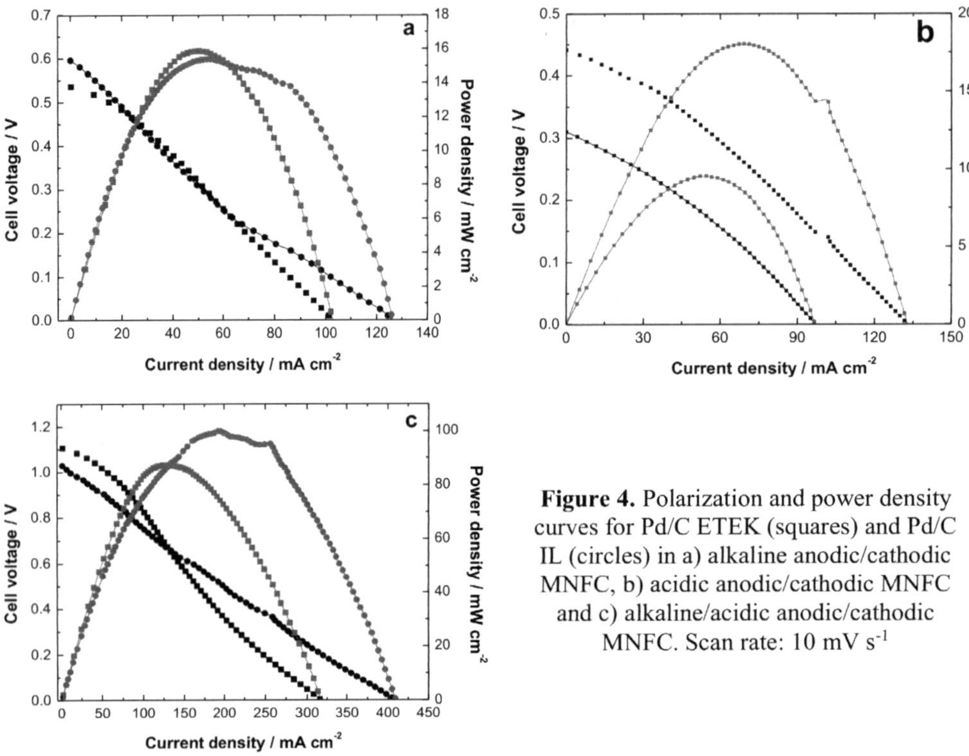

Figure 4. Polarization and power density curves for Pd/C ETEK (squares) and Pd/C IL (circles) in a) alkaline anodic/cathodic MNFC, b) acidic anodic/cathodic MNFC and c) alkaline/acidic anodic/cathodic MNFC. Scan rate: 10 mV s^{-1}

4. Conclusions

Commercial Pd/C and Pd/C synthesized using ionic liquids were successfully used as electrocatalysts in a nanofluidic fuel cell with flow through electrodes. Pd/C IL showed better activity in an alkaline environment showing higher performance in alkaline than in the acidic MNFC because of an enhancement of the ethanol reaction kinetic at high pH. The MNFC which employed commercial Pd/C showed no significant change with changes in pH. The mix of streams alkaline (anode)/acidic (cathode) resulted in power densities up to 4-fold higher than using equal electrolytes. This can be related to the mobility of protons produced in the anode toward the cathode and their further reduction forming multiple oxidant species (oxygen from the saturated aqueous solution, from air and protons). The use of ionic liquids as a reaction medium resulted in a Pd/C material which showed a higher cell performance than commercial Pd/C in the mixed streams despite that it has a particle size almost 3-fold bigger.

References

[1] Kjeang E, Djilali N and Sinton D 2008 *J. Power Sources* **186** 353
[2] Choban E R, Spendelow J S, Gancs L, Wieckowski A and Kenis P J A 2005 *Electrochim. Acta* **50** 5390
[3] Soloveichik 2014 Beilstein *J. Nanotechnol.* **5** 1399
[4] An L, Zhao T S and Li Y S 2015 *Renew. Sust. Energ. Rev.* **50** 1462
[5] Cohen J L, Westly D A, Pechenik A and Abruña H D 2005 *J. Power Sources* **139** 96
[6] Duan H, Wang D and Li Y 2015 *Chem. Soc. Rev.* **44** 5778
[7] Dimitratos N, Kiely C J and Hutchings G J 2013 General Introduction to the Field of Eviromental Catalysis: Green Catalysis with Supported Gold and Gold Bimetallic Nanoparticles *Enviromental Catalysis Over Gold-Based Materials* (RSC Catalysis Series No. 13) ed G Avgouropoulos and T Tabakova (London: RSC) pp 2.
[8] Carrera-Cerritos R, Guerra-Balcázar M, Fuentes-Ramírez R, Ledesma-García J and Arriaga L G 2012 *Materials* **5** 1686

Intermediate-temperature operation of solid oxide fuel cells (IT-SOFCs) with thin film proton conductive electrolyte

T Kariya[1, 2], K Uchiyama[3], H Tanaka[1], T Hirono[2], T Kuse[2], K Yanagimoto[2], M Henmi[4], M Hirose[4], I Kimura[4], K Suu[4] and H Funakubo[1]

[1] Department of Innovative and Engineered Materials, Tokyo Institute of Technology, 4259 Nagatsuta-cho, Midori-ku, Yokohama 226-8502, Japan

[2] Research and Development Canter, Sanyo Special Steel Co., Ltd, 3007, Nakashima, Shikama-ku, Himeji, Hyogo 672-8677, Japan

[3] Department of Creative Engineering, National Institute of Technology, Tsuruoka College, Inooka Sawada 104, Tsuruoka, Yamagata 997-8511, Japan

[4] Institute of Semiconductor & Electronics Technologies, ULVAC, Inc., 1220-1 Suyama, Susono, Shizuoka, Japan 410-1231, Japan

uchiyama@tsuruoka-nct.ac.jp

Abstract. A novel solid oxide fuel cell (SOFC) structure, which is fabricated on a Pd-plated porous stainless steel substrate, was proposed for low-temperature SOFC operation. The surface of the substrate was covered with Pd layer without any pores, which reduces the difficulty of depositing thin film electrolyte on the porous substrate. A 1.2-μm thick proton conductive $Sr(Zr_{0.8}Y_{0.2})O_{3-\delta}$ (SZYO) layer and the cathode of a 100-nm thick $(La_{0.6}Sr_{0.4})(Co_{0.2}Fe_{0.8})O_{3-\delta}$ (LSCF) layer were deposited on the Pd-plated substrates by the pulsed laser deposition (PLD) method. The low temperature operations at 400 and 450 °C were demonstrated with proposed SOFC cells.

1. Introduction

Solid oxide fuel cells (SOFCs) are considered to be one of the candidates for future energy sources because of its high electricity generating efficiency. However, its high operating temperatures of 800-1000 ℃ brings higher apparatus cost and difficulty in its quick start-up, which prevents enlarging its applications, especially for the home uses. Thus, achieving operating temperatures below 600 ℃ is considered to be a key for solving these issues.

To achieve the lower operating temperatures, the reduction of the electrolyte thickness is quite effective. However, thinning the electrolyte requires a substrate that must possess fuel gas permeability. Thus, porous substrates are usually used, which makes depositing thin films electrolyte quite difficult.

In 2005, Ito *et al.* reported good SOFC operations below 600 ℃ with using a thin electrolyte less than 1 μm in thickness [1]. They used palladium (Pd) foils as a substrate to avoid the difficulty of the deposition. However, the cost of the substrate is expensive and it is not realistic for the commercially available SOFCs.

In order to resolve this issue, we have proposed to use a Pd plated porous stainless-steel instead of using solid Pd as a substrate [2,3]. Figure 1 shows a schematic of our proposed SOFC cell structure. As the top surface of the substrate is fully covered with the planer Pd, we can use a variety of deposition techniques of oxide thin films, such as pulsed laser depositions (PLD) method, Sputtering method, and Sol-gel method, etc with a lower substrate cost than that of Ito *et al.*'s. We adopted PLD method for fabricating proton conductive electrolyte of $Sr(Zr_{0.8}Y_{0.2})O_{3-\delta}$ (SZYO) and cathode layer of $(La_{0.6}Sr_{0.4})(Co_{0.2}Fe_{0.8})O_{3-\delta}$ (LSCF) on the Pd-plated substrate

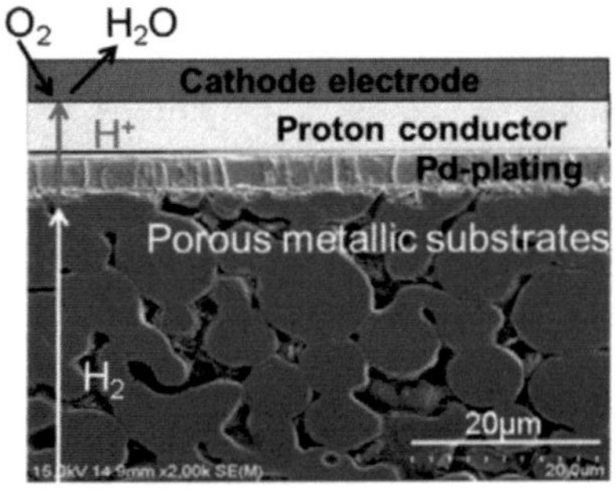

Figure 1. Schematic image of proposed new SOFC structure

In this study we describe the fabrication procedure of our novel SOFCs and its low temperature operations at 400 and 450 ℃.

2. Experimental procedure

2.1 Fabrication of SOFC cells

A fabrication condition of Pd plated porous stainless substrates is described elsewhere [3]. A pressurized hydrogen (H_2) gas was used to measure H_2 permeability.

SZYO and LSCF layers were deposited by PLD method, using an excimer laser with a wavelength of 248 nm. Target materials of SZYO and LSCF used in this study were prepared by spark plasma sintering. The SZYO and LSCF films were deposited on $Pt/SiO_2/Si$ substrate (abridged as Pt/Si sub. in this paper) for evaluating crystallinity and conductivity and Pd-plated stainless steel substrate to fabricate the SOFC structure.

2.2. Characterization

Crystallinity of the deposited films was evaluated by X-ray diffractions (XRD) with θ-2θ scans. Deposition conditions of PLD method were optimized by the XRD to obtain single phase of SZYO and LSCF with high crystallinity.

Arrhenius plots for proton conductivity of the cell were obtained from measuring alternating-current impedance in dry and wet Ar atmospheres by 4194A (Agilent Technologies) impedance analyser. Conditions of alternating-current impedance measurement were as follows: measurement frequency of 100 to 400 MHz, constant current of 20 mA, humidification temperature of 40 °C for wet Ar.

The single-cell performances, current density - cell voltage and current density - power density curves, were measured by an evaluation system, TOYO-Auto PEM (TOYO Corporation) installed by TOYO TFT integrated software. Both sides of the sample cell were connected to TOYO-Auto PEM system by Pt paste and Pt wire. The power generation properties were evaluated under the following test conditions: cell temperature at 400 °C and 450 °C, cathode humidification temperature of 25 °C, anode humidification temperature of 25 °C, hydrogen feed rate of 0.15 L/min, and air feed rate of 0.10 L/min.

3. Results and discussion

3.1 H_2 permeability

Figure 2 shows the H_2 permeability of the Pd-plated substrate in comparison with a 200 μm-thick Pd sheet. A larger permeability of the substrate than that of Pd sheet was observed at 300 and 400 °C. This was brought from the thinner Pd layer of 5-15 μm in the substrate. On the contrary, a lower permeability of the substrate at higher temperatures of 500 and 600 °C was observed. This might come from the degradation of the Pd layer because of diffusion of alloy elements from the stainless steel. We need more improvement of the substrates to suppress Pd layer degradation at higher temperatures.

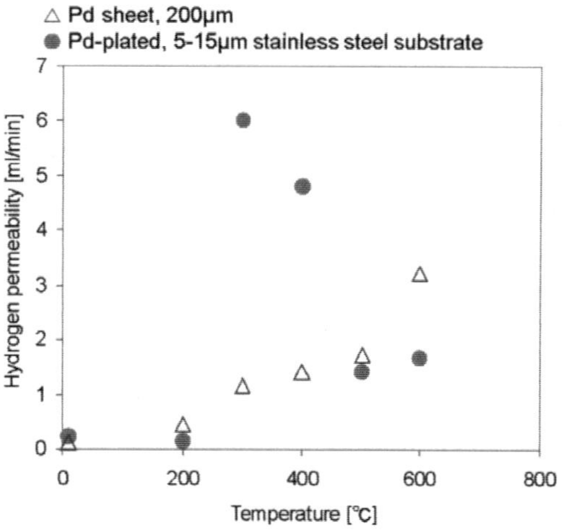

Figure 2. H_2 permeability of Pd-plated substrate and 200 μm-thick Pd sheet vs. temperature

3.2 Optimization of PLD conditions

As for the PLD deposition, we examined oxygen (O_2) partial pressure, laser frequency, and laser-output density, and substrate temperatures. Figure 3 is the one of the data for optimizing PLD deposition conditions. SZYO films were deposited on $SrRuO_3$ with a different laser frequency with 0.53 J/cm^2 of laser power density at 400 °C of the substrate temperature. As can be seen in Figure 3, only 8 Hz of PLD laser frequency can produce the single phase of SZYO. At the other laser frequencies, the films showed SrY_2O_3 phase despite they possess stoichiometric composition of SZYO.

In the same manner, we examined deposition conditions of SZYO and LSCF and the optimized the conditions. The optimized conditions for SZYO were as follows: oxygen partial pressure of 50 mTorr, laser frequency of 8 Hz, laser-output density of 0.53 J/cm$_2$, and substrate temperature of 400 °C. On the other hand, the optimized deposition conditions for LSCF were oxygen partial pressure of 50 mTorr, laser frequency of 4 Hz, laser-output density of 0.53 J/cm^2, and substrate temperature of 650 °C.

3.3 Proton conductivity of PLD deposited SZYO

Proton conductivity of SZYO deposited on $Pt/SiO_2/Si$ substrate by the PLD method was evaluated in dry and wet Ar atmospheres in comparison with the sputtered SZYO reported by M. Arab Pour Yazd *et al.*[4] (Figure 4).

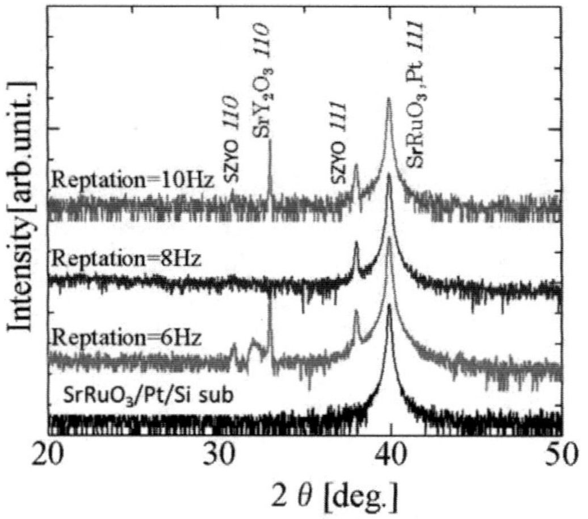

Figure 3. XRD patterns of PLD deposited SZYO thin films with different laser frequencies

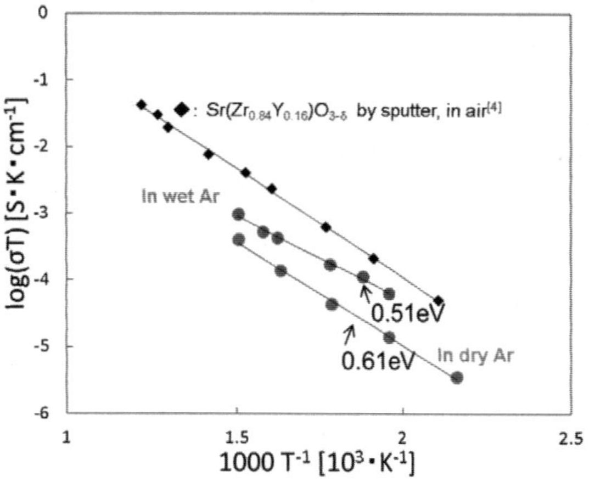

Figure 4. Arrhenius plots of conductivity of PLD deposited SZYO thin films in comparison with sputtered SZYO [4]

The conductivity of derived SZYO is smaller than that of the sputtered one; however, the activation energy in dry atmosphere is almost comparable. We need more improvements of the proton conductivity to achieve better SOFC property.

3.4 Cell performance at intermediate temperatures

Figure 5 is the current density-output voltage and current density-power density characteristics of the SOFC cell operated at 400 and 450 °C. We have successfully obtained power generating characteristics as low as 400 °C. However, its maximum power densities were about 1.2 and 2.4 mW/cm^2 at 400 and 450 °C, respectively, which are two decade smaller than that of commercialized SOFCs. The origin of the low power density might be the cathode of LSCF. As the LSCF later was deposited by the PLD method, the deposited LSCF film was solid and has no pores. This means oxygen (O_2) can be supplied only by the diffusion of the O_2 ions through the LSFC layer. The insufficient supply of the O_2 brings suppressing a reaction between H_2 and O_2, which results in less power generation.

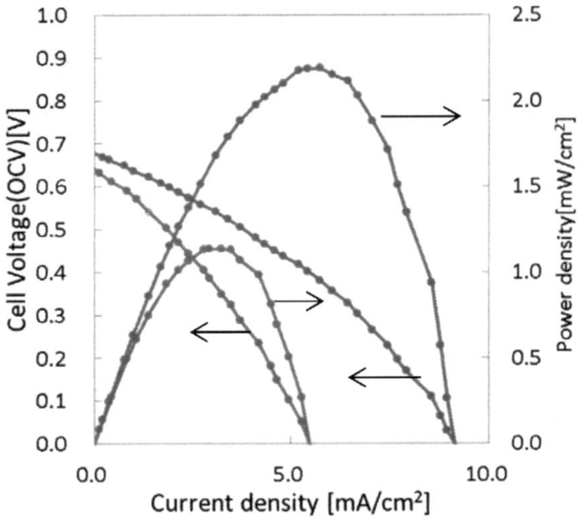

Figure 5. Power generation property of fabricated SOFC operating at 400 (blue lines) and 450 °C (red lines).

In general, porous cathode was generally used in the conventional SOFCs to avoid this issue. However, the processing temperature of fabricating the conventional cathode is above 1000 °C, which is too high for our substrate. We are trying to fabricate porous cathode with a lower processing temperatures to obtain better SOFC operations.

4. Conclusion

We successfully demonstrated low temperature operations at 400 and 450 °C. The SOFC was fabricated on porous stainless steel substrate with Pd, SZYO, and LSCF as an anode, a proton conductive electrolyte, and a cathode, respectively. A solid Pd layer was fabricated on a porous stainless steel substrate by plating. Solid SZYO and LSCF films were deposited on the Pd-plated substrate by the PLD method. The substrate has good H_2 permeability at 300 and 400 °C; however it degraded at above 500 °C.

The fabricated SOFC cell showed low temperature operations at 400 and 450 °C. The power generation densities of the cell was 1.2 and 2.4 mW/cm^2 at 400 and 450 °C, respectively, which is two decades smaller than that of the conventional SOFCs. We consider the origin of this low power generation is insufficient O_2 supply through the cathode.

Acknowledgements

This work is partially supported by JST ALCA and JST A-STEP, Japan. One of the author, K. Uchiyama, also thanks to JSPS KAKENHI Grant number 26420284 for the support.

References

[1] Ito N, Iijima M, Kimura K and Iguchi S, 2005 *J. Power Sources* **152** 200-203

[2] Uchiyama K, Isse Y, Hori Y, Nishida T, Uraoka Y, Kariya T and Yanagimoto K 2010 Proc. 6th Thin-Films Materials and Devices Meeting (Kyoto) 100228102

[3] Kariya T, Tanaka H, Hirono T, Kuse T, Yanagimoto K, Uchiyama K, Henmi M, Hirose M, Kimura I, Suu K and Funakubo H 2016 *J. Alloys Compounds* **654** 171-175
[4] Arab Pour Yazd M, Briois P, Georges S and Billard A 2009 *Solid State Ionics* **180** 1246-1251

A Miniature Radial-Flow Wind Turbine Using Piezoelectric Transducers and Magnetic Excitation

H Fu and E M Yeatman

Department of Electrical and Electronic Engineering, Imperial College London, Exhibition Road, London, SW7 2AZ, UK

E-mail: h.fu14@imperial.ac.uk

Abstract. This paper presents a miniature radial-flow piezoelectric wind turbine for harvesting airflow energy. The turbine's transduction is achieved by magnetic "plucking"of a piezoelectric beam by the passing rotor. The magnetic coupling is formed by two magnets on the beam's free end and on the rotor plate. Frequency up-conversion is realized by the magnetic excitation, allowing the rotor to rotate at any low frequency while the beam can vibrate at its resonant frequency after each plucking. The operating range of the device is, therefore, expanded by this mechanism. Two arrangements of magnetic orientation have been investigated, showing that the repulsive arrangement has higher output power. The influence of the vertical gap between magnets was also examined, providing guidance for the final design. A prototype was built and tested in a wind tunnel. A peak power output of 159 μW was obtained with a 270 kΩ load at 2.7 m/s airflow speed. The device started working at 3.5 m/s and kept operating when the airflow speed fell to 1.84 m/s.

1. Introduction

Airflow energy is a ubiquitous energy source in many situations, including ventilation systems, mining tunnels and moving vehicles, where wireless sensor nodes are often installed [1–3]. Hence, airflow energy harvesting is a potential solution to autonomously power these nodes, instead of using conventional batteries which have a limited lifetime [4].

Miniaturization of wind turbines is one research direction in airflow energy harvesting [5, 6]. Howey *et al.* demonstrated a centimeter-scale electromagnetic wind turbine with a high energy density [7]. However, numerous components were required in this device, which undermined its further miniaturization potential and increased the fabrication complexity. Priya *et al.* developed a piezoelectric windmill with a simpler structure [8], but the mechanical excitation might damage the brittle piezoelectric materials and the size of the device was undesirable for wireless sensing applications. Contactless excitation methods have been used in other areas. Pillatsch *et al.* employed magnetic coupling in a non-harmonic motion energy harvester, in which the piezoelectric beam can be actuated without physical contacts [9]. By taking advantage of this feature, in this paper, a miniaturized wind turbine with piezoelectric transducers and a contactless plucking mechanism has been designed with high reliability and simplicity.

2. Design and Operating Principle

The schematic of the harvester is presented in Figure 1 and Figure 2. The turbine rotor is supported by a rolling bearing mounted on a shaft that is fixed on the rear casing. The rear casing is stationary with the front casing enclosing the turbine rotor. There is a slot, which is designed to hold the piezoelectric beam using a screw, on the rear casing. Two magnets, which form the magnetic coupling, are installed in the harvester: one on the piezoelectric beam's free end (tip magnet in Figure 1) and the other on the rotor plate (primary magnet in Figure 1).

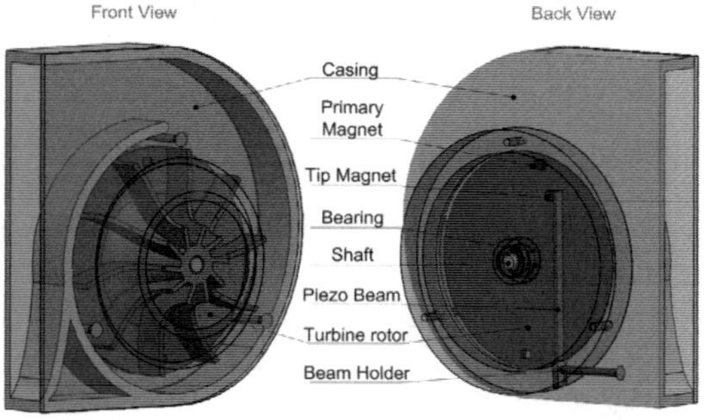

Figure 1. Front and back view of the piezoelectric airflow energy harvester.

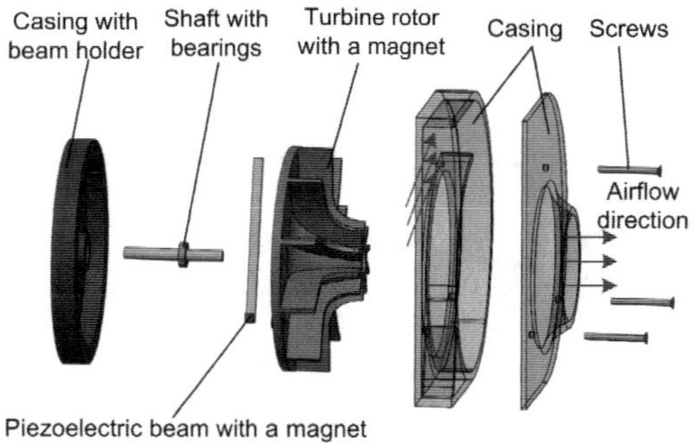

Figure 2. Exploded view of the piezoelectric airflow energy harvester.

Airflow, which enters the turbine radially and exits axially, activates the turbine rotor. The rotor plucks the piezoelectric beam once per rotation cycle by magnetic coupling, providing non-contact excitation. Frequency up-conversion is also implemented by the magnetic excitation, allowing the rotor to rotate at low frequencies while the beam can keep operating at its resonant frequency, enabling the device to have a wide working range. In addition, the non-contact excitation method avoids the mechanical impact on the piezoelectric beam, which enhances the reliability of the device.

3. Optimization of Magnetic Coupling

The magnetic coupling is a critical design factor, as the power output of the beam and the start-up airflow speed of the harvester are directly determined by the magnetic coupling force between the magnets. In order to understand the influence of magnetic coupling on the performance of the piezoelectric beam, an equivalent experimental set-up was built as shown in Figure 3.

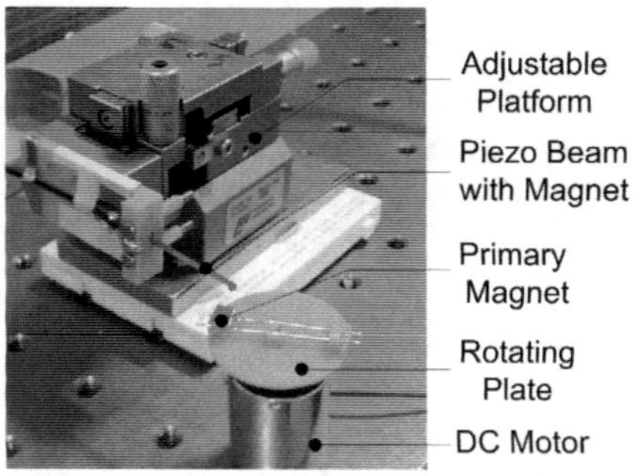

Figure 3. Experimental set-up to optimize the magnetic coupling.

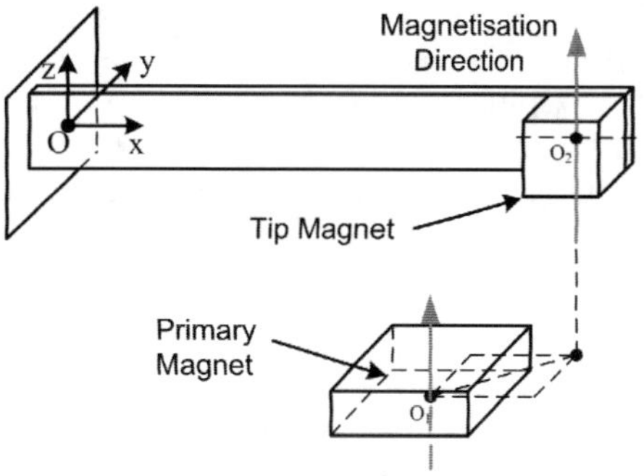

Figure 4. Magnetisation direction of magnets (attractive arrangement).

The piezoelectric beam is installed on an adjustable platform so that the gaps between the two magnets in 3 dimensions can be tuned accurately by the micrometers on the platform. A DC motor is used as an analogy to the rotating turbine. A plate with the same diameter as the turbine rotor is mounted on the motor's rotor with a magnet fixed on it. The magnetisation direction is in the vertical direction as shown in Figure 4. The sizes of these components are listed in Table 1.

Table 1. Dimensions of components.

Item	Dimension
Rotor Magnet	5 mm × 4 mm × 1 mm
Beam Magnet	1 mm × 1 mm × 1 mm
Piezoelectric Beam	26.5 mm × 1.5 mm × 0.3 mm

Two arrangements of the magnetic orientation were investigated against rotational frequency of the primary magnet. The repulsive arrangement has better performance at high frequencies with larger energy output (Figure 5). The influence of the vertical gap between the magnets is also illustrated in Figure 6, which shows the variation of output voltage against the vertical gap. A 2 mm gap was chosen for the final design as a compromise between the start-up airflow speed and the output power.

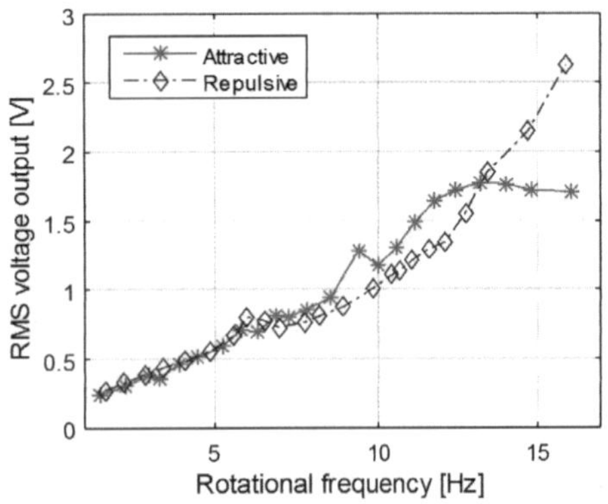

Figure 5. RMS output voltage for attractive and repulsive arrangements of magnets versus rotational speed.

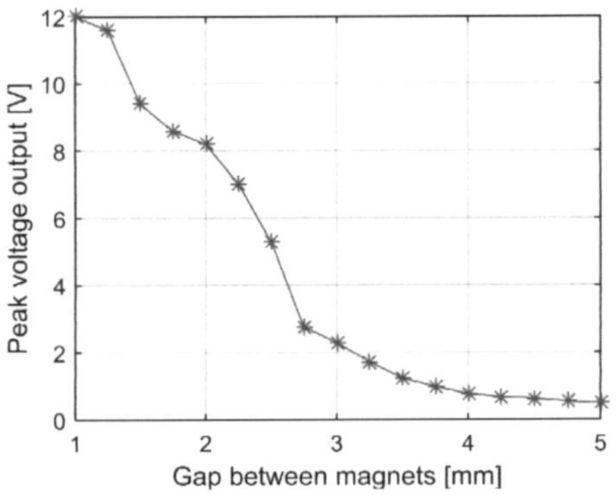

Figure 6. Influence of the vertical gap between magnets on output voltage of the beam.

4. Prototype and Experimental Set-Up

Rapid prototyping was used to fabricate the turbine rotor and casing. Other components were purchased as off-the-shelf products, including cuboidal magnets, piezoelectric beams and rolling bearings. A prototype was assembled as shown in Figure 7(a). The overall size of the device is 47 mm × 42 mm × 25 mm.

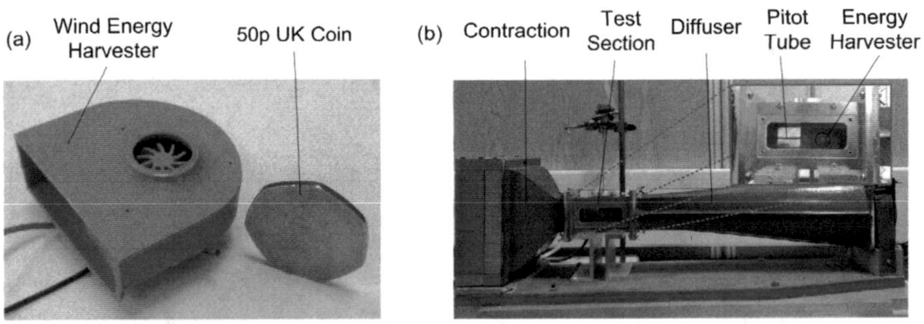

Figure 7. Prototype (a) and experimental set-up (b).

The harvester was tested in the wind tunnel shown in Figure 7(b). This wind tunnel is an open loop wind tunnel with continuously variable airspeed control. It is comprised of four major components: contraction, test section, diffuser and fan. Airflow enters the tunnel from the contraction section and exits from the diffuser. The performance of the harvester was measured in the test section, whose dimension is 100 mm × 85 mm × 85 mm. A pitot tube used to measure the airflow speed was installed in parallel with the harvester in the test section, and the reading of airflow speed was received from a micro manometer.

5. Results

Figure 8 shows the successful implementation of the concept of frequency up-conversion by this device. The turbine rotor rotates at 11 Hz which can be derived from the excitation frequency of the beam. The beam operates in a free-damped-vibration form at its resonant frequency, i.e. 133 Hz after each plucking. The vibration frequency is unaffected by the excitation frequency due to the plucking behaviour.

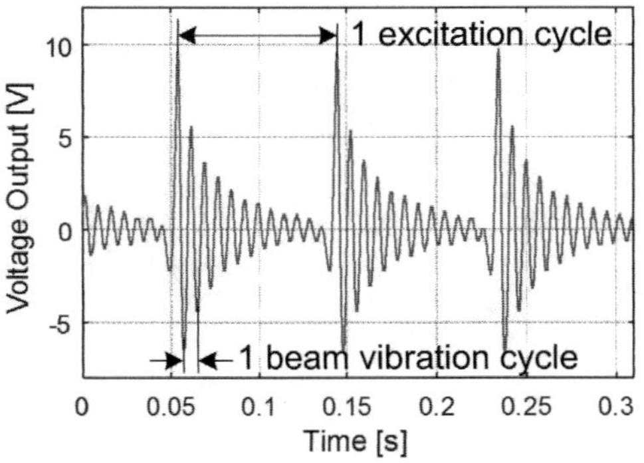

Figure 8. Output voltage with a 1.8 MΩ load at 2.7 m/s, showing the implementation of frequency up-conversion.

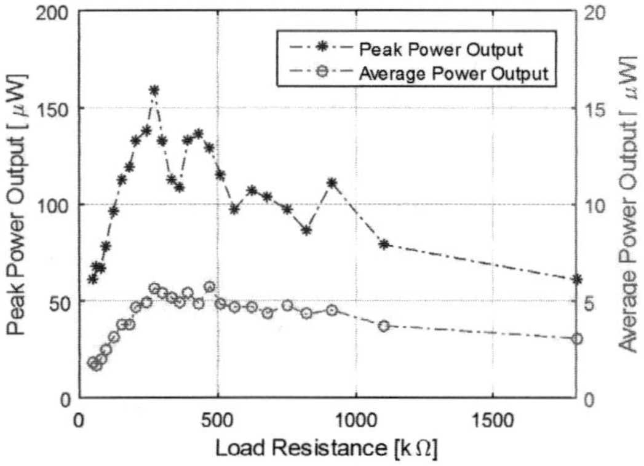

Figure 9. Output power against load resistance at airflow speed 2.74 m/s.

The device has also been tested with different loads and airflow speeds. Figure 9 illustrates the peak and average output power against differing load resistances at 2.7 m/s. A peak power output of 159 μW was obtained with a 270 kΩ load. The fluctuation of output power is caused by the instability of the shaft and the variation of airflow speed. Figure 10 depicts the rotational frequency of the turbine rotor and the output peak power against airflow speed. The device started working at 3.5 m/s and kept operating when the airflow speed fell to 1.84 m/s. The start-up airflow speed is quite high compared to the operating range of the harvester. A self-regulating concept [10] will be implemented in future work to decrease the start-up airflow speed.

In addition, the output power at high airflow speeds decreases sharply due to the high plucking frequency which makes the output of the piezoelectric beam between two excitation cycles overlap with each other, as illustrated in Figure 11. The output voltage is undermined due to the cancelling effect of the overlapping. In order to maintain the output power at high airflow speeds, the rotational speed of the rotor should be controlled without rising with the increase of the airflow rate. Hence, more energy should be consumed from the turbine rotor in other ways. Solutions to expand the operating range will be investigated in future work.

6. Conclusions and Future Work

This paper presents a miniaturized piezoelectric wind turbine for airflow energy harvesting. Magnetic excitation is employed to pluck the beam without any physical contact, enhancing the

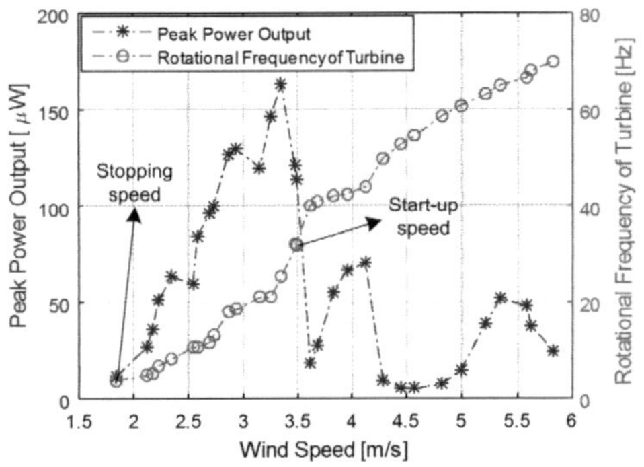

Figure 10. Power output versus airflow speed with a 270 kΩ load.

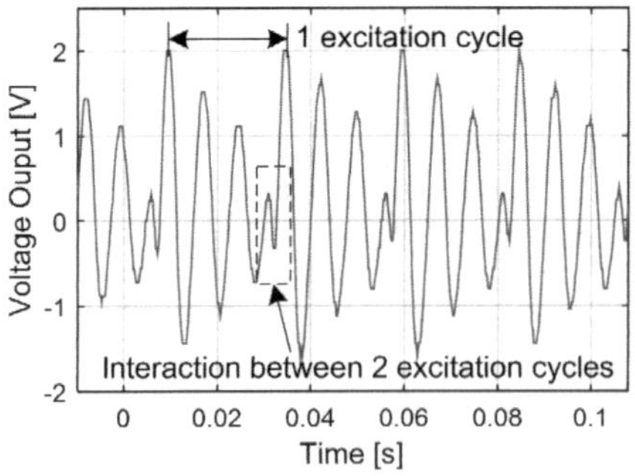

Figure 11. Voltage output at 3.61 m/s, showing the interaction of the beam vibration between two excitation cycles.

reliability of the device. Frequency up-conversion is implemented using the magnetic plucking method, which allows the beam to operate irrespective of the rotational frequency of the turbine rotor.

Two magnet arrangements were investigated, showing that the repulsive arrangement had a better performance at high airflow speeds. The influence of the vertical gap between magnets was also examined. The peak voltage output had a sharp drop with the increase of the vertical gap, which means the torque needed for start-up is decreased as well.

A prototype was fabricated and tested in a wind tunnel. A output power of 159 μW was obtained with a 270 kΩ load at 2.7 m/s. The device started working at 3.5 m/s and kept operating when the airflow speed fell to 1.84 m/s. Future work will be focused on decreasing the start-up airflow speed and improving the power output at high airflow speeds.

Acknowledgments
The work is supported by the China Scholarship Council (CSC) and the Department of Electrical and Electrical Engineering, Imperial College London. We would also like to thank Prof. Andrew Holmes for use of apparatus and valuable suggestions.

References
[1] Briand D, Yeatman E and Roundy S 2015 *Micro Energy Harvesting* Advanced Micro and Nanosystems (Wiley-VCH)
[2] Matiko J W, Grabham N J, Beeby S P and Tudor M J 2014 *Measurement Science and Technology* **25** 1–25
[3] Mellor P, Burrow S, Sawata T and Holme M 2005 *Industry Applications, IEEE Transactions on* **41** 551–556
[4] Yeatman E M 2008 *Proceedings of the Institution of Mechanical Engineers, Part C: Journal of Mechanical Engineering Science* **222** 27–36
[5] Mitcheson P D, Yeatman E M, Rao G K, Holmes A S and Green T C 2008 *Proceedings of the IEEE* **96** 1457–1486
[6] Beyaz M, Hanrahan B, Feldman J and Ghodssi R 2012 *Micro Electro Mechanical Systems (MEMS),IEEE 25th International Conference on* pp 1209–1212
[7] Howey D A, Bansal A and Holmes A S 2011 *Smart Materials and Structures* **20** 1–12
[8] Priya S, Chen C T, Fye D and Zahnd J 2005 *Japanese Journal of Applied Physics* **44** L104–L107
[9] Pillatsch P, Yeatman E M and Holmes A S 2014 *Sensors and Actuators A: Physical* **206** 178 – 185
[10] Gu L and Livermore C 2010 *Applied Physics Letters* **97** 1–3

A MEMS turbine prototype for respiration harvesting

U Goreke[1], S Habibiabad[1], K Azgin[2] and M I Beyaz[3,4]

[1]Graduate School of Natural and Applied Sciences, Middle East Technical University, Ankara, Turkey
[2]Department of Mechanical Engineering, Middle East Technical University, Ankara, Turkey
[3]Department of Electrical and Electronics Engineering, Antalya International University, Antalya, Turkey

Abstract. The design, manufacturing, and performance characterization of a MEMS-scale turbine prototype is reported. The turbine is designed for integration into a respiration harvester that can convert normal human breathing into electrical power through electromagnetic induction. The device measures 10 mm in radius, and employs 12 blades located around the turbine periphery along with ball bearings around the center. Finite element simulations showed that an average torque of 3.07 μNm is induced at 12 lpm airflow rate, which lies in normal breathing levels. The turbine and a test package were manufactured using CNC milling on PMMA. Tests were performed at respiration flow rates between 5-25 lpm. The highest rotational speed was measured to be 9.84 krpm at 25 lpm, resulting in 8.96 mbar pressure drop across the device and 370 mW actuation power.

1. Introduction

Energy harvesting from humans is becoming a promising alternative for powering portable electronic devices [1-2]. A number of harvesters have been reported so far, taking advantage of various body activities and transduction mechanisms. To give examples, thermal generators located around the human skin convert temperature difference into electricity, and tend to deliver nanowatt-level power due to low temperature gradients achievable around the body under comfortable conditions [3-4]. Piezoelectric materials allowed for power generation from high mechanical stress available around the body joints and shoe insoles [5-6]. Electromagnetic generators have been reported to output as high as 5 W utilizing knee motion [7]. In addition to these body activities, normal respiration also poses great potential for energy harvesting. Detailed studies on human breathing showed that watt-level power is available in the form of fluid flow [8-10], which indicates that milliwatts can be generated with a harvester having a few percent efficiency.

In recent works, different respiration harvester designs have been reported based on linear displacement magnets [11] and vibrating structures [12]. However, these designs limited the output power to microwatts due to the inefficient use of airflow. The most efficient technique for converting fluidic flow into electricity involves the use of turbomachinery and electromagnetic induction as in

[4] Corresponding author, Email: mibeyaz@antalya.edu.tr

wind turbines. Accordingly, we have dedicated our efforts to developing a MEMS respiration harvester with a turbine-based architecture that can output several milliwats for powering portable electronic devices and biosensors as well as for recharging batteries. This work focuses on the development of the first turbine prototype, and reports our findings on turbine performance.

2. Design and simulations

The general turbine structure in Figure 1 measures r_{out} = 10 mm in radius and 1.3 mm in thickness. Turbine blades are defined around the turbine periphery with a radial length of $r_{out} - r_{in}$ = 2.5 mm. A test package is designed to tightly encapsulate and actuate the turbine with tangential airflow through rectangular openings (Figure 1b). Stainless steel ball bearings (Ø = 1 mm) are located at r_b = 2.3 mm, and embedded in the grooves etched on both the turbine and the package to enable turbine rotation. The area between the ball bearings and turbine blades are reserved for permanent magnet integration to facilitate electromagnetic induction on stators that are currently under development.

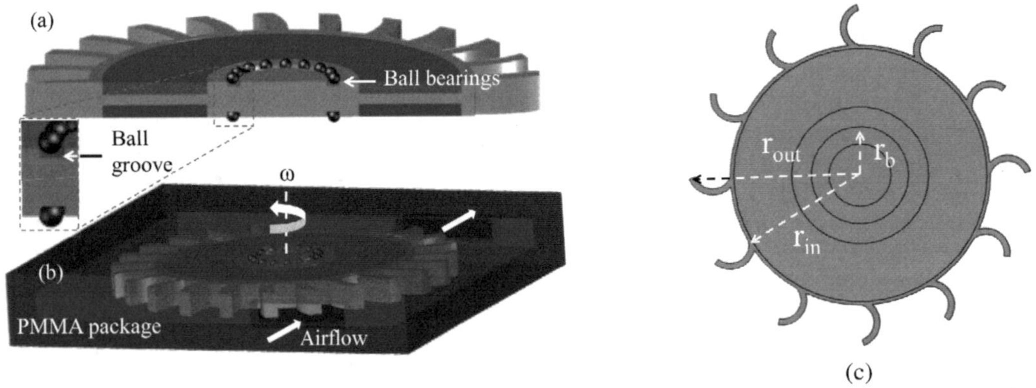

Figure 1. Schematic view of the turbine, (a) cut-away view showing ball bearings, shaded volume is reserved for magnet integration, (b) packaged device, (c) final turbine design with relevant dimensions

Simulations were performed using the Fluid Structure Interaction interface of the COMSOL software for a range of blade numbers to maximize the induced torque and rotational speed for a given actuation flow rate. Figure 2 shows a snapshot of the fluid velocity profile around the turbine at a normal breathing rate of 12 lpm, from which the induced torque is calculated. Based on this study and considering the manufacturing challenges, the number of blades for this first turbine prototype was determined to be 12, leading to 3.07 μNm torque on the turbine.

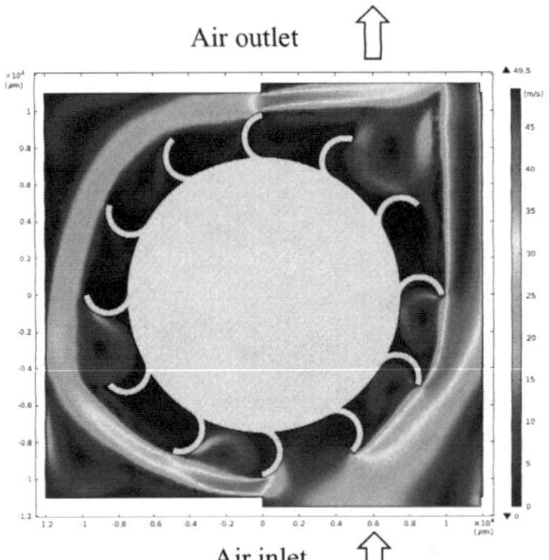

Figure 2. Simulation result showing the air velocity profile around the turbine at 12 lpm

3. Fabrication and test results

The turbine and a test package were fabricated using CNC milling with 50 μm precision on PMMA substrates. The package is made in two pieces, and encapsulates the turbine with four screws at the corners. Metal sheets are placed in between the two package pieces for adjusting the gap and eliminating the normal load on the turbine. A rubber gasket is used for sealing the package. A picture of the fabricated device is shown in Figure 3. The turbine along with other components of the harvester are currently being manufactured using microfabrication technology.

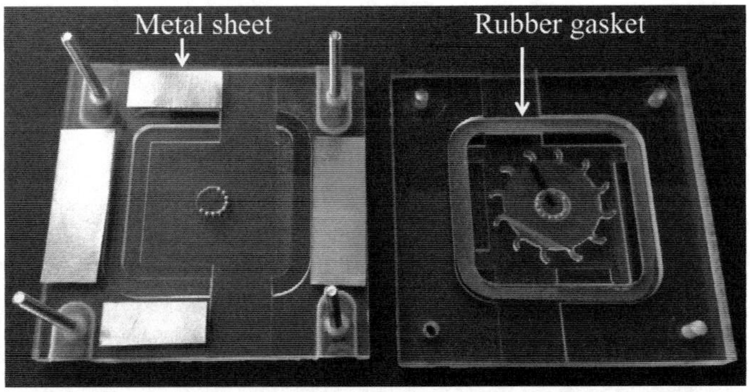

Figure 3. Picture of a fabricated turbine and package

A test setup was built to investigate the turbine performance in the range of normal breathing flow rates. In this setup, the turbine was actuated using pressurized nitrogen as the gas source. The flow rate and pressure drop across the turbine were monitored using a flow meter that can perform both measurements. A black marking was created on the turbine, and the rotational speed was measured by observing this marking during actuation using an in-house-built stroboscope. Figure 4a plots the variation of the rotational speed with respect to flow rate, exhibiting a linear correlation. The change in the slope below 10 lpm is attributed to manufacturing artifacts. Figure 4b shows the corresponding pressure drops across the device with a rather nonlinear relationship. The highest speed was measured to be 9.84 krpm obtained at the maximum actuation flow rate of 25 lpm, which resulted in 8.96 mbar pressure drop and 370 mW actuation power.

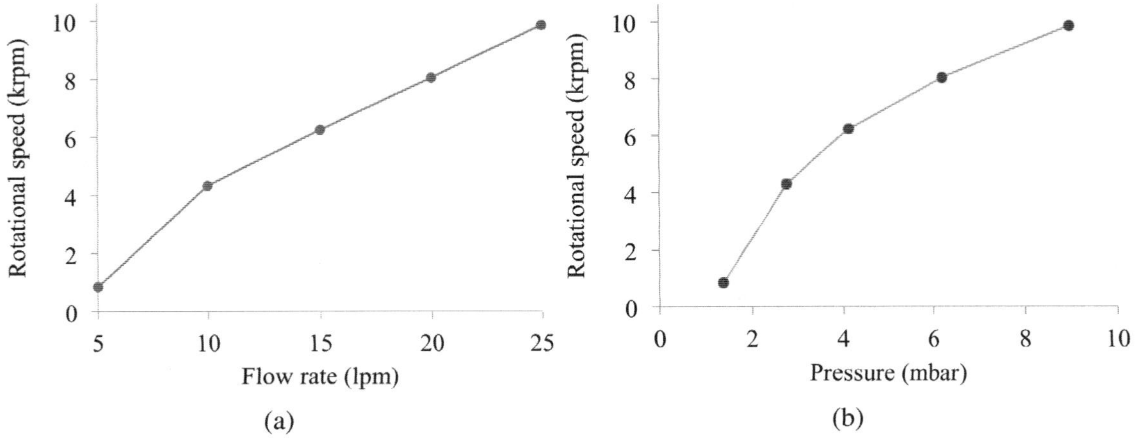

Figure 4. Rotational speed versus (a) actuation flow rate, and (b) pressure drop across the turbine

4. Conclusion

The development of a turbine prototype for respiration harvesting is reported. The turbine design is composed of 12 peripheral blades and ball bearings located around the center. Simulations on fluid velocity profile showed that 3.07 μNm torque is induced on the turbine at 12 lpm flow rate. The device and a test package were manufactured using CNC milling on PMMA substrates. A maximum rotational speed of 9.84 krpm was achieved at 25 lpm with a resulting pressure drop and actuation power of 8.96 mbar and 370 mW, respectively. The speed performance is expected to improve after the microfabrication of the turbine as well as the complete MEMS harvester. The turbine prototype reported here will enable the development of a MEMS respiration harvester that can be integrated into portable electronic devices, and provide milliwatt-level output power. The turbine design presented in this work can also be used in flow sensing applications.

5. References

[1] Paradiso J A and Starner T 2005 *IEEE Pervas. Comput.* **4** 18-27
[2] Mitcheson P D, Yeatman E M, Rao G K, Holmes A S and Green T C 2008 *P. IEEE* **96** 1457-86
[3] Lossec M, Multon B and Ahmed H B 2013 *Energ. Convers. Manage.* **68** 260-5
[4] Jo S E, Kim M K, Kim M S and Kim Y J 2012 *Electron. Lett.* **48** 1015-16
[5] Yang R, Qin Y, Li C, Zhu G and Wang Z L 2009 *Nano Lett.* **9** 1201-5
[6] Klimiec E, Zaraska W, Zaraska K, Gasiorski K P, Sadowski T and Pajda M 2008 *Microelectron. Reliab.* **48** 897-901
[7] Donelan J M, Li Q, Naing V, Hoffer J A, Weber D J and Kuo A D 2008 *Science* **319** 807-10
[8] Starner T 1996 *IBM Syst. J.* **35** 618-29
[9] Gupta J K, Lin C H, and Chen Q 2009 *Indoor Air* **19** 517-25
[10] Gupta J K, Lin C H, and Chen Q 2010 *Indoor Air* **20** 31-39
[11] Delnavaz A and Voix Jeremie 2012 *Proc. IECON 2012* 984-88
[12] Sun C, Shi J, Bayerl D J and Wang X 2011 *Energ. Environ. Sci.* **4** 4508-12

Acknowledgments

This work was supported by Scientific and Technological Research Council of Turkey (TUBITAK) under project number 113E197, and in part by European COST Action MP1303.

Energy Harvesting from Human Motion Using Footstep-Induced Airflow

H Fu[1], R Xu[2], K Seto[2], E M Yeatman[1] and S G Kim[2]
[1] Department of Electrical and Electronic Engineering, Imperial College London, Exhibition Road, London, SW7 2AZ, UK
[2] Department of Mechanical Engineering, Massachusetts Institute of Technology, Cambridge, MA 02139, USA

E-mail: h.fu14@imperial.ac.uk

Abstract. This paper presents an unobtrusive in-shoe energy harvester converting foot-strike energy into electricity to power wearable or portable devices. An air-pumped turbine system is developed to address the issues of the limited vertical deformation of shoes and the low frequency of human motion that impede harvesting energy from this source. The air pump is employed to convert the vertical foot-strike motion into airflow. The generated airflow passes through the miniaturized wind turbine whose transduction is realized by an electromagnetic generator. Energy is extracted from the generator with a higher frequency than that of footsteps, boosting the output power of the device. The turbine casing is specifically designed to enable the device to operate continuously with airflow in both directions. A prototype was fabricated and then tested under different situations. A 6 mW peak power output was obtained with a 4.9 Ω load. The achievable power from this design was estimated theoretically for understanding and further improvement.

1. Introduction

Wearable or portable devices are becoming increasingly popular in our daily lives, including smart watches, health monitors, radio frequency identification tags and emergency locators [1, 2]. However, they are generally powered by batteries that have a limited lifetime. Replacement or recharging is needed regularly, which is inconvenient in some situations. Harvesting wasted energy from human motion to power these devices is a potential alternative to conventional batteries, enabling the devices to be autonomous and fit-and-forget [3].

Extracting energy from foot strike has drawn significant interest [4-6], as the energy from foot-strike during walking is a plentiful and readily tapped wasted energy source [4]. However, it is always in a low frequency range and the deformation of shoe heel is quite limited, which exposes power limits for foot-strike energy harvesting. One interesting development to address these issues is a rotary arm converting the vertical linear motion of footsteps into rotation with a gearbox employed to magnify the rotational speed [5]. However, the rotary arm affects the shoe's comfort, and the gear train increases the manufacturing complexity and system unreliability. Moro *et al.* proposed a piezoelectric shoe energy harvester [6]. Although the structure of the device is simpler, its performance is limited due to the low operating frequency and the restricted volume of piezoelectric material. In this paper, a flexible air pump is employed to convert the vertical footstep motion into airflow, and a miniature wind generator is developed to generate electricity from the airflow at high frequencies. This design overcomes the difficulties in foot-strike energy harvesting mentioned above with a simpler structure and an enhanced power output.

2. Design and Operating Principle

The schematic of the device is presented in Figure 1. It consists of an air bulb pump and a miniature wind turbine. The pump converts the vertical distortion of shoes into airflow, and the turbine generates electrical energy from the passing airflow. The pump cover is constructed from latex rubber, and a wave spring is adopted to restore the shape of the pump after each strike. In order to enhance the speed and pressure of the airflow, an amplification mechanism is achieved by designing a shrunken outlet on the pump. Its sectional area is 500 times smaller than that of the pump, which means the airflow velocity is magnified up to 500 times on entry to the plastic tube. The airflow then passes through the wind turbine and energy is collected from a miniature DC motor as a generator attached to the turbine impeller. This design allows the turbine rotor to operate at a higher frequency than that of human motion. Thus, frequency up-conversion is achieved by the air pump-wind turbine system, addressing the issue of low frequency of footstep and boosting the power output.

The device is installed in the shoe heel, as shown in Figure 2. Due to the miniaturized size, the harvester can be embedded in a shoe heel unobtrusively. The support provided by the wave spring is stiff enough to ensure the wearer's comfort. When a footstep is applied to the air cushion, the pump is deformed and air in the pump is pushed forward to the turbine. When the force is removed, the pump shape is restored by the spring force, and a reversed airflow is generated in the device, further activating the turbine.

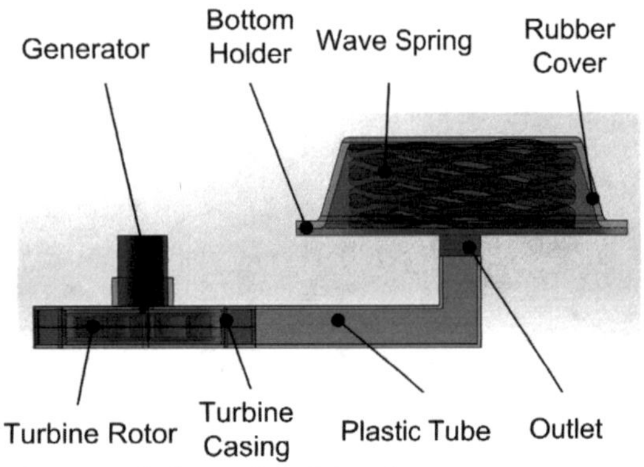

Figure 1. Schematic of the shoe energy harvester.

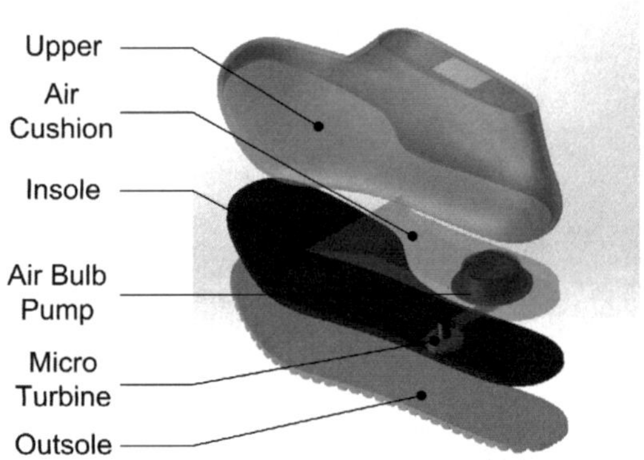

Figure 2. Installation and composition of the harvester in a shoe.

The inlet and outlet of the turbine is designed specially to ensure the airflow direction to be the same during and after each foot strike (Figure 3). The rotational direction of the turbine rotor, therefore, remains uniform during the entire motion cycle, which enables the continuous operation of the device during human motion.

3. Prototype and Experimental Set-up

A prototype of the harvester was built, as illustrated in Figure 4. An air pump was mounted on a sheet of stainless steel with the dimension of 120 mm × 60 mm × 13 mm. It is smaller than the foot heel of a shoe of size 9 (UK). A miniaturized wind turbine was fabricated using rapid prototyping with ABS materials. These two parts were connected by a plastic tube whose cross-sectional area (~Ø4 mm) was 500 times smaller than that of the pump. The overall dimension of the turbine was Ø12 mm × 6 mm, and its transduction was achieved by a DC motor with the dimension of Ø7 mm × 15 mm. The air pump was actuated by normal footsteps from a person weighing 65 kg.

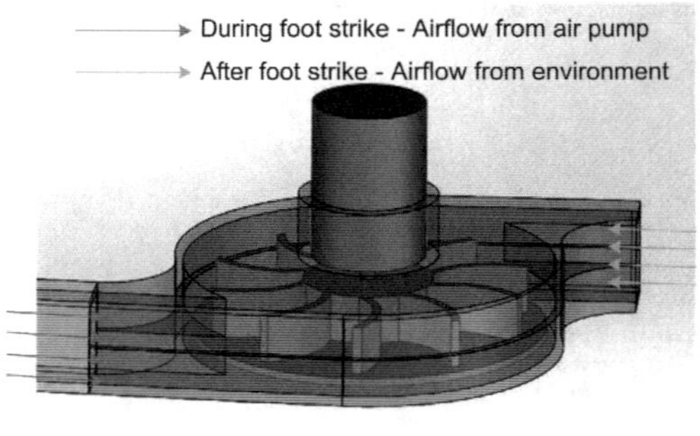

Figure 3. Design of the wind turbine with two generation cycles per foot strike.

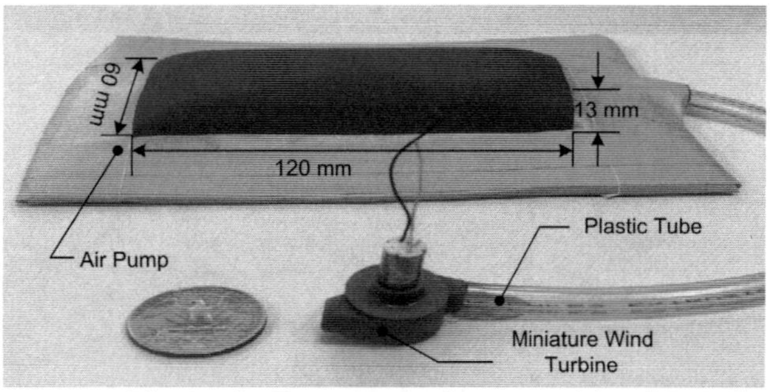

Figure 4. Experimental set-up with a 25¢ US coin as reference.

4. Results

The device was tested under different situations. Figure 5 shows the open-circuit output voltage during walking and running. In walking situations (1 Hz), the turbine rotates intermittently, whereas it can rotate continuously during running (4 Hz). The continuous motion also proves the successful implementation of the inlet and outlet design (Figure 3) for the uniform rotational direction of the turbine during and after each foot strike.

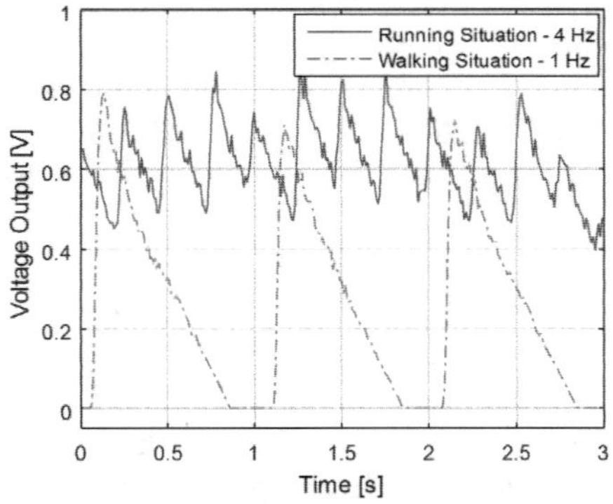

Figure 5. Open-circuit voltage output for different motion situations.

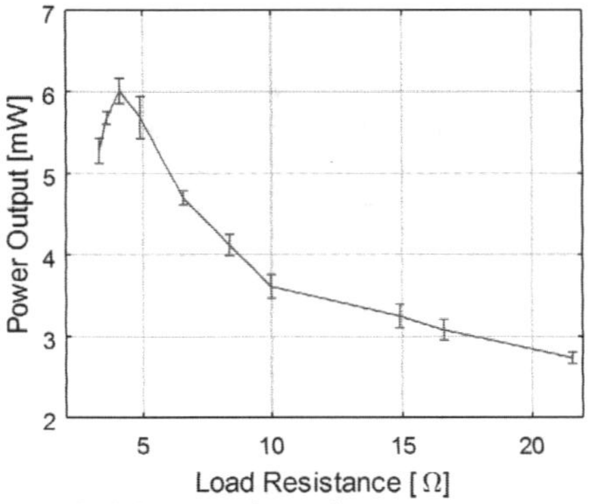

Figure 6. Measured peak power output for a single footstep versus different load resistances.

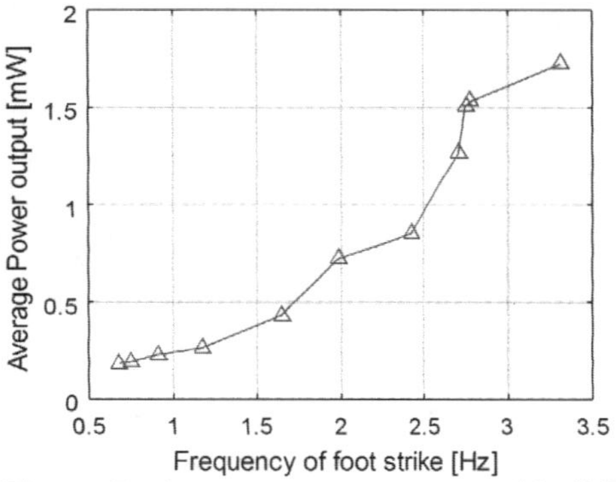

Figure 7. Average power output with different footstep frequencies.

Figure 6 illustrates the measured peak power output for a single footstep with varying load resistances. A 6 mW power output was obtained with a 4.9 Ω load which is identical to the motor's internal resistance. Figure 7 depicts the power output against footstep frequency. The average power output is on the mW scale, which is enough to power certain wireless sensors monitoring the condition of a human body. The device was finally embedded into a running shoe as shown in Figure 8. More tests will be carried out on a treadmill in future work.

Figure 8. An integrated shoe energy harvester.

5. Achievable Energy in Theory

In order to understand the potential achievable power from this mechanism, the upper limit and the theoretically achievable power of using this specific method are estimated analytically, showing the potential improvement of the power output and the unimpeded nature of the harvester to human motion due to the marginal energy extracted.

5.1. Upper Limit of Power Output

The upper limit is estimated without considering the specific operating mechanism. It is defined as the work done on an enclosed air pump during a foot strike. Assuming it is an isotropic process and the temperature, T, is constant, the energy needed to compress the air by a ratio of λ can be calculated using the following equation.

$$W = \int p dv = nRT \int dv/v = nRT \ln(\lambda) \tag{1}$$

where p is the air pressure, v is the volume of the air, n is the number of moles of gas present and R is the ideal gas constant.

Assuming that the striking happens in a gradual process, the momentum of human motion can be neglected. Therefore, for a person weighing m, the compressing ratio can be calculated using

$$\lambda = v/v_0 = p_0/p = \frac{p_0}{mg/A_a + p_0} = \frac{1\times10^5}{637/\left(7.2\times10^{-3}\right)+1\times10^5} = \frac{1}{1.87} \tag{2}$$

where m is the mass of a person, g is the gravitational acceleration, A_a is the sectional area of the air pump and p_0 is the initial pressure of the air in the air pump.

In the test, the weight of the person was 65 kg. The maximum power that can be obtained from each foot strike by the air pump for a compressing and releasing process is

$$W = nRT \ln\left(\frac{mg/A_a + p_0}{p_0}\right) = \frac{0.1}{22.4}\times 8.31\times 293.2\times \ln\left(\frac{1}{1.87}\right) = -6.55 \text{ J} \tag{3}$$

where the negative sign means the work done on the air.

For a walking frequency of 1 Hz, the maximum power obtained by the air pump is

$$P = W/T = 6.55\times 1 = 6.55 \text{ W} \tag{4}$$

The estimated upper limit of power output using this transduction is similar to the estimation in [7].

5.2. Theoretical Achievable Energy of This Mechanism

For this specific design, the compressed and accelerated airflow in the pump is used to drive an electromagnetic turbine. Assuming the airflow exiting the wind turbine at an increased speed with a modest pressure, the energy collected from the foot strike can be calculated from the kinetic energy of the airflow which is

$$E = \frac{1}{2} m_a v_a^{\ 2} \tag{5}$$

where m_a is the mass of the air and v_a is the velocity of the airflow exiting the turbine. In practice, wind turbines has a theoretical maximum power coefficient C_p of 0.593 [8]. According to the result from Figure 5, the working duration of each footstep on the air pump is around 0.1 s and the displacement of the air pump is about 10 mm. Hence, the velocity of the airflow entering the turbine can be up to 50 m/s considering the amplification factor (δ=500).

The energy obtained from each foot strike on the specific air pump mentioned above is

$$E = \frac{1}{2} C_p \cdot m_a \cdot (\delta \cdot v_0)^2 = \frac{1}{2} \cdot 0.593 \cdot (0.12 \times 0.06 \times 0.013 \times 1.29) \cdot (500 \cdot 0.1)^2 = 0.0895 \text{ J} \tag{6}$$

The theoretically achievable power from 1 Hz foot strikes is 89.5 mW. Therefore, the efficiency of the current design at 1 Hz is ~0.28% (0.25/89.5). It indicates that the comfort or feel of shoes is not affected due to the marginal energy extracted, and that the power output can be increased by an order of magnitude or more with appropriate design. Apparently, the theoretically achievable power highly depends on the amplification factor that is adjustable in the design process. Further investigation will be conducted to optimize the structure.

6. Conclusions

In this paper, an in-shoe energy harvester is developed to harness foot-strike energy. An air-pumped turbine system is designed to solve the difficulties of harvesting energy from foot strike, including the limited vertical distortion and the low motion frequency. An amplification mechanism for airflow speed is achieved by designing the air pump and turbine inlet with different cross-section areas. The turbine is activated by airflow and operates with a high frequency, enabling the foot-strike energy to be harvested with a higher frequency. The turbine casing is designed specifically to enable the device to operate continuously with airflow from both directions.

A prototype was fabricated and then tested under different situations. A 6 mW peak power output was obtained with a 4.9 Ω load. The achievable power from this design was estimated theoretically, showing the marginal energy extracted from human motion and the potential improvement of the power output. Future work will be on improving the efficiency of the system and exploring potential applications.

Acknowledgements

The work is supported by the China Scholarship Council (CSC), MIT-SUTD International Design Center and the Department of Electrical and Electrical Engineering, Imperial College. The authors would also like to thank the Imperial-MIT Global Fellows Programme for the travelling support.

References

[1] Mitcheson P D, Yeatman E M, Rao G K, Holmes A S and Green T C 2008 *Proceedings of the IEEE* **96** 1457-86
[2] C -Y Sue and N -C Tsai 2012 *Applied Energy* **93** 390-403
[3] Pillatsch P, Yeatman E M and Holmes A S 2014 *Wearable and Implantable Body Sensor Networks (BSN), 2014 11th International Conference on,* pp 6-10
[4] Paradiso J A and Starner T 2005 *Pervasive Computing, IEEE* **4** 18-27
[5] Hayashida J Y 2000 MIT Media Lab
[6] Moro L and Benasciutti D 2010 *Smart Materials and Structures* **19** 115011
[7] Starner T and Paradiso J A 2004 *Low-power Electronics Design* 1-35
[8] Briand D, Yeatman E M and Roundy S 2015 *Micro Energy Harvesting* Advanced Micro and Nanosystems (Wiley-VCH)

Optimization of a microfluidic based electromagnetic energy harvester for shoe insoles

M M Rahman[1], **R Atkin**[2] **and H Kim**[1, 2]
[1]ECE department, University of Utah, Utah, USA.
[2]SOLFIRE, Salt Lake City, Utah, USA.

E-mail: mahbubur.rahman@utah.edu

Abstract. This paper reports improved performance of the 4th generation microfluidic based energy harvester by finding global optimization among various geometric parameters, resulting in the increase of power density by 6.89 times. Specifically, the power output was optimized by varying diameters and spans of a coil at different frequencies. To verify the optimization, a custom testing platform was constructed, which mimicked the periodic linear movement caused by a human foot. The final device produced total power of 455.77mW from a volume of 20x3.74x0.75cm^3, resulting in a power density of 8.13mW/cm^3 that was identified as one of the highest power densities among human-body-induced vibration based energy harvesters.

1. Introduction

Despite remarkable progress in micro-scale energy harvesting devices, the conversion efficiency, in comparison to the enormous amounts of ambient energy available, still remains low, resulting in relatively low power output densities, i.e. 100μW/cm^3. Such low conversion efficiencies could be attributed to limitations in energy storing capability imposed by tiny dimensions in conventional micro-scale energy harvesting devices.

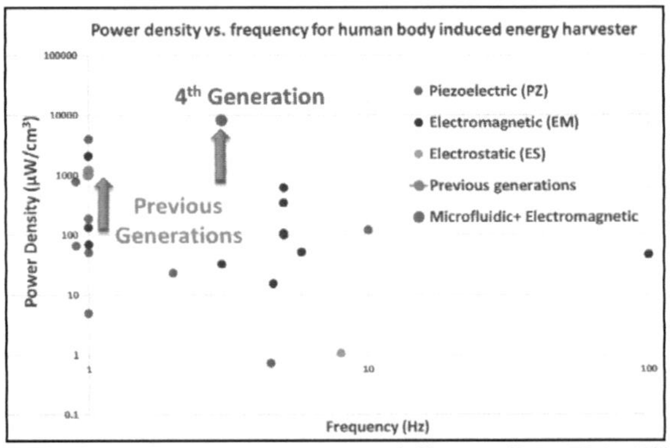

Figure 1: Comparison among different (PZ, EM, ES) human body induced vibration based energy harvesters and power densities compare to this and previous energy harvesters.

As one potential solution to enhance energy storing capability, a concept of microfluidic energy conversion was suggested in our previous prototypes [1, 2]. The concept intended to de-amplify the input displacement into a small device, thus enabling larger accommodation of ambient energy under given dimensions. Previously-reported hydraulic energy harvesters demonstrated the proof-of-concept results [1] and multiple-channel systems [2]; however, they produced only moderate power densities of 1.18mW/cm^3 [1] and 1mW/cm^3 [2] insufficient for numerous practical applications. This paper reports the enhancement of power densities of the 4th generation microfluidic-based energy harvester by optimizing design parameters, such as coil spans and diameters, to produce sufficient amounts of power density (8.13mW/cm^3) from foot movements.

2. Microfluidic Electromagnetic Energy Harvesting System

The microfluidic electromagnetic energy harvesting device mainly consists of two reciprocal diaphragms placed respectively at the heel and toes where maximum pressure is alternately exerted from human walk (Figure 2). This alternating pressure onto two diaphragms moves magnets between the chambers under diaphragms, producing electromagnetic energy in association with coils around the channels.

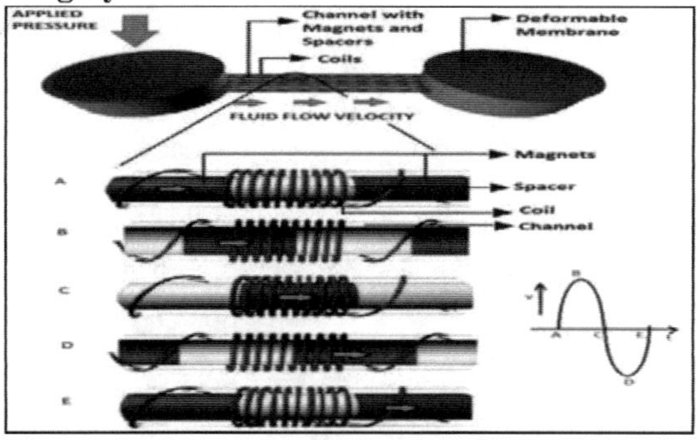

Figure 2: Structure of a microfluidic based electromagnetic energy harvesting device.

3. Design Optimization

In the microfluidic energy harvester, the key principle in energy generation is electromagnetic power generation. As a magnet passes through a surrounding coil due to hydraulic pressure, it induces alternating e.m.f. in the coil. Such a phenomenon can be theoretically modelled as below [6]:

$$P = \frac{M^2 A_g^2 \omega^2 B_L^2 R_L}{\left[\left(K(R_L + R_C) - M\omega^2(R_L + R_C)\right) - DL_C\omega^2\right]^2 + [D\omega(R_L + R_C) + (B_L^2\omega + KL_C\omega - ML_C\omega^3)]^2} \quad (1)$$

where power P is generated across matching load R_L for a system of magnet with mass M, electromagnetic coefficient B_L, spring constant K moving with acceleration A_g with diffusion coefficient D and actuation frequency ω through a surrounding coil with coil resistance R_c and inductance L_c. Diffusion coefficient could be assumed close to zero for the a fluidic system where magnets are moving without any friction.

To optimize the power generation, four primary parameters of coil turn numbers (N_c), acceleration (A_g), spacing of span (h_c), and coil diameters (D_c) were varied. First, the voltage output was theoretically computed by utilizing a unit magnet and a unit coil and then verified by COMSOL simulation (Figure 3). The simulation results showed that a permanent magnet of 1.48T moving into 40 turns of 140µm diameter coil produced the output peak-to-peak voltage of 32mV or a power density of 1.2675µW/cm^3 with the matching load of 0.9Ω. Next, the voltage output was maximized by varying the number of turns (N_c) and coil diameters (D_c). For a fixed span distance (h_c), the number of turns and coil diameters were traded-off. Then, the span distance as well as acceleration was varied while monitoring the power output changes.

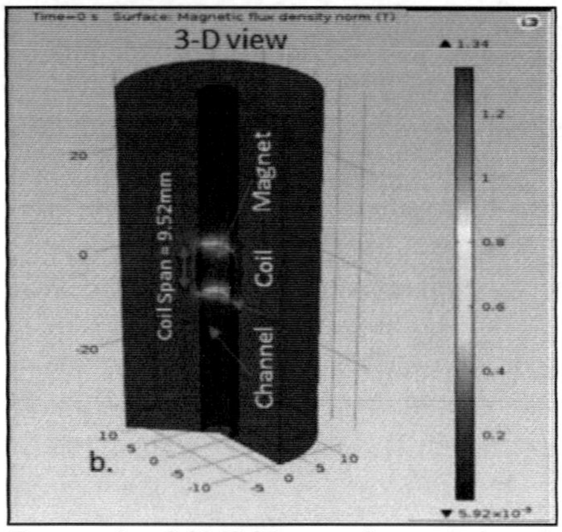

Figure 3: COMSOL simulation of coil power generation for a magnet moving through the channel surrounded by the coils and its magnetic flux leakage on coil.

4. Experimental Methodology

To realize the periodic acceleration caused by human foot motions, an acrylic energy harvester testing platform was constructed with a stepper motor as the main driving force (Figure 4). The platform consisted of a rotating crank arm and a rod that were precisely-aligned on top of a flat platform that housed multiple channels. The rotating crank connected the stepper motor to the rod by converting a rotational movement into a linear motion with variable acceleration. Figure 5 details the illustrations and the corresponding equations to relate angular speed of a motor with linear acceleration. Resultantly, simple control over the angular motor rotation was capable of mimicking a wide range of variable accelerations. For example, the rotation speeds of 54rpm (1Hz), 104rpm (2Hz), and 154rpm (3Hz) produced a maximum acceleration of 1.316ms^{-2}, 5.262ms^{-2}, and 11.83ms^{-2}, covering the conventional range of human foot acceleration during walking. Further variations were achieved by utilizing different arm lengths of the crank from 10mm to 30mm and different crank radii from 10mm to 20mm.

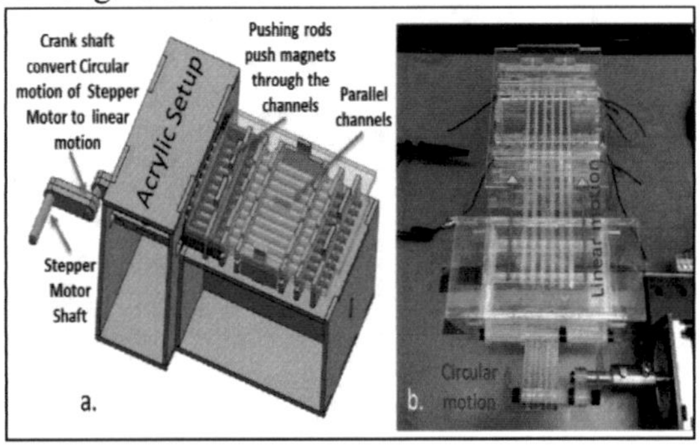

Figure 4: Device setup of a stepper motor converting the circular motion to linear motion by a crank shaft and using the push rod to move the magnets through the channel.

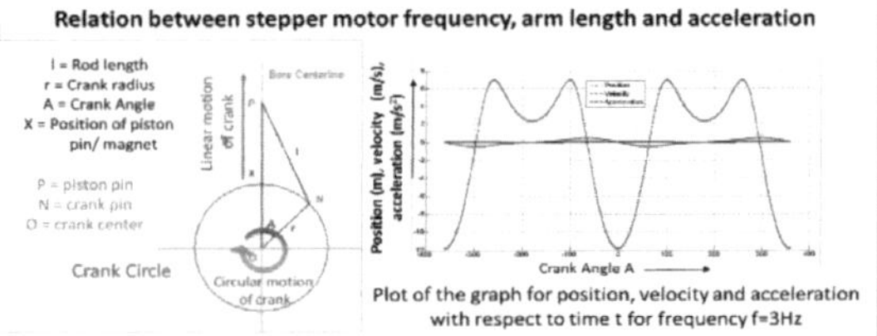

$$x = r\cos A + \sqrt{l^2 - r^2\sin^2 A}$$

$$x' = -r\sin A - \frac{r^2 \sin A\cos A}{\sqrt{l^2 - r^2\sin^2 A}}$$

$$x'' = -r\cos A - \frac{r^2(\cos^2 A - \sin^2 A)}{\sqrt{l^2 - r^2\sin^2 A}} - \frac{r^4 \sin^2 A\cos^2 A}{(\sqrt{l^2 - r^2\sin^2 A})^3}$$

Frequency f_0 (Hz)	Arm Length x or l (mm)	Acceleration $\lvert x\_max \rvert$ (ms^{-2})
1	30	-1.31 to 0.77
2	30	-5.26 to 3.10
3	30	-11.83 to 6.98

Figure 5: Motor frequency vs. linear acceleration of channel relationship and range of acceleration for a specific frequency and arm length of acrylic setup.

Utilizing the testing platform, the actuation frequency ω was varied from 1Hz to 3Hz by controlling the rotation speed per minute of the stepper motor for different arm distances (10mm and 20mm) and magnet/spacer combinations (1 magnet or 2 magnets/1 spacer), while the output voltage V_{p-p} and the power density for a single coil span of 9.52mm were monitored. Additionally, various coil diameters were tested with respectively physically-possible numbers of turns in the fixed span distance of 9.52mm with an actuation frequency of 3Hz: 63µm(151turns), 127µm(74turns), 160µm(59turns), 202µm(47turns), 255µm (37turns), 329µm(29turns) and 405µm(23turns). COMSOL simulation was also performed for verification.

5. Results

The COMSOL simulation (Fig.6) showed that (1) increasing actuation frequencies from 1 to 3 Hz linearly increased the power density by 9 times from 0.1237µW/cm^3 to 1.114µW/cm^3; (2) increasing arm lengths from 10mm to 20mm produced proportionally increase of 4.84 times in power density from 1.114 µW/cm^3 to 5.392µW/cm^3; and (3) doubled number of magnets from 1 to 2, as expected, increased power density by 4.83 times from 4.456 µW/cm^3 to 21.56 µW/cm^3. In combination, a total enhancement of 174.29 times was achieved. Such simulation results were closely matched to experiments as below.

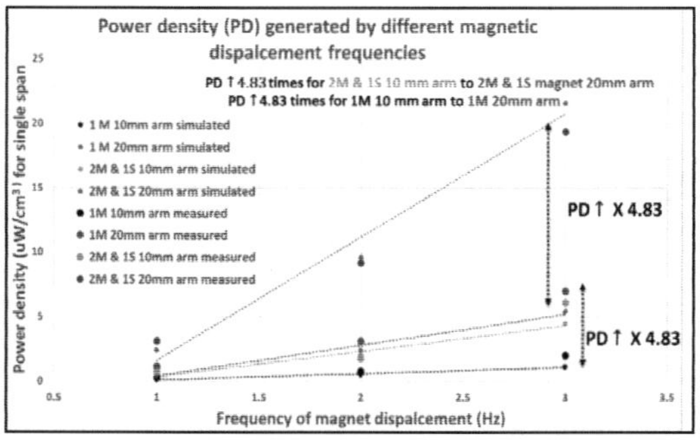

Figure 6: COMSOL simulation of power density for 4 different magnet and spacer combinations with different arm distances and frequencies to verify experimental data from device setup.

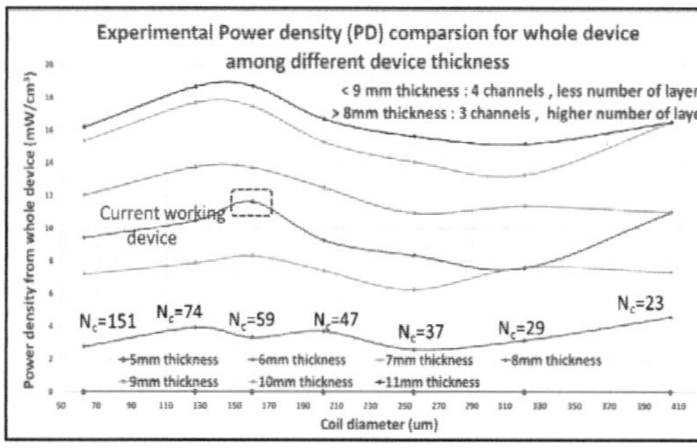

Figure 7: Optimization of power density by experiment for different coil diameters and their respective physically possible turn number in a fixed span (9.52mm) while also varying device thickness from 5mm to 11mm.

Experimental measurement showed that power output was optimized at different wire diameters depending on the restrictions on the device thicknesses. Figure 7 summarizes such optimization results. First, each diameter of coil determined the maximum discrete number of turns to fit in a given span length of 9.52 mm. For example, coil diameters of 63μm, 127μm, 160μm, 202μm, 255μm, 321μm and 405μm could be wound at 151, 74, 59, 47, 37, 29, and 23 times within a single span. Second, each diameter of coil also determined the layer numbers under the device thickness restrictions. For a device thickness of 8 mm, coil diameters of 63μm, 127μm, 160μm, 202μm, 255μm, 321μm and 405μm could form layers of 17, 8, 7, 5, 4, 3, and 3. Resultantly, the optimized power outputs were found at different coil diameters depending on the target device thicknesses. For a device thickness of 8mm (purple line), the optimized power density was measured at a coil diameter of 160μm as 11.6318mW/cm^3, while another device with a thickness of 6mm (red line), the maximum power density was measured at a coil diameter of 405μm as 4.5866mW/cm^3.

The optimized designs significantly improved the total power generation by 12.31 times experimentally, in comparison to previous generation (Fig.8). The final device produced total power of 455.77mW from a volume of 20x3.74x0.75cm^3, resulting in a power density of 8.13mW/cm^3 that was identified as one of the highest power densities among human-body-induced vibration based energy harvesters.

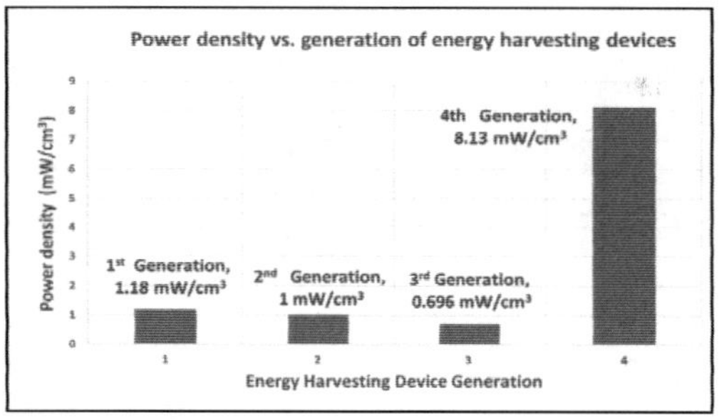

Figure 8: Power density comparison for different generation of energy harvesters.

6. Conclusion

The 4th generation microfluidic based electromagnetic energy harvester has improved the power density 6.89 times compare to the previous generations by enhancing power extraction with the accommodation of larger input force. The device was fabricated and evaluated in terms of power density with previously reported generations in figure 8. The optimized harvester produced total power of 455.77mW with a power density of 8.13mW/cm^3 from a volume of 20x3.74x0.75cm^3 resulting in 6.89 times enhancement in power density, compare to the previous generation devices.

7. References

[1] Bramhanand A and Kim H 2011 Micro Fluidic Energy harvesting system for high force and large deflection accommodation *PowerMEMS* 74-77
[2] Bramhanand A, Rahman M M, Bae Y and Kim H 2012 Hybrid Energy Harvesting System based on Microfluidics with interface circuitry for shoe-embedded application *Hilton Head*
[3] Kymissis J, Kendall C, Paradiso J and Gershenfeld N 1998 Parasitic power harvesting in shoe *ISWC* 132-139
[4] Buren T V, Mitcheson P D, Green T C, Yeatman E M, Holmes A S and Troster G 2006 Optimization of inertial micropower generators for human walking motion *IEEE Sens. J.* **6** 28- 38
[5] Saha C R, O'Donnell T, Wang N and P. McCloskey 2008 Electromagnetic generator for harvesting energy from human motion *Sensors and Actuators* **A 147** 248– 253
[6] Elvin N G, Elvin A A 2011 An experimentally validated electromagnetic energy harvester *Journal of Sound and Vibration* 330 2314–2324

Acknowledgments
The authors disclose that they have financial conflict of interests to Solefire. The authors would also like to thank Moses Noh for helping with the AutoCAD design and processing of the acrylic materials.

Hierarchical cellulose-derived carbon nanocomposites for electrostatic energy storage

V Kuzmenko[1,2], A M Saleem[1,3], A Bhaskar[1], H Staaf[1], V Desmaris[1,3] and P Enoksson[1,2,6]

[1] Department of Microtechnology and Nanoscience, Chalmers University of Technology, Kemivägen 9, Gothenburg, SE-41296, Sweden
[2] Wallenberg Wood Science Center, Chalmers University of Technology, Kemivägen 4, Gothenburg, SE-41296, Sweden
[3] Smoltek AB, Regnbågsgatan 3, Gothenburg, SE-41755, Sweden

E-mail: peter.enoksson@chalmers.se

Abstract. The problem of energy storage and its continuous delivery on demand needs new effective solutions. Supercapacitors are viewed as essential devices for solving this problem since they can quickly provide high power basically countless number of times. The performance of supercapacitors is mostly dependent on the properties of electrode materials used for electrostatic charge accumulation, i.e. energy storage. This study presents new sustainable cellulose-derived materials that can be used as electrodes for supercapacitors. Nanofibrous carbon nanofiber (CNF) mats were covered with vapor-grown carbon nanotubes (CNTs) in order to get composite CNF/CNT electrode material. The resulting composite material had significantly higher surface area and was much more conductive than pure CNF material. The performance of the CNF/CNT electrodes was evaluated by various analysis methods such as cyclic voltammetry, galvanostatic charge-discharge, electrochemical impedance spectroscopy and cyclic stability. The results showed that the cellulose-derived composite electrodes have fairly high values of specific capacitance and power density and can retain excellent performance over at least 2 000 cycles. Therefore it can be stated that sustainable cellulose-derived CNF/CNT composites are prospective materials for supercapacitor electrodes.

1. Introduction

The growing interest in long-lasting energy storage devices leads to the necessity of using more efficient carbon electrode materials [1]. Flexible sheets made of carbon nanofibers (CNFs) with their well-interconnected pores, high mechanical and electrochemical stability are prospective electrode materials for electrostatic energy storage devices such as supercapacitors. Their mesoporosity is of great importance, as it allows unrestricted ion diffusion into electrode cavities. Therefore, CNFs sustain a substantial uptake of electrolyte solution and enable high ion conductivity. In addition to that, the freestanding nature of CNF material allows using it without a polymeric binder [2, 3].

[6] To whom any correspondence should be addressed.

However, the main drawbacks of the CNF material are its relatively low specific surface area and electrical conductivity, which leads to low values of specific capacitance. Functionalization with chemical vapor deposited carbon nanotubes (cvdCNTs) should improve the mentioned flawed properties of CNFs making CNT/CNF composites more suitable for supercapacitors [4].

Previous studies have shown beneficial impacts of cvdCNTs on the electrochemical performance of carbon nanocomposite electrodes in supercapacitors [5-7]. Yet, composites in the aforementioned studies consisted of unsustainable components, while efficient use of carbonaceous electrodes derived from sustainable renewable resources is still very limited for supercapacitors. As demand on carbon nanostructures continues to grow, renewable resources should be accounted as an alternative to the currently most used CNF precursors: coal tar pitch and synthetic polymers. The biopolymer cellulose is a vast source that could be used for the synthesis of CNFs [8]. In this study, new cellulose-derived freestanding hierarchical carbon nanocomposite materials with prospective properties for electrostatic energy storage were evaluated as electrodes that potentially can be used for novel on-chip supercapacitors [9].

2. Experimental

2.1. Fabrication of composite electrodes

The composite CNF/CNT electrodes were produced via chemical vapor deposition of CNTs on top of cellulose-derived CNFs. Initially, CNF sheets were made by three consecutive steps of cellulose acetate electrospinning (17 wt% solution of the polymer in 2:1 solvent ratio of acetone and dimethylacetamide), cellulose regeneration (in 0.1 M water solution of NaOH) and carbonization (in a quartz tube furnace with N_2 flow by heating up to 800 °C with the heating rate of 5 °C/min) according to [8]. Subsequently, CNTs were thermally deposited on CNF substrates at 700 °C for 10 min using acetylene as a carbon source, 2 nm thick iron layer as a catalyst and hydrogen as a carrier gas (in AIXTRON Nanoinstruments Black Magic 2-inch machine).

2.2. Material characterization

The morphology of the composite materials was observed using high resolution scanning electron microscopy (SEM, Leo Ultra 55 FEG SEM, Zeiss) in a secondary electron mode at an acceleration voltage of 3 kV. The surface area of the materials was measured using the Brunauer-Emmett-Teller (BET) nitrogen adsorption method, and mesopore size distribution was quantified by the Barett-Joyner-Halenda (BJH) method using an adsorption isotherm (TriStar 3000 V6.04 A surface area and pore analyzer). The samples were degassed under vacuum at 225 °C for 4 h prior to the measurements. The electrical conductivity of the materials was evaluated using a four-point probe system (Parameter Analyzer-Keithley 4200-SCS).

2.3. Electrochemical analysis

Electrochemical performance was measured in a supercapacitor cell consisting of a symmetrical two-electrode system with the nanocomposites as working electrodes, electrospun cellulose as a separator, and 6 M aqueous solution of KOH as an electrolyte. The working electrodes and separators were cut to circular area of 0.5 cm^2 to fit the current collectors. Before starting the measurements the electrodes were immersed into the electrolyte solution for 24 h. Electrochemical measurements were performed with Gamry Reference 3000 potentiostat/galvanostat/ZRA and data were analyzed with Gamry Echem Analyst. A voltage range between □-0.5 V and 0.5 V was used for CV (cyclic voltammetry) measurements at five different scan rates (5, 10, 20, 100 and 200 mV s^{-1}). GCD (galvanostatic charge-discharge) tests were performed at four different current densities (0.75, 1, 1.5 and 2 A g^{-1}). EIS (electrochemical impedance spectroscopy) was done at open circuit potential with amplitude of 5 mV over a frequency range from 100 kHz to 10 mHz. An electrochemical stability tests were performed by cyclic charge-discharge (CCD) for 2 000 cycles with a current density of 1 A g^{-1}.

3. Results and discussion

3.1. Morphology and surface properties

CNF mats with a thickness of 25-40 mm were obtained via carbonization of electrospun cellulosic precursors. The mats consist of fibers with 50-250 nm diameters. The continuous fibers are randomly oriented and have smooth topography (figure 1a, bottom). The morphology of the composite material is rather different (figure 1a, top). After chemical vapor deposition the bigger CNFs were densely covered with much smaller CNTs (1-20 nm tube diameters) thus forming a hierarchical nanocomposite material.

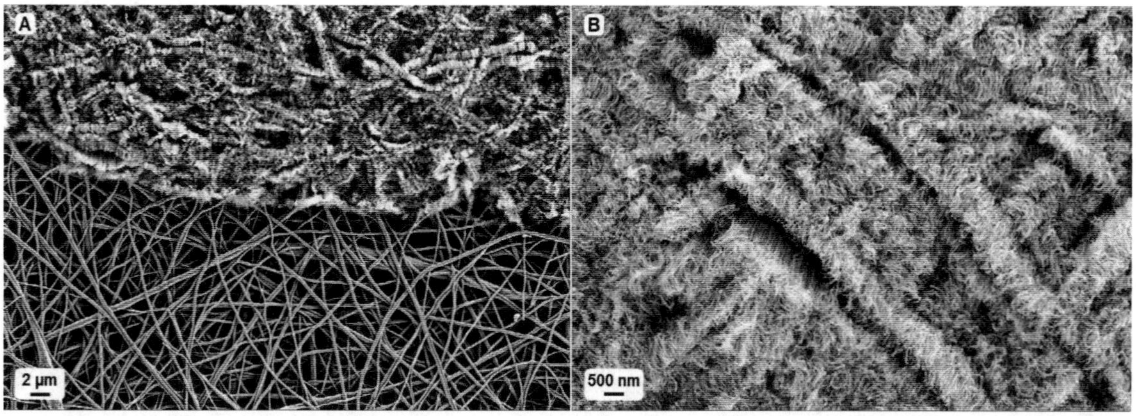

Figure 1. SEM images of the electrode material: A. The interface between pristine CNF region (bottom) and CNF region after deposition of CNTs (top); B. Deposited CNTs on top of CNFs.

Cellulose-derived CNF sheets have great mechanical and electrochemical stability along with valuable mesoporosity. On the one hand, mesoporous CNFs uptake the substantial amount of electrolyte ions and enable these ions to diffuse quickly in and out of electrode cavities [2], while on the other hand, much smaller CNTs increase surface area of the material, i.e. introduce microporosity (figure 2), and improve its electrical conductivity (table 1). These contributions from both components of the resulting composite material are favorable for supercapacitor high power and energy performance [4].

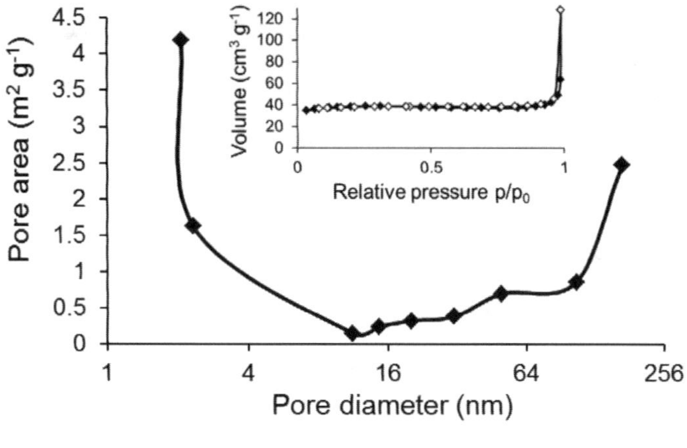

Figure 2. Nitrogen adsorption/desorption isotherms (inset) and pore size distribution (main image) of the electrode material.

Table 1. Properties of cellulose-derived carbon electrode materials.

Sample	Surface area ($m^2 g^{-1}$)	Micropore area ($m^2 g^{-1}$)	Electrical conductivity ($S cm^{-1}$)	Capacitance ($F g^{-1}$)	Energy density ($W h kg^{-1}$)	Power density ($W kg^{-1}$)	Capacitance retention (%)
CNF	45	32	4.2	12.8	0.31	432	88.7
CNF/CNT	131	103	69.4	22.9	0.99	2470	96.6

3.2. Electrochemical performance

Various electrochemical measurements were used to evaluate the performance and behavior of the electrode materials.

Figure 3A shows dependence of capacitance on a scan rate for the composite CNF/CNT electrode. CV curves have moderately rectangular shape, which indicates an electrical double layer (EDL) capacitive behavior [10]. Distortion of the rectangular shape at higher scan rates happens because of the lack of time for electrolyte ions to penetrate completely inside the electrode, whereas for lower scan rate the ions get to the inner surface of electrode providing higher accumulative charge [2]. The comparison of the CV curves of two different CNF-based nanostructured materials is presented in figure 3B. It is clear seen that the CNF/CNT composite electrodes show higher values of specific capacitance in comparison to the pure CNF electrodes.

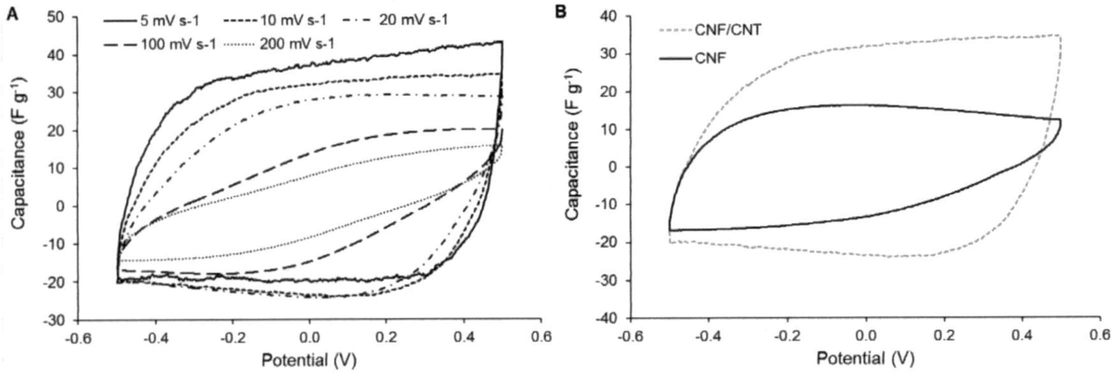

Figure 3. A. CV curves of the CNF/CNT composite electrode at different scan rates. B. CV curves of the composite and pure CNF electrodes at 10 mV s^{-1} scan rate.

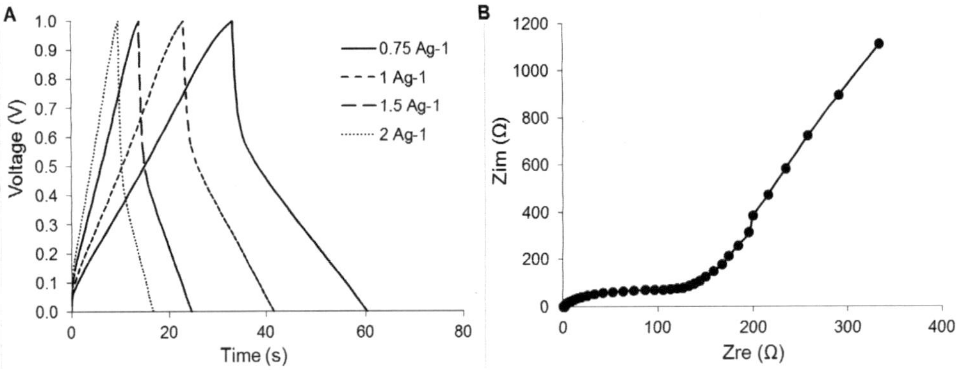

Figure 4. A. GCD curves of the electrode material at different current densities. B. Nyquist impedance plot of the electrode material.

GCD curves show almost symmetrical shape specifying EDL behavior as well (figure 4A). From GCD test power and energy density values were found to be reasonably higher for the composite electrode materials. Fast current-voltage response proves that electrode material have high power density [11].

According to EIS equivalent series resistance of the system is defined as an initial intercept of the plot with the X-axis in the high frequency region (figure 4B). Resistance value is only 0.57 Ohm, which is beneficial for effective performance of a supercapacitor. The verticality of the Warburg line at medium frequency validates sufficient pore accessibility for electrolyte ion diffusion [6].

Moreover, the electrodes retained 96.6% of the initial capacity after 2 000 charge-discharge cycles, which is a very good stability for an energy storage device such as supercapacitor as it has to deliver the harvested energy through quick charging and discharging many times [12].

4. Conclusions

Overall, functionalization of CNF led to the composite materials with higher capacitance values. This positive effect can be explained by few factors. First of all, the deposition of CNTs on top of the CNF substrate improved electrical conductivity (electrode's ability to transfer charges), while CNTs also increased surface area (electrode's ability to uptake electrolyte and accumulate charges). Hence, the hierarchical cellulose-derived CNF/CNT nanocomposites showed their great potential as electrode materials for electrostatic energy storage.

Acknowledgments

The Wallenberg Wood Science Center funded by Knut and Alice Wallenberg Foundation, the Vinnova CarPolCap and the EU Smart-MEMPHIS projects are greatly acknowledged for their financial support.

5. References

[1] Béguin F, Presser V, Balducci A and Frackowiak E 2014 *Adv. Mater.* **26** 2219–51
[2] Gao K, Shao Z, Li J, Wang X, Peng X, Wang W and Wang F 2013 *J. Mater. Chem.* A **1** 63–7
[3] Fang J, Niu H T, Lin T and Wang X G 2008 *Chin. Sci. Bull.* **53** 2265–86
[4] Deng L, Young R J, Kinloch I A, Abdelkader A M, Holmes S M, De Haro-Del Rio D A and Eichhorn S J 2013 *ACS Appl. Mater. Interfaces* **5** 9983–90
[5] Zhou Z, Wu X F and Fong H 2012 *Appl. Phys. Lett.* **100** 023115
[6] Gao Y, Pandey G P, Turner J, Westgate C R, Sammakia B 2012 *Nanoscale Res. Lett.* **7** 651–8
[7] Yang Z Y, Zhao Y F, Xiao Q Q, Zhang Y X, Jing L, Yan Y M and Sun K N 2014 *ACS Appl. Mater. Interfaces* **6** 8497–504
[8] Kuzmenko V, Naboka O, Gatenholm P and Enoksson P 2014 *Carbon* **67** 694–703
[9] Saleem A M, Göransson G, Desmaris V and Enoksson P 2015 *Solid-State Electron.* **107** 15–9
[10] Madhu R, Sankar K V, Chen S M and Selvan R K 2014 *RSC Adv.* **4** 1225–33
[11] Kuzmenko V, Naboka O, Staaf H, Haque M, Göransson G, Lundgren P, Gatenholm P and Enoksson P 2015 *Mater. Chem. Phys.* **160** 59–65
[12] Davies A and Yu A 2011 *Can. J. Chem. Eng.* **89** 1342–57

A liquid-activated textile battery for wearable biosensors

X Liu[1] and P B Lillehoj[1]

Department of Mechanical Engineering, Michigan State University, East Lansing, MI, USA

E-mail: lillehoj@egr.msu.edu

Abstract. Wearable biosensors is an emerging field which offers great potential for many applications including human health monitoring, environmental sensing and bioagent detection. An important requirement for these systems is the need for robust, lightweight batteries which can be easily integrated with wearable materials (i.e. textiles). In this paper, we demonstrate a new textile Ag-Al battery which is activated by liquids. Experiments were performed to optimize the battery performance by altering the device parameters, such as the electrode/electrolyte area and electrolyte concentration. Based on these results, we developed a single-cell battery that can produce a voltage of 1.3 V. By connecting two cells in series, we could power a 1.6 V LED for up to 30 minutes.

1. Introduction

In the past decade, the field of wearable sensors has made substantial progress [1]. Much of the work in this area is focused on human health applications, mainly, monitoring physiological parameters such as heart rate [2], respiratory rate [3] skin temperature [4], and blood pressure [5]. Recently, the concept of 'wear and forget' [6] was introduced which aims to design wearable sensors that don't compromise user comfort or distract them from routine activities. Toward this end, researchers have been developing sensors fabricated directly onto flexible substrates including plastic films [7], wearable textiles [8] and even temporary tattoos [9]. While these devices demonstrate the feasibly of creating flexible sensors, a major challenge that remains is powering these systems, which is currently achieved using batteries.

Conventional portable batteries (NiCd, NiMH, Li-ion) typically consist of metallic electrodes, separators and electrolytes housed within a metal or aluminium casing, and these rigid components hinder their integration with flexible materials. Hence, new materials, such as carbon nanotube/polyaniline composites [10] and carbon nanotube films [11], have been explored for the development of flexible energy storage devices. Flexible batteries based on textiles have also been demonstrated which utilize solid polymer electrolytic layers [12, 13]. While promising, these batteries still pose limitations for wearable biosensor systems since they require additional circuitry for start/stop operation and are not suitable for aqueous conditions.

Alternatively, liquid-activated paper batteries have been demonstrated [14, 15] where the application of an aqueous solution causes the device to turn 'on'. This approach enables on-demand power generation and can simplify overall device design. These batteries could provide voltages of 1.21 V with a 1 kΩ load and 1.3 V with an open circuit, respectively. To improve the durability and integration with wearable materials, we demonstrate a new liquid-activated textile Ag-Al battery. Our textile battery can be fabricated using inexpensive materials and methods, which can significantly lower overall device costs. Additionally, this battery can be activated using different types of liquids, broadening its versatility for various biosensing applications.

2. Materials and Methods

2.1. Designing the battery

The electrochemical cell consists of three layers of cotton/polyester textile sandwiched between two metal foil electrodes (Ag and Al), which serve as the anode and cathode (Figure 1). Two layers of textile are impregnated with silver nitrate ($AgNO_3$) and aluminum chloride ($AlCl_3$) respectively, to serve as dry electrolytes, and one layer is impregnated with sodium nitrate ($NaNO_3$) to serve as a salt bridge. The cell is sheathed between two additional layers of fabric, which are held together using double-sided tape. When liquid is applied to the cell, the dry electrolytes become hydrated and a electrochemical reaction is generated. The chemical reaction for the battery at the anode (oxidation) and cathode (reduction) are represented by equations (1) and (2), respectively:

$$Ag \rightarrow Ag^{+} + e^{-} \quad (1)$$

$$Al^{3+} + 3e^{-} \rightarrow Al \quad (2)$$

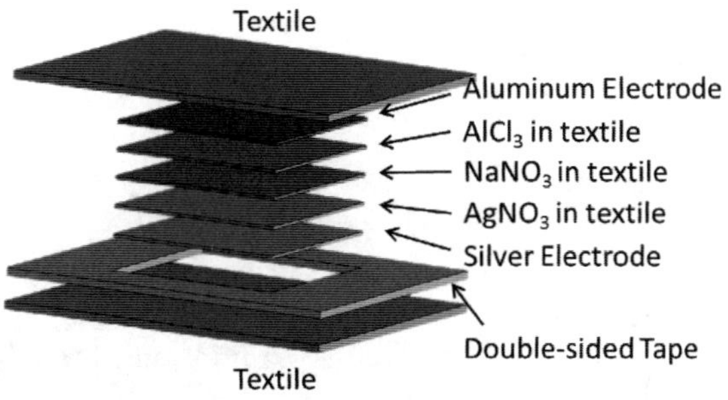

Figure 1. Schematic illustration of the textile battery.

2.2. Materials

Al foil, Ag foil, $AlCl_3$, $NaNO_3$ and $AgNO_3$ were purchased from Sigma-Aldrich (St. Louis, MO). Cotton/polyester fabric was purchased from Jo-Ann Fabrics (Hudson, OH) and double-sided tape was obtained from Intertape Polymer Group (Marysville, MI). Deionized (DI) water was generated using a Barnstead Smart2Pure water purification system and phosphate buffered saline (PBS) was purchased from Thermo Scientific.

2.3. Fabricating the battery

Ag and Al electrodes, cotton/polyester fabric and double-sided tape were cut into 6 mm × 6 mm pieces. A square hole was cut in the double-sided tape to provide access for the activation liquid. The fabric pieces were soaked in separate solutions of $AlCl_3$, $NaNO_3$ and $AgNO_3$ for 1 minute and air dried overnight at room temperature. The fabric and electrodes were stacked together, as shown in Figure 1, and sheathed between two 18 mm × 12 mm pieces of fabric. Prior to assembly, the metal electrodes were sanded to remove any surface oxidation.

2.4. Experimental procedures

For electrical measurements, wires were soldered onto the anode and cathode of the electrochemical cell and connected to a Fluke 87-V digital multimeter using alligator clips. To activate the cell, 20 μL of liquid was applied to the middle of the cell using a pipette. Measurements were performed after ~30 seconds, which was sufficient time for the liquid to fully hydrate the dry electrodes and activate the battery. The data was averaged over 3 individual measurements and plotted as the mean ± standard deviation (SD).

3. Results and Discussions

Experiments were carried out to optimize the performance of the battery by studying the influence of various battery parameters on the maximum output current. For all of these studies, the electrode area was fixed at 36 mm^2 and the salt bridge was fabricated from fully saturated $NaNO_3$ solutions.

3.1. Influence of electrode/electrolyte area on battery performance

As shown in Figure 2, we observed a linear correlation between the electrolyte area and the short circuit current for electrolyte areas from 18 mm^2 to 209 mm^2. For electrolyte areas larger than 36 mm^2, there was no significance increase in the current. These results suggest that the electroactivity of the cell is affected by the ratio of the electrode/electrolyte surface area where a ratio of ≤ 1 is needed to maximize the electrochemical reactions occurring between electrodes and the electrolytes. For these studies, the concentration of $AgNO_3$ was 3 M.

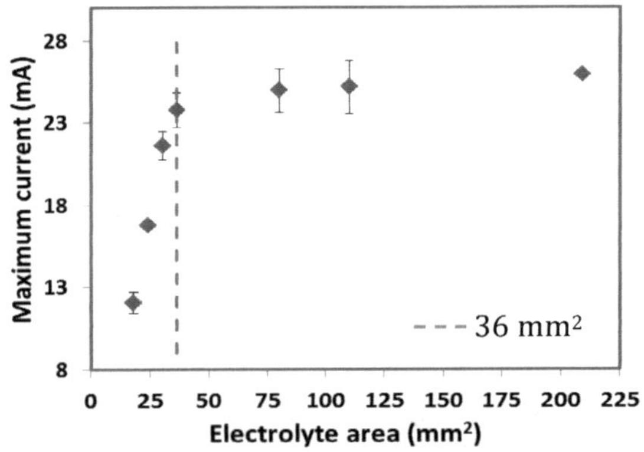

Figure 2. Maximum electrical current as a function of electrolyte area for a single cell battery activated using DI water. The electrode area is 36 mm^2 and the $AlCl_3$ electrolyte concentration is 1M. Each data point represents the mean ± SD of three measurements.

3.2. Influence of electrolyte concentration on battery performance

Experiments were also carried out to in investigate the effect of the electrolyte concentration on the short circuit current of the electrochemical cell. For these studies, the concentrations of $AgNO_3$ and $AlCl_3$ were fixed at a 3:1 ratio to balance the reaction rate (during the reaction, Ag loses one electron while Al gains three electrons) and maximize the efficiency of the half-reaction. As shown in Figure 3, our results show a linear correlation between the electrical current and the concentration of $AlCl_3$ from 0.5 M to 1.5 M. For $AlCl_3$ concentrations above 1.5 M, the output current levels off, which could be due to the saturation of the electrolyte species. This could also be attributed to issues we encountered when fabricating the $AgNO_3$ dry electrolyte where $AgNO_3$ solutions at concentrations > 4.5 M became very viscous, limiting its abilitv to be absorbed into the textile.

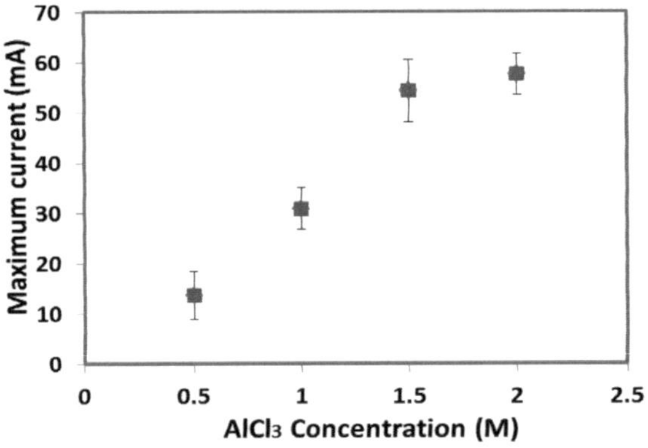

Figure 3. Maximum electrical current as a function of $AlCl_3$ concentration for a single cell battery activated using DI water. The electrode and electrolyte areas are 36 mm^2. Each data point represents the mean ± SD of three measurements.

3.3. Influence of liquid conductivity on battery performance

The impact of liquid conductivity on the battery performance was briefly studied by activating the battery using three different types of liquids: DI water (0.0055 mS/m), tap water (5-50 mS/m), and PBS (~1,600 mS/m). As shown in Figure 4, no noticeable difference in the output current was observed for these three types of liquids. These results indicate that the liquid conductivity has a minimal effect on the performance of the battery, suggesting that this platform could be useful for various biosensing applications which utilize different types of aqueous samples.

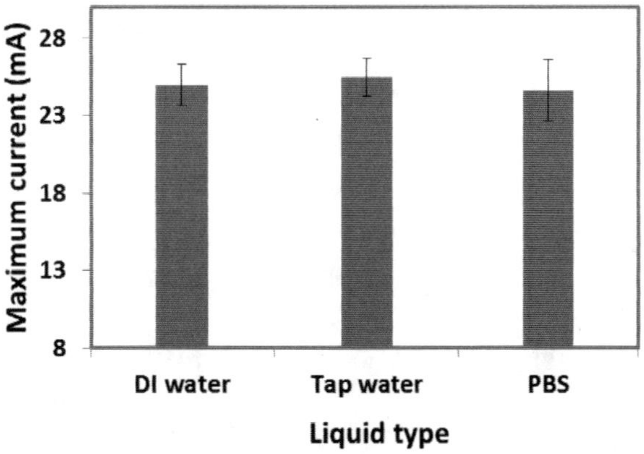

Figure 4. Maximum current from a single cell battery activated using different types of liquids. The electrode and electrolyte areas are 36 mm^2. Each bar represents the mean ± SD of three measurements.

Based on the optimized parameters obtained from the studies above, our single cell battery can produce a maximum open circuit voltage of 1.3 V. A unique feature of our liquid-activated battery is that it enables on-demand power generation where current is generated only when the electrolyte is hydrated. When the liquid has fully evaporated, the chemical reactions within the cell cease and the battery shuts 'off'. The battery can simply be reactivated (at 1.3 V), by applying more liquid to the cell. This process can be repeated several times which demonstrates the multiuse capability of this platform. To demonstrate the functionality of this technology, we connected two cells in series and used it to power a 1.6 V LED, which could remain lit for up to 30 minutes (Figure 5). Typically, biosensing measurements last tens of seconds to a couple of minutes [16], therefore, our textile battery can provide enough power for multiple bioassay measurements or continuous temporal sampling over long durations.

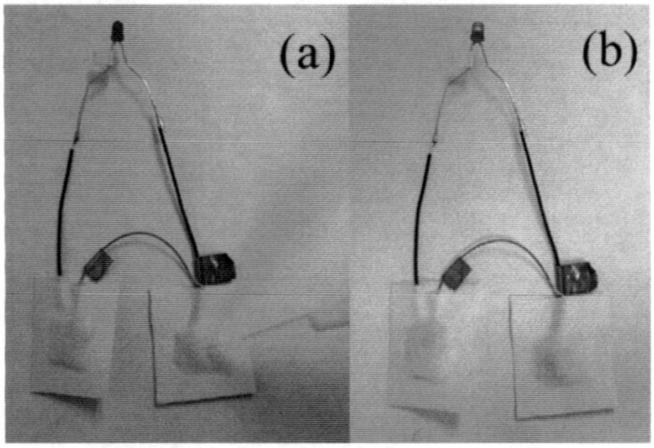

Figure 5. A two-cell textile battery connected to a LED before (a) and after (b) activation using 20 μL of DI water.

4. Conclusions

In this paper, we introduce a new liquid-activated Ag-Al textile battery which can produce a maximum open circuit voltage of 1.3 V. By connecting two cells in series, this battery could power a 1.6 V LED for up to 30 minutes. Studies to optimize the battery revealed that its performance is influenced by the electrode/electrolyte area and the electrolyte concentration, but not by the conductivity of the activation liquid. One of the unique features of this textile battery is its mode of activation, which is initiated by simply applying a liquid sample to the cell. This approach enables on-demand power supply where the battery turns 'on' only when a liquid sample is applied, and turns 'off' when the liquid has fully evaporated. This capability makes our textile battery suitable for various biosensing applications requiring multiple measurements or continuous temporal sampling over long periods of time. Furthermore, this battery can be easily integrated with sensors fabricated on textiles and fabrics, thereby minimizing the overall size, weight and complexity of wearable sensor systems.

Acknowledgments
This work was support by the National Science Foundation CAREER award (ECCS-135056).

References

[1] Gaikwad A M, Arias A C and Steingart D A 2015 *Energy Technology* **3** 305-28
[2] Paradiso R, Loriga G and Taccini N 2005 *IEEE Trans Inf Technol in Biomed* **9.3** 337-44
[3] Johnston W S and Mendelson Y 2004 *Conf Proc IEEE Eng Med Biol Soc* **2** 5388-91
[4] Jung S, Taeksoo J and Vijay K V 2006 *Smart Mater Struct* **15** 1872-76
[5] Teng X F, Zhang Y T, Poon C C Y and Bonato P 2008 *IEEE Rev* **1** 62-74
[6] Tharion W J, Buller M J, Karis A J and Mullen S P 2007 *Proc Hum Fact Ergon Soc Annu Meet* **51** 1006-10
[7] Cao Q, Kim H, Pimparkar N, Kulkarni J P, Wang C, Shim M, Roy K, Alam M A and Rogers J A 2008 *Nature* **454** 495-500
[8] Yang Y L, Chuang M C, Lou S and Wang J 2010 *Analyst* **135** 1230-34
[9] Jia W, Bandodkar A J, Ramirez G V, Windmiller J R, Yang Z, Ramirez J, Chan G and Wang J 2013 *Anal Chem* **85** 6553-60
[10] Meng C, Liu C, Chen L, Hu C and Fan S 2010 *Nano Lett* **10** 4025-31
[11] Hu L, Wu H, Mantia F L, Yang Y and Cui Y 2010 *ACS Nano* **4** 5843-48
[12] Wei D, Cotton D and Ryhänen T 2012 *Nanomaterials* **2** 268-74
[13] Bhattacharya R, Kok M M D and Zhou J 2009 *Appl Phys Lett* **95** 223305
[14] Lee K B 2005 *J Micromech Microeng* **15** S210-14
[15] Thom N K, Yeung K, Pillion M B and Phillips S T 2012 *Lab Chip* **12** 1768-70
[16] Windmiller J R and Wang J 2013 *Electroanalysis* **25** 29-46

Design, Fabrication, and Testing of Silicon-integrated Li-ion Secondary Micro Batteries with Interdigital Electrodes

K Hoeppner[1,3], M Ferch[1], A Froebe[2], R Gernhardt[2], R Hahn[2], P Mackowiak[1], B Mukhopadhyay[2], S Roder[2], I Saalhofen[2], and K-D Lang[1,2]

[1]Berlin Institute of Technology, Research Center of Microperipheric Technologies, Gustav-Meyer-Allee 25, 13355 Berlin, Germany
[2]Fraunhofer IZM, Gustav-Meyer-Allee 25, 13355 Berlin, Germany

katrin.hoeppner@tu-berlin.de

Abstract. This paper reports the design, fabrication, and testing of silicon-integrated lithium ion secondary micro batteries with a side-by-side electrode setup. Cavities separated by narrow silicon spacers served as containments for two interdigitally arranged electrodes and were etched into <110>-Si by wet chemical etching. The etched silicon battery containments were passivated by a layer of SiO_x/Si_xN_y. Al current collectors were applied by sputtering and back etching. A volumetric micro dispenser served to fill the cavities with slurries of the active materials - lithium cobalt manganese oxide ($Li_y(Ni_{1/2}Co_{1/5}Mn_{3/10})O_2$) as the cathode and lithium titanate ($Li_4Ti_5O_{12}$) as the anode material. Filling with electrolyte, encapsulation, and electrochemical characterization of the finished cells took place in an Ar-filled glove box. The fabricated batteries with IDE show considerably lower impedances than cells with single side by side electrodes and are capable of constant current loads up to 10 C. A linear capacity loss rate of <0.1 % per cycle was observed over 30 full cycles at 0.2C.

1. Introduction

The fabrication of silicon integrated micro batteries appears to be of interest for applications where a small volume, high capacity energy supply is needed. Examples are small energy autarkic sensors, e.g. implantable sensors for medical applications [1][2], miniaturized sensor networks [3], active smart labels, or MEMS. Such applications cannot be supplied by state of the art solid state batteries due to their limited specific capacity per area which usually accounts for values of around 100 μAh/cm² only [1][3]. As the size of the battery decreases, the amount of passive packaging material in proportion to active material will increase. For very small systems this is reducing the energy density significantly. Therefore, the direct integration of the battery into the chip or package is crucial for highly miniaturized applications.

2. Design

Previously, we investigated a battery design with two electrodes lying side by side, separated by a silicon spacer (figure 1a) [4]. As compared to a conventional setup with stacked electrodes [5], such a setup facilitates assembly and encapsulation considerably since no polymer separator is needed and

[3] To whom any correspondence should be addressed.

both electrodes can be contacted from downside. However, because of the comparatively high cell impedance due to the long paths of Li-ion movement, the fabricated cells have shown a rate capability that was limited to < 5C. In this study we therefore developed the side-by-side electrode concept further to a design with narrower and interdigitally arranged electrodes (figure 1b).

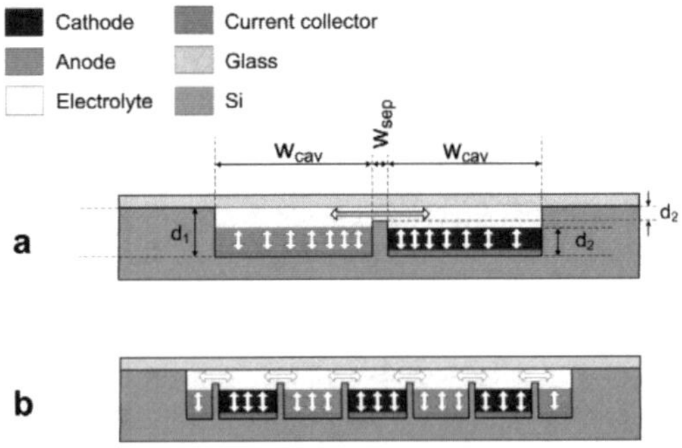

Figure 1. Schematic profiles of battery cells with a side-by-side electrode setup **a**: previous design, **b**: design with IDE (arrows indicate paths of Li-ion movement)

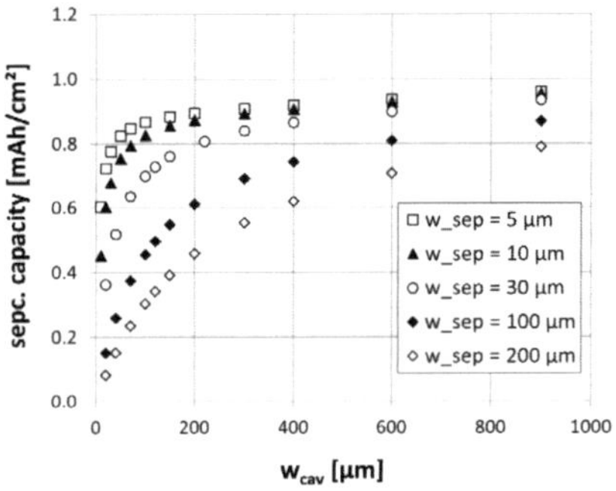

Figure 2. Specific capacity of microbatteries with IDE as function w_{cav} and w_{sep} (d_3~160 µm; reference area is footprint of IDE without package; capacities extrapolated from practically achieved values)

In an IDE setup the width of the separator (w_{sep}) must be minimized to limit losses in specific capacity per area, especially if the cavities are very narrow (figure 2). However, since the active electrode materials undergo volume changes during cycling, the feasible aspect ratio of w_{sep} : (d_1-d_2) is limited to provide for sufficient mechanical stability. Furthermore, the dimension of w_{sep} and d_2 determine the shape of the separator tip, which in turn influences the process of dispensing the electrode materials. Separators that have flattened tips are beneficial for a reliable separation of the electrodes (see section 3.2). Based on these findings and a preliminary investigation on the etch behavior at the separator (see section 3.3), we chose to fabricate IDE with cavities of 600 and 900 µm in width and 100 µm wide silicon spacers. The clearance above the silicon spacer was set to 80 µm.

3. Fabrication

3.1. Fabrication of the electrode containments

Silicon chips structured as shown in figure 1b were fabricated from a p(B) <110>-silicon substrate by applying two steps of wet chemical etching. In the first etching step, two cavities separated by a Si-spacer of full height are formed, whereas the second etching step gives rise to a clearance above the Si-spacer as well as a tapering of its edges. The substrates were passivated by a double layer of SiO_x/Si_xN_y. A 1 µm thin film of Al was applied by sputtering and structured by back etching to serve as current collector for the cathode and the anode, respectively. A more detailed description of the single processing steps can be found in [4]. Figure 3a shows electrode containments ready for dispensing the active electrode materials.

3.2. Deposition of active materials and encapsulation

The active materials for cathode and anode were prepared from slurries of commercially available powders. Lithium cobalt manganese oxide ($Li_y(Ni_{1/2}Co_{1/5}Mn_{3/10})O_2$, NCM) was used as the cathode and lithium titanate ($Li_4Ti_5O_{12}$, LTO) as the anode active material (both from MTI Corp.), polyvinylidene difluoride (PVDF) as a binder, N-methyl-2-pyrrolidone (NMP) as a solvent, and carbon black as a conductive agent. The slurries were applied to the cavities by two proceeding steps of volumetric micro dispensing (Ultimus IV, Nordson) and the solvent removed by appropriate heat treatment. Depending on their individual consistency, the electrode slurries are likely to ascend the walls of the etched cavities due to capillary action during dispensing, especially at the slanted edges of the tips of the silicon spacers that feature a rough, disintegrated surface. The separation of the electrode slurries is mainly based on edge effects. Thus, for a reliable separation of the electrodes, separators that have flattened tips (four edges) are necessary (figure 4).

The battery containments with the applied electrodes were introduced into an Ar-filled glove box and sealed and filled with electrolyte as described in [4]. The bond interface adds up some 20 µm to the final dimension of d_2. We used 1M $LiPF_6$ dissolved in a mixture of ethylene carbonate and dimethylcarbonate (EC:DMC, 1:1wt%) as the electrolyte.

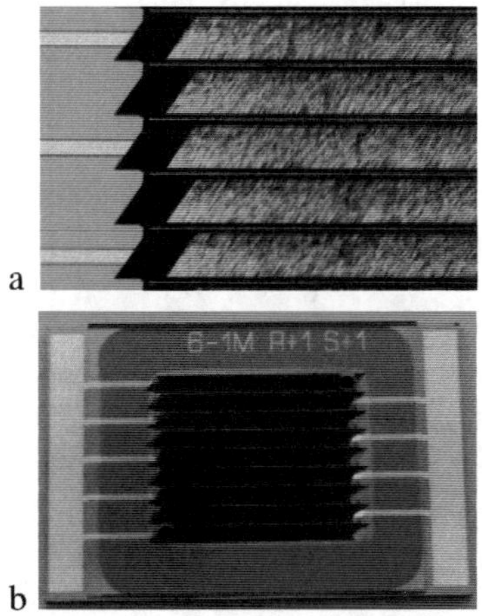

Figure 3. a: Electrode containment (w_{cav} 600 µm, w_{sep} 100 µm; d_1 300 µm, d_2 80 µm) **b:** after encapsulation (chip size 10x15 mm²).

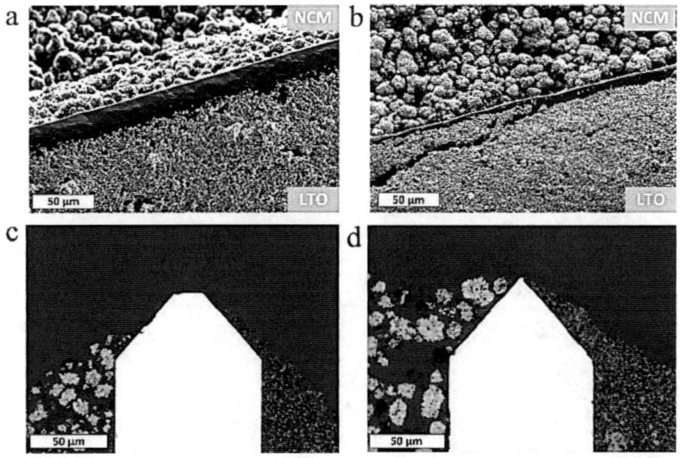

Figure 4. Separator region of cells with applied electrodes and w_{sep} = 100 µm (top: SEM picture, bottom: cross section polish). **a,c:** separator with flattened tip ($w_{plateau}$=20 µm). **b,d:** sharp separator tip.

3.3. Deduction of design rules for shaping the separator tip

As discussed above, it is crucial to precisely control the shape of the separator. The unmasked exposure of convex edges to KOH etching solution is known to result in a rounding of these edges. We simulated and observed this effect in our previous study [4]. The sloped edges at the separator tip form an angle α of around 40° with the vertical {111} side walls of the cavities. Depending on the etch time, there is a critical minimum separator width below which these sloped edges will run into each other and separator tips with acute angles be formed. This process results in a growth of the clearance d_2 above the separator tip.

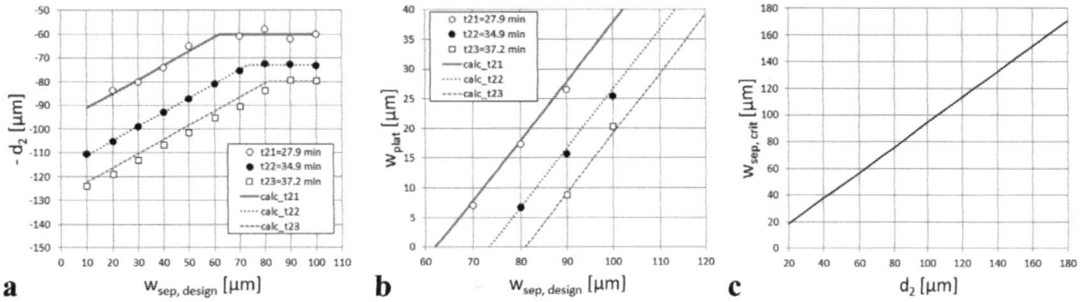

Figure 5. Etching characteristics at the separator tip. **a:** dimension of clearance d_2 as function of separator width w_{sep} and duration t_2 of the second KOH etch step **b:** width of the plateau at the separator tip (w_{plat}) as function of w_{sep} (markers: measurement, lines: calculation using equation 1 and an etch rate of 0.77 µm/min at the separator edges). **c:** critical dimension of w_{sep} as function of d_2.

We measured d_2 in dependence on w_{sep} and the duration of the second KOH etch step by using a profilometer. For wider separators, a plateau parallel to the wafer surface is still visible on the topside of the separator. In this case, the dimension of d_2 is consistent with the expected etch rate of <110>-Si. For smaller w_{sep}, however, d_2 was found to nearly linearly increase with a decrease in w_{sep} (figure 5a).

The orientation of the tapered edges at the separator tips does not correspond to a specific high-index crystal plane. However, a macroscopic etch rate for an etch attack perpendicular to the disintegrated slopes at the separator edges could be determined using geometrical data of separators with a distinct plateau, as given in figure 5b. We found this rate to account for some 0.77 µm/min for our etching conditions (80°C, 33% KOH). Using this rate and the observed angle α the slanted edges of the separator form with the {111} planes of the cavity side walls, a theoretically expected clearance d_2 can be calculated according to equation (1).

$$d_2 = \tan(90-\alpha)\ \frac{w_{sep}}{2} - \frac{d_{etch\perp slope}}{\sin(90-\alpha)} \qquad (1)$$

The output of equation 1 is in very good agreement with the measured data for the first two immersion times (t_{21}, t_{22}) during etch step 2. For the longest time, t_{23}, however, we measured slightly higher values of d_2 than expected by calculation (figure 5). This difference can be explained by an increased etch rate at the acute angles of the separator tips, giving rise to a rounding of the newly formed convex edges. As an important output, the critical separator width, $w_{sep,crit}$, from which on acute angles are formed can be calculated in dependence on the desired clearance d_2 above the separator tip (equation 2, figure 5c).

$$w_{sep,crit} = 2\, r_{etch,slope}\, t_2(\sin(90-\alpha))^{-1} \qquad (2)$$

4. Testing and Evaluation

We tested the fabricated batteries using a multichannel battery test system (CTS-LAB, Basytech GmbH). To investigate the rate capability the fabricated cells were cycled between 2.0 and 2.9 V with CC/CV-charge steps at a rate of 0.2C and increasing discharge rates ranging from 0.2 to 10 C. For each discharge rate six complete cycles were run through. The reference capacity for calculating the C-rates was determined during the first cycle after formation at a discharge rate of 0.2C. At this rate the value basically equaled the nominal capacity of the cathode material (~130 mAh/g) as measured with standard laboratory test cells in EC:DMC (1:1wt) / 1M $LiPF_6$ vs. Li(0).

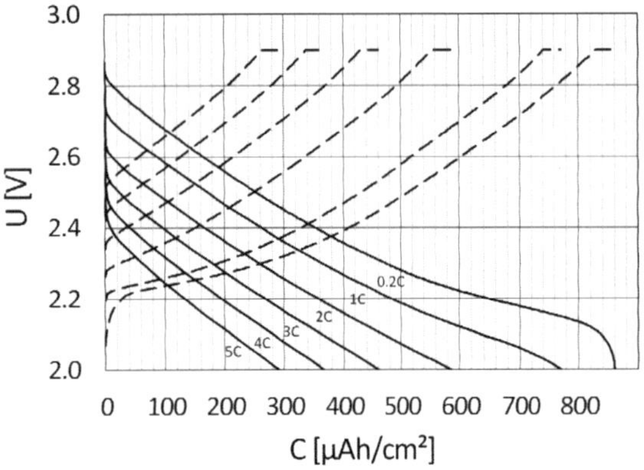

Figure 6. Cell voltage as function of extracted / inserted capacity per active area for micro battery with IDE during CC discharge at different loads and CC-CV charge at 0.2C (w_{cav}: 600 µm, w_{sep}: 100 µm, d_1: 300 µm, d_2: 100 µm, d_3: 180µm)

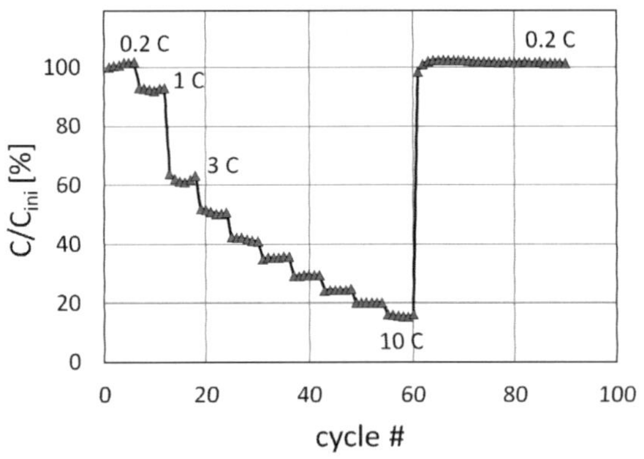

Figure 7. Capacity retention as function of cycle number and C-rate (rate capability test 0.2C to 10C followed by cycling performance test with 30 additional full cycles at 0.2C) (w_{cav}: 600 µm, w_{sep}: 100 µm, d_1: 300 µm, d_3: 170µm)

Figure 6 shows the discharge characteristics of a micro battery with 600 µm wide IDE digits at different CC loads until 5C. Cells of this geometry were capable of constant current loads of 10 C at some 15% capacity retention (figure 7). A comparison between micro batteries with single side-by-side electrodes and micro batteries with IDE of the same geometry (w_{cav}, w_{sep}, d_1, d_2,d_3) shows that by implementing an IDE design the rate capability can be improved significantly (figure 8). To evaluate the cycling stability each rate capability test was directly followed by 30 full cycles at a charge and discharge rate of 0.2C, respectively. The linear capacity loss rate over these 30 cycles accounted for < 0.1 % per cycle for all investigated IDE cells.

Yoshima et al. recently proposed a similar micro battery design with interdigitally arranged side-by-side electrodes (20 µm digits, 20 µm spacers) structured by micro injection of electrode slurries into dry film resist channels and subsequent lift-off of the resist [6]. The cells were capable a rate of 20 C at 60 % capacity retention. For the present micro batteries, some 60% of the discharge capacity at 0.2 C could be retained only at much lower rates of about 2 to 3C. However, the feasible specific capacity of the present cells (up to 850 µAh/cm², figure 6) is much higher than for the cells in [6]. That means, in terms of current and capacity per area, the present battery even outperforms [6] for moderate current loads, whereas for higher current loads a design as proposed in [6] is favorable (figure 9).

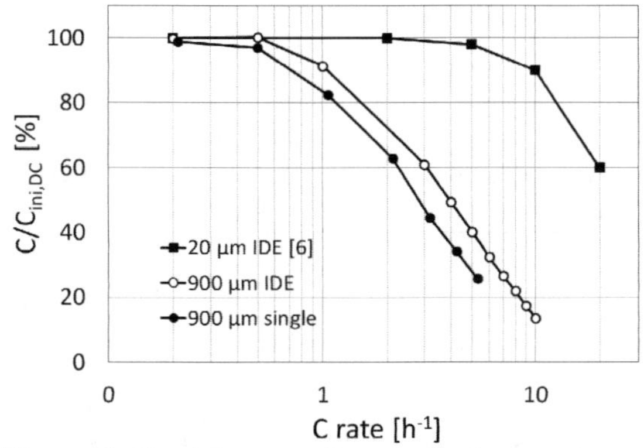

Figure 8. Capacity retention vs. discharge rate

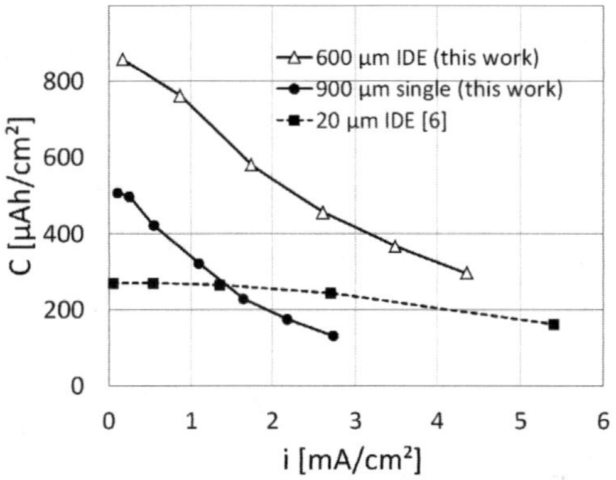

Figure 9. Capacity density vs. current density

5. Conclusion

The present micro battery concept combines state-of-the-art lithium ion battery materials and Si-processing. We hereby seek to overcome the limitations of thin film and 3D micro batteries considering specific capacity per area and cost. For packaging reasons it is of high advantage to have a co-planar arrangement of both electrodes. The fabricated micro batteries with IDE are capable of higher current loads than cells with single side-by-side electrodes [4]. The feasible specific capacity of the present cells is in the range of 600 to 900 µAh/cm² (0.2C) and hence considerably higher than for thin film batteries. The choice of the dimension of the separator width is crucial not only for realizing a high specific capacity per area but also for the reliable separation of the electrode slurries. The critical separator width, $w_{sep,crit}$, from which on acute angles are formed and the necessary etch time to generate a certain clearance d_2 can be calculated based on equation 1 and 2, respectively. A linear capacity loss rate of <0.1 % per cycle was observed over 30 full cycles at 0.2C CC-CV charge and CC discharge, respectively.

References

[1] Albano F et al 2008 *Journal of Power Sources* **185** 1524–1532
[2] Belleville M et al 2010 *Microelectronics Journal* **41(11)** 740-745
[3] Hahn R et al 2014 *Proceedings GMM-Workshop Energieautarke Sensorik* **7** 33-38
[4] Hoeppner K et al 2013 *J Phys: Conf Ser* **476(1)** 012086 DOI:10.1088/1742-6596/476/1/012086
[5] Marquardt K et al 2010 *Microsyst Technol* **16** 1119-1129 DOI: 10.1007/s00542-009-0954-7
[6] Yoshima et al 2012 *Journal of Power Sources* **208** 404-408

An *in situ operando* MEMS platform for characterization of Li-ion battery electrodes

H Jung[1,2], K Gerasopoulos[2], X Zhang[1] and R Ghodssi[1,2*]
[1] Department of Electrical and Computer Engineering, University of Maryland, College Park, MD 20742, USA
[2] Institute for Systems Research, University of Maryland, College Park, MD 20742, USA

[*]E-mail: ghodssi@umd.edu

Abstract. This paper presents an *in situ operando* approach that allows characterization of lithium-ion battery electrodes. A MEMS sensor is designed and integrated with a commercially available Raman spectroscope to enable monitoring the stress and structural changes in the electrodes. An interferometric method with an enhanced image processing algorithm is applied for analyzing the crystal phase-dependent stress changes – contributing to higher sensitivity compared to a previously reported technique – while the structural changes are monitored using Raman spectroscopy. New capabilities of our platform are highlighted, allowing visual observation of crystal phase-dependent structural changes in the electrode. Simultaneous characterization of the stress and structural changes are achieved concurrently *in situ operando*. The results show excellent agreement with previous literature reports on both phenomena.

1. Introduction

Li-ion battery (LIB) electrodes react with Li ions through an intercalation or alloying mechanism. These mechanisms involve continuous volume expansion/contraction-induced stress and structure/phase changes in LIB electrodes, which lead to gradual deterioration of the battery performance. Therefore, there is a need for real-time monitoring platforms capable of providing simultaneous information on the stress and structural evolutions during battery operation.

Different *in situ* analysis techniques have been utilized in order to understand the mechanical changes in LIB electrodes. Among these, a multi-beam optical sensor (MOS) was shown to be an effective method for characterizing Li-ion battery thin film electrodes [1]. This method enables measurement of important mechanical parameters such as average stress, biaxial modulus and fracture energy for both silicon thin film and commercial electrodes. While the technique has proven to be beneficial for mechanical characterization, limited capability to directly access the electrode hinders its potential integration with other methods for providing structural information.

In parallel, a significant amount of effort has been put into analyzing structural changes in LIB electrodes during battery operation. *In situ* nuclear magnetic resonance (NMR) spectroscopy is used to study detailed kinetics of the electrochemical lithiation and delithiation reactions in nm-sized silicon-based anodes over multiple cycles [2]. Compared to the NMR technique, which relies on bulky

* To whom any correspondence should be addressed.

external equipment and complex electrochemical cells, Raman spectroscopy is a valuable tool for analyzing structural and chemical variations in LIB electrodes in a relatively simple fashion. Nevertheless, these methods are restricted to providing only structure or crystal phase (from now on 'phase') evolutions during lithium cycling.

In this work, an *in situ operando* approach is presented, which enables characterization of stress changes using interferometry in parallel with structural changes *via* Raman spectroscopy in a single experiment. The V_2O_5 LIB electrode is used as a model electrode in this work. Additionally, visual observation of phase-dependent asymmetric changes in the electrode is achieved.

2. Design and Fabrication

2.1. Design

The MEMS sensor consists of a Fabry-Perot (FP) cavity and a battery cavity separated by a flexible membrane (figure 1a). The reflective surface of the membrane forms one side of the FP interferometer, while a bonded glass wafer forms the other. The reverse side of the flexible membrane houses a battery cavity coated with the active battery material (V_2O_5). When the battery cavity is exposed to lithium-conducting electrolyte, and metallic Li foil is incorporated as a counter electrode, this device operates a half-cell Li-ion battery. During electrochemical battery operation, the V_2O_5 electrode experiences expansion and contraction as a result of Li-ion insertion/extraction, inducing stress changes in the electrode. These stress changes cause deflection in the flexible membrane (figure 1b), changing the interference patterns accordingly. These pattern changes are recorded and analyzed for stress change characterization. The structure evolutions in the electrode are monitored using Raman spectroscopy which can simultaneously probe the electrode surface through the intervening thin or transparent layers (figure 1b).

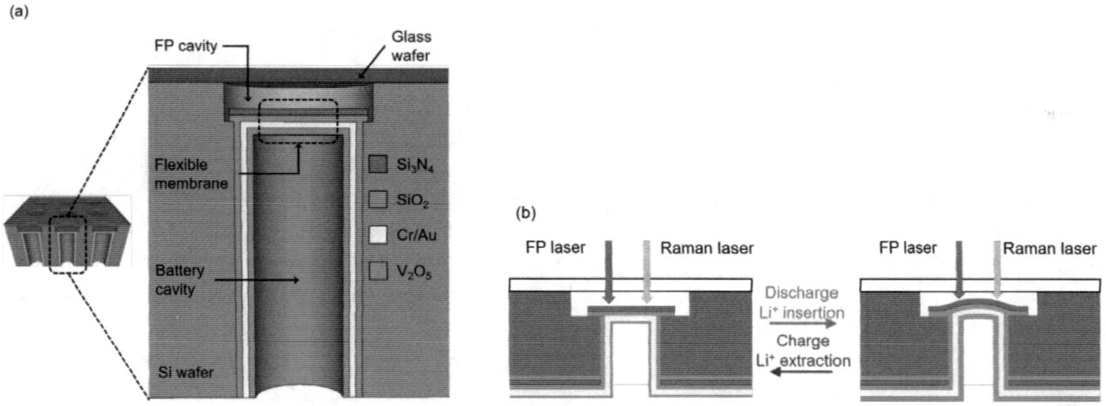

Figure 1. (a) 3D and cross-section schematic of MEMS sensor and (b) schematic representation of the sensing mechanism.

2.2. Fabrication

A double-side polished silicon wafer is used as the substrate for the MEMS sensor fabrication. First, the FP cavity (10 µm deep) is etched using deep reactive-ion etching (DRIE). A 300 nm thick layer of SiO_2 is thermally grown followed by the deposition of a 700 nm thick layer of Si_3N_4 on both sides of the wafer using low pressure chemical vapor deposition. The battery cavity (288 µm deep) is formed by DRIE on the wafer backside, aided by the SiO_2 etch-stop layer. The diameter of the battery cavity (150 µm) determines the diameter of the membrane. A glass wafer is anodically bonded to form the optical FP cavity. The cross-section SEM image (figure 2a) shows well-defined optical and battery cavities separated by the membrane. An additional SiO_2 passivation layer (50 nm) is deposited followed by Cr/Au (5 nm/15 nm) thin films to form the current collector. Finally, the battery cavity is coated with a thin-film V_2O_5 (135 nm) electrode using atomic layer deposition (ALD) (figure 2b). On finishing the fabrication process, the MEMS sensor device is packaged in a modified coin cell for testing. The complete fabrication process and device packaging are described in detail in our previous work [3].

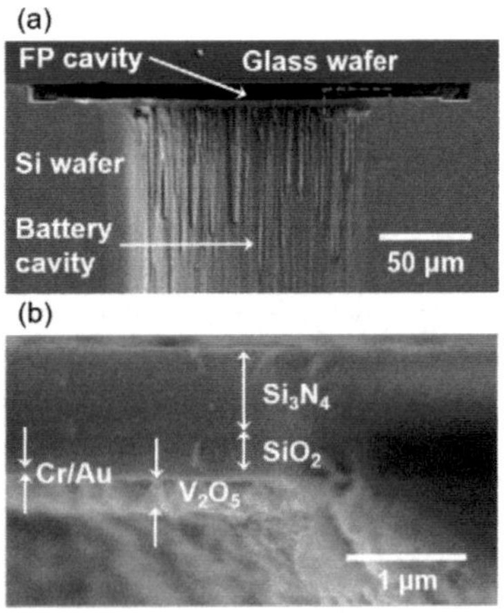

Figure 2. (a) Cross-section SEM of the membrane and (b) enlarged view highlighting the well-defined multi-layer structure.

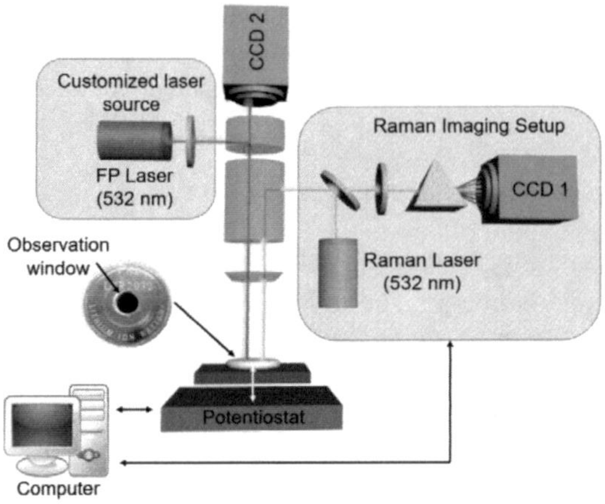

Figure 3. Simplified schematic of the experimental setup showing the coin cell under test relative to the Raman microscope.

3. Experimental set-up

A modified coin cell with an optical window is devised to facilitate the experiment during battery operation (figure 3). The assembled coin cell is connected to a potentiostat (Compactstat, Ivium Technologies) and placed under a Raman microscope (Yvon Jobin LabRam Aramis, Horiba, Ltd.) for testing. The white-light illumination source of the Raman microscope is replaced with a customized laser source (532 nm) for measuring the membrane deflection by utilizing the FP interferometry, while the Raman system laser (532 nm) is used as the excitation source for Raman spectroscopy measurements. By employing FP interferometry, the same microscope used for the Raman microspectroscopy analysis can be used to collect information on the membrane deflection, thus enabling multi-modal, real-time monitoring in a unified setup. All these measurements are conducted in air under normal ambient conditions.

4. Results

4.1. Qualitative stress change characterization

Previously, utilizing a similar sensing platform, the stress changes were characterized by monitoring the radius change of one of the concentric rings in the interference patterns [3]. However, the magnitude of the radius change can be quite small when the stress changes in the electrode are not large enough, which can lead to losing subtle or critical stress changes in the electrode. To overcome this limitation, the stress changes in the electrode are characterized instead by analyzing the interference intensity change at the center of the membrane, which is the part of the membrane most sensitive to stress changes. The intensity change induced by the electrode underneath the membrane is calculated based upon the comparison between a pristine electrode and an electrode under electrochemical testing as shown in figure 4a. The correlation between electrochemical data and changes in interference pattern intensity for the membrane during the first electrochemical cycle is shown in figure 4b. Upon lithium insertion (Discharge), the V_2O_5 generally experiences increasing tensile stress, and the opposite trend is observed during the charge process. Breaking these general trends are two abnormal regions (figure 4b, region I and II), which correspond to elongation/contraction in the lattice constant (*c*) of the crystalline V_2O_5 during the ε- to δ-phase and δ- to ε-phase transformations [4]. These smaller abnormal changes are not detected when the fringe radius monitoring method [3] is applied to the same experimental data used in this work, which demonstrates the enhanced sensitivity of the center intensity monitoring method.

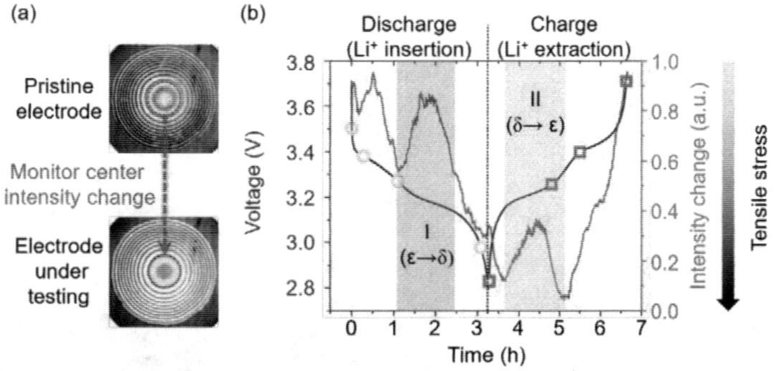

Figure 4. (a) Interference patterns from pristine electrode and electrode under test indicating center intensity change and (b) correlation between the intensity change and discharge-charge curve of the first cycle.

4.2. Structural changes

Simultaneously, *in situ* Raman spectra are collected and representative data at various points along the discharge-charge curve (indicated by blue circles and red squares in figure 4b) are plotted (figure 5). Initially, nine distinct peaks associated with the V_2O_5 electrode are observed. It has been reported that the Raman intensity change at 145 cm^{-1} indicates the level of disorder between V_2O_5 layers [4]. As the level of disorder increases during lithium insertion, the Raman peak intensity at 145 cm^{-1} diminishes and shifts to 155 cm^{-1} (figure 5a). Conversely, during lithium extraction, the Raman peak at 153 cm^{-1} shifts back to 145 cm^{-1} and the intensity of the peak is progressively recovered (figure 5b), indicating the V_2O_5 lattice is fully restored. Progressive decrease and full recovery in the intensity of the other eight peaks, as well as the Raman shift at 996 cm^{-1}, verify the high electrochemical reversibility of the electrode [4].

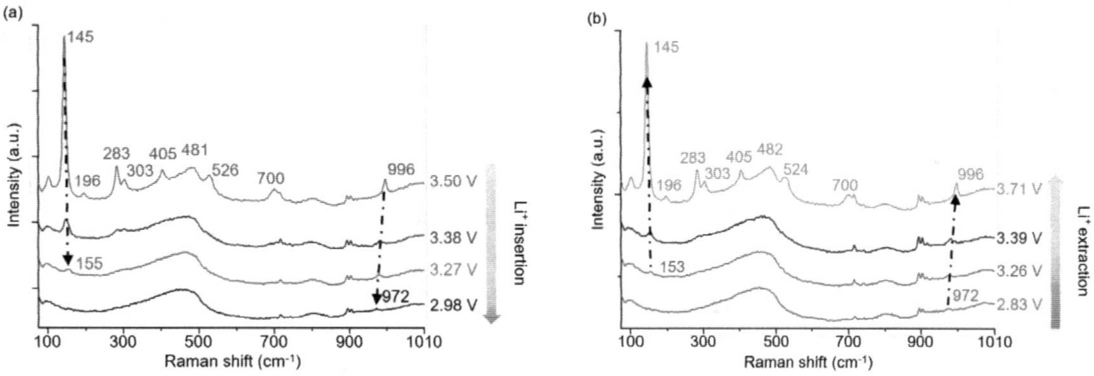

Figure 5. (a) *In situ* Raman spectra collected during first discharge (cyan circles in Figure 5(b)) and (b) charge (magenta rectangles in Figure 5(b)) cycle.

4.3. Visual observation

Visual observation of phase dependent structural changes has a distinct advantage compared to other characterization methods since it allows monitoring of significant structural evolutions in the electrode by merely observing the series of interference pattern images recorded during the experiment. The visual observation of the crystal phase-dependent structural changes in the electrode is shown in figure 6. The interference patterns remain symmetric in the first discharge-charge cycle (figure 6a, region A corresponding to figure 4b), which are the expected interference patterns from a symmetric circular membrane. However, operating over a larger voltage regime in the second cycle, more drastic phase changes can be indirectly visualized through the asymmetric interference patterns themselves (region B), which then reversibly transition back to symmetric patterns (region C) depending on the phase transitions in the V_2O_5 electrode [4].

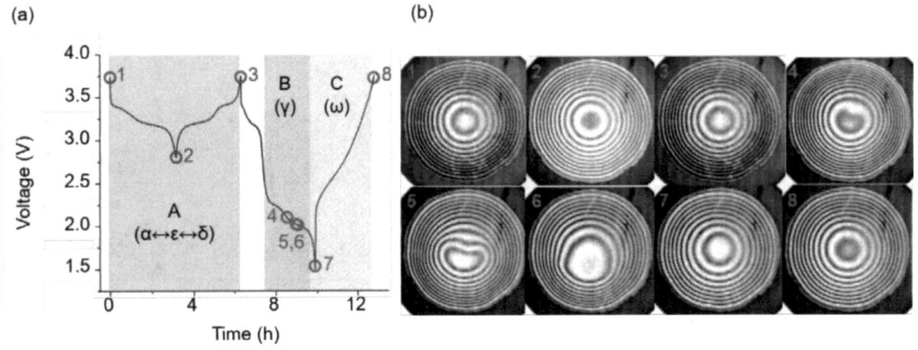

Figure 6. (a) Discharge-charge cycles annotated with the relevant V_2O_5 phase transitions and (b) series of interference images of the circular membrane collected along the discharge-charge curve.

5. Summary

The work presented here demonstrates the capability of the *in situ operando* approach for characterization of LIB electrodes. Stress changes in V_2O_5 electrodes are characterized by monitoring the center intensity changes of the interferometric patterns, yielding higher sensitivity compared to the previously reported method. At the same time, reversible structure changes in the V_2O_5 electrodes are observed by the Raman spectroscopy measurements. These stress and structural evolutions are monitored simultaneously in a unified set-up enabled by the MEMS sensor. Thus, this MEMS platform facilitates simultaneous investigation of electrochemically-driven stress and structural changes in LIB electrodes, and can be leveraged as a discovery platform for various thin-film electrode materials in the field of battery electrode characterization.

Acknowledgement

This research was funded by Nanostructures for Electrical Energy Storage, an Energy Frontiers Research Center funded by the U.S. Department of Energy, Office of Science, Office of Basic Energy Sciences. The authors would like to acknowledge the staff at the Maryland Nanocenter.

6. References

[1] Sethuraman V A, Chon M J, Shimshak M, Srinivasan V and Guduru P R 2010 *J. Power Sources* **195** 5062-5066
[2] Ogata K, Salager E, Kerr C J, Fraser A E, Ducati C and Morris 2014 *Nat. Commun.* **5** 3217 (11pp)
[3] Pomerantseva E, Jung H, Gnerlich M, Baron S, Gerasopoulos K and Ghodssi R 2013 *J Micromech Microeng* **23** 114018 (10pp)
[4] Baddour-Hadjean R and Pereira-Ramos J –P 2010 *Chem. Rev.* **110** 1278-1319

Journal of Physics: Conference Series **660** (2015) 012066 doi:10.1088/1742-6596/660/1/012066

Thermoelectric Energy Conversion: Materials, Devices, and Systems

Gang Chen

Department of Mechanical Engineering
Massachusetts Institute of Technology
Cambridge, MA 02139

This paper will present a discussion of challenges, progresses, and opportunities in thermoelectric energy conversion technology. We will start with an introduction to thermoelectric technology, followed by discussing advances in thermoelectric materials, devices, and systems. Thermoelectric energy conversion exploits the Seebeck effect to convert thermal energy into electricity, or the Peltier effect for heat pumping applications. Thermoelectric devices are scalable, capable of generating power from nano Watts to mega Watts. One key issue is to improve materials thermoelectric figure-of-merit that is linearly proportional to the Seebeck coefficient, the square of the electrical conductivity, and inversely proportional to the thermal conductivity. Improving the figure-of-merit requires good understanding of electron and phonon transport as their properties are often contradictory in trends. Over the past decade, excellent progresses have been made in the understanding of electron and phonon transport in thermoelectric materials, and in improving existing and identify new materials, especially by exploring nanoscale size effects. Taking materials to real world applications, however, faces more challenges in terms of materials stability, device fabrication, thermal management and system design. Progresses and lessons learnt from our effort in fabricating thermoelectric devices will be discussed. We have demonstrated device thermal-to-electrical energy conversion efficiency ~10% and solar-thermoelectric generator efficiency at 4.6% without optical concentration of sunlight (Figure 1) and ~8-9% efficiency with optical concentration. Great opportunities exist in advancing materials as well as in using existing materials for energy efficiency improvements and renewable energy utilization, as well as mobile applications.

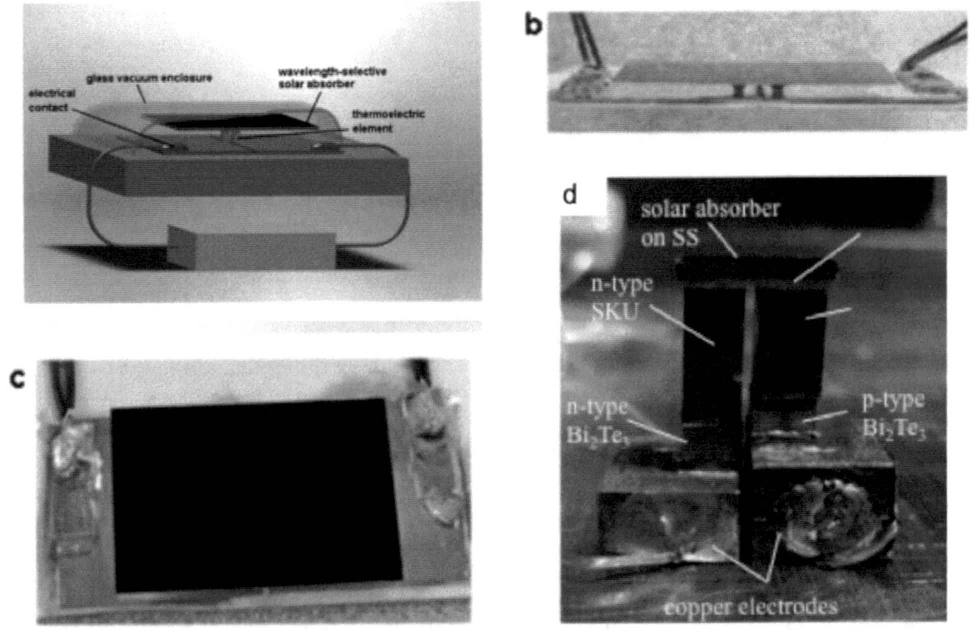

Figure 1 Solar thermoelectric generators (STEG) (a) a flat-panel STEG consisting of an absorber and a pair of thermoelectric legs enclosed in a vacuum chamber, (b) and (c) are cross-sectional and top views of a device, and (d) a segmented thermoelectric generator with optical concentration.

Process for integration of energetic porous silicon devices

Matthew H Ervin, Brian Isaacson and Louis B Levine
US Army Research Laboratory, 2800 Powder Mill Road, Adelphi, MD 20783 USA

Matthew.H.Ervin.civ@mail.mil

Abstract. We have developed new procedures for etching porous silicon (PSi) which allow PSi devices to be more easily integrated with other devices used for controlling or utilizing the output of the PSi devices. The output from energetic PSi devices may be used for MEMS actuation or energy harvesting applications. Initial proof-of-concept energy harvesting with a macro-scale, piezo cantilever has been demonstrated. Of the 2 etch processes developed, the sacrificial electrode process is the simplest, but it produces an inhomogeneous PSi thickness across the wafer and introduces surface topography due to electropolishing. By using the anchored electrode method, which incorporates a dielectric layer, more controllable etch depths and patterned devices are obtained. However, a proximity effect is observed where features closer to the electrode etch more rapidly. A simple voltage divider model can be used to predict these relative etch depths, but more work is required to develop a quantitative model.

1. Introduction

Porous silicon (PSi) devices have been investigated for a number of applications including chem/biosensors, optoelectronics, and solar cells [1] as well as for energetic devices such as micro-thrusters [2]. Here, we describe a metal-assisted etch process [1, 3] that will facilitate integration of PSi with other devices. The difficulty in such integration stems from the aggressive etch solution (3:1 HF:ethanol with 2.4% H_2O_2). The new process, where the metal electrode and the Si to be etched are on the same side of the wafer, permits etchant sensitive devices on the other side of the wafer to be protected during the etch. This is important as it provides significantly more flexibility for integrating PSi with other devices on the same wafer or chip.

PSi is most commonly formed by anodic etching of a silicon (Si) wafer in an etch solution containing a high concentration of HF [1]. We have focused on using a metal-assisted etch, also referred to as a galvanic etch, which is a simplified etch procedure that does not require a potentiostat or external wires to the wafer. In this approach, a metal, typically platinum (Pt) is deposited on 1 side of a Si wafer, and this metallized wafer is immersed in a HF, hydrogen peroxide (H_2O_2), and ethanol etch solution. The H_2O_2 reacts with hydrogen ions from the HF at the catalytic metal surface to become water molecules while liberating electron-hole (h+) charge carriers into the metal/wafer. These h+ charge carriers traverse the wafer to the other side where they induce the exposed Si to react with the HF to produce a soluble dihydrogen silicon hexafluoride product [3]. In this arrangement, the HF and H_2O_2 concentrations control the electrochemical current, and therefore reaction rate, without any external instrumentation. In addition, the Pt to Si surface area ratio also affects the etch rate, as the etch current is proportional to the area of exposed Pt, and the area of exposed Si determines the resultant etch current density. If the current density at the Si surface becomes too large, the etch transitions from producing PSi to an electropolishing regime where the Si is completely etched away [1]. Two approaches were taken for using same-side Pt electrodes. In the first approach, Pt electrodes were deposited directly on the Si wafer to be etched. This approach is termed the sacrificial Pt electrode approach since the PSi etch etches the Si underneath the Pt, eventually releasing the Pt from the wafer. In the second approach, termed the anchored Pt electrode approach, the Pt electrodes are deposited onto a Si wafer that is covered with a silicon nitride (Si_3N_4) layer which is patterned to expose the Si to the etchant, and to allow electrical contact between the Pt electrodes on top of the Si_3N_4 and the Si.

2. Experimental

In this study, 1–10 Ω p+ Si wafers were used as they produce nanoporous Si, which is desired for our energetic PSi applications. These wafers have 500 nm of Si_3N_4 on both sides as received from Rogue Valley Microdevices. In the case of the sacrificial Pt electrodes, the Si_3N_4 is removed from 1 side of the wafer while, in the case of anchored Pt electrodes, photolithographically defined windows are etched in the Si_3N_4 to expose Si areas for etching and for electrical contact to the Pt. These Si_3N_4 etches are carried out using a Unaxis VLR 700 reactive ion etch system.

After the Si_3N_4 has been removed as desired, the wafer is etched for 2 min in 6:1 H_2O:HF to remove any native oxide prior to being sputter cleaned for 30 s in a Unaxis Clusterline 200 sputter deposition system. Without breaking vacuum, the samples are then transferred to the Pt sputter chamber where 170 nm of Pt is sputter deposited at a sample temperature of 250 °C. This Pt layer is then patterned using photolithography and a 4Wave Inc. 4W-PSIBE ion mill etch system. The resulting wafers, which have patterned Pt electrodes and exposed Si, are then etched with a HF, ethanol, and peroxide mixture. The ethanol reduces the viscosity/surface tension of the etch solution so that it can penetrate into the small pores being formed. A typical etch composition is 3:1 HF:ethanol with 2.4% peroxide.

The etch results were characterized using scanning electron microscope (SEM) cross-section micrographs obtained with a Hitachi S4500 SEM. Etch depth measurements were also made, after removing the PSi with a 1 molar potassium hydroxide (KOH) etch, using a Wyko NT1100 optical profilometer. In order to measure the combustion rates of the PSi, bridge wires were photolithographically deposited onto the wafers, prior to the PSi etching, using a lift-off approach with metal deposited using a CHA Industries e-beam evaporator [2]. In order to burn the PSi an oxidizer is applied to the PSi in the form of a 3 M sodium perchlorate in methanol solution. After drying for 30 min, the PSi is ignited by passing 20 V/1 A of current across the bridge wire. This is done while synchronously triggering a Photron Devices Inc, high-speed camera used to measure the PSi burn rate.

An energy harvesting experiments was performed by mounting PSi device, made using backside electrodes [2], onto a macro-scale piezoelectric cantilever model V22B obtained from Mide. Upon ignition of the PSi, the output of the piezo was put into a Mide model EH20W rectifier circuit with the output stored on a capacitor. The piezo signal and the charging of the capacitor were monitored with an oscilloscope.

3. Results

3.1. Sacrificial Pt Electrodes

The sacrificial Pt approach was investigated because it is the simplest approach to same-side Pt electrodes as it requires only a Pt deposition and patterning process before the PSi etch. Figure 1 is a drawing showing the sample structure and associated etch process. The sacrificial Pt electrodes were designed to be at least as wide as the desired etch depth so that the Pt would not be completely undercut until the desired etch depth had been achieved. In practice, the PSi etching under the Pt edges enables the Pt to detach from the Si surface. As a result, the Pt peels back from the etch-front producing greater lateral than vertical etching as shown in figure 2. This delamination results from current crowding [4] at the edge of the Pt/Si interface which increases the local etch current density into the electropolishing regime [1] which removes all of the Si. This Pt delamination and electropolishing limits the PSi thickness, results in PSi thickness inhomogeneity, and produces surface topography. It may be possible to reduce the deleterious electropolishing by reducing the etch solution concentrations and the Pt:Si ratios. However, this is likely to adversely affect the quality of the PSi for many applications, and the PSi thickness will still be non-uniform across the sample surface.

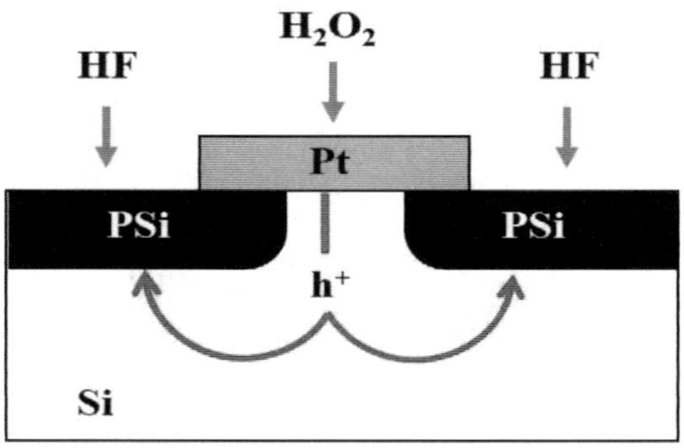

Figure 1. Cross-sectional diagram of the sacrificial electrode PSi etch showing that the etch undercuts the Pt electrode.

Figure 2. SEM cross-sectional image showing a Pt electrode peeling back from the PSi etch-front resulting in greater lateral etching under the Pt than the vertical etching into the Si surface.

3.2 Anchored Pt Electrodes

In order to achieve thicker PSi films along with better control over the PSi pattern on the wafer, the anchored Pt approach was developed. In the this approach, the Pt is deposited onto Si_3N_4 which is patterned to expose the Si to be etched, and to allow electrical contact between the Pt and Si, as shown in figure 3. The Si_3N_4 performs 2 important functions for this etch approach. As before, its resistance to the PSi etchant allows it to be used to define where the PSi will be etched. In addition, its dielectric nature prevents electropolishing induced delamination at the edge of the Pt electrodes.

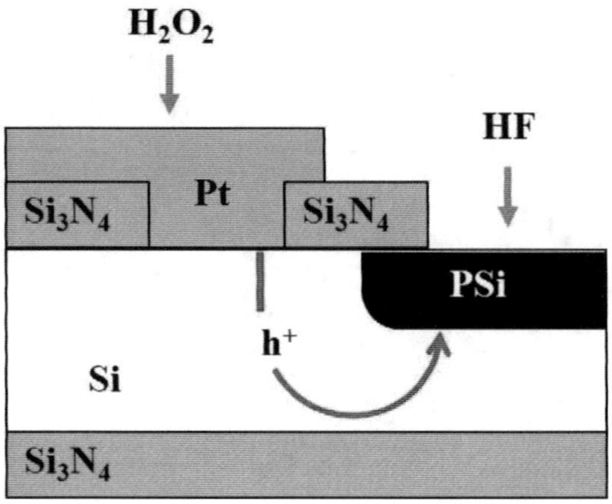

Figure 3. Cross-sectional drawing of the anchored electrode structure and etch process showing the PSi etch undercutting the Si_3N_4, and that the Pt is isolated from the PSi etch front.

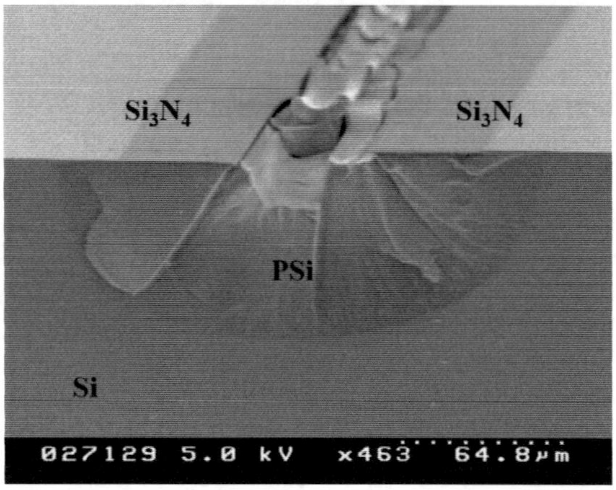

Figure 4. SEM cross-sectional image showing signs of electropolishing early in the etch, with better PSi formed as the etch-front area grows, thus lowering the etch current density.

Even away from the Pt contact, electropolishing of the exposed Si can still occur if the H_2O_2 concentration is too high or if the Pt:Si ratio is too large. Figure 4 shows an example of a line of etched

PSi where the initial etching was near the electropolishing regime leading to broken and isolated pieces of PSi. As the etch progressed, the area of Si at the etch-front increased so that the etch current density decreased to below the electropolishing regime so that better quality PSi was formed.

There can also be variations in etch rates as a function of distance from the Pt electrode. This proximity effect results in features closer to the Pt having different PSi etch depths and properties from features further from the Pt. A quantitative measure of the proximity effect was made by etching a number of lines at various distances from a Pt electrode. Figure 5a-b shows plots of etch depths for 5 lines at different distances from Pt electrodes. These depths were measured using optical profilometry after removing the PSi. It can be seen that there is a greater proximity effect when the PSi lines are close to the Pt electrode (figure 5a). In this case, the etch depths range from 22.4 to 11.8 microns when the lines were from 2.1 to 3.9 mm from the Pt contact point, respectively. In the second sample (figure 5b), the etch depths vary from 7.3 to 6.5 microns for lines 20.1 to 21.9 mm from the Pt contact point, respectively. The additional resistance between the Pt electrode and the etched lines in this sample results in more uniform etching as well as reduced etch rates.

A simple voltage divider model serves as a first-order approximation for this effect, and it could be used to design electrodes for minimizing or utilizing differences in etch depth across the wafer. Normalizing the voltage divider model output to match the etch depth of the line closest to the Pt electrode produces the results shown in figure 6 for the sample showing more variation in etch depth. Figure 6 shows a less than perfect fit between the model and experimental results. More experimental measurements need to be made before a more accurate model can be developed.

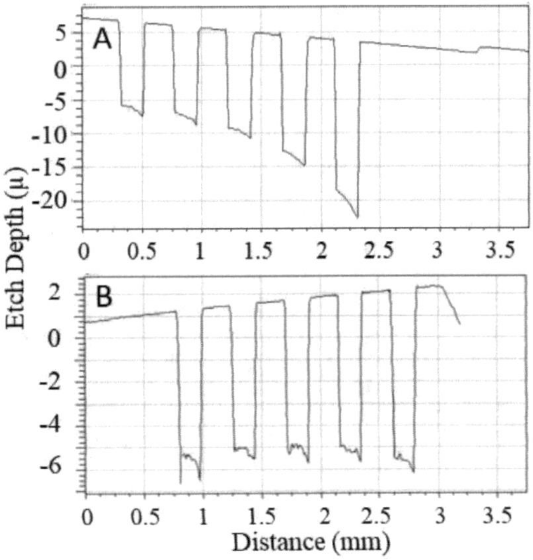

Figure 5. Optical profilometry of PSi line depths at various distances from a Pt electrode (which are to the right of the plots). (A) lines etched at 2.1, 2.55, 3.0, 3.45, and 3.9 mm from their electrode. (B) lines etched at 20.1, 20.55, 21.0, 21.45, and 21.9 mm from their electrode.

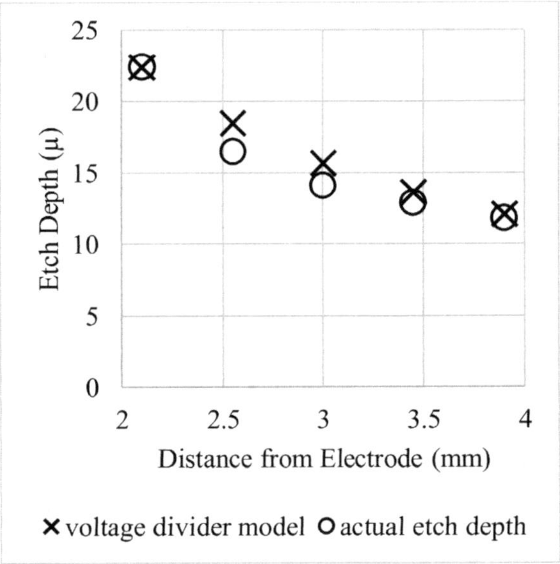

Figure 6. Plot of etch depth versus distance from the electrode data from figure 5a compared to predictions using a simple voltage divider model scaled to the deepest etch

Long straight line devices were made in order to characterize the burn properties of PSi produced with anchored Pt same-side electrodes. By filming the burning of a 0.5-mm-wide, 23-mm-long, and approximately 18-micron-thick PSi device, a flame speed of 3.6 m/s was measured when using sodium perchlorate as the oxidizer. This is comparable to similar PSi etched using the standard backside electrode process, which produces m/s to km/s burn rates [5].

Since the energy density of PSi is very high, it would be useful if it could be converted to electrical energy. A proof of concept energy harvesting demonstration was performed by mounting a 2 mm diameter PSi device [2] onto a macro-scale piezo cantilever. Upon ignition, the force deflected the cantilever which oscillated producing an AC signal which was rectified and used to charge a capacitor (see figure 7). By using force measurements and equations for strain energy in a cantilevered beam, we can estimate that amount of harvestable mechanical energy for these devices is on the order of 1 μJ, much lower than the expected 250 μJ of mechanical energy from the PSi [2]. While initial system efficiencies are low, there is significant opportunity to improve them. For example, the need to excite the proper oscillation modes in the cantilever was observed. This may be done through tuning the burn rate of the PSi to match the harmonics of the cantilever, and by placing the sample at the harmonic nodes of the cantilever.

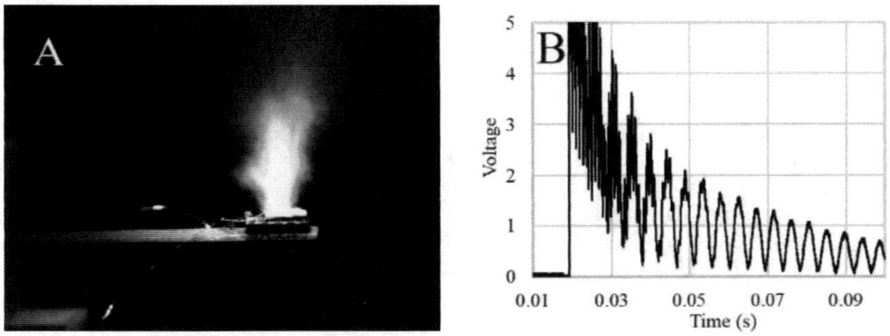

Figure 7. (A) Photograph of a PSi device being ignited while attached to a piezoelectric cantilever. (B) Output of cantilever during oscillation.

4. Conclusions

We have developed new procedures for etching PSi while potentially only exposing 1 side of the sample to the etch solution. This allows PSi devices to be more easily integrated with existing devices on a wafer or chip. These other devices may be used for initiating or utilizing the mechanical force, gases, light, or thermal energy from energetic PSi devices for MEMS actuation or energy harvesting applications. Of the 2 processes developed, the sacrificial electrode process is the simplest, but it produces an inhomogeneous PSi thickness across the wafer and introduces surface topography due to electropolishing. The anchored electrode approach, which includes an etch-resistant dielectric layer, produces more controllable etch depths and makes patterned devices possible. One complication is a proximity effect is where features closer to the electrode etch more rapidly. However, a simple voltage divider model can be used to predict the relative etch rates, and therefore, develop electrode patterns that will produce the desired results. Proof-of-concept energy harvesting with a macro-scale, piezo cantilever has been demonstrated.

5. References

[1] Korotcenkov G and Cho B K 2010 Silicon porosification: state of the art *Crit Rev in Solid State and Mater Sci.* **35** 153-260

[2] Churaman W A, Currano L J, Morris C J, Rajkowski J E and Bergbreiter S 2012 The first launch of an autonomous thrust-driven microrobot using nanoporous energetic silicon *J. of Microelectromech. Sys.* **21** 198-205

[3] Li X and Bohn P W 2000 Metal-assisted chemical etching in HF/ H2O2 produces porous silicon *Appl. Phys. Lett.* **77** 2572-2574

[4] Murrmann H and Widmann D 1969 Current crowding on metal contacts to planar devices *IEEE Trans. On Electron Dev.* **16** 1022-1024

[5] Piekiel N W, Morris C J, Churaman W A, Cunningham M E, Lunking D M and Currano L J 2015 Combustion and material characterization of highly tunable on-chip energetic porous silicon *Propellants Explos. Pyrotech.* **40** 16-26

Journal of Physics: Conference Series **660** (2015) 012069 doi:10.1088/1742-6596/660/1/012069

Photonic Crystal Enabled Thermophotovoltaics for a Portable Microgenerator

Walker R. Chan[1], Veronika Stelmakh[1,2], Christopher M. Waits[3], Marin Soljacic[4], John D. Joannopoulos[1,4] and Ivan Celanovic[1]

[1] Institute for Soldier Nanotechnologies, Massachusetts Institute of Technology, 77 Massachusetts Avenue, Cambridge, MA 02139, USA
[2] Department of Electrical Engineering, Massachusetts Institute of Technology, 77 Massachusetts Ave., Cambridge, Massachusetts 02139, USA
[3] Sensors and Electron Devices Directorate, 2800 Powder Mill Rd., US Army Research Laboratory, Adelphi, Maryland 20783, USA
[4] Department of Physics, Massachusetts Institute of Technology, 77 Massachusetts Ave., Cambridge, Massachusetts 02139, USA

E-mail: wrchan@mit.edu

Abstract. This work presents the design and characterization of a first-of-a-kind millimeter-scale thermophotovoltaic (TPV) system using a metallic microburner, photonic crystal emitter, and low-bandgap photovoltaic (PV) cells. In our TPV system, combustion heats the emitter to incandescence and the resulting thermal radiation is converted to electricity by the low bandgap PV cells. Our motivation is to harness the high specific energy of hydrocarbon fuels at the micro- and millimeter-scale in order to meet the increasing power demands of micro robotics and portable electronics. Our experimental demonstration lays the groundwork for developing a TPV microgenerator as a viable battery replacement.

1. Introduction

For durations greater than one day, the energy demands of micro robotics and portable electronics are too large to be comfortably supplied by batteries and yet too small to warrant a gasoline or diesel generator. Microgenerators promise to fill the 1–100 W range in a compact form factor by extending the high specific energy of hydrocarbon fuels into the millimeter scale. Hydrocarbon fuels have specific energies close to 12.8 kWh/kg whereas state of the art rechargeable batteries are closer to 180 Wh/kg. Thus, even a relatively inefficient generator can significantly exceed the specific energy of batteries. To this end, researchers have explored several possible energy conversion routes for providing next-generation portable power: micro mechanical heat engines, fuel cells, thermoelectrics, and thermophotovoltaics [1, 2].

In this work, we present a thermophotovoltaic (TPV) approach to a hydrocarbon-fueled microgenerator. TPV is the conversion of heat-to-electricity via the thermal emission of photons and their subsequent absorption and conversion to electricity by low bandgap photovoltaic (PV) cells. In our TPV system, shown in Fig. 1(a), propane is burned in a microburner to generate heat which brings a photonic crystal emitter to incandescence and the resulting thermal radiation drives a suitable low bandgap PV cell. This approach is extremely appealing for microgenerators

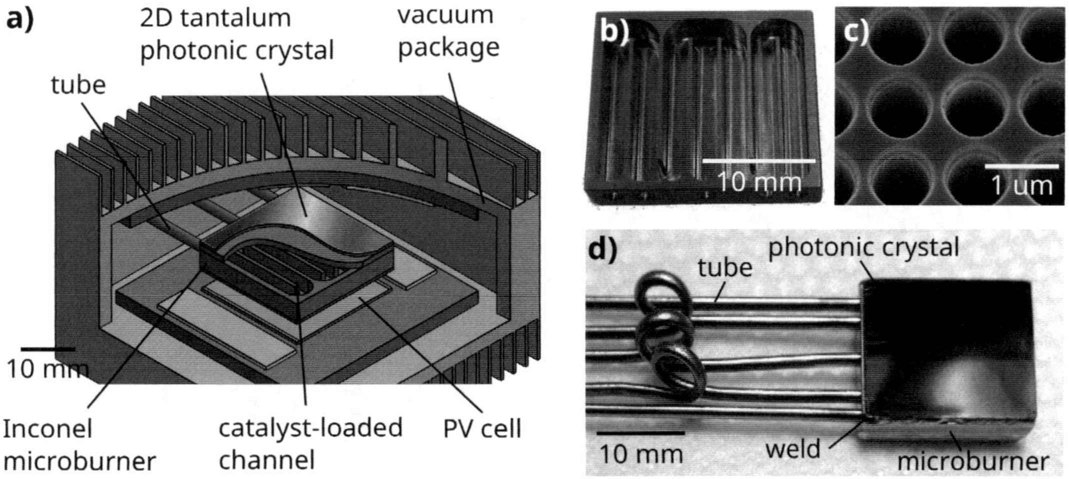

Figure 1. (a) A labeled CAD model of the conceptual TPV microgenerator based on our experimental setup comprised of an Inconel microburner to convert fuel into heat, a photonic crystal emitter to convert the heat into spectrally confined infrared radiation, and PV cells to convert the infrared spectrum into electricity. (b) A photograph of microburner interior with visible channels. (c) A SEM micrograph of the photonic crystal depicting etched cavities. (d) The microburner with attached tubes and photonic crystal.

because of its fully static conversion process, multifuel operation, and high specific power relative to other static conversion processes.

Unfortunately, the demonstrated fuel-to-electricity efficiency of millimeter-scale TPV systems has traditionally been limited to a few percent because of the need for high temperature material performance and synchronization between chemical, thermal, optical, and electrical domains [3–6]. One of the fundamental challenges to realizing a high efficiency is matching the radiated spectrum to the quantum efficiency of the PV cell. The PV cell can convert in-band radiation (photon energies above its bandgap) reasonably efficiently to electricity but out-of-band (photon energies below its bandgap) are wasted. Thus, a selective emitter with high in-band emissivity and low out-of-band emissivity is required. In this work we present a new approach to TPV enabled by a photonic crystal selective emitter, capable of near perfect spectral control, that promises to realize a high fuel-to-electricity efficiency. The low achieved efficiency is not a fundamental limitation of TPV: fuel-to-electricity efficiency of 30% should be achievable with this approach [7].

2. System demonstration

Although considerable technological barriers still need to be overcome to reach full performance, we have performed a robust experimental demonstration that validates the theoretical framework and the key system components. Our system was comprised of an Inconel microburner, 2D photonic crystal, and low bandgap PV cells.

The microburner, shown in Fig. 1, was a 20×20×3 mm chip with an internal catalyst-loaded serpentine channel suspended on tubes that doubled as fluidic connections [8]. Unmixed propane and oxygen were flowed coaxially through the inlet tubes which contained an inner capillary in a tube-in-tube configuration. Upon entering the channel, the gases mixed and reacted on the alumina supported platinum catalyst washcoated on the walls to generate heat. The exhaust gases exited through an outlet tube. The channel width and total length were designed for

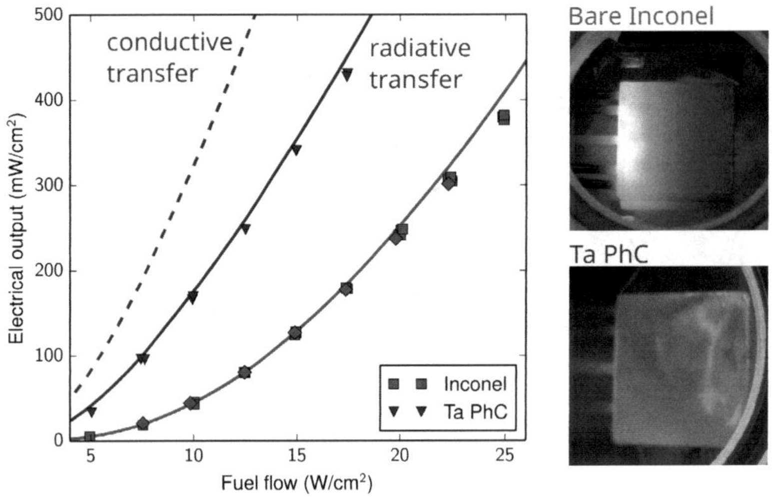

Figure 2. Fuel-to-electricity experimental results obtained from the TPV system. A microburner with and without a photonic crystal emitter, shown to the right, was characterized. Electrical power was scaled for a full set of cells.

complete combustion at the target fuel flow: the timescale for diffusive transverse transport of unreacted gases to catalyst sites on the walls was much shorter than the residence time in the channel. The microburner was fabricated from Inconel, which offers high temperature oxidation resistance, by machining and welding. The tubes were welded into holes drilled into the edge of the machined microburner, then the exposed channels were sealed by welding on a cover.

The 2D photonic crystal, shown in Fig. 1(c), was comprised of a square array of cylindrical cavities defined by interference lithography and deep reactive ion etching in a polished tantalum substrate [9]. The thermal emission spectrum was selectively enhanced through the introduction of cavity resonances in a wavelength range controlled by the geometry of the structure. Outside of the enhanced region, the thermal emission approached that of the metallic substrate. The present photonic crystal, which was only optimized for emission at normal incidence, achieved a hemispherically averaged in-band emissivity of $\varepsilon_{in} = 0.59$ and out-of-band emissivity of $\varepsilon_{out} = 0.16$. With dielectric-filled cavities, the in-band emissivity is predicted to increase to $\varepsilon_{in} = 0.92$ while maintaining the same out-of-band emissivity [10].

The microburner and photonic crystal were integrated by electron beam welding, as shown in Fig. 1(d). The hot side assembly was partially surrounded with either InGaAsSb [11, 12] or InGaAs [13] cells which were maintained at 20°C with a chilled water loop. The distance between the emitter and cells was approximately 1 mm. The microburner, emitter, and cells were contained in a vacuum chamber (5×10^{-5} Torr base pressure) to prevent oxidation of the photonic crystal and convective losses. A CaF window replaced cells on one side (the reported electrical output was scaled for a full set of cells on the front and back of the microburner) in order to allow for infrared temperature measurement and optical preheating.

To ignite the microburner, it was heated to approximately 400°C with a halogen lamp. Above that temperature, the propane kinetics over the catalyst were sufficient for autothermal operation, and the lamp was shut off. The fuel and oxygen flows were increased in small increments, maintaining an equivalence ratio of $\phi = 1.5$. Once steady state was reached, the electrical power output (at the maximum power point) was recorded and is reported in Fig. 2. Initial trials with a welded photonic crystal fell short of expectations because the weld was only made around the perimeter resulting in poor heat transfer between the microburner and photonic crystal. Measured electrical output was consistent with pure radiative heat transfer between the microburner and emitter. We are currently brazing the photonic crystal to the microburner for improved heat transfer. The electrical output for conductive heat transfer between the microburner and photonic crystal is shown in Fig. 2.

3. Towards a TPV microgenerator

In order to move from this bench-top demonstration to a portable microgenerator, we need to redesign the microburner and develop vacuum packaging. The first challenge is developing a vacuum package to maintain a vacuum level sufficient to prevent convective losses and degradation of the photonic crystal. While difficult, vacuum packaging has been successfully demonstrated in MEMS devices and incandescent and fluorescent light bulbs.

The second challenge is to realize an air-breathing microburner capable of processing conventional liquid fuels (gasoline, diesel, and JP-8 military logistics fuel). Transitioning from a gaseous single component, light hydrocarbon fuel to a liquid multi-component, heavy hydrocarbon fuels will require the microburner to compensate for overall longer diffusion times in combustion dominated by surface reactions, fuel vaporization or atomization, and potentially increased carbon formation due to the sulfur and aromatic content of the fuels. The transition from pure oxygen to air-breathing will require the microburner to compensate for a higher flow velocity that both decreases residence time and increases the exhaust heat loss. Excess enthalpy through exhaust heat recirculation holds promise to counteract the variable thermodynamic properties, longer diffusion times, higher flow velocity, and provide heat for vaporization by pre-heating the fuel and air using heat exchanged from the exhaust. These challenges in microburner design are not insurmountable and can leverage the extensive knowledge in microchannel heat exchanger technology.

4. Conclusion

We performed an initial proof-of-concept demonstration of a TPV system as a step towards a TPV microgenerator that could fill the 1–100 W gap between batteries and conventional generators by extending the high specific energy hydrocarbon generators to the millimeter scale. Our TPV demonstration used an Inconel microburner, 2D photonic crystal emitter, and low bandgap TPV cells. Our initial results indicated that TPV can be used in a high specific energy microgenerator, and with modest improvements (improved microburner-emitter heat transfer, filled cavity photonic crystal) greatly improved fuel-to-electricity efficiency is possible. Furthermore, we presented a path towards a TPV microgenerator.

Acknowledgments

The authors would like to thank Christine Wang of MIT Lincoln Laboratory for providing the InGaAsSb cells used in this work and Ben Wilhite of Texas A&M University for performing computation fluid dynamic simulations and assisting with the characterization of the microburner. Photonic crystal fabrication was supported as part of the Solid-State Solar-Thermal Energy Conversion Center (S3TEC), an Energy Frontier Research Center funded by the U.S. Department of Energy (DOE), Office of Science, Basic Energy Sciences (BES), under Award No. DE-SC0001299 / DE-FG02-09ER46577. This work was additionally supported by the Army Research Office through the Institute for Soldier Nanotechnologies under Contract No. W911NF-13-D-0001 and the Micro Autonomous Systems and Technology Collaborative Technology Alliance under Contract No. 892730.

References

[1] Committee on Soldier Power/Energy Systems 2004 *Meeting the Energy Needs of Future Warriors Committee* (National Academies Press) ISBN 0309533449 ISBN 0309533449
[2] Mitsos A 2009 *Microfabricated power generation devices: design and technology* (Weinheim: Wiley-VCH) ISBN 978-3527320813 ISBN 978-3527320813
[3] Doyle E, Shukla K and Metcalfe C 2001 Development and Demonstration of a 25 Watt Thermophotovoltaic Power Source for a Hybrid Power System Tech. Rep. TR04-2001 National Aeronautics and Space Administration National Aeronautics and Space Administration
[4] Fraas L M, Avery J E and Huang H X 2003 *Semiconductor Science and Technology* **18** S247 **18** S247
[5] Wenming Y, Siawkiang C, Chang S, Hong X and Zhiwang L 2005 *Journal of Micromechanics and Microengineering* **15** S239 **15** S239
[6] Chan W R, Bermel P, Pilawa-Podgurski R C N, Marton C H, Jensen K F, Senkevich J J, Joannopoulos J D, Soljacic M and Celanovic I 2013 *Proceedings of the National Academy of Sciences* **110** 5309–5314 **110** 5309–5314
[7] Zenker M and Heinzel a 2001 *IEEE Transactions on Electron Devices* **48** 367–376 ISSN 00189383 ISSN 00189383
[8] Chan W R, Wilhite B A, Senkevich J J, Soljacic M, Joannopoulos J and Celanovic I 2013 *Journal of Physics: Conference Series* **476** 12017 **476** 12017
[9] Stelmakh V, Rinnerbauer V, Geil R D, Aimone P R, Senkevich J J, Joannopoulos J D, Soljačić M and Celanovic I 2013 *Applied Physics Letters* **103** 0–4 ISSN 00036951 ISSN 00036951
[10] Yeng Y X, Chan W R, Rinnerbauer V, Joannopoulos J D and Celanovic I 2013 *Optics Express* **21** 2879–2891 ISSN 1094-4087 ISSN 1094-4087
[11] Dashiell M W, Beausang J F, Ehsani H, Nichols G J, Depoy D M, Danielson L R, Talamo P, Rahner K D, Brown E J, Burger S R, Fourspring P M, Jr W F T, Baldasaro P F, Wang C A, Huang R K, Connors M K, Turner G W, Shellenbarger Z A, Taylor G, Li J, Martinelli R, Donetski D, Anikeev S, Belenky G L and Luryi S 2006 *Electron Devices, IEEE Transactions on* **53** 2879–2891 ISSN 0018-9383 ISSN 0018-9383
[12] Chan W R, Huang R, Wang C A, Kassakian J, Joannopoulos J D and Celanovic I 2010 *Sol. Energy Mater. Sol. Cells* **94** 509–514 **94** 509–514
[13] Wilt D, Wehrer R, Palmisiano M, Wanlass M and Murray C 2003 *Semiconductor Science and Technology* **18** S209–S215 ISSN 0268-1242 ISSN 0268-1242

Autoparametric Resonance Systems for Vibration-Based Energy Harvesters

L. Kurmann[1], D. Hoffmann[2], B. Folkmer[2], Y. Manoli[2,3], P. Woias[4], R. Anderegg[1]

[1]University of Applied Sciences and Arts Northwestern Switzerland, School of Engineering, Institute of Automation, Klosterzelgstrasse 2, 5210 Windisch, Switzerland
[2]Hahn-Schickard, Wilhelm-Schickard-Str. 10, 78052 Villingen-Schwenningen, Germany
[3]Fritz Huettinger Chair of Microelectronics, Department of Microsystems Engineering – IMTEK, University of Freiburg, Georges-Köhler-Allee 102, 79110 Freiburg, Germany
[4]Laboratory for the Design of Microsystems, Department of Microsystems Engineering – IMTEK, University of Freiburg, Georges-Köhler-Allee 102, 79110 Freiburg, Germany

E-mail: lukas.kurmann@fhnw.ch

Abstract. Motivation for this paper is the creation of a new kind of (vibration) kinetic energy harvester systems that can effectively transfer environmental mechanical vibrations into electrical energy over a wider frequency bandwidth than conventional devices. This paper presents a potential improvement in the 1DoF vibration transducer class and examining therefore analytically the behavior of such systems using strong nonlinear springs. Then a new 2DoF class of vibration transducer is presented having a strong nonlinear characteristic which is well suited for autoparametric resonance vibrations.

1. Introduction

Energy Harvesting is a technology for capturing non-electrical energy from ambient energy sources, converting it into electrical energy and storing it to power wireless electronic devices. The process of capturing mechanical energy such as shocks and vibrations is a particular field of energy harvesting requiring specific types of energy harvesting devices, so called kinetic energy harvesters. Conventional, first generation types of such transducers can harvest mechanical vibration energy effectively only in a narrow frequency window. Over time many different types of systems have been analytically characterized, designed and tested. Most of these systems show only small improvements with respect to their bandwidth. None of those systems can transfer mechanical vibration power into electrical energy over a wide frequency band. The main requirements of such a kinetic harvester system shall be a simple mechanical structure as well as a wide vibration frequency range for which the system can transfer very effectively environmental mechanical vibrations into electrical energy.

Chapters 2. and 3. deal with 1DoF vibration harvester systems using nonlinear springs. This idea has been proposed recently, by using springs with nonlinear hardening or softening characteristics; some propositions made in [1] and [2] lead to the duffing equation (cubic displacement of path proportional to the restoring force). Based on the duffing oscillator we create in these two chapters, an analytical model and measure its accuracy using numerical solutions. Then we explore the behavior of such 1DoF energy harvester systems by experimenting with larger nonlinear of springs.

In chapters 4. and 5. we deal with 2DoF vibration harvester systems and propose a new kind of vibration energy harvester, capable of autoparametric resonance. A dimensionless system is modeled and numerical solutions show the stability and energy transfer of such a transducer system.

2. Investigation in 1DoF Systems

In this chapter we deal with analytical and numerical simulations of vibration energy harvesters with one single degree of freedom (1DoF). The inertia of the oscillating proof mass results in a relative displacement when the frame experiences acceleration. Once the proof mass is displaced from its equilibrium position, the restoring force of the suspension starts to act on the proof mass and oscillations occur. The oscillations of the proof mass can be damped by a suitable transducer mechanism and thus kinetic energy is converted into electrical energy [1].

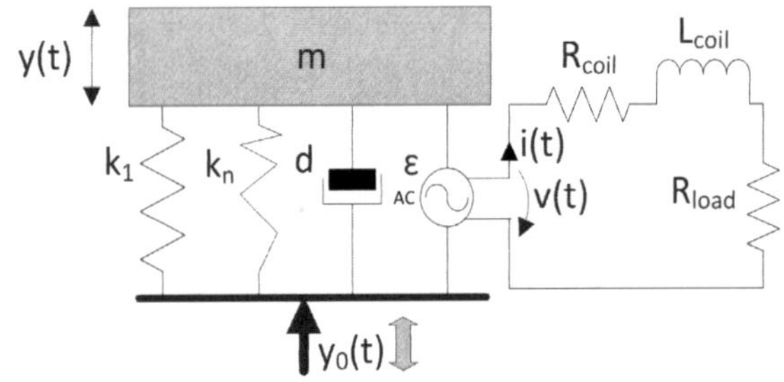

Figure 1. System model elements: a nonlinear oscillator with mass m_1, a damping factor d_1, a linear k_1 and a nonlinear spring k_n. An external basepoint excited harmonic force with amplitude $y_0(t)$; electromagnetic coupling elements: inductance L, resistive load R. System coordinates: mechanical displacement $y(t)$ of mass m, current $i(t)$ through the inductor, and the voltage $v(t)$ generated with the coupling factor ε.

2.1. Lumped parameter model of the electromagnetic 1DoF system

The model comprises a mechanical and an electrical domain which interacts with each other through a coupling element ε, see also system elements in Figure 1. System parameters are the oscillating mass m, the springs k_1 and k_n with the degree of nonlinearity n and the damping d. The amplitude of the ambient vibration is A and its frequency ω; assuming the system is excited by a harmonic function:

$$\ddot{y}_0 = \frac{d^2}{dt^2}(A\cos\omega t) = -A\omega^2\cos(\omega t) \tag{1}$$

The equation of motion for such a system is given by

$$-m\,g - k_1(y_1 - y_0) - k_n(y_1 - y_0)^n - d\,(\dot{y}_1 - \dot{y}_0) + \varepsilon\, i = m\,\ddot{y}_1 \tag{2}$$

Changing equation by adding $-m\,\ddot{y}_0$ and substituting resulting differences with y, $\dot{y}$ and $\ddot{y}$, e.g. $y = y_1 - y_0 + c_0, \dot{y} = \dot{y}_1 - \dot{y}_0, \ddot{y} = \ddot{y}_1 - \ddot{y}_0$ we obtain a DE (of accelerations):

$$\ddot{y} + \frac{d}{m}\dot{y} + \frac{k_1}{m}y + \frac{k_n}{m}y^n + \frac{\varepsilon}{m}\,i = \ddot{y}_0 = -A\omega^2\cos(\omega t) \tag{3}$$

Introducing the coil to obtain a voltage for an electrical resistive load, Faraday's law of induction needs to be applied (see also [1] for a more detailed treatment); equation for the generated voltage ($R = R_{coil} + R_{load}$):

$$v(t) = \varepsilon\,\dot{y} = L\frac{di}{dt} + R\,i \tag{4}$$

The set of equation (3) and (4) models the system of motion. The parameter ε is the electromagnetic coupling constant (transduction factor) $[Vs/m]$, and is measured for the electromagnetic current i $[A]$ that is generated upon velocity of deflection y, assuming having resistive load R in the electrical system domain. To obtain a more general solution, the equations of motion are written in the dimensionless form:

$$u'' + 2\delta u' + u + \beta u^n + \kappa\lambda\,\zeta = -\Omega^2\cos(\Omega\tau) \tag{5}$$

$$\kappa\, u' = \zeta' + \lambda\, \zeta \tag{6}$$

Using following parameters for nondimensionalization:

$$y_0 = A; i_0 = \frac{\varepsilon\, y_0}{R}; \omega_0{}^2 = \frac{k_1}{m}; \beta = \frac{k_n}{k_1}\, y_0{}^2; \tau = t\, \omega_0; \delta = \frac{d}{2\, m\, \omega_0}; \Omega = \frac{\omega}{\omega_0}; \kappa = \frac{\varepsilon^2}{L\, d\, \omega_0}; \lambda = \frac{R}{L\, \omega_0}; \tag{7a}$$

$$path\; u(\tau) = \frac{y(t)}{y_0}\; and\; current\; \zeta(\tau) = \frac{i(t)}{i_0} \tag{7b}$$

In the equation system of (5) and (6), δ is the damping ratio and β the nonlinearity constant together with the exponent n in u^n indicating the system's degree of nonlinearity. A well-known case with n=3 describes the duffing oscillator. The parameter κ is called transduction factor (and assumed to be constant in this analytical treatment) and the load matching constant λ. In case of resistive load matching, the resistance is equal to the apparent impedance of the inductance:

$$\lambda = \frac{|Z(L)|}{L\, \omega_0} = \frac{\omega\, L}{L\, \omega_0} = \Omega \tag{8}$$

For a nonlinear system it is often sufficient to use a limited set of harmonics to obtain a sufficient accurate solution for the system, see also [5]; here only the fundamental frequency is considered:

$$u(\tau) = u_C\, cos(\Omega\tau) + u_S\, sin(\Omega\tau) \tag{9a}$$

$$\zeta(\tau) = \zeta_C\, cos(\Omega\tau) + \zeta_S\, sin(\Omega\tau) \tag{9b}$$

The ansatzfunction (9a) is inserted in equation (6) and its analytical solution is obtained:

$$\zeta(\tau) = e^{-\lambda\tau} c_1 - \frac{\Omega}{\lambda^2+\Omega^2}\left((u_S\lambda + u_C\Omega)\, cos(\tau\Omega) + (-u_C\lambda + u_S\Omega)\, sin(\tau\Omega)\right) \tag{10}$$

Here only the nonhomogeneous solution is of interest, as for large τ only this solution will remain. Inserting again (9a) and its derivative in (10) yield the relation between displacement and voltage:

$$\zeta(\tau) = \frac{\lambda}{\lambda^2+\Omega^2} u' - \frac{\Omega^2}{\lambda^2+\Omega^2} u \tag{11}$$

With this relation, equation (5) can be rewritten:

$$u'' + 2\delta u' + u + \beta u^n + \underbrace{\frac{\kappa\lambda^2}{\lambda^2+\Omega^2}}_{d_{EM}} u' - \underbrace{\frac{\kappa\lambda\Omega^2}{\lambda^2+\Omega^2}}_{k_{EM}} u = -\Omega^2\, cos(\Omega\tau) \tag{12}$$

Energy harvesters are typically characterized by the three different parameters, the open-circuit voltage $\lambda \to 0$ ($d_{EM} = 0, k_{EM} = 0$) the short circuit current $\lambda \to \infty$ ($d_{EM} = \kappa, k_{EM} = 0$) and the maximum power output (load matching) $\lambda \to \Omega$ ($d_{EM} = \kappa/2, k_{EM} = -\kappa\Omega/2$). These three load cases correspond to the terms of d_{EM}, k_{EM} in equation (12). In theory such considerations are reasonable, as for many electrical circuits load matching is used. For electromagnetic harvesters, the influence of the reactance is below 10% of the resistance and optimal load is obtained of adding coil resistance plus the quadratic transduction factor divided by the parasitic mechanical damping d, see also [1].

3. Comparison of frequency response for electromagnetic 1DoF vibration harvesters with different degrees of nonlinearity

Equation (12) gives a general relation for basepoint excited (nonlinear) electromagnetic harvester systems; if β is zero, no nonlinear spring is present and we deal with a linear transducer system. If β is nonzero, we examine nonlinear springs with nonlinearity exponent $n = 3$ (duffing case) and $n = 5$.

3.1. Frequency response with nonlinear spring n=3 (duffing case)

A sufficient accurate approximation for equation (12) can be obtained using the harmonic balance (HB) method. As already stated, we neglect the influence of higher harmonics and assume the ansatzfunction (9a); the cubic term u^3 is approximated by:

$$u^3 = \tfrac{3}{4} \approx \tfrac{3}{4}(u_C^2 + u_S^2)\, u = \tfrac{3}{4}\, \hat{u}^2 u \tag{13}$$

Two linear equations for sine and cosine balance term are formed following the HB procedure. An implicit relation between displacement amplitude and excitation frequency can here be obtained:

$$\hat{u}^2\left(\left(-\Omega^2 + 1 + \tfrac{3}{4}\beta\, \hat{u}^2 - \tfrac{\kappa\lambda\Omega^2}{\lambda^2+\Omega^2}\right)^2 + \left(2\delta\,\Omega + \tfrac{\kappa\lambda^2}{\lambda^2+\Omega^2}\right)^2\right) = 1 \tag{14}$$

As proposed by Neiss et al in [4], we also limit our investigation of the maximum power output to weakly coupled systems, i.e. $\kappa \leq 4\delta\Omega$ and follow also by using a quadratic approximation of the frequency around the linear resonance frequency by applying:

$$\left(2\delta\,\Omega + \frac{\kappa}{2}\right)^2 \approx \left(\delta + \frac{\kappa}{4}\right)(\kappa + 4\,\delta\,\Omega^2) \tag{15}$$

By therefore replacing the damping and stiffness and inserting (15) into (14) we obtain:

$$\hat{u}^2\left(\left(-\Omega^2 + 1 + \tfrac{3}{4}\beta\, \hat{u}^2 - \tfrac{\kappa}{2}\right)^2 + \left(\delta + \tfrac{\kappa}{4}\right)(\kappa + 4\,\delta\,\Omega^2)\right) = 1 \tag{16}$$

This simplified implicit frequency response is now a quadratic polynomial which can be elegantly solved for Ω^2:

$$\Omega_{1,2} = \tfrac{1}{2}\Big(3\,\hat{u}^2\beta + 4 - 8\delta^2 - 2\kappa - \delta\kappa \pm \tfrac{2}{\hat{u}^2}\Big(4\,\hat{u}^2 - 3\hat{u}^6\beta\delta(4\delta+\kappa) + \hat{u}^4(4\delta+\kappa)\big(4\delta(-1+\delta^2) + (-1+\delta(2+\delta))\kappa\big)\Big)^{1/2}\Big)^{1/2} \tag{17}$$

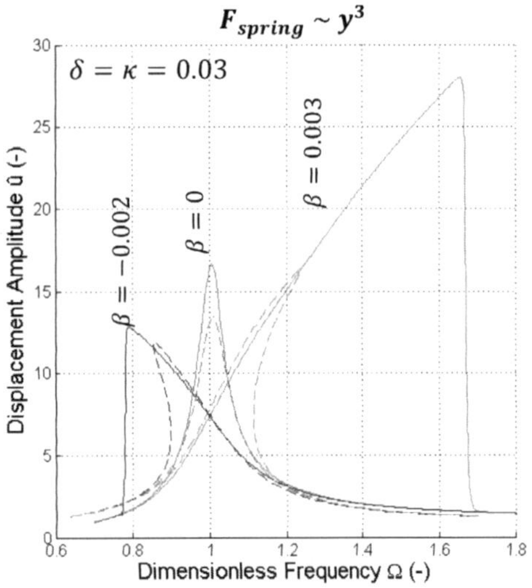

Figure 2. Exemplary frequency responses linear (grey), softening (blue) and hardening (green) using numerical simulation (solid lines) and analytical solution (dashed lines).

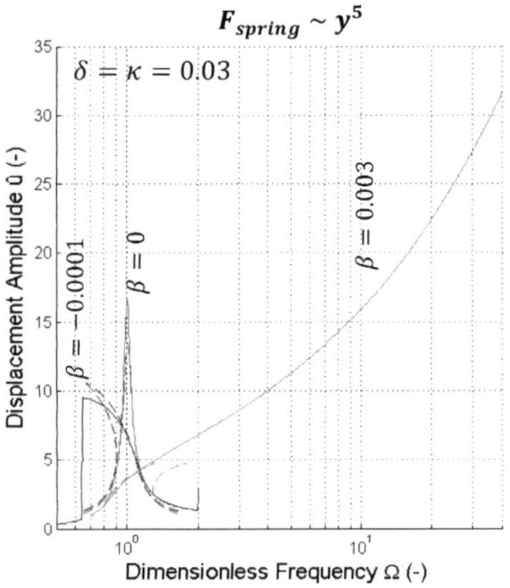

Figure 3. Exemplary frequency responses linear (grey), softening (blue) and hardening (green) using numerical simulation (solid lines) and analytical solution (dashed lines).

A selection for the analytic solution from equation (17) along with the numerical solutions obtained by solving the coupled DE system in (5) and (6) is shown in Figure 2. The solution errors between analytical and numerical analysis show that the backbone curve deviation is small (in the range of ca. <1% over a large parameter range). The amplification errors for the analytical solutions are larger, best matches at ca. 3%. Jump-up and jump-down point identification, maximum power output and bandwith calculations are not presented, as the scope of the paper is not limited to this special nonlinear case.

3.2. Frequency response with nonlinear spring n=5

As stated for the duffing case (n=3) we approximate equation (12) again by using the harmonic balance (HB) method and neglecting again the influence of higher harmonics assuming the ansatzfunction (9a); the stiffness term with u^5 is approximated by:

$$u^5 \approx \frac{5}{8}\hat{u}^4\, u \tag{18}$$

Using again the same approach as presented in chapter 3.1. leads to the analytical simplified implicit frequency response; its quadratic solution for Ω yields:

$$\Omega_{1,2} = \frac{1}{2}\left(\frac{5u0^4\beta}{2} - 2\big(-2+\kappa+\delta(4\delta+\kappa)\big) \pm \frac{\sqrt{2}}{u0^2}\left(u0^2\left(8 + u0^2(4\delta+\kappa)\big(-2\kappa+\delta(-8-5u0^4\beta+8\delta^2+2(2+\delta)\kappa)\big)\right)\right)^{1/2}\right)^{1/2} \tag{19}$$

Figure 3 shows exemplary frequency responses for the analytic solution from equation (19) together with the numerical solution of the coupled DE system. For the hardening case, numerical simulation show that the displacement amplitude is growing indefinitely, until it reaches so large amplitudes that the system becomes instable ($\Omega > 148$ @ $\hat{u} \approx 150$) and the jump-down phenomenon occurs therefore very late; the analytical solution shows no jump-down phenomenon, as the amplitude reaches a complex solution for frequencies $\Omega > 1$. Furthermore in the analytical solution no such amplitude amplification can be observed. It is important to note for all numerical simulations that the generated basepoint excitation is a chirp signal that is slowly ramped up for hardening systems, e.g. $\beta > 0$ and slowly ramped down for softening systems ($\beta < 0$); if a quasi-static excitation is used for each frequency (always with initial conditions set to zero), no large amplitude amplifications can be observed.

The analytical frequency response grows with a higher power. For the backbone deviation in the u^5 solution is ca. 3%; the amplification deviation is larger, best matches at ca. 5%. These mismatches can be attributed to the fact of having only the fundamental harmonics introduced for an acceptable manageable analytical expression. Making a spring with u^3 is feasible, see also [1]; it needs further investigation whether a spring with u^5 can be realized without introducing too much of damping!

4. Investigation in 2DoF Systems

In this chapter we deal with models and numerical simulations of vibration energy harvesters with two degrees of freedom (2DoF). Such systems are capable of autoparametric resonances, e.g. they stabilize their frequency response in a limited window of excitation, see also [6] and [7]. Figure 4 shows one configuration of a new basepoint excited vibration harvester system. The corresponding electrical network is deliberately drawn disconnected, as the transducer can be placed between $y_0$and y_1 or $y_0$and y_2 or $y_1$and y_2.

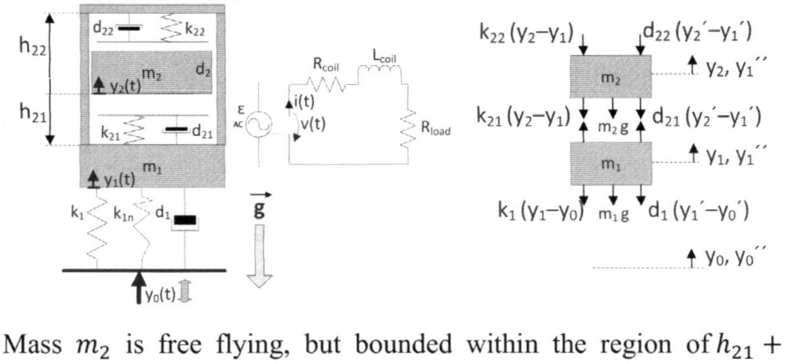

Figure 4. Lumped parameter model of the electromagnetic 2DoF vibration energy harvester; the electrical network showing the transducer unit can be placed (in theory) at three locations: between $y_0$and y_1 or y_0 and y_2 or $y_1$and y_2. On the right hand side the cut free system.

Mass m_2 is free flying, but bounded within the region of $h_{21} + h_{22}$. If the drawn spring $k_{21,22}$ elements are present, mass m_2 has soft impact, otherwise hard impact behavior. The mechanical damping elements $d_{21,22}$ are always present. The input vibration is y_0 and like in the 1DoF system, a linear and also a nonlinear spring might be present, plus the unwanted mechanical dampings d_1 plus the sketchily drawn system wall damping d_2.

4.1. Lumped parameter model of the electromagnetic 2DoF system

The lumped parameter model is shown in Figure 4. The proposed harvester can be thought as 2DoF upgrade of the 1DoF model shown in Figure 1 using the same nomenclature as already introduced in chapter 2. (with additionally on top of mass m_1 a second bounded in $h_{21} + h_{22}$ free flying mass m_2). Following the same nondimensionalization procedure as shown in chapter 2.1. will lead to the three dimensionless coupled differential equations for the system dynamic:

$$u_1'' + \frac{g}{y_0\omega_0^2} - s_a(\lambda_{K1}\Delta u + 2\delta\lambda_{D1}\Delta u') - s_b(\lambda_{K2}\Delta u + 2\delta\lambda_{D2}\Delta u') - s_c(\lambda_{K1}(\Delta u - \rho) + 2\delta\lambda_{D1}\Delta u') \quad = -\Lambda(\tau) \qquad (20)$$

$$u_2'' + \frac{\lambda_M\, g}{y_0\omega_0^2} + s_a(\lambda_{K1}\Delta u + 2\delta\lambda_{D1}\Delta u') + s_b(\lambda_{K2}\Delta u + 2\delta\lambda_{D2}\Delta u') + s_c(\lambda_{K1}(\Delta u - \rho) + 2\delta\lambda_{D1}\Delta u') + \kappa\lambda_E\,\zeta = 0 \qquad (21)$$

$$\kappa\, u' = \zeta' + \lambda_E\,\zeta \qquad (22)$$

with following supplementary equations for (21) and (20):

$$\Delta u = u_2 - u_1;\ \Delta u' = u_2' - u_1';\ s_a = \sigma(-\Delta u);\ s_c = \sigma(\Delta u - \rho);\ s_b = 1 - s_a - s_c \qquad (23)$$

and with the inserted basepoint excitation $A\cos(\Omega\tau)$ the stimulation Λ can be written as:

$$\Lambda(\tau) = u_1 + \beta\, u_1^n - A\,cos(\Omega\tau) + 2\delta(u_1' + A\,\Omega\,sin(\Omega\tau)) \qquad (24)$$

The activation of according stiffness and damping depending on freefall (s_b), lower impact (s_a) and upper impact (s_c) are switched on and off via only path depending Heaviside functions.

5. Comparison of frequency response for electromagnetic 2DoF vibration harvesters

Frequency response of the three dimensionless coupled differential equations (20), (21) and (22) are depicted in Figure 5, using numerical simulation tools. The nonlinearity of the spring k_{1n} is cubic (using a weak nonlinearity with $\beta = 0.03$), e.g. exponent $n = 3$ in the stimulation function Λ.The springs $k_{21} = k_{22}$ are linear, the damping for impact $d_{21} = d_{22}$ kept also equal (see also Figure 4), resulting in dimensionless constants $\lambda_{K1} = \lambda_K = 20$ and $\lambda_{D1} = 10^4$. The free fall damping $\lambda_{D2} = 1$ is weak (free fall stiffness $\lambda_{K2} = 0$) and the proportional factor for masses m_1 and m_2 is set to $\lambda_M = 1/3$. In the electrical domain the dimensionless constants are the transduction factor κ (also assumed to be constant) and the resistive load is assumed to be weak $\lambda_E = 0.05$. For the upper mass m_2 bounded path limits are set to h_{21}=0 and h_{22} dimensionless transformed constant set large to $\rho = 50$ (resulting in no upper impact).

Exciting the system with a frequency $\Omega < 4$ results in a linear frequency response (phase-plot in detail A), as for such frequencies the resulting acceleration for m_2 is too weak (in comparison to the gravitational acceleration) and therefore it will not lift off. Increasing the frequency further, m_2 starts intermittently to lift off, this is shown in phase-plots details B-E. Especially interesting is the behavior of mass m_2 in interval $0.8 < \Omega < 1.5$ (phase-plot details C and D are similar) – over this large interval, we have an auto-stabilization frequency effect of mass m_2 (autoparametric resonance).

6. Conclusions

New analytical models for the set of DE given in (5) and (6) have been created using a well-studied nonlinear exponent ($n = 3$) and a new ($n = 5$) for 1DoF vibration energy harvesters leading to solutions presented in (17) and (19); these analytical models provide a qualitative estimation of the nonlinear frequency response (compared to the numerical solutions). The larger the exponent grows, the larger the estimated analytical error becomes. Furthermore we showed, that introducing a very strong nonlinear spring with exponent $n = 5$, we stay almost indefinitely in resonance for hardening systems (Figure 3).

We then proposed and modeled a new class of 2DoF vibration harvester systems and showed a first none exhaustive analysis of the energy transfer using a dimensionless model, showing autoparametric resonance for both masses m_1 and m_2 over a large frequency range (see also Detail C and D, Figure 5 and chapter 5. for further details).

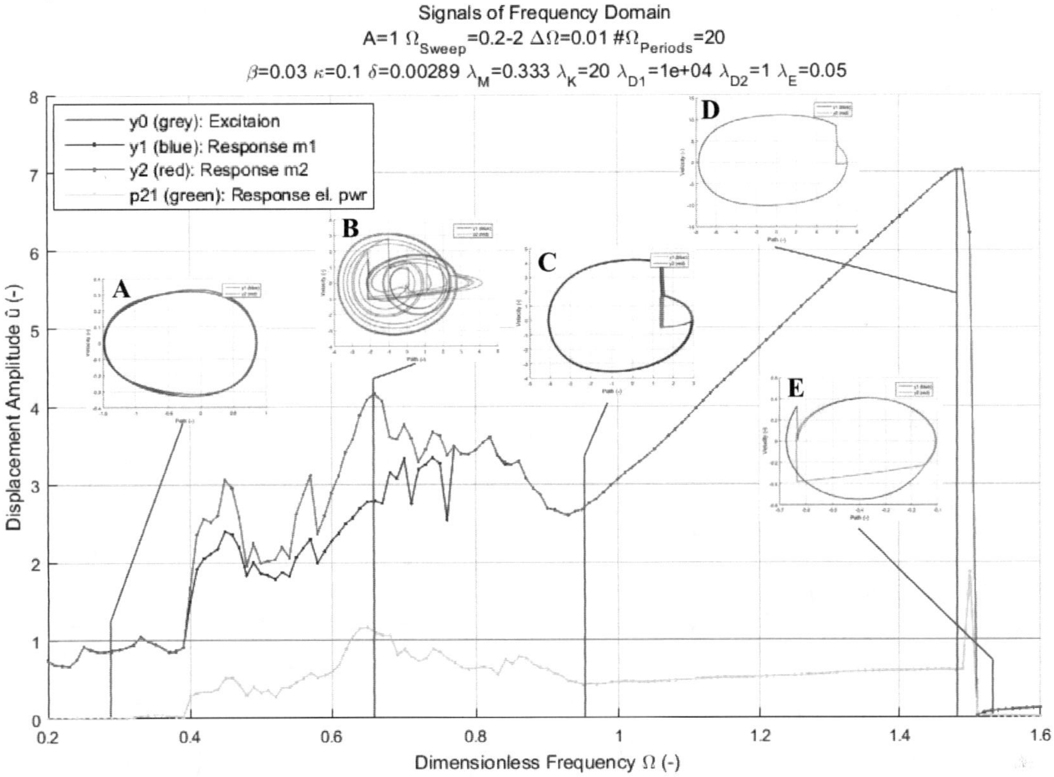

Figure 5. Dimensionless frequency responses y_1 (blue) and y_2 (red) and electrical power (green) having weak electrical load (see also chapter 4.) using an excitation chirp signal y_0 (grey).

References

[1] Spreemann, Manoli, Electromagnetic Vibration Energy Harvesting Devices, Springer, ISSN1437-0387

[2] D. Hoffmann et al, „Erarbeitung von Grundlagen zum Entwurf von Vibrationswandlern mit verbesserter Frequenzbandcharakteristik zur Steigerung der effektiven Energiewandlung“ http://www.hahn-schickard.de/en/projects-publications/publications/#read, 2014 Abschlussbericht IGF-Vorhaben-Nr: 16910 N

[3] F. Goldschmidtboeing, M. Wischke, C. Eichhorn and P. Woias, “Parameter identification for resonant piezoelectrical energy harvesters in the low- and high-coupling regimes”, Journal of Micromechanics and Microengineering, 2011

[4] Neiss et al, “Analytical model for nonlinear piezoelectrical energy harvesting devices”, Smart Mater. Strcut. 23, 2014

[5] Aiharal, “Analysis of Steady State Vibration for Piecewise Linear Systems Subjected to Periodic Excitation”, Dissertation, Tokyo Metropolitan University, 2013

[6] Lukas Kurmann, Roland Anderegg, Sandro Wiedmer, Henner Rüschkamp, “Control of weak and strong nonlinear dynamics in a novel milling cutting process”, IEEE Conference on Mechatronics and Automation in Chengdu, China, 2012

[7] Roland Anderegg, Lukas Kurmann, Sandro Wiedmer, “Mechatronics of activated milling process for coal mining”, International Conference of Mechatronic, Linz, Austria, 2012

Multi-physics modelling approach for oscillatory microengines: application for a microStirling generator design

F. Formosa[1], L.G. Fréchette[2]
1 SYMME, Université Savoie Mont Blanc, Annecy 74000, France,
2 Université de Sherbrooke, Département de génie mécanique, 2500 boul. Université, Sherbrooke, Québec J1K 2R1, Canada

fabien.formosa@univ-smb.fr

Abstract. An electrical circuit equivalent (ECE) approach has been set up allowing elementary oscillatory microengine components to be modelled. They cover gas channel/chamber thermodynamics, viscosity and thermal effects, mechanical structure and electromechanical transducers. The proposed tool has been validated on a centimeter scale Free Piston membrane Stirling engine [1]. We propose here new developments taking into account scaling effects to establish models suitable for any microengines. They are based on simplifications derived from the comparison of the hydraulic radius with respect to the viscous and thermal penetration depths ($\delta_v = \sqrt{2\mu/(\bar{\rho}\omega)}$ and $\delta_k = \sqrt{2k/(\bar{\rho}C_p\omega)}$ respectively).

1. Introduction

Multi-physics coupling is a critical issue for microengines design. On the one hand, the electrical analogy is a useful strategy as it allows multiple physics to be described in a common language. On the other hand, micromachines can be usually divided into elementary components/functions (transducers, membrane/piston, thermal insulation…) and geometries (heat exchangers ducts, chambers…). Apart from the microturbines, they present common features: oscillatory operation, membranes, piezoelectric transducers [2]. Considering the gas flow, linear thermoacoustics theory [3] offers a framework for small harmonic variations that defines equivalent electrical models in which the pressure is the potential and the fluid flow is the current. Thus, we can develop the model of each component of an engine from a simplified 1D scheme of the fluid flow along the fluid flow axis ($\vec{x}$). Moreover, the assumed simplified geometry consists in two parallel plates for which in plane dimension along one direction is much greater than the transverse dimension ($\vec{y}$). For preliminary design, the presented approach is deemed accurate enough as long as it has been successfully run for a centimeter scale Free Piston membrane Stirling engine [1].

The variations of the pressure $p(x,t)$, the volume $V(t)$, the temperature $T(x,y,t)$ and the velocity $u(x, y, t)$ are assumed to be small with respect to their mean value over a cycle which are noted $\bar{p}$, $\bar{V}$, $\bar{T}$ respectively (the mean velocity is zero).

The equilibrium equations (1-3) for the mass, the momentum and the energy conservations respectively are the starting point of the ECE developments:

$$\frac{1}{\gamma r\bar{T}}\left(\frac{\partial p}{\partial t} - (\gamma-1)\Phi\right) + \bar{\rho}\frac{\partial u}{\partial x} = 0 \qquad (1)$$

$$\bar{\rho}\frac{\partial u}{\partial t} = -\frac{\partial p}{\partial x} + \frac{1}{2}\mu\frac{\partial^2 u}{\partial y^2} \tag{2}$$

$$\bar{\rho}C_p\frac{\partial T}{\partial t} = \frac{\partial p}{\partial t} + k\frac{\partial^2 T}{\partial y^2} \tag{3}$$

In which Φ is the thermal heat flux per unit volume, k the thermal conductivity, γ is the ratio of specific heats and C_p the thermal capacity.

2. Developments of electrical equivalent models for basic components

2.1 Chambers

Chambers are defined as 0D volumes such as unique pressure, temperature and density stand for the state of the gas within the chamber. Figure 1a presents the generic schematic of the studied geometry. The adiabatic assumption is usually retained when the reaction time of the heating process is much longer than the period of oscillation. Besides, technological constraints may lead to thick walls to endure the large pressure solicitation. Considering microengines, the downsizing induces large heat transfer so the isothermal assumption is also of relevance and considered here.

From eq. (1), the adiabatic and isothermal assumptions lead to the ECE model of figures 1b and 1c respectively. The expression of the capacitors are given in table 1.

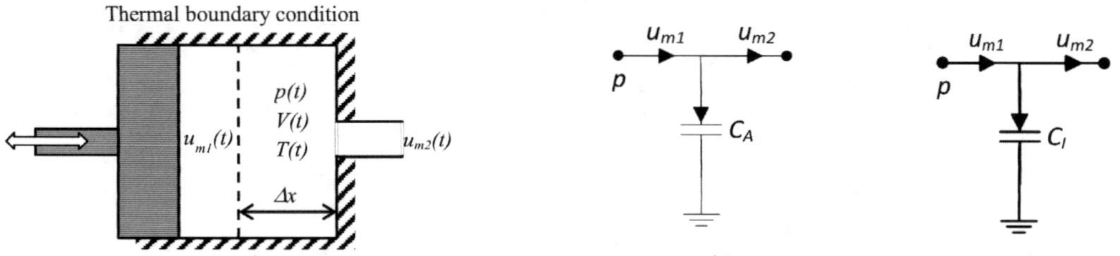

a) b) c)

Figure 1. a) basic scheme for the study of compression/expansion chamber, b) ECE model for adiabatic chamber c) ECE model for isothermal chamber

Table 1. Expressions for the electrical components for chambers and heat exchangers

	Boundary	Electrical circuit equivalent components	
		$r_h < \delta$	$r_h > \delta$
Chamber	Isothermal	$C_I = \Delta x/\bar{p}$	
	Adiabatic	$C_A = \Delta x/\gamma\bar{p}$	
Heat exchanger	Isothermal	$R_{vS} = 3\mu\,\bar{\rho}\frac{\Delta x}{{r_h}^2}$ $C_{IS} = \Delta x\,(2-\gamma)/(\gamma\bar{p})$	$R_{vL} = \frac{1}{2}\,\bar{\rho}\,\omega\frac{\delta_v}{r_h}\Delta x$, $L_{vL} = \bar{\rho}\left(1+\frac{1}{2}\frac{\delta_v}{r_h}\right)\Delta x$ $C_{IL} = \Delta x/(\gamma\bar{p})$, $R_{ILk} = 2\bar{p}r_h\gamma/(\omega\delta_k\Delta x(\gamma-1))$
	Generic heat exchange	$R_{vS} = 3\mu\,\bar{\rho}\frac{\Delta x}{{r_h}^2}$ $C_{HS} = \Delta x\left(1+2(\gamma-1)\frac{1}{\omega}\frac{h}{Nu}\frac{\bar{T}}{\bar{\rho}C_p}\frac{r_h}{\delta_k}\right)/(\gamma\bar{p})$	$R_{vL} = \frac{1}{2}\,\bar{\rho}\,\omega\frac{\delta_v}{r_h}\Delta x$, $L_{vL} = \bar{\rho}\left(1+\frac{1}{2}\frac{\delta_v}{r_h}\right)\Delta x$ $C_{HL} = \Delta x/(\gamma p_m)$, $R_{HLk} = C_p\bar{T}\bar{p}\bar{\rho}\gamma/(h\Delta x(1-\gamma))^*$ * high heat transfer coefficient

2.2 Heat exchangers

The heat exchangers model has to take into account viscous dissipation along the $\vec{x}$ axis as well as various standard thermal boundary conditions: isothermal walls (T_w = cste) and generic heat transfer for which the mean thermal heat flux is $\bar{\Phi} = h/r_h(\bar{T} - T_{ext})$ and the instantaneous heat flux is $\Phi = h/r_h\,T$.

Figure 2a presents an architecture of a micro-heat exchanger [4]. From its dimensions, assuming air as the working fluid, the thermal and viscous penetration depths are plotted as a function of the operating frequency in the range [100 Hz, 4000 Hz] (figure 2b). Dashed dotted horizontal lines are the upper and lower bounds of the considered width of the channel from [4]. It can be inferred from figure 2b that both small and large hydraulic radius to penetration depths ratio cases have to be considered.

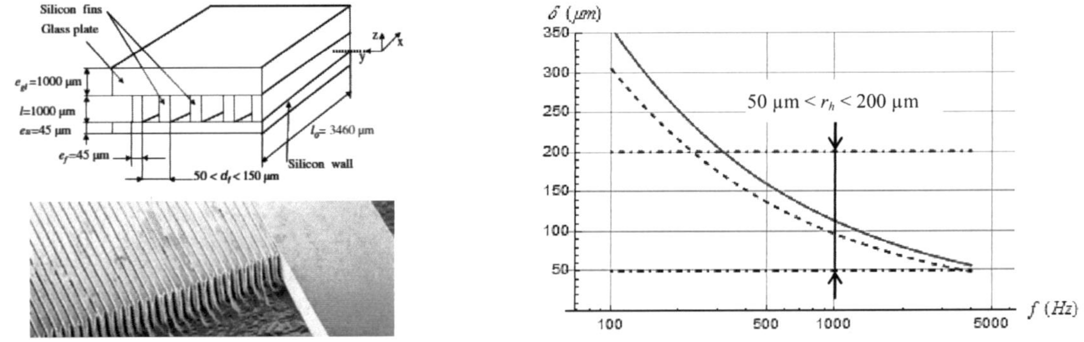

a) b)

Figure 2. a) LIGA technology micro-heat exchanger, b) evaluation of δ_v and δ_k

For isothermal wall conditions, we obtain the ECE models shown in figure 3a and 3b for small and large penetration depth to hydraulic radius ratio respectively. For generic heat transfer, we obtain the scheme of figure 3c-3d. Table 1 gives the values for the electrical components.

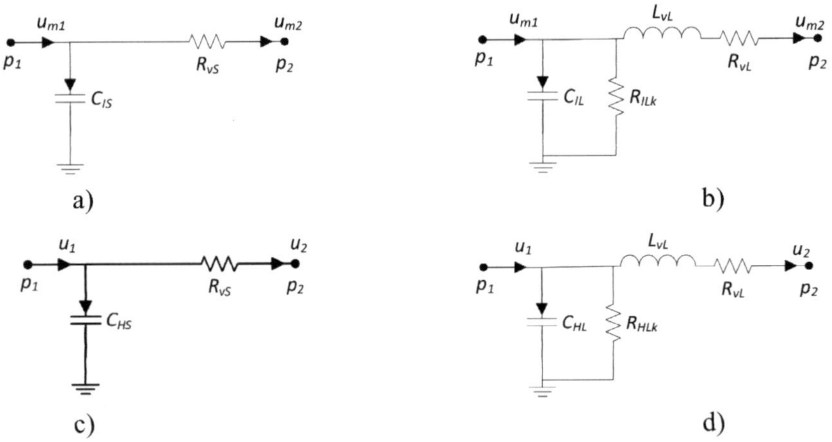

Figure 3.ECE for heat exchangers considering isothermal walls: a) $r_h < \delta$ and b) $r_h > \delta$; generic heat transfer: c) $r_h < \delta$ and d) $r_h > \delta$

2.3- Membrane and electromechanical convertor

Oscillating membranes are usually preferred compared to sliding pistons as they are compatible with planar process and provide perfect sealing with controlled dissipation. Yet, their dynamic behavior has to be accurately modelled including potential large deformation effects. Indeed, micro-engines are free pistons mechanical arrangement so the steady state operation stroke arises from nonlinear effects. Large deformation analytical membranes models are developed in the literature and their integration in the electrical model framework has been demonstrated in [1]. The associated ECE model is presented in figure 4b.

As an example of an integrated piezoelectric the single layered PVDF membrane such as shown in figure 4a can be used for electromechanical conversion. Large deformation effects induces tension stress which produces electrical charge on the electrodes [5]. The ECE model of the piezoelectric membrane including large deformation effect is given in figure 4c in which C_0 is the actual electrical capacitance of the PVDF membrane and R_l the actual electrical resistance load.

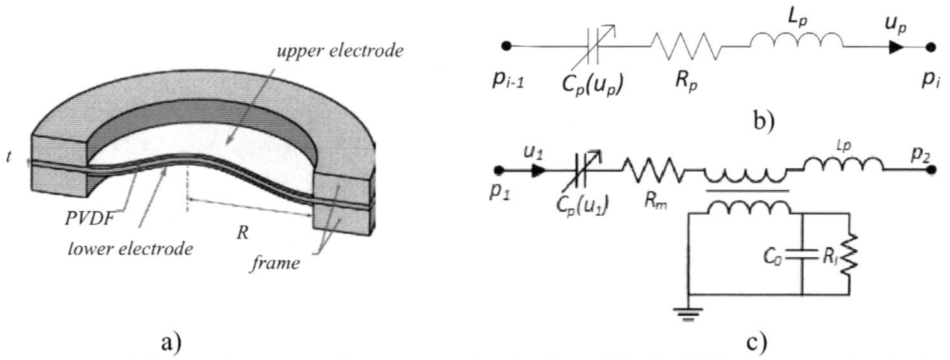

Figure 4. a) PVDF membrane [5], b) ECE mechanical model, c) ECE electromechanical model

3. Design test cases: Multiphase micro-Stirling engine

This modeling approach has been used to design a miniature planar free-piston Stirling generator for low temperature waste heat recovery (figure 5). It consists of three coupled elementary phases, each composed of: *i)* a membrane (∅ 5 mm, thickness h = 40 µm) which acts as a piston and transducer integrating a piezoelectric planar spiral, *ii)* an expansion chamber ($T_e \approx$ 120°C), *iii)* a compression chamber ($T_c \approx$ 30°C) linked by *iv)* a regenerator.

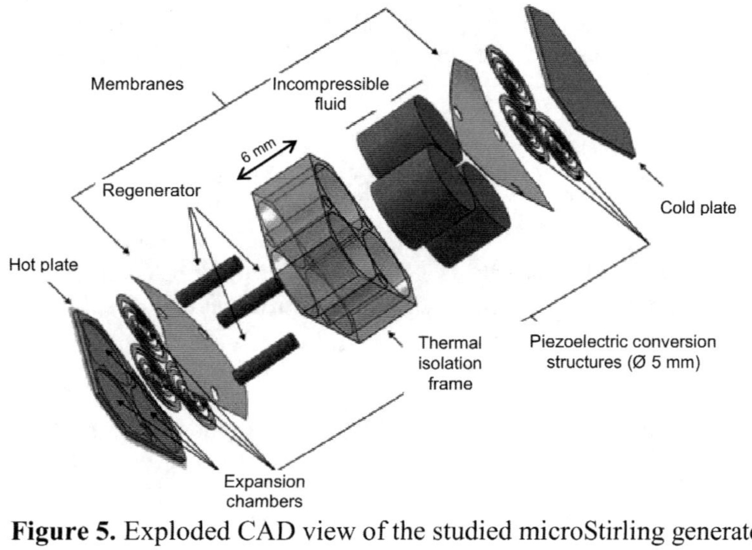

Figure 5. Exploded CAD view of the studied microStirling generator

The associated ECE model for a single phase is shown in figure 6a. The relationship between the pressures from one engine to the other (p^e_{i-1}, p^e_i) and fluid flow rate (u_{i-1}, u_i) can then be easily established [1]. The condition of operation are a 120° phase angle between the variables and a magnitude equals to unity. Therefore, its operation is simulated and the output electrical power derived. Figure 6b and 6c present the evolution of the electrical power and the membrane flexion amplitude as a function of the electrical load, R_l. As the load increases, the power rises (figure 6b) but the harvested energy induces a higher damping, hence reducing the amplitude (figure 6c) and the operating frequency. Design of a microStirling was also proposed with power generation in the range 60-80 mW/cm^2.

These results underline the ability of the proposed ECE modeling approach for the design and optimization of oscillating microengines, accounting for coupled multi-physics and dynamic behavior.

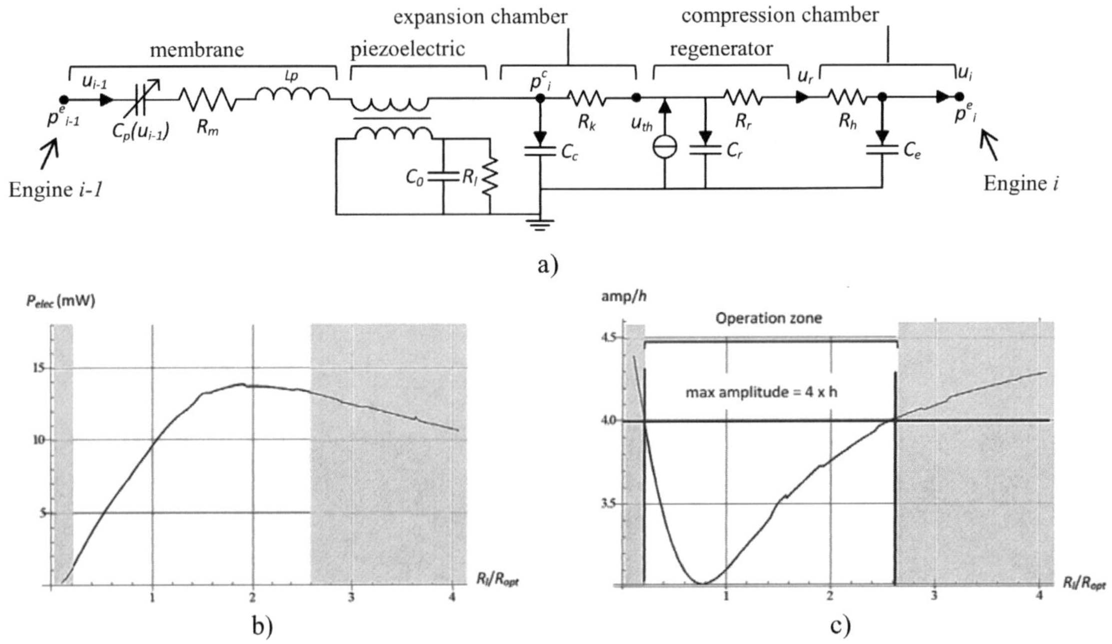

Figure 6. a) Equivalent model for one phase, b) Evolution of the output power with respect to the load resistance, evolution of the stroke with respect to the electric charge

4. Conclusion and perspectives

A novel modelling approach dedicated to oscillating microengines has been proposed. The availability of a model for the main microengine's components allows new concept to be theoretically evaluated and designed. As the phase change thermodynamic cycles prove to be especially relevant for power density, future works will aim at the development and validation of micro-evaporator and micro-condenser modelling in the spirit of the electrical circuit equivalent (ECE) approach.

Acknowledgment

This work has been funded by the French National Research Agency (ANR) under grant no. ANR-12-SEED-0005-01 (MISTIC Project).

References

[1] F. Formosa, A. Badel, and J. Lottin, "Equivalent electrical network model approach applied to a double acting low temperature differential Stirling engine.," *Energy Convers. Manag.*, vol. 78, pp. pp. 753–764, 2014.

[2] S. Percy, C. Knight, S. McGarry, A. Post, T. Moore, and K. Cavanagh, *Thermal Energy Harvesting for Application at MEMS Scale*. Springer-Verlag New York, 2014.

[3] G. W. Swift and S. L. Garrett, *Thermoacoustics: A Unifying Perspective for Some Engines and Refrigerators*, vol. 113, no. 5. 2003.

[4] P. Nika, Y. Bailly, and F. Guermeur, "Thermoacoustics and related oscillatory heat and fluid flows in micro heat exchangers," *Int. J. Heat Mass Transf.*, vol. 48, no. 18, pp. 3773–3792, Aug. 2005.

[5] F. Huet, F. Formosa, and A Badel, "New nonlinear vibration energy harvesters based on PVDF hybrid fluid diaphragm," *J. Phys. Conf. Ser.*, vol. 557, p. 012088, 2014.

Power density improvement of bi-conductive polymer membrane fuel cells by optimization of its internal resistances

S. Hamel, L.G. Fréchette

Institut Interdisciplinaire d'Innovation Technologique (3IT), Université de Sherbrooke, 3000 Boulevard Université, Sherbrooke, J1K 0A5, Québec, Canada

Simon.hamel@usherbrooke.ca

Abstract. This paper reports the development of a new set of equations that defines the internal resistances in bi-conductive membrane (BCM) fuel cells and allows optimization of its critical dimensions and power density. Based on these equations, a design approach is proposed and fabrication steps that limit the power density of the final device are identified. The equations developed in this study can be used for any configuration and shape of holes and current collectors. It demonstrates that by improving fabrication techniques used in a previous study, total internal resistance of the BCM fuel cell could be reduced by 80%, showing that this type of fuel cell could potentially deliver very high power density compared to standard PEMFCs.

1. Introduction

Because of their potentially higher energy density and their instantaneous recharge, polymer electrolyte fuel cells (PEMFCs) are slowly making their way into portable electronic device markets [1]. To become a widespread technology, cost and low energy density issues of PEMFC should however be addressed. To replace the bulky packaging generally used with PEMFCs that significantly reduce their volumetric power density, a new approach to integrate all the necessary components of a fuel cell system on a single layer has been successfully developed in a previous work [2].

Bi-conductive membrane (BCM) fuel cells include on the same layer the current collection, mechanical support and ionic conductive paths and have showed three times better volumetric power density than PEMFC systems with standard packaging (Figure 1). Despite these encouraging results, the areal power density of BCM fuel cells is about 3 times lower than for PEMFCs because of the higher internal resistances generated by this new design. Identifying and quantifying these resistances would allow a better comprehension of BCM fuel cell system and would ease their optimization, thereby increasing the final power density of the device.

In this work, the internal resistances inherent to the BCM fuel cell design are identified and the related analytical equations are developed. This allows the evaluation of the total internal resistance of a BCM fuel cell as a function of its critical dimensions (size of the holes (r_H), spacing (S), radius of patterns (r_P) and membrane thickness (t); Figure 1) and guide the efforts to improve the device.

2. Internal resistances of Bi-conductive membrane (BCM)

To compare easily BCM fuel cells and traditional PEMFCs independently of their area, resistances of fuel cells are evaluated in terms of *area specific resistance*, $\tilde{R}$ ($\Omega\cdot cm^2$), which when multiplied by the current density per area of electrode J ($A\cdot cm^{-2}$) result in the voltage drop in the cell.

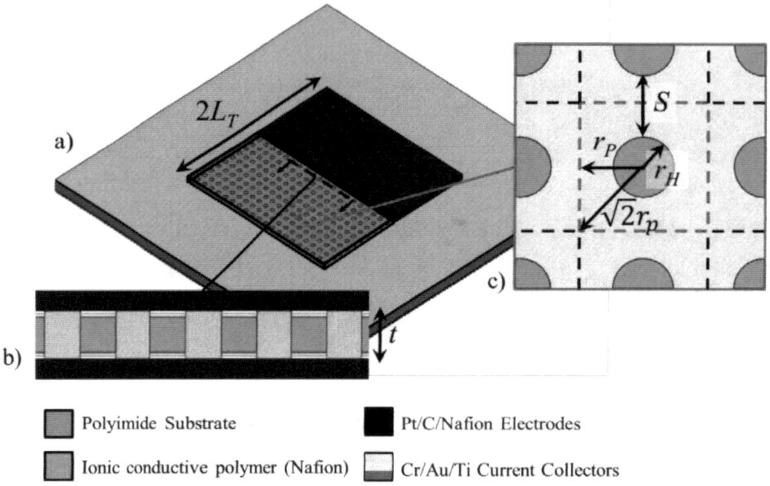

Figure 1 - a) Isometric b) Cross-sectional and c) Close up views of BCPM fuel cell. L_T : Half width of the etched area, t : Thickness of the membrane, r_H : Radius of holes, r_P : Radius of patterns, S : Spacing

In a typical stack configuration, electrons are collected with thick metal grids placed on top of the electrodes while the protons can cross the Nafion membrane which occupies all the volume between the two electrodes. This leads to negligible lateral movements of the electrons and ions in the catalyst layers, hence uniform current generation. In this configuration, the reaction resistance in the catalyst layer ($\tilde{R}_{th}$) as well as the ionic resistance of the Nafion membrane ($\tilde{R}_{IM}$) are the main sources of voltage losses in the cell (Figure 2a). In the BCM configuration, a part of the electrons and protons must overcome a lateral electrical ($\tilde{R}_{EE}$) or ionic ($\tilde{R}_{IE}$) resistance of catalyst layer, respectively, in addition of $\tilde{R}_{th}$ before the reaction. The effective resistance of the catalyst layer $\tilde{R}_E$ must take into account these lateral and reaction resistances. Also, the ionic resistance of the membrane $\tilde{R}_{IM}$ is increased since an important part of the section between the two catalyst layers is occupied with a non ionic conductive substrate (polyimide). Finally, the lateral displacement of the electrons in the integrated current collectors is also subject to a resistance ($\tilde{R}_{EC}$) (Figure 2b).

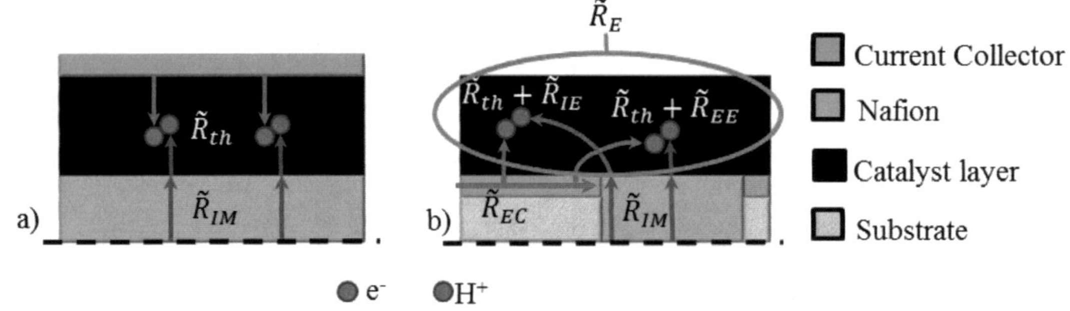

Figure 2 – Area specific resistances in a) half PEMFC with a typical stack configuration and b) half BCM fuel cell.

2.1. Effective resistance of the catalyst layer ($\tilde{R}_E$)

The electrical and ionic lateral resistances of the catalyst layer ($\tilde{R}_{EE}$ and $\tilde{R}_{IE}$) generate a non-linear drop of current density away from the junction between the Nafion hole and integrated current collector (ICC) regions. It can be quantified by evaluating the efficiency of the catalyst layer, E_E, defined as the ratio between the current that a BCM catalyst layer can provide ($I_{P\text{-}BCM}$) on the current that would have been generated by a standard PEMFC ($I_{P\text{-}PEMFC}$) with the same area and $\tilde{R}_{th}$ (Eq. (1)).

$$E_E = \frac{I_{P-BCM}}{I_{P-PEMFC}} \tag{1}$$

For PEMFCs, the generated current can be easily evaluated with Eq. (2) where θ_b is the lost of voltage in the catalyst layer and A its area.

$$I_{P-PEMFC} = \frac{\theta_b}{\tilde{R}_{th}} A \tag{2}$$

For a BCM, because of $\tilde{R}_{EE}$ and $\tilde{R}_{IE}$, a distributed resistance model must be solved to estimate $I_{P\text{-}BCM}$. To find an analytical expression for $I_{P\text{-}BCM}$, the catalyst layer is modelled as a series of distributed resistances on each side of the junction between the Nafion hole and ICC regions (Figure 3a). Also, as an approximation, the square shape of the ICC surrounding the Nafion hole is first considered as a circle shaped current collector of radius $\sqrt{2}r_P$ (Figure 3b). Produced current is estimated independently in each region of the unit cell and then added up to evaluate total $I_{P\text{-}BCM}$.

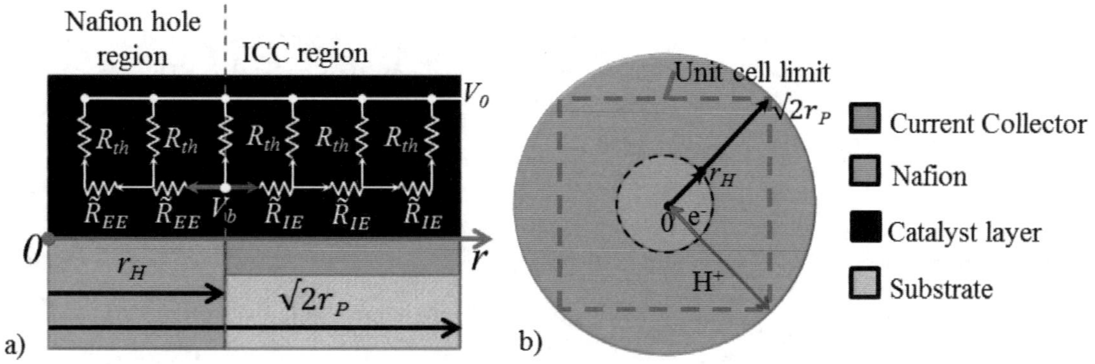

Figure 3 - Top view of BCM (without catalyst layer) illustrating the geometrical limits of the studied zones in the analytical model. b) Distributed resistances model in a cross-sectional view of a BCM.

$\tilde{R}_{EE}$ and $\tilde{R}_{IE}$ vary in function of radial position r because of the radial shape of the unit cell and are proportional to the respective sheet resistance (R_{SE} and R_{SI}) of the catalyst layer. For this work, R_{SE}, have been evaluated to 500Ω/sq with a 4 probes measurement on commercial catalyst layers (Ion power inc., 0.3 mg Pt/cm^2). Also, from the work of Boyer et al. [3], the ionic sheet resistance of the electrodes R_{SI} is evaluated to approximatively 10^4Ω/sq.

Current produced in the Nafion hole region (from $r = 0$ to $r = r_H$) is first solved numerically using the distributed resistance model. It was found that the resulting current produced over the Nafion hole region can be approximated with Eq. (3) if r_H is under 500µm.

$$I_P(0{:}\,r_H) = \pi r_H^2 \int_0^{r_H} \frac{V(r)}{\tilde{R}_{EE}(r)} \approx \pi r_H^2 \frac{\theta_b}{\tilde{R}_{th} + \frac{R_{SE}}{9} r_H^2} \tag{3}$$

For the ICC region, by analogy with heat transfer equations for radial fins, an analytical solution can be determined using modified Bessel functions of the first (I) and second (K) kind of zero and first order [4] :

$$I_P\left(r_H{:}\,\sqrt{2}r_P\right) = \frac{2\pi r_H \theta_b}{R_{SI}\lambda} \cdot \frac{K_1\left(\frac{r_H}{\lambda}\right) I_1\left(\frac{\sqrt{2}r_P}{\lambda}\right) - I_1\left(\frac{r_H}{\lambda}\right) K_1\left(\frac{\sqrt{2}r_P}{\lambda}\right)}{K_0\left(\frac{r_H}{\lambda}\right) I_1\left(\frac{\sqrt{2}r_P}{\lambda}\right) + I_0\left(\frac{r_H}{\lambda}\right) K_1\left(\frac{\sqrt{2}r_P}{\lambda}\right)} \tag{4}$$

where

$$\lambda = \sqrt{\frac{\tilde{R}_{th}}{R_{SI}}} \tag{5}$$

With Eqs. (3) and (4), an analytical solution for the relative efficiency of the catalyst layer, E_E, that includes only the geometrical and resistive terms is defined (Eq. (6)).

$$E_E = \frac{2r_H\lambda}{r_P^2} \cdot \frac{K_1\left(\frac{r_H}{\lambda}\right) I_1\left(\frac{\sqrt{2}r_P}{\lambda}\right) - I_1\left(\frac{r_H}{\lambda}\right) K_1\left(\frac{\sqrt{2}r_P}{\lambda}\right)}{K_0\left(\frac{r_H}{\lambda}\right) I_1\left(\frac{\sqrt{2}r_P}{\lambda}\right) + I_0\left(\frac{r_H}{\lambda}\right) K_1\left(\frac{\sqrt{2}r_P}{\lambda}\right)} + \frac{1}{1 + \frac{R_{SE}}{9\tilde{R}_{th}} r_H^2} \left(\frac{r_H}{\sqrt{2}r_P}\right)^2 \tag{6}$$

The equivalent area specific resistance of one electrode $\tilde{R}_E$ can therefore be evaluated (Eq. (7)):

$$\tilde{R}_E = \frac{\tilde{R}_{th}}{E_E} \tag{7}$$

2.2. *Ionic resistance through the membrane ($\tilde{R}_{IM}$)*

In a BCM fuel cell, part of the surface is used by the mechanical support thereby reducing the area where the ions can cross from the anode to the cathode and increasing $\tilde{R}_{IM}$. This resistance is dependant of the ionic resistivity of the Nafion itself ρ_{Nafion} (~10Ω·cm [5]), thickness of the membrane, t, and the porosity of the current collector, C (Eq. (8)).

$$\tilde{R}_{IM} = \frac{\rho_{Nafion} \cdot t}{C} \tag{8}$$

Ionic resistance through the BCM is therefore proportional to its thickness and inversely proportional to porosity of the ICC, which is also the area covered with Nafion that can be evaluated for the presented configuration with Eq. (9).

$$C = \frac{\pi}{4}\left(\frac{r_H}{r_P}\right)^2 \tag{9}$$

2.3. *Electrical lateral resistance in the ICCs ($\tilde{R}_{EC}$)*

When the electrons reach the ICCs, they must be directed to the periphery by the metallic thin film before going in the external circuit. To find $\tilde{R}_{EC}$, the same model of distributed resistance as for Eq. (3) is used, but a correction factor K_{EC} must be introduced to take into account the large number of etched holes that increase the resistance significantly (Eq. (10)).

$$\tilde{R}_{EC} = K_{EC} \cdot \frac{R_{SC}}{\tilde{R}_{th}} \frac{L_T^2}{9} \tag{10}$$

The value of K_{EC} is evaluated numerically by comparing the loss of voltage of a square porous layer to a plain round one of radius L_T with uniform current. This process has been done for a wide variety of porosity and holes radius. It has been found that for ratio of r_H/L_T under 0.05 (e.g. for a 1 cm^2 fuel cell, $r_{H\text{-}Max}$ = 250μm), K_{EC} factor is constant for a given porosity. In this case, the correction factor can be expressed solely as a function of porosity C (Eq. (11)):

$$K_{E1} = 35.6C^3 - 26.2C^2 + 10.5C + 0.4 \text{ if } \frac{r_H}{L_T} \leq 0.05 \tag{11}$$

2.4. *Performance index and fabrication limitations*

To facilitate design of BCM fuel cells, the total resistance of a standard fuel cell can be compared to the total resistance of the new configuration of the same dimension resulting in its performance index, P_I (Eq. (12)).

$$P_I = \frac{\tilde{R}_{IM-PEMFC} + 2\tilde{R}_{th}}{2\tilde{R}_E + 2\tilde{R}_{EC} + \tilde{R}_{IM}} \tag{12}$$

This index gives a relative performance of the selected dimensions and allows quick comparison of a wide range of configurations. In this work, a commercial Ion power Inc. fuel cell with a 50μm thick

NR-212 Nafion membrane and catalyst loading of 0.3 mg Pt/cm^2 is used as a references in the evaluation of P_I. By extracting data from its polarization curves performed at 40°C and saturated flows conditions, it was established that its total resistance was equal to 0.3Ω·cm^2.

P_I depends on the critical dimensions r_H, S and t which are limited by available technologies and materials, and greatly influence optimization of the BCM. For example, in the preceding work of Hamel et al. [2], r_H was restricted to a minimum of 10μm due to the precision of the chosen etching technique and S to 50 μm to avoid delamination of the ICCs during the fabrication process. Furthermore, a minimal ratio of 0.5 between r_H and t was needed to ensure a complete filling of the holes.

With these fabrications limitations, the optimal P_I for a variety of critical dimensions can be estimated. Figure 4 illustrates the variation of P_I as a function of r_H and C for a Kapton sheet of 50.8μm or 25.4μm thick. The red dotted lines represent the superior limit of P_I for a given S and the dark regions are the possible critical dimensions depending on fabrication limitations. With fabrication limitations stated in [2], the possible combinations are limited to porosities under 0.2 and the best theoritical P_I is 0.4 for r_H = 25μm and S = 50μm. Incidently, these dimensions have been selected in [2], with an experimental P_I of 0.31, relatively close to the predicted value.

Figure 4 also shows achievable P_I if S can be reduced to 25μm and $(r_H / t)_{max}$ increased to 1. These improvement could increase the maximum value of P_I from 0.4 to 0.65 while keeping t = 50.8μm, but could be increased to 0.72 is using a thinner Kapton membrane (25.4μm). It demonstrates that achievable improvement in fabrication techniques could lower the total internal resistance of BCM fuel cell by 80%.

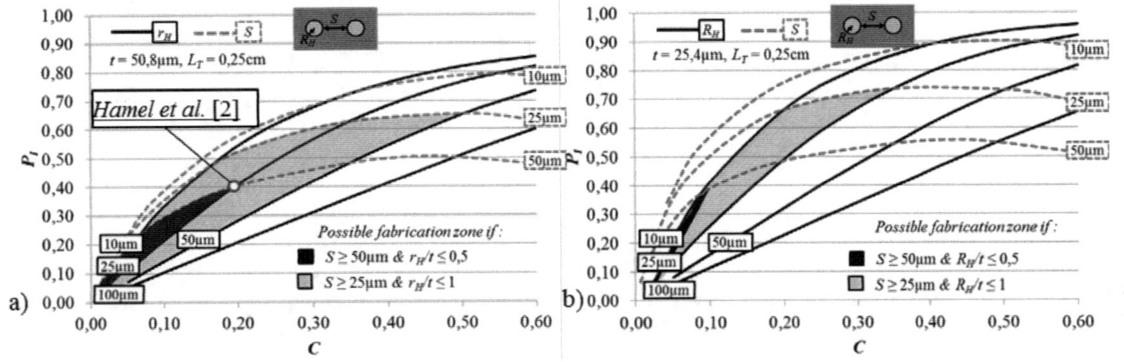

Figure 4 - Performance index of the BCM fuel cell design in function of critical dimensions and fabrication limitations for a) t = 50.8μm, b) t = 25.4μm, and L_T = 0.25cm.

3. Conclusion

The set of equations developed in this paper give simple guidelines to design a BCM fuel cell and limit its internal resistances. It has been shown that fabrication limitations greatly affect the choice of dimensions and should be taken into account at the beginning of the design process.

Simple improvements to the fabrication techniques presented in a preceding work [2] could lead to an areal current density of the BCM of 80%, increasing even more the dominance of this type of fuel cell when it comes to volumetric power density.

References

[1] S.-G. Kim and S.-J. Lee, "A review on experimental evaluation of water management in a polymer electrolyte fuel cell using X-ray imaging technique," *J. Power Sources*, vol. 230, pp. 101–108, May 2013.

[2] S. Hamel, T. Tsukamoto, S. Tanaka, and L. G. Fréchette, "Microfabrication of a Polymer Based Bi-Conductive Membrane for a Polymer Electrolyte Membrane Fuel Cell," *J. Phys. Conf. Ser.*, vol. 476, p. 012109, Dec. 2013.

Flexible solid-state fabric based supercapacitor

S Yong[1], J R Owen[2], M J Tudor[1], S P Beeby[1]

[1]Electronic &Electrical Engineering Group, Electronics and Computer Science, University of Southampton, Southampton, SO17 1BJ, UK
[2]Electrochemistry, Chemistry, University of Southampton, Southampton, SO17 1BJ, UK

E-mail: sy2g11@soton.ac.uk

Abstract. This paper reports details of the design, fabrication and characterisation of a solid-state fabric supercapacitor device. The proposed supercapacitors were based on fabric electrodes fabricated with low cost carbon materials via a spray coating technique. The two layer supercapacitors achieved a specific capacitances of 10.6 $F.g^{-1}$, area capacitance of 71.8 $mF.cm^{-2}$ and maintained 99% of the initial capacitance after cycling the device for more than 15000 times

1. Introduction
Wearable electronics can be defined as the integration of electrical system in clothing and accessories. Powering such system could benefit from a flexible solid-state energy storage device with high energy and power density. A supercapacitor also known as ultra-capacitor, is an electrochemical energy storage device that can provide both high energy density and power density, supercapacitor have been commonly used in smart sensor networks [1], energy harvesting powered systems [2] and power backup systems [3]. This paper presents a detailed fabrication process for a flexible solid-state supercapacitor device based on non-hazardous materials and polyester-cotton fabric substrates.

Previously carbon fabric (CF) [7] or carbon nanotube (CNT) [4] have been typically used in fabric supercapacitor electrodes. CF has the advantages of good mechanical stability, relative low cost in comparison with CNT, flexible, wearable and good electrical conductivity. Jin. H et.al [7] implemented a solid-state flexible supercapacitor using functionalized CF; this device achieved an aerial capacitance of 155.8 mF.cm-2 with an operating voltage up to 1.6 V. Cheng .H [4] reports of a supercapacitor electrode made by weaving CNT coated graphene woven fabric film, the scalable electrode achieves very high specific capacitance of 200.4 $F.g^{-1}$ (0.98 $mF.cm^{-2}$). Activated carbon, is another potential carbon material for supercapacitor electrode, it has large specific surface area, low electric resistivity and relatively low cost in comparison with other materials like CF and CNTs. Previously M. Cowell et.al [5] presented a flexible supercapacitor made with stencil casted polymer-carbon electrodes and gel- electrolyte. The symmetrical capacitive cells achieved 43 $mF.cm^{-2}$.

In comparison with conventional rigid supercapacitor, the supercapacitor used in wearable electronics needs to be lightweight, flexible, solid-state and made with non-hazardous gel-electrolyte and electrode. Previously, researchers have used hazardous substances in their solid-state supercapacitor, for example acid [6] in the gel electrolyte or corrosive oxide material [8] in the electrode. In wearable electronics application, supercapacitor made with hazardous materials requires extra care when packaging reduces the potential for skin irritation. Therefore although such supercapacitor may demonstrate electrochemical performance the use of hazardous materials mean they are not the best opinion for wearable electronics.

Supercapacitors store energy with two types of capacitive mechanisms: the electrical double layer capacitance (EDLC) and redox pseudocapacitance. The EDLC is based on the capture and release of electrostatic charge that occurs at the electrode and electrolyte solution interface. It is constructed using two electrical double layer interfaces, each having opposite polarity with respect to the electrolyte solution. Conventionally a charge separator containing an aqueous electrolyte will be placed in between the two supercapacitor electrodes to prevent the electrical short circuits. In solid-state supercapacitor, a gel electrolyte will replace the charge separator to prevent the short circuits, and provide free ions to forms the double layer interfaces. Therefore, in a solid-state supercapacitor it is not essential to place another layer of charge separator in between the two supercapacitor electrodes [9].

This work demonstrates an energy storage device for wearable electronics by fabricating a solid-state fabric electric double layer supercapacitor with non-hazardous materials. The fabric electrodes were achieved via the spray coating process with various inexpensive carbon powders and poly-cotton fabric substrates that are commonly used in the clothing industry. The supercapacitor was fully tested with a gel electrolyte in order to study their operation and stability as a function of cycle rate and cycle number.

2. Device fabrication

In order to create a permanently conductive carbon-coated fabric, carbon powders are mixed with a liquid binder that provides the adhesion between the carbon powders and fabrics to form a carbon ink. The carbon ink will be spray coated on selected area of the poly-cotton fabric substrate. In this work the carbon powder was mixed with binder Poly (vinyl alcohol-co-ethylene), dimethyl sulfoxide solvent and a surfactant. The carbon powders used in this work are activated carbon powder Kuraray Yp-80F and carbon black powder Shawinigan Black. Kuraray Yp-80F, has high purity, a particle size between 5-25 μm in diameter and an effective surface area of 2200 $m^2.g^{-1}$. Shawinigan Black has a mean particle size of 42 nm in diameter and an effective surface area of 75 $m^2.g^{-1}$.

In the next step the carbon ink was spray coated on both side of the fabric samples with a mask and spray gun connected to an air compressor. The fabric substrate in this work is polyester cotton material that has both the advantages of polyester and cotton material. During the spray coating process, carbon vapor penetrates into the selected part of the fabric substrate and adheres to the yarns of fabrics uniformly. After the sample is fully cured, the carbon powder bonds to the fabric substrate via the polymer binder; it introduces a large, conductive surface area that forms an electrical double layer on contact with the electrolyte. The photograph of the spray coated carbon fabric electrode on the poly-cotton fabrics can be seen in figure 1 (a)

Journal of Physics: Conference Series **660** (2015) 012074 doi:10.1088/1742-6596/660/1/012074

Figure 1. (a) Black carbon fabric electrode on original blue fabric substrate. **(b)** SEM photograph of carbon fabric electrode

Figure 1 (b) shows the SEM micrograph of the carbon particles adhered to the fibre to make a conductive network in the fabric substrate. The circular fabric electrode shown in figure 1 (a) will be cut out to assemble the supercapacitor cell. In this study the area of each poly-cotton fabric electrode is 0.785 cm^2 with thickness about 350 µm, each single piece of fabric electrode has a weight of 21.9 $mg.cm^{-2}$ before coating and carbon materials account for 13.4% of the final electrode weight.

The gel electrolyte was prepared by mixing ammonium salt, vinyl- alcohol polymer and water. The gel electrolyte solution was put under vacuum for 2 hours to reduce the amount of air bubbles inside the solution. In the next step, the two pieces of fabric electrodes were dipped into the gel electrolyte under vacuum for 20 minutes. This step enhances the wettability between the fabric electrodes and the polymer electrolyte. Finally, the two fabric electrodes containing the wet gel electrolyte were compressed together and cured in oven for 2 hours at 60 °C.

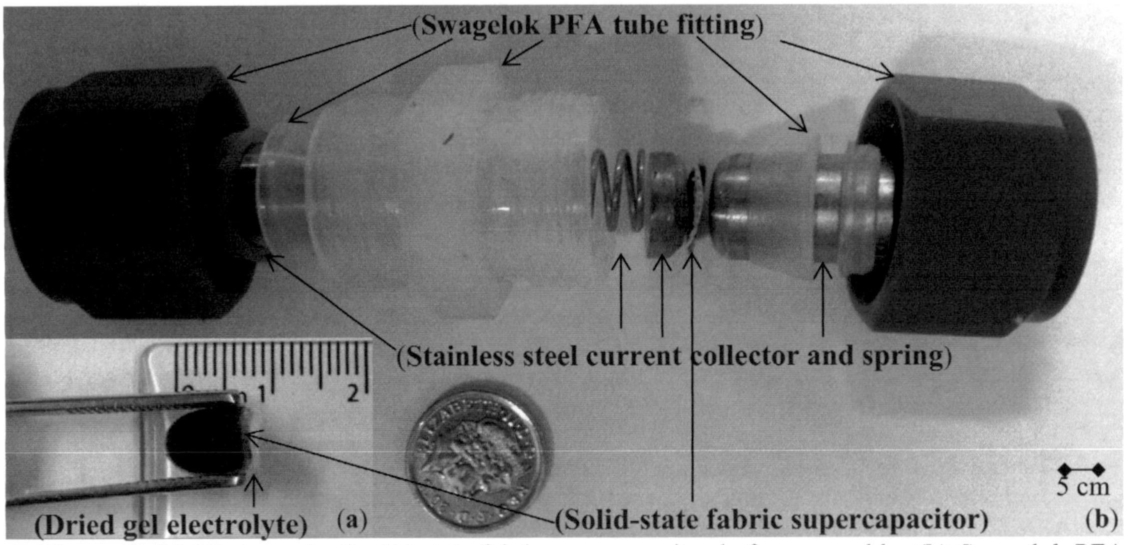

Figure 2. (a) Photograph of solid-state fabric supercapacitor before assembly. **(b)** Swagelok PFA tube fitting used to test the fabric supercapacitor before closing the two fitting caps

Figure 2 (a) shows the dried solid-state fabric supercapacitor is flexible after curing, The white layer in figure 2 (a) is the dried polymer electrolyte that had flowed out from the fabric electrodes. As shown in figure 2 (b) the assembled supercapacitor device is compressed from the top to bottom electrodes using spring loaded stainless steel current collectors and spring housed within a Swagelok PFA tube fitting for testing.

3. Results and discussion

The Solid-state fabric supercapacitor was tested using a VMP2 potentiostat/galvanostat (Biologic, France). The encapsulated supercapacitor was characterised by impedance spectroscopy (EIS), cyclic voltammetry (CV) at different scan rates and galvanostatic cycling (GC). and 4.

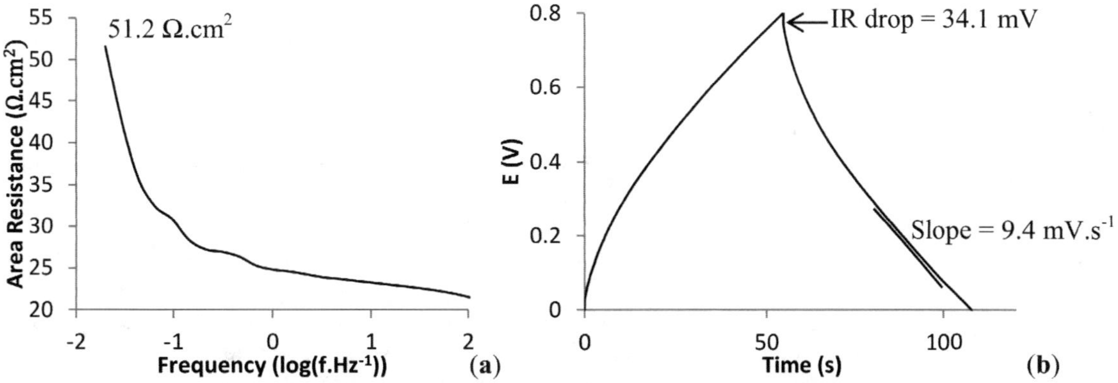

Figure 3. (a) Bode plot of device from 100 Hz to 20 mHz, extract from EIS test with equivalent series resistance (ESR) marked. **(b)** Galvanostatic cycling (GC) tests (0.675 mA.cm^{-2})

The ESR of the supercapacitor was determined from the EIS test from the bode plots extracted from the EIS response at 20 mHz. In the GC test, the ESR of supercapacitor was calculated by measuring the ohmic drop or IR drop and the test current. Figure 3 (a) shows at low frequency (20 mHz) the ESR of the solid-state fabric supercapacitor is 51.2 Ω.cm^2 This ESR is due to the electrolyte and electrode materials and the diffusion resistance. From figure 3 (b) The ESR result from the EIS test is very close to the ESR (50.5 Ω.cm^2) calculate from the IR drop and test current in the GC test. According to figure 3 (b) the mass specific capacitance is found to be 10.6 F.g^{-1}, the area capacitance 71.8 mF.cm^{-2}.

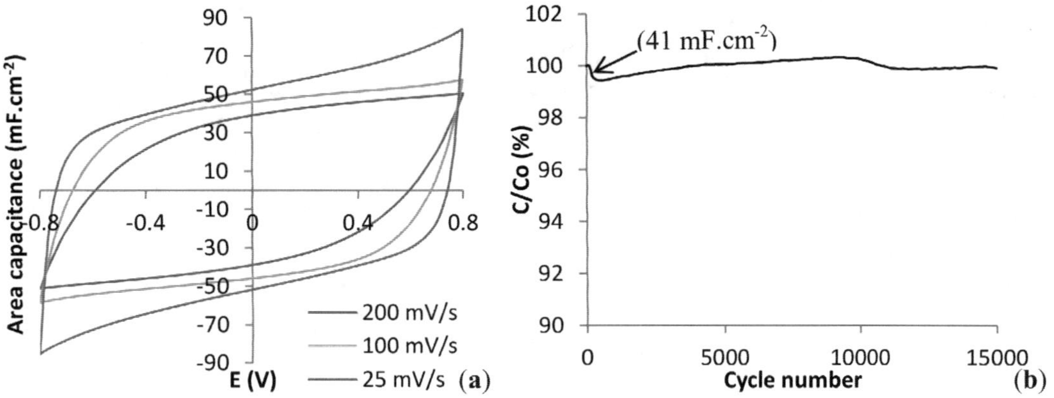

Figure 4. (a) CV test of the device between +/- 0.8 V at the scan rate of 200, 100, 25 mV.s^{-1}, **(b)** stability of device over 15000 cycles. C_o is the initial area capacitance (41 mF.cm^{-2}) of the device measured from cycle 1 of the CV test between +/- 0.8 V at the scan rate of 200 mV.s^{-1}.

As shown in figure 4 (a), the CV curves indicate that the device is electrochemically stable at the scan rates from 25 to 200 mV, and area capacitance values vary from about 40 to 80 mF.cm^{-2} depending on the different scan rate. Figure 4 (b) shows the device stability with a CV response. The overall variation of less than 1% correlates with the small temperature changes in the test lab. The high device stability indicates excellent adhesion of the carbon powders forming a continuous conducting network, which is due the use of the spray coating technique and vacuum impregnation process with poly ethylene vinyl alcohol binder and gel electrolyte. The spray coating process results in the blend of activated carbon and carbon black powder being spread uniformly over the fabric samples and infiltrate into the poly-cotton fabrics and fill up the fabric yarns to form a conductive network. The vacuum impregnation process ensures the gel electrolyte get in contract with maximum amount of the electrode materials to forms an effective double layer interface during characterisation.

4. Conclusion

This report presents a two layer supercapacitor on a fabric substrate. The non-hazardous solid-state fabric supercapacitor implemented in this work achieves a mass specific capacitance of 10.6 F.g^{-1}, area specific capacitance of 71.8 mF.cm^{-2}, a low normalized ESR of 50.5 Ω.cm^{-2} and achieves good cycling stability of less than 1% capacitance variation over 15000 cycles. In comparison with other works, These result are not as good as formal aqueous supercapacitor, however the individual carbon electrode can be sprayed on selected area of the fabrics, the proposed supercapacitor is non-hazardous, fully wearable, scalable, inexpensive and made without an individual charge separator. This work offer a practical approach for achieving a low cost and reliable energy storage supercapacitors in fabrics for wearable applications. Future work will include improving the formulation of the carbon and gel electrolyte ink so they can be spray painted into single piece of fabric to forms a solid-state fabric supercapacitor. The final device could see use applications in a wide range of wearable electronic systems like energy harvesters, medical sensors and a wide range of personal wearable electronics.

5. Acknowledgment

The authors thank the EPSRC for supporting this research with grant reference EP/1005323/1.

References

[1] Zhu D, Beeby S, Tudor J and Harris N. 2011 Sensors and Actuators A Physical 169 **2** 317-25
[2] Beeby, S and Zhu, D. 2011 Materials Science Foundations 197-218.
[3] Sahay K. and Dwidevi B. 2009 J. Elec. Systems **5** 8
[4] Cheng H, Dong Z, Hu C, Zhao Y, Hu Y and Qu Y, 2013 Nanoscale **5** 3428–34
[5] Cowell M, Winslow R, Zhang Q. 2014 Journal of Physics: Conference Series **557** 012061
[6] Zang X, Li X, Zhu M, Li X, Zhen Z, He Y, Wang K, Wei J, Kang F and Zhu H. 2015 Nanoscale **7** 7318
[7] Jin H, Peng Z, Tang W and Chana H. 2014 RSC Advanes **4** 33022
[8] Chodankar N, Dubal D, Gund G, Lokhande C. 2015 Electrochimica Acta **165** 338-47
[9] Rosi M, Iskandar F, Abdullah M and Khairurrijal M. 2014 Int. J. Electrochem Sci **9** 4251 –56

Multiphysics simulations of nanoarchitectures and analysis of germanium core-shell anode nanostructure for lithium-ion energy storage applications

T Clancy[1], J F Rohan[1]

[1]Tyndall National Institute, University College Cork, Lee Maltings, Cork, Ireland

Email: james.rohan@tyndall.ie

Abstract. This paper reports multiphysics simulations (COMSOL) of relatively low conductive cathode oxide materials in nanoarchitectures that operate within the appropriate potential range (cut-off voltage 2.5 V) at 3 times the C-rate of micron scale thin film materials while still accessing 90% of material. This paper also reports a novel anode fabrication of Ge sputtered on a Cu nanotube current collector for lithium-ion batteries. Ge on Cu nanotubes is shown to alleviate the effect of volume expansion, enhancing mechanical stability at the nanoscale and improved the electronic characteristics for increased rate capabilities.

1. Introduction

Portable personal electronic devices rely on batteries for energy storage and operation. Energy provision and storage are well recognised issues for integrated energy storage on chip as the device dimensions decrease, more functionality is added or devices become more autonomous. Lithium-ion batteries are a mature technology and have a high gravimetric and volumetric capacity, which makes them a leading contender for integration with microelectronic devices. Thin-film solid state batteries are being developed for such devices as they have excellent cycle life and can be processed on silicon substrates. The solid state electrolyte ensures that they are intrinsically safe and utilize standard packaging and fabrication processes. In thin-film solid state batteries the cathode thickness is limited to micrometers (<5μm) in a 2D geometry due to the low conductivity and slow transport of ions in the solid state materials. Current commercial thin-film solid state batteries are appropriate for some uses but do not meet the need for developing applications that require more energy and power per area. To meet this increased energy and power density demand, nanoarchitectures and higher specific energy electrode materials need to be developed for lithium-ion batteries.

The geometry and size of the electrodes play an important role in the electrochemical reaction and ion transport in a lithium battery. 3D architectures allow for more electrode surface area to be in direct contact with the electrolyte and theoretically increase the power density. Advances in micro and nano fabrication techniques have allowed for more creativity in battery design. Multiphysics simulations using mathematical models to describe the electrochemical reactions in 2D and 1D porous lithium-ion batteries were first developed by Newman et al. [1]. The Li^+ ion transport in the active material and electrolyte are modeled using Fick's second law and concentration solution theory respectively. Recent works have used Newman's models and applied them to 3D nanoarchitectures to gain insight into the electrochemical reactions [2].

3D nanoarchitectures have also been shown to alleviate the effect of volume expansion, enhancing mechanical stability at the nanoscale [3]. High capacity anode materials such as Si (theoretical capacity 4200 mAh/g) and Ge (1623 mAh/g) are now realistic replacements for the conventional graphite anode (372 mAh/g). Both materials undergo a volume expansion of up to 300% which causes low cycle life in its bulk state due to delamination from the current collector. Ge has many advantages over Si as an anode material for high power applications with 400 times higher rate of Li^+ diffusion at room temperature and 10,000 times the electrical conductivity [4, 5]. They both have a natural oxide on the outermost layer which results in the formation of an inactive solid electrolyte interface (SEI) during the first charge causing a lower coulombic efficiency [6]. Reported 3D geometries of Ge electrodes to date have been nanowires [7], nanotubes [8] and direct deposition onto the 3D current collector [9].

In this work, multiphysics simulations based on COMSOL modules to compare relatively low conductive cathode oxide materials in solid state thin-film micro, nanowire and core-shell nanowire battery geometries. The aim is to gain an insight on the affect the geometry has on battery performance. This work also focuses on the improving the performance of a Ge electrode as an anode material for lithium-ion batteries. The method used DC sputtering of the Ge onto a Cu nanotube array that acted as a current collector. Initial testing of this Cu nanotube core/Ge shell anode array has shown excellent cycle stability and rate capability.

2. Multiphysics simulations

The multiphysics simulations using COMSOL software are based on standard thin-film solid state lithium-ion battery materials: $LiCoO_2$ cathode, Li metal anode and a 1M electrolyte. Non-porous electrodes are used so only Li^+ ion transportation through the electrode/electrolyte boundary areas is considered. The discharge current is measured in C-rate where 1C is a current required to fully discharge the battery in 1 hour.

COMSOL lithium-ion battery and transport of diluted species modules are used to model the thin-film solid state lithium-ion battery. These modules use predefined mathematical equations that are based on the work of Newman [1]. The following assumptions are made:

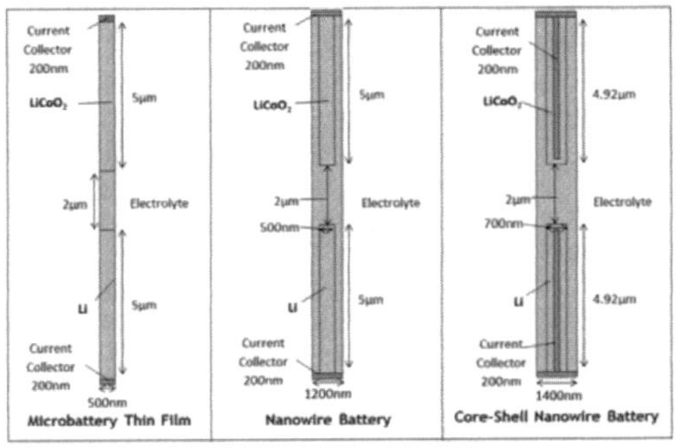

Figure 1. Microbattery thin-film, nanowire battery and core-shell nanowire battery geometries used in simulations.

1. Side reactions are neglected.
2. Volume changes in the electrode are neglected.
3. Atomic movement is described by diffusion where the active material particles are assumed to be one solid non-porous electrode.
4. Diffusion coefficients and conductivities are assumed to be constant within their respective regions in the battery.
5. Electroneutrality is assumed in the electrolyte
6. Anodic and cathodic transfer coefficients are set equal, $\alpha_a = \alpha_c = \alpha = 0.5$, so the exchange current density is calculated from equation (1)

$$i_{0\_pos} = k_{pos}(c_{max} - c)^{\alpha_c} c^{\alpha_a} c_{Li}^{\alpha_a} \quad (1)$$

The geometries used in the simulation are shown in figure 1. All three geometries have an out of plane thickness of 100 μm and volume of $2.5x10^{-16}$ m^3. The microbattery thin film geometry comprises 5 μm thick electrodes separated by 2 μm of electrolyte. The nanowire battery geometry is composed of a nanowire electrode with a height of 5 μm and diameter of 500 nm, separated by 2 μm and 350 nm either side of the nanowires filled with electrolyte. The base of the nanowire electrodes are connected to 200 nm conductive current collector. The core-shell nanowire geometry is 200 nm diameter conductive nanowire current collector that is covered in electrode material. The electrodes are separated by 2 μm and have 350 nm either side of the electrode which is filled with electrolyte. The core-shell nanowire electrodes have an overall height and diameter of 4.92 μm and 700 nm respectively.

Table 1. Parameters used in the simulations.

Symbol	Description	Value
$c_{Li\ init}$	Initial Li^+ concentration in electrolyte	1000 mol m^{-3}
c_{init}	Initial Li^+ concentration in cathode	24400 mol m^{-3}
c_{max}	Maximum Li^+ concentration in cathode	51600 mol m^{-3}
D	Diffusion coefficient for Li^+ in cathode	$5x10^{-13}$ m^2 s^{-1}
D_{Li}	Diffusion coefficient for Li^+ in electrolyte	$7x10^{-11}$ m^2 s^{-1}
$conduct_{pos}$	Conductivity of cathode	$1x10^{-6}$ S cm^{-1}
$conduct_{neg}$	Conductivity of anode	$1.05x10^5$ S cm^{-1}
$conduct_{electroylte}$	Conductivity of electrolyte	$1x10^{-6}$ S cm^{-1}
k_{pos}	Rate constant charge transfer of cathode	$1.27x10^{-6}$ A m^{-2} (mol $m^{-3})^{-1.5}$
$i_{0\ neg}$	Exchange current density of anode	85 A m^{-2}
t_o	Transference number	0.5
α	Transfer coefficient	0.5
T	Temperature	298.15

The mesh used for this study was an extremely fine edge mesh on the electrode/electrolyte boundaries while the mesh for the remaining geometry was extra fine free triangular mesh. A parametric sweep was used to vary the discharge C-rate. The time dependent study was between 0 and 3600 s with a relative tolerance of 1e-4. The stop condition was timestep <1e-10.

3. Experimental

The design and fabrication of the Cu nanotube core/Ge shell anode is shown in figure 2. Cu nanotubes were fabricated by electrodeposited in 21 mm diameter AAO (Anodised aluminum oxide membranes, Whatman, 60 μm thick, 250-300 nm pore diameter and 10^9 pores cm^{-2}) template with a 700nm Ag seed layer as described in an earlier publication [10]. Briefly, a Cu backing layer was electrodeposited to the Ag-conducting side of the template to give the Cu nanotubes more support once the template was removed. Cu Nanotubes were deposited by using 0.24 M $CuSO_4.5H_2O$ (Sigma Aldrich), 1.8 M H_2SO_4 (Sigma Aldrich), 400 ppm PEG (Sigma Aldrich) and 120 ppm NaCl (Sigma Aldrich) electrolytic bath. A potentiostat (CH instrument 660C) was used to apply a constant current of 40 mA for 900 sec in the two electrode setup with Cu foil as the anode and the AAO template as the cathode. The AAO template was dissolved in a 1M NaOH (Sigma Aldrich) solution for 1 hr, washed with DI water and dried in air. Ge was deposited onto the surface of the Cu nanotube array using a 99.99% pure Ge target (Kurt J. Lesker) and was DC-sputtered (Quorum Q300T D Dual) at a pressure of $1x10^{-2}$ mBar. The sputtering current used was 90 mA for 11 min.

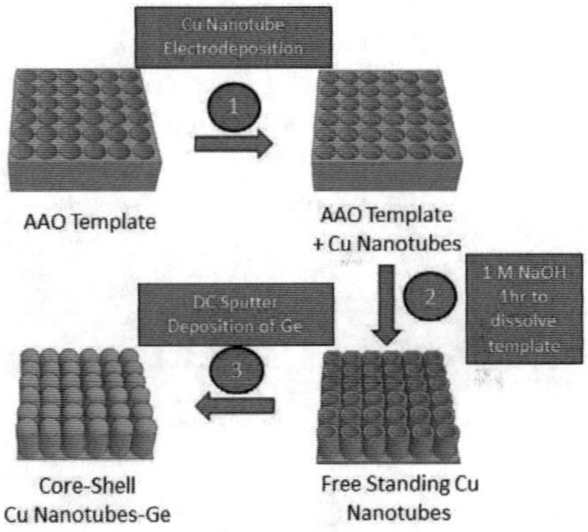

Figure 2. Schematic of the design and fabrication of Cu nanotube core/Ge shell anode. (1) Cu nanotube electrodeposition; (2) AAO template dissolved in 1 M NaOH; (3) DC sputtering of Ge onto the surface of the Cu nanotubes

The structure and the morphology of the samples were analysed (FEI Nova 630 Nano-SEM) coupled with an energy dispersive X-ray (EDX) (Hitachi S4000) and an X-ray powder diffraction (XRD) (Philips PW3710-MPD with Cu K_α radiation, l = 1.54056 Å, at 45 kV (40 mA), and data was analyzed using Philips X'Pert XRD software).

Electrochemical measurements of the Li^+ capacity were performed by cyclic voltammetry (CV) and galvanostatic cycling tests using a potentiostat (Bio-logic VSP) at various scan rates and discharge/charge currents, respectively. A two electrode cell setup of lithium foil 0.25 mm thick (Sigma Aldrich) acted as counter and reference and the Cu nanotube core/Ge shell anode as the working electrode in 1 M $LiPF_6$ in ethylene carbonate/diethyl carbonate (50:50 volume) (Sigma Aldrich) electrolyte assembled in an argon-filled glove box (M. Braun LABstar Glove Box) with O_2 and H_2O maintained below 0.1 ppm.

4. Results and discussion

The electrochemical activity of the battery geometry was evaluated by the development of concentration profiles and corresponding discharge curves. The potential cut-off is 2.5 V. The COMSOL simulation uses the diffusion coefficient and conductivity values of a solid state electrolyte. The conductivity value of the cathode oxide material is characteristic of a cathode without any conductive additives. Simulated battery discharge capacity at various C-rates for the microbattery thin-film, nanowire and core-shell nanowire geometries are presented in figure 3. The microbattery thin-film curve is characteristic of thin-film solid state batteries due to the internal resistance from the low conductivity of the cathode oxide material and electrolyte. This internal resistance causes a large voltage drop at high discharge currents forcing the potential to drop below the cut-off voltage of 2.5 V. The simulations suggest the implementation of nanoarchitectures such as nanowires and core-shell nanowires for all solid state batteries could increase the discharge rate up to 25C and 30C with a maximum charge/discharge of the cell capacity in 2 minutes.

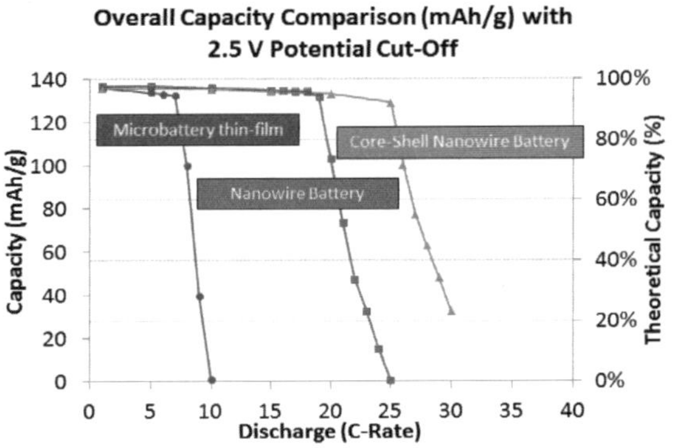

Figure 3. Galvanostatic C-rate discharge comparison for microbattery thin-film, nanowire battery and core-shell nanowire battery geometries.

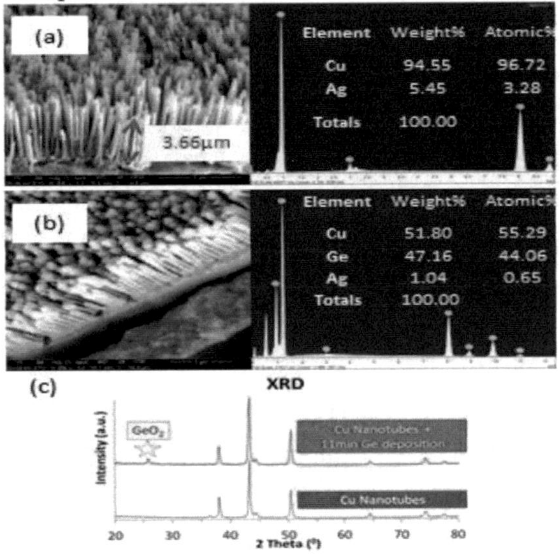

Figure 4. (a) SEM/EDX of electrodeposited Cu nanotubes; (b) SEM/EDX of Ge DC sputtered on Cu nanotube array; (c) XRD of Ge on Cu nanotubes and Cu nanotubes.

Figure 4a is a SEM and EDX image of the electrodeposited Cu nanotubes current collector. The image shows electrodeposited hollow Cu nanotubes, perpendicular to the Ag seed layer and have a smooth surface. The SEM and EDX image of the Cu nanotubes with 11 min DC sputtering of Ge is shown in figure 4b. The EDX confirms the presence of Ge and Cu as the prevalent elements while the SEM illustrates the formation of a uniform layer of Ge on top of the Cu nanotubes from the increase in diameter and rough surface. The XRD analysis shown in figure 4c suggests a natural oxide GeO_2 formed on the surface the Ge. The weight of Ge deposited on the Cu nanotubes that is used for the electrochemical analysis is calculated by measuring the thickness deposited on a planar Cu substrate, the density of Ge 5.323 g/cm^3 and the anode area exposed to the electrolyte in the electrochemical cell.

The electrochemical performance of the Cu nanotubes core/Ge shell was studied by CV and galvanostatic cycling. Figures 5a,b show a CV of cycles 36-40 at a scan rate 0.10 mV/s and the charge/discharge capacities of a range of CVs at different scan speeds vs cycle numbers 1-45 respectively. The cathodic peaks at 0.61 V, 0.31 V and 0.05V during the charging cycling are associated with the formation of the Li-Ge alloy while the broad anodic peak at 0.49 V is associated to the de-alloying of Li_xGe to Ge. The large initial charge capacity of 2300 mAh/g can be linked to the formation of the SEI layer due to the native oxide of GeO_2 identified from the XRD. The CVs of cycles 36-40 show excellent overlap which indicates little or no degradation in performance. Galvanostatic cycling was performed on the same sample over a range of current. Figure 5c,d shows the galvanostatic cycles 81-85 at a current of 120 μA and the charge/discharge capacities of the galvanostatic cycles performed over a range of currents vs cycle numbers 66-100. The galvanostatic cycles show great reproducibility and Coulombic efficiency above 90%.

The exceptional cycling stability over a range of scan rates and current densities highlights the robustness and advantages associated with the 3D core-shell nanoarchitecture permitting the Ge to expand and contract without detaching from the Cu nanotube current collector.

5. Conclusion

In summary, nanowire and core-shell nanowire battery geometries have shown greater rate capabilities compared to thin film micro-batteries using COMSOL multi-physics simulations. These results indicate that nanowire and core-shell nanowire battery geometries can potentially help meet the demand of for energy and power per unit footprint. Cu nanotube core/Ge shell anode design and fabrication can alleviate the volume expansion during high-rate cycling for lithium-ion batteries.

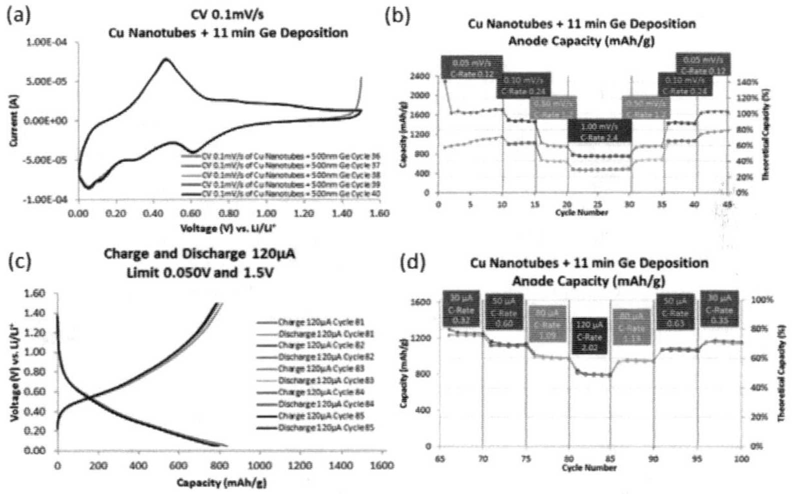

Figure 5. (a) CV at 0.10 mV/s of cycles 36-40; (b) Charge/discharge capacities of CVs vs cycle number; (c) Galvanostatic cycling at 120 μA for cycles 81-85; (d) Charge/discharge capacities of galvanostatic cycles vs cycle number.

Acknowledgments

The authors would like to acknowledge the financial support from SFI Grant number: 12/IP/1722, Nanomaterials design and fabrication for energy storage.

References

[1] Doyle M, Fuller T F and Newman J 1993 *J. Electrochem. Soc.* **140** 1526-33
[2] Zadin V, Kasemagi H, Aabloo A and Brandell D 2010 *J. Power Sources* **195** 6218-24
[3] Hasan M, Chowdhury T and Rohan J F 2010 *J. Electrochem. Soc.* **157** A682-A88
[4] Graetz J, Ahn C C, Yazami R and Fultz B 2004 *J. Electrochem. Soc.* **151** A698-A702
[5] Wang D, Chang Y-L, Wang Q, Cao J, Farmer D B, Gordon R G and Dai H 2004 *J. Am. Chem. Soc.* **126** 11602-11
[6] Seng K H, Park M-h, Guo Z P, Liu H K and Cho J 2013 *Nano Lett.* **13** 1230-36
[7] Kennedy T, Mullane E, Geaney H, Osiak M, O'Dwyer C and Ryan K M 2014 *Nano Lett.* **14** 716-23
[8] Park M-H, Cho Y, Kim K, Kim J, Liu M and Cho J 2011 *Angew. Chem. Int. Edit.* **50** 9647-50
[9] Wang J, Du N, Zhang H, Yu J and Yang D 2012 *J. Mater. Chem.* **22** 1511-15
[10] Chowdhury T, Casey D P and Rohan J F 2009 *Electrochem. Commun.* **11** 1203-06

C-axis tilted AlN films for vibration energy harvesters

H. H. Nguyen, M. Hara, H. Oguchi, H. Kuwano
Department of Nanomechanics, Tohoku University, Japan
E-mail: hung@nanosys.mech.tohoku.ac.jp

Abstract. In this study, we reported on the sputtering growth of the c-axis tilted AlN films, which may have higher electromechanical coupling coefficient k_{31} than [0001]-oriented AlN, on the Si (100) substrates in the aim of high output power vibrational energy harvesters (VEHs). We implemented the incident angle deposition in a wide range of temperature from room temperature (RT) to 650°C. At 420°C the c-axis tilted 22°, which is the largest tilt angle reported for the AlN films grown by incident angle deposition. In addition, by changing growth temperature, we succeeded in preparing c-axis tilted AlN films with different tilt angle. Scanning electron spectroscopy (SEM) revealed that the c-axis tilted films have tilted columnar structures which prevents cracks and pinholes going straight through the films. This feature may help avoiding electrical short between electrodes fabricated on top and bottom of the AlN films in VEHs.

1. Introduction

In recent years much attention has been paid on an AlN thin film as an alternative piezoelectric material to Lead Zirconate Titanate (PZT) in vibrational energy harvesters due to more concern to the environment and human health that asks society to restrict the use of lead in industry. Accordingly researchers have searched for lead-free piezoelectric materials for vibrational energy harvesters (VEHs) [1-3], among lead-free piezoelectric materials developed so far, an AlN (wurtzite structure) film has been considered as one of the most promising materials for vibration energy harvesters because of its high figure of merit (FOM) of output power density [4, 5]. Furthermore AlN, a semiconductor process compatible materials, can be easily integrated into silicon devices [4, 6].

In general people use [0001]-oriented AlN films which have c-axis oriented normal to the substrate surface because electromechanical coupling coefficient (k_{31}) is large along this crystal orientation [10] and output power of the VEHs can be increased. However, k_{31} of the [0001]-oriented AlN is still smaller than that of PZT films [4, 6-9]. Thus it is important to further increase k_{31} of the AlN thin films, and tilting c-axis of the [0001]-oriented can be one effective approach [11, 12]

There have been several reports of c-axis-tilted AlN thin films aiming for optoelectronics applications. For example, Rodriguez-Navarro *et al.* have deposited AlN on a glass substrate using RF reactive sputtering by incident angle deposition where they arranged a substrate and a target so as to set incident angle of the deposition flux to be 45°, and obtained films with ~15° tilted c-axis [13]. In addition, Dellas and Harper obtained films with 8° tilted c-axis [14] using the similar method. However, to further increase k_{31} to catch up with PZT, more tilt angle is desired. Moreover no one investigated k_{31} of the c-axis-tilted AlN thin films.

In this study we aim to grow AlN films with the c-axis tilt angle of more than 15° by RF sputtering. In addition we prepare AlN films with different c-axis tilt angle for the future investigation of k_{31} of the c-axis tilted AlN films and fabrication of high output power VEHs.

2. Experiment

Figure 1 shows a schematic of the sputtering set up for incident angle deposition of the AlN films. A substrate holder and an Al target were arranged so as to set incident angle of deposition flux to be 45° from substrate normal. During film growth Ar and N_2 are used as reactive gases. N_2 was introduced into the chamber through the ion gun to make nitrogen plasma to facilitate AlN formation [15]. N_2/Ar gas flow rate of 3 was fixed to obtain AlN stoichiometry according to the previous investigation shown in Figure 2. Other sputtering parameters were summarized in Table 1. Prior to each AlN film deposition, silicon substrates were cleaned by standard cleaning processes of Radio Corporation of America (RCA1 and RCA2). The Al target was pre-sputtered for 10 minutes in Ar and for 5 minutes in the mixture of Ar and N_2. Thicknesses of the films were measured by a surface profiler. Field emission scanning electron microscopy (FE-SEM, Hitachi SU-70) was used to perform morphological analysis. The composition was confirmed by using the EDX spectroscopy. An atomic force microscopy (AFM, Nikon TE2000-U) was used to investigate the surface roughness. X-ray diffractometer equipped with a two dimensional detector (2D-XRD) was employed to study crystal quality and c-axis tile-angle. Tilt angles were calculated from 2D-XRD diffraction images using software named General Area Detector Diffraction System (GADDS).

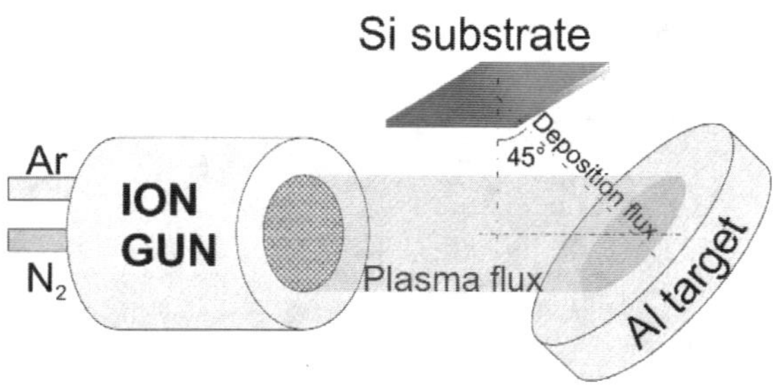

Figure 1. Schematic of deposition

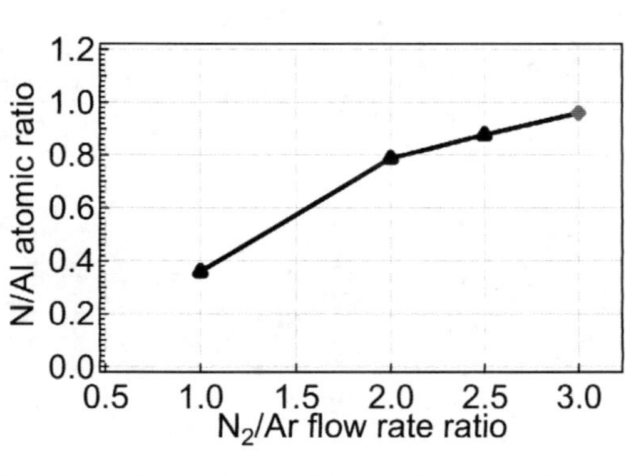

Figure 2. Dependence of Al and N ratio in the thin film deposited at RT as a function of Ar and N_2 flow rate.

Table 1. Deposition conditions

Substrate temperature	RT – 650°C
Back pressure	$1.5x10^{-7}$ Toor
Depositing pressure	$1.2x10^{-3}$ Toor
RF power	140 W
Deposition time	1 h
Film thickness	600 nm
Ar gas flow rate	2.5 sccm
N_2 gas flow rate	7.5 sccm
Accelerated voltage	1200 V
Beam current	43 mA

3. Results and discussions

XRD measurements verified that we have succeeded in growing c-axis tilted AlN thin films. Figure 3a shows a 2D-XRD diffraction image of the AlN film grown at 650°C. Except a diffraction peak from Si, only peaks from (0002) and (0004) plane of AlN were observed almost symmetrically along the chi axis, indicating that film has [0001] orientation. With decreasing growth temperature, the (0002) peak gradually moves down along the chi axis. This clearly indicates that the c-axis tilted from the substrate normal. Tilt-angles were determined as Chi angle of the (0002) peak centre (figure 3), and found to be 6° for 560°C (Figure 3b) and 22° for 420°C (Figure 3c). For the films grown at lower than 300°C (figure 3d and 3e) the tilt angle could not be determined due to poor crystallinity, though the c-axis likely tilt more from the substrate normal at lower growth temperature.

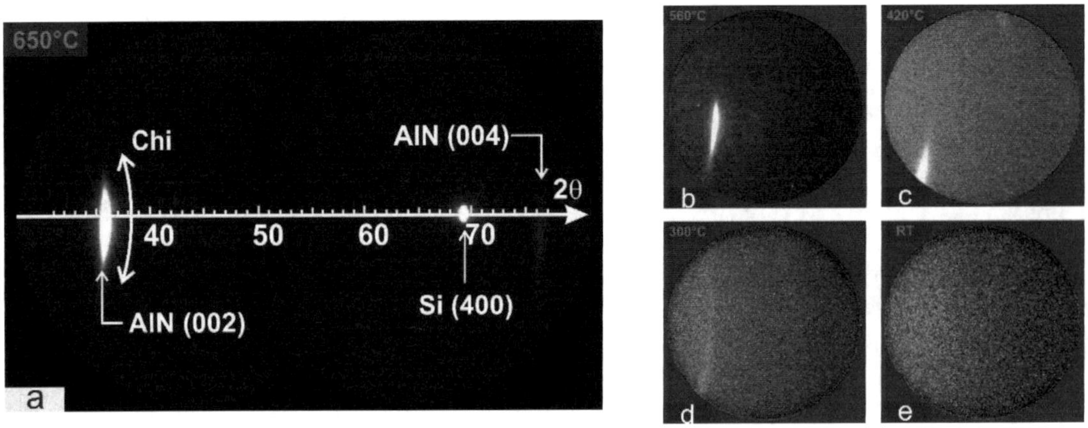

Figure 3. 2D-XRD patterns of AlN films deposited at various temperature.

To understand the mechanism determining the tilt-angle of the c-axis, we conducted FE-SEM. A cross-sectional FE-SEM image of the film grown at 420°C (tilt-angle: 22°) shows a columnar structure with tilt-angle of 43° (Figure 4b), which is close to the incident angle of the deposition flux. In contrast, the columnar structure for 650°C (tilt-angle: ~ 0°) does not tilt (Figure 4a). These facts indicate that tilting of the c-axis can be obtained by growing crystal along the tilted direction using the incident angle deposition if growth temperature is low enough to suppress atomic migration.

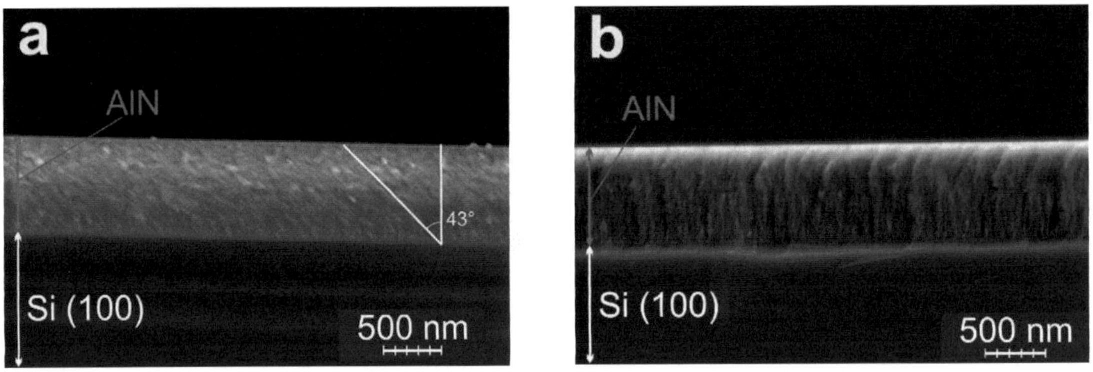

Figure 4. Cross-sectional FE-SEM images of AlN thin film deposited at (a) 420°C and (b) 650°C

SEM and AFM measurements for the film surface further confirmed our hypothesis. Figure 5 shows SEM micrographs of the thin films grown at different temperatures. The size of the islands decreases with increasing growth temperature from ~50 nm for 300°C to ~10 nm for 620°C. The root mean square roughness R_{RMS} of the films deposited at 300°C, 420°C and 650°C determined by AFM measurements are 3.89 nm, 1.15 nm and 1.09 nm, respectively. These observation consistently indicates that more atomic rearrangement occurs at higher growth temperature.

Finally we point out that incident angle deposition can be an appropriate growth method to suppress electrical short between top and bottom electrodes that is often a problem reported for AlN films in micro devices [15, 18, 19]. Because the tilted columnar structures are stacked, films do not allow cracks and pin-holes to go straight through the film. This unique geometry will prevent connection between top and bottom electrodes, which are usually fabricated using physical vapour deposition methods with line-of-sight deposition.

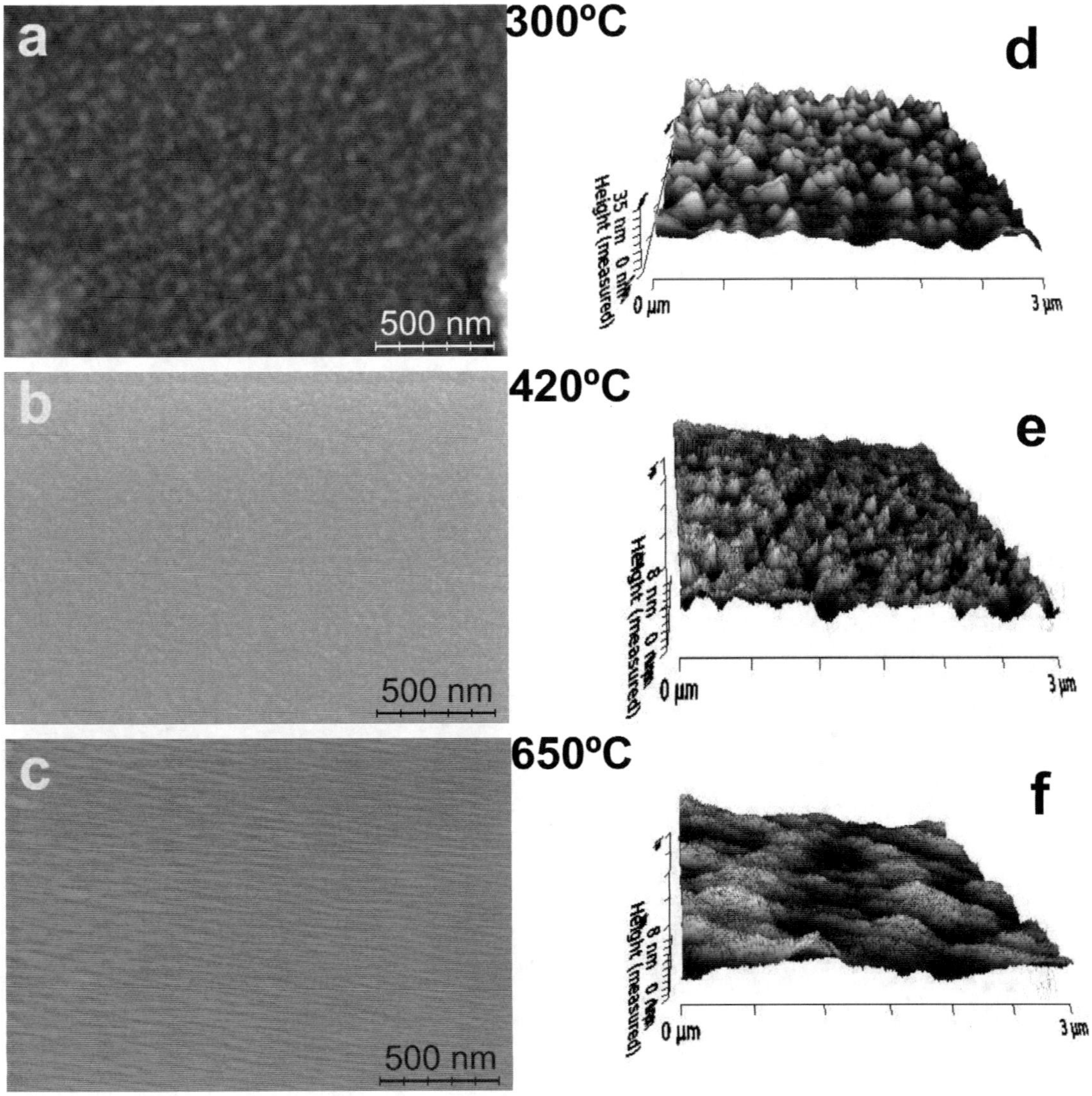

Figure 5. plane view FE-SEM and AFM images of AlN thin films deposited at (a, d) 300°C, (b, e) 420°C and (c, f) 650°C

4. Conclusion

In summary, we have succeeded in growing the AlN films with most tilted c-axis ever among the films grown by incident angle deposition. The results of this study revealed that one key parameter controlling the tilt angle of the c-axis is the growth temperature. For the incident angle deposition with incident flux angle of 45°, the tile angle was successfully changed from 0° for 420°C to 22° for 650°C. This study also indicated that incident angle deposition may be an appropriate method to avoid electrical short through the AlN films by avoiding formation of cracks and pin-holes straight through the films.

References

[1] Le VM, Hara M, Kuwano H. 2013. J. J. Appl. Phys. **52** 07HD8
[2] Springer Science 2008 *Piezoelectric MEMS*, ed Bassiri-Gharb N, chapter 20: Materials and Devices, pp. 413

[3] Hara M. and Kuwano H. 2012 Jpn. *J. Appl. Phys.* **51** 07GC11
[4] Matloub R., Hadad M., Mazzalai A., Chidambaram N., Moulard G., Sandu C. S., Metzger Th., and Muralt P. 2013, *Appl. Phys. Let.* **102** (**15**) 152903
[5] Sakurai T, Yamada O, Miyamoto Y. 2006. Mat. Sc. Eng.: A **415** 40
[6] Nathan J, Lynette K, Alan M. 2013. *Smart Mater. Struc.* **22** 115033
[7] Tasnádi F, Alling B, Höglund C, Wingqvist G, Birch J, et al. 2010. *Phys. Rev. Let.* **104** 137601
[8] Akira N, Mari A, Tsuyoshi Y, Satoru M, Masanori U, et al. 2015. *J. J. Appl. Phys.* **54** 07HD1
[9] Zhang J, Cao Z, and Kuwano H 2011 *Jpn. J. Appl. Phys.* **50** 09ND18.
[10] Trolier-McKinstry S, Muralt P. 2004. *J. Electr.* **12** 7
[11] Wang Q, Cao Z. and Kuwano H. 2012 *Proc. PowerMEMS2012* 327
[12] Wang Q, Oguchi H, Hara M, Kuwano H. *Micro Electro Mechanical Systems (MEMS), 2014 IEEE 27th International Conference*, 636
[13] Rodríguez-Navarro A, Otañno-Rivera W, García-Ruiz JM, Messier R, Pilione LJ. 1997. *J. Mat. Res.* **12** 1689
[14] Dellas NS, Harper JME. 2006. *Thin Sol. Fil.* **515** 1647
[15] Nguyen HH, Oguchi H, Kuwano H. 2014. *Journal of Physics: Conference Series* **557**:012047
[16] Deng R, Muralt P, Gall D. 2012. *J. Vac. Sc. Tech.* A **30** 051501
[17] Jochen MS, Suzanne R, William DS, Allan M. 2000. *Journal of Physics D: Applied Physics* **33** R173
[18] Kar JP, Bose G, Tuli S. 2005. *Mat. Sc. Sem. Proc.* **8** 646
[19] Kumada T, Ohtsuka M, Fukuyama H. 2015. *AIP Advances* **5** 017136

Characterization of Electret Based on Inorganic-organic Nanocomposite Using Fluoropolymer and Silica Nanoparticles

M Suzuki[1], M Shimokizaki[1], T Takahashi[1], Y Yoshikawa[2], and S Aoyagi[1]

[1] Faculty of Engineering Science, Kansai University, 3-3-35 Yamate-cho, Suita, Osaka564-8685, Japan

[2] ROHM Co., Ltd., 21 Saiin Mizosaki-cho, Ukyo-ku, Kyoto 615-8585, Japan

E-mail: m.suzuki@kansai-u.ac.jp

Abstract. An A novel electret based on inorganic-organic nano composite using fluoropolymer and silica nanoparticles was developed in this study. CYTOP® is used to fabricate the nanocomposite electret, which is one of fluoropolymer. Three kinds of silica nanoparticles dispersed in methyl ethyl ketone were employed. Each type of nanoparticles was mixed in the CYTOP or stuck between three layers of CYTOP. Then, negative charge was implanted by corona discharge method. The initial surface potential of the nanocomposite electret was higher than that of a control electret made of pure CYTOP. Additionally, time stability of those was also better than that of control electret. However, above mentioned properties of the mix-typed electret was worse than that of stuck-typed electret, because of discharging through aggregates composed of the nanoparticles.

1. Introduction

Recently, vibrating energy harvesters (VEHs), which generate electrical power from ambient vibration, have attracted much attention [1-5], because it is prospective power supply for wireless sensors [1, 2]. Since main frequency of ambient vibration is usually on the order of a few to several tens of hertz, electret-based electrostatic VEH are more suitable as power generator using ambient vibration than magnet-based VEHs and piezoelectric-based VEHs [3]. Electret is dielectric material of which surface potential s stable based on electrical charges trapped in itself. Previously, many type of electret-based VEH were developed. We also developed an electret-based VEH consists of a fixed electret on bade electrode and a vertically vibrated counter electrode with high dielectric constant (high-*k*) plate [4, 5]. Theoretical output electrical power of electret-based VHE is generally proportional to square of surface potential of electret. Therefore, improvement of surface potential of electret is important to enhance output of electret-based VEH. Conventionally, fluorocarbon polymers [1-5] were frequently used as electret in VEHs because they keep higher concentration of electrical charge than other materials. CYTOP® is famous fluorocarbon polymer as an electret, of which surface potential and its time stability are better than other fluorocarbon polymer [6]. In previous work, we demonstrate that surface potential of an electret made of organic-inorganic nanocomposite materials, in which $BaTiO_3$ nano-particles are mixed in CYTOP, is higher than that made of pure CYTOP. In this study, it was investigated the surface potential of electrets made of CYTOP based nanocomposite materials when the silica nanoparticles is mixed in CYTOP.

2. Fabrication of nano/micro textured electret

In this study, CYTOP (provided by Asahi Glass Co. Ltd., Model: CTL-809-M) solved by a fluorinated organic solve materials, and three kinds of silica nanoparticles (provided by Nissan Chemical Industries, Ltd., Model: MEK-AC-2140Z, MEK-AC-4130Y, MEK-AC-5130Z) dispersed in methyl ethyl ketone (MEK) were employed. Grain sizes of the nanoparticles are shown in Table 1.

In this study, CYTOP (provided by Asahi Glass Co. Ltd., Model: CTL-809-M) solved by a fluorinated organic solve materials and three kinds of silica nanoparticles (provided by Nissan Chemical Industries, Ltd.,) dispersed in methyl ethyl ketone (MEK) were employed. Grain sizes of the nanoparticles are shown in Table 1.

Additionally, two types of electret structures were employed, one of which is "mix-typed electret," and the other of which is "stack-typed electret." In fabrication of the mix-typed electret, CYTOP and silica nanoparticles are mixed by hand, followed by dispersion of the mixed particles using ultrasonic homogenizer (NIHONSEIKI KAISHA LTD., model:US-150T). Then, the mixture was spin coated on a thermal oxidized silicon substrate, and baked to form an electret. The thickness of CYTOP layer and SiO_2 layer are 5.4 μm and 0.8 μm, respectively. In fabrication of the stack-typed electret, CYTOP were spin-coated in three separate times (Thickness of each layer is 1.7 μm, total thickness is 5.4 μm). The nanoparticles dispersed in MEK were coated between the CYTOP layers. Schematic of the electret sample is shown in Fig. 1. After fabricating the two types of electrets, negative charge was implanted into the electrets by corona discharge method as shown in Fig. 2.

Table 1. Specification of Silica nano-particles

Model ID	SiO_2 Ratio (w%)	Dispersion medium	Diameter of particles (nm)
MEK-AC-2140Z	40	methyl ethyl ketone	10 – 20 (typically 20)
MEK-AC-4130Y	30	methyl ethyl ketone	40 – 50 (typically 50)
MEK-AC-5130Z	40	methyl ethyl ketone	70 – 100 (typically 70)

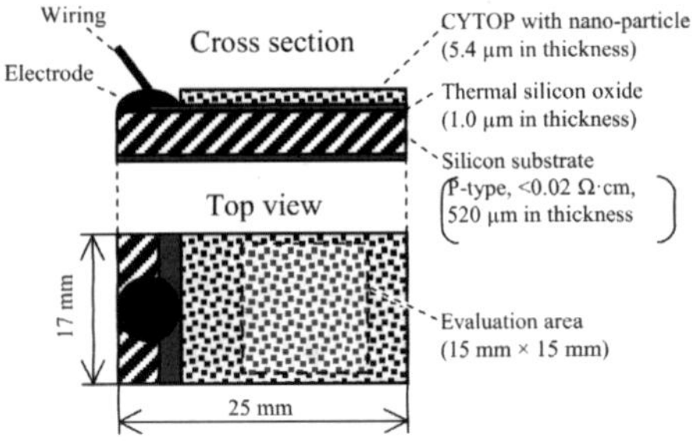

Figure 1. Fluorpolymer (CYTOP) electret with silica nano-particles which is fabricated on the thermal oxidation layer of a low-resistive silicon substrate (SiO_2). The Si substrate also acts as a lower electrode for the electret.

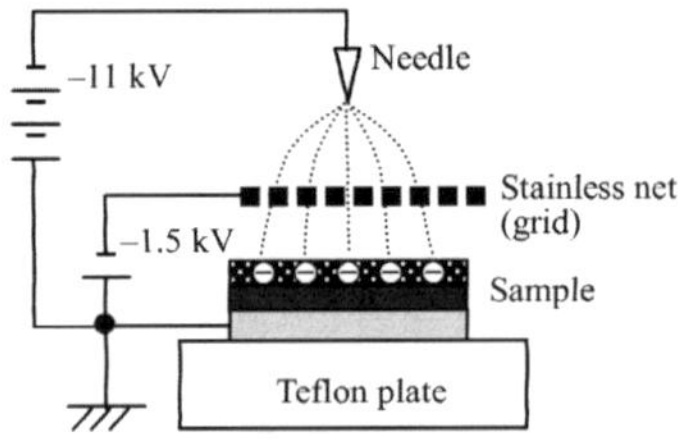

Figure 2. Schematic of setup for corona discharge to implant electron to the fabricated nanocomposite film.

3. Measurement results and discussion

The initial surface potential of CYTOP mixed or stacked with silica nanoparticles was higher than that of a control electret made of pure CYTOP (5.4 μm in thickness). Additionally, time stability of those was also better than that of control electret, as shown in Figs. 3 and 4. Particularly, above mentioned properties of the stack-typed electret was better than that of CYTOP mixed with $BaTiO_3$ nanoparticles which was reported at PowerMEMS2014 [8]. As a result of the observation of mix-typed electret by an SEM, it was found that there are many projections which may be derived from aggregate of nanoparticles. We consider that aggregates of nanoparticles were formed in nanocomposite when the CYTOP and the silica nanoparticles were mixed, and implanted charges were discharged through the aggregates, as shown in Fig.5.

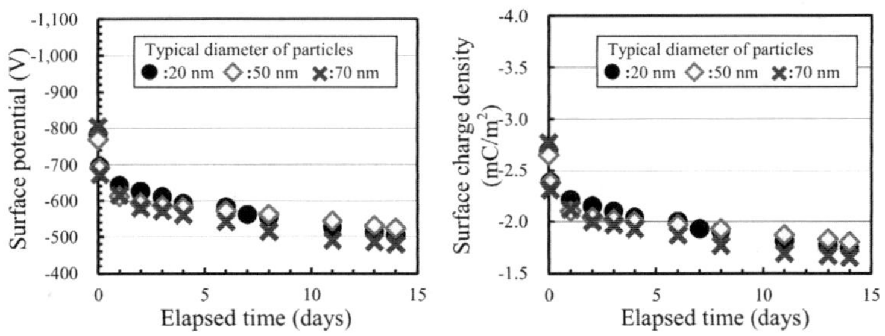

Figure 3. Time degradation of surface potential of electrets made of mixture of CYTOP and silica nano-particles.

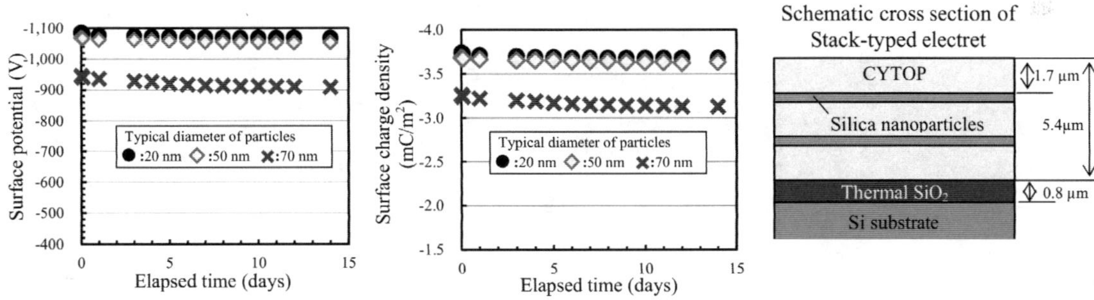

Figure 4. Time degradation of surface potential of electrets made of stacks of CYTOP and silica nano-particles.

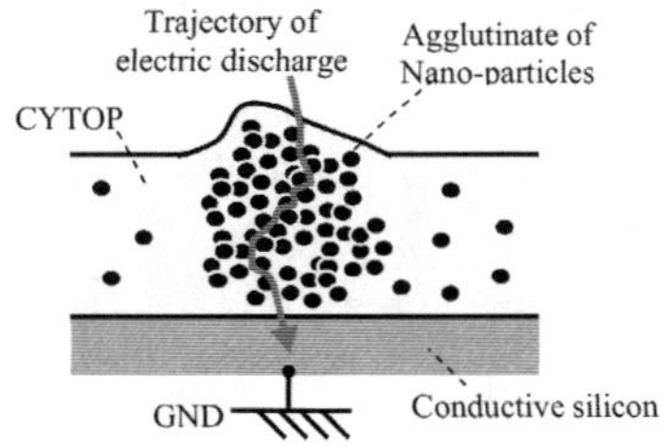

Figure 5. Schematic illustration describing how electric charge discharges through the aggregate of nano-particles.

4. Acknowledgments
This work was supported in part by a grant of "Strategic Research Foundation Grant-aided Project for Private Universities": Matching Fund Subsidy of MEXT (Ministry of Education, Culture, Sport, Science, and Technology, Japan), 2015-2019. This work was supported in part by JSPS (Japan Society for the Promotion of Science) KAKENHI (26870728). Silica nanoparticles used in this study were provided by Nissan Chemical Industries, Ltd.

5. References

[1] S. P. Beeby, M. J. Tudor, and N. M. White 2011 Meas. Sci. Technol., 17, 175-195
[2] K. Matsumoto, K. Saruwatari, Y. Suzuki 2011 in Digest Tech. PowerMEMS'11, 134-137
[3] T. Tsutsumino, Y. Suzuki, N. Kasagi, and Y. Sakane 2006 in Digest Tech. IEEE Internat. Conf. MEMS'06 98-101
[4] T. Takahashi, M. Suzuki, T. Nishida, Y. Yoshikawa, and S. Aoyagi, 2012 in Digest Tech. MEMS'12 1265-1268
[5] T. Takahashi, M. Suzuki, T. Nishida, Y. Yoshikawa, and S. Aoyagi 2013, in Digest Tech. MEMS'13 877-880
[6] Y. Sakane, Y. Suzuki, N. Kasagi Journal of Micromechanics and Microengineering 2008 18 104011(6 pp)
[7] K. Kashiwagi, K. Okano, N. Tanabe, Y. Sera, T. Miyajima, Y. Morizawa, Y. Sakane, Y. Hamatani, F. Nonaka, A. Asakawa, and Y. Suzuki 2011 in Proc. IEEE ISE14, 71-72.
[8] M. Suzuki, T. Takahashi, and S. Aoyagi 2014 Journal of Physics, Conference Series 557 (Power MEMS 2014), 012062 (5Page).

Characterization of Piezoelectric PDMS-Nanoparticle Composites

C J Borsa[1], M Mionić Ebersold[2,3,4], P Bowen[3], P-A Farine[1] D Briand[5]
Ecole Polytechnique Fédérale de Lausanne (EPFL)
[1]Electronics and Signal Processing Laboratory EPFL Neuchâtel, Switzerland
[2]Department of Radiology, University Hospital Lausanne (CHUV) and University of Lausanne (UNIL), Switzerland
[3]Center for Biomedical Imaging (CIBM), Lausanne, Switzerland
[4]Powder Technology Laboratory, EPFL Lausanne, Switzerland
[5]Sensors, Actuators and Microsystems Laboratory, EPFL Neuchâtel, Switzerland

E-mail: christopher.borsa@epfl.ch

Abstract. In this work, the novel fabrication and characterization of elastomeric piezoelectric nanocomposites are explored. Fabrication methods explored herein utilize ball milled barium titanate powder dispersions, along with double walled carbon nanotubes which are dispersed in toluene though the use of an ultrasonic probe. Test devices are then constructed with electrodes made from evaporated gold on polyimide foils and protective dielectrics of pristine PDMS. Two different device construction methods are explored utilizing both direct contact bonding and plasma bonding of the active composite layers to the dielectric/electrode. Test samples are evaluated through the use of a dedicated Berlincourt type piezoelectric d_{33} meter.

1. Introduction

With the ever shrinking power consumption of modern electric devices and sensors, the feasibility of autonomous sensor nodes powered exclusively from renewable energy harvesters is ever increasing. For powering sensor nodes for use in concealed environments such as in textiles or on the skin, piezoelectric generators that utilize kinetic energy are advantageous as they do not require direct sunlight or thermal differentials needed for thermoelectric or photovoltaic generation methods. One of the challenges in traditional resonant piezoelectric generators is that they optimally produce power at frequencies in the range of hundreds of Hertz[1]. While there are many strategies for frequency down converters, researchers are now looking towards the new paradigm of composite piezoelectric energy harvesters that are capable of low frequency power generation.

While there have been many advances in piezoelectric composite development over the past several years, most authors reported generator outputs generators of different areas, and under different strain conditions. Herein we synthesize piezoelectric elastomeric composites composed of polysiloxane (PDMS), double walled carbon nanotubes (DWCNT) and barium titanate ($BaTiO_3$) and characterize test structures using a Berlincourt type d_{33} meter for which the static and dynamic clamping forces can be precisely controlled. Utilizing a controlled characterization method as described above, we compare the differences that can be seen by directly adhering vs plasma bonding of the active piezoelectric layer to foil electrodes comprised of gold evaporated on polyimide.

2. Fabrication

When synthesising the piezoelectric active layer, the following stock materials are utilized. Barium titanate particles consisting of a tetragonal crystal structure and a nominal diameter of 200nm were purchased from US Research Nanomaterials Inc, DWCNT's with an average diameter of 3.5nm's and length of 1-10 microns were purchased from Nanocyl S.A., Dow Sylgard 184 was utilized as a matrix material. Fabrication begins with the de-agglomeration of the $BaTiO_3$ particles. De-agglomeration is carried out through attrition milling for one and a half hours in ethanol with a phosphoric ester dispersant (BYK W – 9010). The size distribution of the nanoparticles before and after ball milling was characterized using a Malvern Mastersizer laser diffraction instrument, and it can be seen on Figure 1 that the majority of the bulk micron sized agglomerates can be reduced within 1.5 hours of milling.

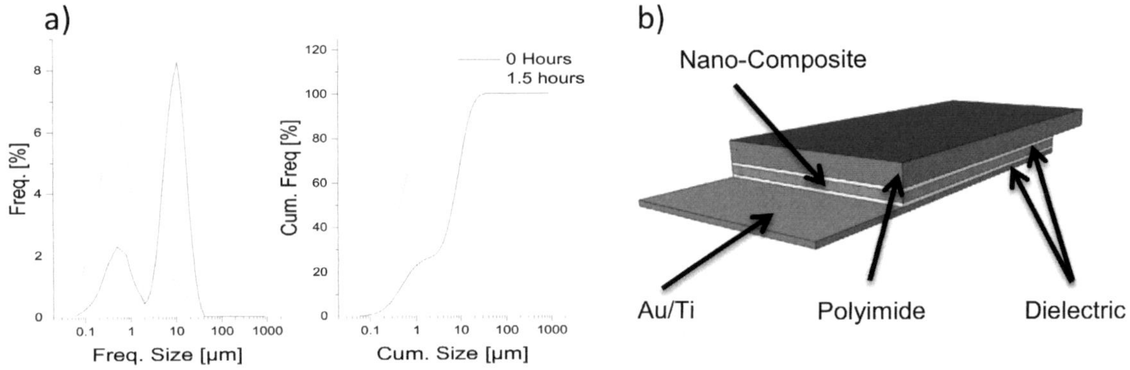

Figure 1. a) Frequency and cumulative size distribution of $BaTiO_3$ in ethanol before and after ball milling. b) Schematic overview of piezoelectric nanocomposite test sample.

After milling a solvent exchange to toluene was performed by ultracentrifugation of the $BaTiO_3$/ethanol mixture, then removal of the supernatant and addition of toluene, this mixture was then added to uncured PDMS resin. Separately, DWCNT's in toluene were given a 10 minute ultrasonic treatment, the $BaTiO_3$/PDMS mixture was then added to the DWCNT/toluene mixture while stirring. Throughout the piezoelectric composite literature it is common to use carbon allotropes around 1 wt%, whereas pristine CNT composites we expect appreciable changes in electrical conductivity and Young's modulus at ranges of 0.0025 wt% [2]. With this in mind it was decided to use 0.001 wt% DWCNT's and 10 wt% $BaTiO_3$ to explore compositions with smaller inert carbon loading.

Assembly of the piezoelectric test structures begins with the addition of PDMS curing agent followed by bar casting of 250µm piezoelectric composites membranes on a PET foil. The electrodes were fabricated by first evaporating titanium/gold (10nm/100nm) on 125µm polyimide foil. Dielectric layers of 15µm pristine PDMS were then directly cast over the gold electrodes, these dielectrics protect the piezoelectric nanocomposite layer from electric breakdown while poling. Finally assembly of the test samples were then carried out by both contact adhesion, and O_2 plasma bonding of the active layers to the PDMS dielectrics. Bonding of PDMS layers by first exposing both surfaces to O_2 plasma before contact leads to the formation of highly reactive silanol groups at the PDMS surface[3]. These silanol groups subsequently form permanent bonds when brought into contact with oxides or neighboring PDMS molecules. After assembly, all samples are poled at a voltage of 2500V and a temperature of 150°C for three hours.

3. Characterization

As the development of composite piezoelectric energy harvesters is still in its early phases, standardized testing procedures have yet to be adopted. In order to address the issue of a standardized test we utilize a Berlincourt type piezoelectric d_{33} meter which can precisely deliver both static and dynamic forces to the test sample that is seen in Figure 4. The piezoelectric d_{33} coefficient can be calculated by dividing the output charge signal created from the piezoelectric sample by the dynamic force that was applied Initial piezoelectric characterization was carried out on a piezoelectric composite containing 1 wt% CNT's, and 10 wt% $BaTiO_3$ and with a 0.25N dynamic force. After measuring the initial response seen in Figure 2, the leads from the sample to the charge amplifier used to measure the piezoelectric output signal were switched, and an otherwise identical measurement was carried out. As can be seen in Figure 3, the polarity of the piezoelectric response was reversed. Such behavior is indicative piezoelectric materials, which will have distinct positive and negative poles. Similar *"Switching Polarity Tests"* are widely used to confirm the presence of a piezoelectric behavior in experimental materials[4].

It has been known that increasing the static preloading will lower the charge output in soft materials and thus the d_{33} coefficient[5]. Such behaviour was seen during initial testing of piezoelectric composite samples and dynamic and static loading of 5 and 3 Newton's were selected as such setting provided the clearest output charge signals. PVDF control samples with a known d_{33} of 22 pC/N were first measured in order to compare experimental outputs to know piezoelectric materials.

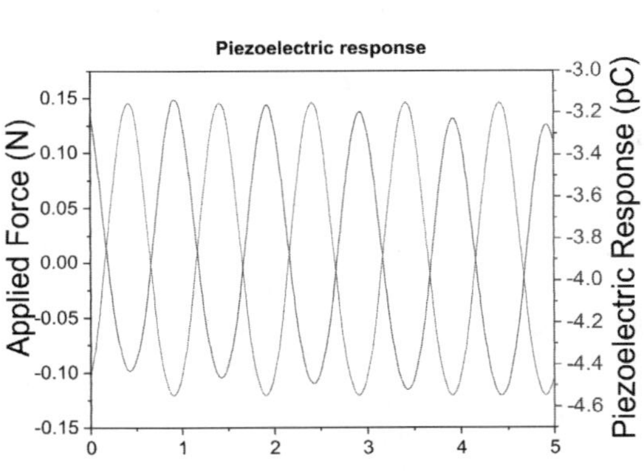

Figure 2. Initial piezoelectric response.

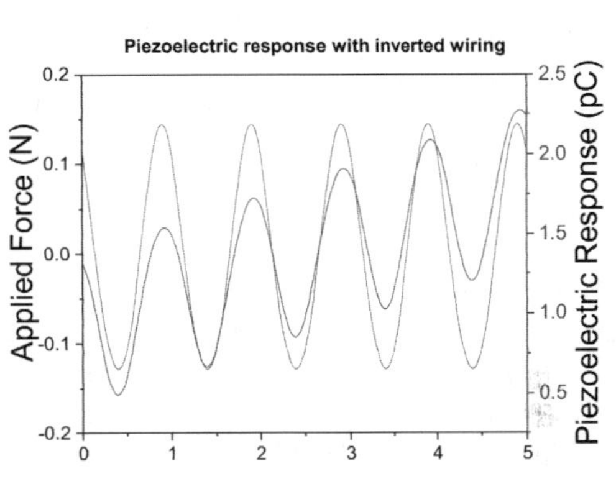

Figure 3. Piezoelectric response with opposite leads connected.

Figure 4. Custom built d_{33} meter.

Figure 5. Piezoelectric composite test sample complete with gold/polyimide electrodes and wire leads.

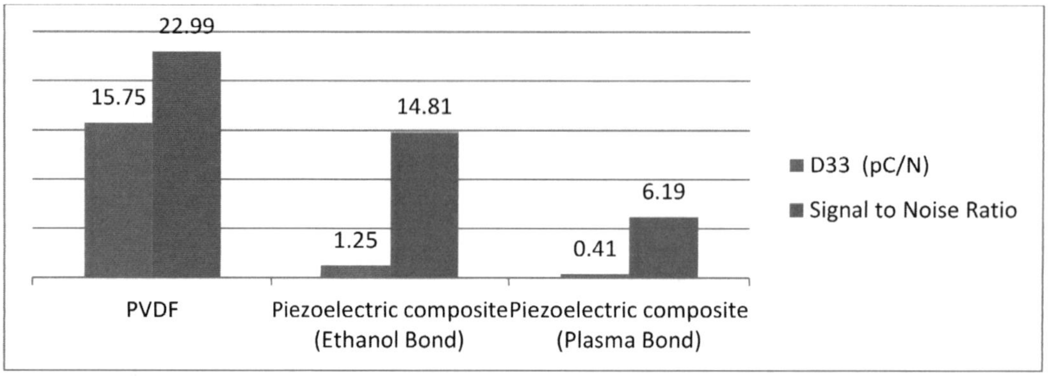

Figure 6. Measured d_{33} coefficients and signal to noise ratios.

Figure 6 shows both the average piezoelectric coefficients the signal to noise ratios for both the test samples which utilize plasma bonding and contact bonding assembly methods, as well as the PVDF control sample. It is immediately evident that the plasma bonding the active layer to the dielectric/electrodes hinders charge generation when compared to the simply utilizing direct contact methods. At present, it is difficult to pinpoint the exact cause as to why the plasma bonded samples provide less charges, but one could speculate that the cracking and increased surface roughness known to be induced from oxygen plasma treatment[6] ultimately serves to resist charge flow more than direct contact adhesion to the electrodes.

4. Conclusion

In the following work we illustrate the raw materials processing, device fabrication and characterization of a piezoelectric $BaTiO_3$/DWCNT/PDMS nanocomposite. Utilizing traditional Berlincourt piezoelectric characterization methods we can conclude that the use of O_2 plasma bonding in device fabrication hinders charge generation when compared to non-permanent contact bonding of the active nanocomposite layer to the protective dielectric and electrode.

Acknowledgements

This research has been funded by Nano-Tera.ch, a program of the Swiss Confederation, evaluated by the Swiss National Science Foundation. Authors would like to extend a warm thank you to Prof. Dragan Damjanovic of the EPFL Ceramics Laboratory for his extensive advice, as well as the use of his equipment.

References

[1] A. V. Quintero, N. Besse, P. Janphuang, R. Lockhart, D. Briand, N. F. de Rooij, *Smart Mater. Struct.* **2014**, *23*, 045041.

[2] Y. Y. Huang, E. M. Terentjev, *Adv. Funct. Mater.* **2010**, *20*, 4062.

[3] S. Bhattacharya, A. Datta, J. M. Berg, S. Gangopadhyay, *J. Microelectromechanical Syst.* **2005**, *14*, 590.

[4] K.-I. Park, S. Xu, Y. Liu, G.-T. Hwang, S.-J. L. Kang, Z. L. Wang, K. J. Lee, *Nano Lett.* **2010**, *10*, 4939.

[5] M. Stewart, M. G. Cain, in *Characterisation Ferroelectr. Bulk Mater. Thin Films* (Ed: M.G. Cain), Springer Netherlands, **2014**, pp. 37–64.

[6] H. Hillborg, J. F. Ankner, U. W. Gedde, G. D. Smith, H. K. Yasuda, K. Wikström, *Polymer* **2000**, *41*, 6851.

Highly Methanol Tolerant Cathode Based on PtAg for Use in Microfluidic Fuel Cell

J. C. Abrego-Martínez[1], L. H. Mendoza-Huizar[2], J. Ledesma-Garcia[3], L. G. Arriaga[1], F. M. Cuevas-Muñiz[1*]

[1]Centro de Investigación y Desarrollo Tecnológico en Electroquímica, Parque Tecnológico Querétaro S/N, Pedro Escobedo 76703, Querétaro, Mexico

[2]Área Académica de Química, Universidad Autónoma del Estado de Hidalgo, Carretera Pachuca-Tulancingo Km 4.5, Mineral de la Reforma 42184, Hidalgo

[3]División de Investigación y Posgrado, Facultad de Ingeniería, Universidad Autónoma de Querétaro, Cerro de las campanas S/N, Santiago de Querétaro 76010, Mexico

*E-mail:fcuevas@cideteq.mx

Abstract. In the present work, a methanol tolerant cathode catalyst was developed through the electrodeposition of bimetallic nanoparticles PtAg on a glassy carbon electrode using potentiostatic and voltammetric techniques. The resulting particles were characterized by Scanning Electron Microscopy (SEM) and electrochemical techniques. According to the SEM study the synthetized nanoparticles have semi-spherical shape and their average diameter is 100 nm. This material shows catalytic activity to the oxygen reduction reaction with high tolerance to methanol.

1. Introduction

The development of microdevices has encouraged the quest for power sources able to operate for long time with a low operation cost [1]. In this sense, microfluidic fuel cells (μFFCs) are an attractive alternative for power supplies because can operate using co-laminar flow of fuel [2-5] and oxidant electrolytes without a physical barrier to separate the two half-cells. The main components and functions related to the electrochemical reactions are located in a single microfluidic channel. In this microfluidic channel, the mixing of two parallel streams is minimal, nonetheless, if a reactant crossover exists, the performance of the cell is affected [6]. This situation has motivated the development of new catalysts that allow using them in conjunction with a single reactants stream in order to simplify behavior and cell design.

The purpose of this work is developing a PtAg/C bimetallic catalyst, prepared by electrodeposition (a rapid and low cost technique), with high tolerance to methanol in order to apply it in a flow methanol μFFC.

2. Experimental

2.1. Materials

Electrodeposits of Pt and Ag were carried out by chronoamperometry technique from two aqueous solutions, the first one containing 5 mM K_2PtCl_4 + 0.5 M H_2SO_4, while the second one contains 5 mM $AgNO_3$ + 0.5 M H_2SO_4. The plating baths were prepared using analytic grade reagents with deionized water. The working electrode was a glassy carbon tip provided by BAS™ with 0.071 cm^2 of area. The exposed surface was polished with alumina 0.05 µm. A graphite bar was used as counter electrode. The reference electrode was an $Hg/HgSO_4$ electrode. All the potentials here reported are referred to the SCE. The electrochemical experiments were carried out with a VSP Biologic potentiostat/galvanostat.

2.2. Synthesis of PtAg/GC catalyst

In order to study the electrochemical behavior of platinum and silver, cyclic voltammetry was carried out in the 1.35 to -0.05 V potential range at different scan rates. PtAg/GC catalyst was obtained in two consecutive steps. In the first one, Pt particles were synthetized by applying a pulse of 0.27 V during 10 seconds from solution 1. After that, the electrodes were washed with deionized water and then introduced in the solution 2 in order to proceed with the second step of the synthesis. Electrodeposition of Ag onto Pt particles was carried out by applying a potential pulse of 0.54 V during 30 seconds from solution 2. Pt/GC particles were also synthetized, with the only purpose of compare them with the PtAg/GC particles, following the procedure above described. The electrodeposits were characterized by SEM and EDS using a JEOL JSM 6510LV.

2.3. Electrochemical characterization and evaluation of catalysts

Performance of modified electrodes was studied in both acidic (0.5 M H_2SO_4, J. T. Baker 99.7%) and basic (0.3 M KOH, J. T. Baker 87%) media using cyclic voltammetry. Evaluation of Pt/GC and PtAg/GC materials in the ORR was conducted in basic media under oxygen atmosphere in hydrodynamics conditions. The catalysts were tested in absence and presence of different concentrations (up to 5 M) of methanol. Linear sweep voltammetry at a scan rate of 5 mV s^{-1} was used for the evaluation.

3. Results and discussion

3.1. Electrochemical deposition of PtAg particles on glassy carbon

A typical cyclic voltammogram of 5 mM K_2PtCl_4 in 0.5 M H_2SO_4 is shown in Fig. 1. During the potential scan in cathodic direction, 3 peaks, A, B and C were recorded at 0.767, 0.415 and 0.080 V respectively. Peak A is associated with hydrogen adsorption process and peak B to the hydrogen reduction at underpotential conditions. Peak C is related to Pt reduction. During the inverse potential scan, peaks D and E may be associated with the dissolution of platinum deposited at the direct scan.
Also, in Fig. 1 it is shown the cyclic voltammogram of 5 mM $AgNO_3$ + 0.5 M H_2SO_4. During the cathodic scan, a peak is observed at 0.435 V, associated with the reduction of silver. In the inverse potential scan, the peak G is associated with the dissolution of silver. Additionally, it is shown the voltammogram obtained only from supporting electrolyte; note that there are not any electrochemical signals. Therefore, the peaks previously discussed are due to Pt and Ag.

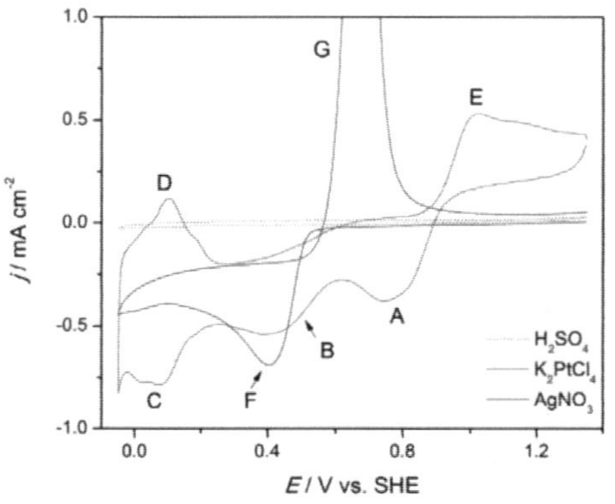

Figure 1. Cyclic voltammetry for platinum and silver precursors ($\nu = 50$ mV s^{-1})

3.2. Characterization of Pt/GC and PtAg/GC particles

Figure 2a and 2b show SEM micrographs of the Pt/GC and PtAg/GC deposits. In both cases, spherical particles (around 100 nm) are observed. The figure 2c shows the distribution of Pt (green) and Ag (red) in the PtAg/GC sample. It is also possible to observe the presence of both metals homogeneously distributed onto the glassy carbon surface.

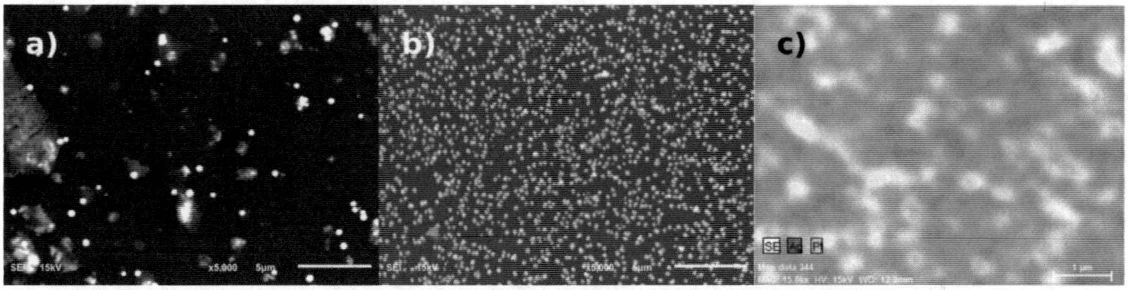

Figure 2. Scanning electron micrography of a) Pt/GC and b) PtAg/GC at 5,000X c) PtAg/GC element mapping at 15,000 X

3.3. Electrochemical characterization of Pt/GC and PtAg/GC catalyst

Figure 3 shows the cyclic voltammograms of the Pt/GC and PtAg/GC electrodes. For Pt/GC, in H_2SO_4 0.5 M, observe the reduction of platinum oxide at 0.75 V and the hydrogen adsorption/desorption zone at a potential value lower than 0.3 V, but these signals disappear in the PtAg/GC system. This can be attributed to the masking induced by the silver covering. This phenomenon is also observed in alkaline media (KOH 0.3 M). However, it is important to mention that the peaks recorded were associated with formation and reduction of silver oxides.

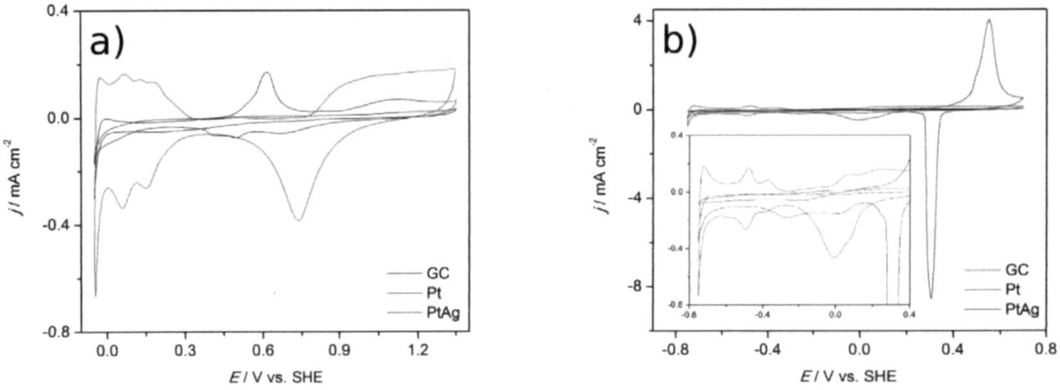

Figure 3. Cyclic voltammograms for bare GC, Pt/GC and PtAg/GC a) acid media and b) alkaline media.

3.4. *Evaluation of Pt/GC and PtAg/GC catalysts in the ORR*

The activity of synthetized catalysts towards oxygen reduction reaction was studied by linear sweep voltammetry in hydrodynamic conditions. Polarization curves were obtained from 0.3 M KOH in an oxygen-saturated atmosphere in the rotation rate range [100 - 2500] rpm. The results for Pt/GC and PtAg/GC are shown in the figure 4a and 4b, respectively. The onset potential of the ORR for Pt is 0.93 V vs. RHE, while the onset potential for PtAg/GC is 0.80 V vs. RHE. This negative shift is due, probably to the Ag layer on Pt, which reduces its catalytic activity. From the polarization curves, the disk-limiting current density was used to construct the Levich plot [7], shown in Fig 4c. The theoretical plots for the transfer of two and four electrons for the ORR were calculated using the Levich equation (Eq. 2).

$$i_L = 0.62nFAD^{2/3}\omega^{1/2}\nu^{-1/6}C_0 \qquad \text{Eq. 2}$$

where i_L is the limiting current (A), n is the total number of electrons transferred, F is the Faraday constant (96485.34 C mol^{-1}), A is the area of the electrode (cm^2), D is the diffusion coefficient (cm^2 s^{-1}), ω is the rotation rate (rad s^{-1}), ν is the kinematic viscosity (cm^2 s^{-1}) and C_0 is the concentration in bulk (mol cm^{-3}). By comparison of the experimental and theoretical plots, it is demonstrated that both catalysts show consistent behavior with a four-electron transfer mechanism.

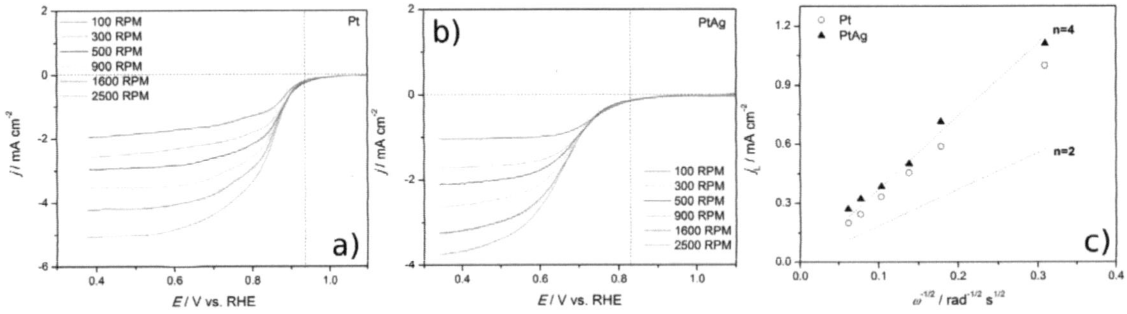

Figure 4. Oxygen Reduction Reaction on a) Pt, and b) PtAg. c) Levich plot for both catalysts.

3.5. Oxygen Reduction Reaction in presence of Methanol.

The electrochemical behavior of the catalysts for the ORR in presence of methanol was also evaluated in a rotating rate of 1600 RPM. The methanol concentration used are 0.1, 0.5, 1.0, 2.0 and 5 M. The evaluation was carried out through polarization curves, which are shown in Figure 5. In absence of methanol, both catalysts promote the ORR. However, when methanol is added, in the case of Pt/GC a decrease in the cathodic current is observed, due to the competition between ORR and methanol oxidation reaction (MOR) processes [5], resulting in anodic currents. Also, it can be observed that at the higher methanol concentration, it is obtained the larger negative shift in the onset potential of the ORR, becoming up to 200 mV in comparison to the system where methanol is not added. On the other hand, the PtAg/GC catalyst keeps its catalytic activity towards the ORR regardless of the concentration of methanol. No anodic currents are observed with the PtAg/GC catalyst, which means that no methanol oxidation reaction occurs. Additionally, the shift on the onset potential of the ORR is 20 mV when the highest concentration of methanol is added.

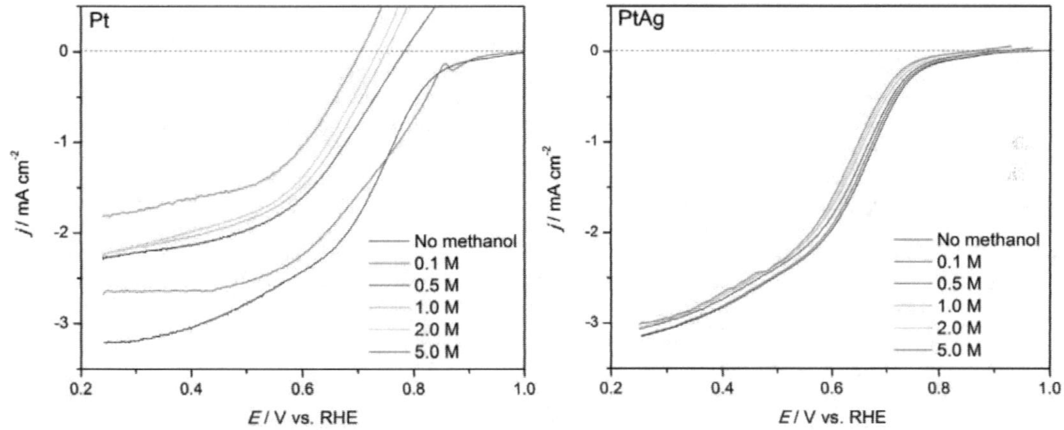

Figure 5. Oxygen reduction in presence of Methanol on Pt/GC and PtAg/GC at 1600 rpm.

Conclusions

A bimetallic PtAg/GC material was obtained by electrochemical technique. This material shows a high methanol tolerance, above 5.0 M in alkaline media. Hence, this material is a promising candidate, in order to avoid the crossover effect in methanol fuel cell, and for using in microfluidic fuel cell.

Acknowledgement

Abrego-Martínez acknowledges to CONACYT for a postgraduate scholarship. Mendoza-Huizar gratefully acknowledges financial support from the Universidad Autónoma del Estado de Hidalgo.

References

[1] Choban E R, Markoski L J, Wieckowski A and Kenis P J A, 2004 *J. Power Sources* **128** 5462.
[2] Moreno-Zuria A, Dector A, Cuevas-Muñiz F M, Esquivel J P, Sabaté N, Ledesma-García J, Arriaga L G and Chávez-Ramírez A U, 2014 *J. Power Sources* **269** 783.
[3] López-González B, Dector A, Cuevas-Muñiz F M, Arjona N, Cruz-Madrid C, Arana-Cuenca A, Guerra-Balcázar M, Ledesma-García J and Arriaga L G, 2014 *Biosens. Bioelectron* **62** 221.
[4] Choi B, Nam W, Chung D Y, Park I, Yoo S J, Song J C and Sung Y, 2015 *Electrochim. Acta* **164** 235.
[5] Asteazaran M, Cespedes G, Bengió S, Moreno M S, Triaca W E and Castro Luna A M, 2015 *J. Applied Electrochem.* In press DOI:10.1007/s10800-015-0845-9.
[6] Kjeang E, 2013 *Microfluidic Fuel Cells and Batteries*, Kjeang E, New York, Springer, Chapter 2.
[7] Treimer S, Tang A, Johnson D C, 2002 *Electroanalysis* **3** 165.

Photonic Crystal Emitters for Thermophotovoltaic Energy Conversion

Veronika Stelmakh[1,2], Walker R. Chan[1], Michael Ghebrebrhan[3], Marin Soljacic[4], John D. Joannopoulos[1,4] and Ivan Celanovic[1].

[1] Institute for Soldier Nanotechnologies, Massachusetts Institute of Technology, 77 Massachusetts Ave., Cambridge, MA 02139, USA

[2] Department of Electrical Engineering, Massachusetts Institute of Technology, 77 Massachusetts Ave., Cambridge, Massachusetts 02139, USA

[3] U.S. Army Natick Soldier Research, Development, and Engineering Center, 1 4th Ave., Natick, MA 01760, USA

[4] Department of Physics, Massachusetts Institute of Technology, 77 Massachusetts Ave., Cambridge, Massachusetts 02139, USA

E-mail: stelmakh@mit.edu

Abstract. This paper reports the design, fabrication, and characterization of 2D photonic crystal (PhC) thermal emitters for a millimeter-scale hydrocarbon TPV microgenerator as a possible replacement for batteries in portable microelectronics, robotics, etc. In our TPV system, combustion heats a PhC emitter to incandescence and the resulting radiation is converted by a low-bandgap TPV cell. The PhC tailors the photonic density of states to produce spectrally confined thermal emission that matches the bandgap of the TPV cell, enabling high heat-to-electricity conversion efficiency. The work builds on a previously developed fabrication process to produce a square array of cylindrical cavities in a metal substrate. We will present ongoing incremental improvements in the optical and thermo-mechanical properties, the fabrication process, and the system integration, as recently combined with fabrication using novel materials, such as sputtered coatings, to enable a monolithic system.

1. Introduction

The increasing power demands of communication equipment, sensors, micro robotics platforms, and portable electronics has driven recent interest in micro- and millimeter-scale thermophotovoltaic (TPV) generators. Thermophotovoltaic energy conversion is a solid-state high-temperature heat-to-electricity conversion scheme with no moving parts that allows for scalable energy production with high specific energy from a variety of heat sources. In TPV systems, thermal radiation from a heat source at high temperature drives a suitable low-bandgap photovoltaic cell (PV). The heat can be produced by hydrocarbon combustion, ideal for lightweight, high specific energy portable power sources [1–3]; by radioisotope decay, ideal for space missions and remote missions requiring power sources with long lifetimes and low maintenance [4]; or by concentrated solar radiation [5]. A selective emitter between the source and the cell enables high heat-to-electricity conversion efficiency by enhancing in-band (convertible) radiation and suppressing out-of-band (nonconvertible) radiation that would otherwise be wasted.

The high temperature requirements of TPV applications constrain the choice for the selective emitter. Multilayer stacks [6], cermets [7], and other heterogeneous platforms are subject to thermo-mechanical stresses and chemical reactions at material interfaces that are initiated at elevated temperatures. Even homogeneous material platforms can degrade at high temperature: radius of curvature driven surface diffusion shortens the lifetime of complex structures such as 3D PhCs [8,9]. Our unique approach, the refractory metal 2D PhC consists of a square periodic array of holes etched into the low-emissivity polished metallic substrate. The PhC is fabricated in a single material, with a simple geometry that minimizes degradation by diffusion [10] and meets the requirements of TPV systems offering high spectral selectivity and high-temperature stability.

This 2D PhC emitter has been fabricated on single-crystal tungsten (W) substrates [11,12], on polycrystalline Ta and Ta-W alloy substrates [13,14], as well as sputtered Ta coatings [15]. The emissivity of the substrate material is selectively enhanced to near that of a blackbody emitter by the resonant cavity modes resulting from the etched holes. The spectral range of enhancement can therefore be easily tailored to the specific system needs by tuning cavity geometry to match the PhC to different low-bandgap TPV cells. In this paper, we present an overview of the advances that have been made over the past decade specifically in the design, materials, and fabrication of these 2D metallic PhCs that enable the high heat-to-electricity system efficiency of TPV systems.

2. Advances in Design, Optimization, and Simulation

Our chosen PhC design consists of a 2D square array of cylindrical holes etched in a large area metallic surface. This relatively simple design allowed us to simultaneously achieve near-blackbody emittance at short wavelengths as well as emittance almost as low as a polished metal at long wavelengths, with a sharp cutoff separating the two regimes [12]. The high emittance at short wavelengths is achieved by matching for each resonance the radiative rate to the absorptive rate. This is assisted by the fact that both the radiative and absorptive rates increases with mode number and hence frequency. At long wavelengths the emittance can be shown to be due to a surface area weighted effective impedance created by the holes [16].

The spectral emissivity of the PhC at normal incidence was initially determined by MEEP via finite-difference time-domain (FDTD) numerical methods [17]. However its high computational requirements limited its application in determining the optimized hemispherical emissivity. Thus, for quicker estimation, a mode matching formalism was developed matching the radiation fields at the boundary of free space and the cylindrical cavities via expansion in the basis of a small number of cavity modes for shorter wavelengths and utilizing a surface area weighted impedance for longer wavelengths [16]. Recently, rigorous coupled wave analysis methods (RCWA) [18] were also made available to simulate the full hemispherical performance relatively quickly. In all simulation methods, the material properties of the substrates were taken into account using a Lorentz-Drude model fitted to the elevated temperature emissivity to capture the optical dispersion of the substrate at high temperature. Nonlinear optimization determined the period, cavity radius, and height with bounds based on fabrication constraints [19]. The figure of merit used in the optimization was spectral selectivity defined as the ratio of power emitted in the desirable wavelength range to the total power at a given temperature. It has been verified that the most significant influence on the spectral properties in the range of interest is given by the cavity radius, whereas the period mainly influences the spectral position of the diffraction limit.

3. Advances in Substrate Materials

The emitter of a TPV system must withstand high temperatures for long operational lifetimes, be simple and relatively low-cost to fabricate, and have spectrally selective emission over a large

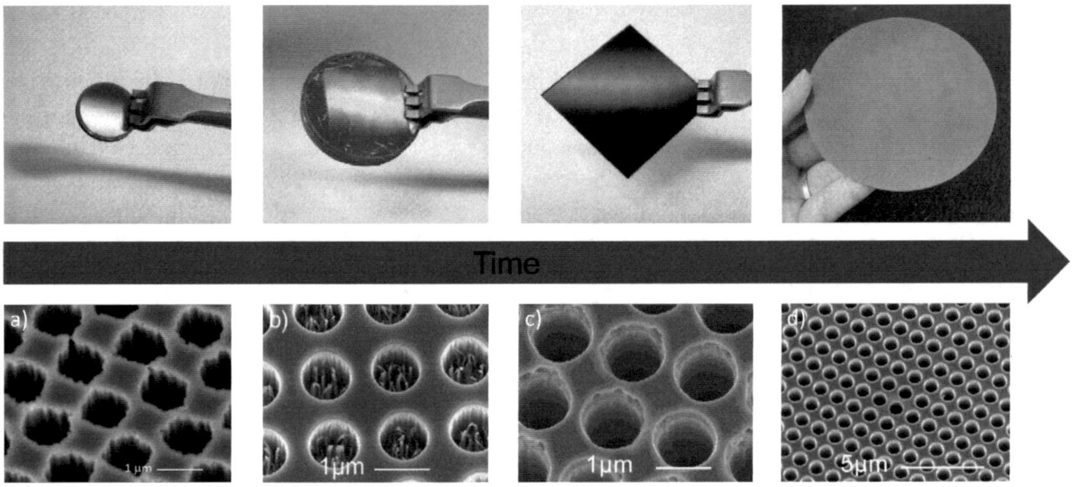

Figure 1. The evolution over time of the substrates, from single crystal W to polycrystalline Ta, and fabrication, from wet etching to RIE, of our 2D PhC, as described in Sections 3 and 4.

uniform area. Refractory metals, such as Ta and W, are most suitable for TPV applications due to their high melting point, low vapor pressure, low infrared emissivity, and ability to be etched. *Single-crystal tungsten* (W) was initially chosen for its optical and thermo-mechanical properties and relative availability [11, 12]. Tungsten was difficult to integrate into a system because it was difficult to machine and weld and because high quality, low cost, large area W was not available. *Polycrystalline tantalum* (Ta) is machinable, weldable and is available as high-purity, large area, cold rolled sheets, and allowed a direct system integration path [13]. *Polycrystalline Ta3% W* combined better thermo-mechanical properties of W with the more compliant material properties of Ta [14]. Substrates were cut, lapped, and polished to an optical finish. Some were pre-annealed to stabilize grain structure growth.

Metallic coatings were investigated in parallel as an alternative route to system integration. In initial studies, *evaporated Ta coatings* on Si were found to have low film density and high surface roughness, resulting in low reflectance in the near-infrared wavelength range [20] and thick *sputtered W coatings* on Si delaminated at high temperatures [21]. On the other hand, thick *sputtered Ta coatings* on polished Inconel 625, a readily available low-cost nickel-chromium-based superalloy used in many high-temperature applications, were found to be suitable substrates for high-temperature energy conversion applications [15]. The PhC fabricated in the Ta coating showed little deterioration of either the optical or thermo-mechanical properties after annealing for 1h at temperatures up to 1100°C and 24h at 900°C with the HfO_2 thermal barrier coating.

4. Advances in Fabrication Methods

As shown in Fig. 2, the fabrication process begins with by defining a periodic pattern in photoresist (PR). The requirement for large area photonic crystal excluded methods such as electron beam lithography. The requirement for low cost metallic substrates excluded contact lithography where surface defects and roughness can result in large inaccuracies in the pattern. Thus *interference lithography* was used, which provides excellent pattern coherence across large areas of exposure and resolution comparable to stepper and electron-beam lithography. Two orthogonal exposures in a Mach-Zehnder setup generated the pattern. In order to prevent standing-wave induced vertically sinusoidal walls in the PR resulting from reflections from the hard-mask layer, a numerically optimized anti-reflection coating (ARC)/SiO_2/PR stack was used

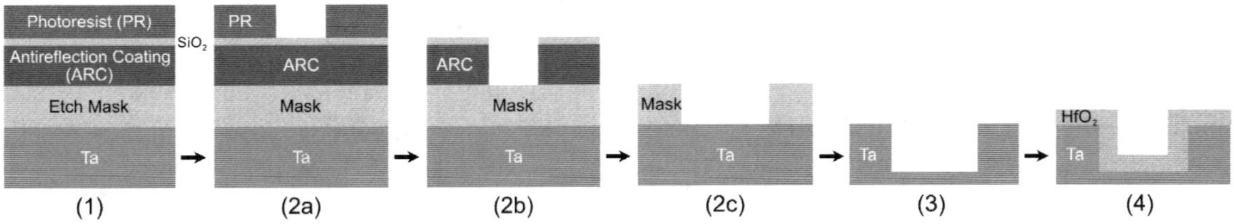

Figure 2. A general process flow for the fabrication of our 2D PhC involves (1) pattern generation, (2) pattern transfer into a hard mask by etching, (3) pattern transfer into the metallic substrate by etching, and (4) surface passivation for thermal stability.

where the SiO_2 served as a protection layer for the ARC while etching the PR.

Using *reactive ion etching* (RIE), the pattern was transferred through the tri-layer stack into the hard mask. Originally a chromium (Cr) hard-mask was used, patterned by wet chemical etching [11]. Initial results were acceptable however it was found that improvements to the optical properties (emittance) of the PhC could be achieved by fabricating a more precise structure, using dry etching. Reactive ion etching of the Cr hard-mask improved uniformity and accuracy of the cavities, facilitating large-area photonic crystal fabrication [22]. Switching from a Cr hard-mask to a *plasma-enhanced chemical vapor deposition* (PECVD) dense SiO_2 mask improved the performance of the PhC in the long wavelength. Indeed, residual Cr on the surface of the Ta decreased the reflectance whereas SiO_2 was completely removed by hydrofluoric acid.

Deep reactive ion etch (DRIE) process for Ta using a Bosch process was developed, which enabled $\sim$8 µm deep cavities with an aspect ratio of 1:8, with very steep and smooth sidewalls. The thermal emitters fabricated by this method showed unprecedented spectral selectivity, enhancement of the emissivity below cut-off approaching unity, and a sharp cut-off between the high emissivity region and the low emissivity region, while maintaining the low intrinsic emissivity of bare Ta above the cut-off wavelength.

One of the greatest challenges of TPV is operating at such high temperatures, which impose strict requirements on the materials used. To address the problems of surface diffusion and surface reactions at high temperatures, a thin surface protective coating of HfO_2 was deposited by *atomic layer deposition* (ALD), which acts as a *thermal barrier coating* and diffusion inhibitor due to its lower surface diffusion [23, 24]. In a number of studies, we observed the degradation of Ta samples without such a coating even after 3h at 900°C due to the formation of tantalum carbide on the surface, resulting in an increased emissivity at long wavelengths. In contrast, those with the coating have consistently showed no degradation of either the optical or structural properties for up to 144h at 900°C.

5. Conclusion

Selective emitters are an enabling component of a TPV system. Even a simple multilayer stack PhC has been shown to double the fuel-to-electricity conversion efficiency versus a greybody emitter for a millimeter-scale TPV system [3]. A similarly large increase in efficiency is predicted for a 2D PhC. Using our 2D PhC, a TPV microgenerator can thus greatly exceed the performance of batteries. This paper presented an overview of the advances made over the past decade specifically in the design, materials, and fabrication of 2D metallic PhCs that enable high heat-to-electricity system efficiency of TPV systems, in particular our hydrocarbon TPV microburner.

Acknowledgments
The authors would like to thank: J. Daley and T. Savas at NSL (MIT), and R. Geil at the University of North Carolina at Chapel Hill for assistance with fabrication of the PhC. H.C. Starck for Ta and Ta3%W substrates; and Southwest Research Institute for Ta sputtering. Fabrication was done in part at NSL at MIT and at CNS at Harvard University, a member of the National Nanotechnology Infrastructure Network (NNIN), supported by the National Science Foundation (NSF) under NSF Award No. ECS-0335765. Fabrication was supported as part of the Solid-State Solar-Thermal Energy Conversion Center (S3TEC), an Energy Frontier Research Center funded by the U.S. Department of Energy (DOE), Office of Science, Basic Energy Sciences (BES), under Award # DE-SC0001299 / DE-FG02-09ER46577 and by the U.S. Army Research Laboratory and the U.S. Army Research Office through the Institute for Soldier Nanotechnologies, under Contract # W911NF-13-D-0001.

References

[1] Doyle E, Shukla K and Metcalfe C 2001 Development and Demonstration of a 25 Watt Thermophotovoltaic Power Source for a Hybrid Power System Tech. Rep. TR04-2001 National Aeronautics and Space Administration
[2] Bitnar B, Durisch W and Holzner R 2013 *Applied Energy* **105** 430–438 ISSN 03062619
[3] Chan W R, Bermel P, Pilawa-Podgurski R C N, Marton C H, Jensen K F, Senkevich J J, Joannopoulos J D, Soljacic M and Celanovic I 2013 *Proceedings of the National Academy of Sciences* **110** 5309–5314 ISSN 0027-8424
[4] Crowley C J, Elkouh N a, Murray S and Chubb D L 2005 *AIP Conference Proceedings* **746** 601–614 ISSN 0094243X
[5] Lenert A, Bierman D M, Nam Y, Chan W R, Celanovic I, Soljačić M and Wang E N 2014 *Nature Nanotechnology* **9** 1–5 ISSN 1748-3387
[6] Narayanaswamy A and Chen G 2004 *Physical Review B - Condensed Matter and Materials Physics* **70** ISSN 01631829
[7] Kennedy C E 2002 *NREL Technical Report* 1–58
[8] Fleming J G, Lin S Y, El-Kady I, Biswas R and Ho K M 2002 *Nature* **417** 52
[9] Lin S Y, Moreno J and Fleming J G 2003 *Applied Physics Letters* **83** 380–382 ISSN 00036951
[10] Peykov D, Yeng Y X, Celanovic I, Joannopoulos J D and Schuh C a 2015 *Optics Express* **23** 9979 ISSN 1094-4087
[11] Celanovic I, Jovanovic N and Kassakian J 2008 *Applied Physics Letters* **92** 193101
[12] Yeng Y X, Ghebrebrhan M, Bermel P, Chan W R, Joannopoulos J D, Soljacic M, Celanovic I, Soljacic M and Celanovic I 2012 *Proceedings of the National Academy of Sciences* **109** 2280–2285 ISSN 0027-8424
[13] Rinnerbauer V, Ndao S, Yeng Y X, Chan W R, Senkevich J J, Joannopoulos J D, Soljačić M and Celanovic I 2012 *Energy & Environmental Science* **5** 8815 ISSN 1754-5692
[14] Stelmakh V, Rinnerbauer V, Geil R D, Aimone P R, Senkevich J J, Joannopoulos J D, Soljacic M, Celanovic I, Soljačić M and Celanovic I 2013 *Applied Physics Letters* **103** 0–4 ISSN 00036951
[15] Stelmakh V, Peykov D, Chan W R, Senkevich J J, Castillo R, Coulter K, Wei R, Joannopoulos J D, Soljačić M and Celanovic I 2015 *Journal of Vacuum Science & Technology A* submitted
[16] Ghebrebrhan M, Bermel P, Yeng Y X, Celanovic I, Soljacic M and Joannopoulos J D 2011 *Physical Review A - Atomic, Molecular, and Optical Physics* **83** 1–6 ISSN 10502947
[17] Oskooi A F, Roundy D, Ibanescu M, Bermel P, Joannopoulos J D and Johnson S G 2010 *Computer Physics Communications* **181** 687–702 ISSN 00104655
[18] Liu V and Fan S 2012 *Computer Physics Communications* **183** 2233–2244 ISSN 00104655
[19] Johnson S G The NLopt nonlinear-optimization package URL `http://ab-initio.mit.edu/nlopt`
[20] Rinnerbauer V, Ndao S, Xiang Yeng Y, Senkevich J J, Jensen K F, Joannopoulos J D, Soljacic M, Celanovic I and Geil R D 2013 *Journal of Vacuum Science & Technology B: Microelectronics and Nanometer Structures* **31** 11802 ISSN 21662746
[21] Stelmakh V, Rinnerbauer V, Geil R D, Aimone P R, Senkevich J J, Joannopoulos J D, Soljačić M and Celanovic I 2013 *Applied Physics Letters* **103** ISSN 00036951
[22] Araghchini M, Yeng Y X, Jovanovic N, Bermel P, Kolodziejski L A, Soljacic M, Celanovic I and Joannopoulos J D 2011 *Journal of Vacuum Science & Technology B: Microelectronics and Nanometer Structures* **29** 61402
[23] Nagpal P, Josephson D P, Denny N R, DeWilde J, Norris D J and Stein A 2011 *Journal of Materials Chemistry* **21** 10836 ISSN 0959-9428
[24] Arpin K A, Losego M D and Braun P V 2011 *Chemistry of Materials* **23** 4783–4788 ISSN 08974756

Fabrication of a rotary carbon nanotube bearing test apparatus

E H Cook[1], M S Weinberg[1], Z S Spakovszky[2] and D J D Carter[1]

[1] Draper Laboratory, 555 Technology Square, Cambridge MA, 02139, USA
[2] Massachusetts Institute of Technology, 77 Massachusets Ave., Cambridge MA, 02139, USA

Email: ecook@draper.com

Abstract. Carbon Nanotubes (CNTs) are attractive elements for bearings in Micro-Electro-Mechanical Systems (MEMS), because their structure comprises nested shells with no bonding and sub-nanometre spacing between them, enabling relative motion with low friction and wear. A reliable bearing technology is critical to bringing rotating MEMS machines from laboratory demonstrations to common use. We report here the design and fabrication of a test rotor, a testing apparatus and testing attempts, and integration of CNTs with MEMS. The device improves on existing CNT bearing demonstrators by establishing a vertical bearing orientation (enabling superior rotor balance and speed, and drive mechanism placement flexibility) and a manufacturable process (employing CNTs grown in place by chemical vapour deposition (CVD)). The main outstanding challenge to demonstrating rotation is available CVD CNT quality.

1. Motivation

Micro- and Nano- electro-mechanical Systems (MEMS and NEMS) have proven effective at providing functionality similar to their larger-scale counterparts, with significant savings in cost, weight, size, and power consumption. Examples of commercially successful devices include pressure sensors, accelerometers, and inkjet printer heads. However, devices that feature moving contacts (such as micro mirror arrays or relays) are rare, and no widespread devices include rubbing surfaces. The tribological phenomena of interfacial friction and wear are responsible [1, 2] for the absence of any commercially successful devices requiring sliding motion. To overcome these issues it will be necessary to reduce friction at the interfaces, and especially to control the mechanisms that cause friction to increase over the device lifetime, with novel bearing technologies.

Recognizing the critical importance of bearing development for the future of MEMS, many have invested substantially in bearing research. From simple bushings, with parts rubbing directly on each other [3], to macro-inspired designs such as ball-bearings [4], to more complex systems such as gas [5] or liquid film bearings [6], a diverse array of approaches has arisen. While most of these solutions offer improved performance over bearing-less designs, they often come at the cost of fabrication complexity or design restrictions, and further performance improvements are desirable (figure 1).

Carbon nanotube (CNT) based bearings are a promising approach, as CNTs have long been known to exhibit low sliding resistance between adjacent walls [7] and several experimental demonstrations have been accomplished [8-13]. Other experimental work has shown that friction performance can decrease after "self-repair" of the CNT undergoing motion [14]. These encouraging properties can be attributed to the energetically favorable and chemically stable structure of CNTs and the inability of debris to accumulate in the 0.34 nm gap between adjacent walls. Still, a quantitative understanding of the friction properties within the bearing system remains elusive [15]. Therefore, it is critical to develop a robust test article that can be used to measure CNT friction and to validate the existing friction models.

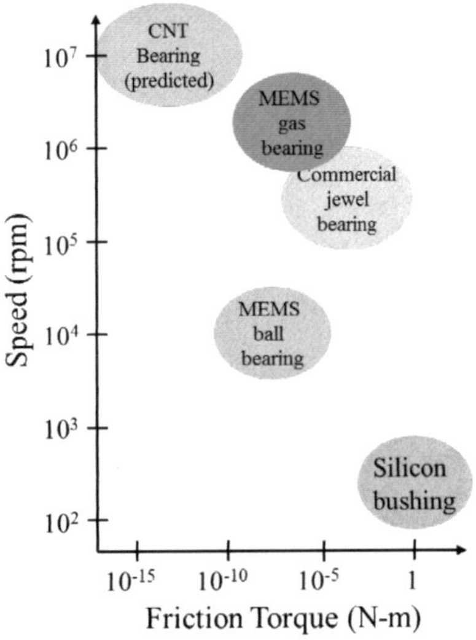

Figure 1. Performance characteristics of several small bearing technologies.

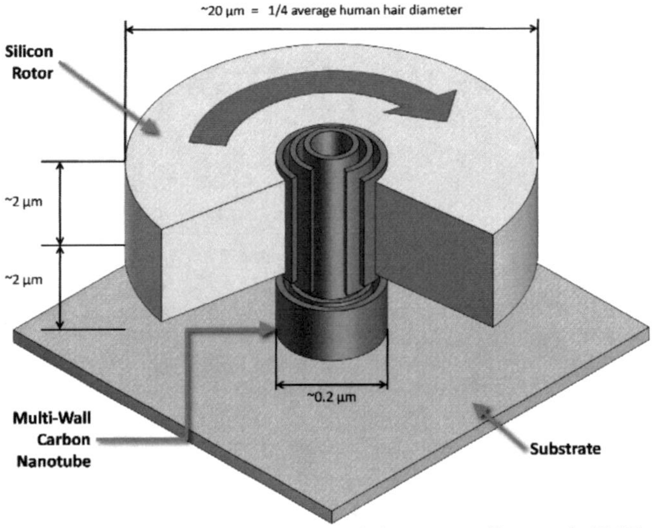

Figure 2. The rotor is supported by a cantilevered CNT, with multiple concentric shells. Cut-away for clarity.

2. Applications

The device presented here (figure 2) paves the way for a host of micromechanical devices. Any rotating mechanical device would be a candidate for miniaturization with CNT bearings. Power devices include turbines, compressors, reactant pumps, and active cooling systems. Propulsion devices such as wheels and propellers could enable micro robots. Bio-mems examples include pumps, valves, fluid shear stress control, viscometers, and centrifuges.

A few applications in particular stand to benefit significantly, such as flywheel energy storage. The energy storage density of a flywheel is limited by the strength of the flywheel material, and independent of volume. Stored energy density in silicon flywheels is comparable to batteries, but could be discharged faster. The CNT bearing could make such flywheels feasible by enabling high-speed rotation, and by providing a low-friction support that reduces unwanted energy dissipation.

The CNT bearing also has promise is a microscale rotating gyroscope. The bearing's very low "wobble" (angular displacement of the axis of rotation, in the CNT case set by the inter-layer spacing of 0.34 nm) would reduce the dominant cause of errors in dynamically tuned gyroscopes. The high stiffness would enable high-speed operation, for improved bandwidth and sensitivity. Also, low friction suggests that the drive power to operate the gyro and associated errors could be very low.

3. Test apparatus design

Previous work showing rotary CNT bearings [8, 9] has used CNTs scattered randomly on a substrate, so that they are horizontally oriented. The devices must be made by painstakingly hand-selecting CNTs and patterning features by e-beam lithography around each one, and furthermore must be planar in shape. For the present device, a vertically oriented CNT was chosen (figure 2), enabling axisymmetric geometry to be defined lithographically.

The basis of the design is the well-known Stodola rotor model [16]. This model considers the overhung rotor to consist of a rigid disk supported on a shaft, which is treated as a classical beam element. The rigid rotor assumption is justified since the silicon rotor is two orders of magnitude thicker than the CNT axle. A more critical assumption is that the CNT can be treated in a continuum fashion, using a classical beam model. Some [17] have pointed out that molecular scale structures do not have truly uniform mechanical properties because of the quantized nature of the atoms. Furthermore, geometric features such as cross-sectional area are not well-defined for atoms. However, the CNTs used in this device are relatively large (50 or more walls) so that geometric ambiguities are less important. Furthermore, experimental investigations [18] have indicated that for large tubes, mechanical properties that do not vary from tube to tube can be established with some degree of accuracy and repeatability. The Young's modulus value from that study (1.28 TPa) was calculated from beam bending of CNTs, and is used to model CNT beam bending in the present model.

Within the envelope defined by the failure limits of the CNT and rotor, based on assumed rotor misalignments of 1 μm (from lithography) and 10° (from observed CNTs), an optimal design was selected by maximizing coast-down time, or the time the rotor would take to come to rest if given an initial spin rate. Maximum coast-down time provides the best possible precision in measuring the friction in the rotor via a coast-down experiment. Assuming a constant friction of 0.85 MPa [9], the coast-down time is proportional to both the maximum possible speed and the disk inertia, so that the maximum possible diameter of 20 μm was selected, taking into account the deflection and stress constraints, as well as fabrication limits for the CNT and the rotor disk itself. This corresponds to a disk thickness of 2 μm, a CNT length of 4 μm, and an expected coast-down time of about 1 ms, from a maximum speed of 3.2 Mrpm.

To actuate the rotor, electrostatic rotational motors [3, 19, 20] were considered, and would likely be the obvious choice for eventual devices based on the bearing. They provide bi-directional actuation, uniform torque coverage, 360° actuation at high angular resolution, and precise electronic control. However, they introduce complexity, so it was decided to actuate the device using jets of air directed at blades attached to the rotor.

4. Fabrication

After thoroughly exploring the interactions of established MEMS processes with CNTs [21], a fabrication method leveraging these techniques was developed to build a CNT bearing test prototype.. Fabrication begins with an isolated, vertically oriented CNT (Figure 3a), grown by CVD [22, 23]. Isolated nanotubes are obtained [24] by patterning the metal catalyst particles (via e-beam liftoff) to a size comparable to the CNT diameter. CNTs had diameters of 10-30 nm and lengths of 2-4 μm. With the CNT in place, a 2 μm PECVD $SiO2$ layer was deposited at 1 Torr and 380°C to act as the sacrificial layer. The $SiO2$ coats the CNT conformally. While it is possible to achieve release with a timed etch, in practice it was determined that better results can be obtained by securing the polysilicon rotor layer directly to the CNT, by removing the $SiO2$ around the upper portion of the CNT. This was accomplished by spin-coating a 0.5 μm layer of photoresist, which coats the flat substrate but lets the CNT protrude (Figure 3b). BOE was then used to strip the oxide from this exposed portion of the CNT (figure 3c). The polysilicon rotor layer was then deposited via LPCVD.

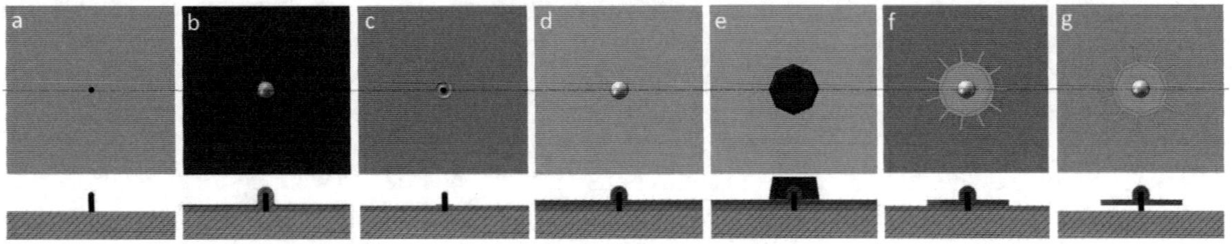

Figure 3. The fabrication process for the rotor includes CVD sacrificial and rotor layers on top of a vertical CNT, e-beam and optical lithography, and wet release etch.

A two-step lithography procedure was used to define the rotor so that a thick resist could cover the CNT protrusion, while a thin resist provided the necessary resolution. Two metal layers (Cr and Ti) were first sputtered on the sample (figure 3d). A thick negative resist (Futurrex NR9-3000P, 5 μm) was applied, and an optical microscope was used to do the exposure at 100x magnification, 3 minute exposure, 15 μm spot size. Contact lithography cannot be used because of the protruding CNTs. After etching the first metal layer with this thick resist mask (figure 3e), the rotor itself was then patterned using e-beam lithography (with 0.5 μm maN-2403) and the other metal layer was etched using that mask. Using the double-metal mask, the rotor was etched with a cryogenic SF6/O2 ICP-RIE process that provides smooth vertical sidewalls at this etch depth (figure 3f). Finally, the rotors were released in BOE (figure 3g). Transmission microscopy was used to monitor the progress of the release etch undercut front. The completed device is shown in figure 4.

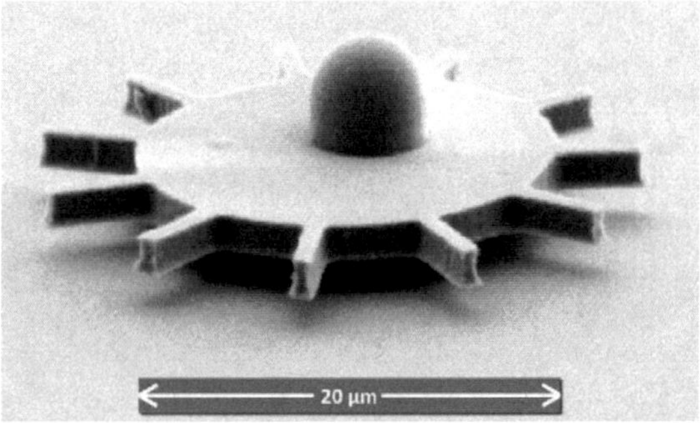

Figure 4. A completed device in SEM. The central "dome" is the location of the CNT axle. The fins are for actuation via an external air jet.

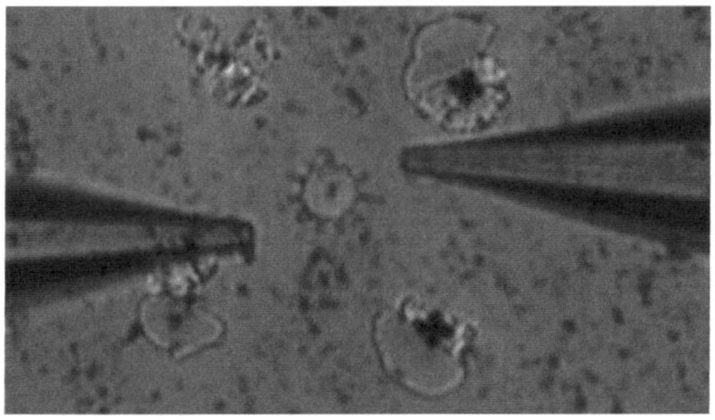

Figure 5. The test setup is shown. The conical objects are the micro capillary tubes, blowing air onto the vanes of the rotor (center).

5. Testing

CNT rotors were tested by blowing air onto the rotor vanes, providing a torque. A microelectronics probe station was converted to interface with glass micropipettes (available from a medical equipment supplier) instead of electronic probes. These micropipettes with a tip orifice diameter of 5 μm can direct air at the vanes of the prototype device from opposite sides, providing a balanced torque. Given the small size of the pipettes and the sub-micrometer precision of the probe station manipulators, the tips of the pipettes can be brought within one blade length of the blades providing a jet of air to impinge on the blades (as seen in figure 5). The air to the micropipettes is supplied by a syringe pump, which provides air at flow rates from under a microliter per minute to milliliters per minute.

In order to observe any rotation, the microscope was fitted with a video recorder. To date, a successful test has not been performed. Although air has been confirmed to be flowing from pipettes (by a water submersion test), and hence a torque is applied, no motion of the rotor has been observed. The suspected reason is that the CVD-grown CNTs used in construction of prototype devices thus far have exhibited poor crystalline quality (figure 6).

To address the crystalline quality issue, we have also considered the use of arc-discharge CNTs (figure 7), such as those used in previous demonstrations [8, 9]. These CNTs are produced at much higher temperatures than CVD CNTs, (4000°C vs. 600°C) and therefore are annealed into the more energetically favourable concentric tube structure [25]. However, arc-discharge CNTs are available as individual tubes and must be manipulated and attached to the substrate before being incorporated into the device. We have used a SEM/FIB tool with integrated nanomanipulation probes to pick-and-place CNTs. The tool uses in-situ FIB-enhanced amorphous carbon deposition to weld CNTs to the probe and to the substrate, and then FIB cutting to detach. While it is possible to move and attach CNTs in this manner, it is difficult to achieve a vertical orientation because of the limited viewing angle.

Further work in improving the available CNT quality, either by manipulation of high-quality tubes into place or improving the quality of CVD-grown tubes, is required to achieve a functional prototype.

Acknowledgements
The authors gratefully thank ZF Ren, T Paudel, G McMahon and DZ Wang of Boston College for CVD-CNT samples, and assistance with TEM characterization and SEM/FIB manipulation of CNTs.

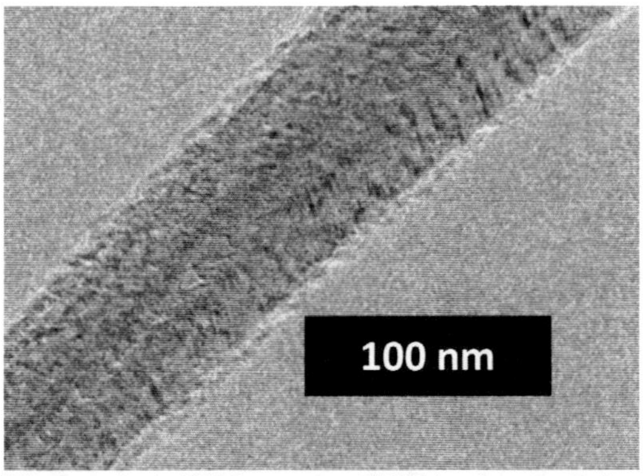

Figure 6. A CVD-grown CNT is shown. It is thicker than a typical arc-discharge tube, and includes many structural defects.

Figure 7. An arc-discharge CNT is shown. The concentric rings are visible as lines in this high resolution TEM image. The structure is close to ideal.

References

[1] Romig A D, Dugger M T and McWhorter P J 2003 *Acta Mater.* **51** 5837-66
[2] Flater E E, Corwin A D, de Boer M P and Carpick R W 2006 *Wear* **260** 580-93
[3] Mehregany M, Bart S F, Tavrow L S, Lang J H, Senturia S D and Schlecht M F 1990 *Sensor. Actuat. A-Phys.* **21** 173-9
[4] Ghalichechian N, Modafe A, Beyaz M I and Ghodssi R 2008 *J. Microelectromech. S.* **17** 632-42
[5] Lang J ed 2010 *Multi-Wafer Rotating MEMS Machines* (New York: Springer)
[6] Chan M L, *et al.* 2012 *Sensor. Actuator. A-Phys.* **177** 1-9
[7] Charlier J-C and Michenaud J-P 1993 *Phys. Rev. Lett.* **70** 1858-61
[8] Fennimore A M, Yuzvinsky T D, Han W-Q, Fuhrer M S, Cumings J and Zettl A 2003 *Nature* **424** 408-10
[9] Bourlon B, Glattli D C, Miko C, Forro L and Bachtold A 2004 *Nano Lett.* **4** 709-12
[10] Cumings J and Zettl A 2000 *Science* **289** 602-4
[11] Yu M-F, Yakobson B I and Ruoff R S 2000 *J. Phys. Chem. B* **104** 8764-7
[12] Akita S and Nakayama Y 2005 *e-Journal of Surface Science and Nanotechnology* **3** 86-93
[13] Akita S and Nakayama Y 2003 *Jpn. J. Appl. Phys.* **42** 4830-3
[14] Kis A, Jensen K, Aloni S, Mickelson W and Zettl A 2006 *Phys. Rev. Lett.* **97** 025501-4
[15] Cook E H, Buehler M J and Spakovszky Z S 2012 *J. Mech. Phys. Solids* **61** 652
[16] Childs D W 1993 *Turbomachinery Rotordynamics: Phenomena, Modeling, and Analysis* (New York: Wiley-Interscience)
[17] Yakobson B and Avouris P 2001 *Carbon Nanotubes: Synthesis, Structure, Properties, and Applications* ed Dresselhaus M S, Dresselhaus G and Avouris P (New York: Springer-Verlag) pp 287-327
[18] Wong E W, Sheehan P E and Lieber C M 1997 *Science* **277** 1971
[19] Bart S F, Mehregany M, Tavrow L S, Lang J H and Senturia S D 1992 *IEEE T. Electron Dev.* **39** 566-75
[20] Tavrow L S, Bart S F and Lang J H 1991 *Proc. Int. Conf. Solid-State Sensors and Actuators* (San Francisco, CA: IEEE) 877-81
[21] Cook E H and Carter D J D 2011 *J. Vac. Sci. Technol. B* **29** 06FE04
[22] Ren Z F, Huang Z P, Xu J W, Wang J H, Bush P, Siegal M P and Provencio P N 1998 *Science* **282** 1105-7
[23] Tu Y, Huang Z P, Wang D Z, Wen J G and Ren Z F 2002 *Appl. Phys. Lett.* **80** 4018-20
[24] Ren Z F, *et al.* 1999 *Appl. Phys. Lett.* **75** 1086-8
[25] Dai H 2001 *Carbon Nanotubes: Synthesis, Structure, Properties, and Applications* ed Dresselhaus M S, Dresselhaus G and Avouris P (New York: Springer-Verlag) pp 29-53

Green piezoelectric for autonomous smart textile

E Lemaire[1], C J Borsa, D Briand
Ecole Polytechnique Fédérale de Lausanne (EPFL)
The Microsystems for Space Technologies Laboratory, Neuchâtel, Switzerland

E-mail: etienne.lemaire@epfl.ch

Abstract. In this work, the fabrication of Rochelle salt based piezoelectric textiles are shown. Structures composed of fibers and Rochelle salt are easily produced using *green* processes. Both manufacturing and the material itself are really efficient in terms of environmental impact, considering the fabrication processes and the material resources involved. Additionally Rochelle salt is biocompatible. In this *green* paradigm, active sensing or actuating textiles are developed. Thus processing method and piezoelectric properties have been studied: (1) pure crystals are used as acoustic actuator, (2) fabrication of the textile-based composite is detailed, (3) converse effective d_{33} is evaluated and compared to lead zirconate titanate ceramic. The utility of textile-based piezoelectric merits its use in a wide array of applications.

1. Introduction

Textile-based sensors [1, 2] and piezoelectronics are have been recently utilized in the development of smart textiles for medical applications [3]. In this work, the development of a textile-based piezoelectric material based on large area, low-cost manufacturing is addressed. We first report on the actuation of cantilevers using pure crystals demonstrating the efficient and well-known [4] piezoelectric character of the pure material: Rochelle salt. Then the fabrication of the composite material composed of polypropylene (PP) fibers and Rochelle salt is described. Finally, the converse piezoelectric effect of the composite material is demonstrated and compared with a standard lead zirconate titanate (PZT) ceramic of similar size. Complementing that, impregnated PP textile has been used for the piezoelectric actuation of a PET-based cantilever.

2. Context

Sodium potassium tartrate tetrahydrate, $KNaC_4H_4O_6 \cdot 4H_2O$ also called Rochelle or Seignette salt is a ferroelectric and piezoelectric material that does not containt rare elements, therefore being intrinsically environmental friendly. Rochelle salt is the oldest and has been for a long time the only known ferroelectric and piezoelectric material [5, 6]. Nowadays Rochelle salt is used in chemistry for different reactions especially in organic synthesis [7] and it is massively produced as a food additive (E337) also called cream of tartar [8].

In order to demonstrate piezoelectricity of Rochelle salt (called PiezoSalt in this work), we used pure millimeter sized crystals as a piezoelectric actuator of a paper-based cantilever (Figure 1). A driving voltage was applied to the crystal and the cantilever displacement was measured using a laser

[1] To whom any correspondence should be addressed.

vibrometer. The dynamic first resonant mode of the paper cantilever exhibited an out-of-plane amplitude higher than 200 pm. The decrease of the phase from 0 to -90° indicated that the out-of-plane displacement was the consequence of the driving signal applied to the crystal. During this experiment acoustic waves produced by the crystal were audible.

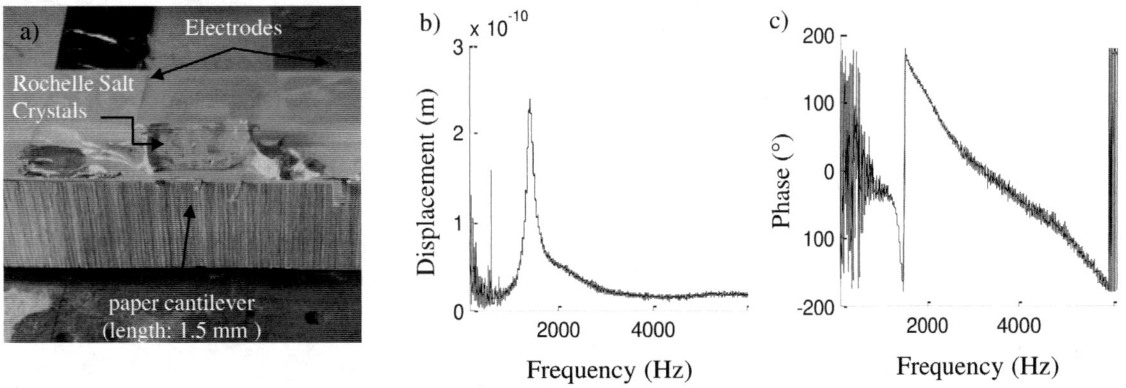

Figure 1. (a) Array of paper cantilevers of several lengths actuated by acoustic waves produced by a Rochelle crystal fixed on a paper sheet. (b) Module and (c) Phase of the displacement of the cantilever (length 1.5 mm).

This experiment illustrates (1) the piezoelectric behavior of pure Rochelle salt crystals, (2) its ability to produced easily acoustic waves that could drive a cantilever-based structure, (3) a micro-actuator concept simply built-up with paper, Rochelle salt and aluminum. The next step was to functionalize some material with this piezoelectric crystal. Therefore this micro-actuator concept was tested again in section 3.4 with the composite PiezoSalt.

3. Fabrication and characterization

In this section we discuss how we manufacture the textile-base piezoelectric. Then, some micrograph highlights the polycrystalline structure obtained in between the fibers. Finally, the composite is actuated at low voltage and the out-of-plane displacement produced is measured. A Comparison between PZT ceramic and textile-base piezoelectric is performed in order to evaluate the converse effective d_{33}. Additionally the impregnated textile is used to actuate a PET cantilever demonstrating the micro-actuator concept within the textile.

3.1. Fabrication

Polypropylene (PP) based absorbent tissue usually provided for cleaning floors and surfaces has been used as matrix. The thickness of the initial tissue is 3 mm. The textile has been impregnated into Rochelle salt solution and dried. Figure 2 illustrates the PP sample before and after impregnation. Stretchable silver-plated nylon supplied by Sparkfun were used as electrode and fixed on the PP textile before impregnation. After impregnation the stretchable conductive textile (cf. conductive textile on Figure 2) is naturally attached to the impregnated PP textile due to the presence of the salt.

Impregnation and drying procedure were repeated two times in order to increase the loading of Rochelle salt inside the matrix. A Rochelle salt composite material has been formed using an absorbent textile composed of polypropylene (PP) fibers. The resultant PiezoSalt Textile is hard and stiff while the initial absorbent tissue was soft and compressible. The thickness of the impregnated textile is about 10% thicker than the initial tissue.

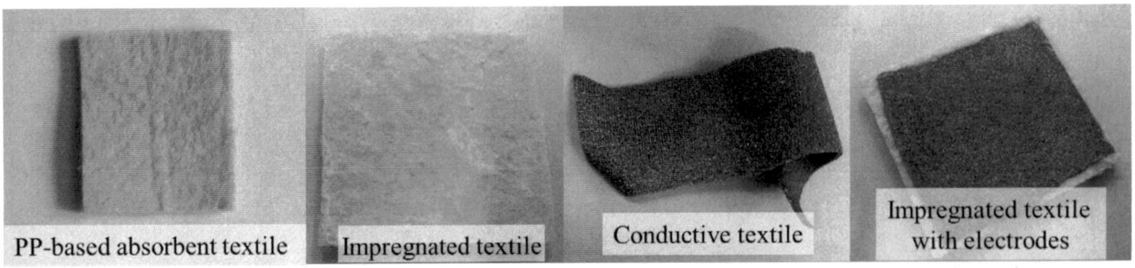

Figure 2. Image of the absorbent textile, the conductive textile and the samples after impregnation and drying.

3.2. Material structure

The initial absorbent tissue is formed of random arrangement of polypropylene fibers as shown is Figure 3a. Micrographs have been taken using the Hirox KH-8700 digital microscope. After the impregnation, the fibers network is trapped into a polycrystalline structure clearly visible on the micrograph of Figure 3b.

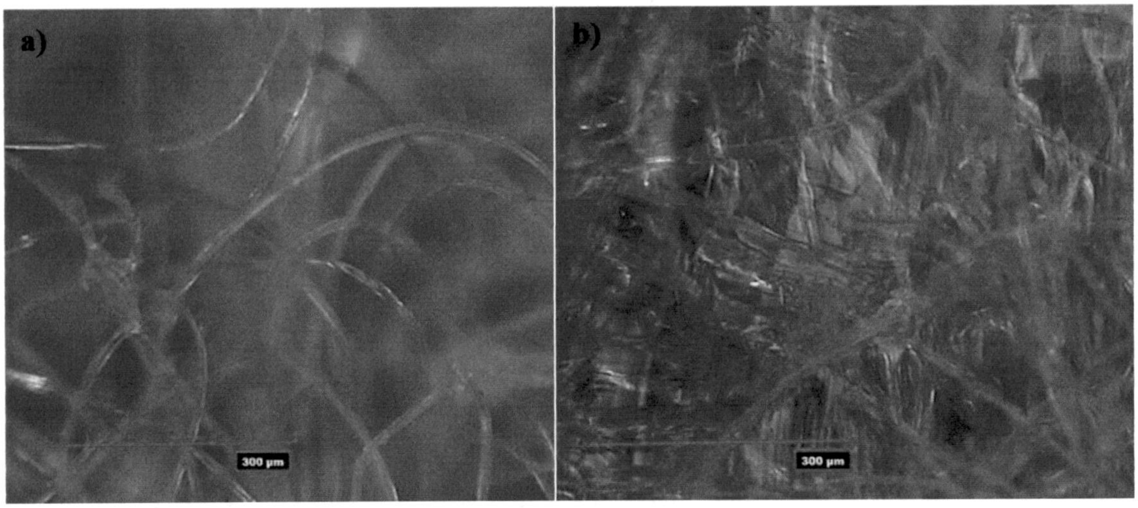

Figure 3. (a) Network of PP fibers composing the absorbent textile. (b) Network of PP fibers impregnated with solid Rochelle salt polycrystalline structure.

3.3. Evaluation of the converse piezoelectric effect

The evaluation of the converse piezoelectric effect has been done by comparing the Rochelle salt impregnated textile samples (3 mm thick) and a standard PZT ceramic from PI (ref. PIC181, 1 mm-thick). The first easiest test was an audio sweep. Both materials emitted audible acoustic waves when applying a 3V periodic chirp defined from 10 Hz to 10 kHz. The exact same actuating conditions have been applied to the samples when placed under a laser vibrometer (MSA-400 from Polytec). Out of plane complex displacement of both piezoelectric devices were measured on the top planar electrode of the samples, the resulting spectrums being presented in Figure 4.

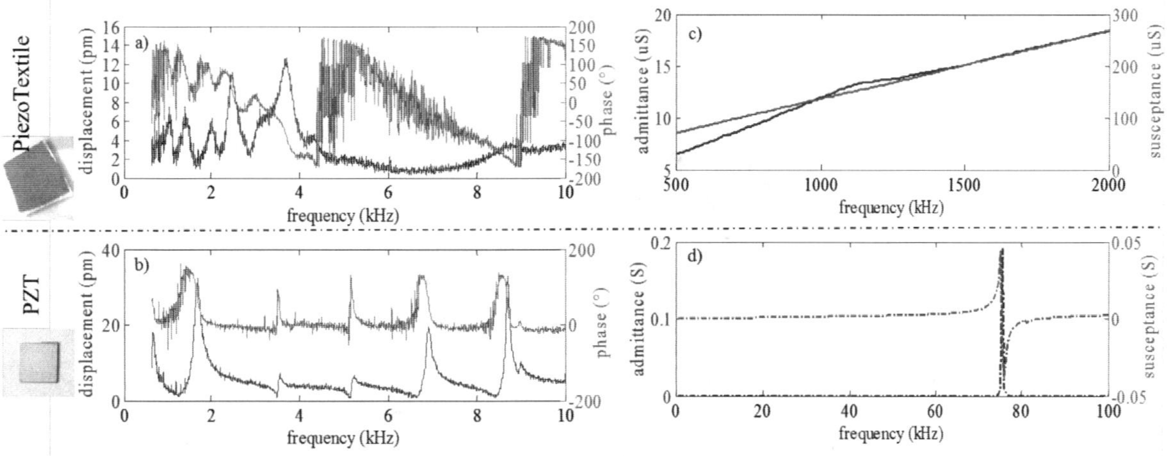

Figure 4. Displacement module and phase of a) PiezoSalt Textile and b) PZT samples measured using a laser vibrometer. Admittance and susceptance of c) PiezoSalt Textile and d) PZT samples measured using an impedance analyser.

The comparison of displacement spectrums shows that the maximum displacement measured, in the range 1 to 10 kHz, was 14 pm for the PiezoSalt Textile while displacement reached 29 pm for the PZT sample. We could also notice that the sharpness of resonant peaks is deprecated in case of PiezoTextile due to the difference in the stiffness of the materials. Conductance spectrums shown in Figure 4-c and –d highlighted also an important difference in the electromechanical coupling factor: a narrow bandwidth indicating a high coupling factor (which was the case for PZT but not for PiezoSalt Textile). Uncalibrated effective d_{33} and coupling factor k obtained for the PZT and the PiezoTextile are presented in Table 1.

Table 1. Uncalibrated effective d_{33} and coupling factor k respectively obtained by laser vibrometer and impedance analyser.

	Effective d_{33} (pm/V)	k_p
PiezoTextile	45	0.01
PZT	94	0.5

The PiezoTextile does not have the excellent and unbeatable piezoelectric properties of PZT. Nevertheless, these electromechanical tests have demonstrated high piezoelectric properties for a textile-based material.

3.4. Micro-actuator concept

A microcantilever composed of PET coated with aluminium has been fixed on top of the PiezoTextile. The dynamic first out-of-plane resonant mode of the cantilever has been measured. A drive oscillating voltage of 5V actuated the impregnated PP textile which induced the phase-shifted displacement of the PET cantilever. Module and phase of the cantilever displacement recorded with a laser vibrometer is shown on Figure 3. Any phase-shifted displacement and neither any sound production could be recorded on a non-impregnated reference device.

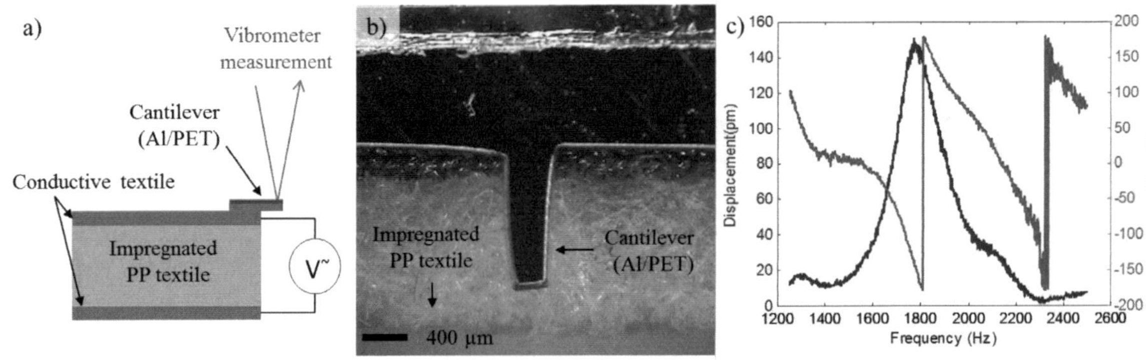

Figure 5. (a) Piezoelectric textile actuating Al/PET cantilever (drive voltage of 5V). (b) View of a cantilever fixed on the top electrode of the actuator. (c) Module and phase of the cantilever displacement induced by piezoelectricity.

The microcantilever has been successfully actuated piezoelectrically with the PiezoTextile. This micro-actuator concept works identically that the one presented in Figure 1. Therefore Rochelle salt could stay piezoelectrically active when impregnated into the polypropylene-based textile.

4. Conclusion

In this work, the piezoelectric behavior of Rochelle salt crystal has been highlighted within a textile matrix composed of polypropylene fibers. We developed a method to functionalize textile with this piezoelectric material. Electromechanical properties of the piezoelectric textile have been evaluated. Finally the converse piezoelectric effect existing into the composite material has been demonstrated. A PET microcantilever could be successfully actuated piezoelectrically with the PiezoSalt Textile.
This concept easily manufacturable could find application into many fields. Smart textile, large area functionalized surfaces, integrated sensors and actuators or even harvesting surfaces are some of the numerous potential applications that can be envisioned at low cost. In the future, wearable smart devices will be developed out of this PiezoSalt Textile.

5. References

[1] Morris, D., Coyle, S., Wu, Y., Lau, K. T., Wallace, G., & Diamond, D. (2009). Bio-sensing textile based patch with integrated optical detection system for sweat monitoring. *Sensors and Actuators B: Chemical, 139*(1), 231-236.
[2] Yang, Y. L., Chuang, M. C., Lou, S. L., & Wang, J. (2010). Thick-film textile-based amperometric sensors and biosensors. *Analyst, 135*(6), 1230-1234.
[3] Carpi, F., & De Rossi, D. (2005). Electroactive polymer-based devices for e-textiles in biomedicine. *Information Technology in Biomedicine, IEEE Transactions on, 9*(3), 295-318.
[4] Unsworth, J. (1992, January). Piezoelectricity and piezoelectric materials. In *Key Engineering Materials*, Vol. 66, pp. 273-310.
[5] Busch G., 1987, Early History of Ferroelectricity. *Ferroelectrics*; *74*; 267-28.4
[6] Kanzig W., 1987, History of Ferroelectricity 1938-1955. *Ferroelectrics*:*74*; 285-291.
[7] Fieser, L. F.; Fieser, M., *Reagents for Organic Synthesis*; Vol.1; Wiley: New York; 1967, p. 983.
[8] Furia, T. E. (1973). *CRC handbook of food additives* (Vol. 1). CRC Press.

Nonlinear electromagnetic energy harvesters fabricated by rigid-flex printed circuit board technology

Yi Chiu[1], Hao-Chiao Hong[1], Wei-Hung Hsu[2]
[1] Department of Electrical and Computer Engineering, National Chiao Tung University, Hsin Chu, Taiwan, R.O.C.
[2] Green Energy and Environment Research Laboratories, Industrial Technology Research Institute, Hsin Chu, Taiwan, R.O.C.

E-mail: yichiu@mail.nctu.edu.tw

Abstract. In this paper, a wideband electromagnetic energy harvester designed and fabricated by commercial rigid-flex PCB technology is demonstrated. The rigid FR-4 boards are used for mechanical frames and coil winding whereas the flexible polyimide film is used for mechanical springs and mass platforms. The total dimension of the device is $20 \times 20 \times 2$ mm^3. The internal coil resistance is 15 Ω. In vibration tests, nonlinearity can be observed even at 0.1 g vibration level due to the spring hardening effect. The peak frequency was increased as the vibration level increased. The effective bandwidth was increased from 6 Hz at 0.1 g to 21 Hz at 0.5 g and 27 Hz at 1 g, respectivel, due to the hysteresis effect. For a matched load and 1 g vibration at 240 Hz, the maximum output power is 24.5 nW, corresponding to a power density of 31 nW/cm^3.

1. Introduction

Energy harvesters are typically cm-scale devices fabricated by micromachining technology [1] or by assembling components with cm to sub-mm dimensions [2, 3]. The fabrication or assembly cost can be high and impede wide application of the emerging technology. In comparison, printed circuit boards (PCB) is a mature and lost-cost fabrication technology suitable for cm-scale devices. Its design rules are typically 100 μm and can be used for most harvesting devices. Among different types of energy harvesters, electromagnetic harvesters have the advantage of simple structures, simple material requirement, and low output impedance. PCB technology is particularly attractive for fabricating electromagnetic harvesters since fine structures and gaps are seldom used in these devices. Furthermore, PCB can be used for coils winding and as mounting platforms for permanent magnets. Both rigid PCB [4, 5] and flexible PCB [3] have been used to fabricate energy harvesters. In [4], springs, magnet platforms, and support frames were fabricated in standard FR-4 PCB by CNC machining. A copper wire wound coil was bonded to a glass substrate to pick up the flux change and generate power during vibration. In [2, 6-8], various polymer films such as PE, PI/Kapton, and PDMS, were used as suspension membranes for the permanent magnet to achieve low resonance frequency. However, these films had to be attached to another rigid frame during device assembly. This extra assembly process can increase fabrication cost and reduces device reliability. Both flexible polyimide (PI) films and rigid FR-4 boards used in the literature are common and well known substrates in the PCB industry. They can be laminated together to form rigid-flex PCB. In [9], rigid-flex PCB were used to fabricate an integrated piezoelectric/electromagnetic harvester. However, the flexible layer in

[9] was used mainly to support the PZT bimorph only. In this paper, an electromagnetic energy harvester fabricated by the rigid-flex PCB technology is demonstrated. The rigid boards are used for mechanical frames and coil winding whereas the flexible polyimide is used for mechanical springs and mass platform. The device is designed and fabricated by using commercial PCB technology, thus low cost and good reliability can be ensured.

2. Device principle and design

The schematic of the proposed harvester is shown in Figure 1. The 50-μm thick flexible polyimide layer is sandwiched between top and bottom rigid FR-4 boards. Springs (1×2.5 mm^2) and central platforms (ϕ 5 mm) are fabricated in the flexible PI layer. After the rigid-flex structures are fabricated, two cylindrical permanent magnets are attached to the central platform. When subject to external vibration, the magnets move in the vertical z direction and an electromotive force (emf) V is induced in the coil fabricated in the rigid boards. The PCB layout is shown in Figure 2. Both the top and bottom rigid boards have 3 layers of coil windings designed with 5-mil line/space design rules. Each layer has 11 turns. The two sets of coils can be connected externally in parallel to increase the output current or in series to increase the output voltage.

The output voltage V is determined by Faraday's law of electromagnetic induction,

$$V = -n\frac{\mathrm{d}\Phi}{\mathrm{d}t} = -n\frac{\mathrm{d}\Phi}{\mathrm{d}z}\dot{z},$$

where Φ is the total magnetic flux enclosed by the coils and z is the position of the magnets with respect to the coils. To maximize the output voltage, the spatial derivative of the total flux Φ should be maximized. In finite element simulation, it was found that $\mathrm{d}\Phi/\mathrm{d}z$ reached maximum when the coils were placed at the same plane as the top surface of the magnets. Therefore the rigid FR-4 boards are designed so that the coil routing are laminated in boards with minimum thickness (0.1 mm) and a thick dummy board is used to adjust the position of coils to align with the magnet surfaces, as shown in Figure 3(c).

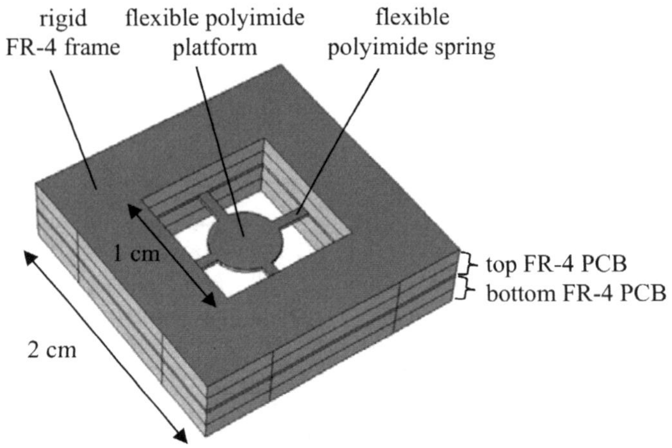

Figure 1. Schematic of harvester fabricated by rigid-flex PCB technology.

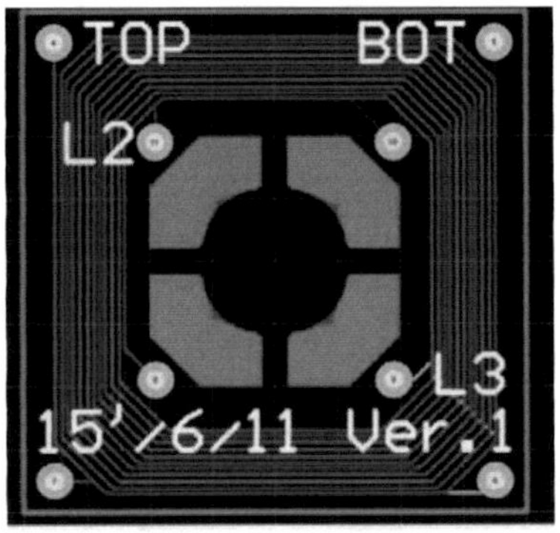

Figure 2. PCB layout.

3. Fabrication and measurement

The device was fabricated by a commercial PCB manufacturer, as shown in Figure 3. First, the rigid and flex boards were prepared by standard PCB processes (Figure 3(a)). The copper thickness of the inner and outer coils were 17.5 μm and 35 μm, respectively. Then the center hole was drilled in the rigid boards and the spring/platform shapes were punched in the flexible boards (Figure 3(b)). Next, the three boards were laminated (Figure 3(c)) and through holes were drilled and plated to connect coil windings in different layers (Figure 3(d)). Figure 4 shows a fabricated device with two 2-mm thick NdFeB magnets with 0.3-T surface flux density attached to the central platform. The total dimension of the device is $20 \times 20 \times 2$ mm^3 and the center hole in the rigid boards are 10×10 mm^2. The measured internal coil resistance is 15 Ω.

In vibration tests, the harvester was excited by a shaker and a commercial accelerometer was used to monitor the vibration level. The harvester output was connected to an oscilloscope and a dynamic signal analyzer; the total input impedance of the instruments was 500 kΩ. Figure 5 shows the output voltage vs. driving frequency for various vibration levels. Due to the spring hardening effect [1, 2, 4] of the miniaturized springs, nonlinearity and hysteresis was observed even at 0.1 g vibration. The effective bandwidth was increased from 6 Hz at 0.1 g to 21 Hz at 0.5 g and 27 Hz at 1 g, respectively. Thus the bandwidth can be increased by 3 to 4 times at high vibration levels. The peak output frequency was also shifted as the vibration level increased, as shown in Figure 6. A load resistor was connected to the harvester and the output power at 240 Hz and 1 g vibration is shown in Figure 7. For

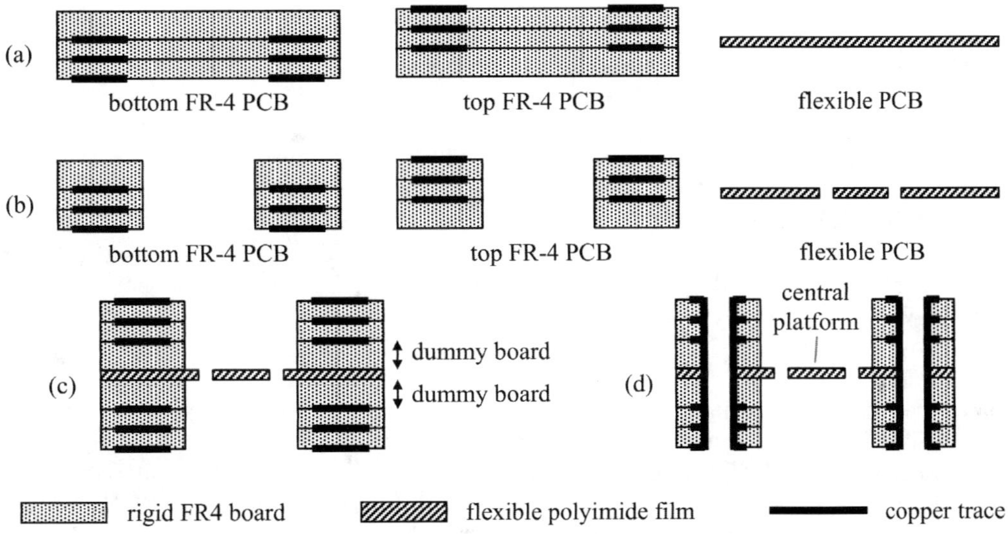

Figure 3. Fabrication processes, (a) preparation of rigid PCB and flexible polyimide films, (b) drilling of rigid boards and punching of flexible boards, (c) lamination, (d) through hole drilling and plating

Figure 4. Fabricated device with magnets attached.

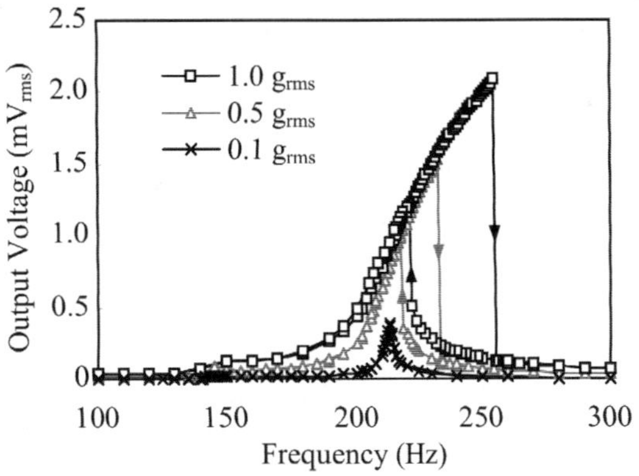

Figure 5. Output voltage vs. frequency for various vibration levels.

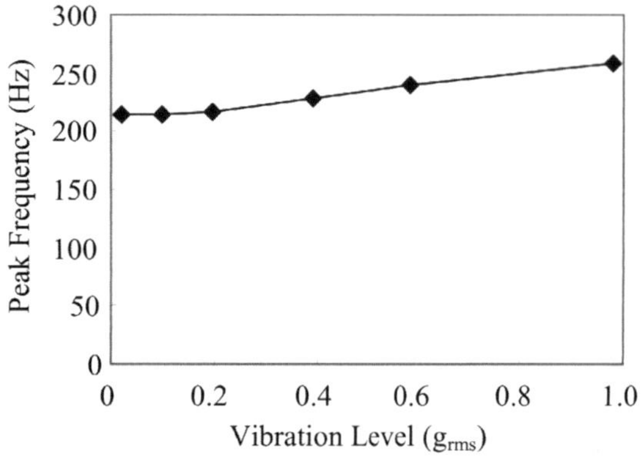

Figure 6. Peak frequency shift.

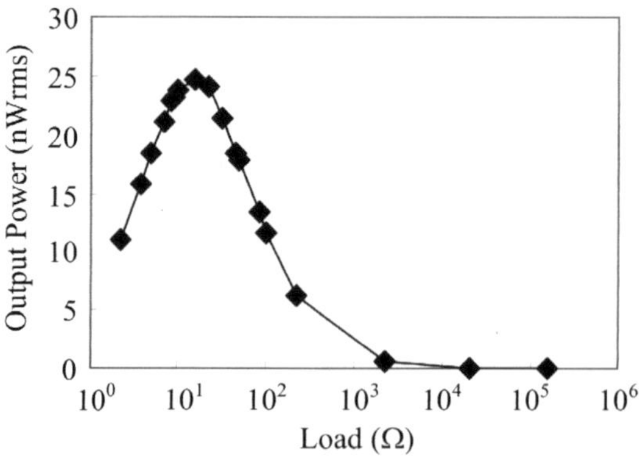

Figure 7. Output power vs. load at 240 Hz and 1 g vibration.

the matched load of 15 Ω, the maximum power was 24.5 nW, corresponding to a power density of 31 nW/cm^3, which was comparable to the reported value in literature [1].

4. Conclusion

This paper demonstrates an electromagnetic energy harvester designed and fabricated by the commercial rigid-flex PCB technology. The simple device structure design and mature manufacturing technology reduce production cost. Due to the nonlinear spring effect, the bandwidth was increased by 3 to 4 times at high vibration levels. Therefore the proposed device can be employed in wideband applications. For a matched load, a maximum output power of 24.5 nW, corresponding to a power density of 31 nW/cm^3, was achieved at 240 Hz and 1 g vibration.

Acknowledgments

The project was supported in part by the Ministry of Science and Technology, Taiwan, R.O.C., (MOST 103-2221-E-009-218 and MOST 104-2221-E-009-168-MY2) and the Bureau of Energy, Taiwan, R.O.C., (E455CU4400). The authors were grateful to the use of facilities at the Chip Implementation Center, the National Center for High-performance Computing, the National Nano Device Laboratory, and the National Chiao Tung University Nano Facility Center, Taiwan, R.O.C.

References

[1] H. Liu, et al., "An in-plane approximated nonlinear MEMS electromagnetic energy harvester," *J. MEMS*, 23 (3), 2014, pp. 740-749.
[2] F. Khan, et al., "Nonlinear behaviour of membrane type electromagnetic energy harvester under harmonic and random vibrations," *Microsyst. Technol.*, 20, 2014, pp. 1323-1335.
[3] Y. Chiu, et al., "Stacked electret energy harvesting system fabricated with folded flexible printed circuit board," *Proc. PowerMEMS 2012*, 2012, pp. 149-152.
[4] D. Mallick, et al., "A nonlinear stretching based electromagnetic energy harvester on FR4 for wideband operation," *Smart Mater. Struct.*, 24, 2015, 015013 (14 pp).
[5] G. Hapitoglu, et al., "FR4-based electromagnetic energy harvester for wireless sensor nodes," *Smart Mater. Struct.*, 19, 2010, 015022 (11 pp).
[6] C. B. Williams, et al., "Development of an electromagnetic micro-generator," *IEE Proc. Cirsuits Devices Syst.*, 148 (6), 2001, 337-342.
[7] C. Serre, et al., "Linear and non-linear behavior of mechanical resonators for optimized inertial electromagnetic microgenerators," *Microsyst. Technol.*, 15, 2009, 1217-1223.
[8] D.-A. Wang, et al., "Electromagnetic energy harvesting from wind induced vibration," *Microelectronics J.*, 41, 2010, 356-364.
[9] M. C. Wurz, et al., "Highly-integrated energy harvesting device for rotational applications utilizing quasi-static piezoelectric and electromagnetic generators," *Proc. SPIE*, Vol. 8688, 2013, 86881L-1-86881L-8.

Novel Fabrication Process for Micro Thermoelectric Generators (μTEGs)

U Pelz, J Jaklin, R Rostek, M Kröner, P Woias

University of Freiburg, Department of Microsystems Engineering - IMTEK, Lab for Design of Microsystems, Georges-Köhler-Allee 102, 79110 Freiburg, Germany

Email: uwe.pelz@imtek.de

Abstract. A cost effective bottom-up process for the fabrication of micro thermoelectric generators (μTEGs) was developed. It is based on a novel fabrication method involving a selectively sacrificial photoresist for the sequential galvanostatic electrodeposition of thermoelectric materials. The use of an industrial pick and placer (P&P) for dispensing the second photoresist allows for accurate and flexible μTEG designs. The process makes use of Ordyl® as a negative dry film photoresist template and sequential lamination steps for shaping all thermoelectric legs and contacts. All structures of the μTEG are generated in one photoresist multi-layer – this represents the most significant advantage of the process. The process uses a minimum of clean room processing for the preparation of pre-structured substrates for electrodeposition and therefore provides a cost-effective, highly flexible fabrication platform for research and development.

1. Introduction

Smart energy autonomous micronodes (SEAM) are one example of autonomous wireless sensor systems which have experienced increased interest during the last decade. Application areas for these systems range from medical healthcare to automotive industry. As one application, a consortium of workgroups and institutions around the University of Freiburg is - among other topics - investigating new ways of detection, analysis and treatment of diseases connected to the brain, e.g. epilepsy or Parkinson's disease [1]. In this context implantable SEAMs have potential applications such as brainwave analysis or the autonomous neurostimulation of targeted areas within the brain. All of these applications require a reliable energy supply which shall be provided via application of a micro thermoelectric generator. Micro thermoelectric generators (μTEGs) are miniaturized TEGs and therefore able to convert thermal energy into electrical energy through the Seebeck effect. They consist of alternating *n*- and *p*-type semiconductor thermolegs placed thermally in parallel and electrically in series. The specific application of the μTEG presented in this paper is a system for transcranial energy harvesting (through the skull bone) as an energy supply for implanted SEAMs. The goal is a long lasting, maintenance-free, low energy supply without the requirement of a battery replacement.

Several different μTEG fabrication processes have been published already. The processes presented by Lim et al. [2], as well as the one by Glatz et al. [3], make use of low-cost electrodeposition to generate the required thermolegs. However, a multi-step stripping and recoating process (Ref. [2]) and the need to restructure the lower contacts at the end of the electrochemical deposition process (Ref. [3]) must be performed in a clean room. The commercial process developed by German company Micropelt uses plasma sputtering to form the thermolegs, which is carried out in a clean room environment as well [4]. Other μTEG build-up processes involve e.g. special photosensitive glass molds for aerosol deposition [5], or magnetron sputtering [6]. There are few examples of specific low-cost μTEGs; Madan et al. used a precision dispenser to print thermoelectric components of generators [7] and Shin et al. recently published a fabrication process via electrodeposition [8].

The aim of the fabrication process described here was to reduce manufacturing complexity compared to existing processes, enabling the production of cheaper μTEGs with a minimal dependence on cleanroom technology while maintaining the ability to use very small structures in the future. The process therefore makes use of a cost effective galvanostatic electrodeposition technique developed in our labs [9–11] to generate *n*- and *p*-type thermolegs directly in pre-structured molds. The use of an industrial pick and placer (P&P) equipped with a dispensing unit allows selective sealing of pre-structured molds with photoresist and enables sequential deposition of *n*- and *p*-type thermolegs. The μTEG structures are pre-fabricated in a clean room using Ordyl® as a negative dry film photoresist. After this initial step no further clean room processing is required.

2. Fabrication Process

The fabrication process was developed for a μTEG consisting of seven thermocouples (TCs) and has been presented at the International Conference on Thermoelectrics ICT 2015, in Dresden (Germany) [12]. This process has been adapted and improved for a μTEG consisting of 71 TCs. The fabrication process for a 7 TC μTEG is shown in figure 1 and is described here for completeness. Steps **a** to **f** were carried out in a clean room environment on wafer level. The subsequent steps (**g** to **m**) were performed on diced individual chips of 15.8 x 24.8 mm size (leg ø 300 μm). High resolution printed transparency masks (Koenen GmbH, Ottobrunn, Germany) were used for UV photolithography.

Alumina wafers (525 μm thick, diameter 100 mm) were used as a substrate. A 100 nm Au seed layer was thermally evaporated on top of a 10 nm Ti adhesion layer (figure 1**a**). A positive photoresist (AZ® 1518, MicroChemicals, Ulm, Germany) was used as a mask for wet chemical etching (aqueous KI solution followed by 1 % HF) to structure the seed layer and interconnected bottom contacts (figure 1**b**) that are used as plating electrodes and internal connections for the μTEG. A 17 μm thick Ordyl® laminate dry film photoresist with high chemical and thermal stability (Ordyl® SY-355, Elga Europe, Italy) was laminated on top of the substrate and structured using a maskaligner (MA6, Karl SUSS, Garching, Germany) to form a template (figure 1**c**) for Au electroplating of the lower contact pads (figure 1**d**, carried out outside the clean room). Subsequently, a second 55 μm thick Ordyl® layer was laminated on top to form a template for the μTEG thermolegs (figure 1**e**). This was followed by lamination of a third, 17 μm thick Ordyl® layer, structured to form a template for the upper contact structures (figure 1**f**). Both Ordyl® layers were developed simultaneously and hard baked at 150 °C to guarantee a high chemical stability. To allow a sequential galvanostatic electrodeposition of *n*- and *p*-type legs a P&P (SAM42, Amadyne, Bühl, Germany) with a mounted dispenser was used to apply positive photoresist drops (AZ® 1518, MicroChemicals, Ulm, Germany) locally as sealing for alternating cavities (figure 1**g**). The second photoresist was exposed to UV light 3 times for 10 s with 30 s intervals in between, and was subsequently hard baked at 60 °C for 20 min. All open cavities were filled with *n*-type Bi_2Te_3 using the pulsed electroplating process (200 ms pulse length; 10 % duty cycle; mean current density 3.3 mA/cm²) described in reference [9]. The deposition was carried out in acidic Bi/Te electrolytes as described in [9-11]. The second photoresist was selectively stripped with isopropanol (figure 1**h**) before sealing the previously deposited Bi_2Te_3 thermolegs the same way with the P&P as done with the other cavities (figure 1**i**). The deposition of the *p*-type material (Cu) into the new open molds was carried out the same way as described in [9]. The second set photoresist mask drops was removed by immersing the μTEG in isopropanol (figure 1**i**). A thin Au layer was sputtered on the sample surface and polished to ensure metal coverage inside the molds for the upper contacts only (figure 1**l**). Galvanostatic electrodeposition of Cu was used once more to generate the upper contacts (figure 1**m**). The Ordyl® template was preserved and acted as structural support for the TEG.

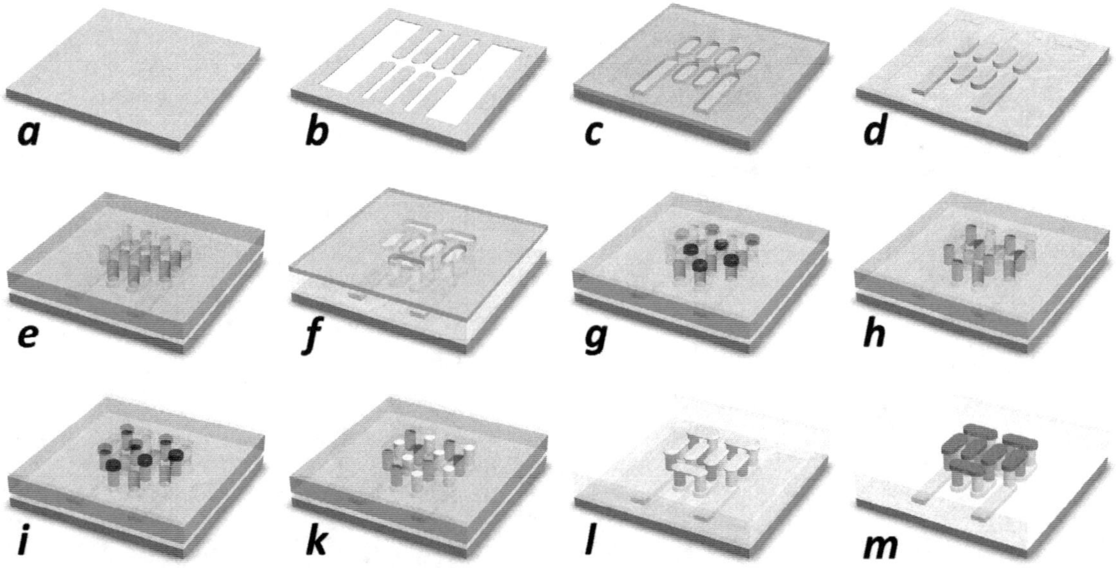

Figure 1. Steps of the multi-lamination process shown for a μTEG with 7 TCs (all heights are increased 10x for better visibility) – description is given in the text.

3. Results and Discussion

The complete process is discussed elsewhere [12]; in this paper only modifications to the original process are discussed. Compared to the previously published process the generator design was modified to have 71 TCs (figure 2) instead of 7, aiming for higher output voltage. Additionally, two larger single couples (not visible) were added to the deposition template which can be used for characterization purposes. For depositions of the μTEG structures only, the large couples are covered by adhesive tape, preventing any deposition during structure plating. The fabrication of the new 71-TCs generator is shown in figure 2. The key step, the sequential electrodeposition of *n*- and *p*-type thermoelectric material, is depicted in figure 2**b-e**, showing the photoresist pattern produced by the P&P. Modifications to the process included the change of the second photoresist from the negative resist AZ® nXT125 (MicroChemicals, Ulm, Germany) to a positive resist. The original negative photoresist has a high chemical stability and was therefore a good choice for electrolyte treatment. Due to the high stability, however, it could only be removed using acetone, which also attacked the Ordyl®-template over time possibly resulting in destroying the electroplating template. The new positive photoresist (AZ® 1518, MicroChemicals, Ulm, Germany) is less viscous and therefore easier to apply but, most important, can easily be removed by isopropanol alone. After deposition, the photoresist is exposed to UV light 3 times for 10 s with 30 s intervals in between. This step destabilizes the photoresist and makes it easily dissolvable in the solvents. The subsequent hard bake results in an increased stability towards the electrolytes, which is sufficient for deposition times up 24 h. The subsequent removal of the sealant does not involve long treatment in acetone, but rather short exposure to isopropanol. This prevents damage to the Ordyl®-template. Copper was chosen as p-type material because of the easy deposition process. Due to the very weak thermoelectric properties of Cu, the TCs in this μTEG are actually uni-legs with the Bi_2Te_3 accounting for the thermoelectric properties and the Cu just creating the electrical connection. In future iterations of the μTEG, Cu will be replaced by other materials such as pure Te or $(Bi,Sb)_2Te_3$. These deposition process are currently still under investigation. As shown in figure 2**a** the bottom contacts are fully interconnected to provide a uniform electrode for the electrodeposition process. For the finalized μTEG (figure **2g**), these interconnects, together with the Ordyl®-template in those areas, were removed via laser ablation to electrically separate the individual TCs.

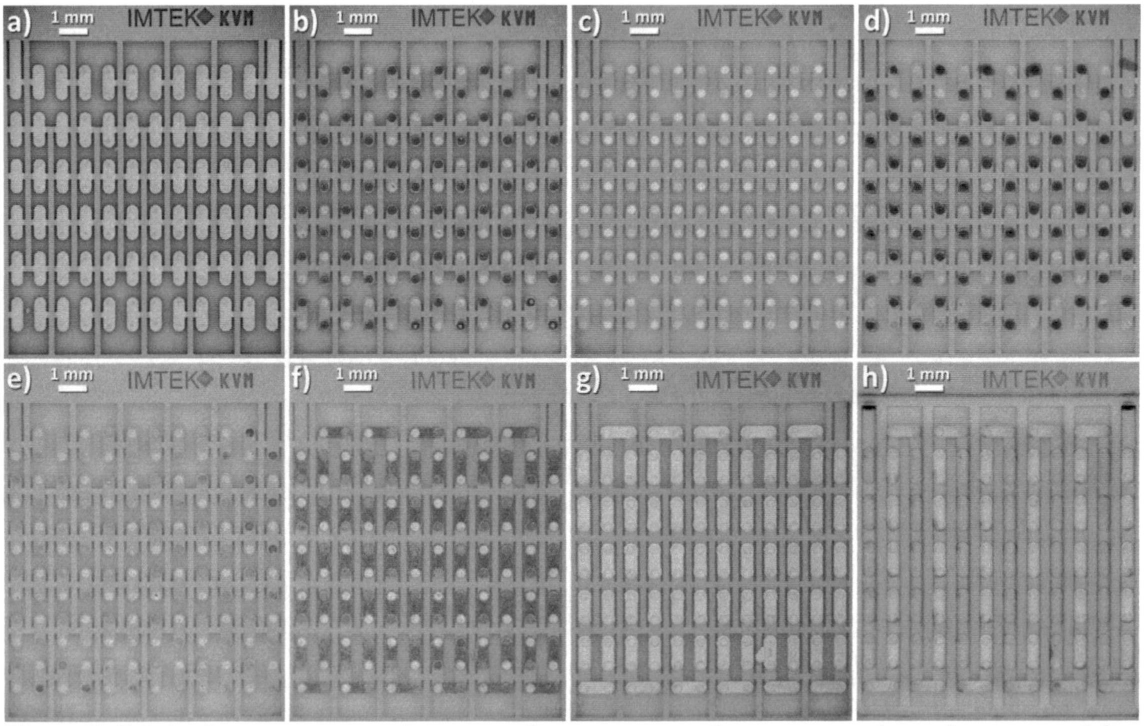

Figure 2. Fabrication process as a microscope photo sequence: a) pre-structured template prepared for electrodeposition with Au contact pads and interconnects; b) half of the molds were covered by the second photoresist and *n*-type Bi_2Te_3 was deposited into the open molds; c) the other molds were opened with isopropanol; d) the deposited Bi_2Te_3 thermolegs were covered with a photoresist layer and Cu was deposited into *p*-type cavities; e) both thermoleg-types were opened; the Ordyl® layer is still in place; f) after Au sputtering, and surface polishing, the upper contact molds are filled with a Au seed layer; g) electrodeposition of the Cu upper contacts; h) the Au interconnects as well as the Ordyl® layers on top are removed by a laser.

Some measurements on the new μTEG have been performed but far more will be carried out in the near future. All measurements in table 1 were carried out prior to the upper contact deposition on as-deposited samples. The electrical resistance of μTEG 1 and 3 shows low values with deviations comparable to the previous 7-TC μTEGs from reference [12]. The other two μTEGs appear to have a contact problem because measurements on different thermolegs yielded a variety of values. An improvement of the electrical properties is expected after a temperature treatment at 200 °C for several hours. The Seebeck voltage values are even higher compared to those for the previous μTEG [12].

Table 1. Measurement of Seebeck voltage and electrical resistance on 4 individual μTEGs; averaged values from 3 independent measurements.

Measurement of 3…	μTEG 1 … Bi_2Te_3/Cu TCs	μTEG 2 … Bi_2Te_3 legs only	μTEG 3 … Bi_2Te_3 legs only	μTEG 4 … Bi_2Te_3 legs only
Seebeck voltage in μV/K with sD	57,7±1,52	57,0±2,10	51,6±0,56	53,2±0,52
R electric in Ω with sD	0,244±0,134	8,418±13,943	0,003±0,0003	32,539±37,505

4. Conclusion and Outlook

In the present work, a low cost approach to the fabrication of µTEGs was shown. The process exhibits two key features: the use of a multi lamination technique based on Ordyl® dry film photoresist as a chemically stable template, and the use of a P&P with a second, positive photoresist to cover individual pre-structured cavities and allow sequential electrodeposition of two different thermoelectric materials. The high flexibility of the process provides a perfect platform for research and development on thermoelectric materials as well as µTEGs as systems themselves. Initial measurements of the Seebeck voltage and the electrical resistance show promising values. In upcoming work the characterization of the µTEG will be finished and accompanied by simulations.

Acknowledgement

The authors thankfully acknowledge the financial support by BrainLinks-BrainTools cluster of excellence funded by the German research foundation (DFG, grant number EXC 1086).

References

[1] University of Freiburg *web presence of the BrainLiks BrainTools cluster* www.brainlinks-braintools.uni-freiburg.de last time accessed September 21, 2015

[2] Lim J R, Snyder G J, Huang C-K, Herman J A, Ryan M A and Fleurial J-P 2002 Thermoelectric microdevice fabrication process and evaluation at the Jet Propulsion Laboratory (JPL) *Proc. of ICT'02. 21st Int. Conf. on Thermoelectrics,* Long Beach, CA, USA, 25-29 Aug. 2002 (IEEE) pp 535-9

[3] Glatz W, Schwyter E, Durrer L and Hierold C 2009 *J. Microelectromech. Syst.* **18** 763–772

[4] MicroPelt *web presence of the company MicroPelt* www.micropelt.com last time accessed on July 16, 2015

[5] Kouma N, Nishino T and Tsuboi O 2013 *J. Micromech. Microeng.* **23** 114005

[6] Fan P, Zheng Z-H, Li Y-Z, Lin Q-Y, Luo J-T, Liang G-X, Cai X-M, Zhang D-P and Ye F 2015 *Appl. Phys. Lett.* **106** 73901

[7] Madan D, Wang Z, Chen A, Winslow R, Wright P K and Evans J W 2014 *Appl. Phys. Lett.* **104** 13902

[8] Shin K J and Oh T-S 2015 *J. Electron. Mater.* **44** 2026

[9] Roth R, Rostek R, Cobry K, Kohler C, Groh M and Woias P 2014 *J. Microelectromech. Syst.* **23** 961

[10] Rostek R, Kottmeier J, Kratschmer M, Blackburn G, Goldschmidtboing F, Kröner M and Woias P 2013 *J. Electrochem. Soc.* **160** D408

[11] Rostek R, Sklyarenko V and Woias P 2011 *J. Mater. Res.* **26** 1785

[12] Pelz U, Jaklin J, Rostek R, Thoma F, Kröner M and Woias P 2015 Fabrication Process for Micro Thermoelectric Generators (µTEGs) *J. Electr. Mater.* accepted for publication

Journal of Physics: Conference Series **660** (2015) 012085 doi:10.1088/1742-6596/660/1/012085

Powering autonomous sensors with miniaturized piezoelectric based energy harvesting devices operating at very low frequency

G. Ferin, C. Bantignies, H. Le Khanh, E. Flesch, and A. Nguyen-Dinh
VERMON, 180 rue du general Renault, 37000 Tours, France

E-mail: g.ferin@vermon.com

Abstract. Harvesting energy from ambient mechanical vibrations is a smart and efficient way to power autonomous sensors and support innovative developments in IoT (Internet of Things), WSN (Wireless Sensor Network) and even implantable medical devices. Beyond the environmental operating conditions, efficiency of such devices is mainly related to energy source properties like the amplitude of vibrations and its spectral contain and some of these applications exhibit a quite low frequency spectrum where harvesting surrounding mechanical energy make sense, typically 5-50Hz for implantable medical devices or 50Hz-150Hz for industrial machines. Harvesting such low frequency vibrations is a challenge since it leads to adapt the resonator geometries to the targeted frequency or to use out-off band indirect harvesting strategies. In this paper we present a piezoelectric based vibrational energy harvesting device (PEH) which could be integrated into a biocompatible package to power implantable sensor or therapeutic medical devices. The presented architecture is a serial bimorph laminated with ultra-thinned (ranging from 15µm to 100µm) outer PZT "skins" that could operate at a "very low frequency", below 25Hz typically. The core process flow is disclosed and performances highlighted with regards to other low frequency demonstrations.

1. Introduction

Piezoelectric MEMS is one of the smartest energy harvesting solution which can efficiently convert mechanical energy into electrical energy. These devices can indeed exhibit various shapes and principle of operations but their intrinsic quality lies on the piezoelectric component used to perform such a conversion and the affordable energy source. Piezoelectric materials for MEMS can be deposited using physical or chemical processes on various substrates. Such deposition processes are quite reliable and can be considered as mature technologies since it is widely used in standard inertial sensors. Piezoelectric MEMS also offers a high integration degree through common microelectronic interconnection and packaging solutions. As a result, piezoelectric MEMS are considered as an essential component in electromechanical energy harvesting and has been tested through numerous important and serious studies. [1]

However, piezoelectric MEMS based on thick or thin films cannot address every vibrational energy harvesting applications. At very low frequency, typically between 25 and 50Hz, it starts to be difficult to properly design a performant piezoelectric MEMs. At low frequency, the geometrical dimension became larger and are directly related to the overall beam stiffness which is mainly driven by the substrate mechanical properties and thickness. Thinning silicon or glass substrate below 50µm become challenging in terms of yields and costs limiting the interest for such nice and sophisticated MEMS

construction. There is two main strategies enabling to overtake these limitations, either you design a larger bandwidth vibrating devices that mechanically convert low frequencies into higher frequencies (like beam bashing systems) where piezoelectric MEMs exhibit a better Figure of Merit or you can use very thinned bulk piezoelectric material and even integrate it under a MEMS shape [2].
In this paper, we propose to assess the performances of such a thinned bulk piezoelectric material to harvest mechanical vibrations around 25Hz corresponding to existing vibration in human tissues close to the heart (valve noise) for a relatively low acceleration amplitude, typically around 0.1G where G is the gravitational constant. We will address manufacturing and performance assessment processes to finally conclude on the obtained performances and viability of such approaches.

2. Manufacturing process flow

2.1. Vertical Architecture

Choosing bulk piezoelectric materials instead of deposited piezoelectric material offers to liberate the design from certain MEMS constraints like achievable piezoelectric film thickness, substrate thickness and its mechanical impact, i.e. material type. The most important one is the vertical cantilever architecture which can be a unimorph like most of the Piezoelectric MEMS, a bimorph (see Figure 1) or a multilayered architecture. In this paper, we have decided to focus our efforts on a bimorph architecture electrically connected in serial. We have not yet investigated parallel piezoelectric outer layer connections because at such low vibration amplitude it would be difficult to pass over rectifiers threshold values but it is technologically doable and would only require few developments.

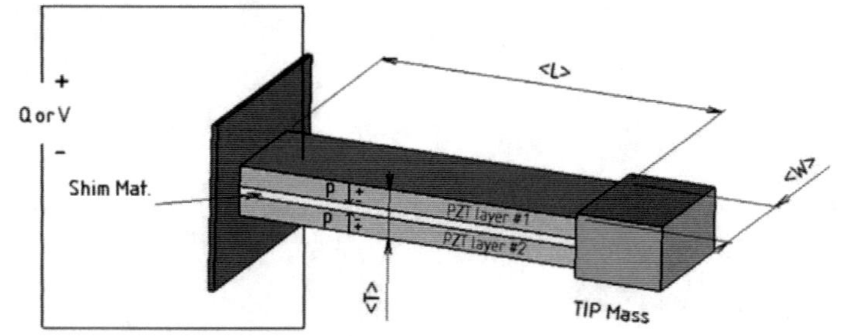

Figure 1 : Schematic of a clamped/free serial bimorph architecture for energy harvesting

2.2. Design

Clamped free cantilever beams is without any doubt the most known mechanical structure [3]. Its vibrating behaviour has been widely described using standard or more sophisticated computation tools. However, such a structure dedicated to low frequency energy harvesting using d31 piezoelectric coefficient push us into a complex world where tip displacement can be up to ten times the overall structure thickness generating both mechanical and piezoelectric non-linearities.

Table 1 : Bimorph architecture description

	Piezo outer layer	Inner shim	Piezo outer layer
Type	Hard PZT	Copper	Hard PZT
Thickness	45µm	10µm	45µm
Electrode	Gold plated		Gold plated
Beam length		25mm	
Beam width		4mm	

More, to keep a relatively small device that could be implemented within an implantable medical device, we decided to limit the beam length to 25mm, leading us to grind down PZT layer between 80µm and 20µm according to the internal shim layer thickness. As a result, we have defined the optimized vertical architecture as disclosed in Table 1.

Based on this first architecture definition, a manufacturing process flow has been defined and iteratively optimized. It has to be noticed that obtaining performant bulk PZT material with a thickness ranging between 20µm and 80µm is already challenging and can be directly compared with PZT thick films. A French research program called LAUREAT is currently assessing both reliability and aging of such piezoelectric structures.

2.3. Manufacturing process flow

A very simple approach has been used. As a first step, each piezoelectric layers are bonded at room temperature using conductive epoxy on a copper foil through advanced chemical and mechanical adhesion improvement processes. Then each face are grinded down and polished to their final thickness. The metallisation is sputtered and mechanically patterned to enable both interconnection and tip mass assembly.

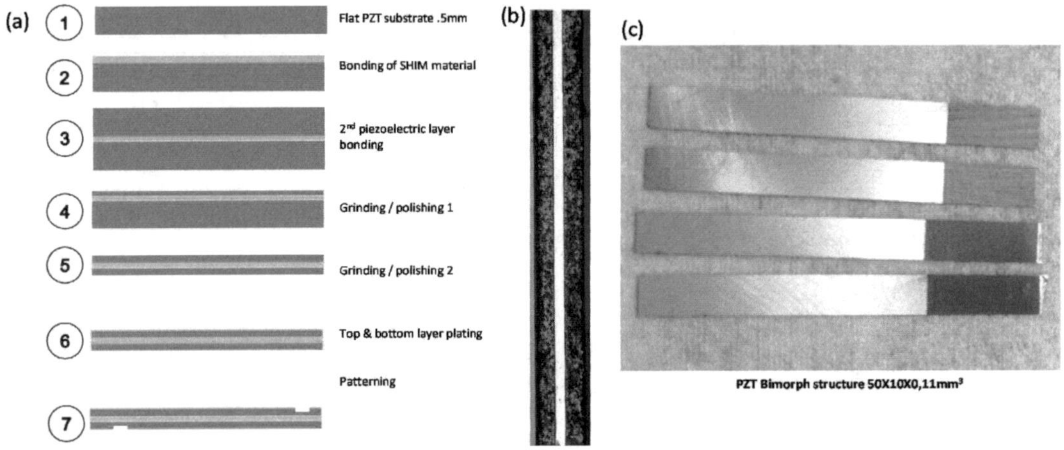

Figure 2 : Manufacturing process flow is disclosed in (a), a microscope photo (X50) of the bimorph cross cut (b) and four bimorphs presenting unplated areas for tip mass assembly (c).

Our process flow, based on seven simple individual steps, is presented Figure 2. A cross section obtained with an optical microscope is also presented and a more general view of the final bimorph beams as well. The process flow is quite reproducible with an accuracy of the thickness over the bimorph surface around 2µm. Internal stresses trapped within the layered structure during the assembly are quite low since no visible deflection has been detected. A Panel of 40x40 square millimeters can be processed. Increasing the panel dimensions should lead to considerably reduce costs and increase the yield.

3. Performance assessment

3.1. Measurement Setup

We also present a fully programmable uniaxial harmonic test bench (presented Figure 3) which enables us to dynamically assess the performances even at non-linear mechanical solicitations. Instead of using a sophisticated shaker, we have built around a loud speaker (SPH-275C from MONACOR) a custom shaker where the membrane is actuated using a waveform generator from TEKTRONIX (AG3001), the displacement is applied to a grip tool through a piston fixed on the membrane where cantilever beam is mechanically clamped on one end at a monitored pressure force ensuring a good clamping reproducibility. A tip mass is fixed on both side of the free end of the beam using cyanoacrylate glue.

An accelerometer ADXL 337 from Analog Devices, with a maximal resolution of 300mV/G is used to monitor the vibration amplitude and spectral contains and generated voltages are acquired through a pure resistive load with an oscilloscope TEKTRONIX TDS3052B.

Figure 3 : Uniaxial harmonic vibrating bench enabling the output power measurement on an optimized resistive loaded.

A computer drive the whole acquisition chain enabling an automated scan of the output power for given vibration frequencies and amplitudes at an optimized resistive load determined at the maximum frequency thanks to a custom programmable decade box ranging from 1Ohm to 10KOhms. The internal oscilloscope impedance is taken into account. Through this automated bench, it is possible to get output voltages with regards to vibration frequency, amplitude and parallel resistive load.

3.2. Performances

Using the test bench described above, we have measured with two different tip masses, the RMS output power for an acceleration of 1m.s-^{2} at, when possible, the resonant (blue curve) and antiresonant (red curve) time frequencies.

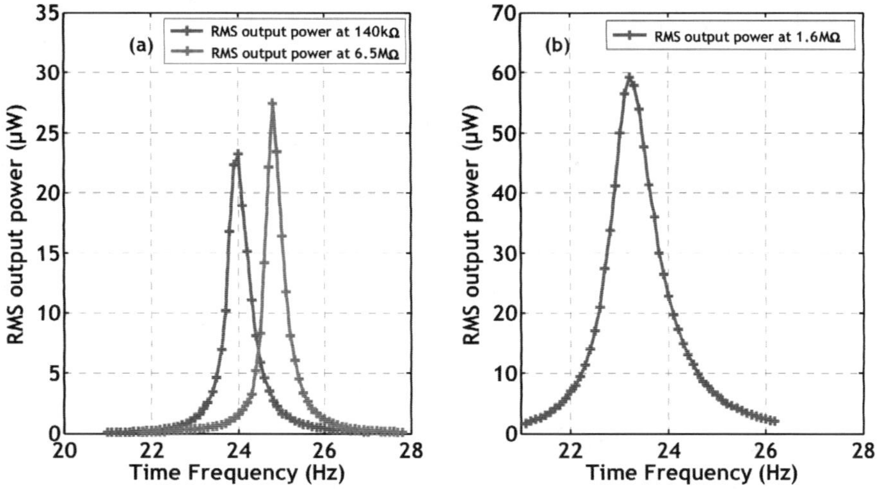

Figure 4 : RMS output power measured for a clamped free cantilever beam of : 4mmx36.2mm with a tip mass of 4mmx7.5mm of 0.5g measured for a resistive load optimized at the resonant (in blue) and antiresonant (in red) frequencies (a), 4mmx25mm square millimeters with a tip mass of 4mmx7.5mm of 3.7g measured for an optimized resistive load (b). Both measurements have been made for a uniaxial acceleration of $1m.s^{-2}$ along the gravitational direction.

For the device equipped with a heavier tip mass (3.7g), the mechanical quality factor become very low and it is no more possible to get distinct antiresonant and resonant frequencies. The optimized resistive load is also lower and the maximum output voltage is about 15 Volts pick to pick at 0.1G of acceleration which is already high compared to diode threshold value of a classical rectifier circuit.

4. Conclusion

Our optimized thinned-bulk bimorph PEH device has been fully characterized and has exhibited a RMS output power of up to 60μW at $1m.s^{-2}$ of acceleration at 23.2Hz along the gravitational direction for a 4X25 square millimetre footprint. The resulting RMS output power density measured at $1m.s^{-2}$ is up to $6.6mW/cm^3/g^2$ (assuming a capsule of $6x30x5mm^3$) which is very promising and already competitive to previous works [4] using same design strategy and can already serve batteryless implanted medical devices.

Beyond the performance demonstration, such bimorph architectures can be considered as an alternative to more traditional MEMS energy harvesting devices enabling to address lower frequency applications with large displacements. Another key should lies on the durability of such a construction and a study has been launched to monitor piezoelectric and mechanic constant during mechanical cycling.

References

[1] Erturk, A. and Inman, D.J., 2011, Piezoelectric Energy Harvesting, Wiley, Chichester, UK

[2] Ethem Erkan Aktakka, Member, IEEE, Rebecca L. Peterson, Member, IEEE, and Khalil Najafi, Fellow, IEEE, "High Stroke and High Deflection Bulk-PZT Diaphragm and Cantilever Micro Actuators and Effect of Pre-Stress on Device Performance", JOURNAL OF MICROELECTROMECHANICAL SYSTEMS, 2014.

[3] Jan G. Smits, Susan I.Dalke, "THE CONSTITUENT EQUATIONS OF PIEZOELECTRIC BIMORPHS", 1989 ULTRASONICS SYMPOSIUM.

[4] A MICRO INERTIAL ENERGY HARVESTING PLATFORM WITH SELF-SUPPLIED POWER MANAGEMENT CIRCUIT FOR AUTONOMOUS WIRELESS SENSOR NODES", E. E. AKTAKKA, AND K. NAJAFI, IEEE JOURNAL OF SOLID-STATE CIRCUITS, VOL. 49, NO. 9, SEPTEMBER 2014.

Sub-millimeter arbitrary arrangements of monolithically micro-scale electrical double layer capacitors

Karolina U Laszczyk[1], **Kobashi Kazufumi**[1, 2], **Shunsuke Sakurai**[1, 2], **Atsuko Sekiguchi**[1, 2], **Don N Futaba**[1, 2], **Takeo Yamada**[1, 2] **and Kenji Hata**[1, 2]

[1]Technology Research Association for Single Wall Carbon Nanotubes (TASC), Central 5, 1-1-1 Higashi, Tsukuba, Ibaraki, 305-8565, Japan
[2]National Institute of Advanced Industrial Science and Technology (AIST), CNT-Application Research Center, Central 5, 1-1-1 Higashi, Tsukuba, Ibaraki, 305-8565, Japan

E-mail: kenji.hata@aist.go.jp

Abstract. We report the investigation on the reproducibility of micro-scale electric double layer capacitors (micro-EDLCs). The micro-EDLC components were fabricated parallel using photolithography, wet and dry processing. Electrodes of the micro-EDLCs are highly dense packed Single Wall Carbon Nanotubes (SWCNTs) that form a mesh structure. The micro-EDLCs are connected 1-10 in series and in parallel being unified electrical circuits to tune the capacitance and the operational voltage. To confirm the reproducibility of the cells as well as the yield we performed electrochemical measurements in order to define the performance uniformity between cells strings and individual cells connected in a string. For 1-10 cells in series and in parallel the trends for the capacitance and operational voltage satisfied electro-physics rules governing cells addition. However, the measurements of the individual cells in a string revealed the significant performance discrepancy that might result in a shorten life cycling of a circuit.

1. Introduction

The trend in portable electronic devices have advanced by scaling down the electrical components, such as transistors, resistors and inductors, with the exception of Al electrolytic capacitors (AECs) which have remained on the centimeter-scale for the past 50 years. The obstacle in size shrinking lies in a restricted accessible area of the Al electrodes. Contrary, supercapacitors, based on the electric double layer to store the energy, offer energies approaching the level of batteries. However, their power is minor compared to the AECs due to the limited ions diffusion in an inner pore network of the activated carbon (AC) electrodes and a separator placed between an anode and a cathode. Recently, use of nanocarbonous materials (CNTs, graphene, carbon onion) together with fine resolution techniques resulted in micro-supercapacitors with significantly improved speed because of shorten the ions diffusion path. The first has attained the improved accessible surface area for the electrodes, while the latter has been used to form planar, interdigital structures with an empty space instead of the separator. Parallel to the high power performance (two-order higher than supercapacitors) the micro-supercapacitors benefit of high capacitance, hence energy, becoming the potential candidate to replace the bulky AECs [1,2].

Recently, we developed the scalable and compatible process for the Integrated Energy Device Circuit (IEDC) that performance was similar to a target AEC but its volume differed 1 to 1000th in the size [2]. The IEDC was composed of the individual, similar micro-EDLCs. Each micro-EDLC worked as an independent electrochemical cell yet all were connected into an electrical circuit. By taking the advantage of electro-physics rules governing the cells addition we managed to tune the operational voltage and the capacitance.

However, the cells strings to be ultimately attractive and viable for electronic applications must demonstrate the performance uniformity for the individual micro-EDLCs. One of the important aspect of this issue is the reproducibility and yield of the micro-EDLCs fabricated in a one process flow.

In this paper we report the comparison of co-fabricated on-chip circuits consist of monolithically integrated micro-EDLCs connected in parallel, in series, and individual cells in circuits about their patternability and electrochemical performance. Finally, we did a comparable study (section 3) and discussion (section 4) of obtained results. Based on the performance deviation among the micro-EDLC in a circuit we could estimate the major advantages and drawbacks of used technology. These important points represent some of the fundamental steps towards the on-chip large-scale integration of the IEDCs

2. Integrated Energy Device Circuits

The 1-10 in series and in parallel samples were fabricated following the fabrication process described in [2].

2.1. CNT electrodes

The mesh electrodes consist of dense packed single wall Super Growth CNTs (SGCNT) forest grown by water assisted CVD process [3]. The SGCNT forest is characterized by high impurity and Specific Surface Area (SSA, > 1000 gcm^{-3}). To make the uniform and thick film, we developed the homogenous, high viscous, 0.5 wt.-% suspension of SGCNTs in *1-Methyl-2-pyrrolidinone* (NMP). The paste was used for a paint-like coating on the substrate area with patterned electrical pads and cell barriers for an electrochemical isolation. Next, the coated paste was dried to get the CNT film free of a solvent. Finally, the micro-meter thick interdigitated electrodes were released.

3. Results

All of the electrochemical measurements were done using two electrodes set-up of VMP3 (Bio-logic Instruments, US). The thickness of the CNT electrodes was determined by a 2D surface profiler Dektak 150 (Veeco Instruments, US). The determination of *iR* drop, calculation of the capacitance were performed the same style as it was reported in [2].

3.1. 1-10 cells in series and in parallel

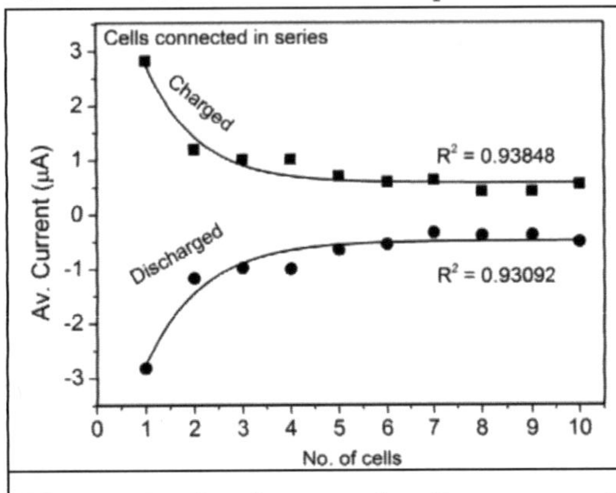

Figure 1. Graph: no. of cells connected in series vs. av. current.

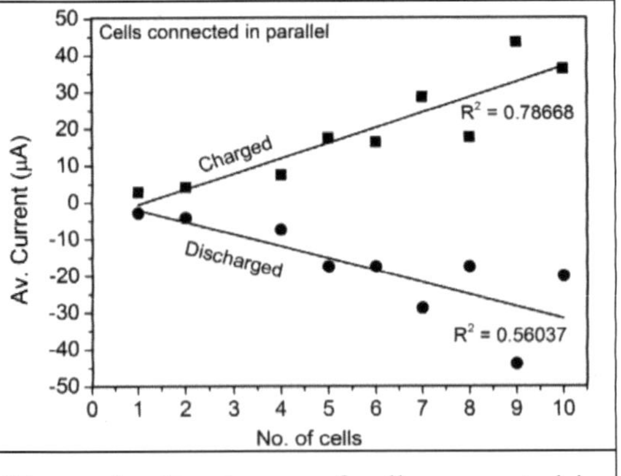

Figure 2. Graph: no. of cells connected in parallel vs. av. current.

Previously we reported the strings of 1-10 cells in series and in parallel that trend lines of the operational voltage and capacitance satisfied the electro-physics rules governing cells addition [2].

From that point we assumed the uniformity of the cells is approved. However, this time we decided to take a detail look on the performance characterization. Figure 1 and 2 present the charged and discharged currents level for 1-10 cells in series and in parallel, respectively, obtained from cyclic voltammetries (CVs). The charge-discharge currents for particular connections showed mirror responses except for 10 cells in parallel connection. The current values for the anode and the cathode were -20.04 μA and +36.34 μA, respectively. In all cases the maximum divergence from the trend lines equals to +/-0.35 μA and +/-13.00 μA for in series and in parallel connections, respectively. The divergences could be a consequence of: 1) a non-uniform pore network, 2) various electrodes volume, 3) the leakage current reflected in an *iR* drop, 4) various equivalent series resistance (*ESR*), which include the contact resistance between CNT electrodes and metal electric pads, 5) the ohmic resistance of the electrical pads.

3.2. Individual cells combined in a string

In order to understand the origins of the divergence in the current responses, reported in section 3.1, we did measurements of individual micro-EDLCs being in a string. Not all of the 10 cells in a string were possible to measure; cell no. 2 and 3 have a short circuit between the CNT cathode and anode. Nonetheless, the other cells presented in figure 5, due to independent leads, were characterized.

First, we checked how much the CNT electrodes differentiate in a thickness obtained from a 2D surface profiler, Average Height Step (ASH, figure 3). We assumed the electrodes areas are invariant

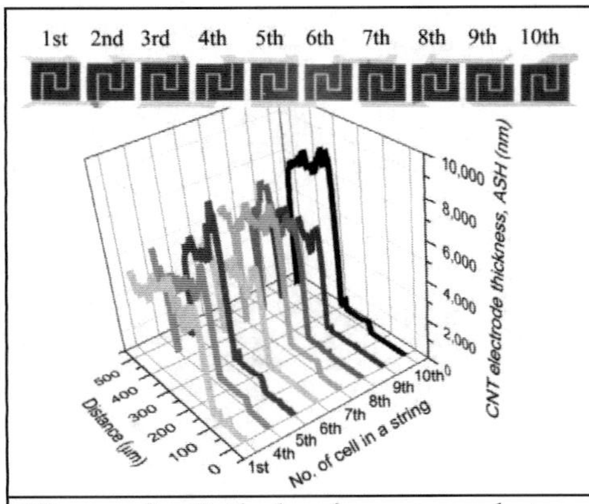

Figure 3. Optical microscope photography: cells in a string (up); graph: distance vs thickness of electrode (down).

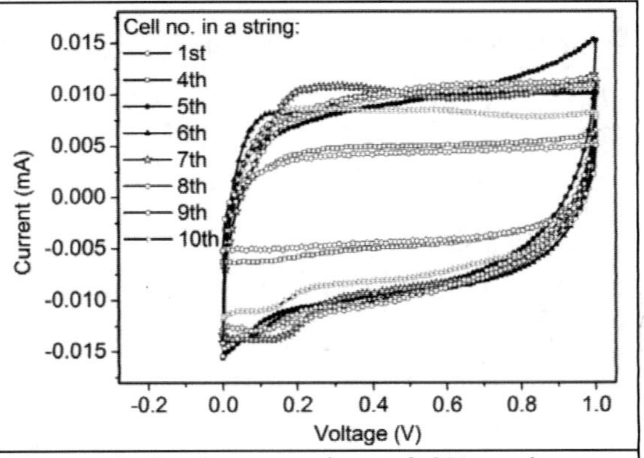

Figure 4. Graph: comparison of CVs: voltage vs current conducted at 500 mV/s.

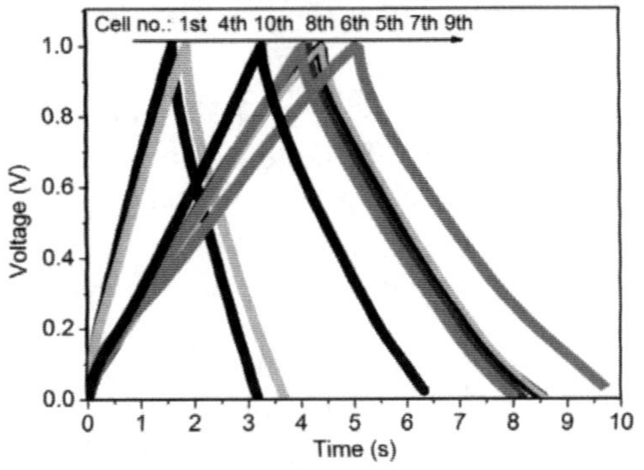

Figure 5. Graph: comparison of CCCDs: time vs voltage conducted at 0.2 mAcm^{-3}.

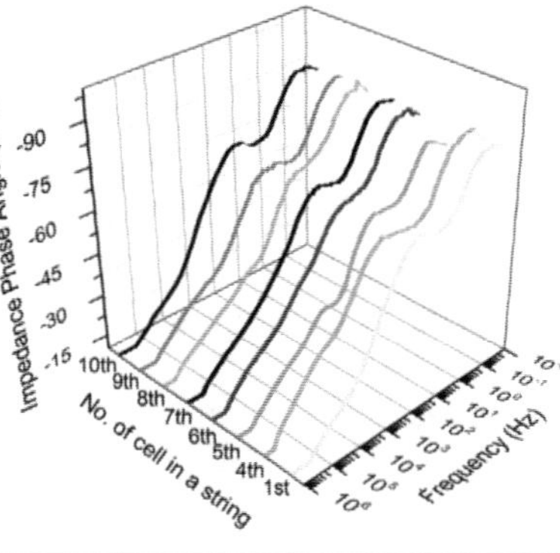

Figure 6. Graph: comparison of impedance spectras: frequency vs impedance phase angle.

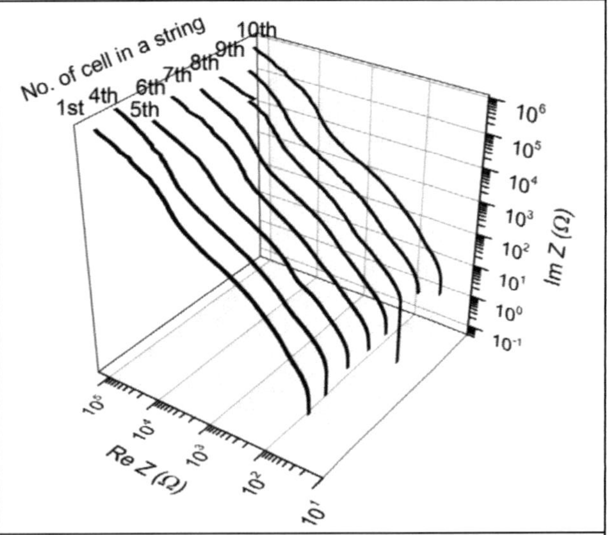

Figure 7. Graph: comparison of Nyquist Plots: Re*Z* vs Im*Z*.

due to the photolithography technique we used for patterning. The electrode thickness of cell no.1 and 4 (~3.3 μm) attained about 60% of the thickness of the other six measured cells (4.5-6.4 μm, table 1). The thickness results was used to estimate the total volume of the CNT electrodes.

The surface roughness (*Ra*) of the electrodes are in the range of 390-905 nm (table 1). Further, from CVs (figure 4), CCCDs (figure 5) we got the various responses of the individual micro-EDLCs to deduce the capacitances and *iR* drops. Additional, we performed the Potential Electrochemical Impedance Spectroscopy (PEIS) at 0V with 10 mV amplitude in order to confirm the impedance angle response vs frequency, a time relaxation constant, τ (figure 6), and *ESR* (figure 7). All the data were summarized in table 1 to give a comparative summary of individual micro-EDLCs in a string.

Table 1. Comparative summary of the performance of individual cells in a string.

[a] *ESR* was taken from PEIS Nyquist Plot for ReZ at ImZ=0.

Cell no.	Electrode thickness, ASH (nm)	*Ra* (nm)	Electrode volume ($x10^{-5}cm^3$)	Capacitance based on CV (μF)	Capacitance based on CCCD (μF)	*iR* (V)	τ (s)	ESR^a (Ω)
1st	3286.97.	679.48	1.57	7.86	7.24	0.12	0.002	27
4th	3392.15	389.71	1.62	8.76	8.01	0.14	0.001	32
5th	5695.76	582.19	2.71	16.82	16.39	0.11	0.013	28
6th	6418.25	780.64	3.06	17.86	16.98	0.08	0.005	28
7th	5021.89	561.97	2.39	17.50	17.50	0.09	0.005	30
8th	5129.20	502.22	2.44	17.39	16.98	0.12	0.011	35
9th	5474.81	720.52	2.61	17.54	20.12	0.11	0.005	N/A[b]
10th	4516.61	904.49	2.15	8.76	13.51	0.11	0.002	27

[b] ImZ $\neq$0.

4. Discussion

The purpose of the evaluation of monolithic integrated micro-EDLCs was to determine the factor that has the major influence on the cells uniformity. At this stage we would like to confirm the level of uniformity of our IEDC by comparative study of cells' performances being monolithic connected in a string.

We found the discrepancy in capacities among micro-EDLCs being in the range of (7.24-20.12) μF from CCCD tests, and (7.86-17.86) μF from CVs (table 1). We relate it to the unstable CNT suspension used to form the mesh electrodes; within short time (hours) since the dispersion process had been finished, we observed the separation of the CNTs/solvent and the solvent. Moreover, the separation was present during the formation of the CNT film resulting in diversified CNTs distribution. Thus, the variation in the CNT film thickness hence electrodes volumes were unavoidable (figure 3 and table 1). The origins of the mechanism standing behind this issue is under the investigation. We suppose it might cause some local CNT agglomerations that disallow the isolation of polarized electrodes within one cell, as it happened in case of cells no.2 and 3 (short circuit) decreasing the yield of working cells.

Additional *iR* drops (figure 5 and table 1) were estimated to be in range of (0.09-0.14) V. Here, *iR* drop is assigned to the inner pore network of CNT mesh electrodes. For operational voltage equals to 1 V it is about 1% difference. From the impedance spectra graphs we could notice analogous energy dissipations for cells (figure 6) and line profiles related to diffusion impedance elements (figure 7). The *ESRs* from the Nyquist Plots (figure 7 and table 1) were resembling (27-35) Ω what means the formation of contacts between the CNT mesh and metal pads had a similar structure. All of these results indicates the identical inner pore network among cells, though the recalled separation phenomena took place.

The impedance phase angle -90 deg was kept up to 0.2 Hz (figure 6) and is not applicable for the electronics applications. However, we believe by the modification of a pore size in the CNT film and an electrode thickness we could enhance the frequency response.

Towards the reproducible IEDCs, which possess reliable long life cycling, we need comparable electrodes in their volumes. One of the solution is to focus on obtaining the uniform thick film by improvement of the coating technique.

References

[1] El-Kady M L and Kaner R 2013, *Nat. Comm.* 4:1475, 1-4.

[2] Wu Z.-S, Parvez K, Feng X and Mullen K 2013, *Nat. Comm.* 4:2487, 1-4.

[3] Laszczyk K U, Kazufumi K, Sakurai S, Sekiguchi A, Futaba D N, Yamada T and Hata K 2015, *Adv. En. Mat. in press*

[4] Hata K, Futaba D N, Mizuno K, Namai T, Yumura M and Iijima S, 2004, *Science* 306, 1362-1364.

Acknowledgements

This presentation is based on results obtained from a project commissioned by the New Energy and Industrial Technology Development Organization (NEDO).

Low power electronic interface for electrostatic energy harvesters

Tra Nguyen Phan, Mehdi Azadmehr, Cuong Phu Le and Einar Halvorsen

Department of Micro- and Nanosystem Technology, Buskerud and Vestfold University College, Raveien 215, 3184 Borre, Norway

E-mail: Einar.Halvorsen@hbv.no

Abstract. This paper presents design and simulation of a power electronic interface circuit for MEMS electrostatic energy harvesters. The designed circuit is applicable to highly miniaturized electrostatic harvesters with small transducer capacitances below 10 pF. It is based on comb-drive harvesters with two anti-phase capacitors that are connected as charge pumps and uses a flyback-path scheme. Controlled activation and deactivation of sub-circuits, some by help of clocking, were exploited to reduce power consumption down to 1.03 μW. Net power generation can be achieved with as low initial voltage as 3.0 V.

1. Introduction

Harvesting energy from the environment is not a new concept and has been widely explored in various applications such as wireless sensor nodes [1], structural health monitoring [2] or implantable biomedical devices [3]. Among ambient energy sources, kinetic energy is abundant and therefore an attractive alternative motivating a number of studies [4, 5]. Commonly used mechanisms for energy conversion in kinetic energy harvesters are electromagnetic, piezoelectric and electrostatic transduction [6]. Capacitive electrostatic harvesters are attractive due to their suitability for microfabrication and their potentially small chip sizes. These harvesters suffer from correspondingly small variable capacitances, limited by standard fabrication processes and device size, which are obstacles for effective energy conversion. To design an effective micro energy harvesting system, not only the generator design and fabrication should be optimized, but also the power electronic interface for the generator. The electronic interface circuitry in a micro energy harvester system must be highly effective and power conservative as it is supplied by the harvester system. Due to the limited power available in the microscale electrostatic generators, designing effective power electronic interface is especially challenging.

There have been published several interesting studies on the interface circuits for capacitive transducers, especially for those with large capacitances. Meninger *et al.* [7] proposed a synchronous capacitive energy harvester consisting of two active switches to charge and discharge a variable capacitor through an inductor. Complexity of control circuitry and power dissipation for synchronization are main concerns for this architecture. Miyazaki *et al.* [8] improved this topology by proposing timing-capture scheme instead of programmable controller with delay locked loop (DLL) feedback. A, by now, classic circuit topology was proposed by Yen *et al.* [9] in which a charge pump was used with a flyback circuit to store and transfer extracted energy.

This topology was further investigated by Galayko *et al.* [10], focusing on design of the control circuit to optimize the flyback switching sequence. Other architectures used voltage-constrained topology by Torres *et al.* [11], and modified storage capacitor in the charge pump part by Kempitya *et al.* [12].

In this work, an interface circuit for a MEMS electrostatic energy harvester with transducer capacitance range below ~10 pF is presented. The interface circuitry is based on the aforementioned flyback technique and automatically detects the switching instants for optimal power transfer through the flyback path without using any external control signals. A circuit alternative to the comparators that are used in other implementations of this architecture is investigated. The need to power the circuitry by extracted power only and the limited available power due to a highly miniaturized MEMS harvester makes low power consumption of the circuit a main design driver. All the simulations presented were made with Tanner Tools using models for the high-voltage AMS035 CMOS process.

2. Power electronic interface

2.1. Overview

The circuit topology is an adaption of Yen's circuit [9] to anti-phase capacitive structure with double, instead of single charge pumps. Figure 1 shows the block diagram of the whole circuit which consists of three main elements: double charge pump, flyback path and switch controller. In this topology, optimization of switching sequence in the flyback has significant effect on harvested power from electrostatic transducer. Thus, the control circuitry presented is designed to detect these optimal switching instants and then turn on/off the switch correspondingly.

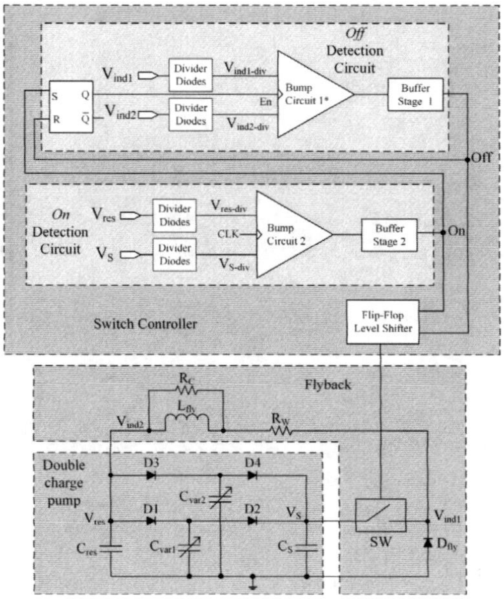

Figure 1. Block diagram of proposed circuit.

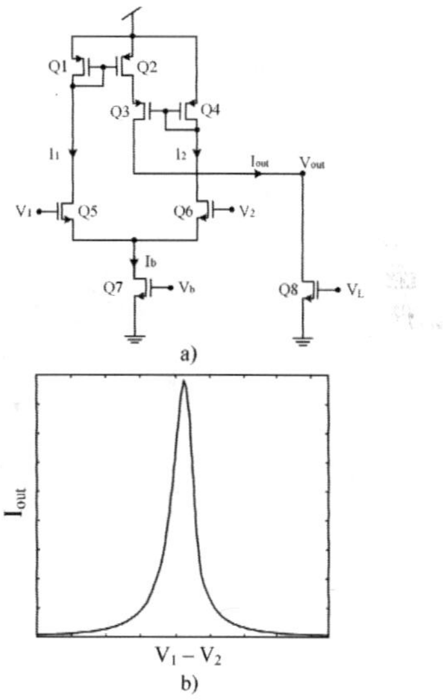

Figure 2. *a*) Bump circuit. *b*) Input-output characteristics of bump circuit. (After Delbruck [13])

2.2. Double charge pump

The double charge pump is used because two capacitors varying with opposite of phases are found in many electrostatic harvesters. Operation of the double charge pump is similar to the case of one pumping capacitor except that it exploits the entire mechanical cycle by transferring charges into the store capacitor C_s during both half cycles. The voltage V_s on the store capacitor after n cycles of vibration is

$$V_{s,n} = V_{res}[(1 - \frac{C_{max}}{C_{min}})(\frac{C_s}{C_{min} + C_s})^{2n} + \frac{C_{max}}{C_{min}}] \tag{1}$$

where V_{res} is the voltage across the reservoir capacitor and C_{max}/C_{min} are the maximum and minimum capacitances of the transducer during one vibration cycle.

2.3. Flip-flop level shifter

As the voltage across the harvesting capacitor and the flyback switch are much higher than the voltage controlling the switch, a level shifter is needed to match these two. A flip-flop is also needed to hold the switch state after receiving the *on* or *off* signal at the input of level shifter. Dudka *et al.* [14] proposed a flip-flop level shifter that fulfills these requirements and has a zero static power consumption.

2.4. Circuit detecting on event

Assume $V_s = V_{res} = V_0$ at startup when $t = 0$ and define $V_1 = V_s\ (t = t_1)$ as the optimal V_s at which the switch should be turned on. Similar to the power obtained from a single transducer as calculated in [10], the power harvested from the two anti-phase transducers here is

$$P = \frac{C_s(1 + \frac{C_s}{C_{res}})V_0^2}{T} \ln \frac{C_s}{C_{min} + C_s} \frac{(1-\theta)^2}{\ln \frac{\theta-\alpha}{1-\alpha}} \tag{2}$$

where T is the vibration period, $\theta = V_1/V_0$ and $\alpha = C_{max}/C_{min}$. There is an optimal value $\theta = \theta_{opt}$ at which the power is maximum. It is given by

$$2 \ln \frac{\theta_{opt} - \alpha}{1 - \alpha} = \frac{\theta_{opt} - 1}{\theta_{opt} - \alpha}. \tag{3}$$

To detect when V_s reaches the optimal value and the switch should be turned on, V_s is first divided by θ_{opt}, then a bump circuit [13], as shown in figure 2, is used to trigger switching when the obtained value is equal to V_{res}. The bump circuit is a voltage similarity detector which produces a signal when the two inputs are equal. A diode-connected voltage divider is used to downscale the voltage to a suitable level at the input of bump circuit.

2.5. Circuit detecting off event

When the switch is turned on, the circuit has C_s, L_{fly} (including parasitics) and C_{res} in a series configuration. In order to optimize the power transfer, the switch should be turned off when the inductor current reaches maximum value, i.e. when voltage across L_{fly} is equal to zero. To detect this instant, a bump circuit is connected to the two terminals of the inductor. The bump circuit's output voltage goes high when its two inputs are equal. In order to reduce power consumption, this bump circuit is activated only during the on-time of the switch. The bump circuit deactivates itself after it detects the *off* event. The bias voltage V_b of the bump circuit is used for activation/deactivation.

3. Simulation results

In the simulations, we adopt transducer parameters of the small-scale electrostatic harvester prototype published in a previous work [15]. The transducer is a comb-drive capacitance with a maximum capacitance C_{max} = 8.62 pF and a minimum capacitance C_{min}=2.16 pF, ignoring parasitic capacitances. The operating frequency of the harvester is f_0 =650 Hz. The optimal value θ_{opt}=3.15 is then found from (3).

3.1. Basic operation

Figure 3 a) demonstrates the basic operation without the flyback path. When the capacitances vary, V_s increases to a saturated value as the charge is built up on C_s. With the flyback path, V_s is prevented from saturation by discharging it through flyback path as illustrated in figure 3 b) where the switch is periodically turned on by an external signal.

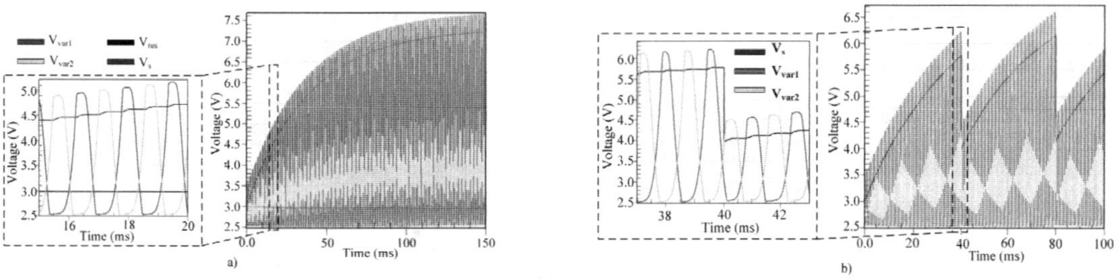

Figure 3. Voltage waveforms across capacitors in double charge pump circuit: *a*) without flyback path and *b*) with flyback path and periodically closing switch.

3.2. With control circuitry

We now consider operation with the control circuitry described in the previous section. We let the *on*-detection bump circuit stay enabled all the time as it is unknown when it needs to do its comparison, except for the short duration when the switch is already on. Figure 4 shows that the pulse *off* occurs when V_{ind1} approaches V_{ind2}. The *on* detection circuit detects when V_s/V_{res} reaches 3 and then generates the pulse *on* as shown in figure 5. However, the power consumption of the control circuitry in this configuration is 97.2 μW. It is too high to be supported by the harvested power so an external source was used in the simulation. Most of the consumption is in the second bump circuit which is enabled all the time.

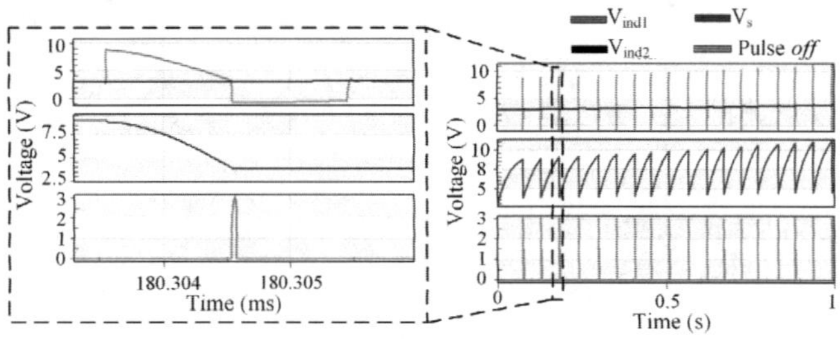

Figure 4. The switch is turned *off* when V_{ind1} approaches V_{ind2}.

3.3. Clocking technique for lowering power consumption

Reduced power consumption can be obtained by clocking the enable signal of the second bump circuit, as shown in figure 1. By choosing a small duty cycle, the circuit is kept mostly in off-mode. The reduced power comes at the expense of accuracy in the timing of the switch, but is controlled by clock frequency. With a duty cycle of 0.005%, the power consumption of the control circuitry is reduced to 1.03 μW which is comparable to that of the circuit in [16].

With an initial voltage of 3.0 V, the system power-budget is balanced with a small margin of 8.3 nW as shown in figure 6. We estimate the power required to drive the second bump circuit's

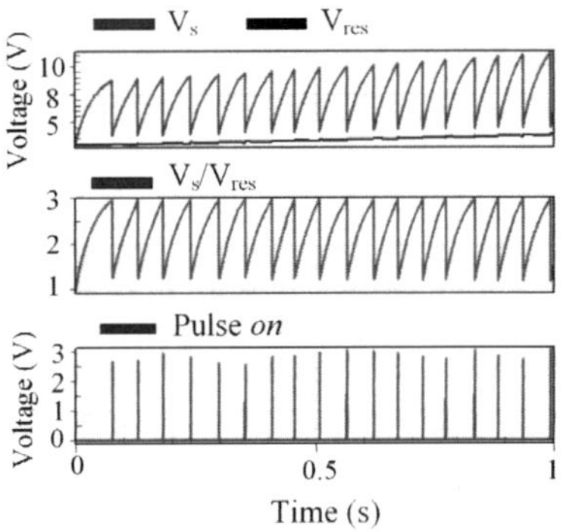

Figure 5. The switch is turned *on* when $V_s/V_{res} = 3$.

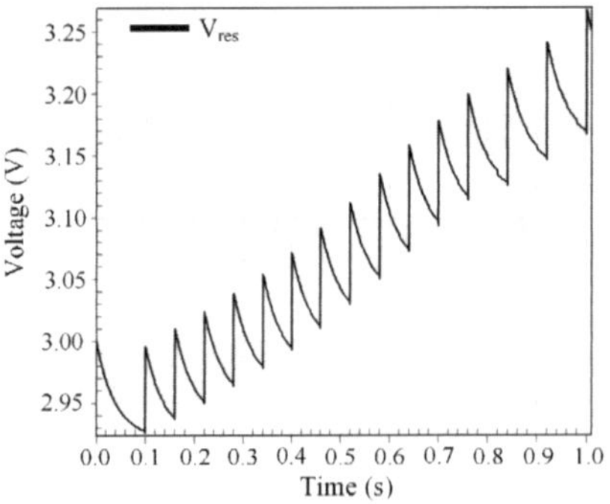

Figure 6. Voltage waveform on C_{res} of the designed circuit when using harvested energy to power control blocks.

enable port to about 150 pW. The clock generator was not simulated, but state-of-the-art low power oscillators can have a power consumption in the nW-range [17] and could therefore be driven by the small power margin found here. Hence, the capacitance variation and initial voltage here constitute a break-even case for operation of this circuit.

4. Conclusions

A low power electronic interface for MEMS electrostatic energy harvesters was designed using bump circuits. The design consists of double charge pumps and a flyback switch with an associated switch controller. The controller automatically detects the optimal switching instant solely based on internal voltages. The power consumption of the entire circuitry was reduced by a factor about 94 by using clocking. Simulations showed that with an initial bias voltage of 3.0 V, the interface circuitry could be powered by a harvester with less than 10 pF capacitance without need for external bias sources.

Acknowledgment

This work was supported by the Research Council of Norway through Grant no. 229716/E20.

References

[1] Roundy S J 2003 Ph.D. thesis The University Of California, Berkeley
[2] Musiani D, Lin K and Rosing T 2007 *Proc. IEEE IPSN* 390
[3] Deterre M, Risquez S, Bouthaud B *et al.* 2013 *J. Phys. Conf. Series* **476** 012039
[4] Mitcheson P D, Green T C, Yeatman E M and Holmes A S 2004 *J. Microelectromech. Syst.* **13** 429
[5] Beeby S P, Tudor M J and White N M 2006 *Meas. Sci. Technol.* **17** 175
[6] Mitcheson P, Yeatman E, Rao G, Holmes A and Green T 2008 *Proc. IEEE* **96** 1457
[7] Meninger S, Mur-Miranda J *et al.* 2001 *IEEE Trans. VLSI Syst.* **9** 64
[8] Miyazaki M, Tanaka H *et al.* 2003 *Proc. Int. Symp. Low Power Electron. Design* 193
[9] Yen B and Lang J H 2006 *IEEE Trans. Circuits Syst. I* **53** 288
[10] Galayko D, Pizarro R, Basset P and Paracha A 2007 *Proc. IEEE BMAS Conf* 115
[11] Torres E O and Rincon-Mora G A 2010 *IEEE J. Solid-State Circuits* **45** 483
[12] Kempitiya A, Borca-Tasciuc D and Hella M 2013 *IEEE Trans. Ind. Electron.* **60** 5639
[13] Delbruck T 1991 *IJCNN-91* **1** 475
[14] Dudka A, Basset P, Cottone F, Blokhina E and Galayko D 2013 *J. Phys. Conf. Series* **476** 012127
[15] Le C and Halvorsen E 2012 *J. Micromech. Microeng.* **22** 074013
[16] Dudka A, Galayko D and Basset P 2013 *Proc. DTIP* 1
[17] Denier U 2010 *IEEE Trans. Circuits Syst.* **57** 1973

Thermal Design of a Thermoelectric Micro-Generator

S Hama[1], T Yabuki[1], L Tranchant[1], K Miyazaki [1, 2]

[1]Department of Mechanical and Control Engineering, Kyushu Institute of Technology, 1-1 Sensui-cho, Tobata-ku, Kitakyushu, Fukuoka 804-8550, Japan.

[2]e-mail: miyazaki@mech.kyutech.ac.jp

Abstract. In this study, we fabricated micro thermoelectric power generator using free-standing film substrate, and we evaluated the performance of the generator from the standpoint of thermoelectric performance and thermal design. We fabricated a SiN_x free-standing film substrate about 5 μm thick on Si wafer, using MEMS processes. Then, we prepared for both p and n type of bismuth telluride thermoelectric thin films by using a coaxial type vacuum arc evaporation method, and annealed for one hour at 573 K. As an electrode, Cu was deposited using a vacuum deposition method. We fabricated the thermoelectric power generator of 5 mm x 5 mm using a shadow mask for the patterning. The fabricated generator can create temperature difference of 22.3 K due to its high thermal resistance of the structure when the heat source temperature is 373 K. The exergy of the thermoelectric device is up to 7 %. Therefore, the generator can convert about 0.4 % of thermal energy into electric energy, even though the material performance is low with ZT = 0.28. The conversion efficiency is much higher than that of the conventional Π type thermoelectric module. It was possible to get higher performance by the thermal design, which is a more simple way than an improvement of ZT.

1. Introduction

Micro-generators have been intensively investigated for energy harvesting using the light, vibration and thermal energy [1]-[3]. Electricity can be generated from the thermal energy directly by thermoelectric materials without any moving parts even at room temperature. The stable operation with long life time is expected, and the thermoelectric micro-device is one of important energy harvesting technologies. The performance is evaluated in a dimensionless figure of merit, ZT, which is defined as $ZT = S^2\sigma T/\kappa$, where S is the Seebeck coefficient, σ is the electrical conductivity, κ is the thermal conductivity, T is the absolute temperature. The relatively low ZT is the critical issue for the practical use at the present stage, although high ZT is reported by using nano-structures, such as nano-crystalline, superlattices, nano-porous and so on [4]-[7]. Here, we intend to improve the performance of the micro-generator by the thermal design.

We fabricated thermoelectric micro-generator of Bi_2Te_3 thin films on SiN_x free-standing film as the substrate. We fabricated a SiN_x free-standing film with 5 μm thick from Si wafer by using standard micro-fabrication processes [8],[9]. The schematic cross-sectional view of a thermoelectric micro-generator is shown Fig.1. Thermal resistance of the thermoelectric legs on the free standing film is very high due to their tiny cross-sectional area. The whole surface of the device is cooled by natural convection with high thermal resistance in in-plane direction. Large temperature difference in thin film thermoelectric device can be realized. The output power of the device and thermoelectric properties of consisting materials are measured. We confirmed the validity of the thermal design, and

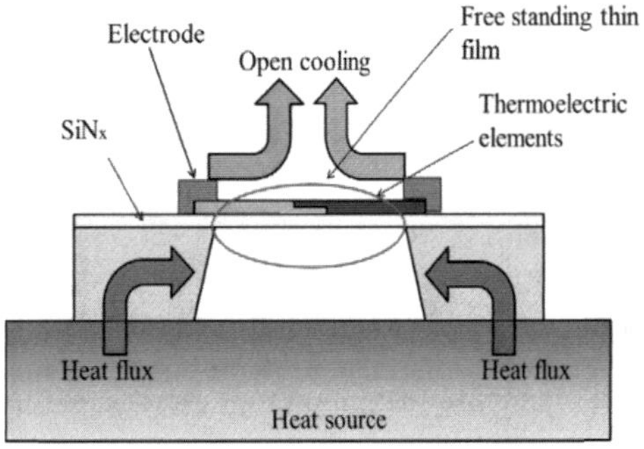

Fig.1 The schematic cross-sectional view of a thermoelectric micro-generator

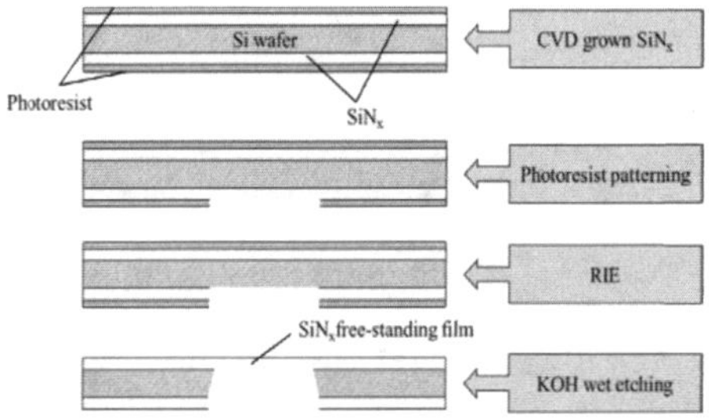

Fig.2 Microfabrication processes for a free-standing substrate of SiN_x

evaluated the performance of the device using exergy. We discuss the thermal design of a thermoelectric micro-generator.

2. Experiment

We fabricated a SiN_x free-standing film substrate about 5 μm thick from Si wafer using MEMS processes as shown in Fig.2. We deposited SiN_x on both sides of the Si wafer. The top side SiN_x film will be a free-standing thin film, and the back side is a as passivating layers for the etching process by PECVD (Plasma-Enhanced Chemical Vapor Deposition). The part of SiN_x film of the back side is selectively removed by RIE (Reactive-Ion Etching). The Si substrate without SiN_x film is removed by wet-etching using KOH. The SiN_x free-standing film with 4mm x 4mm was made.

Bismuth telluride is well known as the best thermoelectric material at room temperature. We used a coaxial type vacuum arc evaporation method to make Bismuth telluride thin films [10],[11] , because the stoichiometric alloy can be made although there are many thin film methods, such as flash evaporation [12],[13], molecular beam epitaxy (MBE) [6],[14], metaloganic chemical vapor deposition (MOCVD) [15],[16], co-sputtering [17],[18] and so on. In addition, coaxial type vacuum arc evaporation method is very simple structure, consisting only a vacuum chamber, an evaporation source, and a substrate holder. We prepared both p and n type of bismuth telluride thermoelectric thin films, and annealed them for one hour at 573 K. We made both p-type and n-type bismuth telluride thin film on 2mm x 2mm substrate. The distance between substrate and target is 150 mm. The chamber was evacuated to 4.0×10^{-3} Pa by a turbo molecular pump, and the discharge voltage set to 80 V at a frequency of 1 Hz. The deposited thin films were annealed in high-purity Ar mixed with 5 % H_2 gas by an electric furnace. Cu was deposited for the electrode of the micro-generator using a vacuum deposition method. The chamber was evacuated to 3.0×10^{-3} Pa by oil diffusion pump. The electric current of 80 A was applied to a tungsten filament with Cu, and the distance between substrate and tungsten filament was 200 mm. We fabricated the thermoelectric power generator of 5 mm x 5 mm using a shadow mask for the patterning.

The generator is shown in Fig.3. When the Si wafer is heated, SiN_x thin film deposited on Si wafer is heated, however the center of free-standing thin film is not well heated due to the high thermal resistance of the thin film. The large temperature difference can be realized, and the output voltage is proportional to the temperature difference. The schematic diagram of the measurements for the output voltage of the generator is shown in Fig.4. The bottom of the generator fixed on the hot plate, and measured the output voltage. Temperature at the center of SiN_x free-standing thin film was measured by an infrared radiation thermometer. The thermoelectric properties were measured such as Seebeck coefficient by each methods steady-state temperature difference method, electrical conductivity by four probe method, and thermal conductivity by periodically heating method [19].

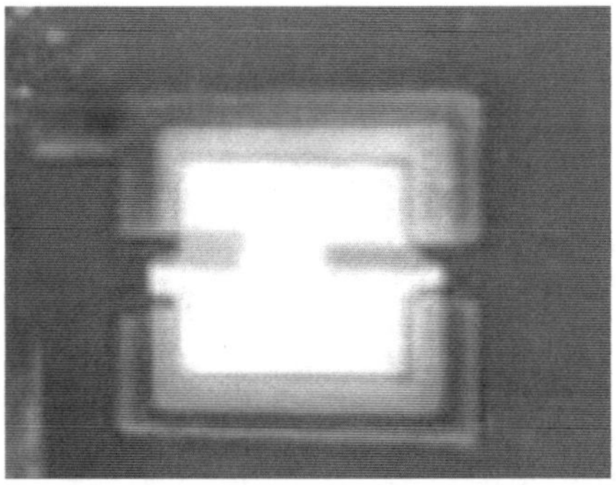

Fig.3 Photograph of a fabricated in-plane thermoelectric generator on a SiN_x free standing film.

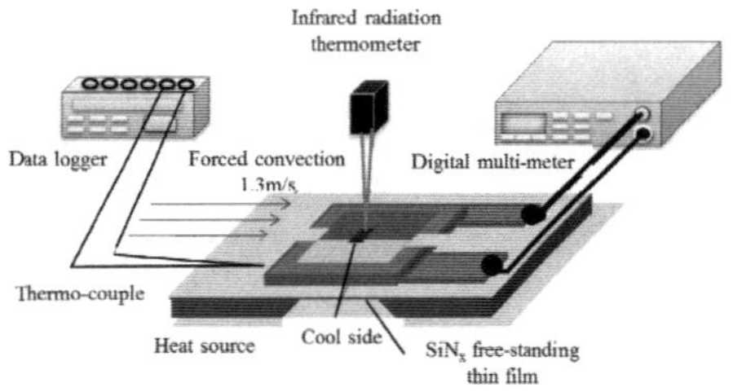

Fig.4 The schematic diagram of the measurements for the output voltage of the generator

3. Result and discussion

We measured the thermoelectric properties of both p and n type bismuth telluride as shown in Table 1. The Seebeck coefficient was 169.1 μV/K (p-type) and -119.2 μV/K (n-type). Electrical conductivity was 311 S/cm (p-type) and 844 S/cm (n-type). Thermal conductivity was 1.51 $Wm^{-1}K^{-1}$ (p-type) and 1.23 $Wm^{-1}K^{-1}$. Power factor (σS^2) was calculated as 8.89 $\mu Wcm^{-1}K^{-2}$ (p-type) and 12.00 $\mu Wcm^{-1}K^{-2}$ (n-type), *ZT* was 0.22 (p-type) and 0.36 (n-type). The output voltage of the generator at room temperature while heating at the bottom of the substrate is shown in Fig.5 (a). Maximum output power of the generator, *P*, can be evaluated by $P = V^2 / R$. Here, R is the internal resistance, *V* is the output open circuit voltage. It was shown in Fig.5 (b). The measurement setup is described in Fig.4. The output voltage was increased with heat source temperature and maximum is 3.57 mV at a heat source temperature of 403 K. Similarly, the maximum output power was increased. When the heat source temperature was 373 K, a temperature difference at the generator of 22.3 K, the output voltage was 2.63 mV, an internal resistance of 18.3 Ω, and the calculated maximum output power was 94.5 μW. The calculated conversion efficiency is only 0.4 % by energy balance at the generator. On the other hand, the conversion efficiency can be calculated by the following equation,

$$\eta = \left(1 - \frac{T_{gc}}{T_{gh}}\right) \frac{\sqrt{1+ZT} - 1}{\sqrt{1+ZT} + \frac{T_{gc}}{T_{gh}}} \qquad (1)$$

T_{gc} is the cold temperature of the generator, and T_{gh} is the hot temperature of the generator [20]. T is the average temperature, $(T_{gh}+T_{gc})/2$. If the non-dimensional figure of merit, *ZT* is improved, the conversion efficiency will be increased. T_{gh} and T_{gc} will be determined by the structure of the generator. The thermal resistance between hot junction and cold junction of the generator is huge in the present study, therefore T_{gc} is relatively low. The calculated first term is equivalent to the exergy of the device, and it is up to 6.4 %. The efficiency η is 0.40 % although the *Z* is only 0.22. Both T_{gc} and T_{gh} of the conventional Π type thermoelectric generator can be calculated under the assumption of the same volume of the thermoelectric material as that of the present study. The calculated exergy is only 1.1 %. If we need the same efficiency as that of the present in-plane generator by using the conventional Π type structure, *ZT* should be 3.3. Here, we demonstrated that the thermal design of a generator is very important as well as the thermoelectric material performance. The exergy of the device should be improved by thermal design to enhance the device efficiency.

Table1. The thermoelectric properties of both p and n type bismuth telluride

thick[nm]	type	S[μV/K]	ρ [S/cm]	P.F[$\mu Wcm^{-1}K^{-2}$]	κ [$Wm^{-1}K^{-1}$]
1700	p	169.1	311	8.89	1.51
	n	-119.2	844	12.00	1.23

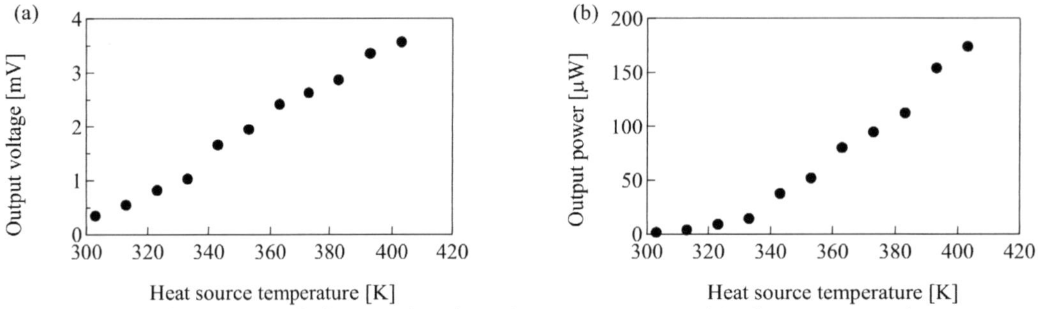

Fig.5 The performance of thermoelectric micro-generator : (a) the output voltage at room temperature while heating at the bottom of the substrate (b) maximum output power evaluated by the internal resistance.

4. Conclusion

We fabricated an in-plane thermoelectric micro-generator using SiN_x free-standing film with 5 μm thickness on Si wafer. We deposited both p and n type bismuth telluride thin films by a coaxial type vacuum arc evaporation method using shadow masks for patterning. As an electrode, Cu was deposited using a vacuum deposition method. The dimensions of both type bismuth telluride are 2 mm (length) x 2 mm (wide) x 1.7 μm (thickness). We measured their thermoelectric properties and output voltage at room temperature while heating at the bottom of the substrate. The output voltage was 2.63 mV with 22.3K temperature difference at 373K. The large temperature difference can be realized by the in-plane structure. The high exergy of the generator was calculated to be 6 %. As a result, 0.4 % of thermal energy was converted into electrical energy, even though the material performance *ZT* is only 0.28. We demonstrated that it was possible to get higher performance by the thermal design, which is a more simple way than an improvement of *ZT*.

References

[1] Priya S and Inman D J 2009 *Energy Harvesting Technologies* Springer pp. v-vi
[2] Harb A. 2011 *Renewable Energy* **36** 2641-2654
[3] Kim H S, Kim J H, and Kim J 2011 *IJPEM* **12** 1129-1141
[4] Xie W, Tang X, Yan Y, Zhang Q and Tritt T M 2009 *J. Appl. Phys.* **105** 113713
[5] Poudel B, Hao Q, Ma Y, Lan Y, Minnich A, Yu B and Ren Z 2008 *Science* **320** 634-638
[6] Harman T C, Tayor P J, Walsh M P and Laforge B E 2002 Science **297** 2229
[7] Kashiwagi M, Hirata S, Harada K, Zheng Y, Miyazaki K, Yahiro M and Adachi C 2011 *Appl. Phys. Lett.* **98** 023114
[8] Kurosaki J, Yamamoto A, Tanaka S, Cannon J, Miyazaki K and Tsukamoto H 2009 *J. Electron. Mater.* **38** 1326
[9] Yamamoto A, Hagino H, Hashimoto Y and Miyazaki K 2012 *J. Electron. Mater.* **41** 1799-1804
[10] Gasenkova I V and Tochitsky E I 1995 Proceedings of XIV International Conference on Thermoelectrics **29**
[11] Uchino M, Kato K, Hagino H and Miyazaki K 2013 *J. Electron. Mater.* **42** 1814-1819
[12] Takashiri M, Shirakawa T, Miyazaki K and Tsukamoto H 2007 *J. Alloys Compd.* **441** 246
[13] Takashiri M, Miyazaki K and Tsukamoto H 2008 *Thin Solid Films* **516** 6336

[14] Inoue T and Miyazaki K 1999 *Therm. Sci. Eng.* **7** 95-101
[15] Venkatasubramanian R, Colpitts T, Watko E, Lamvik M and El-Masry N 1997 *J. Cryst. Growth* **170** 817
[16] Al Bayaz A, Giani A, Al Khalfioui M, Foucaran A, Pascal-Delannoy F and Boyer A 2003 *J. Cryst. Growth* **258** 135
[17] Böttner H, Schubert A, Schlereth K H, Eberhard D, Gavrikov A, Jägle M, Ku¨hner G, Künzel C, Nurnus J and Plescher G 2004 *J. Microelectromech. Syst.* **13** 414
[18] Noro H, Sato K and Kagechika H 1993 *J. Appl. Phys.* **73**, 1252
[19] Hatta I 1990 *Int. J. Thermophys* **11** 293-303
[20] Rowe D M 1995 *CRC Handbook of Thermoelectrics* CRC pp.19-25

On the performance of gas-lubricated bearing-rotor system in microengine

Xiaoli Wang

Beijing Institute of Technology, Beijing, China

xiaoli_wang@bit.edu.cn

Abstract. This paper reports a comprehensive analysis of the performance for gas-lubricated bearing-rotor system in silicon microengine. The generalized lubrication model and dynamics model are developed and the effects of micro fabrication defects, temperature and viscous friction force on both steady-state and dynamic characteristics of bearing system are investigated. The results can serve as a useful tool for the design of silicon microbearing- rotor systems.

1. Introduction

Gas-lubricated bearings are the best candidates to support the rotating members in Power MEMS devices such as micro-engines, due to their unique advantages of low friction losses, near-zero wear and no contamination through lubricant leakages compared with their rolling-element or oil-lubricated counterparts. However, the implementation of gas bearings encounters challenges because of their characteristics which are different from their conventional ones as follows: the very high rotating speed and temperature to achieve high power density, the super thin gas film between rotors and static structures, and the relative large fabrication uncertainties according to the imperfection of the fabrication technology [1-3]. In fact, as the rotating speed is supper high, the compressibility of the gas film force is very significant, and the gas film force is strongly nonlinear. The forces and moments of the system depend not only on the instantaneous kinematical state but also upon the history of the motion, which results in coupling of the lubrication analysis with the dynamics of the rotor motion. Moreover, tightened by the constraints of the micro fabrication progress (DRIE, etc.), the length-to-diameter ratio (L/D) of the silicon-based micro gas journal bearing is about 0.05～0.1, which is one order of magnitude lower than in conventional gas journal bearings. For the conventional gas journal bearing-rotor systems, the viscous friction force is much lower compared with the gas film force and usually ignored in the research of the system [4]. However, in the ultra-short micro journal bearing-rotor systems, the order of magnitude of the viscous friction force is the same as that of the gas film force. Many researchers have studied the dynamic characteristics of micro gas bearings [5-7].Unfortunately, the influence of the viscous friction force on the nonlinear dynamic behavior has not been studied yet. This work is undertaken to provide insight into the effects of micro fabrication defects, temperature and viscous friction force on both steady-state and dynamic characteristics.

2. Mathematical analysis

2.1. Lubrication model

Figure 1 is a three-dimensional schematic drawing of a silicon-based microengine in which the rotor is supported axially by a pair of spiral-grooved micro gas hydrodynamic thrust bearings, and radially by a hydrodynamic journal bearing. When operating the film thickness h lies in the range of $0.5\mu m \leq h \leq 20\mu m$, resulting the Knudsen number Kn is about 0.0032～0.1291, which means the gas film is rarefied and the continuous flow model is no longer valid. With the assumptions of laminar and isothermal flow and the ignorance of the inertia item, the non-dimensional governing compressible form of the MGL model is [8]

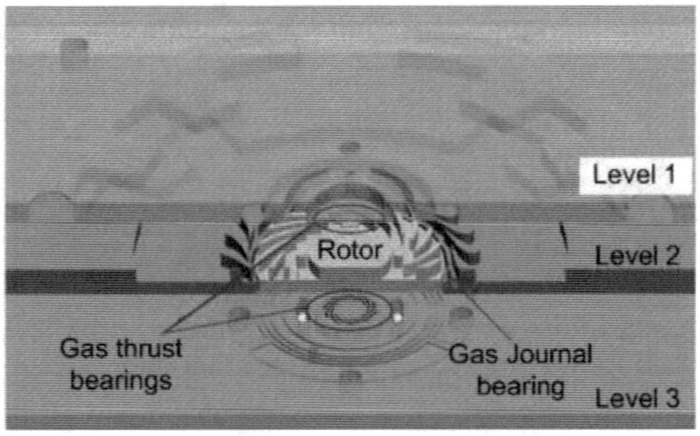

Figure 1. Schematic of the bearing-rotor system for micro gas turbine engine.

$$\frac{\partial}{\partial\theta}(PH^3Q\frac{\partial P}{\partial\theta})+\frac{\partial}{\partial\xi}(PH^3Q\frac{\partial P}{\partial\xi})=\Lambda\frac{\partial}{\partial\theta}(PH)+2\Lambda\frac{\partial}{\partial\tau}(PH) \tag{1}$$

where H=1- $\varepsilon\cos(\theta-\varphi)$ is the non-dimensional film thickness, $P = p/p_a$ is the non-dimensional gas film pressure, $\tau=\omega t$ is the non-dimensional time, Λ is the bearing number, Q is the non-dimensional flow rate coefficient which is related to the inverse Knudsen number D_K [8]. In terms of continuous flow model, Q equals 1 at any film thickness. θ and ξ are the circumferential and axial coordinates, respectively. The non-dimensional gas film force, $F_g=f_g/(P_aLD)$, and the attitude angle φ can be obtained as

$$\begin{cases} F_{gn} = \dfrac{R}{2L}\displaystyle\int_{-\frac{L}{D}}^{\frac{L}{D}}\int_0^{2\pi} P\cos\theta \mathrm{d}\theta\mathrm{d}\xi \\ F_{g\tau} = \dfrac{R}{2L}\displaystyle\int_{-\frac{L}{D}}^{\frac{L}{D}}\int_0^{2\pi} P\sin\theta \mathrm{d}\theta\mathrm{d}\xi \\ F_g = \sqrt{F_{gn}^2 + F_{g\tau}^2} \\ \varphi = \arctan\left(-\dfrac{F_{g\tau}}{F_{gn}}\right) \end{cases} \tag{2}$$

The sheer stress at the rotor surface arising from the viscosity of the gas is

$$\tau_h = \frac{Hcp_a}{2R}\frac{\partial P}{\partial\theta} + \frac{\mu\omega R}{cH} \tag{3}$$

The non-dimensional viscous friction force, $F_f=f_f/(P_aLD)$, can then be calculated as

$$\begin{cases} F_{fx} = \dfrac{R}{2L}\displaystyle\int_{-\frac{L}{D}}^{\frac{L}{D}}\int_0^{2\pi} \tau_h\cos\theta \mathrm{d}\theta\mathrm{d}\xi \\ F_{fy} = \dfrac{R}{2L}\displaystyle\int_{-\frac{L}{D}}^{\frac{L}{D}}\int_0^{2\pi} \tau_h\sin\theta \mathrm{d}\theta\mathrm{d}\xi \\ F_f = \sqrt{F_{fx}^2 + F_{fy}^2} \end{cases} \tag{4}$$

2.2. Rotor dynamic model

The non-dimensional kinematic equations of motion of the rotor is

$$\begin{cases} M\ddot{X} = F_{ex} + F_{gx} + F_{fx} + M\tilde{\rho}\sin\tau \\ M\ddot{Y} = F_{ey} + F_{gy} + F_{fy} + M\tilde{\rho}\cos\tau \end{cases} \quad (5)$$

where F_e stands for non-dimensional external force. Eq. (1) and Eq. (5) are not independent of each other, causing the coupling of the kinematic equation and the lubrication model.

3. Results and discussion

3.1. Effects of microfabrication defects on the performance of microbearings

The pressure distribution comparison among the bearings with different degrees of taper can be seen in Figure 2. As shown, the clearance of the tapered bearing with $\Delta c \neq 0\mu m$ is not constant axially, thus the pressure distribution is not symmetrical about the axial mid-plane where ξ=0. Besides, the pressure peak value of the tapered bearing drops and the greater effects will happen when Δc becomes larger.

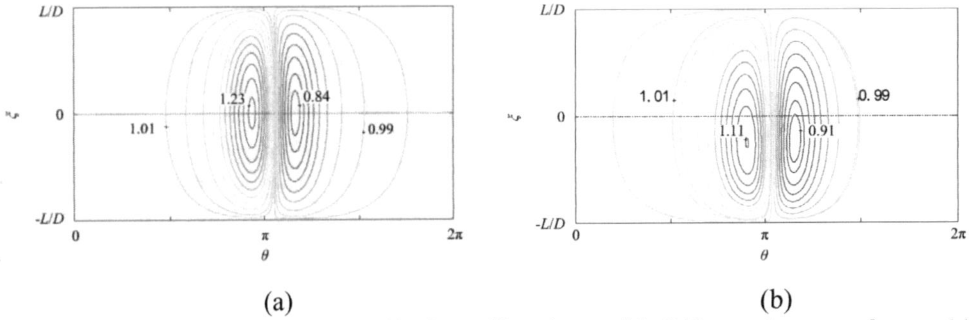

Figure 2. Non-dimensional pressure distribution of bearings with different degrees of taper (a) Δc = 0μm and (b) Δc = 2μm

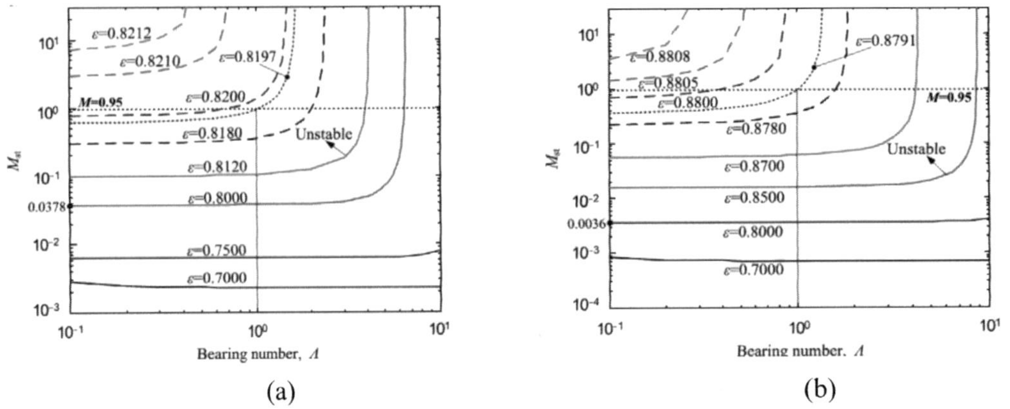

Figure 3. Comparison of M_{st}-Λ chart among bearings with different degrees of taper (a) Δc = 0μm and (b) Δc = 2μm

The M_{st}-Λ charts are denoted in Figure 3, where the dotted lateral lines (i.e., M=0.95) represent the present rotor mass value. When the rotor mass is fixed, the turning point of eccentricity to be stable or unstable is different for bearings with different degrees of taper. For example, when M=0.95, Λ=1, the transition occurs at about 0.8197 for a no-tapered bearing. However, when clearance of 2μm difference exists, the transitional eccentricity becomes about 0.8791. Thus the tapered bearing needs to achieve a higher eccentricity to realize the stable operation. On the other hand, at the same eccentricity and bearing number, the tapered bearing needs to have a smaller rotor mass to operate stably. For example, to running stably in a range of bearing number from 0.1 to 10 at ε=0.8, the non-dimensional rotor mass has to be less than 0.0378 for a no-tapered bearing, whereas the value is 0.0036 for a

Δc=2μm bearing. In conclusion, a tapered bearing is more liable to contact with the static structure or become unstable.

3.2. *Effects of temperature on the performance of microbearings*

As discussed above, the increase of the temperature could cause the increase of molecular mean free path and thus cause the rarefaction effect, which might decrease the load capacity, to become more important. However, on the other hand, the gas viscosity may increase with the increase of the temperature, and higher gas viscosity implies larger load capacity. The simulation results show that the temperature variation range from 300K to 1600K could be divided into two parts, as shown in Figure 4. The left part is the viscosity effect domain region (300K～753K) and the other one is the rarefaction effect domain region (753K～1600K). In the first part, the influence of temperature on gas viscosity is dominant and the load capacity becomes larger as temperature increases. In the other part, the influence of temperature on rarefaction effect is more important and the inverse trend can be found as temperature goes high.

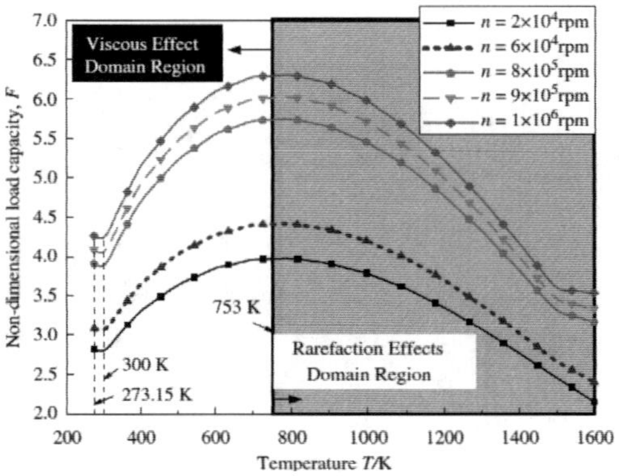

Figure 4. Curves of non-dimensional load capacity versus temperature at different rotational speeds.

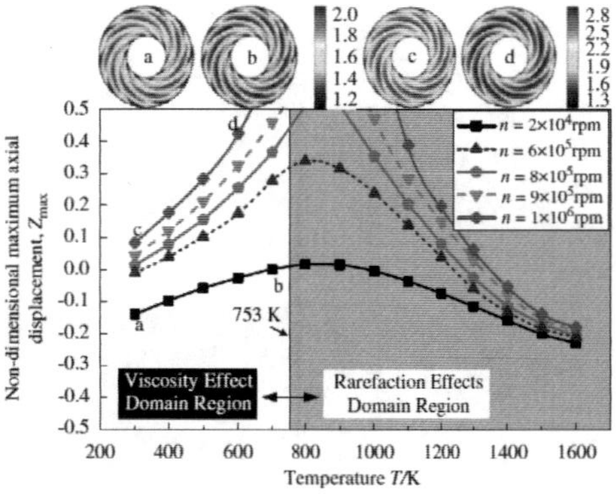

Figure 5. Maximum axial displacement at different values of temperature

The maximum axial displacement in the history of the transient response of the rotor, Z_{max}, is an important parameter to characterize the dynamic behavior of the micro bearing system since $h_{min} = 0.5 - Z_{max}$ represents the minimum distance between the rotor and the bearing pad. Figure 5 shows the curves of Z_{max} versus temperature at different rotational speed. The contour graphs of pressure distribution corresponding to points of a, b, c and d are also displayed at the top of Figure 5. When the rotational speed go beyond the value of 6×10^5 rpm, as can be seen, the stability of the thrust bearing system becomes weaker when the operating temperature lies in the range from 700K to 1000K, which is the junction zone between the viscosity domain region and the rarefaction domain region. In this case, the excessive deviation of the rotor position from the original equilibrium position (Z=0) will causes the system to crash.

3.3. *Effects of viscous friction on the performance of microbearings*

In this section, the nonlinear dynamic characteristics for both conditions of with and without the viscous friction force considered are studied and compared. Here the bearing number is selected as the bifurcation parameter and increases from 2.0 to 14.0, at m=10.0mg, $\varepsilon 0$=0.85 and L=330μm. The bifurcation diagrams without or with the consideration of the viscous friction force are shown in Figure 6(a) and (b) respectively. When ignoring the viscous friction force, as the increase of the bearing number, the system performs period-1 synchronous motion firstly when Λ is less than 2.7, and then bifurcates into quasi-periodic motion or multi-periodic motion. The system motion turns back to synchronous when Λ lies between 3.3 and 7.6 and then shifts between quasi-periodic motion and multi-periodic motion once again after Λ=7.6. Quite different phenomenon could be observed by taking the viscous friction force into account. As plotted in Figure 6 (b), the system exhibits synchronous motion at low bearing numbers, and loses its stability at 7.1, from which a period-doubling phenomenon occurs. As Λ increases over 9.2, the system shifts back to 1-periodic motion, and then performs quasi-periodic motion over the bearing number interval $10.0 \leq \Lambda < 10.7$. After that, the rotor center motion transits through the following behaviors: 3T—T—3T—quasi.

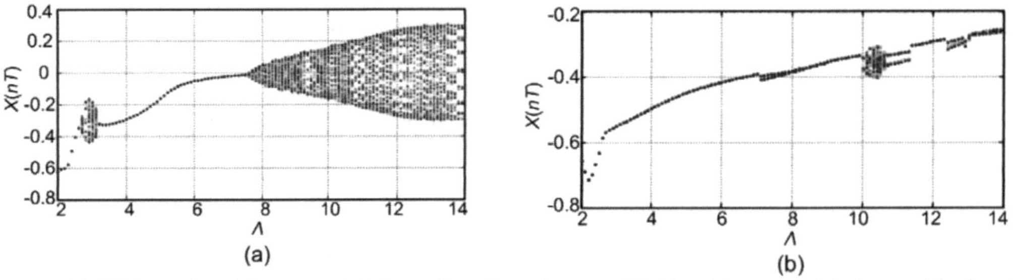

Figure 6. Bifurcation diagrams (a) ignoring the viscous friction force and (b) considering the viscous friction force

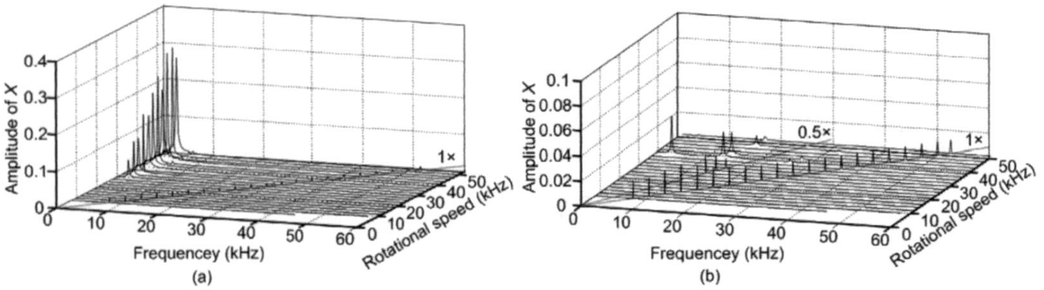

Figure 7. Waterfall plots of the system response in the x-direction (a) ignoring the viscous friction force and (b) considering the viscous friction force

From the waterfall plots of the system responses for the two conditions illustrated in Figure 7 (a) and (b), one can see that a low-frequency large-amplitude self-excited whirl motion occurs and dominates as the rotational speed increases when ignoring the viscous friction force. However, for the other case, the synchronous motion prevails in the whole rotational speed range with some narrow extents of whirl motion, and moreover, the amplitude of the motion is rather smaller than the case shown in Figure 7 (a). The reason for this phenomenon is that for ultra short gas journal bearing-rotor systems, the viscous friction force absorbs the energy of the whirl motion and thus decreases the whirling amplitude and a more stable rotor motion then appears.

4. Conclusions

The work presented here differs from previous work in the specific effects of micro fabrication defects, viscous friction force and temperature on both static and dynamic characteristics of gas-lubricated bearing-rotor system in silicon microengine, and the results can serve as a useful tool for design of silicon microbearing-rotor systems.

References

[1] Epstein A H 2004 *ASME J. Eng. Gas Turbines Power* **126** 205–26
[2] Teo C J, Liu L X, Epstein A H and Spakovszky Z S 2009 *ASME J. Tribol.* **131** 032001
[3] Shan X C, Murakoshi Y and Wang Z F 2004 *Proc. of SPIE.* **5276** 428–33
[4] Rashidi R, Ardeshir K M and Firooz B N 2010 *Nonlinear Dyn.* **61** 783–802
[5] Piekos E S and Breuer K S 1999 *ASME J. Tribol.* **121** 604–09
[6] Yang P, Zhu K Q and Wang X L 2009 *Tribol. Int.* **42** 71–76
[7] Liu R, Wang X L and Zhang X Q 2012 *ASME J. Tribol.* **134** 022201
[8] Fukui S and Kaneko R 1981 *ASME J. Tribol.* **10**, 253–62

A compact architecture for passively-switched energy harvesters

Tian Liu and Carol Livermore

Department of Mechanical and Industrial Engineering, Northeastern University, USA

Email: tianliu2010@gmail.com, livermore@neu.edu

Abstract. This paper presents the design and experimental characterization of a compact parallel-beam architecture for low-frequency energy harvesters that switch passively among dynamical modes to extend their operational range. Two beams interact to generate power; a driving beam couples into low frequency vibrations, and a higher frequency generating beam outputs power upon impact by the driving beam. The system switches between modes in which the driving beam bounces off the generating beam (coupled motion) and modes in which the driving beam passes the generating beam (plucked motion). The compact structure is realized by mounting the generating beam within a U-shaped driving beam on a single support. A flexible tip is mounted inside the driving beam's U shape to enable robust interactions. This new architecture reduces system volume by 80% compared with an earlier model that has the same resonance frequency, but it also changes the flexible tip's role in the contact dynamics. The flexible tip is experimentally tailored to optimize performance. The harvester generates power over the measured range of acceleration from 0.2 g to 2 g and driving frequency from 5 Hz to 20 Hz. With one tip design, the harvester offers peak power of 0.267 mW with plucked operation covering 32% of the tested range. With a second tip design, the harvester offers a lower peak power of 0.036 mW with plucked operation covering 73% of the tested range.

1. Introduction

Vibrational energy harvesters output electrical energy from ambient vibration via transduction mechanisms such as magnetics, electrostatics and piezoelectrics [1-2]. Power output and bandwidth are perennial challenges. Resonant operation can increase power output, but resonant harvesters' powers decline significantly when running off of their resonances. To increase harvesting bandwidth, mechanical tuning methods may be used to tune a harvester's resonance to match the ambient driving frequency, including active [3-4] and passive tuning [5-7] of harvesters' resonances. Nonlinear systems also offer increased operational bandwidth [8].

Low frequency energy harvesting, such as for extracting energy from human motions, is another challenge. Frequency up-conversion has proved to be an effective method of coupling low ambient frequencies into higher generating frequencies [9-12]. In frequency up-conversion, a driving element with low resonance frequency interacts with a generating element with high resonance frequency, thereby transferring mechanical energy from lower frequency to higher frequency motions.

Previously we have reported a passively-switched energy harvester with increased operational range that has a measured maximum average power of 1.56 mW under 0.5 g at its resonance of 7 Hz [13]. The harvester comprises a low frequency driving beam and a higher frequency generating beam that are positioned facing each other with overlapping beam tips. The harvester passively switches between modes of operation in which the driving beam bounces off the generating beam (coupled-motion mode) and modes of operation in which the driving beam plucks the generating beam by moving from above it to below it or vice versa (plucked mode). Each mode has its own resonance frequency. The resulting multiple system resonances enable increased operational range. However, the size of the previous passively-switched, low-frequency harvester is too large for many compact, portable applications. Direct down-scaling does not provide an attractive solution because smaller harvesters typically have higher resonance frequencies.

This paper proposes and experimentally demonstrates a new, 80% smaller, parallel-beam architecture for an energy harvester that allows it to couple robustly into 10 Hz-range vibrations while maintaining passive switching among dynamical modes for increased operational range. As shown in figure 1a, the harvester includes a U-shaped plastic cantilever with a proof mass at the tip as the driving beam. A piezoelectric cantilever identical to that of [13] is mounted inside the driving beam's U shape to serve as the generating beam. Unlike the harvester of [13], the driving beam and generating beam are mounted together on a single support so that they are in parallel and are in the same plane. The driving beam's U shape lets the generating beam move freely through the driving beam. A flexible tip mounted inside the driving beam's U shape overlaps the tip of the generating beam.

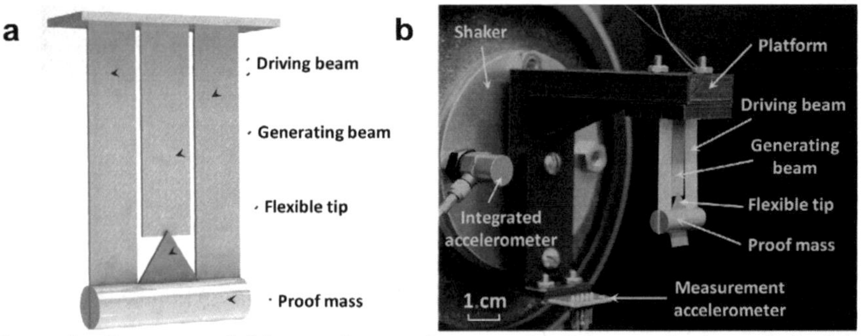

Figure 1. (a) Schematic diagram and (b) experiment photograph of the passively-switched harvester with the new parallel-beam architecture.

For small or off resonance accelerations, the driving beam bounces off the generating beam in coupled motion mode. After the driving beam impacts the generating beam, the two beams move together for about half of an oscillation cycle before separating. The characteristic frequency of this mode reflects a combination of both beams' resonance frequencies [11]. For large or on resonance accelerations, the driving beam deflects the generating beam until their tips pass and the generating beam is plucked. The characteristic frequency of this mode is dominated by the driving beam's own resonance. Mixed mode occurs when the beams alternate between coupled-motion and plucked mode.

The parallel-beam architecture fundamentally changes the flexible tip's role in the dynamics. For facing beams as in [13], the lengthwise overlap between the driving and generating beams decreases as the beams deflect. If the initial overlap is small enough, the tip of the driving beam eventually passes the generating beam. For the parallel architecture, deflecting the beams increases their lengthwise overlap, making it harder for the driving beam to pass the generating beam. The design of the flexible tip is therefore critical for the new architecture to undergo plucked harvesting as well as coupled motion harvesting.

2. Experiment

2.1. Experimental Setup

Figure 1b shows the experimentally-implemented harvester. Two bronze half-cylinders comprise the 8.76 g proof mass at the tip of the U-shaped ABS driving beam. A PZT bimorph cantilever (Piezo System, Inc.) inside the driving beam's U shape serves as the generating beam. The harvester's dimensions and parameters are listed in table 1. The generating and driving beams are epoxied on a common 3D printed support.

Unlike the rectangular flexible tip in the previous passively-switched harvester [13], the present tips are shaped as isosceles triangles. The tips have two thirds of the stiffness of a rectangular cantilever with the same base and length dimensions and material, but the triangular shape reduces the actual impact area between the flexible tip and the generating beam by more than 90%.

Table 2. Mechanical and electrical specifications of the driving beam and generating beam.

	Driving beam	**Generating beam**
Material	ABS plastic with proof mass	PZT-5A
Length (mm)	36	28.5
Width (mm)	6 on each side, 12 total	6.4
Thickness (mm)	0.53	0.51
Effective mass (g)	8.76	0.174
Young's modulus (GPa)	2.24	66
Measured stiffness (N/m)	17.2	578
Capacitance C (nF)	Not applicable	7.4

Two variations of the triangular tip design are used, in each case with the same driving beam and generating beam. Both tips are isosceles triangles with a base of 6.8 mm and a length of 5.7 mm. One flexible tip is made of two layers of VWR labelling tape (0.29 mm thick total), and the other is made of three layers of the same tape (0.43 mm thick total). The experimentally-estimated stiffness values of the two flexible tips are 20 N m-1 (two-layer) and 37.5 N m-1 (three-layer). Because the driving beam and generating beam are mounted in the same plane, the flexible tip initially touches the generating beam and is deflected by generating beam's thickness. The overlap between the flexible tip and the generating beam can be adjusted to vary the distribution of dynamical modes.

The energy harvester is mounted on a vibration testing system (LW139.151-30, Labworks Inc.) and tested under sinusoidal excitation. The acceleration is displayed in real time by the system's integrated accelerometer. A second accelerometer (ADXL203CE, Analog Devices) mounted on the harvester's base measures the acceleration output. The synchronized generating beam voltage and acceleration are recorded. For both tips, the harvester's dynamics and power output are measured from 0.2 g to 2 g and from 5 Hz from 20 Hz with a load resistance matched to the coupled-motion mode. Coupled-motion dynamics are identified when the driving beam always bounces off the generating beam on one side in steady state. Plucked dynamics are identified when the driving beam repeatedly plucks the generating beam in steady state. Mixed dynamics are identified when coupled-motion and plucked dynamics co-exist in steady state. When the driving beam undergoes large deflections greater than about 30 mm, the generating beam has the potential to break; these conditions are avoided and no data are recorded. The inclusion of motion stops in future harvesters may reduce the risk of breakage.

2.2. Experimental results

Figure 2 shows the experimentally-observed dynamics vs. frequency and acceleration for harvesters with the two tip designs. The lengthwise overlap of the tip on the generating beam is similar for the two harvesters. For the more flexible, two-layer tip, plucked dynamics initiate at 0.1 g when the driving frequency is near the driving beam's resonance of 7 Hz. For the stiffer, three-layer tip, plucked dynamics initiate at 0.5 g when the driving frequency is near the driving beam's resonance. Under all measured operating conditions, the system exhibits either coupled-motion, plucked, or mixed dynamics. Unlike in [13], there are no conditions under which the driving beam fails to impact the generating beam. The successful production of power across the entire measured range reflects the co-planar (zero-gap) mounting of the driving and generating beams. For both tips, the system passively switches between coupled-motion and plucked harvesting with mixed mode between them.

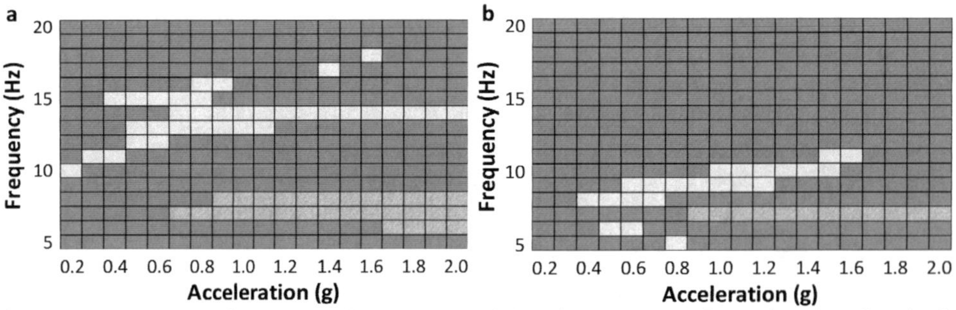

Figure 2. Measured maps of dynamical mode vs. driving frequency and acceleration for the harvester with (a) a two-layer triangular flexible tip and (b) a three-layer triangular flexible tip. Green, blue and yellow represent coupled-motion, plucked, and mixed harvesting, respectively. Gray represents no measurement (to prevent overdeflection).

The harvester with the stiffer, three-layer tip has a much smaller range of plucked dynamics (32% of the tested range) than the harvester with the two-layer tip (73% of the tested range). The stiffer tip deforms less under a given load, making it harder for the driving beam to pass the generating beam. For the two-layer tip, plucked and mixed modes appear at lower accelerations near the system's characteristic frequencies than they do off of the system's characteristic frequencies. These frequencies reflect the driving beam's resonance (7 Hz), the coupled-motion resonance (11 Hz), and the first harmonic of the driving beam's resonance (14 Hz). For the three-layer tip, plucked and mixed modes only appear near the driving beam's resonance because the other characteristic frequencies are not powerful enough to drive plucked modes with the stiffer tip in the tested acceleration range.

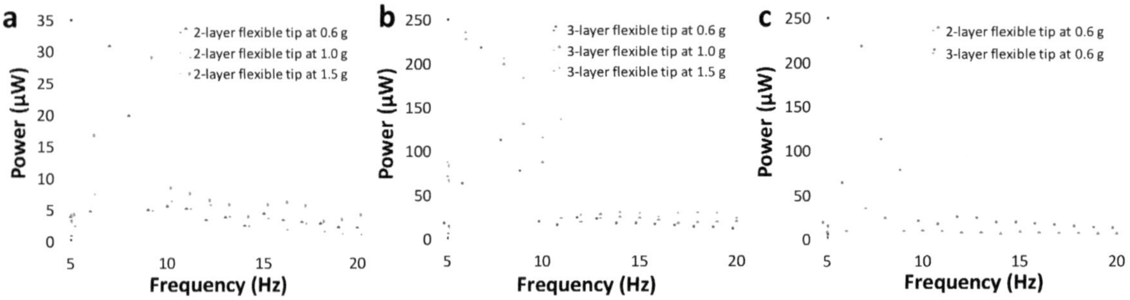

Figure 3. Plots of average power vs. driving frequency at a driving acceleration of 0.6 g (blue), 1.0 g (orange), and 1.5 g (gray) for (a) the harvester with a two-layer triangle flexible tip and (b) a three-layer triangle flexible tip, and (c) comparison of average power for the harvester with a two-layer triangle flexible tip (green) and a three-layer triangle flexible tip (red) at the same acceleration of 0.6g. Missing data points in (a) and (b) were not measured to prevent overdeflection.

Average power outputs from harvesters with two-layer and three-layer tips are compared in figure 3. Power output is calculated by averaging instantaneous power from stable dynamics over time. Figures 3(a) and 3(b) plot average power vs. driving frequency for harvesters with two-layer and three-layer flexible tips at accelerations of 0.6 g, 1 g and 1.5 g. Under 0.6 g, power outputs from both harvesters peak at the driving beam's 7 Hz resonance; output at other frequencies is small. Under 1 g and 1.5 g, the frequencies near the driving beam's resonance are omitted to avoid breakage, but peak broadening and the emergence of subsidiary peaks at other frequencies are evident. For the two-layer tip, additional power peaks appear near the coupled-motion and second harmonic frequencies, albeit at reduced amplitude because coupled-motion and mixed dynamics output less power than plucked dynamics. For the stiffer three-layer tip, the consistent coupled-motion dynamics at higher frequencies prevents the appearance of significant power peaks at the secondary resonances. Figure 3(c) compares the overall power level of the two- and three-layer tips at 0.6 g. The harvester with the stiffer tip generates higher power in both plucking and coupled-motion modes because it transfers more energy from the driving beam to the generating beam. However, the stiffer tip also suffers a narrower range of plucked harvesting and a larger chance of the tip becoming wedged with the generating beam.

3. Conclusion

The new, parallel-beam architecture presented here for passively-switched energy harvesters enables harvesters that are 5X more compact than their predecessors [13] to convert human-scale low frequency accelerations into effective power generation across a wide range of frequencies and accelerations. The wide operational range is achieved by the same kind of passive switching among dynamical modes that is used in [13], but the present unique architecture of a generating beam embedded within a co-planar U-shaped driving beam enables the harvester's dramatic reduction in size. The detailed design of the flexible tips by which the driving and generating beams interact is a key consideration for the parallel-beam design. A triangular tip minimizes the opportunity for the driving and generating beams to become wedged together, and the system's dynamics and power output vary enormously with tip stiffness. The harvester with a stiffer tip produces higher powers of up to 0.267 mW; the harvester with the more flexible tip outputs a peak power of 0.036 mW. The harvester with a stiffer tip requires higher accelerations to switch from coupled-motion to plucked harvesting, so that only 32% of the operating range is in the plucked harvesting region, In contrast, the harvester with a flexible tip undergoes plucked harvesting over 73% of its operational range. In the future, motion-stops may deter breakage under the driving beam's large resonant deflection.

References

[1] Saadon S and Sidek O 2011 A review of vibration-based MEMS piezoelectric energy harvesters *Energy Conversion and Management* **52** 500-504
[2] Harb A 2011 Energy harvesting: State-of-the-art *Renewable Energy* **36** 2641-54
[3] Huang S and Lin K 2012 A novel design of a map-tuning piezoelectric vibration energy harvester *Smart Materials and Structures* **21** 085014
[4] Eichhorn C, Tchagsim R, Wilhelm N and Woias P 2011 A smart and self-sufficient frequency tunable vibration energy harvester *J. Micromech. Microeng.* **21** 104003
[5] Gu L and Livermore C 2010 Passive self-tuning energy harvester for extracting energy from rotational motion *Applied Physics Letter* **97** 081904-1
[6] Gu L and Livermore C 2012 Compact passively self-tuning energy harvesting for rotating applications *Smart Materials and Structures* **21** 015002
[7] Liu T and Livemore C 2012 Passively tuning harvesting beam length to achieve very high harvesting bandwidth in rotating applications *Proceedings of PowerMEMS 2012* pp. 492-495
[8] Harne R and Wang K 2013 A review of the recent research on vibration energy harvesting via bistable systems *Smart materials and structures* **22** 023001
[9] Kulah H and Najafi K 2008 Energy scavenging from low-frequency vibrations by using frequency up-conversion for wireless sensor applications *IEEE Sensor Journal* **8** 261-268
[10] Pozzi M and Zhu M 2011 Plucked piezoelectric bimorphs for knee-joint energy harvesting: modelling and experimental validation *Smart Materials and Structures* **20** 055007
[11] Gu L and Livermore C 2011 Impact-driven, frequency up-converting coupled vibration energy harvesting device for low frequency operation *Smart Materials and Structures* **20** 045004
[12] Pillatsch P, Yeatman E M and Holmes A S 2014 A piezoelectric frequency up-converting energy harvester with rotating proof mass for human body applications *Sensors and Actuators A: Physical* 206 178-185
[13] Liu T, St Pierre R and Livermore C 2014 Passively-switched energy harvester for increased operational range *Smart Materials and Structures* **23** 095045

A low-frequency hybrid energy harvester with high output performance

Y Xia, W Liu, T Chen, Z Yang, P Wang, H Liu* and L Sun

Robotics and Microsystems Center, Jiangsu Provincial Key Laboratory of Advanced Robotics, and Collaborative Innovation Center of Suzhou Nano Science and Technology, Soochow University, Suzhou 215123, China

Email: hcliu078@suda.edu.cn

Abstract. Harvesting energy from ambient environment as a power source has attracted a wide attention now. However, it meets with difficulties of achieving high output performance under low excitation frequency. This paper presents a low-frequency hybrid energy harvester with high output performance based on frequency-up-conversion (FUC) approach. Under the strike of electromagnetic harvesting suspension structure, the PVDF cantilevers can operate at their resonant frequency with larger amplitude. The power output and energy conversion efficiency can be improved significantly at low frequency excitation. For an excitation acceleration of 0.5g, the peak voltage output of electromagnetic part can reach to 1.9 V at driving frequency of 23.6 Hz. In addition, the peak voltage of PVDF cantilever increases from 30 mV to 3 V and the working frequency increase from 23.6 Hz to 60.6 Hz by using FUC approach.

1. Introduction

Vibration energy harvesting attracts a wide attention at present for the application on power supply of microelectronics [1-2]. Many vibration energy harvesting devices based on different conversion mechanisms have been proposed [3-5]. In order to achieve high power efficiency, the natural frequency of an energy harvester should be designed to match with the environment vibration source [6-7]. Research shows that frequencies of most vibration sources in the environment are below 200 Hz [8]. The existing harvesters with high energy output always work at high frequency while devices with low working frequency cannot achieve high power output. It becomes a great challenge to design an energy harvester with high energy output and low working frequency. The key solutions are to improve the energy output density and energy conversion efficiency of devices. Wang's group has developed a variety of triboelectric nanogenerators based on vertical contact-separation model, sliding model, and single-electrode model. The highest energy density is up to 45 mW/cm^3 [9-11]. Tang et al proposed a low-frequency vibration piezoelectric MEMS generator by bonding bulk PZT on an SOI structure [12]. The power density is 28.9 mW/cm^3 under the excitation acceleration of 1g. To obtain higher energy conversion efficiency at low frequency, Kulah and Najafi [13] presented an electromagnetic power generator based on FUC system, which can scavenge energy efficiently from low-frequency external vibrations. Liu et al [14] also proposed two piezoelectric energy harvesting systems based on FUC mechanism. The peak power density can reach to 159.4 μW/cm^3 at a low operating frequency of 25 Hz.

In this study, we demonstrated a low-frequency hybrid energy harvester with high output performance based on FUC approach. Under an external excitation, the electromagnetic harvesting suspension structure is able to impact with four high-frequency PVDF harvesting cantilevers with large amplitude, the output performance of PVDF cantilevers can be improved significantly. At the same time, the electric voltage across the coil is also generated from the electromagnetic part.

2. Device and working principle

The schematic diagram of the fabricated harvesting device is shown in figure 1, which consists of electromagnetic and piezoelectric parts with general dimensions of 7 cm×7 cm×4.3 cm. The main components are winding coil, permanent magnet, spring structure beam, excitation piece and four pieces of PVDF cantilevers. The spring structure beam is processed from a circular copper with the thickness of 0.8 mm. A permanent magnet with the diameter of 40 mm and height of 4 mm is attached on the one side of the spring structure beam, while excitation piece is attached on the other side. The winding coil of 0.09 mm in diameter and 200 m in total length is tightly wound surrounding the hollow cylinder. Four commercial PVDF cantilevers with the size of 41 mm×16 mm are distributed below the excitation piece. The photograph of hybrid energy harvester is shown in figure 2(a) and the components are shown in figure 2(b), (c), (d) and (e).

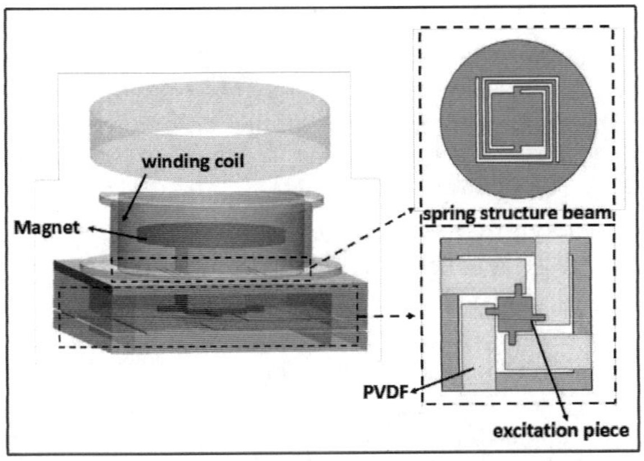

Figure 1. Schematic drawing of the hybrid energy harvester.

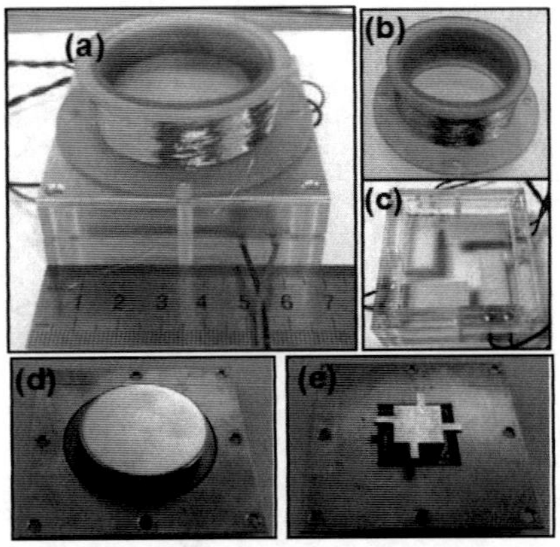

Figure 2. Photograph of the harvester and its components.

Figure 3 shows the sectional view of the device and the vibration state of inner components. As the harvesting device is driven by an external excitation, both the electromagnetic and piezoelectric parts generate electrical power. For electromagnetic part, the working principle is based on law of electromagnetic induction. Under the external excitation, the spring structure beam will drive the permanent magnet to vibrate at vertical direction. Magnetic flux crossing the coil changes and electrical voltage is generated.

As to piezoelectric part, the commercial PVDF cantilevers with high resonant frequency are chosen as energy harvester for the reason that the generated power is theoretically proportional to the cube of the operation frequency [15]. However, the output performances of high-frequency PVDF cantilevers are relatively low under the low driving frequency. In order to improve the power output, an excitation piece connecting with spring is used to impact against the PVDF harvesting cantilevers. As shown in figure 3(b), the excitation piece vibrates in response to external excitation until it impacts with PVDF cantilevers located beneath. As the excitation piece separates from PVDF cantilevers as shown in figure 3(c), the released cantilevers will vibrate damply at their higher resonant frequency with large amplitude before the next impact occurs. Therefore, the AC voltage output of PVDF cantilevers can be improved by an order of magnitude.

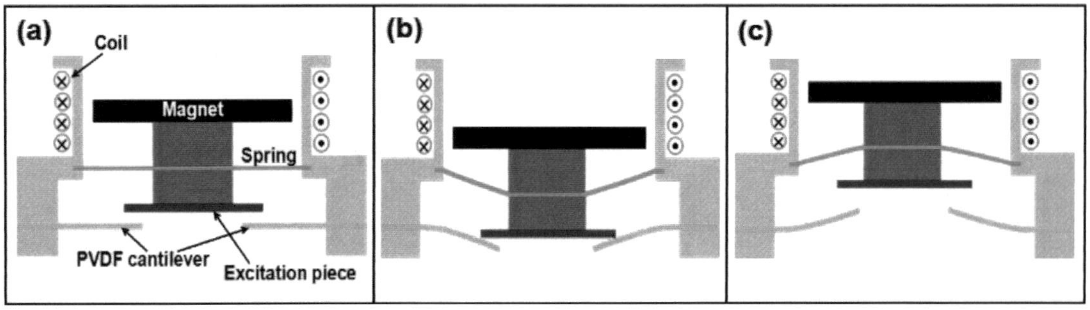

Figure 3. Illustration of the operation process of the energy harvester

3. Experiment and results

The assembled device is tested by a vibration control system. The experimental result in figure 4(a) shows the output performance of electromagnetic harvesting part in frequency domain. For an excitation acceleration of 0.5g, the peak voltage output across the coil can reach to 1.9 V at resonant frequency of 23.6 Hz. The voltage waveform of the electromagnetic part in time domain is also shown in the inset figure of figure 4(a). The output performance of PVDF harvesting cantilever in frequency domain is shown in figure 4(b). The maximum output voltage of a single PVDF cantilever can reach to 3 V with the strike of excitation piece at the excitation frequency of 23.6 Hz (point **A**). After that, the second peak voltage output of 0.9 V will occur when it reaches to the resonant frequency of 60.6 Hz (point **B**). The inset figure of figure 4(b) shows the voltage waveform of the PVDF harvesting cantilever at resonant frequency of 60.6 Hz at point **B**. Figure 5 shows the voltage waveforms of PVDF with (point **A**) and without the strike of excitation piece at excitation frequency of 23.6 Hz. The peak voltage increases from 30 mV to 3 V and the working frequency increase from 23.6 Hz to 60.6 Hz under the strike of excitation piece. It can be easy to get a conclusion that the output performance of PVDF harvesting cantilever is improved significantly by using FUC approach. In addition, there are four cantilevers integrated in the energy harvester and the electromagnetic part will also generate electrical power. Therefore, the total output performance of this device is quite high.

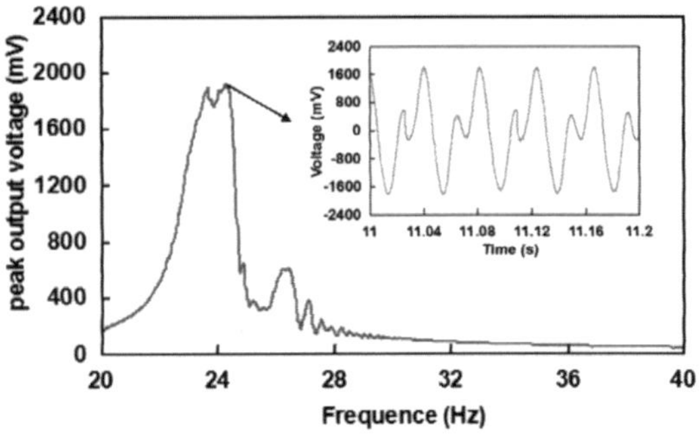

Figure 4. (a) The voltage output of the electromagnetic harvesting structure in frequency and time domain at 0.5g.

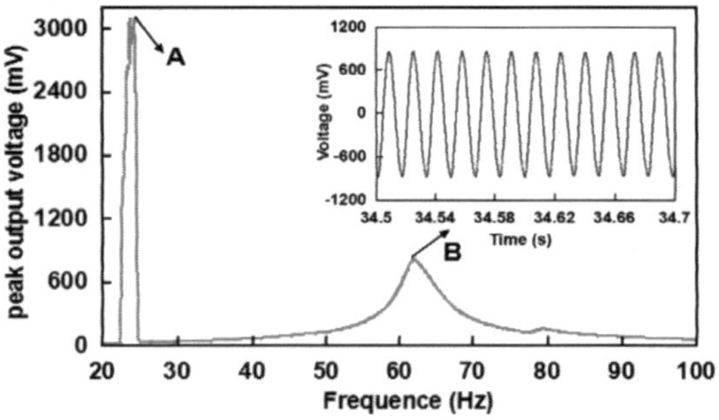

Figure 4. (b) The voltage output of the PVDF harvesting cantilever in frequency and time domain (point **B**) at 0.5g.

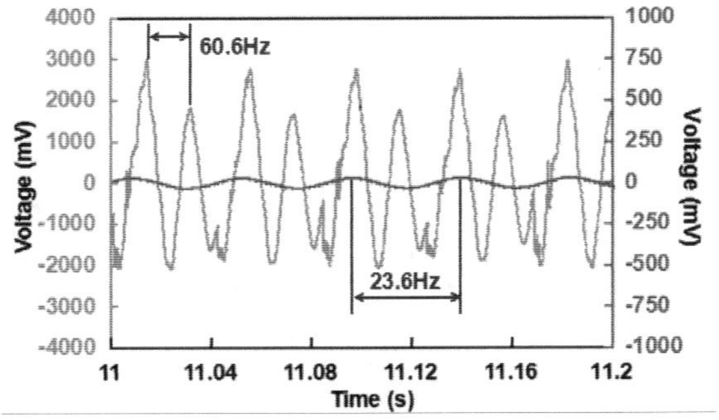

Figure 5. The voltage waveforms of PVDF cantilever with and without the strike of excitation piece at excitation frequency of 23.6 Hz

4. Conclusion

In this paper, we proposed a hybrid energy harvester based on FUC approach, which can achieve high output performance under low excitation frequency. An excitation piece is designed to impact with PVDF cantilevers, resulting in a significant improvement of electrical output. For an excitation acceleration of 0.5g, the peak voltage output of electromagnetic and piezoelectric parts can reach to 1.9 V and 3 V, respectively, at the driving frequency of 23.6 Hz.

References

[1] Beeby S P, Tudor M J and White N M 2006 *Meas. Sci. Technol.* **17** R175-R195.
[2] Zhang X S, Han M D, Meng B and Zhang H X 2015 *Nano Energy* **11** 304-322.
[3] Khaligh A, Zeng P, and Zheng C, 2010 *IEEE Trans. Ind. Electron.* **57** 850-860.
[4] Glynne-Jones P, Tudor M J, Beeby S P and White N M 2004 *Sens. Actuators, A.* **110** 344-349.
[5] Zhang X, Han M, Wang R, Zhu F, Li Z, Wang W and Zhang H 2013 *Nano Lett.* **13** 1168-72.
[6] Williams C B and Yates R B 1996 *Actuat. A-Phys.***52** 8-11.
[7] Dhakar L, Tay F and Lee C 2015 *J. Microelectromech. S.* **24** 91-99.
[8] Roundy S, Wright P K, and Rabaey J 2003*Comput. Commun.* **26** 1131-44.
[9] Bai P , Zhu G, Zhou Y S, W S, Ma J, Zhang G and Wang Z L 2014 *Nano Res.* **7** 990-997.
[10] Tang W, Zhang C, Han C B and Wang Z L 2014 *Adv. Funct. Mater.* **24** 6684-90.
[11] Xie Y, Wang S, Niu S, Lin L, Jing Q, Yang J, Wu Z and Wang Z L 2014 *Adv. Mater.* **26** 6599-6607.
[12] Tang G, Liu J Q, Yang B, Luo J B, Liu H S, Li Y G, Yang C S, He D N, Dao V D, Tanaka K and Sugiyama S 2012 *J. Micromech. Microeng.* **22** 065017.
[13] Kulah H and Najafi K 2008 *IEEE Sens .J.* **8** 261-268.
[14] Liu H, Lee C, Kobayashi T, Tay C J and Quan C 2012 *Sens. Actuators, A.* **186** 242-248.
[15] Beeby S P,Tudor M J and White N M 2006 *Meas. Sci. Technol.* **17** 175-195.

A non-linear 3D printed electromagnetic vibration energy harvester

P Constantinou and S Roy[1]
Micropower Systems and Nanomagnetics Group, Micro-Nano Systems Centre, Tyndall National Institute, Lee Maltings, Dyke Parade, Cork, Ireland

E-mail: saibal.roy@tyndall.ie

Abstract. This paper describes a novel electromagnetic energy harvester that exploits the low flexural modulus of ABS and comprises of a nonlinear mechanism to enhance the generated power and bandwidth. The device is printed using desktop additive manufacturing techniques (3D printing) that use thermoplastics. It has a 'V' spring topology and exhibits a softening spring non-linearity introduced through the magnetic arrangement, which introduces a mono-stable potential well. A model is presented and measurements correspond favourably. The produced prototype generates a peak power of approximately 2.5mW at a frame acceleration of 1g and has a power bandwidth of approximately 1.2→1.5Hz and 3.5→3.9Hz during up and down sweeps respectively. The device has a power density of 0.4mW/cm^3 at a frame acceleration of 1g and a density of 0.04mW/cm^3 from a generated power of 25μW at 0.1g.

1. Introduction

With the advent of the Internet of Things there will be a large quantity of wireless sensor nodes that would be required to operate autonomously. To achieve this, energy harvesting offers an alternative source of energy to traditional battery sources, which have a finite capacity, requiring regular maintenance. This paper presents a topology that can harness mechanical vibrations and is fabricated using a 3D printer exploiting Fused Deposition Modelling (FDM). Additive manufacturing or 3D printing promises to be a disruptive technology and currently, in comparison to traditional (subtractive) manufacturing techniques, it is more cost effective to manufacture a small to medium number of devices and is capable of manufacturing unique and complicated products where traditional techniques cannot. A 3D object is printed as a series of 2D patterns fused on top of each other. Here, a heated plastic is extruded and deposited in thin layers to form the 3D topology sent to the printer. FDM typically uses thermoplastics to construct the object being printed and of the available thermoplastics, ABS (Acrylonitrile Butadiene Styrene) and PLA (Polylactic Acid) are the most common. ABS has a flexural modulus of 2.1GPa and PLA has a higher value of 3.8GPa. However, their material properties can be potentially manipulated with the method of printing. For example, a structure can be printed with a certain fill density or print orientation. For example if an ABS structure is printed in the ZX axis (Cartesian axis) the structure will have a lower flexural modulus (1.65GPa) to when the structure is printed in the XZ axis (2.1GPa) [1]. Better geometric resolution than that achieved using FDM can be achieved using alternative 3D printing technologies, such as stereolithography, however, this may increase the cost of fabrication. Furthermore in [2], where

[1] corresponding author

microsterolithography is used, it is discussed that it is possible to 'tune' the material's mechanical properties such that devices can be developed to operate at certain frequencies.

In [3] a spring is printed using ABS (FDM (assumed)) and is used as part of an electromagnetic energy harvester that can harness vibrations from an insect. The folded beam structure supports a magnet arrangement and between the magnets coils fabricated on PCB are placed. Through a coil, a load power of 0.9mW is reported for a frame excitation of 0.61g (25.8Hz). Alternative 3D printed technologies such as inkjet printing [4] has been used to fabricate a non-linear energy harvester. The system comprises of two PMMA (Polymethyl methacrylate) folded back cantilevers, with their tips facing each other. At the tips magnets are placed such that as they move they actuate each other. A wide (displacement) bandwidth of approximately 24Hz is demonstrated at frequencies under 40Hz. The transduction mechanism is of the form of piezoelectric patches, and a reported total power of 18.45μW at a frequency of 30Hz is demonstrated. Projection microstereolithography has been used to manufacture energy harvesters [2]. Here a UV curable resin (HDDA – Hexanediol diacrylate) was used to manufacture four helical springs arranged in a 2 x 2 pattern, all supporting a single magnetic mass. Each spring had a diameter of 1010μm and a thickness of 163.5μm. The magnetic mass passed through a coil and a generated power of approximately 2.114μW was generated at a frame acceleration of 0.23g at a resonant frequency of 61Hz.

Of late, though some of the literatures (as discussed above) have described topologies that utilise 3D printing to fabricate vibrational energy harvesters (while one utilises a non-linear mechanism in a piezoelectric transduction) this paper describes a novel electromagnetic vibration energy harvester that exploits the low flexural modulus of ABS, in a non-linear configuration, to increase the system bandwidth and power simultaneously at low acceleration and device volume.

2. Topology configuration and model

The topology is as shown in Figure 1. The suspension consists of two ABS printed beams in the form of a 'V'. The base of the 'V' is attached to the base and a support platform is attached to the opposite end. The platform holds two magnets whose magnetization directions (black arrows) are anti-parallel and face two magnets fixed to the frame with the same magnetisation direction. Between the magnets is a coil, attached to the frame and the magnetic flux distribution is shown in figure 2(a).

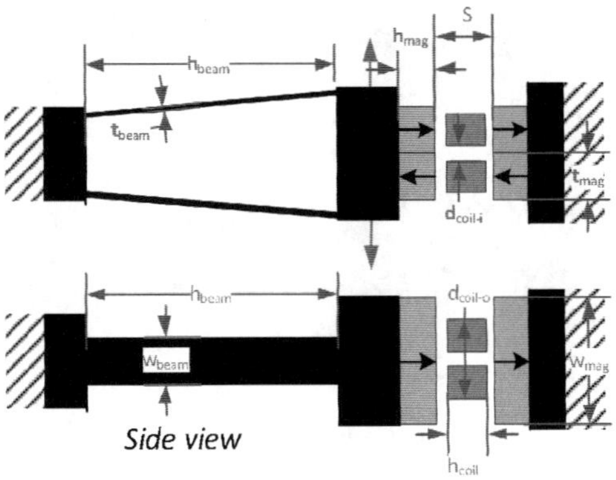

Figure 1: Topology.

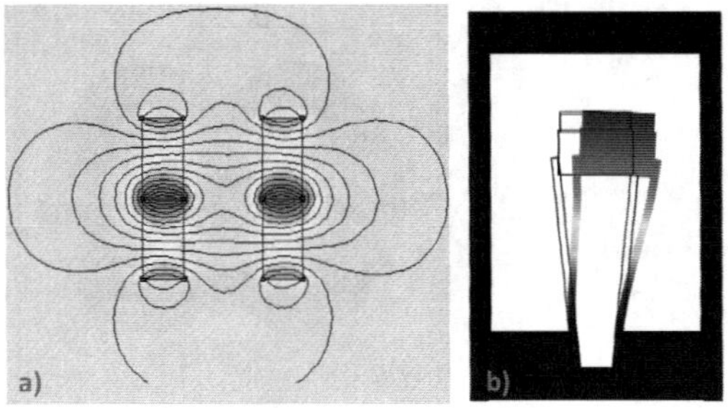

Figure 2 (a) Flux distribution (b) First eigen mode.

When the frame is excited the suspended magnets on the platform move (indicated by red arrows) and the coil experiences a change in magnetic flux linkage resulting in an induced current through a load attached to the coil. As a result of the restoring force (f_{res}), the system will resonate. The system is modelled through:

$$m\ddot{z(t)} + \left(c_m + c_e(z(t))\right)\dot{z(t)} + f_{res}(z(t)) = -m\ddot{y(t)} \;\; \& \;\; c_e(z(t)) = \frac{(k_e(z(t)))^2}{\sqrt{(R_C+R_L)^2+(\omega L_C)^2}} \qquad (1)$$

where m is the equivalent mass of the beam, support and magnets; z is the relative displacement from the centre position (neutral position); c_m is the viscous mechanical damping; c_e is the equivalent electrical damping; k_e is the electromagnetic coupling coefficient; R_L is the load resistance; R_C is the coil resistance; L_C is the coil inductance; ω is the excitation frequency; and y is the frame excitation displacement.

The restoring force comprises both suspension and magnetic components. As a result of the magnetic components the restoring force is non-linear. This is such that one pair of magnets move relatively to the other, which are fixed to the frame. As a consequence the device exhibits the performance of that with a mono-stable potential well.

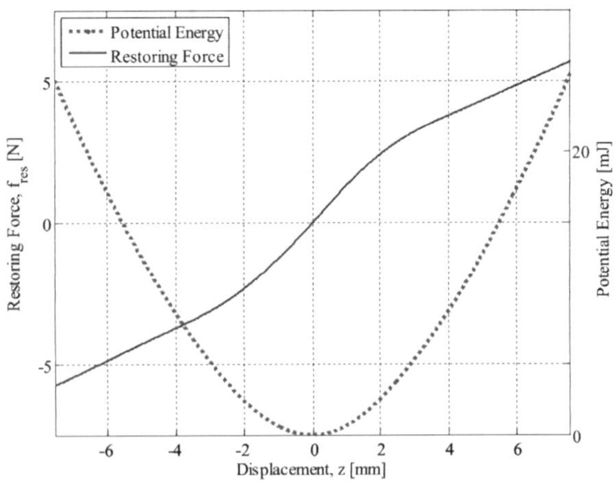

Figure 3: Restoring force and corresponding potential well profiles.

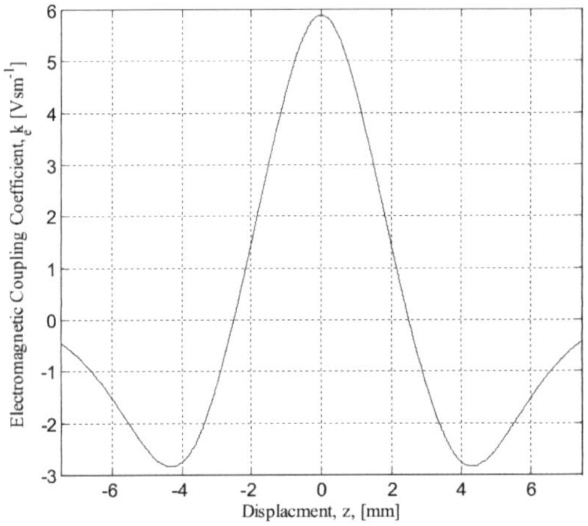

Figure 4: Electromagnetic coupling coefficient.

Table 1: Prototype Parameters [mm].

Property	Symbol	Value
Beam Height	h_{beam}	17.8
Beam Thickness	t_{beam}	0.8
Beam Width	w_{beam}	4
Coil Outer Diameter	$d_{coil\text{-}o}$	6.5
Coil Inner Diameter	$d_{coil\text{-}i}$	1.15
Coil Height	h_{coil}	2
Magnet Width	w_{mag}	8
Magnet Thickness	t_{mag}	4
Magnet Height	h_{mag}	2
Separation between magnets	S	4

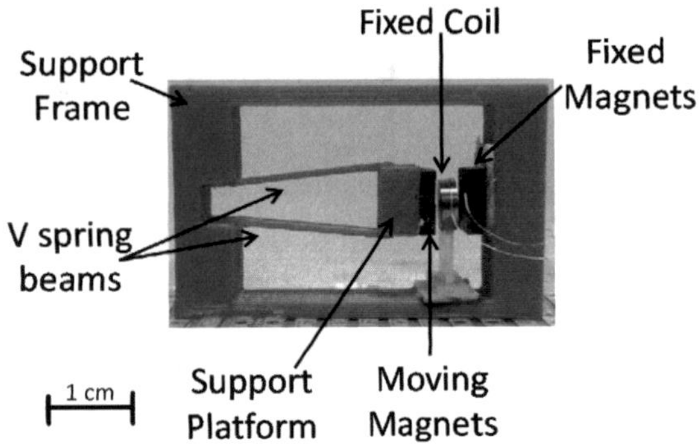

Figure 5: Photograph of prototype.

The linear component of the restoring force is determined, using FEA, where the mode utilised is the first in-plane mode (figure 2(b)). For the prototypes beam thickness an eigen-frequency of 117 Hz is determined using a flexural modulus of 2.1GPa. However due to the FDM process the beam thickness can vary, changing the eigen frequency. Figure 3 shows the restoring force profile as a function of displacement and its corresponding potential well profile.

The electromagnetic coupling coefficient determines how well the electrical and mechanical domains of the system are linked and hence how much energy can be transferred between the domains. It is given by Faraday's law of electromagnetic induction, where the induced voltage, is given by:

$$v_{oc}(t) = -\dot{z}(t)N\frac{d\phi_{avg}(z(t))}{dz} = -\dot{z(t)}k_e(z(t)) \quad (2)$$

where ϕ_{avg} is the average flux through the coil and *N* is the total number of turns in the coil. The corresponding electromagnetic coupling coefficient for the topology is as shown in Figure 4. Here, the system operates around the neutral position ($z = 0$) where the electromagnetic coupling coefficient is at a maximum (and the device is at the bottom of the potential well). A maximum occurs at this point due to the rapid change in magnetic flux direction, due to the anti-parallel configuration.

3. Experimental and model evaluation

The topology shown in Figure 1 (and Figure 5) was fabricated using a Stratasys Mojo 3D printer, which uses FDM technology. The thermoplastic that was used was ABS and NdFeB permanent magnets (N35, 1.17T) were glued to the frame and support platform. A coil ($N = 7060$ and gauge is 25μm, Recoil Ltd., UK) was fixed to the frame and placed between the magnets. The coil had a measured resistance and inductance of 3.287kΩ and 196mH respectively. Further details of the prototype's physical properties can be found in Table 1. The prototype was evaluated on an LDS455 dynamic shaker, as depicted in Figure 6. Measurements were in the form of up and down frequency sweeps in the range of 100Hz to 200Hz, at a rate of 1Hz/s.

The measured and calculated open circuit frequency responses, at a frame acceleration of 1g are shown in Figure 7 below and show that quality factors, *Q*, in the range of $Q = 100 - 120$, yield good correlation between measured and calculated responses. This Q factor range corresponds to an equivalent mechanical damping ratio of $\zeta_m = 0.00417 - 0.005$ as $Q = 1/(2\zeta_m)$. As can be seen from the figure, it is clear that the frequency response is of a softening spring type, and the bandwidth of the device increases between the up and down sweeps and demonstrates hysteresis. The measured peak frequencies (at which maximum occurs) of both the up and down sweeps correspond to that determined by the model and differ slightly, possibly due to the variability in the printed beam thickness. Table 2 below lists both the calculated and measured peak frequencies of the topology at different frame accelerations. As is evident, the peak frequencies decrease with the frame acceleration, however, the downward sweep peak frequency generally occurs at a lower value than that of the upward sweep. The open circuit measured/calculated bandwidths for the up and down sweeps are respectively 2.6/2.4 Hz and 6.5/4Hz.

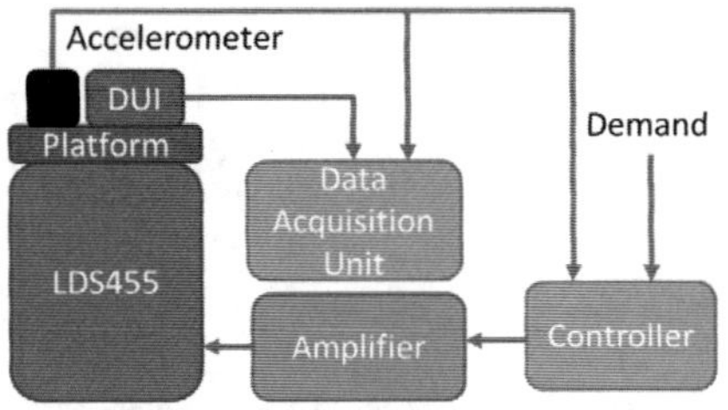

Figure 6: Block diagram of experimental setup (DUI = Device Under Investigation).

Table 2: Open Circuit Resonant Frequency

Frame Acceleration	0.1g	0.5g	1g
Up – calculated	153.3	152.7	151.7-151.8
Up- measured	150	150	149
Down – calculated	152.3-152.4	151.4-151.7	148-149.4
Down- measured	150	149	145

Figure 8 shows the peak power determined from frequency sweeps for both the model and that generated from the prototype at frame acceleration of 1g. Similar curves are seen for frame accelerations of 0.1g and 0.5g. Under these conditions, the generated peak powers for the up and down sweeps (average of up and down values) are: 25μW, 0.7mW and 2.5mW for frame accelerations of 0.1g, 0.5g and 1g respectively. These correspond to power densities of 0.04mW/cm^3, 0.12mW/cm^3, and 0.4mW/cm^3 respectively from a device with an active volume of 6cm^3. The power bandwidth was determined from the measurements as between 1.2→1.5Hz and 3.5→3.9Hz for the up and down sweeps respectively, for the maximum power points in figure 8 below. The corresponding peak frequencies were determined from the measurements as approximately 146-149Hz, with that of the down sweep slightly lower than that of the up sweep. From the simulations it is observed that the peak frequency, at which the generated power frequency response occurs at a maximum changes with the load. At 1g it varies by 1-2Hz (depending on Q factor) and varies by a lower amount (~<0.5Hz) at lower frame accelerations.

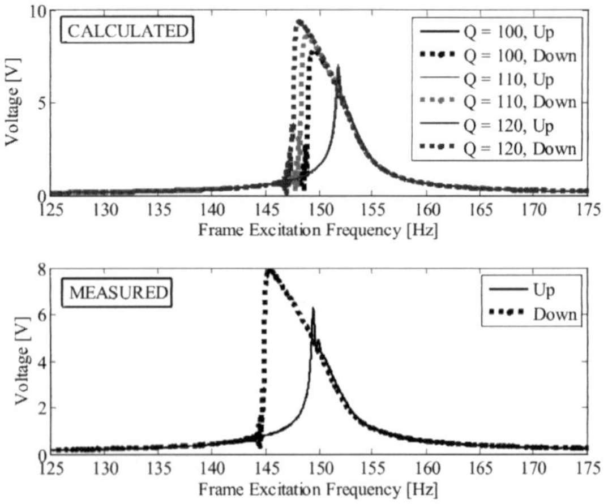

Figure 7: Open Circuit Frequency Response at 1g.

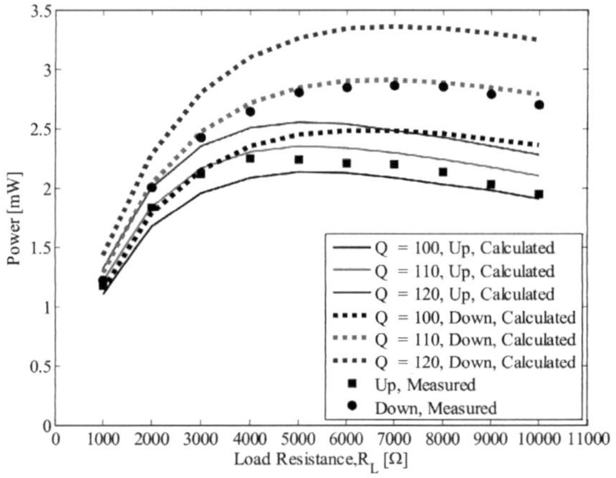

Figure 8: Peak power for different loads at 1g.

4. Summary

This paper reports a non-linear vibration energy harvester that has successfully been fabricated using 'desktop' 3D printing, demonstrating its potential to be exploited with the advent of the Internet of Things. The topology comprises an electromagnetic vibration energy harvester that exploits the low flexural modulus of ABS, in a non-linear configuration, to increase the bandwidth and power at low acceleration (25μW observed at 0.1g) and device volume. It has a mono-stable non-linearity and a softening spring response. The model and measurements are in agreement and demonstrate a peak power of 2.5mW at 1g. The up/down sweep maximum peak power frequencies lie in the range of 146-149Hz and a power bandwidth of 1.2→1.5Hz and 3.5→3.9Hz respectively are observed at 1g. This yields a power density of approximately 0.4mW/cm^3 at 1g.

Acknowledgements

This work is financially supported by Science Foundation Ireland (SFI) Principal Investigator (PI) project on 'Vibrational Energy Harvesting' grant no. SFI-11/PI/1201.

References

[1] Stratasys 2014 ABSplus P430 datasheet.

[2] Baker E, Reissman T, Zhou F, Wang C, Lynch K and Sun C 2012 Microstereolithography of three dimensional polymeric springs for vibration energy harvesting *Smart Materials Research* **2012**

[3] Chang C S, Yaul M F, Dominguez-Garcia A, O'Sullivan F, Otten M D and Lang H J 2009 *Proc. PowerMEMS* Washington DC USA pp. 57-60

[4] Wu X and Lee W D 2015 *Proc. IEEE MEMS* Estiril Portugal pp1110-1113

A Silicon Disk with Sandwiched Piezoelectric Springs for Ultra-low Frequency Energy Harvesting

J. Lu[1], L. Zhang, T. Yamashita, R. Takei, N. Makimoto, T. Kobayashi
Research Center for Ubiquitous MEMS and Micro Engineering (UMEMSME)
National Institute of Advanced Industrial Science and Technology (AIST)
Namiki 1-2-1, Tsukuba, Ibaraki, Japan 305-8564

E-mail: jian-lu@aist.go.jp

Abstract. Exploiting the sporadic availability of energy by energy harvesting devices is an attractive solution to power wireless sensor nodes and many other distributed modules for much longer operation duration and much lower maintenance cost after they are deployed. MEMS energy harvesting devices exhibit unique advantageous of super-compact size, mass productivity, and easy-integration with sensors, actuators and other integrated circuits. However, MEMS vibration energy harvesting devices are rather difficult to be used practically due to their poor response to most of the ambient vibrations at ultra-low frequency range. In this paper, a micromachined silicon disk with sandwiched piezoelectric springs was successfully developed with resonant frequency of 15.36~42.42 Hz and quality factor of 39~55 for energy harvesting. Footprint size of the device was 6 mm × 6 mm, which is less than half of the piezoelectric cantilevers, while the device can scavenge reasonably high power of 0.57 μW at the acceleration of 0.1 g. The evaluation results also suggested that the device was quite sensitive as a sensor for selective monitoring of vibrations at a certain frequency.

1. Introduction

In recent years, the rapid market growth, along with many great challenges, in Internet-of-Things (IoT), wearable healthcare and medical devices, industrial 4.0, and wireless sensor networks (WSN) is becoming one of the major drivers for the R&D of innovative semiconductor devices, various MEMS sensors and actuators, as well as integration techniques [1][2]. The ubiquitous applications of above system may require thousands of sensor nodes, actuators, or other modules to be deployed arbitrarily and operate autonomously for collecting enough data in real-time, cooperatively pass their data through network to a main location, and then execute instruction efficiently. Among the challenges to be coped with, power source to each sensor nodes, actuators, or modules is most essential for high system flexibility and low maintenance cost [3]. The research on power source may create huge new opportunities for manufacturers in near future.

Compared to many other alternative power sources such as thin-film and printed batteries, small flexible photovoltaics panels and thermoelectric modules, MEMS energy harvesting devices [4]-[6] exhibit unique advantageous of super-compact size, mass productivity, and easy-integration with sensors, actuators, and other integrated circuits. Especially to piezoelectric lead zirconate titanate

[1] To whom any correspondence should be addressed.

(PZT) cantilever, although the application fields so far mainly focused on low power devices, it still remains in critical area of interest due to its excellent electrical-mechanical conversion coefficient [7]. PZT cantilever is also believed promising as a self-powered sensor for selectively monitor of vibrations from a machine or infrastructures, if the cantilever is well designed for a certain target frequency [8]. However, since most of the ambient vibrations occur in ultra-low frequency range, i.e. 1-20 Hz for human and animals, <100 Hz for machines, it is rather difficult to obtain an enough high power density and device output from ambient vibrations by PZT cantilever for its poor responses. Increase cantilever's footprint size of up to 1 cm × 1 cm may dramatically reduce its resonant frequency to less than 30 Hz [9-11], but at the cost of process expenses and integration difficulties.

In this paper, a micromachined silicon disk with piezoelectric springs was successfully developed with ultra-low resonant frequency of 15.36~42.42 Hz and quality factor (*Q*-factor) of 39~55. The dependence of resonant frequencies, *Q*-factor, and initial deformation on device structure was investigated both theoretically and experimentally. The results reveal that the device can scavenge reasonably high power of 0.57 μW at the acceleration of 0.1 g, while footprint size of the device was less than half of most piezoelectric cantilevers. Moreover, the potential applications of above devices for not only energy harvester but also self-powered vibration sensor were investigated and discussed herein.

2. Experimental

2.1. Device design and fabrication

Fig.1 shows SEM image of the device. A silicon disk, which is supported by 3 springs symmetrically, was designed as the proof mass with the thickness of up to 400 μm. Sandwiched piezoelectric PZT thin film was fabricated on the springs to harvest energies when the disk was actuated along the out-of-plane direction. The PZT thin film can be also used as self-powered sensing elements to monitor vibrations of a machine or infrastructure. For the pursuit of low resonant frequency, finite elements analysis (FEA) was carried out, and then thickness of the spring and the PZT film was set at 3 μm and 1.9 μm, respectively. Width of the springs was 100 μm or 200 μm, and diameter of the disk was 2 mm, 2.5 mm, or 3 mm to investigate their performances at different resonant frequencies.

Fig. 2 shows flowchart of the fabrication process, which was started by using silicon in insulator (SOI)

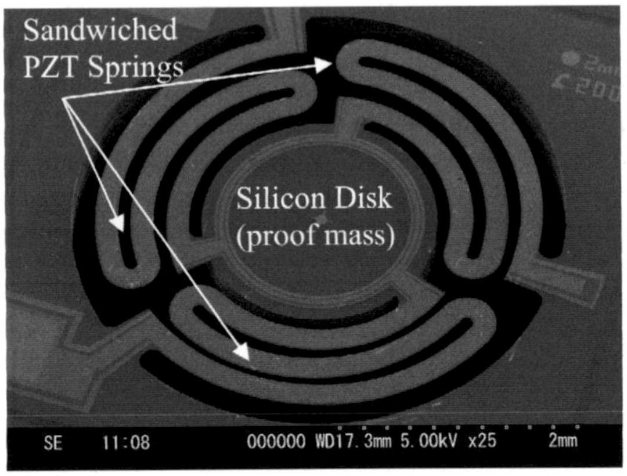

Figure 1. SEM image of the developed silicon disk (diameter: 2 mm) with sandwiched piezoelectric springs (width: 200 μm).

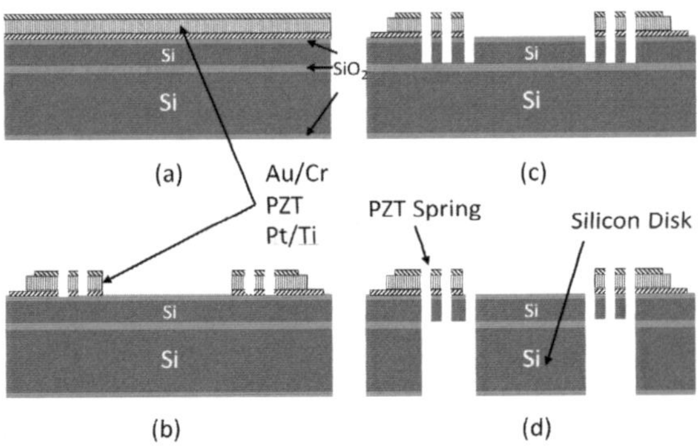

Figure 2. Flowchart of the device fabrication process.

wafer with a 3 μm-thick active silicon layer. The PZT and its electrodes were first deposited on the SOI wafer by sol-gel process and sputter, respectively (fig.2 (a)). Then, the PZT and the electrodes were patterned by wet etching and dry etching in turn (fig.2 (b)), followed by inductively coupled plasma reactive ion etching (ICP-RIE) (MUC-21, Surface Technology System Co., Ltd) of active silicon layer and reactive ion etching (RIE) (RIE-10NRS, SAMCO Co., Ltd) of SiO_2 box layer (fig.2 (c)) to create the springs. Finally, the silicon disk was released from backside of the wafer by ICP-RIE.

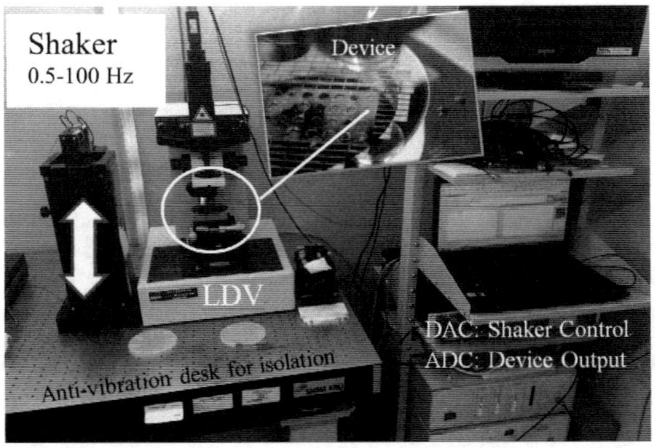

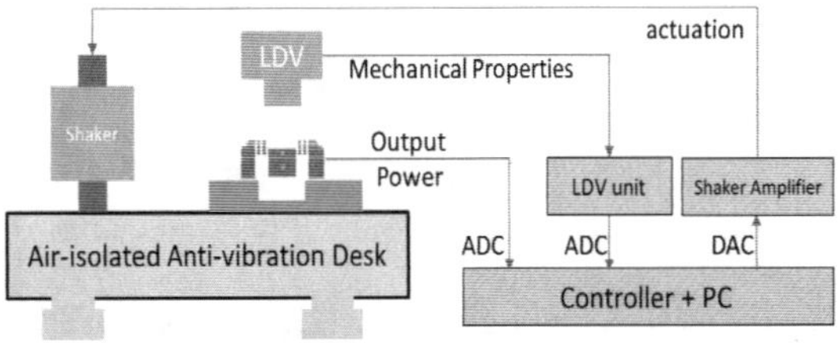

Figure 3. Photo of the evaluation system (left) and its configurations (upper-right).

2.2. Device evaluation

The initial deformation of the device was measured by using a high-precision non-contact depth measuring microscope (DH2-150, UNION Optical Co., Ltd). Then, the device was set on an air-isolated anti-vibration table to avoid any effects from ambient environment. A program controllable mechanical shaker was used for device actuation at a certain frequency and acceleration. Configurations and photo of the evaluation system was shown in Fig. 3. Laser Doppler velocimetry (LDV) (LT7901, Graphtec Co.) was used to evaluate resonant frequencies and *Q*-factor of the device. Electrical output current and voltage of the device was measured by a controller equipped with 24-bits AD convertor.

3. Results and Discussion

The resonant peaks of the device with 100 μm-wide PZT springs and different disk diameters were summarized in Fig. 4 for comparisons. It can be found that ultra-low resonant frequency of 15.36 Hz, which was believed by the disk vibration in out-of-plane direction, was obtained when the disk diameter was 3 mm. *Q*-factor of the disk was calculated as 39, indicating that except for energy harvesting, the device was quite sensitive for the monitoring of vibrations with frequency of less than 20 Hz. In the calculation, *Q*-factor was defined as the ratio of the cantilever resonant frequency to the bandwidth of the resonant peak at -3 dB power.

Fig. 4 also reveals that when the disk diameter was 2.5 mm and 2 mm, resonant frequency of the device was slightly increased to 21.70 Hz and 34.11 Hz, respectively. *Q*-factor of the devices was then increased from 39 to 55 with the decrease of disk diameter. As shown in Fig. 5, the device with 200 μm-wide PZT springs shows very close performances with resonant frequency of 19.0 Hz, 27.19 Hz, and 42.42 Hz when disk diameter was 3 mm, 2.5 mm and 2 mm. *Q*-factor of the devices was increased from 41 to 54 with the decrease of disk diameter. Clearly, the developed device exhibits mechanical characteristics similar to other piezoelectric energy harvesters, but the footprint size was less than half of PZT cantilevers [11] and the volume was less than 10 % of the piezoelectric ceramics [12]. Those issues are favourable for manufacturers to reduce production cost.

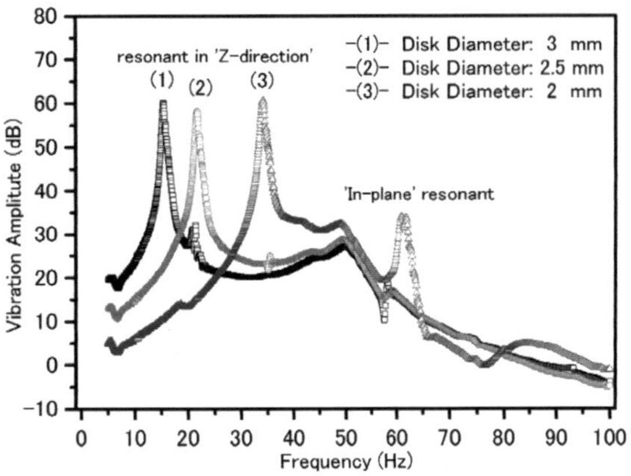

Figure 4. Measured device responses at different frequencies and disk diameters. The width of the piezoelectric spring was 100 μm.

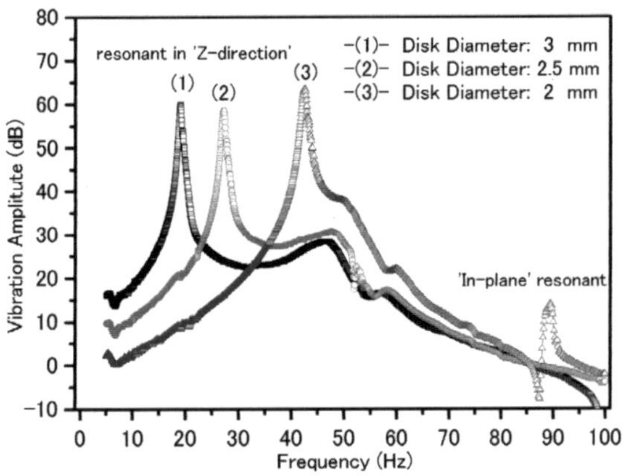

Figure 5. Measured device responses at different frequencies and disk diameters. The width of the piezoelectric spring was 200 μm.

It worth noticing that resonant peak at around 50 Hz can be found from all the devices in both Fig. 4 and Fig. 5. It is believed due to the resonant frequency of the anti-vibration desk. Besides, sub-peaks can be observed in the device with disk diameter of 2 mm, probably due to 'in-plane' vibration because 'smaller' disk is more 'un-stable' by actuation. Although corresponding resonant frequency of those sub-peaks was a bit high (60 Hz when spring width was 100 μm; 90 Hz when spring width was 200 μm), it might be an promising solution to 'flatten' the resonant peak with lower Q-factor for better energy harvesting from ambient vibrations.

Table 1. Summery of the measured mechanical properties of the devices.

Disk Diameter (mm)	2		2.5		3	
Spring Width (μm)	100	200	100	200	100	200
Resonant Frequency (Hz)	34.11	42.42	21.70	27.19	15.36	19.00
Q-factor	55	54	46	50	39	41
Initial Deformation (μm)	208	175	640	413	1525	688

Most of the device suffers from package difficulties and reliability because of the proof mass [11] [13]. As shown in Table 1, initial deformation of the disk was a bit large when disk diameter was 3 mm, but it is acceptable if the ratio of disk size to initial deformation is considered. Table 1 also indicated that the initial deformation can be dramatically reduced by increasing of the spring width or decreasing of the disk diameter, but at the cost of higher resonant frequencies. By applying residual stress in sandwiched PZT springs may greatly reduce the device initial deformation as well as to improve performance of the device. Moreover, it is believed that many other piezoelectric thin films, i.e. aluminium nitride (AlN), can be used as the energy harvesting or sensing elements for a better process compatibility and process flexibility. These works are undergoing and the results will be presented in our future publications.

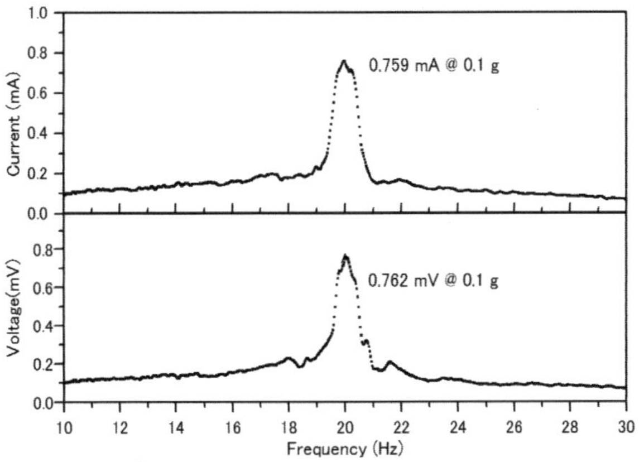

Figure 6. Measured voltage and current output of the device with disk diameter of 3 mm and spring width of 200 μm, when the acceleration was 0.1 g. The peak power was calculated as 0.57 μW.

Fig. 6 shows measured electrical output from the sensor when the acceleration was 0.1 g. The peak power was calculated as 0.57 μW, close to that of PZT cantilever [11]. The output power is expected can be further improved by higher acceleration of up to 0.2~0.3 g, which exists in most of the machines, infrastructures, as well as ambient environments. Fig. 6 also suggested that this device may selective response to a certain frequency, not only as an energy harvester but also as a vibration sensor.

4. Conclusions

In this paper, we have presented a silicon disk with PZT springs for energy harvesting applications. The results demonstrated that ultra-low resonant frequency of < 20 Hz and reasonably high power output of > 0.5 μW can be achieved by the device with footprint size of 6 mm × 6 mm. Further research and device optimization will enable the practical applications of this device in IoT, WSN, etc.

Acknowledgements

This work has been conducted as one of the research topics of the New Energy and Industrial Technology Development Organization (NEDO) project "Utility Core Monitoring System development (UCoMS)".

References

[1] Stankovic J. A. 2014 *IEEE Internet of Things Journal* **1** 3
[2] Patel S, Park H, Bonato P, Chan L, and Rodgers M 2012 *Journal of NeuroEngineering and Rehabilitation* **9** 1
[3] Dewan A, Ay S U, Karim M N, Beyenal H 2014 *Journal of Power Sources* **245** 129
[4] Wang F, Hansen O 2014 *Sensors and Actuators A: Physical* **211** 131
[5] Saadon A, Sidek O 2010 *Energy Conversion and Management* **52** 500
[6] Usman A, Mukhtar M, and Mohammad U R 2014 *Proc. 2014 Int. Conf. Emerging Technologies (Islamabad)* (IEEE) p 142
[7] Caliò R, Rongala U B, Camboni D, Milazzo M, Stefanini C, Petris G, and Oddo C M 2014 *Sensors* **14** 4755
[8] Elvin n G, Lajnef N, Elvin A A 2006 *Smart Mater. Struct.* **15** 977
[9] Liu H C, Tay C J, Quan C, Kobayashi T, and Lee C K 2011 *Journal of Microelectromechanical Systems* **20** 1131
[10] Xie L H, Cai M J 2014 *Appl. Phys. Lett.* **105** 143901
[11] Liu H C, Lee C K, Kobayashi T, Tay C J, and Quan C G 2012 *Microsystem Technologies* **18** 497
[12] Colin M, Basrour S, Rufer L, Bantignies, and Aguyen-Dinh A 2013 *Journal of Physics: Conference Series* **476** 012133
[13] Le C P, and Halvorsen E 2012 *J. Micromech. Microeng.* **22** 074013

A vibration powered wireless mote on the Forth Road Bridge

Yu Jia[1,2], Jize Yan[2], Tao Feng[2], Sijun Du[2], Paul Fidler[2], Kenichi Soga[2], Campbell Middleton[2] and Ashwin A Seshia[2]

[1]Faculty of Science and Engineering, Thornton Science Park, University of Chester, Chester CH2 4NU, UK

[2]Department of Engineering, University of Cambridge, Trumpington Street, Cambridge CB2 1PZ, UK

E-mail: yu.jia.gb@ieee.org

Abstract. The conventional resonant-approaches to scavenge kinetic energy are typically confined to narrow and single-band frequencies. The vibration energy harvester device reported here combines both direct resonance and parametric resonance in order to enhance the power responsiveness towards more efficient harnessing of real-world ambient vibration. A packaged electromagnetic harvester designed to operate in both of these resonant regimes was tested *in situ* on the Forth Road Bridge. In the field-site, the harvester, with an operational volume of ~126 cm^3, was capable of recovering in excess of 1 mW average raw AC power from the traffic-induced vibrations in the lateral bracing structures underneath the bridge deck. The harvester was integrated off-board with a power conditioning circuit and a wireless mote. Duty-cycled wireless transmissions from the vibration-powered mote was successfully sustained by the recovered ambient energy. This limited duration field test provides the initial validation for realising vibration-powered wireless structural health monitoring systems in real world infrastructure, where the vibration profile is both broadband and intermittent.

1. Introduction

Harvesting ambient kinetic energy holds the promise of realising decentralised power generation for the electronic systems at the point of application. Example applications include structural health monitoring of civil infrastructural assets such as bridges, railways and tunnels by using wireless sensor networks (WSN). However, majority of the vibration energy harvesting (VEH) systems reported in the literature are designed for single sinusoidal frequency vibration sources [1], while real vibration environments in these applications tend to be broadband intermittent or of rapidly varying frequency content.

Common frequency broadening techniques in the literature [2] include arraying of multiple harvesters each at a slightly different frequency (at the cost of overall power density), mechanical frequency tuning (actuation power is required), electrical frequency tuning (moderate tuning range) and various other nonlinear vibrational approaches such as Duffing oscillators (moderate broadening) and bi-stable structures (design complexity). At the core, these techniques still involve the direct excitation of a classic linear (or weakly nonlinear) resonator.

The adjustment of the quality factor for a given linear resonator can only maximise either the power peak or the frequency bandwidth. Therefore, a compromise between peak power

Journal of Physics: Conference Series **660** (2015) 012094 doi:10.1088/1742-6596/660/1/012094

and operational frequency bandwidth needs to be made while designing for a specific vibration profile. On the other hand, a parametrically excited resonator has been demonstrated as a viable solution to noticeably enhance both the power output as well as the operational bandwidth due to its fundamentally different instability phenomenon [3].

This paper reports a packaged vibration energy harvester designed to operate in both direct resonant and parametric resonant regimes. The harvester prototype, integrated with a power conditioning circuit, has been demonstrated to successfully power a wireless mote on the Forth Road Bridge.

2. Method and apparatus

A packaged electromagnetic harvester based on the design of a previously reported auto-parametrically excited vibration energy harvester [4] is shown in figure 1. The package volume was approximately 300 cm^3 while the operational volume of the harvester was 126 cm^3. The electromagnetic transducer comprised of a coil with a wire resistance of 4 kΩ and two pairs of neodymium iron boron magnets. The experimentally matched load resistance, at which maximum power can be extracted, was in the range of 4 to 5 kΩ for both direct and parametric resonant peaks.

The resonant power amplitudes of the prototype, when driven into both direct and parametric resonance regimes, compare favourably to a commercial electromagnetic VEH counterpart of comparable size (135 cm^3) [5] as can be seen in figure 2. The parametric resonant response onsets upon attaining an initiation threshold excitation amplitude (<0.2 g_{rms}) and then rapidly outperforms its direct counterpart.

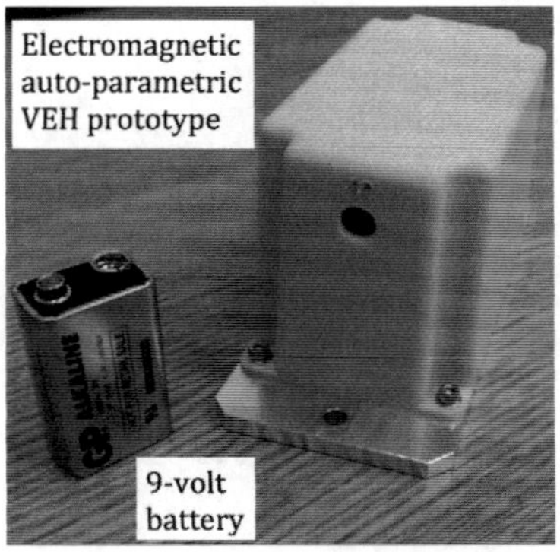

Figure 1: Auto-parametric vibration energy harvester prototype (right) next to a 9 V battery (left).

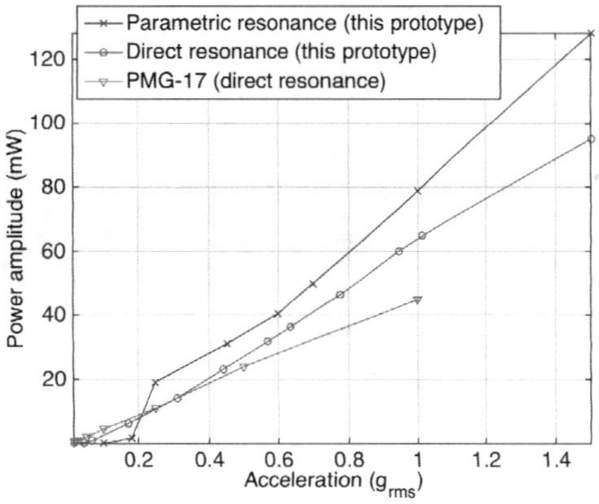

Figure 2: The parametric and direct resonant power output, compared to PMG-17 [5].

The natural frequency is in the vicinity of 13.2 Hz and auto-parametric resonance is observed at approximately twice of this frequency as illustrated in figure 3. It can also be noted that at the same acceleration level in figure 3, the parametric resonant peak is both higher and broader. Table 1 summaries the performance of the reported prototype and commercial harvester. The metrics employed are two of the most widely used in the literature; namely, power density normalised against acceleration squared (N.P.D) [1] and a figure of merit derived from N.P.D normalised against operational bandwidth divided by resonant frequency (F.O.M) [6].

Limited duration *in situ* field-testing of the prototype was carried out at various locations under the deck of the Forth Road Bridge, a suspension bridge in Scotland with a main span of 1006 m and services an average traffic of ~60 thousand vehicles per day [7]. Lateral bracings were identified as the primary points of interest (figure 4), where abundance of traffic-induced

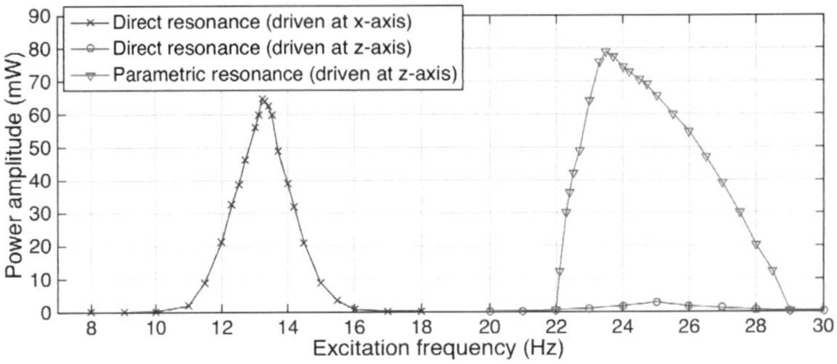

Figure 3: Frequency domain power response when the device is driven at 1 g_{rms}.

Table 1: Comparison of the power performance of the auto-parametric prototype when driven into direct resonance and parametric resonance, and a commercial VEH. All are subjected to 1.0 g_{rms} (13.9 ms^{-2}) of acceleration at their resonant frequencies. N.P.D denotes normalised power density [1] and F.O.M represents figure of merit [6].

Device	Power (mW)	Frequency (Hz)	-3dB band (Hz)	Volume (cm^3)	N.P.D ($\mu Wcm^{-3}m^{-2}s^4$)	F.O.M
Parametric	78.9	23.5	4.5	126	3.24	0.62
Direct	64.8	13.2	2.0	126	2.66	0.40
PMG-17	45	110	2.0	135	1.73	0.03

(a) Forth road bridge

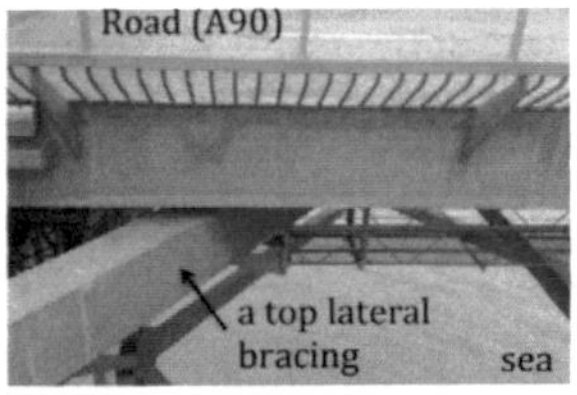

(b) Bracing below deck

Figure 4: A top lateral bracing and cross girder at Forth Road Bridge were used to test VEH-WSN.

vibration is suitable for VEH and stems the motivaton for structural health monitoring of the dynamically stressed structures.

A 6-stage charge pump circuit to amplify the raw AC voltage of the harvester, an off-the-shelf power conditioning circuit (LTC3588-1), a 5 mF storage supercapacitor and an in-house ultra-low power WSN mote (based on Atmel Lightweight Mesh) were integrated with the harvester (figure 5) for the trial test. The mote was programmed a transmission rate of once per minute and consumes 11 μW average power (measured value). No external sensors were used for this particular trial and the vibration powered mote transmits a preset message to a battery powered mote, in order to demonstrate successful transmission. Figure 6 illustrates the attachment (magnetically) of the harvester kit onto one of the locations on the Forth Road Bridge. An acceleration data logger was also used to record the vibration experienced by the harvester, for follow up lab-based characterisation tests.

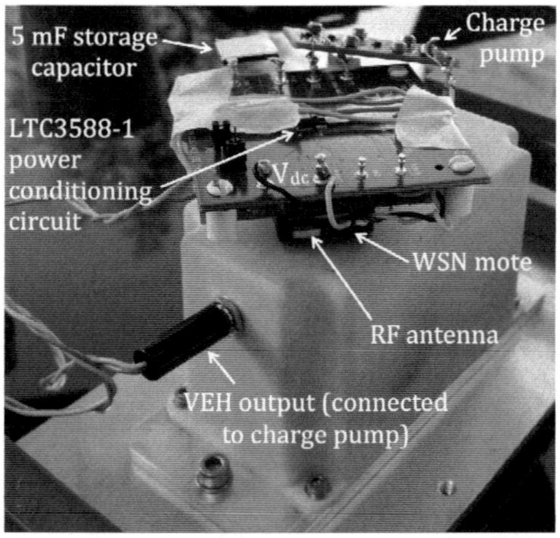

Figure 5: Vibration powered wireless mote system used for the field site trial.

Figure 6: Photograph of the deployment trial testing using the VEH prototype powering a WSN mote. This particular location illustrates testing on the cross girder underneath the deck of the Forth Road Bridge.

3. Results

The harvester produced >1 mW average raw AC power when attached to one of the top lateral bracings at a particular orientation. The eventual conditioned power delivered to the load suffered due to the poor efficiency of the current conditioning and power management circuit. Nonetheless, the conditioned power achievable on the bridge was more than sufficient to sustain the power budget of the wireless mote purely from the intermittent traffic induced vibration energy in the site. Table 2 summarises the power values attained at some of the locations. The location-dependency of power output was due to the localised nature of the vibration.

Table 2: The estimated average power values generated by the harvester prototype at various locations on the Forth Road Bridge. Typical traffic conditions were assumed for the day of measurement. Net power represents remaining average DC power after accounting for the average power consumption of the wireless mote.

Location and orientation	Active frequency range (Hz)	Raw AC power (μW)	Conditioned power (μW)	Net power (μW)
Cross girder vertical	10 to 30	160	32	+20
Top lateral vertical	10 to 30	800	174	+160
Top lateral horizontal	7 to 26	1050	315	+300

The measured vibration data from the field site was used to program a mechanical shaker in the lab in order to experimentally simulate the vibration conditions from the bridge. Reduced-amplitude profile of the measured data was used, coupled with only single axis excitation achievable by the shaker, the simulated bridge vibration produced in the lab was conservative. Figure 7 represents the raw power response and figure 8 demonstrates a transmitting wireless mote powered by the harvester prototype driven with the above-mentioned vibration conditions.

As the voltage across the supercapacitor is charged to 4.0 V, a regulated DC voltage of 2.5 V is supplied to the wireless mote. As can be seen from figure 8, apart from the large initial energy drain required to initialise the wireless network, a steady and continuous rising voltage across the supercapacitor can be seen. Therefore, net power gain was achieved despite the energy drain from wireless transmissions of once per minute (denoted by the red circles).

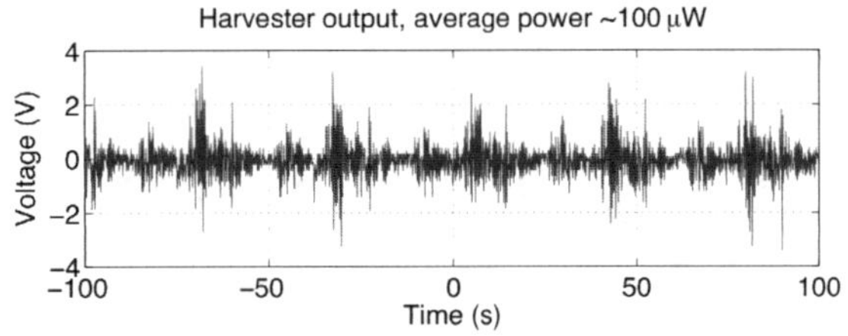

Figure 7: Lab-simulated experimental testing using recorded vibration profile (conservative values) from the Forth Road Bridge.

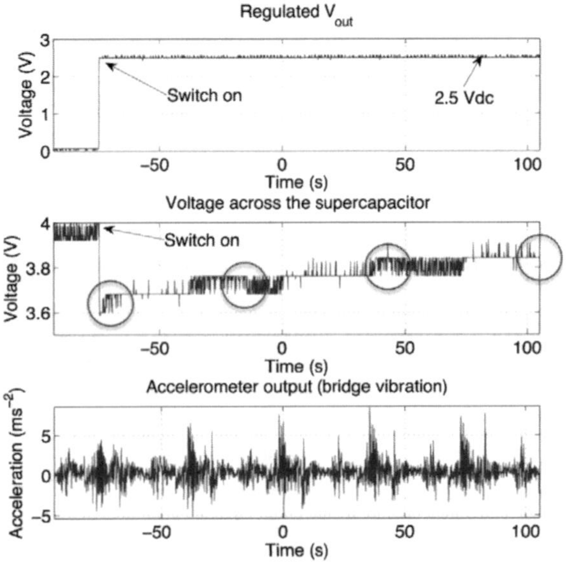

Figure 8: VEH-powered WSN mote using recorded bridge vibration profile driven by a shaker in the lab. Red circles indicate transmission events. Net gain in power can be observed for a transmission rate of once per minute.

4. Conclusion and future work

A packaged auto-parametric vibration energy harvester, integrated with power conditioning circuit and a wireless mote, was demonstrated *in situ* on the Forth Road Bridge. Over 1 mW average AC power was generated at certain locations. Despite the poor efficiency of the current power conditioning circuitry, ~0.3 mW of conditioned DC power delivered to a wireless mote was more than sufficient to successfully sustain wireless transmissions to another battery powered mote (average power consumption of the mote: 11 μW).

Further work involves enhancing the robustness of the harvester prototype, improving the efficiency of the power conditioning circuitry, further minimising the power requirement of the WSN mote and incorporating sensor systems onto the vibration powered mote in order to realise long term deployment trials at field-sites.

Acknowledgement

This work was supported by EPSRC [EP/L010917/1]. We would also like to thank Barry Colford and the Forth Estuary Transport Authority for providing access to the Forth Road Bridge.

References

[1] S Priya and D Inman, 2009, *Energy Harvesting Technologies*, (New York: Springer US)
[2] D. Zhu, M.J. Tudor and S.P. Beeby, 2010, *Meas. Sci. Technol.*, **21**(2)
[3] Y Jia, J Yan, K Soga and A A Seshia, 2014, *Smart Mater. Struct.* **23**
[4] Y Jia and A A Seshia, 2014, *Sens. Actuators A.*, **220**, 69-75
[5] Perpetuum - Products, 2008, PMG-17, URL: `http://perpetuum.com/products.asp`
[6] R Andosca *et al.*, 2012, *Sens. Actuators A.*, **178**, 76-87
[7] Forth Road Bridge, URL: `https://www.forthroadbridge.org/home`

Acoustic energy transmission in cast iron pipelines

Michail E Kiziroglou, David E Boyle, Steven W Wright and Eric M Yeatman
Electrical and Electronic Engineering, Imperial College London, SW7 2AZ, UK

E-mail: m.kiziroglou@imperial.ac.uk

Abstract. In this paper we propose acoustic power transfer as a method for the remote powering of pipeline sensor nodes. A theoretical framework of acoustic power propagation in the ceramic transducers and the metal structures is drawn, based on the Mason equivalent circuit. The effect of mounting on the electrical response of piezoelectric transducers is studied experimentally. Using two identical transducer structures, power transmission of 0.33 mW through a 1 m long, 118 mm diameter cast iron pipe, with 8 mm wall thickness is demonstrated, at 1 V received voltage amplitude. A near-linear relationship between input and output voltage is observed. These results show that it is possible to deliver significant power to sensor nodes through acoustic waves in solid structures. The proposed method may enable the implementation of acoustic - powered wireless sensor nodes for structural and operation monitoring of pipeline infrastructure.

1. Introduction

Monitoring the condition and operation of infrastructure in industrial plant, transportation and distribution networks are important application areas for emerging Internet of Things technologies. This application is important because it can reduce operation and maintenance costs, improve service availability and introduce reliable safety and security mechanisms. It can also enable new services, including centralized industrial control and intelligent control of resource networks. The implementation of such systems is increasingly becoming economically beneficial, for a variety of applications, in terms of utilization, reliability and insurance, but also in environmental and infrastructure security terms.

The development, installation and operation of continuous monitoring systems for infrastructure such as aircraft engines [1], urban trash collection [2], oil and gas pipelines [3] and intelligent transportation systems [4] has progressed rapidly in recent years largely due to the maturity of wireless communications. A challenge in the broad adoption of permanent monitoring installations is powering the sensor nodes involved, especially when deployed in large numbers and/or at inaccessible locations. For this purpose, considerable attention has been attracted to the development of low-power electronics and high-density batteries for wireless sensor nodes. Ultimately, a maintenance-free permanent monitoring system requires energy autonomous wireless nodes, which is achievable by harvesting energy from the local environment [5], using a radioactive power source [6], or wireless power delivery [7, 8]. For the latter, power can be delivered in various ways, the most commonly proposed in recent publications being via inductive coupling and via vibrational waves. Other possible means include electromagnetic radiation, light, heat and low-frequency motion and vibration.

Here, acoustic energy transmission through metal pipes is studied and assessed as a method for powering sensor nodes located on the outer pipe surface. This is relevant to monitoring applications involving pipeline structures, such as gas, oil, water, and sewage networks, and also industrial environments such as plant, production and mining sites.

2. System concept

The objective of the proposed system is to remotely power wireless sensor nodes that are installed on the surface of a pipe, by transmitting a vibration wave along the pipe walls; shown in in Fig. 1. The system consists of a vibration exciter, which induces a transverse, longitudinal or shear vibration wave into the pipe structure. The vibration propagates down the pipe through the pipe walls, the contents and the surrounding material. Each wireless sensor node is equipped with a vibration harvester, which can be inertial or direct-force-based, depending on the installation requirements. The received energy is transduced into electrical form and, subsequently, it is either used directly or accumulated in intermediate storage. In this work the exciter and sensor/harvester are piezoelectric transducers.

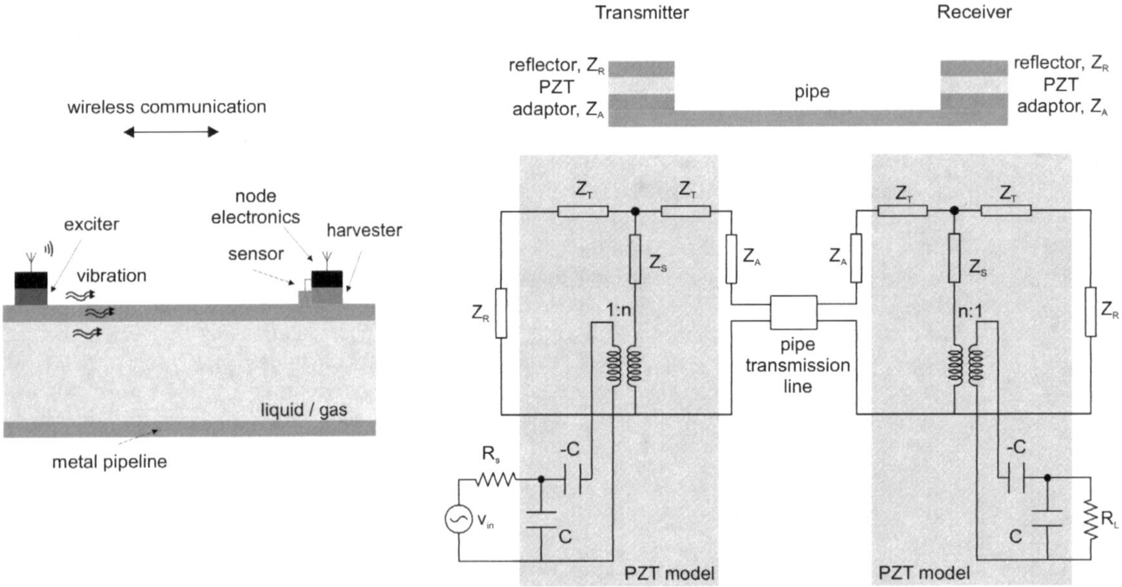

Figure 1: Left: Conceptual description of the proposed powering method. Right: Model of acoustic power transmission based on the Mason equivalent circuit.

3. Theoretical background

Acoustic properties of pipeline structures have been rigorously studied in developing methods for structural characterization, and especially for the non-invasive measurement of inner corrosion in metal pipes [9]. Driven by a large commercial market, this work has been extended to include simulation and analytical tools, providing a strong theoretical and experimental background [10, 11].

In support of the discussion of the experimental results in this paper, an outline of a simple lumped-element modelling method, based on the Mason model [12] for piezoelectric transducers and a T equivalent circuit for acoustic transmission lines is given below. It is noted that in this model, as a first approximation for vibration transmission at low frequencies, the pipe structure is treated as a bulk material. The corresponding lumped-element circuit is shown in Fig. 1 (right).

The case of using identical transducers for the transmitter and the receiver is analysed, as illustrated in the right-top of Fig. 1. Each transducer consists of a PZT crystal with area A, thickness τ, dielectric constant ε, bulk modulus K and density ρ. An adaptor layer is used between the crystal and the pipe, while a reflector layer is used on top, allowing for acoustic impedance matching. The pipe structure is treated as a bulk material, with non-dispersive wave propagation. This means that a constant PZT sound speed $\upsilon_{PZT} = \sqrt{K/\rho}$ can be assumed for all frequency values. The specific acoustic impedance quantity, denoted as Z in Fig. 2, is defined as the ratio of force over velocity, with unit N / (m/s). The Mason model specific impedances, for zero attenuation, are $Z_T = jZ_0\tan(\Gamma\tau/2)$, $Z_S = -jZ_0\csc(\Gamma\tau)$, $Z_0 = \rho \cdot \upsilon \cdot A$ and $\Gamma = \omega/\upsilon_{PZT}$, where ω is the wave angular velocity [12, 13]. The transformer factor is $n = g \cdot C \cdot K$, with unit A / (m/s), where g is the voltage constant of the piezoelectric material, with unit (V/m) / Pa, and C is the capacitance of the PZT device. The reflector and adaptor specific impedances are denoted as Z_R and Z_A respectively. The pipe can be modelled as a T network. A detailed analysis of modelling of PZT devices using the Mason and the KLM equivalent circuits can be found in [12, 13].

4. Experimental results

To evaluate the feasibility of the proposed concept, an experimental system was constructed. A schematic of the setup is shown in Fig. 2 (left). Two identical transducers are used as the transmitter and the receiver, each consisting of an APC International Ltd PZT 850 disk with diameter 48 mm and thickness 7.9 mm, a 60 × 60 × 20 mm aluminium reflector on top and an aluminium adaptor of the same size with a concave side on the bottom. The density, dielectric and voltage constants of the PZT 850 material are 7600 kg/m^3, 1900 and 26.8 mV/m respectively [14]. The two transducers were mounted on a 1.8 m long, Ø 118 mm cast iron pipe, with 8 mm wall thickness, at various distances, using adjustable straps (red in Fig. 2, right).

The electrical impedance of the PZT disk was characterized by I-V measurements, using a 10 V amplitude signal in the range between 1 kHz and 1 MHz. The current was monitored through the voltage drop on a 50 Ω resistance connected in series with the transducer. The response of an un-mounted PZT disk is presented in Fig. 3 (left). Characteristic resonance-anti-resonance peaks that correspond to vibration resonance, expressed by the tangential terms in the Mason model, are observed at various frequencies. In order to identify their origin, the calculated impedance of the PZT is also plotted for comparison, for axial and radial vibration. For this calculation, the model outlined in Section 3 was used, using the nominal values of the PZT devices, with a sound speed of 4600 m/s and a corresponding bulk modulus of 168 GPa. The un-mounted PZT device corresponds to zero forces at its surfaces, and hence short-circuited terminals in the Mason model. The comparison illustrates that the lower frequency resonances correspond to the radial vibration modes of the disk, with $\tau = 46$ mm, corresponding to 196 kHz, while the larger frequency resonance, corresponds to the axial mode, with $\tau = 7.9$ mm, corresponding to 1.19 MHz. For reference, the case of a hardbound PZT, such that no lattice motion is possible at the edges is also plotted in Fig. 3 (left). This corresponds to open circuit conditions in the Mason model, giving an impedance of $1/\omega C$.

The effect of mounting on the electrical impedance of the PZT disk is studied in Fig. 3 (right), where the responses of an un-mounted disk, a disk attached to the reflector and adaptor parts, and a pipe-mounted disk (shown in Fig. 3, right) are compared. As expected, the additional impedance induced by the aluminium brackets reduces the intensity of the PZT disk resonance effects. This serves as an indication of power delivery to the pipe and can be used as a measure of mounting quality, or to identify a successful device mounting during the installation of transmitter and receiver devices.

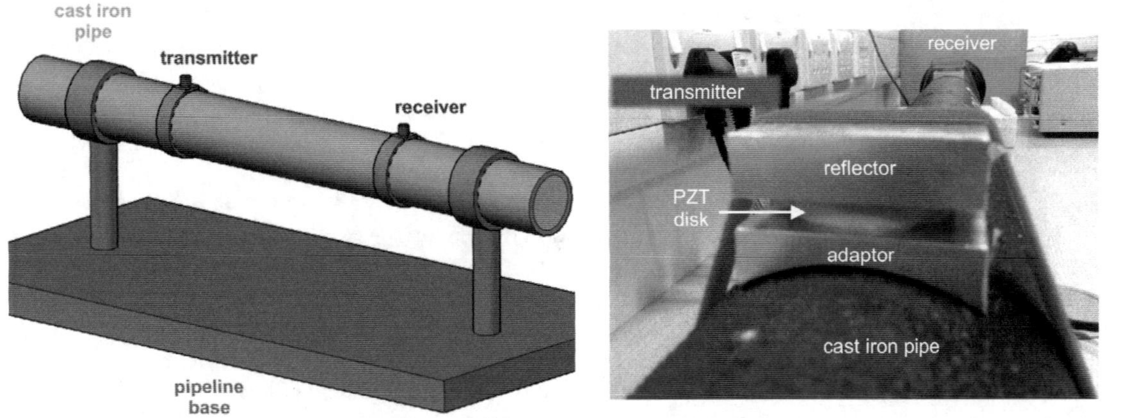

Figure 2: Schematic (left) and image (right) of the acoustic power transmission evaluation setup.

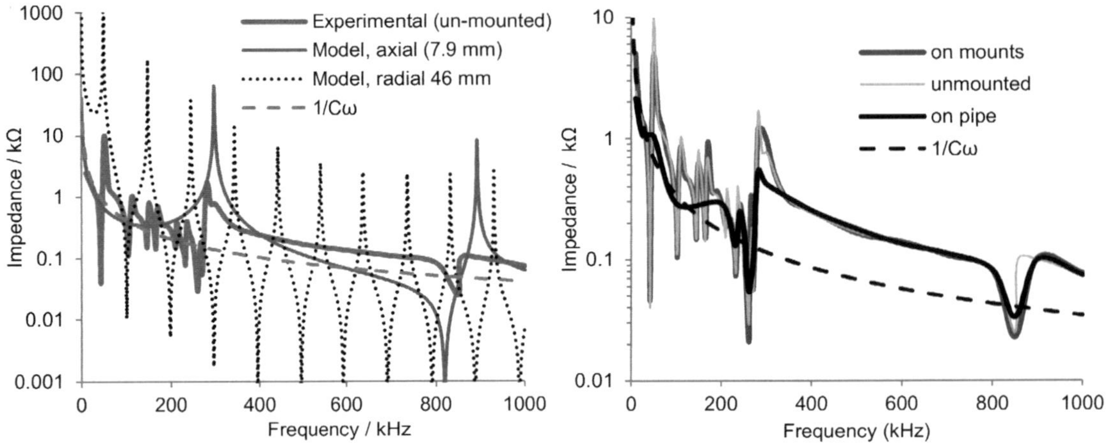

Figure 3: Left: Calculated and measured electrical response of the PZT transducer without mounting. Right: Measured electrical response in different mounting states.

Power transfer experiments were carried out using a function generator in the 1 kHz – 100 kHz frequency range, with a Falco WMA-300 high voltage amplifier to drive the transmitter PZT disk, with voltage amplitude up to 150 V. The distance between the transmitter and the detector was 1 m. A resistive load R_L was connected directly to the receiver PZT output. The optimum R_L value depends on the operation frequency. The level of received power is also frequency dependent. In the particular set up used, peak received power was observed at 48.5 kHz, with $R_L = 1.5$ kΩ. This is demonstrated in the measured transmitted and received signals shown in Fig. 4 (left). The transmission signal is a 150 V, 48.5 kHz sinusoidal signal, pulse-modulated at a frequency of 200 Hz and 25% duty cycle. The amplitude of the sinusoidal voltage amplitude measured on R_L was 1 V, corresponding to a received RMS power of 0.33 mW for a continuous drive signal.

The received voltage was measured as a function of the applied voltage amplitude at the transmitter. The results are shown in Fig. 4, right, both for the open-circuit voltage, and the 1.5 kΩ load. A linear relationship between input and output is observed, at good approximation, with a gradient of 2.5 mV/V. The ratio of output voltage on R_L to the open circuit voltage remains constant in the transmission voltage range between 50 V and 150 V, at around 55%. The corresponding power delivery range that is demonstrated in this experiment is between 20 μW and 100 μW. This is less than the 0.33 mW demonstrated in the experiment of Fig. 4, left, because during the voltage sweep, emphasis was given to consistency of mounting conditions rather than to achieving maximum power. In practice, the 0.33 mW output was demonstrated by pressing down the receiver to achieve better contact. In voltage sweep, this was avoided for consistency. This power delivery level is adequate for the requirements of low power sensor nodes and the corresponding voltage is sufficient to drive commercial power management systems designed for energy harvesting devices.

6. Conclusion

In this paper, a new method for remote powering of sensor nodes installed on structures such as metal pipes was proposed. A simple theoretical analysis framework was drawn and an experimental validation setup was designed and built. Acoustic delivery of 0.33 mW RMS power through a pipeline at a distance of 1 m was demonstrated. This power range may be adequate for various application scenarios, such as corrosion monitoring of pipeline networks, where the data acquisition scheduling permits duty cycled node operation. Optimization of the acoustic mode selection, transmitter and detector designs, and practical mounting, are expected to enable a practical solution to the challenge of powering sensor nodes that are permanently installed in remote locations.

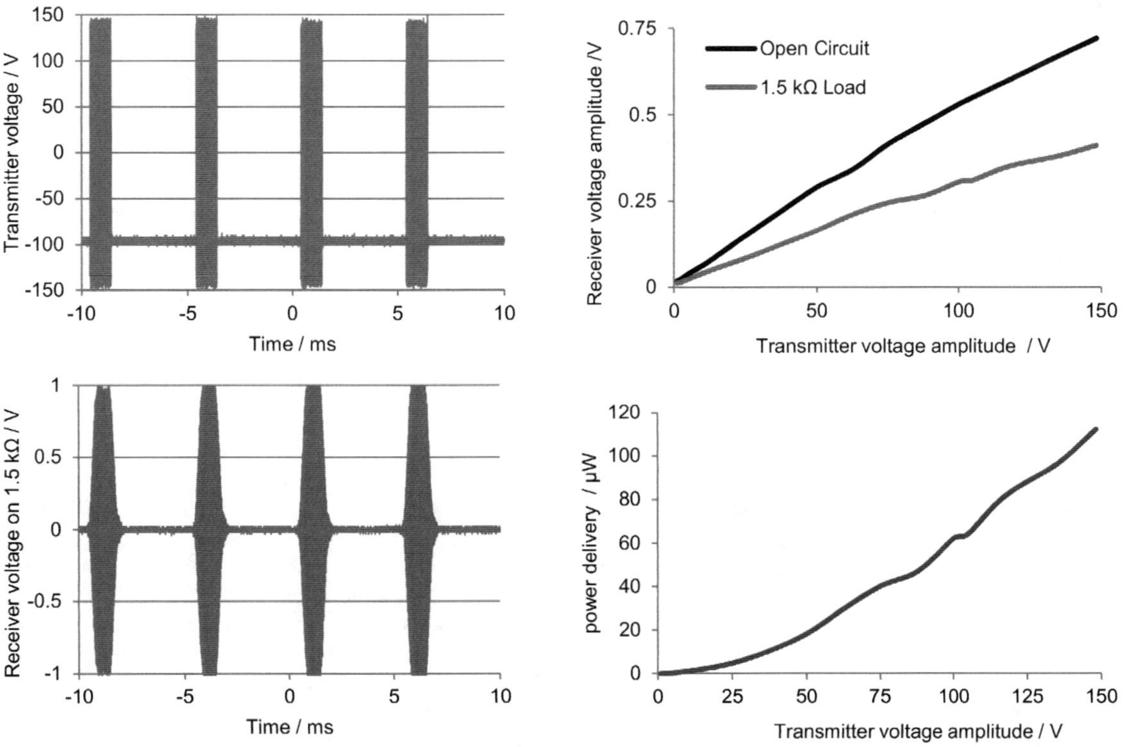

Figure 4. Left: Transmitter and receiver voltage at 48.5 kHz, at a distance of 1 m, demonstrating a received power of 0.33 mW. The transmission is pulsed to show the acoustic delay between the signals. Right: Received vs transmitted voltage (top) and corresponding delivered power (bottom) at 48.5 kHz, at a distance of 1 m.

References

[1] N. Waters, "Engine Health Management," *Ingenia,* 2009.
[2] *Big Belly Solar* 2014 http://www.cityofboston.gov/publicworks/cleanliness/litter.asp
[3] F. Cegla and J. Allin, *Oil and Gas Pipelines*, ed: John Wiley & Sons, Inc., 2015, pp. 571-8.
[4] Barrero F, Guevara J A, Vargas E, Toral S and Vargas M. 2014 *Comp. Stan. Interf.* **36** 300-11
[5] Briand et al 2015 *Micro Energy Harvesting* **12**: John Wiley & Sons.
[6] Olsen L C, Cabauy P and Elkind B J 2012 *Physics Today,* **65** 35-8.
[7] Kurs A, Karalis A, Moffatt R, Joannopoulos J D, Fisher P Soljačić M 2007 *Science,* **317**, 83-6
[8] Denisov A and Yeatman E M, 2010 *Int. Conf. on Body Sensor Networks (BSN)* 84-9.
[9] Silk M G and Bainton K F, 1979, *Ultrasonics* **17** 11-9.
[10] Cawley P, Cegla F and Stone M 2013 *J. Nondestr. Eval.* **32** 156-63
[11] Long R, Lowe M and Cawley P 2003 *Ultrasonics* **41** 509-19
[12] Sherrit S, Leary S P, Dolgin B P and Bar-Cohen Y 1999 *Ultrasonics Symp.* **2** 921-6
[13] O'Donnell et al 1981 *Methods in Experimental Physics*. **19**, Academic Press 29-65
[14] APC materials 2015 http://www.americanpiezo.com/apc-materials/piezoelectric-properties.html

All dispenser printed flexible 3D structured thermoelectric generators

Z Cao, J J Shi, R N Torah, M J Tudor, S P Beeby

Printed Electronics and Materials Lab, Electronics and Computer Science, University of Southampton, Southampton, SO17 1BJ, UK

E-mail: z.cao@soton.ac.uk

Abstract. This work presents a vertically fabricated 3D thermoelectric generator (TEG) by dispenser printing on flexible polyimide substrate. This direct-write technology only involves printing of electrodes, thermoelectric active materials and structure material, which needs no masks to transfer the patterns onto the substrate. The dimension for single thermoelectric element is 2 mm × 2 mm × 0.5 mm while the distance between adjacent cubes is 1.2 mm. The polymer structure layer was used to support the electrodes which are printed to connect the top ends of the thermoelectric material and ensure the flexibility as well. The advantages and the limitations of the dispenser printed 3D TEGs will also be evaluated in this paper. The proposed method is potential to be a low-cost and scalable fabrication solution for TEGs.

1. Introduction

Flexible thermoelectric generators (TEGs) have the potential for use in generating electricity from heat flux to power small devices in wearable systems [1][2]. Flexible TEGs are becoming more attractive for wearable health and environmental monitoring because of the combination of comfort, silent operation ability to supply power over a long time period [3].

The basic module of a TEG is a thermocouple, which consists of two thermolegs, one thermoleg being n-type and the other p-type material. When a heat flux passes through a thermocouple from one p/n junction to another, a voltage and current will be generated with the electrical power dissipated in a load resistor R_L. This arises from the Seebeck effect [4]. In order to fabricate the TEGs onto flexible substrates, dispenser and inkjet printing technologies have been introduced to realise thermocouples on Kapton [5][6]. However, only planar structured thermocouples have been printed with the final TEG being coiled up to form a cylinder that harvests the temperature gradient along its length. These planar structures are less efficient than vertically fabricated thermocouples [7], such as those found in conventional rigid thermoelectric devices.

This work shows the use of the dispenser printing process to fabricate 3D thermoelectric structures on Kapton substrate. Both n and p types of thermocouples have been dispenser printed layer by layer to form the cube structure. A polymer based structure layer was also been printed with the same height to physically support the top electrodes, therefore enabling a fully 3D dispenser printed, vertical TEG to be realised.

2. Experiment

The setup for dispenser printing includes a 3 axis stage (Sigma Koki SGSP26-50), a stage controller (Sigma Koki SHOT-204MS), a dispenser (Musashi ML-808 FX COM-CE) and an air pump (JUN-AIR

OF302). The dispenser printer used in this experiment was a pneumatic type dispenser printer. This dispenser machine can print pastes with a wide range of viscosity (100 – 10000 cP) by varying the nozzle size, air pressure and dispensing time.

For a working TEG the thermocouples should be connected electrically in series, which requires both the top and the bottom of the thermoelectric material cubes to be connected to electrodes. This 3D structure is illustrated in figure 1.

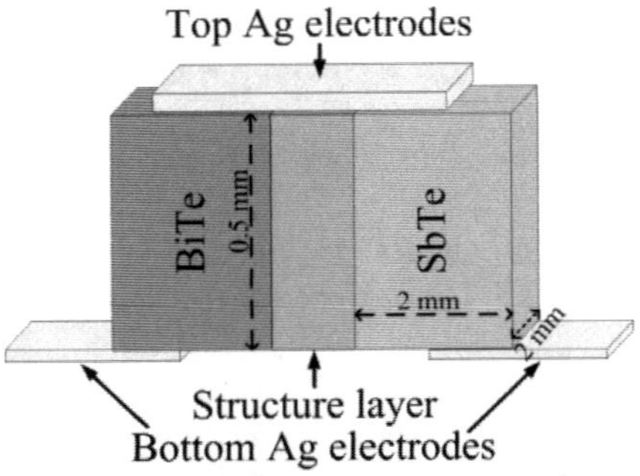

Figure 1. Illustration of a 3D-structured dispenser printed BiTe/SbTe thermocouple.

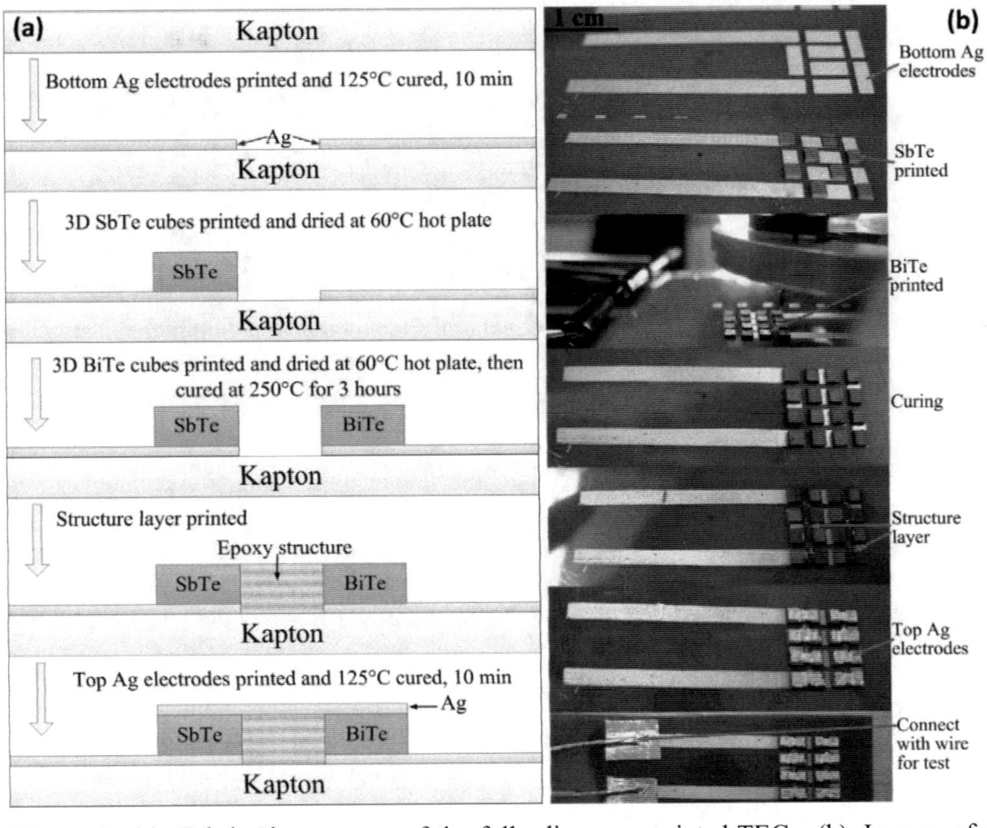

Figure 2. (a): Fabrication process of the fully dispenser printed TEGs; (b): Images of a fully dispenser printed TEGs with 8 thermocouples in different fabrication processes

The thermoelectric materials used here were bismuth telluride (BiTe) and antimony telluride (SbTe) for n-type and p-type thermolegs, respectively. The compositions of the dispenser printable pastes were the same as the screen printable pastes from previous work [8]. The thinner ratios in both pastes were modified to meet the required viscosity level. The bottom silver (Ag) electrodes can be printed onto the substrate, while a structure layer is required in order to support the top electrode layer. The dispensable silver used here was Johnson Matthy s-020 silver polymer conductive ink. Kapton was used as the substrate. The ultraviolet (UV) curable epoxy OG675 from EPO-TEC was used as the structure layer material. This UV curable single component epoxy has a pot life over 3 hours and fast curing time of 2 seconds, which makes it suitable for dispenser printing. Also, a non-thermally cured structure layer minimises the oxidation of the thermoelectric materials.

Firstly, the Ag paste was dispenser printed onto the Kapton substrate and cured. Then, the SbTe and BiTe pastes were dispenser printed onto the Ag electrodes in the required patterns. The stage was heated up to 60°C using a Peltier plate. The printed patterns were dried using the heated stage in order to build up the thickness of the cubes. The dimensions of individual cubes was 2 × 2 × 0.5 mm. The distance between adjacent cubes was 1.2 mm. The thickness of the bottom silver electrodes was 10 μm. The stage controller has a resolution of 1 μm, which meets the requirements of these dimensions. The fabrication processes are illustrated in figure 2-(a). The curing of thermoelectric materials were processed in nitrogen to prevent oxidation. The images of a dispenser printed TEG sample at different stages of the process are shown in figure 2-(b).

In this experiment, 3 samples with the same thermoelectric material cubic size but different number of thermocouples were dispenser printed. The matrix of the cubes were 1×2, 2×2 and 4×4 for sample (a), sample (b) and sample (c) respectively. The flexibility of the samples are shown in figure 3.

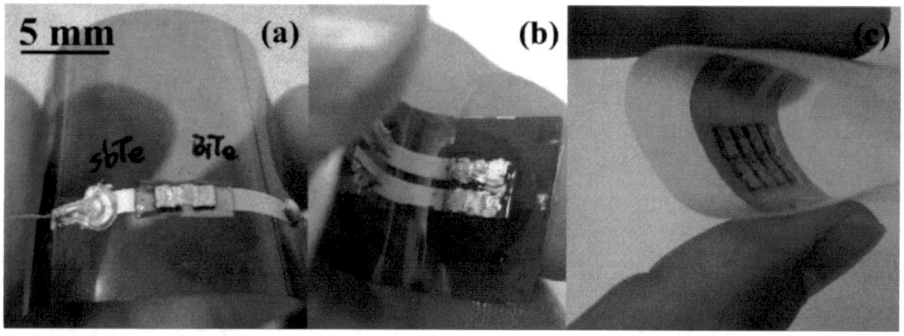

Figure 3. Flexibility demonstration of 3 fully dispenser printed TEGs, sample (a): 1 thermocouple with 1×2 cubes, sample (b): 2 thermocouples with 2×2 cubes, sample (c): 8 thermocouples with 4×4 cubes.

3. Measurement and results

For dispenser printed 3D TEGs, the heat flux must be perpendicular to the substrate. The temperatures on each side of the vertical sample was measured from the hot and cold source it is attached to. In figure 4-(a), aluminium (Al) blocks were used as heat conductors to allow the temperature to be measured, while also supplying a temperature gradient.

The 3D TEGs were placed between two Al blocks. The thermal conductivity of the Al is 205 W/(m·K), which is 5 times higher than the ceramic plates used on the Peltier. The high thermal conductivity ensures a uniform temperature distribution over the surface of the TEGs. Figure 4-(b) and (c) show the temperature distribution over these 2 Al blocks, which were measured by the Testo 875 thermal imaging camera. The probes of the thermometer were put in the middle of the Al blocks by drilling a hole into the centre, and sealing with thermal grease.

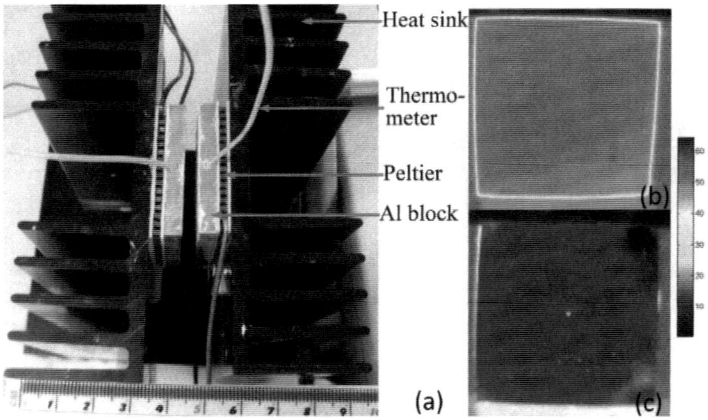

Figure 4. The Seebeck voltage and power output measurement setup (a), and the temperature distribution over the Al blocks, (b): hot side. (c): cold side.

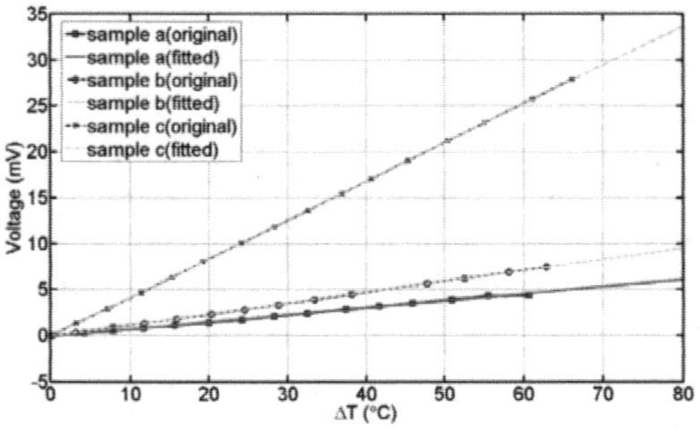

Figure 4. Generated Seebeck voltages of dispenser printed TEGs increase with the temperature gradient.

Table 1. The Seebeck coefficient α, resistance R and power output for single BiTe/SbTe thermocouple

Sample name	(a)	(b)	(c)
α (μV/K) from Seebeck voltage measurement	23.56	59.8	52.9
R (Ω)	844	1260	2276
Pmax (nW) at ΔT=20°C	1.54	0.25	0.14

From the above figure, it can be seen that the output voltage increased linearly with an increase in the temperature gradient. At a temperature gradient of 60°C, a single thermocouple can generate a voltage of 4.5 mV, 3.6 mV and 3.2 mV calculated from sample (a), sample (b) and sample (c) respectively, which are all lower than the 16.6 mV generated by a single planer BiTe/SbTe thermocouple [8]. The actual temperature gradient is difficult to maintain in the experiment setup due to the close proximity of the two Al blocks. In addition, the vertically fabricated device is on a polyimide substrate, which has a low heat conductivity. This will limit the accuracy of the Seebeck voltage and power test.

The calculated α value from both the Seebeck voltage and power measurement along with the resistance of the single thermocouple for these three samples are shown in table 1. The voltage measurement over the load resistor shows the maximum power per single thermocouple among sample a, b and c is 1.5 nW at a temperature difference of 20 °C, which is much lower than the 48 nW for the planar thermocouple [8]. The resistance of a single dispenser printed thermocouple is much higher than the screen-printed planar one (160 Ω in reference [8]).

An experiment was carried out to find which part in the TEG contribute the highest resistance. 8 printed thermocouples with bottom silver electrodes are shown in figure 5, the whole resistance was 480 kΩ. The same BiTe and SbTe pastes, same silver paste and same curing condition were applied in this experiment.

The resistance measured using a multimeter is shown in table 2. The resistance of the junctions contributed 85% of the entire resistance. The resistance was measured using a Wayner Kerr 4300 LCR Meter. The applied AC voltage had a frequency of 1 kHz.

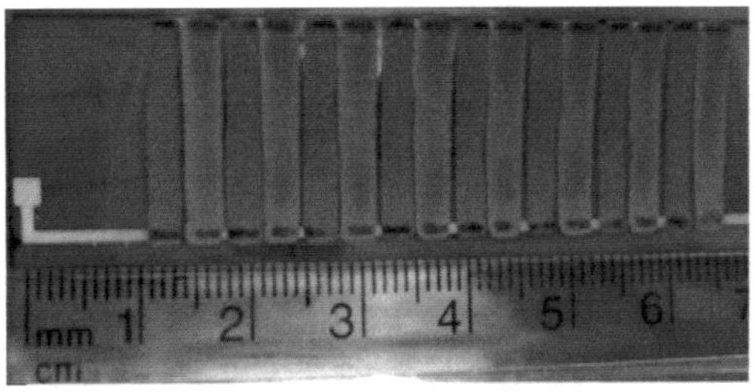

Figure 5. Printed BiTe/SbTe thermocouples on silver electrodes

Table 2. Resistance of different functional parts in the above 8 planar thermocouples

Sample name	NO.s	range	average
BiTe legs	8	1.4 ~ 2.0 kΩ	1.6 kΩ
SbTe legs	8	8 ~ 26 Ω	16.3 Ω
Junctions	15	411 Ω ~ 9.6 kΩ	4.9 kΩ

4. Discussion

In this paper, using dispenser printing technology to print vertical TEGs was investigated. The dispenser printer demonstrated the ability to build up 3D-structured thermocouples. However, the time spent on printing the cubes increased with the number of the cubes. With the desired resolution in this work, printing two cubes in sample (a) took nearly 100 minutes while printing the 16 cubes in sample (c) took over 700 minutes. The printing time can be decreased by applying a dispenser printer with multiple nozzles to build up the thickness of the repeated pattern [9].

The reason behind the dispenser printed samples having a higher resistance than screen-printed samples, comes from the contact resistance of the junction between silver electrodes and the thermoelectric materials. In addition, from figure 5, the overlap areas of the silver patterns and the thermoelectric materials patterns show a different colour from the other parts. A potential reason is that there was a chemical reaction between the silver electrode patterns and the uncured thermoelectric materials during the curing process.

In conclusion, dispenser printing is able to fabricate TEGs with 3D structures. However, the mechanism behind the high contact resistance needs to be further investigated. Evaporating and sputtering could be alternative solutions to deposit the conductive electrodes.

5. Acknowledgment

The authors thank the EPSRC for supporting this research with grant reference EP/I005323/1.

References

[1] Kim S L, Choi K, Tazebay A and Yu C 2014 *ACS Nano* **8** 2377–86
[2] Kim M-K, Kim M-S, Lee S, Kim C and Kim Y-J 2014 *Smart Mater. Struct.* **23** 105002
[3] Misra V, Bozkurt A, Calhoun B, Jackson T, Jur J, Lach J, Lee B, Muth J, Oralkan O, Ozturk M, Trolier-McKinstry S, Vashaee D, Wentzloff D and Zhu Y 2015 *Proc. IEEE* **103** 665–81
[4] Beeby S and White N 2010 *Energy Harvesting for Autonomous Systems* ed S Beeby and N White (Norwood: Artech House)
[5] Chen a, Madan D, Wright P K and Evans J W 2011 *J. Micromechanics Microengineering* **21** 104006
[6] Lu Z, Layani M, Zhao X, Tan L P, Sun T, Fan S, Yan Q, Magdassi S and Hng H H 2014 *Small* **10** 3551–4
[7] Glatz W, Schwyter E, Durrer L, Hierold C and Member S 2009 *Microelectromechanical Syst.* **18** 763–72
[8] Cao Z, Koukharenko E, Tudor M J, Torah R N and Beeby S P 2013 *J. Phys. Conf. Ser.* **476** 012031
[9] Khan A, Rahman K, Kim D S and Choi K H 2012 *J. Mater. Process. Technol.* **212** 700–6

An improved switching control law for the optimized synchronous electric charge extraction circuit

Weiqun Liu[1], Adrien Badel[2], Fabien Formosa[2], Congzhi Liu[1],Guangdi Hu[1]

[1]Department of Vehicle Engineering, School of Mechanical Engineering, Southwest Jiaotong University, China
[2]Laboratoire SYMME, Université Savoie Mont Blanc, Annecy, F74000 France

E-mail: weiqunliu@home.swjtu.edu.cn
Abstract. Nonlinear switching interface circuits are considered as an efficient way to improve the performance of vibration energy harvesters. Among the various approaches, OSECE (Optimized Synchronous Electric Charge Extraction) exhibits satisfying properties: simple switching strategy, good performance in low coupling cases and low load dependency. However, the overdamping induced by the voltage inversion at maximal points leads to performance degeneration in high coupling cases. This paper presents an improved switching control law for the OSECE technique. The new OSECE_PT (OSECE with switching Phase Tuning) technique presented here is to let the switches act ahead or after the maximal point with a phase tuning. Theoretical analysis and numerical simulations show that the OSECE_PT technique can improve the power performance effectively and preserves desired load independence properties.

1. Introduction

Autonomous WSN (Wireless Sensor Network) with the self-regenerated power source is expected in great demand. It allows the WSNs to get rid of the restriction of the battery and prolong the working life to a new record. Vibration energy harvesting is one of the most promising solutions to be embedded in WSN nodes to realize the power regenerating ability [1]. In order to ensure the robustness of the WSN node, the VEH (Vibration Energy Harvester) is required to provide a considerable level of power output under variable environmental vibrations. It contains two aspects: First, the vibration generator part of the VEH has to be wideband [2, 3]; Second, the interface circuit must be capable of extracting the electric power with high performance for different load situations.

Due to the standard circuit's poor performance in low coupling cases and its high load dependence, nonlinear switching interface circuits [4-5] are considered as an efficient way to improve the performance of vibration energy harvesters. Among the various approaches, OSECE presented in figure 1 (a) shows satisfying properties: simple switching strategy, good performance in low coupling cases and low load dependency [6]. However, the overdamping induced by the voltage inversion at maximal points leads to performance degeneration in high coupling cases when the ratio of the extracted energy in a period exceeds the optimal point [7, 8]. A new controller introducing switching pauses has been proposed for the overdamping effect of the SSHI (Synchronous Switching Harvesting on Inductor) circuit with a constant load in high coupling cases [9].

In this paper, an improved switching control law for the OSECE named as OSECE_PT is proposed. Analytical and numerical analysis shows that the OSECE_PT approach can keep the optimal power performance in high coupling cases and low load dependence at the same time.

2. Improved control law for OSECE

2.1. OSECE circuit

Before starting the discussion of the interface circuit, the general electromechanical model of a linear piezoelectric generator can be written as:

$$\begin{cases} M\ddot{u} = F - Ku - \mu\dot{u} - \alpha V \\ I = \alpha\dot{u} - C_0\dot{V} \end{cases} \tag{1}$$

in which M is the inertial mass, F the excitation force, u the displacement, K the stiffness, μ the damping coefficient, α the piezoelectric force factor, V the piezoelectric voltage, I the current out of the piezo element and C_0 the piezo element's capacitance.

When the OSECE circuit is used for extracting the electric energy of the generator's piezo element as shown in figure 1 (a), the operation principle is explicated as follows: As the displacement u and the in-phase piezoelectric voltage V reach the positive maximum point , switch S_1 is closed while S_2 is kept open. A LC oscillation starts and the energy is returned to C_0 with the voltage V inverted. Once V reaches $-V_{DC}/m$ (m: turn ration of the transformer; $V_{DC:}$ load voltage), the remained energy in L_1 goes to the load side. Subsequently, the piezoelectric voltage decreases with u. As V arrives at the negative maximum, the status of the switches are inverted (S_1 open, S_1 closed). A similar process as half period 1 is started.

Due to the presented inverted piezoelectric voltage as an offset for the charge accumulation in the piezoelectric, more power can be harvested in low coupling cases. However, it imposes an overdamping effect and reduces the available power in high coupling cases. This performance degeneration trend is clearly seen from the normalized max power curve versus the coupling figure k^2Q_m (k^2: electromechanical coupling level, Q_m: mechanical quality factor of the generator) in figure 3.

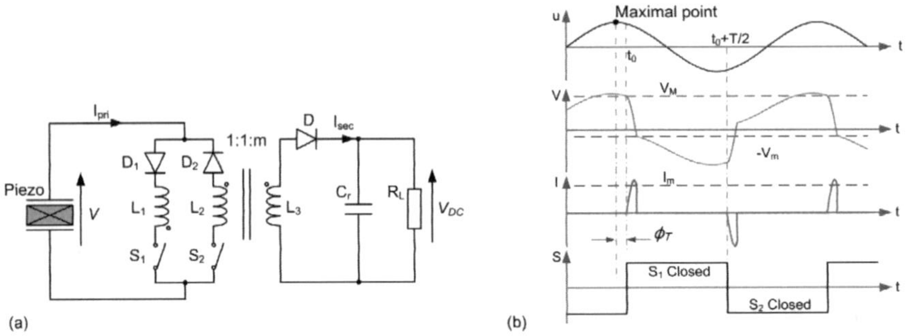

Figure 1. (a) OSECE circuit; (b) Phase tuning operation and waveforms of OSECE_PT circuit.

2.2. OSECE_PT technique

The new OSECE_PT (OSECE with switching Phase Tuning) technique presented here is to let the switches act ahead or after the maximal point with a phase tuning value ϕ_T as indicated by the waveforms in figure 1 (b). During the period between the maximum voltage point and the switching point, the energy stored on the piezo element is returned to dynamic system. Consequently, the energy extracted from the piezoelectric transducer can be controlled by altering ϕ_T.

Following the approach in [6], the extracted energy in each switching operation can be expressed as:

$$E_{extr} = \frac{V_{DC}^2}{R_L}\frac{\pi}{\omega} = \frac{1}{2}(C_0 V_M^2 \sin^2\phi_0)e^{-\frac{\phi_0}{Q_l}} \tag{2}$$

where ϕ_0 is the end phase of the LC oscillation, Q_l is the quality factor of LC circuit, ω is the excitation frequency and other variables defined in figure 1. We also have:

$$V_m = \frac{V_{DC}}{m} = V_M \cos\phi_0 e^{-\frac{\phi_0}{Q_I}} \tag{3}$$

$$V_M = V_m + \frac{2\alpha u_m}{C_0}\cos\phi_T \tag{4}$$

$$\int_{t_0}^{t_0+\frac{T}{2}} F\dot{u}dt = \mu \int_{t_0}^{t_0+\frac{T}{2}} \dot{u}^2 dt + \frac{1}{2}C_0(V_M^2 - V_m^2) \tag{5}$$

Eq. (4) is obtained by integrating the electric part of eq. (1) for a half period between two switching actions (*I=0*) while eq. (5) is from energy balance relationship. Using eqs.(2)-(5), we can write the power expression as:

$$P_{OSECE\_PT} = \frac{2(\alpha\cos\phi_T)^2}{\pi C_0\omega}\frac{\sin^2\phi_0 e^{-\frac{\phi_0}{Q_I}}}{\left[1+\cos\phi_0 e^{-\frac{\phi_0}{2Q_I}}\right]^2}\frac{F_M^2}{\left[\mu + \frac{4(\alpha\cos\phi_T)^2}{\pi C_0\omega}\frac{1-\cos\phi_0 e^{-\frac{\phi_0}{2Q_I}}}{1+\cos\phi_0 e^{-\frac{\phi_0}{2Q_I}}}\right]^2} \tag{6}$$

F_M is the amplitude of the excitation force. When normalized by the maximum available power $P_{max}=F_M^2/8\mu$ [10], we have the normalized expression:

$$\bar{P}_{OSECE\_PT} = \frac{\sin^2\phi_0 e^{-\frac{\phi_0}{Q_I}}}{\left[1+\cos\phi_0 e^{-\frac{\phi_0}{2Q_I}}\right]^2}\frac{\pi k^2 Q_m \cos^2\phi_T}{\left[\frac{\pi}{4} + k^2 Q_m\cos^2\phi_T\frac{1-\cos\phi_0 e^{-\frac{\phi_0}{2Q_I}}}{1+\cos\phi_0 e^{-\frac{\phi_0}{2Q_I}}}\right]^2} \tag{7}$$

It is interesting to find that k^2Q_m is multiplied by an additional coefficient $cos^2\phi_T$ when comparing with the normalized expression of OSECE in [6]. It is inferred that tuning the switching phase ahead or after the original maximum voltage (displacement) amount to decrease the coupling level k^2 by a factor of $cos^2\phi_T$. In this way, the figure k^2Q_m can be controlled at the optimal level avoiding to overdamp the system responses.

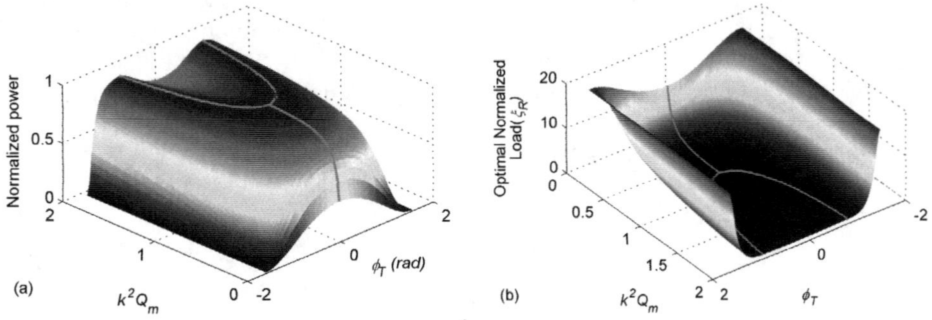

Figure 2. (a) Normalized maximal power versus k^2Q_m and ϕ_T with optimal power points highlighted in pink; (b) Corresponding optimal normalized load.

The power performance of the OSECE_PT is plotted in figure 2 (a) with the optimal load selected ($Q_I=15$). The optimal power points for different k^2Q_m are highlighted by the pink line. Clearly, when k^2Q_m is less than the optimal level $\pi/4$, the optimal switching phase is always at zero. In this case, the OSECE_PT technique is identical to the OSECE. when k^2Q_m approaches $\pi/4$, there are two available optimal switching phases which are symmetrical about zero. It hints that tuning ahead or after the maximum voltage point gives exactly the same results. The optimal normalized load resistance

($\xi_R=2R_L\omega C_0/\pi$) is also plotted versus k^2Q_m and ϕ_T in figure 2 (b) with the values corresponding to the optimal power points marked by the pink lines as well. The load resistance goes small as k^2Q_m increases and approaches to zero with $k^2Q_m>\pi/4$.

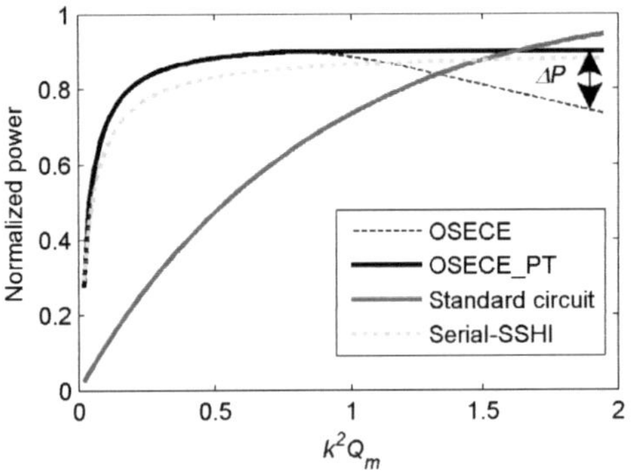

Figure 3. Normalized max power comparison between OSECE_PT and other approaches .

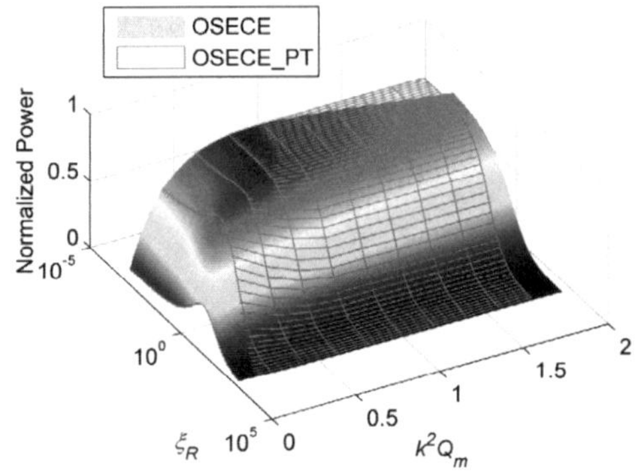

Figure 4. Normalized power versus different load and coupling cases.

Figure 3 presents the optimal normalized power for several often used approaches. The OSECE_PT approach has the best performance for low k^2Q_m values as the OSECE technique and maintains the best performance for high k^2Q_m values. A performance boost of ΔP is observed when compared with OSECE. Moreover, the load dependence of OSECE_PT is also investigated as shown in figure 4 with OSECE for comparison. Obviously, the desired weak load dependence is preserved.

3. Numerical validation

3.1. LTSpice Model

In order to validate the performance of the OSECE_PT technique, an approach of utilizing the equivalent network instead of the mechanical dynamic system is used. Consequently, the harvester system composed of a linear piezoelectric generator and the OSECE_PT circuit can be represented by the Spice model in figure 5. The switching control signal generation part which provides a square waveform with a phase shift to the piezoelectric voltage is not shown here. With the model realized in the LTspice© software, numerical simulations are performed for different k^2Q_m and load resistance cases (turn ratio m=4). In order to simplify the analysis, the diode in the circuit is assumed as an ideal unidirectional component with zero forward voltage and very small conducting resistance.

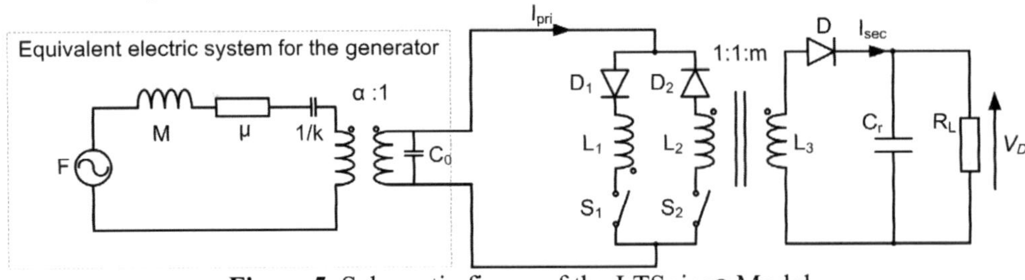

Figure 5. Schematic figure of the LTSpice© Model.

3.2. Results

By varying the switching phase and the load, simulations are repeated to search the optimal power points and the corresponding phase and load. Figure 6 (a) presents the maximum normalized power obtained from the simulation. The numerical results are in good agreement with the theory. The deviation is due to the inadequate simulation time and the decreased Q_I in the low coupling cases. As can be seen, the performance is stabilized around the optimal value after the coupling is greater than a certain value. The optimal switching phase is also obtained with only the phase-lag (ϕ_T>0) branch considered in figure 6 (b). The numerical results confirms the theory well again.

With the optimal phase determined, the load dependence of OSECE_PT is studied by changing the load in a wide range. Figure 6 (c) shows the available normalized power for different cases. As predicted in theory, the weak load dependence is observed. Moreover, the load independence can be further improved by increasing the turn ratio of the transformer in the circuit [6].

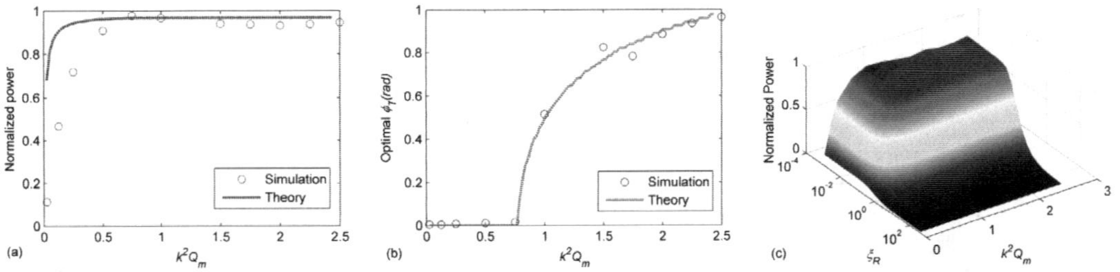

Figure 6. Numerical simulation results for the OSECE_PT technique: (a) Maximum normalized power; (b) Optimal switching phase ϕ_T; (c) Normalized power for different load and k^2Q_m.

4. Conclusion

In this paper, an improved switching control law for the OSECE technique is presented. It is to let the switches act ahead or after the maximal point with a phase tuning. Theoretical analysis and numerical simulations show that the OSECE_PT technique can improve the power performance effectively and preserves desired load independence properties. This is an outstanding advantage over the standard circuit, especially in the wideband energy harvesting applications where the load matching is difficult to achieve. Another important advantage is the expected bandwidth widening effect as presented in [11] for SECE circuit. It is understood that the OSECE_PT technique can greatly extend the application area to all coupling level cases while the tuning realization is easily feasible with an autonomous circuit as shown in [12].

References

The work is supported by the NSFC (National Science Foundation of China) under grant No.51505395, the key basic research project of Sichuan province in China under grant No. VQ21SS1102Y15006 and the Fundamental Research Funds for the Central Universities of China (No. A0920502051514-5).

References

[1] Beeby S P, Tudor M J and White N M 2006 *Meas. Sci. Technol.* 17(12) R175-R195.
[2] Liu W Q, Badel A, Formosa F, Wu Y and Agbossou A 2013 *Smart Mater. Struct.* 22(3) 035013.
[3] Liu W, Formosa F, Badel A, Wu Y and Agbossou A 2014 *Sensor. Actuat. A-Phys.* 216 106-105.
[4] Guyomar D and Lallart M *Micromachines* 2(2), 274–294 ,2011.
[5] Lefeuvre E, Lallart M, Richard C and Guyomar D Piezoelectric Ceramics (Sciyo, 2010) Chapter 9 165–184.
[6] Wu Y, Badel A, Formosa F, Liu W and Agbossou A E 2013 *J. Intel. Mat. Syst. Str.* 24(12) 1445-1458.
[7] Liu W Q, Badel A, Formosa F, Wu Y and Agbossou A 2013 *Smart Mater. Struct.* 22(12) 125038.
[8] Liu W Q, Badel A, Formosa F, Wu Y and Agbossou A 2013 *Journal of Physics: Conference Series* 476 (Proc. PowerMEMS 2014) 012115.
[9] Makihara K, Yamamoto Y, C. Horiguchi, H. Sakaguchi and K. Fujimoto, *Journal of Physics: Conference Series,* 557 (Proc. PowerMEMS 2014) 012023.
[10] Lefeuvre E, Badel A, Richard C, Petit L and Guyomar D 2006 *Sensor. Actuat. A-Phys.* 126(2) 405-416.
[11] Badel A and Lefeuvre E, *Journal of Physics: Conference Series,* 557 (Proc. PowerMEMS 2014) 012115
[12] Boisseau S, Gasnier P, Gallardo M and Despesse G *Journal of Physics: Conference Series* 476 (Proc. PowerMEMS 2013) 012080.

Analysis of Magnetic Plucking Configurations for Frequency Up-Converting Harvesters

T Xue and S Roundy
University of Utah, Department of Mechanical Engineering, 1495 E 100 S, Salt Lake City, UT 84112, USA

E-mail: tiancheng.xue@utah.edu

Abstract. Magnetic plucking applies the strategy of frequency up-conversion in inertial energy harvesting when the energy source, such as human motion, only provides excitations with very low and irregular frequencies. This paper presents an analysis of three different magnet configurations to achieve magnetic plucking based on a three-dimensional analytical cube permanent magnet model: direct repulsive configuration, orthogonal configuration and indirect repulsive configuration. Simulation and experimental results indicate that the indirect repulsive configuration generates the largest tip displacement given the pratical constraints in designing a wearable energy harvester. We have implemented this configuration in a wrist-worn rotational energy harvester to pluck multiple piezoelectric beams. Other configurations, however, can potentially be advantageous in applications with alternative constraints.

1. Introduction

Energy harvesting for wearable wellness sensors could provide the potential for continuous health monitoring by eliminating the need to replace or recharge batteries manually. The inherent limitation of utilizing human motion as the source for inertia energy harvesting is that it only provides excitations with very low and irregular frequencies. Frequency up-conversion is a commonly used strategy to tackle this issue by transforming the low-frequency input motion into high-frequency resonance of the transducer. Plucking is one technique which applies such a strategy. Eccentric rotor-based wearable energy harvesters have been demonstrated in previous endeavors either using magnets [1][2] or pins [3][4] to pluck piezoelectric beams. Magnetic coupling provides better reliability since it can be designed contact-free. Usually magnets are arranged to pluck the beam in the direction of magnet motion with either a direct repelling or an attractive configuration [5], i.e. in-plane plucking. While the in-plane plucking introduces a stronger jump phenomenon in the cantilever beam, the out-of-plane deflection configuration provides the capability of fabricating multiple beams in the harvester on a single substrate. Furthermore, mechanical design complexity is reduced and space utilization is increased which ultimately improves power output. However, if the magnets are simply aligned, as in the direct repulsive configuration in Figure 1, the overall thickness of the harvester will grow. Additionally, in this configuration the opposing magnets can get stuck in a side by side orientation, i.e. there is a pull-in effect. Thus exploring alternative magnet configurations with the ability to trigger out-of-plane plucking as well is worthwhile.

Among all the possible combination of orientations, we present two alternative magnet configurations: the orthogonal and the indirect repulsive configuration, which not only have demonstrated the capability of triggering plucking but also are free of the pull-in effect. As shown in

Figure 1, the direct repulsive configuration is listed alongside for later comparison as we have applied this orientation in our previous harvester prototyping with off-the-shelf components [2]. This paper will introduce the modelling, experimental validation and finally an implementation of the alternative magnetic configuration in a frequency up-converting energy harvester.

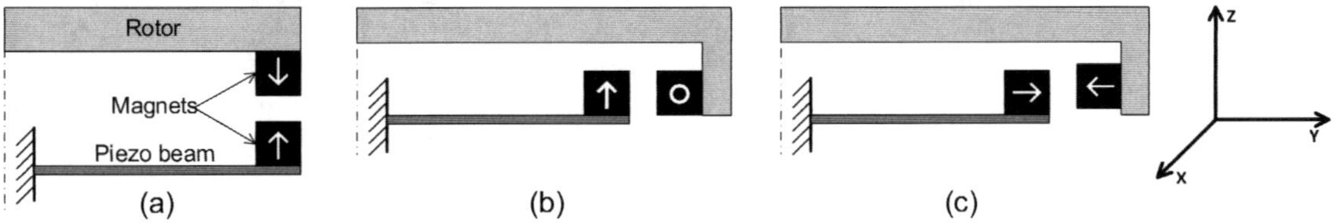

Figure 1. Magnet configurations: (a) Direct Repulsive Configuration, (b) Orthogonal Configuration, (c) Indirect Repulsive Configuration.

2. Modelling of the Magnetic Interaction

The magnetic interaction model is a vital component in the coupled system model to accurately predict the actuation to the beam. In previous studies on magnetic plucking, the Gilbert model or the Ampère model is often applied which assumes the permanent magnet as either a magnetic charge or magnetic dipole. As a result, the force between magnets follows an inverse squared or cubic relationship with respect to the distance, which is a valid assumption when the size of magnets are negligible compared to the distance between them. However when devices are miniaturized as in our case, the size of magnets are often found to be on the same order of magnitude as the distance. In addition, torques are usually neglected, which could potentially provide a nontrivial excitation. The finite element method is a common approach to a more accurate approximation, which serves well as a validation, whereas the analytical expression provides fast calculation and integration with the dynamic simulation. Our study applies the distributed Gilbert model [6] to obtain the forces and torques between cube magnets analytically. The permanent cube magnet is modelled as two surfaces of evenly distributed magnetic charges. The total potential energy between two cube magnets shown in Figure 2 is given by:

$$E=-\frac{JJ'}{2\pi\mu}\int_{-L}^{L}\int_{-L}^{L}\int_{-L}^{L}\int_{-L}^{L}\left((\alpha+X-x)^2+(\beta+Y-y)^2+\gamma^2\right)^{-\frac{1}{2}}dxdydXdY$$
$$+\frac{JJ'}{4\pi\mu}\left(\begin{array}{l}\int_{-L}^{L}\int_{-L}^{L}\int_{-L}^{L}\int_{-L}^{L}\left((\alpha+X-x)^2+(\beta+Y-y)^2+(\gamma+2L)^2\right)^{-\frac{1}{2}}dxdydXdY\\+\int_{-L}^{L}\int_{-L}^{L}\int_{-L}^{L}\int_{-L}^{L}\left((\alpha+X-x)^2+(\beta+Y-y)^2+(\gamma-2L)^2\right)^{-\frac{1}{2}}dxdydXdY\end{array}\right) \quad (1)$$

Where the J and J' are the magnetic polarizations for each cube magnet of length $2L$, μ is the permeability of the media and the origin of the O'XYZ frame is located at (α, β, γ) with respect to the Oxyz frame (see Figure 2). The distance between two magnets d is defined as the length OO'. The forces and torques between two magnets are the derivative of the potential energy with respect to each direction. A similar formula can be derived for the case where magnetic polarizations are perpendicular as shown in [7].

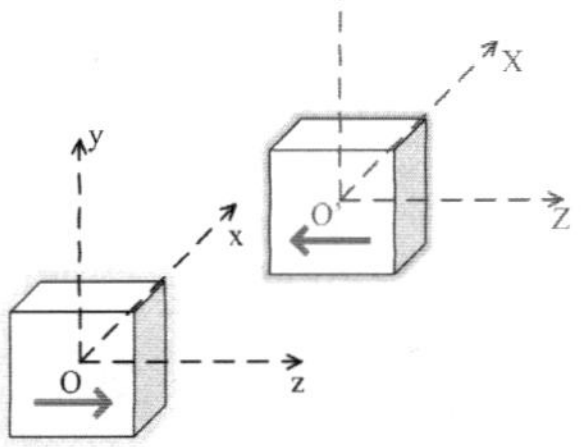

Figure 2. Schematic of the generic magnetic configuration.

We use a static force profile to analyze the actuation mechanism of different magnetic plucking configurations. The force profile is given as a function the displacement normalized by the length of magnet. As illustrated in Figure 1, the beam is assumed to be rigid such that the magnet attached is considered static and does not move. The moving magnet on the rotor is assumed to follow a linear path as the small angles between the opposing surfaces of the magnets are neglected. Note that the model is created within the size of a wrist-worn device. The length of the cube permanent magnet (N42) is 1.59 mm (1/16 inch). A 1 mm gap between magnets is given to all the configurations for comparison as in the direct repulsive configuration such space is required to avoid the pull-in. In the alternative configurations the gap can be further reduced to increase the coupling. Particularly in the indirect repulsive configuration, a 0.5 mm offset is applied to induce the transverse force. We used the Radia software developed by ESRF to provide a numerical validation for the analytical model, which is plotted in the corresponding figures as well. The torque components are neglected as their contribution to the dynamics of the beam are approximately one order of magnitude smaller than the forces.

In general, the transverse force F_z and the axial force F_y (directionality shown in Figure 1) are of interest to the plucking. The transverse force directly deflects the beam while the axial force softens the beam and provides a bending torque when the beam is already deflected. The lateral force F_x is irrelevant in terms of plucking. As shown in Figures 3 to 5, the transverse force exists among all the configurations and its magnitude is the greatest in the direct repulsive configuration given the same gap between the magnets. The indirect repulsive configuration benefits from a large axial force while the orthogonal configuration operates at a pure deflecting mode, i.e. only the transverse force is generated. Although the direct repulsive configuration seems to be the optimal mode to trigger the plucking based on the static force profile alone, as mentioned earlier, it requires a larger gap to avoid the pull-in effect and a larger space for beam deflection since the magnets are aligned in the direction of device thickness, which makes this configuration less appealing for a wearable device as it is size-sensitive. There is no possibility for the magnets to touch each other in the alternative magnetic configurations, thus we are able to implement a smaller gap between magnets which results in a larger transverse force. The offset in the indirect repulsive configuration can be optimized to increase the transverse force as well. As shown in Figure 6, the transverse force is in the optimal range and becomes less sensitive to the change in offset when the offset is in the range of 50% to 80% of the size of magnet.

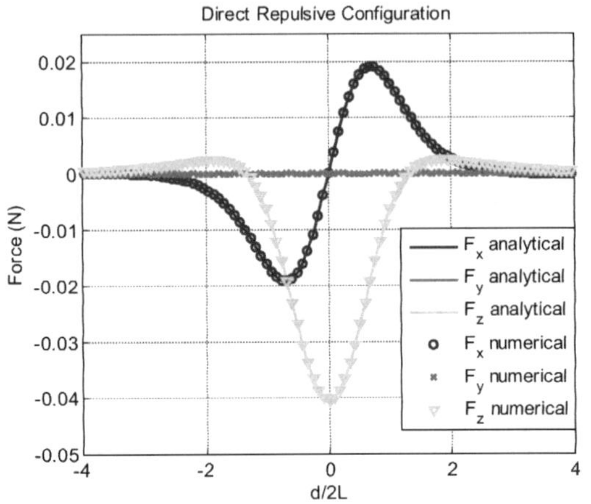

Figure 3. Static force profile of the direct repulsive configuration (gap = 1 mm).

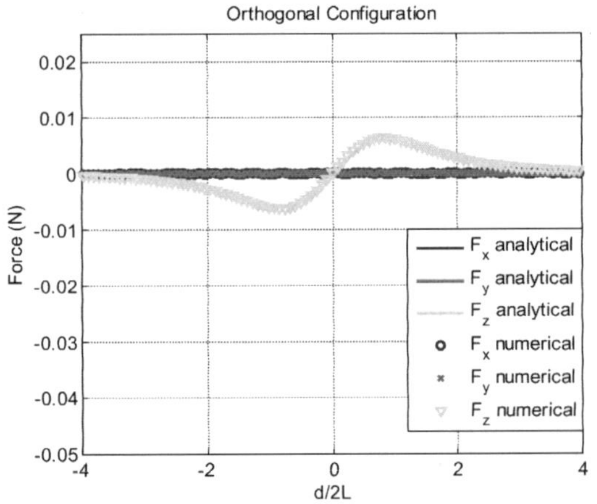

Figure 4. Static force profile of the orthogonal configuration (gap = 1 mm).

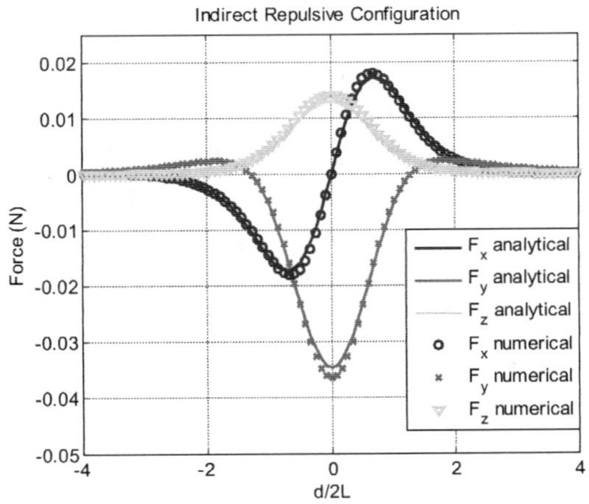

Figure 5. Static force profile of the indirect repulsive configuration (gap = 1 mm, offset = 0.5 mm)

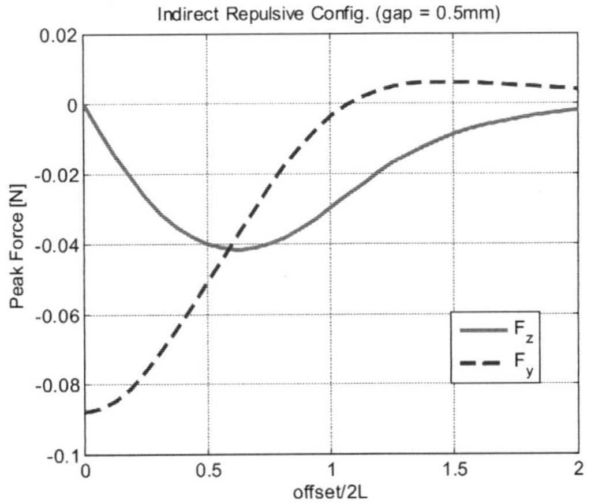

Figure 6. Peak transverse (F_z) and axial force (F_y) as a function of the normalized offset in indirect repulsive configuration.

3. Experimental Validation

We created a simple cantilever beam structure with magnet attached on the tip to corroborate our model since the beam tip displacement is the direct result of the magnetic interaction. The scale is slightly increased compared to the previous model parameters for a more convenient measurement with a laser displacement sensor. A permanent cube magnet (N50) of 2 mm in length is glued to the tip of beam with corresponding orientation. The cantilever beam a brass reinforced piezoelectric bending actuator made by Piezo Systems with the dimension of 0.38 mm (t) × 3.2 mm (w) × 30 mm (l). The gap between magnets for the direct repulsive configuration is 2 mm as the pull-in occurs with a smaller gap while we can safely lower the gap in the other two configurations to 0.8 mm for a larger transverse force. The offset in the indirect repulsive configuration falls in the optimal range found in Figure 6.

Figure 7 illustrates our preliminary results compared with simulated profile. The indirect repulsive configuration outperforms other two configurations with the largest displacement when practical constraints are considered, i.e. the gap and the offset allowed in each configuration. While the shape of the direct repulsive and orthogonal configurations matches calculations well, the model seems to calculate a smaller tip displacement for the indirect repulsive configuration. The discrepancy is due to the softening effect caused by the axial force neglected in the beam model. In terms of designing a frequency up-converting harvester, all of these configurations are capable of plucking the beam as long as the magnet is moving fast enough such that the frequency of the magnetic force profile is well above the natural frequency of the beam.

4. Implementation of the Alternative Magnetic Configuration

As mentioned the main purpose of choosing out-of-plane plucking with alternative magnetic configuration is to allow the use of multiple beams fabricated on a single substrate and assembled with a simple fixture. We have designed and fabricated a wrist-worn eccentric rotor-based piezoelectric energy harvester shown in Figure 8 implementing the indirect repulsive magnetic configuration to pluck 6 bimorph piezoelectric beams. The thin-film piezoelectric beams are fabricated on a nickel substrate. A peak to peak output voltage of 5 volts has been achieved from a single unimorph beam across a 47 kΩ load under the excitation of intense walking when the device is worn on wrist. This will ultimately result in a total power output of over 40 µW given 6 working bimorph beams, which is enough to power a wearable wellness sensor continuously.

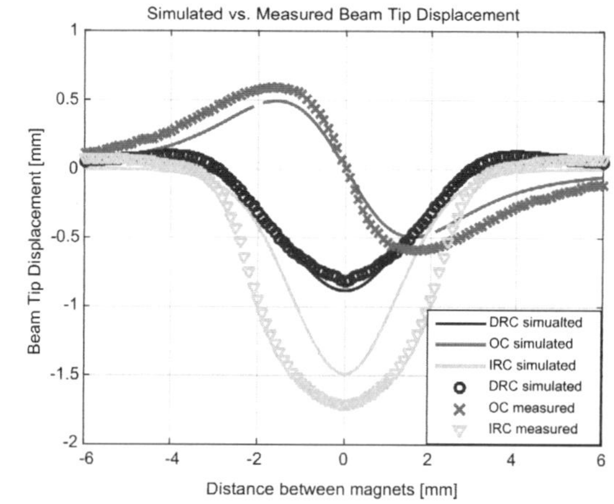

Figure 7. Simulated vs. measured beam tip displacement in different configurations (Direct Repulsive Configuration, Orthogonal Configuration and Indirect Repulsive Configuration).

Figure 8. Photo of the rotational energy harvester implementing the indirect repulsive magnet configuration to pluck multiple piezoelectric beams.

5. Conclusions

This paper has explored alternative configurations to achieve the magnetic plucking in frequency up-converting harvesters. Simulation and experimental results indicate that the indirect repulsive configuration is preferred for a size (and especially thickness) sensitive device such as a wearable energy harvester, which has been demonstrated by a wrist-worn piezoelectric harvester prototype. In a broader sense, however, these configuration can be implemented in other applications. For instance, the direct repulsive configuration should be considered when the device thickness is not a primary concern; the orthogonal configuration can potentially be utilized in precision sensing or actuation due to its pure actuation in the transverse direction.

6. Acknowledgement

This work was supported by the National Science Foundation ASSIST Nanosystems ERC under Award Number EEC-1160483. The authors would also like to thank the Trolier-McKinstry Group and Jackson's Electronics Research Group from Pennsylvania State University for cooperation in fabricating the rotational harvester prototype.

References

[1] Pillatsch P, Yeatman E M and Holmes A S 2013 *J. Phys. Conf. Ser.* **476** 012010
[2] Xue T, Ma X, Rahn C and Roundy S 2014 *J. Phys. Conf. Ser.* **557** 012090
[3] Pozzi M and Zhu M 2011 *Smart Mater. Struct.* **20** 055007
[4] Lockhart R, Janphuang P, Briand D and de Rooij N F 2014 *2014 IEEE 27th Int. Conf. Micro Electro Mech. Syst.* 370–3
[5] Pillatsch P, Yeatman E M and Holmes A S 2014 *Smart Mater. Struct.* **23** 025009
[6] Akoun G and Yonnet J 1984 *IEEE Trans. Magn.* **MAG-20** 1962–4
[7] Allag H and Yonnet J 2011 *LDIA 2011 8th Int. Conf. Linear Drives Ind. Appl.* **2011** 1–5

Band widening of piezoelectric vibration energy harvesters by utilizing mechanical stoppers and magnets

T. Maeguchi[1], **A. Masuda**[1], **H. Katsumura**[2], **H. Kagata**[2] **and H. Okumura**[2]

[1] Kyoto Institute of Technology, Matsugasaki, Sakyo-ku, Kyoto 606-8585, Japan
[2] Panasonic Corporation, Kadoma 571-8506, Japan

E-mail: m4623044@edu.kit.ac.jp

Abstract. This paper presents a design of a piezoelectric hardening-type nonlinear vibration energy harvester which has widened resonance band while maintaining the same peak performance at the resonance frequency as that of the reference linear harvester. To this end, a pair of mechanical stoppers and a pair of repulsive magnets are introduced in this study. An experimental prototype device is designed by using a stainless steel-based piezoelectric cantilever, and numerical simulations and experiments are conducted to examine the validity of the presented design strategy. It is concluded that using the magnets to shift the resonance peak toward the lower frequency and using stoppers to expanding the resonance band toward the higher frequency can broaden the resonance band effectively maintaining the peak response. The damping due to the contact of the tip mass with the stopper is one of the key parameters which should be as small as possible to enhance the band widening effect.

1. Introduction

There has been increasing interests in vibration energy harvesting technologies that aim to retrieve discarded kinetic energy from environment as the power source of wireless sensors. Many researchers have proposed resonance-type devices with a linear oscillator, in which the natural frequency of the oscillator is matched with the dominant frequency of the vibration source. In this type of energy harvesters, the mechanical Q factor of the oscillator is designed as large as possible to maximize the energy harvesting performance. The larger Q factor, however, bounds resonance in a narrower frequency band, therefore, the harvested power decreases significantl if the dominant frequency of the source moves out of this band. This trade-off is recognized as one of the most significan issues in resonance-type vibration energy harvesters [1] because the actual vibration sources present in the environment are likely to have fluctuation in their frequency components. A number of efforts to pursue wider-band vibration energy harvesters have been reported in the literature [1, 2]. Among them, the concept of introducing a nonlinear oscillator to widen the resonance frequency band has attracted increasing attention. A variety types of nonlinear oscillators including hardening, softening, and bistable oscillators are reported in literature.

One of the most important requirements in the practical design of nonlinear wide-band vibration energy harvester is to widen the resonance band while remaining the peak performance at the specifi target frequency. Theoretical analysis using Krylov-Bogoliubov averaging method has indicated that it is required to make the backbone curve, i.e., the equivalent natural frequency of the nonlinear oscillator, to reach the maximum possible stroke at the resonance frequency of the reference linear oscillator [3]. In this study, a pair of mechanical stoppers and a pair of repulsive magnets are introduced to try to achieve this requirement. Numerical studies and experiments using a cantilever oscillator with a thin piezoelectric bimorph are carried out to clarify the feasibility of the idea of tailoring the restoring force of the oscillator by introducing these mechanical elements.

2. Band-widening by introducing nonlinear oscillator

Let us consider a cantilever oscillator consisting of a piezoelectric bimorph and a tip mass. Suppose that there are a pair of stoppers which limits the motion of the tip mass in a certain stroke. Also, suppose a pair of repulsive magnets, one of which is fi ed at the end of the tip mass, while the other is fi ed on the base to induce axial compressive force in the cantilever at the neutral position. Assuming that the cantilever always deforms in a single mode shape (in the lowest vibration mode), the governing equations of the cantilever is derived as

$$m\ddot{x}(t) + c\dot{x}(t) + kx(t) + F_{stopper}(x) + F_{magnet}(x) + \theta v_p(t) = f(t) \tag{1}$$

and

$$C_p v_p(t) = \theta x(t) - q(t) \tag{2}$$

where $x(t)$, $v_p(t)$, $f(t)$, and $q(t)$ are the relative displacement of the tip mass, the voltage across the piezoelectric layers, the external force given by a base excitation, and the electrical charge, respectively. The constants m, c, k, θ, and C_p are the equivalent mass of the cantilever including the tip mass, the mechanical damping coefficient, the equivalent stiffness of the cantilever in the short circuit configuration the electromechanical coupling factor, and the clamped capacitance of the piezoelectric layer, respectively.

The force function $F_{stopper}$ is an additional restoring force induced by the collision of the tip mass with the stopper, while the function F_{magnet} is a negative restoring force induced by the repulsive force between the magnets. Considering that these forces are well localized at the position of the stoppers and around the neutral position, respectively, they give a hardening effect to the oscillator by harden the oscillator at the stroke limits by $F_{stopper}$, and by soften the oscillator around the neutral position by F_{magnet}. The basic concept of this study is to tailor the oscillator's restoring force as desired by adjusting the parameters of these mechanical components, i.e., the stopper position, stopper material, specificatio of the magnets, and the gap between the magnets.

The piezoelectric layer in this study is assumed to be connected directly to a pure resistance R without any interface circuits, thus, the boundary condition for the piezoelectric voltage and charge is given by

$$v_p(t) = R\dot{q}(t) \tag{3}$$

The instantaneous harvested power, which is evaluated as the power consumption at the load resistance R, is given by

$$P(t) = R\dot{q}(t)^2 \tag{4}$$

Let us suppose the sinusoidal base excitation with a frequency of ω. Assuming sinusoidal variation of all the state variables, the equation of motion of the cantilever is expressed as follows by utilizing equivalent linearization method (or equivalently, via Krylov-Bogoliubov averaging method):

$$m\ddot{x}(t) + c\dot{x}(t) + K_e(x_0)x(t) + \theta v_p(t) = f_0 \cos \omega t \tag{5}$$

where K_e is the equivalent stiffness in function of the displacement amplitude x_0.

The design objective of the nonlinear restoring force in this study is to achieve widening of the resonance band of the harvester while remaining the peak performance compared with a reference linear harvester. This requires

$$K_e(x_{\max}) = k_r \tag{6}$$

and

$$K_e(0) = k_0 < k_r, \qquad \frac{dK_e}{dx_0} \geq 0 \tag{7}$$

where k_r, $x_{\max}$, and k_0 are the stiffness of the reference linear oscillator, its maximum displacement at the resonance peak for a specifi level of excitation, and the stiffness of the nonlinear oscillator linearized at the neutral position, respectively. Equation (6) is the requirement to maintain the peak performance, while equation (7) is for the band widening by hardening.

3. Numerical and experimental studies

3.1. Piezoelectric bimorph cantilever

The piezoelectric bimorph cantilever used in this study is made with screen-printing technique to form a thick fil of piezoelectric ceramics(PZT) directly on a stainless steel substrate [4]. The thickness of the PZT layer is 30 μm and the thickness of the based stainless layer is 0.2 mm. Different from a conventional piezoelectric element which is easily damaged when undergoes large tensile deformation, this element withstand such deformation because of its large residual compression stress (approx. 500 MPa). Also, it shows remarkable long life under repetitive plucking test over 10 million times. Other specification of the bimorph are listed in table 1.

3.2. Experimental Setup

Figure 1 shows a schematic illustration of an experimental setup. A tip mass of 28.6 g made of copper was attached at the end of the bimorph cantilever. The resulting short circuit natural frequency was 30.3 Hz. At the end of the stroke of the tip mass, a pair of hemisphere stoppers were mounted on fin stages which enable to adjust the stoppers' vertical position. A pair of neodymium permanent magnets with a diameter of 3 mm and a height of 2 mm were adhered on the tip mass and on the base with the same polarity facing each other. The whole setup was mounted on a vertical shaker and the displacement of the tip mass and the base were measured by laser displacement sensors. A sinusoidal excitation was applied to see the steady-state response of the harvester. The input acceleration was controlled as 0.5 m/s^2 during the numerical simulations and the experiments.

Table 1. Specificatio of piezoelectric bimorph.

Effective length of substrate	25	[mm]
Width of substrate	14	[mm]
Thickness of substrate	0.2	[mm]
Density of substrate	7900	[kg/m^3]
Young's modules of substrate	200	[GPa]
Capacitance of piezoelectric layer	22.5	[nF]

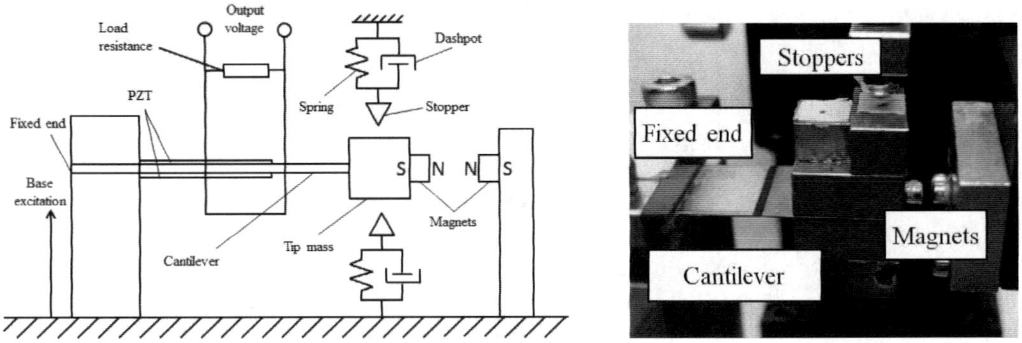

Figure 1. Experimental setup.

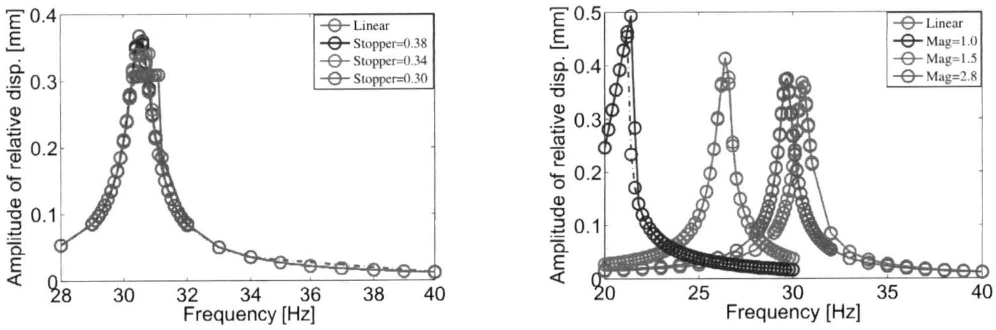

(a) Effect of stoppers. (b) Effect of magnets.

Figure 2. Numerical results showing the effect of independent use of stoppers and magnets.

First, the stoppers and magnets were removed, and the load resistance was determined so that the output power of the linear harvester is maximized. The resultant values of the resistance were 100 k Ω for the numerical simulations and 90 kΩ for the experiments.

In the following numerical calculations, the stopper force was modeled as an extra set of spring and dashpot as presented in the figure The spring constant was set to 100 times of that of the linear oscillator, and the damping coefficient was determined so that the damping ratio ζ_2 during the collision of the tip mass with the stopper was to be 0.35 which was identifie from the experiment. The magnetic force was modeled by following the formulation presented by Stanton et al. [5].

4. Results and discussions

4.1. Effect of independent use of stoppers and magnets

Figure 2 (a) shows the numerically calculated steady-state displacement amplitude for the case with stoppers at 0.38 mm, 0.34 mm, and 0.30 mm, compared with the case without stoppers and magnets (linear). From these results, one can figur out that the introduction of the stoppers folds the backbone curve, i.e., the center line of the resonance peak, toward the higher frequency, and the band widening effect becomes stronger when the stopper location becomes closer.

4.2. Effect of magnets

Figure 2 (b) shows the numerically calculated steady-state displacement amplitude for the case with magnets with the gap of 1.0 mm, 1.5 mm, and 2.8 mm, compared with the case without stoppers and magnets (linear). These plots indicate that the effect of introducing the magnets is to shift the backbone curve toward the lower frequency and add a slight effect of hardening. The frequency shift effect becomes stronger when the gap between the magnets becomes closer.

4.3. Effect of simultaneous use of stoppers and magnets

Figures 3 and 4 show the combination effect of the stoppers and magnets. Figure 3 shows the numerically calculated steady-state displacement amplitudes while figur 4 shows the experimental results of the steady-state displacement amplitudes and the averaged harvested power. In figur 3 (a), ζ_2, the damping ratio while the tip mass contacting with the stopper, is set to 0.35 which is identifie from the experimental results shown in figur 4. These results suggest that using the magnets to shift the resonance peak toward the lower frequency and using stoppers to expanding the resonance band toward the higher frequency can broaden the resonance band effectively maintaining the peak response. Figure 3 (b), in which ζ_2 is decreased to 0.01, implies that, to enhance the band widening, it is effective to reduced the damping due to the contact of the stopper. This may be realized by appropriately designing the shapes and materials of the contacting part of the stopper and the tip mass as well.

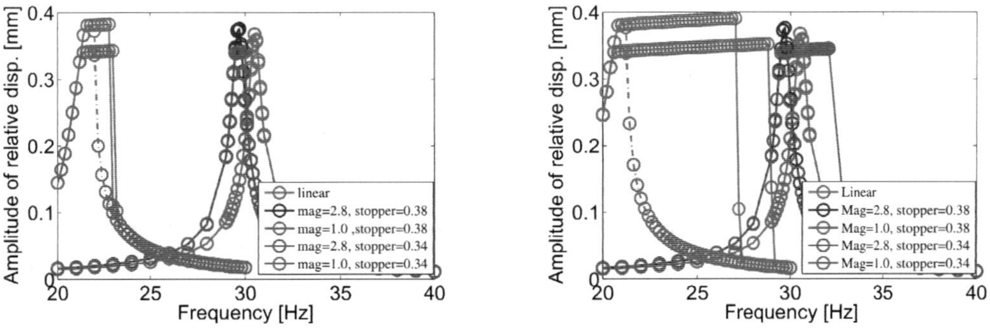

(a) ζ_2=0.35 identifie from experiment. (b) ζ_2=0.01.

Figure 3. Numerical results showing the effect of simultaneous use of stoppers and magnets.

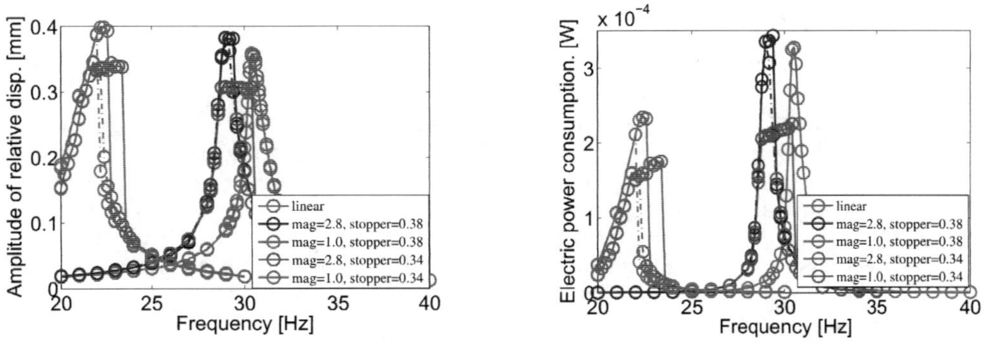

Figure 4. Experimental results showing the effect of simultaneous use of stoppers and magnets.

5. Conclusion

The feasibility of the band widening of vibration energy harvester by introducing stoppers and magnets have been verifie in this paper. The numerical and experimental studies also suggested the importance of suppressing the energy loss during the contact of the tip mass with the stopper to enhance the band widening effect.

6. Reference

[1] Tang L, Yang Y and Soh C K 2010 Toward broadband vibration-based energy harvesting, *Journal of Intelligent Material Systems and Structures* 21(18), pp 1867–1897.

[2] Zhu D,Tudor M J and Beeby S P 2010 Strategies for increasing the operating frequency range of vibration energy harvesters: a review, *Measurement Science and Technology*, Vol. 21, 022001, pp 1-29.

[3] Masuda A, Senda A, Sanada T, and Sone 2013 A Global stabilization of high-energy responce for a Duffing-type wideband nonlinear energy harvester via self-excitation and entrainment, *Journal of Intelligent Material Systems and Structures*, Vol.24, No.13, 1589–1612.

[4] Oishi A, Okumura H, Katsumura H and Kagata H 2014 The power generation characteristics and durability of a vibration power generation device using piezoelectric thick fil formed directly by screen printing on a stainless steel substrate, *Journal of Physics: Conference Series, PowerMEMS 2014*, 557-1.

[5] Stanton S C, McGehee C C and Mann B P 2010 Nonlinear dynamics for broadband energy harvesting: Investigation of a bistable piezoelectric inertial generator, *Physica D*, Vol. 239, No. 10, pp 640-653.

PowerMEMS 2015 IOP Publishing
Journal of Physics: Conference Series **660** (2015) 012100 doi:10.1088/1742-6596/660/1/012100

Batteryless wireless transmission system for electronic drum uses piezoelectric generator for play signal and power source

H. Nishikawa, A. Yoshimi, K. Takemura, A. Tanaka, and T. Douseki

Ritsumeikan University, Graduate School of Science and Engineering.
1-1-1, Noji-Higashi, Kusatsu, Shiga, 525-8577 Japan

rfnext@nethome.ne.jp

Abstract. A batteryless self-powered wireless transmission system has been developed that sends a signal from a drum pad to a synthesizer. The power generated by a piezoelectric generator functions both as the "Play" signal for the synthesizer and as the power source for the transmitter. An FM transmitter, which theoretically operates with zero latency, and a receiver with quick-response squelch of the received signal were developed for wireless transmission with a minimum system delay. Experimental results for an electronic drum without any connecting wires fully demonstrated the feasibility of self-powered wireless transmission with a latency of 900 μs.

1. Introduction

One problem with the plethora of electronic instruments in the music world today is all the wires and cables that need to be transported, plugged in, and managed. A conventional electronic drum set employs wires to connect the sensing pads to the synthesizer unit. The wires complicate setup and look messy, and the difficulty of assembly and disassembly limits use of such a drum set to one fixed place. A wireless system, on the other hand, would provide portability and greatly expand the range of venues where electronic drums could be used in place of acoustic ones. A batteryless wireless FM transmitter, which uses the output of a piezoelectric generator both as a power source and to produce an analog signal, has been developed. An FM receiver with an ultrafast squelch that operates with almost zero latency helps reduce the system delay in a wireless electronic drum (Fig.1).

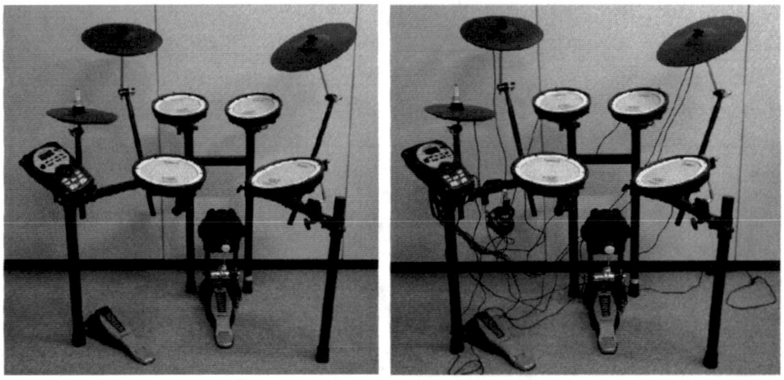

(a) (b)

Figure 1. (a) New wireless drum system and (b) Conventional electronic drum system with connecting wires.

2. Basic Theory

The drum pad of an electronic drum contains a piezoelectric sensor [1]. When the pad is hit with a drumstick, the vibration propagates to the piezoelectric sensor, which produces an electrical signal. In a conventional system, the signal is sent over a wire to a synthesizer, which produces the appropriate sound.

In our newly developed wireless system, a piezoelectric generator was mounted in a Styrofoam housing, covered with a protective silicone sheet (to protect it from strikes with a drumstick), and placed in a drum pad (Fig. 2). When the pad is struck, the force is conducted to the generator, and electrical energy is produced. The generated signal drives a transmitter, which sends a modulated signal wirelessly to a receiver in the synthesizer. This is in contrast to a conventional system, which sends the signal to the synthesizer over a wire (dotted line in Fig. 3).

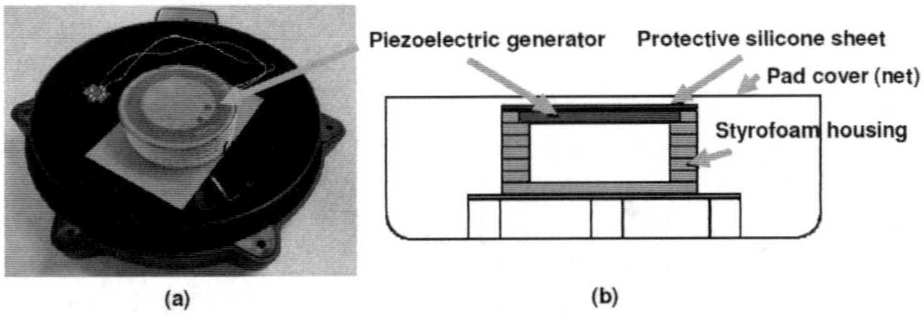

Figure 2. Electronic drum pad with piezoelectric generator: (a) photograph and (b) cross-sectional diagram.

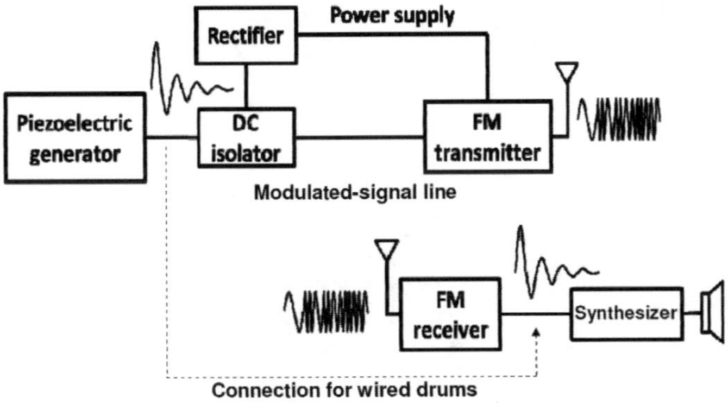

Figure 3. Block diagram of self-powered wireless electronic drum

The energy produced by the piezoelectric generator is used both for the "Play" signal sent to the synthesizer and as a power source [2] for the wireless transmitter to support batteryless operation. A full-wave-rectifier circuit with a diode bridge is placed between the generator and the transmitter. Since the ground of the transmitter is taken to be the DC reference, the negative swing of the generator output is inverted to a positive swing by the rectification of the diode bridge. Thus, DC isolation is required between the drum signal and the power circuitry so that a proper signal with both positive and negative polarities can be sent to the transmitter for modulation. In this study, that was accomplished by placing a transformer with appropriate coupling capacitors between the generator and transmitter (Fig. 4).

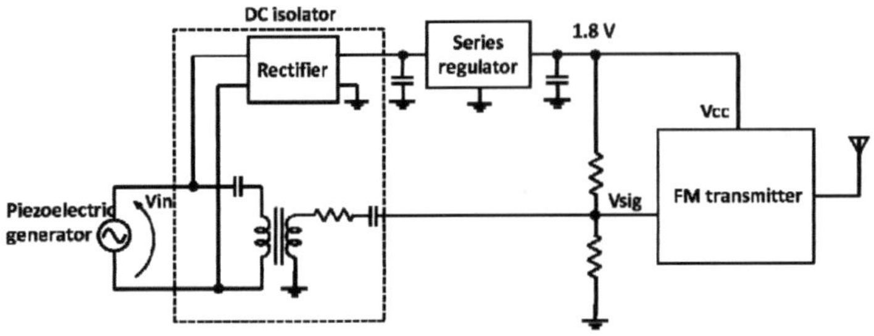

Figure 4. Block diagram of DC isolation, which drives power generated by piezoelectric generator into power source Vcc) and analog signal for transmission (Vsig).

For percussion instruments, such as drums, the maximum acceptable latency in the signal transmission process has been empirically estimated to be 1 ms. A digitally modulated radio signal is widely used for wireless microphones [3][4] and other equipment because there is no significant distortion where latency is not a major concern. For a drum system, however, a small latency has a higher priority than good sound quality; and thus, an analog FM modulation architecture, which theoretically provides zero latency, was selected for our system. Although the latency is zero in the wireless process, the charge-up time of the rectification in the transmitter and the squelch response in the receiver are the main components of the total delay of the system (Fig. 5). The values of the decoupling and filtering capacitors in the rectification and regulation circuit of the transmitter (in Fig. 4) must be reduced to the minimum values needed to obtain a quick charge-up time. A high-speed squelch is required to obtain a short enough recovery time from the shut-down mode in the FM receiver (Fig.6).

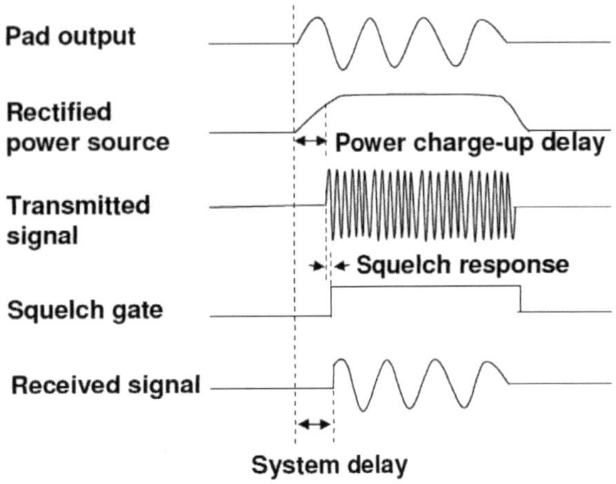

Figure 5. Relationship among pad output, rectified power source, transmitted signal, squelch response, and received signal.

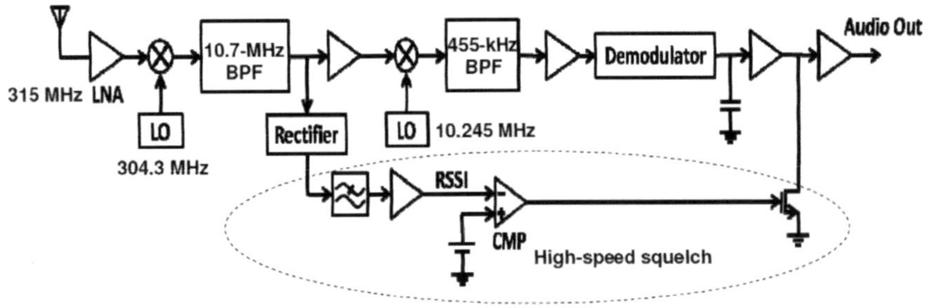

Figure 6. Block diagram of FM receiver with high-speed squelch.

3. Fablication and experiments

The FM transmitter, which includes DC isolation, rectifiers, and a regulator, is constructed on a small printed circuit board measuring 2.3 cm × 3.5 cm. A plug on the transmitter board connects the input line to the drum pad. All of the receiver functions, including the quick-response squelch, are implemented on a printed circuit board measuring 8.7 cm × 5.6 cm (Fig. 7).

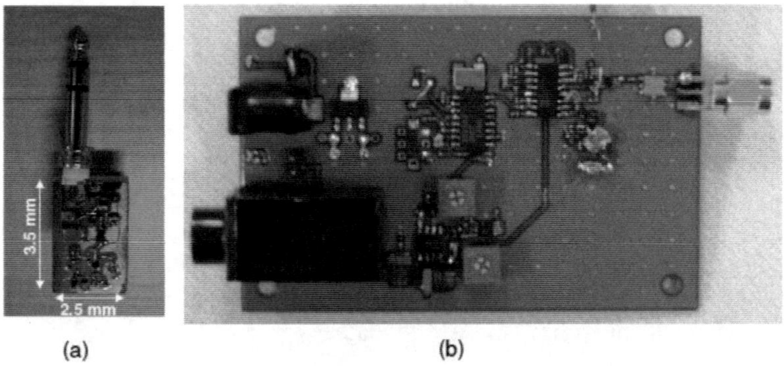

Figure 7. Photographs of (a) 315-MHz batteryless FM transmitter, and (b) receiver board with embedded high-speed squelch.

The rectified regulated DC power was measured at the output of the regulator, to which a 1.2-kΩ resistor was connected to simulate the load of the FM transmitter. The required charge-up time to reach a stable 1.8-V DC level was 500 μs. The output of the regulator remained 1.8 V for more than 2 ms, which is enough power for the transmitter to maintain oscillations and modulation (Fig. 8).

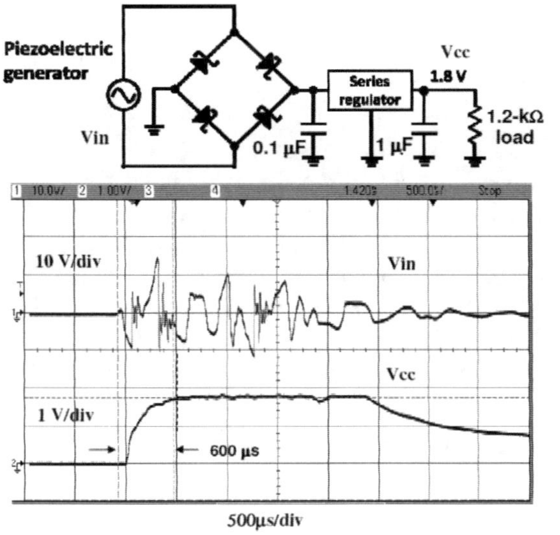

Figure 8. Measured waveforms: input from piezoelectric generator (Vin), and rectified DC (Vcc).

The response time of the squelch in the FM receiver was measured (Fig. 9). The squelch gate opens right after a 315-MHz carrier is supplied to it. The squelch response time is almost zero.

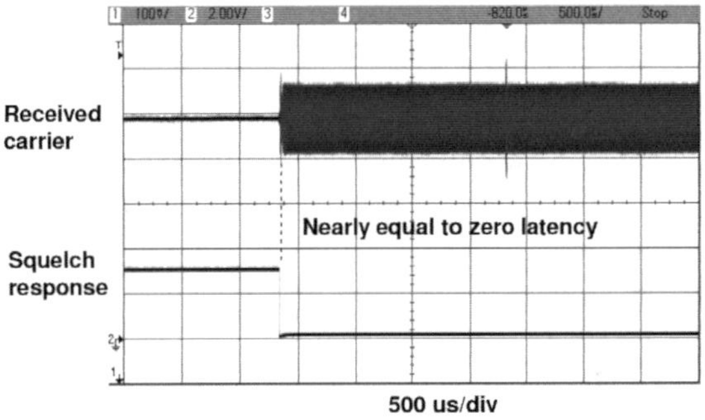

Figure 9. Measured waveform of received carrier (top) and squelch response (bottom).

Experiments were performed by attaching the transmitter to a drum pad and connecting the receiver to a synthesizer (Fig. 10a). The transmitter input and the receiver output were monitored on an oscilloscope, and the waveforms (Fig. 10b) were found to be in good agreement. The latency was 900 μs, which meets the target specification of 1 ms. Thus, it has been demonstrated that a batteryless FM transmission system is applicable to an electronic drum, and that the latency is sufficiently small.

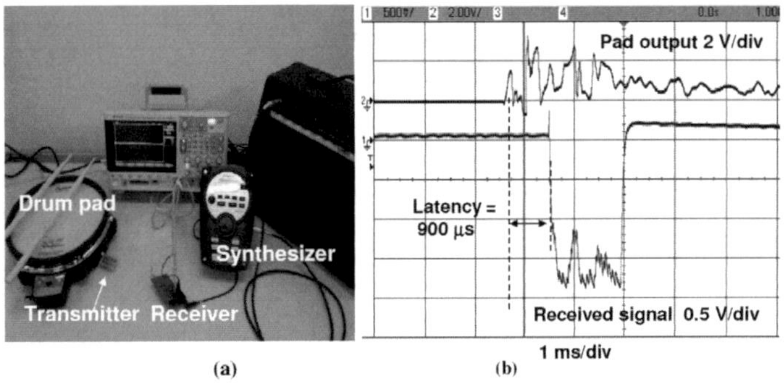

Figure 10. (a) Setup (drum pad, transmitter, receiver, synthesizer) for evaluating self-powered FM transmission system for electric drum, (b) Measured waveform of pad output and signal received at synthesizer after wireless transmission.

4. Conclusion

A batteryless self-powered wireless transmitter has been developed that sends a signal from a drum pad to a synthesizer. The power generated by a piezoelectric generator functions both as the "Play" signal for the synthesizer and as the power source for the wireless transmitter. A DC isolation technique based on a transformer and coupling capacitors provides efficient power division. The system employs direct FM modulation so that it theoretically provides zero latency in the wireless transmission process. The latency of the whole system, including power rectification, the startup of the FM transmitter, and the receiver response, is 900 μs, which is small enough for an electronic drum.

References

[1] Electronic Drums [Online]. Available at http://pwc.theclarkwebsite.com/piezos.php
[2] S. Wang, et.al., "Energy harvesting with piezoelectric drum transducer", Applied Physics Letters 90, 113506,
[3] Wireless Microphone System [Online]. Available at http://www.jstage.jst.go.jp/article/ieiej/31/11/31_858/_pdf
[4] Digital Signal Processing in RF Applications [Online]. Available at http://cas.web.cern.ch/cas/Sweden2007/Lectures/Web-versions/Schilcher-1.pdf pp. 1-3, 2007

Journal of Physics: Conference Series **660** (2015) 012101 doi:10.1088/1742-6596/660/1/012101

Bottom-up silicon nanowire-based thermoelectric microgenerators

D Dávila, R Huber and C Hierold

Micro and Nanosystems, Department of Mechanical and Process Engineering, ETH Zurich, Tannenstrasse 3, 8092, Zurich, Switzerland

E-mail: diana.davila@micro.mavt.ethz.ch

Abstract. In this work, bottom-up intrinsic crystalline Si nanowire arrays in combination with top-down microfabrication techniques and a vertical device architecture have been proposed to develop an all-silicon nanostructured thermoelectric generator. To fabricate this device, a suitable vertical integration of Si NWs on patterned microstructures, which define the thermoelectric legs of the generator, has been achieved by bonding top and bottom silicon structures through nanowires. The process has been proven to be a feasible approach that employs a regrowth process of the nanowires for bonding purposes.

1. Introduction

Thermoelectric generators (TEGs) are able to convert heat in electrical energy by exploiting the intrinsic properties of the materials that compose them. TEGs possess the potential to become the power generating core of zero-power systems such as sensor networks working in hazardous or remote environments. These prospective applications have led to the development of a broad range of thermoelectric materials with the aim of enabling high performance thermoelectric devices. Among these materials, low-dimensional materials have received particular attention due to the possibility of enhancing the thermoelectric properties of bulk materials through nanostructuring effects, which allow filtering out the low-energy electrons by scattering phonons more effectively than electrons becoming suitable for thermoelectric applications [1-2]. In this way materials such as silicon, considered an inefficient thermoelectric material ($ZT \approx 0.01$) due to its high thermal conductivity, have shown to be a promising material [3-4] that could be integrated in thermoelectric devices.

In this work, silicon nanowires (Si NWs) have been vertically integrated in microfabricated structures to act as the thermoelectric material of a full silicon-based micro thermoelectric generator (μTEG). A thermoelectric generator is composed by *p*-type and *n*-type material legs, connected electrically in series and thermally in parallel. Figure 1a depicts the basic concept of a single element or thermocouple of the μTEG presented in this work in which arrays of *p*- and *n*-type SiNWs substitute the thermoelectric legs of a standard TEG device. Achieving different doping types of silicon nanowires within the same substrate entails technological constrains when working with bottom-up nanowires, specifically when *in situ* doping is performed. Therefore, the device concept of this work focuses on the growth of nanowires on two independent substrates and on the bonding of these two structures through an additional growth process step to achieve a complete μTEG (Figure 1b).

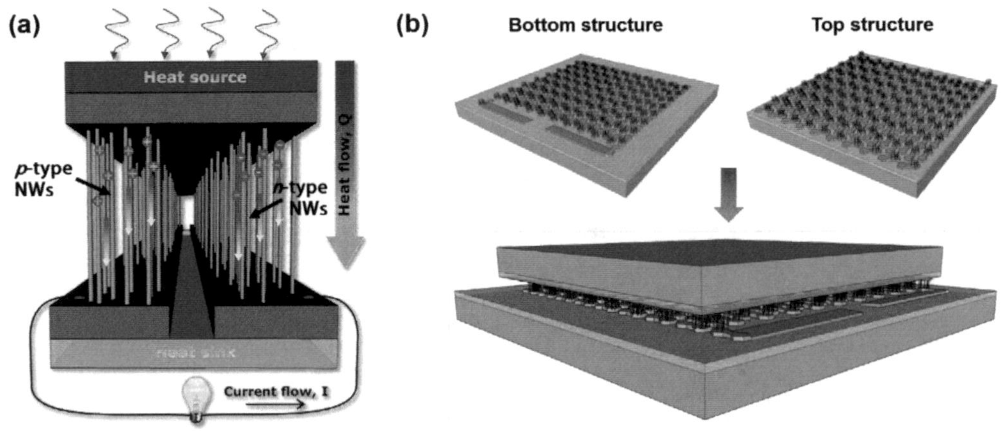

Figure 1. Schematic illustration of the device concept. (a) Thermoelectric elements consisting of *p*-type and *n*-type silicon nanowires arrays are electrically connected in series and thermally in parallel for power generation (Seebeck effect). (b) Once the Si NW legs have been separately grown on two different samples, the top structure is flipped and aligned with the bottom structure with the help of a silicon holder.

2. Experimental

Two different substrates (for the top and bottom components) were patterned with defined hexagonal Si structures that shape the thermoelectric legs of the device (Figure 2a). For this purpose, silicon on insulator (SOI) wafers with a 20 µm-thick *p*-type Si (111) device layer and a 0.5 µm-thick buried silicon oxide layer over a 250 µm-thick (100) silicon handle wafer were employed. The device layer of the SOI substrates was structured with a deep reactive ion etching (DRIE) process to define the thermoelectric legs. A thermal oxidation of the substrates was then performed to cover all the silicon exposed areas with an oxide layer and to avoid in this way the exposure of unwanted silicon areas during the further deposition of the nanoparticles employed for the growth of the nanowires. Metal contacts were patterned by means of a reactive ion etching (RIE) process that exposed the areas defined for this purpose followed by the sputtering and lift-off of a tungsten layer. A plasma-enhanced chemical vapor deposition (PECVD) oxide layer was used to passivate the metal contacts. The silicon areas for the selective growth of silicon nanowires were then exposed through a RIE process. At this point, the deposition of the metal catalyst required for the synthesis of silicon nanowires by means of the Vapor-Liquid-Solid (VLS) mechanism took place. The VLS process allows for high-density epitaxial growth of nanowires on free silicon surfaces using catalytic Au nanoparticles as mediators. The Au nanoparticles were deposited by means of the galvanic displacement method, which allows to control the nanoparticle density and diameter distribution. This highly selective method enables the deposition of nanoparticles only on conducting or semiconducting surfaces and the subsequent growth of silicon nanowires only on this (silicon-exposed) areas.

In order to deposit the gold nanoparticles, a solution, consisting of water droplets (reversed micelles) stabilized by a surfactant and emulsified in oil, was prepared. The gold clusters contained in the micelles of the solution are reduced only at the exposed silicon surfaces through a redox reaction. Silicon nanowires were grown on these substrates in a low pressure chemical vapor deposition (LPCVD) furnace at 630°C using a mixture of SiH_4, H_2 and HCl gases at a pressure of 9 mbar. It is important to mention that the results presented in this work correspond only to intrinsic silicon nanowires.

Highly dense arrays of silicon nanowires with diameters of 84.4 ± 24.7 nm and lengths of 16.7 ± 0.9 µm were selectively grown on two independent substrates for a period of 90 minutes. Figure 2b shows a tilted image of an individual hexagonal thermoelectric leg in which silicon nanowires have been selectively grown.

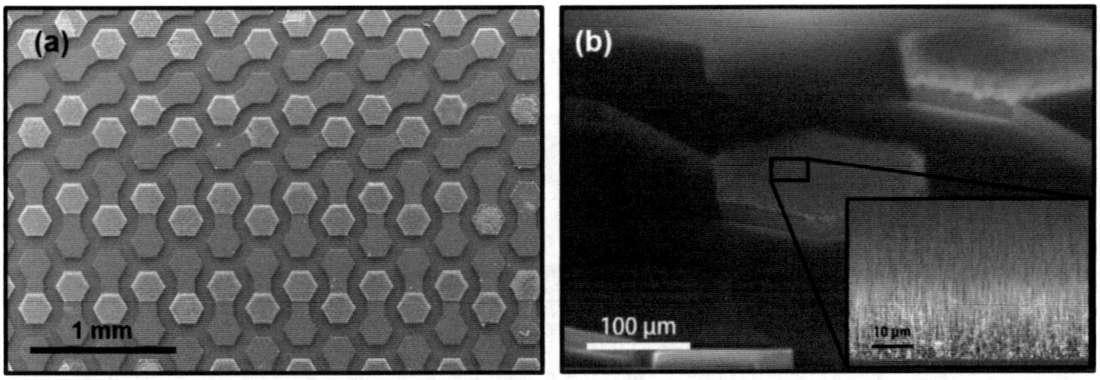

Figure 2. (a) SEM image showing the hexagonal design of the thermoelectric elements comprising the device before bonding the top and bottom structures. (b) Highly dense and ordered arrays of Si NWs have been selectively grown on defined Si structures.

In order to obtain a functional TEG structure, the SiNW arrays, which act as the thermoelectric legs, must provide electrical and mechanical connection between the top and bottom substrates. These two substrates will in turn connect the thermoelectric legs electrically in series and will be exposed respectively to a high and a low temperature source, thus connecting them thermally in parallel.

The ability of SiNWs to connect two silicon surfaces has been previously reported for horizontal architectures in which two pre-fabricated silicon electrodes or structures are bridged through a large number of highly directional nanowires [5-7]. In the case of TEG devices, a vertical architecture that allows a good thermoelectric leg integration and an efficient exploitation of naturally occurring thermal gradients would be desired. The vertical bonding of the two, top and bottom, structures was achieved using a microfabricated silicon holder that served to align the samples and bond them through an additional nanowire growth process step. Since the silicon nanowires on the samples have been synthesized using a bottom-up technique (i.e. the VLS process) the catalyst gold nanoparticle, which remains on top of the nanowires, allows to resume the growth as depicted in figure 3. The Au nanoparticle is responsible for the passage of silicon from vapor phase (SiH_4), trough liquid phase (in the Au-Si alloy), to solid phase (in the NW's body). Since the nanoparticles stay on top of the wires, it is reasonable to expect that, by reintroducing an array of already grown SiNWs in the LPCVD system and by recreating the same growth conditions, the growth of the NWs can be resumed and the their length further increased. Therefore, the vertical nanowire arrays of the thermoelectric legs continue to grow reaching the opposite surface and achieving an epitaxial contact between the two structures.

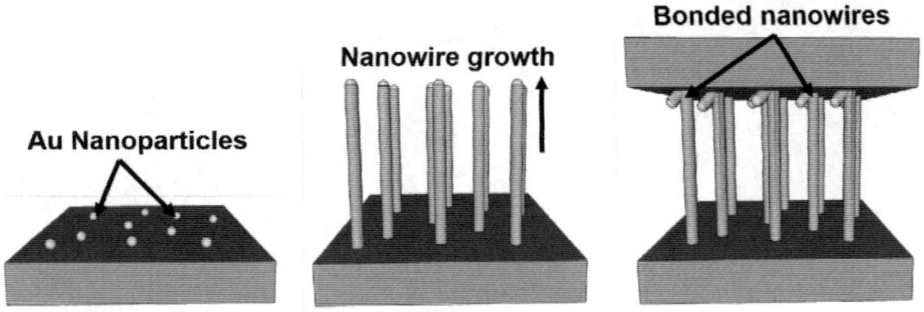

Figure 3. Schematic illustration of the bonding mechanism. Vertically aligned arrays of silicon nanowires grown on a silicon substrate bond with a top silicon surface.

3. Results

To prove the existence of a regrowth mechanism, silicon nanowires were grown on (111) Si substrates for a period of 60 minutes after which the samples are removed from the LPCVD and analyzed using SEM imaging. Then these samples are reintroduced in the LPCVD chamber, and the same growth procedure is reapplied. SEM images after this process reveal that, as expected, the silicon nanowire arrays continue to grow in the <111> direction (figure 4).

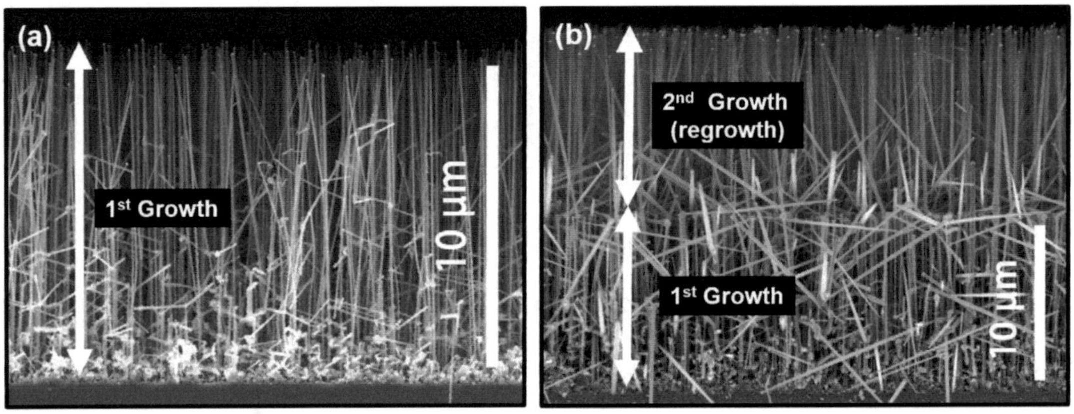

Figure 4. SEM images showing the feasibility of resuming the growth of silicon nanowires. (a) A single growth of silicon nanowires yielded a length of 12.4 ± 0.5μm for a 60 minutes period growth. (b) The same sample is reintroduced in the LPCVD chamber and the same growth procedure is performed on the sample. A homogeneous second regrowth process yielding a total length of 24.8 ± 0.7 μm is achieved.

The alignment of bonded structures was verified by manually detaching the top and bottom samples of the device. As expected, SiNWs successfully bond the two substrates. SEM inspection confirms that the nanowires grow from the bottom to the top substrate (and vice versa) providing linkage between the two structures. As shown in figure 5, silicon nanowires break where they have reached and contacted the opposite substrate, whereas misaligned nanowires, lacking of a top structure where to bond and make contact, are left intact.

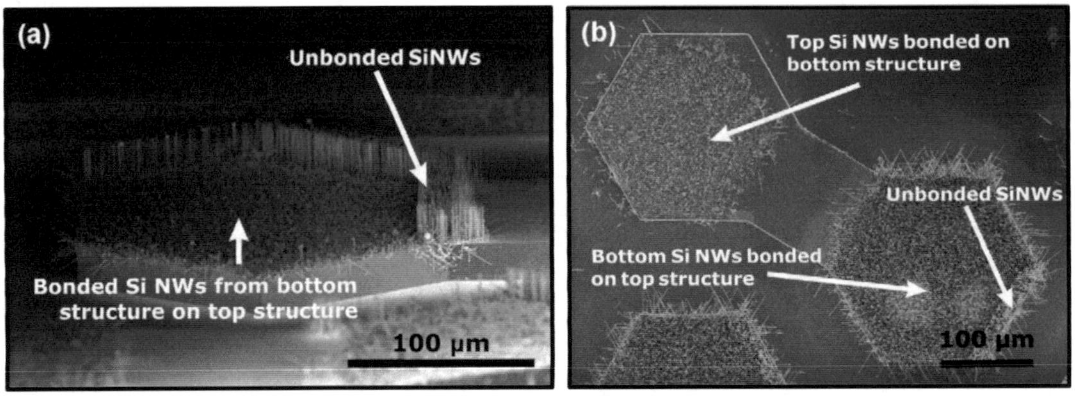

Figure 5. SEM images showing the areas where the top and bottom patterns have been aligned and bonded. The Si NWs on these areas have broken when detaching the samples, whereas the misaligned NWs, which have not been bonded with the opposite structure, have continued to grow. A minimum misalignment of 30 μm has been achieved using a microfabricated silicon holder.

Devices with 276 hexagonal elements connected through intrinsic silicon nanowires have been fabricated yielding a total electrical resistance of 4.4 ± 0.53 kΩ at room temperature. Although further efforts have to be performed to reduce the internal resistance of the device through doping od the nanowires, these preliminary results together with the SEM analysis performed after detaching the samples prove the feasibility of integration of two different types of silicon nanowires on future thermoelectric microgenerators and provide the basis of a fabrication process to produce a functional device.

4. Conclusions

In this work, the feasibility of a thermoelectric microgenerator with a vertical structure in which the thermoelectric legs consist of silicon nanowires has been proven. Selective growth of VLS-synthesized silicon nanowires has been achieved on defined silicon structures that shape the legs of the thermoelectric device.

The regrowth of silicon nanowires was successfully carried out by applying the same growth conditions for a second time in samples with arrays of silicon nanowires. The results obtained allow to exploit this mechanism to assemble two different substrates on which nanowires have been already grown and doped separately. By vertically growing silicon nanowires in two separate silicon substrates, flipping one of these substrates over the other and performing a second growth step, the possibility of bonding a top and a bottom sample was demonstrated.

Integration of *p*- and *n*-type silicon nanowires and further characterization to ensure a good electrical connection of the NWs and a good thermal connection of the substrates to the thermal interfaces is needed to confirm the feasibility of this design for thermoelectric applications.

Acknowledgments

D Dávila would like to thank the financial support granted by the ETH Zurich Postdoctoral Fellowship Program supported under the FP7 European Union Program. Furthermore, the authors would like to thank the cleanroom operation teams of the Binnig and Rohrer Nanotechnology Center (BRNC) and of the Frontiers in Research: Space and Time (FIRST) facilities. In addition, we would like to thank U Drechsler for helpful discussions.

References

[1] Hicks LD and Dresselhaus MS 1993 *Phys. Rev. B* **47** 12727
[2] Feser JP, Sadhu SJ, Azeredo BP, Hsu KH, Ma J, Kim J, Seong M, Fang NX, Li X, Ferreira PM, Sinha S and Cahill DG 2012 *Jour. App. Phys.* **112** 114306
[3] Boukai AI, Bunimovich Y, Tahir-Kheli J, Yu JK, Goddard WA and Heath JR 2008 *Nature* **451** 168
[4] Hochbaum AI, Chen R, Delgado RD, Liang W, Garnett EC, Najarian M, Majumdar A and Yang P 2008 *Nature* **451** 163
[5] Islam MS, Sharma S, Kamins TI and Williams RS 2004 *Nanotechnology* **15** L5
[6] Islam MS, Sharma S, Kamins TI and Williams RS 2005 *Applied Physics A* **80** 1133
[7] Dávila D, Tarancón A, Calaza C, Salleras M, Fernández-Regúlez M, San Paulo A and Fonseca L 2012 *Nano Energy* **1** 812

Broadband electrostatic device for power harvesting

N S Yuksek[1], Z C Feng[2], and M Almasri[1]
[1] University of Missouri, Electrical and Computer Engineering, Columbia, MO, USA
[2] University of Missouri, Mechanical and Aerospace Engineering, Columbia, MO, USA

almasrim@missouri.edu

Abstract. This paper introduces a prototype energy harvester device with integrated MEMS capacitive plate and two impact oscillators for transferring energy from low frequency structural vibration with varying mechanical spectra to a vibration of a high resonance frequency cantilever. The use of the two impact oscillators not only harvested energy at low frequencies but also had demonstrated exceptionally sufficient and optimum dynamic responses to a broad frequency bandwidth between 13 Hz and 39 Hz, the bandwidth covering wide range of residual vibrations in structures and systems, without reduction in output power. The device was designed with a MEMS capacitor fixed at the free end of an aluminium cantilever clamped at one side, with a high resonance frequency of 605 Hz matched with the single-cavity capacitor, and two cantilevers made of Al sheet with low resonance frequencies of 18 Hz and 25 Hz. The results clearly demonstrates the device's ability for frequency up-conversion and harvesting power on a wide range from 13 Hz to 39 Hz at 1g excitation.

1. Introduction

Inertial forces are widely used in MEMS power harvesters, charge pumps and voltage step-up converters for energy conversion from unwanted ambient vibrations into electrical energy with high enough voltage level to power up wireless sensors used in construction, security, aviation and biometry. Autonomous wireless sensor networks can be achieved with vibration power harvesters for structural health monitoring in smart buildings and structures. The majority of MEMS energy harvesters are designed to operate at resonance with narrow bandwidth [1, 2]. However, in the real environment, the vibration frequency varies with time, resulting in significant power reduction for resonance operations. Nonlinear energy harvesting devices have been shown to be capable of operating in a broader frequency range and harvesting more power than resonant devices, making them more appropriate for real life applications. To enhance the operational frequency range, researchers have employed solutions by tuning the resonance frequency to match the ambient vibration or broadening the operational bandwidth of the energy harvester [3, 4]. The resonance frequency was tuned by adjusting the stiffness of suspension beams, changing the size of the proof mass or by bidirectional tuning [3, 5]. Bi-stable springs have been used to harden or soften the suspensions to increase the bandwidth of the wideband MEMS electrostatic energy harvester utilizing nonlinear springs [6, 55], electromagnetic induction energy harvesters [7, 56], a bistable piezoelectric inertial generator [8, 57], a wideband micro-electromechanical energy harvesters [9, 58]. Various impact mechanisms were attempted but still resulted in a relatively narrow bandwidth with low power [10-13]. Frequency-up conversion techniques have been used increase the power harvester's efficiency at low frequency. For example, a periodic single impact between a low frequency driving beam resonator and a single or double high frequency piezoelectric generating beam resonators.

Device Design and Fabrication

The power harvesting device harvests power with a wide bandwidth by transferring the power from low frequency vibration to the high frequency vibration through impact oscillators. The device uses oscillators with low resonance frequencies to convert the energy to a high resonance frequency power harvester through oscillations resulting from intermittent physical contacts between a low frequency oscillator and the power harvester. This technique will enable energy transfer from low frequency structural vibration to power harvester whose natural frequency lies outside the frequency ranges of typical mechanical vibration that the device is designed to harvest energy from. In addition, harvesting low frequency energy is not restricted to the narrow band at the resonance frequency of the impact oscillator. Due to the nonlinearity of the impact oscillator, it has a much broader bandwidth, from its natural frequency to several times of its natural frequency. If the forcing frequency is above the natural frequency when the oscillator impacts on a stiffer structure, it results in a higher "equivalent stiffness" thus a much broader bandwidth than in the absence of impact.

We have designed and built a hybrid macro scale wideband electrostatic power harvester prototype for low frequency vibration (Figure 1). The device consists of a single-cavity MEMS capacitor fixed at the free end of an aluminum cantilever clamped at one side with a length, width and thickness of 63 mm, 13 mm, 3 mm, respectively, with a high resonance frequency of 605 Hz matched with the single-cavity capacitor, and two cantilevers with a length of 34 mm and 50 mm, width of 9.5 mm, 8 mm, and thickness of 0.5 mm, respectively, made of Al sheet vibrating at low excitation frequency, each with a thick impact mass attached at its unclamped end. The low-frequency cantilevers were attached on top of the clamped side of the high frequency cantilever.

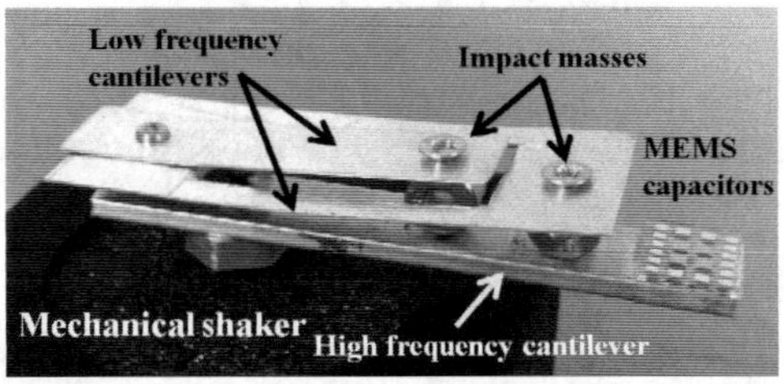

Figure 1. A photo of the macro-scale wideband electrostatic power harvester

One of the two low frequency cantilevers on the macro device was designed to resonate at 18 Hz while the other resonates at 25 Hz within the range of ambient vibration frequency. The metallic impact masses were used in order to reduce the cantilever's resonance frequency and improve the momentum transfer to the high frequency cantilever. When the power harvester was subjected to a low frequency vibration using a mechanical shaker, the two low frequency cantilevers responded by vibrating at low frequencies with at least one of them vibrating close to its resonance frequency, and thus their thick metallic masses made impacts with the high resonance frequency cantilever repeatedly at two locations. This has caused it along with the MEMS capacitor to oscillate with decaying amplitude at its resonance frequency. After the impact occurs, the absorbed energy by the two low-frequency cantilevers were transferred to the high resonance frequency cantilever and to the single-cavity MEMS capacitor, and then to the electrical domain.

The MEMS capacitors were fabricated successfully using surface micromachining technologies on top of a substrate (Figure 2). Cr and Au layers are sputter-deposited on glass substrate with thicknesses of 50 nm and 150 nm, respectively and patterned. A thin layer of NR-9 3000PY negative photoresist is used for the SiO_2 dielectric lift-off process. The first mold for the anchors was prepared with a 10 μm thick layer of AZ P4620 photoresist. The same photoresist layer functions as a sacrificial layer. The photoresist sacrificial layer is cured in a convection oven with an optimized temperature 120 C for 90 minutes and followed by the deposition of another Cr/Au seed layer. A second photoresist mold was patterned on top of the second seed layer to create the top movable plate along with the suspension beams. The mold was then patterned to provide the movable plates with square holes. The moving plate and suspension beams are grown on the second seed layer by electroplating Ni for a thickness of 10 μm. Magnified optical microscopy image of electroplated Ni plate can be seen in Figure 3. After desired thickness of the material is deposited, the top mold is washed away carefully by solvents. The seed layer is then removed by etching Au and Cr consequently.

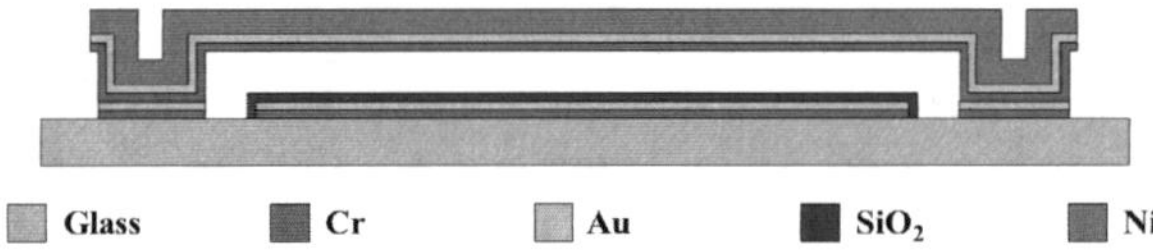

Figure 2. Schematics of the MEMS variable capacitor fabrication process by surface micromachining on a glass substrate.

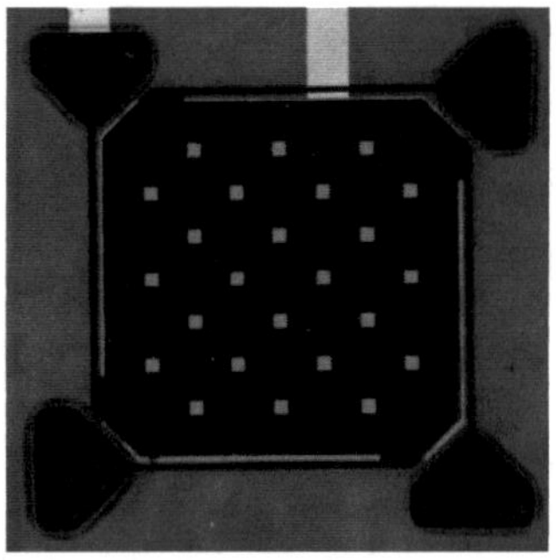

Figure 3. Optical image of the magnified view of the device after removing the Au and Cr layers. After this step, bottom electrode and trace lines can be seen through transparent photoresist layer.

2. Device Testing

Prior to assessing the fabricated devices, a simple electrical circuit was built to extract the output power generated by the MEMS power harvester. The power source performance was characterized by placing it on a controllable vibration shaker with adjustable vibration frequency and acceleration. The device was connected to a load resistor and the output voltage was measured across a load resistor connected in series with the MEMS capacitor. The voltage was monitored on a PC using LabVIEW and a data acquisition cable through an oscilloscope.

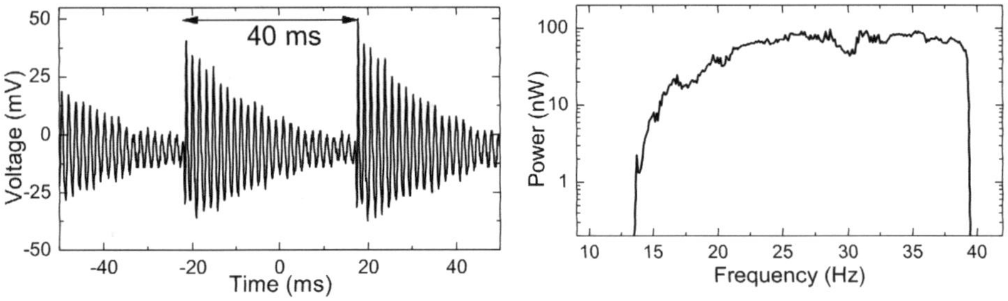

Figure 4. a) Time domain plot of the output voltage across the load resistor. b) The RMS output power as a function of frequency for 1 g excitation acceleration.

An analog accelerometer was used to determine the acceleration at various vibration conditions, and thus calibrate the power harvesting device. The power harvesting devices were characterized by fixing the device on a rigid stage on the shaker. AC signals of different frequencies were applied to the shaker, and the resulting output voltages were measured across a load resistor connected in series with MEMS capacitor, using an oscilloscope and an AC voltmeter. To determine the resonance frequency of the high frequency cantilever, the excitation frequency was swept from 100 to 1 kHz, and the generated output voltages were measured using a Multimeter. The resonance frequency was observed around 605 Hz for low frequency excitation conditions (Figure 4.a). A typical time domain plot of the measured output AC voltage across a load resistor (100 kΩ) when the device was exposed to an ambient vibration at 25 Hz is shown in Figure 4.a. The plot clearly demonstrates a nonlinear behavior with exponential decay at its resonance frequency of 605 Hz, and the peak to peak signal corresponds to a low frequency vibration of 25 Hz.

This clearly indicates the device ability for frequency up-conversion. To estimate the response bandwidth, the devices were first characterized by sweeping the excitation frequency from 5 Hz to 50 Hz with 1 g of constant acceleration, while the output voltage was measured for each excitation frequency across the load resistor. Initially, with a small acceleration of 1 g, only one impact cantilever with a resonance frequency of 18 Hz responded and made impacts with the high frequency cantilever up to 21 Hz. As the frequency increased further, both cantilevers started to make impact with the high frequency cantilever. As the frequency increased further, the low frequency cantilever stopped hitting the high frequency cantilever beam, and only the second cantilever beam kept hitting the high frequency cantilever beam up to 39 Hz. This has resulted in a wide bandwidth response from 13-39 Hz at 1 g (Figure 4.b). Qualitatively speaking, the oscillator responded with a variable resonance frequency depending on the degree of impact. Therefore, the device had an optimal response over a much broader low frequency. It is noted that two impact masses were enough to cover the frequency range from 13 to 39 Hz at 1 g with a maximum power of 96.2 nW.

References

[1] E. O. Torres and G. A. Rincón-Mora, "Electrostatic energy-harvesting and battery-charging CMOS system prototype," IEEE Trans. Circuits Syst., vol. 56, no. 9, pp. 1938-1948, 2009.

[2] T. Von Buren, P. D. Mitcheson, T. C. Green, E. M. Yeatman, A. S. Holmes, and G. Troster, "Optimization of inertial micropower generators for human walking motion," IEEE Sensors J., vol. 6, pp. 28-38, 2006.

[3] D. Zhu, M. J. Tudor, and S. P. Beeby, "Strategies for increasing the operating frequency range of vibration energy harvesters: a review," Meas. Sci. Technol., vol. 21, no. 2, pp. 1-29, 2010.

[4] S. Roundy and Y. Zhang, "Toward self-tuning adaptive vibration based micro-generators," Proc. SPIE, pp. 373–384, 2005.

[5] V. R. Challa, M. G. Prasad, Y. Shi, and F. T. Fisher, "A vibration energy harvesting device with bidirectional resonance frequency tunability," Smart Mater. Struct., vol. 17, no. 1, 2008.

[6] D.S. Nguyen, E. Halvorsen, G. U. Jensen, and A. Vogl, "Fabrication and characterization of a wideband MEMS energy harvester utilizing nonlinear springs," J. Micromech. Microeng., vol. 20, no. 12, 2010.

[7] B. P. Mann and N. D. Sims, "On the performance and resonant frequency of electromagnetic induction energy harvesters," J. Sound Vib. vol. 329, pp. 1348-1361, 2010.

[8] S. C. Stanton, C. C. McGehee, and B. P. Mann, "Nonlinear dynamics for broadband energy harvesting: investigation of a bistable piezoelectric inertial generator," Physica D: Nonlinear Phenomena, vol. 239, pp. 640-653, 2010.

[9] S. D. Nguyen, E. Halvorsen, and I. Paprotny, "Bistable springs for wideband microelectromechanical energy harvesters," Appl. Phys. Lett., vol. 102, 2013.

[10] S. Moss, A. Barry, I. Powlesland, S. Galea, and G. P. Carman, "A low profile vibro-impacting energy harvester with symmetrical stops," Appl. Phys. Lett., vol. 97, no. 23, 2010.

[11] K. Ashraf, M. H. Md Khir, J. O. Dennis, and Z. Baharudin, "A wideband, frequency up-converting bounded vibration energy harvester for a low frequency environment," Smart Mater. Struct., vol. 22, no. 2, 2013.

[12] B. P. Mann and N. D. Sims, "Energy harvesting from the nonlinear oscillations of magnetic levitation," J. Sound Vib., vol. 319, pp. 515-530, 2009.

[13] N. S. Yuksek, Z. C. Feng, and M. Almasri. "Broadband electromagnetic power harvester from vibrations via frequency conversion by impact oscillations," *Appl. Phys. Lett.*, vol. 105.11 (2014): 113902.

Reliability study of Piezoelectric Structures Dedicated to Energy Harvesting by the Way of Blocking Force Investigation

S Maaroufi[1,2], F Parrain[1], E Lefeuvre[1], B Boutaud[2], R Dal Molin[2]
[1]Univ Paris-Saclay – CNRS, Institut d'Electronique Fondamentale, Centre Scientifique d'Orsay, bât. 220, 91405 Orsay Cedex, France
[2]SORIN CRM: Advanced research department, 92140 Clamart, France

E-mail: seifeddine.maaroufi@u-psud.fr

Abstract. In this paper we propose an approach to study the reliability of piezoelectric structures and more precisely energy harvesting micro-devices dedicated to autonomous active medical implants (new generation pacemakers). The structure under test is designed as a bimorph piezoelectric cantilever with a seismic mass at its tip. Good understanding of material aging and mechanical failure is critical for this kind of system. To study the reliability and durability of the piezoelectric part we propose to establish a new accelerated methodology and an associated test bench where the environment and stimuli can be precisely controlled over a wide period of time. This will allow the identification of potential failure modes and the study of their impacts by the way of direct mechanical investigation based on stiffness and blocking force measurements performed periodically.

1. Introduction

Recent advances in microfabrication and biotechnology have allowed the development of a wide variety of implantable miniaturized systems. Their application can range from monitoring and diagnostics to localized treatments. All these systems include, among other things, electronics parts and require an energy source to be supplied. Even if the use of a battery is still possible, the replacement of it (and more generally the whole system) is not a common and simple procedure especially in the case of a leadless pacemaker that is directly implanted within a heart cavity of the patient (Figure 1). In this case energy harvesting is an interesting alternative to traditional batteries making the targeted systems fully autonomous and allowing further miniaturization and enhanced functionalities.

Academic and industrial studies have shown through different proofs of concept the viability of this approach, particularly for leadless pacemakers [1]. Different principles/architectures were developed and evaluated in the literature. As example, some of them use electrostatic transduction [2] [3], blood pressure variations [4] [5] or piezoelectric stacks as the targeted system developed by SORIN CRM (Cardiac Rhythm Management) and studied in this paper.

In all these cases, maximizing the power density invited to use different materials to their full potential. However, the long term reliability of these devices/materials is largely unknown today, and detailed study is needed to pass from proof concepts to demonstrators and finally to industrialization. Indeed, the knowledge of the lifetime materials, components and assemblies is necessary to ensure proper long-term operation of the implant, and optimize device performance. In the field of medical implants like pacemakers, their long-term reliability is crucial because it could impact patient survival.

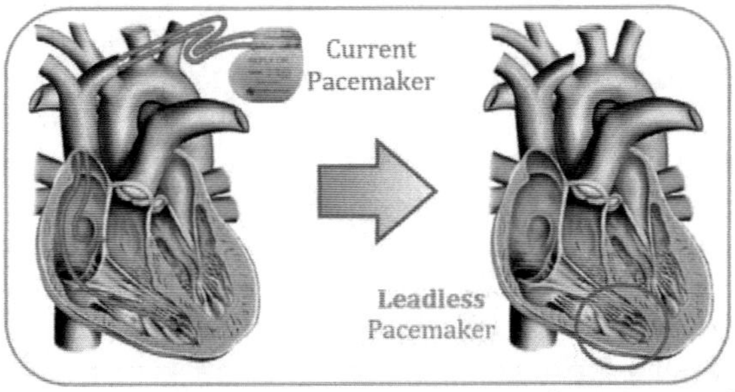

Figure 1. Current Pacemaker versus Leadless Pacemaker.

To investigate electro-mechanical properties and so the reliability of a piezoelectric-based device over its lifetime, several approaches are possible. The most common is certainly impedance spectroscopy that allows extracting from the measurement of impedance/admittance over frequency the main electro-mechanical parameters (resonant frequencies, associated Q factors, coupling coefficient [6]). Nevertheless, due to excitation amplitude generally provided by the commercial impedance meters and the underlying formalism (small-signal hypothesis), this method doesn't allow to investigate large signal behaviour and so possible piezoelectric/mechanical nonlinearities.

In our case we propose an original massively parallel aging test bench (up to 36 structures could be simultaneously tested) where the device under test is actuated through its base by the way of a shaker. Thanks to a custom system associated to each structure under test, it makes possible to extract the structure stiffness measuring the reaction force for a given movement but also the blocking force for a given excitation voltage. That allows to extract periodically electro-mechanical parameters of the device under test (mainly Young's modulus, piezoelectric coefficient d_{31}) and so the evolution of them to monitor material aging or ruptures (cracks and/or delamination).

2. Studied structure architecture

Aside from the associated circuitry and the seismic mass (a few grams in tungsten) that allows the reduction of the system resonant frequency (~10 Hz) to tune the bandwidth over the vibration environmental spectrum, the energy harvester is mainly composed of a rectangular-shaped bimorph with a low thickness with respect to its others in-plane dimensions (a few millimeters wide and about 20m in length, see Figure 2).

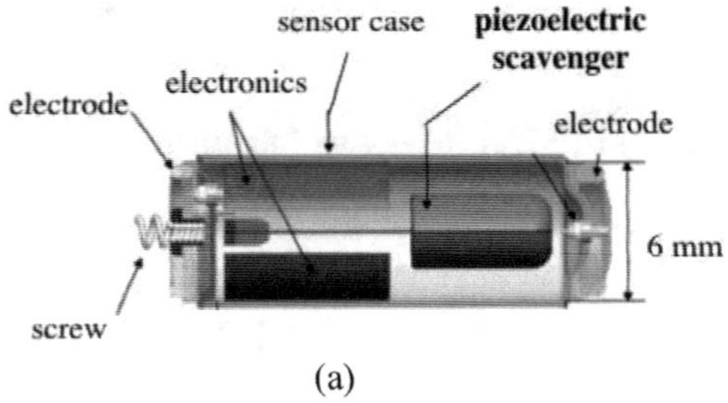

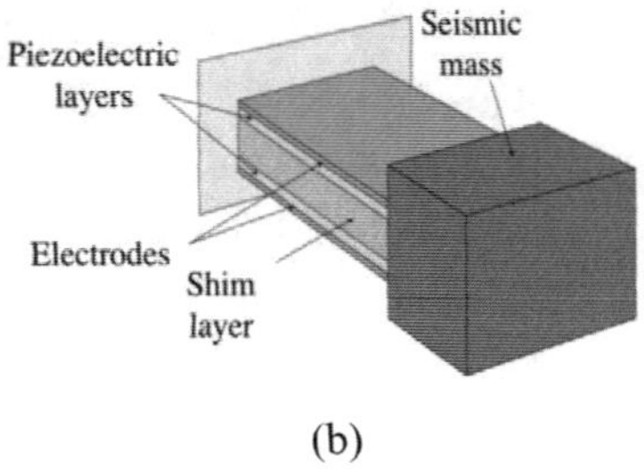

Figure 2. Leadless pacemaker architecture (a). Piezoelectric scavenger (b).

The structure consists of a stack of different layers: two ceramic PZT (Lead Zirconate Titanate) films are bonded onto a structural metallic layer (shim). Finally both sides are metalized to collect the generated electrical charges and to make possible wire bonding. In the final application, the structure will operate by inertia effect during the heartbeat. Successive constraints elongation and compression of PZT layers will provide electrical energy by piezoelectric coupling.

3. Accelerated aging test bench

In this paper the proposed test bench mainly consists of a shaker that can actuate simultaneously up to 36 piezoelectric energy scavengers samples for long time periods (up to 10^9 cycles). The shaker will mimic same mechanical deformation and the same electrical loading as in a real use but at much higher frequency (from 100 to 200Hz). By this way, the ageing is expected to be accelerated by a factor proportional to the frequency ratio. The scavenger operation frequency in real environment is of nearly 10Hz, so the aging is expected to be accelerated by a factor in the range of 10 to 20. If necessary, the aging may be largely increased using large deflection. To be under realistic conditions, the structures under test will be placed in a controlled environment to maintain the temperature at 37°C (98.6 °F) which is the working temperature of implanted devices. To control the induced mechanical deformation under imposed displacement the shaker movement is continuously measured by the way of a laser vibrometer system.

So as to impose the mechanical deformation corresponding to the first flexion mode of a cantilever beam as in real use, the tip of the scavengers (without the seismic mass always omitted during testing) will be maintained vertically fixed by a custom "pinch system" nevertheless leaving free the rotation in both directions thanks to sphere-plan hertzian contacts ensured by a calibrated spring as shown in Figure 3.

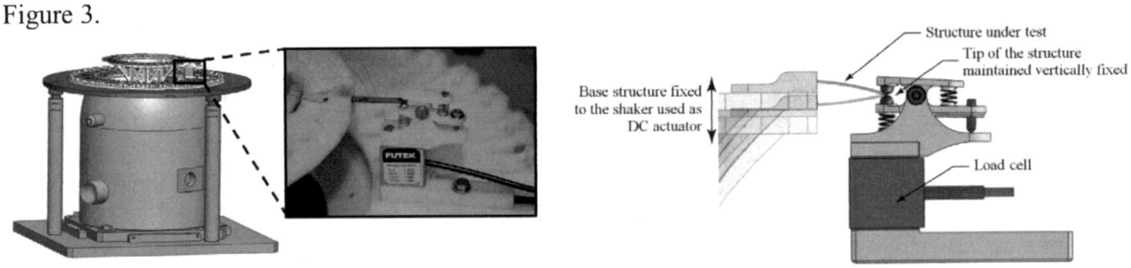

(a) (b)

Figure 3. (a) Global view of the proposed test bench that can test simultaneously up to 36 piezoelectric scavengers. (b) Pinch system that allows to maintain vertically fixed the tip of the structure under test.

The majority of the mechanical parts make up that system were realized by the way of stereolithography that allows to obtain resistant and light elements. This is well suitable for those in movement (attached to the shaker rod) where the weight bounds the maximal frequency for a given amplitude (typical amplitude of 2mm peak to peak). The two beads used for hertzian contacts have a diameter of 3mm and are made of treated and rectified stainless-steel that provides an excellent wear resistance minimizing surface damage at the sample level (electrodes).

4. Electro-mechanical investigation methodology

Throughout the test the accelerated aging cycle at "high" frequency is periodically stopped (every couple of hours) to monitor the health of the structures performing DC characterization. Thus in the case of the proposed test bench the device under test could also be moved through its base by the way of the shaker operating at DC thanks to a custom power amplifier. Indeed standard power amplifiers generally associated with this kind of equipment has a cut-off frequency of a few Hz prohibiting DC actuation. As in AC long-term excitation, the amplitude of the DC movement is precisely controlled via a laser vibrometer system having a repeatability of about 100nm (Keyence vibrometer). As shown in Figure 3, for each structure, the pinch system used to maintain vertically fixed the tip of the structure is itself mounted onto a 50gf load cell (LSB200 from Futek) that makes possible to extract the structure stiffness measuring the reaction force for a given base displacement (in open or short circuit) but also the blocking force for a given excitation voltage (in this case the based is maintained in a fixed position). The measurement of the stiffness gives us

informations about the Young's modulus of the different materials but also about the mechanical integrity of the structure (cracks, delamination …). In other hand the blocking force is directly related to the transverse piezoelectric coefficient d_{31} as shown in [7].
The load cell is associated to a custom readout electronics based upon an AD7780 converter from Analog Devices (24-bit resolution). According to the datasheet, this sigma delta (Σ-Δ) ADC has a p-p resolution for 18.8 effective bits (PGA gain set to 128V/V, 10SPS update rate). Considering the sensitivity of the load cell (2mV/V at rated load), the detection limit is better than 1mgf (~10μN).
The main drawback of the used pinch system is to induce a restoring couple attributable to the hertzian contacts. This has as effect to modify the effective stiffness of the structure. Thus, in the case of small deflections the restoring couple M could be expressed as:

$$M(\alpha) = -\frac{F_B.(2.R+e).tan(\alpha)}{cos(\alpha)} \approx -F_B.(2.R+e).tan(\alpha) \quad \mathbf{(1)}$$

Here α, e, F_B, R are respectively the tip angle deflection with respect to the horizontal, the thickness of the scavenger, the pinch force and the beads diameter. Using the stiffness matrix of Euler-Bernoulli beam (scavenger is equivalent to a thin cantilever) [8], it's possible to calculate the effective stiffness k_{eff}:

$$k_{eff} = k_0.\left[1 - \frac{9}{4}.\frac{F_B.(2.R+e)}{L^2.k_0 + 3.F_B.(2.R+e)}\right]^{-1} \quad \mathbf{(2)}$$

Here k_0 is the initial stiffness of the scavenger (without pinch system) and L the structure free length between the anchor and the hertzian contact point. The last expression clearly shows that the effective stiffness k_{eff} increases with respect to the pinch force F_B.
Others contributions such as the stiffness of the association of the load cell topped by the pinch system could affect the mechanical measurements. Nevertheless the stiffness of the whole system has been experimentally evaluated to more than 2.5kN.m^{-1}. This value is to compare to the stiffness of the scavengers that is equal to about 350N/m^{-1} that allows us to conclude that the influence of it is more than negligible.

5. Experimental results

A number of preliminary tests were carried out using damage-free scavengers from Piezo Systems Incorporated. A set of measurements of the reaction force for a given base displacement and the blocking force for a given excitation voltage is shown in Figure 4.

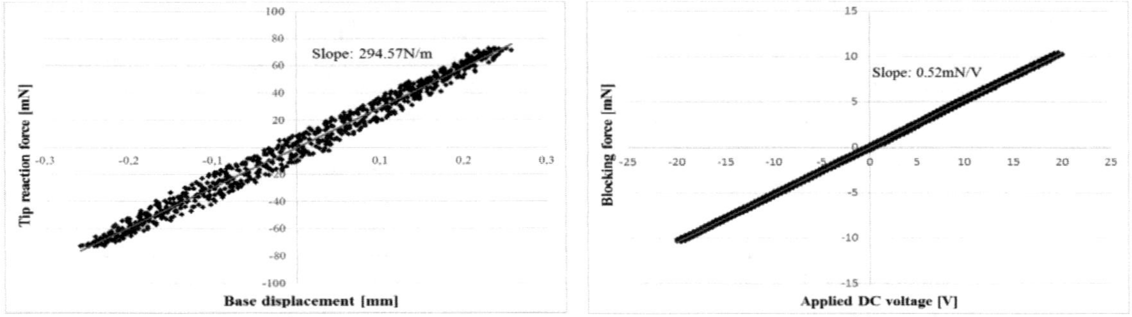

(a) (b)

Figure 4. (a) Tip reaction force versus base displacement. (b) Blocking force versus applied voltage.

The scavenger stiffness and the blocking force coefficient have been extracted by linear regression and compared to the theoretical values calculated using the material parameters provided by the manufacturer. All these data are summarized in the table 1 below:

Table 1. Experimental values compared to theoretical ones.

Parameters	**Theory**	**Measurements**	**Variation**
Scavenger stiffness [N.m^{-1}]	359	294±10	19.7%
Blocking force coefficient [mN.V^{-1}]	0.618	0.52±0.05	17.1%
Transverse piezoelectric coefficient d_{31} [pC/N]	-320	-329±40	2.7%

The different curves have been plotted following several loading cycles and exhibit a slight hysteresis behaviour. This could be explained, amongst others, by drift and possible stiction between the steel beads and the structure surface. Concerning the theoretical variation of the stiffness with respect to the pinch force, this phenomenon cannot be verified hidden by the uncertainty concerning the structure free length that plays a crucial role in the calculation (the stiffness is proportional to L^{-3}).

6. Conclusion

This paper presents a massively parallel test bench that allows to study the reliability of up to 36 energy harvesting micro-devices. Here the mechanical integrity of the structures but also the piezoelectric material aging could be monitored by the way of direct mechanical investigation based on stiffness and blocking force measurements. This especially allows us to investigate large signal behaviour unlike impedance spectroscopy commonly used for this kind of system as reported in the literature. However the presented method doesn't allow to obtain the Young's modulus of each material but just the value of the effective one related to global structure. Similarly impedance spectroscopy is still necessary if the knowledge of the electrical permittivity is needed.

Notwithstanding this, preliminary test results on undamaged structures show that experimental and expected values are close together demonstrating the pertinence and the efficiency of this testing/characterization methodology. Our future work will focus on the analysis of the impact of the pinch system especially on the sample surface that could be damaged by sphere/plan contact. A particular effort will be also made to automate the different measurements so as to study, for this time, the long term reliability of a complete set of 36 structures.

7. References

[1] Deterre M Lefeuvre E and Dufour-Gergam E 2012 *Smart Materials and Structures* 21

[2] Mitcheson P D Green TC and Yeatman E M 2007 *Microsystem Technologies* 13 1629

[3] Meninger S Mur-Miranda J O Amirtharajah R Chandrakasan A P and Lang J H 2001 *IEEE Transactions on Very Large Scale Integration (VLSI) Systems* **9** 64

[4] Horowitz S B Sheplak M Cattafesta L and Nishida T 2006 *Journal of Micromechanics and Microengineering* 16 174

[5] Iizumi S Kimura S Tomioka S Tsujimoto K Uchida Y Tomii K Matsuda T and Nishioka Y 2011 *Eurosensors XXV Procedia Engineering* **25** 187

[6] Barsoukov E and Macdonal J R *Impedance Spectroscopy Theory, Experiment, and Applications, Second Edition* (John Wiley & Sons, Inc., Publication)

[7] Dunsch R and Breguet J M 2006 *Sensors and Actuators A* **134** 436

[8] Kwon Y W and Bang H C *The Finite Element Method Using MATLAB, Second Edition* (CRC Mechanical engineering series)

Acknowledgments

The authors acknowledge the support of the French Agence Nationale de la Recherche (ANR) under reference ANR-14-CE17-0010 (LAUREAT project).

Design and manufacture of perpendicular bi-stable cantilever for vibrational energy harvesting on the basis of stochastic resonance

M Kawano, Y Zhang, R Zheng, K Nakano and B Kim
Institute of Industrial Science, The University of Tokyo, 4-6-1 Komaba, Meguro, Tokyo, 153-8505, Japan

E-mail: kawano-m@iis.u-tokyo.ac.jp,

Abstract. This paper describes extremely simple configuration of novel vibrational energy harvester, which can harness low frequency (less than 5 Hz, such as various environmental vibrations) over a broad frequency band for the first time. A design that utilizes a phenomenon called stochastic resonance can give significantly enhanced vibration mode for increasing efficiency, and simple bi-stable cantilever with tip mass installed a basement vertically fulfils the requirements for stochastic resonance. We fabricated bi-stable cantilever with tip mass and validated whether the cantilever could be used as an effective low frequency vibration energy harvester. In the experiment, when a 1 Hz periodic force and environmental noise vibration were applied, stochastic resonance occurred. The amplitude of the energy harvester increased over tenfold (over 30 mm).

1. Introduction

Recently, the idea called "Internet of Things [1]" is obtaining a great deal of attention and the number of wireless sensors has been growing exponentially. In this situation, there will be too many sensors to charge or to exchange in our society, and the field of energy harvesting is gathering attention as a solution of sensors' power source problem. Energy harvesting is a method for scavenging power from environmental energy sources like solar energy, thermal energy and vibrational energy instead of batteries. However, the solar power is unavailable in dark environment and thermal energy is unavailable in cold environment. By contrast, vibrational energy is more robust for condition changes. Therefore, more and more researcher are focusing on vibrational energy harvesting.

To convert vibrational energy to electrical energy, there are three types of vibrational energy harvesting. First type is electromagnetic induction energy harvesting. The electromagnetic induction type requires large magnets, so its structure is likely to become bulky. Second type is electrostatic energy harvesting. This type has difficulty in controlling gaps between electrodes in order that the gaps are small and electrodes does not touch each other. Third type is piezo energy harvesting. Energy density of this type is three times larger as compared to the two counterparts [2].

In previous studies, cantilevers with piezo films are utilized for piezo vibrational energy harvesting frequently. However, there are two major problems in using piezo films on cantilevers [3]. First, ambient vibration frequency is much lower than the resonant frequency of standard energy harvesters imposed by harvester's stiffness. Second, electro-mechanical conversion efficiency decreases promptly when environmental vibration frequencies deviates from resonant frequencies of energy harvesters due to their narrow frequency band operation design.

The objective of this work is to design and manufacture a new piezo energy harvester that can convert low frequency vibrations to electrical energy over a wide frequency band.

2. Proposed energy harvester

To fulfil the research objective, we proposed to use a remarkable phenomenon called stochastic resonance. When stochastic resonance occurs, amplitude of bi-stable oscillators increases [4]. To make stochastic resonance occur, three conditions need to be fulfilled: double-well potential, a periodic force and environmental noise vibration. A cycle of stochastic resonance is shown in figure 1. In this cycle, first, a periodic force reduces the potential barrier height. Second, oscillator state jumps to the other well, being triggered by environmental noise vibration. When this cycle periodically repeats, the oscillator fluctuates between two wells at the resonant point.

Various methods have been proposed to build double-well potential systems, such as the use of permanent magnets [5]. When permanent magnets are utilized to build double-well potential system, fabrication process becomes complex and miniaturization of devices becomes difficult. In this work, we proposed to use vertical bi-stable cantilevers because its structure is simple to fabricate and the miniaturization is possible. Figure 2 shows the structure of perpendicular bi-stable cantilever where v and u are respectively the horizontal and vertical displacement of the tip mass, s is the distance of the beam from the basement, v_{p} and u_{p} are respectively the horizontal and vertical displacement at the point where the distance of the beam is s, x and y are respectively the horizontal and vertical displacement of the basement and M_{t} is the tip mass. In this study, vertical vibration $\ddot{y}$ is taken into account for the first time in addition to horizontal vibration [6]. By considering vertical vibration and horizontal vibration, vibrations in all directions could be simulated.

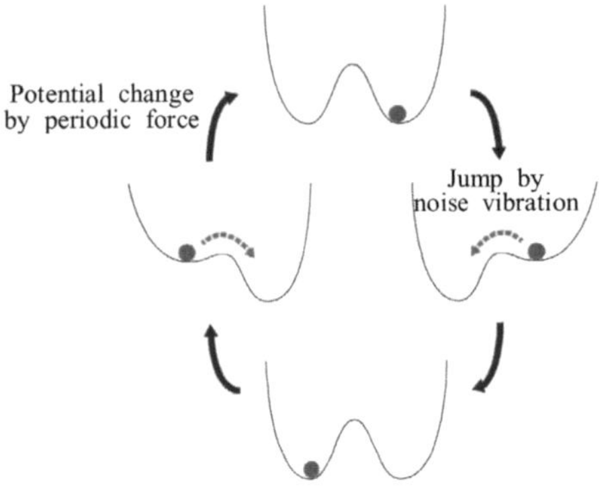

Figure 1. The cycle of stochastic resonance.

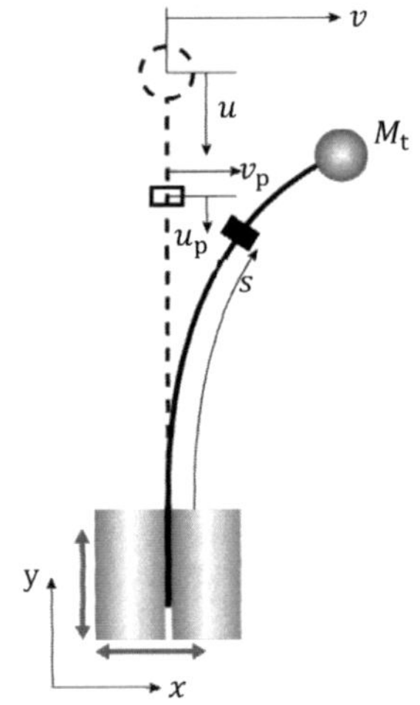

Figure 2. Mechanical system of cantilever.

The kinetic energy of this cantilever's system T is defined by

$$T = \frac{1}{2}\rho A \int_0^L \left[\left(\dot{v}_{\mathrm{p}}(s,t) + \dot{x}\right)^2 + \left(\dot{u}_{\mathrm{p}}(s,t) - \dot{y}\right)^2\right] \mathrm{d}s + \frac{1}{2}M_{\mathrm{t}}[(\dot{v} + \dot{x})^2 + (\dot{u} - \dot{y})^2] + \frac{1}{2}I_{\mathrm{t}}\dot{\phi}^2 \quad (1)$$

where ρ is density of the beam, A is the cross section area of the beam, L is the length of the beam, x is the horizontal displacement of the basement, w is the vertical displacement of the basement, I_{t} is moment of inertia of the end mass, ϕ is the slope of the tip mass and the dot means time-derivative.

The position potential energy of the system Π is defined by.

$$\Pi = \frac{1}{2}EI\int_0^L\left(\kappa(s,t)\right)^2\mathrm{d}s - \rho Ag\int_0^L\left(u_\mathrm{p}(s,t)-y\right)\mathrm{d}s - M_\mathrm{t}g(u-y) \tag{2}$$

where E is Young Modulus of the beam, κ is the beam's curvature and g is acceleration of gravity.

Simplifying Lagrange's equation, equation of motion of the system is obtained as in equation (3).

$$\begin{aligned}[M_\mathrm{t} + \rho AN_1 + I_\mathrm{t}N_5^2 + (\rho AN_3 + M_\mathrm{t}N_4^2 + I_\mathrm{t}N_5^4)v^2]\ddot{v} + (M_\mathrm{t}N_4^2 + \rho AN_3 + I_\mathrm{t}N_5^4)v\dot{v}^2 \\ +(EIN_6 - N_9\rho Ag - N_4M_\mathrm{t}g + 2EIN_7v^2)v \\ = -(\rho AN_2 + M_\mathrm{t})\ddot{x} + (\rho AN_9 + M_\mathrm{t}N_4)v\ddot{y} + N(t)\end{aligned} \tag{3}$$

where N_1-N_9 are constants and $N(t)$ is environmental noise vibration.

3. Results

In this section, the simulation and experiment are described. The simulation and experiment results are discussed at the end of this section.

3.1. Simulation

Optimization of the parameters is conducted under the following conditions. First, tip mass must be over the mass where beam starts buckling. Second, the range of equilibrium position is set between 1 mm and 50 mm. Based on these conditions, tip mass and beam thickness were determined as 7.2 g and 100 µm, respectively. Then, three numerical simulations were implemented as shown in figure 3 (a), (b) and (c) and parameter setting in simulation is described in table 1. As figure 3 (a) and (b) shows, when the energy harvester is excited only by ambient noise or periodic force, tip mass of the beam oscillates around only one equilibrium position. By contrast, when periodic force and environmental noise vibration oscillate the vertical bi-stable cantilever, the beam's tip mass is able to oscillate between two equilibrium positions, increasing the beam oscillation considerably. Therefore, the amplitude of the tip mass increases more than twofold (more than 20 mm).

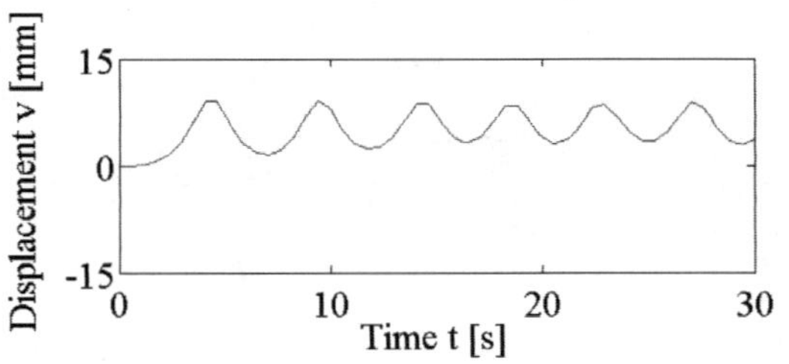

(a)

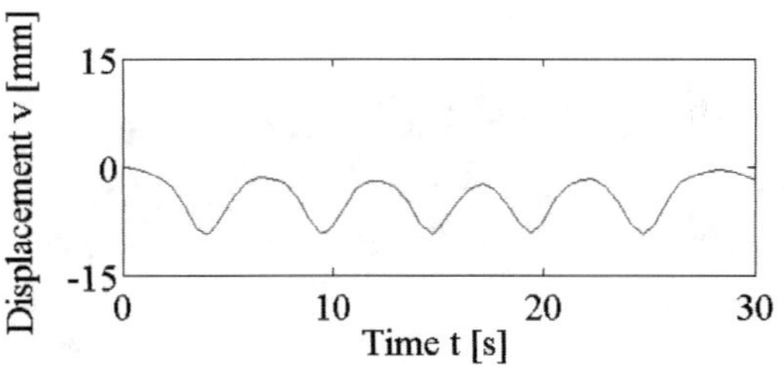

(b)

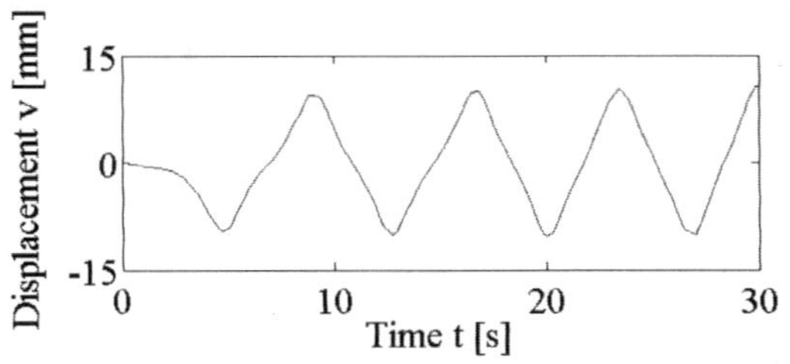

(c)

Figure 3. Simulation results. Three types of simulations are conducted: (a) excited by only environmental noise vibration, (b) excited by only a periodic force and (c) excited by both of them.

Table 1. Parameter setting in simulation.

Length of the beam (mm)	Width of the beam (mm)	Tip mass (g)	Thickness of the beam (µm)	Young Modulus (GNm^{-2})	Density of the beam (kgm^{-3})
50	4	7.2	100	197	7930

3.2. Experiment
In this subsection, experimental setting and results are described.

3.2.1. Experimental setting. An energy harvester is designed to utilize stochastic resonance and fabricated. Figure 4 shows fabricated energy harvester. Specification of the energy harvester is showed in table 2. This energy harvester is the second version because the first version of the energy harvester was fabricated without consideration of the vertical vibration mode of actual vibration generator. Furthermore, the second version of the tip mass is triangular prism type because the first version of the tip mass which is quadrangular type was likely to rotate in addition to vertical vibration. The beam and the tip mass were made of stainless steel (SUS304) and carbon steel (S45C), respectively. The energy harvester and a laser sensor (IL-100, Keyence, Japan) are installed on a vibration generator (m060, IMV, Japan). Amplitude of the beam was measured at 20 mm above the basement. A periodic force's frequency is 1 Hz, respectively. Directions of the periodic force and environmental noise vibration are vertical directions.

Table 2. Specifications of energy harvester.

Length (mm)	Width (mm)	Tip mass (g)	Thickness of the beam (μm)
50	4	7.2	100

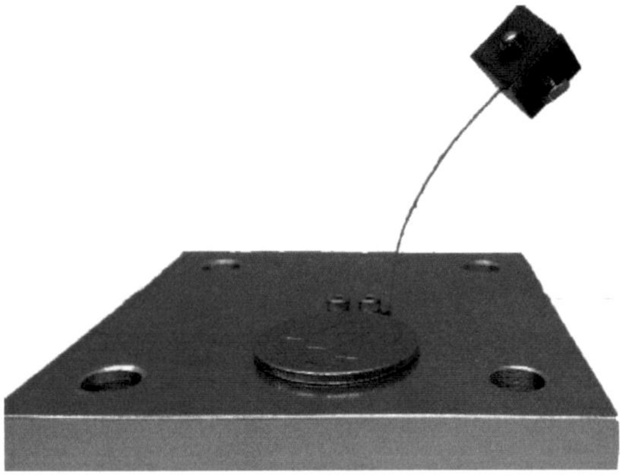

Figure 4. Fabricated energy harvester. The length and width of the energy harvester is 50mm and 4mm, respectively. The tip mass is 7.2 g.

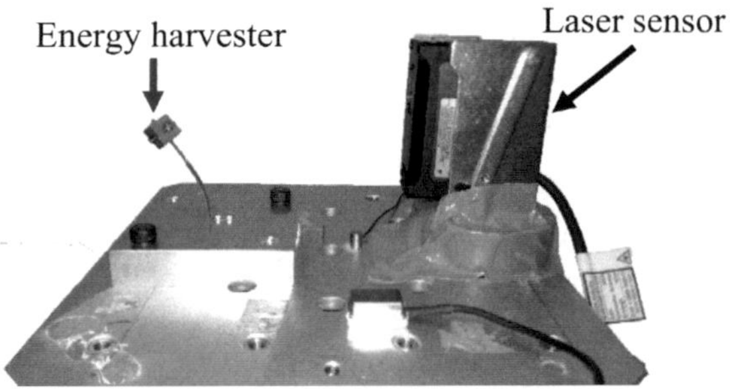

Figure 5. Experimental setting. Energy harvester and laser sensor are installed on vibration generator.

3.2.2. Experimental results. Figure 6 (a) shows the result when only the periodic force was applied and Figure 6 (b) shows the result when only the environmental noise vibration is applied. In both cases, the oscillator vibrates around only one equilibrium as seen in simulation, so the amplitude is about 3 mm. On the other hand, Figure 6 (c) shows the result when both the periodic force and the ambient noise vibration are applied, and the oscillator can move between two equilibrium positions, so stochastic resonance occurred and the amplitude of the tip mass increased over tenfold (over 30 mm).

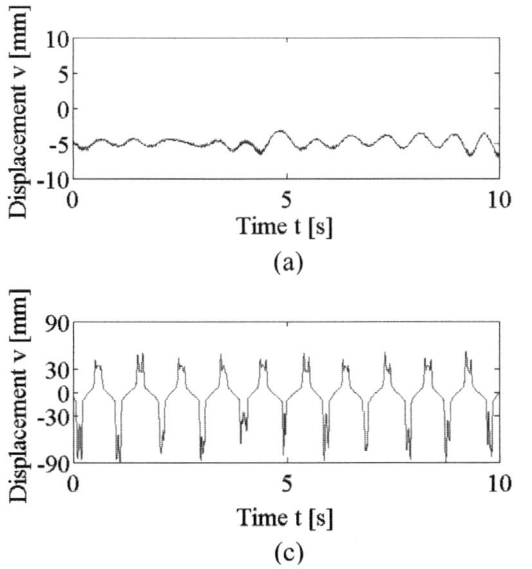

Figure 6. Experimental results. Three types of experiments were conducted: (a) excited by only environmental noise vibration, (b) excited by only a periodic force and (c) excited by both of them.

3.3. Discussion

The disturbance of wave showed in figure 6 (c) is due to tip mass's blocking laser sensor light. To eliminate this disturbance, the point that will be measured by the laser sensor should be lowered to less than at 20 mm above the basement. In the simulation and experiment, only when the periodic force and environmental noise vibration excited the energy harvester, the tip mass could vibrate between two equilibrium positions. These results verified stochastic resonance's occurrence.

4. Conclusion

In conclusion, this paper designed and demonstrated a new energy harvester that can be used for low frequency vibration energy harvesting over a wide frequency band. Our current effort is to attach a piezo film to the energy harvester in order to generate electrical power. For future work, we will downsize the energy harvester and the energy harvester will be tested in real environment such as on bridges and cars.

References

[1] Hecht J 2014 *Nat. Phys.* **10** 538
[2] Priya S 2007 *J. Electroceramics* **19** 167-84
[3] Eichhorn C, Tchagsim R, Wilhelm N and Woias P 2011 *J. Micromech. Microeng.* **21** 104003
[4] Benzi R, Parisi G, Sutera A and Vulpiani A 1982 *Tellus* **34** 10-6.
[5] Nakano K, Cartmell M P, Hu H and Zheng R 2014 *J. Mech. Eng.* **60** 314-20
[6] Friswell M I, Ali S F, Adhikari S, Lees A W, Bilgen O and Litak G 2012 *J. Int. Mat. Sys. Stc.* **13** 1505-21

Dynamic study for performance improvements of a thermo-mechanically bistable heat engine

J. Boughaleb[1,2,4], A. Arnaud[1], S. Monfray[1], P.J. Cottinet[3], S. Quenard[2], G. Pitone[4], F. Boeuf[1], D. Guyomar[4], T. Skotnicki[1]

[1]STMicroelectronics, 850 rue Jean Monet – 38920 Crolles, France
[2]CEA-LITEN, 17 rue des Martyrs – 38054 Grenoble Cedex 9, France
[3]LGEF, INSA Lyon, 8 rue de la Physique – 69621 Villeurbanne Cedex, France
[4]Delta-concept, 6 rue de Chamechaude, 38360, Sassenage, France

E-mail: jihane.boughaleb@st.com

Abstract. This paper focuses on a thermal study of a thermal energy harvester based on the coupling of a bimetallic strip heat engine with a piezoelectric membrane for wasted heat scavenging. Such a harvester is dedicated to power autonomous systems such as wireless sensor nodes. For a better understanding of the working principle of the system, it is compulsory to have a good understanding of the thermal specificities and phenomenon taking place inside the harvester. Attention is consequently focused on the thermal modeling of the harvester in static mode using the equivalence between the electrical and thermal quantities. This first modeling step allowed the improvement of the thermal properties inside the system by increasing the thermal gradient across it. However, the bimetal being the active part of the system has not been taken into account in this model and shadow zones persisted regarding the bimetal operation windows as a function of its snapping temperatures and hysteresis. To overcome this, a dynamic model is proposed in this paper taking into account the bimetal as a switched capacitance alternatively in contact with the hot source and the cold surface. This last model completed the static one by predicting the bimetal's operation windows in function of its intrinsic properties and the operation range evolution in function of the snapping temperature first and then in function of the bimetal thermal hysteresis. Moreover, experimental measurements enable to validate the proposed model and to point out the most powerful bimetals for scavenging higher amounts of power.

1. Introduction

Energy harvesting technologies have known a significant expansion during the last decades and researchers are still focusing on this area as being a clean and promising technology field. This increased development has been allowed by the need for wireless sensors. Internet of things, smart buildings, industrial and health monitoring these are some domains where sensor networks are widely used. However, the impractical aspect of wired systems and the problems related to the power supply by short-lifetime batteries are raised for those applications. This brought attention on ambient energy harvesters and explains the great interest in energy harvesting devices over the last years. This paper presents a study conducted on a thermal energy harvester previously presented in [1,2,3] and based on a two-step conversion mechanism. At first a thermo-mechanical conversion insured by a bimetallic strip occurs and it is then coupled to an electromechanical transduction ensured by a piezoelectric membrane. The thermal energy harvester is presented in Fig. 1.a. The bimetallic strip used in this harvester is characterized by two equilibrium positions as shown in Fig. 1.b. This makes it keeps on oscillating between the cold and the hot reservoirs. For an efficient energy harvesting, the bimetal should be set in a significant thermal gradient to ensure its thermal instability and to increase its cycling frequency. To ensure a good functioning of the bimetal, a static and a dynamic models are established.

(a)

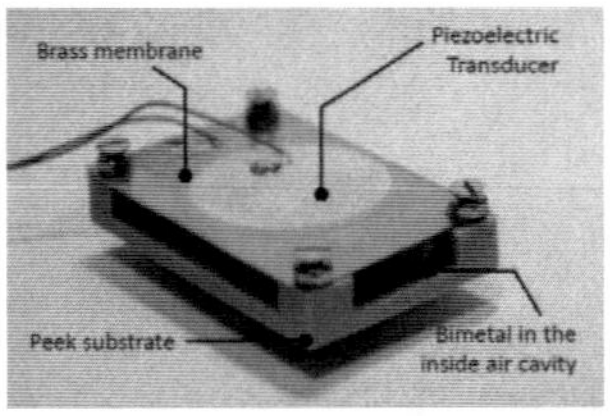

(b)

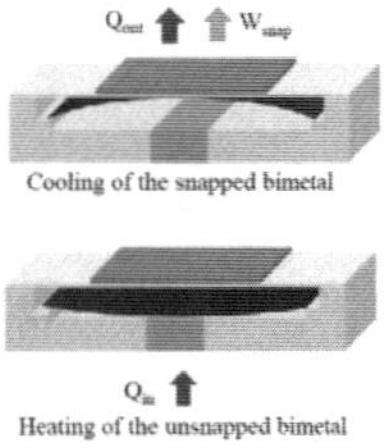

Fig. 1. a. Image of the thermal energy harvester, **b.** scheme of the harvester when the bimetal is at its upper position then at its lower one.

2. Bimetals experimental characterization

The heat engine presented in this work uses a shell-like bimetal made of two materials with a mismatch in their coefficients of thermal expansion. This mismatch makes them bend when subject to temperature variations. It is made of Invar (Fe-Ni 36%) acting as the low coefficient of thermal expansion coefficient (CTE) layer and of NC4 (Fe-Ni 22%-Cr 3%) acting as the high CTE layer**.** To point out the most powerful heat engines, bimetals of different thermal hystereses from 3K to 15K are characterized. First a force sensor is needed to measure its delivered force to the piezoelectric transducer then a vibrometer is used to estimate its snapping velocity and thus its available mechanical power. When the bimetal snaps, these parameters are measured thanks to the experimental setup described in Fig. 2.b and Fig. 2.c.

(a)

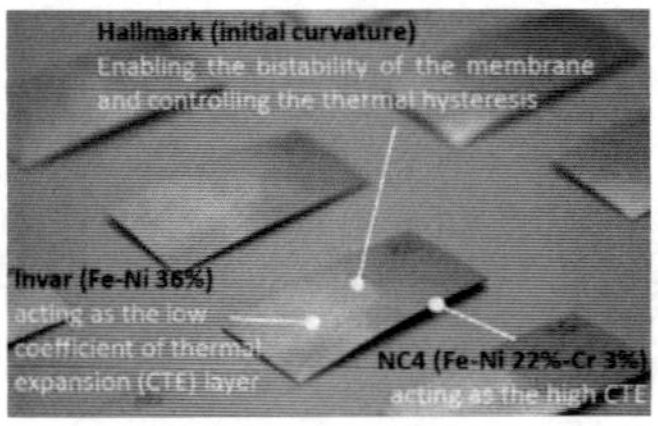

(b)

(c)

Fig. 2. a. Bimetals in the inside air cavity in Fig. 1.a, Experimental setup for the bimetals characterization: **b.** force captor, **c.** vibrometer

The results are shown in Fig.3. These graphs show that high bimetal hysteresis improves the force and the mechanical power, in comparison with lower hystereses [4,5].

(a)

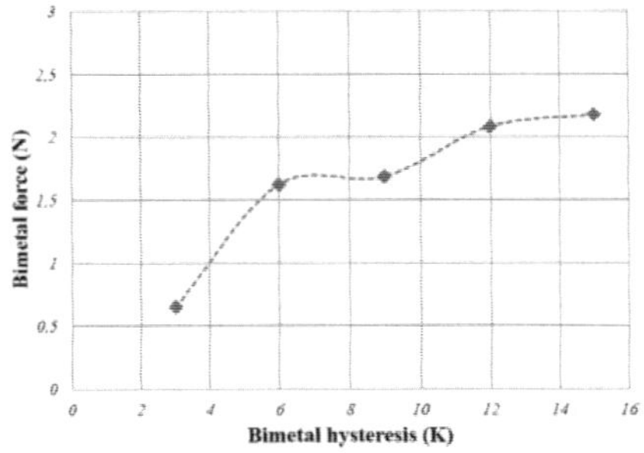

(b)

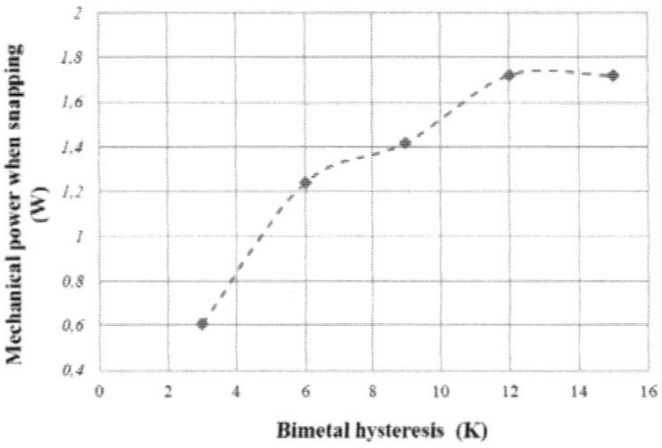

Fig. 3. a. Bimetals average force in function of their thermal hysteresis, **b.** Bimetals instantaneous mechanical power when snapping up as a function of the bimetal thermal hysteresis

3. Thermal modelling of the harvester in static mode without bimetal

The key point of this harvester is its ability to maintain an important thermal gradient across it. At first and before any thermal improvements, the thermal specificities of the device were not sufficient to allow bimetals with hystereses higher than 3K oscillate between the hot and the cold sources. A thermal model in static mode was then developed in [6] for the device in Fig. 4.a taking into account all the processes by which heat is transferred inside the structure. This model provided the main guidelines for the optimization step. A new design was then proposed and allowed to increase the temperature difference across the harvester by at least a factor 2. The comparison of the thermal properties of the initial harvester with the optimized one is shown in Fig. 4.b. Experimental data show good agreements with the analytical one confirming this model's validity.

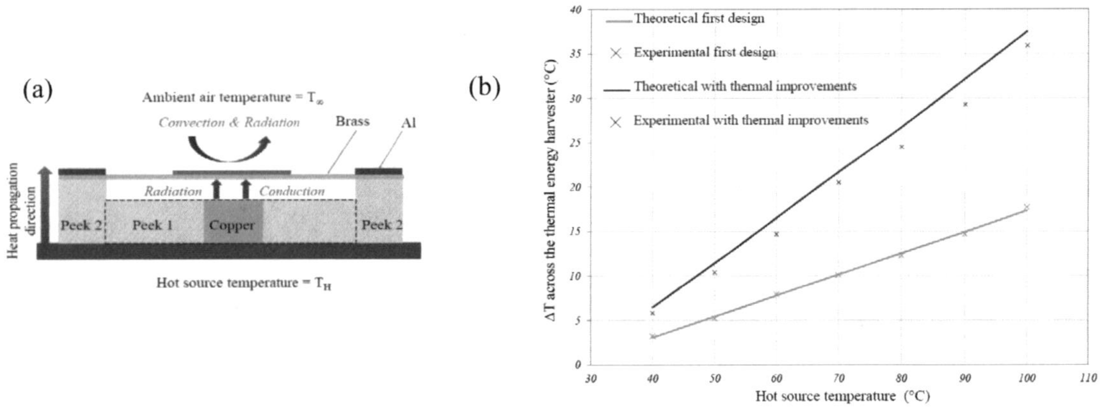

Fig. 4. a. Schematic representation of the harvester with the different heat transfer process occurring through it, **b.** Thermal properties of the harvester before and after the thermal management improvement

4. Bimetal impact on the device thermal behaviour

The thermal model in static mode allowed thermal optimization of the structure. However, this model did not take into account the oscillating bimetal. What made the dynamic study become compulsory was the shift observed experimentally between the operation window of a bimetal when it is directly put onto a hot plate to oscillate freely and the one when it is mounted in a harvester. In Fig. 5, this shift is clearly illustrated for a bimetal having a snapping temperature of 70°C and a snapping back one of 67°C. Once mounted in the harvester, it only starts oscillating between the cold and hot source at 84°C and stops at 102°C. To close this gap, a dynamic model taking into account the bimetal became necessary.

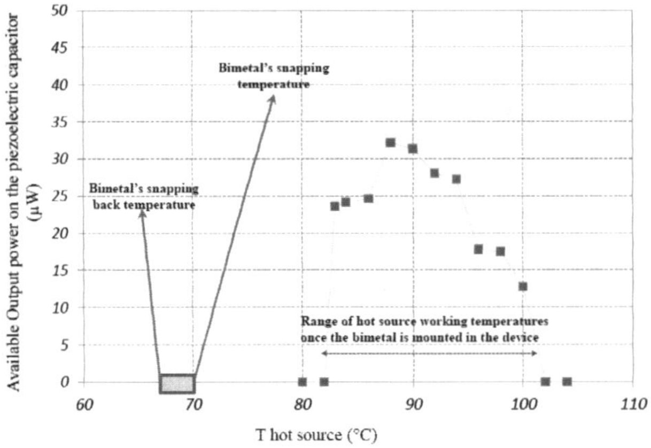

Fig. 5. Range of hot source functioning temperatures for a 3°C hysteresis bimetal (67°C-70°C)

5. Dynamic model of the harvester taking into account the bimetal

To estimate the bimetal snapping frequency, the amount of harvested energy and the bimetals operation windows in function of their hystereses and their snapping up and down temperatures, a dynamic model of the harvester is proposed in Fig.6. The bimetal is represented by a switched capacitance [4][7] with two switches to model its contact with the hot source (Fig. 6.a) and with the cold surface (Fig. 6.b).

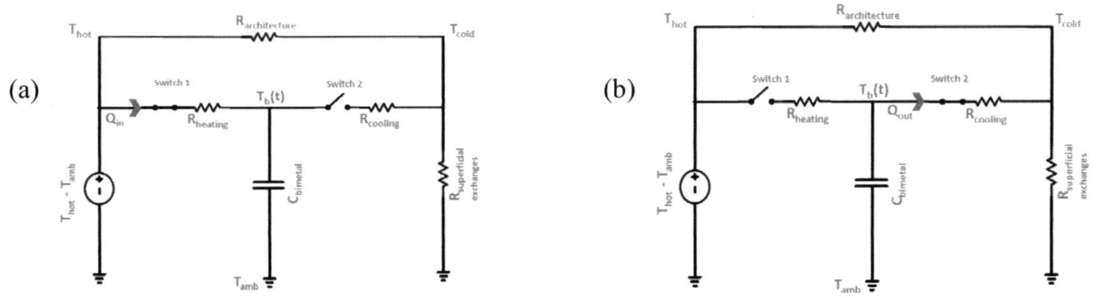

Fig. 6. a. Equivalent circuit of the harvester in a dynamic mode when the bimetal is at its lower state, **b.** Equivalent circuit of the harvester in a dynamic mode when the bimetal is at its upper state

Using these thermo-electrical circuits, one can consequently obtain the expressions of the time spent by the harvester in the lower and upper positions. This leads to the calculation of the hot source operation window by means of the calculation of the minimal hot source temperature for the bimetal to start oscillating $\mathbf{T_{hot}^{min}}$ and the maximal hot source temperature for the bimetal to be stuck in its upper position $\mathbf{T_{hot}^{max}}$.

$$T_{hot}^{min} = \frac{2.T_s^2 - 2.T_{sb}^2 - T_s.T_{amb} + T_{sb}.T_{amb}}{3.(T_s - T_{sb})} \quad (1)$$

$$T_{hot}^{max} \geq 2.T_{sb} - T_{amb} \quad (2)$$

$$\text{Hot source operation window} = T_{hot}^{max} - T_{hot}^{min} \quad (3)$$

Where $\mathbf{T_{amb}}$ is the ambient air temperature, $\mathbf{T_s}$ and $\mathbf{T_{sb}}$ respectively the snapping and snapping back temperatures.
Simulations of the hot source operation windows are performed and compared with the experimental data for the same bimetals in Fig. 7.a allowing the validation of the dynamic model. Simulations of the hot source operation windows as a function of the bimetals hysteresis are presented in Fig. 7.b revealing that lowering the hysteresis allows to widen the operation window of the harvester.

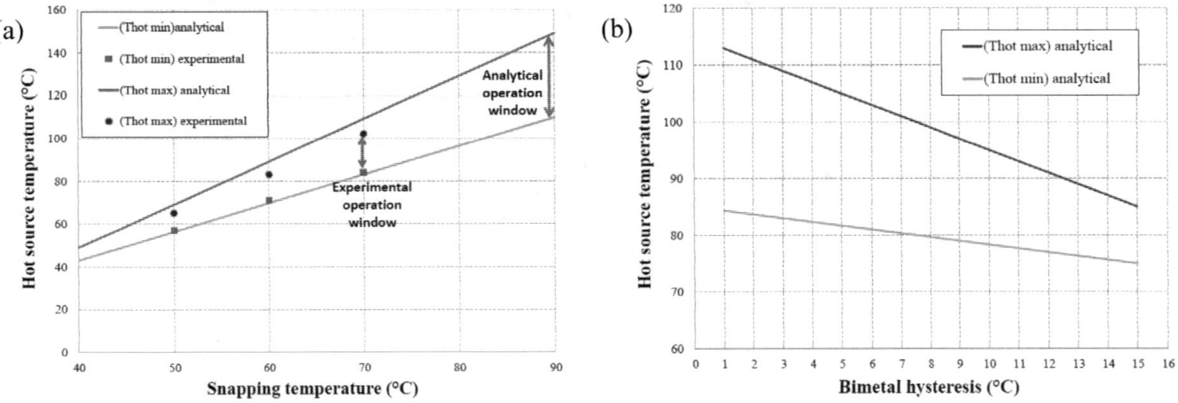

Fig. 7. a. Comparison of the experimental operation windows with the analytical one (bimetals hysteresis=3K), **b.** Effect of the bimetal hysteresis on the operation window (Ts=70°C for all the bimetals)

To complete our study, we simulated the snapping frequency of various bimetals. This study is led in two cases: first for bimetals having the same snapping temperature with different thermal hystereses (Fig. 8.a), then for bimetals having the same thermal hysteresis and different snapping temperatures (Fig. 8.b). Fig. 8.a presents the evolution of the cycling frequency as a function of hot source temperature for bimetallic membranes having the same snapping temperature. This figure shows that bimetals with low hysteresis can work on wider hot source temperature windows. Fig. 8.b presents the evolution of the cycling frequency as a function of hot source temperature for bimetallic membranes having different snapping temperatures and the same thermal hysteresis. It reveals that it is more interesting to work on high hot source temperatures as for a fixed thermal hysteresis, bimetals snapping up on 100°C have wider operation windows than bimetals snapping at 70°C. This work completes the mechanical characterization of bimetals and gives the main guidelines for a bimetal choice depending on the use case: if one needs more power, it is interesting to choose bimetals with high hysteresis but the hot source operation window will be lower. Contrary, if one needs a bimetal that is able to oscillate on a wide hot source temperatures, it is more interesting to choose bimetals with the lowest hystereses.

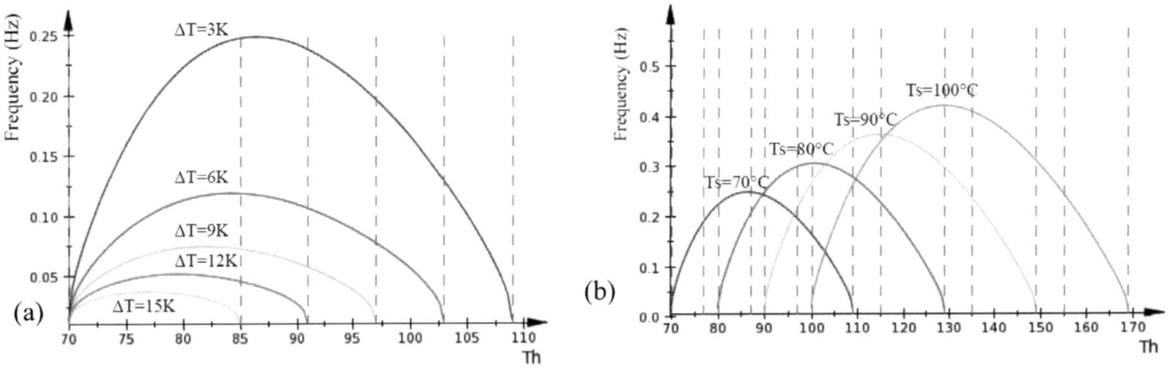

Fig. 8. a. Frequency evolution in function of the hot source temperature (Ts=70°C for all the bimetals), **b.** Frequency evolution in function of the hot source temperature (ΔT=3K for all the bimetals)

6. Conclusion

This paper presents the thermal optimization of a thermal energy harvester based on bistable bimetallic strips and piezoelectric materials and a dynamic model taking into account the bimetal. This allows predicting the temperature operation ranges and their relations to the bimetals hystereses and snapping temperatures. This gives the main guidelines for the bimetal choice depending on the requirement specifications of the application. The established models are then validated using experimental measurements that fit well with the simulation results.

7. Acknowledgments

This work was funded by the Fonds Unique Interministériel (FUI), through the HEATec project.

8. References

[1] T. Skotnicki, 2009, Dispositifs de conversion d'énergie thermique en électricité In: France Patent FR2951873 (A1).
[2] Arnaud A *et al.*, 2013, Proc. PowerMEMS, (London, UK, 3-6 december 2013), pp. 295-299.
[3] Boughaleb J *et al.*, 2014 , Proc. PowerMEMS, (Japan, Awaji Island , 18-21 november 2014), pp. 457-461.
[4] S K T. Ravindran *et al.*, 2011, Appl. Phys. Lett. 99, 104102.
[5] A. Arnaud et al, 2015, Journal of micromechanics and microengineering, accepted manuscript for publication.
[6] J. Boughaleb et al., 2015, Smart Mater. Struct. 24 (2015) 085025.
[7] E. Lurie *et al.*, Eneg convers manage., Volume 51, Issue 6, June 2010, pp. 1202–1209.

Electrical characterization of a buckling thermal energy harvester

E Trioux[1,2], L Rufer[1,2], S Monfray[3], T Skotnicki[3], P Muralt[4] and S Basrour[1,2]

[1] Université Grenoble Alpes, TIMA Laboratory, F-38031 Grenoble, France
[2] CNRS, TIMA Laboratory, F-38031 Grenoble, France
[3] STMicroelectronics, Crolles, France
[4] Ceramics Laboratory, EPFL, Lausanne, Switzerland

E-mail: emilie.trioux@imag.fr

Abstract. This paper presents the electrical characterizations of a novel concept for thermal energy harvesting at micro scale. The devices presented here are based on a two-step transduction combining thermo-mechanical and piezoelectric conversion. The piezoelectric layer is directly integrated into a buckling bilayer plate made of aluminium and aluminium nitride. For the first time, we have characterized the structures electrically and we have investigated their output power during the buckling. Firstly, we have used an insulating tip to make the plate buckle in order to have an estimation of the output power due to piezoelectric contribution only, and to eliminate any pyroelectric contribution that might be present during the thermal actuation. Then, we heated up the structure and we collected the output signal with an instrumentation amplifier in order to measure the voltage generated during the buckling. The output power during the mechanical and the thermal buckling is compared in the paper.

1. Introduction

The development of harvesting systems is growing steadily these last years. They are harvesting all kinds of surrounding energies in order to power autonomous wireless sensor networks. The piezoelectricity is commonly used for mechanical energy harvesting under the form of vibrations (human body movements, household appliances, industrial machines…) [1, 2]. Devices based on the Seebeck effect are the most popular to scavenge thermal energy, and a lot of efforts are made to downscale them [3, 4]. We have originally presented an alternative to thermoelectric materials for micro-scale thermal harvesting applications in [5, 6]. These devices combine piezoelectric transduction and thermal harvesting and are fabricated using CMOS compatible processes. In this work, the devices are characterized electrically and for the first time the output power furnished during the thermal buckling of the bilayer structure is measured.

The working principle of the harvester is based on a two-step transduction. A bilayer plate initially curved downward is put in contact with a hot source. The difference of thermal expansion coefficients (CTE) of the two materials is causing thermal stresses in the films that are leading to a mechanical instability resulting in the buckling. The plate is passing suddenly from a downward to an upward curvature. The large displacements due to the buckling are converted into electric charges by the piezoelectric layer. After the change of curvature, the plate is not anymore in contact with the hot source. It is cooling down by natural convection, and for the same reasons as before, the change in temperature of the device is generating thermal stresses that are leading to the inverse buckling of the structure. The curvature being again downward, the plate comes again in contact with the hot point and the thermal cycle is closed. The regular oscillations of the plate are generating electrical peaks that could be considered for powering a low-consumption system.

In this contribution, the output power of the structure is characterized. The buckling is firstly induced mechanically by a tip pushing the curved plate. Such a process helps to distinguish the expected piezoelectric signal from eventual pyroelectric charges due to the change in temperature. Then, the buckling is induced thermally, by putting the device onto a hotplate. The temperatures of snap up and snap down are measured as well as the output power, which will be compared to the one obtained during the mechanical buckling.

2. Mechanical actuation

The structure presented here are composed of a seed layer of AlN 200nm thick, patterned Pt electrodes of 100nm and the two active materials of the bilayer plate: 0,5μm of AlN as the piezoelectric layer and 0,5μm of Al as the metallic layer with high CTE. The layers are deposited by sputtering on a Si (100) wafer with 1μm thermal oxide on both sides. The structures are liberated from the backside by a Bosch process. A top view SEM picture of a 1400μm by 700μm plate is shown in Figure 1 as well as its cross-section in Figure 2. This structure will be characterized this paper. The combination of the 150MPa tensile stress in the top AlN layer, -400MPa compressive stress in the active AlN layer and 230MPa tensile stress in the AlN seed layer result in the initial downward curvature of the structure. The detailed fabrication process is presented in [5].

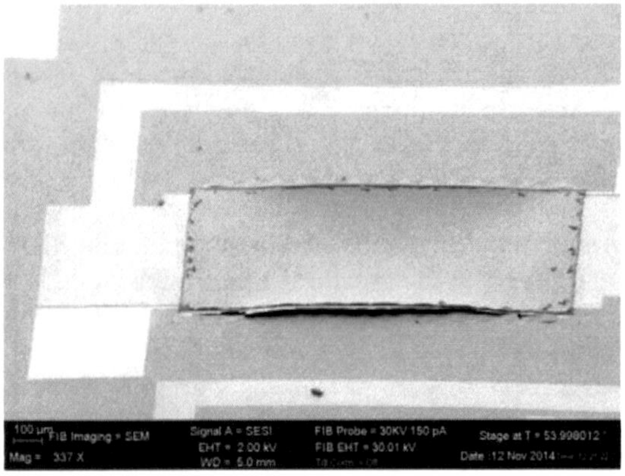

Figure 1. SEM image of a 1400*700 harvester

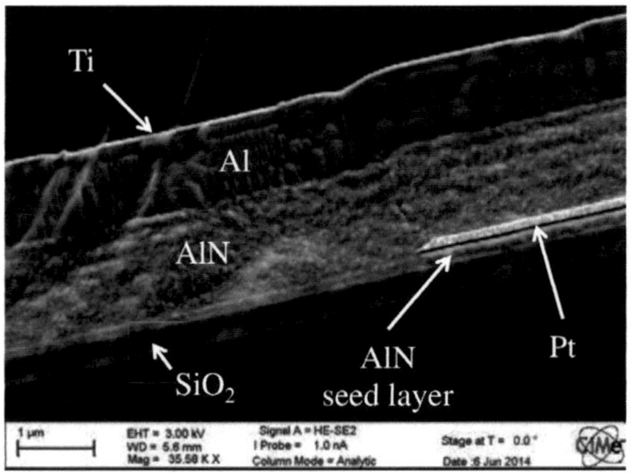

Figure 2. SEM cross-section of a device

In order to actuate mechanically the membrane, it was necessary to build a setup (Figure 3 and Figure 4), which is able to reach the backside of the plate, and to give a controlled micrometric displacement. An insulating tip with a diameter of 70μm is then mounted on a piezoelectric multilayer stack, which is able to provide a 30μm displacement at frequencies below 100Hz. Thanks to a camera placed on the front side, we can see the deformation of the plate when the tip enters in a contact. The tip is pushing until the plate buckles. At the beginning, the plate is following the tip displacement. When a critical force applied by the tip is reached, the plate becomes mechanically unstable and snaps up. As the temperature of the plate is the ambient temperature, it has only one stable position, which is the initial position. This means that when it buckles up, it does not reach a more stable position, so the plate returns back in contact with the tip.

The output signal is collected by an instrumentation amplifier with a gain of 1. The load resistance placed in parallel to the input of the amplifier is optimized to obtain the maximum output power of the tested plate and it is found to be at 1MΩ (Figure 5). This value is then taken for the following measurements. The same plate is then actuated by the tip in order to cause its buckling. In Figure 6, the dashed line represents the displacement of the tip. The maximum voltage corresponds to the maximum elongation of the piezoelectric stack while the lower voltage corresponds to its minimum. The positive peak is the piezoelectric signal of the device when it buckles upward and the negative peak the signal when it buckles back downward.

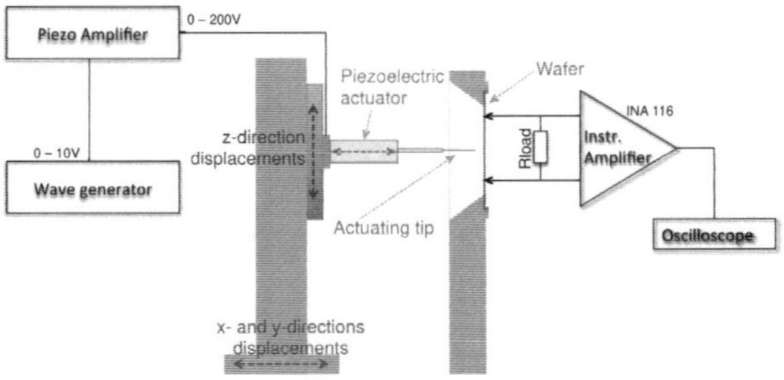

Figure 3. Schematic of the actuation setup

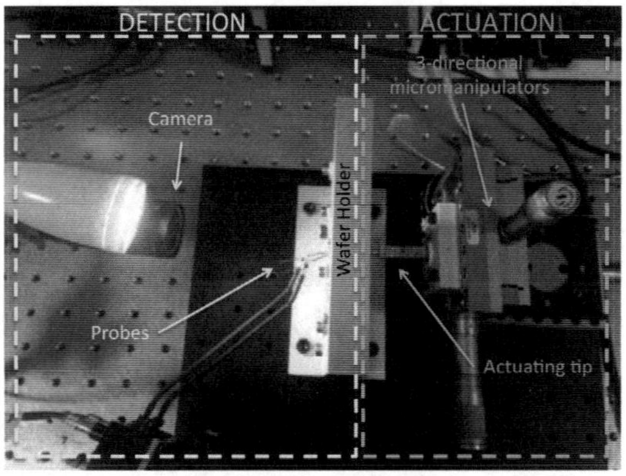

Figure 4. Image of the setup

The instantaneous power of one peak p_{int} is related to its energy w_{int} by the following formula:

$$p_{\text{int}} = \frac{1}{\Delta t}\int_{\Delta t} v_{out}^2 / R_{load} = w_{\text{int}} / \Delta t \tag{1}$$

where v_{out} is the output voltage, R_{load} the load resistance and Δt the time period over which the integration is done. The average power generated by the device at a given actuation frequency is calculated by the product of the energy per peak, W_{int}, and the actuation frequency. For instance when the plate is actuated at 20Hz, the mean instantaneous power density is of 48nW.cm^{-2}, and the corresponding mean energy density per peak is 3,6pJ.cm^{-2}. The average power density, 0,7nW.cm^{-2}, is then much lower than the instantaneous power density. Increasing the actuation frequency would also increase the average output power density of the structure. Measurements at 30Hz were performed, and with a relatively unchanged instantaneous power density (41nW.cm^{-2}), the average power density increased to 1,9nW.cm^{-2}.

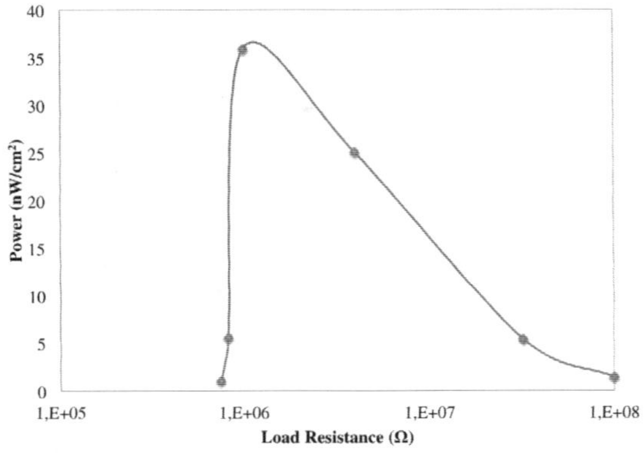

Figure 5. Output power in function of the load resistance

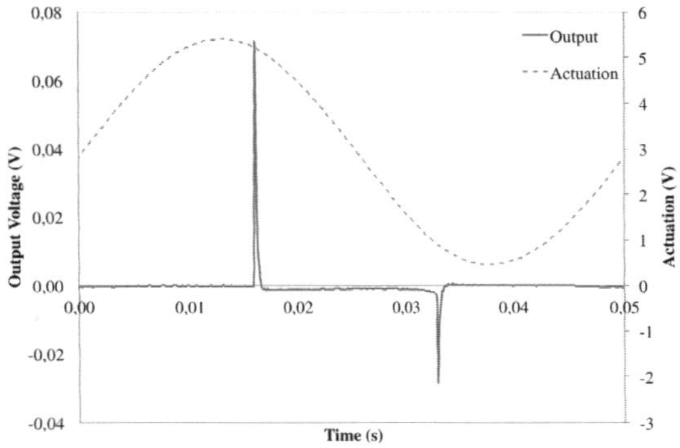

Figure 6. Mechanical buckling of the plate at 20Hz

3. Thermal actuation

With a mechanical actuation, we have obtained the pure piezoelectric signal as a result of the device buckling. To actuate thermally the plates, we put them on a hotplate and the temperature was slowly increased. The structure was heated mostly by conduction through the substrate. A thermocouple was measuring the temperature of the surface of the hotplate. The same structure tested before was buckling up at 83°C and buckling down at 62°C. A top view of the structure is shown in inset of the Figure 7 when buckling up (a) and when buckling down (b) which is similar to the initial shape of the device.

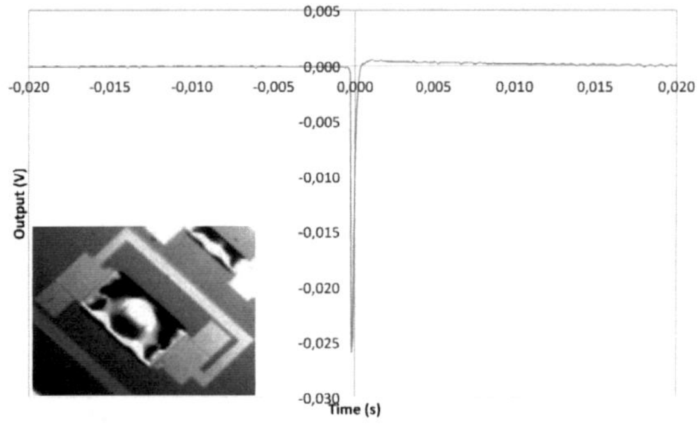

(a)

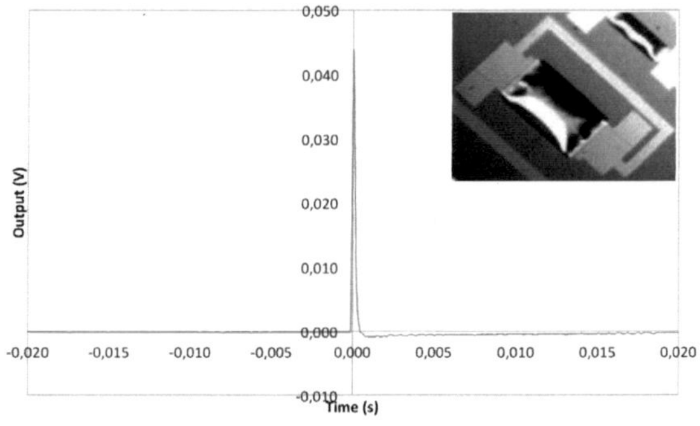

(b)

Figure 7. Output signal for thermal buckling up at 83°C (a) and thermal buckling down at 62°C (b) with in inset images of the device after buckling up and after buckling down respectively.

The corresponding instantaneous power of the buckling up and buckling down peaks is calculated in the same manner as for the mechanical actuation. For the snap up, the device generated 18nW.cm^{-2} while for the snap down, the power was 49nW.cm^{-2}. The peaks are of the opposite sign compared to the mechanical actuation because the positive and negative inputs of the amplifier were inverted. The same value is found for the snap down as for the mechanical actuation, which is what was expected. On the other hand, the instantaneous power during the snap up is much lower than for the mechanical actuation. This can be explained by the pyroelectric charges that are acting against the piezoelectric charges. The average power density is not relevant in this case because of the time necessary for the hot plate cools down to lower the temperature and cause the snap back of the plate. The time between the snap up and down is then very long and does not correspond to the real thermal cycle of the system at which the system is able to work.

4. Conclusion

This paper presents the first electrical characterization of the mechanical and thermal buckling of a micro thermal energy harvester based on a bilayer structure. The instantaneous output power in both cases is measured around 48nW.cm^{-2} for the mechanical actuation, value that is confirmed in the snap back of the thermal actuation. The instantaneous power during the thermal snap up is 4 times lower, which might be due to pyroelectric charges that are acting against the piezoelectric signal. Moreover, the thermal hysteresis of the tested device is of 20°C, which is relatively high. We are currently working on different geometries that might reduce this hysteresis to make the system more adapted to future applications.

References

[1] Mitcheson P D, Green T C, Yeatman E M and Holmes A S 2004 *J. MEMS* **13** 429
[2] Kim S-G, Priya S and Kanno I 2012 *MRS Bulletin* **37** 1039
[3] Hourdakis E and Nassiopoulou A 2013 *Sensors* **13** 13596
[4] Moser A, Erd M, Kostic M, Cobry K, Kroener M and Woias P 2012 *J. Electr. Mat.* **41** 1653
[5] Trioux E, Monfray S, Skotnicki T, Basrour S and Muralt P 2014 *J. Phys.: Conf. Ser.* **557** 012087
[6] Arnaud A, Monfray S, Boughaleb J, Trioux E, Boeuf F, Cugat O and Skotnicki T 2014 *J. Phys.: Conf. Ser.* **557** 012015

Journal of Physics: Conference Series **660** (2015) 012107 doi:10.1088/1742-6596/660/1/012107

Electrostatic Energy Harvester Utilizing High Density of Electrode for Higher Output Power

K Minami, N Miwatani, K Kanda, T Fujita, and K Maenaka

Graduate School of Engineering, University of Hyogo
2167 Shosha, Himeji 671-2280, Japan

E-mail: er14q043@steng.u-hyogo.ac.jp

Abstract. In this study, we report an improvement of output power from an electret type vibration energy harvester. Typical crossing-area change harvester has a stripe-shaped electret and counter electrode for making the capacitance change. In order to improve space efficiency, the counter electrodes are divided and arraignment with the same pitch of the electret. We investigate that adjoining the counter electrodes, the fringing effect is decreased and the capacitance change between the electrodes is larger than the conventional design from FEM analysis. The output power of 2.5 μW and 5.3 μW are obtained in each kind of counter electrode with the applied acceleration of 3 G at 350 Hz, which is about 2 or 4 times as high as our previous work.

1. Introduction

Wireless sensor networks (WSNs) are attractive for a tire pressure monitoring system (TPMS) or a human healthcare system. A power supply for the system needs long lifetime. Energy harvesters are expected as a power source for these systems [1][2]. Mechanical vibration sources are abundant in the environment and easy to use for conversion to electrical energy. We study an electrostatic vibration energy harvester, which transfers charge on the counter electrodes (CE) by capacitive induction from an electret material with electrical charge. Typical crossing-area change electrostatic energy harvester has a stripe-shape electret and the CE for making capacitance change. In order to obtain the larger capacitance change, the pitch of CE and the electret pitch are same in general (Fig. 1a) [3]. However, this design shows small capacitance change because of a large fringing effect. In this study, we demonstrate novel device with doubled number i.e. half pitch size of the CE for improving output power. The novel electrostatic energy harvester with high-density electrodes was fabricated by MEMS batch process. The preliminary measurement results are also described.

2. FEM analysis

In order to improve space efficiency and obtain the large capacitance change, the CE is divided to two parts, CE-N (negative) and CE-P (positive) with the same pitch of electret as shown in Fig. 1(b).The capacitance between the electrodes is calculated by FEM analysis (ANSYS 14.0) with considering fringing effect. The FEM analysis model for conventional and high-density (HD) CE model is same as Fig. 1. A 2.5 μm thick electret polymer CYTOP (CTL-809M, Asahi glass Corp.,) and 0.5 μm thick bottom electrodes (BGE-N, BGE-P; abbreviation of buried grid electrode-negative and positive) are fabricated on a 1 μm thick SiO_2 isolation layer on a 500 μm thick Si substrate. The all the dimensions are as same as our previous electrostatic energy harvester.

Figure 2 shows the vector plots of the electric field directions at half pith displacement of CEs. The BGE-N and the BGE-P are set to the -200 V and GND (0 V), respectively. The vector arrow direction depicts an electric field direction, length and color shows intensity. From the results, the diffusion of electric field from the BGEs is smaller for the HD CE structure. Figure 3 shows the capacitance change between the CE-N and the BGE-N on conventional and HD CE design. Since the diffusion of electric field from the BGEs is reduced by adjoining CE, the fringing effect is decreased. Thus the capacitance change for the HD structure is twice as large as conventional design and the offset caused by the fringing effect was reduced.

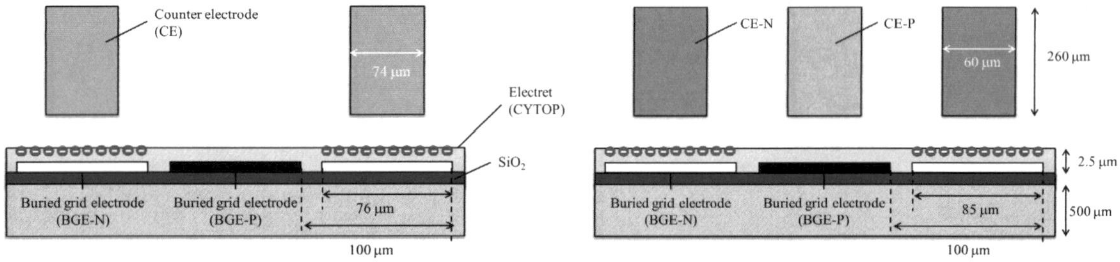

Figure 1. FEM analysis model for (a) general, and (b) high-density of counter electrode.

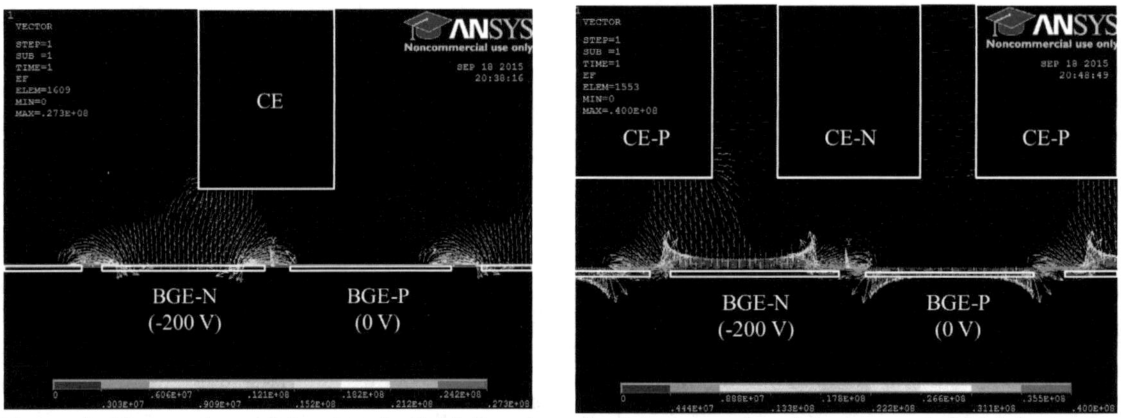

Figure 2. Vector plots of the electric field directions for (a) conventional design, and (b) high-density counter electrode. The charged voltage BGE-N and BGE-P are -200 V and 0 V at BGE-P.

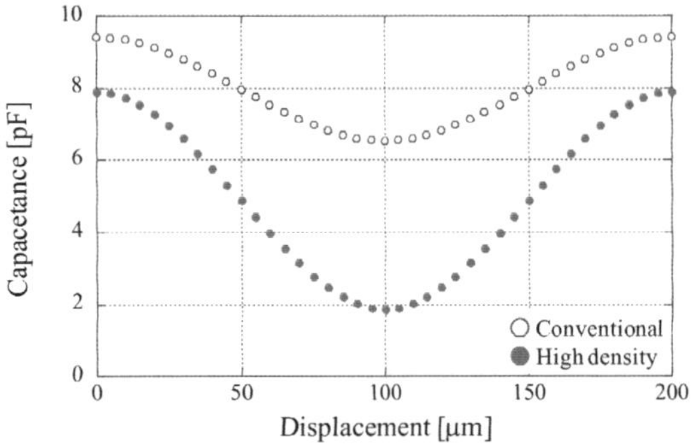

Figure 3. The capacitance change between the CE-N and BGE-N for conventional and high-density structure

3. Device design and fabric process

3.1 Device design

Figure 4 shows the cross-sectional schematic diagram of our electrostatic energy harvester. The harvester consists of a Si counter electrode, an electret with base electrodes placed on a mass-spring structure, and cover glass. The Si counter electrode functions harvesting counter electrode and charging grid. In order to prevent pull-in, the Si grid also has micro pillars for spacer. Figure 5 shows the fabrication process of the energy harvester. The shape comparison for conventional and HD CEs is shown in Fig. 6. In the novel design, CE is divided into an interdigital shaped CE-N and CE-P.

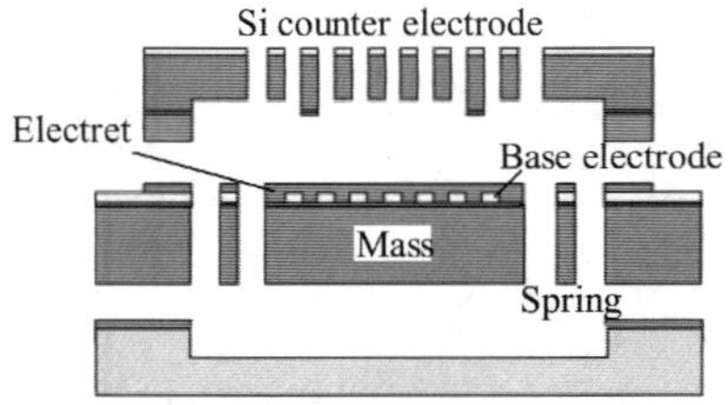

Figure 4. The concept of our electrostatic energy harvester [4].

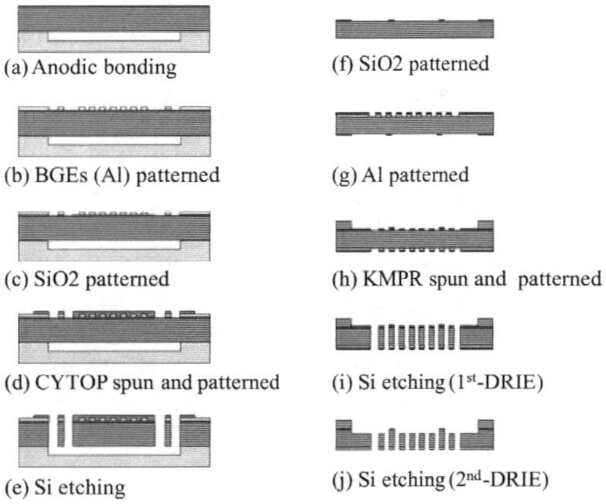

Figure 5. Fabrication process of the energy harvester.

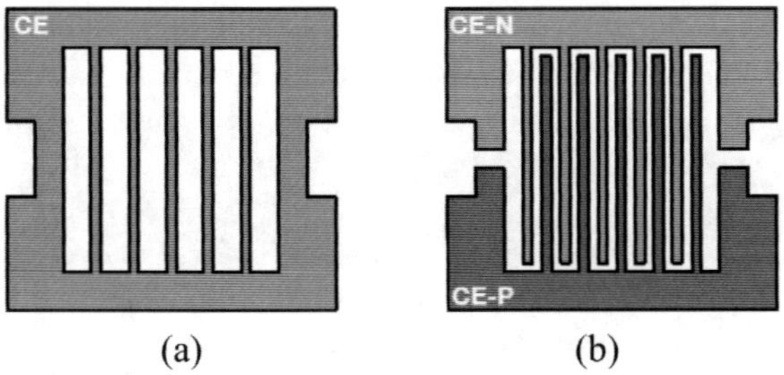

Figure 6. The conceptual diagram of counter electrode for (a) conventional, and (b) slit high-density device.

3.2 Fabrication Process

An electret part, Cr was sputtered on a glass wafer (Pylex code 7740; Corning Inc., USA) and patterned. The glass wafer was etched by BHF with the Cr pattern as a mask. Then, the glass wafer was bonded to 500 μm thick Si wafer with 1 μm thick SiO_2 in anodic bonding (Fig. 5a). BGEs made of Al are sputtered on the Si substrate and patterned with the 85 μm width 15 μm intervals (Fig. 5b). Then, the SiO_2 film was patterned by CF_4 plasma etching (Fig. 5c). A 2.5 μm thick CYTOP was spun on and patterned by O_2 plasma etching. Then, the mass-spring structure was formed by DRIE.

The counter electrode is fabricated monolithically on a p-type Si wafer (<0.1 Ω·cm resistivity, 300 μm thick). The SiO_2 is patterned for the pillars and etch mask of 2nd-DRIE (Fig. 5f). The sputtered Al on Si surface is patterned (Fig. 5g). In order to improve the connectivity Al and Si are sintered at 460 °C for 30 min. Positive type photoresist (KMPR 1005; Microchem Corp., USA) is formed as the spacer for the air gap (Fig. 5h). The counter electrodes are fabricated in the shape of interdigital by 1st-DRIE (Fig. 5i). The air gap and the pillars are fabricated by 2nd-DRIE (Fig. 5j). Figure 7 shows the 4-inch wafer with CE-N and CE-P shape after 1st-DRIE process.

The electret wafer and counter electrodes were bonded by CYTOP heat-pressing method at 150 °C. An estimated output current is anti-phased for CE-N and CE-P. Therefore, the counter electrodes should be divided to individual electrodes after a dicing process. Finally, the device was charged by corona discharging method. The dimension of the entire device is 13 × 12 × 1.2 mm^3. Figure 8 shows the fabricated harvester compared with coin battery (CR2032).

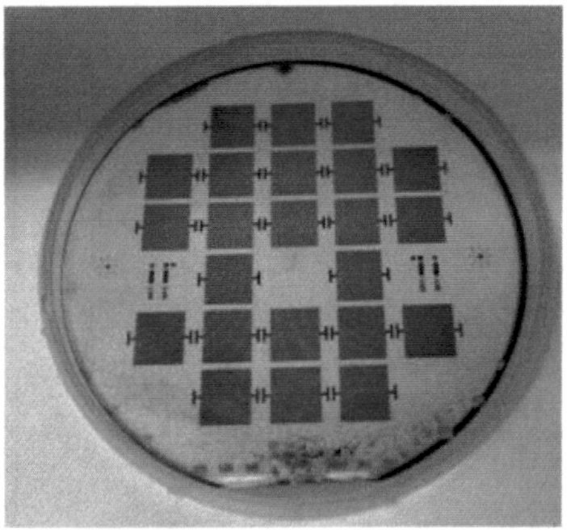

Figure 7. The structural of counter electrode after DRIE on 4-inch wafer.

Figure 8. The comparison of the electrostatic energy harvester and coin battery. (CR2032)

4. Harvesting experiment

Our electrostatic energy harvester was charged with the grid setup voltage of -200 V at the charge process. The device is attached on the shaker (ET-136; Labworks Inc., USA) and vibrated at 350 Hz. The acceleration of 3 G is sufficiently large to ignore the electrostatic force, which is caused by charged electret. Figure 9 shows the output waveform for CE-N and CE-P connected with 1 MΩ load resistance. The output waveform of individual CE-N and CE-P is anti-phase. Assuming the actual harvesting form the differential of CE-N and CE-P, the peak-to-peak voltage is twice as large as each one. The output frequency of our harvester is larger than the applied acceleration frequency because the amplitude of the vibration is larger than the electrode pitch. Utilizing HD CE, the output voltage is higher for improving the space efficiency.

Figure 10 shows the measurement output power for CE-N and CE-P; CE-N or CE-P is connected to a variable resistor with ultra high-impedance buffer (AD549; Analog Device Inc., USA) and the other is connected to GND. The simulated output power, which is calculated by the equivalent circuit model [5] that is shown in Fig. 10. The maximum output power for CE-N and CE-P are 2.5 μW and 5.3 μW at 5 kΩ and 7 kΩ load resistance, respectively. The measurement result of the harvester corresponds with the simulated one. Since the capacitance change between the electrodes is lager, the output power of the device is higher than conventional design. The output power will be 4 times as high as the result utilizing the bipolar electret, which charges not only negative but also positive charge in our previous work [6].

We evaluate a harvesting system by combing with a power management IC and our harvester on a SPICE circuit simulator (LTspice IV; Linear Technology Corp., USA). Figure 11 shows the harvesting system connected with LTC3588-1 and our novel harvester with HD CE and bipolar charged. The charged voltage of the harvester is ±200 V. From the result, the voltage after the power management of 1.8 V will be obtained after 35 s intervals. It is 5 times faster than our previous work.

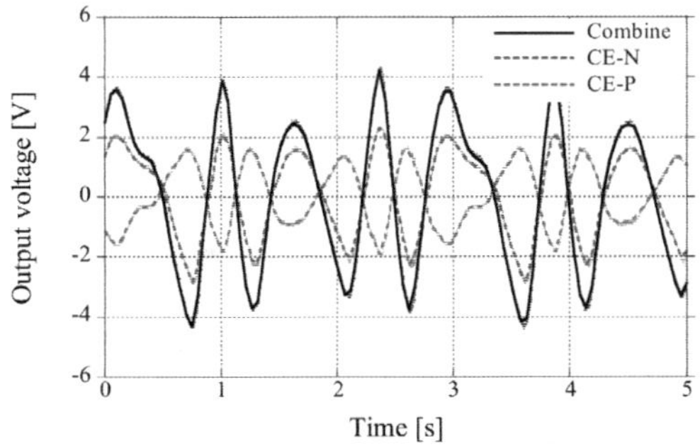

Figure 9. Output waveform of CE-N, CE-P, and Combing.

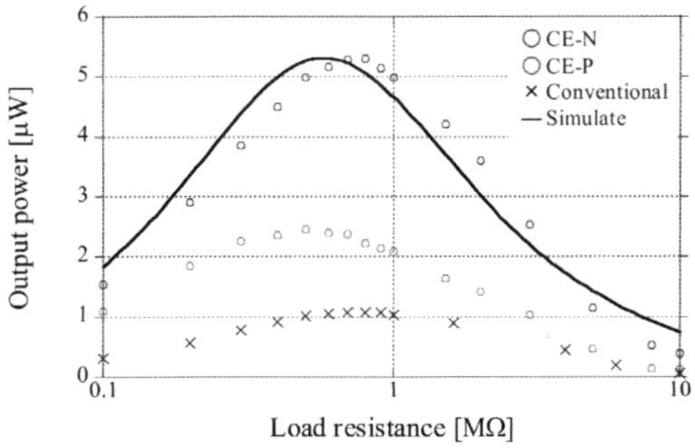

Figure 10. Output power vs. load resistance.

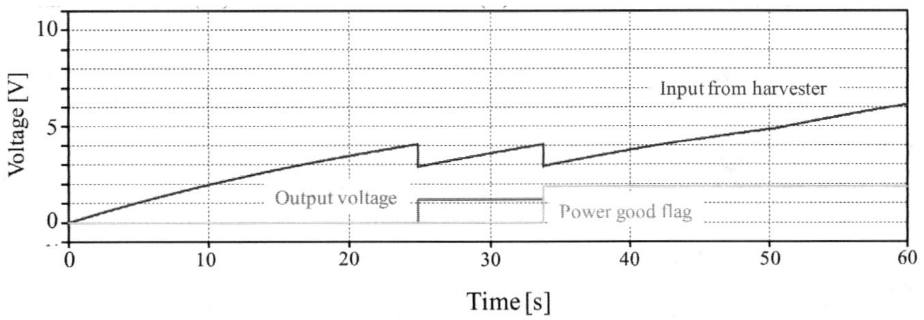

Figure 11. The power management output by LTC3588-1. The charged voltage is ±200 V.

5. Conclusion

In this study, an improvement of output power from an electrostatic energy harvester is investigated. Utilizing double number of counter electrode, the capacitance change between the electrodes is larger from FEM analysis. The electrostatic energy harvester was fabricated by MEMS batch process. The output power of the high-density device was 2.5 μW at CE-N and 5.3 μW at CE-P , which is 2 or 4 times as large as conventional design. The device connected with LTC3588-1 will output the management voltage of 1.8 V at 35 s intervals from the circuit simulator.

References

[1] Altena G, Renaud M, Elfrink R, Goedbloed M H, Nooijer C de and Schaijk R van 2013 *Proc. of PowerMEMS2013* 371-375
[2] Mitcheson PD, Yeatman EM, Kondala G Rao, Holmes AS, and Green TC 2008 *Proc. of IEEE* **96** 14577-1486
[3] Marboutin C, Suzuki Y, and Kasagi N 2007 *Proc. of PowerMEMS2007* 141-144
[4] Fujita T, Fujii K, Onishi T, Kanda K, Higuchi K, Maenaka K 2011 *Proc. of Eurosensors* **25** 733-736
[5] Minami K, Fujita T, Sonoda K, Miwatani N, Kanda K, Maenala K 2014 *Proc. of PowerMEMS2014* 208-212
[6] Fujita T, Onishi K, Fujii K, Katsuma K, Kanda K, Higuchi, Maenaka K 2012 *Proc. of PowerMEMS2012* 436-439

Energy harvester using contact-electrification of magnetic fluid droplets under oscillating magnetic field

D Kim and K-S Yun
Department of Electronic Engineering, Sogang University, Seoul, Korea

E-mail: ksyun@sogang.ac.kr

Abstract. This paper reports a fluidic-based energy harvester generating electric power through contact-electrification of ferrofluid droplets, which will allow the power generation using oscillating magnetic field without vibration of any mechanical structure such as membrane or cantilever. The proposed device consists of top and bottom plates with a conducting electrode coated with a hydrophobic layer and water-based ferrofluid droplet. The contact area between the ferrofluid and the solid surface is changed according to the magnetic field applied by a magnet, which generates AC output power by contact electrification at the ferrofluid–solid interface.

1. Introduction

When the surface of a solid material makes contact with an aqueous solution such as water, an electrical double layer (EDL) is formed at the interface by the interaction of surface charges and their counter ions in aqueous solution [1-3]. Recently, a number of research results regarding electric energy harvesting by using the EDL mechanism has been reported. Lin et al. presented energy harvesting devices based on contact electrification between water and a hydrophobic polytetrafluoroethylene (PTFE) surface and the corresponding EDL formation at the interface [4, 5]. In addition, it has been demonstrated that electric energy can be harvested by mechanical modulation of the EDL capacitors formed by the contact of water with the conducting plates [6, 7]. However, those studies used droplets falling on a substrate from some height or mechanical vibration of the substrate itself to create continuous dynamic contact conditions, which hinder the packaging of the devices by limiting their application to a restricted area.

In this paper, we propose and demonstrate an energy harvester using ferrofluid droplets that are deformed by an external magnetic field and consequently provide a dynamically changing contact area at the interface, resulting in continuous charge flow by contact electrification.

2. Proposed device and operational principle

Ferrofluids are categorized as oil-based and water-based fluids according to the base media composing them. When a water-based ferrofluid is in contact with a solid surface, an EDL is formed at the interface. Figure 1 shows the power generation mechanism of the proposed energy harvester using a ferrofluid and the oscillating external magnetic field. The device consists of top and bottom plates with a conducting electrode covered by a hydrophobic layer and water-based ferrofluid droplet. The ferrofluid droplet is dispensed by a pipette. The ferrofluid droplet in this experiment is positively charged while separating from dispenser [8]. So, the electrode surface becomes negatively charged to maintain the balance of the electrical potential. In the initial state, no current flows because there is no potential difference between the top and bottom electrodes, as shown in figure 1(a). When the magnet moves close to the bottom plate, the ferrofluid is pulled downward, and finally separated from the top plate, as shown in figure 1(b) and figure 1(c). During this operation, the contact area with the ferrofluid decreases on the top plate and increases on the bottom plate. As a result, electrons move from the top electrode to the bottom electrode through the load resistance until electrical equilibrium is achieved. In figure 1(d), the ferrofluid returns to its initial shape to maintain the balance of surface tension on the hydrophobic bottom surface and makes contact with the top plate when the magnetic field applied to the ferrofluid decreases as the magnet moves away from the bottom plate. During this step, the positive charges on the ferrofluid draw the electrons from the bottom electrode to the top electrode, reversing the direction of current flow. Continuous output power can be generated by the reciprocal oscillating motion of the external magnet.

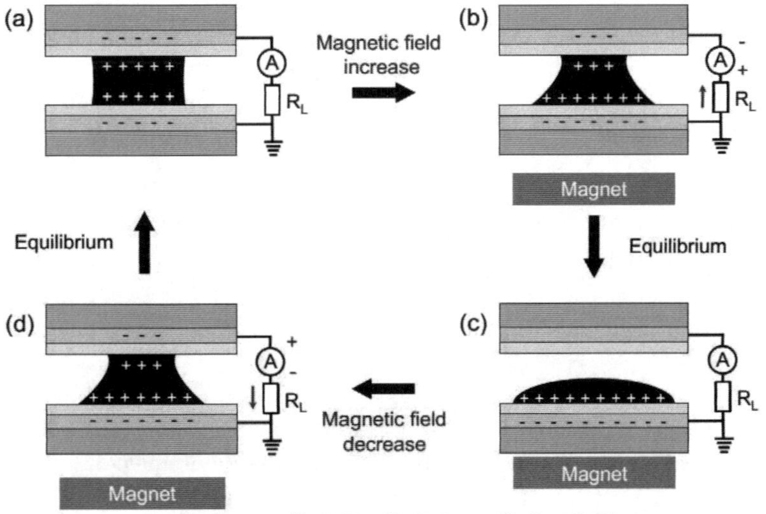

Figure 1. Power generation mechanism of the proposed energy harvester.

3. Fabricated device and experimental setup

In this experiment, we used gold as electrode material as shown in figure 2. The top and bottom electrodes were coated with Teflon (AF 1600) to provide hydrophobic surfaces and prevent stiction of the ferrofluid during operation. A ferrofluid droplet was placed between the two plates. The gap distance between the top and bottom plates was determined by acrylic spacers. The fabricated energy harvester was mounted on a z-axis stage. A cylindrical neodymium magnet with diameter was fixed on a computer-controlled linear actuator and placed under the device.

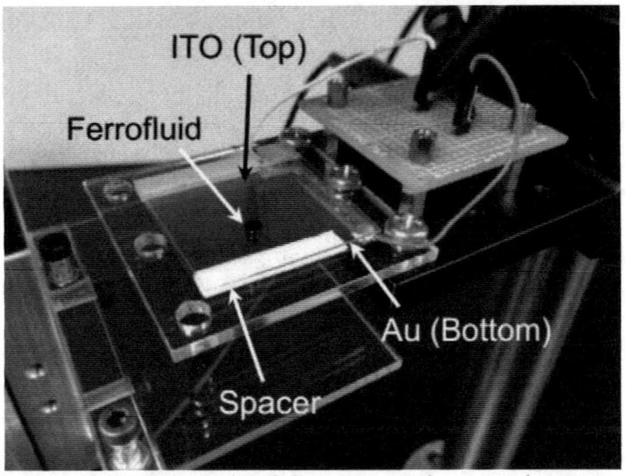

Figure 2. Photograph of the proposed energy harvester.

4. Experimental results

Figure 3 shows the profile of the open-circuit output voltage from the device with a 1.8-mm gap between the top and bottom plates. Positive and negative output signals were observed as the magnet moved up and down. Furthermore, it is clearly shown that the magnitude of the output voltage increases with the actuation frequency because it is determined by the change in the charge flow with time.

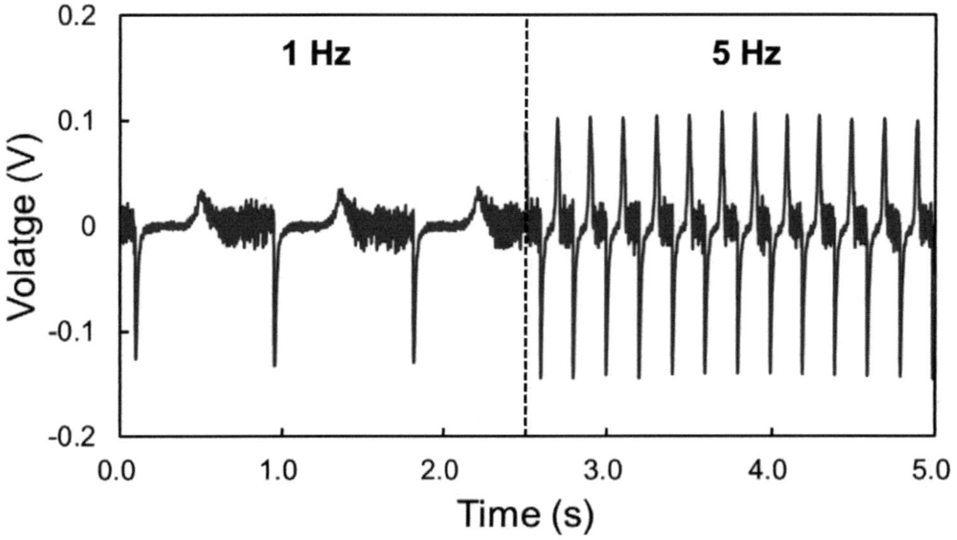

Figure 3. Typical profile of the output voltage at oscillation frequencies of 1 and 5 Hz.

5. Conclusions

In conclusion, we have proposed an energy harvester using contact electrification with a water-based ferrofluid operated by an oscillating magnetic field. The modulation in the contact area between the ferrofluid and the electrodes and resultant power generation were successfully demonstrated. One important advantage of the proposed device is that it eliminates the need to vibrate the device itself to actuate the ferrofluid droplet and to harvest electric power because the shape of the ferrofluid inside the device can be modulated by using an external magnetic field.

References

[1] Grahame D C 1947 The electrical double layer and the theory of electrocapillarity *Chem. Rev.* **41** 441-501

[2] Parsons R 1990 Electrical double layer: recent experimental and theoretical developments *Chem. Rev.* **90** 813-826

[3] Schoch R B, Han J and Renaud P 2008 Transport phenomena in nanofluidics *Rev. Mod. Phys.* **80** 839-883

[4] Lin Z-H, Cheng G, Lee S, Pradel K C and Wang Z L 2014 Harvesting water drop energy by a sequential contact-electrification and electrostatic-induction process *Adv. Mater.* **26** 4690-4696

[5] Lin Z-H, Cheng G, Lin L, Lee S and Wang Z L 2013 Water–solid surface contact electrification and its use for harvesting liquid-wave energy *Angew. Chem. Int. Ed.* **52** 12545-12549

[6] Moon J K, Jeong J, Lee D and Pak H K, 2013 Electrical power generation by mechanically modulating electrical double layers *Nature comm.* **4** 1487

[7] Kwon S-H, Park J, Kim W K, Yang Y, Lee E, Han C J, Park S Y, Lee J and Kim Y S, 2014 The effective energy harvesting method from natural water motion active transducer *Energy Environ. Sci.* **7** 3279-3283

[8] Choi D, Lee H, Im D J, Kang I S, Lim G, Kim D S and Kang K H 2013 Spontaneous electrical charging of droplets by conventional pipetting *Sci. Rep.* **3** 2037

Experimental verification of load resistance switching for global stabilization of high-energy response of a nonlinear wideband electromagnetic vibration energy harvester

T. Sato, A. Masuda and T. Sanada

Department of Mechanical and System Engineering, Kyoto Institute of Technology, Matsugasaki, Sakyo-ku, Kyoto, Japan

E-mail: m5623113@edu.kit.ac.jp

Abstract. This paper presents an experimental verificatio of a self-excitation control of a resonance-type vibration energy harvester with a Duffing-type nonlinearity which is designed to perform effectively in a wide frequency range. For the conventional linear vibration energy harvester, the performance of the power generation at the resonance frequency and the bandwidth of the resonance peak are trade-off. The resonance frequency band can be expanded by introducing a Duffing-type nonlinear oscillator in order to enable the harvester to generate larger electric power in a wider frequency range. However, since such nonlinear oscillator can have multiple stable steady-state solutions in the resonance band, it is difficult for the nonlinear harvester to maintain the high performance of the power generation constantly. The principle of self-excitation and entrainment has been utilized to provide the global stability to the highest-energy solution by destabilizing other unexpected lower-energy solutions by introducing a switching circuit of the load resistance between positive and the negative values depending on the response amplitude of the oscillator. It has been experimentally validated that this control law imparts the self-excitation capability to the oscillator to show an entrainment into the highest-energy solution.

1. Introduction

Vibration energy harvesting is a micro-power generation technique using a mechanical oscillator with a certain transduction mechanism which is designed to resonate with the environmental vibration source to collect unused or wasted mechanical energy and convert it to electric energy. The conventional vibration energy harvester with a single-degree-of-freedom linear oscillator, however, has a narrow bandwidth because of the power-bandwidth trade-off. One of the most promising nonlinear approaches is Duffing-type harvesters that uses a nonlinearly spring oscillator to widen the resonance frequency band [1, 2, 3, 4, 5]. This idea is an attempt to develop a single-degree-of-freedom energy harvester which can respond largely in a wide frequency range by folding the resonance peak toward the higher (lower) frequency direction by utilizing a hardening (softening) spring instead of the linear one. This type of nonlinear oscillator, however, can have two or more stable solutions with different amplitude in the resonance band when it is subjected to a sinusoidal excitation. It is difficult to keep the response on the highest energy branch because initial conditions determines which solution the response converges. In most literature, this difficulty is recognized as a fatal limitation of this type of nonlinear energy harvesters [1]. In the previous study [6] presented by the authors, a concept of a nonlinear energy harvester in which only the highest energy solution is global stabilized has been proposed. This is done by destabilizing other possible stable solutions by introducing a switching control circuit that switches the load resistance

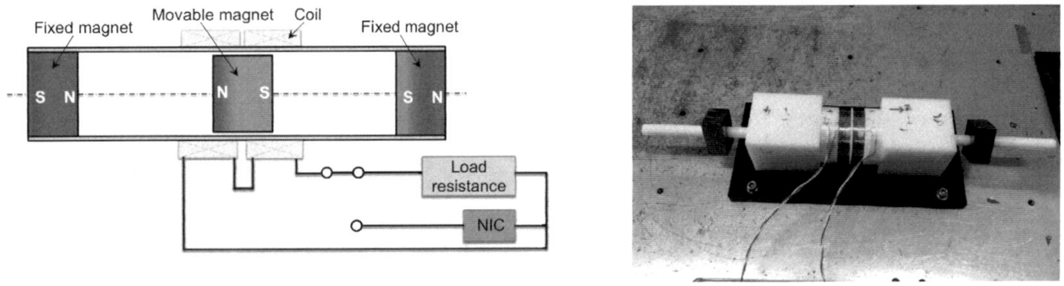

Figure 1. Schematic illustration and photograph of prototype harvester.

between positive and negative values depending on the response amplitude. In the article [6], the validity of the concept was examined by theoretical and numerical analyses, but experimental studies have not been presented. This study aims to conduct an experimental verificatio of the proposed concept by using a conceptual prototype of the proposed harvester. The response of the harvester subjected to the sinusoidal excitation is examined for the cases of with and without the load resistance switching.

2. Prototype design

2.1. Nonlinear oscillator

The prototype of the nonlinear vibration energy harvester investigated in this study is a single-degree-of freedom oscillator with an electromagnetic transducer followed by a load circuit as schematically illustrated in figur 1. It consists of a plastic cylinder with a diameter of 0.041 m and a length of 0.046 m, which has a pair of permanent magnets mounted at both ends and a movable permanent magnet enclosed inside which is guided in the axial direction by a pair of linear bearings. The polarities of the magnets are set such that the movable magnet receives repulsive forces from the fi ed magnets. A copper wire is wound outside the cylinder to form a 500-turn solenoid coil followed by another 500-turn coil in the reverse direction. This arrangement allows the movable magnet not only to perform as an inertial mass of a nonlinear hardening oscillator but also to as an electromagnetic transducer. The transducer is connected to the load circuit, through which the electric power induced by the transducer is transmitted and delivered to the output load. For simplicity, it is assumed that all the circuits used in this study are represented by an equivalent load resistance R serially connected to the internal resistance of the transducer R_g; hence, the output power of the harvester can be evaluated via the power consumption at the resistance R. The photograph of the prototype harvester is also shown in figur 1.

The equation of motion of the prototype harvester is given by

$$m\ddot{x}(t) + \left(c + \frac{\Phi^2(x(t))}{R + R_g}\right)\dot{x}(t) + F(x(t)) = -mu_a \cos \omega t \tag{1}$$

where x is the relative displacement of the mass and u_a is the amplitude of the input acceleration, further, m and c denote the inertial mass and the mechanical damping coefficient, respectively. The functions $\Phi(x)$ and $F(x)$ are the electromechanical force factor, and the restoring force of the magnetic nonlinear spring, as functions of the relative displacement, respectively.

The restoring force $F(x)$ is modeled by following the approximated formulation presented by Vokoun et al. [7] as

$$F(x) = \frac{\pi\mu_0 M^2 r^4}{4} \sum_{j,k=0}^{1} (-1)^{j+k} \left[\frac{1}{(d - x + jh_{fix} + kh_{mov})^2} - \frac{1}{(d + x + jh_{fix} + kh_{mov})^2} \right] \tag{2}$$

where μ_0, M, r, d, h_{fix}, and h_{mov} are the permeability of vacuum, saturation magnetization of the magnets, radius of the magnets, gap between the fi ed magnet and the movable magnet when the latter is in the

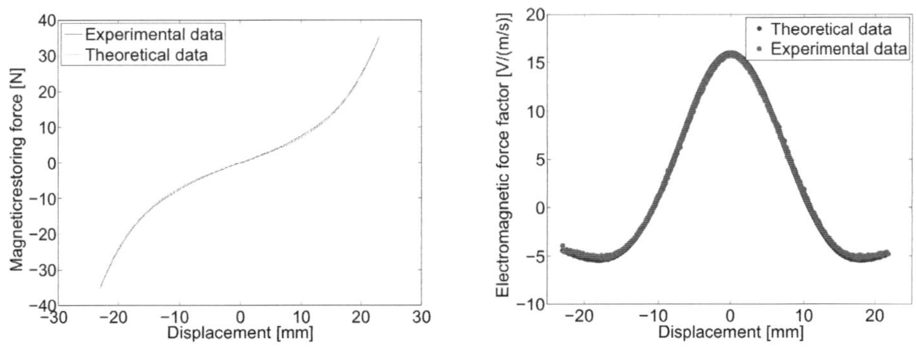

Figure 2. Experimentally identifie functions.

neutral position, thickness of the fi ed magnet, and thickness of the movable magnet, respectively. The electromagnetic force factor $\Phi(x)$ is formulated as

$$\Phi(x) = -\frac{\pi D_c^2}{4} B_{\mathrm{rmov}} \sum_{n=1}^{N} \left[\frac{D_{\mathrm{mov}}^2}{\left\{4\left(x + \frac{h_{\mathrm{mov}}}{2} - x_n\right)^2 + D_{\mathrm{mov}}^2\right\}^{3/2}} - \frac{D_{\mathrm{mov}}^2}{\left\{4\left(x - \frac{h_{\mathrm{mov}}}{2} - x_n\right)^2 + D_{\mathrm{mov}}^2\right\}^{3/2}} \right] \tag{3}$$

All the parameters in equations (2) and (3) were experimentally identified The identifie functions are plotted in figur 2 compared with the experimental data.

2.2. *Load switching circuit*

Since the nonlinear oscillator described above has strong hardening effect, it has two stable steady-state solutions in the resonance band. Because the initial conditions determine which solution emerges, it is difficult for this nonlinear harvester to maintain the high performance of the power generation constantly. In the study done by Masuda et al. [6], the principle of self-excitation and entrainment has been utilized to provide the global stability to the highest-energy solution by destabilizing other unexpected lower-energy solutions. They proposed to introduce a switching circuit of the load resistance between positive and the negative values depending on the response amplitude of the oscillator as follows:

$$R = \begin{cases} R_{positive} & (a \geq a_\theta) \\ R_{negative} & (a < a_\theta) \end{cases} \tag{4}$$

where a is the displacement amplitude of the oscillator, and a_θ is a threshold value. $R_{positive}$ is the load resistance, while $R_{negative}$ is a negative resistance which performs as a energy source to excite the oscillator. This control law imparts the self-excitation capability to the oscillator to perform an entrainment into the highest-energy solution.

In this study, the negative lord resistance is implemented by using a negative impedance converter (NIC) circuit. The response of the oscillator is measured by a laser displacement sensor, which is fed to the control circuit with a microprocessor (mbed LPC1768) which calculates the amplitude of the displacement to perform the load resistance switching as given by equation 4. The switching circuit is controlled by a switching circuit consisting of FET semiconductor switches.

3. Experimental verificatio

An experimental verificatio of this concept of global stabilization was carried out using an experimental setup depicted in figur 3. The prototype harvester described above was mounted on a single axis

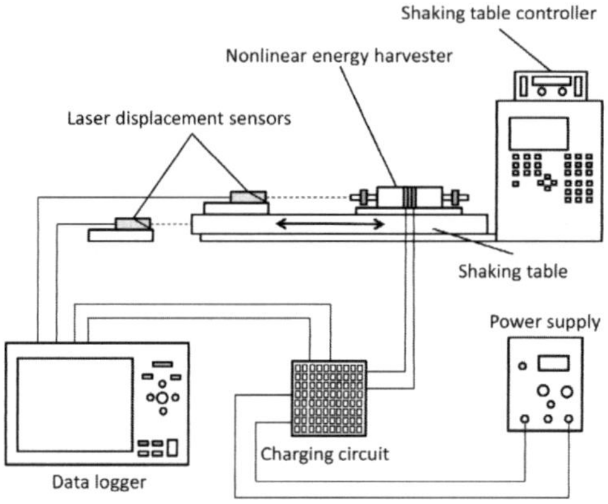

Figure 3. Experiment setup.

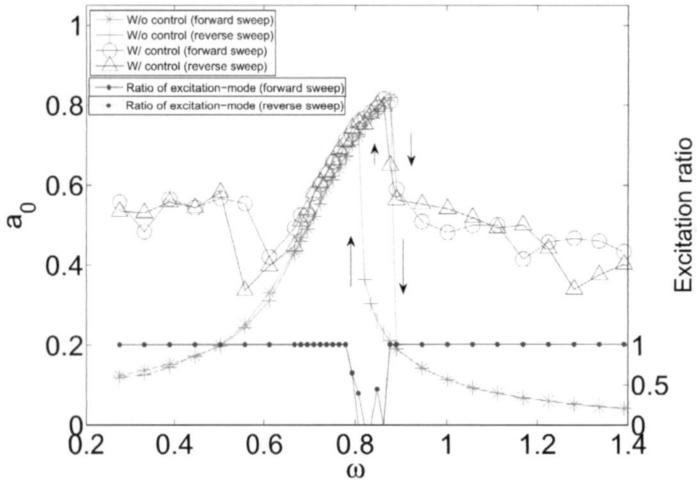

Figure 4. Experimental result.

horizontal shaking table and the relative displacement response of the harvester was measured by a laser displacement sensor mounted on the table. The shaking table was excited by a sinusoidal swept wave from 2 to 10 Hz with an acceleration amplitude fi ed at 0.3 G. The positive and negative resistances were set to 100 Ω and -75 Ω, respectively. The threshold value used in the switching control was set to 74 % of the full stroke.

Figure 4 shows the experimentally obtained resonance curve of displacement responses with and without switching control plotted in a dimensionless form. From the figure one can see that there is a hysteresis of the resonance curve in the case without control. This is because there are two stable solutions in this frequency band. In contrast, the resonance curve with control in the figur shows that the presented prototype harvester successfully destabilized the low energy solution, which provides the global stability to the highest-energy solution. The right axis of the plot is an excitation ratio, which is define as a duty ratio of negative resistance. The frequency band where the excitation ratio completely vanishes is the band where the oscillator is ideally entrained into the high energy solution, while in other frequency band the load resistance is switched to negative intermittently to excite the oscillator. This suggests that the switching control should be activated only in the band of the entrainment to prevent undesirable loss of energy.

4. Conclusion

In this study, an experimental verificatio of a self-excitation control of a resonance-type vibration energy harvester with a Duffing-type nonlinear energy harvester which is designed to perform effectively in a wide frequency range is presented. The experimental results show that the prototype switching circuit between positive and negative load resistances and its control law as a function of the displacement amplitude have successfully destabilized the undesirable low energy solution and provided the global stability to the highest energy solution.

5. Reference

[1] Tang L, Yang Y and Soh C K 2010 Toward broadband vibration-based energy harvesting, *Journal of Intelligent Material Systems and Structures* 21(18), pp 1867–1897.

[2] Soliman M S M , Abdel-Rahman E M , El-Saadany E F and Mansour R R 2008 A Wideband Vibration-Based Energy Harvester, *Journal of Micromechanics and Microengineering* 18(11), p 115021.

[3] Ferrari M, Ferrari V, Guizzetti M, And'o B, Baglio S, and Trigona C 2009 Improved energy harvesting from wideband vibrations by nonlinear piezoelectric converters, *Procedia Chemistry* 1(1), pp 1203 – 1206.

[4] Stanton S C, McGehee C C, and Mann B P 2009 Reversible hysteresis for broadband magnetopiezoelastic energy harvesting, *Applied Physics Letters* 95(17), p 174103.

[5] Karami M A and Inman D J 2010 Nonlinear hybrid energy harvesting utilizing a piezo-magneto-elastic spring, *in Proceedings of SPIE*, 7643, p 76430U.

[6] Masuda A, Senda A, Sanada T, and Sone A 2013 Global stabilization of high-energy responce for a Duffing-type wideband nonlinear energy harvester via self-excitation and entrainment, *Journal of Intelligent Material Systems and Structures*, Vol.24, No.13, pp 1589–1612.

[7] Vokoun D, Bleleggia M, Heller L, Heller L, Sittner P 2009 Magnetostatic interactions and forces between cylindrical permanent magnets, *Journal of Magnetism and Magnetic Materials* 321(22) pp 3758–3763.

Fabrication of Scalable Indoor Light Energy Harvester and Study for Agricultural IoT Applications

M. Watanabe[1], A. Nakamura[1], A. Kunii[2], K. Kusano[2] and M. Futagawa[3]
[1] Business Development Project, LAPIS Semiconductor Co., Ltd. Yokohama, JAPAN
[2] Device Technology Division, LAPIS Semiconductor MIYAGI Co., Ltd. Furukawa, JAPAN
[3] Electrical and Electronic Engineering, Shizuoka University, Hamamatsu, JAPAN

E-mail: watanabe424@mnf.lapis-semi.com

Abstract. A scalable indoor light energy harvester was fabricated by microelectromechanical system (MEMS) and printing hybrid technology and evaluated for agricultural IoT applications under different environmental input power density conditions, such as outdoor farming under the sun, greenhouse farming under scattered lighting, and a plant factory under LEDs. We fabricated and evaluated a dye-sensitized-type solar cell (DSC) as a low cost and "scalable" optical harvester device. We developed a transparent conductive oxide (TCO)-less process with a honeycomb metal mesh substrate fabricated by MEMS technology. In terms of the electrical and optical properties, we achieved scalable harvester output power by cell area sizing. Second, we evaluated the dependence of the input power scalable characteristics on the input light intensity, spectrum distribution, and light inlet direction angle, because harvested environmental input power is unstable. The TiO_2 fabrication relied on nanoimprint technology, which was designed for optical optimization and fabrication, and we confirmed that the harvesters are robust to a variety of environments. Finally, we studied optical energy harvesting applications for agricultural IoT systems. These scalable indoor light harvesters could be used in many applications and situations in smart agriculture.

1. Introduction

In the IoT era, many sensor nodes with integrated energy harvesters were installed in many places and used in many applications. Since over one trillion sensors [1] per year is the scale required, facility usage and cost effectiveness is an important factor for practical energy harvesting applications. Indoor lighting power sources exist everywhere, although conventional silicon-type solar cells cannot be used in all applications because of their photoelectric conversion properties. Recently, next-generation solar cells such as dye sensitized solar cells (DSC) [2], organic thin films [3], and perovskite type [4] technologies have progressed, and they are expected to be applied for harvesting device relays because of their robustness to environmental lighting conditions and cost effective manufacturing processes.

"Scalability" is a keyword for lighting harvester devices. One type of scalability is output power scalability, for example output power linearity and voltage regulation flexibility, which depends on the external dimensions or design layout. For area scalability, DSCs commonly use transparent conductive oxide (TCO) substrates, but TCO is a high-cost material and has high sheet resistance and low transparency, so there is trade-off between efficiency and cell area. The other type of scalability is input power scalability, i.e., the dependence on environmental conditions such as light power intensity, wavelength, and incident angle. These two types of scalability are very important to expand IoT energy harvesting applications.

There are many IoT sensing applications and many environmental conditions in each case. IT agricultural sensing is one representative application, and cultivated fields are exposed to a variety of environment lighting conditions from outdoor farming under the sun to indoor plant factories under LED lighting.

In this paper, scalable indoor light energy harvesting devices were fabricated by MEMS and printing hybrid technology, and harvester devices were evaluated under a variety of environmental conditions for agricultural IoT sensing applications.

2. Experiment

Fabrication was performed as follows. We used paste to fabricate a nano-porus TiO_2 layer and a Ru-based metal complex for the dye sensitizer. TiO_2 layer patterning was performed by a room-temperature nanoimprint process. The substrate metal mesh material was a titanium-based film.

The devices were irradiated by 100 mW/cm^2 with a light source simulating AM 1.5 global solar radiation. The current-voltage curves of the harvesters were measured with a semiconductor characterization source meter system at room temperature with masks under illumination. The shape analysis was performed with scanning electron microscopy (SEM; Hitachi, S-5000), the electrical properties were characterized by the sheet resistance, and the optical properties were measured with a spectrophotometer (Eiko-seiki, PSL 100).

3. Results and discussion

3.1 Device fabrication

First, we developed the cell material and process. DSC harvesters are mainly fabricated via screen printing. Figure 1 shows the conventional structure with a TCO substrate. We have developed TCO-less [5] process with a micro mesh metal structure, as shown in figure 2. Figure 3 is the typical fabrication result for a titanium-based metal mesh, which has low sheet resistance and high visible light transparency.

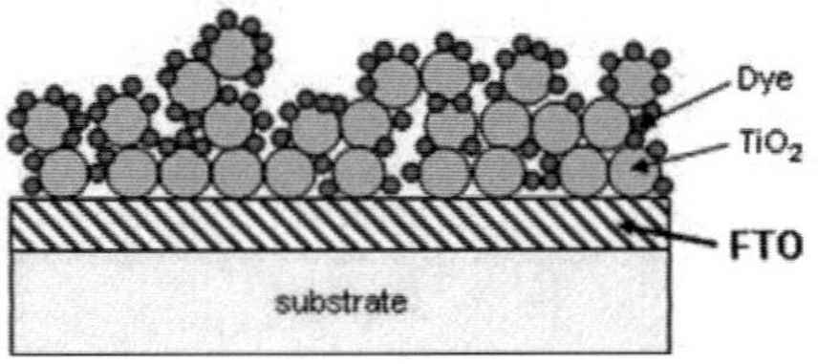

Figure 1. Conventional DSC structure

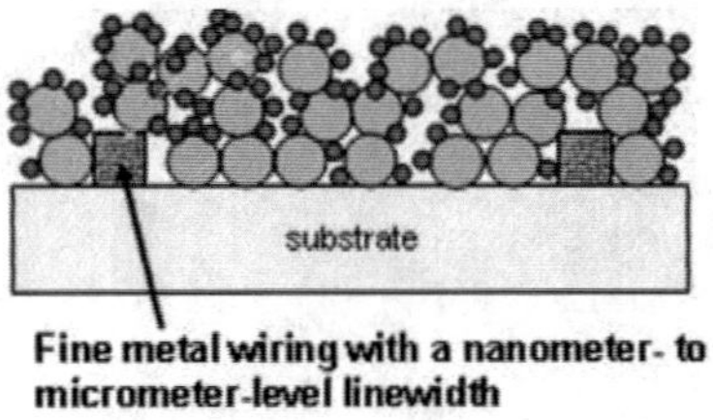

Figure 2. TCO-less structure

Journal of Physics: Conference Series **660** (2015) 012110 doi:10.1088/1742-6596/660/1/012110

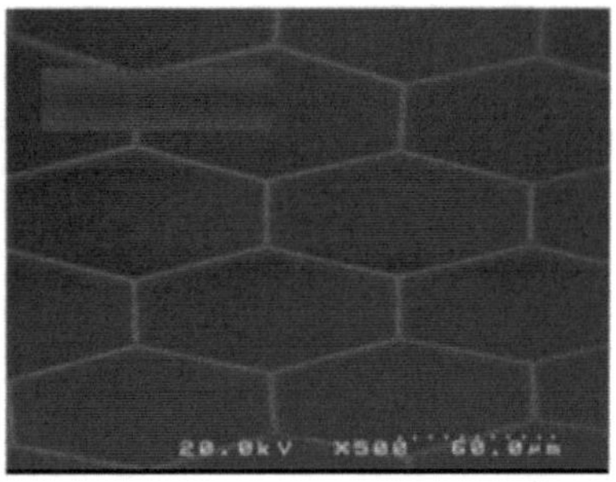

Figure 3. Fabrication of honeycomb mesh

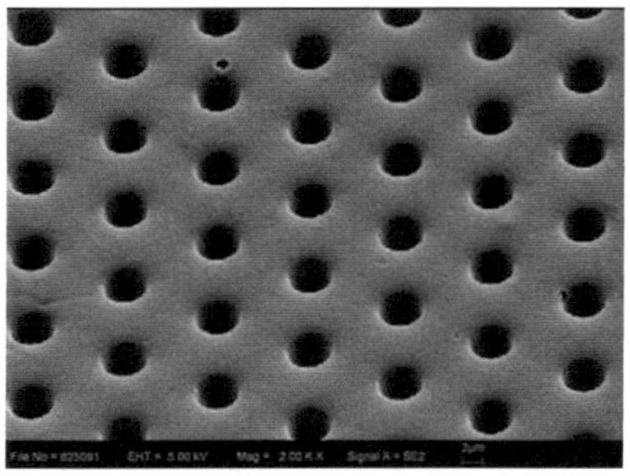

Figure 4. Hole array nanoimprinting

We fabricated TiO_2 electrodes and patterned a nanometer- to micrometer-scale structure that was designed for optical optimization and fabrication dye absorption process tact time. Figure 4 shows the hole array pattern fabricated by nanoimprint technology. The imprint process was carried out on coated TiO_2 paste at room temperature.

Second, we designed and fabricated the harvester module for a wearable thin film with a substrate thickness of 200 μm (figure 5a), a module for mobile use that was 40 mm by 40 mm (figure 5b), and a module for power generation that had a large area of 120 mm by 120 mm (figure 5c).

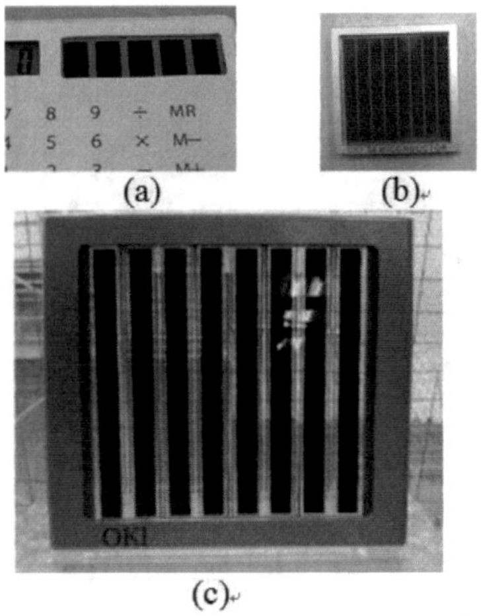

Figure 5. Module prototype of (a) a thin film module, (b) a 40 mm by 40 mm mobile use module, and (c) a 120 mm by 120 mm large area module.

3.2 I-V characteristics and scalability

The I-V output characteristics of chips with a variety of cell areas were estimated by cell fabrication and compared to the properties of a 5 mm square small test cell chip. Figure 6 shows cell area scalability with the micro honeycomb mesh structure developed. The photo-electron conversion efficiency had a constant value that depended on the cell area. Therefore, we confirm that the cell output power scalability does not depend on cell size.

Typically, the DSC output voltage is approximately 0.6 to 0.8 V per cell. Therefore, for output voltage scaling, we designed a module layout with cells connected in series. The harvester module showed scalable linear output properties that depended on the cell number, as shown in figure 7. Thus, we designed many module types for many application specifications.

Figures 8 and 9 show the inlet lighting condition dependence of the output power density. We confirmed that the DSC harvester could demonstrate photon-electron conversion, even in the low-light intensity condition below 100 lux, and we also checked other light sources, a florescence light and an OLED, that indicated the same trend as the LED. According to the results for the incident light intensity and inlet angle dependency, the harvester developed is robust to environmental conditions.

Furthermore, this system could also be used for high-power-consumption applications, such as wireless power supply systems for smart phones [6]. Therefore, harvester output power scaling was accomplished by the hybrid fabrication technologies presented here.

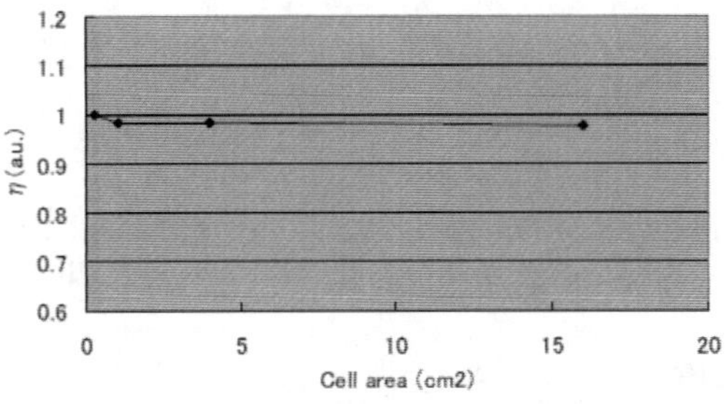

Figure 6. Cell area scalability with a micro mesh number

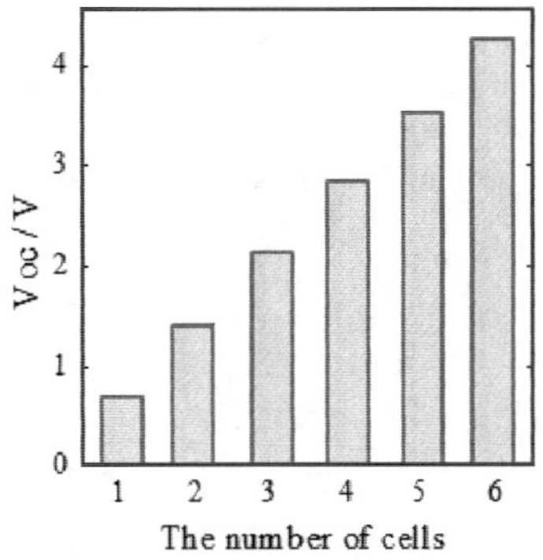

Figure 7. V_{oc} dependence on cell

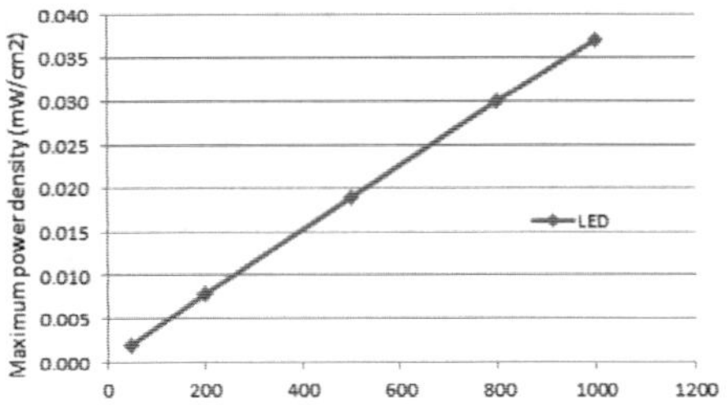

Figure 8. Maximum output power density

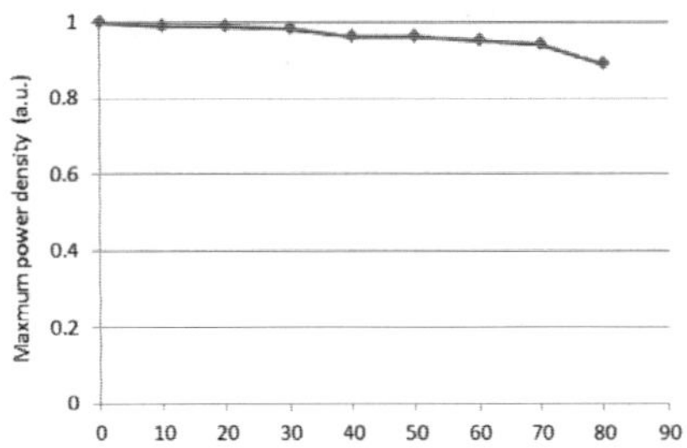

Figure 9. Light incident angle dependency

3.3 Study for agricultural IoT applications

IoT environmental sensing is used widely in outdoor and indoor environments. In an IT agricultural sensor network, sensor nodes are installed in various environmental lighting conditions such as those for outdoor cultivation, plastic light shield greenhouse farming, and indoor plant factories. We evaluated the environmental luminance intensity at a height of 30 cm from the ground. Illumination intensities were 80,120 lux at an outdoor cultivation farm, 40,280 lux at a greenhouse farm, and 17,500 lux in a plant factory on sunny days. On cloudy and rainy days or during morning and evening, the light intensity decreased to 1/10 to 1/100 of the intensity level under sunny daytime conditions. According to these results, the harvesters developed could cover the almost every agriculture environment application, even under conditions of low illumination intensity.

4 Conclusion

A scalable indoor light energy harvester was fabricated by MEMS and printing hybrid technology. It was evaluated for agricultural IoT applications under different environmental input power density conditions. We developed a TCO-less process with a honeycomb metal mesh substrate and TiO_2 hole array pattern fabricated by MEMS technology and evaluated a DSC module as a low-cost and "scalable" optical harvester device. This scalable indoor light harvester is robust to a variety of environmental conditions, and thus it could be used in many applications and environmental situations in IoT smart agriculture.

5 Acknowledgement

The authors would like to thank Professor Toyota of Tokyo University of Agriculture and Technology for agricultural application study and Professor Hirai of Osaka Prefecture University for the nanoimprint process.

References

[1] Trillion sensor universe web site, www.tsensorssummit.org/
[2] B. Oregon and M. Gretel, Nature 353 (1991) 737.
[3] C.W.Tang et al., Appl. Phys.Lett. 48 (1986) 183.
[4] A.Kojima, T.Miyasaka et al., J. Am. Chem. Soc. 131 (2009) 6050.
[5] M. Watanabe, H. Fujimaki et al., 25th EUPVSEC, 1DV4.5 (2010) 784.
[6] M.Watanabe, K Takeno et al., IEICE-EE, 2017-17, 110, 218 (2010) 7.

Finite element analysis of combined magnetoelectric-electrodynamic vibration energy converter

Sonia Bradai[1,2], Slim Naifar[1,2] and Olfa Kanoun[1]

[1]Chair for Measurement and Sensor Technology, Technische Universität Chemnitz, Reichenhainer Straße 70, 09126 Chemnitz, Germany
[2]Laboratory of Electromechanical Systems, National Engineering School of Sfax, University of Sfax, Route de Soukra km 4, 3038 Sfax, Tunisia

E-mail: sonia.bradai@s2013.tu-chemnitz.de

Abstract. In this paper we report on the design and optimization of a novel combined vibration energy harvester based on the use of electrodynamic and magnetoelectric (ME) principles to increase the energy outcome of an electrodynamic harvester without significantly increasing its size. Thereby the most important aspect is the dependence of magnetic flux variation on design parameters, as is it is the decisive effect for energy conversion. Magnetic circuit form and magnetization are optimized for maximizing energy outcome. We conclude that better magnetic flux variation is reached for a magnetic circuit formed with two magnets stacked one within the other using the same magnetization. Results illustrate that the use of combined converter enables to enhance the performance of simple electrodynamic or ME converter.

1. Introduction

Harvesting energy from natural sources to power wireless devices present a big interest for research. In particular, due to the availability of vibration energy, harvesting energy from vibration sources is investigated based on electrostatic, electrodynamic [1, 2], piezoelectric [3, 4] and magnetoelectric [5, 6] principles. Developed converters have some advantages but present also some limitations such as the size, energy density and system reliability.

To improve vibration energy converter performance, different solutions are developed such as the enhancement of the electronic to improve the converter performance [7]. Also, the use of multi-harvesters based on the same principle [8, 9] are investigated which enable to improve the working frequency for the converter but lead to a bulky system.

Another interesting solution is the use of combined converter based on different principles especially electrodynamic, magnetoelctric and piezoelectric [10, 11]. This method enables to enhance the energy outcome for the system and its life time. Until now, combined converters based on electrodynamic and piezoelectric present some limitations due to the aging problem of the piezoelectric material receiving an external excitation and the needed volume to realize such solution. For the combination of electrodynamic and ME principles, developed system based on cantilever beam enables to improve the converter energy outcome but still have some limitations such the size of the whole system and life time because of using the cantilever beam.

In this context, this work presents an alternative combined vibration converter based on electrodynamic and ME principles. The developed system is formed with a magnetic spring, magnetic circuit, magnetoelctric transducer and coil. This work is devoted to evaluate the behavior of the combined vibration energy converter using finite element analysis. It is organized in three main sections. In section 1, the developed converter design is presented. In section 2, a finite element analysis model is investigated to optimize the ME transducer behavior. At the end, comparison for the combined converter with single ME or electrodynamic converter is developed.

2. Design of the combined energy converter

Combined converter design based on electrodynamic and ME principles is proposed as shown in Figure 1.(a). The electrodynamic principle is based on the relative movement between magnets and coil which leads to generate energy based on Faraday's Law. For the ME principle, it consists on the placement of a ME transducer formed with two magnetostrictive (MS) layers and one piezoelectric in between, in a variable magnetic field environment. Due to the presence of vibration, the changing of the magnetic flux leads to a deformation on the MS layers. Receiving this stress, piezoelectric layer placed between two MS layers generates energy. In the developed concept, the magnets movement is ensured by the use of magnetic spring principle for better robustness.

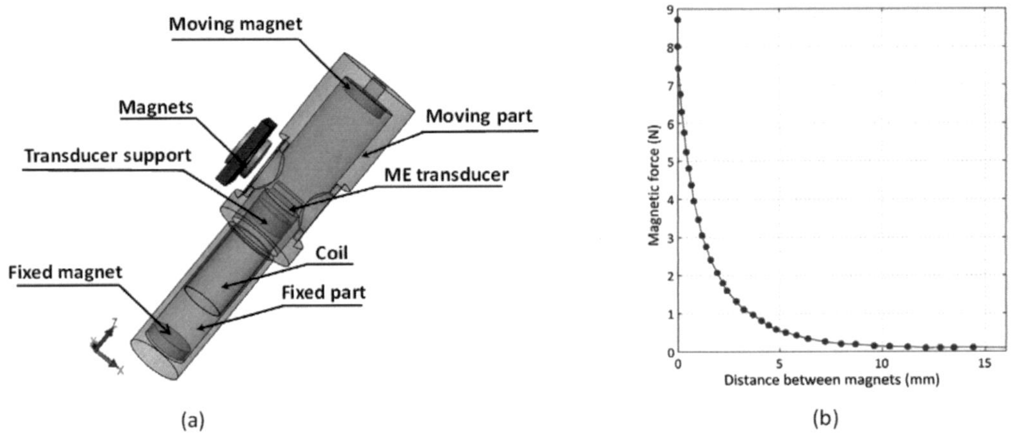

Figure 1. (a) Proposed design for the electrodynamic-magnetoelectric converter, (b) Magnetic Force for the magnetic spring

The system consists on the use of magnetic spring formed with two magnets with the size of 8 mm of diameter and 2 mm of thickness placed in repulsive direction to transmit the external vibration source. The top magnet is attached to the moving part. Magnetic circuit formed with cylindrical magnets for the ME transducer is also fixed to the moving part. Second magnet, ME transducer and coil are attached to the fixed part.

For the magnetoelctric transducer, rectangular geometry is selected based on previous realized work. It is formed with two magnetostrictive layers with the size of 6 mm of length, 4 mm of width and 1 mm of thickness and one piezoelectric layer in between with the size of 7 mm of length, 4 mm of width and 1 mm of thickness.

Magnetic circuit form, size, polarization and positions are optimized to get better performance from the converter. The housing coil is placed surrounding the ME transducer with the size of 8 mm of outer diameter, 25 mm of length and 0.5 mm of thickness. The total volume for the converter is 20 cm^3. Static position for the moving part is determined by the calculation of magnetic force for the magnetic spring depending on the distance between the magnets as shown in Figure 1.(b).

3. Simulation results and discussions

3.1. Definition of the finite element model

Figure 2 presents the developed finite element model for this study. The magnetic circuit is defined using cylindrical magnets. In between, we define the composite formed by two MS layers and one PZT layer. NdFeB and copper materials are used for the magnets and coil

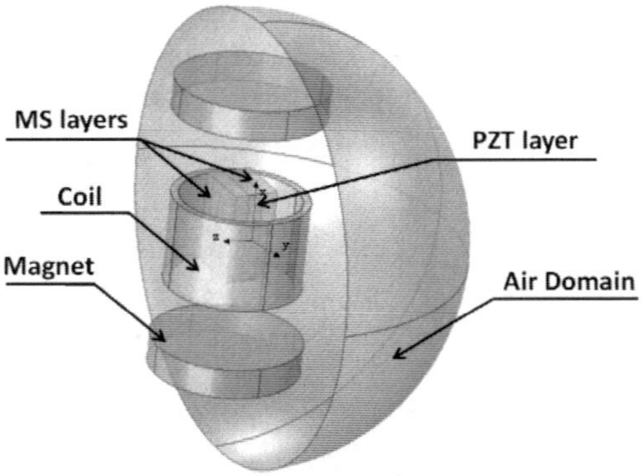

Figure 2. Defined finite element model for the converter

respectively. For the magnetic transducer, Terfenol-D and PZT-5H materials are used for the MS and piezoelectric layers respectively. Terfenol-D is used due to its high MS performance and its nonlinear behavior is defined using HB curve [12] from which the magnetic field H is determined relative to the magnetic flux density value. The MS layers are polarized along their length and PZT along the thickness which enables to get better performance for the MS layers. The induced voltage for the converter is related directly to the variation of magnetic field through the MS layers and coil. For that, we concentrate on the evaluation of magnetic field variation relative to different parameters such as magnetic circuit form, size and magnetization.

3.2. Optimization of magnetic circuit

The MS layers behavior is evaluated depending on different parameters. We start with the evaluation of the suitable magnetic circuit by comparing three possibilities, which consist on the use of one disc magnet (Figure 3.(a)), two separated discs magnets from each MS layer side (Figure 3.(b)) or two magnets stacked one within the other (Figure 3.(c)). The magnets size is fixed to 8 mm of diameter and 1 mm of thickness for the use of one disc magnets and to 4 mm of diameter and 1 mm of thickness for the use of two separated discs magnets. For the use of two stacked magnets, the influence of the ring magnet size is studied. Simulations are conducted for

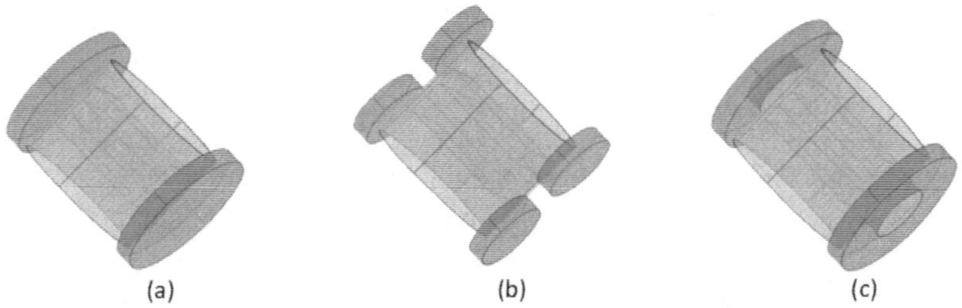

Figure 3. Proposed magnetic circuit for the converter (a) one disc magnet (b) two discs magnets (c) two magnets stacked one within the other

an inner disc magnet with the size from 1.6 mm to 7.8 mm. Results show that the use of two separated discs magnets leads to the decrease of the magnetic flux density to 0.6 T as maximum (Figure 4.b) comparing to the use of only one disc magnet (Figure 4.(a)). Using two magnets stacked one within the other, same behavior is reached than the use of one disc magnet with a more higher magnetic flux density level.
We conclude that the use of ring magnet with 8 mm of inner diameter, 3.2 mm of outer diameter and 1 mm of thickness and an inner disc with the size of 3.2 mm of diameter and 1 mm of thickness enables to reach better magnetic flux density level. This is due to the concentration of magnetic flux density lines which are surrounding the MS layers.

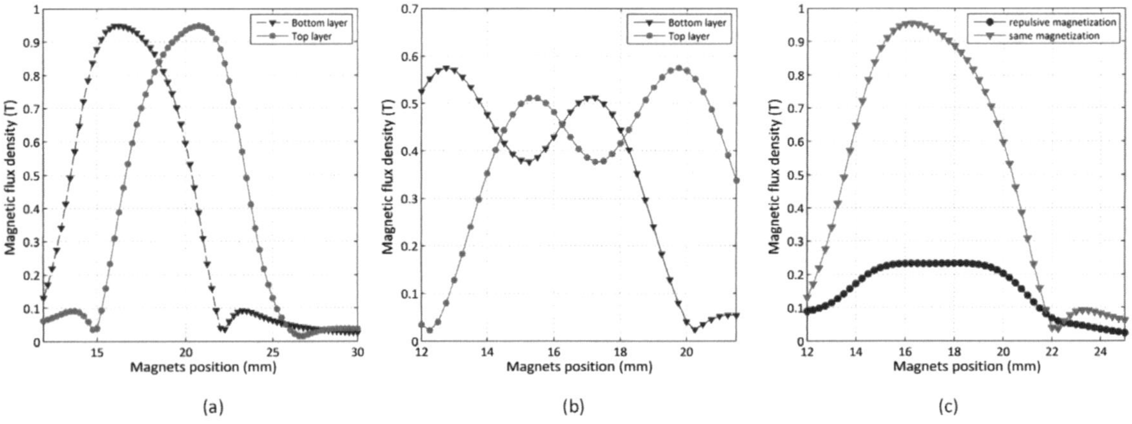

Figure 4. Variation of magnetic flux density for bottom and top MS layers relative to the magnets position along z axis (a) one disc magnet (b) two discs magnets (c) two magnets stacked one within the other for one layer with same and repulsive magnetization

Taking into account the obtained results for the magnetic flux density variation through the bottom and top MS layers, the static position for the magnets is fixed to 18.5 mm along z axis. Figure 4.(c) presents the influence of the magnets polarization on the magnetic flux density variation. The magnet movement is limited to 2 mm. In the case of repulsive direction, magnetic flux density is limited to 0.23 T which is four time less than the reached magnetic flux density using same magnetization. We conclude that the use of two stacked magnets with same magnetization for the magnetic circuit enables to enhance the magnetic flux variation through the MS layers.

3.3. Comparison of combined magnetoelectric-electrodynamic with simple electrodynamic or magnetoelectric converter

After the determination of optimal magnetic circuit and magnetization, influence of the two principles combination is studied. Magnetic field evaluation is illustrated in Figure 5. Results show that combined converter enables to enhance the magnetic field level especially for the case of electrodynamic converter. The magnetic field is increased by 125 Oe through the coil comparing to the case of electrodynamic converter. For the magnetoelectric transducer, a small difference is observed. Therefore, due to the increase of the magnetic field through the combination of the two principles, electrodynamic-magnetoelectric converter is able to reach better performance comparing to the use of single electrodynamic or ME converter.

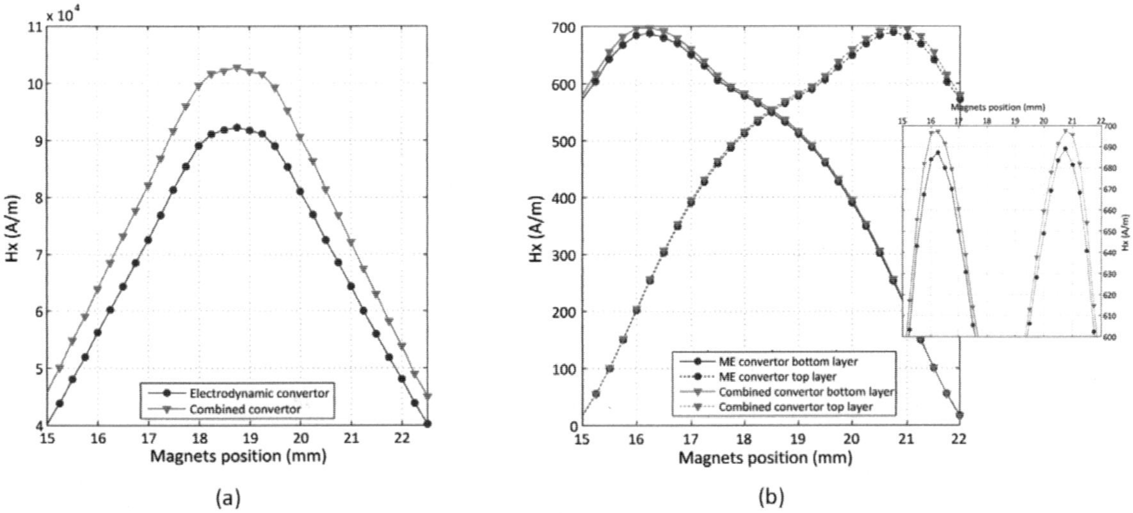

Figure 5. Variation of magnetic flux density through the (a) coil for electrodynamic and combined converters (b) transducer layers for ME and combined converters

4. Conclusion

In this paper, optimization of combined vibration converter based on electrodynamic and ME principles is investigated. Mechanical design for the combined converter composed with magnetic spring, magnetic circuit, one ME transducer and a coil is proposed. For this purpose, the used ME transducer is formed with two MS layers (Terfenol-D) and one PZT layer in between. Through the realized finite element analysis, magnetic flux density variation through the transducer and the coil is evaluated. Results show that the use of magnets with same magnetization enables to reach higher magnetic field variation than repulsive magnetization. Finally, an enhancement of magnetic flux density is reached through the coil and the the ME transducer while combining the two principles comparing to a simple electrodynamic or ME converter.

References

[1] Bradai S et al, 2014 *International Instrumentation and Measurement Technology Conference Proceedings* 1152-1156.
[2] Morais R et al, 2011 *Sensors and Actuators A: Physical* **172** 259-258
[3] Khaligh A et al, 2010 *IEEE transactions on industrial electronics* **57** 850-860.
[4] Gao J et al, 2012 *J. Appl. Phys.* **112** 104101.
[5] Naifar S et al, 2014 *International Instrumentation and Measurement Technology Conference Proceedings* 1157-1162.
[6] Qiu J et al, 2015 *J. Appl. Phys.* **117** 17E705.
[7] Zhu D et al, 2012 *J. Smart Mater. Struct.* **21** 105039.
[8] Dibin Z et al, 2010 *Measurement Science and Technology* **21** 022001.
[9] Zhu D et al, 2014 *J. Phys. Conf. Series* **476** 1-5.
[10] Tadesse Y et al, 2009 *J. of Intelligent Material Systems and Structures* **20** 625-632.
[11] Qiu J et al, 2015 *J. Appl. Phys.* **117** 17A331.
[12] http://www.etrema.com/terfenol-d/.

Frequency up conversion approach to scavenge mechanical energy from an electromagnetic digital actuator

Linjuan Yan[1], Adrien Badel[1], Laurent Petit[2], Fabien Formosa[1]

[1] SYMME, Université Savoie Mont Blanc, Annecy 74000, France
[2] Roberval, Université Technologique de Compiègne, Compiègne 60200, France

adrien.badel@univ-smb.fr

Abstract. This paper reports the practical design and experimental testing of a piezoelectric energy reclamation system. The presented system aims at scavenging energy from a miniaturized electromagnetic digital actuator for additional function such as obtains the discrete position information for enhanced reliability. Based on frequency up conversion technique, and the considered actuator dynamical responses, a piezoelectric energy harvester has been experimentally evaluated. In plane integration is the next step.

1. Introduction

The integration of several actuators to realize a smart surface is a potential route to develop flexible conveyors. Compared to analog actuators, the digital ones have the features of relatively simple control loop, low cost and low energy consumption. Although feedback sensor is not mandatory in multi-stable digital actuators design, the implementation of sensors can improve the reliabilities of such actuators. In order to prevent extra energy supply and wires for this additional sensing function, energy harvesting design for powering sensing system is a promising approach.

The final target of this study is a fully integrated miniaturized self-powered position-sensing device for multi-stable digital actuators. Vibration energy harvesting strategy to scavenge energy from the actuation movement is investigated. The concept is demonstrated on an electromagnetic (EM) digital actuator proposed by Petit et al [1-2] (figure 1). The electromagnetic digital actuator is capable of achieving two-dimensional movement in plane by switching a mobile permanent magnet (MPM) among four discrete positions by generating electromechanical force from related fixed wires carrying proper currents.

Since the position switching movement of the EM digital actuator is a low frequency motion (less than a few hertz) and it could be intermittent due to the practical applications and tasks, the conventional natural frequency matched resonator is not suitable for the VEH design. Therefore, an alternative adaptive harvester design based on frequency up-conversion strategy is investigated. In order to avoid modifying the magnetization distribution of the EM digital actuator, piezoelectric transduction is chosen due to its compactness, lightness and ease of implementation features.

In the literature, several piezoelectric harvesters, which implement a frequency up-conversion strategy, have been reported. It aims at decoupling the excitation frequency and the structural frequency of the VEH, allowing scavenging energy by a high resonance frequency harvester in a low frequency excitation scenario. The frequency up-conversion effect is usually enabled either by contact mechanisms [3-4] or by non-contact interactions [5-6].

2. Energy harvesting concept

Position switch movement of the MPM between two stable positions is used to initiate the deformation of a high frequency VEH device. The proposed approach is similar to the magnetic latching principle introduced in [4]. However, instead of magnet, a small piece of ferromagnetic material is attached on the tip of the VEH to achieve mechanical deformation and to avoid changing the magnetization distribution. Once the restoring force of the VEH is greater than the magnetic force interaction between the MPM and the tip end of the VEH, it is released. Then part of the obtained mechanical energy can be converted to electrical energy by the piezoelectric material and stored in the dedicated circuit.

From the energy transfer point of view (from energy supply to the mechanical energy of the system consisting of the MPM and the VEH, part of the mechanical energy of the VEH to electrical energy), the harvested energy is linked to the energy supplied to drive the EM actuator. The amount of the obtained mechanical energy in the VEH is designed to be much lower than the supplied energy so that negligible supplementary power for actuation is required. From the piezovoltage waveforms of the VEHs system, the position of the MPM can be detected. Although the feature of position detection is theoretically dispensable for the operation of a digital actuator, it allows to improve the reliability and to monitor the actuators array, which is a requested feature for practical implementation.

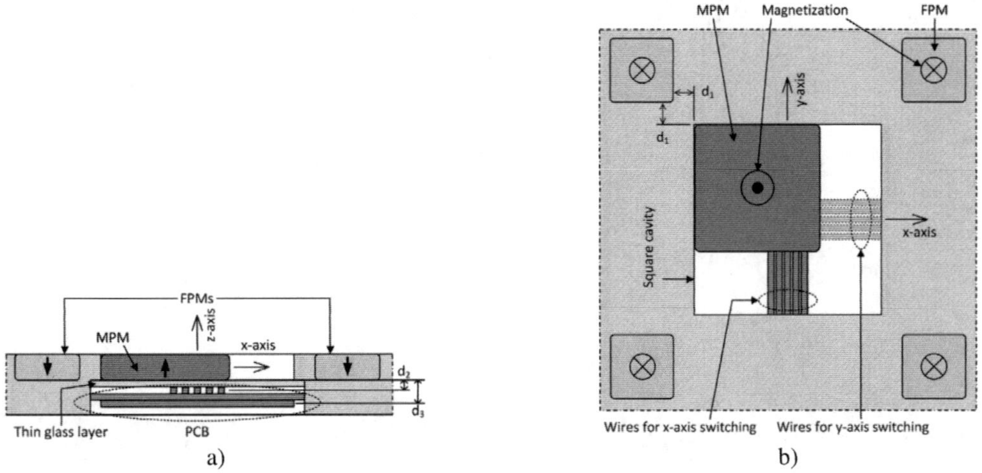

Figure 1. Illustration of the electromagnetic digital actuator: a) side view and b) top view

3. Experimental results

In order to validate the design concept, preliminary experiments are carried out. Simple cantilever beams equipped with a piezoelectric patch are arranged over the EM digital actuator (figure 2) and the tip of each beam connects to the MPM when the latter is at the $(x, y)=(-d_{stroke}/2, d_{stroke}/2)$ discrete position. To control the magnetic attraction between the tip end of the cantilever and the MPM, a media plastic layer is added on the iron piece at the tip of the cantilever beam. By tuning the thickness of this layer, the magnetic force between the MPM and the tip of the cantilever beam can be controlled.

dSPACE real-time solution and a voltage/current converter ([-10V, 10V]) implement the pulse signal of the driving current. A 250 ms current pulse signal is supplied to generate the electromagnetic force for the position switch of the MPM. It should be noted that the out of plane cantilever harvester is only used in this study for the concept validation. In-plane integration will be accomplished in future work. Dynamical waveforms of the VEH in open-circuit are plotted in figure 3.

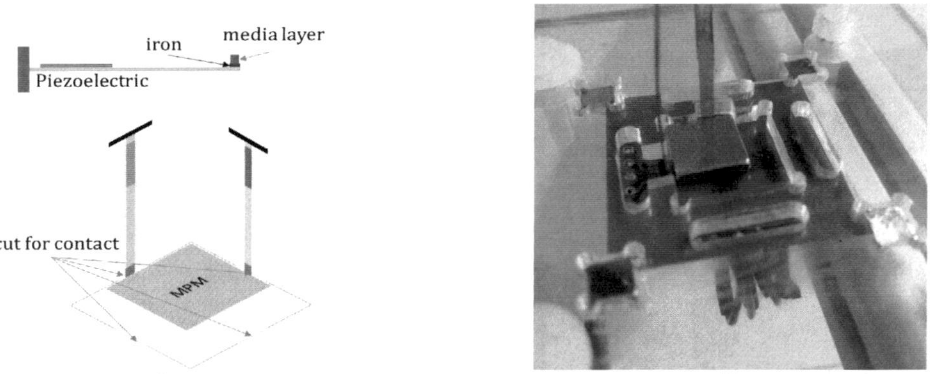

a) b)

Figure 2. a) Illustration of the VEH; b) picture of the system at the discrete position ($-d_{stroke}/2$, $d_{stroke}/2$).

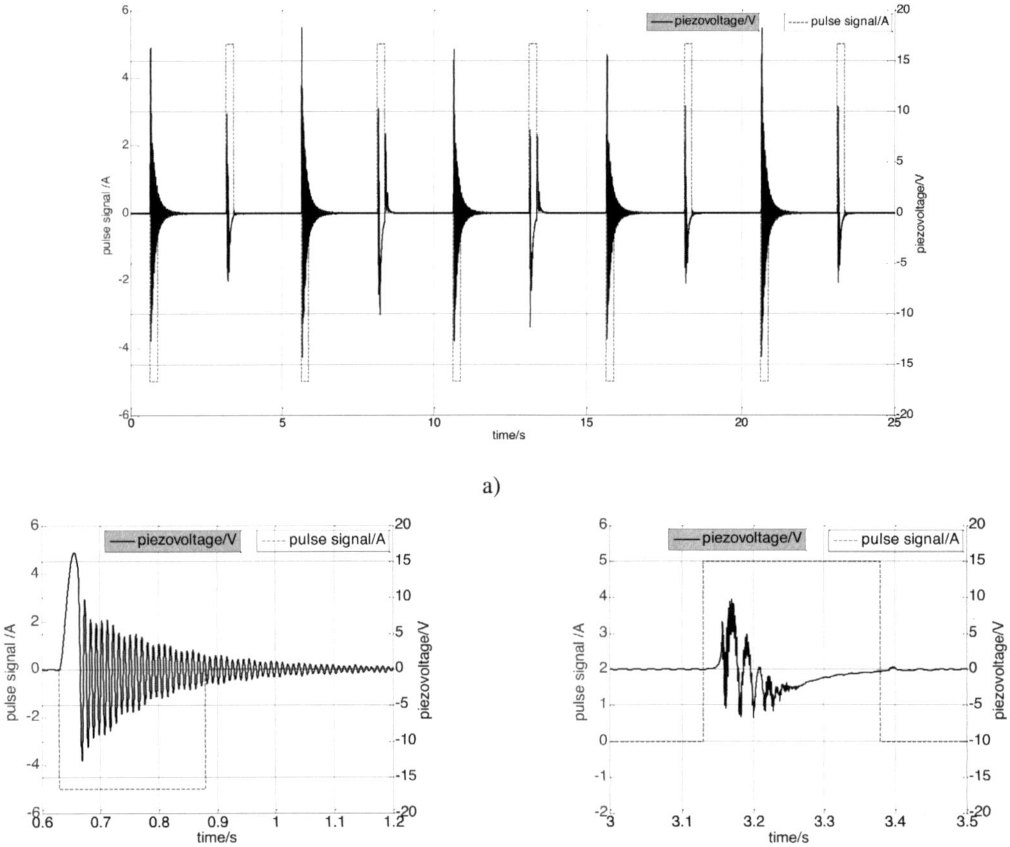

b) release c) contact

Figure 3. Experimental waveforms in open circuit

The experimental results illustrate the release and the contact of the VEH cantilever with the MPM. It is shown that the VEH can successfully separate from the MPM and transfer the potential energy stored in the cantilever beam during its free oscillation. It is also shown that when the MPM goes back and impacts the VEH, a voltage signal can be observed.

By connecting the piezoelectric element to a dedicated circuit, part of the stored potential energy of the cantilever beam can be harvested. A voltage doubler circuit (figure 4) which is one of the simplest yet effective circuits for energy harvesting [7] is taken for this first step of feasibility proof.

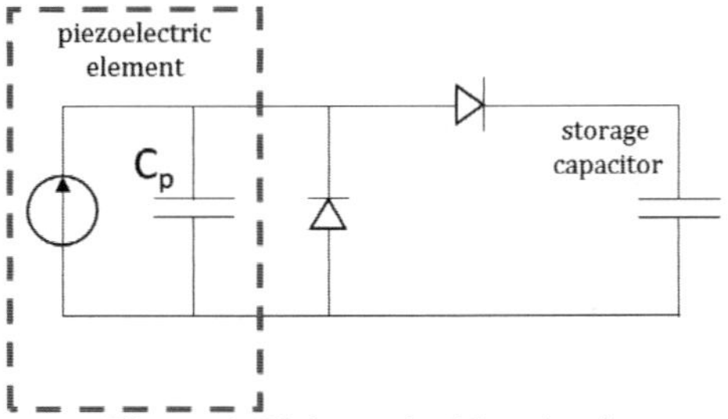

Figure 4. Voltage doubler circuit

The dynamical response for the piezovoltage and the voltage on the storage capacitor (1 μF) are plotted in figure 5. It can be found that around 4.7 μJ (more than 30% of the VEH stored potential energy) is harvested at each release. The small stiffness of the cantilever (k = 9 N/m) is chosen to maintain the proper function of the EM digital actuator. Moreover, the position information of the MPM can clearly be inferred from the waveforms.

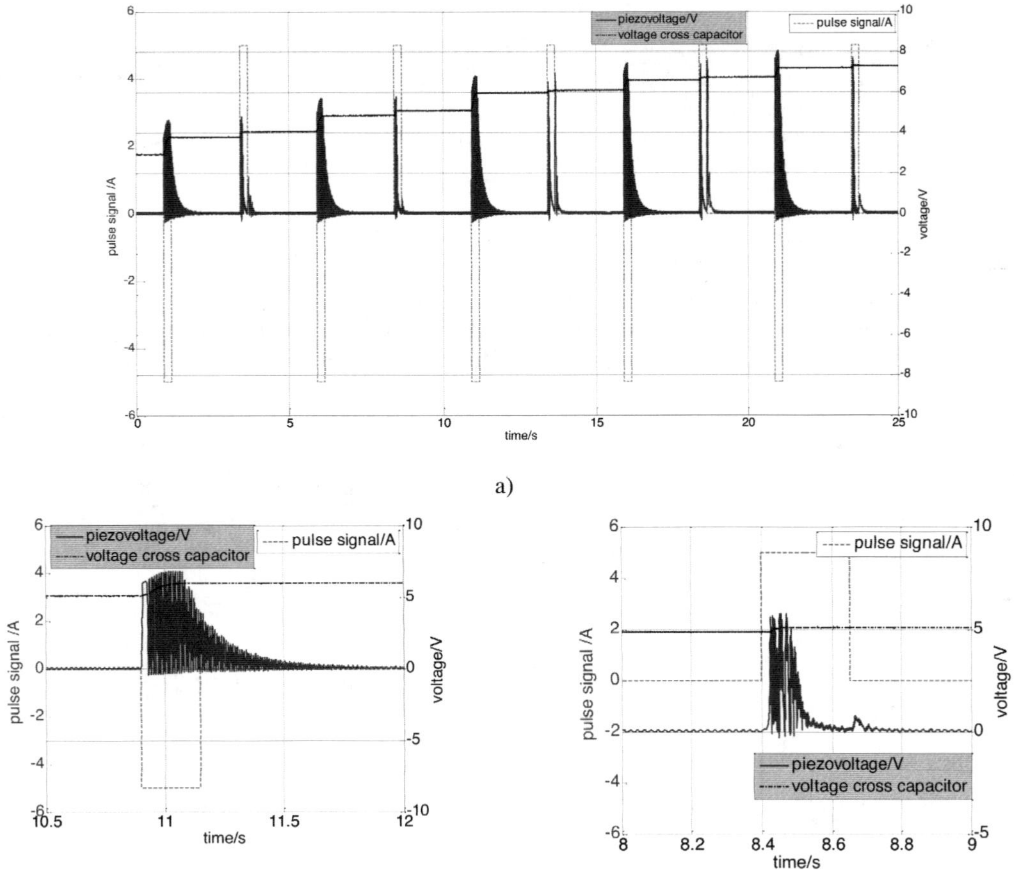

a)

b) release

c) contact

Figure 5. Energy harvesting experimental waveform

4. Conclusion

In this paper, a piezoelectric vibration energy harvester system is designed based on the frequency up conversion principle aiming at harvesting energy from an electromechanical digital actuator and obtaining position information. In order to implement the frequency up conversion technique, a small piece of ferromagnetic material is used for generating the magnetic attraction between the harvester and the mobile permanent magnet, taking advantage of the characteristics of the EM digital actuator.

When the MPM is experiencing an electromagnetic force for position switch, the harvester will move with the MPM towards the aimed discrete position. After reaching a critical position where the restoring force equals the magnetic attraction, the harvester and the MPM separate and the VEH oscillates freely. Thus, mechanical energy can be converted by the piezoelectric materials and further stored on a capacitor. 4.7 μJ harvested energy is obtained in the preliminary experiment with a 16 mN magnetic attraction between the tip end of the harvester and the MPM. When the MPM is close to the aimed discrete position, another harvester is attracted by the MPM and comes into contact. Then they move together until the rest position. Thus, based on the dynamical responses of the energy harvesting system, the position information of the EM digital actuator is obtained from the piezoelectric voltages.

By integrating the proposed VEH system, additional functions such as sensing capability of the electromagnetic digital actuator smart surface will be achieved and its reliability can be eventually improved without adding supplementary power supply and extra wires.

Acknowledgment

This work was supported by the French National Research Agency (ANR) under grant no. ANR-13-BS03-0003-04 (READMI Project) and by the Centre Technique des Industries mécaniques (CETIM).

References

[1] L. Petit, C. Prelle, E. Doré, F. Lamaque and M. Bigerelle, *Four discrete positions electromagnetic actuator: modelling and experimentation*, Mechatronics, IEEE/ASME Transactions, 2010, vol.15, no.1, pp.88-96.

[2] L. Petit, C. Prelle, E. Doré, F. Lamaque and M. Bigerelle, *Optimized design of a four discrete positions electromagnetic actuator*, Proc. IEEE ICM, Malaga, Spain, Apr. 2009, vol.1/2, pp. 509-514.

[3] S. Priya, *Modeling of electric energy harvesting using piezoelectric windmill*, applied physics letters, 2005, 87, 184101

[4] T. Galchev, E. E. Aktakka and K. Najafi, *A piezoelectric parametric frequency increased generator for harvesting low-frequency vibration*, journal of microelectromechanical systems, 2012, Vol. 21, No. 6, pp: 1311-1320

[5] H. T. Luong and N. S. Goo, *Use of a magnetic force exciter to vibrate a piezocomposite generating element in a small-scale windmill*, smart materials and structures, 2012, 21, 025017

[6] S. Jung and K. Yun, *Energy-harvesting device with mechanical frequency-up conversion mechanism for increased power efficiency and wideband operation*, applied physics letters, 2010, 96, 111906

[7] T. Le, J. Han, A. von Jouanne, K. Mayaram, and T. Fiez, *Piezoelectricmicro-power generation interface circuits*, IEEE J. Solid-State Circuits, 2006, vol. 41, no. 6, pp. 1411–1420.

Improved thermal isolation of silicon suspended platforms for an all-silicon thermoelectric microgenerator based on large scale integration of Si nanowires as thermoelectric material

L Fonseca[1], I Donmez[1], M Salleras[1], C Calaza[1], G Gadea[2], J D Santos[2], A Morata[2], A Tarancon[2]
[1] IMB-CNM (CSIC), Campus UAB, E08193 Bellaterra, Spain
[2] IREC, Jardins de les Dones de Negre 1, 2ª pl., E08930 Barcelona, Spain

E-mail: luis.fonseca@imb-cnm.csic.es

Abstract. Special suspended micro-platforms have been designed as a part of silicon compatible planar thermoelectric microgenerators. Bottom-up grown silicon nanowires are going to bridge in the future such platforms to the surrounding silicon bulk rim. They will act as thermoelectric material thus configuring an all-silicon thermoelectric device. In the new platform design other additional bridging elements (usually auxiliary support silicon beams) are substituted by low conductance thin film dielectric membranes in order to maximize the temperature difference developed between both areas. These membranes follow a sieve-like design that allows fabricating them with a short additional wet anisotropic etch step.

1. Introduction

Silicon nanowires (Si NWs) thermoelectric properties are much better than those of bulk silicon [1,2]. Taking advantage of silicon microfabrication techniques and harmonizing the device fabrication with the growth of Si NWs, an all-silicon thermoelectric (micro) generator approach can be attempted. Such microgenerators would be useful as energy harvesting sources to power sensors nodes when a waste heat source is available. While standard thermoelectric modules use exotic materials and their assembling is hard to automate, especially when reducing their size, silicon microtechnology offers the possibility to exploit an abundant material and a mature technology that enables large volume fabrication and also offers an integration route for miniaturized devices. Moreover, silicon micromachining allows fabricating free surfaces or volumes physically coupling the device to the environment and also adds a third dimension to otherwise planar silicon devices. This additional dimension is useful to integrate the 3D nature of bottom-up growth methods of Si NWs.

2. Design and fabrication

Examples of this integration can be found in the literature such as in [3], where a trench is micromachined by DRIE on the silicon device layer of an SOI silicon wafer defining a silicon platform separated from the surrounding silicon rim (figure 1a). The handle wafer is patterned and the silicon under the platform region is removed for thermal isolation. Gold nanoparticles, with appropriate size and density, are selectively deposited in the opposing vertical walls of the silicon platform and rim by means of the galvanic displacement method. These gold nanoparticles act as seeds for the Si NWs, which are then grown in a CVD following the VLS mechanism [4,5]. Si NWs grow preferentially on <111> planes. Vertical <111> walls are found in (110) silicon, so that the device layer of the starting SOI wafers should feature such orientation and the device design must follow the geometrical disposition of the two intersecting <111> family planes.

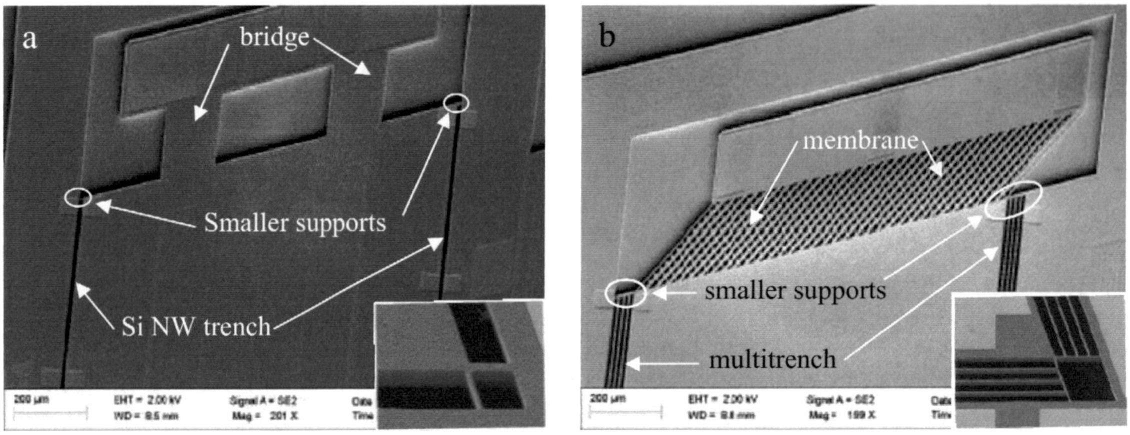

Figure 1. Suspended silicon platforms fabricated in SOI wafers. The platforms are supported by two bulk silicon bridges and other auxiliary bars (a) or by a perforated thin dielectric membrane (b). The later device also shows a set of spacers supported by longer auxiliary bars. The insets show a detail of the corners not shown that feature similar bars.

Once grown, Si NWs connect the silicon platform (the cold area) with the surrounding silicon bulk (the hot area) configuring the silicon leg of a (nanostructured) silicon-metal thermocouple. The length of the nanowires depends on the CVD process time, and it must be adjusted to cover the trench width. Nanowire tapering may occur at prolonged growth times. A given number of intermediate silicon spacers can be defined between the platform and the silicon rim (figure 1b). In this way, nanowires with arbitrary length can be designed/obtained as a multistage concatenation of shorter nanowires in a single controllable CVD process as in figure 2. Longer nanowires connecting heat source and heat sink shall feature higher thermal resistances and lead to a higher temperature drop across them [6]. Typical dimensions of the elements of those devices are 1 mm^2 platforms and 15 μm deep and 10 μm wide trenches (single or multi-trenches).

Figure 2. Silicon nanowires bridging four 10 μm (wide) x 15 μm (deep) trenches defined by three intermediate silicon spacers. These spacers are hold together between the suspended platform and the silicon rim by additional support bars (not shown). The growth of Si NWs into the current structures is still work in progress

In spite of the extra vertical dimension added by silicon micromachining to host the nanowires, the device is still essentially planar and its geometry obeys to the fact that vertical thermal gradients must be transposed in a the laterally assembled device. This is the reason why the suspended platform is needed in the first place: to force the heat flow through the laterally grown nanowires when the device is left to rest onto a hot surface. Material paths, other than the thermoelectric active ones, connecting the platform and the surrounding rim will negatively impact the thermal isolation of the suspended platform diminishing the attainable temperature difference. As shown in figure 1 and figure 2 and their insets, the 'suspended' platform is laterally supported by different features: two main bulk silicon bridges, which also host the electrical connection tracks; small secondary silicon bars at the four corners of the platform keeping it aligned to the silicon rim during Si NWs growth (stress gradients due to the metal and dielectric layers present in the platform may produce a vertical displacement); similar silicon bars in the multi-trench device holding the silicon spacers' ensemble in place.

Once the nanowires are grown, the smaller supporting bridges are no longer useful and they can be eliminated with FIB. This removal has a measurable impact on the thermal response of the platform [7]. However, this is not an option for the two main bridges hosting the metal tracks since they are the physical embodiment of one of the thermocouple legs. The purpose of this work is to find a larger thermal resistance alternative for those bridges. The solution devised consists of building a connecting dielectric membrane in substitution of the two bulky bridges. To do so, a sieve-like structure is defined in the supporting side of the device (figure 1b). The layers present in this region prior to any silicon DRIE are the silicon itself, a LPCDV silicon nitride layer, the patterned metal tracks and a PECVD oxide layer. The sieve holes and the trenches that will accommodate the Si NWs are defined by the same dry etch perforating the two dielectric layers and the silicon device layer of the SOI wafer till reaching the buried oxide (figure 3a). The sides of the sieve cavities are designed in a way their walls are aligned to fast-attacking silicon planes, so that a subsequent KOH wet anisotropic etch will remove laterally the silicon beneath the sieve area. Depending on the design, a few minutes etch may free the membrane, which is kept tight and flat by the tensile stress of the nitride (figure 3b). Metal tracks are defined finger-like to be accommodated in a distributed way on top of some of the nitride strips of the perforated membrane. The <111> vertical walls of the trenches to be covered by the Si NWs are preserved during the anisotropic etch releasing the membrane because such plane is very slowly etched by KOH. In fact, the final condition of the exposed <111> planes is improved because the typical scalloping found after DRIE is smoothed out.

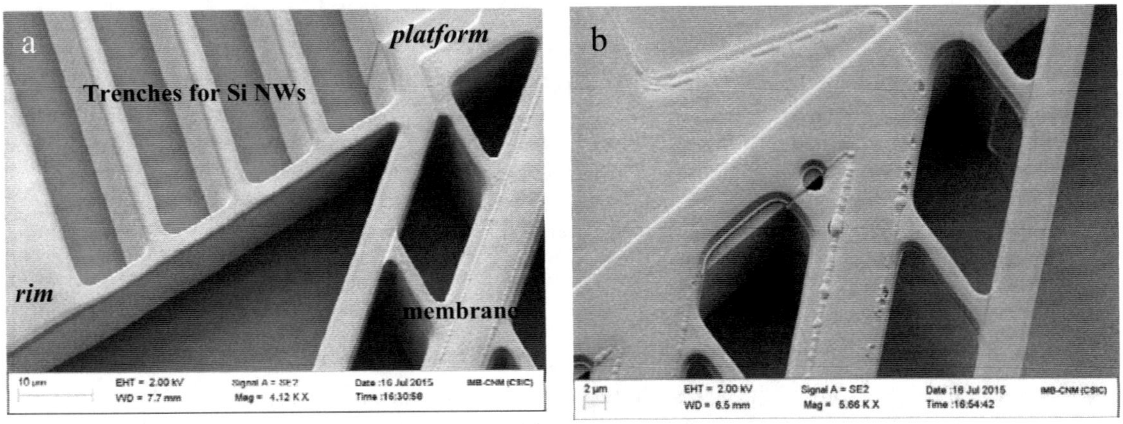

Figure 3. (a) Zoom-in of the perforated membrane and (multi)trench region after DRIE. Hole angles in the membrane are defined so that fast attacking planes are revealed. (b) Result after a KOH anisotropic etch that removes the silicon underneath the membrane.

3. Results

The thermal isolation offered by any physical support will depend on its cross-section, length and on the thermal conductivity of the building material. Silicon is a good thermal conductor so bridges of this material should be narrow and long for good thermal isolation. Since the main support structure should also host the metal leading to one of the terminals of the device, too narrow bridges will increase the internal electrical resistance of the generator. Typical silicon support bridges are 200 μm wide and 200 μm long. The nitride based membrane will offer a better isolation because, although wider (1 mm maximum), it will be ten times thinner (1.5 μm *vs.* 15 μm) than the bridges, and because silicon nitride and oxide have thermal conductivities five to one hundred times smaller than silicon.

In order to assess the thermal isolation provided by both alternatives, an indirect electrical way has been used. Both types of devices have the same built-in heater in their platforms. A similar power has been dissipated in the heaters of devices that only differ in the main support strategy. Those featuring a better thermal isolation will attain a higher platform temperature, which can be assessed by the value of the heater resistance itself (once its TCR is known). An example of such characterization is shown in figure 4. It can be seen that for any dissipated power the device with a membrane support reaches higher platform temperatures, so it can be concluded that such an approach will offer larger power densities in future generators implementations. Moreover, in the particular case shown in figure 4, the membrane length was half of the bridge length, so the membrane device occupy less area, which would also influence positively in the attainable power density.

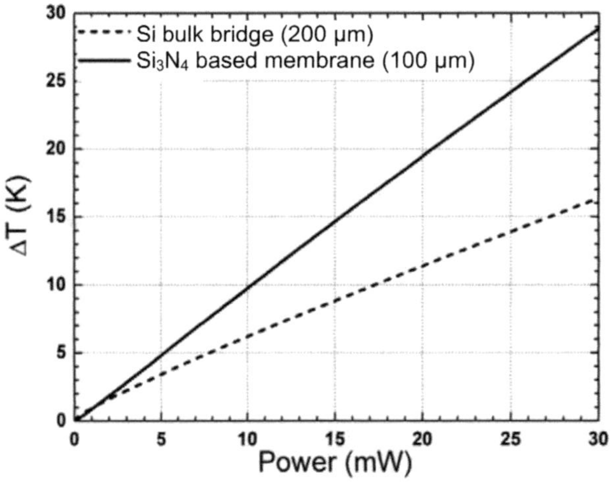

Figure 4. Platforms supported by dielectric membranes reach a higher temperature than the supported by bulk silicon bridges when the same power is dissipated by a built-in heater. This points to a much improved thermal isolation, even though in this particular example the membrane length (100 μm) was just half of the silicon bridges one. Power density will be thus further improved.

Acknowledgements

This work has been supported by the EU FP7-NMP-2013-SMALL-7, SiNERGY (Silicon Friendly Materials and Device Solutions for Microenergy Applications), under contract n. 604169, the Spanish Ministry of Economy and Competitiveness (TEC-2010-20844) and the ''Generalitat de Catalunya'' (Advanced Materials for Energy Network (XaRMAE), 2009-SGR-440). C. Calaza and A. Tarancón would like to thank the financial support of the Ramón y Cajal postdoctoral program of the Spanish Ministry of Economy and Competitiveness.

References

[1] Hochbaum A I, Chen R, Delgado R D, Liang W, Garnett E C, Najarian M, Majumdar A and Yang P 2008 *Nature* **451** 163–7

[2] Boukai A I, Bunimovich Y, Tahir-Kheli J, Yu J K, Goddard III W A and Heath J R 2008 *Nature* **451** 168–71

[3] Davila D, Tarancon A, Calaza C, Salleras M, Fernandez-Regulez M, San Paulo A and Fonseca L 2012 *Nanoenergy* **1** (6), 812-9

[4] Paulo A S, Arellano N, Plaza J A, He R, Carraro C, Maboudian R, Howe R T, Bokor J and Peidong Y 2007 *Nano Letters* **7**, 1100-4

[5] Gadea G, Morata, A, Santos J D, Davila D, Calaza C, Salleras M, Fonseca L and Tarancon A 2015 *Nanotechnology* **26** (19), 195302

[6] Davila D, Tarancon A, Calaza C, Salleras M, Fernandez-Regulez M, San Paulo A and Fonseca L 2013 *J. Electron. Mat.* **42** (7), 1918-25

[7] Donmez I, Salleras M, Calaza C, Santos J D, Gadea G, Morata A, Dávila D, Tarancon A and Fonseca L Proc. *SPIE 9517, Smart Sensors, Actuators, and MEMS VII; and Cyber Physical Systems, 95172C (21 May 2015)* 11 pages

Maximizing Output Power in a Cantilevered Piezoelectric Vibration Energy Harvester by Electrode Design

Sijun Du, Yu Jia and Ashwin Seshia

Department of Engineering, University of Cambridge, Trumpington Street, Cambridge CB2 1PZ, UK

E-mail: sd672@cam.ac.uk; yu.jia.gb@ieee.org; aas41@cam.ac.uk

Abstract. A resonant vibration energy harvester typically comprises of a clamped anchor and a vibrating shuttle with a proof mass. Piezoelectric materials are embedded in locations of high strain in order to transduce mechanical deformation into electric charge. Conventional design for piezoelectric vibration energy harvesters (PVEH) usually utilizes piezoelectric material and metal electrode layers covering the entire surface area of the cantilever with no consideration provided to examining the trade-off involved with respect to maximizing output power. This paper reports on the theory and experimental verification underpinning optimization of the active electrode area of a cantilevered PVEH in order to maximize output power. The analytical formulation utilizes Euler-Bernoulli beam theory to model the mechanical response of the cantilever. The expression for output power is reduced to a fifth order polynomial expression as a function of the electrode area. The maximum output power corresponds to the case when 44% area of the cantilever is covered by electrode metal. Experimental results are also provided to verify the theory.

1. Introduction

For the purpose of vibration energy harvesting, cantilevers with a piezoelectric layer sandwiched between two electrode layers, and resting on on top of a substrate, are widely used due to its simplicity and moderately high power density [1, 2, 3], as shown in figure 1. Currently, top and bottom electrode layers usually cover the entire piezoelectric layer to enable transduction of mechanical response to electrical power [4]. However, due to the distribution of strain in the piezoelectric layer while vibrating, the volumetric strain is higher near the clamped end and very little near the free end of the cantilever [5]. Due to the non-uniformly distributed strain along axis x, there should be an optimal value for the area of electrode, or in other words the active piezoelectric area. In this paper, the optimal area of active piezoelectric layer for a maximum power output is calculated from the Euler-Bernoulli beam theory and the result is experimentally verified by a MEMS scale cantilevered PVEH.

2. Theoretical derivation

In this section, the optimal area of electrode layer is theoretically derived for a maximum power from a cantilevered energy harvester. Figure 2 shows the structure of a cantilever with some parameters for calculation. The length, width, thickness of the piezoelectric and substrate layers are L, H, W and h respectively. It is assumed that the width of the electrode layer is also W, but its length starts from the clamped end is a variable x, which is the value that we aim to find to maximize the power output.

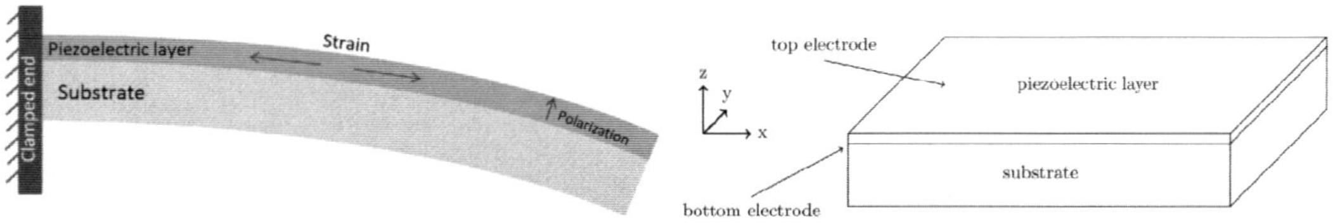

Figure 1. Cantilevered PVEH without proof mass

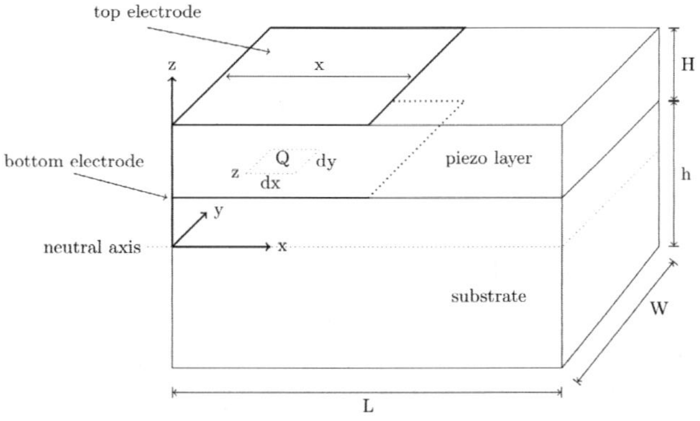

Figure 2. Schematic of a cantilevered PVEH for calculation

The calculation starts from the Euler-Bernoulli Beam Theory [6], which gives an approximate relation between displacement along z-axis for a specific point of beam at (x) and the applied external force, This equation is given by equation 1 [7]:

$$EI\frac{d^4\omega(x)}{dx^4} = q(x) \tag{1}$$

In the equation 1, the parameters E and I represent the Young's modulus and Second moment of area of the entire cantilever respectively; $\omega(x)$ is the displacement (m) of a point at x, and $q(x)$ is the external excitation force per unit length (N/m). Assuming that the excitation force is $F = F_0 \sin\omega_0 t$ and the force is uniformly distributed along x-axis, so we have:

$$q = \frac{F}{L} = \frac{F_0}{L}\sin\omega_0 t \tag{2}$$

By integrating the equation 1 and applying the Dirichlet Boundary Conditions (at the clamped end: $\omega' = 0$ and $\omega = 0$; at the free end: $\omega''' = 0$ and $\omega'' = 0$), we have:

$$\omega(x) = \frac{1}{24}Ax^4 - \frac{1}{6}ALx^3 + \frac{1}{4}AL^2x^2 \tag{3}$$

where $A = \frac{q}{EI}$. For a symmetrical bending, the tensile stress experienced by the beam can be expressed as $\sigma_{(x,y,z)} = \frac{Mz}{I}$, where M is the bending moment which is given by $M = -EI\frac{d^2\omega(x)}{dx^2}$, I is the second moment of area, so we have the stress given by:

$$\sigma_{(x,y,z)} = -zE\frac{d^2\omega(x)}{dx^2} = -z\frac{q}{I}(\frac{1}{2}x^2 - Lx + \frac{1}{2}L^2) \tag{4}$$

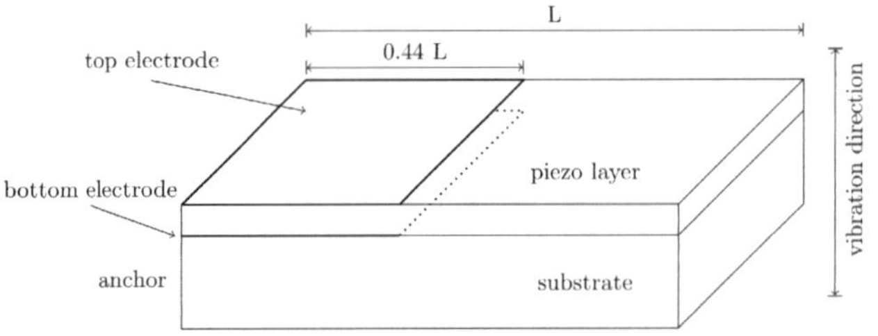

Figure 3. Schematic of a cantilevered PVEH with optimal electrode length

So the amount of charge generated by the strain is expressed as:

$$Q_{(x,y,z)} = d_{31}\sigma_{(x,y,z)} = -zd_{31}\frac{q}{I}(\frac{1}{2}x^2 - Lx + \frac{1}{2}L^2) \tag{5}$$

This is the charge generated per area $dxdy$ at location z, as shown in figure 2. The total surface charge can be calculated by integrating equation 5.

$$Q_{total} = -d_{31}\frac{F_0}{L}\frac{W(h+H)}{2I}(\frac{1}{6}x^3 - \frac{1}{2}Lx^2 + \frac{1}{2}L^2x)\sin\omega_0 t \tag{6}$$

When the cantilever is vibrating at its natural frequency, the equivalent electrical circuit for the cantilever can be equivalent to a current source I_P connected with a capacitor C_P and a resistor R_P in parallel [8]. The generated power by the harvester is the power consumed by the internal impedance (capacitor and resistor in parallel) from the current source. The current source I_P can be calculated from the derivative of charge to time:

$$i_p = \frac{dQ_{total}}{dt} = i_0\cos\omega_0 t \;\; (with\; i_0 = -d_{31}\frac{F_0\omega_0}{L}\frac{W(h+H)}{2I}(\frac{1}{6}x^3 - \frac{1}{2}Lx^2 + \frac{1}{2}L^2x)) \tag{7}$$

The internal impedance can be deduced from the capacitance and resistor in parallel:

$$Z_p = C_p//R_p = \frac{\frac{R_p}{2\pi C_p}}{R_p + \frac{1}{j\omega_0 C_p}} = \frac{\rho}{1 + j\omega_0\varepsilon_r\varepsilon_0\rho}\frac{H}{xW} \tag{8}$$

By applying the second moment of area $I = \frac{W(h+H)^3}{12}$, It can be seen that the generated power by the harvester is:

$$\Rightarrow P_0 = B(\frac{1}{36}x^5 - \frac{1}{6}Lx^4 + \frac{5}{12}L^2x^3 - \frac{1}{2}L^3x^2 + \frac{1}{4}L^4x)\left(B = d_{31}^2F_0^2\omega_0^2\frac{18H}{WL^2(h+H)^4}\frac{\rho}{1+j\omega_0\varepsilon_r\varepsilon_0\rho}\right) \tag{9}$$

Equation 9 gives the expression of the generated power by the harvester. From the plot of this expression shown in figure 5, the output power is at its maximum value when $x \approx 0.44L$. From the expression of stress along x-axis in equation 4, it can be found that the stress at $x = 0.44L$ is around 31% of the maximum stress value.

3. Experiment

In order to experimentally verify the theoretical calculation of the active area for optimal output power, a MEMS-scale cantilevered harvester without tip mass is fabricated. The size of the cantilever is 3.5 mm × 3.5 mm and the top electrode is split into 8 segments as shown in figure 4 (left). From region 1 to region 8, they sequentially occupy 20%, 10%, 10%, 10%, 10%,

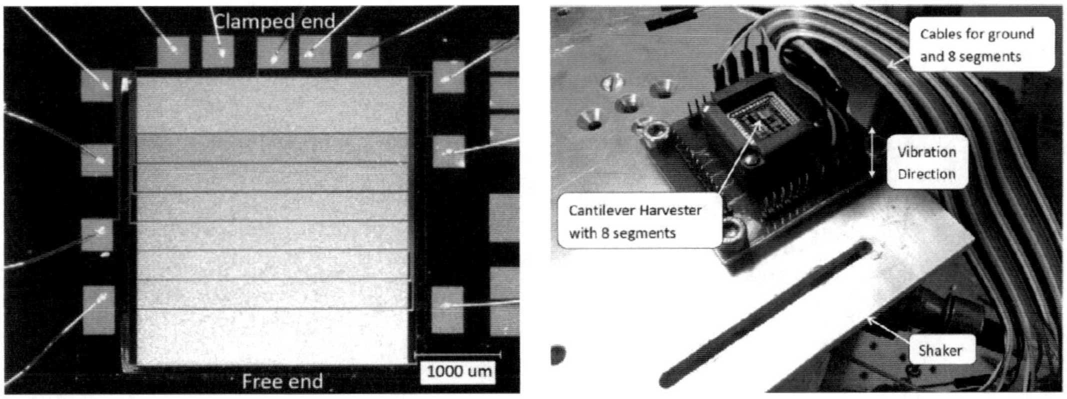

Figure 4. Optical micrograph of MEMS cantilever with 8 electrode regions (left) and the experimental setup (right)

Table 1. Experimental output power comparison of cantilever with 8 regions (frequency: 1208 Hz, acceleration: 0.5 g)

Electrode area	Measured capacitance (nF)	Matched load resistor (kΩ)	Measured Power (nW)
0%	0	-	0
20%	0.464	280	140.01
30%	0.858	160	180.63
40%	1.128	115	214.07
50%	1.401	95	222.01
60%	1.673	75	213.16
70%	1.945	65	199.51
80%	2.217	55	189.55
100%	2.689	50	153.6

10%, 10% and 20% respectively, of the total length of the cantilever. The device in the figure contains 12 electrode pads where there are 8 pads for 8 regions and 4 pads for ground. In order to minimize the effect of capacitance of ground pads, only one of them is used in experiments.

The MEMS device to be tested is clamped in a chip socket, which is fixed on a shaker, see figure 4 (right). The natural frequency of the cantilever is 1208 Hz and the peak input acceleration is around 0.5 g. Experiments are performed with gradually increased top electrode area by adding regions from region 1 to 8.

For each active electrode area, the load resistor is varied to determine the value that matches the internal impedance. Table 1 shows the measured results and the figure 5 illustrates how the output power varies with different active electrode area (comparison of theoretical results and experimental results). By fitting the 8 points with a polynomial trend line, the maximum value can be found at around 48%. The discrepancy with respect to the theoretical value (44%) (solid line shown in figure 5)can be attributed to the parasitic capacitance in the ground pad and also due to fabrication tolerances.

4. Conclusion

A theoretical calculation and experimental verification in this paper are performed to find an optimal active piezoelectric layer area for maximizing output power of a cantilevered PVEH. The results show that maximizing active area does not always increase output power; in the

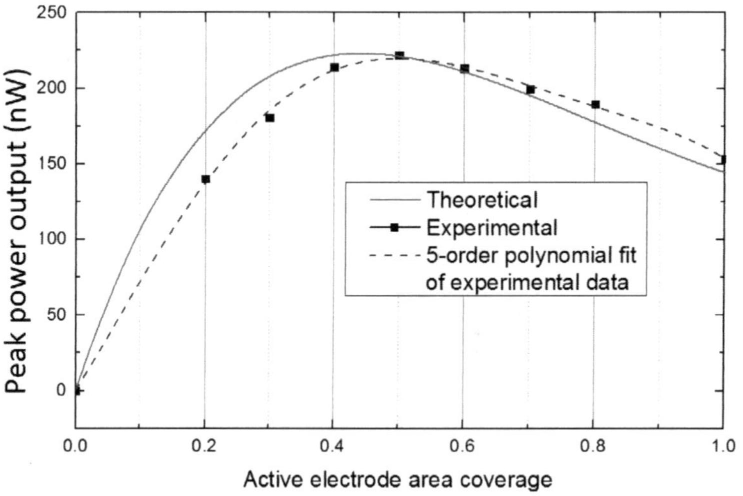

Figure 5. Measured peak output power of the cantilevered PVEH compared to model predictions

contrast, power can be reduced if the low-strain area is covered. For designing a piezoelectric vibration energy harvester at either macro-scale or MEMS-scale, the active layer does not necessarily need to cover the entire piezoelectric layer. This work shows that the piezoelectric area corresponding to less than 31% of the maximum bending stress should not be included in the active electrode region. This design approach can also be applied to other structural topologies and mode shapes for PVEHs.

References

[1] S. R. Anton and H. A. Sodano, "A review of power harvesting using piezoelectric materials (2003–2006)," *Smart materials and Structures*, vol. 16, no. 3, p. R1, 2007.

[2] S. P. Beeby, M. J. Tudor, and N. M. White, "Energy harvesting vibration sources for microsystems applications," *Measurement Science and Technology*, vol. 17, no. 12, p. R175, 2006.

[3] A. G. A. Muthalif and N. H. D. Nordin, "Optimal piezoelectric beam shape for single and broadband vibration energy harvesting: Modeling, simulation and experimental results," *Mechanical Systems and Signal Processing*, vol. 54-55, no. 0, pp. 417–426, 2015.

[4] M. Stewart, P. M. Weaver, and M. Cain, "Charge redistribution in piezoelectric energy harvesters," *Applied Physics Letters*, vol. 100, no. 7, p. 073901, 2012.

[5] P. D. Mitcheson, E. M. Yeatman, G. K. Rao, A. S. Holmes, and T. C. Green, "Energy harvesting from human and machine motion for wireless electronic devices," *Proceedings of the IEEE*, vol. 96, no. 9, pp. 1457–1486, 2008.

[6] S. Priya and D. J. Inman, *Energy harvesting technologies*, vol. 21. Springer, 2009.

[7] W. Thomson, *Theory of vibration with applications*. CRC Press, 1996.

[8] N. G. Elvin, "Equivalent electrical circuits for advanced energy harvesting," *Journal of Intelligent Material Systems and Structures*, 2014.

MEMS based Nonlinear Monostable Electromagnetic Vibrational Energy Harvester for Wider Bandwidth

D Mallick[1], A Amann[2], S Roy[1]*
[1]Micropower Systems & Nanomagnetics Group, Micro-nano-systems Center, Tyndall National Institute, Lee Maltings, Dyke Parade, Cork, Ireland.
[2]School of Mathematical Science, University College Cork, Cork, Ireland.

E-mail: saibal.roy@tyndall.ie

Abstract. This paper reports a wideband vibrational energy harvesting scheme using a MEMS based nonlinear electromagnetic transducer. The nonlinearity is incorporated in the proposed device through the stretching strain in addition to the bending of the fixed-guided configured beams of the designed structure. The thin spring structure is fabricated on Silicon-On-Insulator substrate with device layer thickness of 50 m. The MEMS spring structure is packaged and characterized with wire wound copper coil (NE1) and micro fabricated double layer copper coil (NE2) for comparison. Measurement results show that ~80 Hz half power bandwidth is obtained for the fabricated devices with maximum load powers of 2.8 W (NE1) and 0.4 W (NE2) respectively at 0.5g which improves the 'power-bandwidth gain' to one of the highest among reported works.

1. Introduction

With advancement of low power electronics, the speculated 'Internet of Things (IoT)' has opened up a window to visualize a connected world of tiny, autonomous Wireless Sensor Nodes (WSNs), which would transform our surrounding into an intelligent and responsive environment. To enable the 'fit-and-forget' operation of the WSNs, scavenging electrical energy from ambient vibrations has gained major attention. Nonlinear oscillation based Vibrational Energy Harvesters (VEHs) [1-8] have received enormous attention recently due to their inherent ability to improve the off-resonance performance compared to a resonant system. In this work, we have developed stretching strain based nonlinear monostable VEH which is much compact and easy to implement at MEMS scale.

2. Design and FEM Analysis

We have designed our devices in such a way that the nonlinearity appears due to stretching of the spring arms in addition to the bending strain in a fixed-fixed beam configuration. The resultant spring force of such beam, which consists of two parts, one due to bending (F_b) and the other due to stretching (F_S) has been calculated analytically and is given by [4]

$$F = F_b + F_s = -\frac{Y}{L^3} W d^4 \left[\frac{d_f}{d} + \frac{18}{25}\left(\frac{d_f}{d}\right)^3\right] \quad (1)$$

* Corresponding Author

Where, L, W, d and Y are length, width, thickness and Young's modulus of the beam and d_f is the beam tip deflection. For large amplitude oscillations (d_f>d), the stretching terms become significant and nonlinearity appears. We have designed the device in such a way to exploit this stretching based nonlinearity that results in wideband response. The thin spring configuration on silicon potentially brings the different fundamental modes of the designed structure close together. As shown in Fig. 1, the first three modes of the proposed device are 406 Hz (out-of-plane vibration), 425 Hz (torsional) and 512 Hz (torsional in different direction) respectively as obtained from the FEM analysis in COMSOL Multiphysics. The closely spaced resonance modes interact between themselves and enhance the otherwise wideband output frequency response due to high nonlinear stiffness of the device.

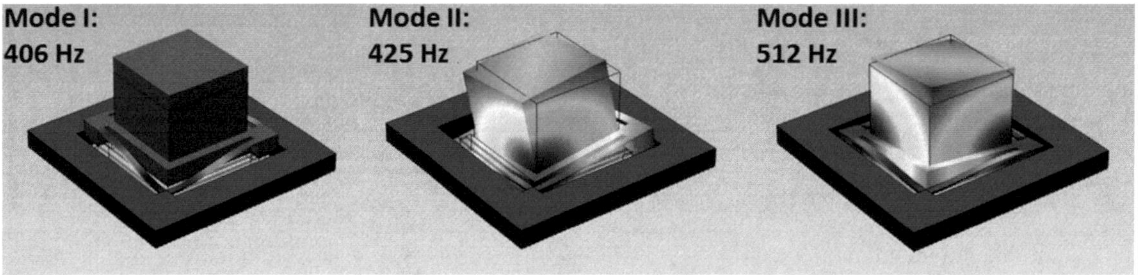

Figure 1. First three vibrational modes of the MEMS nonlinear spring structure.

3. Fabrication

The designed spring structure is fabricated on a Double Side Polished (DSP) <100> Silicon-on-Insulator (SOI) wafer with device layer thickness of 50 m, 3 m buried oxide (BOX) and a handle layer of 450 m using a double mask process. The device layer defines the thickness of the spring arms. The fabrication starts by thermally growing oxide layers on front and back side of the wafer which act as the mask layer during front and back etch. The front side is then patterned and developed using a first mask and the exposed oxide sites are etched using Plasma Enhanced Reactive Ion Etching (PERIE) to reach the device layer silicon. Same is performed in the back side of the wafer using a second mask. The device layer silicon is etched using Deep Reactive Ion Etching (DRIE) till the BOX layer to define the thin spring structure. Lastly, the handle layer is etched from the back using DRIE to release the MEMS structure. The microscopic images of the fabricated spring structure are shown in Fig. 2.

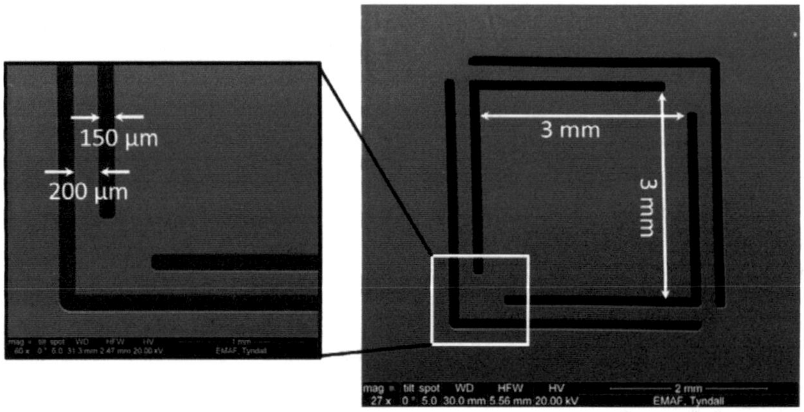

Figure 2. SEM image of the Nonlinear Spring fabricated on SOI substrate.

The double layer electroplated copper coil is fabricated on a separate Silicon substrate [9]. The fabrication process starts by sputtering Ti/Cu (20/200 nm) seed layer on silicon. The substrate is then patterned and the first layer of copper is electroplated and resist is stripped off. Using a second mask, via layer is patterned and copper is again electroplated to fill the via. Next, SU-8 insulation layer is spun to isolate the bottom copper tracks from the top layer. Similar seed layers as the bottom is sputtered on SU-8. The corresponding layer is patterned again, top layer copper is electroplated and the resist is stripped off. Finally another layer of SU-8 is spun to provide passivation to the structure. The track width, the inter-track gap and the height of each track of the coil is 10 m each. The square coil has an outer dimension of 2.8×2.8 mm^2. The Microfabricated coil is shown in Fig. 3(a) which has 144 turns with the coil resistance of 192 Ω.

The micro-fabricated spring structure and coil are packaged together to form the MEMS based nonlinear electromagnetic energy harvesting device. A small NdFeB block magnet ($2.5 \times 2.5 \times 2$ mm^3) is epoxy bonded onto the movable paddle which provides mass (9.83×10^{-5} Kg) to the oscillator. The device is characterized at different level of integration to report a comparative study. First, the nonlinear spring structure is packaged with a wire wound copper coil having comparable dimensions to the micro coil. The wire wound coil with 3mm outer diameter, 0.5 mm thickness and 25 m wire diameter, has 560 turns and a coil resistance of 140 Ω. The packaged MEMS devices [Fig. 3(b)] have a volume of 0.14 cm^3.

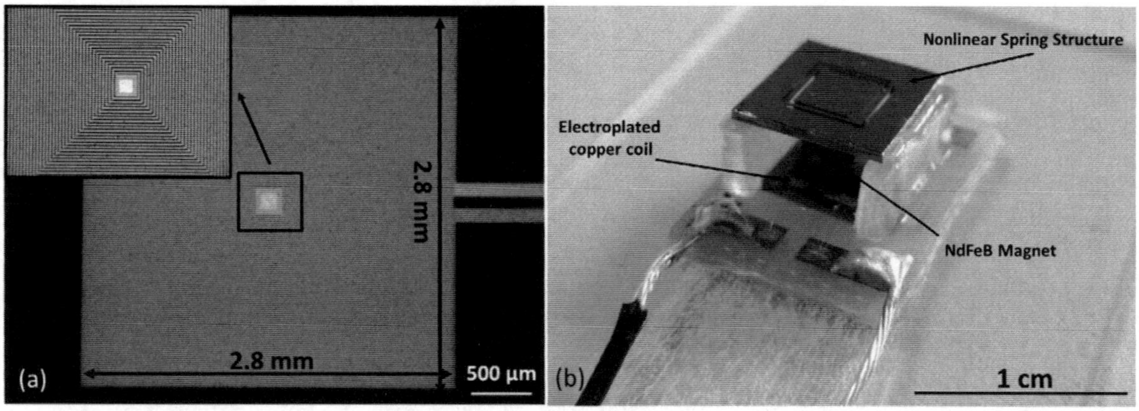

Figure 3. (a) Microscopic image of the micro-fabricated coil. (b) Packaged device (NE2) on chip for testing.

4. Characterization & Discussions

The two prototypes, one with wire wound coil (NE1) and the other one with the micro coil (NE2) are tested under swept sine input vibrations under different acceleration levels (0.1g, 0.2g and 0.5g; g=9.8 m/s^2). The electrical damping due to the small size magnet and coil assembly is found to be smaller than the parasitic damping. Thus the optimized load for both the prototypes are equal to the respective coil resistances i.e., for NE1 it is 140 Ω and for NE2 the same is 190 Ω respectively. The load power vs frequency response of the prototypes under different acceleration levels are shown in Fig. 4 as the frequency is swept in the forward direction. As mentioned in section 2, the first three vibration modes of the designed spring are relatively close. The large deflection, stretching strain nonlinearity will be stronger for the first out-of-plane vibration mode. However since the other two modes (torsions in two opposite directions) are also out-of-plane, the nonlinear contribution due to those two modes also adds up. Consequently, very wideband response is achieved and the corresponding wideband nature escalates with the input acceleration. At 0.5g, the peak load power of 2.8 W is obtained with NE1 with a half power bandwidth of 78 Hz. Peak load power of 0.4 W is obtained with prototype NE2 along with a bandwidth of 82 Hz. The slight fall of the bandwidth of NE1 can be due to the comparatively larger electrical damping produced from the wire wound coil with large turn numbers than a micro coil.

As a figure of merit of wideband energy harvesters, the concept of Power Integral is introduced before [6]. It is the area under the load power-frequency curve at a fixed acceleration level and is defined as ($\boldsymbol{P_f = \int P_L . df}$) where P_L is the load power and f is the frequency. In MEMS based systems, the power is normally significantly reduced due to the scaling of the active material whereas the bandwidth increases due to the larger nonlinearity because of the thinner spring structure. In the following Table, we provide a comparative study of some of the reported MEMS energy harvesting devices along with our work. The efficient nonlinear spring design in our device produces sufficient wideband response, which results in large power integral values.

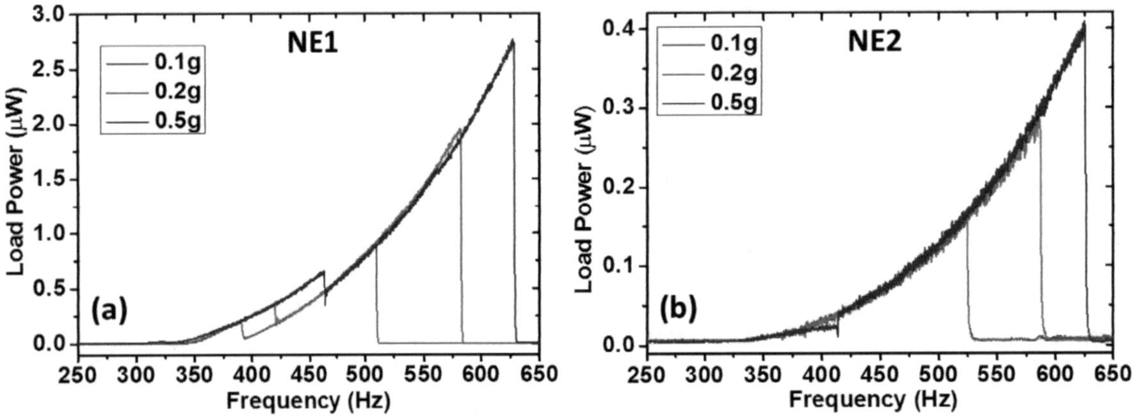

Figure 4. Load power vs frequency response of the two prototypes under different accelerations for respective optimized load resistances: (a) NE1 for 140 Ω load and (b) NE2 for 190 Ω load.

Table 1. Comparison of Power Integral for different MEMS nonlinear energy harvesters.

Reference	Type of Transduction	*Volume (cm³)*	Acceleration (g)	*Power Integral (W-Hz)*
7	EM	0.032	3g	1.8
8	PZ	0.016	0.6g	2.8
10	ES	0.042	1g (35V bias)	164
11	EM	0.158	1g	0.02
NE1	EM	0.14	0.5g	840
NE2	EM	0.14	0.5g	123

5. Conclusions

In summary, this paper reports a bandwidth widening scheme using a MEMS based monostable nonlinear electromagnetic vibration energy harvesting device. The nonlinearity is incorporated in the proposed device through the stretching strain in addition to the bending of the fixed-guided configured beams of the designed structure. The thin spring structure is fabricated on SOI substrate with device layer thickness of 50 m. The device is tested with wire wound copper coil (NE1) and micro fabricated double layer copper coil (NE2) for comparison. Testing results show that ~80 Hz half power bandwidth is obtained for the fabricated devices with maximum load powers of 2.8 W (NE1) and 0.4 W (NE2) respectively at 0.5g which enhances the 'power-bandwidth gain' to one of the highest among reported works.

Acknowledgement

This work is financially supported by Science Foundation Ireland (SFI) Principal Investigator (PI) project on 'Vibration Energy Harvesting' -grant no SFI-11/PI/1201. The authors would like to thank the support of Central Fabrication Facilities (CFF) at Tyndall National Institute for the MEMS fabrications. The authors would also like to thank Peter Constantinou for his help in the micro coil fabrication. One of the authors D. Mallick is supported from UCC Strategic Research Fund Award.

References

[1] Cottone F, Vocca H, Gammaitoni L, Phys. Rev. Lett., **102**, 080601 (2009).
[2] Erturk A, Hoffmann J, Inman D J, Appl. Phys. Lett., **94**, 254102 (2009).
[3] Stanton S C, McGehee C C, Mann B P, Physica D, **239**, 640-53 (2010).
[4] Mallick D, Amann A, Roy S, Smart Mater. Struct., **24**, 015013 (2015).
[5] Podder P, Amann A, Roy S, Sens. Actuators A: Phys., **227**, 39-47, (2015).
[6] Mallick D, Amann A and Roy S, Interplay between electrical and mechanical domains in a high performance nonlinear energy harvester, Smart Mater. Struct. (Under Review).
[7] Liu H, Qian Y, Wang N, Lee C, J. Microelectromech. Syst., **23**, 740-749 (2014).
[8] Liu H, Lee C, Kobayashi T, Tay C J, Quan C, Smart Mater. Struct. **21**, 035005 (2012).
[9] Podder P, Constantinou P, Mallick D, Roy S, Silicon MEMS bistable electromagnetic vibration energy harvester using double-layer micro-coils, PowerMEMS, 1-4 December, 2015, Boston, Massachusetts, USA.
[10] Basset P, Galayko D, Cottone F, Guillemet R, Blokhina E, Marty F, Bourouina T, J. Micromech. Microeng., **24**, 035001 (2014).
[11] Liu H, Koh K H, Lee C, Appl. Phys. Lett., **104**, 053901 (2014).

MEMS Narrow Gap Electromagnetic Harvester with Mitigation of Curvature Distortion

K Yamaguchi, T Fujita, K Kanda and K Maenaka
Graduate School of Engineering, University of Hyogo
2167 Shosha, Himeji, Hyogo 671-2280, Japan

E-mail: er14a049@steng.u-hyogo.ac.jp

Abstract. The narrowing air-gap between coil and magnet of electromagnetic energy harvester is one of the most important parameter to improve the generated power. Because of the sputtering 15 μm thickness NdFeB/Ta multilayer film with high temperature annealing, the large curvature distortion is occurred on Si substrate. In this paper we demonstrate a compensation method for residual stress on the sputtered NdFeB magnetic film on Si wafer. By using both-side sputtering method, we reduced the residual stress, which reduces the reliability of MEMS fabrication and performance of harvesters, from 144 MPa to 9.4 MPa. The electromagnetic type energy harvester with narrow gap of 20 μm was successfully fabricated and evaluated.

1. Introduction

Recently the wireless sensor network systems for automotive e.g. TPMS (Tire Pressure Monitoring Systems) [1] and for human monitoring systems [2] are collecting lot of attentions. These systems are typically composed of sensors, MPU, battery and wireless transmitter. The battery i.e. power source without maintenance or replacing is most vital part of the systems.

The vibration energy harvesting has various method and mechanism of power generation. We studied the electromagnetic harvester that consists of coil and magnet, which has some features e.g. (1) large harvesting current by low output coil impedance, (2) no pull-in problem unlike to electrostatic type harvester and (3) simple and easy to fabricate structure.

In order to improve a generated power, it is required that the rapid and large transient change in magnetic flux density passing through a coil. Accordingly, a strong magnet and large stroke range of vibration are required. However these are both difficult in MEMS (micro electromechanical systems) fabrication. In the previous work, we have already succeeded to establish the combination of Si MEMS structure and NdFeB/Ta multilayer magnet film [3]. Then we demonstrated the electromagnetic harvester with corrugated shape magnet and serpentine coil, which generates 2.67 mV of electromotive force (EMF) and 7.6 nW of power at constant amplitude vibration of 100 Hz, 300 $\mu m_{p\text{-}p}$. The result was obtained from the optimized design on 30 μm air-gap [4]. Designing the optimal dimensions for electromagnetic type generating maximum power, it is essential that the equivalent and the improvement of both the attenuation coefficient of electrical and mechanical with resonant state. In order to obtain a large electrical attenuation coefficient, a narrow gap and a miniaturization of effective generation areas are required, however it is the issue that large curve in substrate is occurred by high temperature annealing process of magnet sputtering.

In this paper, for the narrow gap harvester, we demonstrate the improvement method of mitigating the internal stress by deposition magnet film to both sides of substrate. In addition, we fabricate the prototype harvester with 20 μm of narrow gap.

2. Electromagnetic harvester

2.1. Device structure and principle

The electromagnetic harvester consists of magnet, coil and mass-spring structures (Fig. 1). The chip size including frame, effective power generation areas i.e. coil and magnet are 15×12×1 mm^3, 10 ×10 mm^2, respectively. The coil structure has the serpentine shaped [5] with Au on a Si mass that supported by Si linear spring. In the opposing section, the magnet array is made by NdFeB/Ta multilayer sputtering on the corrugated Si. The magnet film with 40 layers and total 15 μm thickness is deposited on a Si structure with trench depth of 300 μm (Fig. 2). The corrugate shaped magnet array produces the periodical distribution of magnetic flux density through the coil. The interval of coil line is designed with synchronizing a change of magnetic flux. Therefore when a mass vibration causes magnetic flux changing through the coil surface, and then EMF will be generated on the coil.

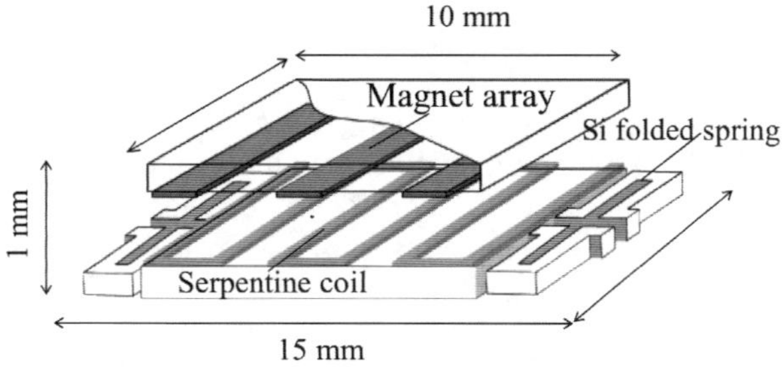

Figure 1. Schematic of harvester

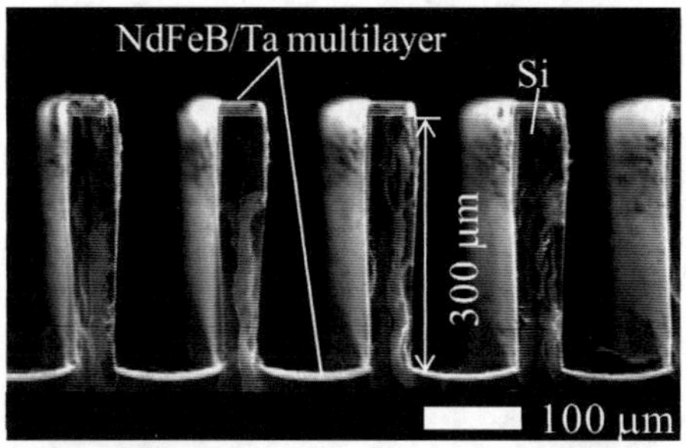

Figure 2. Cross-sectional SEM image of magnet on corrugated Si. Thickness of magnets are top of 15 μm, wall of 6 μm and bottom of 3 μm.

2.2. Estimation of power generation with Lorentz force consideration

The electromagnetic type energy harvester is belonging to the velocity damped resonant generator (VDRG) model (Fig. 3). The Lorentz force $F(\mathrm{t})$ is proportional to vibration velocity $\dot{x}(t)$ of coil mass. In the VDRG model, the maximum power will be generated when the coil mass vibrates at the resonant state and the electrical attenuation coefficient, c_e i.e. Lorentz force equals to the mechanical attenuation coefficient, c_m [6]. $c_m = m\omega/Q_m$ can be designed by changing the mass-spring parameters i.e. coil mass: m, resonant angular frequency: ω and mechanical Q-factor: Q_m. c_e is decided by effective power generation areas i.e. magnet and coil structure. c_e is depicted by following equations:

$$c_e = \frac{F(\mathrm{t})}{\dot{x}(t)} = \frac{WL^2}{4\rho SP(L+P/2)} \Delta B_y(x)^2 \qquad (1)$$

$$(\Delta B_y(x) = B_y(x) - B_y(x+P/2))$$

, where W, L, ρ, S, P and ΔB_y are length of coil, width of coil, resistivity, sectional area of coil, magnet pitch and peak-to-peak of magnetic flux density for vertical direction, respectively (Figs. 4 and 5). In order to increase c_e and keep the device dimension, miniaturizing a magnet pitch and narrowing a gap are required simultaneously. The calculated value of the magnet pitch and gap vs. c_e is shown in

Fig. 6. The magnet pitch has an optimal value P_{opt} that can generate maximum power, and P_{opt} is also proportional to the gap. In previous work, the previous work has W and L of 10 mm, ΔB of 0.09 T, P of 140 μm, ρ of 2.21 × 10^{-8} Ωm and S of 20 ×6 μm^2, therefore c_e is estimated to 7.1×10^{-6}. The motion of coil mass on VDRG model is following equation:

$$m\ddot{x}(t) + (c_e + c_m)\dot{x}(t) + kx(t) = -m\omega^2 \ddot{y}(t) \tag{2}$$

, where k and y are mechanical spring constant and position of external vibration, respectively. The generated power is calculated by following equation:

$$Power(t) = c_e \dot{x}(t)^2 \tag{3}$$

, where $\dot{x}$ is vibration velocity of coil mass. Assuming the applied vibration of 100 Hz, 300 $\mu m_{p\text{-}p}$, designed harvesters with 30 μm, 20 μm and 10 μm gap can generate 0.009 μW, 0.026 μW and 0.12 μW in the estimation.

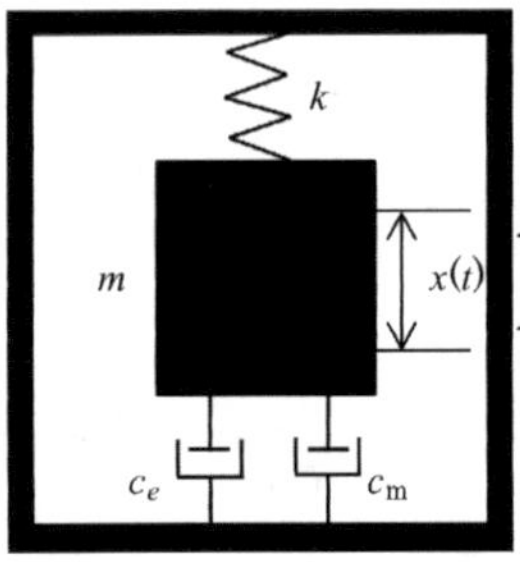

Figure 3. VDRG model

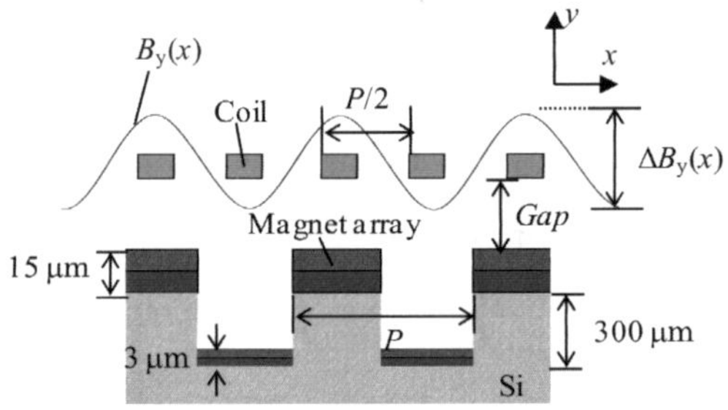

Figure 4. Analytical model of effective power generation area. The magnet on wall is outside a consideration because easy axis of magnetization is differ from the top and the bottom.

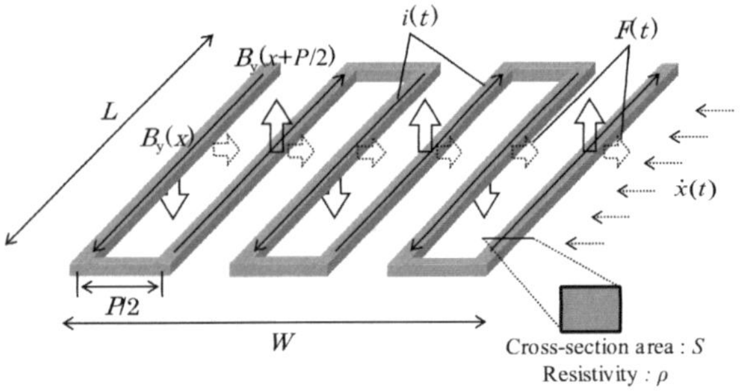

Figure 5. Overview of serpentine coil

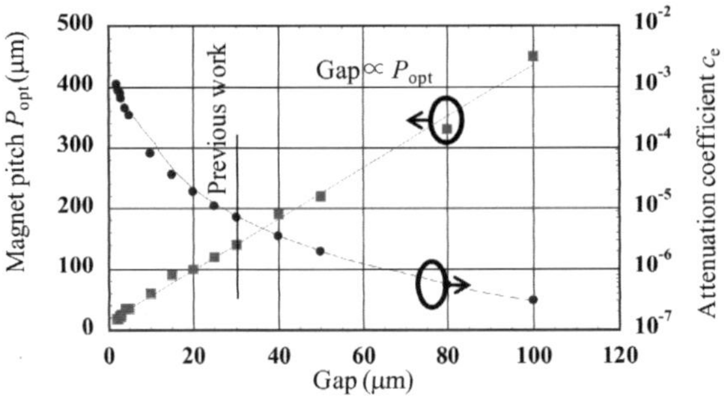

Figure 6. Electrical attenuation coefficient and optimal magnet pitch for each gap. The optimal pitch is proportional with gap.

3. Fabrication of narrow gap device

3.1. Improvement of internal stress

As mentioned previously, the narrow gap the larger generated power. In order to narrow the gap, the biggest problem is a curvature distortion of the Si substrate with magnet film. Our 15 μm NdFeB/Ta multilayer film shows maximum radius of curvature 5.00 m i.e. it has residual stress of 144 MPa after

600 °C annealing (Fig. 7a). The 200 μm camber on 4 inch Si wafer that caused by the residual stress is serious obstacle for MEMS fabrication. A deposition at room temperature and applying high temperature post-annealing could control the residual stress, however the magnetic characteristic was extremely declined. There also is the optimum combination of gas and pressure for sputtering.

Thus, we try to deposit the magnetic film on opposite side of the Si wafer for compensating the curvature distortion. By depositing the similar magnetic film of 10 layers on the reverse surface, the distortion is compensated from 5.00 m to 57.9 m curvature radius. The residual stress is successfully reduced from 144 MPa to 9.4 MPa (Fig. 7b).

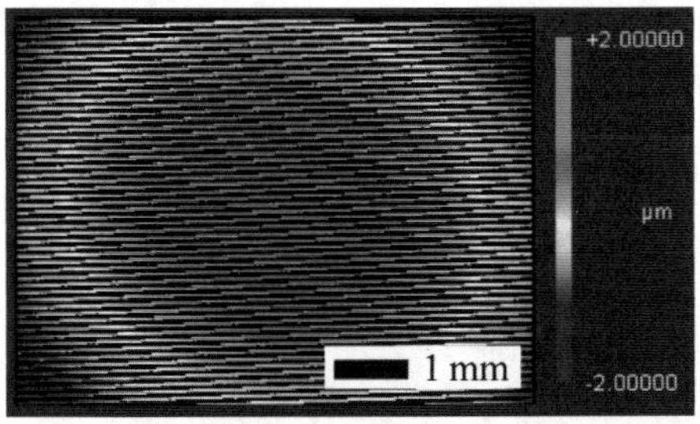

(a) Sputtering magnet on corrugate Si side. The magnet is 15 μm thickness and 40 layers at 600 °C.

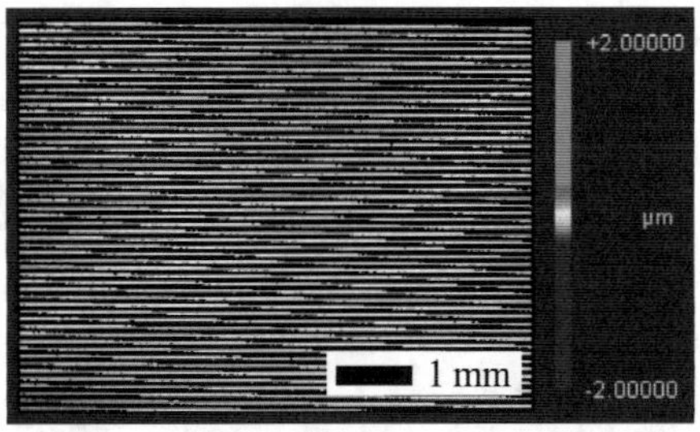

(b) Sputtering magnet on both-side. The magnet are 15 μm thickness and 40layers at 600 °C on corrugate side and 4 μm thickness and 10 layers at 600 °C on back side.

Figure 7. Surface profile of corrugate shaped magnet

Regarding the coil wafer, there is no distortion because thin Au/Cr is deposited at room temperature and 6 μm thickness Au coil line was electroplated, though (Fig. 8). From the measured curvature radius of magnet and coil wafer, we confirmed to narrow the gap to 20 μm for MEMS batch fabrication with 4 inch wafer.

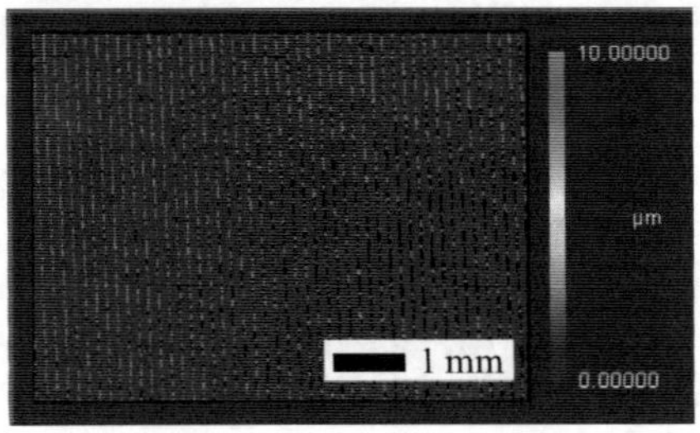

Figure 8. Surface profile of serpentine coil

3.2. Fabrication of harvester with narrow gap

We fabricated the prototype of harvester with 20 μm gap (Fig. 9a) from the result of reducing the residual stress. The Au coil line was fabricated on the Si mass-spring wafer by Au electroplating and deep-RIE (MCU-21-090; Sumitomo Precision Product Co.,) (Fig. 9b). The NdFeB/Ta multilayer (40 layers) is sputtered on miniaturized corrugate shaped Si surface and additional 10 layers are sputtered on the back side for compensation (Fig. 9c). In order to regulate the gap to 20 μm, a negative type photo resist (KMPR; Nippon Kayaku Co.,) was used for spacer. The mass-spring structure is not packaged for optical measurement from the bottom side. We measured the displacement of coil mass during changing vibration frequency with sweeping up condition by using laser Doppler vibrometer (MLD-103A; Neoark Co.,). The frequency response of mass vibration amplitude and phase at 0.03 G

(G: gravitational acceleration) and 0.07 G of applied acceleration are shown in Fig. 10. The measured resonant frequency and mechanical Q-value are 215 Hz and 715, respectively. The larger acceleration causes a collision over the maximum travel range of 180 μm_{p-p}, however we confirmed that the 20 μm narrow gap device could vibrate with high Q-factor.

(a) Overall view of harvester (b) Coil (c) Magnet

Figure 9. Electromagnetic harvester with 20 μm gap

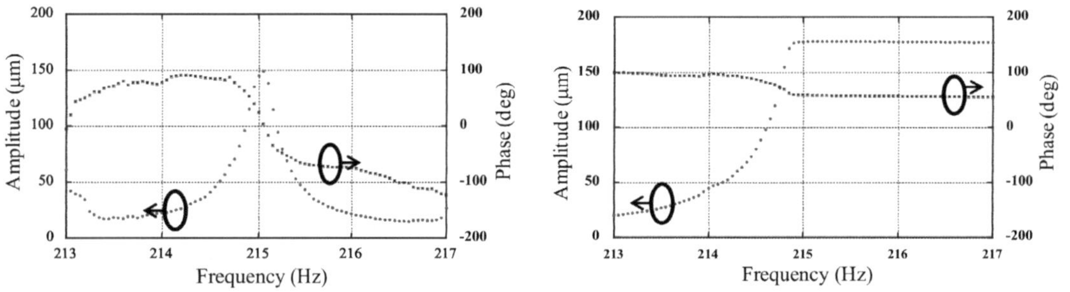

(a) 0.03 G of application acceleration (b) 0.07 G of application acceleration

Figure 10. Vibration test with sweeping up vibration frequency.

4. Conclusion

We demonstrated the improvement method of curvature radius of the NdFeB/Ta sputtered wafer. During high temperature annealing for sufficient magnetic performance, the magnet film has large curvature distortion. By applying the additional sputtering on the Si back side, we could reduce the residual stress from 144 MPa to 9.4 MPa. In addition the electromagnetic harvester with 20 μm narrow gap was successfully fabricated and cleared the vibration test.

5. References

[1] Grygier P et al. 2001 *An Evaluation of Existing Tire Pressure Monitoring Systems* (Washington D C : NHTSA)

[2] Maenaka K et al. 2007 *Proc. of The 4th Int. Conf. on Networked Sensing Systems* (Braunschweig) pp 47-52

[3] Uehara M et al. 2009 *J. Magn. Soc. Jpn.* **33** 227-231

[4] Yamaguchi K et al. 2014 *J. Phys.: Conf. Ser.* **557** 012056

[5] Tanaka Y et al. 2013 *J. Phys.: Conf. Ser.* **476** 012095

[6] Mitcheson P D et al. 2004 *J. Microelectromechanical systems* **13** 429-440

Multilayer ferroelectret-based energy harvesting insole

Z Luo[1], D Zhu[1,2] and S P Beeby[1]

[1]Electronics and Computer Science, University of Southampton, Southampton, SO171BJ, UK
[2]School of Computing, Electronics and Maths, Faculty of Engineering, Environment and Computing, Coventry University, Coventry, CV15FB, UK

E-mail: z.luo@soton.ac.uk

Abstract. This paper reports a flexible energy harvesting insole made of multilayer ferroelectrets, and demonstrates that this insole can power a wireless signal transmission. We have previously studied the energy harvesting characteristics of single and 10-layer ferroelectrets under compressive forces with quantified amplitudes and frequencies. In this work, we fabricate a flexible insole using multilayer ferroelectrets, and increase the number of layers from 10 up to 80, then use this insole to harvest energy from footsteps. We use this insole to power a commercial ZigBee wireless transmitter, and successfully demonstrate that an 8-bit data transmission can be solely powered by the energy harvested from this insole for every 3 to 4 footsteps. It confirms the anticipation from our previous work that the multilayer ferroelectrets are capable of powering the start-up and transmission of a low-power chipset, and shows a potential of using this energy harvesting insole in wearable applications.

1. Introduction

Ferroelectret is a thin and flexible cellular polymer foam that generates electric pulses under mechanical force [1,2], can be used as both a sensor and an energy harvester. When used as an energy harvester, the electric pulses from deforming the ferroelectret can charge a capacitor and power the electronics. Our previous study [3] has investigated the energy harvesting ability of porous polypropylene (PP), which is one of most researched ferroelectrets [1, 4, 5]. The PP ferroelectret is a polymer film with thickness of only 70 to 80 μm, and Young's modulus of about 1 MPa in the thickness direction [6, 7]. This soft polymer nature and small thickness make it suitable to be used in an insole for harvesting power while maintaining comfort. However, a single layer of PP ferroelectret can only produces 1.12 μJ of energy under 800 N of compressive force [3]. This output is not enough to power any commercial low-power chipsets at present. Therefore, we designed a multilayer structure of ferroelectret to improve its output power. Using an electrodynamic instrument to apply a waver function of compressive forces, we demonstrated that a 10-layer PP ferroelectret can produce more than 10 times of energy compared to a single layer [3].

Our previous work anticipates that the multilayer ferroelectret could be capable of powering the start-up and transmission a low-power chipset [3], in this work we fabricate a ferroelectret insole and integrate it with a wireless transmitter, to show that a wireless transmission can be powered solely by the energy harvesting insole that harvests power from footsteps. This insole is made by inserting a multilayer ferroelectret into the heel. The tested multilayer samples includes 10-layer, 20-layer, 30-layer, 40-layer and 80-layer. By directly walking on this energy harvesting insole and using the generated electrical pulses to charge a capacitor, it shows that an 80-layer ferroelectret insole can produce more than 100 μJ of energy from every footstep. A prototype has been made by connecting this 80-layer insole with a power conditioning circuit and a commercial ZigBee wireless transmitter. It shows that for every 3 to 4 footsteps, the transmitter gains sufficient energy from the insole and is able to send an 8-bit wireless signal to its receiver, which is within 8 meters from the source.

2. Experimental

The PP ferroelectret films were commercially purchased from Emfit Ltd. These commercial samples have been charged by the manufacturer. They were sheets in the size of 230 mm × 210 mm, with thickness of 70 μm. These films were further cut into testing samples with size of 60 mm × 70 mm. Ag electrodes were printed on both sides of the samples using a screen printer (Dek 248, Dek Printing Machines Ltd). The Ag printing solution was silver Fabinks TC C40001 from Smart Fabric Inks Ltd. The electrode has an area size of 50 mm × 60 mm. The samples after electrode printing were cured in an oven at a temperature of 50°C for 10 minutes.

The multilayer ferroelectret was fabricated by stacking layers of electroded PP ferroelectrets with bonding films in between each two layers. These bonding films were used both for insulation and bonding to improve the synchrony of the output electrical signals. The stacked ferroelectret layers were then connected in parallel. A schematic diagram of the fabrication process is shown in Figure 1.

The multilayer ferroelectret was directly inserted into the heel of a shoe. In the test a person with weight of 80 kg walked wearing with this shoe. The generated pulses from the energy harvesting insole were used to charge a capacitor, and the capacitor voltage was monitored by a data logger (KUSB3102, Keithley Instruments Ltd). The wireless transmission powered by the energy harvester was demonstrated using a self-built power conditioning circuit and a ZigBee RF transceiver module.

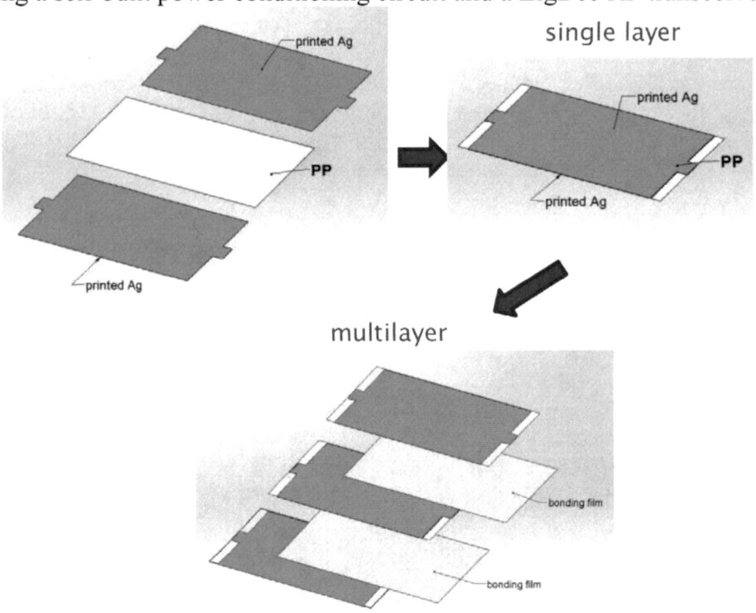

Figure 1. Fabrication process of multilayer ferroelectret

3. Results and discussion

3.1. Energy output from the ferroelectret insole

Insole samples with 10-layer, 20-layer, 30-layer, 40-layer and 80-layer of PP ferroelectrets are tested in this work. During the walking test, the electrical pulses generated from the ferroelectret insole are rectified and used to charge a 2.2 μF capacitor. The voltage of the capacitor during charging is recorded by the data logger and shown in Figure 2a. From Figure 2a, it can be clearly observed that one footstep results in an increase of the voltage, then the voltage remains no change in the interval before next footstep. This stepping shape of the charging curve shows the change in the capacitor voltage from one footstep. Using the equation $E = \frac{1}{2}CV^2$, where C is the capacitance of the capacitor, V is the voltage of the capacitor, the energy E in the capacitor is calculated and shown in Figure 2b. Similarly, the change in charged energy from each footstep is observed. It shows that for every single footstep, the 10-layer, 20-layer, 30-layer, 40-layer and 80-layer ferroelectret insoles charge 19.8 μJ, 31.9 μJ, 40.2 μJ, 65.6 μJ and 100.9 μJ of energy respectively into a capacitor. The 80-layer insole consists 8 times of ferroelectret layers compared to the 10-layer one, but the generated energy of the former is only 5.1 times of the latter. This shows that the generated energy does not increase linearly with the number of layers. This is due to the fact that the stresses distributed between each stacked layer are uneven, which the stress sustained by the ferroelectret layers in the bottom of the insole decreases as the number of layers increase. This has been shown by both the simulation and experimental results in our previous work [3]. It is also worth mentioning that ferroelectret insole can lose more than 50% of efficiency during capacitor charging. Therefore, the energy generated by the insole are in fact higher than the one that estimated by the capacitor charging in Figure 2.

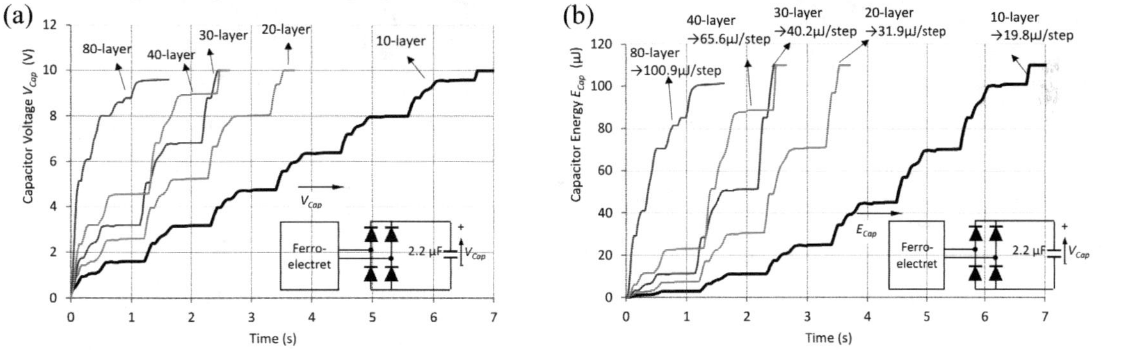

Figure 2. (a) Voltage charging curve of a 2.2 μF capacitor charged by 10, 20, 30, 40 and 80 layers of ferroelectrets. (b) Energy charging curve of a 2.2 μF capacitor charged by 10, 20, 30, 40 and 80 layers of ferroelectrets

3.2. Wireless transmission powered by the ferroelectret insole

It is estimated that the commercial ZigBee wireless transmitter consumes more than 100 μJ of energy per transmission. Since the 80-layer insole generates about 100 μJ of energy per footstep when charging a capacitor, it is selected to be tested for powering the transmission in this work. The energy generated from the ferroelectret insole will need to be regulated first before powering the transmitter. The circuit diagram of the power conditioning circuit is shown in Figure 3. In this circuit the output of the ferroelectret energy harvester is rectified by a bridge rectifier and used to charge two 4.7μF electrolytic capacitors. Once the voltage on both capacitors reach 5.7V, the voltage regulator MIC5231 is enabled and kept on until the input voltage is dropped below its output voltage of 2.75V. The threshold voltage of the voltage detector can be changed if different operation voltage range is required. The output of the voltage regulator is then used to power the ZigBee transmitter. The actual circuit layout is shown in Figure 4.

Figure 5 shows the ferroelectret insole connecting the power conditioning circuit and the ZigBee transmitter. The wireless transmission is monitored by a receiver within 8 meters from the transmitter. A LED on the receiver will turn off when an 8-bit data is received. When walking on the ferroelectret insole, the LED light turns off for every 3 to 4 footsteps, indicating a wireless transmission has been powered by the insole. This result also suggests that the harvested energy loses more than 50% during the power conditioning, since 300 to 400 μJ of energy is generated from the insole for 3 to 4 steps, only about 150 μJ is fed into the transmitter.

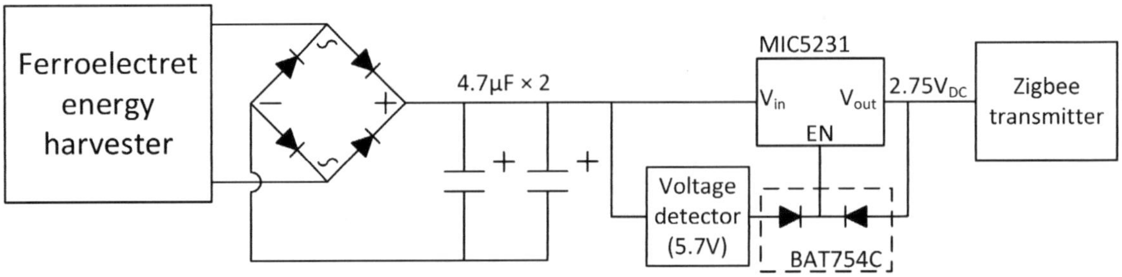

Figure 3. Circuit diagram of the power conditioning circuit

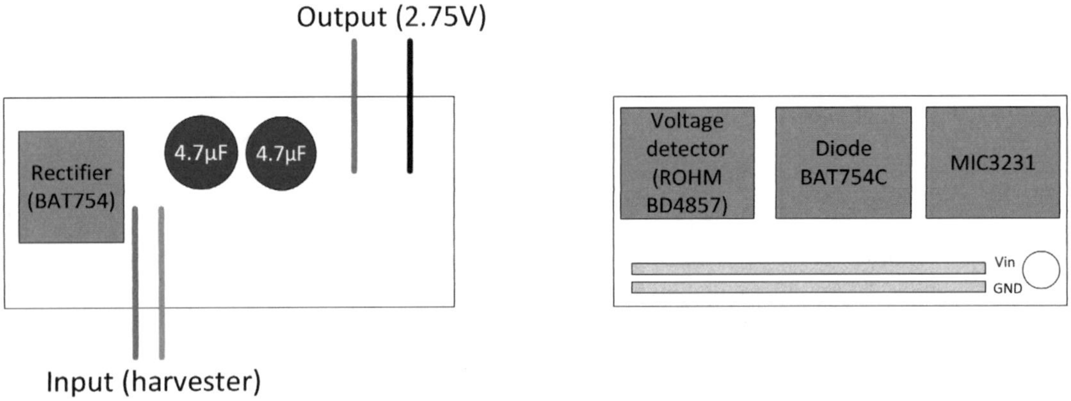

(a) top side (b) bottom side

Figure 4. Layout of the power conditioning circuit

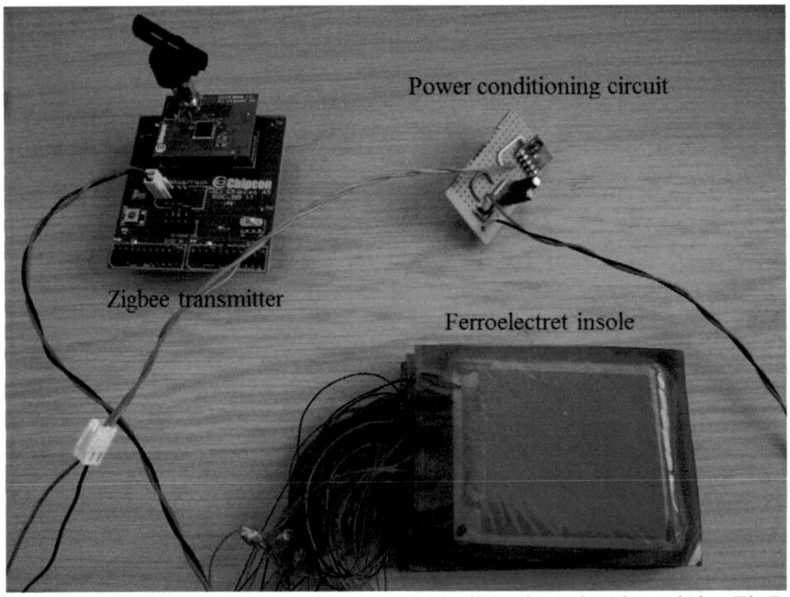

Figure 5. Ferroelectret insole connecting to the power conditioning circuit and the ZigBee transmitter

4. Conclusion and future work

This paper reports the design and test of an insole energy harvester made of multilayer ferroelectrets, and demonstrates the wireless transmission that powered by this insole. Experimental results shows that an 80-layer ferroelectret insole can generate more than 100 μJ of energy per footstep when charging a capacitor. This insole is used to power a commercial ZigBee transmitter. It is demonstrated that for every 3 to 4 footsteps, the energy harvested from the insole is able to power an 8-bit wireless data transmission. The next step forward will be to improve the energy output from the ferroelectret insole and reduce the energy loss during power conditioning

5. Reference

[1] Hillenbrand J and Sessler G M 2000 Piezoelectricity in cellular electret films *IEEE Trans. Dielectr. Electr. Insul.* **7** 537-42

[2] Neugschwandtner G S, Schwodiauer R, Bauer-Gogonea S, Bauer S, Paajanen M, and Lekkala J 2001 Piezo-and pyroelectricity of a polymer-foam space-charge electret *J. Appl. Phys.* **89** 4503-11

[3] Luo Z, Zhu D, Shi J, Beeby S P, Zhang C, Proynov P and Stark B 2015 Energy harvesting study on single and multilayer ferroelectret foams under compressive force *IEEE Trans. Dielectr. Electr. Insul.* **22** 1360-68

[4] Bauer S, Gerhard-Multhaupt R and Sessler G M 2004 Ferroelectrets: Soft electroactive foams for transducers *Phys. Tod.* **57** 37-43

[5] Paajanen M, Lekkala J and Kirjavainen K 2000 ElectroMechanical Film (EMFi) — a new multipurpose electret material *Sens. Actuat. A: Phys.* **84** 95-102

[6] Zhang X, Sessler G M and Wang Y 2014 Fluoroethylenepropylene ferroelectret films with cross-tunnel structure for piezoelectric transducers and micro energy harvesters *J. Appl. Phys.* **116**, 074109 (8pp)

[7] Neugschwandtner G S, Schwödiauer R, Vieytes M, Bauer-Gogoena S, Bauer S, Hilenbrand J, Kressmann R, Sessler G M, Paajanen M and Lekkala J 2000 Large and broadband piezoelectricity in smart polymer-foam space-charge electrets *Appl. Phys. Letts.* **77**, 3827-29

Acknowledgments

This work was performed under the SPHERE IRC funded by the UK Engineering and Physical Sciences Research Council (EPSRC), Grant EP/K031910/1.

Passively-Switched, Non-Contact Energy Harvester for Broad Operational Range and Enhanced Durability

William Z Zhu[1] and Carol Livermore[1]

[1] Department of Mechanical and Industrial Engineering, Northeastern University, Boston, MA 02115 USA

E-mail: zhu.wi@husky.neu.edu, livermore@neu.edu

Abstract. Impact-based vibrational energy harvesters that switch passively among dynamical modes to best match the ambient conditions and perform frequency up-conversion to maximize power output can increase the operational range over which harvesters output useful power. The disadvantage of impact-based harvesting is that it can lead to premature system failure. This paper presents modelling and experimental validation of a harvester with passively-switched dynamics in which a magnetic non-contact interaction replaces the impact-based interaction of previous systems for greater robustness. A low-frequency driving beam couples to a higher-frequency generating beam via tip magnets. Simulations predict that the non-contact interaction drives a diversity of dynamical modes in which the tip of the driving beam variously is repelled from and passes the tip of generating beam. Experiments validate the predicted dynamical behaviour of the system. Output power levels are similar to predictions, with a power of 903 μW on the primary resonance at 1 g acceleration.

1. Introduction

Vibrational energy harvesting can increase the operation time of wireless sensors by converting mechanical energy from ambient accelerations into electric power [1]. One challenge of vibrational energy harvesting is the balance between resonant operation (for increased output power) and wide bandwidth (for increased operational range). Solutions to this challenge include nonlinear harvesting [2] and both active and passive tuning of harvesters' resonance frequencies, for example as in [3, 4]. For human-powered systems, a second challenge lies in capturing the low frequency accelerations that are typical of human motion while maintaining sufficient power output. It is possible to address both challenges by using separate but interacting driving and generating elements that couple into low frequencies and output power at high frequencies, respectively. Such systems may be driven by mechanical impact or non-contact means, for example as in [5, 6]. An important design consideration for such systems is robustness; large amplitude motions must not harm either the driving or the generating elements.

One example of a wide-bandwidth harvester that converts low frequency excitations into higher frequency motions of the generating element is in [7]. In [7], a driving beam and a generating beam that overlap only at their tips can interact in a few different dynamical modes, each of which has its own characteristic frequency. The harvester passively switches among modes in response to changes in ambient conditions, and impact between the beams up-converts the low ambient frequency to a higher generating frequency. However, because the beams interact via impact, the generating beam is expected to break when the driving beam undergoes very large amplitude motions.

This paper presents modelling and experimental demonstration of a new kind of passively-switched harvester in which the inter-beam contact interaction of [7] is replaced by a magnetic non-contact interaction for greater durability. Despite the non-contact interaction, the system is still designed to switch passively among dynamical modes in response to changing ambient conditions. The low-frequency driving beam and the higher frequency piezoelectric bending actuator (the generating beam) are arranged in a facing configuration. Permanent magnets on the tips of the beams repel each other, so that deflection of the driving beam induces deflection in the generating beam.

One of the primary dynamical modes of [7] occurs when the tip of the driving beam passes from above the tip of the generating beam to below the tip of the generating beam and back. In [7], these passing dynamics are described as the driving beam 'plucking' the generating beam. Unlike the case of [7], here the magnetic interaction drives the generating beam to deflect away from the driving beam first before the tips pass, and again after the tips pass, potentially increasing the strength of the plucking. The second main dynamical mode of [7] occurs when the driving beam interacts with the generating beam, but the driving beam does not pass the generating beam and always remains on the same side of it (e.g. above the generating beam). These modes of motion can have different characteristic frequencies, thereby broadening the bandwidth. One goal of the present research is to characterize the role of these passing and non-passing modes of operation in the non-contact system.

2. Modelling and simulation

Figure 1(a) shows the passively switched, non-contact magnetic harvester modeled as a single degree of freedom lumped parameter system. The model includes the mass, stiffness, and damping of each beam, as well as the external vibration. The magnetic force is represented as a nonlinear stiffness. Power extraction from the generating beam is modeled as electrical damping and is included in the total damping of the generating beam. Although damping is not an exact description of piezoelectric energy conversion, it provides a rapid simulation tool for investigating the system's dynamics.

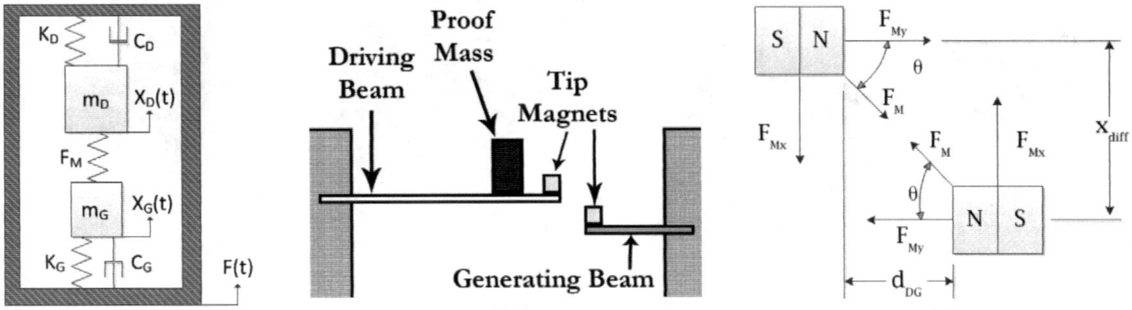

(a) (b) (c)

Figure 1. Representation of (a) model and (b) schematic of vibrational harvester and (c) magnetic force.

The equation for the harvester's equivalent 1D lumped parameter system is

$$[m\ddot{x} + c\dot{x} + kx]_{D,G} = F(t) + F_m(x_D, x_G), \quad (1)$$

where m, c, and k are the mass, damping, and stiffness of the beam, and the subscripts D and G denote either the driving or generating beam. The total damping term for the generating beam, c_G, accounts for both viscous damping and electrical extraction from the PZT to a load. Modelling electrical extraction as damping is described in [8]. The function $F_m(x_D, x_G)$ denotes the magnetic force between the driving and generating beams as a function of the deflection of the beams in the x direction. A full analytic equation of magnetic force between two non-identical permanent magnets would require resource-intensive calculations. Instead, the magnetic force is approximated using a least squares method as in [9]. The least squares approximation is

$$F_m(x_D, x_G) = \propto_{DG} \frac{d_{DG}^2(x_D - x_G + y_{DG})}{\sqrt[2/3]{d_{DG}^2(x_D - x_G + y_{DG})^2}}, \tag{2}$$

where d_{DG} and y_{DG} are the horizontal and vertical separations between the tip magnets at rest and α_{DG} is the magnetic force at that horizontal separation. The values used for α_{DG} are found in [10].

The harvester's dynamics are integrated using the Runge-Kutta method in MATLAB (ode45). The simulation calculates parameters for failure detection due to over deflection of the PZT beam as well as voltage, power, and converted energy over time. Table 1 lists the parameters used in the simulation.

Table 1: List of parameters for model

	Driving Beam	**Generating Beam**
Material	ABS plastic	PZT-5A
Mass m (g)	9.6	0.217
Total Damping Coefficient (c)	0.0300	0.0235
Stiffness k (Nm^{-1})	14.1	445
Capacitance C (nF)	N/A	3.5

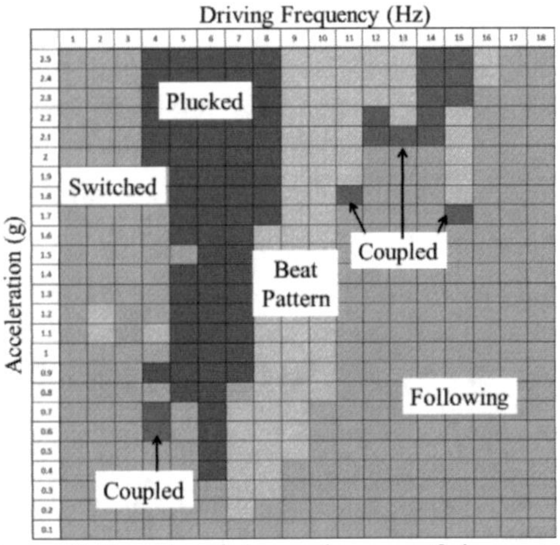

Figure 2. Predicted dynamic map of the system

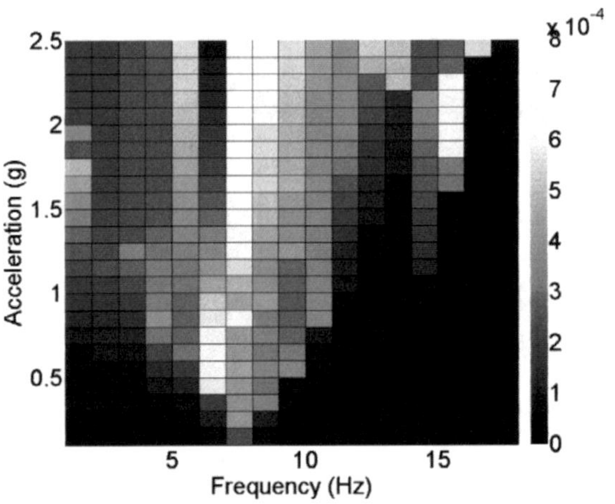

Figure 3. Predicted average power map

Figure 2 shows the predicted dynamics map of the harvester. Two types of passing dynamics are observed, referred to here as switched and plucked dynamics. In plucked dynamics, the driving beam passes the generating beam, leaving the generating beam free to ring down and generate power. In switched dynamics, the driving beam passes the generating beam but remains so close to the generating beam after passing that its magnetic force impedes the free ring-down of the generating beam, limiting the output power. Two types of non-passing dynamics are observed, referred to here as following and coupled dynamics. In coupled dynamics, the driving beam is repelled by the magnetic force so that it appears to "bounce off" of the generating beam without actually touching it. After the driving beam moves away, the generating beam undergoes free ring-down. In following dynamics, the driving beam is again repelled by the magnetic force, but it never moves away far enough for the generating beam to undergo a free ring-down. In some regions, passing (typically plucked) and non-passing (typically coupled) dynamics are interleaved in a beat pattern (also shown in figure 2). Passing dynamics occur around the primary and secondary resonances and/or at high accelerations, whereas non-passing dynamics occur around the secondary resonance, off resonance, and/or at low accelerations. Figure 3 shows the predicted power vs driving acceleration and driving frequency, with substantial power contributions at the secondary resonance. Powers of greater than 100 μW are typically produced in regions where passing dynamics are exhibited. The peak power shifts from 6 Hz to 7 Hz as the driving acceleration is increased. Figure 4 plots the predicted average power vs driving frequency for driving accelerations of 0.5g, 1.0g, 1.5g, and 2.0g. A single power peak is observed at 0.5g but broadens at higher accelerations to a pronounced two-peak structure with significant power output at both the primary and secondary resonances.

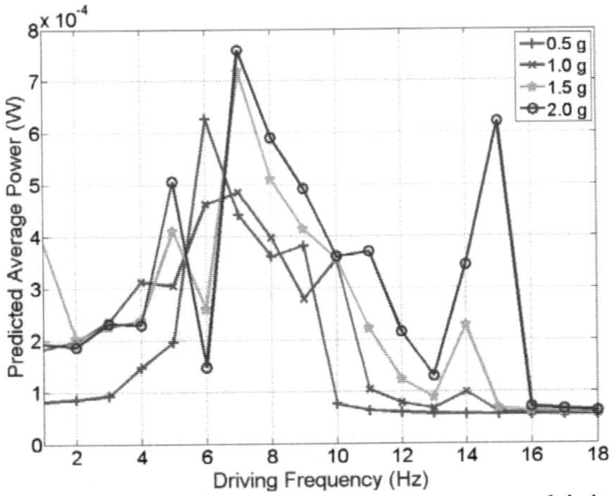

Figure 4. Predicted average power vs driving acceleration for driving accelerations of 0.5g, 1.0g, 1.5g, and 2.0g.

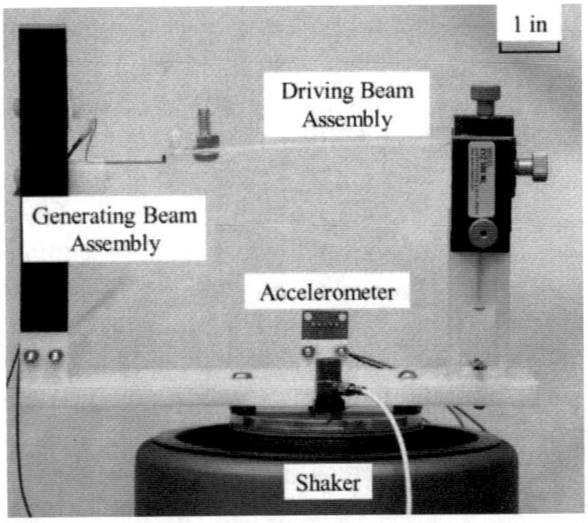

Figure 5. Experimental setup of vibrational harvester system

3. Experiments

Figure 5 shows the experimental setup of the system. The harvester base is made from 3D printed plastic (fused deposited polylactide resin). Attached to the base is an accelerometer (Analog Devices model ADXL 203EB). The harvester is driven by a closed loop feedback shaker table (Labworks Inc., model ET-139). A 10 g proof mass and a 4.76 mm diameter, 0.79 mm thick N52 grade neodymium magnet are mounted at the tip of an acrylonitrile butadiene styrene plastic driving beam so that the magnetization direction is parallel to the neutral axis of the driving beam. The generating beam is a 31.75 mm long, 3.175 mm wide, 0.51 mm thick, two layer, brass reinforced, series poled PSI-5A4E bending actuator (Piezo Systems, INC. T220-A4-103X) with a tip-mounted 3.175 mm square, 1.59

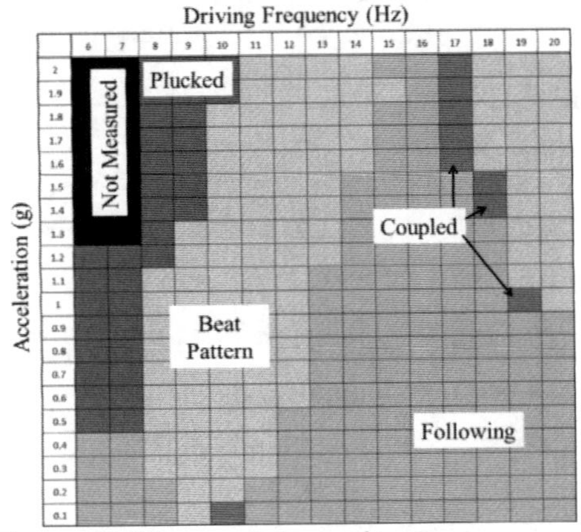

Figure 6. Observed dynamics map.

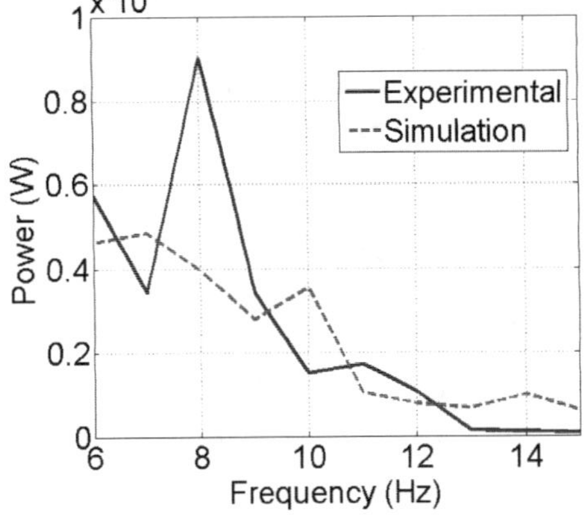

Figure 7. Measured and predicted average power vs driving frequency for a driving acceleration of 1.0g

mm thick N42 grade neodymium magnet. The generating beam is connected to an impedance matched load resistor, and the acceleration and the voltage across the load are recorded. The average power is calculated by numerically integrating instantaneous power over time and normalizing by time.

Figure 6 maps the observed dynamics. Driving frequencies of 5 Hz and below and high accelerations at 6 Hz and 7 Hz were omitted due to technical limitations of the test platform. However, the system itself is successfully operated at all other frequencies and accelerations, including at high accelerations on resonance. This is in contrast to the system of [7], in which the potential for damage upon impact limited the harvester's safe operational range. The experiments show similar dynamics to the predictions of figure 2. The robust two-peak structure observed of figure 3 is shifted to slightly lower frequencies than in the simulations, reflecting the effects of fabrication variations. Plucked and beat pattern dynamics are observed over a greater frequency range, and switched dynamics are not observed. These results indicate that the real system undergoes energetic plucking and beat pattern dynamics more readily than the simulated system. The discrepancy reflects the limitations of the approximations used in the simulation, in particular the simulation's approximation of the dynamics as one-dimensional rather than two-dimensional.

Figure 7 plots the measured and predicted average power vs driving frequency for a driving acceleration of 1.0 g. The measured amplitude of the peak output power (903 μW) is similar to but somewhat exceeds the predicted amplitude (486 μW). The measured and simulated half-power bandwidths cannot be directly compared because the driving beam impacts the harvester base on resonance, reducing the output power. The measured range at 1.0 g does not encompass the secondary resonance at which following dynamics give way to higher power output dynamics, so the anticipated secondary power peaks are not visible.

4. Conclusion

The present simulations and experiments demonstrate that a magnetically-interacting, non-contact harvester can switch passively among dynamical modes to adapt to changing ambient vibrational conditions. Unlike the case of [7], in which the highest amplitude regions of the operational range had to be avoided to prevent breakage, the present harvester is able to operate at large accelerations on resonance without damage. Measured and predicted powers are similar, with a measured power of 903 μW on the primary resonance at 1 g acceleration.

References

[1] Harne R L and Wang K W 2013 *Smart Mat. and Struct.* **22(2)** 023001
[2] Cammarano A, Neild S A, Burrow S G and Inman D J 2014 *Smart Mat. and Struct.* **23(5)** 055019
Zhu D, Tudor M J and Beeby S P 2010 *Meas. Sci. and Tech.* **21(2)** 022001
[3] Eichhorn C, Goldschmidtboeing F and Woias P 2009 *J. of Micromech. and Microeng.* **19(9)** 094006
[4] Liu T and Livemore C 2012 *Proc. of PowerMEMS 2012* pp. 492-495
[5] Gu L and Livermore C 2011 *Smart Mat. and Struct.* **20(4)** 045004
[6] Kulah H and Najafi K 2008 *Sens. J. IEEE* **8(3)** 261-268
[7] Liu T, St Pierre R and Livermore C 2014 *Smart Mat. and Struct.* **23(9)** 095045.
[8] Inman D J and Erturk A. 2011 *Piezoelectric Energy Harvesting Modelling and Application* (NJ, USA: John Wiley & Sons) pp 68-73
Wang Q-M, Xiao hong D, Xu B and Cross L E 1999 *J. of App. Phys.* **85(3)** 1702-1712
[9] Al-Ashtari W, Hunstig M, Hemsel T and Sextro W 2012 *Smart Mat. and Struct.* **21(3)** 035019
[10] Zhu W Z 2015 M.Sc. thesis (Northeastern University)

Piezo-magnetic energy harvesting from movement of the head

Aidin Delnavaz and Jérémie Voix

École de technologie supérieure, Montreal, Quebec, Canada

E-mail: aidin.delnavaz@ens.etsmtl.ca

Abstract. This paper reports the design, modeling, optimization and testing of the piezo-magnetic energy harvester that is capable of converting non-harmonic movement of the human head into electricity. The rolling magnet and doubly-clamped piezoelectric configuration of the device makes the energy harvesting from small-amplitude and low-frequency movements of the human head efficient. In addition, the device can inconspicuously be integrated with the glasses. the experimental results show that the energy harvester device could deliver the maximum instantaneous power of 0.5 µW to the impedance matched resistive load.

1. Introduction

Wearable technologies are being interwoven with the daily life. As they are becoming smarter and smaller, new efficient and cost effective powering solutions are needed. Although primary batteries have been the traditional source of power for mobile systems, energy harvesting is emerging in recent years. Energy can be harvested from almost any environment, however, since a wearable device is attached to the human body, one of the most readily available sources is body movement.

Head movement contains some kinetic energy that could be used to fully or partially power electronic head-mounted devices such as hearing aids, wearable EEG systems, handsfree bluetooth devices and optical eyewear displays. Although human head is not as dynamic as other human parts such as arms and legs, it could be highly advantageous to power head-mounted-devices for the obvious reason of proximity. Nodding or shaking of the head typically occur during conversations and it is estimated that such movements can generate a mean power of 30 µW [1]. This amount of power would be sufficient to power a wearable sensor node [2].

Electromagnetic and piezoelectric transduction methods are two most competitive approaches for kinetic energy harvesting. Electromagnetic generators benefit from bulk magnets that make them suitable for low-frequency energy harvesting [3, 4]. On the other side, most piezoelectric generators have the advantages of simple mechanical structure and less peripheral components [5, 6]. The energy harvesting device presented in this paper is a combination of both transduction methods to combine their advantages. It is composed of a flexible piezoelectric composite in a doubly-clamped structure actuated by magnetic repulsive force. Piezoelectric energy harvesting with magnetic coupling has been previously used in a rotational configuration to harness human body power [7]. In contrast, the piezo-magnetic energy harvester proposed in this paper is in a rolling configuration; resulting in larger magnet velocity and stronger magnetic force. This configuration suits low-frequency and small-amplitude applications like movement of the head.

2. Energy harvesting device

The energy harvesting device is composed of a doubly-clamped piezoelectric fiber composite (PFC) beam and two permanent magnets. The PFC beam (Advanced Cerametrics, NJ, USA) comprises unidirectionally aligned piezoceramic micro-fibers with interdigitated electrodes. One of the Neodymium magnets (Amazing Magnets, CA, USA) is bonded to the PFC beam while another one is rolling along the beam length as shown in Figure 1. The rolling magnet moves back and forth while the device is tilting. The magnets are positioned such that like poles face each other and hence each time the moving magnet passes in front of the PFC magnet, the beam is deformed by the repulsive magnetic force and an electric current flows in the load circuit. The energy harvesting device is mounted on a pair of eyeglasses as shown in Fig. 2 to become wearable.

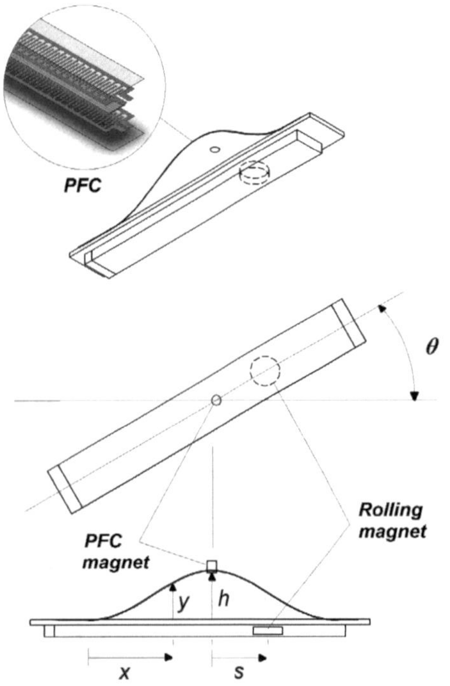

Figure 1. Schematic representation of the energy harvesting system

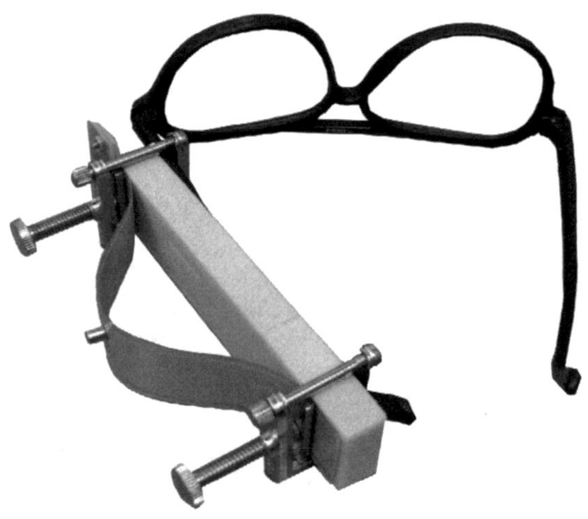

Figure 2. Piezoglasses energy harvester

3. System modeling

Fitting a beam of length L_0 in a gap of width l, where $L_0 > l$ necessitate a curvature to the beam. The equilibrium equation of an axially loaded beam is [8]:

$$y'''' + n^2 y'' = 0 \tag{1}$$

where y describes the deflection of the beam, $n^2 = P/EI$, and P, E and I are the axial load, the Young's modulus and the beam's moment of inertia respectively. Considering the clamped-clamped boundary conditions, the equation of the buckled beam in the first mode can be written as

$$y = \frac{h}{2}\left(1 - \cos\frac{2\pi x}{l}\right) \tag{2}$$

where l is the distance between the beam supports and h is the deflection of the center of the beam; $y(l/2) = h$. Using the formula for calculating the arc length of the curve, the length of the deflected beam can be derived as

$$L = \int_0^l \sqrt{1 + y'^2}dx = \int_0^l \sqrt{1 + \left(\frac{\pi h}{l}\sin\frac{2\pi x}{l}\right)^2}dx \tag{3}$$

The square term is assumed to be much smaller than unity and hence Eq.(3) can be approximated by the Taylor's series expansion.

$$L \approx \int_0^l \left[1 + \frac{1}{2}\left(\frac{\pi h}{l}\sin\frac{2\pi x}{l}\right)^2\right]dx = \frac{\pi^2 h^2}{4l} + l \tag{4}$$

The potential energy stored in the beam due to the axial deformation can be expressed as

$$V_a = \frac{1}{2}\frac{AE}{L_0}(L - L_0)^2 = \frac{1}{2}\frac{AE}{L_0}\left(\frac{\pi^2 h^2}{4l} + l - L_0\right)^2 \tag{5}$$

where A is the cross section area of the beam. In addition to axial deformation, the bending of the beam can also be quantified as the potential energy.

$$V_b = \frac{EI}{2}\int_0^l (y'')^2 dx = \frac{\pi^4 EI}{l^3}h^2 \tag{6}$$

The repulsive magnetic force has been measured as a function of the separation distance (r) between the magnets used in the prototype mechanism by using a precise scale. This force is found to be proportional to $1/r^3$ and the corresponding potential energy can be calculated by

$$V_m = -\int_\infty^r \frac{k}{r^3}dr = \frac{1}{2}\frac{k}{r^2} \tag{7}$$

in which k is the coefficient of proportionality and $r = \sqrt{s^2 + h^2}$.

The Lagrangian function of the entire system is defined as the subtraction of the kinetic and potential energies.

$$\begin{aligned} L = T - V = \; & \tfrac{1}{2}m\dot{h}^2 + \tfrac{1}{2}M\left(\dot{s}^2 + (s\dot{\theta})^2\right) + \tfrac{1}{2}J\left(\tfrac{2\dot{s}}{\mu d}\right)^2 \\ & - \left[\frac{1}{2}\frac{AE}{L_0}\left(\frac{\pi^2 h^2}{4l} + l - L_0\right)^2 + \frac{\pi^4 EI}{l^3}h^2 \right. \\ & \left. + \frac{1}{2}\frac{k}{h^2 + s^2} + Mgs\sin\theta\right] \end{aligned} \tag{8}$$

The Lagrange's formula $\frac{d}{dt}\left(\frac{\partial L}{\partial \dot{q}_i}\right) - \frac{\partial L}{q_i} = Q_i$ is then applied to drive the equations of motions for two generalized coordinates, h and s:

$$\begin{cases} m\ddot{h} + c\dot{h} + \dfrac{2\pi^2 AEh}{4lL_0}\left(\dfrac{\pi^2 h^2}{4l} + l - L_0\right) + \dfrac{2\pi^4 EI}{l^3}h - \dfrac{kh}{\left(h^2 + s^2\right)^2} = 0 \\ \left(M + \dfrac{4J}{\mu^2 d^2}\right)\ddot{s} - Ms\dot{\theta}^2 - \dfrac{ks}{\left(h^2 + s^2\right)^2} + Mg\sin\theta = 0 \end{cases} \tag{9}$$

Table 1. System parameters

Magnetic force	k	$1.67 \times 10^{-6}\,\mathrm{Nm}^3$
Fixed magnet	m	$2\,\mathrm{g}$
Moving magnet	M	$4\,\mathrm{g}$
	d	$12\,\mathrm{mm}$
	J	$80\,\mathrm{gmm}^2$
PFC	A	$6\,\mathrm{mm}^2$
	I	$0.045\,\mathrm{mm}^4$
	E	$2.44 \times 10^{10}\,\mathrm{Pa}$
	C	$10 \times 10^{-9}\,\mathrm{F}$
	d_{33}	$550 \times 10^{-12}\,\mathrm{C/N}$
Device parameters	l	$10\,\mathrm{cm}$
	R	$660\,\Omega$
Initial conditions	s_0	$5\,\mathrm{cm}$
	h_0	$0.5\text{-}5\,\mathrm{cm}$
	$\dot{s}_0$	0
	$\dot{h}_0$	0

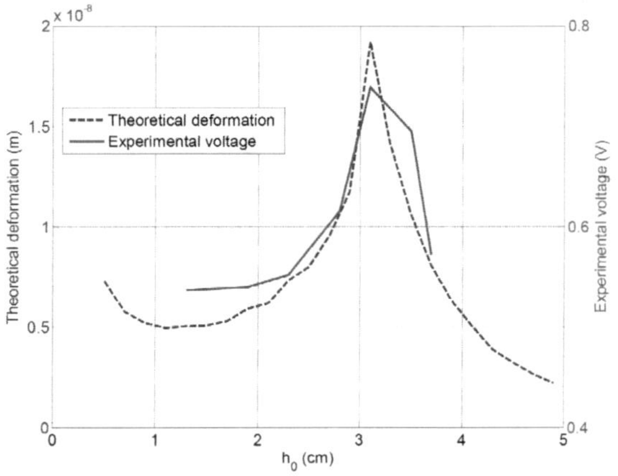

Figure 3. Theoretical deformation and experimental voltage for $\theta = 45°$ and $\dot{\theta} = 0$

4. Model validation

The theoretical model expressed in Eq. 9 is experimentally tuned and validated for $\theta = 45°$ and $\dot{\theta} = 0$ and the results are shown in Figure 3 for different beam apex heights. All the parameters used in the theoretical model are also given in Table 1. According to Figure 3, there is a good agreement between experimental measurements and theoretical predictions. In fact, the maximum measured voltage of the device and the maximum theoretical deformation of the beam occur nearly at the same beam apex height.

5. System optimization

According to the derived equations of motion, Eq. (9), the magnet travel distance (s_0) and the piezoelectric beam apex height (h_0) have significant effects on the performance of the device. The more distance the magnet can travel (large s_0), the larger potential energy it gains and the faster it passes in front of the PFC fixed magnet, resulting in larger magnetic force and more deformation of the beam. However, this parameter is eventually limited by the size of the device. On the other hand, if the beam is too curved (large h_0), the separation distance is too large to produce significant magnetic force and deform the beam. Conversely, if the beam curve is too little (small h_0), the repulsive force component along the direction of the moving magnet is too large to let the moving magnet pass and deform the beam. Therefore, there is an optimum value for the beam apex height (h_0). This optimum value can be identified in Figure 3.

Having validated the theoretical model for the constant θ in Figure 3, Eq. 9 is again used to optimize the device for more realistic scenario and practical conditions in which $\dot{\theta} = \bar{\theta} f$, where $\bar{\theta} = 16.5°$ and $f = 3.7\,\mathrm{Hz}$ and are respectively average amplitude and frequency of the head movements during conversation [1]. The optimum value is then calculated to be $\tilde{h}_0 = 2.5\,\mathrm{cm}$.

6. Experimental results

A prototype is fabricated based on the derived optimum value ($\tilde{h}_0$) and is mounted on a pair of eyeglasses as shown in Figure 2. To determine the device's approximate raw power, the PFC was terminated with the matched resistive load and its voltage was measured for a test-subject wearing the piezo-magnetic glasses while working with the computer as shown in Figure 4. The voltage output of the energy harvester during a ten-minute test is illustrated in Figure 5.

According to the results, the maximum measured voltage is 0.53 V that corresponds to the instantaneous power of 0.5 µW delivered to the impendence matched resistive load.

Figure 4. Experimental setup

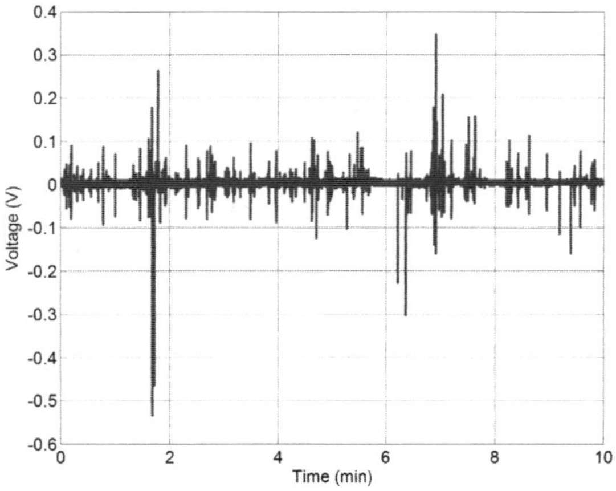

Figure 5. Piezo-magnetic glasses voltage output

7. Conclusions

To the best of the authors' knowledge, the presented device is the first inertial energy harvester with the flexible piezoelectric beam that is actuated by the rolling magnet. The main advantage of the proposed energy harvester is that the velocity of the rolling magnet that excites the beam is a function of its traveled distance and the tilt angle. Therefore, the velocity and hence the kinetic energy of the rolling magnet can be quite considerable even if the host motion is very limited and slow. Consequently, the device can operate with the small-amplitude and low-frequency movements of the head for which other energy harvesting solutions are rarely efficient. Moreover, the proposed energy harvester can be unobtrusive when integrated with the glasses or any other head-mounted devices. The power output of the energy harvester device is currently much far from the expected power, however it can multiply by using heavier rolling magnets and installing more piezoelectric layers in both temples. The piezo-magnetic energy harvester can be ultimately used to power a body sensor network or any other wearable or implantable medical devices in the region of the head.

References

[1] Erich Goll, Hans-Peter Zenner, and Ernst Dalhoff. Upper bounds for energy harvesting in the region of the human head. *IEEE transactions on bio-medical engineering*, 58(11):3097–103, November 2011.

[2] Ruud Vullers, Rob van Schaijk, Bert Gyselinckx, and Chris Van Hoof. Is there a sweet spot for energy harvesting? In *2009 Device Research Conference*, pages 7–8. IEEE, June 2009.

[3] B.P. Mann and N.D. Sims. Energy harvesting from the nonlinear oscillations of magnetic levitation. *Journal of Sound and Vibration*, 319(1-2):515–530, January 2009.

[4] C.R. Saha, T. O'Donnell, N. Wang, and P. McCloskey. Electromagnetic generator for harvesting energy from human motion. *Sensors and Actuators A: Physical*, 147(1):248–253, 2008.

[5] A. Khaligh. Kinetic Energy Harvesting Using Piezoelectric and Electromagnetic Technologies—State of the Art. *IEEE Transactions on Industrial Electronics*, 57(3):850–860, March 2010.

[6] Seok Min Jung and Kwang Seok Yun. Energy-harvesting device with mechanical frequency-up conversion mechanism for increased power efficiency and wideband operation. *Applied Physics Letters*, 96, 2010.

[7] P. Pillatsch, E. M. Yeatman, and A. S. Holmes. A wearable piezoelectric rotational energy harvester. In *2013 IEEE International Conference on Body Sensor Networks, BSN 2013*, 2013.

[8] Mattias Vangbo. An analytical analysis of a compressed bistable buckled beam. *Sensors and Actuators A: Physical*, 69(3):212–216, 1998.

Piezoelectric energy harvesting from heartbeat vibrations for leadless pacemakers

M H Ansari[1, 3] and M Amin Karami[2]

[1] IDEAS Lab, Department of Mechanical and Aerospace Engineering, University at Buffalo (SUNY), Buffalo, New York, USA
[2] Department of Mechanical and Aerospace Engineering, University at Buffalo (SUNY), Buffalo, New York, USA

E-mail: mansari2@buffalo.edu

Abstract. This paper studies energy harvesting from heartbeat vibrations using fan-folded piezoelectric beams. The generated energy from the heartbeat can be used to power a leadless pacemaker. In order to utilize the available 3 dimensional space to the energy harvester, we chose the fan-folded design. The proposed device consists of several piezoelectric beams stacked on top of each other. The size for this energy harvester is 2 cm by 0.5 cm by 1 cm, which makes the natural frequency very high. High natural frequency is one major concern about the micro-scaled energy harvesters. By utilizing the fan-folded geometry and adding tip mass and link mass to the configuration, this natural frequency is reduced to the desired range. This fan-folded design makes it possible to generate more than 10 μW of power. The proposed device does not incorporate magnets and is thus Magnetic resonance imaging (MRI) compatible. Although our device is a linear energy harvester, it is shown that the device is relatively insensitive to the heartrate. The natural frequencies and the mode shapes of the device are calculated. An analytical solution is presented and the method is verified by experimental investigation. We use a closed loop shaker controller and a shaker to simulate the heartbeat vibrations. The developed analytical model is verified through comparison of theoretical and experimental tip displacement and acceleration frequency response functions.

1. Introduction

Unlike traditional pacemakers, leadless pacemakers do not have leads and do not need an open surgery for implantation The required power for a pacemaker is about 1 μW [1]. The main obstacle for the development of leadless pacemakers is the power issue, as its battery takes about 60% of the size of a conventional pacemaker. The size of the conventional pacemaker batteries is too large for leadless pacemakers. The battery size issue hindered the development of leadless pacemakers for 20 years. Recently novel batteries have been developed that make leadless pacemakers realizable [2]. Still, the battery life is the same as traditional pacemakers and it lasts 6 to 7 years [2]. Extraction of leadless pacemakers is very difficult so when the battery is depleted, a new pacemaker has to be implanted. It is shown that our proposed device generates an order of magnitude more power than the nominal power needed for a leadless pacemaker. The small size of the energy harvester and sufficient output

[3] Author to whom any correspondence should be addressed.

power of the device are the advantages of this new design. The size of the device is $2cm \times 0.5\ cm \times 1cm$ (1 cc in volume).

Several studies have been done on energy harvesting from heartbeat and fan-folded structure. None of the groups looked into energy harvesting for leadless pacemaker using fan-folded structure. Karami uses linear and non-linear design for pacemaker energy harvesting [3]. Khattak studies the interesting behaviour of a fan-folded structure [4]. He finds that structures with even beams have repeated resonances and structure with odd beams have double roots coinciding with a cantilevered beam. In another research, Reissman studies the coupling effect of vibration fan-folded energy harvesting structures [5]. In this research, a fan-folded structure is studied. The free vibration modes for a five beam structure are found and it is shown that the generated power of the structure is sufficient to power a pacemaker.

2. Device configuration

The device consists of several horizontal bimorph beams and vertical rigid beams. Each horizontal beam is a bimorph piezoelectric beam, with one brass layer as the substrate and two piezoelectric layers attached on each side. The bimorph beams are connected to each other by rigid links made from platinum. Platinum is chosen for the links due to its high density. One end of the structure is clamped as the other end is free to move (Figure 1 and Figure 2). High natural frequency is one major problem in small energy harvesters. In order to generate sufficient power for a pacemaker, the first natural frequency of the energy harvester should be less than 50 Hz. The fan-fold geometry is a very effective design to reduce the natural frequency. We also utilize the tip mass and the link masses to reduce the natural frequency of our device to 22 Hz.

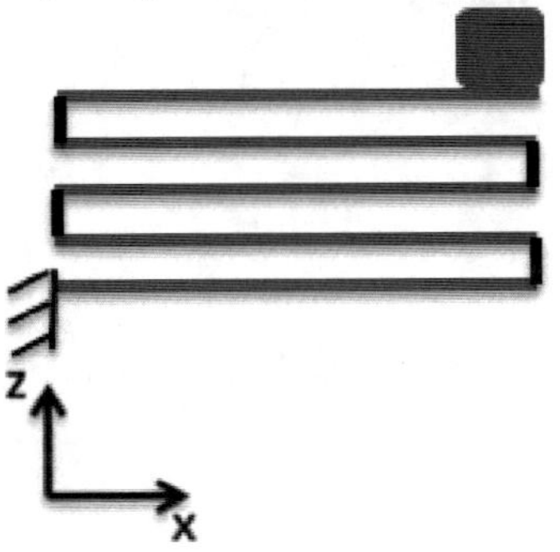

Figure 1. Fan-folded energy harvester with five beams and tip mass

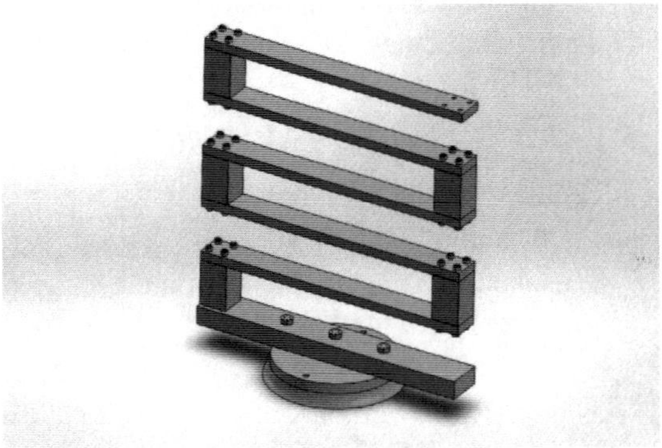

Figure 2. 3D model of a five beam fan-folded structure

3. Modelling and solving the governing equations

The device consists of several uniform composite beams and each beam is modeled with the Euler-Bernoulli beam theory. The structure vibrates due to base excitation. Each beam can bend and deflect which changes the start position of the next beam. The deflection of the beam w_i is a function of the length x, and time t. The index i is the beam number (from 1 to n). The coupled mechanical equation of a beam with tip mass is [6]:

$$\rho A\,\frac{\partial^2 w_{rel}}{\partial\, t^2} + YI\,\frac{\partial^4 w_{rel}}{\partial x^4} = -\alpha\left[\frac{d\delta(x)}{dx} - \frac{d\delta(x-L)}{dx}\right]v(t) - [\rho A + m_l\delta(x - x^*, i - k) + M_t\delta(x - x_{end}, i - n)]\frac{d^2 w_b}{dt^2} \quad (1)$$

Where ρA is the total mass per unit length of the beam. $w_{rel}\,(x,t)$ is the deflection along the z-axis, YI is the equivalent bending stiffness of the composite beam. m_l is the link mass and x^* is 0 (if i is an odd number) or length of the beam, L, (if i is an even number). Index k is from 2 to n. M_t is the tip mass, $\delta(x)$ is the Dirac delta function and α is the coupling term and for parallel connection. To find the vibration mode shapes, we solve the undamped, uncoupled equation. The solution for the free vibration can be shown as a linear combination of all natural motions of the beam [7] (Section 11):

$$w_i(x,t) = \sum_{j=1}^{\infty} \phi_{ji}(x) T_j(t) \tag{2}$$

Where, ϕ_{ij} is j^{th} mode shape of the ith beam, and T_j is the time dependent function. By using the above equations and considering boundary and continuity conditions, the mode shapes and natural frequencies of a fan-folded structure are derived. In the configuration that we have, the beams are connected electrically in parallel. Since the deflection of the beams is opposite of each other in some mode shapes, we use a switch to keep or to reverse the polarity of the generated voltage [8]. These switches decide if the current going to each member should be added to or subtracted from the current in other members. Considering the switches and using equation (1) and equation (2), we have:

$$YI \sum_{j=1}^{\infty} T(t)_j \, \phi_{ji}^{(4)} + \rho A \sum_{j=1}^{\infty} \ddot{T}(t)_j \, \phi_{ji} = -\alpha P_i \left[\frac{d\delta(x)}{dx} - \frac{d\delta(x-L)}{dx}\right] v(t) - [\rho A + m_l \delta(x - x^*, i-k) + M_t \delta(x - x_{end}, i-n)]\ddot{w}_b \tag{3}$$

Where P_i is the switch for the ith member and it is either 1 or -1. The sign of the switches for each mode is decided based on $P_i = Sign(\phi'_{ji}(l) - \phi'_{ji}(0))$. Using equation (3) and Kirchhoff laws for parallel connection we find the expression for the multi-mode power frequency response function. For more details on the mathematical modeling please refer to [9]:

$$\frac{p(\omega)}{{a_b}^2(\omega)} = \frac{1}{2R}\left(\frac{\sum_{j=1}^{\infty}\frac{j\omega \chi_j \gamma_j}{\omega_{nj}^2 + 2\,\zeta\omega_{nj} j\omega - \omega^2}}{\frac{1}{R} + j\omega C_p + \sum_{j=1}^{\infty}\frac{j\omega {\chi_j}^2}{\omega_{nj}^2 + 2\,\zeta\omega_{nj} j\omega - \omega^2}}\right)^2 \tag{4}$$

The tip deflection of the structure and the relative tip acceleration to the base acceleration in the frequency domain are calculated as:

$$\frac{w_{tip}(\omega)}{a_b(\omega)} = \sum_{j=1}^{\infty} \frac{-1}{\omega_{nj}^2 + 2\,\zeta\omega_{nj} j\omega - \omega^2} \times \left(\chi_j \frac{\sum_{j=1}^{\infty}\frac{j\omega \chi_j \gamma_j}{\omega_{nj}^2 + 2\,\zeta\omega_{nj} j\omega - \omega^2}}{\frac{1}{R} + j\omega C_p + \sum_{j=1}^{\infty}\frac{j\omega {\chi_j}^2}{\omega_{nj}^2 + 2\,\zeta\omega_{nj} j\omega - \omega^2}} + \gamma_j\right) W_{jn}(x_{end}) \tag{5}$$

$$\frac{a_{rel}(\omega)}{a_b(\omega)} = \frac{w_{tip}(\omega)}{a_b(\omega)} * (-\omega^2) \tag{6}$$

4. Results for a case study

In this section, we study the generated energy form a 5 beam structure using the heartbeat vibrations as the source of the vibration to the system. The structure consists of five bimorph beams with 2 cm length and 0.5 cm width. PSI - 5A4E PIEZO sheets from PIEZO SYSTEMS, INC. are used as the piezoelectric element. The thickness of the brass layer and the piezoelectric layer are 0.0254 and 0.0127 cm. Each beam is connected to the next beam by a platinum rigid beam. The length of the rigid part is 0.2 cm. The tip mass is 21 grams made from platinum. A minimum safety factor of 4 is considered for the device to avoid fracture at where the beams are connected and where the first beam is clamped.

We then consider the heartbeat vibrations as the base acceleration of the system. To estimate the vibrations in the vicinity of the heart due to the heartbeat, we use the ultrasonic velocity measurements performed by Kanai [10]. Figure 3 shows the heartbeat acceleration in time domain and frequency domain.

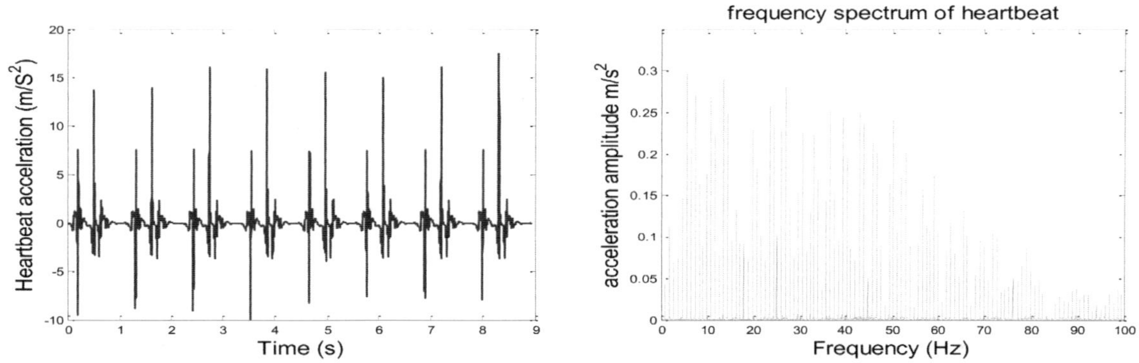

a) **b)**

Figure 3. Heartbeat wave acceleration a) in the time domain and b) Frequency domain

By using equation.(4) and equation.(6) we calculate the generated voltage and power from heartbeat vibrations. The instantaneous power across a $416\ k\Omega$ purely resistive load is plotted in Figure 4 The value of the resistive load matches the resistance having the maximum power in the first natural frequency of the device. The average power for generated electricity in Figure 4 is $10.24\ \mu W$. The power needed for a pacemaker is less than $1\ \mu W$. The first natural frequency of the device occurs at $16.18\ Hz$ which satisfies the condition of having a small natural frequency to generate more power. The assumed damping ratio in this case is 4%. Figure 5 shows different average powers for different heartrate. Although the device is linear, the output power does not change significantly with change in the heartrate.

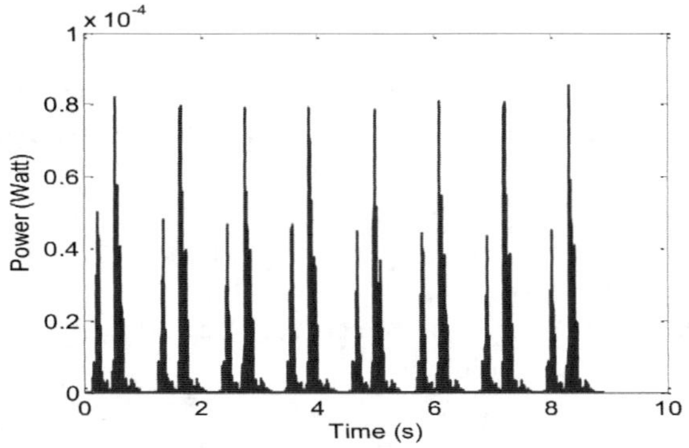

Figure 4. Instantaneous power

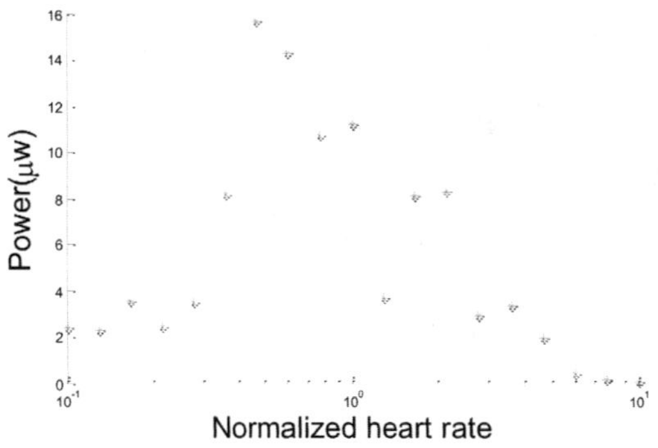

Figure 5. Output power for different heartrates

5. Experimental verification

In this section, we verify the results using experimental data. In order to do the experiment, we use a closed loop shaker and an accelerometer to measure the base acceleration. We use a laser vibrometer to measure the tip velocity of the structure. The structure is made from Aluminium and we use four steel bolts to connect the beams to the links. To verify the method, we do a sine sweep test and plot the relative tip acceleration to the base acceleration in the frequency domain. Figure 6 shows the comparison between the experimental and theoretical results for different number of beams. There is good match between the results which shows that the theoretical method predicts the natural frequency and the mode shapes of the device correctly. In our model we assume that the length of the links is much shorter than the length of the beams. Thus, there is a discrepancy between the theoretical results and the experimental results as the number of the beams increase.

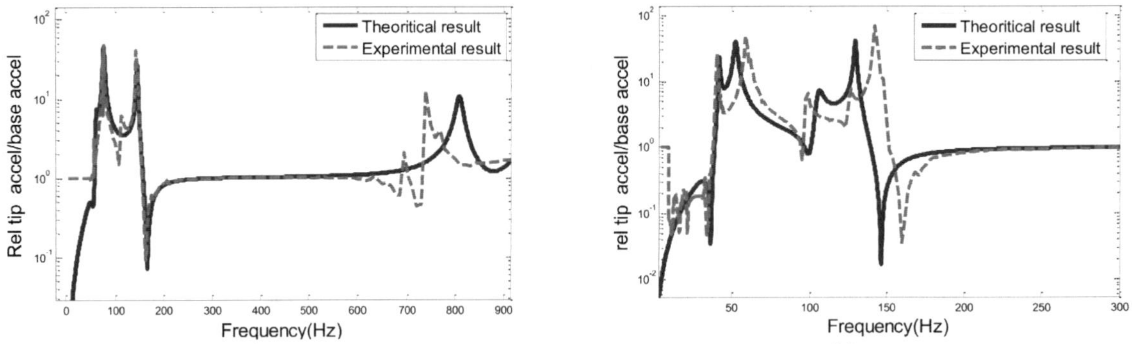

a) **b)**

Figure 6. Analytical predictions versus experimental measurements for three(a) and four(b) beams

6. Conclusion

This article presents a 3D fan-folded design to generate energy from heartbeat vibrations. The natural frequency and mode shapes of the device are found and the results are verified with experimental results. The electro-mechanical equations are solved for this configuration and the energy generated from 9 seconds of heartbeats is calculated. It is shown that the energy harvester generates enough energy to power a pacemaker. The fan-folded geometry and the tip mass makes it possible to have the energy harvester in small size ($2cm$ by $0.5cm$ by $1cm$). Adding the tip mass reduces the natural frequency significantly. The device can be implemented inside the body to generate the electricity needed for leadless pacemakers and makes it possible to have an autonomous pacemaker without the need of a battery.

7. References

[1] O. J. OHM and D. Danilovic, "Improvements in pacemaker energy consumption and functional capability: four decades of progress," *Pacing and clinical electrophysiology,* vol. 20, pp. 2-9, 1997.

[2] http://www.sjm.com.

[3] M. A. Karami and D. J. Inman, "Powering pacemakers from heartbeat vibrations using linear and nonlinear energy harvesters," *Applied Physics Letters,* vol. 100, p. 042901, 2012.

[4] A. Khattak, S. Garvey, and A. Popov, "Repeated resonances in folded-back beam structures," *Journal of sound and vibration,* vol. 290, pp. 309-320, 2006.

[5] T. Reissman and E. Garcia, "Electro-mechanical effect on resonances in fan-folded piezoelectric beam structures," in *CanSmart international workshop on smart materials and smart structures*, 2008, pp. 23-24.

[6] A. Erturk and D. J. Inman, "A distributed parameter electromechanical model for cantilevered piezoelectric energy harvesters," *Journal of Vibration and Acoustics,* vol. 130, p. 041002, 2008.

[7] S. S. Rao, *Vibration of continuous systems*: John Wiley & Sons, 2007.

[8] M. A. Karami and D. J. Inman, "Analytical modeling and experimental verification of the vibrations of the zigzag microstructure for energy harvesting," *Journal of Vibration and Acoustics,* vol. 133, p. 011002, 2011.

[9] M. H. Ansari and M. A. Karami, "Energy harvesting from heartbeat using piezoelectric beams with fan-folded configuration and added tip mass," in *SMASIS 2015*, Colorado, USA, 2015.

[10] H. Kanai, M. Sato, Y. Koiwa, and N. Chubachi, "Transcutaneous measurement and spectrum analysis of heart wall vibrations," *Ultrasonics, Ferroelectrics and Frequency Control, IEEE Transactions on,* vol. 43, pp. 791-810, 1996.

Piezoelectric Vibration Energy Harvester Using Indirect Impact of Springless Proof Mass

S Ju and C-H Ji

Department of Electronics Engineering, Ewha Womans University, Seoul, Korea

E-mail: cji@ewha.ac.kr

Abstract. This paper presents an impact-based piezoelectric vibration energy harvester using freely movable spherical proof mass and MFC (Macro Fiber Composite) beams as piezoelectric cantilevers. When external vibration is applied, a metal sphere moves freely along the channel and collides with both ends of the cavity, which induces the vibration of parallel-connected MFCs and generates electric power. A proof-of-concept device having the form-factor of a wristwatch has been designed and tested. Moreover, spherical proof mass made of different materials has been tested to analyze the relationship between output power, long-term reliability, and audible noise level during operation. Maximum peak-to-peak open circuit voltage of 41.2V and average power of 908.7μW have been obtained in response to a 3g vibration at 17Hz for device with parallel-connected MFC beams.

1. Introduction

Harvesting energy from low frequency vibrations such as human-body-induced motion has been drawing substantial research interest to eventually remove the need for battery replacement in some of the applications, such as implantable bio-sensors and wearable devices [1]. To overcome the major challenges in generating electric power, including low and random frequency of the vibrations, wide variety of non-resonant harvesters have been reported as alternative approaches. Frequency-up-converted electromagnetic transducers and piezoelectric generators using magnetic or impact-based systems are some of the examples [2, 3]. Although utilization in a specific wearable device application would require further research efforts, harvesters having the form-factor of a conventional wristwatch have also been reported, which utilized the asymmetric rotor to convert the external vibration into a rotational motion and generate power [4, 5]. Due to the straightforward operating principle and proven performance in the watch manufacturing industry, utilization of asymmetric rotor as a proof mass in energy harvesting device could potentially be a promising approach, but the improvement of output power and reduction of fabrication cost and complexity need to be accompanied.

Piezoelectric energy harvesters using impact-based power generation mechanisms have been widely researched due to simplicity of the system and effectiveness of energy conversion resulting from high energy density of the material. Despite the advantages, reliability issues in contact regions resulting from high amplitude excitation of proof mass pose major challenges. Previously, we have reported an indirect impact-based energy harvester to avoid the catastrophic failure of contacting surface due to high stress and deflection in conventional impact-based systems [6].

In this research, we present a low frequency vibration energy harvester using impact between single springless proof mass and aluminum housing of monolithic construction, which conveys the vibration to two parallel-connected MFC beams for high output power generation while maintaining a small form-factor.

2. Harvester design

Schematic diagrams of the proposed vibration energy harvester are shown in figure 1. The harvester consists of a straight channel with spherical proof mass in the middle and two piezoelectric cantilevers attached on both sides of the channel. Piezoelectric cantilevers have been prepared by attaching copper blocks at top and bottom side of the free end of each MFC beams. Device housing, whose size was inspired by a conventional wristwatch, has been fabricated with aluminum, considering the ease of fabrication and high restitution coefficient. Previously, we have reported an indirect impact-based energy harvester with multiple proof mass and single piezoelectric cantilever in a circular geometry [7]. Despite some of the technological improvements from the earlier works, output power improvement was limited due to curved geometry. In this research, rectangular housing has been designed for higher power density and parallel-connected MFC beams have been used to improve the output power.

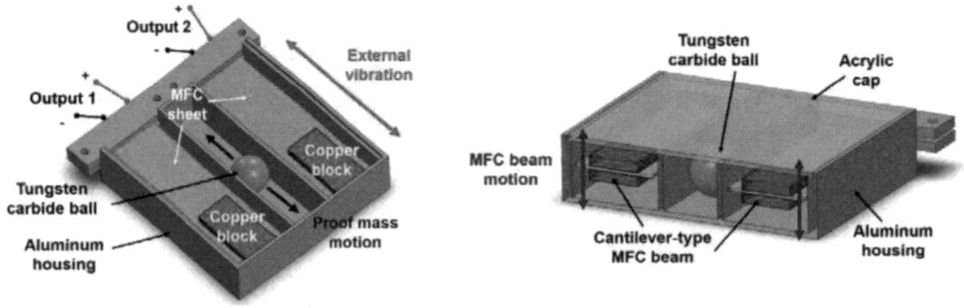

Figure 1. Schematics of the proposed piezoelectric energy harvester using tungsten carbide ball and MFC cantilevers: perspective view (left), cross-sectional view (right).

3. Fabrication

As shown in figure 2, a proof-of-concept harvester has been fabricated by attaching two MFC cantilevers (26x6.5x0.3mm^3 each) on one end of the channel which has 5mm-diameter tungsten carbide proof mass inside. To evaluate the relationship between output power, long-term reliability, and audible noise level of the device during operation, devices with various proof mass materials have been tested (figure 2(c)). Assembled device, which measures 27x26.5x6.5mm^3, has been covered with an acrylic plate and tested using vibration exciter and pendulum test setup.

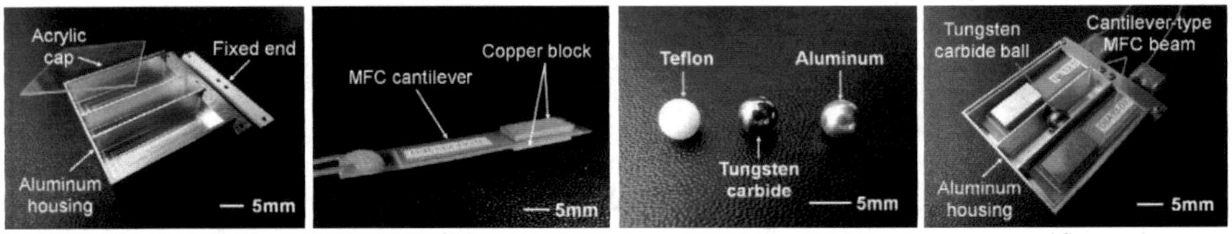

(a) (b) (c) (d)

Figure 2. Assembled device: (a) Housing components, (b) MFC cantilever with copper blocks, (c) three different types of proof mass tested, (d) fabricated piezoelectric energy harvester.

4. Results and discussion

4.1. Vibration exciter test

To test the device performance at various input frequencies, output from the fabricated device has been measured while the device was shaken with vibration exciter. As shown in figure 3, sinusoidal vibration has been applied in vertical direction. Input acceleration of 3g and input frequencies ranging from 10 to 20Hz with 1Hz step have been tested.

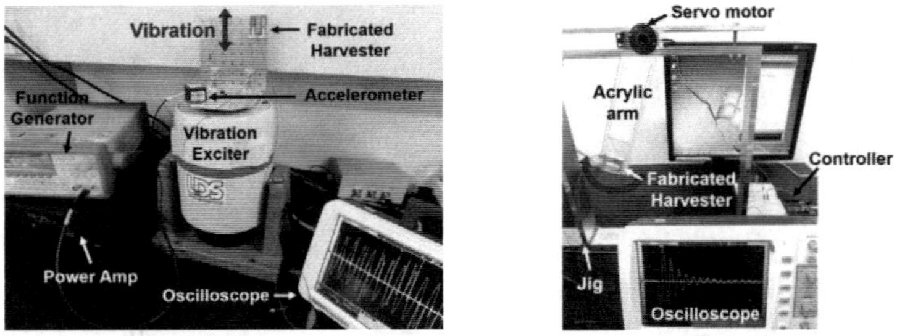

Figure 3. Experimental setup for the vibration exciter test (left) and pendulum test (right).

Output voltage waveforms of each MFC cantilevers have been compared to verify the symmetry (figure 4) and fabricated devices with single MFC and parallel-connected MFC cantilevers have been tested and compared. As shown in figure 5, peak-to-peak open circuit voltage of both devices show a similar trend at various input frequencies at 3g acceleration. Maximum open circuit voltage of 37.8V and 41.2V have been obtained for devices with single and parallel-connected MFCs, respectively.

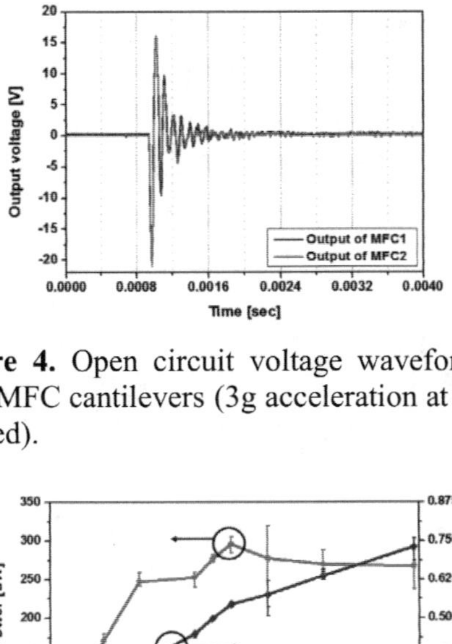

Figure 4. Open circuit voltage waveform of each MFC cantilevers (3g acceleration at 17Hz applied).

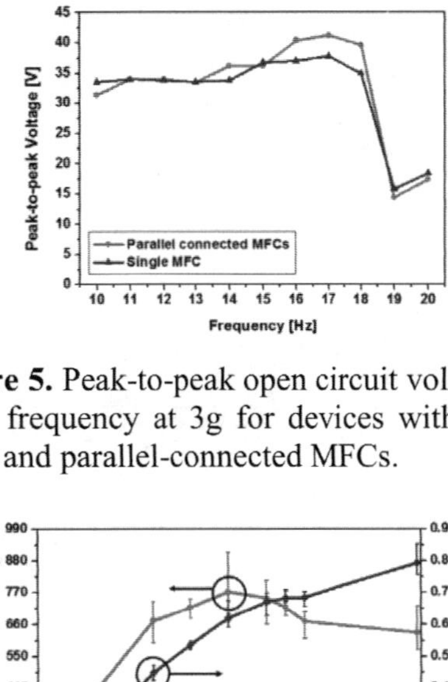

Figure 5. Peak-to-peak open circuit voltage vs. input frequency at 3g for devices with single MFC and parallel-connected MFCs.

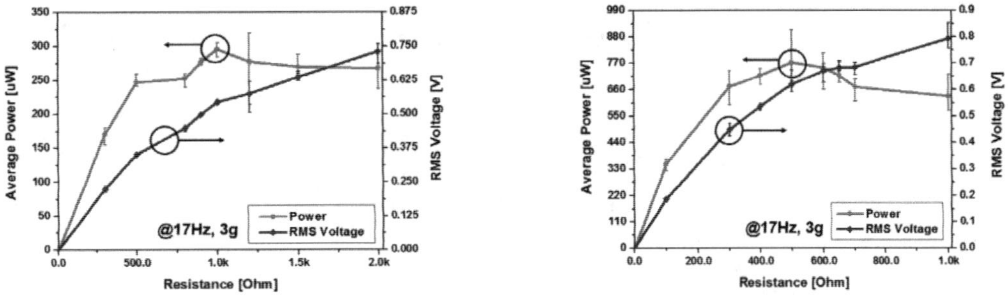

(a) (b)

Figure 6. RMS voltage and average power at various load resistance (3g acceleration at 17Hz applied): (a) device with single MFC cantilever, (b) device with parallel-connected MFC cantilevers.

Figure 6 shows the root-mean-square (RMS) voltage and average power delivered to external resistive load measured with devices having single MFC and parallel-connected MFCs. Input frequencies and accelerations for maximum open circuit voltages have been used in the experiment. At 3g acceleration at 17Hz, maximum output power of 320μW has been obtained at 1.2kΩ using the device with single MFC. For the device with parallel-connected MFCs, maximum output power of 908.7μW has been achieved at 500Ω. Output power of the device with double MFCs was 284% higher than that of the device with single MFC, which can be ascribed to the lower impedance of the device with parallel-connected MFC cantilevers. Moreover, output voltage and power at low frequency vibrations have been improved significantly compared to previous results obtained with device having single MFC beam and metal spheres in two parallel channels [7].

4.2. Pendulum test

To evaluate the device under real-world operating conditions, pendulum test has also been carried out (figure 3). Assembled device with parallel-connected MFCs has been attached to a 20cm-long acrylic swing arm controlled by a servo motor. Output voltage and power have been measured at very low input frequencies ranging from 0.5 to 2.5Hz while varying the swing speed and angle. Open circuit voltage showed an increasing trend as the angular velocity and angle of swing were increased except at swing speed of 60°/sec (figure 7(a)). Maximum open circuit voltage of 42V has been obtained at swing speed of 150°/sec and total swing angle of 120°. RMS voltage and average power measured with 500Ω resistive load at various angular velocities are shown in figure 7(b). Increase of output power due to increased swing speed was more pronounced compared to that resulting from the increase of swing angle. Maximum output power reached 97.8μW at speed of 150°/sec and swing angle of 60°, where the input frequency was twice that of the 120° swing condition. Although average power from pendulum test setup was much lower than that generated with vibration exciter, total amount of power generated at very low frequency vibration, mimicking that of the wrist motion, was relatively higher than expected.

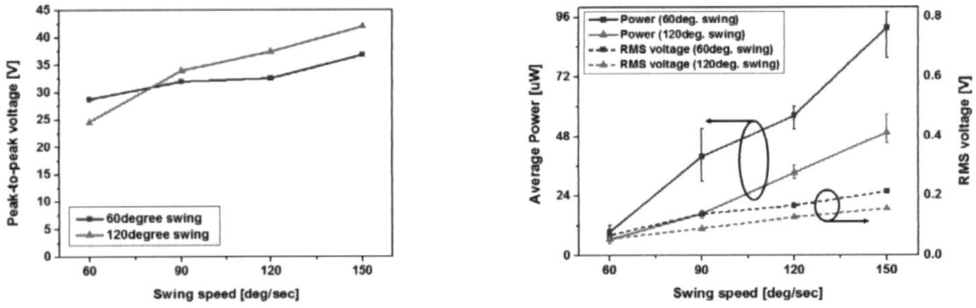

(a) (b)

Figure 7. Pendulum test results: (a) peak-to-peak open circuit voltage at various swing speeds and angles, (b) RMS voltage and average power at various swing speeds and angles.

4.3. Performance analysis

As shown in figure 8, fabricated device has been connected to a simple storage circuit with diode rectifier and capacitor. For device with parallel-connected MFCs, 277seconds of vibration at 3g acceleration at 17Hz was sufficient to fully charge the 1,000μF capacitor and stored energy was 1,549μJ, which was high enough to turn on the LED in storage circuit.

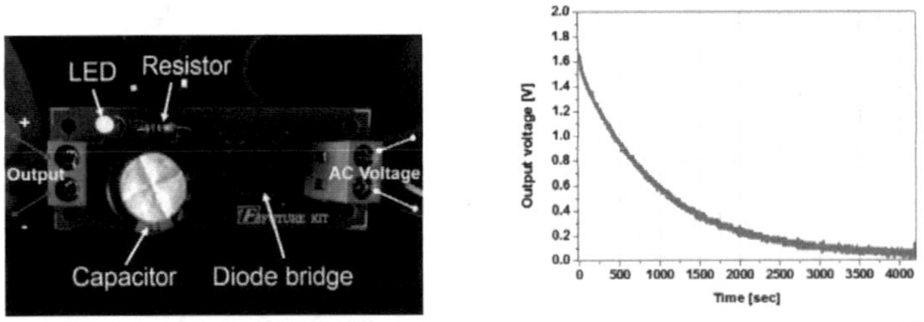

Figure 8. Energy storage test results: circuit with diode rectifier, capacitor and LED (left), discharge waveform (right).

As the output power, long-term reliability, and audible noise level of the device are closely related to the impact between proof mass and channel end, performance of the devices with various proof mass materials have been compared. Cyclic test has been performed with 3g acceleration at 17Hz. As shown in figure 9(a), noticeable drop of output power has been observed with device having tungsten carbide proof mass. However, devices with aluminum and Teflon proof mass showed no sign of power degradation up to 489,600cycles at the cost of reduced initial output power. Measured noise level from device with tungsten carbide and Teflon proof mass were 112dB and 41dB, respectively (figure 9(b)). Aluminum or Teflon proof mass can be an option in terms of enhanced long-term reliability and noise level with trade-off in output power. Device performance can be further improved by appropriate choice of spherical proof mass and housing materials, and optimized device geometry. Further improvement of output characteristics is expected by additional dynamic analysis and optimization of the device geometry at both ends of the cavity. Moreover, position and size of the proof mass attached to the MFC cantilever require optimization to maximize the efficiency of the energy conversion.

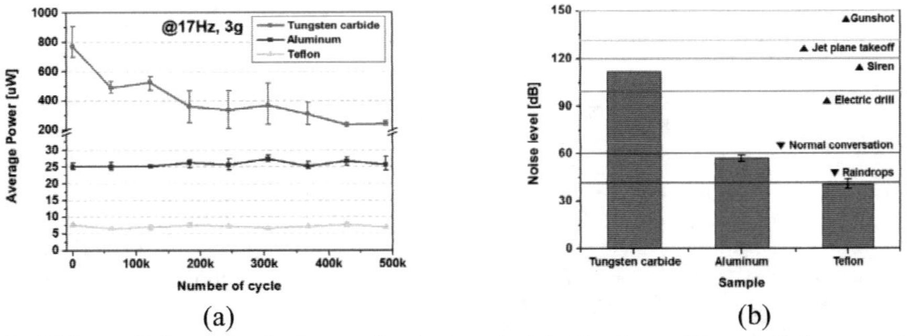

Figure 9. Effect of the spherical proof mass material (device with parallel-connected MFC beams tested): (a) output power variation, (b) audible noise measurement result.

5. Conclusion

We have demonstrated an indirect impact based piezoelectric vibration energy harvester using freely movable spherical proof mass and parallel-connected MFC beams as piezoelectric cantilevers. A proof-of-concept device having the form-factor of a conventional wristwatch has been designed, fabricated and tested successfully in various operating conditions. Maximum peak-to-peak open circuit voltage of 41.2V and average power of 908.7μW have been obtained at 3g acceleration at 17Hz in vibration exciter test. Effect of proof mass material in long-term reliability and audible noise level of the fabricated device have been verified by cyclic testing and noise measurement. Although fabricated device requires further optimization, we have successfully verified the concept of an indirect impact-based piezoelectric energy harvesting device for potential wearable device applications.

6. Acknowledgement

This research was supported by the Pioneer Research Center Program through the National Research Foundation of Korea funded by the Ministry of Science, ICT & Future Planning (2010-0019313), and by the Center for Integrated Smart Sensors as GFP (CISS-2012M3A6A6054204).

References

[1] Galchev T V *et al.* 2011 *J. Micromech. Microeng.* **21** 104005
[2] Bowers B J and Arnold D P 2009 *J. Micromech. Microeng.* **19** 094008
[3] Renaud M *et al.* 2009 *Smar Mater. Struct.* **21** 035001
[4] Pillatsch P, Yeatman E M and Holmes A S 2014 *Sens. Act. A* **206** 178-185
[5] Xue T, Ma X, Rahn C and Roundy S 2014 *J. Phys. Conf. Ser.* **557** 012090
[6] Ju S, Chae S H, Choi Y and Ji C-H 2015 *Sens. Act. A* **226** 126-136
[7] Ju S and Ji C-H 2015 *Proc. Transducers 2015* (Anchorage, 21-25 June 2015) 1913-1916

Remote power transfer using magneto-electric devices

K Sinha[1] and M Tabib-Azar[1, 2]
[1]Department of Electrical and Computer Engineering, University of Utah, Utah, USA
[2]University of California Berkeley, EECS Department, Berkeley, CA 94704

E-mail: azar@berkeley.edu

Abstract. We report remote power transfer using magneto-electric devices. The experiments were performed at room temperature for piezoelectric beam coupled with electromagnet. Neodymium magnet was used as mass loading. We observed the output power of the order +19.3 to -71.1 dB given the gap between the input and output source was varied from 4 mm to 12 mm for the device (21.3 mm× 3.59 mm× 0.57 mm) with best performance at the resonance peak. We tested the device for frequency sweeps of 10-100 Hz and 100-5000 Hz. This enabled us to figure out the output power for the device at resonant frequencies over a wide frequency range. The device has high input impedance (as opposed to coils) and can be miniaturized aggressively to below 100 μm linear dimensions. The piezoelectric beams have much higher quality factors (Q) larger than 1000 while coils have low Qs (~ 20) and the harvesting efficiency is proportional to Q.

1. Introduction

Coils are used to inductively power implantable devices. To reduce the size of the coils and associated components one needs to increase the signal frequency. It is well known that losses increase at higher frequencies and the power transfer efficiency reduces as given by Friis's near-field magnetic formula (1). So it is desirable to have power transfer links operating at low frequencies that at the same time are smaller in size for implant applications. As long as the implantable part of the power link is made very small, the part that is outside the body can be as large as necessary. Here we discuss a device consisting of a piezoelectric beam integrated with a permanent magnet to convert acoustic and magnetic signals to electricity. This approach enables miniaturization of the implantable part of the power link down to micrometer size levels which otherwise is not possible with coils even at high frequencies. Moreover, the power transfer efficiency in resonant circuits also depends on the quality factors of its receiver/transmitter parts and piezoelectric resonant structures can easily achieve Q's in excess of a few 1000 compared to 18 or 20 that can be achieved in the best inductors.

Powering wireless sensors in biomedical and consumer electronics has been a major driving force behind the significant research done in this area. Energy scavenging from radio waves [1-5], vibration [6-9], electric field generated by over-head power line [10-11], thermal gradient and photons has been reported using MEMS devices. Ashebo *et al.* [12] reported a concept to convert vibration energy from the bridges and pavements and obtained ~6 μW output power when subjected to an acceleration of 0.05 m/sec^2. Mongia and Abdelmoneum [13] proposed the conversion of heat energy 300-600 mW in

[2]To whom any correspondence should be addressed. On sabbatical leave from U of Utah.

mobile platform when operated at 35-50 W thereby saving 1% of the input energy, using micromachined thermopile. Marzencki *et al.* [14] reported thin film AlN cantilever micro generator with power output of 0.038 μW from a 0.5 g vibration at 204 Hz resonant frequency. Marzencki *et al.* [15] increased the vibration to 4g at 1368 Hz resonant frequency to generate a power of 1.97 μW. Shen *et al.* [16], reported a PZT cantilever MEMS-based micro generator with power output of 2.15 μW for a vibration of 2 g (g = 9.81 m/s^2) at resonant frequency of 461.15 Hz. Challa *et al.* [17] demonstrated output of 332 μW from a piezoelectric and electromagnetic coupled device.

In this paper we use a piezoelectric beam with a neodymium magnet to use magnetic field variations in addition to vibrations for power harvesting and transfer. It is also possible to add electrically "charged" regions as shown by Pai *et al.* [18] to enable electric fields powering as well. Apart from determining the amount of power available, we have also used isolated piezoelectric material to quantify energy available as a function of frequency. The ultimate aim is to develop a device that can be used to transfer maximum power from wireless devices, eliminating use of low Q MEMS coil. The objective of the paper is to determine the power transfer efficiency over a large frequency range.

2. Device fabrication and set-up

The amount of energy available in the environment is quite large but only a small portion of it can be converted to useful energy. Solar energy [19] alone can provide 15000 $\mu W/cm^2$ power density and 150 $\mu W/cm^2$ during cloudy conditions. This reduces to 6 $\mu W/cm^2$ indoors. The vibration energy on the other hand gives around 100-200 $\mu W/cm^2$, acoustic noise is around 0.003 dB, daily temperature variation can provide 10 $\mu W/cm^2$, and the temperature gradient in the surrounding accounts for 15 $\mu W/cm^2$ for 10 °C gradient. It is thus very important to understand how much energy can be detected and utilized by the energy harvesting devices. To demonstrate the same we used a set-up to excite the best device, which we obtained from section1, for further study. We used the Keysight (Agilent) Technologies 4395A Network/Spectrum/Impedance Analyzer for the study. The device, with the magnets and coils has been shown in the figure 1. The device was subject to different cases for the different frequency sweeps. For the 10-100 Hz sweep the case 1 was when input was applied to coil 1 and output was taken from coil 2. For case 2 the input was given to coil 1 and output was taken from PZT/magnet. In case 3 the input was applied to PZT/magnet and output from taken from coil 2.

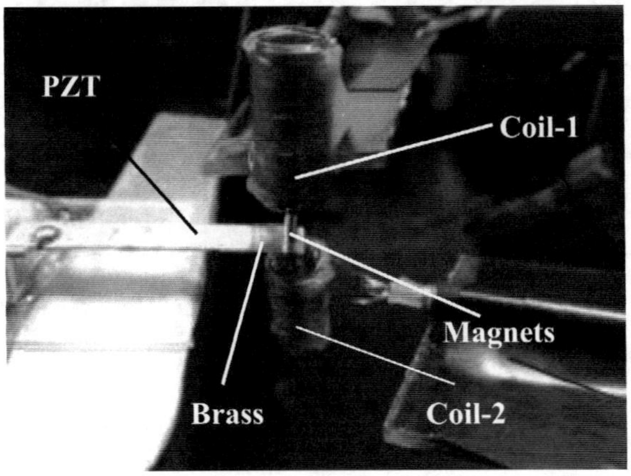

***Fig. 1**: Optical image of device, magnets and the coils used in the measurement of energy transfer. The beam is made of Non-Ferromagnetic material and has a PZT layer on it. The two coils used had 50 turns each.*

For the 100-5000 KHz sweep the cases were similar. The distance was measured from input to output/ (4 mm, 7 mm, and 12 mm). The bandwidth was 10 Hz, input power was 1 dBm (251 mV), input gain (RF Source) was 20 and the output gain (channel B) was 5. Figure 2 below shows the common set-up for all the cases. The input and output changed according to the case used.

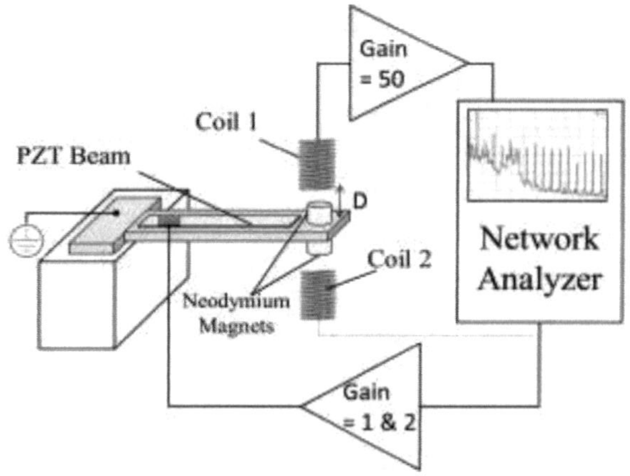

***Fig. 2**: Schematic of the device used for power transfer efficiency measurement. The device was tested under different excitation. (10-100Hz - case1: RF input was applied to coil 1 and output was taken from coil 2. Case 2: RF input to coil 1 and output from PZT. Case 3: RF input to PZT and output from Coil 2.)*

3. Experimental Results

We examined 3 cases. In case 2 we used a coil as the primary and a PZT/magnet as the secondary (Figure 3). In case 3, we reversed the roles of the coil and the PZT/magnet and used the coil as the secondary and the PZT/magnet as the primary (Figure 4). We also used a coil-coil coupling (case 1) through air for comparison (Figure 1). When the PZT/magnet was used as the primary, the vibration of the magnet produced a time dependent magnetic flux density that induced an EMF in the receiving (secondary) coil. Our devices were fabricated using PZT beams (Figure 1) cut from PZT discs. The beams had fundamental resonance frequencies below 100 Hz (Figures 3 and 4) to enable harvesting environmental vibrational energy as well. Higher frequencies (100s of KHz) can be designed if needed for ultrasonic powering and communication.

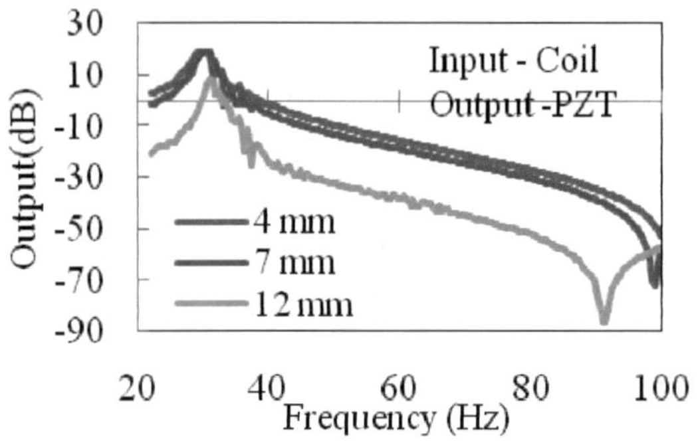

***Fig. 3**: Output spectra with a coil used for excitation (230 mV) and the output was taken from the PZT/magnet.*

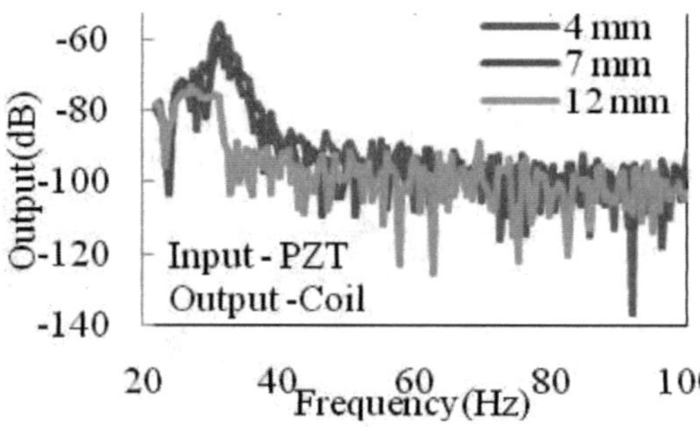

***Fig. 4:** Output spectra with PZT/magnet used for excitation and the output was taken from the coil.*

The frequency sweep from 100-5000 Hz was very informative as it gave multiple peaks for higher frequencies. The gap range in this case was only taken as 4 mm, 7 mm and 12 mm between the input and the output source. The bandwidth was set to 10 Hz and the input power was 251 mV for all the cases.

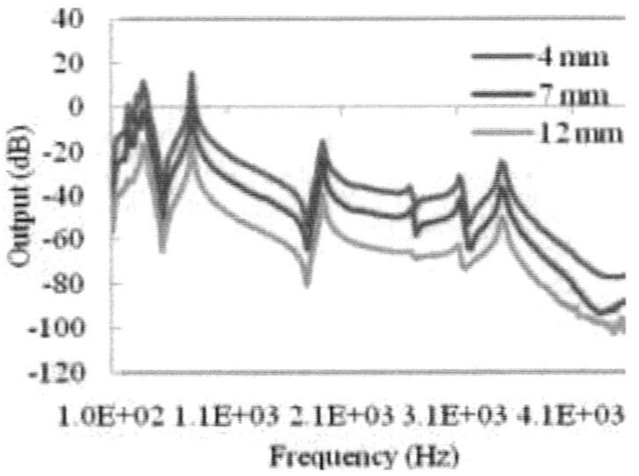

***Fig. 5**: The same as Fig. 3 but with higher frequency range.*

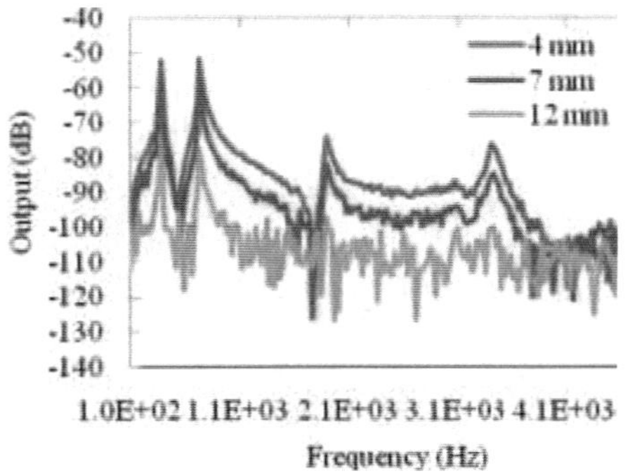

***Fig. 6**: The same as Fig. 4 but with higher frequency range.*

The device had multiple peak resonances and that were reproducible in all the cases at ~230 Hz, ~370 Hz, ~737 Hz, ~1890 Hz, and ~3430 Hz. The best response was 15 dB at 786 Hz for case 2 at 4 mm gap and -51.9 dB at 737 Hz for case 3.

4. Calculations and discussion

The power transfer efficiency is higher (~50 %) at lower frequencies (100 Hz) as given by the Friis's near-field magnetic field power formula:

$$P_{RX(H)} = \frac{P_{TX}G_{TX}G_{RX}}{4}\left(\frac{1}{(kr)^2}+\frac{1}{(kr)^4}\right), \quad (1)$$

Where P_{RX} is the received and P_{TX} is the transmitted powers, G_{TX} is the transmitter magnetic dipole gain, G_{RX} is the receiver dipole gain, $k = 2\pi/\lambda$ is the wave number with λ the wavelength, and r is the distance between the transmitter and the receiver.

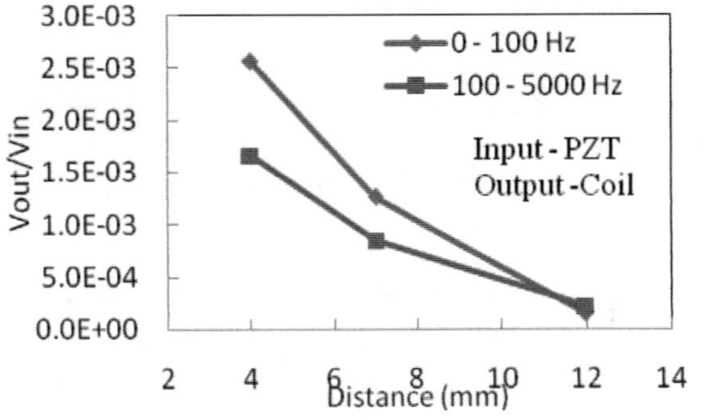

(a)

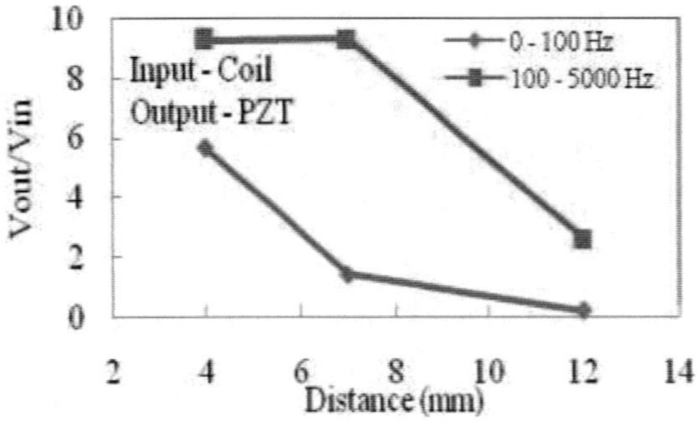

(b)

***Fig. 8**. Voltage transfer efficiency (averaged over different bandwidths) as a function of distance for a) excitation applied to PZT- magnet beam and coil produced the output and b) excitation applied to coil and PZT-magnet produced output.*

The results for power transfer over different frequency ranges are shown in Figures 8a and 8b. Three different measurements were performed: a) coils were used to transmit and receive power, b) a coil generated time varying magnetic signals and the PZT-magnet was the receiver, and c) PZT-magnet generated the magnetic signal and a coil received it. The first case is not shown here but it can be used to communicate signals from the implantable device. The best result was obtained when we used a coil to generate the magnetic signal and the PZT-magnet to receive it. We note that it is more convenient to micro fabricate high-Q piezoelectric beams than coils that tend to be very lossy at micro-scale. Our work reported here introduces the possibility of using a PZT-magnet beam for communicating with implantable devices as well as powering it efficiently.

5. Conclusion

A new resonant power transfer/harvesting device consisting of a primary coil and a secondary PZT/magnet is discussed. The best power transfer efficiency (52%) was achieved at 786 Hz for output taken from the PZT beam at 4 mm gap between the coil and the beam. For the 10-100 Hz 19.3 dB (54% efficiency) was obtained at ~30 Hz for 2 mm gap between the input and output source. The receiving part of power transfer/harvester device discussed here can be fabricated with very small dimensions unachievable for coils. It can also have high quality factor improving the power transfer efficiency and can be used to harvest mechanical as well as magnetic field fluctuations in the environment.

References

[1] Kotani K, Bando T, and Sasaki Y 2012 *Proceedings of the 2012 IEEE Asian solid state circuits conference (A-SSCC)* p 329-32
[2] Aminov P, Agrawal J P 2014 *IEEE 64th Electronic components and technology conference (ECTC)* p 1838-41
[3] Lu Y, Wang W, Zhang Z, Huang A, Lau V K N 2014 *IEEE Global communications conference. GLOBECOM* p 923-8
[4] El-Anzeery *et al.* 2012 *IEEE International power engineering and optimization conference* p 209-13
[5] Bait-Swailam M M, Nadir 2014 *International journal of communications* **8** p 1-7
[6] Triplett A and Quinn D 2009 J*ournal of intelligent material systems and structures* **20** p 1959-1967
[7] Wasa K, Matsushima T, Adachi H, Kanno I, Kotera H 2012 *Journal of microelectromechanical systems* **21** no. 2
[8] Fakhzan M N, Muthalif A G A, 2011 *4th International Conference on Mechatronics (ICOM)* **7**
[9] M.N Fakhzan, Asan G. A. Muthalif 2013 *Mechatronics* **23** 1 p 61-6
[10] Kim H, Choi D, Gong S, and Park K 2014 *Electronics Letters* **50** 17 p 1236-8
[11] He W, Li P, Wen Y, Zhang J, Lu C and Yang A 2013 *Review of Scientific Instruments* **84** 10 p 105004 (9 pp.)
[12] Ashebo D B, Tan C A, Wang J, Li G 2008 *SPIE Proceedings* **6934** *Progress in NDE*
[13] Mongia R and Abdelmoneum M *2010 Proceedings 2010 International Conference on Energy Aware Computing (ICEAC 2010)* p 4 pp.
[14] Marzencki M, Basrour S, Charlot B, Colin A, Valbin L 2005 *DTIP'05, Montreux, Switzerland*
[15] Marzencki M, Ammar Y, and Basrour S 2008 *Sensors and Actuators a- Physical* **145** p 363-370
[16] Shen D, Park J H, Ajitsaria J, Choe S Y, Wikle H C, and Kim D J 2008 *Journal of Micromechanics and Microengineering* **18**
[17] Challa V R, Prasad M G and Fisher F T 2009 *Smart Materials and Structures* **18** n 9 p 095029 (11 pp.)
[18] Pai P and Tabib-Azar M 2013 *IEEE MEMS* 572-575
[19] Külah H and Najafi K 2008 *IEEE SENSORS JOURNAL* **8**

Silicon MEMS bistable electromagnetic vibration energy harvester using double-layer micro-coils

P Podder, P Constantinou, D Mallick and S Roy*

Micropower Systems and Nanomagnetics Group, Micro-Nano Systems Centre, Tyndall National Institute, Lee Maltings, Dyke Parade, Cork, Ireland

E-mail: saibal.roy@tyndall.ie

Abstract. This work reports the development of a MEMS bistable electromagnetic vibrational energy harvester (EMVEH) consisting of a silicon-on-insulator (SOI) spiral spring, double layer micro-coils and miniaturized NdFeB magnets. Furthermore, with respect to the spiral silicon spring based VEH, four different square micro-coil topologies with different copper track width and number of turns have been investigated to determine the optimal coil dimensions. The micro-generator with the optimal micro-coil generated 0.68 micro-watt load power over an optimum resistive load at 0.1g acceleration, leading to normalized power density of 3.5 kg.s/m^3. At higher accelerations the load power increased, and the vibrating magnet collides with the planar micro-coil producing wider bandwidth. Simulation results show that a substantially wider bandwidth could be achieved in the same device by introducing bistable nonlinearity through a repulsive configuration between the moving and fixed permanent magnets.

1. Introduction

With the progress in CMOS integrated circuit technology, the power consumption of individual electronic devices has been lowered to the level of few micro-watts. These advancements in low power electronics have facilitated the drive towards the development of the 'internet of things (IoT)' in the forthcoming decades, comprising billions of low power autonomous sensor modules connected to the internet via wireless networks. However, one of the major roadblocks towards the deployment of the IoT is the lack of appropriate energy sources to power the individual sensor modules for long period. Ambient mechanical vibrations, which are abundant in the modern urbanized landscape, provide an untapped energy source which can be harvested to power wireless sensor nodes for autonomous operation [1-3]. Nonetheless, most of the ambient vibrations are either random in nature, or have frequencies spread across a considerably wide range, which are difficult to harvest using linear oscillator based vibration energy harvesters. Additionally the power output of the micro-fabricated VEH devices utilizing piezoelectric transduction is often too low to power sensor modules. An alternative solution to this problem would be the adoption of electromagnetic transduction [4, 5] and nonlinear bi-stable oscillator based VEH device [6, 7], which typically has a wider operational bandwidth. In this paper we present the design, fabrication and experimental validation of a MEMS electromagnetic VEH device utilizing double-layer planar micro-coils. Four different types of planar micro-coils, each with different number of turns and copper track width within the same area, have been investigated to experimentally determine the optimum coil layout. The micro-VEH generated 0.68 μW power at 0.1g acceleration for the optimum coil and resistive load configuration.

* Corresponding author

2. Design and simulation of micro-EMVEH

The proposed micro electromagnetic vibrational energy harvester comprise of silicon spring, tiny NdFeB magnets and double-layer planar copper micro-coil. The silicon spring and micro coils are fabricated separately using standard MEMS fabrication process and all components are assembled to form the micro-EMVEH. As illustrated in figure 1(a), bistable nonlinearity is incorporated in the simulation by means of repulsive interaction between a pair of additional magnets placed on the frame. Figure 1(b) exhibits the simulated stress profile in the silicon spring structure when the central pad is deflected by 1mm. The maximum stress (750 MPa) is shown to be well below the yield stress of silicon (7000 MPa). The numerically simulated frequency responses for two different micro-coils, type 2 and 3 (see Table 1) are illustrated in figure 1(c). It is shown that while the linear configuration of the device produces resonant peak in the frequency response, the nonlinear bistable configuration achieves a hysteretic response with wider bandwidth.

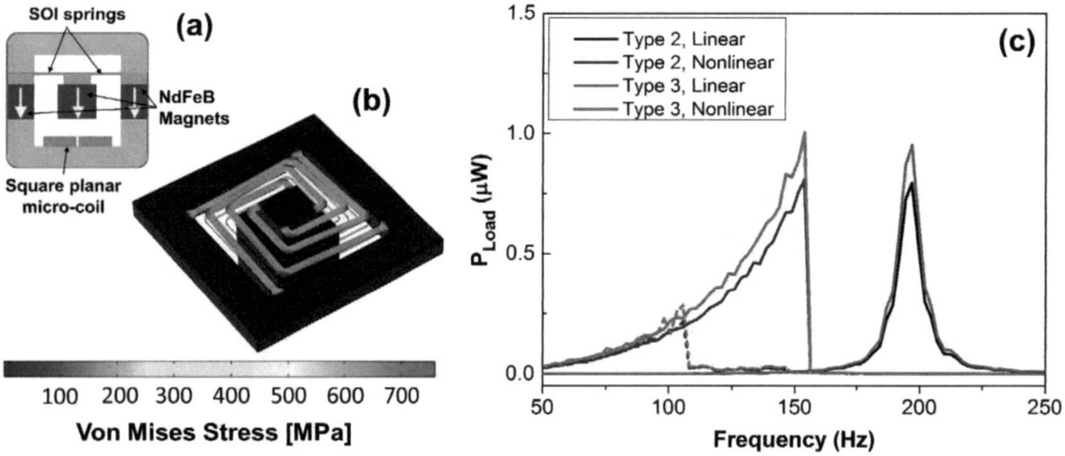

Figure 1. (a) Schematic diagram of the proposed micro-EMVEH, (b) mechanical simulation showing the von Mises stress distribution and (b) frequency responses for coil types 2 and 3. Solid lines represent the forward sweep and dashed lines represent the reverse sweep plots respectively.

3. Fabrication of micro-EMVEH

The micro-fabricated EMVEH consists of three principal components, the silicon spring, the planar micro-coil and the miniaturized NdFeB magnets. The silicon springs and micro-coils are fabricated separately using silicon MEMS process and post-fabrication, these two components together with the miniaturized magnets are assembled to form the complete micro-EMVEH system.

3.1. Fabrication of silicon spring

The silicon springs (Figure 2(a, b)) are fabricated on double-side polished 0.5mm thick SOI (silicon-on-insulator) wafer with SOI, buried oxide and handle wafer thicknesses of 50μm, 3μm and 450μm respectively. The fabrication process begins with transferring the spring patterns onto the SOI layer using photolithography and etching the spring patterns using DRIE process. In the next step, the back side (handle wafer) is etched using DRIE to release the springs. Following release of the springs, the delicate spring structures are protected by sputtered aluminium (0.2 μm) and sprayed photoresist. The wafers are then diced with individual die size of 7mm × 7mm and a movable central paddle (1.1mm × 1.1mm) supported by four spiral springs (50μm thick, 0.2mm wide). Finally, the protective aluminium and sprayed photoresist layers are etched to obtain the individual springs.

3.2. Fabrication of double-layer planar copper micro-coil

The double layer planar micro-coils (Figure 2(c, d)) are fabricated using photolithography and electro-deposition of copper on sputtered seed layers. The two coil layers are separated by an insulating SU8 layer and connected at the centre through a via which traverses through the insulation layer. The thicknesses of the bottom and top electroplated copper coil layers are 15μm and 12.5μm respectively. The overall area of the coils is 2.8mm × 2.8mm where the total number of turns, for both layers and the corresponding measured resistances are listed in Table 1. A passivation layer of SU-8 was used to protect the coil from shorting or damaging due to contaminants. The two ends of the copper coils are terminated in two copper pads to provide contact with PCB through wire bonding or soldering.

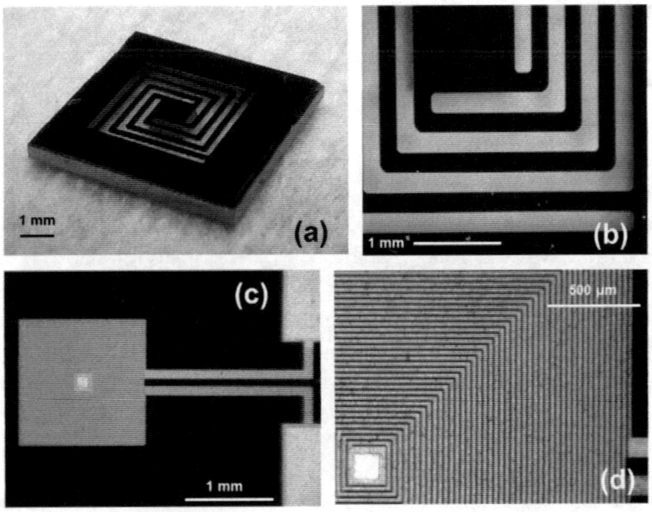

Figure 2. Fabricated (a, b) silicon-on-insulator spring structure and (c, d) planar micro-coil

3.3. Assembly of the silicon micro-EMVEH

The silicon spring and double-layer planar micro-coils together with the miniaturized NdFeB magnets are assembled to form the complete micro-EMVEH system. The planar micro-coil is bonded using epoxy adhesive on an FR4 PCB board (10mm × 15mm) with copper pads onto which the copper coil pads are wire-bonded. An NdFeB magnet (2mm × 2mm × 2mm) is bonded on one side of the movable central paddle of the silicon spring structure. The silicon spring with the NdFeB magnet is aligned above the centre of the copper coil such that the magnetic flux lines pass through the coil. Two Perspex spacers are positioned in between the coil and frame of the spring such that the vertical gap between the micro-coil plane and the magnet is 0.5mm. The magnet can move vertically (out-of-plane) and induce current in the micro-coil due to change of flux-linkage.

Table 1. Micro-coil parameters

Coil	Copper Track width (μm)	Inter Track gap (μm)	No. of turns	Resistance (Ohm)
Type 1	15	10	104	57.7
Type 2	12.5	12.5	104	95.6
Type 3	10	10	130	127.7
Type 4	10	8	144	156

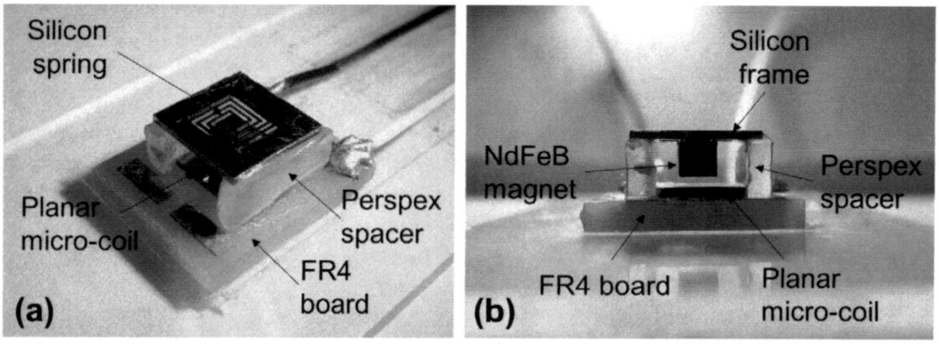

Figure 3. Fully assembled micro-EMVEH (a) perspective view showing the silicon spring (b) side view showing the NdFeB magnet suspended above the planar micro-coil.

4. Experimental results and discussions

Four different devices, each with the same magnet and silicon spring configurations but different micro-coils are fabricated and tested. The parameters of the different coils are given in Table 1. In the first set of experiments the optimum load resistances of the individual devices were obtained for a linear configuration of the device. It is observed that the optimum resistive load is very close to the resistances of the individual coils (Figure 4). This is due to the fact that the weak magnetic flux linkage between the magnet and coils result in weak electromagnetic coupling coefficient, leading to low electromagnetic damping in comparison to the mechanical damping which has negligible effect on the optimum load.

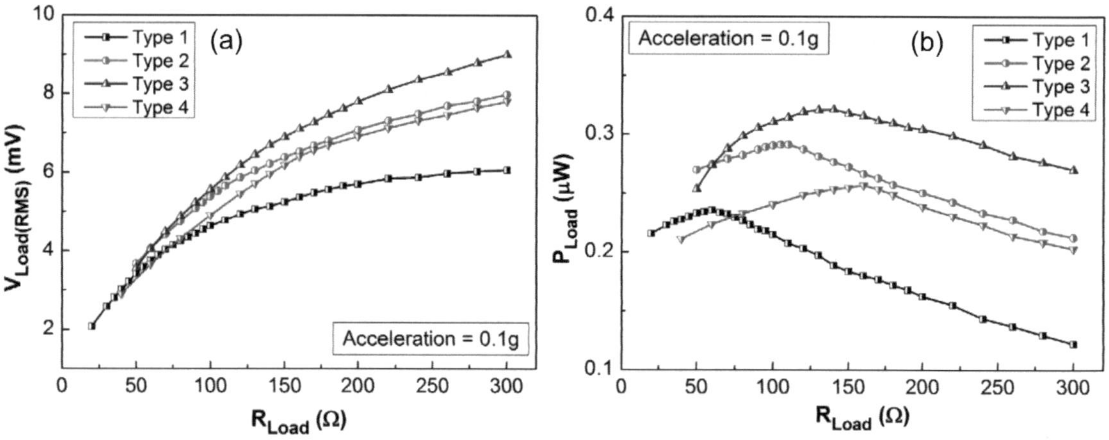

Figure 4. Variations of (a) RMS load voltage and (b) Load power with load resistance for different types of micro-coil at 0.1g acceleration.

The RMS load voltage and the load powers of the different devices at 0.1g acceleration are shown in figure 4. At 0.1g acceleration, the device with Type 3 micro-coil generates the highest power (0.32 μW) while the device with Type 1 micro-coil produces the smallest power (0.24 μW). Experimentally, the Type 3 micro-coil is the optimum for this device as this configuration generates the highest voltage and power across the optimum load among all the configurations, which could be due to higher packing factor and optimum coil geometry. The experimental test set up, comprising of electromagnetic shaker, power amplifier, accelerometer and computer controlled vibration controller is schematically illustrated in figure 5(a). The output voltage responses of the micro-EMVEH are recorded using a digital storage oscilloscope and stored in a computer.

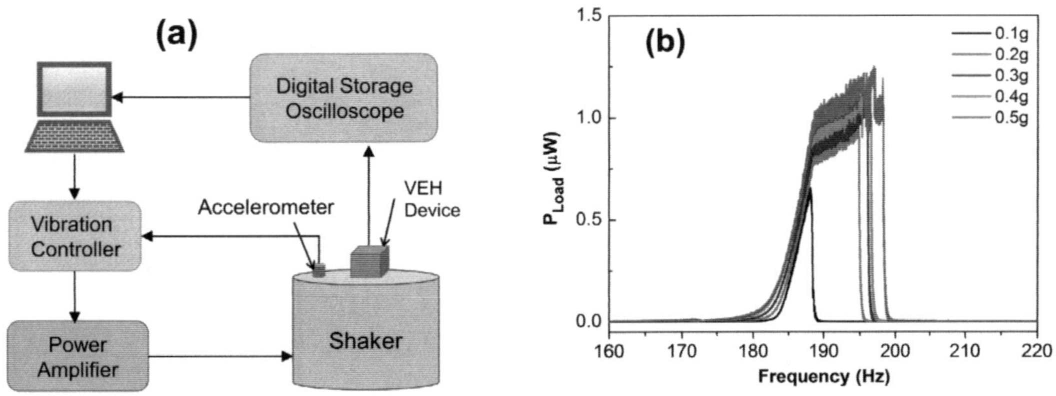

Figure 5. (a) Vibration test set up. (b) Load power vs frequency responses for different acceleration.

The experimental frequency response plots for the device with coil type 3 for different accelerations are shown in figure 5(b), which exhibits nonlinear behaviour due to stretching of the spiral beams, producing 0.68 μW power at 0.1g acceleration. As the acceleration is increased, the magnet starts to collide with the coil and the peaks of the frequency responses for accelerations 0.2g-0.5g are flattened, leading to wider bandwidth of 8-12 Hz. This type of impact induced bandwidth widening has been explored at macro scale [8, 9] for electromagnetic systems and at micro scale [10, 11] for electrostatic transduction systems previously. It is evident from figure 5(b) that the mechanical impact phenomenon can also be exploited for bandwidth widening at micro-scale using electromagnetic transduction.

5. Conclusions

This work describes the design, fabrication and experimental validation of a nonlinear electromagnetic micro-power generator. Four different types of double-layer planar micro-coils are tested at the same vibrational acceleration (0.1g) and the power generated in the optimum coil is 0.68 μW. Furthermore, it is observed that at higher vibrational accelerations, the movable magnet collides with the planar coil, resulting in wider bandwidth in comparison to a non-impact linear configuration. This work forms the background for further development in the device topology by incorporation of magnetically induced bistable nonlinearity and investigation of its effect on the device performance.

Acknowledgements

This work is financially supported by Science Foundation Ireland (SFI) Principal Investigator (PI) project on 'Vibrational Energy Harvesting' grant no. SFI-11/PI/1201. The authors also would like to acknowledge the help provided by the Central Fabrication Facility (CFF) and their staff in Tyndall National Institute for silicon MEMS fabrication.

References

[1] N Mitton et al, EURASIP Journal on Wireless Communications and Networking 2012, 2012:247.
[2] S Roundy, Journal Of Intelligent Material Systems And Structures, Vol. 16, October 2005.
[3] S P Beeby et al, Journal of Micromechanics and Microengineering 17 (2007) 1257–1265.
[4] S Kulkarni et al, Journal of Applied Physics, 99, 08P511, 2006.
[5] S Kulkarni et al, Sensors and Actuators A: Physical, Volumes 145–146, July–August 2008, Pages 336–342.
[6] F Cottone et al, Phys. Rev. Lett., vol. 102, p. 080601, 2009.
[7] M Ferrari et al, Sens. Actuators A, Phys., vol. 162, pp. 425–431, 2010.
[8] M. S. M. Soliman et al, J. Micromech. Microeng., vol. 18, p. 115021, 2008
[9] P Podder et al, Mechatronics, IEEE/ASME Transactions on, (DOI: 10.1109/TMECH.2015.2451016), 2015.
[10] C. P. Le et al, J. Micromech. Microeng., vol. 22, p. 074013, 2012.
[11] P. Basset et al, J. Micromech. Microeng., vol. 24, p. 035001, 2014.

SMART DESIGN SELFTUNING PIEZOELECTRIC ENERGY HARVESTER INTENDED FOR GAS TURBINES

L G H Staaf, E Köhler, M Soeiro, P Lundgren and P Enoksson

Department of Microtechnology and Nanoscience, Chalmers University of Technology, 412 96 Gothenburg, Sweden

Corresponding author: v96staaf@chalmers.se

Abstract. Piezoelectric energy harvesting on a gas turbine implies constraints like high temperature tolerance, size limitation and a particular range of vibrations to utilise. In order to be able to operate under these conditions a harvester needs to be small and efficient and to respond to the appropriate range of frequencies. We present the design, simulation and measurements for a clamped-clamped coupled piezoelectric harvester with a free-sliding weight which adds self-tuning for improved response within the range of vibrations from the gas turbine. We show a peak open circuit voltage of 11.7 V and a 3dB bandwidth of 12 Hz.

1. Introduction

The coming of the internet of things requires that we now develop alternative energy sources to replace or support today's and tomorrow's batteries. Alternative energy sources such as energy harvesters have been focused by many researchers during the past years [1]. Energy harvesters convert ambient energy in our surrounding, like solar irradiation, wind, heat and mechanical vibrations into electric energy. For a gas turbine the main ambient energies are heat and mechanical vibrations. The thermal energy is concentrated to certain areas within the gas turbine, while vibrations are more or less available everywhere, making piezoelectric energy harvesting a viable option. The main challenge for piezoelectric energy harvesting is to maintain a sufficient power output over a broad bandwidth; much research has been done on broadening the harvesting bandwidth [1-2]

In this paper we report on the design, simulation and measurements of a piezoelectric harvester with self-tuning for wider bandwidth and coupled piezoelectric cantilevers to maintain a high power output by extended strain distribution. The harvester is designed to meet the gas turbines specific conditions on size, temperature and frequency. Previous work [3] has shown that by utilizing a distributed stress pattern over the whole cantilever the complete area of the piezoelectric cantilever is used and not only the clamped end, thereby yielding a higher power output. In previous work [4] this distributed stress pattern was observed for a cantilever with one end fixed and the other end coupled to a second (top) cantilever (which showed a stress pattern similar to a single cantilever with one end fixed). In the present design both cantilevers are clamped and coupled at each end, hence using all available piezoelectric capacity to optimize the power output. To extend the bandwidth, passive self-tuning is introduced by a free sliding weight. In previous work a free sliding proof mass has been used on a thin fixed-fixed beam. When the proof mass slides to one end of the thin beam it can get stuck there even if the frequency is changed [5-6]. To avoid this problem we have made our middle beam thicker, in order to lower or remove the impact of the natural frequencies that might force the weight to one end. The movement range of the proof mass has also been limited to prevent the occurrence of it getting stuck at one end.

2. Simulation

The harvester was modelled in the simulation tool COMSOL. Figure 1 shows a schematic image of the harvester; it consists of two piezoelectric cantilevers that are clamped to an attachment at one end.

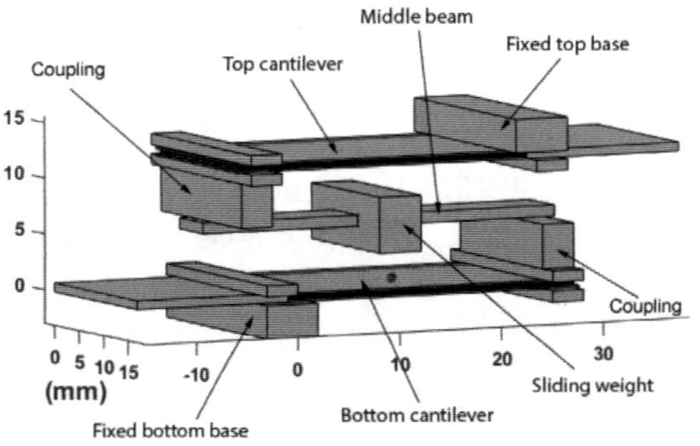

Figure 1. Schematic image of the harvester setup in COMSOL.

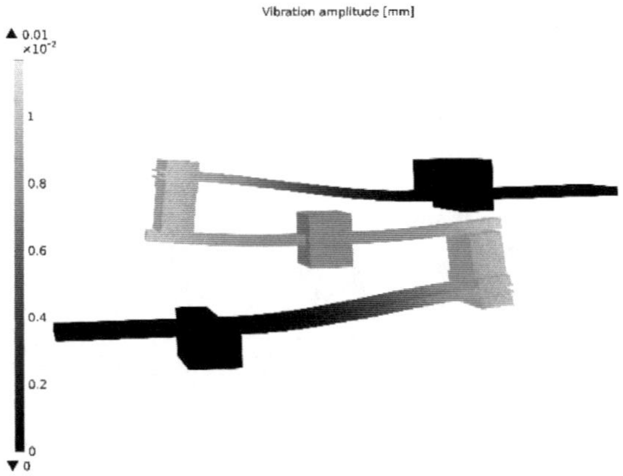

Figure 2. The mode shape at resonance frequency at 380 Hz from simulation

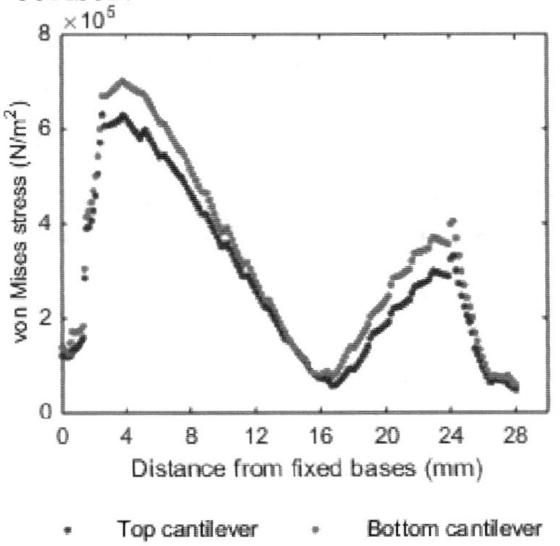

Figure 3. Simulated distributed stress over top and bottom cantilever.

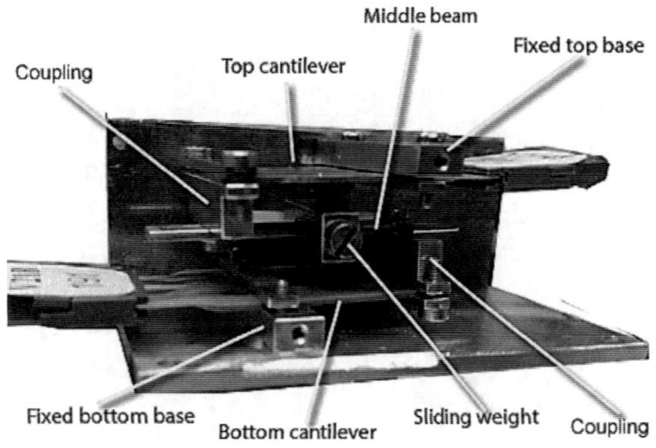

Figure 4. The setup of the harvester, at measurement 1, tuned to 373 Hz

On the other end the cantilevers are attached to a coupling that connects the two cantilevers by an aluminium beam where a sliding weight is placed.

The targeted frequency range is 370 - 380 Hz with self-tuning from the weight. The simulated harvester is tuned to 380 Hz, with a fixed weight at the centre of the middle beam, and the simulation provides the required length of the cantilevers on the real harvester. In figure 2 the mode shape of the resonance frequency mode at 380 Hz is shown. From figure 2 we can conclude that the cantilever attachments and the couplings are not colliding and thus not creating any collision-induced damping.

In figure 3 the stress pattern for the cantilevers is simulated. Both cantilevers have similar stress patterns: higher at the clamped end and less at the coupled end. As shown in previous work [3] the coupling is similar to clamping and a higher open circuit voltage for the piezoelectric cantilever is reached by it utilizing a larger area of stress.

3. Experimental setup

The characterization of the harvester is done on a sinusoidal excited shaker table. The cantilevers electrodes are connected to multimeters measuring the open circuit voltage over the frequency span 330-410 Hz. In figure 4 the harvester assembly is shown. The front wall is removed for a better view from the side. In measurement 1 (M1), figure 4, the cantilevers are 24 mm long and the length of the middle beam is 16 mm. The coupling weight is in total 3.87 g, where the sliding weight is 0.87 g. The active middle beam is 1 mm thick, 3 mm broad and 16 mm long. The natural frequency of the middle beam is 2974 Hz, which is far above our targeted frequency range of 370 – 380 Hz.
In measurement 2 (M2), figure 5, the top and bottom cantilever lengths are 26 mm and 28 mm respectively. The middle beam is 1 mm thick, 3 mm wide and 27.5 mm long. The natural frequency of the middle beam is 1006 Hz, still above our targeted frequency range of 370 – 380 Hz.

4. Result

Table 1. Comparison of measurement 1 (M1) and measurement 2 (M2).

	Primary mode (Hz)	*Open circuit voltage (V)*	3 dB open circuit voltage (V)	*3 dB bandwidth (Hz)*
M1 16 mm beam, sliding weight	373	3.12	2.2	12
M1 16 mm beam, fixed weight	378	3.26	2.3	8
M2 27.5 mm beam, loose sqrew	373	11.2	7.9	10
M2 27.5 mm beam, extremely loose screw	372	11.71	8.28	12
M2 27.5 mm beam, fixed weight	378	11.5	8.13	8

4.1 Measurement 1

The first measurement series was made with 370 - 380 Hz as target frequency. The setup was manually tuned to a resonance frequency of 373 Hz. Even though the movement of the proof mass was visibly much more limited than expected, a small effect could be seen when comparing to a completely fixed weight. In figure 6 the measured result (M1) for the total output with non-fixed and fixed weight with a 16 mm long middle beam is shown for the primary frequency mode of 373 Hz with a 3 dB bandwidth of 12 Hz. The measured open circuit voltage output peaks at 3.1 V and 2.2 V at the 3 dB border. The self-tuned harvester has nearly as high voltage output as the harvester with the fixed weight, but shows a near 50% increase in bandwidth as shown in table 1.

4.2 Measurement 2

From the first measurement we could conclude that the non-fixed weight can increase the bandwidth compared with a fixed weight. Shorter cantilevers are harder to tune than longer, therefore new simulations were performed with longer cantilevers. For the second measurement three parameters were changed, two based on simulation results and one mechanical (a more reliable weight fixture). The middle beam was set to be as long as possible (27.5 mm) due to the current setup geometric constraints as shown in figure 5. In this setup the top cantilever is 26 mm long and the bottom one is 28 mm long. The target frequency was still 370 - 380 Hz; with longer cantilevers and longer middle beam the harvester showed resonance at 372 Hz.
In the second measurement the new weight fixture screw was tested for two different settings; *loose* (-½ screw turn from fixed condition) and extremely loose (unscrewed). For both settings, the weight could be seen to move to different positions when the frequency was changed. By slowly increasing the frequency we could see that after 2-3 seconds at a specific frequency, the weight moved to another

position, where it henceforth remained if the frequency was kept constant. At the start of a measurement sweep the weight was positioned at the middle of the beam. When the frequency reached 357 Hz it moved to the left end (refer to figure 5) and at 368 Hz it started to slide back and forth up to 372 Hz from when it remained stable at the left side. When the frequency sweep was downwards the sliding weight positioned itself accordingly at the same frequency as on the sweep upwards. In figure 6 the open voltage output for the screw settings are compared; it is inferred that a loose screw setting seems preferable.
Compared to the first measurement we can conclude that the second setup had approximately 3.5 times higher open circuit voltage output of 11.7 V (table 1) for the same volume, which might occur from that the natural frequency of the middle beam which is more close to the resonance frequency of the harvester on measurement 2 than on measurement 1. Hence with a lower natural frequency for the middle beam a higher impact on the open circuit voltage might occur. Thinner middle beams with lower natural frequency have to be tested for further evidence of this theory. The bandwidth was the same 3 dB bandwidth range of 12 Hz as in measurement 1.

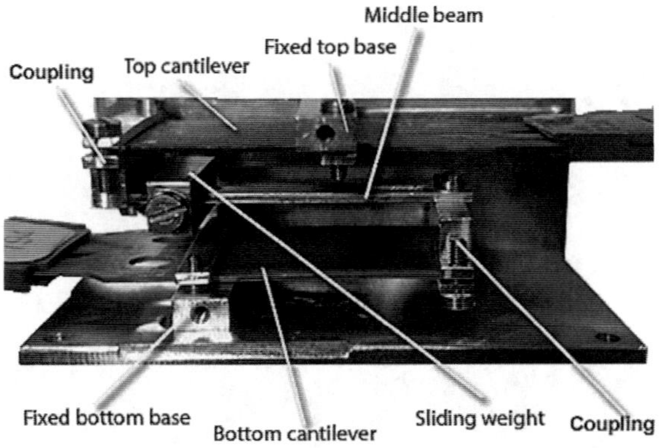

Figure 5. Measurement 2 setup with 27.5 mm middle beam and longer top (26 mm) and bottom (28 mm) cantilevers compared with measurement 1 setup.

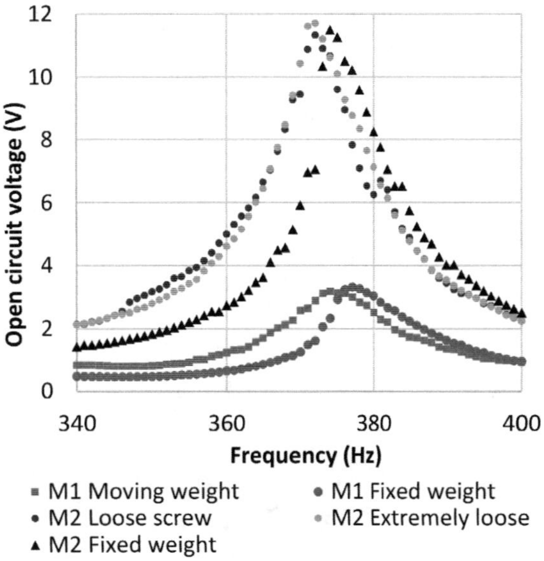

Figure 6. Second measurement (M2) open circuit voltage output with different screw tightness and compared with measurement 1 (M1) which had a 16 mm middle beam and 3.5 times lower output.

5. Conclusion

Our second generation of coupled harvester displayed a bandwidth of 12 Hz and an open circuit voltage of 11.7 V, which is 3.5 times higher, with the same volume, than our first harvester implementation. The higher open circuit voltage might occur from that the natural frequency of the middle beam is closer to the resonance frequency of the harvester on the second measurement. Future harvester improvements are foreseen when tuning 1) the proof mass weight, 2) the stiffness of the couplings and 3) the thickness of the middle beam. Despite improvements remaining to be implemented, the tested harvester demonstrates the bandwidth broadening impact of a passively sliding weight and high open circuit voltage which looks promising for our gas turbine application.

Acknowledgment

The European project STARGATE is greatly acknowledged for their financial support.

References

[1] A E Kubba and K Jiang, 2014, Efficiency enhancement of a cantilever-based vibration energy harvester, *Sensors 2014, 14, 188-211, DOI:10.3390/s140100188*

[2] H Wu, L Tang, Y Yang and C K Soh, 2014, Development of broadband nonlinier two degree of freedom piezoelectric energy harvester, *Journal of Intelligent Material Systems and Structures, vol 25(14) 1875-1889*

[3] L G H Staaf, E Köhler, D Parthasarathy, P Lundgren, P Enoksson, 2015, Simulation and experimental demonstration of improved efficiency in coupled piezoelectric cantilevers by extended strain distribution, *Sensors and Actuators A: Physical Volume 229, Pages 136–140*

[4] L G H Staaf, E Köhler, P Lundgren, P Enoksson, 2014, Modelling and experimental verification of more efficient power harvesting by coupled piezoelectric cantilevers, *PowerMEMS 20141742-6588). Vol. 557,*

[5] C G Gregg, P Pillatsch, P K Wright, 2014, Passively Self-tuning Piezoelectric Energy Harvesting System, *PowerMEMS 2014, Journal of Physics: Conference Series 557*

[6] P Pillatsch, L M Miller, E Halvorsen, P K Wright, E M Yeteman and A S Holmes, 2013, Selftuning behavior of a clamped-clamped beam with sliding proof mass för broadband harvesting, *Journal. of Phyis Conference Series 476 (2013) 012068*

Study on Tire-attached Energy Harvester for Low-speed Actual Vehicle Driving

Y. Zhang[1], R. Zheng[1], T. Kaizuka[1], D. Su[1], K. Nakano[2]

[1] Institute of Industrial Science, The University of Tokyo, 4-6-1, Komaba, Meguro, Tokyo, 153-8505, Japan
[2] Interfaculty Initiative in Information Studies, The University of Tokyo, 4-6-1, Komaba, Meguro, Tokyo, 153-8505, Japan

Email: zysgluck@iis.u-tokyo.ac.jp, topzrc@iis.u-tokyo.ac.jp, tkaizuka@iis.u-tokyo.ac.jp, sudx@iis.u-tokyo.ac.jp, knakano@iis.u-tokyo.ac.jp

Abstract. This study reports a tire-attached energy harvester, in which a cantilever beam pasted piezoelectric film and magnets with the same polarity are fabricated as a bistable vibrating system, for low-speed actual-vehicle driving. As the wheel rotates, the energy harvester is subjected to the noise produced from the interaction between the paved road and the rotating tire, and tangentially gravitational force as a periodic input can be applied to achieve the occurrence of stochastic resonance. Stochastic resonance can significantly stimulate the response of the bistable vibrating system, and therefore enhance the energy harvesting efficiency.

1. Introduction

Several researchers have developed bistable energy harvesters, which performs a single degree of freedom behavior in a certain direction to exploit stochastic resonance on a vibrator [1]. In this approach, a novel system presented can be excited to vibrate under the condition of double degrees of freedom by the rotating motion; thereby, the periodic force is autonomously self-excited from a combination of gravitational effects and the rotational degree of freedom without consuming additional energy.

Moreover, this energy harvesting model is exposed under low-speed actual noise instead of typical Gaussian noise [2], ensuring that this novel bistable device has the ability to perform a real behaviour for the realization of occurrence of stochastic resonance. Furthermore, comparison with other researches related to nonlinear bistable systems applied on the tire environment [3], with consideration of the theory of stochastic resonance in this research [4], the vibrating responses become significantly strong to generate more available energy, where the energy captured from not only the road ambient, but also its gravitational acting, especially at the lower frequency filed when one potential well motion of nonlinear bistable system is dominant under the only periodical force excitation.

Based on experiment results, when the angular frequency of tire revolution is 34.5 rad/s, which is close to the value theoretically predicted for the stimulation of stochastic resonance, the system can maintain the inter-well motion going all the time, and the displacement responses maintain the highest level. Hence, it is demonstrated that the prototype of the proposed energy harvester can be efficiently utilized under the low-speed actual-vehicle driving circumstances.

2. Mathematic model

A new energy harvester model is presented for adjusting Kramers rate [5] to the corresponding angular velocity of automobile tires by tuning the distance d between two magnets, when tire wheel revolves at different speeds, as shown in Figure 1. In order to eradicate all passive centrifugal force effect on the stiffness of the cantilever beam, the tip mass of the proposed energy harvester is mounted on the center axis of the tire wheel. By using a dipole mode [6], the analytic expression of the resultant restoring force of magnetic mass and cantilever beam in the tangential direction is derived and can be expressed as

$$F(x)=\frac{\mu_0 v^2}{4\pi}\left(\frac{3[M_{cy}\left(dM_{fx}+xM_{fy}\right)+M_{cx}\left(dM_{fy}+3xM_{fx}\right)]}{\left(d^2+x^2\right)^{5/2}}-\frac{15x\left(xM_{fx}+dM_{fy}\right)\left(xM_{cx}+dM_{cy}\right)}{\left(d^2+x^2\right)^{7/2}}\right)-kx\,, \quad (1)$$

where μ_0 is the permeability of free space, v is the volume of the magnets, and $M_f=\left(M_{fx},M_{fy}\right)$, $M_c=\left(M_{cx},M_{cy}\right)$ are the magnetization amplitudes of permanent magnet attached on frame and the magnetic tip mass, k is the linear coefficient of the beam, and x is the deflection of the tip magnet.

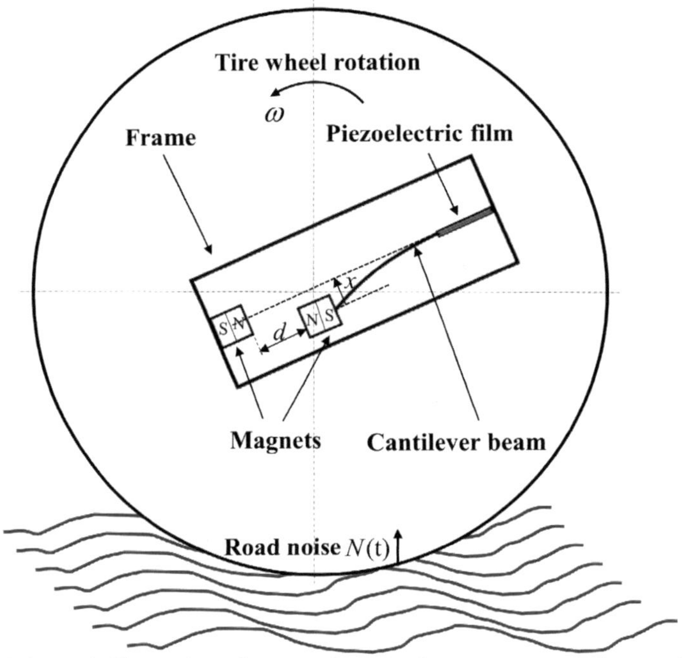

Figure 1. Schematic illustration of proposed energy harvester mounted on a rotating tire.

The equation (1) can be expanded into a Taylor series, calculated around $x=0$ and truncated as

$$F(x)=\left(\mu v^2\frac{9M_{cx}M_{fx}-12M_{cy}M_{fy}}{4\pi d^5}-k\right)x-\left(\mu v^2\frac{75M_{cx}M_{fx}-90M_{cy}M_{fy}}{8\pi d^7}\right)x^3\,. \quad (2)$$

From the reference [3], the duffing equation of tire rotation system can be acquired as

$$m\ddot{x}+c\dot{x}-ax+bx^3=N(t)+G\sin\omega t\,, \quad (3)$$

where m is the mass of the magnetic tip mass, c is the viscous damping coefficient of the beam, and $N(t)=\sqrt{2D}\xi(t)$ is the noise excitation from the road. By means of the rotation of tire the periodic force G will be produced from the effect of gravity on the end magnet, a is the combinational linear coefficient of the beam and magnets, and b is the nonlinear coefficient between two magnets. Thereby, the nonlinear restoring force can be given as

$$F_M' = -ax + bx^3 . \tag{4}$$

Therefore, corresponding to Eq. (2), Eq. (5) and (6) are obtained as follows:

$$a = \frac{\mu v^2 \left(\frac{3\left(3M_{cx}M_{fx} + M_{cy}M_{fy}\right)}{d^5} - \frac{15M_{cy}M_{fy}d^2}{d^7} \right) - 4\pi k}{4\pi} , \tag{5}$$

$$b = \frac{\mu v^2 \left(75M_{cx}M_{fx} - 90M_{cy}M_{fy}\right)}{8\pi d^7} . \tag{6}$$

Substituting Eq. (5) and (6) into Eq. (7) Kramers rate equation in the case of weak friction as

$$f = \frac{\left[\mu v^2 \left(\frac{3\left(3M_{cx}M_{fx} + M_{cy}M_{fy}\right)}{d^5} - \frac{15M_{cy}M_{fy}d^2}{d^7} \right) - 4\pi k\right]}{4\sqrt{2}\pi} \exp\left(- \frac{\left(4\pi k - \mu v^2 \left(\frac{3\left(3M_{cx}M_{fx} + M_{cy}M_{fy}\right)}{d^5} - \frac{15M_{cy}M_{fy}d^2}{d^7} \right)\right)^2}{\frac{8\pi D \mu v^2}{d^7}\left(75M_{cx}M_{fx} - 90M_{cy}M_{fy}\right)} \right) , \tag{7}$$

Hence, Kramers rate formula is derived as a function of the distance between two magnets.

3. Road noise measurement and simulation study

The smooth paved road and EV car (Coms, ZAD-TAK30-DS, TOYOTA AUTO BODY Co., Ltd) are taken into consideration as the experimental environments (see Figure 2). A wireless acceleration sensor (MVP-RF3-J, MicroStone Corporation) is installed on the inside of the front-wheel in where the road noise can be gauged vertically (see Figure 3). When the vehicle travels on the smooth paved road at the low speeds, the density of the accelerations almost keep the same level by the analyses of the measured signals. Hence, the acceleration measured around the speed of 20 km/h can be considered as the simulation condition. Based on Kramers rate, the stochastic resonance is easy to occur at the angular frequency which is smaller than 43rad/s.

Figure 2. Test car travelling on paved road.

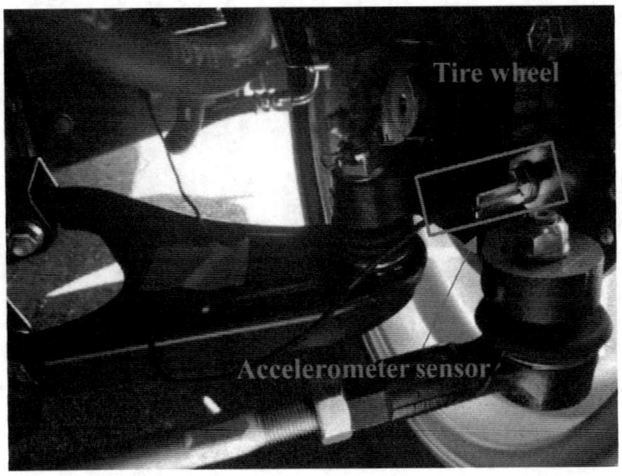

Figure 3. The road noise measurement on front-wheel.

From the implemented simulation results as depicted in Figure 4, under the input signals of gravitational acceleration and measured road noise, the responses of cantilever beam become strong owing to the inter-well motion between two stable positions, when the angular frequency of tire revolution is 38 rad/s, which is close to the value theoretically predicted for the stimulation of stochastic resonance, the system can maintain the inter-well motion going all the time, and the displacement and velocity responses maintain the highest level, among the five cases as shown in Figure 4(c). The availably estimated power reaches maximum of 3.5 mW with the effective value of

1.02 mW as depicted in Figure 5. Hence, it is demonstrated that the probability of the proposed energy harvester is verified under the low-speed actual-vehicle driving circumstances.

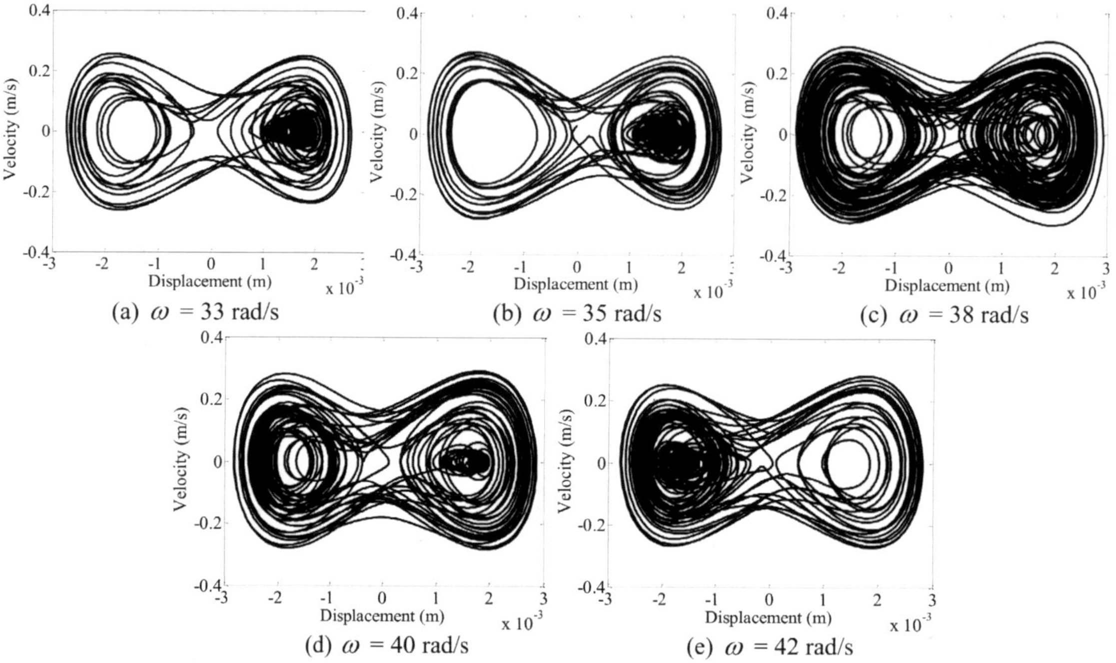

Figure 4. Displacement and velocity responses under paved road noise excitation and self-provided rotational force at different rolling angular frequencies: (a) 33 rad/s, (b) 35 rad/s, (c) 38 rad/s, (d) 40 rad/s, and (e) 42 rad/s.

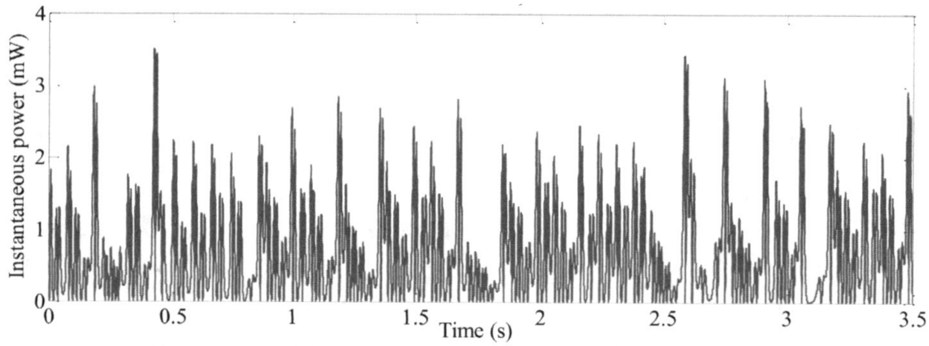

Figure 5. Availably estimated power under stochastic resonance.

4. Experimental investigation

A shaker (F-08000BDH/SLS16/Z02, EMIC Corporation) that can generate periodic vibration and noise vibration under the range of amplitude 100 mm_{P-P} and frequency 0 Hz – 2000 Hz is utilized to reappear the measured real-world road noise excitation as shown in Figure 6. The energy harvester installed on the shaker consists of a cantilever beam that has dimensions of 42 mm × 10 mm × 4 mm, and an adjustable screw rod that enables it to be optimal application for the automotive field through adjusting the distance between two magnets as described in figure 7.

By means of analysing the power spectral density of the measured road acceleration, it shows that the energy of paved road noise excitation is trapped in the frequency bandwidth of 0 Hz – 200 Hz, and the simulated road noise from the shaker is consistent with the real-world road noise by using a composite filter as compared in figure 8. Finally, as shown in figure 9, under both the gravitational effect and real-world road noise condition, the response of cantilever beam can be enhanced due to the occurrence of stochastic resonance at the angular frequency of 34.5rad/s.

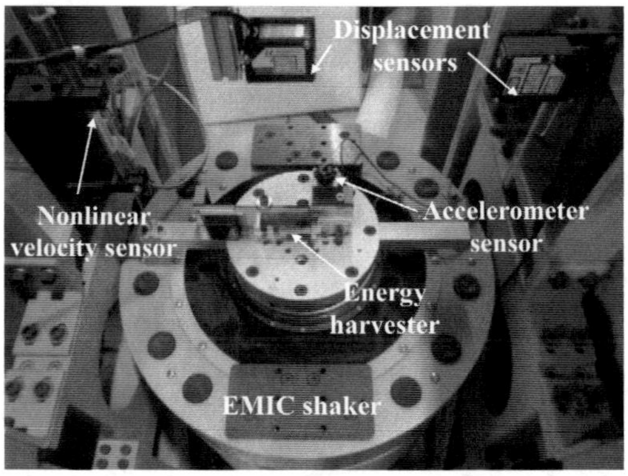

Figure 6. Overview of the experimental setting.

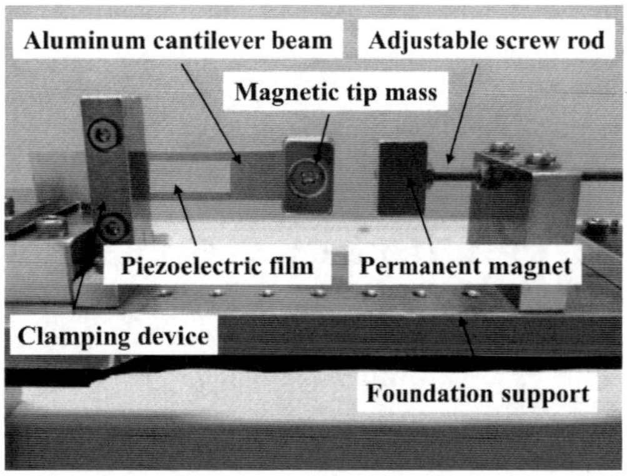

Figure 7. Fabricated energy harvester.

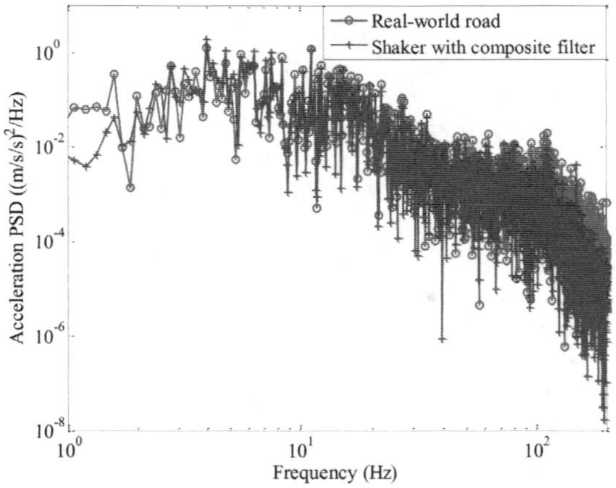

Figure 8. Regeneration of real-world road noise.

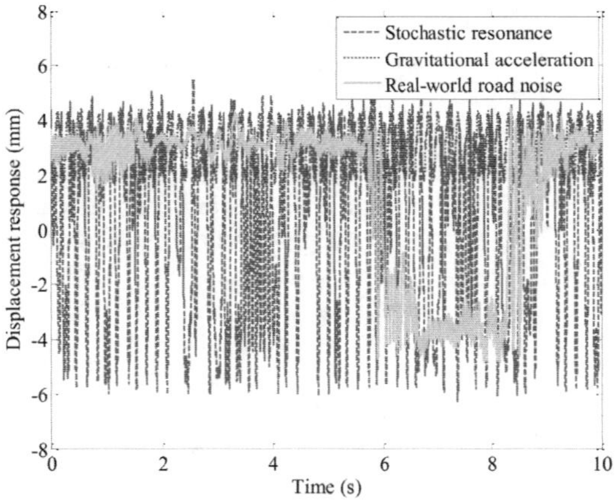

Figure 9. Responses under three conditions.

5. Conclusion

In this study, a miniature energy harvester is proposed and fabricated to apply to low-speed vehicle tires. The simulation study is investigated under the case of real-world road noise excitation and gravitational effect, and it is verified that the feasibility of stochastic resonance can enhance the vibration response. Meanwhile, the effective power that can be generated is of the order of 1.02 mW. By using a shaker to accurately regenerate the vibrational noise from tire rotation, the occurrence of stochastic resonance is validated at the low-speed actual-vehicle driving circumstances.

References

[1] R. Zheng, K. Nakano, H. Hu, D. Su, M.P. Cartmell, "An application of stochastic resonance for energy harvesting in a bistable vibrating system", *J. Sound Vib*, vol 333, pp. 2568–2587, 2014.

[2] Y. Zhang, R. Zheng, K. Nakano, "Feasibility of energy harvesting from a rotating tire based on the theory of stochastic resonance", *J. Phys: PowerMEMS2014,* vol 557, pp. 12097–12101, 2014.

[3] S. Roundy, J. Tola, "Energy harvester for rotating environments using offset pendulum and nonlinear dynamics", *Smart Mater. Struct,* vol 23, pp. 105004–105014, 2014.

[4] P. Hänggi, P. Talkner, M. Borkovec, "Reaction Rate Theory: Fifty Years After Kramers", *Rev. Mod. Phys,* vol 62, pp. 251–342, 1990.

[5] B.P. Mann, B.A. Owens, "Investigations of a nonlinear energy harvester with a bistable potential well", *J. Sound Vib,* vol 329, pp. 1215–1226, 2010.

Vibration Energy Harvester with Bi-stable Curved Beam Spring Offset by Gravitational Acceleration

Koki Yamamoto[1], Takayuki Fujita[1], Adrien Badel[2], Fabien Formosa[2], Kensuke Kanda[1] and Kazusuke Maenaka[1]
[1]Graduate School of Engineering, University of Hyogo, 2167 Shosha, Himeji, Hyogo, Japan
[2]SYMME Laboratory, Universite Savoie Mont Blanc, F7400, Annecy, France

E-mail: er14h053@steng.u-hyogo.ac.jp

Abstract. We developed MEMS bi-stable spring for vibration energy harvester (VEH), which consists of intrinsically curved shape spring and gravitational acceleration. By applying the gravitational acceleration, the curved beam is offset to the gravity direction. It will make more symmetrical bi-stable motion and the symmetry is improved from 3.3 to 65.4%. We proposed that the combination between curved beam and gravity acceleration for decreasing snap-through acceleration. From the analytical result, we investigate the combination can effective to use for decreasing of snap-through force. We also fabricated the prototype device by using MEMS fabrication process. The frequency response for horizontal direction and the acceleration response for vertical direction are measured. The acceleration response shows that the gravitational acceleration improves the symmetry of snap-through force.

1. Introduction

Recently, vibration type energy harvesters (VEHs) are widely noticed as an alternating battery for various sensing systems. Our group have been studying the human sensing system [1] that can effectively use for health monitoring and risk prediction etc. For human sensing systems, the VEHs should be suitable for human motion. The human motion was consisted by combination of lower frequency and pink noise band vibration [2]. In order to collect the sufficient mechanical energy, we have developed the bi-stable spring system that can scavenge low frequency vibration by bi-stable snap-thorough motion with nonlinear behaviours [3]. Generally it was known that the snap-through is happened when the external force exceed the snap-through force. On typical type of bi-stable mechanism, snap-through force depends on the curvature of clamped-clamped beam, and it can be controlled by compressed axis force [4]. However, it was difficult to develop the compression force in MEMS (micro electromechanical systems). In regard to bi-stable beam in MEMS technology, we proposed and designed the novel bi-stable mechanism by using intrinsically curved beam for snap-through motion [5]. In our previous work, curved beam spring was designed for low acceleration operation, however it still required more than 1 G (G; gravitational acceleration) acceleration. By introducing the parallel type curved beam, we achieved lower acceleration of 0.24 G for snap-through In this paper, we propose the additional gravitational acceleration to improve the symmetry for intrinsically curved beam spring.

2. Snap-through motion with gravitational acceleration

2.1. Bi-stable operations on difference structures

The snap-through motion is decided by the applying force (acceleration), the curvature and the geometry of clamped-clamped beam spring. Figure 1 shows the comparison of (a) typical bi-stable spring model, (b) identically curved spring model and (c) curved beam with gravitational acceleration. The typical type shows good symmetry for snap-through forces i.e. Force1 and Force2 are equal. Since the curved spring with curvature of δ was formed and adjusted by applying in-plane force to the straight beam (Fig. 1a). The Force1 or Force2 exceed the snap-through force, the beam jumps from one stable position to another and vice versa, when the device is vibrated along to horizontal direction. However, it is difficult for MEMS harvester applying the in-plane force to the spring after the fabrication. Thus the curved spring was intrinsically designed and fabricated with curvature of δ. In that case, the snap-through force should be asymmetry i.e. Force2 is greater than Force1 (Fig. 1b). In order to improve the symmetry of snap-through force, gravitational acceleration was useful for applying the bias force. Fig. 1(c) shows similar curved beam but the beam is vibrated along to the vertical direction. Although a proof mass was omitted in the simplified model, the mass is connected on the center of the beam. By using the gravitational acceleration, the snap-through force can be balanced by adjusting weight of proof mass.

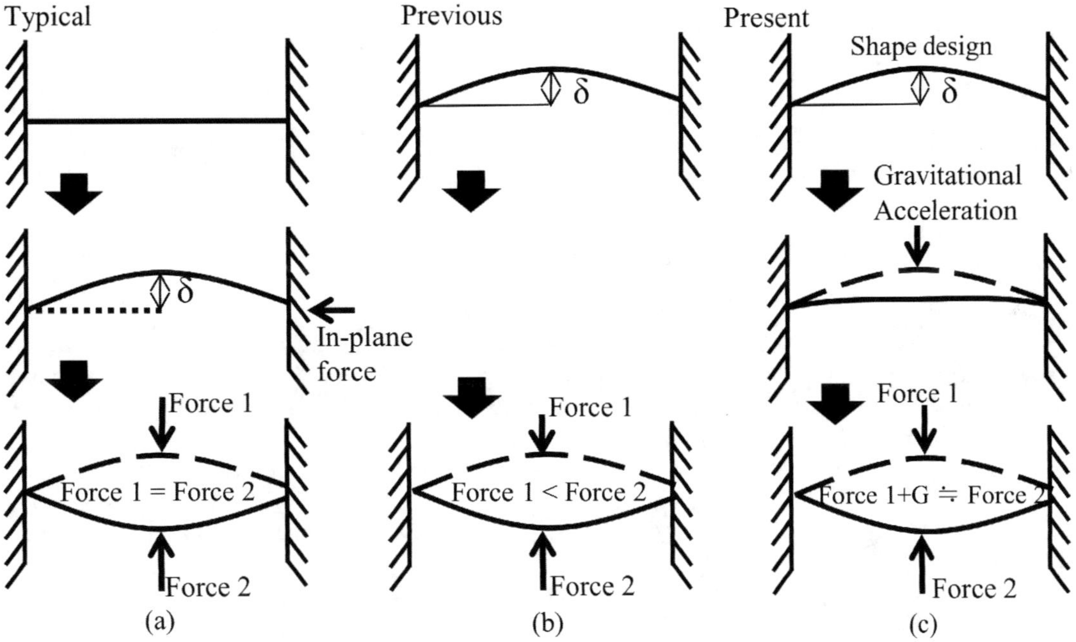

Figure 1. The comparison of snap-through motion
for conventional spring and intrinsically curved beam spring

2.2. Spring characteristics and static energy analysis

The spring characteristic for force vs. displacement (*F-X* curve) and static energy for three types of bi-stable structures are evaluated. The *F-X* curves were obtained by FEM simulator (Ansys Ver.15.5) with static structure analysis. By applying the force on the center of beam, the spring jumps from position A to B as shown in Fig. 2. Before snap-through motion, the beam has softening characteristic by non-linearly beam. As described before, the typical type shows good symmetry of *F-X* curve. However, the previous type shows asymmetry curve with force offset. The present type that has curved beam with gravitational acceleration improves the symmetry of *F-X* curve.

Furthermore, the static energy was also calculated from the *F-X* curve. The static energy also shows the symmetry of three types of structure. The elastic energy of the typical type is completely symmetric. The energy for the previous type is asymmetric and it has no energy stable position. The difference of the static energy P_1 and P_2 can be used as an index of symmetry. In present type, the index of symmetry is improved by applying gravity with optimized proof mass.

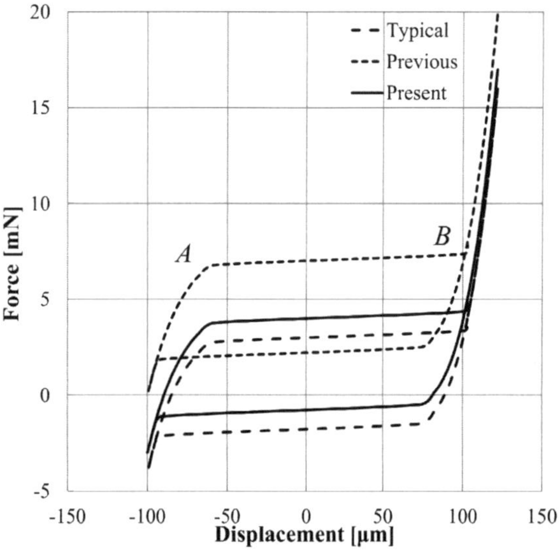

Figure 2. *F-X* curves from FEM buckling analysis.

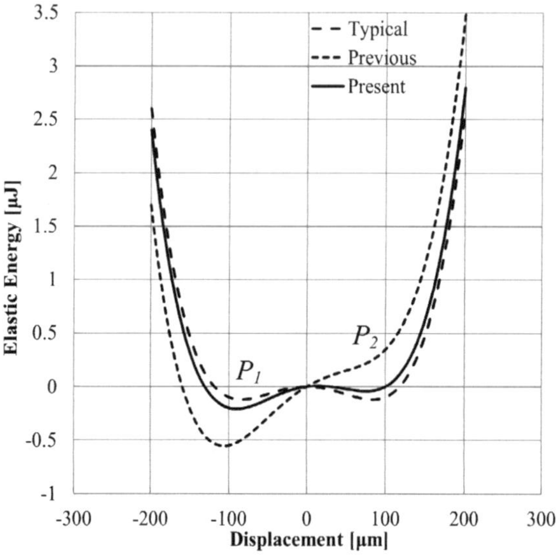

Figure 3. Elastic energy comparison for displacement.

3. Fabrication of curved beam in MEMS

3.1. Prototyping for design and fabrication process evaluation

For preliminary evaluation, we fabricated the test prototype for spring design. The structure was fabricated on the n-type single crystal silicon with thickness of 525 µm. The proof mass and curved support spring were fabricated by deep-RIE process as shown in Fig. 4. The 15 mm length beam supports the 12 mm × 15 mm proof mass. From the calculated proof mass weight, the gravitational acceleration generates an offset force of 2.13 mN to the curved beam. The upper left figure shows close up of beam partand. It's surface profile measured by a Scanning White Light Interferometer (New View 7300; Zygo Corporation).We confirmed the spring width shows 40 µm as same as designed value.

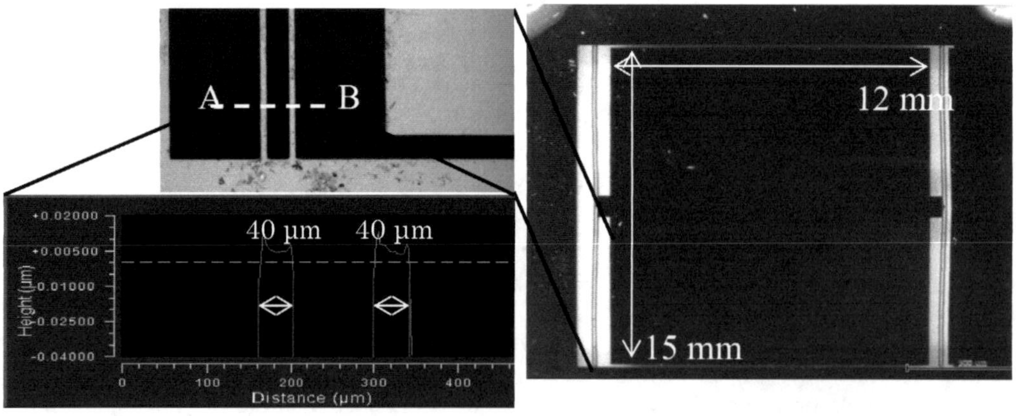

Figure 4. The Photograph of Silicon mass structure and surface profile result

4. Experimental result

Vibrational measurements are carried out by a shaker (M060;IMV Corporation). Firstly, we measured the frequency response without gravity acceleration by horizontal shaker. The figure5 (a) shows just a resonant frequency of the curved beam in each amplitude of acceleration. The maximum displacement points are decreased as increasing the applied acceleration. This softening spring effect is usual characteristics for bi-stable spring structure before the snap-through phenomenon. The bi-stable should response represented hard spring response. We supposed that the spring was used for dozen hertz. However in low frequency band, we could not get enough displacement so the external acceleration should be increased for low frequency band.

Secondly, we measured the acceleration response with applying the gravity acceleration at 28Hz. The response shows the suddenly increasing the displacement. Because, the snap-through phenomenon was happed at 0.28 G. In spite of the small acceleration and low frequency, we could get snap-through phenomenon. We sure that the gravitational acceleration was controlled by weight of proof mass.

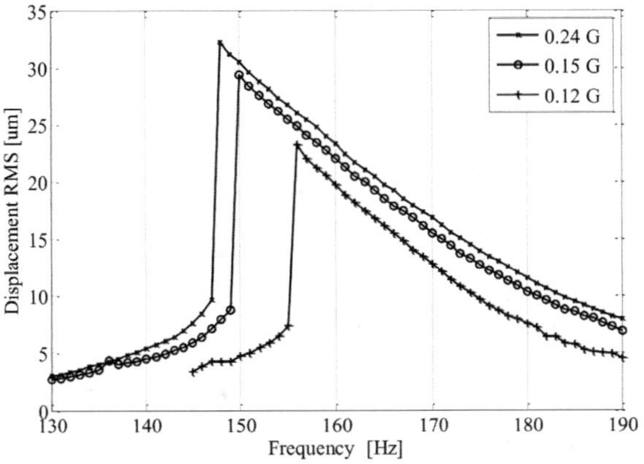

Figure 5. Frequency response of curved beam without gravitational acceleration

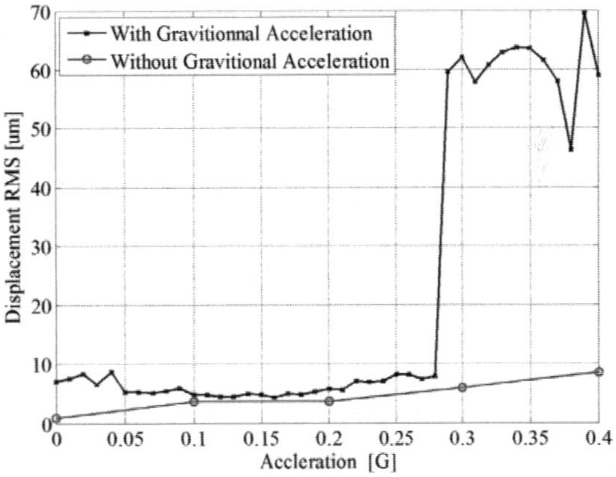

Figure 6. Acceleration response with gravitational acceleration@ 28Hz

5. Conclusion

We presented the combination about gravity acceleration and curved beam system. In FEM analysis, the gravitational acceleration can improves the symmetric of bi-stable characteristic. We fabricated the prototype by MEMS process. Finally the prototype has good advantage with applying acceleration. We sure that this good advantage was generated by improvement of symmetrically.

6. References

[1] Maenaka, K., Masaki, K., & Fujita, T. "Application of Multi-Environmental Sensing System in MEMS Technology-Monitoring of Human Activity. In Networked Sensing Systems", INSS'07. Fourth International Conference on (pp. 47-52). IEEE.

[2] Neri, Igor, et al,"A real vibration database for kinetic energy harvesting application",Journal of Intelligent Material Systems and Structures 23.18 (2012)

[3] W Q Liu et al, "Novel piezoelectric bistable oscillator architecture for wideband vibration energy harvesting", Smart Mater. Struct. 22 (2013) 035013

[4] Liu, W. Q., et al. "Wideband energy harvesting using a combination of an optimized synchronous electric charge extraction circuit and a bistable harvester". Smart Materials and Structures 22.12

[5] K. Yamamoto et al., "A design of Mechanical Frequency Converter Linear and Non-linear Spring Combination for Energy Harvesting" Smart Mater.Struct. (2014) J. Phys.: Conf. Ser. 557 012065

Which is better, electrostatic or piezoelectric energy harvesting systems?

A D T Elliott[1], L M Miller[2], E Halvorsen[3], P K Wright[4] and P D Mitcheson[1]

1 Department of Electrical and Electronic Engineering, Imperial College London, London, UK
2 Alphabet Energy, Eden Landing Rd, Hayward, CA, USA
3 Buskerud and Vestfold University College, Grønland 58, 3045 Drammen, Norway
4 University of California, Berkeley, CA, USA

E-mail: alwyn.elliott07@imperial.ac.uk

Abstract. This paper answers the often asked, and until now inadequately answered, question of which MEMS compatible transducer type achieves the best power density in an energy harvesting system. This question is usually poorly answered because of the number of variables which must be taken into account and the multi-domain nature of the modelling and optimisation. The work here includes models of the mechanics, transducer and the power processing circuits (e.g. rectification and battery management) which in turn include detailed semiconductor models. It is shown that electrostatic harvesters perform better than piezoelectric harvesters at low accelerations, due to lower energy losses, and the reverse is generally true at high accelerations. At very high accelerations using MEMS-scale devices the dielectric breakdown limit in piezoelectric energy harvesters severely decreases their performance thus electrostatics are again preferred. Using the insights gained in this comparison, the optimal transduction mechanism can be chosen as a function of harvesting operating frequency, acceleration and device size.

1. Introduction

Vibrational energy harvesters are often considered as an alternative to batteries for wireless sensors as once installed the harvester does not require regular maintenance. However the choice of transducer (e.g. electrostatic or piezoelectric) to use to convert the mechanical vibration into electrical energy isn't always obvious due to variation in dominance of various loss mechanisms and material constraints. A comparison of the full system models for both electrostatic and piezoelectric energy harvesters is therefore required in order to select the correct transduction method in a given situation.

Electrostatic energy harvesters extract power by using mechanical vibrations to separate charged plates, causing work to be done against the electrostatic attraction. Many examples of MEMS scale electrostatic energy harvesters have been modelled and fabricated [1, 2]. They can be divided into two main types, constant charge, and constant voltage. These harvesters have then been modelled for optimisation in [3, 4], however [3] assumes an optimal resistive load is connected to the output therefore ignoring the constraints due to the low power conditioning circuit required, and [4] ignores analysis of the circuit efficiency. Instead the constant voltage model presented in [5] is prefered and used here as it generates a fully parametrised model. Note the constant charge implementation is ignored as its performance was shown to be worse than constant voltage in [5].

Piezoelectric energy harvesters extract power by inducing a voltage across the piezoelectric material when mechanical stress from a vibration is applied. Examples of piezoelectric energy harvesters can be found in [6], however fully modelling a piezoelectric energy harvester system is more difficult as the type of power conditioning circuit and the limitations these impose have to be taken into account. In [7], the single-supply pre-biasing (SSPB) circuit was shown to outperform all other power conditioning circuits, however the models used did not take into account the effect of the damping force on the transducer. The full system model presented in [8] was therefore chosen for comparison as the same model parametrisation as in [5] is used.

This paper reveals when to use an electrostatic or piezoelectric energy harvesting system for a given harvesting operating frequency, acceleration and device size. The paper first presents the models and algorithms for each transduction method. The maximum power generation and system effectiveness at 100 Hz operating frequency for both transducers are then compared enabling an engineer to choose an optimal transduction mechanism for a given application.

2. Background

Electrostatic and piezoelectric energy harvesters have been modelled using the mass-spring-damper model and electrical equivalent circuits for each has been derived previously [9, 7]. Both the electrostatic and piezoelectric energy harvesters have been shown to preferable be operated as Coulomb damped resonant generators (CDRGs) [10, 11]. The advantage of this is the electrical damping force may be set by applying a fixed voltage making it very easy to control and set to the optimal value.

2.1. Electrostatic model

Figure 1 (a) shows the physical structure of the constant voltage electrostatic energy harvester. The top and bottom electrodes (E_t and E_b) alternatively mesh together with the mass electrode, E_m, as the applied mechanical excitation oscillates the mass vertically. For more details see [5].

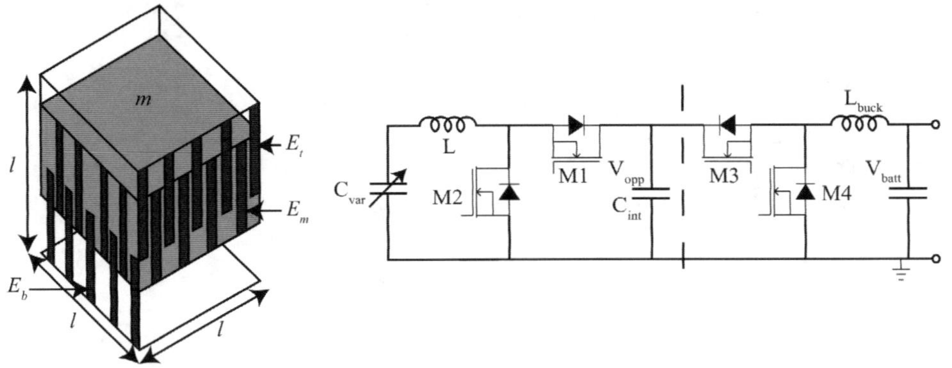

(a) Constant voltage physical configuration (b) Constant voltage power conditioning circuit

Figure 1. Mechanical configuration and power conditioning circuit for constant voltage electrostatic energy harvester based on [5].

The power conditioning circuit required to implement the constant voltage configuration is shown in Figure 1 (b). Switches M3 and M4 act as a boost circuit to prime the intermediate capacitor, C_{int}, to the optimal voltage (V_{opp}). The voltage on C_{var} is then increased by pulsing

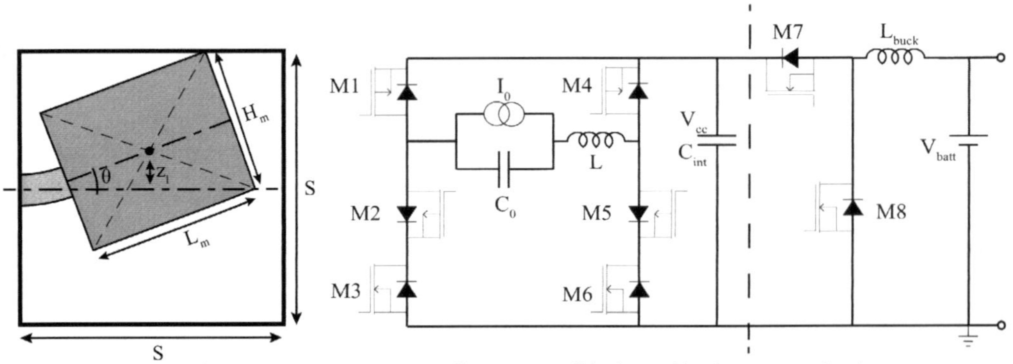

Figure 2. Mechanical structure and power conditioning circuit for piezoelectric energy harvesting system [8].

switch M1 until the voltage on C_{var} is equal to V_{opp}. The electrodes move apart, causing almost all the charge to be transferred from C_{var} to C_{int} through $M1$. As the electrodes move back together again M1 is opened and M3 and M4 are operated as a buck circuit to transfer the extracted energy to the storage element, V_{suppply} [5].

2.2. Piezoelectric model

The mechanical structure of the piezoelectric transducer is assumed to be a simple cantilever beam with a proof mass on the end (Figure 2 (a)). The beam consists of a piezoelectric thin film with silicon and oxide structural layers. The displacement of the mass is calculated to ensure that the deflection of the beam does not cause the mass to collide with side of the box.

The most compact circuit capable of operating a piezoelectric energy harvester as a CDRG is the single-supply pre-biasing (SSPB) circuit (Figure 2 (b)) [12]. This circuit transfers charge from an intermediate storage capacitor, C_{int}, to the piezoelectric layer on the cantilever when it reaches its extreme points of travel in order to pre-bias it through a set of switches (e.g. M1, M5, M6). This pre-bias charge induces a Coulomb-damping force which acts in the opposite direction to the motion of the cantilever, therefore increasing the work done. As the cantilever deflects to the opposite extreme point, more charge is induced. The pre-bias and induced charge are then returned back to C_{int} through the same set of switches used to pre-bias (e.g. M1, M5, M6) before the pre-bias is applied with the opposite polarity through the other set of switches (e.g. M4, M2, M3). Switches M7 and M8 are used to transfer extracted energy from C_{int} to the storage battery, V_{batt}, to maintain the optimal voltage on C_{int}.

2.3. System effectiveness

A harvester's maximum theoretically available power at a given acceleration (A_{input}), with a volume of side length S, at an operating frequency (ω_{input}) and proof mass density (ρ_{mass}) [10] is

$$P_{\mathrm{max}} = \frac{1}{16}\rho_{\mathrm{mass}} A_{\mathrm{input}} \omega_{\mathrm{input}} S^4. \tag{1}$$

The effectiveness of a harvesting system can thus be calculated by dividing the final output power, P_{out}, of the harvesting system by the maximum available power:

$$\eta_{\mathrm{system}} = \frac{P_{\mathrm{out}}}{P_{\mathrm{max}}}. \tag{2}$$

3. Model Implementations

Several assumptions were made about both systems to help constrain the problem for comparison. Firstly the displacement of the mass was not allowed to exceed the given volume as this would cause damage to device. The inductor is assumed to occupy half the given volume of the device. The proof mass was assumed to be made of gold as it is compatible with MEMS manufacturing techniques and has a high density (19320 kg m^{-3}). The 1.5 V battery, used to store the extracted energy, is assumed the same size and hence excluded as part of the system volume. All semiconductor devices are assumed to be small enough as to occupy negligible space.

3.1. Electrostatic algorithm

The electrostatic model algorithm operates as follows. First the operating frequency, ω_{input}, and vectors for side length, S, and acceleration, A_{input}, are defined. The optimal operating voltage for given S and A_{input} is then found. If this voltage exceeds semiconductor device breakdown the pre-charge voltage is reduced. Numerical solutions for η_{system} using different numbers of semiconductor cells are then calculated. The optimal number of cells which maximises η_{system} is thus selected. The algorithm is repeated with the next value of S and A_{input}. For more detail on the algorithm see [5].

3.2. Piezoelectric algorithm

The piezoelectric model algorithm is similar to the electrostatic model and operates as follows. First the operating frequency, ω_{input}, and vectors for side length, S, and acceleration, A_{input}, are defined. Beam length and thickness pairs are then found which resonate within 1 % of ω_{input}. Using these beam definitions, the transducer parameters are calculated. The optimal pre-bias voltage required for CDRG operation and the end voltage after beam deflection are then calculated. If the end voltage exceeds piezoelectric dielectric breakdown, the displacement of the mass is decreased. Initial numerical solutions for the circuit's parameters are then calculated for different circuit inversion efficiencies. The circuit parameters are then iteratively updated to allow for charge sharing, leakage and conduction losses in the SSPB and buck circuits. The net power and system effectiveness are finally calculated. The algorithm is repeated with the next value of S and A_{input}. For more detail on the algorithm see [8].

4. Results

The comparison of the two systems was performed at 100 Hz with the volume constrained between 1 mm and 15 mm and the acceleration varied between 0.01 ms^{-2} to 100 ms^{-2} using a gold proof mass. The semiconductor device breakdown voltage was assumed to be 1.5 kV and the piezoelectric dielectric breakdown voltage was assumed to be 750 V.

Figure 3 compares the power output and system effectiveness for both energy harvesting systems. It can be clearly seen that the electrostatic system is preferred at low accelerations over the piezoelectric system as the energy losses in the piezoelectric devices at low power levels severely reduces the output power. As mechanical acceleration is increased, the electrostatic system requires increasingly high biasing voltages, causing the semiconductor device on-state resistance to increase in order to block the higher voltage, resulting in increased conduction losses and a preference for the piezoelectric system. However at very high accelerations the power output of the piezoelectric system is severely limited as the dielectric breakdown voltage in MEMS scale devices forces the mass displacement to be decreased thus not utilising the maximum deflection possible, therefore the electrostatic devices are preferred in this case.

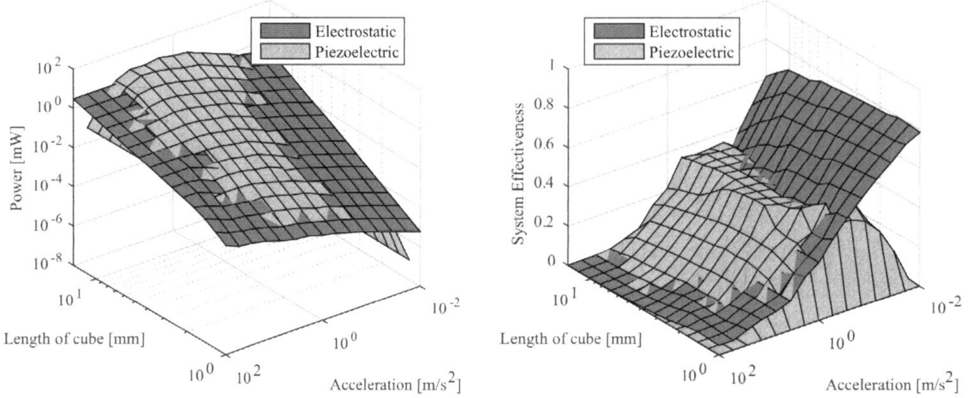

Figure 3. Comparison of the power output and system effectiveness between electrostatic and piezoelectric MEMS scale energy harvesting systems.

5. Conclusion

A full system model comparison of electrostatic and piezoelectric energy harvesters has been presented for the first time. It is clear that at very low accelerations, electrostatic devices are preferred as they have lower losses. As the damping force and therefore required optimal biasing voltage increases, piezoelectric devices have better performance due to increased conduction losses than electrostatic devices, however at very high accelerations MEMS scale piezoelectric material suffer from voltage breakdown thus electrostatic devices should be used. Future work therefore should examine larger devices where bulk piezoelectric material can be used as it has a higher breakdown voltage. A comparison with electromagnetic resonant vibrational energy harvesting systems should also be conducted in order to compare all three technologies.

Acknowledgements

The authors would like to thank the California Energy Commission and Research Council of Norway grant number 229716 and for their support, and EPSRC for their sponsorship of this project through a doctoral training award at Department of Electrical and Electronic Engineering, Imperial College London.

References

[1] Suzuki Y 2011 *IEEJ Transactions on Electrical and Electronic Engineering* **6** 101–111 ISSN 1931-4981
[2] Le C P and Halvorsen E 2012 *Journal of Micromechanics and Microengineering* **22** 074013
[3] Mitcheson P D, Reilly E K, Toh T, Wright P K and Yeatman E M 2007 *J. Micromech. Microeng* **17** S211
[4] Guillemet R, Basset P, Galayko D and Bourouina T 2010 *Procedia Engineering* **5** 1172 – 1175 ISSN 1877-7058 eurosensor {XXIV}
[5] Mitcheson P and Green T 2012 *Circuits and Systems I, IEEE Transactions on* **59** 3098–3111 ISSN 1549-8328
[6] Kim H, Kim J H and Kim J 2011 *INT J PRECIS ENG MAN* **12** 1129–1141 ISSN 1229-8557
[7] Dicken J, Mitcheson P D, Stoianov I and Yeatman E 2012 *Power Electronics, IEEE Transactions on* **27** 4514–4529 ISSN 0885-8993
[8] Miller L M, Elliott A D T, Mitcheson P D, Halvorsen E, Paprotny I and Wright P K UNDER REVIEW *IEEE Transactions on Circuits and Systems - Part 1*
[9] Mitcheson P D and Toh T T 2010 Power management electronics *Energy Harvesting for Autonomous Systems* (Norwood, MA: Artech House, Inc) chap Chapter 6
[10] Mitcheson P, Green T, Yeatman E and Holmes A 2004 *Microelectromechanical Systems, Journal of* **13** 429–440 ISSN 1057-7157
[11] Miller L M, Mitcheson P D, Halvorsen E and Wright P K 2012 *Applied Physics Letters* **100** 233901
[12] Dicken J, Mitcheson P D, Elliott A D T and Yeatman E M 2011 Single-supply pre-biasing circuit for low-amplitude energy harvesting application *PowerMEMS* (Seoul, South Korea) pp 46–49

A novel study of the kinetics of external hierarchical nanostructures in methanol fuel cell

M Al-Halhouli[1], J Kieninger[2], O Yurchenko[1], G Urban[1,2]
[1]Freiburg Materials Research Center, FMF, University of Freiburg, Germany
[2]Department of Microsystems Engineering, IMTEK, University of Freiburg, Germany

E-mail: mohammad.halhouli@fmf.uni-freiburg.de

Abstract. For the purpose of a direct methanol fuel cell (DMFC), this research investigates the kinetics of methanol oxidation in a porous layer consisting of external hierarchical nanostructures through 2D COMSOL simulation. Three different lengths of nanowires (*L*) were considered in simulations. The investigation showed that specific activity was reversely proportional to nanowires length and roughness factor (*Rf*). However, the current density increased by increasing *Rf*. Although the current density in case of L = 200 nm and 500 nm was identical with respect to *Rf*, there was a slight deviation when L = 1000 nm due to the changing from kinetic to diffusion controlled regime, which was identified by investigation of Thiele modulus. The catalytic efficiency for L = 1000 nm dropped to 50% at Rf = 140, whereas the high efficiency with no mass-transport limitation was achieved by shorter nanowires. Therefore, increasing *Rf* within the simulation range resulted in increasing the total catalytic activity but simultaneously decreasing the specific activity because of the decrease in pore accessibility and catalytic efficiency of nanostructures.

1. Introduction

The high surface-to-volume ratio of nanoporous electrodes results in high catalytic surface area. [1] Therefore, there is an increasing number of nanostructures that were reported for several catalytic reactions. Hierarchical nanostructures represent one of the most significant nanostructures generating a nanoporous layer. [2] Generally, they consist of two sub-structures, macrostructures and micro-/mesostructures. [3] According to our definition, hierarchical nanostructures can be divided into two types, internal and external. The internal structures consist of macropores that work as a distributor of diffused species to the interconnected micro-/mesopores. [3] On the other hand, the external structures consist of macrostructures that grow on the electrode surface and serve as a base for the interconnected nanowires.

According to Butler Volmer equation (BV), the overall catalytic activity of the nanostructures is influenced by the reaction kinetics and the concentration of active species at the active surface. [4] The catalytic activity defines the number of reactions at the active site and the number of active sites on the nanostructures [5]. Therefore, surface *Rf*, which is defined as the ratio between the total nanoscopic catalytic surface area and the geometrical surface area [6], influences the catalytic activity through its influence on the number of active sites. In addition, the concentration of active species influences the reaction rate at the active sites due to local concentration factor in BV. [4] Accordingly, pore accessibility is an additional factor that influences the overall catalytic activity.

Specific activity (A/cm^2) and current density (A/cm^2) are experimental quantities, mostly used for determining surface catalytic activity. Specific activity, which defines the average current per the unit of electrochemical active surface area [7, 8], measures the catalytic activity based on the pore accessibility only. However, catalytic activity based one current density includes *Rf* and pore accessibility. On the other hand, Thiele modulus and effectiveness factor are numerical values that can evaluate the catalytic activity and determine the regime of reaction kinetics. Electrode kinetics according to Thiele modulus is classified as one of two regimes, diffusion controlled (D-controlled) or reaction kinetic controlled (k-controlled). [9, 10]

Unlike many former researches that have investigated internal hierarchical structures, like reference [11], this paper investigated the kinetics in a porous layer that consists of external hierarchical nanostructures through 2D COMSOL simulation. Beside the investigation of specific activity and current density based on BV, mass-transport and pore accessibility was also examined based on Thiele modulus. Simulations were performed for the purpose of a direct methanol fuel cell.

2. Methods

2.1. Reaction kinetics

The kinetic of methanol (Red) oxidation at the anode was investigated in our simulations by considering a reaction that releases one electron and one proton, i.e., oxidizing agent (Ox). The kinetics of reactions occurring on the electrode surface is influenced by the electrode potential. [4, 12] The current-potential characteristics were determined according to BV equation, which predicts the currents resulting from the overpotential on the electrode [12]:

$$j = \frac{j_0}{C}\left[C_{Ox}\exp\left(\frac{-\alpha F\eta}{RT}\right) - C_{Red}\exp\left(\frac{1-\alpha F\eta}{RT}\right)\right] \tag{1}$$

where

$$j_0 = Fk^0 C_{Ox}^{1-\alpha} C_{Red}^{\alpha} = Fk^0 C \tag{2}$$

where j is the current density (A/m^2), F is the Faraday constant (C/mol), j_0 is exchange current density (A/m^2), C is the reference concentration at which the bulk concentrations of oxidizing and reducing agents are equal (mol/m^3), C_i is the concentration at the electrode surface of species i (mol/m^3), i is either Ox or Red, α is the charge transfer coefficient, k^0 is the standard heterogeneous rate constant (m/s), η is the overpotential (V), R is the gas constant (J/(mol K)) and T absolute temperature (K).[4]

2.2. Thiele modulus

Thiele modulus ϕ_n and the effectiveness factor η are simple terms to analyze the catalytic activity and the diffusional resistance of dispersed species in a porous electrode. [13] The lower Thiele modulus leads to better diffusion profile. For each nanostructure, Thiele modulus was derived based on the following equation[9, 10]:

$$\varphi_n = L \cdot \sqrt{\frac{k}{D}} \tag{3}$$

where

$$k = k^0 \frac{A_r}{V_D} \tag{4}$$

L is the diffusion distance (m), D is the diffusion coefficient (m^2/s), V_D is the volume available for diffusion of molecules (m^3), and A_r is the area of the surface of reaction (m^2). Likewise, the effectiveness factor is defined in the following equation:

$$\eta = \frac{\tanh(\varphi_n)}{\varphi_n} \tag{5}$$

2.3. Simulation software
Finite element simulator COMSOL Multiphysics® was used for studying and comparing the catalytic activity of the proposed external hierarchical nanostructures. The catalytic activity was investigated in 2D simulation model after mapping the diffusion coefficient from 3D simulation model.

2.4. Geometry of nanostructures
The geometry of investigated nanostructures is shown in figure 1. The hierarchical nanostructure contains one macrostructure (gray) and many nanowires with a variable length (L) as microstructure (black). The length of nanowires and the diameter of the macrostructure adjust Rf. The length of nanowires was set during the simulation to 200 nm, 500 nm and 1000 nm.

In figure 1, the gray and black colors represent the solid-phase and the blue color represents the void-phase. The reaction surface, at which the oxidation of methanol takes place, is the outer surface of the solid phase. Each Rf and each L had a stand-alone 2D component in one simulation file. In each component, one nanostructure was placed at the bottom of a rectangular simulation model of width 20 µm and height 10 µm. Moreover, the top edge of the rectangle was set as an inflow boundary condition, the side edges were set as symmetry boundaries, and the bottom edge was set as a wall.

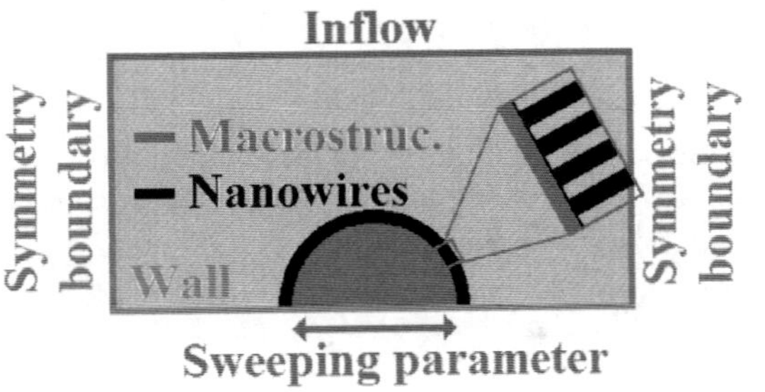

Figure 1. The model of half-cell of DMFC and boundary conditions of simulation.

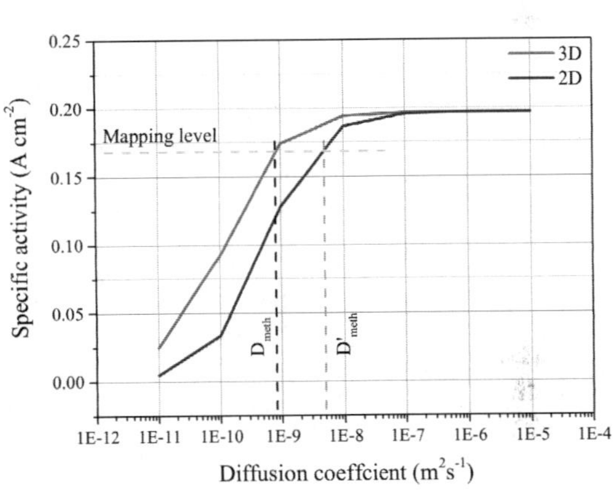

Figure 2. Diffusion constant from 3D to 2D mapping.

3. Results and discussion

3.1. Diffusion coefficient mapping
Because of computational limitations in 3D models, 2D simulation models were chosen to investigate the catalytic efficiency of proposed geometries and variations. Unfortunately, there is no 2D cross-section that achieves symmetry in the 3D geometry. Therefore, it was not possible to map the results of 2D model to the results of 3D model based on geometrical aspects. Thus, specific activity was considered as a mapping criterion because it involves the average rate of mass transport of active species to the active sites and the average molecule turnover rate at each active site. Figure 2 shows the specific activity of 2D and 3D simple external non-hierarchical nano-structures (not shown in this paper) in a wide range of diffusion coefficients. Considering the diffusion coefficient of methanol $D_{meth} = 0.84e^{-9}$ m^2/s [14], the equivalent diffusion coefficient of methanol in 2D for the equivalent specific activity is $D'_{meth} = 5e^{-9}$ m^2/s. The found value of D'_{meth} was used for all simulations in this paper.

3.2. Kinetics of external hierarchical nanostructures

The specific activity, as shown in figure 3, is inversely proportional to the length of nanowires. Additionally, specific activity is also inversely proportional to *Rf* of surface. Therefore, when the nanostructures achieve an equal *Rf* of surface, the one with shorter nanowires length features higher specific activity. Moreover, the specific activity is related to the concentration at the direct vicinity of nanostructures surface because both quantities are influenced by molecules turnover rate at surface and mass transport rate to active sites. However, the current density is directly proportional to *Rf* at all nanowires lengths (figure 4). Only a slight deviation in the curve of L = 1000 nm at higher *Rf* is observed. The shorter nanowires show larger current density, especially at high *Rf* values. Unlike the specific activity, although the average turnover rate decreases at high *Rf,* the number of catalytic active sites increases and therefore the total turnover of molecules increases, which results in a larger current density. It is import to note, even though the shorter nanowires at specific *Rf* address higher total turnover, the high *Rf* values are only achieved by long nanowires.

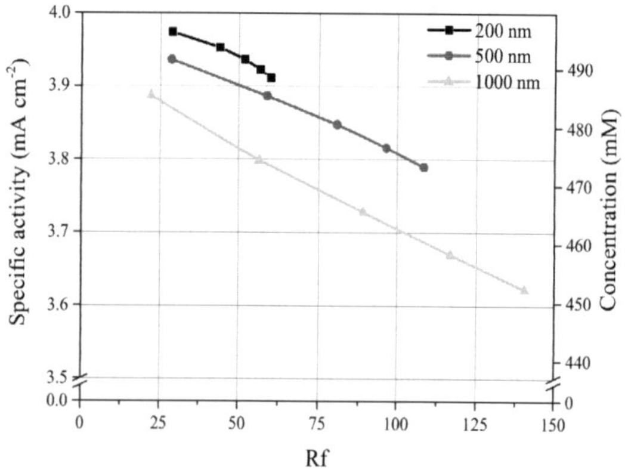

Figure 3. Specific activity of different nanowires length with respect to the surface roughness factor.

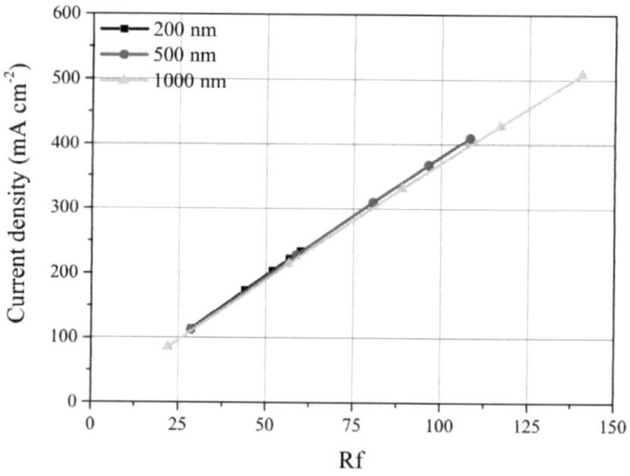

Figure 4. Current density of different nanowires length with respect to the surface roughness factor.

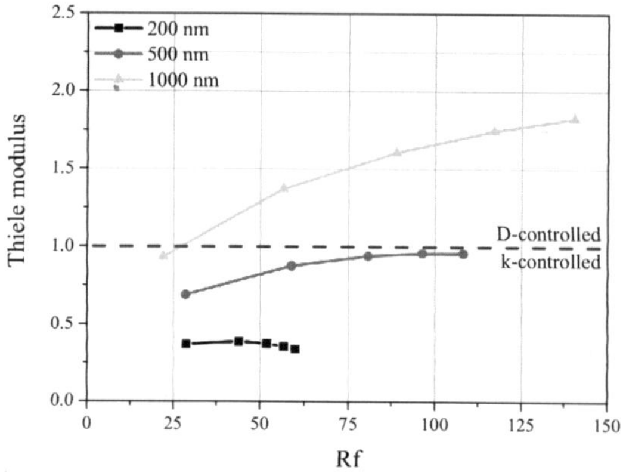

Figure 5. Thiele modulus of different nanowires length with respect to the surface roughness factor.

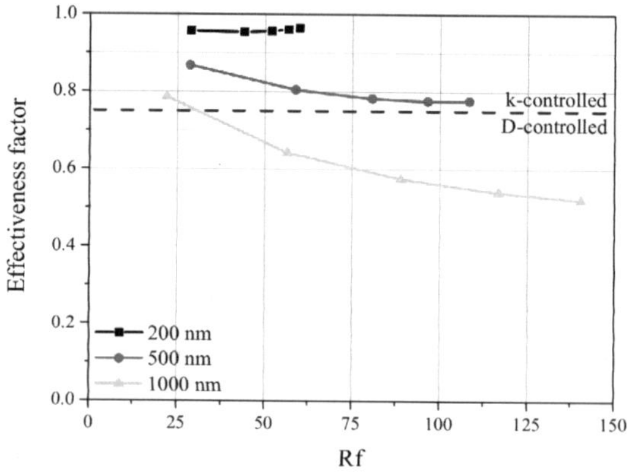

Figure 6. Effectiveness factor of different nanowires length with respect to the surface roughness factor.

The current densities (figure 4) in case of L = 200 nm and 500 nm with respect to *Rf* are more identical compared to the current density in case of L = 1000 nm because of the kinetics regime. According to Thiele modulus (figure 5), the kinetic regime in case of L = 1000 nm is limited by diffusion, while there are no mass-transport limitations due to diffusion when L = 500 nm and 200 nm. The catalytic efficiency (figure 6) for L = 1000 nm drops to 50% at Rf = 140, whereas the high efficiency with no mass-transport limitation is achieved by the shorter nanowires. When L = 200 nm, the effectiveness factor is close to 1, which means perfect mass transport to all active sites regardless the value of Rf. Likewise, when L = 500 nm, the kinetic regime stays at k-controlled region and high mass transport efficiency is achieved. Therefore, increasing Rf, within the simulation range, results in increasing the total catalytic activity but simultaneously decreasing the specific activity because of decreasing in pore accessibility and catalytic efficiency of nanostructures.

4. Conclusions

Our investigations showed that 3D simulation and 2D simulation of external hierarchical nanostructures are asymmetrical. Therefore, the variation in 3D catalytic activity was compensated by the diffusion coefficient in 2D simulations, which delivered the specific activity of methanol in 3D simulations. It was shown that specific activity was inversely proportional to the length of the nanowires and Rf of the surface. This inverse proportion was a result of pore accessibility, which was confirmed by calculating Thiele modulus and effectiveness factor. The nanostructures with longer nanowires and their higher density resulted in lower pore accessibility as well as lower specific activity because of larger diffusional resistance. It was determined that in the simulated Rf range, the nanostructures with nanowires having L = 200 nm and 500 nm showed a kinetic controlled regime, while the nanostructures, which have nanowires with L = 1000 nm, showed diffusion controlled regimes. Nevertheless, the number of the catalytic actives sites increased at large values of Rf. Therefore, it was confirmed that the current density increased by enlarging the surface roughness.

Acknowledgments

We gratefully acknowledge the International Graduate Academy (IGA) of the University of Freiburg for the financial support of the PhD thesis of M.Sc. M. Al-Halhouli through LGFG scholarship.

References

[1] Wang J 2005 *Analyst* **130** 421
[2] Christopher M. A., Parlett, Wilson K and Lee A F 2013 *Chemical Society reviews* **42** 3876–93
[3] Xiao S, Xiao F, Hu Y, Yuan S, Wang S, Qian L and Liu Y 2014 *Scientific reports* **4** 4370
[4] Bard A J and Faulkner L R 2000 *Electrochemical methods and applications* 2nd edn (New York, London: Wiley-Interscience)
[5] Nørskov J K, Studt F, Abild-Pedersen F and Bligaard T 2015 *Fundamental concepts in heterogeneous catalysis* (Hoboken, New Jersey: Wiley)
[6] Yi J L and Jae Y P 2011 *J. Korean Phys. Soc.* **58** 1505
[7] Boreskov G K 1957 *Zhu. Fiz. Khim.* **31** 937–51
[8] Dzis'ko V A 1974 *Russ. Chem. Rev.* **43** 435–51
[9] Thiele E W 1939 *Ind. Eng. Chem.* **31** 916–20
[10] Davis M E and Davis R J 2003 *Fundamentals of chemical reaction engineering (McGraw-Hill chemical engineering series)* (Boston: McGraw-Hill)
[11] Prachayawarakorn S and Mann R 2007 *Catalysis Today* **128** 88–99
[12] Zoski C G 2007 *Handbook of electrochemistry* 1st edn (Amsterdam, Boston: Elsevier)
[13] Janardhanan V M and Deutschmann O 2011 Computational Fluid Dynamics of Catalytic Reactors *Modeling and Simulation of Heterogeneous Catalytic Reactions* ed O Deutschmann (Weinheim, Germany: Wiley-VCH Verlag GmbH & Co. KGaA) pp 251–82
[14] Cussler E L 2009 *Diffusion: Mass transfer in fluid systems* 3rd edn (Cambridge, New York: Cambridge University Press)

Stretchable glucose biofuel cell with wirings made of multiwall carbon nanotubes

Yusuke Fujimagari, and Yasushiro Nishioka*
Department of Precision Machinery Engineering, College of Science and Technology, Nihon University, Funabashi, Chiba 274-8501, Japan

E-mail: nishioka@eme.cst.nihon-u.ac.jp

Abstract. In this study, we fabricated a flexible and stretchable glucose-biofuel cell with wirings made of multi wall carbon nanotube (MWCNTs) on a polydimethylsiloxane substrate. The biofuel cell investigated consists of a porous carbon anode (area of 30 mm^2) modified by glucose oxidase and ferrocene, and a cathode (area of 30 mm^2) modified by bilirubin oxidase. The anode and the cathode were connected with the MWCNT wirings. The maximum power of 0.31 μW at 76.6 mV, which corresponds to a power density of 1.04 $\mu W/cm^2$, was realized by immersing the biofuel cell in a phosphate buffer solution with a glucose concentration of 100 mM, at room temperature.

1. Introduction

Recently, enzymatic biofuel cells (BFCs) that use glucose in a human body to produce electricity have been of special interest. This is because they work under mild conditions (room temperature, neutral pH and atmospheric pressure), which make them amenable to use in the human body [1-12]. However, when these devices are used in the human body, the flexibility and stretchability of BFCs are among their most important features. Recently, as a potential candidate for a flexible BFC to be used in vivo, Fukushi et al. reported a flexible glucose BFC on a polyimide (PI) sheet because PI offers good biocompatibility, with high thermal stability and flexibility [13]. In addition, Miyake et al. fabricated flexible BFCs with a structure of sandwiched d-fructose dehydrogenase in carbon fabric (CF) [14, 15]. Although these devices were flexible, they were not stretchable. The fabrication of stretchable electrodes and wiring is as necessary as bending flexibility to realize wearable BFCs. Fujimagari et al. recently reported a stretchable glucose BFC using wirings made of silver nanowires [16].

In contrast, carbon nanotube (CNTs) have drawn much attention because they are both electrically conductive and stretchable [17-26]. Polydimethylsiloxane (PDMS) is a material that has been extensively investigated for use in many biodevices because it is biocompatible, flexible, and stretchable [27-30]. For example, Lu et al. fabricated a strain sensor using metal wiring patterns transferred on a stretchable PDMS substrate [31]. We considered that a stretchable wiring for flexible and stretchable BFCs can be realized by utilizing CNTs embedded in a PDMS substrate. However, there has been no report on creation of a stretchable glucose BFC using CNT wirings. In this study, we fabricated a glucose BFC with flexible, stretchable wirings made of multi wall carbon nanotube (MWCNT) on a PDMS substrate.

2. Experimental

2.1. Power Generation Mechanism

The BFC investigated in this research consists of a porous carbon anode modified by glucose oxidase (GOD) and ferrocene, and a cathode modified by bilirubin oxidase (BOD). Figure 1 shows the power generation principle. Electrons removed from glucose by GOD are transferred to the anode through the ferrocene mediator, and protons are released from glucose at the same time. The electron flow though the resistance, and are delivered to the cathode where oxygen is reduced by these electrons and protons. The generated power can be diverted through an external circuit connected between the anode and the cathode.

2.2. Design of wiring and electrodes

Figure 2 shows the layout of the cathode, anode and MWCNT wirings. Both the anode and cathode had a width of 3 mm and a length of 10 mm. The gap between the anode and cathode was 1 mm. The anode and cathode were deposited on MWCNTs with a width of 3 mm and length of 25 mm. The areas of the anode and the cathode were 3 x 10 mm^2.

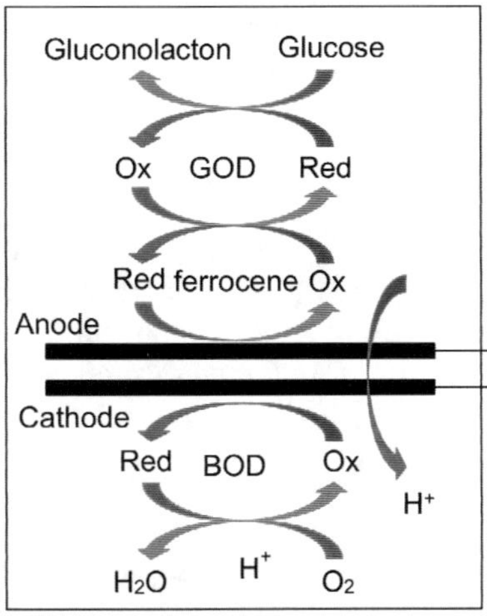

Figure 1. Possible mechanism of power generation.

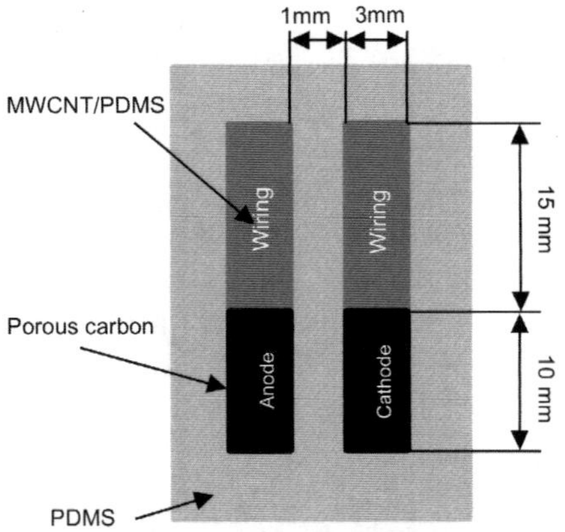

Figure 2. Design of stretchable wiring and electrodes.

2.3 Wiring fabrication processes

Figure 3 shows the MWCNT wiring fabrication processes. At first, a PI film (Du Pont-Toray Co., Ltd.) with a thickness of 50 μm was cleaned by acetone and ethanol, and place a mask made of poly-ethylene-terephthalate film with the wiring patterns with the areas of 3 x 30 mm^2 and the depth of 100 μm [Fig. 3(a)]. 10 mg of MWCNT (Nanocyl) and 136.4 mg of a PDMS precursor (Dow Corning Toray Co. Ltd.) was mixed with the addition of 13.6 mg of a curing agent (Dow Corning Toray Co. Ltd.), and the resultant paste was filed in the wiring patterns of the mask using a screen printing technique. The mask was removed after the solidification of the MWCNT/PDMS mixture. The wiring patterns were cured at 100 °C for 20 min. Another PDMS layer was formed on the top of the wiring patterns on the PI sheet, and the PDMS layer was solidified at 100 °C for 20 min to form the MWCNT wiring embedded in the PDMS substrate [Fig. 3(b)]. Finally, the PI sheet was peeled off [Fig. 3(c)]. A similar stretchable wiring using carbon black was also fabricated using a similar processes using a mixture of carbon black and PDMS. The thickness of the resultant porous carbon electrode was approximately 100 μm.

2.4 Fabrication of the carbon-modified electrodes

The surfaces of the MWCNT wirings film were coated by screen-printing with carbon paste (carbon black) and polyvinylidene fluoride (PVDF, KUREHA, Inc.) mixed ink, which was then cured at 80 °C for 1 h.

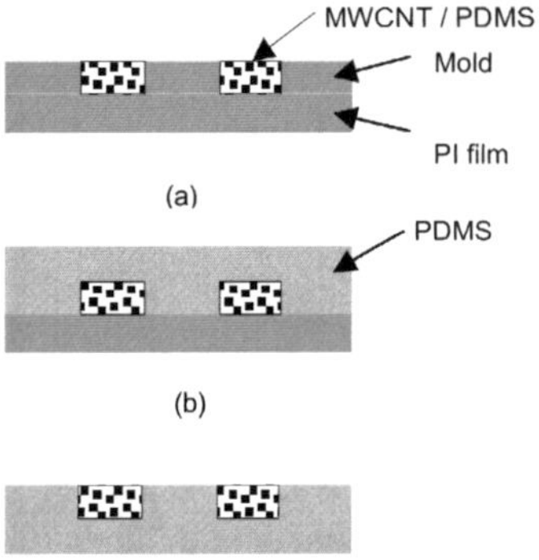

Figure 3. Wiring fabrication processes.

2.5 Fabrication of the enzyme-modified electrodes

A small amount of GOD (0.05 mg) was mixed with 30 µl of phosphate buffer solution (PBS, 50 mM, pH 7.0). Then, 0.05 mg of the ferrocene mediator was mixed with 30 µl of PBS. Then, 10 µl of the resulting solutions was dropped onto the porous carbon anode, and left to dry for 90 min. Similarly, 0.05 mg of BOD was mixed with 30 µl of PBS. Then, 10 µl of the resulting solution was dropped onto the porous carbon cathode and left to dry for 120 min. As a reference, carbon electrodes were also fabricated using similar conditions to those previously reported.

2.6 Measurements

Changes in resistance of the MWCNT or carbon black wirings were measured by stretching the wirings. Each wire was stretched in 2-mm intervals using a jig until it broke. Electrochemical measurements were performed in an air-saturated, 50 mM, 100 mM and 200 mM phosphate buffer solution (PBS) of glucose, at room temperature. The power measurement was performed with the wiring part stretched and the porous carbon electrodes unstretched. The power W generated was evaluated by measuring the cell voltage while varying the external load resistance between 0 and 2.7 MΩ using a source meter (Keithley 2400, Source Meter). The following equation was used to calculate W:

$$W = V^2 / R \qquad (1)$$

where R is the load resistance, and V is the generated voltage measured between the terminals of the BFC. The power density was calculated by dividing W by the area of the anode or cathode, which was 30 mm^2.

3. Results and discussion

Figure 4 shows the resistance change of the wirings with a length of 30 mm, a width of 3 mm and a thickness of 100 µ as a function of elongation. The resistance of a carbon black wiring was also shown as a reference. The resistance change of a carbon black wiring was also shown as a reference. The maximum elongations of MWCNT and carbon black were 10 mm and 30 mm, respectively. The resistance of the MWCNT wirings increased from 267 Ω to 601 Ω, and that of the carbon black wiring increased from 44 kΩ to 55 kΩ at the elongation of 10 mm. The smaller resistance of the MWCNT wiring compared to that of the carbon black wiring may be due to the wired structure of MWCNTs [26].

Figure 5 shows the relationship between the power density and the output voltage of the BFCs with different wirings. The smaller power density of the BFC with the carbon black wirings was considered due to the larger resistance of the wirings. The power density of the glucose BFC with the MWCNT wirings exhibited the maximum of 1.04 μW/cm^2 at 76.6 mV, and that of the porous carbon wiring exhibited 0.123 μW/cm^2 at 44.3 mV.

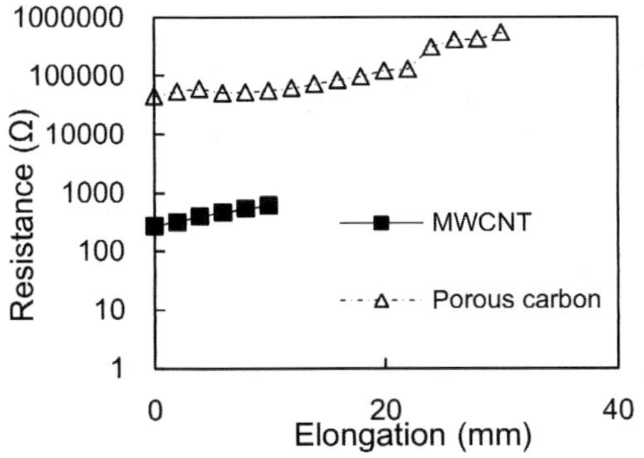

Figure 4. Change of resistance of MWCNT and porous carbon wirings as a function of elongation.

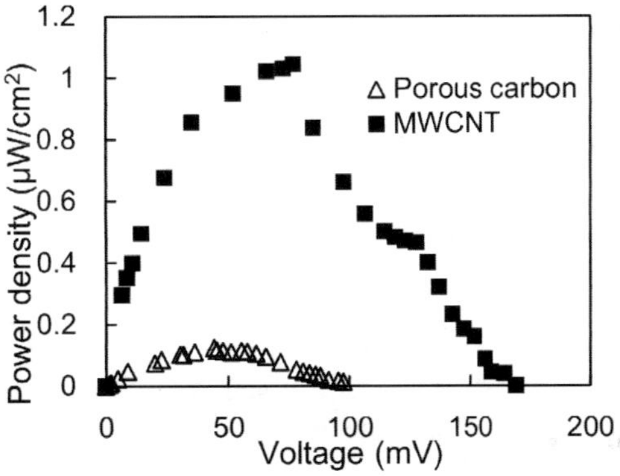

Figure 5. Relationships between power density and output voltage for BFCs with MWCNT and porous carbon wirings.

Figure 6 shows the relationships between the power density and the output voltage for the BFC measured in glucose solutions with different glucose concentrations of 50, 100, and 200 mM. The power densities for the BFCs in 50, 100, and 200 mM were 0.70 μW/cm^2 at 65 mV, 1.04 μW/cm^2 at 76.6 mV, and 1.07 μW/cm^2 at 80 mV, respectively. The power density was saturated above 100 mM, and this may suggest that the rate of glucose oxidation at the anode was limited by electron transfer processes at the anode.

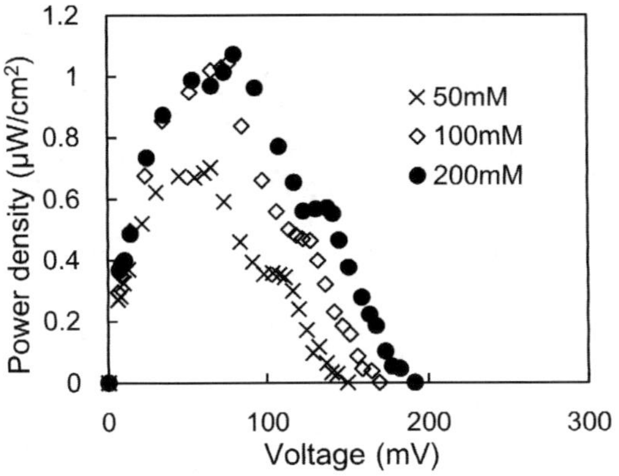

Figure 6. Relationships between power density and output voltage for BFCs in glucose solutions of 50, 100, and 200 mM.

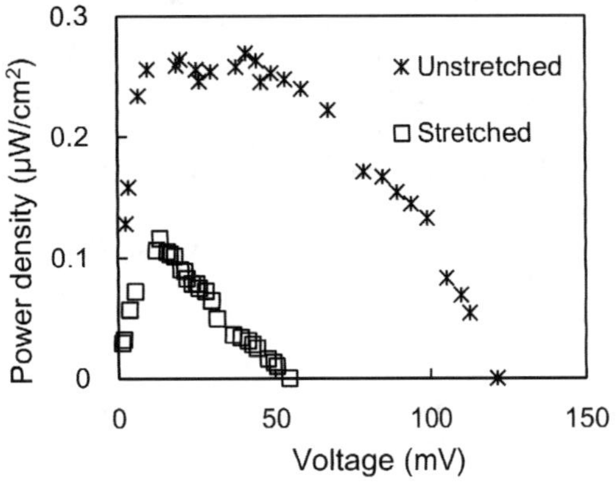

Figure 7. Relationships between power density and output voltages for BFCs without elongation and with elongation of 2.5 mm in 100 mM glucose solution.

Figure 7 shows a comparison of the relationships between the power density and the output voltage without stretching and being stretched at the elongation of 2.5 mm measured in the glucose solution of 100 mM. The power density without elongation was 0.26 μW/cm^2 at 44.2 mV, and that with the elongation of 2.5 mm was 0.12 μW/cm^2 at 15.7 mV. This reduction of the power density under elongation was due to the increased resistance. The power density without elongation in Fig. 8 was smaller than that of Fig. 7. This is due to the fact the resistance was increased by the contact resistance between the jig and the BFC electrodes.

4. Conclusion

In this study, we fabricated a flexible, stretchable glucose BFC with the MWCNT wirings on a PDMS substrate. The BFC investigated consists of a porous carbon anode (area of 30 mm^2) modified by GOD and ferrocene, and a cathode (area of 30 mm^2) modified by BOD. The anode and the cathode were connected the with the MWCNT wirings. The maximum power of 0.31 μW at 76.6 mV (corresponding with a power density of 1.04 μW/cm^2) was realized by immersing the BFC in a PBS with a glucose concentration of 100 mM, at room temperature.

5. Acknowledgment

This research was conducted in supports of the staffs in the Micro Functional Device Research Center of the College of Science and Technology, Nihon University.

References

[1] Zebda A, Renaud L, Cretin M, Innocent C, Ferrigno R, and Tingry S 2010 *Sens and Actuators B: Chemical* **1** 44
[2] Brunel L, Denele J, Servat K, Kokoh K B, Jolivalt C, Innocent C, Cretin M, Rolland M and Tingry S 2007 *Electrochem. commun.*, **9** 331
[3] Zou Y, Xiang C, Yang L, Sun L X, Xu F, and Cao Z, 2008 *Int. J. Hydrogen Energy*, **33** 4856.
[4] Togo M, Takamura A, Asai T, Kaji H and Nishizawa M 2008 *J. Power Sources*, **178** 53
[5] Wei X and Liu J 2008 *Front. Energy Power Eng. China*, **2** 1
[6] Kamitaka Y, Tsujimura S, Setoyama N, Kajino T and Kano K 2007 *Phys. Chem. Chem. Phys.*, **9** 1793
[7] Mano N, Mao F, Shin W, and Heller A 2003 *Chem. Commun.* **2** 518
[8] Katz E, Bückmann A F and Willner I 2001 *J. Am. Chem. Soc.* **123** 10752
[9] Mano N, Mao F and Heller A 2003 *J. Am. Chem. Soc.***125** 6588
[10] Cinquin P, Gondran C, Giroud F, Mazabrard S, Pellissier A, Boucher F, Alcaraz J P, Gorgy K, Lenouvel F, Mathé S, Porcu P and Cosnier S 2010 *PLoS One* **5** 1
[11] Zebda A, Cosnier S, Alcaraz J P, Holzinger M, Goff A Le, Gondran C, Boucher F, Giroud F, Gorgy K, Lamraoui H and Cinquin P 2013 *Sci. Rep.* **3** 1516
[12] Zebda A, Gondran C, Goff A Le, Holzinger M, Cinquin P and Cosnier S 2011 *Nat Commun.* **2** 370
[13] Fukushi Y, Koide S, Ikoma R and Akatsuka W 2013 *J. Photoplym. Sci. Technol.* **26** 303
[14] Miyake T, Haneda K, Yoshino S and Nishizawa M 2013 *Biosens. Bioelectron.* **40** 45
[15] Haneda K, Yoshino S, Ofuji T, Miyake T and Nishizawa M 2012 *Electrochim. Acta* **82** 175
[16] Fujimagari Y, Fukushi Y and Nishioka Y 2015 *J. Photoplym. Sci. Technol.* **28** 357
[17] Ding H, Feng T, Chen Y and Sun Z 2012 *Appl. Surf. Sci.* **13** 5191
[18] Niu Z, Dong H, Zhu B, Li J, Hng H H, Zhou W and Xie S 2013 *Adv. Funct. Mater.*7 1058
[19] Lee D H, Kim J E, Han T H, Hwang J W, Jeon S, Choi S Y and Kim S O 2010 *Adv. Mater.* **11** 1247
[20] Xu F, Wang X, Zhu Y, and Zhu Y 2012 *Adv. Funct. Mater.* **6** 1279
[21] Zhang Y, Sheehan C J, Zhai J, Zou G, Luo H, Xiong J, and Jia Q X. 2010 *Adv. Mater.* 28 3027
[22] Lee J H, Lee K Y, Gupta M K, Kim T Y, Lee D Y, Oh J, and Kim S W 2014 *Adv. Mater.* **5** 765
[23] Liang H W, Guan Q F, Song L T, Yao H B, Lei X, and Yu S H 2012 *NPG Asia Materials* **6** e19.

[24] Mi H Y, Li Z, Turng L S, Sun Y and Gong S 2014 *Mater. Des.*, **56** 398
[25] Cai L, Li J, Luan P, Dong H, Zhao D, Zhang Q, and Xie S 2012 *Adv. Funct. Mater.* **24** 5238
[26] Yamada T, Hayamizu Y, Yamamoto Y, Yomogida Y, Izadi-Najafabadi A, Futaba D N, and Hata K *Nature nanotechnology* 2011 **5** 296
[27] Hocheng H, Chen C M, Chou Y C and Lin C H 2010 *Microsyst Technol,* **16** 423
[28] Siu C P and Chiao M 2008 *J. Microelectromech S* **17** 1329
[29] Kudo H, Sawada T, Kazawa E, Yoshida H, Iwasaki Y and Mitsubayashi K 2006 *Biosens. and Bioele.* **22** 558
[30] Chen C Y, Chang C L, Chien T F and Luo C H 2013 *Sens. and Actuators A: Physical* **203** 20
[31] Lu N, Lu C, Yang S, and Rogers J 2012 *Adv. Funct. Mater.* **19** 4044

Evaluation of alcohol dehydrogenase and aldehyde dehydrogenase enzymes as bi-enzymatic anodes in a membraneless ethanol microfluidic fuel cell

J Galindo-de-la-Rosa[1], N Arjona[2], L G Arriaga[3], J Ledesma-García[1] and M Guerra-Balcázar[1]*

[1] Facultad de Ingeniería, División de Investigación y Posgrado, Universidad Autónoma de Querétaro, Querétaro, Qro., C.P. 76010, México.
[2] Centro de Investigación y Desarrollo Tecnológico en Electroquímica, Unidad Tijuana, Tijuana, B.C., C.P. 22444, México.
[3] Centro de Investigación y Desarrollo Tecnológico en Electroquímica, Pedro Escobedo, Qro., C.P. 76703, México.

E-mail: minbalca@yahoo.com.mx

Abstract. Alcohol dehydrogenase (ADH) and aldehyde dehydrogenase (AldH) enzymes were immobilized by covalent binding and used as the anode in a bi-enzymatic membraneless ethanol hybrid microfluidic fuel cell. The purpose of using both enzymes was to optimize the ethanol electro-oxidation reaction (EOR) by using ADH toward its direct oxidation and AldH for the oxidation of aldehydes as by-products of the EOR. For this reason, three enzymatic bioanode configurations were evaluated according with the location of enzymes: combined, vertical and horizontally separated. In the combined configuration, a current density of 16.3 mA cm^{-2}, a voltage of 1.14 V and a power density of 7.02 mW cm^{-2} were obtained. When enzymes were separately placed in a horizontal and vertical position the ocp drops to 0.94 V and to 0.68 V, respectively. The current density also falls to values of 13.63 and 5.05 mA cm^{-2}. The decrease of cell performance of bioanodes with separated enzymes compared with the combined bioanode was of 31.7% and 86.87% for the horizontal and the vertical array.

1. Introduction

The use of immobilized enzymes onto the electrode surfaces in fuel cells has been of great interest, the major issues to overcome in membraneless microfluidic fuel cells are the increase of power density of such devices and find strategies for the enzymatic immobilization [1-3]. Ethanol has been recognized as the most promising alcohol for use in the energy conversion area due to its physicochemical properties, including high energy density (8.0 kWh^{-1} kg^{-1}) and the multiple pathways to obtain it [4-5]. Alcohol dehydrogenase (ADH) and aldehyde dehydrogenase (ALDH) are the principal enzymes responsible for metabolism of ethanol in humans, catalyzing the conversion of ethanol to acetaldehyde and then to acetate, respectively [6-8]. The ADH system can be regarded as major detoxifying machinery for alcohols and aldehydes. The conversion of ethanol to acetaldehyde coupled with the reduction of NAD^+ to NADH results in an altered NAD^+/NADH ratio and is responsible for many metabolic effects of ethanol [9]:

$$CH_3CH_2OH + NAD^+ = CH_3CHO + NADH + H^+$$

Immobilization is the key factor to enhance the activity of enzymes, especially if two enzymes are immobilized in the same electrode. In this way, with a good immobilization, it is possible to improve almost all the catalytic properties of enzymes: stability, activity, specificity, selectivity, and the inhibition reduction [10]. In this work, alcohol dehydrogenase and aldehyde dehydrogenase were used as catalysts for the ethanol oxidation and the acetaldehyde oxidation (as byproduct of ethanol reaction) respectively, in order to improve the obtaining electrical energy. For this reason, three enzymatic arrangement on Toray paper electrodes were tested: a physical mixture dispersed on the electrode, and the enzymes horizontal and vertically separated. The effect of the electrode array was analyzed on the cell performance of an ethanol membraneless microfluidic fuel cell.

2. Experimental

2.1 Electrodes preparation

Three catalytic inks were prepared, one for alcohol dehydrogenase (EC 1.1.1.1 initial activity ≥300 units/mg protein from *Saccharomyces cerevisiae*), other for aldehyde dehydrogenase (EC 1.2.1.5 initial activity ≥2 mg protein from *Saccharomyces cerevisiae*) and also a third, which involve the mixture of both enzymes. The catalytic ink was prepared using 5 μL of Nafion (Sigma Aldrich), 10μL of glutaraldehyde, 1000 μL of deionized water, 2 mg β-nicotinamide adenine dinucleotide sodium salt (NAD^+, Sigma Aldrich) and 20 mg of tetrabutylammonium bromide (Sigma Aldrich) per milligram of enzyme for the first two enzymes individually and per milligram of the mixture of enzymes (half of alcohol dehydrogenase and half aldehyde dehydrogenase).

The bioanodes consisted of two 2.5×30 mm (wide and long) Toray porous paper electrodes (EC-TP1-060T for combined immobilization of enzymes across the electrode and the vertical configuration) and two 1.25 ×30mm Toray porous paper electrodes (EC-TP1-060T for horizontal configuration). These carbon paper slides were covered with 100 μL of 5:100 v/v% Nafion-water and dried with hot air. After, theses slides were employed for the methylene blue electro-polymerisation using a Biologic VSP Potentiostat/Galvanostat, by cyclic voltammetry from -1 to 1 V vs. SCE for 15 cycles at 0.05 V s^{-1} scan rate. The electrolytic bath consisted of 0.05 M methylene blue. The inks were then sonicated for 20 min and mixed by vortex for other 15 min. Three configurations for the immobilization were performed: a) the first corresponded to the use of the combined enzymatic mixture. This was done placing 100 μL of the catalytic ink over the entire surface of the electrode. b) In the second, two smaller electrodes were used and 50 μL of the alcohol dehydrogenase ink was depositing onto the surface of one them and 50μL of the aldehyde dehydrogenase ink on the other electrode. c) A slide was vertically divided in two sections. In the superior section were deposited 50 μL of alcohol dehydrogenase ink and 50 μL of ink enzyme aldehyde dehydrogenase in the bottom section (Figure 1). All the electrodes with deposited ink were dried at room temperature.
The Pt-based abiotic cathode was prepared using commercial Pt/C (30% E-TEK) as the catalyst and carbon nanofoam (2.5×30 mm, Marketech®) as the electrode. Briefly, 120 μL of isopropyl alcohol and 14 μL of Nafion were added per milligram of catalyst and mixed for 30 minutes. The catalytic ink was deposited on the carbon nanofoam surface using the spray technique until the electrode weight increase 1 mg.

2.2 Hybrid microfluidic fuel cell evaluation

The design of the microfluidic fuel cell [11] consists of three main sections, which were constructed using non-sophisticated techniques. The first section consisted in two supporting plates made of Poly-(methyl methacrylate) (PMMA). These were patterned using a computer numerical control (CNC). The second section consists in a homemade silicone elastomer film (Silastic®, Dow Corning, prepared using an Elcometer® Film Applicator with a final thickness of 200 μm) used as both, gasketing and as cell channel structure. The third section consisted in the core of the device, which is the bioanodes and cathode electrodes. The hybrid microfluidic fuel cell tests were performed by injecting 0.1 M ethanol N_2-saturated (4 U. P. D. Praxair) as fuel and oxygen from air as oxidant within two aqueous streams that contained phosphate buffer (pH 8.86) and 0.3 M KOH on the anodic and the cathodic side respectively and used as supporting electrolytes. The flow rate was 50 μL min^{-1} for both streams. Figure 1 shows the configurations of the bioanodes used in microfluidic fuel cells: a) which have the two enzymes immobilized along the electrode, b) having two thin electrodes where the first from right to left is immobilized enzyme alcohol dehydrogenase and the second with aldehyde dehydrogenase enzyme, and c) where the electrode is divided into two, which are immobilized on *top* alcohol dehydrogenase and bottom aldehyde dehydrogenase.

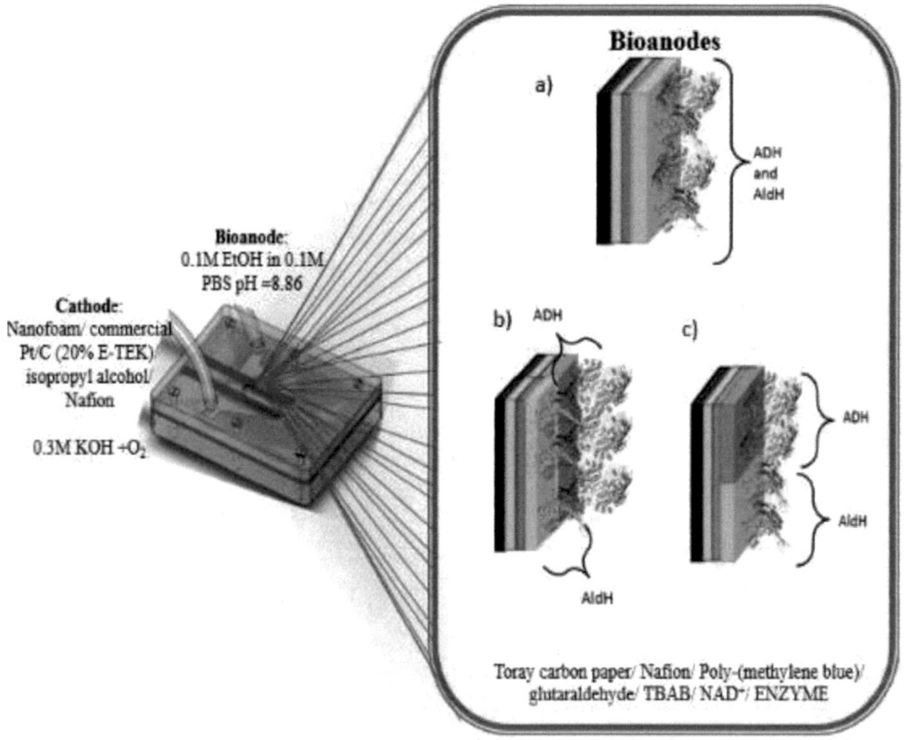

Figure1. Schematic representation of the configuration of enzymatic bioanodes: a) combined, b) horizontally separated and c) vertically separated.

3. Results and discussion

3.1 Microfluidic fuel cell performance

The performance of the bi-enzymatic bioanodes toward the EOR in a membraneless ethanol hybrid microfluidic fuel cell is illustrated in Figure 2 and the obtained results for the three hybrid microfluidic fuel cells are presented in Table 1. The hybrid MFC with the mixture of enzymes exhibited a cell voltage of 1.141 V and a maximum current density of 16.3331 mA cm^{-2} (Fig. 2). In contrast, the hybrid MFC with horizontal bioanode configuration exhibited an open circuit voltage of 0.6795V and a current density of 5.0735 mA cm^{-2}. The power density for the hybrid MFC with the vertical configuration was 13.6176 and 4.8158 mW cm^{-2}, respectively. The drop in performance exhibited by the vertical electrode array can be related –according with the behavior of the curve-to a deficiency in both, the electron transference between enzymes and a poor mass transport due to the electrode configuration. In the case of the vertical array, the lower performance compared with the homogeneous enzyme mixture are related to ohmic issues, such as problems with electron transference. The stability tests were performed by chronoamperometry (Fig. 3). The electrode configuration related to a horizontal array showed the poorest stability, the current density fell over the entire period of time that the experiment was carried out. The combined enzyme mixture and the vertical enzymatic array showed good stability over time. The vertical array exhibited a constant current density of 5 mA cm^{-2} and the combined array of 7.5 mA cm^{-2}.

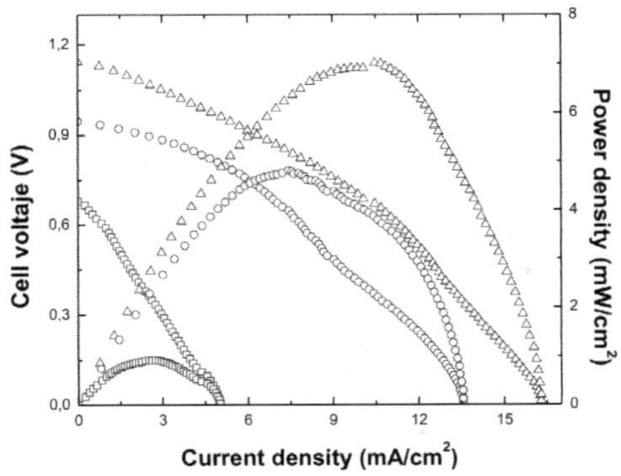

Figure2. Polarization and performance curves of the device using three enzymatic bioanodes arrays: combined (blue), horizontally separated (green) and vertically separated (black).

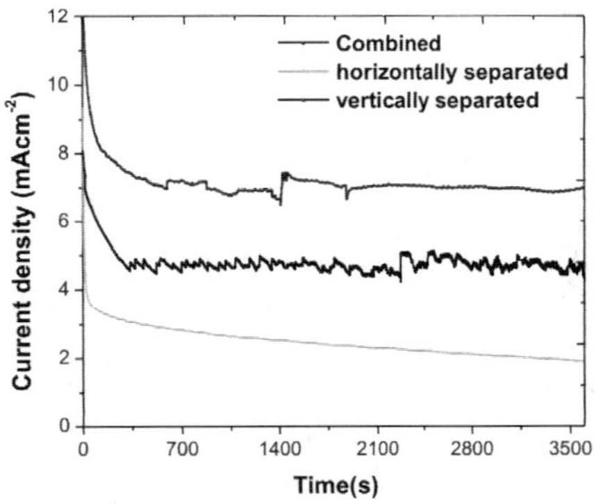

Figure3. Chronoamperometry using 0.02 V as potential at maximum current density.

Table1. Performance of microfluidic fuel cells witch bi-enzymatic bioanodes: Alcohol dehydrogenase and aldehyde dehydrogenase.

Bioanode configuration	Current density (mA/cm^2)	Open circuit potential (V)	Power density (mW/cm^2)
Enzyme combined	16.3331	1.141	7.0289
Enzyme horizontally separated	5.0735	0.6795	0.9186
Enzyme vertically separated	13.6176	0.9466	4.8158

3.2 Bioanode SEM analysis

Scanning electron microscope analysis of the anode corresponding to the combined enzymatic array immobilized on the Toray paper surface was made (Fig. 4). This electrode was selected because of its high performance in the microfluidic fuel cell. SEM images (Fig. 4a and 4b) were acquired to show the successful use of a covalent binding method to immobilize the alcohol dehydrogenase and aldehyde dehydrogenase enzymes. The micrograph of naked Toray electrode is shown in Fig. 4a, left. Meanwhile, the Toray electrode with the bioanode array (Nafion, methylene blue, glutaraldehyde, TBAB, NAD$^+$, Nafion) is shown in Fig. 4b right. It is clearly observed from micrograph at the right position that, some material is deposited over the entire surface of the electrode which could indicate that, through electro-polymerisation, ADH y AldH enzymes can be adhered to the carbon-based surface.

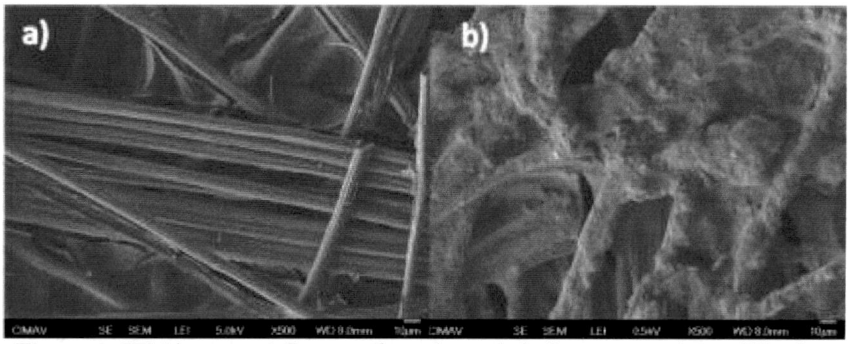

Figure 4. SEM images: a) Toray carbon electrode before enzymatic modification and b) electrode with ADH y AldH immobilized

Conclusion

The immobilization method as well as the enzymatic arrangement in multi-enzymatic electrodes have a significant effect on the performance of microfluidic fuel cells. In this work, three electrode configurations were proposed in which the enzymes were immobilized on the electrode surface. The homogeneous mixture between both enzymes showed the highest power density (7.0289 mW cm-2) herein reported. This performance, compared with the other electrode configuration, was related to an enhancement in the electron transference probably because of the short distance between them which allowed a better conduction of charge from both electro-oxidation reactions. Also, the order of magnitude (milliwatts) of the cell performance was achieved with the use of three-dimensional electrodes as flow-through electrodes. In addition, a covalent binding immobilisation method based on Nafion, poly-(methylene blue), glutaraldehyde, TBAB and NAD^+ was used to enhance the durability of the MFC.

Acknowledgements

The authors thankfully acknowledge to national council of science and technology (Conacyt) for the support through SEP-CONACYT-2012-01-17992

References

[1] Bullen R A, Arnot T C, Lakeman J B and Walsh F C 2006 *Biosens. Bioelectron* **21** 2015-1045
[2] Akers N L, Moore C M and Minteer S D 2005 *Electrochim. Acta* **50** 2521–2525
[3] Minteer S D, Liaw B Y and Cooney M J 2007 *Curr. Opin. Biotech.* **18** 228–234
[4] Carrera-Cerritos R, Fuentes-Ramírez R, Cuevas-Muñiz F M, Ledesma-García J and Arriaga L G 2014 *J. Power Sources* **269** 370-378
[5] Ermete A 2007 *J. Power Sources* **170** 1-12
[6] Staab C A, Alander J, Morgenstern R, Grafström R and Höög J-O 2009 *Chem. Biol. Interact.* **178** 29-35
[7] Dolega Anna 2010 *Coord. Chem. Rev.* **254** 916-937
[8] Staab C A, Hellgren M, Höög J-O 2008 *Cell. Mol. Life Sci.* **65** 3950
[9] Lieber C S 1999 Alcohol *Clin. Exp. Res.* **23** 991
[10] Joonmok S, Gha-Young K, and Seung-Hyeon M 2011 *J. Electroanal. Chem.* **653** 14 20
[11] Ortiz-Ortega E, Goulet M A, Lee J W, Guerra-Balcázar M, Arjona N, Kjeang E, Ledesma-García J and Arriaga L G 2014 *Lab Chip* **14** 4596 4598

A Solar Cell Powered Adaptive Charging Circuit for CMOS Integrated Micro Fuel Cells

C. Moranz[1], H. Ghafarian M.[1], K. Ylli[2] and Y. Manoli[1,2]

[1]University of Freiburg - IMTEK, Freiburg, [2]Hahn-Schickard, Villingen-Schwenningen

E-mail: Christian.Moranz@imtek.uni-freiburg.de

Abstract. This paper presents an autonomous interface circuit which uses solar cells to automatically recharge chip integrated micro fuel cell accumulator arrays. These accumulators comprise a fuel cell for powering systems and a hydrolysis cell for charging the integrated hydrogen storage. The charging current is continuously monitored and the interface circuit automatically maximizes and controls the charging current. The solar cells powering the charging process are projected to be part of the chip package. The presented system, in combination with a previously developed voltage regulator and integrated sensors or actuators enables the implementation of fully integrated energy autonomous systems.

1. Introduction

Various efforts in the last years aim for fully integrated autonomous microsystems [1, 2, 3]. While miniaturization of the system size is resulting in a reduction of power consumption, a reliable miniaturized fully integrated power supply is a common issue. Commonly, batteries are used as energy source. However the limited life time, maintenance requirements and the challenge of scaling down the size lead to new power supply strategies.

On one hand, energy harvesting is one of the promising approaches to overcome these drawbacks. Nevertheless, the variable available environmental energy requires a temporary energy storage to ensure a sustainable power supply.

On the other hand, integrated micro fuel cells have been a novel approach [1, 4], providing a CMOS-integrated versatile power supply. However, the life time of this power supply is limited by the storage capacity of the fuel cells. In order to enable a virtually unlimited lifetime of autonomous systems, an electrical recharging of the fuel cells enabled by a hydrolysis cell was proposed [5].

Therefore, we present an adaptive interface circuit which enables the usage of solar cells as energy harvesting devices and fuel cell accumulators [5] as temporary energy storage. The presented interface circuit allows for a charge current control and maximization independent of the envrironmental conditions. This interface circuit in combination with the solar cells and the fuel cell accumulators, results in an autonomous power supply with virtually unlimited lifetime without any additional external power source.

2. Fuel cell accumulator and solar cell as power supply

A cross section of the fuel cell accumulator (FCA) presented in [5] is shown in Fig. 3. It comprises a fuel cell to power a connected system and the hydrolysis cell to allow for an electrical recharge

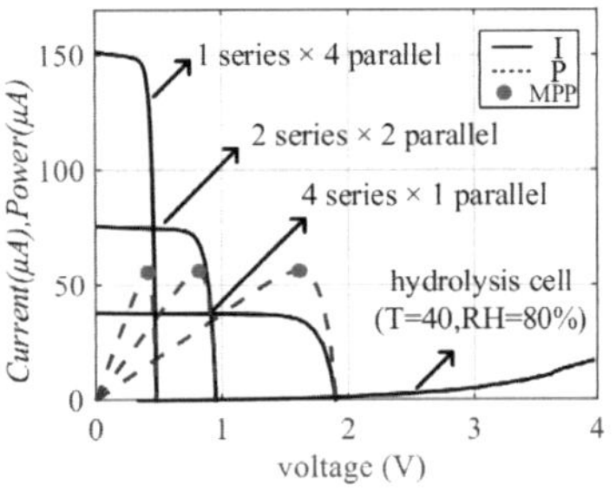

Figure 1. Typical voltage to current characterics of the hydrolysis cell presented in [5] and a solar cell module with different configurations.

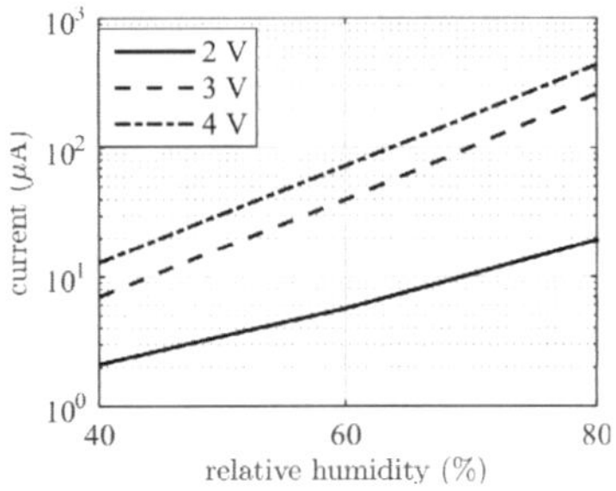

Figure 2. Depdendency of the charging current on the charge voltage and ambient relative humidity

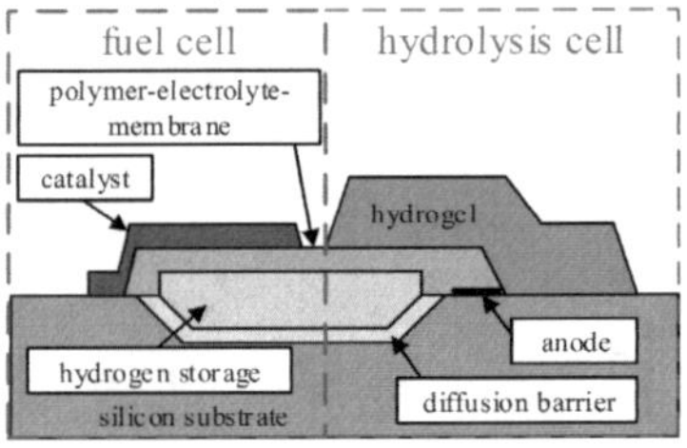

Figure 3. Cross section of the fuel cell accumulator proposed in [5]

of the hydrogen storage. The fuel cell converts the chemical energy stored as hydrogen in the palladium hydrogen storage and oxygen from the ambient air into electrical energy and releases water. In the hydrolysis cell, a hydrogel collects water from the ambient air which can then be electrically split into hydrogen and oxygen. The hydrogen then diffuses into the palladium storage while the oxygen is released into the ambient air.

A typical voltage to current characteristic of the hydrolysis cell at a temperature of 40 °C and a relative humidity of 80 % with an area of 3.15 mm^2 together with the different solar cell configurations with an area of 4 cm^2 is shown in Fig. 1. If a voltage higher than approximately 1.5 V is applied on the hydrolysis cell, the current is positive which means the FCA is being charged.

The voltage to current characteristic of the solar cell with an area of 4 cm^2 is modeled with a two diode model [6] and is shown in Fig. 1. It shows a typical open circuit voltage of approximately 0.6 V and a voltage of approximately 0.4 V at the maximum power point (MPP). In addition, this graph shows different configurations of solar cells while the area of all configurations is kept constant. If the area of each single solar cell is lowered but the number of series connected solar cells is increased, the open circuit voltage increases while the maximum output current decreases. Connecting more than 4 solar cells in series is not advisable due to resulting low area of each of the solar cells while the boundary losses increase, lowering the overall output power.

Connecting one of the solar cells directly to the hyrolysis cell is not advisable because of the low charge current and the operation of the solar cell far off its MPP. Moreover, due to the strong dependence of the electical characteristic of the FCA [1] and the solar cell, a direct connection between these two would not provide an optimized charging time. The dependency of the FCA charging current can be seen in Fig. 2. It shows that the charging current can change by a factor of approximately 10 at equal hydrolysis voltage.

In addition, the charging current of the hydrolysis cell is not allowed to exceed a specific maximum current in order not to irreversible damage the FCA. Therefore, we propose the application of a charge pump based interface circuit, monitoring and maximizing the charge current.

3. Charging System

The proposed power management circuit as shown in Fig. 4 fulfils three tasks: Maximize the charging current independent of input power (e.g., sunlight intensity) and ambient conditions (e.g., temperature, humidity), track the state of charge and set an over-current limit.

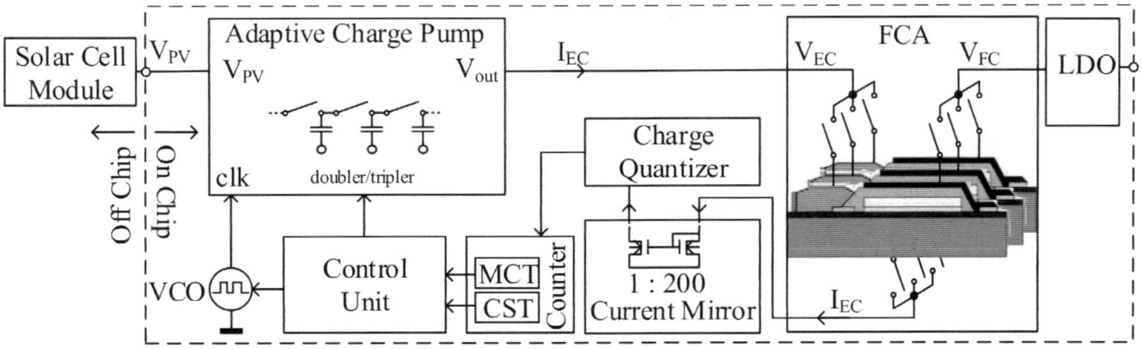

Figure 4. Proposed system with a solar cell module attached to the chip cover, the adaptive charging circuit, the rechargeable fuel cell array and a variable low drop-out voltage regulator (LDO).

The energy transfer stage is an adaptive 6 phase charge pump [7] with two modes of operation. In *doubler* mode, it acts as voltage doubler, in mode *tripler* it acts as a voltage tripler while the number of stacked capacitors is adjusted accordingly to two and three. Additionally, its operation frequency can be varied by a voltage controlled oscillator (VCO). In order to maximize the charge current, a maximum current tracking unit (MCT) monitors the charging current and modifies the operation mode and frequency of the adaptive charge pump. Therefore, with an active current mirror, a fraction of the charging current is mirrored into the charge quantizer. If a fixed amount of charge is delivered to the quantizer, the MCT counter is incremented. After a predefined charge period, the counter value is stored and compared to the previously stored counter value. An increase or decrease of the amount of the charge stored within one period is detected. This comparison is the input for a hill climbing algorithm, adjusting the VCO frequency and the charge pump operation mode accordingly. In addition, the output signal of the charge quantizer is also connected to the charge state tracking counter (CST) in order to track the charge transferred during a single recharge cycle. If this counter reaches a predefined value, representing a fully charged FCA, it is reset and the next empty fuel cell in the array is to be connected to the charging circuit.

3.1. Adaptive Charge Pump

The circuit schematic of the 6 phase adaptive charge pump (CP) is illustrated in Fig. 5(a). The 6 phases driving the charge pump are generated by a dedicated non-overlaping phase generator [7]. To differentiate the left and the right side of the circuit, each component has subscript of A and B, respectively. In order to simplify the explanation, just one half of the CP is discussed next and the subscript is not mentioned in the explanation. The other half of the CP operates with a phase shift of 180°.

The circuit schematic consists of 3 closely coupled charge pumps. The main charge charge pump, consits of M_2, M_4 and CA. The gates of M_2 and M_4 are driven by boosted clock signals, generated with unloaded charge pumps much smaller than the main charge pump. Using a smaller unloaded extra charge pump has the advantage of a swingless driving signal, enhancing the on-resistance of M_2 and M_4 and lowering the charge pump losses.

The key component which makes this charge pump adaptive is the capacitance array CA with the corresponding switches as shown in 5(b). In the mode *doubler*, the two flying capacitors C_1 and C_2 are connected in parallel in both charging and discharging period. However, in the *tripler* mode, the discharge period, C_1 and C_2 are connect in series as shown in Fig. 5(b). In the charging period, both parallel capacitors C_1 and C_2 are charged up to V_{PV} via M_2. In the discharging period, by applying V_{PV} to the bottom of CA, the top of CA would be connected via M_4 to V_{out}. Depending on the mode of operation *doubler* or *tripler* , the boosted voltage on CA is $2 \cdot V_{PV}$ and $3 \cdot V_{PV}$, respectively.

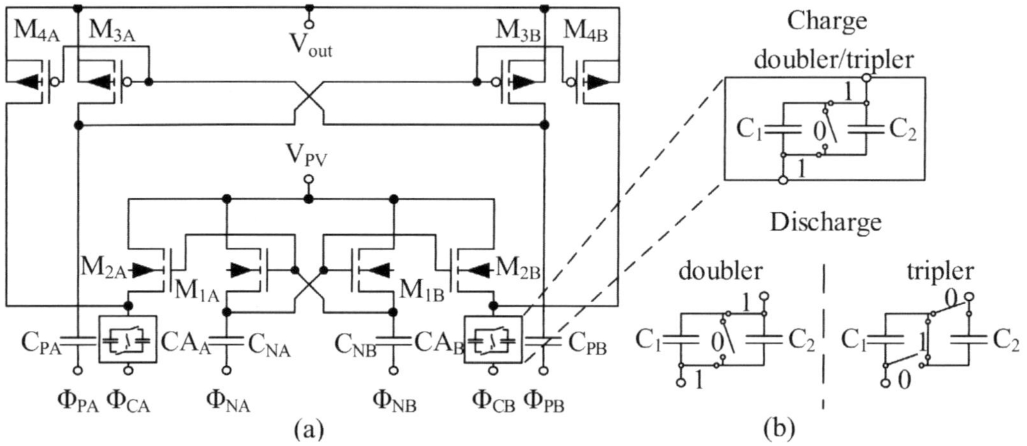

Figure 5. Circuit schematic of the adaptive charge pump (a) and illustration of the flying capacitor array (CA) switch matrix in charge and discharge state (b).the bulk of all NMOS is connected to ground.

4. Results

Fig.6 shows the simulated autonomous operation with start-up and restart at 12 ms of the system. Due to the clocked power transfer, there is a ripple on V_{PV}, V_{out} and I_{out}. Since the logic of the control system is powered directly from the solar cell, the operating point of the solar cell module also has a voltage ripple with a frequency equal to the operation frequency of the logic unit.

In order to show full system operation under varying ambient conditions, the input power density (light intensity) was changed each 10 ms. Within each change of the input light intensity, the system automatically maximizes the charging current, independent of the environmental conditions. The operation frequency and the operation mode are adopted accordingly.

The efficiency of the system, defined as $\eta = P_{out}/P_{in}$ is shown in Fig.6. P_{out} is the power used to charge the FCA and P_{in} is the power extracted from the solar cell module. The power consumption of the control circuits, the charge quantizer, the corrsponding counters and the clock boosters are all included in the input power. The efficiency of the charge pump and the resulting total system efficiency is limited by the stray capacitance. In this work, a poly over diffusion capacitor is deployed where the $\alpha = \frac{C_s}{C_f} = 0.2$ and C_s is the stray capacitance of the top plate to ground. Therefore, the maximum power efficiency which can be assumed for the deployed charge pump in this work, is between 50 % to 64 % [8].The presented system shows a peak efficiency of 60 % including the losses of all control circuits. Furthermore, a term named as total power extraction efficiency (η_{TPE}) is defined as in equation 1 and is depicted as well with η in Fig.6.

$$\eta_{TPE} = \frac{P_{out}}{P_{in,MPP}} \tag{1}$$

Here Pout is the power used to charge the FCA and $P_{in,MPP}$ is the maximum power which can be extracted theoretically from the solar cell module if this is operated at its maximum power point. Therefore, where η_{TPE} is nearer to the efficiency of the system, it shows that the system is extracting more power from the solar cell module and nearer to the input maximum power point within the P-V curve.

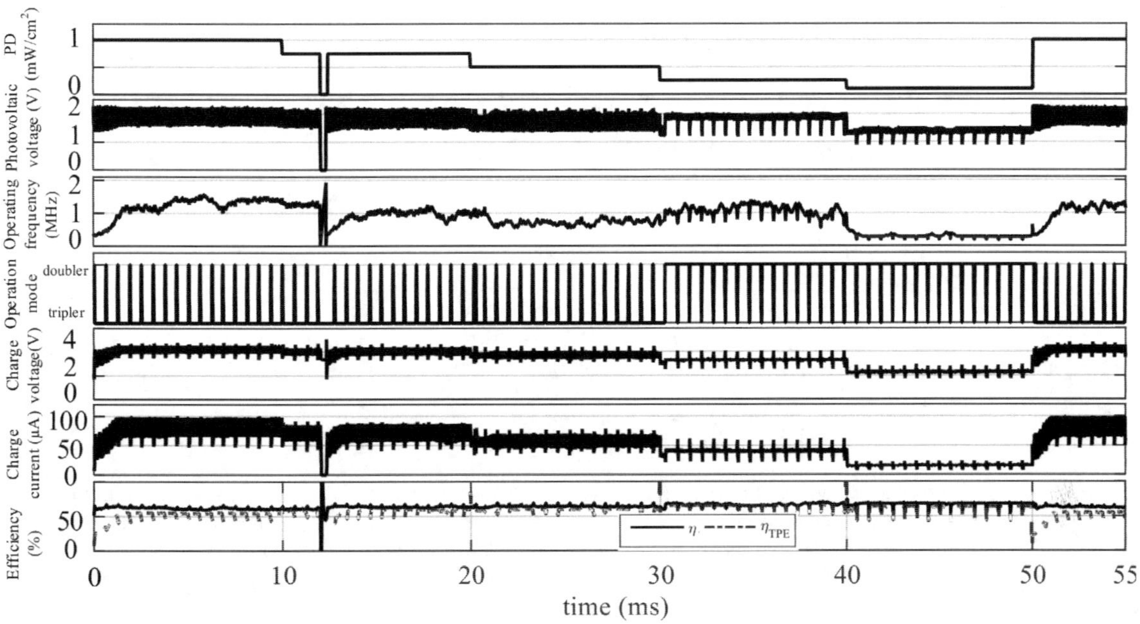

Figure 6. Simulated autonomous system operation showing start-up and the self-adoption of the circuit according to the input power density (PD).

5. Conclusion

This paper presented an autonomous interface circuit which use small scaled solar cells to automatically recharge chip integrated micro fuel cell accumulator arrays. The interface circuit maximizes the charging rate of the fuel cell accumulator by maximizing the charging current. The presented system in combination with a previously developed voltage regulator and additional integrated sensors or actuators resuls in a aunonomous fully integrated system for remote applications.

References

[1] Frank M, Kuhl M, Erdler G, Freund I, Manoli Y and Mller C 2009 *ISSCC 2009 / SESSION 17*

[2] Leicht J, Maurath D and Manoli Y 2012 Autonomous and self-starting efficient micro energy harvesting interface with adaptive mppt, buffer monitoring, and voltage stabilization *ESSCIRC (ESSCIRC), 2012 Proceedings of the* (IEEE) pp 101–104

[3] Leicht J, Amayreh M, Moranz C, Maurath D, Hehn T and Manoli Y 2015 Electromagnetic vibration energy harvester interface ic with conduction-angle-controlled maximum-power-point tracking and harvesting efficiencies of up to 90% *Solid-State Circuits Conference-(ISSCC), 2015 IEEE International* (IEEE)

[4] Frank M, Kuhl M, Erdler G, Freund I, Manoli Y, Muller C and Reinecke H 2010 *Solid-State Circuits, IEEE Journal of DOI - 10.1109/JSSC.2009.2034441* **45** 205–213 ISSN 0018-9200

[5] Zimmermann D, Becker J, Freund I, Muller C and Reinecke H 2013 *Proceedings of the SSI Conference, 2013*

[6] Phang J, Chan D and Phillips J 1984 *Electronics Letters* **20** 406–408

[7] Salem L G and Mercier P P 2014 4.6 an 85%-efficiency fully integrated 15-ratio recursive switched-capacitor dc-dc converter with 0.1-to-2.2 v output voltage range *Solid-State Circuits Conference Digest of Technical Papers (ISSCC), 2014 IEEE International* (IEEE) pp 88–89

[8] Favrat P, Deval P and Declercq M J 1998 *Solid-State Circuits, IEEE Journal of* **33** 410–416

Microwatt-Powered, Low-Cost, Printed Graphene Oxide Humidity Sensors for Distributed Network Applications

Anthony P Taylor[1] and Luis F Velásquez–García[2]*

[1] Edwards Vacuum LLC, Sanborn, NY, 14132, United States
[2] Microsystems Technology Laboratories, Massachusetts Institute of Technology, 77 Massachusetts Ave., Cambridge, MA, 01239, United States

* Email: Velasquez@alum.mit.edu

Abstract. We report the design, fabrication, and characterization of novel conductometric humidity sensors that employ an ultrathin film of graphene oxide (GO) nanoflakes as transducing element. The GO film is deposited with a shadow mask using an electrospray emitter operated at atmospheric pressure; no post treatments–including annealing or doping–were applied to the GO sensors. The Van der Pauw electrode structure used in the sensor had 50 μm linewidths separated by 600 μm and was fabricated with the lift-off metallization technique. In dynamic humidity tests conducted at atmospheric pressure, the sensor tracks the response of a commercial sensor and reacts to changes in humidity in less than 500 ms. There is a quadratic dependence of the relative humidity on the sensor resistance for the relative humidity range between 3% and 63%, with more than a three-fold change in resistance over the range. The power consumption of the sensor is less than 30 μW while drawing 7 μA, and less than 15 μW while drawing 5 μA. Our devices are promising candidates for deployment in a distributed sensor network due to their low cost, small size, and low power consumption.

1. Introduction

Distributed humidity and temperature sensor networks have been investigated for deployment in commercial buildings as part of energy management of heating, ventilating, and air conditioning (HVAC) systems [1]-[2]. For this application, size, power consumption, reliability, and cost are the main driving metrics for sensor selection. Conductometric gas sensors–based on the chemoresistive response of semiconducting metal oxide films–are widely used due to their simplicity, flexibility in production, and broad applicability to many fields [3]. Typically, the adsorption of a gas molecule on the surface of a metal oxide alters surface electronic properties, causing a change in electrical conductivity [4]. Although many metal oxides could be used for gas sensing, only a few show the appropriate combination of adsorption ability, catalytic activity, sensitivity, and thermodynamic stability. These select metal oxides (e.g., SnO_2, TiO_2, and ZnO), however, are the least active from the catalytic point of view [5]; doping with redox-active noble metal nanoparticles, such as Pt, Au, and Pd, is commonly done to enhance conductivity response and gas sensitivity. Unfortunately, noble metals are expensive, thereby precluding their use in low-cost applications.

An appealing alternative material for reactive gas sensing is graphene oxide (GO) because of its high sensitivity to adsorbed surface species and harsh environment compatibility. Thin-film humidity sensors made of GO, and the related 2D material reduced graphene oxide (rGO), have been deposited in aqueous suspensions by drop-casting and airbrush spraying [6]-[7]. However, electrospray printing allows for more precise control of the film properties compared with the other techniques and is compatible with low-cost, large-area manufacturing through emitter multiplexing [8]-[11].

In this work, low-cost GO humidity sensors are reported; the devices have a ~60 nm-thick transducing element fabricated at low temperature (< 65 °C) using electrospray printing of nanoflakes with a shadow mask, and no post treatment to the devices was conducted including annealing and doping. Devices with multiple electrode configurations were fabricated on SiO_2-coated Si wafers using contact photolithography and the lift-off technique. Our devices were characterized as humidity sensors using a home-built environmental chamber at atmospheric pressure while varying the relative humidity (RH) between 3% and 63%; the response was benchmarked using a commercial humidity sensor. Electrospray printed rGO thin film structures have recently been reported for humidity detection [12]. However, our electrospray printed GO humidity sensors are superior to those reported in [12] because our GO sensors respond faster, reaching equilibrium more than 30 times quicker at 50 % RH, and because the change of resistance of our sensors has a dynamic range that is an order of magnitude larger.

2. Description of Fabrication Method

The starting substrate is a 1 cm-wide square piece of single-crystal silicon coated with 500 nm of thermal oxide. First, image-reversal contact lithography is conducted in a spun-coated thin film of photoresist to transfer the layouts of the electrodes. Then, a 100 nm Au film on top of a 10 nm Cr film is deposited everywhere on the substrate using the electron beam evaporation technique. After that, the photoresist is dissolved using acetone, which removes the metal stack everywhere on the substrate except at the bottom of the cavities defined by the lithography, leaving behind the electrode pattern. Finally, the GO layer is electrospray printed on top of the electrodes using a home-built apparatus.

The home-built electrospray printer has a syringe pump that delivers an aqueous dispersion of GO nanoflakes to a blunt hollow stainless steel needle with a 300 μm inner diameter. The starting material is a commercial GO solution (Sigma Aldrich product 777676) with a concentration of 4 mg/ml GO in water. We diluted the starting material using deionized water to a concentration of 40 μg/ml. The needle tip is centered in the plane of a grounded stainless steel annular electrode, and a high-voltage power supply biases the needle at around 3kV to produce droplet emission. A shadow mask is placed over the substrate to be printed, which is on top of a hot plate, and the assembly is mounted on a PC-controlled 3-axis stage above the needle. To perform a film deposition, the substrate is first mounted on the heated stage and brought to temperature for about 5 minutes. The GO nanoflake solution is then loaded into the syringe pump, and the system is primed up to the tip of the emitter. After this, the syringe pump is activated, delivering a flow rate around 1 μl/min, and a high bias voltage is applied to the tip and adjusted to yield a stable Taylor cone. The beam of GO droplets is positioned to the side of the substrate and observations are made to ensure that the emission is stable. Finally, a deposition recipe (i.e., a sequence of commands that move the stage in a pre-established fashion) is run.

Figure 1. Electrospray printed GO films on a dielectric substrate at different separation distances and stage speeds. The third imprint from the right corresponds to an optimal separation distance of 3 cm.

Optimization experiments with the home-built electrospray printer were conducted to deposit interconnected films by keeping constant the flow rate and bias voltage while varying the separation distance, stage speed, number of passes, and surface temperature. From these experiments, the optimum separation distance was found to be 3 cm (Figure 1). From the imprints collected, the beam divergence of the electrospray is estimated at 10° semi-angle, which is in agreement with the literature on cone-jet electrospray sources [13]-[14]. Films deposited at temperatures slightly above room temperature (50 – 64°C) were of better quality than those deposited at room temperature; in particular, heated samples had less thickness variation across the coating and did not show the liquid accumulation at the edges of the imprints created on unheated samples.

Originally, a single shadow mask (stainless steel, 250 μm thick, 1.3 mm ID aperture) was used to create the imprints. However, during operation of the electrospray emitter the GO solution wetted the perimeter of the aperture of the shadow mask, and upon drying and mask removal, freestanding walls of GO nanosheets were left on some areas around the edge of the hole, and patches of the imprint were delaminated from the substrate in other areas. To prevent edge sidewall formation and imprint delamination, a two-layer shadow mask was implemented (two stainless steel sheets, each 125 μm thick, bottom layer 1.3 mm aperture, top layer 1.0 mm aperture, concentrically mounted). With the double shadow mask and under optimized conditions, GO films with average thickness <100 nm were successfully manufactured.

Once fabricated, the chips were placed in standard IC packages with Au wire-bonds (Figure 2, left). The sensor has an active area of about 0.4 mm^2 between the electrodes; the metallization lines underneath the active area are 50 μm wide with 600 μm separation (Figure 2, right). The GO nanoflakes form a connected conductive network between the four electrodes.

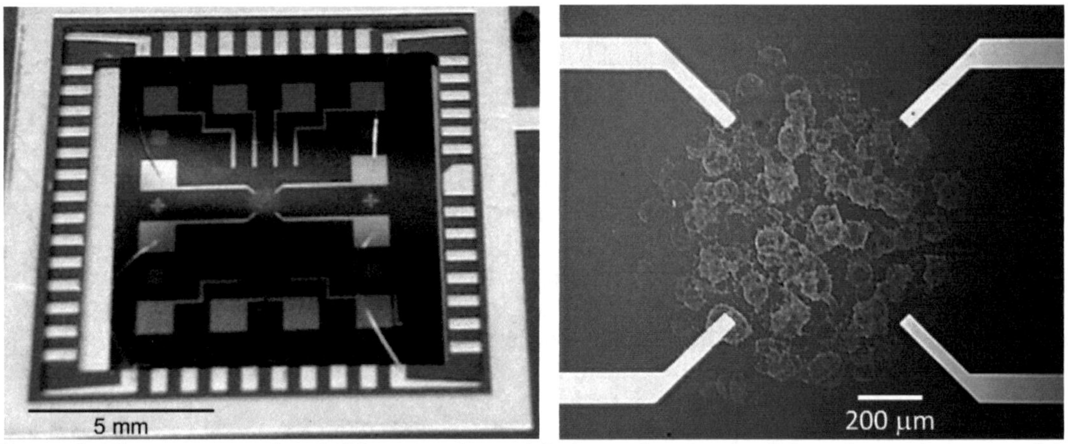

Figure 2. Electrospray-printed GO sensor (greenish dot at the center of the substrate) on top of Au electrodes in the Van der Pauw configuration (left). Close-up of the sensor's active are (right); remnants of the droplets created during electrospray printing are visible.

3. Characterization of the Devices as Humidity Sensors

The packaged sensors were placed inside a custom-built environmental chamber (Figure 3) where the relative humidity was varied between 3% and 63%. Room air was bubbled through a bottle containing water by pulling a house vacuum on the chamber. The relative humidity in the chamber was varied by mixing the moistened room air with dry nitrogen at the inlet and varying the house vacuum draw at the outlet. We were able to set a maximum RH of about 40 % within the chamber while under flow conditions; we obtained RH levels up to around 60% by isolating the chamber and soaking it for about 20 minutes. A commercial humidity sensor (Honeywell HIH-4000) was mounted near the GO sensor for comparison (Figure 3, right).

In the Van der Pauw electrode configuration of the GO sensor, I_1 current was supplied to electrode 1 with a source measuring unit (SMU) Keithley 2612B, electrode 2 was connected to ground, and electrodes 3 and 4 were floating. Voltage readings of the electrodes were logged on the GO sensor and the commercial humidity sensor with a Dataq DI-149, which is an 8-channel data logger and was set to sample data at a rate of 20 Hz. The resistance across pins 3 and 4 was calculated using the expression $R_{1243}= (V_4-V_3)/I_1$ where V_4-V_3 is the bias voltage between electrodes 4 and 3. Through experimentation of the magnitude of I_1 it was determined values in the 5-7 μA range yielded the largest difference in voltage between pins 4 and 3.

We conducted two types of experiments to characterize the change of resistance in the sensors with relative humidity: *dynamic response tests* were the humidity was varied quickly while gas was flowing to the chamber, and *static response tests* where the humidity was raised to highest achievable value attainable by the experimental apparatus to then bled down to the lowest value. In both cases, the boxcar averaging technique with a window size of 0.55 s was used to reduce the signal noise. During the dynamic response tests, the resistance change of the GO sensor (red curves of Figures 4(a)-(c)) and the response of the commercial sensor (black curves of Figures 4(a)-(c)) tracked each other closely as the relative humidity was varied rapidly between 10 and 30%. Figure 4(b), which is a magnified view of a rapid humidity transition zone in Figure 4(a), demonstrates that the signal response time is less than 500 ms for the GO sensors and faster than the signal response of the commercial sensor.

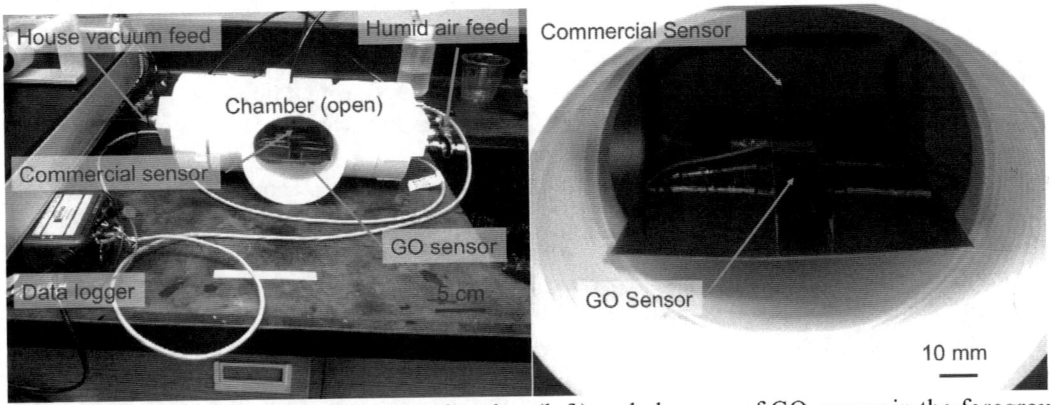

Figure 3. Home-built environmental test chamber (left) and close-up of GO sensor in the foreground and commercial sensor in the background (right).

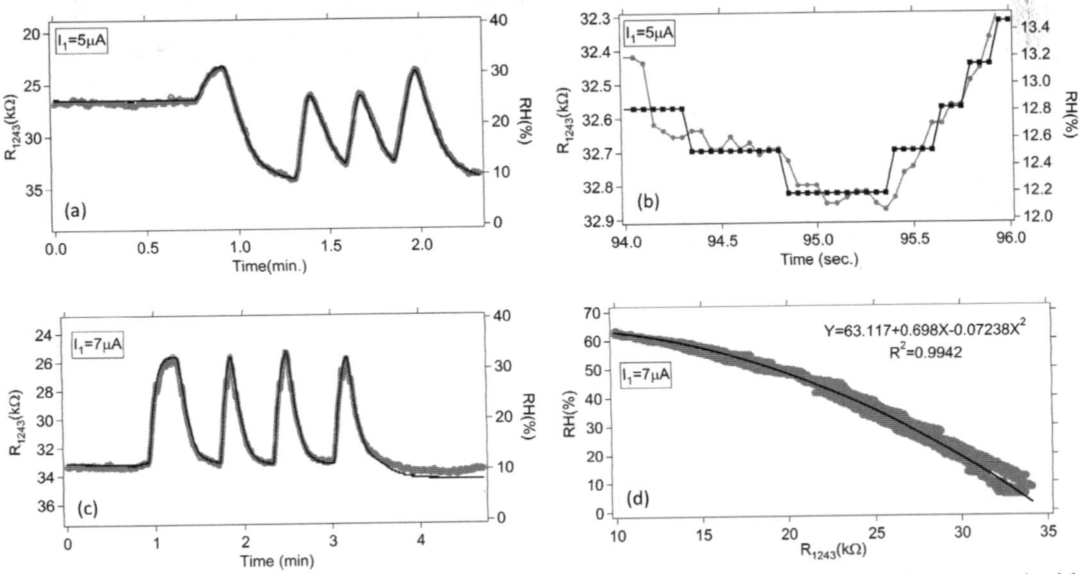

Figure 4. GO sensor response to dynamically changing humidity ((a)-(c) red curves) compared with a commercially available sensor ((a)-(c) black curves) and RH as a function of the sensor response (d).

Analysis of all the data collected during the experiments evidences a quadratic dependence of the relative humidity on the sensor resistance (Figure 4(d)). The scatter in the data falls within R_{1243}+/- 7% and the HIH-4000 has a sensitivity of +/- 5% RH. In addition, more than a threefold change in the GO printed sensor resistance was measured in the 3-63% RH range. The power consumption of the GO printed sensor is estimated at 30 μW or less over the 3-63% relative humidity range while 7 μA are drawn by the device; if the current drawn by the device is lowered to 5 μA, the power consumption drops to less than 15 μW. The decrease in resistance of the GO sensors with increasing relative humidity found in these experiments followed the same behaviour observed for reported sensors made with drop-cast and airbrush sprayed GO thin films [6],[7].

4. Conclusions

Microwatt-powered conductometric humidity sensors were fabricated by shadow mask electrospray printing at atmospheric pressure and slightly elevated temperatures (< 65°C) of an aqueous suspension of GO nanoflakes onto electrode patterns defined by the lift-off metallization technique. No post treatment was required to activate the devices including annealing or doping with metal nanoparticles. The sensor response to humidity followed that of a commercial sensor and there is a quadratic dependence of the relative humidity on the sensor resistance in the 3-63% RH range studied, with more than a threefold change in resistance over this range. The measured decrease in sensor resistance with increasing RH is in agreement with reported GO sensors fabricated by the drop-cast and airbrush spray methods. Our devices are promising candidates for the humidity sensor on the modules dispersed in distributed sensor networks due to their low cost, small size, and low power consumption.

Acknowledgements

The microfabrication of the devices was completed using the facilities at the Microsystems Technology Laboratories of MIT. This work was supported by Edwards Vacuum.

References

[1] Iyenger S S and Brooks R R 2005 *Distributed Sensor Networks* (Boca Raton, FL: Chapman and Hall/CRC Press) p 5
[2] Vacca J R, 2015 *Handbook of Sensor Networking: Advanced Technologies and Applications* (Boca Raton, FL: Taylor and Francis/CRC Press) p 1-4
[3] Henrich V and Cox P 1994 *The Surface Science of Metal Oxides*, (Cambridge, UK: Cambridge University Press)
[4] Wang D et al 2008 *J. Phys. Chem. C* **112** 13499-13509
[5] Korotcenkov G 2007 *Mater. Sci. Eng. B* **139** 1–23
[6] Borini S, White R, Wei D, Astley M, Haque S, Spigone E, Harris N, Kivioja J, and Ryhanen 2013 *ACS Nano,* **7** 11166–11173
[7] Yao Y, Chen X, Zhu J, Zeng B, Wu Z, Li X 2012 *Nanoscale Res. Lett.* **7** 363
[8] Ponce de Leon P, Hill F A, Heubel E V, Velásquez-García L F 2015 *Nanotechnology* **26** 225301
[9] Hill F A, Heubel E V, Ponce de Leon P, Velásquez-García L F 2014 *J. Microelectromech. Syst.* **23** 1237-1248.
[10] Gassend B, Velásquez-García L F, Akinwande A I, Martinez-Sanchez M 2009 *J. Microelectromech. Syst.* **18** 679-694
[11] Velásquez-García L F, Akinwande A I, Martínez–Sánchez M 2006 *J. Microelectromech. Syst.* **15** 1272-1280
[12] Xuriguera E, Monereo O, Varea A, and Cirera A 2014 Flexible reduced graphene oxide gas sensor deposited by electrospray *Proceedings 15th Trends in Nanotech. Inter. Conf. (TNT 2014), Barcelona, Spain, October 27 -31*
[13] Velásquez-García L F, Akinwande A I, Martínez–Sánchez M 2006 *J. Microelectromech. Syst.* **15** 1260-1271
[14] Fernandez de la Mora J 2007 *Annu. Rev. Fluid Mech.* **39** 217– 243

Design of an Inductive Power Transfer System with Flexible Coils for Body-worn Applications

L. R. Clare[1], S. G. Burrow[1], B. H. Stark[1], N. J. Grabham[2], S. P. Beeby[2]
[1] University of Bristol, UK. [2] University of Southampton, UK.

Email: stephen.burrow@bristol.ac.uk

Abstract. This paper describes an IPT (Inductive Power Transfer) system for body worn electronics, and investigates the challenges for an IPT system that arise specifically in this scenario. Principally, these are: highly variable coil coupling through time-varying miss-alignment and coil separation; a requirement that one or both of the coils must be wearable and thus flexible; and proximity to the human body introducing limits on the maximum EM field. The highly variable coupling results in a system that must operate effectively with a large range of received powers, whilst the constraints on the realisation of the coils typically reduce the Q-factor; the human exposure considerations limit both the maximum field strengths that the wearer of a receiver coil might experience, and also the field strengths that a 3rd party might be exposed to, for instance when approaching the transmit coil.

1. Introduction

Wireless power transfer is commonly implemented using inductive coupling to transfer power between transmit and receive coils, this is commonly known as inductive power transfer (IPT) [1], or 'near-field transfer'. Near-field IPT systems operate at frequencies determined by the desire to make both transmit and receive coils electrically resonant, and with power levels and transmission ranges typically of the order of watts over 10's cm range. In this paper IPT for body-worn sensors is considered, and hence the maximum power and maximum coil separation distance is limited by regulations governing the maximum magnetic field strength to which a person might be exposed. In an illustrative scenario the transmit coil is integrated within a chair arm (fig. 1a), and the receive coil is worn on the body, for instance as part of a wrist mounted health sensor (fig. 1b).

The power transfer between the coils of an IPT system is highly dependent on the magnetic coupling between them, which in turn is sensitive to distance and relative orientation of the coils and hence the dynamic body-worn scenario is challenging: if the system is designed for the peak coupling it may 'brown-out' for long periods; if it is designed for the lowest coupling then there will be periods when significant excess power must be dissipated. Additionally when one or both of the coils are flexible, a further degree of freedom is introduced as they can deform, affecting coupling and inductance – key parameters in determining the operation of resonant coils. Flexible coils present their own challenges, particularly as it is more difficult to obtain both flexibility and the low coil resistance required for a high quality factor, which is important in maximising power transfer in a resonant system.

a) Transmit coil b) Receive coil

Figure 1. The IPT scenario showing flexible transmit coil incorporated into furniture and receive coil shown in body-worn electronics housing.

2. System operation

2.1. IPT fundamentals

The doubly-tuned transformer, which forms the basis of most IPT systems, is shown in fig. 2, and is well-described in the literature, e.g. [3,4]. The circuit comprises primary and secondary coils with inductance (L_p, L_s) and parasitic resistance (R_{cp}, R_{cs}) coupled by their mutual inductance (M). Both coils are tuned to resonate at the desired frequency by their respective tuning capacitors: forming the coils into resonant circuits helps to mitigate the effects of reduced magnetic coupling due to higher separations and the absence of a high permeability flux path, compared to iron cored transformers [5].

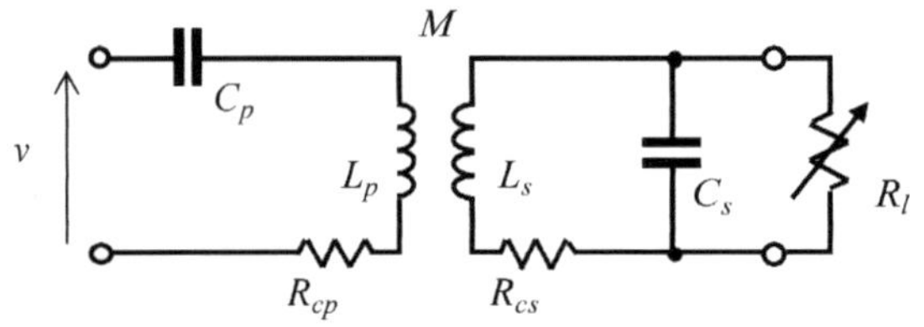

Figure 2. Equivalent circuit of an IPT system.

The resonant circuit on each side could be created by either a parallel or series arrangement, the choice of which depends on driver and load considerations. A parallel primary resonant arrangement requires the driver to support the full coil voltage at a lower current; with a series circuit this is reversed. In addition the series circuit allows class-d drive circuits to be used; a parallel circuit necessitates pure sinusoidal excitation. In both cases, resonant operation significantly reduces the drive VA requirement.

Assuming both primary and secondary are tuned to resonate at the drive frequency, ω, and using standard circuit theory it can be shown that:

$$P_{RL} = \frac{(\omega I_P M Q_S)^2}{R_L} \qquad (1),$$

Where Q_s is the secondary circuit quality factor and M is the mutual inductance, which at critical coupling (peak power) is given by:

$$M = \frac{\sqrt{R_{CP}(R_L{}' + R_{CS})}}{\omega} \quad (2), \text{ and the primary current by: } I_P = \frac{V_P}{R_{CP} + \frac{(\omega M)^2}{R_{CS} + R_L{}'}} \quad (3)$$

$$\text{Where: } R_L{}' = \frac{1}{(\omega C_S)^2 R_L} \quad (4)$$

2.2. *Coupling between coils*

The coupling between primary and secondary coils can be described by a coupling factor, k, a dimensionless number between 0 and 1, defined by equ. 5. Coupling factor is equal to 1 if the secondary coil couples with all of the flux generated by the primary coil.

$$k = \frac{M}{\sqrt{L_P L_S}} = \frac{V_{SEC}}{V_{PRI}}\sqrt{\frac{L_P}{L_S}} \quad (5)$$, where V_{PRI} & V_{SEC} are primary and secondary coil voltages.

The coupling between coils is highly sensitive to separation and alignment, however modelling the variation is not trivial, and mathematical derivations do not yield simple analytical expressions. The majority of authors thus rely on empirical or numerical results and an understanding of qualitative effects. The variation in coupling factor between two pairs of coils, experimentally measured, is shown in fig. 3a, coupling between the two flexible TX coils and a TX/RX pair is shown for contrast; the larger flexible pair give better coupling but would not physically fit in the application.

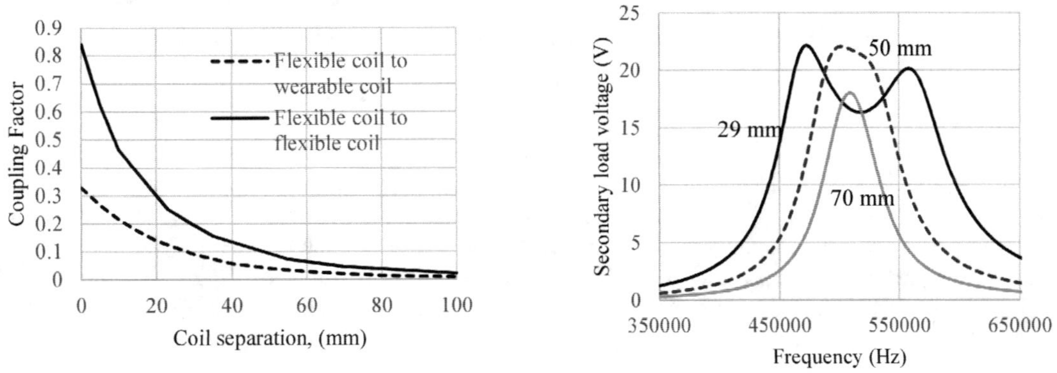

a) Coupling factor b) Frequency response

Figure 3. Coupling factor (k) and load voltage frequency response against coil separation.

A second important effect of a varying coupling factor is seen in the frequency response function of the coupled-tuned system. This effect is illustrated in fig. 3b, and is significant for IPT systems that have variable coupling (as we consider in this paper) since this forms an additional factor modulating the transmitted power with coil separation. This is further illustrated in fig. 4, which shows the locus of the peak in frequency response as the coil separation (hence coupling factor) changes. The bifurcation represents the critical coupling condition, described in section 2.1.

3. System Design

3.1. *Choosing parameters*

Choosing the coil geometry, and electrical parameters for an IPT system is somewhat of an iterative process, constrained by the particular application. Because the coil often must be resonant with a high Q, coils geometry feeds into the choice of operating frequency, but EM emission and human exposure regulations, and topology of drive also strongly influence these design decisions. Often the Q of coils will rise with frequency, indicating efficient operation in the MHz or above region, however higher frequencies require a move from hard-switching drive electronics to resonant approaches.

To ensure human EM exposure limits are not exceeded, the TX coil current in this paper is set such that the resultant magnetic field complies with a Specific Absorption Rate (SAR) of 2W/kg for the set head/trunk by the International Commission on Non Ionizing Radiation Protection (ICNIRP) [6], which, to the authors' knowledge, are the most stringent. At the thermal limit of the flexible coil (2 Amps) the SAR limit is reached at approximately 1MHz.

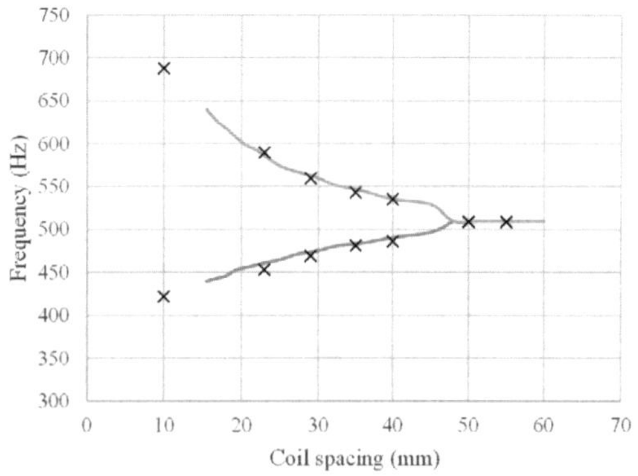

Figure 4. Locus of frequency response peak

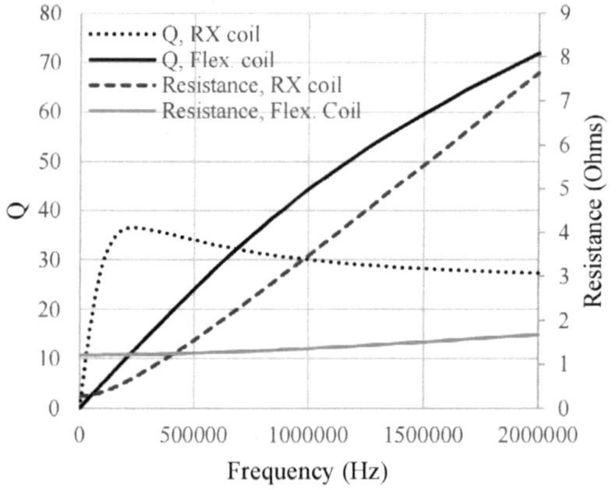

Figure 5. Coil parameters

3.2. Coils

The prototype coil illustrated in fig. 1a, is formed from flexible wire embroidered onto a light canvas backing: 10 turns of 7 x 0.07 mm PTFE-insulated wire, with a diameter of 90mm, produce an inductance of approximately 9.8 μH and a d.c. resistance of 1.2 Ω. The quality factor of the coil i.e. $\omega L/R$, where R is the coil resistance at that frequency, and a.c. resistance is shown in fig. 5

The 12μH rectangular receive coil, fig. 1b, is formed from 10 turns of 36 x 0.04 mm wire, the overall size of 25x47mm is determined by the enclosure for the wrist electronics. The coil is wound to maximize the turn-area product and thereby the induced voltage. The quality factor and resistance of the coil is also shown in fig. 5.

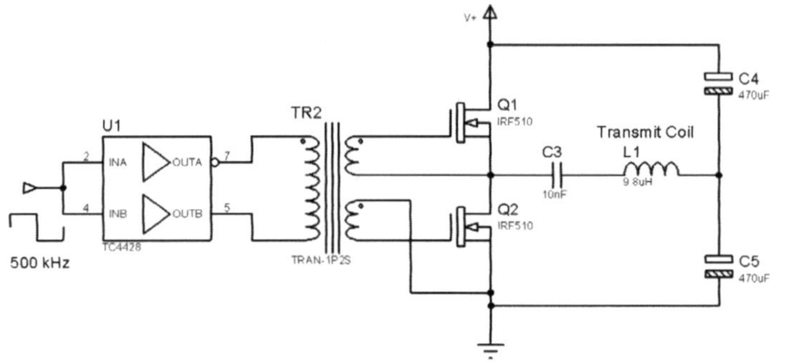

a) Fixed frequency drive

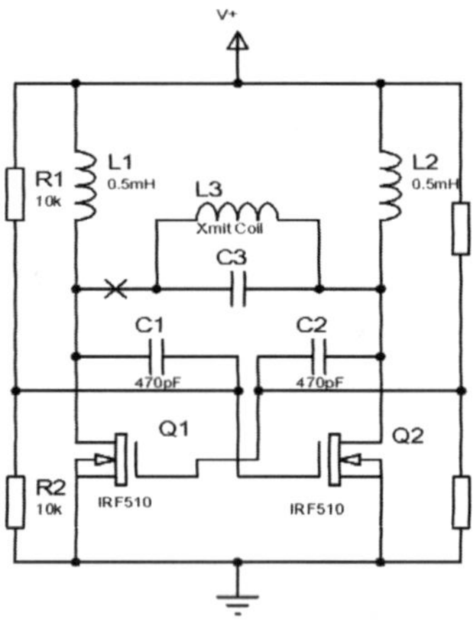

b) Self-Oscillating circuit

Figure 6. Experimental drive circuits

3.3. Drive and load electronics

Two drive topologies are described in this paper: the first is a fixed frequency Class D half-bridge of MOSFET circuit; the second is a self oscillating drive which tracks the peak in frequency response as the coil separation changes. The circuits were tested with several frequencies in the range 0.5-1Mhz, the upper limit being determined by human exposure limit at the peak coil current.

The goal of the self oscillating drive is to maximise the range of coil separations over which the IPT system can usefully operate, but it does so at the expense of being frequency-wild. In both cases the receiver side circuit is a peak rectifier circuit formed from a full bridge of 1N5818 diodes.

4. Experimental results

Fig 7 shows two of the most illustrative test results. In both tests the TX coil drive voltage was set to produce the maximum (2 A) current when the RX coil was at a significant distance; fig. 7a shows the variation in load power as the coils are brought together; fig, 7b shows the primary current variation. Critical coupling (set by load resistance, see equ. 2) occurred a separation distance of 45mm – a compromise to achieve best power transfer over the widest range of separations in the fixed frequency case. The self-oscillating circuit was able to improve upon this performance at close separation.

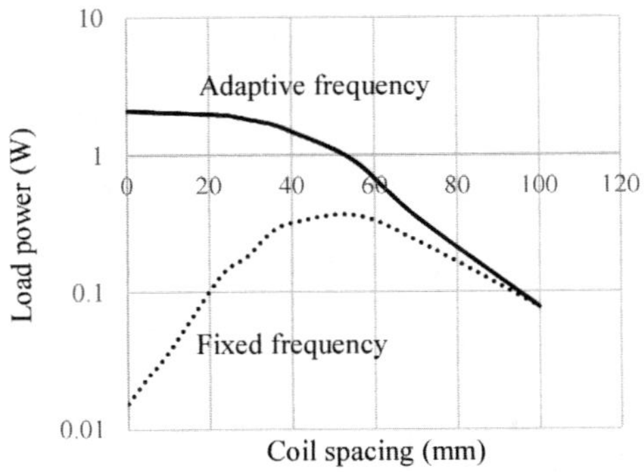

a) Load power

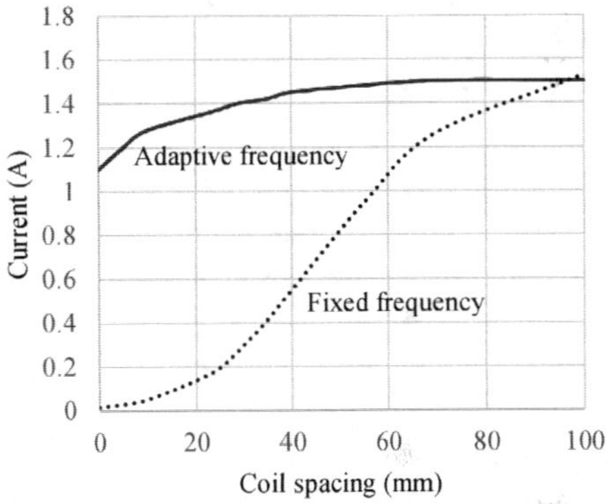

b) TX coil current

Figure 7. Comparison of the system performance over varying separation

5. Conclusions

IPT system-design trades off many design parameters, especially when performance must be persevered against dynamic changes in separation. The adaptive frequency self-oscillating drive shows potential in this respect.

References

[1] Hubregt J. Visser. Aspects of Far-Field RF Energy Transport. Proceedings of the 42nd European Microwave Conference.

[2] S. Y. R. Hui, Wenxing Zhong, and C. K. Lee. A Critical Review of Recent Progress in Mid-Range Wireless Power Transfer, IEEE Transactions on Power Electronics, Vol. 29, No. 9, September 2014

[3] Terman F. E. Electronic and Radio Engineering. McGraw-Hill Book Company Inc. 1955. pp 63-70.

[4] Albert A.L. Radio Fundamentals. McGraw Hill Book Company Inc. 1948. pp 101-108.

[5] General Analysis on the Use of Tesla's Resonators in Domino Forms for Wireless Power Transfer. Wenxing Zhong, Chi Kwan Lee, S. Y. Ron Hui. IEEE Transactions on Industrial Electronics, Vol. 60, No. 1, January 2013

[6] ICNIRP guidelines for limiting exposure to time-varying electric and magnetic fields, Health Physics, vol. 74, no. 4, pp. 494–522, April 1998.

Near field wireless power transfer using curved relay resonators for extended transfer distance

D Zhu[1,3], L. Clare[2], B. H. Stark[2] and S. P. Beeby[1,4]
[1]Electronics and Computer Science, University of Southampton, Southampton, SO17 1BJ, UK
[2]The Department of Electrical and Electronic Engineering, University of Bristol, Bristol, BS8 1TH, UK
[3]School of Computing, Electronics and Maths, Faculty of Engineering, Environment and Computing, Coventry University, Coventry, CV1 5FB

E-mail: spb@ecs.soton.ac.uk

Abstract. This paper investigates the performance of a near field wireless power transfer system that uses curved relay resonator to extend transfer distance. Near field wireless power transfer operates based on the near-field electromagnetic coupling of coils. Such a system can transfer energy over a relatively short distance which is of the same order of dimensions of the coupled coils. The energy transfer distance can be increased using flat relay resonators. Recent developments in printing electronics and e-textiles have seen increasing demand of embedding electronics into fabrics. Near field wireless power transfer is one of the most promising methods to power electronics on fabrics. The concept can be applied to body-worn textiles by, for example, integrating a transmitter coil into upholstery, and a flexible receiver coil into garments. Flexible textile coils take on the shape of the supporting materials such as garments, and therefore curved resonator and receiver coils are investigated in this work. Experimental results showed that using curved relay resonator can effectively extend the wireless power transfer distance. However, as the curvature of the coil increases, the performance of the wireless power transfer, especially the maximum received power, deteriorates.

1. Introduction

Recent developments in printing electronics and e-textiles have seen increasing demand for embedding electronics into fabrics for wearable applications. One of the difficulties with such applications is how to provide a sufficient and reliable power source. Traditional batteries are not an ideal solution for this application because its bulky size makes it difficult to be integrated into fabrics. Furthermore, they have limited capacity and, therefore, require constant maintenance such as replacement or recharging. Research on flexible energy storage devices, such as batteries and supercapacitors, has been reported [1][2] but their capacity and performance to date are relatively poor compared to traditional batteries.

Energy harvesting concerns conversion of ambient energy into electrical energy and can potentially be used to power electronics in wearable applications. Ambient energy sources in this application include human movement, temperature difference and ambient light. However, these energy

[4] To whom any correspondence should be addressed.

harvesting techniques rely significantly on the ambient environment which is unpredictable in wearable applications. Therefore, the amount of energy generated is unpredictable which makes it not a reliable power source for this particular application.

By contrast, active power transfer techniques, such as wireless power transfer [3], are more reliable and suitable for powering wearable devices. Near field wireless power transfer is one of the most promising methods. It operates based on the near-field electromagnetic coupling of coils. This concept can be applied to body-worn textiles by, for example, integrating a transmitter coil into upholstery, and a flexible receiver coil into garments. Flexible textile coils take on the shape of the supporting materials such as garments and, therefore, tend to bend during operation.

The performance of the near field wireless power transfer system is a strong function of the alignment [4] and separation of the inductively coupled coils. Near field wireless power transfer systems can transfer energy over a relatively short distance which is of the same order of dimensions of the coupled coils. Previous work [5] has shown that the energy transfer distance can be increased using flat relay resonators. In this paper, performance of a near field wireless power transfer system that uses a curved relay resonator to extend transfer distance is investigated.

2. Experimental

2.1. Experimental setup

Figure 1 shows the experimental setup. The transmitter coil is driven by a fixed frequency driver. The transmission frequency is set as 250 kHz before other coils are coupled. A relay resonator consisting of a relay coil and a matching capacitor, C1, is placed between the transmitter and receiver coils. The receiver coil is connected in parallel with the tuning capacitor C2. The AC input is rectified using a bridge rectifier. The rectified DC voltage across a variable load resistor R_L, V_L, is measured and received DC power can then be calculated by V_L^2/R_L. V_L was measured across various R_L, thus, the maximum received DC power across the optimal load resistor can be determined.

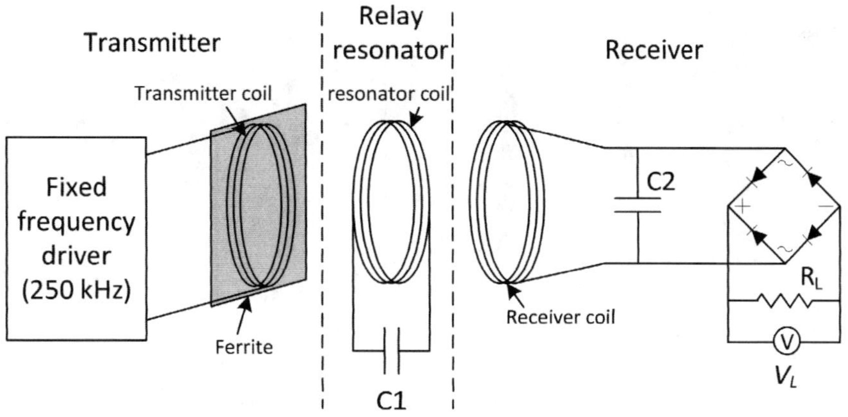

Figure 1. Experimental setup concept.

All coils used in the test are a typical coil used in commercial Qi compatible receivers. Details of the coil can be found in [6]. The inductance of the coil was measured as 7.98 μH without the ferrite back and 10.88 μH with the ferrite back, respectively. The coil is 300-micron thick and therefore is relatively flexible without the ferrite back. It can be bent to make a curved coil. The inductance of the coil reduces as the coil bending curvature increases. When the bending curvature is 0.057 m^{-1}, which is the maximum of all coils used in the test, the coil inductance was measured as 7.017 μH without the ferrite back. Figure 2 shows the practical implementation of the testing circuit.

Figure 2. Practical implementation of the testing circuit.

2.2. Coil arrangement

Figure 3 shows the two coil arrangements that were studied. In both arrangements, the transmitter coils are flat and have ferrite backs. The centres of the transmitter and receiver coils are separated by a distance of d. The relay resonator coil is placed in the middle of the transmitter and receiver coils, i.e. the centre of the resonator coil is $d/2$ apart from both the centres of the transmitter and receiver coils. The resonator and receiver coils in arrangement 1 have lower curvatures than those in arrangement 2.

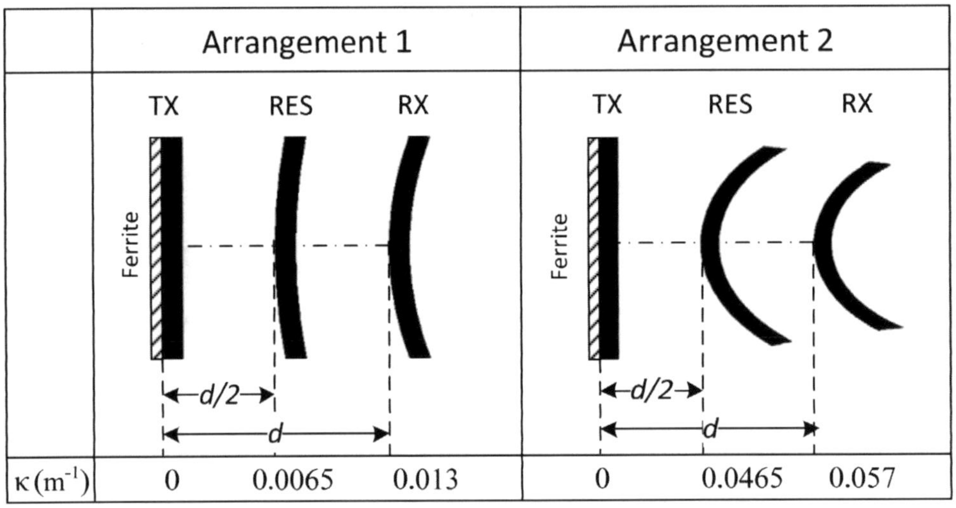

Figure 3. Side view of the two coil arrangements that were tested (TX: transmitter, RES: relay resonator, RX: receiver). Coil bending curvature, κ, is shown beneath the respective coil. Coils in arrangement 1 have lower curvature than those in arrangement 2.

3. Results

3.1. Matching capacitor of the relay resonator

In the test, a variable matching capacitor was first connected to the relay resonator coil to find the optimum value. Figure 4 shows comparisons of DC-DC power transfer efficiency and maximum received power with variations of matching capacitances of the relay resonator in the two coil arrangements. It was found that the DC-DC efficiency reached maximum with a 14.7 nF matching capacitor in both coil arrangements. In terms of maximum received power, arrangement 1 requires a 16.8 nF matching capacitor while arrangement 2 requires 14.7 nF. Therefore, 16.8 nF and 14.7 nF capacitors were connected to resonator coils in arrangements 1 and 2 respectively.

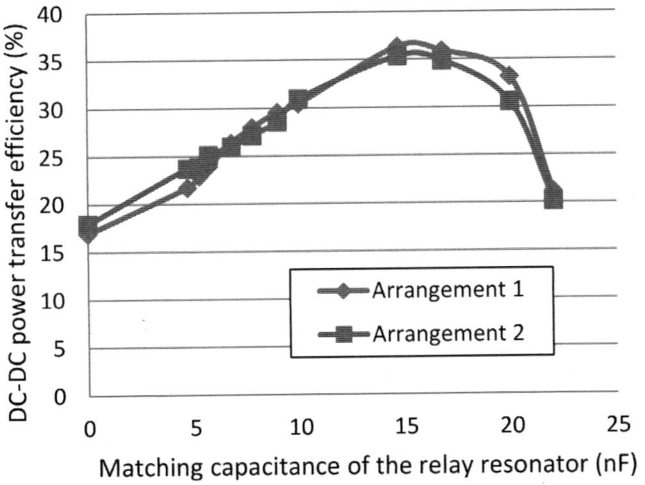

(a) DC-DC power transfer efficiency

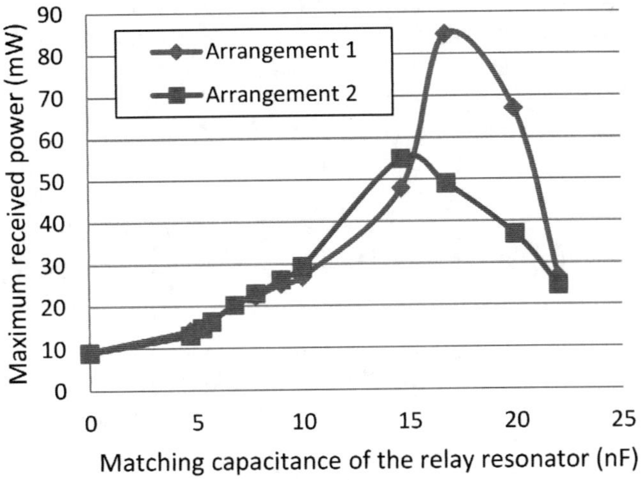

(b) Maximum received power

Figure 4. Comparisons of DC-DC power transfer efficiency and maximum received power with variations of matching capacitance of the relay resonator in the two arrangements.

3.2. Comparisons of experimental results

In the test, three scenarios were considered for each coil arrangement. In the first two scenarios, no relay resonators are used and the TX and RX coils are separated by 10 mm and 20 mm, respectively. In the third scenario, the TX and RX coils are separated by 20 mm and a relay resonator is used as shown in Figure 3. Table 1 compares performance of wireless power transfer in terms of DC-DC power transfer efficiency and maximum received DC power for the three scenarios in each of the two arrangements. An optimum matching capacitor was connected to the receiver coil in all scenarios. The optimum matching capacitor in each scenario was determined using the same method described in Section 3.1.

It was found that using a curved relay resonator can effectively increase both DC-DC power transfer efficiency and the maximum received power over a longer transfer distance. For arrangement 1, the DC-DC power transfer efficiency and the maximum received power are increased by 114.7% and 809.7% by using a relay resonator, respectively. For arrangement 2, these two values are increased by 132.2% and 590.7% respectively by using a relay resonator. The increase in the DC-DC power transfer efficiency is similar in both arrangements while maximum received power is reduced as coil curvature increases.

Table 1. Comparisons of performance of the wireless power transfer system with and without the relay resonator.

	Arrangement 1			Arrangement 2		
Using relay resonator?	No	No	Yes	No	No	Yes
Distance between TX and RX coils, *d*, (mm)	10	20	20	10	20	20
DC-DC power transfer efficiency (%)	46.4	16.9	36.3	41.8	15.2	35.3
Maximum received power (mW)	66.8	9.3	84.6	42.4	7.89	54.5

4. Conclusion

This paper investigates the feasibility of using curved relay resonators to extend the wireless power transfer distance. A basic test platform was built and two coil arrangements with different coil bending curvatures were tested. Experimental results showed that, like using relay resonators with flat coils, curved coil relay resonator can also effectively extend the wireless power transfer distance. However, the performance becomes worse as the coil bending curvature increases.

This technique can potentially be applied to wearable applications, especially those using flexible textile based coils which is likely to bend during operation. Future work include using more curved relay coils to further extend the transfer distance as well as implementing the technique using textile based coils that was demonstrated previously in [4].

5. Reference

[1] Gaikwad A M, Steingart D A, Ng T N, Schwartz D E and Whiting G L 2013 A flexible high potential printed battery for powering printed electronics *Appl. Phys. Lett.* **102** 233302

[2] Yong S, Owen J R, Tudor M J and Beeby S P 2013 Fabric based supercapacitor *Journal of Physics: Conference Series* **476** 012114

[3] Kim K Y 2012 Wireless *Power Transfer - Principles and Engineering Explorations* (Open access, InTech)

[4] Zhu D, Grabham N, Lindsay C, Stark B and Beeby S 2015 Inductive power transfer in e-textile applications: reducing the effects of coil Misalignment *IEEE Wireless Power Transfer Conference, Boulder, US, 13 - 15 May 2015*

[5] Zhong W, Lee C K and Hui S Y R 2013 General Analysis on the Use of Tesla's Resonators in Domino Forms for Wireless Power Transfer *IEEE Transactions on Industrial Electronics* **60(1)** 261-270

[6] WURTH ELEKTRONIK 760308103205 Wireless power charging coil, http://katalog.we-online.de/pbs/datasheet/760308103205.pdf (accessible on 18 September 2015)

Acknowledgments

This work was performed under the SPHERE IRC funded by the UK Engineering and Physical Sciences Research Council (EPSRC), Grant EP/K031910/1. Beeby would also like to acknowledge his EPSRC Fellowship, Grant EP/I005323/1.

A MEMS-based solid propellant microthruster array for space and military applications

A. Chaalane[1,2], R. Chemam[1], M. Houabes[3], R. Yahiaoui[4], A. Metatla[5], B.Ouari[6], N. Metatla[7], D. Mahi[8], A. Dkhissi[9,10] and D. Esteve[10]

[1] Physics Department, LPR laboratory, University of Annaba, BP.12, 23000 Annaba, Algeria
[2] ESIEE-Paris, SEN Department, 2 Bd. Blaise Pascal, 93162 Noisy-le-Grand, France
[3] Department of Electrical Engineering, University of Annaba, 23000 Annaba, Algeria
[4] FEMTO-ST, Department of Micro Nano Sciences and Systems MN2S, 15B ave. des Montboucons, 25030 Besançon, France
[5] Department of Electrical Engineering, University of Skikda, 21000 Skikda, Algeria
[6] Physics Department, University of Tlemcen, BP.119, 13000 Tlemcen, Algeria
[7] Department of Science and Technology, University of Mila, 43000 Mila, Algeria
[8] Department of Electrical Engineering, University of Laghouat, 03000 Laghouat, Algeria
[9] Laboratory for Chemical Technology, Ghent University, 9052 Ghent, Belgium
[10] LAAS-CNRS, NI2S, 7 ave. du Colonel Roche, 31031 Toulouse, France

E-mail: amar.chaalane@esiee.fr

Abstract. Since combustion is an easy way to achieve large quantities of energy from a small volume, we developed a MEMS based solid propellant microthruster array for small spacecraft and micro-air-vehicle applications. A thruster is composed of a fuel chamber layer, a top-side igniter with a micromachined nozzle in the same silicon layer. Layers are assembled by adhesive bonding to give final MEMS array. The thrust force is generated by the combustion of propellant stored in a few millimeter cube chamber. The micro-igniter is a polysilicon resistor deposited on a low stress SiO_2/SiN_x thin membrane to ensure a good heat transfer to the propellant and thus a low electric power consumption. A large range of thrust force is obtained simply by varying chamber and nozzle geometry parameters in one step of Deep Reactive Ion Etching (DRIE). Experimental tests of ignition and combustion employing home made (DB+x%BP) propellant composed of a Double-Base and Black-Powder. A temperature of 250 °C, enough to propellant initiation, is reached for 40 mW of electric power. A combustion rate of about 3.4 mm/s is measured for DB+20%BP propellant and thrust ranges between 0.1 and 3,5 mN are obtained for BP ratio between 10% and 30% using a microthruster of 100 μm of throat wide.

1. Introduction

The idea of powering MEMS has become feasible thanks to propellant integration in micromachined silicon structures. But things are not that easy. At this small scale, heat loss phenomena are so dominant that stifles the chemical reaction of combustion and prevents the reaction-control. This result a non-sustained reaction, not stable over time and not reproducible. Technological solutions are therefore necessary to remove these locks. Micro-combustion is became a vibrant research field and researchers are studying the stability and performance of micro-combustion in different applications such as drug delivery, Swiss roll combustor [1], micro-gas turbine engines [2], micro-thermo-photovoltaic system [3], a free piston knock engine [4], micro-combustor tube [5], radial channel combustors [6], micropropulsion [7, 8, 9] and in various kinds of other micro-combustor [10].

Recently, micro-spacecraft and many other low-cost miniature space systems have got a lot of attention due to the reduction of launch cost and various mission capabilities, and the development of new miniaturized systems is required. The propulsion system must be capable of providing extremely low levels of thrust and impulse, in order to satisfy micro-spacecrafts and miniature space systems propulsion need. However, there is almost no propulsion system appropriate in the market, and the development of new miniaturized systems is required. In this context, many micropropulsion systems based on MEMS technologies have been developed in the world. A micro-thruster is composed basically from three parts: a fuel tank of few cubic millimeters, an igniter and a micro-nozzle. The thrust force is generated by the combustion of fuel stored in a chamber and thus the gas is accelerated through the micro-nozzle. Propellant ignition is achieved by the Joule heating effect thanks to a micro-resistor. The use of solid propellant is an interesting choice because combustion is an easy way to achieve large quantities of energy from a small volume. Moreover, the system is simple, has no moving parts and presents a good efficiency (generated thrust to input power). Furthermore, there could be the perspectives of monolithic integration of electronics in order to realize the command. However, it is difficult to achieve a successful, reproducible and continuum combustion of the propellant in a millimeter scale device due to quenching effect [11]. Within a European project, Rossi et al. [7] fabricated a solid propellant microthruster to deliver thrust impulse of a few tens mN.s and a thrust force ranging from µN to few mN required for micro-spacecraft station keeping needs. It consists of a stack of four micromachined wafers: a chamber, an igniter, a nozzle and a seal closing the structure. Igniters and nozzles were realized on bulk silicon using MEMS technologies; chambers were fabricated in Foturan (photosensitive glass) and the seals were in Pyrex glass. Due to there low thermal conductivity, Foturan and Pyrex were used in order to minimize the thermal cross-talk between neighboring cells. The igniter and the Foturan chamber of 1.5mmx1.5mm of area dimensions were filled with Zirconium Perchlorate Potassium (ZPP) mixture propellant, and the Glycidyle Azide Polymer mixed with Ammonium Perchlorate and Zirconium (GAP/AP/Zr) propellant respectively. This design can no resist very high combustion pressure because the nozzle part is glued to the reservoir using polymer glue. Furthermore, the microthruster is powered from the front side, which lead to some difficulties for a future matrix addressing. Zhang et al. [8] proposed a flexible design of solid propellant microthruster were both reservoir and the nozzle were etched on the same silicon wafer, and its geometry can be changed at the same time using only one step of DRIE. But the use of igniters deposited on bulk material does not allow low electric power consumption.

This paper focuses on our progress toward the development of a micropropulsion array composed of 7 solid propellant utits. Based on a previous work, we propose here a better adapted design, which allows flexibility to obtain a large range of thrust forces. The thrust force can be easily scaled by varying the geometry, the number of micro-thrusters and the type of propellant in order to satisfy the specific space/aeronautic applications. The key point of our design is the concept of initiator of thin membrane. This requires a very low power for ignition but a special care in microfabrication, loading and assembly processes. The design, fabrication process as well as the assembly process are described here in detail. After that, ignition tests and micro-combustion near extinction limit, employing several propellants, are also presented.

2. Solid propellant microthruster design

A microthruster is composed of three layers: a micromachined silicon layer containing initiator on thin membrane, chamber and nozzle, a ceramic (Macor) layer consists of chamber extension glued on the top side of the silicon layer and a glass (Pyrex) layer glued on the silicon bottom side (Fig. 1). This structure has been designed to work with a double base propellant at high combustion pressure (60bar) to reach 1µN to 50mN of thrust forces.A schematic top view of micro-thruster is shown in Fig. 2. Two types of thrusters have been designed in the same silicon wafer: 1.5mm×5.5mm and 1.5mm×3.5mm of $W_c \times L_c$ parameters with membranes of SiO_2/SiN_x in order to have a large range of specific impulsions. From computation, the chamber to throat section ratio must be of 100 to obtain the 60bar with a double base propellant. Three types of nozzle throats (W_t) have been realized: 100µm; 150µm and 200µm. The silicon wafer is a 545µm of thickness.

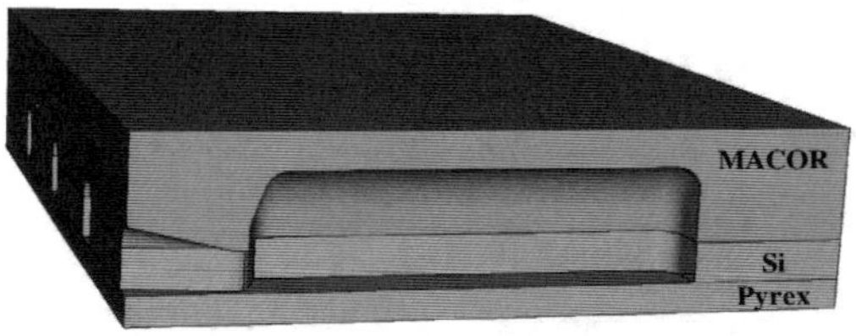

Figure 1. Schematic of the solid propellant structure array.

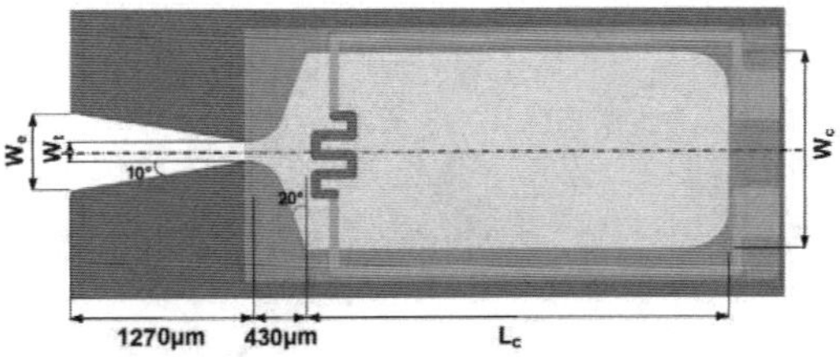

Figure 2. Schematic top view of a single microthruster with the geometric dimensions.

3. Summary of the fabrication and assembly processes

The microthruster process flow is simple but the challenge lies is attaining a highly variable geometry that maintains anisotropy over the full height of the structure. The chambers and the micro-nozzles are designed in the same silicon wafer. We know that large features etch faster than smaller features. This causes a residue of silicon on the nozzle throats when membranes are free. To completely performing nozzle throats, etch time will be extended that causes over etching of SiO_2 and destroy membranes. To provide over etching of the SiO_2/SiN_x thin membrane, we are realized SiO_2 layer thicker than calculation to be slim downed by DRIE during performing the nozzle throats. Fig. 3 shows a SEM picture of the fabricated convergent-divergent nozzle. The throat wide is of 150µm. The etched sidewalls are vertical in spite of the high chamber over nozzle area ratio (about 15). Micro-igniter layer is composed of a thin film polysilicon deposited on SiO_2/SiN_x low stressed membrane and then etched for obtaining the resistor design. Fig. 4 shows a 3D optical microscopy of the polysilicon resistor design. The measured value of resistance is about 500Ω and the heating surface is of 550µm×550µm. A maximum temperature of 100°C is reached for only 15mW and 450°C for 80mW.

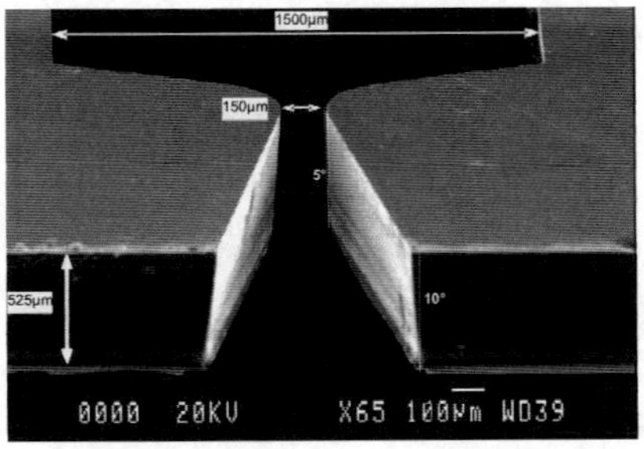

Figure 3. SEM photo of a micro-nozzle etched by DRIE on a 525µm thickness of silicon wafer.

Figure 4. 3D optical profilometry picture of the thin film membrane with polysilicon resistor.

The fabrication and assembly process were described previously in detail [11]. A fabricated but not assembled silicon chip is shown in Fig. 5. It is composed of seven micro-rockets. Each single micro-rocket is separated with air grooves of 250 m width to prevent undesirable ignition from close neighbours. Silicon chip of 545 µm thick can be filled with energetic material and sealed from both top and backside using pyrex-7740 glass. But in this configuration the combustion is not continuum in the silicon chamber, and thus the generated thrust force is not reproducible. This is due to the extinction limit phenomena of the flame caused by heat loss in the silicon surrounding the combustion chamber. For this reason, we sealed the backside of the silicon microthruster array by a machined chip to add a supplementary chamber space. It leads to increase the ratio of chamber section to nozzle throat section, and by consequence increases thrust force. In a previous work, we have seen that silicon supplementary chamber causes unsuitable firing of the close neighbor of the tested chamber [7]. Macor has advantageous mechanical and thermal characteristics: coefficient of expansion very close to the silicon one, very low thermal conductivity (1.46 W/(m ◦C)) and high operating temperature (1000 ◦C). Its drawbacks are: the high weight (2.52 g/cm) and the difficulty of machining.

Figure 5. Photograph of 7x1micromachined but non assembled silicon array

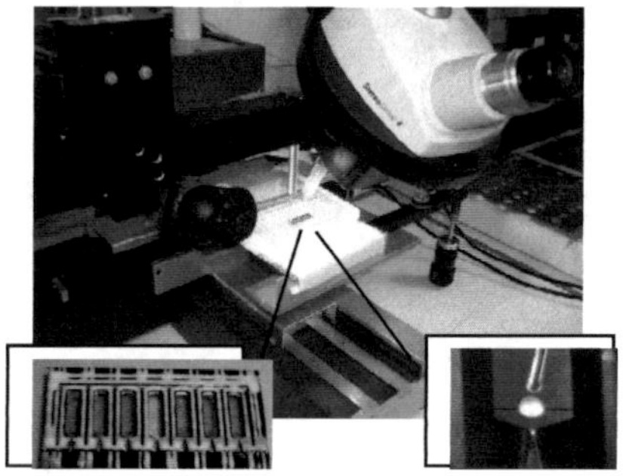

Figure 6. Photograph of the Flip-Chip setup for propellant loading and structure sealing.

There are three steps for the assembly process which are presented in Fig. 6. Firstly, the topside of silicon chip was sealed with Pyrex glass using H70-E epoxy glue. Then the silicon reservoirs were filled of propellant. This propellant was glued on the membrane by means of deposited micro-drops of the H20-E thermal conductive epoxy glue. Finally, the backside of silicon chip was sealed with machined Macor ceramic part using H70-E epoxy glue. Fig. 7 presents a micro-rocket array assembled and ready for test.

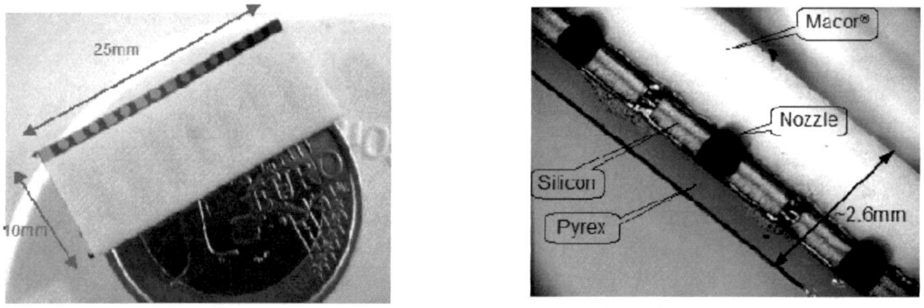

Figure 7. Photograph of a fabricated, filled with propellant and assembled 7x1 microthruster array.

4. Ignition and combustion and thrust investigations

Principle of an ignition test is based on the use of the polysilicon resistor on membrane both as heater part and piezoresistive sensor. The resistor is powered at specified electric current to heat the propellant, by Joule effect, until its ignition temperature. Then, the time of ignition is obtained when the membrane breaks (current =0) due to the gas issue of the starting of the micro-combustion. The voltage at the edges of resistor is measured. Data acquisition (voltage and current) are saved in computer and treated after.

A double base propellant called SD1152 has been used for the ignition and combustion. A preliminary ignition test at atmospheric pressure enables us to find, for each configuration, the best ignition parameters such as minimum electric power and time of application. After that, we apply optimum ignition parameters for performing combustion tests for full assembled structure. Fig. 8 shows a successfully combustion test for an assembled microthruster. First results of combustion tests give a combustion rate of only 1.6mm/s. The combustion stops before the end of reservoir because the SD1152 propellant is not adapted for small sizes. Tests using DB propellant gave a successful ignition but not sustained combustion. Thanks to the home made (DB+x%BP) solid propellant, a complete combustions have been achieved for x more or equal to 10%.

For the thrust force measurement, we used a stand based on capacitive sensor [13]. The resulting thrust over the time is given in Fig. 9. Thrust level and combustion time are reported directly from the thrust curve versus time. The disturbances present in the curve during data acquisition are caused by the material of measurement and the environment air. To these disturbances are added fluctuations present only during the thrust application. They can be explained by inhomogeneity of propellant, down scaling effect (the size of the propellant's grains become considerable compared to the chamber volume) and/or to approach towards the flame extinction limit diameter given of one millimeter by the literature.

Figure 8. Photo of the combustion test of an assembled planar structure of micro-thruster

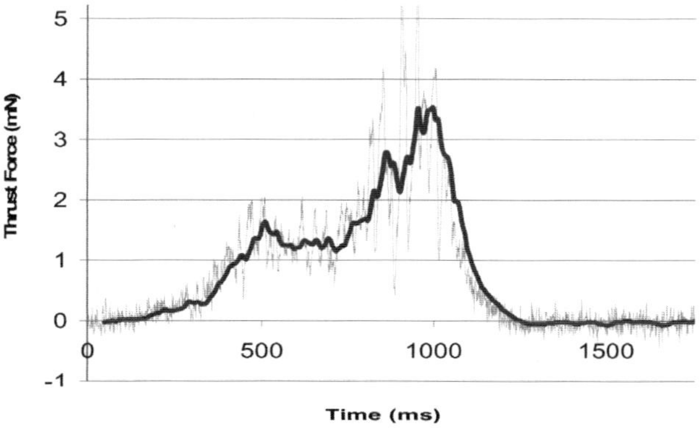

Figure 9. Thrust force vs. Time for thruster of Wt=100μm, loaded of (SD+30%PN) propellant.

5. Conclusion

In this paper, a summary of the fabrication and performance of MEMS solid propellant thruster array have been described. We improved the stability and low electric power consumption of the micro-igniter by using a thin film membrane. We developed a process to fabricate the micro-igniter on a thin dielectric membrane of 2μm thick instead of on bulk silicon or glass. A 0,5μm thick of polysilicon was used as the heater coil and the heating characteristic with the given electric power was estimated. The electric power required to obtain temperatures up to 250 °C, which is the initiation temperature of home made (DB+*x*%BP) propellant, was approximately 40 mW. The proposed concept of micro-igniter provide sufficient heat for the (DB+*x*%BP) propellant and allow the combustion of a large range of propellant because temperature of 450 °C is obtained by only 80 mW of electric power. The micro-igniter, the chamber and the nozzle are realised on the same silicon wafer by the conventional MEMS technology. This study reveals a reproducible ignition process and stable combustion in a millimeter scale device thanks to the DB+*x*%BP propellant. The last one seems more adapted for MEMS than DB propellant. The use of Macor ceramic as chamber results in a better thermal insulation between the thrusters of the array, thus preventing from thermal cross-talk between each thruster. A combustion rate of about 3.4 mm/s is measured for DB+20%BP propellant and thrust ranges between 0.1 and 3,5 mN are obtained for BP ratio between 10% and 30% using microthrusters of 100 μm of throat wide.

6. References

[1] Khandelwal B and Kumar S 2010 *Applied Thermal Engineering* **30** 2718.
[2] Shih H and Huang Y 2009 *Applied Thermal Engineering* **29** 1493.
[3] Yang W.M, Chou S.K, Shu C, Xue H and Lil Z.W 2004 *J.Phys. D: Applied Physics* **37** 1017.
[4] Aichlmayr H.T, Kittelson D.B and Zachariah M.R, 2003 *J. Combustion and Flame* **135** 227.
[5] Junwei L and Beijing Z 2008 *Applied Thermal Engineering* **28** 707.
[6] Kumar S, Maruta K and Minaev S 2007 *J. Micromech. Microeng* **17** 900.
[7] Rossi C et al. 2005 *Sensors and Actuators* A **121** 508.
[8] Zhang K.L, Chou S.K, Ang S.S 2004 *J. Micromech. Microeng*. **14** 785.
[9] Lee J, Kim T 2013 *Sensors and Actuators* A **190** 52.
[10] Khandelwal B and Kumar S 2010 *Applied Thermal Engineering* **30** 2718.
[11] Songqi Hu et al. 2014 *Materials Testing* **56** 399.
[12] Chaalane A, Rossi C and Esteve D 2007 *Sensors and Actuators* A **138** 161.
[13] Orieux S Rossi C Esteve D 2002 *Review of Scientific Instruments* **73** 73.

On designing low pressure loss working spaces for a planar Stirling micromachine

M.-A. Hachey[1], É. Léveillé[1], L.G. Fréchette[1], F. Formosa[2]
[1]Institut interdisciplinaire d'innovation technologique (3IT) Labo. Nanotech. Nanosystèmes (LN2) CNRS UMI 3463, Université de Sherbrooke, Sherbrooke, Canada
[2]Université Savoie Mont Blanc, Annecy, France

Luc.Fréchette@usherbrooke.ca

Abstract. In this paper, research was undertaken with the objective to design low pressure loss working spaces for a Stirling cycle micro heat engine operating from low temperature waste heat. This planar free-piston heat engine is anticipated to operate at the kHz level with mm^3 displacement. Given the resonant nature of the free-piston configuration, the complexity of its working gas' flow geometry and its projected high operating frequency, flow analysis is relatively complex. Design considerations were thus based on fast prototyping and experimentation. Results show that geometrical features, such as a sharp 90° corner between the regenerator and working spaces, are strong contributors to pressure losses. This research culminated into a promising revised working space configuration for engine start-up, as it considerably reduced total pressure losses, more than 80% at Re = 700, from the original design.

1. Introduction

Ongoing research is being undertaken by an international consortium (France-Canada) to produce a novel Stirling micro heat engine, the MISTIC (micro Stirling cluster), which will use low temperature waste heat (150-200° C) as a primary energy source for cogeneration. Three double-acting piston membranes (the double-acting idiom designating that the cold and hot working spaces are simultaneously acting on their respective faces) are set up in a Siemens (daisy chained) configuration [1]. These specially-designed double-acting piston membranes forgo the typical reciprocating pistons, piston rings and mechanical linkages found in Stirling engines for an airtight layered construction of the working spaces. This makes way for enabling the use of both MEMS fabrication and conventional bulk fabrication processes.

Each of the heat engine's three *modules* (a cold and hot piston-membrane face, a cold and hot working space, and the corresponding regenerator) (figure 1) exploit the inertia of the suspended structures to act on the working gas. Thus, cyclic pressure applied between the modules cause the piston-membranes of each module to phase by 120°; i.e., the piston-membranes act as both a piston (produce work with the working gas) and displacer (displace the working gas) over one cycle.

This paper discusses how the MISTIC's working spaces were designed to reduce pressure losses, also called pumping losses, as the working gas is displaced in and out of them. This entails having comprehensive knowledge of the flow characteristics, which is hard to achieve through analytical means because of the scale, oscillating flow and non-traditional geometry [2]. Hence, an experimental approach was favored where the characterization of the pressure losses are first investigated in steady-state unidirectional flow to identify promising configurations. Rapid prototyping, where a set of non-dimensional numbers focusing on certain geometric features, helped iterate working spaces [3]. In this article, working space geometries were trialed in both directions. One of them was found most satisfactory for engine start-up with a given set of mechanical and thermal parameters.

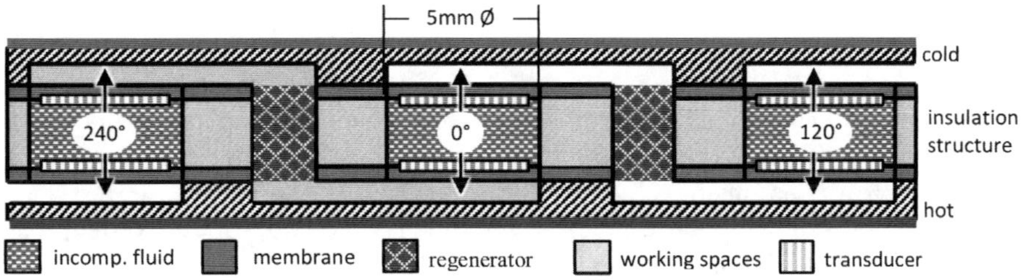

Figure 1. Depiction of a cross-section plane passing through the MISTIC'S three modules

The start-up criterion is currently determined through the use of a specially-developed electrical circuit equivalent (ECE) model elaborated by the Université Savoie Mont-Blanc to analytically characterize free-piston Stirling engines [1]. The model also helped determine the operating frequency of the structure and, consequently, the parameters of this study; given that frequency scales inversely to length, operation is expected to reach at least 2000 Hz with mm^3 displacement. As most structures have a few hundred micrometer diameters, this implicates high flow velocities (some possibly reaching Ma = 0.2 in the original configuration).

2. Designing the MISTIC's working spaces

Each of the MISTIC's three modules comprises cold and hot working spaces. One of the plane walls of these open spaces is in contact with a thermal reservoir. The other plane is occupied by the piston membrane which compresses and/or displaces the working fluid (gas). Given that the exchanger wall and the membrane piston share the same working space, its simplest (and most compact) form is found when most of the shape corresponds to the membrane's perimeter, occupying a 5mm diameter. Moreover, because these piston membranes are anticipated to have a short stroke, the working spaces are required to be thin and the current design expects to have a gap that's approximately 100μm high.

Joined to the working space is the regenerator cartridge that goes through the insulated core that also reaches the opposite working space. The volume contained in all three is the total gas volume contained in one module. Given that the regenerator is at 90° with the working space, local flow phenomena can be problematic in this region and transition effects should be mitigated.

2.1. Design approach

2.1.1. Geometry breakdown. To give some direction to the design process, the working spaces were deconstructed into easily identifiable geometrical elements and non-dimensional groups (NDGs). This enabled us to delineate the changes to the working spaces that were deemed more likely to reduce pressure losses incurred from the gas' displacement. Three NDGs were identified (refer to figure 3):

- NDG_1: ratio of regenerator (1) hydraulic diameter to that of exchanger channel (2)
- NDG_2: ratio of exchanger outlet (3) width to working space diameter (4)
- NDG_3: ratio of exchanger volume (2', 3, 5) over maximum swept volume (2, 4)

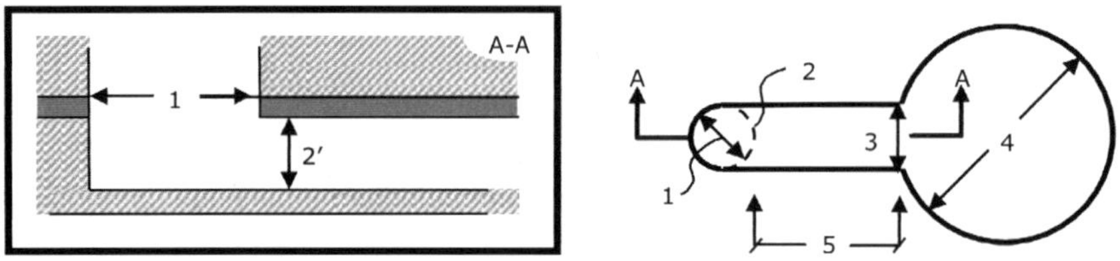

Figure 3. Zones pertaining to the geometries' design and to the formulation of the NDGs.

Given these sets of NDGs, one can surmise that the pressure losses reach a minimum when NDG_1 and NDG_2 are at unity; this would implicate the elimination of the minor losses at zone 3 and the non-recoverable losses from the steep velocity change at the 90° corner. However, increasing NDG_2 increases NDG_3 and subsequently, dead space or volume on which no work is exerted. It should be noted that as trials were being conducted, the regenerator cartridge was found to be more efficiently manufactured by MEMS processes. As such, the regenerator's shape was changed from a cylinder to a rectangular prism, immediately affecting NDG_1 and the working space's shape in subsequent revisions. This resulted in Revision A exploiting three of the four faces of the regenerator interface to increase NDG_1 while slightly increasing NDG_2. Revision B was proposed to maximize the regenerator interface's hydraulic diameter (and NDG_1) by exploiting all four faces of the cartridge at the corner by shaping it similarly as a golden (or logarithmic) spiral. Revision B also maximized NDG_2 as a means to eliminate possible flow separation at the wall at the cost of an increased NDG_3.

Fast prototyping. The experimental trials consisted of reproducing the different working space shapes into a laminate chip. Thus, an LPKF Protolaser U3 was used to cut the working space's shape into a 130µm thick Kapton 500FN131 and mill the two glass caps in which the Kapton was laminated with a hot plate. The Kapton layer included five pressure taps (numbered 1 through 5 in figure 4) to measure the static pressure at the perimeter of the reproduced working space. Given that these pressure taps were set at dominant geometrical features, they would help determine the pressure loss in these areas. In addition, radial relief structures were added in the Kapton's pattern to account for degassing and warping due to thermal dilatation during the device assembly.

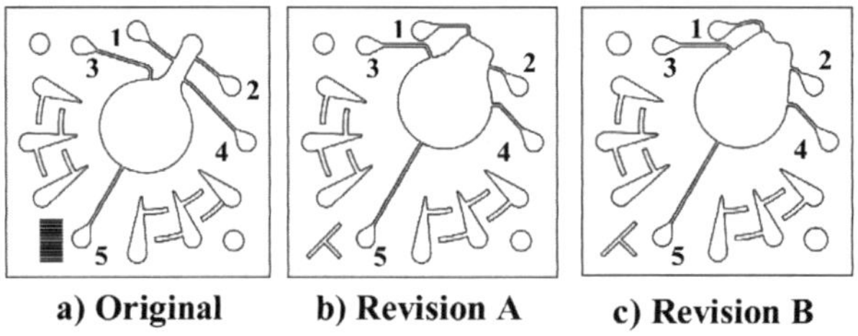

Figure 4. CAD drawings of the three working space configurations pertaining to this paper which are denoted as: a) original design, b) Revision A and c) Revision B. Pressure tap 0 (not shown here) was located in an inlet manifold.

The resulting structure, a 15x15 mm glass-Kapton-glass laminate, could then be easily socketed into a Plexiglas/aluminium packaging. The packaging granted the user access to the chips' pressure taps through conventional fittings and tubing. Lastly, because the flow was to be unidirectional, the regenerator interface and the working space median were exploited as inlets or outlets to the chip, that is, depending on the gas flow's direction (figure 5).

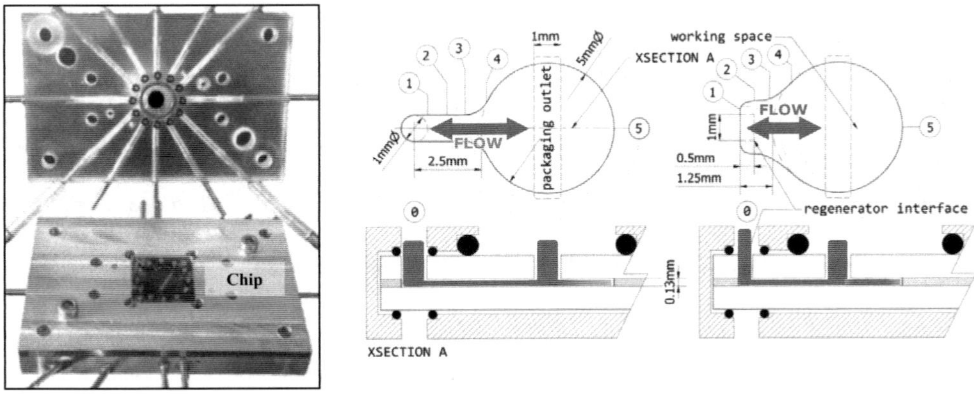

Figure 5. Packaging used in experiments pictured with a chip exposed (left) and a CAD drawing of the packaging's ability to accept the original and Revision A versions of the working space (right).

2.1.2. Experimental setup. With the packaging sealed by O-rings and the tubes secured and tested for integrity, each chip was tested in an incrementing unidirectional flow rate. Static pressure at the taps were recorded to be later processed for determining the pressure losses that correspond to geometrical features and flow rate. The inlet flow rate, controlled (and corroborated) digitally in a range of 100 to 1500 cm^3/min, represented the transient range of a pulsating flow, given the MISTIC's operating parameters and displacement. Meanwhile, the packaging permitted the measurement of the static pressure at taps 1 through 5 by an Omega PX4202 pressure sensor (0-30psig resolution). An additional junction, tap 0, was located in a relatively large inlet manifold; thus, its static pressure was deemed to be the stagnant (total) pressure of the inlet. It is worth of note that all values were recorded relative to the transducer's readout of local atmospheric pressure and inlet flow resistance, i.e. the resistance between tap 1 (or 5 if in reverse flow) and tap 0, was deducted.

3. Results and discussion

Post processing of the static pressures in the general equation for conservation of energy was required to determine pressure losses at each zone (Table 1). These zones were length-wise portions of the working space between the pressure taps. Any non-recoverable losses between two taps were to be found in the general equation by substituting stagnation pressure from one zone to another. This means that kinetic (average velocity) and semi-compressible (local change of density) effects were considered. Pressure at tap 0, being stagnate, corroborated the sum of the calculated stagnation pressure losses.

Table 1. Total pressure losses (Δp_{loss}) calculated for each zone

	Δp_{loss} (Pa)		
Zones	Original	Revision A	Revision B
0→1	394	1473	3563
1→2	3781	1947	*-395*[a]
2→3	1241	1760	129
3→4	2187	57	925
4→5	2003	126	*-307*[a]
total	**9605**	**5361**	**3915**
p_{tap0} (deviation)	**9515** (<1%)	**5171** (3.7%)	**4137** (5.4%)

[a] Negative values are within the range of 0.25% measurement uncertainty (+/-500Pa)

Results show that in a regenerator-to-working-space direction, pressure losses occurred mostly in the 0-1 zone, which corresponds to the 90° angle between the regenerator and the exchanger portion of the working space. The effect of increasing NDG_1 is prominently featured in the left portion of figure 6 where 84% of the pressure drop at Re = 700 was reduced between the original configuration and Revision A. This was likely due from increasing the hydraulic diameter at tap 1 from 200 to 260 µm. Improvement between Revisions A and B was marginal, even if the hydraulic diameter was maximized from 260 µm to 320 µm by way of a spiral relief. The wall's spiral relief proved detrimental to reverse flow, as overall resistance increased for this case in Revision B. Thus, the working space in Revision A was retained for having more favourable (and symmetrical) flow pressure loss in steady state flow.

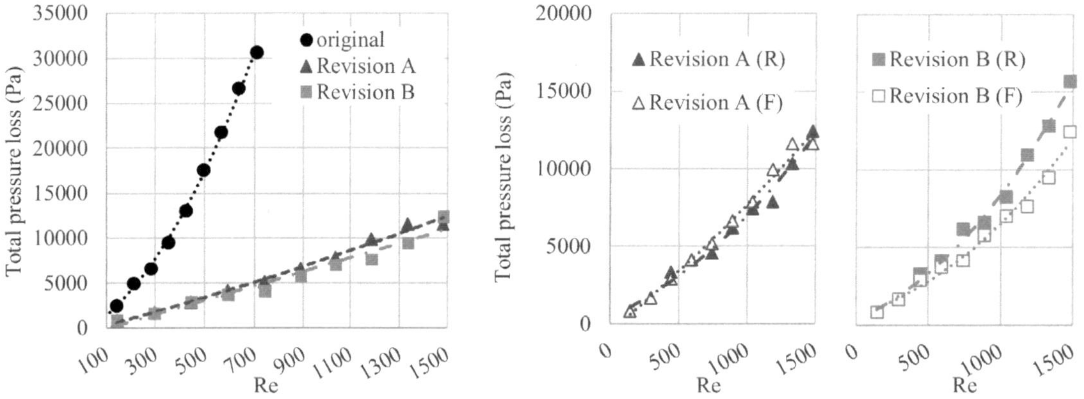

Figure 6. Total pressure loss of the original, Revision A and Revision B configurations in a regenerator-to-working-space flowrate (left) and total pressure loss between Revision A and Revision B according to flowrate direction (right), i.e. regenerator-to-working-space (F) and working-space-to-regenerator (R). Data was processed in relation to tap 1's Re number.

4. Conclusion

This research's objective was to design a working space which would reduce geometry-based pressure losses in the MISTIC's engine modules as they operated at kHz levels. Practical considerations and the use of non-dimensional groups helped to design optimized working spaces for the projected operating conditions. By breaking down the working space geometry into a given set of zones, it was possible to determine which geometrical features were likely to affect the working gas total pressure loss. These findings were further corroborated by the retrofitting of the data into an ECE-based global model of the engine that showed favorable start-up conditions with the configuration that was retained. However, further study is necessary in determining the losses in oscillatory flow and the efficiency of the working space's heat transfer.

Acknowledgement

This is work has been funded by the French National Research Agency (ANR) under grant no. ANR-12-SEED-0005-01 (MISTIC Project)

References

[1] F. Formosa, A. Badel, and J. Lottin, "Equivalent electrical network model approach applied to a double acting low temperature differential Stirling engine," *ECaM* 78, pp. 753–764, 2014.

[2] D. Pfund, D. Rector, A. Shekarriz, A. Popescu, and J. Welty, "Pressure drop measurements in a microchannel," *AIChE Journal*, vol. 46, no. 8, pp. 1496–1507, 2000

[3] M.-A. Hachey, É., Léveillé, J.-F. Dufault, L.G. Fréchette, F. Formosa, "Reducing heat exchanger pressure losses in a microscale Free Piston Stirling Engine", *Proceedings of the JNRSE*, 2015

Energy harvester for use on nearby current-carrying conductors

Duy-Son Nguyen, Zhiwei Wu, and Richard White

University of California, Berkeley, USA

Email: rwhite@eecs.berkeley.edu

Abstract. We report a concept of a new electret-based energy harvester that can extract energy from a nearby conductor carrying a time-varying current. In the harvester, a permanent magnet located near an electret is made move relative to the electret by the magnetic interaction of the circumferential time-varying magnetic field produced by the current flow and the static magnetic field of the permanent magnet. The advantage of this harvester over harvesters that employ piezoelectric elements is that piezoelectric materials are known to fatigue over time, whereas electrets having extremely long lifetimes are possible.

1. Introduction

Energy harvesters that couple to magnetic fields surrounding wires carrying AC current have been presented in previous publications [1, 2] using piezoelectric transduction. However, piezoelectric material has a problem of the material degradation that results to change the resonance of energy harvesters [3]. Recent development of long lifetime electrets [4-6] is attractive to explore electret-based energy harvesters for use nearby the AC-current carrying conductors.

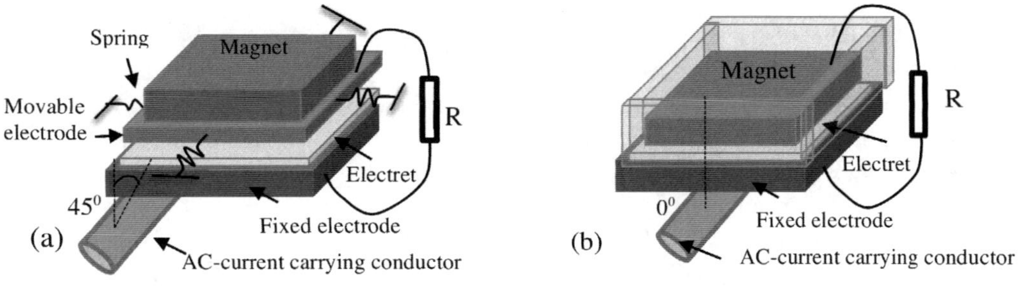

Figure 1: Drawing concept of electret-based energy harvester (EH) near the AC current-carrying conductors. (a) A permanent magnet is mounted on a movable electrode, suspended by springs. The harvester is placed in an orientation of 45^0. (b) A permanent magnet is rested on a rigid surface, inside a cavity, without spring supporting. The harvester is placed an orientation of 0^0.

A proposed electret-based energy harvester (EH) for use nearby the AC-current carrying conductor is depicted in figure 1(a). The structure of the harvester is similar to an out-of-plane energy harvester from ambient motion [7], except for adding a permanent magnet on the movable electrode. The magnet is oscillated mechanically by the circumferential magnetic field around the conductor and actuates the EH. The device is placed in an orientation of 45^0 from the conductor center to obtain the strongest force in the vertical direction on the magnet [1, 2].

The presence of suspension such as cantilever beams or folder-springs could limit the harvester lifetime due to their mechanical fatigue in long term of oscillations. Moreover, more space is required to design the suspended systems. The elimination of suspension seems to be more interested in design of vibration EHs. A propose of EH without suspension is shown in figure 1(b). The magnet is rested on a rigid surface of electret and fixed electrode, inside the cavity. The magnet is also served as a movable electrode of the electrostatic energy harvester.

In this paper, we will report a concept of an electrostatic EH without suspension as illustrated in figure 1(b). The equations of motion and experiment to verify the device concept will be presented.

2. Device modelling

2.1. Forces and torques on the magnet

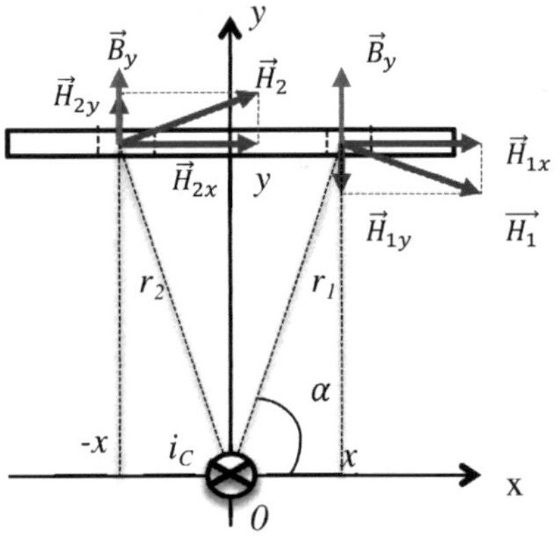

Figure 2: Geometry of element magnets nearby an AC-current carrying conductor i_C, $\vec{H}$ is the circumferential magnetic field from the conductor. $\vec{B}$ is the flux density of the magnet.

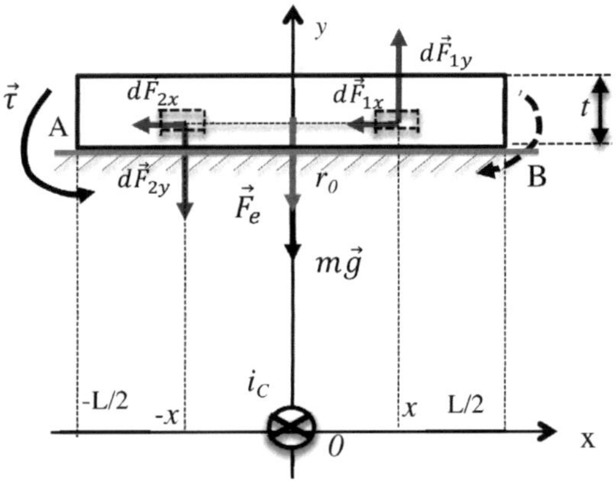

Figure 3: Forces on the magnet. The magnet will rotate anti-clockwise around A when the current i_C is in $-z$ direction, and clockwise around B when i_C flips.

We consider two elements of the magnet at the coordinate of (x, y) and ($-x$, y) (figure 2). The cable along the z-axis carries a current $i_C = I_C \sin(2\pi f t)$ in the $-z$ direction. We assume that the flux density $\vec{B}$ of the permanent magnet is uniform and aligns in the positive y-direction ($\vec{B} = jB_y$). A magnetic field at a distance r from the origin, created by the conductor, has the magnitude of $H = i_C/(2\pi r)$ and the direction perpendicular to the line current and to the position vector r as shown in figure 2. The magnetic forces on the magnet elements are given by

$$dF_y = \frac{B_y i_C}{2\pi} \frac{2xy}{(x^2+y^2)^2}, \quad dF_x = -\frac{B_y i_C}{2\pi} \frac{y^2-x^2}{(x^2+y^2)^2}, \tag{1}$$

The direction of the forces on the elements is shown in figure 3. The forces in y-direction will create a torque and rotate the magnet in the anti-clockwise (and in clockwise when i_C changes to z direction) when the current is larger than the current threshold I_{CT}. Assuming the magnet thickness t is very small compared to its length L, the total magnetic torque due to the magnetic force is approximated by

$$\tau_{magnet} = 2 \iiint_{V/2} x \, dF_y \, dV \approx \frac{B_y i_C}{\pi} \frac{V}{4r_0}, \tag{2}$$

in which V is the magnet volume and r_0 is the distance from the lower edge of the magnet to the cable center. At rest, the torque due to the gravitational force $m\vec{g}$ and the electrostatic force $\vec{F}_e$ will be balanced by the torque of the reaction force from the rigid surface. The magnet starts rotating when the magnetic torque is larger than the torque of the gravitational and electrostatic forces. In the order word, to actuate the magnet, the current in the cable must be larger than the current threshold, given by

$$I_{CT} = \frac{2\pi r_0}{B_y wt}(mg + F_e) \tag{3}$$

where w is the magnet width.

2.2. *Electret-energy harvesting modelling*

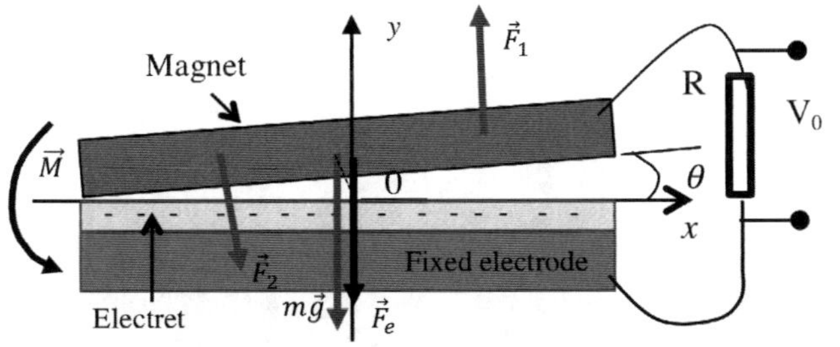

Figure 4: Magnet resting on the rigid plane and rotated an angle of θ under the magnetic torque. $\vec{F}_1$ and $\vec{F}_2$ are the magnetic force on positive and negative sides of x-axis. The movable electrode (magnet) and fixed electrode are connected to the load resistor R.

Assuming the magnet rotates a small angle of θ as shown in figure 4. The capacitance for a small tilt angle of θ is approximated by

$$C = C_0[1 + \frac{\varepsilon_e L}{t_e}\theta] \,, \tag{4}$$

where $C_0 = \varepsilon_0\varepsilon_e({}^{wL}/_{t_e})$ is the initial capacitance (when $\theta = 0$), t_i is the dielectric thickness, ε_0 and ε_e are the absolute permittivity and relative permittivity of the electret, respectively. The equations of motion for the energy harvester can be written as

$$\begin{cases} I\ddot{\theta} + b\dot{\theta} + mg\frac{L}{2}\left(1 - \frac{\theta^2}{2}\right) + \tau_{elec} + \tau_{stopper} = \frac{B_y V}{4\pi r_0} f(\theta) i_C & (5) \\ -\dot{q}R = \frac{\Gamma}{C_0}\theta + \frac{q}{C_0} & (6) \end{cases}$$

Here, I is the moment of inertia of the magnet; b is the damping constant; τ_{elec} is the torque of electrostatic force, and is a function of the charge on the capacitor q and the tilt angle θ; $f(\theta)$ is a function of θ, expressing for the dependence of magnet torque on the tilt angle of θ; R is the load resistor; $\Gamma = V_e C_0({}^{\varepsilon_e L}/_{t_e})$ is the transduction factor; $\tau_{stopper}$ is the stopper torque, using to express the situation of the magnet impacting the rigid surface. In the equations (5) and (6), θ must be positive and $i_C \geq I_{CT}$.

Deriving the closed-forms of τ_{elec} and $f(\theta)$ as well as solving the equations will be addressed in the other work. In this paper, experiments to verify the energy harvester concept will be focused and presented in the next section.

3. Experiment

Figure 5 shows the experimental set up to test the energy harvester. The output of the harvester is connected to a load resistor of 44 MΩ. In this experiment, a DC voltage, stimulating of an electret, is used to bias the harvester. The current in the cable is the 60-Hz power line. The output voltage V_0 in the time and frequency domains is shown in the figure 6. The magnet plate rocks at the frequency of 60 Hz. Since lifting of the magnet on the left and the right have the same effect on the capacitance, V_0 has a large component at the double frequency, 120 Hz.

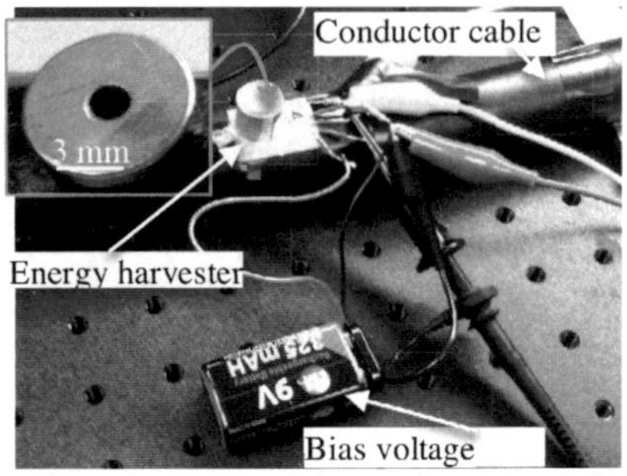

Figure 5: Experimental set up for testing the EH.

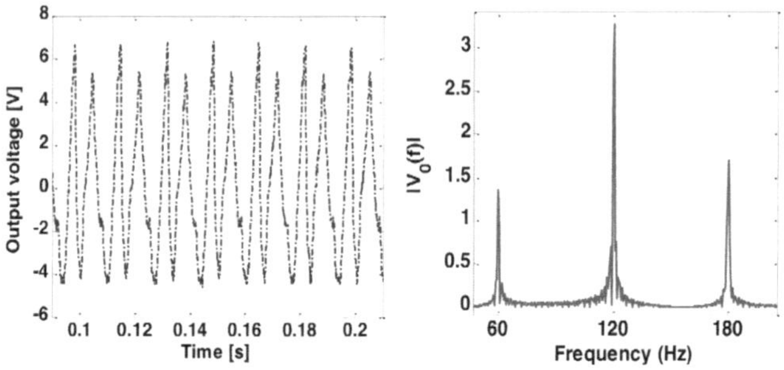

Figure 6: Measurement of the output voltage in the time domain (left) and frequency domain (right), R = 44 MΩ, V_e = 67V. The amplitude of the current in the cable is 30A.

Figure 7 shows the RMS output voltage when increasing the amplitude of the current in the conductor cable. The output voltage slowly increases when the current amplitude is larger than 20 A and sharply increases when reaching the threshold current (Eq. (3)) of 36 A. Beyond 40 A, the RMS output voltage does not increase more. The RMS output voltage when varying the bias voltage is shown in figure 8. In the bias voltage range of experiments, the RMS output voltage increases when the bias voltage increases.

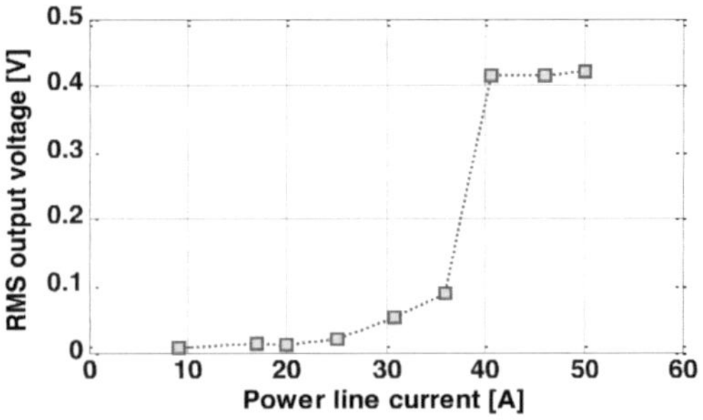

Figure 7: RMS output voltage vs. amplitude current in the conductor cable, R = 44 MΩ, V_e = 9V.

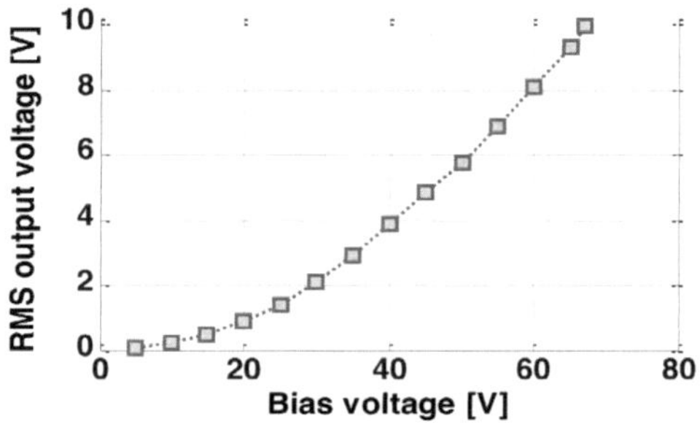

Figure 8: RMS output voltage vs. bias voltage, R = 44 MΩ, the amplitude of the current is 30A.

4. Conclusion

A new concept of an electrostatic energy harvester without suspension for use on nearby an electrical conductor carrying an AC current has been presented. The experimental results of an energy harvester, using external bias voltage, have verified the working principle of the harvester. The RMS output voltage sharply increases when the current amplitude gets 36-40 A, and continuously increases when the bias voltage varies from 5V-67V.

References

[1] E Leland *et al* 2006 *Proc. PowerMEMS*, 165-168
[2] I Paprotny *et al* 2013 *J. IEEE Sensors* **13** 190-201
[3] P Pillatsch *et al* 2014 *J. Physics: Conference Series* **557** 012129
[4] J A Voorthuyzen *et al* 1989 *IEEE Trans. Electrical Insulation* **24** 255-266
[5] S Boisseau *et al* 2011 *Smart Mater. Struct.* **20** 105013
[6] Y Suzuki 2011 *IEEJ Trans* **6** 101-111
[7] H Asanuma *et al* 2014 *J. Phys.: Conf. Ser.* **557** 012124

Mitigation of interfacial silicide reactions for electroplated CoPt films on Si substrates

Ololade D. Oniku and David P. Arnold
Interdisciplinary Microsystems Group, Dept. Electrical and Computer Engineering, University of Florida, Gainesville FL 32611, USA

E-mail: darnold@ufl.edu

Abstract. We report in this paper the influence of film thickness on the material and magnetic properties of electroplated CoPt permanent magnets. Layers of CoPt magnets with film thicknesses ranging from 0.5 μm to 5 μm are deposited into photoresist molds (3.5 mm x 3.5 mm square and 5 μm x 50 μm arrays) on a (100)Si substrate coated with 10 nm/100 nm Ti/Cu adhesion/seed layer. Results show an unexpected reduction in magnetic properties for films below 2 μm thick. This effect is determined to be a consequence of metal-silicide reactions at the substrate interface during annealing leading to the formation of a non-magnetic layer at the interface. Subsequently, a TiN diffusion-barrier layer is added to inhibit the silicide reaction and thereby maintain strong magnetic properties (Hci ~800 kA/m, M_r/M_s = 0.8) in micron-thick electroplated CoPt layers.

1. Introduction

Equiatomic CoPt in the ordered $L1_0$-phase is known to exhibit strong magnetic properties (K_u = 4.9 MJ/m^3, $\mu_0H_{ci} \geq 1$ T, and M_s = 1 T) and excellent corrosion resistance [1]–[3], making it a magnetic material of considerable interest to the PowerMEMS community. In the as-deposited state, CoPt films possess a disordered A1 crystallographic phase, with relatively soft magnetic properties (μ_0H_{ci} < 0.0125 T, M_r/M_s < 0.05). The films typically require a thermal anneal (usually > 500°C) to induce a crystallographic ordering from the disordered A1 phase with a face-centered cubic structure to an ordered $L1_0$ equilibrium phase having a face-centered tetragonal structure. The tetragonal crystal lattice yields high magnetocrystalline anisotropy and plays a key role in yielding the hard magnetic properties.

Because of the advantages offered by electroplating (such as cost efficiency, ease of fabrication and scalability, shape controllability, and general integrability within MEMS processes), the electrodeposition of CoPt permanent magnets has been widely studied as an attractive and practical fabrication technique for various MEMS applications [4]–[12]. However, like many fabrication methods, electroplating is a variation-prone process, in that factors such as plating conditions (pH, current characteristics, temperature, agitation, deposition time), electrolyte composition, and substrate (or seed layer) impact the morphology and overall properties of the plated layer. For example, our group has previously reported the deposition of 3 – 20 μm-thick CoPt magnets and investigated the influence of current density, seed layer, anneal temperature/time on the crystallographic structure and magnetic properties of the plated CoPt magnets [4]–[12]. While most of the above studies have focused on multi-micron thick films, film thicknesses of hundreds of nanometers to a few microns may be desired for certain microsystem device applications. In this work, we study the effects of film thickness variation on the magnetic properties of electroplated CoPt magnets and investigate an unexpected reduction in magnetic properties observed in films with thickness below 2 μm. The deterioration in magnetic properties is identified to be the result of metal-silicide reactions between the metal layers and silicon substrate, which occurs during the post-deposition annealing step. The use of a TiN diffusion barrier is shown to eliminate the silicide reaction and thereby improve the magnetic properties of thinner layers.

2. Experimental Procedure

2.1. Fabrication process

The CoPt magnetic layers are electroplated into photoresist molds, 3.5 mm x 3.5 mm and 5 μm x 50 μm array, on a (100)Si substrate using one of two types of seed layers. Type A samples use a 10 nm Ti adhesion layer and a 100 nm Cu conductive layer. Type B samples use a 25 nm TiN diffusion barrier, followed by 10 nm Ti and 100 nm Cu. The Ti and Cu are DC sputtered, and the TiN is achieved by reactive sputtering of Ti in nitrogen. The electroplating bath consists of 0.1 M of cobalt sulfamate, 0.025 M of diamine dinitrito platinum (II), and 0.1 M of ammonium citrate salts in a 100 mL solution. Films are plated at room temperature, at a pH of 7 using 100 mA/cm^2 currents and subsequently annealed in a forming gas ambient (4% H_2 + 96% N_2) at 700°C with a ramp rate of 20°C/min for 40 min [11]. Figure 1 shows example optical images of CoPt magnets deposited into photo-defined molds. Film thicknesses between ~0.5 μm and ~5 μm are obtained by varying the time of deposition. Table 1 presents the deposition time and corresponding film thickness.

Figure 1: Example top-view optical images of CoPt magnets electroplated into photoresist molds.

Table 1: Deposition times and film thickness of electroplated CoPt.

Sample	Layers	Deposition Time (min)	Film Thickness (μm)
A1	Si/Ti/Cu	5	0.4
A2	Si/Ti/Cu	15	1.3
A3	Si/Ti/Cu	30	2.6
A4	Si/Ti/Cu	45	3.5
A5	Si/Ti/Cu	60	5.2
B1	Si/TiN/Ti/Cu	5	0.4
B2	Si/TiN/Ti/Cu	15	1.3
B3	Si/TiN/Ti/Cu	30	2.6

2.2. Characterization

Energy dispersive x-ray spectroscopy (EDS) was used to confirm the chemical composition (~50:50) of the CoPt films. A FEI Nova NanoSEM 430 scanning electron microscopy was used to image the films and to measure the film thickness. Crystallographic structure analysis was carried out using x-ray diffraction (XRD) patterns from a Panalytical X'Pert Powder diffractometer. In-plane and out-of-plane magnetic measurements were made with an ADE EV9 vibrating sample magnetometer (VSM).

3. Results and Discussion

This section presents experimental results of the material and magnetic characterizations of CoPt samples deposited on Si/Ti/Cu and Si/TiN/Ti/Cu.

3.1. SEM Imaging

Figure 2 shows cross-section SEM images of samples with different thicknesses after annealing. For samples without TiN layer (samples A1–A5), a rough, pulverized appearance of the metal-substrate interface is observed, particularly in the 0.4-μm-thick film. This damaged interface is a consequence of a reaction between the Si substrate and the metal layers at temperatures ≥ 200°C [13], [14] (later confirmed by XRD). It is noted that as the film thickness increases, the effects of the interface reactions become less noticeable and the microstructure of the CoPt film becomes clearly defined. The use of the 25-nm TiN layer (Type B samples) shows a dramatic improvement of the interface, although a new problem arises in the poorer adhesion of the electroplated layer to the TiN barrier layer on Si. The larger area 3.5 mm x 3.5 mm films delaminated from the substrate, but the smaller 5 μm x 50 μm patterned films did not.

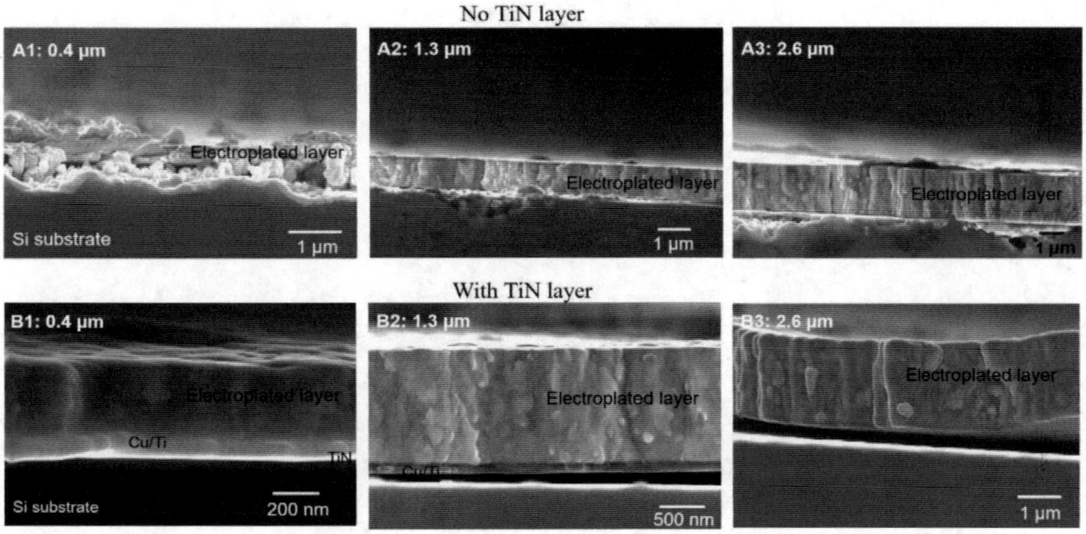

Figure 2: Cross section SEM images of annealed CoPt magnets of various film thicknesses (A1-A3) without TiN layer; and (B1-B3) with TiN diffusion barrier layer.

3.2. Crystallography

Figure 3 shows XRD measurements for the CoPt samples with different film thicknesses. The diffraction patterns confirm the presence of Co- and Cu-silicides in the Type A samples, with strong peak intensities notable for the 0.4-μm-thick sample. The figure also confirms the elimination of the silicides in the Type B samples with the introduction of the TiN barrier layer. Figure 4 shows the $L1_0$ phase volume fraction [9] of samples A1 – A5 (without TiN) and B1 – B3 (with TiN). It is seen here that an increase in the $L1_0$ volume fraction is observed with increasing film thickness for both sets of samples. The low volume fraction obtained in the thinner films is attributed to the formation of metal silicides from the interface reactions inhibiting the progress of the transformation, for the case of the Type A samples.

3.3. Magnetic Properties

Figure 5 presents out-of-plane magnetization curves of the CoPt samples deposited with and without TiN. Very poor magnetic properties are exhibited by the 0.4-μm-thick sample without TiN layer, owing to the large volume fraction of non-magnetic silicides. The magnetic properties improve with thickness suggesting the existence of a "good" magnetic layer on top of a "poor" layer, as illustrated in figure 6. Films at 3.5 μm and above show properties consistent with the previously reported 3 to 20-μm-thick layers [10], [11].

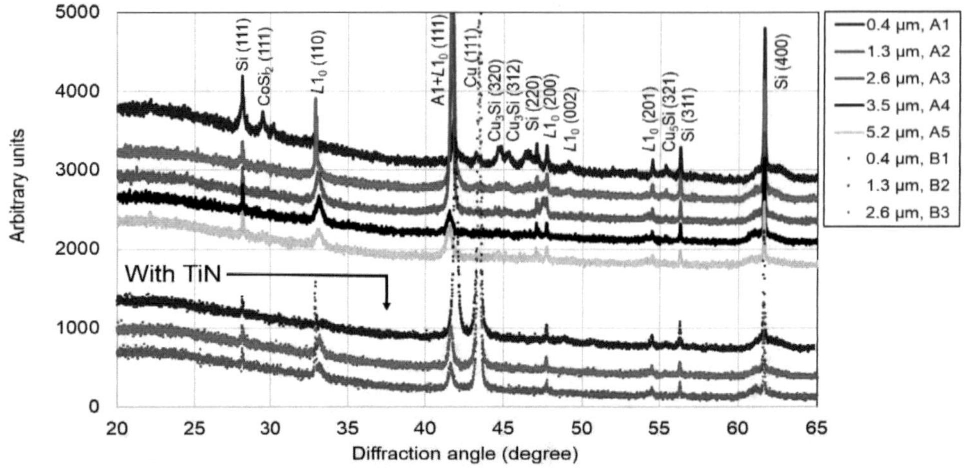

Figure 3: X-ray diffraction patterns of annealed CoPt magnets showing the presence of non-magnetic Cu-silicide and Co-silicide in the Type A samples, with subsequent mitigation of these silicides in the Type B samples.

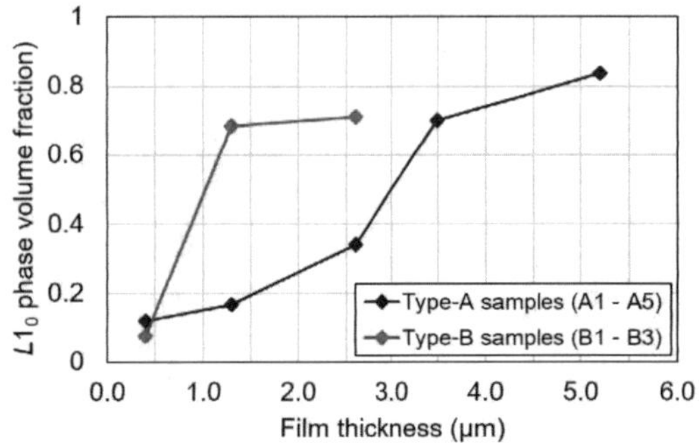

Figure 4: $L1_0$ phase volume fraction of CoPt magnets without (A1 – A5) and with TiN (B1 – B3) for various film thicknesses.

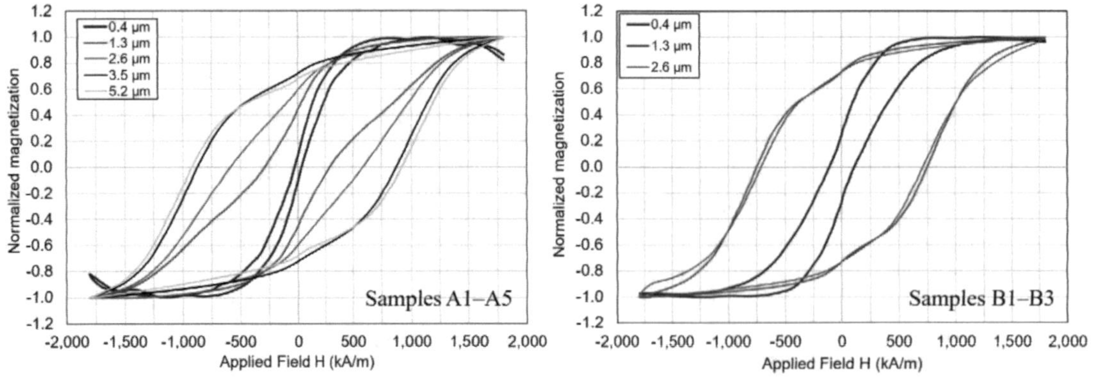

Figure 5: Out-of-plane magnetic hysteresis curves for CoPt samples with various thicknesses (A1-A5) without TiN layer and (B1-B3) with TiN diffusion barrier layer

Substantial improvements in magnetic properties are observed for samples on TiN layer with film thickness of 1.3 μm and above, although the 0.4 μm-thick film still exhibit relatively poor magnetic properties. Lastly, Figure 7 shows a strong correlation between the $L1_0$ volume fraction and the coercivity of the CoPt magnets, the highest volume fraction of 0.83 yielding a high coercivity of close to 900 kA/m.

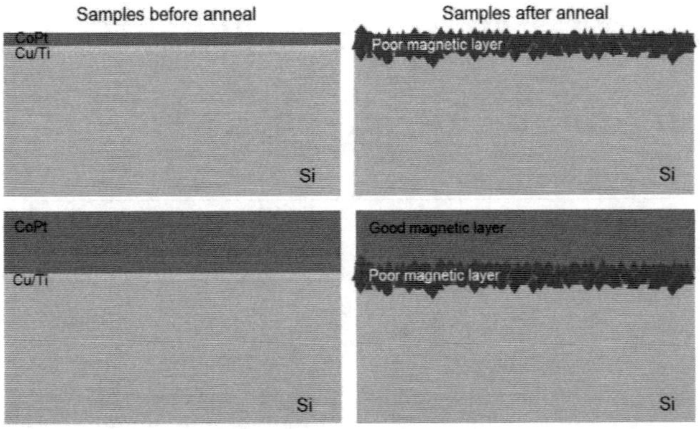

Figure 6: Cartoon image illustrating why magnetic properties improve with increase in CoPt film thickness for samples deposited on Cu/Ti.

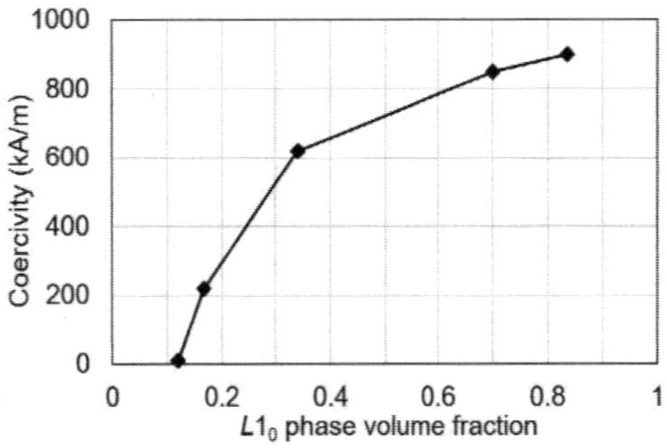

Figure 7: Relationship between coercivity and $L1_0$ volume fraction

4. Conclusion

The mitigation of metal-silicide reactions at the substrate interface of an electroplated CoPt magnetic layer is reported in this paper. More than 4X reduction in coercivity is observed for films with thicknesses less than 2 μm as a result of the silicide reaction. By using a TiN barrier layer, the metal-substrate interface is improved, the silicide reaction is prevented and a coercivity of ~800 kA/m is maintained for micron-thick films.

5. References

[1] A. S. Darling 1963 *Platinum Metals Rev.* Available: http://www.docstoc.com/docs/160242420.
[2] A. Kashyap 2004 *J. Appl. Phys.* **95** 7480
[3] K. Barmak, et. al. 2005 *J. Appl. Phys.* **98** 033904.
[4] T. S. Eagleton, et. al. 2005 *J. Electrochem. Soc.* **152** C27.
[5] K. Ž. Rožman, et. al. 2010 *Thin Solid Films* **518** 1751–55.
[6] F. Wang, et. al. 2004 *Electrochem. commun.* **6** 1149–52.
[7] O. Berkh, et. al. 2008 *Electrochem. Solid-State Lett.* **11** D38.
[8] E. Gileadi, O. Berkh, and Y. Shacham-Diamand 2009 *Electrochem. Solid-State Lett.* **12** D53.
[9] P. L. Wu, X. H. Li, W. Li, H. Y. Sun, Y. Chen, and X. Y. Zhang 2008 *Mater. Lett.* **62** 309–312.
[10] O. D. Oniku, B. Qi, and D. P. Arnold 2014 *J. Appl. Phys.* **115** 17E521.
[11] O. D. Oniku, B. Qi, and D. P. Arnold 2015 *TRANSDUCERS* 596–601.
[12] A. Garraud, O. D. Oniku, and D. P. Arnold 2015 *J. Appl. Phys.* **117** 17C718.
[13] A. Cros, M. O. Aboelfotoh, and K. N. Tu 1990 *J. Appl. Phys.* **67** 3328.
[14] N. Benouattas, A. Mosser, and A. Bouabellou 2006 *Appl. Surf. Sci.* **252** 7572–77.

Acknowledgments
This work was supported in part by the U.S. Army Research Office (W911NF-09-1-0511) and the Multi-functional Integrated System Technology (MIST) Center, an NSF I/UCRC. The authors would also like to thank the technical staff of the UF Nanoscale Research Facility and Major Analytical Instrumentation Center for their assistance with microfabrication and material analyses.

A highly efficient simulation technique for piezoelectric energy harvesters

R Ardito, A Corigliano[1] and G Gafforelli
Department of Civil and Environmental Engineering,
Politecnico di Milano, Piazza Leonardo da Vinci 32, Milan, Italy

E-mail: alberto.corigliano@polimi.it

Abstract. This paper presents a new numerical technique which is aimed at obtaining fast and accurate simulations of piezoelectric beams, used in inertial energy harvesting MEMS. The execution of numerical analyses is greatly important in order to predict the actual behaviour of MEMS device and to carry out the optimization process on the basis of Design of Experiments (DOE) techniques. In this paper, a refined, yet simple, model is proposed with reference to the multi-physics problem of piezoelectric energy harvesting by means of laminate cantilevers. The proposed model is calibrated and validated with reference to 3D finite element analyses.

1. Introduction

The application of piezoelectric materials is continuously increasing, with different possible uses of both "direct" (conversion of mechanical into electric energy) and "indirect" effect. The latter is applied in actuators, e.g. in the case of micropumps [1]; one of the most promising application of "direct" effect is energy harvesting [2]. In recent times, the concept of energy harvesting has been applied to Micro-Electro-Mechanical Systems (MEMS), with similar functioning principles [3], [4]: an additional broadening of applications can be forecast in the next future, with the immediate corollary of a fundamental need for improved computational tools.

In this paper, a simple 1D model is built in order to simulate piezoelectric thin beams and plate harvesters. Starting from the fully coupled 3D constitutive equations of piezoelectricity, appropriate hypotheses are introduced to model strains and stresses so that the 1D model takes into account the 3D effects. The theoretical model is founded on an enrichment of the Euler-Bernoulli kinematic field, with some additional strain contributions in order to introduce some three-dimensional effects: the obtained model is denoted by Modified Transverse Deformation (MTD). The modification is based on the width-to-length ratio and is driven by some considerations on the 3D behaviour of piezoelectric films. Such an enriched model is unprecedented, to the Author's knowledge, and allows the user to obtain reliable results in a negligible time.

The model is applied to some specific examples referred to inertial energy harvesters, see figure 1. A simple resistance R is considered in the electric circuit attached to the piezoelectric element. The numerical solution is obtained making use of a specific discretisation of both the displacement and the voltage fields. The validation is obtained by the critical comparison with the results of full 3D computations.

[1] To whom any correspondence should be addressed.

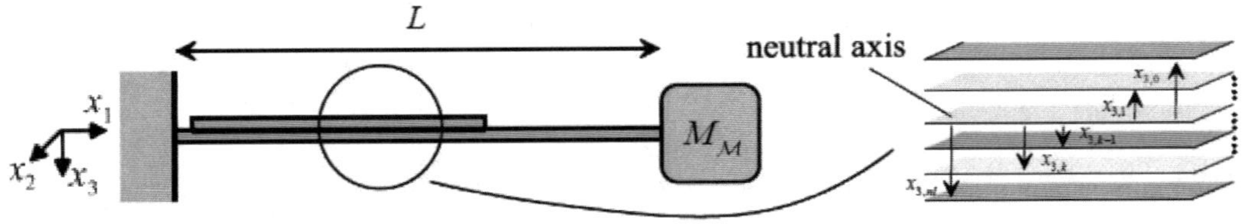

Figure 1. Scheme of a layered cantilever beams, representative of a MEMS harvester.

2. Description of the proposed model

The cantilever harvester is a piezo-laminated beam clamped at one end section and free to oscillate on the other side. The dynamic of the laminated beam can be easily modelled by employing standard beam theories for laminated composites. Herein, Classical Lamination Theory (CLT) [5] is employed and the kinematics of the beam section is described through Euler-Bernoulli hypothesis (i.e. plane sections orthogonal to the neutral axis remain orthogonal to the same after deformation).

The reference system is centred on the neutral axis of the beam and on the built-in end section. The x_1 axis lies along the beam's length, so that it ranges between 0 and L; the x_2 axis is along the beam's width and ranges between $-b/2$ and $b/2$; finally, the x_3 axis is across the beam's thickness h. The Euler-Bernoulli kinematic model, in its original form, depends on the transverse displacement $w(x_1)$ only. In this paper, we consider a suitable modification in order to obtain the Modified Transverse Deformation (MTD) model (further details can be found in [6]). The final model, in terms of the displacement components s_j, reads:

$$s_1 = -x_3 \frac{\mathrm{d}w}{\mathrm{d}x_1}; \quad s_2 = \hat{s}_2(\mathbf{x}, \Lambda); \quad s_3 = w + \hat{s}_3(\mathbf{x}, \Lambda) \tag{1}$$

The additional functions depend on a geometric parameter $\Lambda = L/b$, which represents the transverse slenderness of the beam, i.e. the ratio of the beam's length and the beam's width. The chosen model should fulfil the standard requirements for a beam theory: the in-the-thickness stress must be null, $T_{33}=0$; the in-plane stress must be $T_{22}=0$ at $x_2 = \pm b/2$. The MTD model aims at obtaining a variable response for different transverse slenderness: when $\Lambda \to 0$ the beam is infinitely wide and the strain condition $S_{22}=0$ must be verified; on the other hand when $\Lambda \to \infty$ the beam is extremely narrow and $T_{22}=0$ has to be guaranteed. These features can be obtained if the strain component S_{22} is modified through the multiplication times a "transversal shape function" f_Λ:

$$f_\Lambda = \left[1 - A(\Lambda)\right] \left| \frac{2x_2}{b} \right|^{B(\Lambda)} + A(\Lambda) \tag{2}$$

where A and B depend on two fitting parameters, a_Λ and b_Λ:

$$A = \frac{\Lambda^{a_\Lambda}}{\Lambda^{a_\Lambda} + b_\Lambda}; \quad B = 1 + \frac{1}{\Lambda} \tag{3}$$

3. Governing equations

The governing equations for the piezoelectric problem can be obtained by using the dissipative form of Euler-Lagrange equations:

$$\frac{\mathrm{d}}{\mathrm{d}t}\left(\frac{\partial L}{\partial \dot{q}_i}\right) - \frac{\partial L}{\partial q_i} + \frac{\partial D}{\partial \dot{q}_i} = 0 \tag{4}$$

where D is the dissipation function and L is the Lagrangian function which is given by suitably combining the kinetic energy K, the internal energy E and the external work Y: $L = K - (E - Y)$. An approximate solution is sought starting from some hypotheses on the unknown fields. First, the displacement field is expressed on the basis of a single time-variant parameter W, given a suitable shape function ψ_w:

$$w(x_1) = \psi_w(x_1) W(t) \tag{5}$$

Second, the electric potential is assumed to be linear across the thickness t_p of the piezoelectric layer and constant along the beam length, so that the electric field is uniform: $E_3 = V(t)/t_p$. The electric charge q, collected by the electrodes, is managed by an external circuit, which provides the power supply for the self-powered electronic device. Different schemes of circuitries are investigated in [7]. The harvester provides AC voltage and the simplest solution is the coupling with an external load resistance. The final system of equations reads:

$$\begin{aligned} m\ddot{W} + c_M \dot{W} + k_L W - \Theta_{\chi v} V &= F_{ext} \\ k_E \dot{V} + \Theta_{\chi v} \dot{W} + R^{-1} V &= 0 \end{aligned} \tag{6}$$

4. Numerical results

In order to validate the model developed in section 2, a 3D FE model has been developed with the commercial code ABAQUS. The comparison between the numerical solution of the 3D model and of the proposed beam model is used in order to calibrate the parameters a_Λ and b_Λ, by using some simple static analyses; afterwards, the validity of MTD hypotheses has been checked with reference to quasi-static and dynamic analyses. The comparative analyses have been performed with a 2-layer cantilever (2 μm PZT on 6 μm polysilicon substrate) without tip mass. The beam length is L = 1000 μm, the width is parametric and varies from b = 50 μm to b = 5000 μm. The lower surface of the PZT layer is grounded, from the electric point of view, while the potential on the upper surface is constrained to be uniform in order to reproduce the presence of electrodes. Table 1 reports the material properties of the layers.

Table 1. Material parameters values ($\varepsilon_0 = 8.854 \cdot 10^{-12}$ F/m).

	ρ [g/cm^3]	E [GPa]	ν [-]	e_{31} [N/mV]	e_{33} [N/mV]	ε_{33}^S [ε_0]
PZT	7.70	100	0.30	-12	20	2000
PolySilicon	2.33	148	0.33	0	0	0

4.1. Static analyses

Open circuit static analyses have been performed with a tip load f_c=1 μN/μm at the beam free edge. After the convergence analysis, it has been possible to select the optimal FE mesh for the 3D model, which has been discretised by using brick elements with quadratic displacement fields. The tip displacement and the voltage at the free electrode are evaluated through various 1D models, namely: a classical beam theory, which entails null transverse deformation (NTD); a classical plate theory, which shows null transverse stress (NTS); the MTD theory. The two calibration parameters a_Λ and b_Λ have been computed through fitting on these static analyses. Their final values are a_Λ =0.8 and b_Λ =2.2. The obtained results show that the MTD model correctly describes the variation in stiffness and, most importantly, the variation of the upper electrode voltage, which is strictly connected to the coupling coefficient.

4.2. Quasi-static analyses

Quasi static analyses have been performed; a smooth step load, until the maximum value 1 μN/μm, followed by a constant plateau has been considered. Parametric analyses have been performed varying the load resistance. A peak power generation is obtained for an optimal value of load resistance which inversely depends on the capacitance of the device. All models predict the same optimal load resistance as ABAQUS. The same analysis has been performed for different values of width and the peak power per unit width, computed at the optimal load resistance, has been collected in figure 2a for all models. It is shown that the MTD model correctly predicts the harvester performances for the whole range of length-beam ratios that have been considered. On the other hand, null transverse stress and deformation models always under- and overestimate voltage and power, respectively. As shown by figure 2b the mean error on the power generation can be reduced from barely 30% to less than 5% considering both models. It can also be concluded that wider cantilever harvesters not only recover more power, which is obvious, but also have a higher power density than narrower beams.

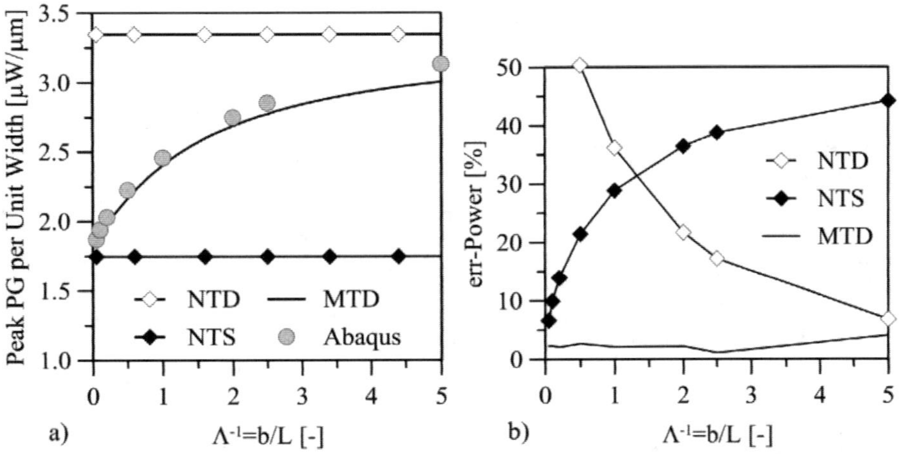

Figure 2. a) Peak power generation at optimal *R* and b) error on power generation.

4.3. Dynamic analyses

Free oscillation analyses have been performed on the same cantilever structure introduced previously. As before, the beam is quasi-statically moved to a certain position but in this case it is suddenly released activating free oscillations. The width is fixed to $b = 200$ μm while the load resistance R and the mechanical quality factor Q are varied in order to carry out parametric analyses.

A first significant result is that while the 1D code takes less than 3 seconds to complete the analysis, ABAQUS requires 5 hours 25 minutes and 23 seconds to produce the same results. Satisfactory agreement is obtained, in spite of a slight mismatch between the oscillation frequencies. Such a difference, which is due to the simplicity of the 1D model, can be appreciated by considering the Fourier transform of the displacement. Moreover, one finds that ABAQUS is able to capture the second eigen-frequency, which produces a small oscillation of the peak values of the responses. This effect originates from higher order resonance modes which are completely neglected in the 1D model since only a single shape function has been employed.

The results of parametric analyses at varying R show that the MTD model better reproduces numerical results than null transverse stress and deformation models. The effect of the mechanical quality factor has been examined. Except for exceptionally low values of Q, the peak power generation remains more or less constant. In spite of this, the mechanical damping has a strong influence on the performances of the harvester in terms of overall harvested energy. In fact, the total energy which is collected is reduced as Q decreases. The energy harvested is also affected by the choice of the resistance. It has to be noted that the peak power generation and the peak energy harvested do not occur at the same load resistance.

5. Conclusions

The main objective of this paper was represented by the implementation of a simulation tool, which merge the simplicity of Classical Lamination Theory with the unavoidable 3D effects due to the finite width of the beam. The answer is represented by an improved theory, so-called Modified Transverse Deformation, which includes the effect of the transverse strain so that boundary conditions and limit stress and strain configurations are recovered. The accuracy of the proposed model has been validated by means of critical comparison with fully 3D analyses, carried out by means of a commercial finite element code. The MTD model allows for considerable computational savings vis-à-vis 3D FE dynamic analyses, with a reasonable degree of accuracy. Once validated, the MTD model can be employed for the characterization and for the optimal design of realistic MEMS harvesters. As an example, a multi-layer model can be introduced in order to include the electrodes, which are usually constituted of different highly conductive materials. The piezoelectric thickness varies between 0.1 and 4 μm; the cantilever width is b = 1000 μm and the beam's length is L = 1000 μm. Figure 3 shows that interesting values of peak power can be obtained by means of an impulsive excitation.

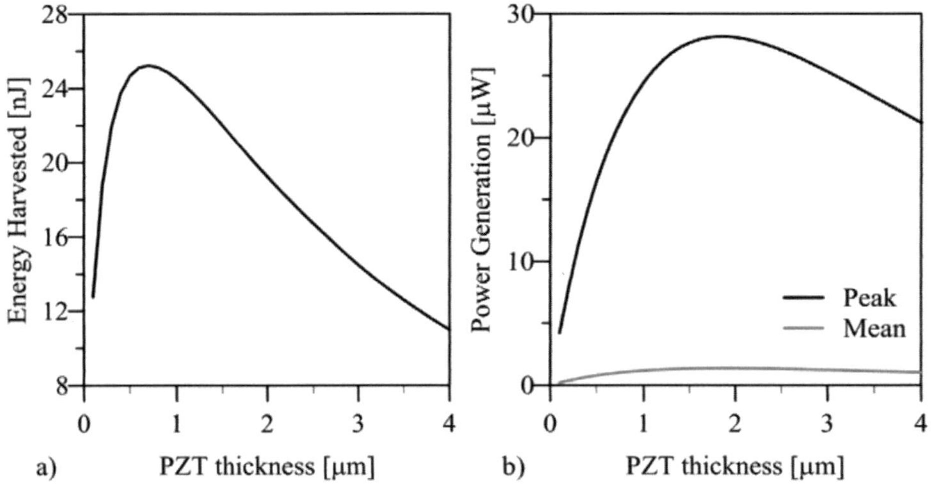

Figure 3. a) Total energy harvested and b) peak and mean power generation as a function of PZT thickness for an impulsive cantilever harvester

6. References

[1] Ardito R, Bertarelli E, Corigliano A and Gafforelli G 2013 *Composite Struct.* **99** 231–40
[2] Roundy S and Wright P K 2004 *Smart Mater. and Struct.* **13** 1131–42
[3] Kim S-G, Priya S and Kanno I 2012 *MRS Bulletin* **37** 1039–50
[4] Jeon Y B, Sood R, Jeong J-H and Kim S-G 2005 *Sens. and Act. A: Phys.* **122** 16–22
[5] Carrera E and Ciuffreda A 2005 *Composite Struct.* **69** 271–93
[6] Ardito R, Corigliano A and Gafforelli G 2015 *Composite Struct.*, **127**, 369–81
[7] Guyomar D, Badel A, Lefeuvre E and Richard C 2005 *IEEE Trans. on Ultras., Ferroel. and Freq. Contr.* **52** 584–94

Acknowledgments

The authors wish to thanks the ENIAC Joint Undertaking, Key Enabling Technology, project Lab4MEMS, Grant no. 325622, for partial funding of this research.